OXAZOLES:
SYNTHESIS, REACTIONS, AND SPECTROSCOPY, PART A

This is the sixtieth volume in the series
THE CHEMISTRY OF HETEROCYCLIC COMPOUNDS

THE CHEMISTRY OF HETEROCYCLIC COMPOUNDS

A SERIES OF MONOGRAPHS

EDWARD C. TAYLOR AND PETER WIPF, *Editors*

ARNOLD WEISSBERGER, *Founding Editor*

OXAZOLES: SYNTHESIS, REACTIONS, AND SPECTROSCOPY

Part A

Edited by

David C. Palmer

Johnson & Johnson Pharmaceutical Research & Development, L.L.C.
Raritan, New Jersey

AN INTERSCIENCE PUBLICATION

JOHN WILEY & SONS, INC.

Copyright © 2003 by John Wiley & Sons, Inc. All rights reserved.

Published by John Wiley & Sons, Inc., Hoboken, New Jersey.
Published simultaneously in Canada.

No part of this publication may be reproduced, stored in a retrieval system, or transmitted in any form or by any means, electronic, mechanical, photocopying, recording, scanning, or otherwise, except as permitted under Section 107 or 108 of the 1976 United States Copyright Act, without either the prior written permission of the Publisher, or authorization through payment of the appropriate per-copy fee to the Copyright Clearance Center, Inc., 222 Rosewood Drive, Danvers, MA 01923, 978-750-8400, fax 978-750-4470, or on the web at www.copyright.com. Requests to the Publisher for permission should be addressed to the Permissions Department, John Wiley & Sons, Inc., 111 River Street, Hoboken, NJ 07030, (201) 748-6011, fax (201) 748-6008, e-mail: permreq@wiley.com.

Limit of Liability/Disclaimer of Warranty: While the publisher and author have used their best efforts in preparing this book, they make no representations or warranties with respect to the accuracy or completeness of the contents of this book and specifically disclaim any implied warranties of merchantability or fitness for a particular purpose. No warranty may be created or extended by sales representatives or written sales materials. The advice and strategies contained herein may not be suitable for your situation. You should consult with a professional where appropriate. Neither the publisher nor author shall be liable for any loss of profit or any other commercial damages, including but not limited to special, incidental, consequential, or other damages.

For general information on our other products and services please contact our Customer Care Department within the U.S. at 877-762-2974, outside the U.S. at 317-572-3993 or fax 317-572-4002.

Wiley also publishes its books in a variety of electronic formats. Some content that appears in print, however, may not be available in electronic format.

Library of Congress Cataloging in Publication Data is available.

Palmer, David C.
  Oxazoles: Synthesis, Reactions, and Spectroscopy, Part A

ISBN 0-471-39494-7

Printed in the United States of America

10 9 8 7 6 5 4 3 2 1

To my wife, Vicki, with love

The Chemistry of Heterocyclic Compounds
Introduction to the Series

The chemistry of heterocyclic compounds is one of the most complex and intriguing branches of organic chemistry, of equal interest for its theoretical implications, for the diversity of its synthetic procedures, and for the physiological and industrial significance of heterocycles.

The Chemistry of Heterocyclic Compounds has been published since 1950 under the initial editorship of Arnold Weissberger, and later, until his death in 1984, under the joint editorship of Arnold Weissberger and Edward C. Taylor. In 1997, Peter Wipf joined Prof. Taylor as editor. This series attempts to make the extraordinarily complex and diverse field of heterocyclic chemistry as organized and readily accessible as possible. Each volume has traditionally dealt with syntheses, reactions, properties, structure, physical chemistry, and utility of compounds belonging to a specific ring system or class (e.g., pyridines, thiophenes, pyrimidines, three-membered ring systems). This series has become the basic reference collection for information on heterocyclic compounds.

Many broader aspects of heterocyclic chemistry are recognized as disciplines of general significance that impinge on almost all aspects of modern organic chemistry, medicinal chemistry, and biochemistry, and for this reason we initiated several years ago a parallel series entitled General Heterocyclic Chemistry, which treated such topics as nuclear magnetic resonance, mass spectra, and photochemistry of heterocyclic compounds, the utility of heterocycles in organic synthesis, and the synthesis of heterocycles by means of 1,3-dipolar cycloaddition reactions. These volumes were intended to be of interest to all organic, medicinal, and biochemically oriented chemists, as well as to those whose particular concern is heterocyclic chemistry. It has, however, become increasingly clear that the above distinction between the two series was unnecessary and somewhat confusing, and we have therefore elected to discontinue *General Heterocyclic Chemistry* and to publish all forthcoming volumes in this general area in *The Chemistry of Heterocyclic Compounds* series.

The chemistry and synthetic applications of oxazoles were first covered in 1986 in an comprehensive volume edited by I. J. Turchi (Volume 45 of *The Chemistry of Heterocyclic Compounds* series). In the meantime, the number of synthetic strategies directed toward oxazole assembly as well as the use of these versatile heterocycles as intermediates, catalytic ligands, and pharmaceutical building blocks has vastly increased. We felt that a supplement and update of oxazole chemistry would be welcomed by the international chemistry community, and we are delighted that Dr. Palmer and his colleagues have accomplished this onerous mission. This volume represents another outstanding service to the organic and heterocyclic chemistry literature that we are pleased to publish within *The Chemistry of Heterocyclic Compounds* series.

Department of Chemistry EDWARD C. TAYLOR
Princeton University
Princeton, New Jersey

Department of Chemistry PETER WIPF
University of Pittsburgh
Pittsburgh, Pennsylvania

Foreword

The subject of heterocyclic chemistry, prior to 1950, had been viewed as the domain of a small group of organic chemists. The perception prevailed that these individuals simply added ingredients together to make a witch's brew, heated it to 150°–250° C, and ultimately isolated a heterocyclic compound. This may be a somewhat exaggerated description of the subject but nevertheless makes the point that up to that time, it was assumed that one needed special training and knowledge to engage in this subject. However, in spite of this, a large number of molecularly distinct heterocyclic compounds were prepared and subsequently found to have highly important uses in medicine, polymers, dyes, and a number of other areas. As biology and biochemistry matured into a true science during the past 50 years, more and more biological phenomena were found to involve heterocyclic systems. This led to an increased appreciation of heterocycles, their chemical properties, and the reactions they undergo. As a result, these ring systems were subsequently regarded as more than a narrow field of chemistry. There is now little need to convince the informed scientific community of the incredible value of heterocyclic compounds.

As organic chemistry entered a new level of sophistication in the 1950s and understanding of chemical reactions was actively pursued, heterocyclic compounds were also included in this exploration and found to play a major role in many important chemical reactions, both as intermediates and as final products.

As this writer predicted in 1974 in a monograph entitled "Heterocycles in Synthesis," these ring systems will not only be crucial to the scientific areas already mentioned above but will also find great importance in the synthesis of all types of organic compounds. In fact, in the current climate, heterocycles and their properties are so well accepted that they pervade all areas of medicine and biology, as well as chemistry.

The present updated volumes relating to oxazoles, oxazolines, and oxazolones are very timely works since these simple five-membered ring heterocycles have contributed much to the knowledge we have acquired in various fields of biological and chemical sciences. For example, we may envision oxazolones as tautomeric derivatives of the old and well-known azlactones, first reported in 1883. They may also be viewed as "cyclized" amino acid derivatives or their dehydro analogs and therefore would be expected as constituents of many biologically active natural products. In addition, these ring systems have shown their versatility in the synthesis of a variety of ligands for metal catalysts, as well as precursors or vehicles to reach many types of functionalized compounds. Furthermore, their chiral counterparts—oxazoles, oxazolines, and oxazolones containing a stereogenic center—have been major players in a very large number of asymmetric syntheses. Hardly a day passes that some journal does not describe the involvement of these chiral, non-racemic heterocycles for preparing an organic compound in very high enantiomeric excess. Thus, oxazoles and their derivatives, whose synthesis, properties, and reactivity are described in the following two volumes, represent an immensely versatile family of heterocyclic compounds for future exploitation by both synthetic and medicinal chemists.

Fort Collins, Colorado A. I. MEYERS
November, 2002

Preface

By far the most comprehensive review of the synthesis and reactions of mononuclear oxazoles and derivatives is *The Chemistry of Heterocyclic Compounds, Volume 45*, edited by I. J. Turchi and published in 1986. This work is the definitive reference for oxazole chemistry through 1983. Subsequently, literally tens of thousands of references appeared in the period 1983–2001 pertaining to this remarkable small ring heterocycle. Oxazoles and derivatives continue to be of great interest and importance in all aspects of synthetic chemistry with applications in medicinal and agricultural chemistry, material sciences, photographic dyes, peptide chemistry, asymmetric catalysis, and polymer chemistry. Indeed, more than 250 reviews focusing on specific aspects of the chemistry and biology of oxazoles, oxazolones, oxazolines, and chiral bis-oxazolines have been published from 1983 to 2001. The continuing interest in oxazoles together with the wealth of new information warrants a second review of this exciting area.

It would require a Herculean effort to prepare a complete discussion and review of every report related to the synthesis, reaction, or application of an oxazole while tabulating every oxazole, oxazolone, oxazoline, and chiral bis-oxazoline prepared and evaluated during the period of 1983–2001. Such an undertaking is beyond the scope of this review. Furthermore, the ease with which electronic databases, including the patent literature, can be searched, the data retrieved, and the information tabulated would render such a project somewhat redundant.

Rather, the intent of the current project is to provide the reader with a discussion and leading examples of significant advances made in the synthesis, reactions, and applications of mononuclear oxazoles, oxazolones, oxazolines, and chiral bis-oxazolines during this time frame. The material focuses on the more recent literature, although an update of the older synthetic literature is included wherever possible. In an effort to be selective, references to relevant reviews of material, not discussed in a chapter, are provided. Completely reduced oxazoles, i.e. oxazolidines as well as benzo-fused derivatives, are outside the scope of this review.

The coverage is similar to that of Volume 45, although the presentation has been changed and the scope has been expanded to include a chapter devoted to the exciting area of chiral bis-oxazolines. The material is presented in nine chapters and two volumes. In some cases, the organization of the individual chapter contents is different from that in Volume 45 to reflect the changing emphasis on newer methodologies and synthetic targets. For example, in Part A, Chapter One contains an expanded section that deals specifically with the synthesis of selected naturally occurring mono-, bis-, and tris-oxazoles to reflect the significant synthetic challenges therein. In addition, the discussion of cycloaddition and Diels-Alder reactions of oxazoles is introduced in Chapter One but is covered in detail in Chapter Three. In Part B, oxazolones are defined by the structure of the individual regioisomer and discussed in Chapters Five, Six, and Seven, respectively. Chapter Eight describes the syntheses and reactions of oxazolines including asymmetric methodology employing monooxazoline ligands. A new chapter, Chapter Nine, was added to include the recent developments in asymmetric synthesis utilizing chiral bis-oxazolines. Discussion of material from the patent literature has been included as an integral part of the volumes. Primary emphasis has been given to general syntheses and reactions. However, reactions that are more limited in scope and yet are singularly unique may also be described.

Tables are included in every chapter. Wherever possible, these contain a variety of selected examples to provide the reader with the scope and limitations of synthetic methods and reactions. However, in some cases a table will contain only the examples reported. No attempt has been made to provide an exhaustive compilation of every oxazole, oxazolone, or oxazoline prepared since 1983.

Part A is devoted specifically to the synthesis, reactions, and spectroscopic properties of oxazoles and encompasses four chapters: Chapter 1—Synthesis and Reactions of Oxazoles; Chapter 2—Spectroscopic Properties of Oxazoles; Chapter 3—Diels-Alder and Cycloaddition Reactions of Oxazoles; and Chapter 4—Synthesis and Reactions of Mesoionic Oxazoles.

Part B is comprised of the following five chapters: Chapter 5—Synthesis and Reactions of 2(3H)-Oxazolones and 2(5H)-Oxazolones; Chapter 6—Synthesis and Reactions of 4(5H)-Oxazolones; Chapter 7—Synthesis and Reactions of 5(2H)-Oxazolones and 5(4H)-Oxazolones; Chapter 8—Synthesis and Reactions of Oxazolines; and Chapter 9—Synthesis and Reactions of Chiral bis-Oxazolines.

Acknowledgments: I thank the authors for their individual contributions and patience through several iterations of the chapters. I am indebted to the library staff at Johnson & Johnson Pharmaceutical Research & Development who secured even the most obscure references in a timely manner. A very special acknowledgment and thanks are due to Dr. Fuqiang Liu for his critical insights, suggestions, comments, and review of individual chapters during preparation of these volumes. The series editors, particularly Professor Ted Taylor, offered many helpful suggestions and guidance. I thank Dr. Darla Henderson and Ms. Amy Romano at John Wiley & Sons for their constant encouragement and support. Special thanks are due to Ms. Shirley Thomas and her staff at John Wiley & Sons for their patience and understanding during preparation of these volumes. Finally, I am deeply thankful to my wife, Vicki, for her continual support, patience, and understanding during this entire project.

The reader may well encounter errors in a work of this magnitude, particularly in one with several thousand structures. I hope such errors will not detract from the overall intent of the volumes. Nonetheless, any errors are the responsibility of the editor.

Johnson & Johnson Pharmaceutical Research & Development, L.L.C.   DAVID C. PALMER
Raritan, New Jersey

Contents - Part A

Abbreviations		xv
1.	**Synthesis and Reactions of Oxazoles** D. C. PALMER and S. VENKATRAMAN	1
2.	**Spectroscopic Properties of Oxazoles** D. LOWE	391
3.	**Oxazole Diels-Alder Reactions** J. I. LEVIN and L. M. LAAKSO	417
4.	**Mesoionic Oxazoles** G. W. GRIBBLE	473
Author Index		577
Subject Index		599

Abbreviations

2-MI	2-methylimidazole
9-BBN	9-borabicyclo[3.3.1]nonane
Ac	acetyl
Acm	acetamidomethyl
AHMHA	4-amino-3-hydroxy-6-methylheptanoic acid
AHPBA	3-amino-2-hydroxy-4-phenylbutyric acid
AIBN	2,2′-azobisisobutyronitrile
Alloc or AOC	allyloxycarbonyl
AMNT	aminomalononitrile *p*-toluenesulfonate
BARF	tetrakis[3,5-bis(trifluoromethyl)phenyl]borate
BDMS	biphenyldimethylsilyl
BINAP	2,2′-bis(diphenylphosphino)-1,1′-binaphthyl
BINOL	[1,1′]binaphthalenyl-2,2′-diol
BINOL-box	3,3′-bis(2-oxazolyl)-1,1′-bi-2-naphthol
Bn	benzyl
Boc	*tert*-butyloxycarbonyl
Boc-Ox	2-oxo-3(2*H*)-oxazolecarboxylic acid *tert*-butyl ester
BOP	benzotriazol-1-yloxytris(dimethylamino)phosphonium hexafluorophosphate
BOP-Cl	*N,N*-bis-(2-oxo-3-oxazolidinyl)phosphonic chloride
BPA	L-4-boronophenylalanine
BPO	dibenzoyl peroxide
Bt	benzotriazol-1-yl or 1-benzotriazolyl
Bz	benzoyl
Cbz	benzyloxycarbonyl
Cbz-Ox	2-oxo-3(2*H*)-oxazolecarboxylic acid benzyl ester
CDI	1,1′-carbonyldiimidazole
CIP	2-chloro-1,3-dimethylimidazolium hexafluorophosphate
cod	cyclooctadiene
Cp	cyclopentadiene
CPTS	collidine *p*-toluenesulfonate
CSA	camphorsulfonic acid
CSI	chlorosulfonylisocyanate
DAST	diaminosulfur trifluoride
dba	dibenzylideneacetone
DBF-box	2,2′-(4,6-dibenzofurandiyl)bis[4,5-dihydro-4-phenyloxazole
DBN	1,5-diazabicyclo[4.3.0]non-5-ene
DBU	1,8-diazabicyclo[5.4.0]undec-7-ene
DCC	*N,N*′-dicyclohexylcarbodiimide

DCE	1,2-dichloroethane
DDQ	2,3-dichloro-4,5-dicyano-1,4-benzoquinone
DEAD	diethyl azodicarboxylate
DECP or DEPC	diethylcyanophosphonate, diethylphosphoryl cyanide
Deoxo-fluor	bis(2-methoxyethyl)aminosulfur trifluoride
DIAD	diisopropyl azodicarboxylate
DIBALH	diisobutylaluminum hydride
DIPEA	diisopropylethylamine
DMAC	dimethylacetamide
DMAD	dimethyl acetylenedicarboxylate
DMAP	4-(dimethylamino)pyridine
DME	1,2-dimethoxyethane
DMF	dimethylformamide
DMI	1,3-dimethyl-2-imidazolidinone
DMPU	1,3-dimethyl-3,4,5,6-tetrahydro-2(1H)-pyrimidinone
DMTO	dimethoxytrityl
DOPA	3,4-dihydroxyphenylalanine
DPPA	diphenylphosphoryl azide
dppb	1,4-bis(diphenylphosphino)butane
DPPC	diphenylphosphoryl chloride, diphenyl phosphochloridate
dppe	1,4-bis(diphenylphosphino)ethane
dppf	1,4-bis(diphenylphosphino)ferrocene
DPPOx	diphenyl-(2-oxo-3(2H)-oxazolyl)phosphonate
dppp	1,4-bis(diphenylphosphino)propane
EDCI or EDAC	1-(3-dimethylaminopropyl)-3-ethylcarbodiimide hydrochloride
ETHP	2-ethyl-1,4,5,6-tetrahydropyrimidine
ETMG	2-ethyl-1,1,3,3-tetramethylguanidine
EVL	ethoxyvinyllithium
FMO	frontier molecular orbital
Fmoc	9-fluorenylmethoxycarbonyl
HATU	O-(7-Azobenzotriazol-1-yl)-N,N,N',N'-tetramethyluronium hexafluorophosphate
HBTU	O-benzotriazol-1-yl-N,N,N',N'-tetramethyluronium hexafluorophosphate
HMPA	hexamethylphosphoric triamide
HMTA	hexamethylenetetraamine
HOAt	1-hydroxy-7-azabenzotriazole
HOBt	1-hydroxybenzotriazole
HOMO	highest occupied molecular orbital
HydrOx	hydroxy-oxazoline
IBCF	isobutyl chloroformate
Im	Imidazole
KHMDS	potassium hexamethyldisilazane, potassium bis(trimethylsilyl)amide

L-(+)-DET	L-(+)-diethyl L-tartrate
LAH	lithium aluminum hydride
LDA	lithium diisopropylamide
LDEA	lithium diethylamide
LiHMDS	lithium hexamethyldisilazane, lithium bis(trimethylsilyl)amide (LHMDS)
LTMP	lithium 2,2,6,6-tetramethylpiperidide
LUMO	lowest unoccupied molecular orbital
m-CPBA	*m*-chloroperoxybenzoic acid (MCPBA)
MeBmt	(2*S*,3*R*,4*R*,6*E*)-3-hydroxy-4-methyl-2-(methylamino)-6-octenoic acid
MEK	methyl ethyl ketone
MIBK	methyl isobutyl ketone
MOM	methoxymethyl
morphoCDI	*N*-cyclohexyl-*N'*-2-(*N*-methylmorpholinio)ethylcarbodiimide *p*-toluenesulfonate
Ms	methanesulfonyl (mesyl)
NaHMDS	sodium hexamethyldisilazane, sodium bis(trimethylsilyl)amide
NBS	*N*-bromosuccinimide
NCS	*N*-chlorosuccinimide
NIS	*N*-iodosuccinimide
NMM	4-methylmorpholine (*N*-methylmorpholine)
NMO	4-methylmorpholine *N*-oxide
NMP	*N*-methyl-2-pyrrolidinone
NOE	nuclear Overhauser effect
Nos	*p*-nitrobenzenesulfonyl (nosyl)
NPM	*N*-phenylmaleimide
PB	Probase
PCC	pyridinium chlorochromate
PDC	pyridinium dichromate
PEG	poly(ethylene)glycol
PET	positron emission tomography
PhosOx	phosphine-oxazoline
Phth	Phthaloyl
Piv	Pivaloyl
PMB	*p*-methoxybenzyl
PPA	poly(phosphoric acid)
PPE	polyphosphate ester
PPL	porcine pancreatic lipase
PPTS	pyridinium *p*-toluenesulfonate
PyBOP	benzotriazol-1-yl-*N*-oxytris(pyrrolidino)phosphonium hexafluorophosphate
PyBroP	bromotris(pyrrolidino)phosphonium hexafluorophosphate
PyrOx	pyridine-oxazoline
RaNi	raney nickel

SelOx	selenide-oxazoline
SEM	2-(trimethylsilyl)ethoxymethyl
SES	2-(trimethylsilyl)ethanesulfonyl
SulfOx	sulfide-oxazoline
TADDOL	α,α,α′,α′-tetraaryl-1,3-dioxolane-4,5-dimethanol
TBAF	tetra *n*-butylammonium fluoride
TBDMS or TBS	*tert*-butyldimethylsilyl
TBDPS	*tert*-butyldiphenylsilyl
TBTU	*O*-benzotriazol-1-yl-*N,N,N′,N′*-tetramethyluronium tetrafluoroborate
TCNE	tetracyanoethylene
TEAHC	tetraethylammonium hydrogen carbonate
TEAP	tetraethylammonium perchlorate
TECM	tandem Erlenmeyer condensation macrolactamization
TEMPO	2,2,6,6-tetramethyl-1-piperidinyl *N*-oxide
TEOF	triethyl orthoformate
TES	triethylsilyl
Tf	trifluoromethanesulfonyl (triflyl)
TFA	trifluoroacetic acid
TFAA	trifluoroacetic anhydride
TFE	2,2,2-trifluoroethanol
THF	tetrahydrofuran
THP	tetrahydropyran-2-yl
TIA	*N,N,N′*-triisopropylacetamidine
TIG	1,2,3-triisopropylguanidine
TIPS	triisopropylsilyl
TMANO	trimethylamine *N*-oxide
TMEDA	*N,N,N′,N′*-tetramethyl-1,2-ethylenediamine
TMG	1,1,3,3-tetramethylguanidine
TMP	2,2,6,6-tetramethylpiperidine
TMS	trimethylsilyl
Tol-BINAP	2,2′-bis(di-*p*-tolylphosphino)-1,1′-binaphthyl
TosMIC	tosylmethyl isocyanide, [(*p*-toluenesulfonyl)methyl] isocyanide
TPAP	tetrapropylammonium perruthenate
Tr	trityl
Troc	2,2,2-trichloroethoxycarbonyl
Ts or Tos	*p*-toluenesulfonyl (tosyl)

OXAZOLES:
SYNTHESIS, REACTIONS, AND SPECTROSCOPY, PART A

This is the sixtieth volume in the series
THE CHEMISTRY OF HETEROCYCLIC COMPOUNDS

CHAPTER 1

Synthesis and Reactions of Oxazoles

David C. Palmer

*Johnson & Johnson Pharmaceutical Research & Development, L.L.C.
Raritan, New Jersey*

Srikanth Venkatraman

*Schering-Plough Research Institute
Kenilworth, New Jersey*

1.1. Introduction
1.2. General Properties of Oxazoles
1.3. Synthesis of Oxazoles
    1.3.1. Oxidation of Oxazolines
        1.3.1.1. Nickel Peroxide
        1.3.1.2. Manganese Dioxide
        1.3.1.3. Copper Salts
        1.3.1.4. Kharasch-Sosnovsky Reaction
        1.3.1.5. Miscellaneous Oxidations
    1.3.2. Rearrangement Reactions
        1.3.2.1. *N*-Acylaziridines
        1.3.2.2. *N*-Acyltriazoles
        1.3.2.3. Isoxazoles and *N*-Acylisoxazolones
    1.3.3. Organometallic Reactions
        1.3.3.1. Rhodium Carbene Additions
        1.3.3.2. Organotellurium Reagents
        1.3.3.3. Organomercury Reagents
    1.3.4. Oxazoles from α-Substituted Ketones
    1.3.5. Oxazoles via Cyclizations
        1.3.5.1. Propargylic Amides
        1.3.5.2. Azides
    1.3.6. Oxazoles from Vinylogous Amides/Enamino Esters
    1.3.7. Oxazoles from Oximes and Hydrazones
    1.3.8. Oxazoles from Imidates/Thioimidates—Cornforth Reaction
    1.3.9. Oxazoles from Isocyanides
    1.3.10. Oxazoles from Vinyl Bromides
    1.3.11. Oxazoles from α-Acyloxyketones
    1.3.12. Robinson-Gabriel and Related Reactions
    1.3.13. Miscellaneous Reactions

The Chemistry of Heterocyclic Compounds, Volume 60: Oxazoles: Synthesis, Reactions, and Spectroscopy, Part A, edited by David C. Palmer
ISBN 0-471-39494-7   Copyright © 2003 John Wiley & Sons, Inc.

1.4. Reactions of Oxazoles
    1.4.1. Electrophilic Reactions
    1.4.2. Nucleophilic Reactions and Hydrolysis
    1.4.3. Reduction
    1.4.4. Oxidation
    1.4.5. Cycloaddition Reactions and Sigmatropic Rearrangements
    1.4.6. Conversion of Oxazoles to Other Heterocycles
    1.4.7. Organometallic Reactions
    1.4.8. Transition Metal Catalyzed Cross-Coupling Reactions
    1.4.9. Trimethylsilyloxazoles
    1.4.10. Oxazolium Salts
    1.4.11. Oxazole Wittig Reagents
    1.4.12. Miscellaneous Reactions
1.5. Oxazole Natural Products
    1.5.1. Group A Stretogramin Antibiotics
    1.5.2. Thiangazole and Tantazole
    1.5.3. Calyculins
    1.5.4. Hennoxazoles
    1.5.5. Diazonamides
    1.5.6. Ulapulides
1.6. Addendum
1.7. Summary
References

1.1. INTRODUCTION

Oxazoles continue to hold a center stage in organic synthesis, although the structure of the first oxazole was reported over a century ago. This is evidenced by the continued growth in the number of research publications and reviews.[1–18] The field of oxazoles is extensive and includes natural products, medicinal chemistry, and materials science. This chapter describes the major developments in this field from 1983 to 2001.

The chapter is divided into the following sections: "Introduction," "Synthesis of Oxazoles," "Reactions of Oxazoles," and "Oxazole Natural Products." The introduction is a brief discussion of the numbering and lists the major ^{1}H, ^{13}C, and ^{15}N resonances of a few selected examples. The reader should consult Chapter 2 for a complete discussion of the spectroscopic properties of oxazoles.

The second section describes the common methods of synthesis. It is important to note that no attempt has been made to describe every monocyclic oxazole synthesized or all synthetic methods. Rather, the most common and useful synthetic methods and some particularly novel methods have been included together with tables of some representative examples. The reader should consult the primary literature for further examples. In addition, this section and the reactions of oxazoles are not necessarily organized in the same manner as in the first volume edited by Turchi. This reflects the changing emphasis on new methods applicable to complex natural product synthesis.

The third section presents important reactions of oxazoles but again no attempt has been made to describe every reaction of an oxazole or every derivative prepared. Here, the aim is to convey the wealth of chemistry available by the selected examples.

The last section includes syntheses of some naturally occurring oxazoles. The choice of these natural products is arbitrary and selected from the recent literature to demonstrate the versatility of oxazoles in synthesis. For more complete discussions of oxazole natural products the reader is referred to specialized reviews.[19–22]

1.2. GENERAL PROPERTIES OF OXAZOLES

Oxazoles are numbered around the ring starting at the oxygen atom (Fig. 1.1) and are designated as 1,3-oxazoles to indicate the position of heteroatoms in the ring.

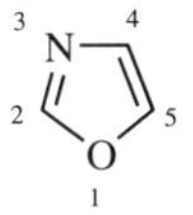

Figure 1.1. Oxazole.

The proton acidities of oxazoles have been theoretically calculated and determined experimentally. The reactivity of oxazoles indicates that the acidity of a hydrogen atom decreases in the order $C(2) > C(5) > C(4)$. This is true in most cases; however, exceptions are well documented and depend on the substitution. The acidity of the hydrogen at C(2) was estimated to be $pK_a \sim 20$ while for oxazole itself the pK_b is reported to be $pK_b \sim 1$.[17,18]

Oxazoles exhibit characteristic resonances in both ^{1}H NMR and ^{13}C NMR spectra. The parent compound displays resonances (δ) between 7.00 and 8.00 in the ^{1}H NMR spectrum, and the presence of substituents can change the chemical shift by up to 1 ppm. The ^{13}C NMR of oxazole displays typical aromatic resonances (Fig. 1.2). The shielding or deshielding effect of C(2) substitution on the C(4) and C(5) resonances is typically < 2 ppm. The ^{1}H, ^{13}C, and ^{15}N chemical shifts of a few selected examples are shown in Figure 1.2.[23–26]

The IR spectrum of oxazole displays absorbances at 1537, 1498, 1326 (ring stretch), 1257 (C-H in plane deformation), 1143, 1080 (ring breathing), and 1045 cm^{-1}.[23–26] In the UV, the λ_{max} of oxazoles depend highly on the substitution pattern. In methanol, the parent ring system has an absorption maximum at $\lambda_{max} = 205$ nm.[17,18]

¹H NMR resonances of selected oxazoles (δ, ppm)

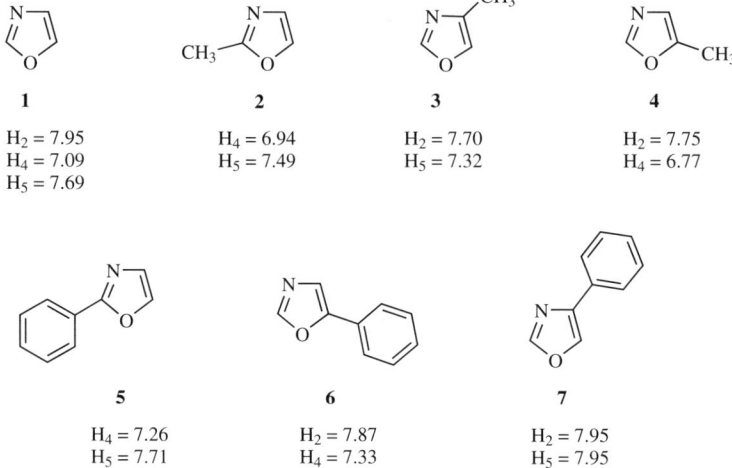

1
$H_2 = 7.95$
$H_4 = 7.09$
$H_5 = 7.69$

2
$H_4 = 6.94$
$H_5 = 7.49$

3
$H_2 = 7.70$
$H_5 = 7.32$

4
$H_2 = 7.75$
$H_4 = 6.77$

5
$H_4 = 7.26$
$H_5 = 7.71$

6
$H_2 = 7.87$
$H_4 = 7.33$

7
$H_2 = 7.95$
$H_5 = 7.95$

¹³C NMR resonances of selected oxazoles (δ, ppm)

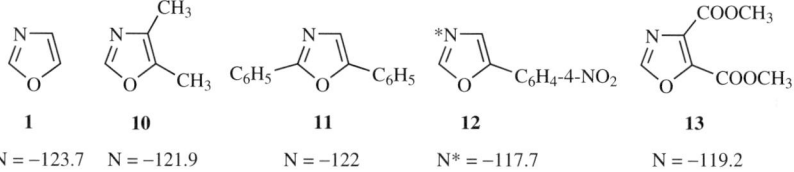

1
$C_2 = 150.6$
$C_4 = 125.4$
$C_5 = 138.1$

8
$C_2 = 151.5$
$C_4 = 126.0$
$C_5 = 141.1$

9
$C_2 = 161.1$
$C_4 = 128.1$
$C_5 = 139.8$

¹⁵N NMR resonances of selected oxazoles (δ, ppm, d₆–DMSO)

1
$N = -123.7$

10
$N = -121.9$

11
$N = -122$

12
$N^* = -117.7$

13
$N = -119.2$

Figure 1.2. ¹H, ¹³C, and ¹⁵N NMR resonances of selected oxazoles.

1.3. SYNTHESIS OF OXAZOLES

1.3.1. Oxidation of Oxazolines

The structural diversity and complexity of naturally occurring oxazoles (see Section 1.5) has fueled a continuing search for mild, efficient methods of oxazole ring construction. One method that has become particularly useful in this context is the oxidation of oxazolines.

The dehydration of β-hydroxyamides is the most widely used method for the synthesis of oxazolines, and the various reagents used for this conversion will not be discussed further in this section. Instead, the reader should consult Chapter 8 and Chapter 9 in Part B for the synthesis and reactions of oxazolines and chiral bis-oxazolines for detailed discussions and leading examples.

A variety of reagents have been evaluated in the search for the direct oxidation of oxazolines to oxazoles. Even though a number of reagents efficiently oxidize activated oxazolines (i.e., oxazolines containing an electron-withdrawing group at the 4 or 5 position), a general, high-yielding method to effect oxidation of unactivated oxazolines has not been described.

1.3.1.1. Nickel Peroxide

Although new methods have been developed recently for the oxidation of oxazolines, oxidation with NiO_2 is one of the oldest and still widely used methods. It was originally reported by Meyers and Evans (Scheme 1.1).[27]

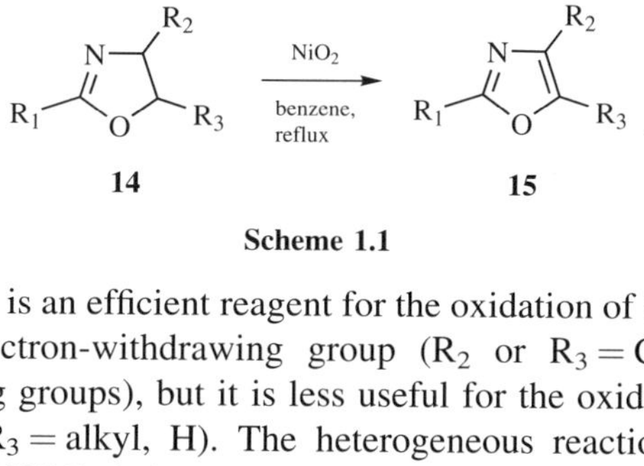

Scheme 1.1

Nickel peroxide[28] is an efficient reagent for the oxidation of activated oxazolines that contain an electron-withdrawing group (R_2 or $R_3 = COOC_2H_5$ or other electron-withdrawing groups), but it is less useful for the oxidation of unactivated oxazolines (R_2 or R_3 = alkyl, H). The heterogeneous reaction is conducted by refluxing an excess of NiO_2 in benzene, and a radical reaction mechanism has been proposed. The application of NiO_2 in the synthesis of eupolauramine[29] is a noteworthy example in which one of the key steps involves the oxidation of the unactivated oxazoline **16** to the 2-aryloxazole **17** in 55% yield (Scheme 1.2).

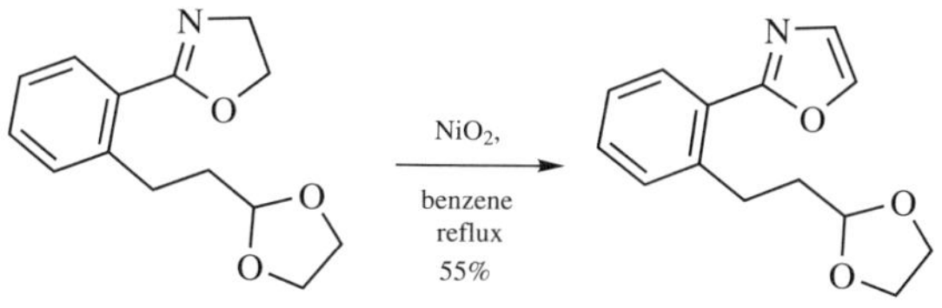

Scheme 1.2

TABLE 1.1. 2,4-DISUBSTITUTED- AND 2,4,5-TRISUBSTITUTED OXAZOLES VIA OXIDATION OF OXAZOLINES

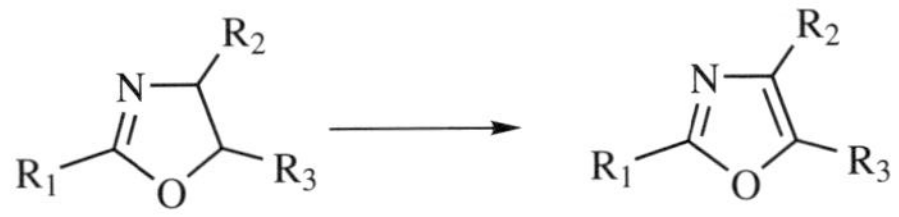

Figure 1.3

Entry	R_1	R_2	R_3	Product	Yield[a] (%)	References
1	C_6H_5	CH_2CN	4-CH_3S-C_6H_4	(2-C_6H_5, 4-CH_2CN, 5-C_6H_4-4-SCH_3 oxazole)	54 (A)	34
2	C_6H_5	$CH_2OCOC_6H_5$	4-CH_3S-C_6H_4	(2-C_6H_5, 4-$CHOCOC_6H_5$, 5-C_6H_4-4-SCH_3 oxazole)	40 (A)	34
3	C_6H_5	$COOCH_3$	H	(2-C_6H_5, 4-$COOCH_3$ oxazole)	55 (B)	32, 33
4	$C_6H_5CH=CH$	$COOCH_3$	H	(2-$C_6H_5CH=CH$, 4-$COOCH_3$ oxazole)	61 (B)	32, 33

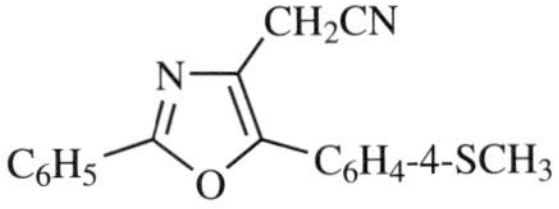

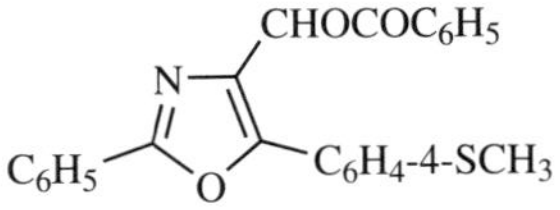

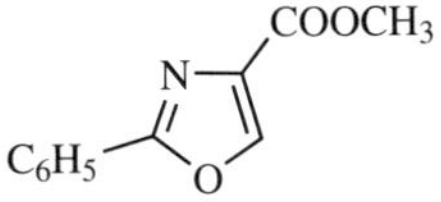

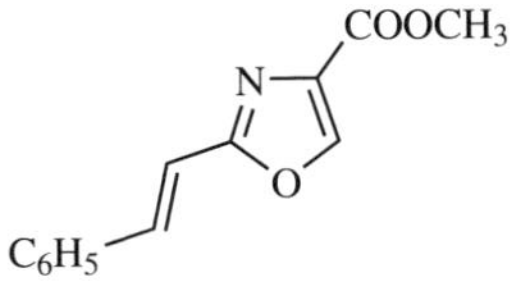

#						
5	H	COOCH$_3$	![structure with NHBoc, CH$_3$]	![oxazole with CH$_3$, BocHN, COOCH$_3$]	51 (C)	35, 36
6	H	COOC$_2$H$_5$	![structure with N$_3$, CH$_3$]	![oxazole with CH$_3$, N$_3$, COOC$_2$H$_5$]	56 (C)	35, 36
7	H	COOCH$_3$	i-C$_3$H$_7$	![oxazole with i-C$_3$H$_7$, COOCH$_3$]	1/76^b (D)	35, 36
8	H	COOCH$_3$	c-C$_6$H$_{11}$	![oxazole with c-C$_6$H$_{11}$, COOCH$_3$]	1/66^b (D)	35, 36

^{a} A, CuBr$_2$/LiBr/CaCO$_3$; B, nickel peroxide; C, CuBr/Cu(OAc)$_2$/C$_6$H$_5$C(O)OOtBu; D, N-Bromosuccinimide.
^{b} Ratio refers to the ratio of **65:66** (see Scheme 1.20).

Pattenden and co-workers[30,31] successfully used nickel peroxide oxidation in the synthesis of the naturally occurring oxazole thiangazole (described in Section 1.5.2). Shioiri's group [32,33] developed a versatile method for the cyclodehydration of β-hydroxyamides that, in combination with NiO_2 oxidation, results in a general synthesis of 2-alkyl(aryl)-4-oxazolecarboxylic acid esters **20** (Scheme 1.3). For example, reaction of **18** (R = OCH_3) with diphenylsulfoxonium triflate, generated in situ, yields oxazolines **19** (R = OCH_3), which are converted to 2-aryl-4-oxazolecarboxylic acid esters **20** (R = OCH_3) using NiO_2. Representative examples of this methodology are shown in Table 1.1, entries 3 and 4.

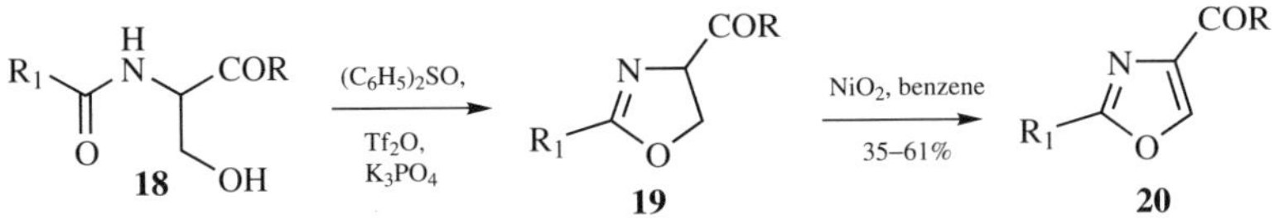

Scheme 1.3

1.3.1.2. Manganese Dioxide

Activated manganese dioxide is a closely related oxidant to NiO_2 but less commonly used. Meguro, Fujita, and co-workers[37] described oxidation of oxazolines using MnO_2 in their synthesis of antidiabetic agents. For example, the serine-derived β-hydroxyamide **21** was cyclized to the oxazoline **22** with polyphosphoric acid. Oxidation of **22** with activated manganese dioxide afforded 2-styryl-4-oxazolecarboxylic acid methyl ester, **23** (Scheme 1.4).

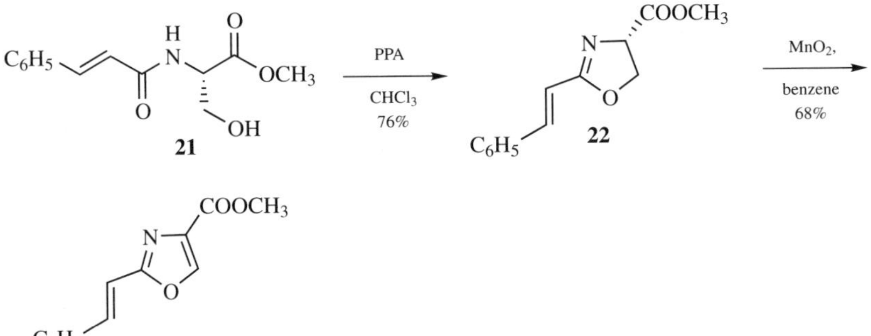

Scheme 1.4

Oxidation of oxazolines using MnO_2 was also employed in the synthesis of a berninamycin A fragment by Shin and co-workers[38–40] and in the preparation of the naturally occurring 2,4-disubstituted oxazole, phenoxan **28**, by Yamamura's group[41] (Scheme 1.5). For berninamycin, the oxazoline **25** was synthesized in 61% yield by Mitsunobu[42] dehydration of the serine-derived amide **24**.

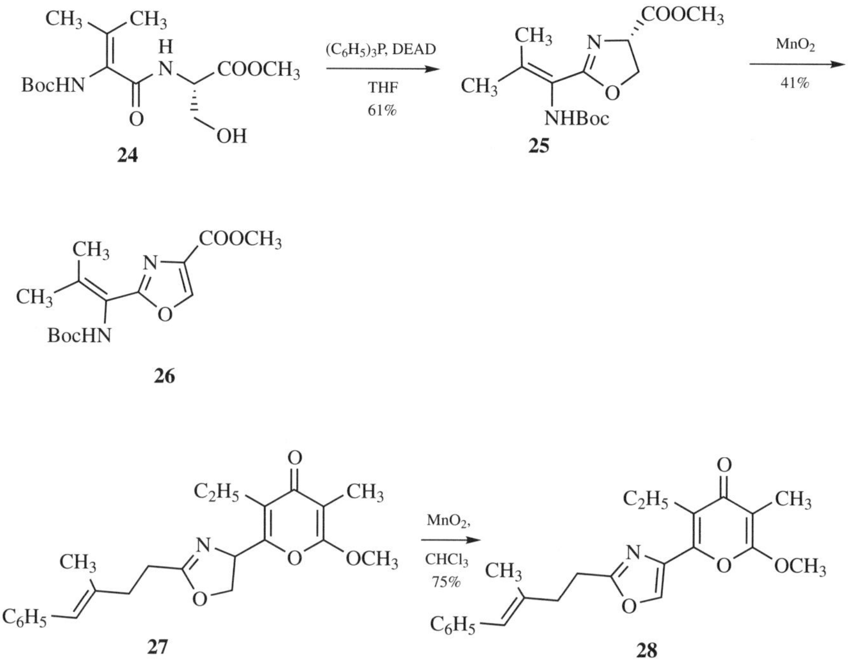

Scheme 1.5

Oxidation of **25** to the 2-substituted 4-oxazolecarboxylic acid methyl ester **26** was accomplished in 41% yield. Similarly, oxidation of **27** yielded phenoxan **28**. It should be noted that yields of MnO_2 oxidations are variable but can be comparable to those obtained using NiO_2. It is likely that both MnO_2 and NiO_2 oxidations are mechanistically similar.

1.3.1.3. Copper Salts

Oxidations employing copper salts are among the newer methods introduced and have proven to be quite useful. For example, $CuBr_2$/DBU/HMTA was used to convert oxazolines to 2-alkyl(aryl)-4-oxazolecarboxylic acid esters or amides (Scheme 1.6).[43,45–47] Barrish and co-workers[43] proposed that the mechanism of this reaction probably involves the formation of a copper enolate **29**, which undergoes an oxidation–reduction with elimination of Cu(I)Br to yield the 2-substituted-4-oxazolecarboxylic acid ester or amide **20**. They believe this redox reaction involves an internal electron transfer process. An alternative mechanism involving elimination of HBr from the 5-bromo analog of **29** could not be ruled out, although there was little evidence to support such a mechanism.

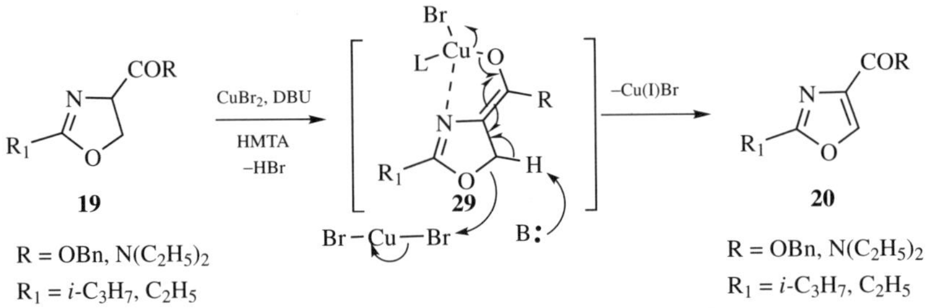

Scheme 1.6

The oxazole-thiazole fragment of the peptide antibiotic microcin B17 was prepared using this methodology (Scheme 1.7).[48] Here, the thiazole hydroxyamide **30** was cyclized to the oxazoline **31** using Burgess reagent,[49] following the procedure of Wipf and Miller.[50] Oxidation of **31**—using CuBr$_2$, DBU, and HMTA—produced the oxazole **32** in good yield.

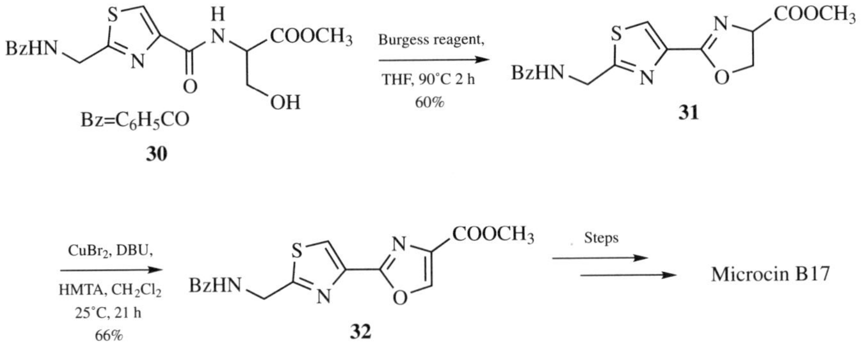

Scheme 1.7

This oxidation protocol was also used to prepare the oxazole fragments of rhizoxin[51] and phorboxazoles.[52] As an example, Leahy and co-workers[51] prepared 2-methyl-4-oxazolecarboxylic acid ethyl ester **34** (R = C$_2$H$_5$), the starting material for the oxazole-diene synthon in their synthetic approach to rhizoxin, in excellent yield (Scheme 1.8). Similarly, Bristol-Myers Squibb workers[45–47] used this methodology to prepare the highly potent, selective, and long-acting thromboxane A$_2$ receptor antagonist **37** in excellent yield (Scheme 1.9).

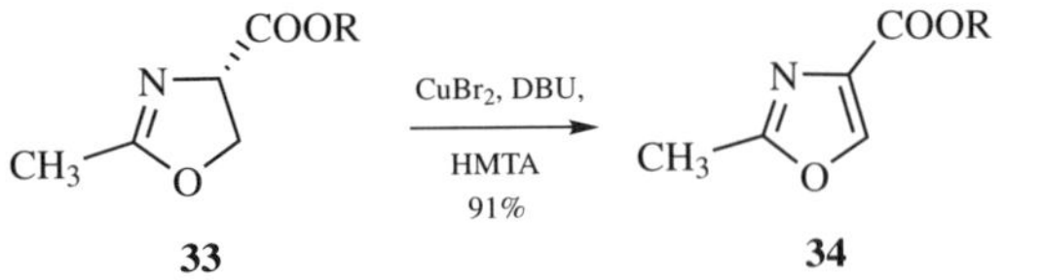

Scheme 1.8

Klein and co-workers[34] developed a similar protocol during their synthesis of new 2,5-diaryl-4-oxazoleacetic acids, evaluated as potential medicinal agents. They

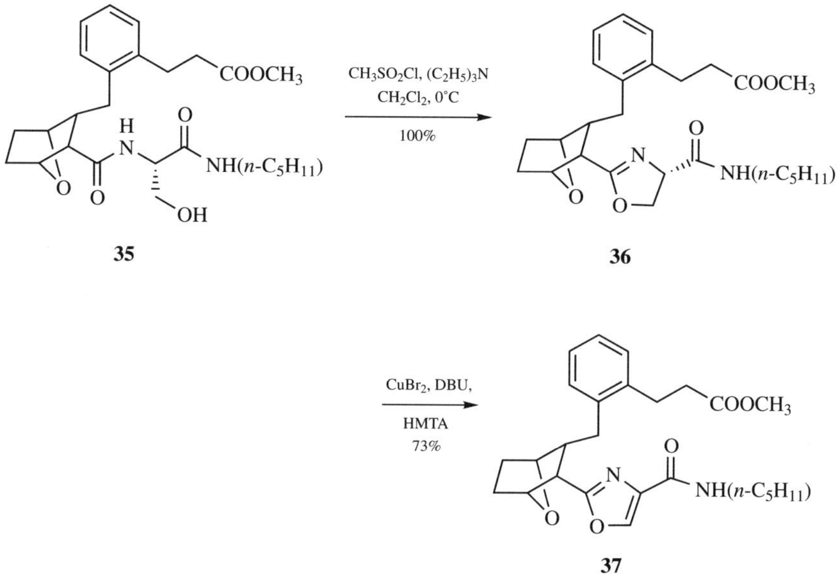

Scheme 1.9

found that oxidation of 2,5-diaryl-4-substituted oxazolines **14** with CuBr$_2$/LiBr/CaCO$_3$ produced 2,4,5-trisubstituted oxazoles **15** in fair to excellent yields (Scheme 1.10). Even thioethers survive the mild reaction conditions (Table 1.1, entries 1 and 2). The authors proposed that a mixture of CuBr$_2$ and LiBr generated bromine in situ, which effected bromination of the oxazoline, since the reaction can also be conducted using Br$_2$ in place of CuBr$_2$.

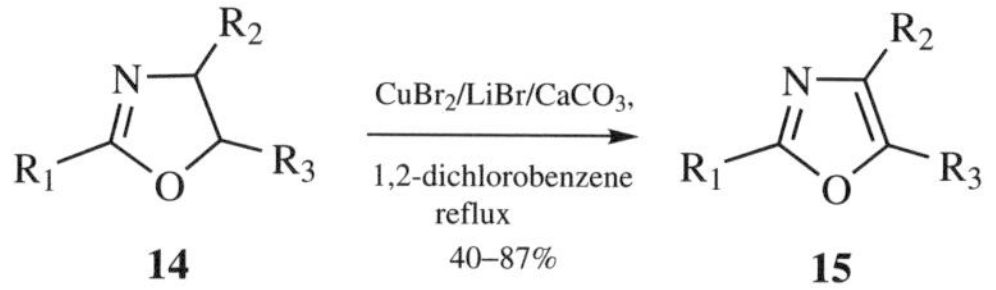

R$_1$, R$_3$ = aryl
R$_2$ = CH$_2$CN or CH$_2$OCOC$_6$H$_5$

Scheme 1.10

The use of the CuBr$_2$/LiBr/CaCO$_3$ protocol has several advantages. No other reagent—including NBS, DDQ, chloranil, or CuBr$_2$/LiBr—produced more than a trace of the desired oxazole. In addition, both NiO$_2$ and activated MnO$_2$ effected oxidation of the sensitive thioether in **14** to the corresponding sulfoxide (Table 1.1, entries 1 and 2). However, a limitation of this methodology occurs with oxazolines containing a potential leaving group, i.e., **14** R$_2$ = CH$_2$X, where X = CH$_3$COO, Cl, Br, I, OH, or OSO$_2$CH$_3$. In these cases, elimination of HX occurs under the reaction conditions to generate the corresponding 4-methyloxazole after a 1,3-hydrogen shift.

1.3.1.4. Kharasch-Sosnovsky Reaction

Meyers and Tavares[35,36] introduced an alternate method to oxidize oxazolines that employs a mixture of Cu(I) and Cu(II) salts together with *tert*-butylperbenzoate (Scheme 1.11, Table 1.1 entries 5 and 6). These reaction conditions are a modification of the original Kharasch-Sosnovsky reaction,[53,54] which employs a mixture of Cu(I) and Cu(II) salts with *tert*-butylhydroperoxide as the oxidant.

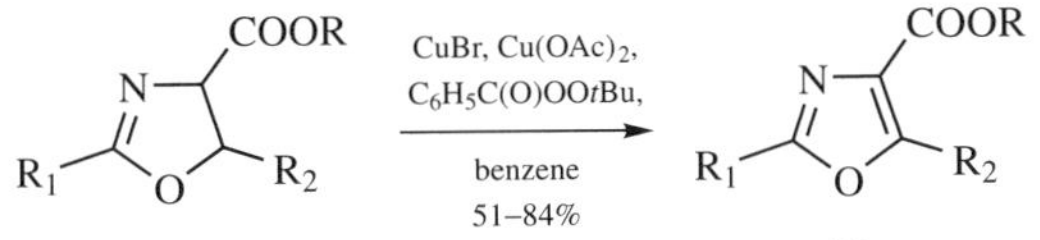

Scheme 1.11

The oxidation is quite effective for oxazolines containing an electron-withdrawing group at the 4 position, e.g., COOR. Significantly, 2-isopropyl and 2-cyclohexyl-4-oxazolecarboxylic acid esters were obtained from the corresponding oxazolines without side chain bromination. In contrast, side chain bromination was nearly the exclusive reaction observed using conventional methodology such as NBS/initiator. A limitation was seen with 4-alkyloxazolines, which afforded only poor yields of the corresponding oxazoles. For example, oxidation of 4-(1-methylethyl)-2-phenyloxazoline **40** with CuBr/Cu(OAc)$_2$ yielded only 10% of 4-(1-methylethyl)-2-phenyloxazole **41**. Most of **40** was recovered, even after prolonged reflux (Scheme 1.12).

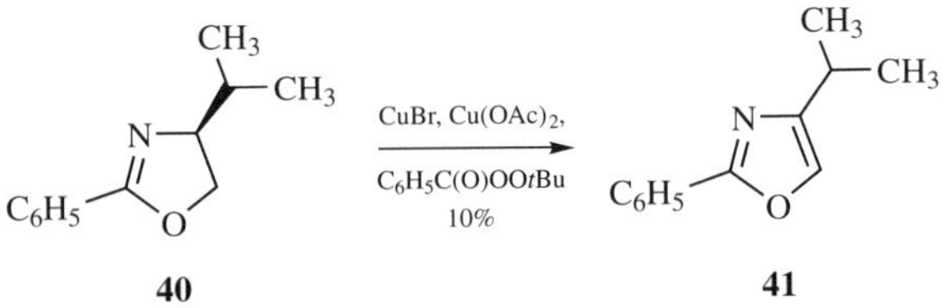

Scheme 1.12

The proposed mechanistic pathway is shown in Scheme 1.13 and is believed to involve the formation of regioisomeric oxazoline radicals, **42** or **44**. For **42**, Cu(II) effects oxidation to the oxazolium ion **43**, which loses a proton to yield **20**. Alternately, ligand transfer to the captodative radical **44** could yield the benzoate **45** from which syn elimination of benzoic acid would also produce **20**. The authors proposed that for an unactivated oxazoline like **40** either radical species analogous to **42** and **44** would not be expected to be as stable, thus explaining the failure of this oxidation methodology in these cases.

This methodology has been widely used in the syntheses of natural products. An example of the application of this reaction in the synthesis of (−)-thiangazole is discussed in Section 1.5.2. Meyers and co-workers[55] applied this method to prepare

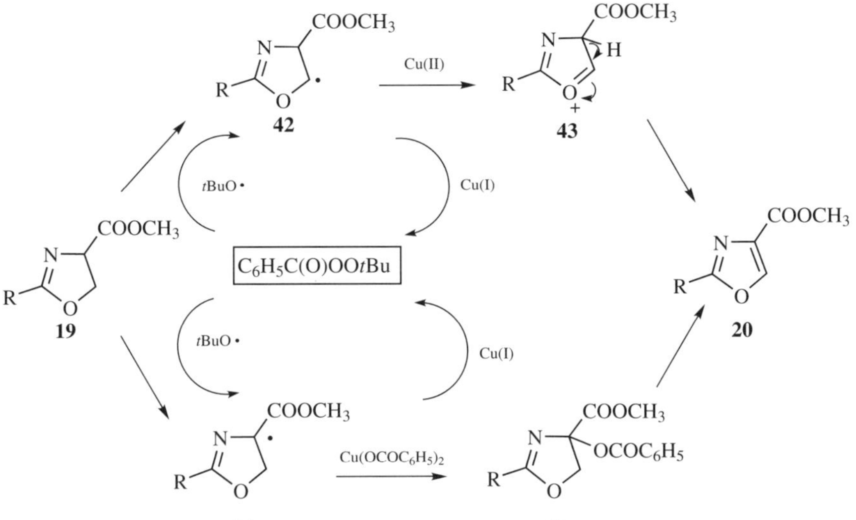

Scheme 1.13

the oxazole fragment **47** used in their total synthesis of bistratamide D. In addition, Pattenden and Chattopadhyay[56] prepared a 2-substituted 4-oxazolecarboxylic acid ester intermediate **49** of ulapualide A using this methodology (Scheme 1.14).

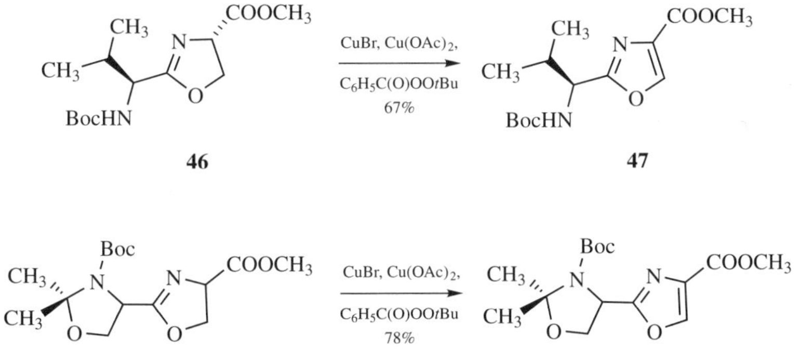

Scheme 1.14

1.3.1.5. Miscellaneous Oxidations

Alternate methods that do not involve a metal-based oxidation have also been described. These methods halogenate an oxazoline with subsequent dehydrohalogenation to effect overall net oxidation. Williams and co-workers[57] used BrCCl$_3$ and DBU to oxidize oxazolines (Scheme 1.15). Reaction of **50** with BrCCl$_3$ in the presence of two equivalents of DBU generates **51** in 87% yield. Similarly, **52**

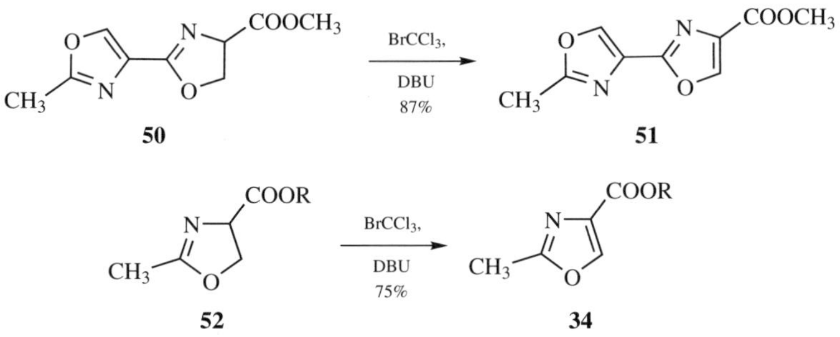

Scheme 1.15

(R = CH$_3$) was converted to 2-methyl-4-oxazolecarboxylic acid methyl ester **34** (R = CH$_3$) in 75% yield.

Meyers's group[55] found this to be the method of choice to prepare the oxazole fragment of bistratamide, wherein **47** was obtained quantitatively from **46** (Scheme 1.16). Jung and co-workers[58,59] employed similar methodology using CCl$_4$ and DBU in their syntheses of oxazole **54** and thiazole **56** fragments of microcin B17 (Scheme 1.17).

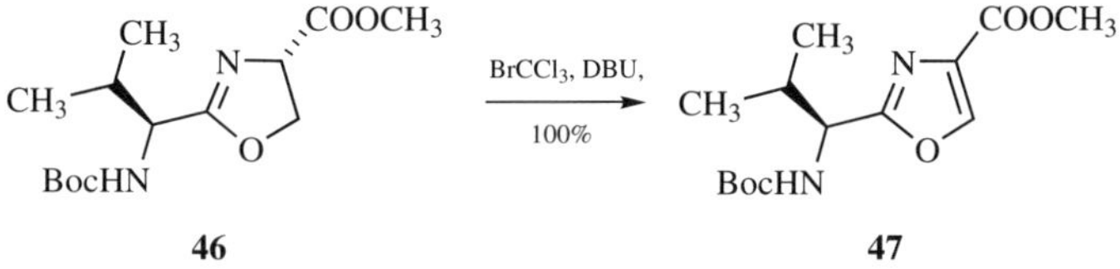

Scheme 1.16

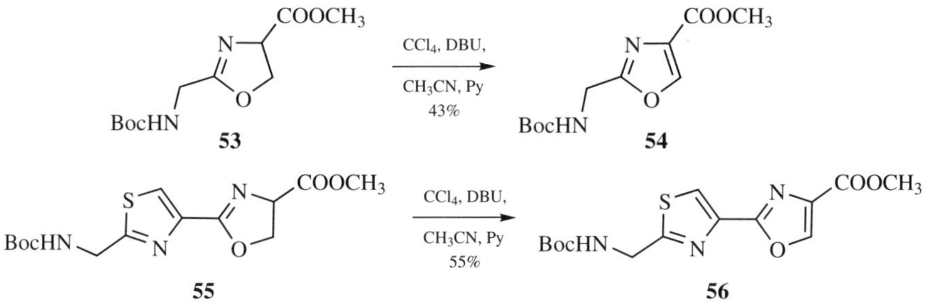

Scheme 1.17

Electophilic iodination using I$_2$ has also been employed to effect net oxidation of oxazolines to 2-substituted 4-oxazolecarboxylic acid esters. Koskinen and co-workers[44,60] prepared an intermediate oxazole fragment of calyculin using this method (Scheme 1.18). Here, the oxazoline **57** was first treated with LiHMDS and TMSCl to protect the carbamate as a silyl amide followed by treatment with KHMDS and iodine to generate the oxazole **58**. Interestingly, the authors[60] also isolated diastereomeric spirocyclic ortho ester aminals **59** in 25 to 30% yield under these reaction conditions.

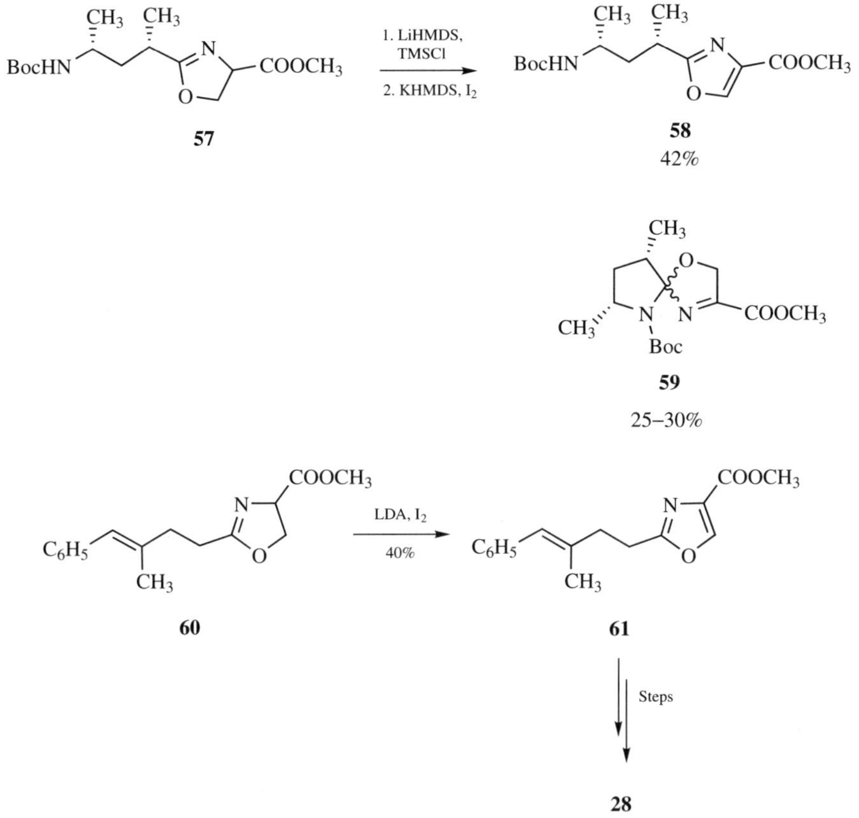

Scheme 1.18

Peña and co-workers[61] applied this methodology to prepare the key oxazole fragment **61** in their synthesis of phenoxan **28** (Scheme 1.18). The oxazoline **60** was not readily converted to **61** using Wipf's[62] protocol of triphenylphosphine/I_2 or using DDQ.[63] However, treatment of **60** with LDA and iodine produced **61** in 40% yield together with 40% recovered starting material.

Balaban and co-workers[64] prepared 2,5-disubstituted oxazoles via side-chain bromination with NBS, followed by dehydrobromination with pyridine. Kashima and Arao[65] described a synthesis of oxazoles using NBS/AIBN to effect oxidation of an oxazoline. In this case, the product was a 5-bromooxazole, e.g., **63** (Scheme 1.19). This study was limited to only three examples of 2-aryloxazoles.

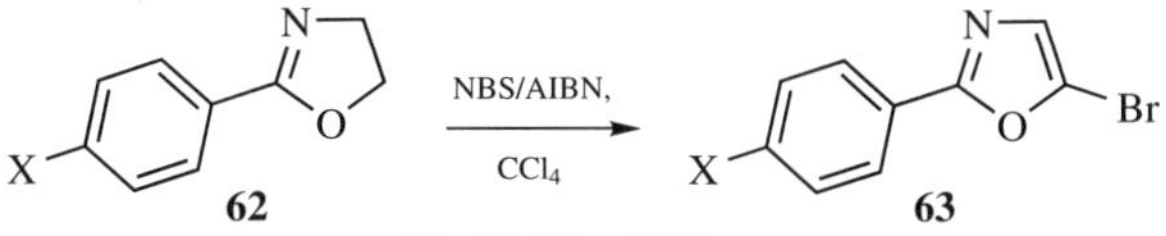

X = H, Cl, t-C_4H_9

Scheme 1.19

Meyers and Tavares[35,36] also investigated radical bromination as a means to effect net oxidation of activated oxazolines. They found these reaction conditions were acceptable for preparing 2-alkyl-4-oxazolecarboxylic acid esters only if the 2-alkyl group was methyl or primary, i.e., **65** $R_1 = R_2 = H$ or $R_1 = n\text{-}C_4H_9$, $R_2 = H$. However, these conditions failed completely if the 2-alkyl group was secondary, e.g., isopropyl or cyclohexyl. In these cases, the desired oxazole **65** was isolated in <1% yield. Instead, the sole product was **66**, the result of oxidation with concomitant side chain bromination (Scheme 1.20, Table 1.1, entries 7 and 8).

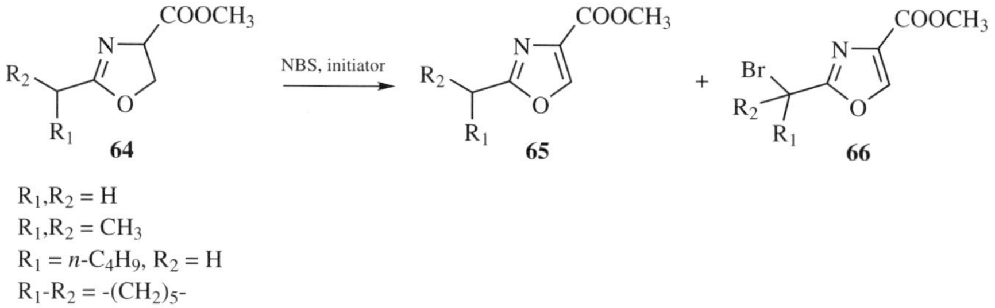

$R_1, R_2 = H$
$R_1, R_2 = CH_3$
$R_1 = n\text{-}C_4H_9, R_2 = H$
$R_1\text{-}R_2 = \text{-}(CH_2)_5\text{-}$

Scheme 1.20

The successful oxidation of oxazolines to oxazoles depends greatly on the nature of the substituents. Several methods have been described that effect this conversion in good to excellent yields for activated oxazolines. However, secondary alkyl groups can suffer free radical bromination (Scheme 1.20). In addition, ring-brominated oxazoles can also be isolated (Scheme 1.19). Despite the progress to date, a general high-yield method to oxidize activated and unactivated oxazolines to oxazoles is still needed.

1.3.2. Rearrangement Reactions

1.3.2.1. *N-Acylaziridines*

Rearrangement of *N*-acylaziridines followed by oxidation is a well-known method of preparing oxazoles. There are several factors that control the regiochemistry of the rearrangement in unsymmetrical aziridines. In the context of their approach to halichondramides, Eastwood and co-workers[66] studied this rearrangement and followed it with NiO_2 oxidation to produce 2,4-disubstituted oxazoles.

If ring expansion of **68** (R_1 = electron-withdrawing group) was affected by a nucleophile then a 2,5-disubstituted oxazoline was produced. However, rearrangement of **68** (R_1 = electron-donating group) under similar conditions affords a 2,4-disubstituted oxazoline **69** (Scheme 1.21).

The authors' initial efforts using NaI or NaBr with **68** ($R_1 = \text{COO}i\text{-}C_3H_7$, $R_2 = C_6H_5$) afforded 31% of the desired **69** ($R_1 = \text{COO}i\text{-}C_3H_7$, $R_2 = C_6H_5$) together with 45% of **67** ($R_1 = \text{COO}i\text{-}C_3H_7$, $R_2 = C_6H_5$) and 14% of the acrylate-elimination

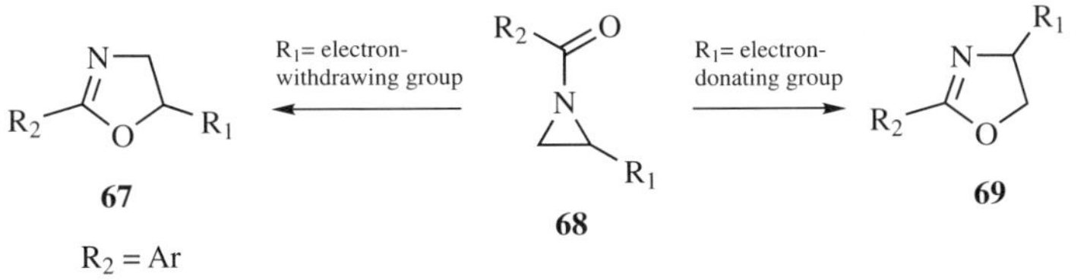

Synthesis of Oxazoles

Scheme 1.21

product. However, if the rearrangement was initiated with triflic acid, the ratio of **69/67** was improved to 5.7:1, with no evidence of the elimination product.

They modified their synthetic strategy and reexamined the original reaction conditions using **68**, wherein R_1 = acetal or ether and $R_2 = C_6H_5$ (Scheme 1.22). Thus an alkene **70** was treated with NaN_3 and ICl to produce a β-iodoazide **71**, which was reduced with $LiAlH_4$, cyclized, and acylated in situ to yield **72**. Rearrangement of **72** with NaI in acetone then gave an oxazoline **73** that was oxidized to the target 2,4-disubstituted oxazole **74**. Examples of this reaction sequence are shown in Table 1.2. Preparation of the *N*-acylaziridines and subsequent rearrangement to the oxazolines proceeds in excellent yield. However, the yields of **74** are limited by the NiO_2 oxidation.

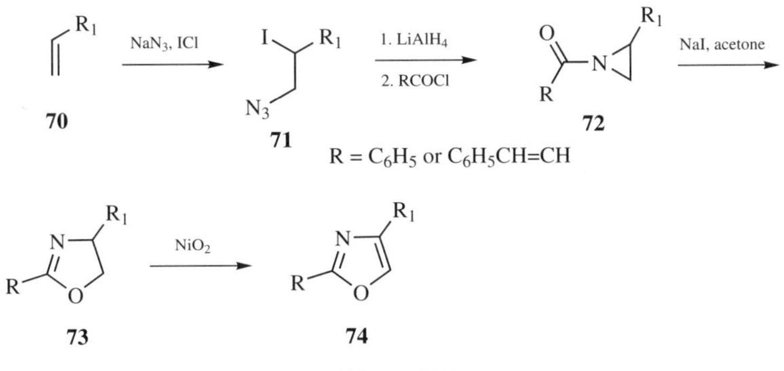

Scheme 1.22

1.3.2.2. N-Acyltriazoles

Both *N*-acyl-1,2,3-triazoles and *N*-acyl-1,2,4-triazoles can rearrange to oxazoles. Williams[67] described a general synthesis of 2-substituted oxazoles based on the rearrangement of *N*-acyl-1,2,3-triazoles. He found that prolonged reaction of the acid chloride **75** with 2-trimethylsilyl-1,2,3-triazole **76** in refluxing toluene (Table 1.3, method A) generated an approximately 1:2 mixture of the expected triazole amide **77** together with the 2-pyridyloxazole **78** (Scheme 1.23). Presumably **78** arises via rearrangement of **77**. Indeed, upon heating, **77** was converted to **78** with loss of nitrogen. Mechanistically, the author proposed acylation of **76**

TABLE 1.2. 2-PHENYL- AND 2-STYRYL-4-SUBSTITUTED OXAZOLES FROM REARRANGEMENT OF N-ACYLAZIRIDINES[a]

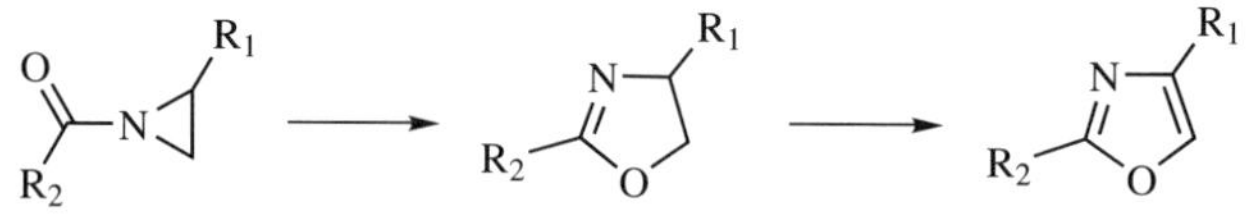

Figure 1.4

$R_2 = C_6H_5$ and $C_6H_5CH=CH$

Entry	R_1	R_2	Yield (%)
1	$CH_2Ot\text{-}C_4H_9$	C_6H_5	36–61
2	$CH(OCH_3)_2$	C_6H_5	36–61
3	$CH(OC_2H_5)_2$	$C_6H_5CH=CH$	70
4	$COOi\text{-}C_3H_7$	$C_6H_5CH=CH$	56[b]

[a] Data from ref. 66.
[b] Triflic acid was used for the rearrangement of the N-acylaziridine.

TABLE-1.3. 2-ALKYL(ARYL)OXAZOLES BY REARRANGEMENT OF N(1)-ACYL-1,2,3-TRIAZOLES[a]

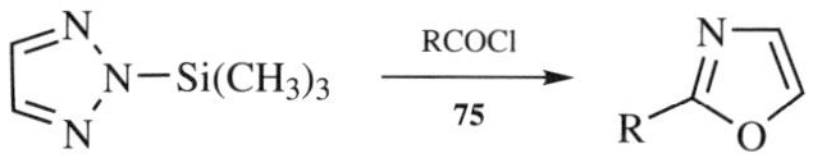

Figure 1.5

Entry	RCOCl	Product	Yield (%)	Method[b]
1	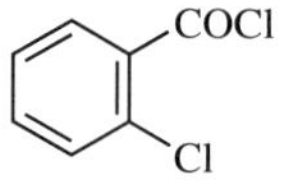 2-chlorobenzoyl chloride	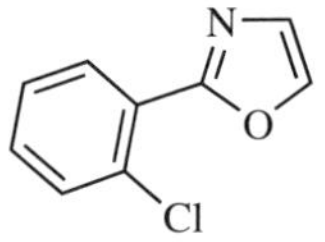 2-(2-chlorophenyl)oxazole	82	A
2	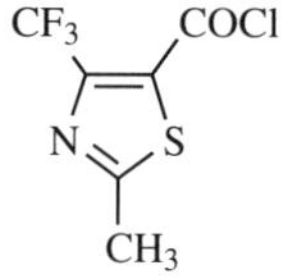 4-COCl-2-CH3-5-CF3-thiazole	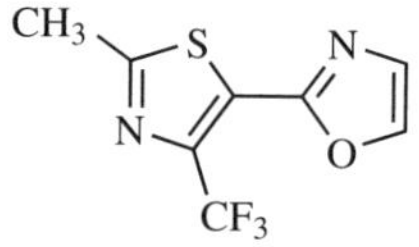 2-(thiazolyl)oxazole	86	B
3	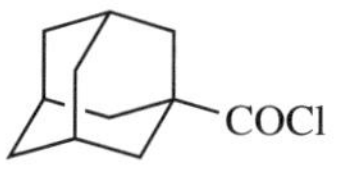 1-adamantanecarbonyl chloride	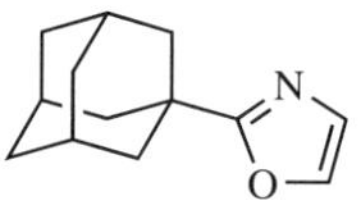 2-(1-adamantyl)oxazole	73	B
4	 ClOC-(CH2)4-COCl	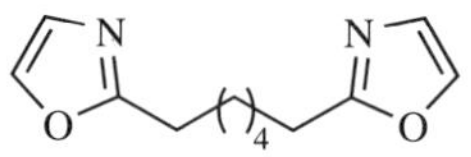 bis-oxazole linked by (CH2)4	37	B

[a] Data from ref. 67.
[b] A, toluene, reflux, 3–7 days; B, sulfolane, 140°–150°C, 2–3 h.

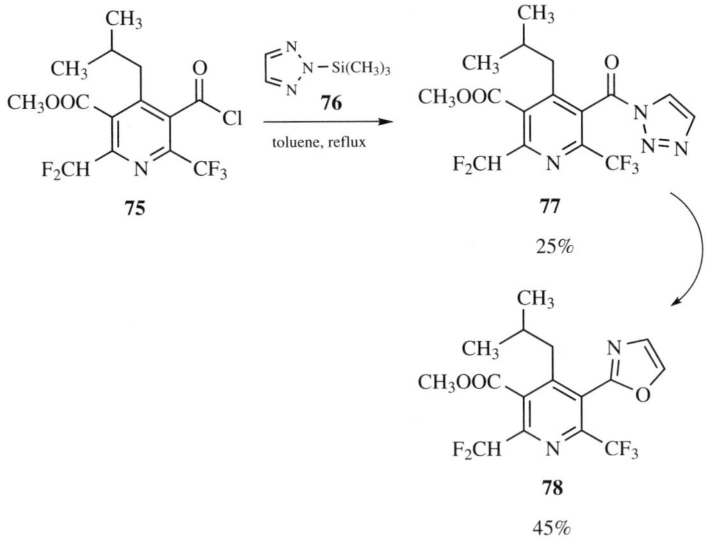

Scheme 1.23

initially gave the $N(1)$-acyl-1,2,3-triazole **79**, which was in equilibrium with the $N(2)$-acyl-1,2,3-triazole **80**. Cleavage of **79** to a zwitterionic intermediate **81**, followed by ring closure with loss of nitrogen, then yielded the 2-substituted oxazole **82** (Scheme 1.24).

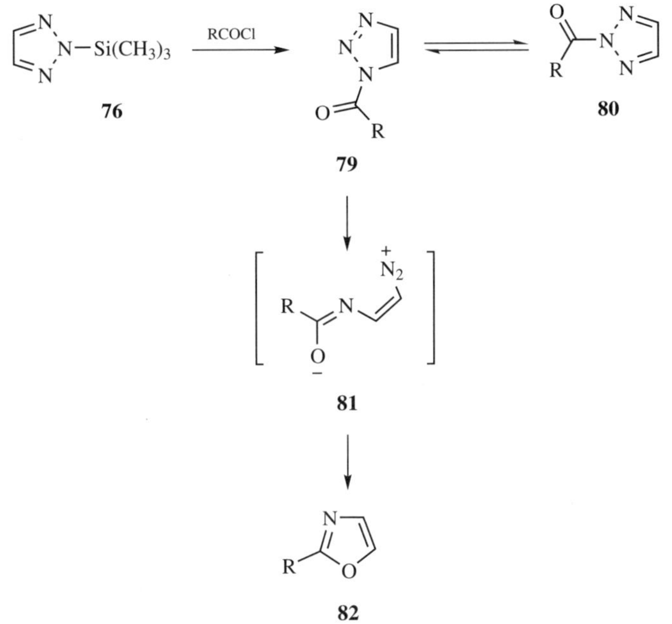

Scheme 1.24

This reaction was extended to prepare other 2-alkyl(aryl)-oxazoles **82** in modest to excellent yields (Table 1.3). The low yield obtained for the bis-oxazole (Table 1.3, entry 4) resulted from difficulties encountered with isolation of the product from sulfolane.

Maquestiau and co-workers[68] investigated thermal rearrangement of N(1)-acyl-1,2,4-triazoles to oxazoles using flash vacuum pyrolysis (FVP). In this case, FVP of an N(1)-acyl-1,2,4-triazole **83** leads to a 5-aryloxazole **86**. It was assumed that a [1,5] sigmatropic shift of the acyl group in **83** produced **84**, followed by loss of nitrogen to yield the diradical **85**, which cyclized to **86** (Scheme 1.25). A limited number of examples were investigated, and the results are shown in Table 1.4. The yields of **86** are excellent for N(1)-aroyl-1,2,4-triazoles but modest in the case of N(1)-acetyl-1,2,4-triazole. Nonetheless, the author noted that this represented a one-step synthesis of a 5-susbstituted oxazole in contrast to the conventional four-step syntheses usually employed.

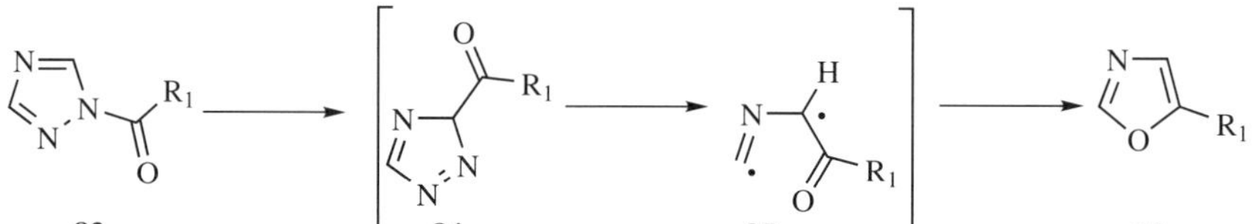

Scheme 1.25

TABLE 1.4. 5-ALKYL(ARYL)OXAZOLES BY FVP OF N(1)-ACYL-1,2,4-TRIAZOLES[a]

Figure 1.6

Entry	R_1	Yield (%)
1	C_6H_5	90
2	4-CH_3-C_6H_4	93
3	CH_3	25[b]

[a] Data from ref. 68.
[b] Approximately 52% of the starting material was recovered.

1.3.2.3. Isoxazoles and N-Acylisoxazolones

Although rearrangement of isoxazoles can be used to prepare oxazoles, unexpected limitations have been reported. Pérez and Wunderlin[69] investigated the FVP of various 5-alkyl-3-methyl-4-nitroisoxazoles **87** as a method for preparing

2,5-disubstituted 4-nitrooxazoles **88** (Scheme 1.26). When the authors subjected **87** ($R_1 = R_2 = H$), **87** ($R_1 = R_2 = CH_3$), or **87** ($R_1 = H$, $R_2 = CH_3$) to FVP at 400°C at 0.1 torr, they isolated 1-nitro-1-cyanoacetone **89** quantitatively, with no evidence for the desired oxazole. In contrast, FVP of 5-ethyl-3-methylisoxazole **92** produced the expected 5-ethyl-2-methyloxazole **93** quantitatively. The authors explained these results via the intermediacy of a 1-azirine **90**. This leads to a nitrile ylide **91** that does not cyclize to **88** due to charge delocalization by the nitro group.

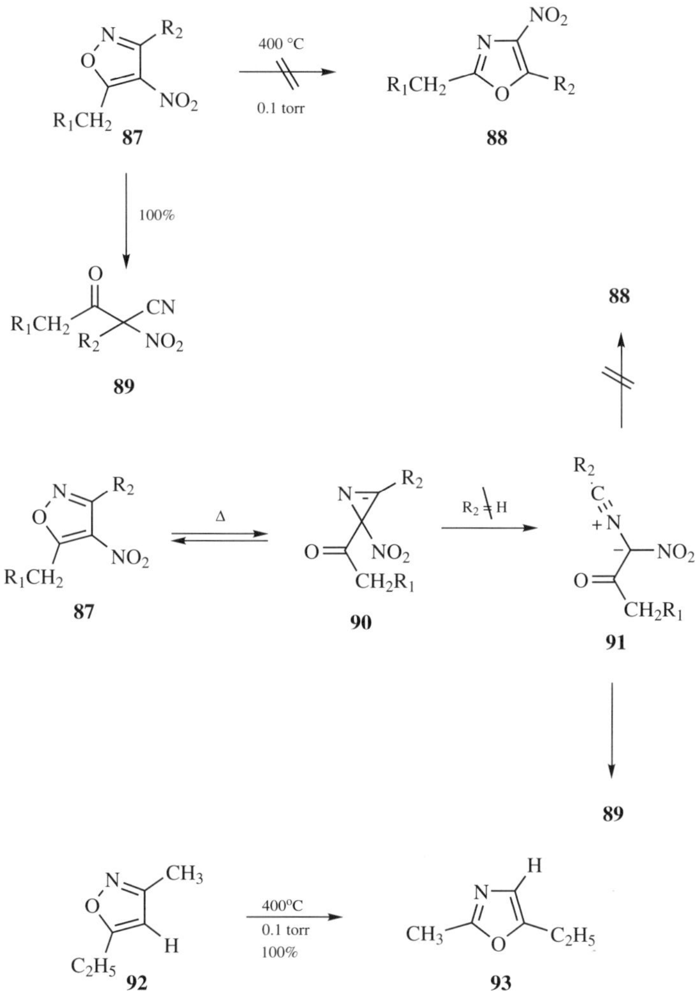

Scheme 1.26

Pascual[70] developed a one-pot synthesis of 2-(alkylamino)-4-cyanooxazoles and 5-alkyl-2-(alkylamino)-4-cyanooxazoles by rearrangement of isoxazole-derived thioureas (Scheme 1.27). Reaction of an isoxazole isothiocyanate **94** with primary amines gave the thioureas **95**, which were dehydrated to carbodiimides **97**

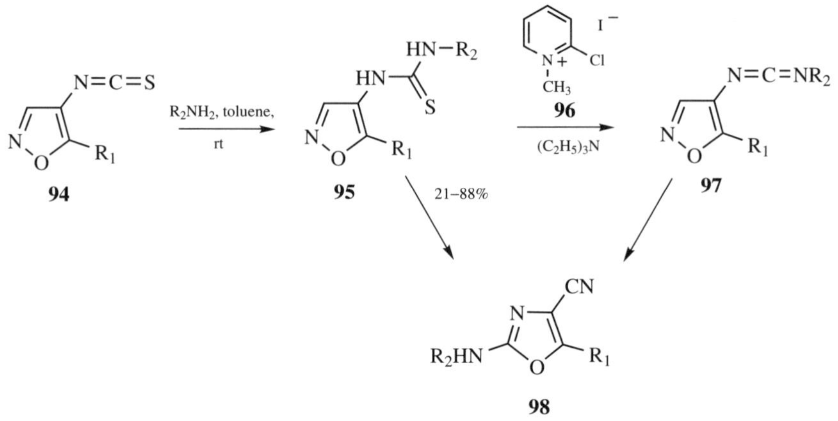

Scheme 1.27

using Mukaiyama's[71] reagent. These somewhat unstable carbodiimides were not isolated but were treated directly with $(C_2H_5)_3N$ to afford the desired 4-cyanooxazoles **98** in excellent yield. Selected examples of **98** prepared via this methodology are shown in Table 1.5.

TABLE 1.5. 2-ALKYLAMINO- AND 5-ALKYL-2-(ALKYLAMINO)-4-CYANOOXAZOLES FROM REARRANGEMENT OF ISOXAZOLE THIOUREAS[a]

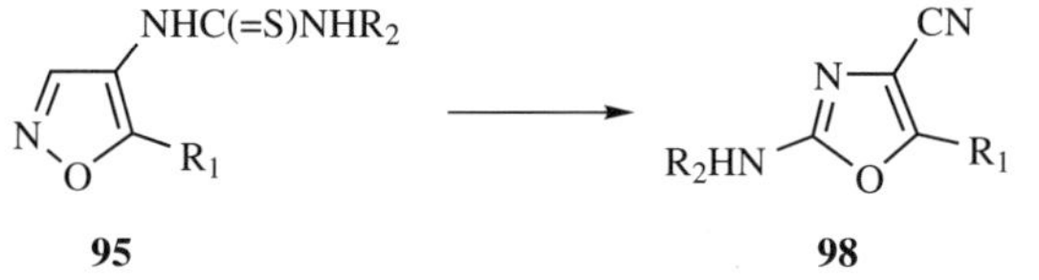

Entry	R_1	R_2	Product	Yield (%)[b]
1	H	t-C_4H_9	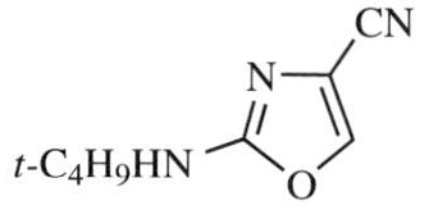	88
2	CH_3	i-C_3H_7	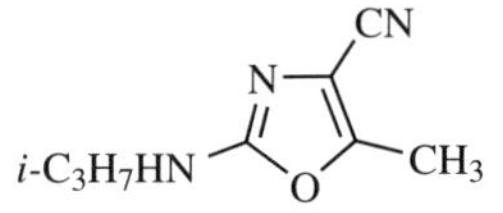	81
3	i-C_3H_7	i-C_3H_7	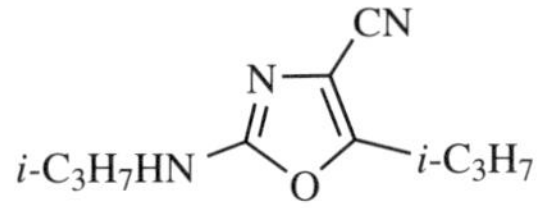	74
4	i-C_3H_7	t-C_4H_9	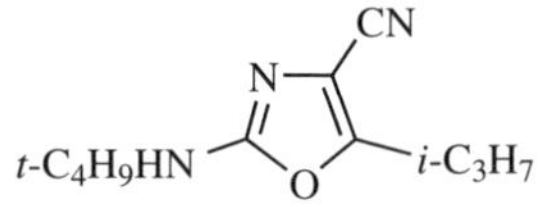	77

[a] Data from ref. 70.
[b] Yield is the conversion of the carbodiimide to the oxazole.

Prager and co-workers[72–74] used the photochemical rearrangement of 2-acyl-isoxazol-5-ones to prepare 2,4-disubstituted, 2,5-disubstituted, and 2,4,5-trisubstituted oxazoles **15**. Both FVP and photolysis of 2-acylisoxazol-5-ones **99** gave the target oxazoles **15** in excellent yields (Scheme 1.28). The authors reported significantly improved yields for oxazoles prepared in this manner vs. preparation from an acyltriazole precursor.[72] Photolysis was superior for **99** (R_3 = electron-withdrawing group), and pyrolysis is the preferred method for all other cases. The generality of this rearrangement is shown by selected examples in Table 1.6.

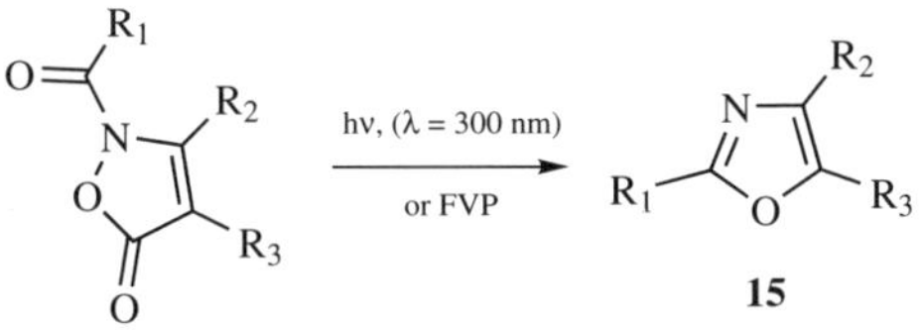

Scheme 1.28

TABLE 1.6. 2,4-DISUBSTITUTED-; 2,5-DISUBSTITUTED-; AND 2,4,5-TRISUBSTITUTED-OXAZOLES FROM REARRANGEMENT OF 2-ACYLISOXAZOL-5-ONES[a]

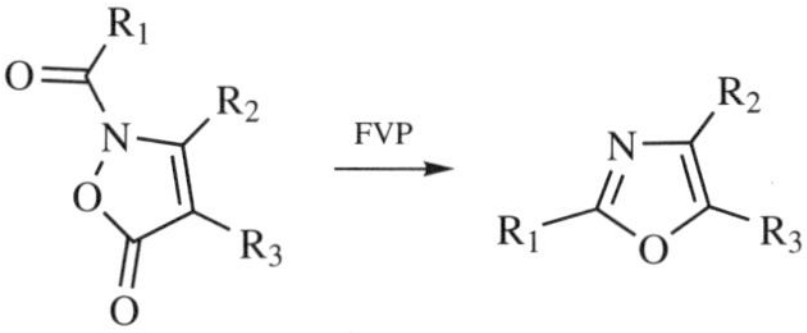

Entry	R_1	R_2	R_3	Product	Yield (%)[b]
1	CH_3	$COOC_2H_5$	CH_3	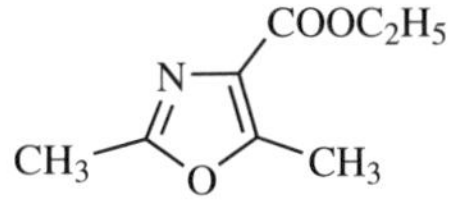	95(40)
2	C_6H_5	$COOC_2H_5$	CH_3	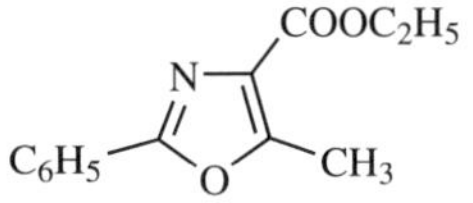	95(60)
3	C_6H_5	C_6H_5	H	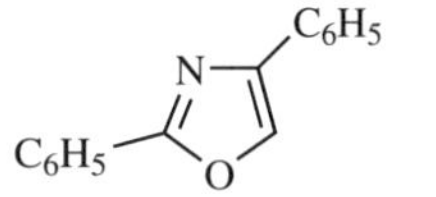	70(24)
4	C_6H_5	H	$COOC_2H_5$	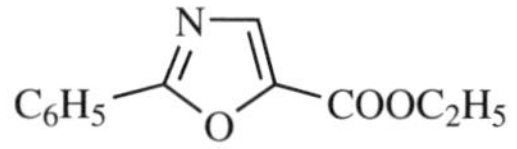	10(85)

[a] Data from ref. 72.
[b] % Yields in parentheses are from photolysis.

Prager and co-workers[73] described an elegant application of this methodology for preparing linear, directly connected tris-oxazoles found in marine natural products (Scheme 1.29). Thus FVP of ethyl-[2-acetyl-2,5-dihydro-4-methyl-5-oxo]-3-isoxazolecarboxylate **100a** (R = Ac) afforded 2,5-dimethyl-4-oxazolecarboxylic acid ethyl ester **101** in 95% yield. Saponification and coupling with **100b** (R = H) gave the bis-oxazole precursor **102**. This was rearranged to **103** in overall yield, and the process was iterated a third time to generate the tris-oxazole **105**, isolated in 10% overall yield.

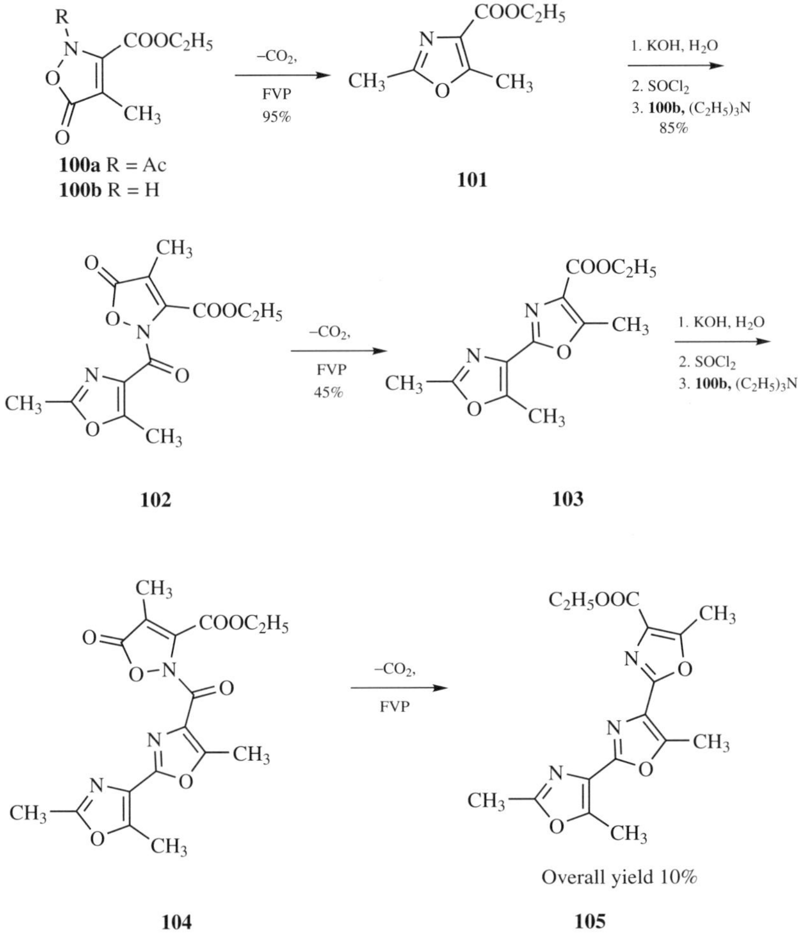

Scheme 1.29

In some cases, mixtures of 2,4-disubstituted and 2,5-disubstituted oxazoles were isolated. For example, pyrolysis of 2-benzoyl-3-phenyl-isoxazol-5-one **106** gave the expected 2,4-diphenyloxazole **110** in 60% yield together with 10% of 2,5-diphenyloxazole **111** and 15% of benzanilide (Scheme 1.30).[73,74] The authors explained these products by initial formation of a singlet iminocarbene **107**. This then

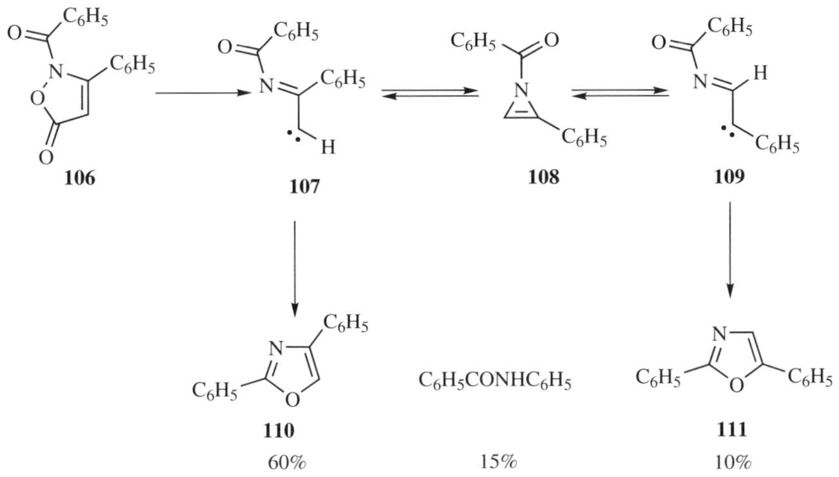

Scheme 1.30

rearranges to the more stable iminocarbene **109** via the *N*-acyl-1*H*-azirine **108**, as shown. The intermediacy of a singlet iminocarbene rather than a triplet diradical was supported by the observation that the presence of triplet quenchers such as 9-methylanthracene or benzophenone had no effect on the photolysis yields.

Further evidence for involvement of an *N*-acyl-1*H*-azirine intermediate is supported by isolation of both 5-bromo-4-methyl-2-phenyloxazole **113** and 4-bromo-5-methyl-2-phenyloxazole **114** from the photolysis of 2-benzoyl-4-bromo-3-methylisoxazol-5-one **112** (Scheme 1.31). In addition, the presence of a 4-halo substituent in **112** seriously limited the synthetic utility of this reaction. Despite extensive ^{13}C and substituent labeling, the authors concluded that further studies were necessary to elucidate fully the mechanism of this rearrangement.

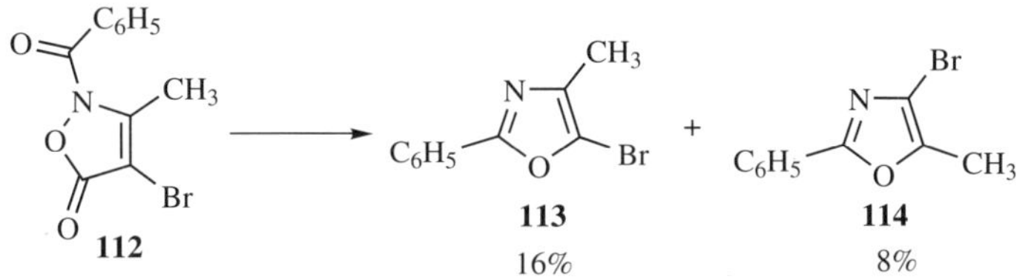

Scheme 1.31

Doleschall and Seres[75] described a novel base-catalyzed isoxazole-oxazole ring transformation. Treatment of ethyl 5-hydroxy-3-(5-methylisoxazol-4-yl)isoxazole-4-carboxylate **115** with KO*t*Bu/DMF produced ethyl 4-cyano-5-methyloxazol-2-yl acetate **120** in 82% yield. Mechanistically, the authors proposed a base-catalyzed Beckmann rearrangement of **115** via the intermediate ketenimine **118** to account for isolation of **120** (Scheme 1.32). The structure of **120** was confirmed by saponification and decarboxylation to 4-cyano-2,5-dimethyloxazole **121**. The generality of this novel rearrangement has not yet been defined.

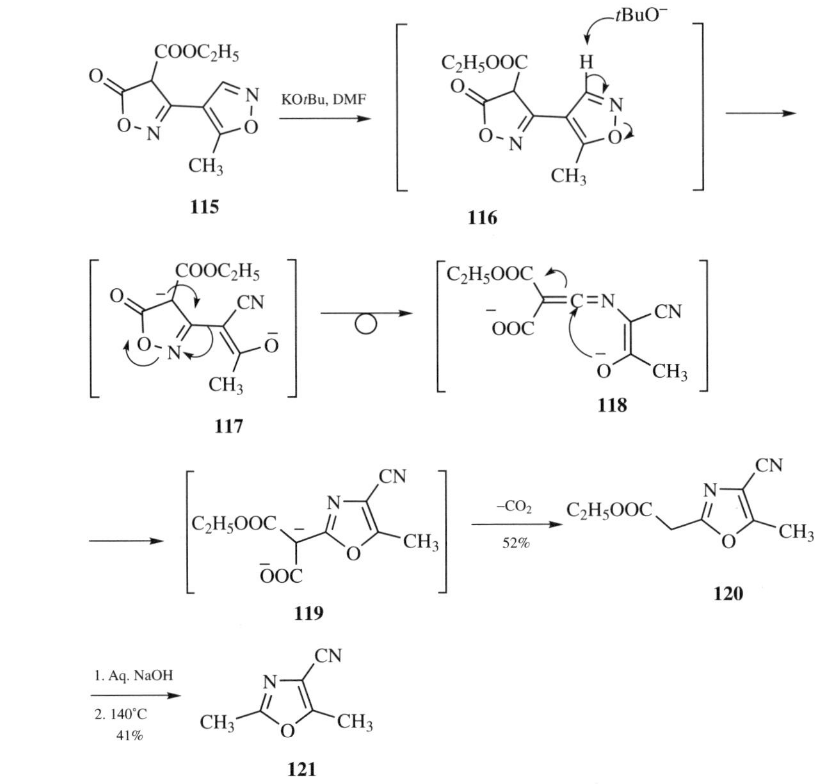

Scheme 1.32

Vivona and co-workers[76] also described an unusual base-catalyzed isoxazole-oxazole rearrangement. Here, treatment of 3-(aroylamino)-5-methylisoxazoles **122** with KOtBu/DMF effected rearrangement to 2-(aroylamino)-5-methyloxazoles **125** (Scheme 1.33). The authors proposed initial deprotonation of **122** to generate **123**,

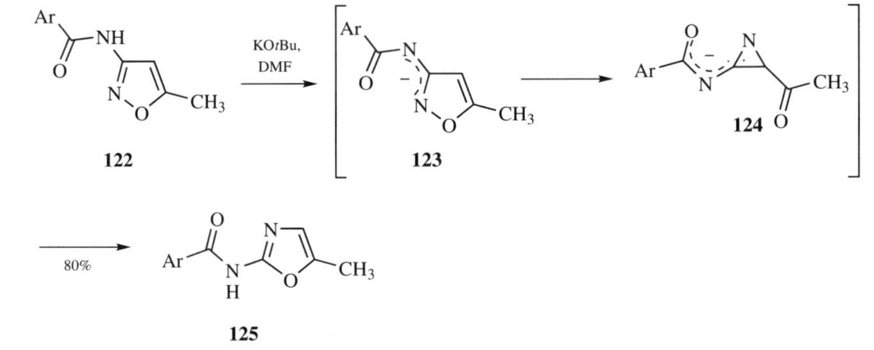

Ar = C_6H_5, 4-CH_3-C_6H_4, 4-CH_3O-C_6H_4

Scheme 1.33

followed by ring contraction to produce the C-aroylamino azirine **124**. Subsequent ring expansion of **124** would then yield **125**. The authors noted that the rearrangement was unique to the combination of KO*t*Bu/DMF. There was no reaction of **122** in a melt, in refluxing DMF, or in the presence of 10% KOH in refluxing ethanol.

Liu and Howe[77] employed a photochemical transformation of 3,5-diarylisoxazoles to prepare 2,5-diaryloxazoles, which were evaluated as plant growth regulators and herbicides. Photolysis of isoxazoles **126** affords the 2,5-diaryloxazoles **127** in good yield (Scheme 1.34). The authors described this as a potentially general method for preparing analogs such as **127**.

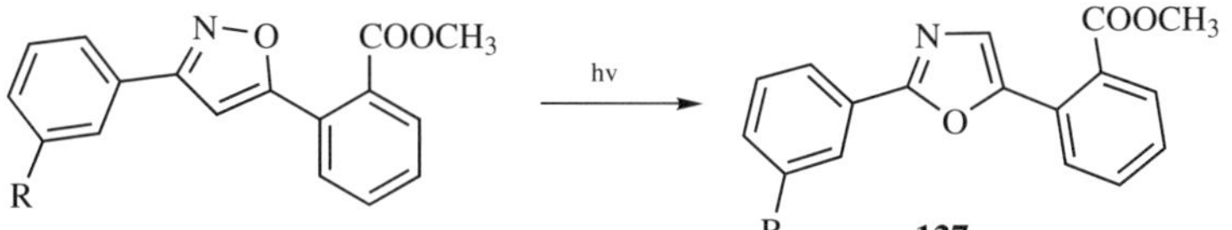

R = H (74%)
R = CF$_3$ (54%)

Scheme 1.34

Nishimoto and co-workers[78] reported a theoretical study on the photochemical rearrangement of isoxazoles to oxazoles. In a later study of the thermal and photochemical transposition of oxazoles, the authors concluded that these rearrangements involve the formation of azirine intermediates.[79]

1.3.3. Organometallic Reactions

1.3.3.1. Rhodium Carbene Additions

Oxazoles can be prepared by the reaction of nitriles with diazocarbonyl compounds in the presence of Lewis acid catalysts or mediated by transition metals.[80,81] Selected examples that illustrate the scope of this chemistry are described in this section. However, the reader should consult both Doyle and Moody's[80] and Helquist and co-workers'[81] papers for excellent compilations of references summarizing other work in this area.

In a model study, Helquist and co-workers[82] described the reaction of dimethyl diazomalonate **128** with benzonitrile to prepare 5-methoxy-2-phenyl-4-oxazolecarboxylic acid methyl ester **129** nearly quantitatively (Scheme 1.35). Several other 2-aryl-5-methoxy-4-oxazolecarboxylic acid methyl esters were prepared analogously. In addition, 2-alkyl(alkenyl)-5-methoxy-4-oxazolecarboxylic acid methyl esters were also prepared, although the yields for aliphatic nitriles were not as good, unless the nitrile was used as solvent.[81,82] Other metal salts—including Rh$_2$(NHAc)$_4$, Cu(OTf)$_2$, Cu(C$_2$H$_5$-acac)$_2$, Rh$_2$(O$_2$CC$_3$H$_7$)$_4$, and Rh$_3$(CO)$_{16}$—were not as effective as Rh$_2$(OAc)$_4$ in this reaction.

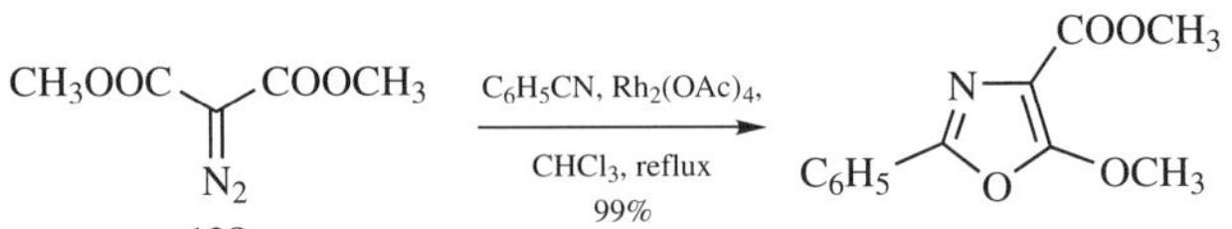

Scheme 1.35

The authors proposed an electrophilic rhodium carbene complex to account for formation of **129**. Thus reaction of **128** with $Rh_2(OAc)_4$ generates the carbene complex **130**, which is trapped by benzonitrile to afford the nitrilium species **131**. Cyclization of **131** through the enolate oxygen then yields **129** (Scheme 1.36).

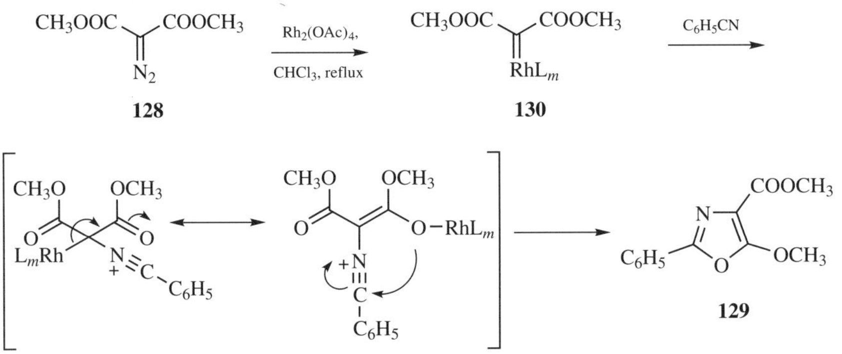

Scheme 1.36

Fukushima and Ibata[83] reported direct evidence for the formation of an acyl nitrile ylide intermediate during $Rh_2(OAc)_4$-catalyzed reactions of α-diazoacetophenones **133** with nitriles in the presence of dimethylacetylene dicarboxylate (DMAD). The authors isolated mixtures of both 5-aryl-2-phenyloxazoles **137** and tetrasubstituted pyrroles **138** (Scheme 1.37). They proposed that **137** arose from the acyl nitrile ylide **136** via 1,5-cyclization, whereas 1,3-diploar-cycloaddition of **136** with DMAD followed by a 1,5-hydrogen shift produced **138**. The authors showed that **138** did not arise via a Diels-Alder reaction of **137** with DMAD.

The yield of **137** and **138** depended on the aryl substitutent. Electron-withdrawing groups (X = Cl, CN, and NO_2) gave a mixture of 6–7:1 of **137/138** in 70–80% yield. In contrast, electron-donating groups (CH_3O and CH_3) gave a mixture of up to 15:1 of **137/138** but only in 34–54% yield. Reaction of benzonitrile with ethyl diazobenzoylacetate **139** gave only 2,5-diphenyl-4-oxazolecarboxylic acid ethyl ester **141**, albeit in very low yield, derived from cyclization with the carbonyl group (Scheme 1.38). There was no evidence for 4-benzoyl-5-ethoxy-2-phenyloxazole **142**.

Xu and co-workers[84] prepared the previously unknown 5-ethoxy-4-(trifluoromethyl)-2-oxazolecarboxylic acid ethyl ester **144** in 90% yield using $Rh_2(OAc)_4$-catalyzed reaction of ethyl 3,3,3-trifluoro-2-diazopropionate **143** with ethyl

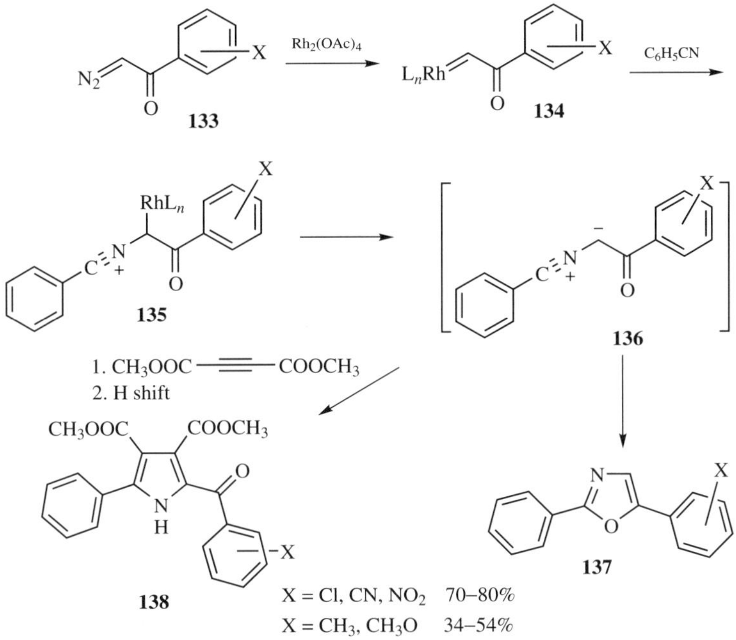

Scheme 1.37

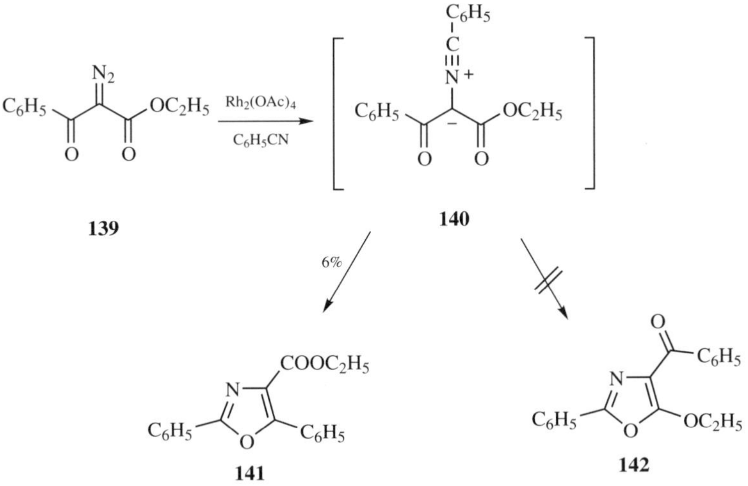

X = Cl, CN, NO₂ 70–80%
X = CH₃, CH₃O 34–54%

Scheme 1.38

cyanoformate (Scheme 1.39). This oxazole ester was a useful starting material for a variety of analogs screened as agrochemicals.

Moody and co-workers[80,85-94] extensively investigated and refined rhodium-catalyzed additions of diazocarbonyl compounds to construct oxazoles. Several significant developments from these studies are described in the following schemes.

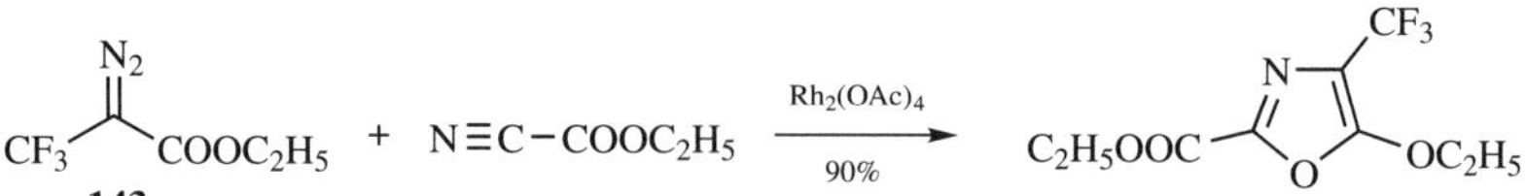

143 **144**

Scheme 1.39

Some noteworthy examples applied to natural product syntheses are (+)-nostocyclamide,[85,89] promothiocin A,[86] and diazonamide.[88,92–94]

Early work from Moody's laboratories[87] extended the rhodium-carbene methodology to include α-diazosulfones (Scheme 1.40). Selected examples of 2-alkyl (aryl)-5-ethoxy-4-(phenylsulfonyl)oxazoles **146** prepared from ethyl phenylsulfonyldiazoacetate **145** and nitriles are shown in Table 1.7.

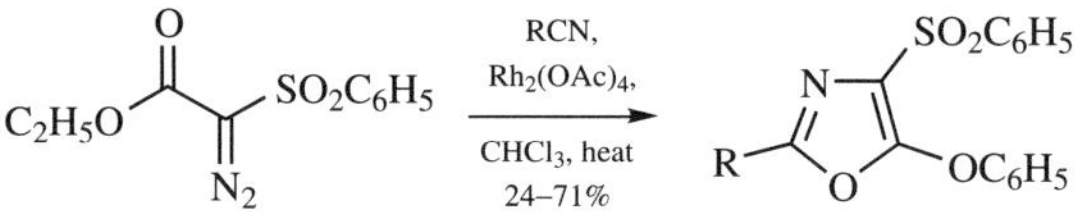

145 **146**

Scheme 1.40

TABLE 1.7. 2-ALKYL(ARYL)-5-ETHOXY-4-(PHENYLSULFONYL)OXAZOLES FROM α-DIAZOSULFONES[a]

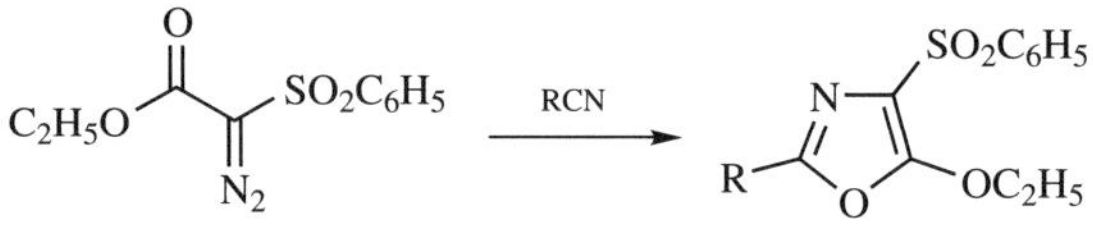

Figure 1.9 **145** **146**

Entry	R	Product	Yield (%)
1	C_2H_5	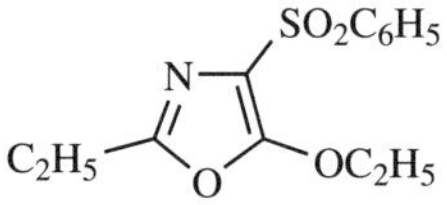	52
2	2-Cl-C_6H_4	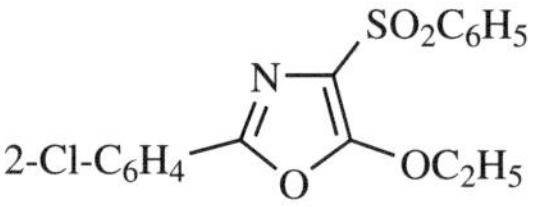	56
3	3-CH_3O-C_6H_4	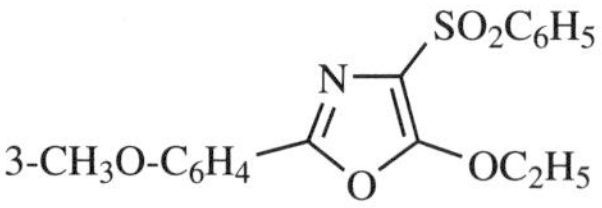	24
4	C_6H_5	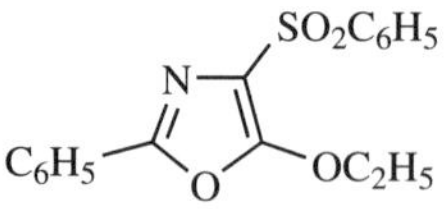	71

[a] Data from ref. 87.

Synthesis of Oxazoles 31

The authors extended this methodology to methyl cyanodiazoacetate in the first use of an α-diazonitrile to prepare a 4-cyanooxazole, which then served as a substrate for a second iteration (Scheme 1.41).[87] Thus benzonitrile reacted with methyl cyanodiazoacetate to produce 4-cyano-5-methoxy-2-phenyloxazole **147**, which was subjected to $Rh_2(OAc)_4$-mediated coupling with dimethyl diazomalonate **128** to furnish the bis-oxazole **148** in only 4% yield. However, using $Rh_2(NHCOCF_3)_4$ in this second coupling afforded **148** in 53% yield. This was a significant advance, since functionalized bis-oxazoles could then be prepared in three steps from the starting diazo compound.

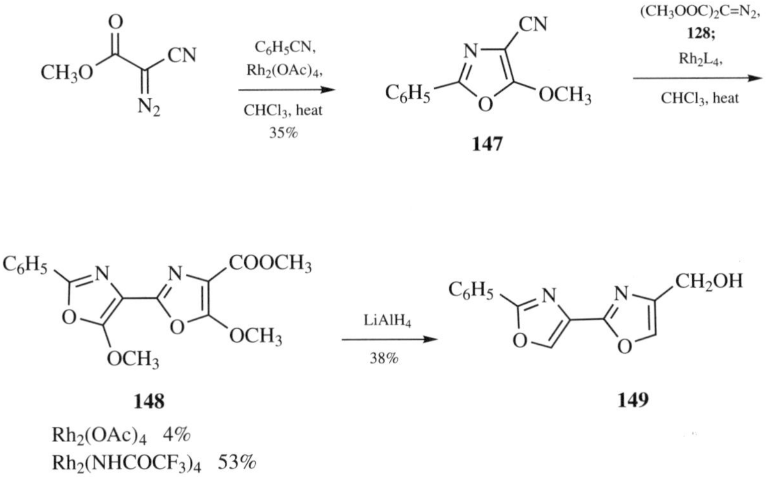

Scheme 1.41

Doyle and Moody[80] investigated the effect of the ligand on the formation of 5-ethoxy-2-ethyl-4-(phenylsulfonyl)oxazole **146** (R = C_2H_5). They isolated comparable yields of the desired oxazole with several ligands (Table 1.8), although the reaction rate was faster using $Rh_2(NHCOCF_3)_4$. The authors found Rh_2(NHCO

TABLE 1.8. EFFECT OF RHODIUM(II) CATALYSTS ON THE SYNTHESIS OF 5-ETHOXY-2-ETHYL-4-(PHENYLSULFONYL)OXAZOLE[a]

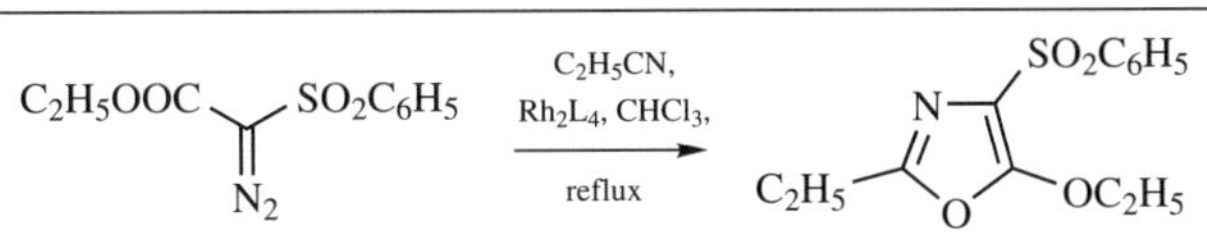

Figure 1.10

Entry	Rhodium Ligand	Yield (%)
1	CH_3COO	52
2	CF_3CONH	64
3	$C_6H_5CH(OH)COO$	68
4	$1\text{-}C_{10}H_7COO$	65
5	1-benzenesulfonylprolinate	44

[a] Data from ref. 80.

$CF_3)_4$ to be a most useful catalyst that also significantly improved the yield of **151** prepared from the α-diazophosphonate **150** (Scheme 1.42).

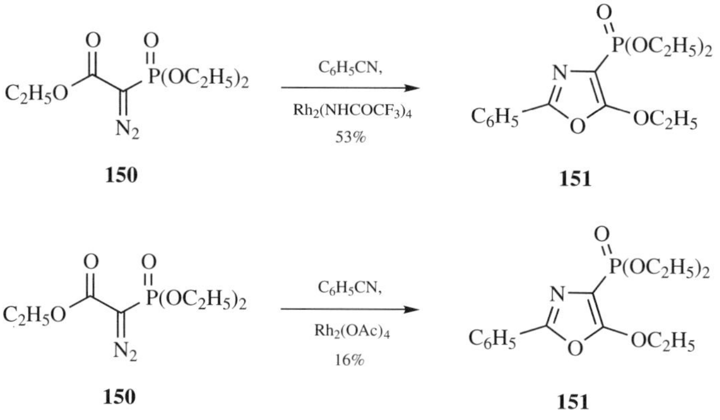

Scheme 1.42

Moody and co-workers[88] adapted their rhodium-carbene methodology to prepare 3-(oxazol-5-yl)indoles **154** as model compounds in their approach to diazonamide. In this case, **152** (R_1 = Boc) was readily converted to the *N*-Boc-diazoacetylindole **153** (R_1 = Boc). Treatment of **153** with $Rh_2(OAc)_4$ in refluxing acetonitrile gave **154** (R = CH_3, R_1 = Boc) in acceptable yield (Scheme 1.43). Preparation of **154** (R = C_2H_5, R_1 = Boc) required the more active catalyst, $Rh_2(NHCOCF_3)_4$. Deprotection of **154** with sodium methoxide produced the naturally occurring alkaloids pimprinine **154** (R = CH_3, R_1 = H) and pimprinethine **154** (R = C_2H_5, R_1 = H).

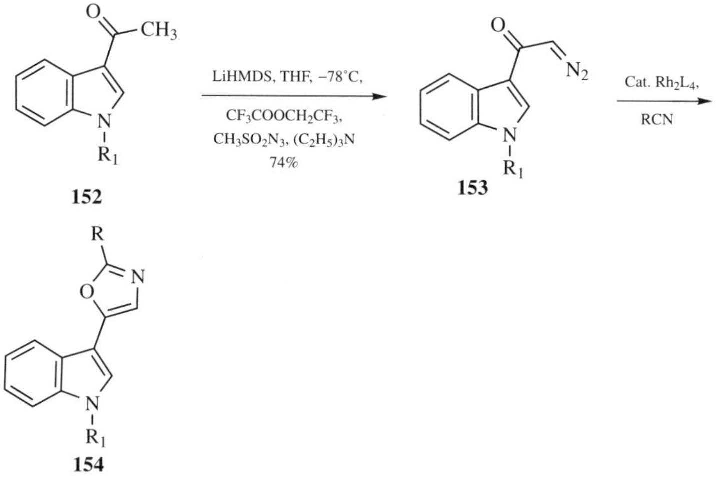

R = CH_3 L = OAc, 40%
R = C_2H_5 L = $NHCOF_3$, 90%

Scheme 1.43

Synthesis of Oxazoles 33

A further refinement in this methodology was reported from Moody's laboratories[90] during investigations leading to phenoxan **155** and muscoride A **156**. The authors' attempts to prepare **158** (R = CH$_3$, t-C$_4$H$_9$) from **157** using diazomalonates gave poor yields. Attempts to employ α-aminonitriles were also unsuccessful. However, a rhodium-carbenoid NH-insertion reaction of a diazomalonate **159** (R$_2$ = OCH$_3$) with an amide was successful and afforded the N-acylamino-β-ketoesters **160**. Cyclodehydration of **160** to the desired 2,5-disubstituted 4-oxazolecarboxylic acid methyl ester **161** was best accomplished using (C$_6$H$_5$)$_3$P and I$_2$ (Scheme 1.44).[62] This methodology was then extended to include amino acid amides requisite to prepare **156**. Representative examples are shown in Table 1.9.

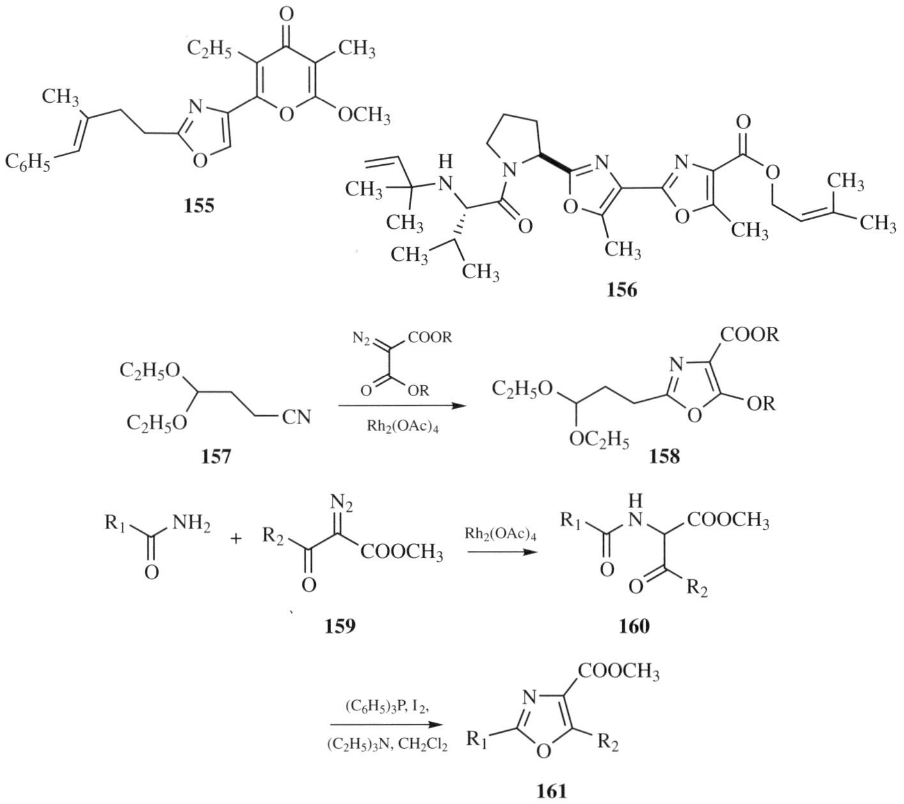

Scheme 1.44

An iterative application of this rhodium-carbenoid NH-insertion reaction with an amide was used in their approach to the bis-oxazole core **166** of muscoride A **156** (Scheme 1.45). N-Carbobenzyloxyproline amide was converted to **162** in good yield using methyl diazoacetoacetate. Cyclodehydration of **162** then gave the 2,5-disubstituted-4-oxazolecarboxylic acid methyl ester **163**, which was converted to the amide **164** uneventfully. Repetition of the rhodium-carbenoid NH-insertion reaction with methyl diazoacetoacetate gave **166** after cyclodehydration. This

TABLE 1.9. 2,5-DISUBSTITUTED-4-OXAZOLECARBOXYLIC ACID METHYL ESTERS FROM RHODIUM-CARBENOID NH INSERTION REACTIONS[a]

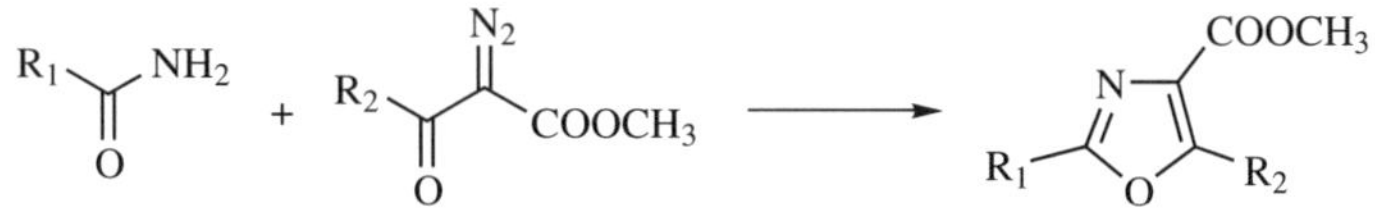

Figure 1.11

Entry	R_1	R_2	Product	Yield (%)[b]
1	$(C_2H_5O)_2CH(CH_2)_2$	OCH_3	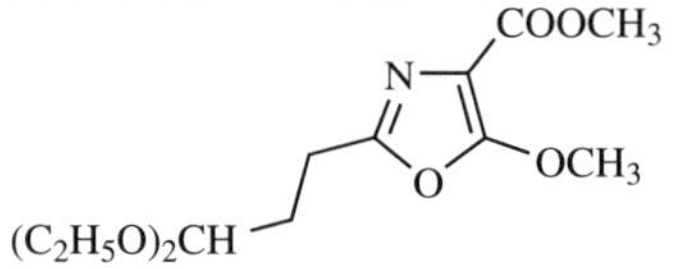 (oxazole with $(C_2H_5O)_2CH$ at 2-position, OCH_3 at 5-position, $COOCH_3$ at 4-position)	68
2	CbzNHCH(i-C_3H_7)	CH_3	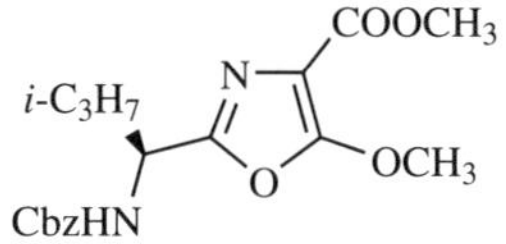 (oxazole with i-C_3H_7/CbzNH at 2-position, OCH_3 at 5-position, $COOCH_3$ at 4-position)	63
3	CbzNHCHCH$_3$	CH_3	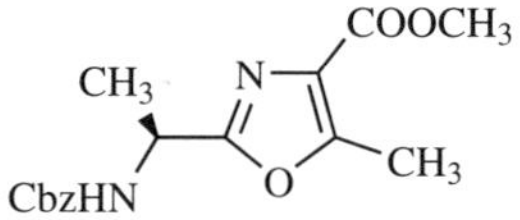 (oxazole with CH_3/CbzNH at 2-position, CH_3 at 5-position, $COOCH_3$ at 4-position)	44
4	CbzNHCH$_2$	CH_3	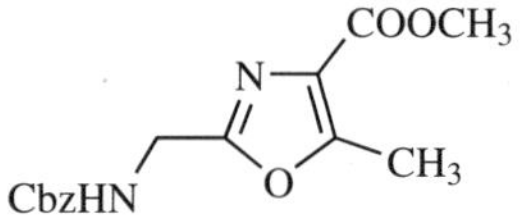 (oxazole with CbzNH at 2-position, CH_3 at 5-position, $COOCH_3$ at 4-position)	40

[a] Data from ref. 90.
[b] Yields are for two steps: N-acylamino-β-ketoester formation and cyclodehydration.

methodology was also successfully applied for the synthesis of (+)-nostocyclamide[85,89] and promothiocin A.[86]

Gogonas and Hadjiarapgolu[95] described the synthesis of oxazoles via a Rh$_2$(OAc)$_4$-catalyzed [3 + 2]-cycloaddition reaction of phenyliodonium salts with nitriles (Scheme 1.46). The starting phenyliodonium salt **167** was prepared from dimedone and diacetoxyiodobenzene in 95% yield. Reaction of **167** with nitriles in the presence of Rh$_2$(OAc)$_4$ gave fused 2-alkyl-, 2-aralkyl-, or 2-aryloxazoles **168** in modest yields. It is noteworthy that the reaction conditions do not require an inert atmosphere. The mechanistic details of this reaction are not yet clear. Selected examples are shown in Table 1.10.

The continuing development of solid-phase synthesis and combinatorial chemistry has led to solid-phase oxazole syntheses with a minimum of purification. Iso and co-workers[96] generated α-(trimethylsilyl)diazoketones on a Wang resin and employed rhodium-catalyzed diazo transfer methodology to prepare oxazoles (Scheme 1.47). Reaction of the resin-bound benzoyl chloride **169** with (trimethylsilyl)diazomethane gave the corresponding α-(trimethylsilyl)diazoketone **170** in excellent yield. Treatment of **170** with an aryl nitrile in the presence of catalytic Rh$_2$(OAc)$_4$ then furnished a resin-bound 2,5-diaryl-4-(trimethylsilyl)oxazole **171**.

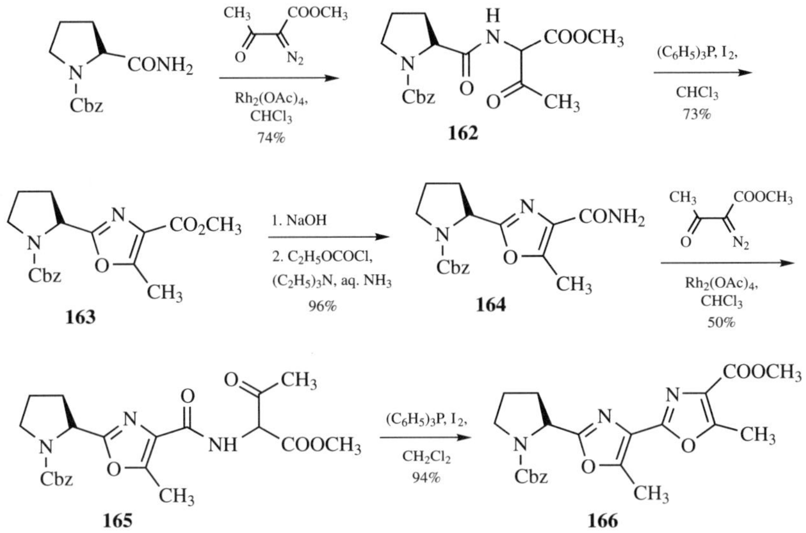

Synthesis of Oxazoles

Scheme 1.45

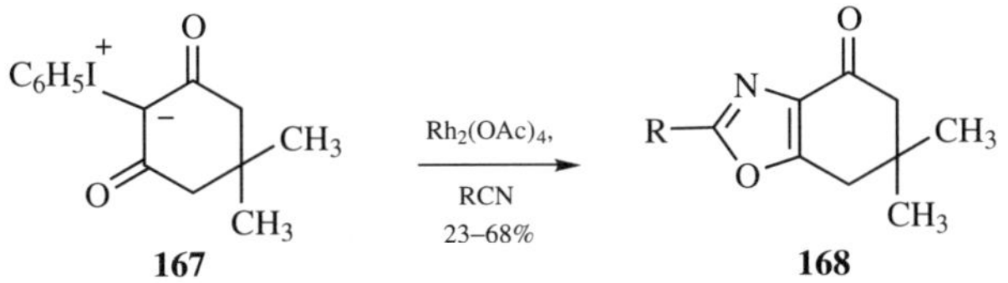

Scheme 1.46

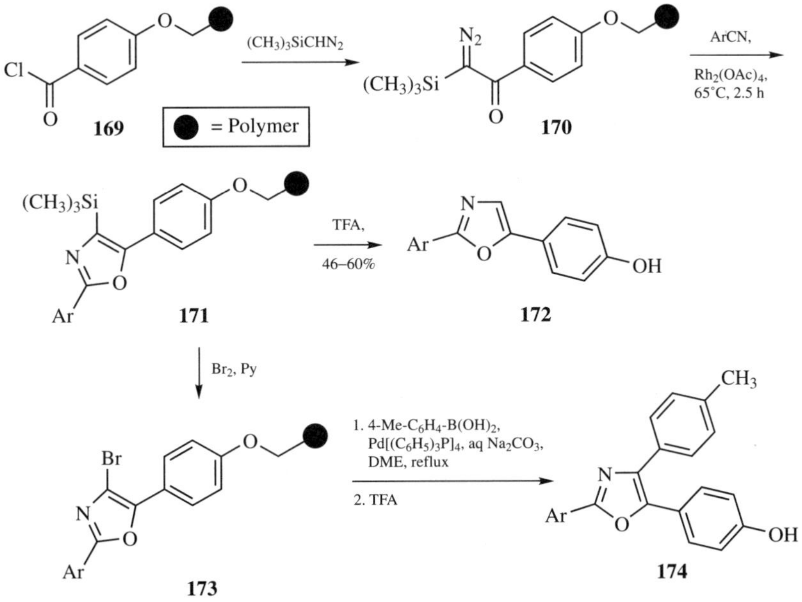

Scheme 1.47

TABLE 1.10. 2-ALKYL(ARYL)-6,7-DIHYDRO-6,6-DIMETHYL-4(5H)-BENZOXAZOLONES FROM IODONIUM SALTS AND NITRILES[a]

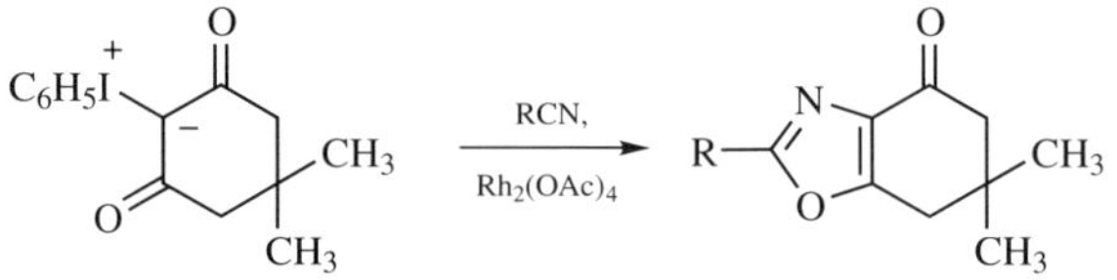

Figure 1.12   167 → 168 (RCN, Rh$_2$(OAc)$_4$)

Entry	RCN	Product	Yield (%)
1	CH$_3$CN	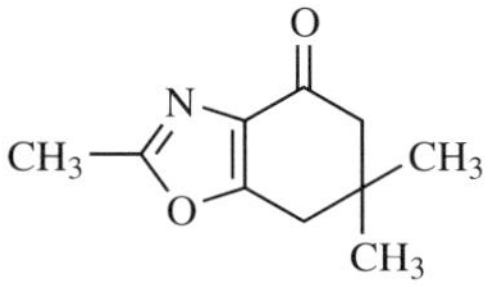 (2-CH$_3$ benzoxazolone)	68
2	C$_6$H$_5$CH$_2$CN	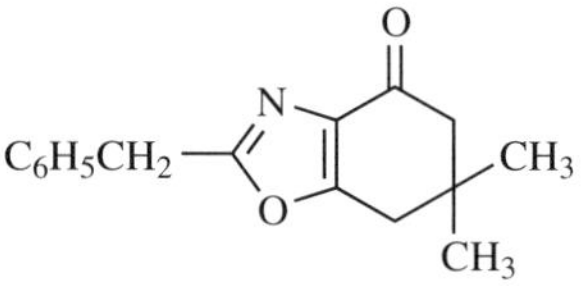 (2-C$_6$H$_5$CH$_2$ benzoxazolone)	23
3	C$_6$H$_5$CN	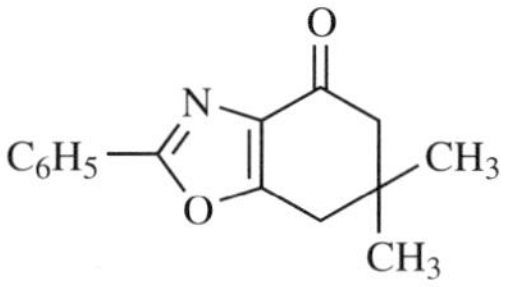 (2-C$_6$H$_5$ benzoxazolone)	28
4	4-CH$_3$C$_6$H$_4$	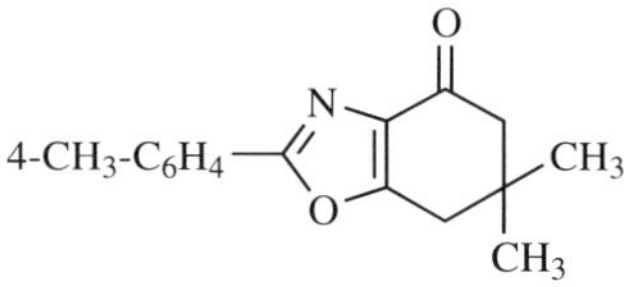 (2-(4-CH$_3$-C$_6$H$_4$) benzoxazolone)	42

[a] Data from ref. 95.

Cleavage of **171** with TFA was accompanied by desilylation to afford 2-aryl-5-(4-hydroxyphenyl)oxazoles **172**. Bromodesilylation of **171** yielded the corresponding resin-bound 4-bromo-2,5-diaryloxazole **173**, which was used in a Suzuki coupling[96a] to prepare 5-(4-hydroxyphenyl)-4-(4-methylphenyl)-2-phenyloxazole **174** (Ar = C$_6$H$_5$). Selected examples are shown in the Table 1.11.

Ducept and Marsden[97] described a general synthesis of 5-ethoxy-2-substituted 4-(triethylsilyl)oxazoles **176** from the rhodium(II)octanoate-catalyzed diazo-transfer reaction of ethyl (triethylsilyl)diazoacetate **175** and nitriles (Scheme 1.48). The reaction conditions tolerate a wide variety of functional groups on the nitrile, including alkyl, aryl, heteroaryl, vinyl, carbonyl, silyloxy, and dialkylamino. Desilylation of **176** with TBAF afforded the corresponding 2-alkyl(aryl)-5-ethoxy-oxazoles **177**, which are normally inaccessible from diazoesters using conventional rhodium-carbene methodology. The authors noted that carbonyl groups in the 2 position of **176** are not compatible with TBAF deprotection.

TABLE 1.11. 2-ARYL-5-(4-HYDROXYPHENYL)OXAZOLES FROM RESIN-BOUND α-(TRIMETHYLSILYL)DIAZOKETONES AND NITRILES[a]

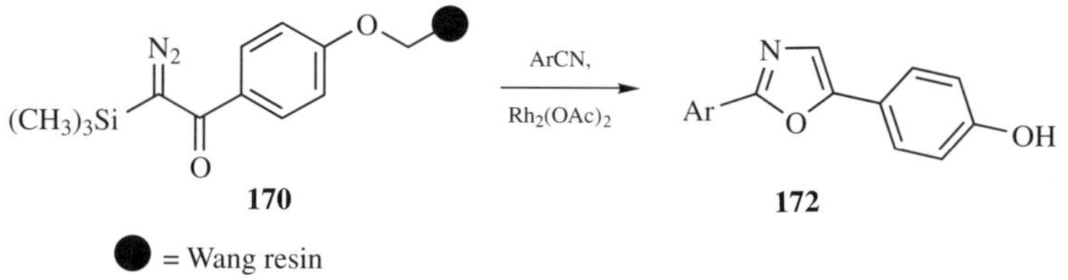

Figure 1.13

Entry	ArCN	Product	Yield (%)
1	C_6H_5CN	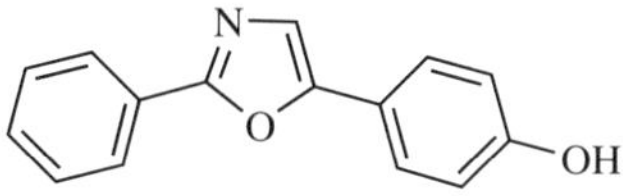 2-phenyl-5-(4-hydroxyphenyl)oxazole	66
2	4-Cl-C_6H_4CN	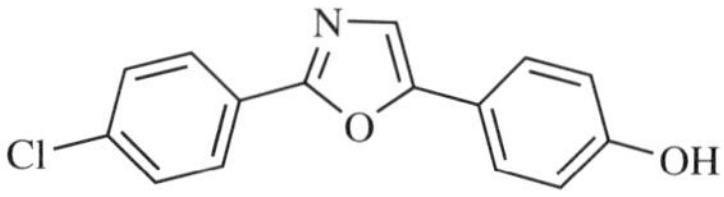 2-(4-chlorophenyl)-5-(4-hydroxyphenyl)oxazole	46
3	1-NaphthylCN	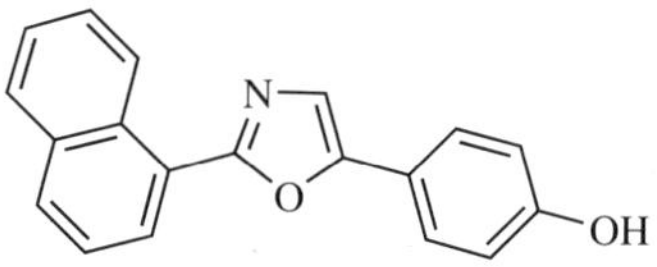 2-(1-naphthyl)-5-(4-hydroxyphenyl)oxazole	60

[a] Data from ref. 96.

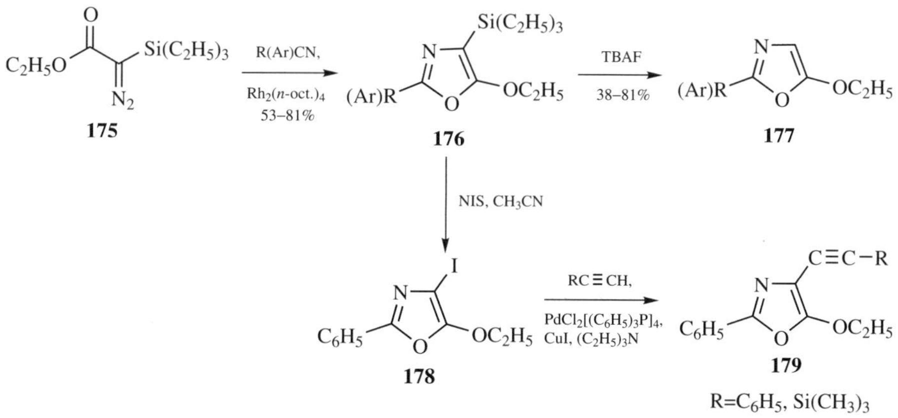

Scheme 1.48

Treatment of **176** (Ar = C$_6$H$_5$) with *N*-iodosuccinimide affords 5-ethoxy-4-iodo-2-phenyloxazole **178**, from which the 4-alkynyl analogs **179** were obtained via Sonogoshira couplings.[98]

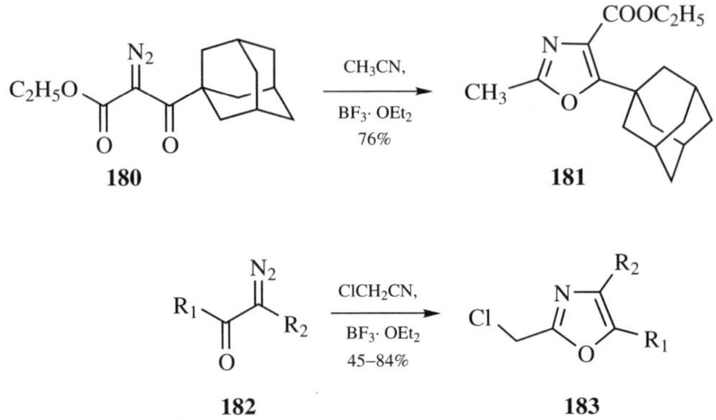

Scheme 1.49

Lewis acids efficiently catalyze the transfer of diazoketones and diazoesters to amides and nitriles to furnish 2,4,5-trisubstituted oxazoles (Scheme 1.49). In particular, Eguchi and co-workers[99] found that BF$_3$•OEt$_2$ was the best catalyst for reaction of the adamantyl diazoketoester **180** with acetonitrile. Attempts to prepare **181** using Rh(II)acetate or photolysis were unsuccessful. Similarly, Ibata and Isogami[100] described a general synthesis of 5-aryl-2-(chloromethyl)oxazoles **183** from a BF$_3$•OEt$_2$-catalyzed addition of α-diazoacetophenones **182** with chloroacetonitrile. The yields were quite respectable (Table 1.12).

1.3.3.2. Organotellurium Reagents

Ogura and co-workers[101] reported a general synthesis of 4,5-disubstituted-2-methyloxazoles **190** from attempted amidotellurinylation of internal acetylenes (Scheme 1.50). Normally, terminal acetylenes undergo an *E*-stereoselective

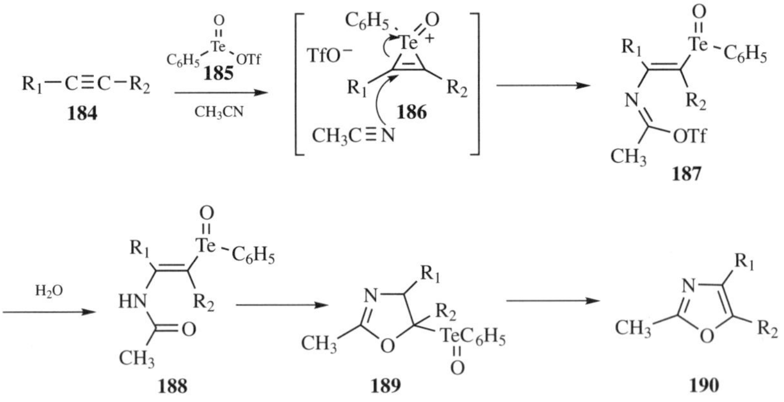

Scheme 1.50

TABLE 1.12. 5-ARYL-2-(CHLOROMETHYL)OXAZOLES FROM $BF_3 \cdot OEt_2$ CATALYZED REACTION OF DIAZOKETONES AND DIAZOESTERS WITH CHLOROACETONITRILE[a]

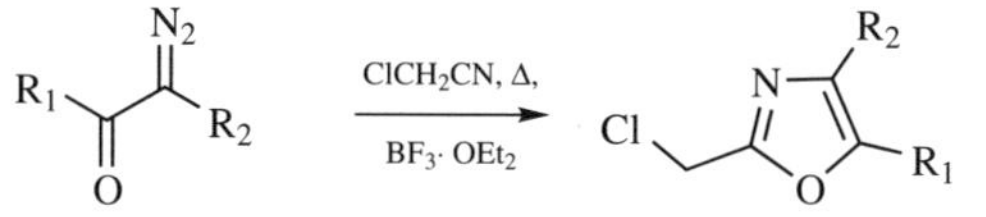

Figure 1.14

Entry	R_1	R_2	Product	Yield (%)
1	4-NO_2-C_6H_4	H	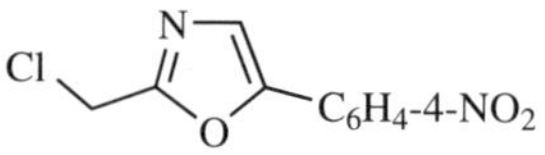 Cl-CH₂-oxazole-C_6H_4-4-NO_2	64
2	3-NO_2-C_6H_4	H	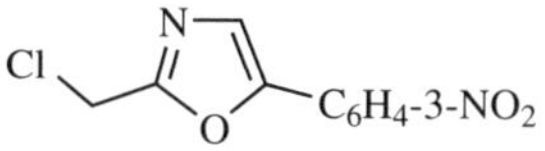 Cl-CH₂-oxazole-C_6H_4-3-NO_2	64
3	C_6H_5	H	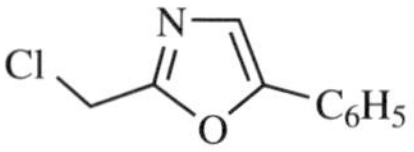 Cl-CH₂-oxazole-C_6H_5	84
4	CH_3O	4-NO_2-C_6H_4	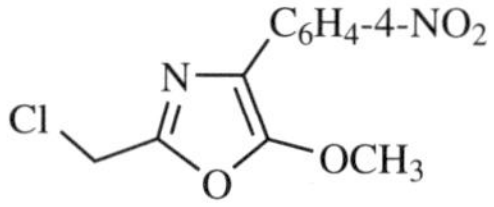 Cl-CH₂-oxazole(4-C_6H_4-4-NO_2, 5-OCH_3)	45

[a] Data from ref. 100.

amidotellurinylation when reacted with a nitrile and benzenetellurinyl triflate **185**. However, reaction of an internal acetylene **184** with **185** in acetonitrile gave 4,5-disubstituted-2-methyloxazoles **190** in good yield. Mechanistically, the authors proposed that **185** added to **184** to produce **186**. Ring opening of **186** by the nitrile or by an O-trifloyl imidate gives the E-amidotellurinylation product **187**. Hydrolysis of **187** affords the acetamide **188**, which cyclizes to the oxazoline **189**. Loss of $C_6H_5Te(O)H$ from **189** then yields **190**. This method appears to be a general approach for preparing a variety of trisubstituted oxazoles. Selected examples of 4,5-disubstituted-2-methyloxazoles are shown in Table 1.13.

1.3.3.3. Organomercury Reagents

Lee and Song[102] prepared 2,4-diaryl- and 5-alkyl-2,4-diaryloxazoles **191** via a microwave-assisted reaction of ketones and nitriles with mercury(II) tosylate

TABLE 1.13. 4,5-DISUBSTITUTED-2-METHYLOXAZOLES FROM INTERNAL ACETYLENES, ACETONITRILE, AND BENZENETELLURINYL TRIFLUOROMETHANESULFONATE[a]

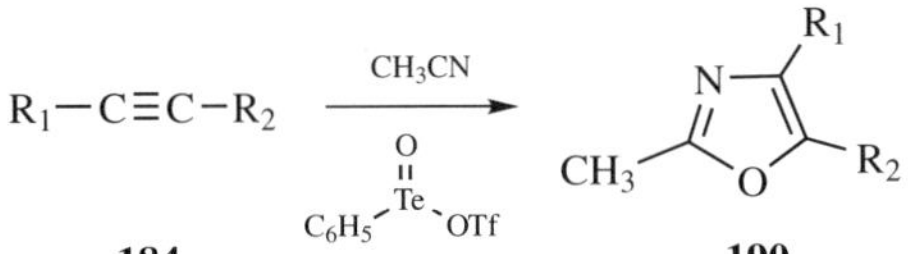

Figure 1.15: $R_1-C\equiv C-R_2$ (184) + CH_3CN + $C_6H_5-Te(=O)-OTf$ (185) → 2-methyl-4-R_1-5-R_2-oxazole (190)

Entry	R_1	R_2	Product	Yield (%)
1	C_6H_5	C_6H_5	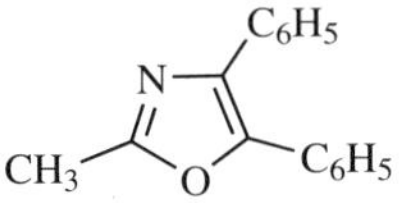 2-CH$_3$, 4-C$_6$H$_5$, 5-C$_6$H$_5$ oxazole	75
2	n-C$_3$H$_7$	n-C$_3$H$_7$	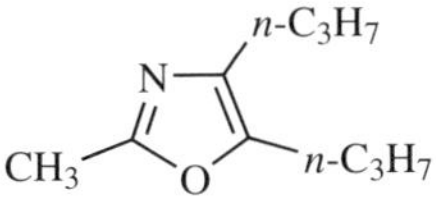 2-CH$_3$, 4-n-C$_3$H$_7$, 5-n-C$_3$H$_7$ oxazole	72
3	C_6H_5	CH$_3$	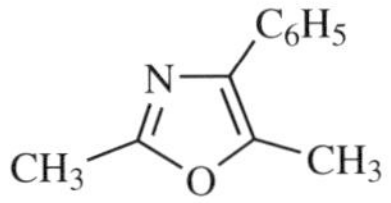 2-CH$_3$, 4-C$_6$H$_5$, 5-CH$_3$ oxazole	44
4	C_6H_5	C_2H_5	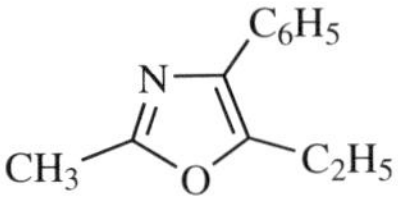 2-CH$_3$, 4-C$_6$H$_5$, 5-C$_2$H$_5$ oxazole	57

[a] Data from ref. 101.

(Scheme 1.51). This is the first report of direct conversion of ketones to oxazoles in the absence of a solvent. The authors found that the enol content of the ketone was critical. Indeed, acetophenones required addition of *p*-toluenesulfonic acid to afford good yields of **191**. Copper(II) triflate was ineffective as a catalyst.

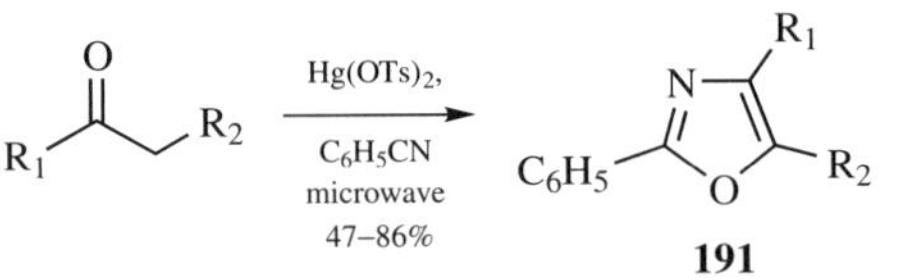

$R_1-C(=O)-CH(R_2)$ + C_6H_5CN →[Hg(OTs)$_2$, microwave, 47–86%] 2-C$_6$H$_5$, 4-R_1, 5-R_2 oxazole (**191**)

Scheme 1.51

For example, reaction of propiophenone and benzonitrile with mercury(II) tosylate affords 2,4-diphenyl-5-methyloxazole **196** (Scheme 1.52). Mechanistically, the authors proposed initial formation of a di(α-arylalkanoyl)Hg(II) species **192** from reaction of the enol tautomer of propiophenone with mercury(II) tosylate. Reaction of **192** with benzonitrile generates **193**, which collapses to the oxirane **194** with loss of mercury. Rearrangement of **194** and loss of

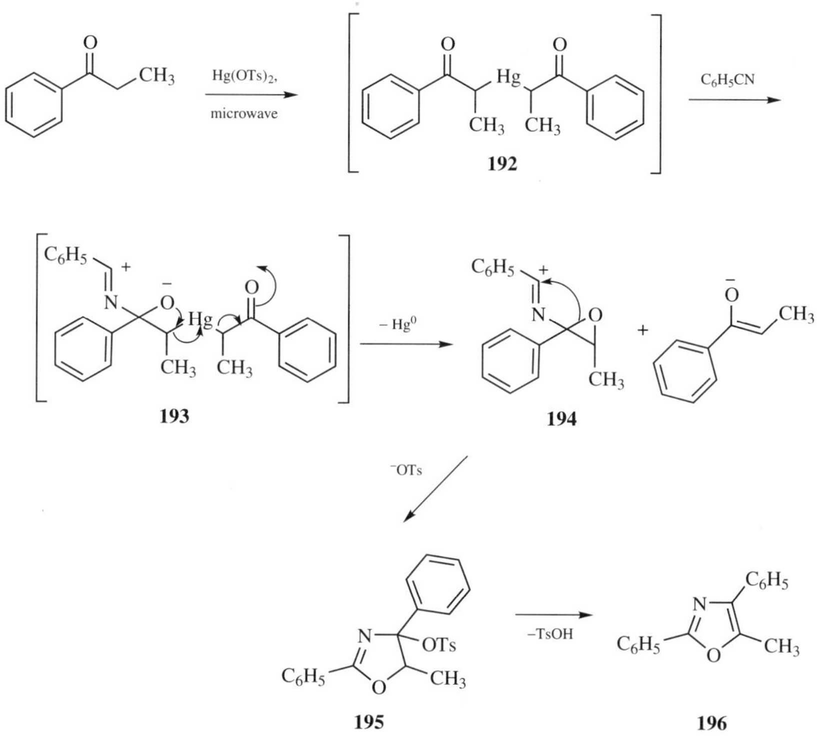

Scheme 1.52

toluenesulfonic acid then affords 2,4-diphenyl-5-methyloxazole **196**. Additional examples are shown in Table 1.14.

1.3.4. Oxazoles from α-Substituted Ketones

Cyclization of an amide or urea with an α-haloketone is a well-established and still widely used method to prepare oxazoles. Both 2,4-disubstituted oxazoles **197** and 2-amino-4-substituted oxazoles **198** are readily available from this methodology (Scheme 1.53).[8] The commercial availability of the starting materials and the

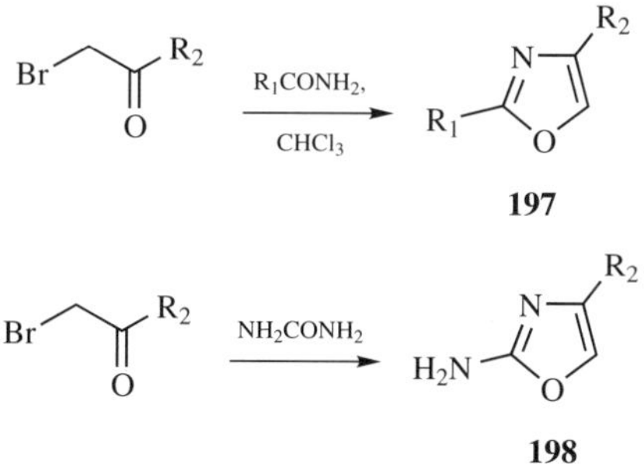

Scheme 1.53

TABLE 1.14. MICROWAVE ASSISTED SYNTHESIS OF 2-PHENYL-4-SUBSTITUTED OXAZOLES AND 4,5-DISUBSTITUTED 2-PHENYLOXAZOLES FROM ARYL KETONES, BENZONITRILE, AND MERCURY(II)TOSYLATE[a]

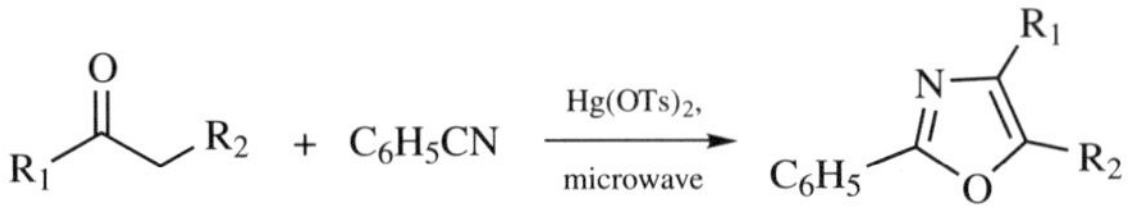

Figure 1.16

Entry	R_1	R_2	Product	Yield (%)
1	C_6H_5	H	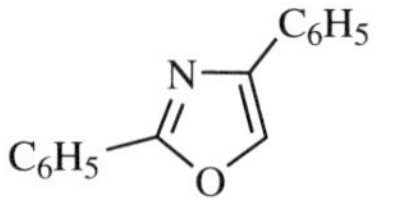	51
2	$4\text{-Cl-}C_6H_4$	H	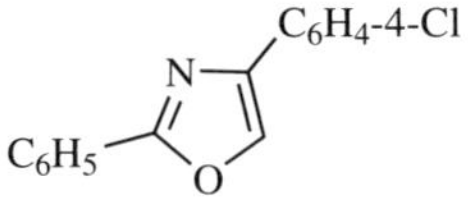	50
3	$4\text{-Cl-}C_6H_4$	CH_2CH_3	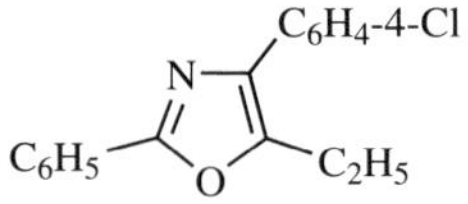	71
4	$COOC_2H_5$	H	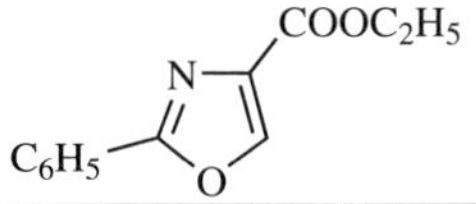	47

[a] Data from ref. 102.

simplicity of the reaction makes this an efficient method for parallel synthesis or in combinatorial reactions, despite the widely varying yields. Several examples that illustrate this method are discussed in this section.

Meguro and co-workers[37] used this method, among others, to prepare a variety of potential antidiabetic agents. For example, cyclization of cyclohexanecarboxamide with ethyl 4-chloroacetoacetate gave 2-cyclohexyl-4-oxazoleacetic acid **199**, albeit in poor yield (Scheme 1.54). Similarly, Ohkubo and co-workers[103] prepared 4-(nitrophenyl)-2-phenyl-5-oxazolecarboxylic acid ethyl esters **200a** and **200b** as precursors to potential cerebral protective agents.

Kelly and Lang[104] found cyclization of an α-haloketone with amides to be a useful strategy for preparing model oxazoles for the synthesis of dimethylsulfomycinate. Refluxing 2-(bromoacetyl)pyridine with methacrylamide or 4-methoxybenzamide gave 2-(2-propenyl)-4-(2-pyridyl)oxazole **201** and 2-(4-methoxyphenyl)-4-(2-pyridyl)oxazole **202** (Scheme 1.55).

Malamas and co-workers[105] adapted Meguro and Fujita's[37] methodology to prepare a series of 2-aryl-5-methyl-4-oxazoleacetic acid methyl esters **203**, which

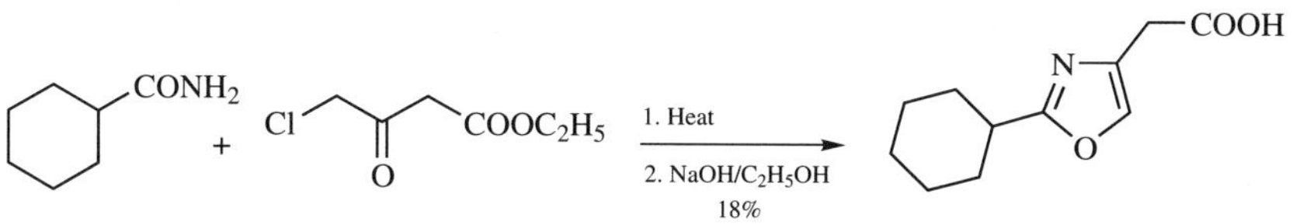

199

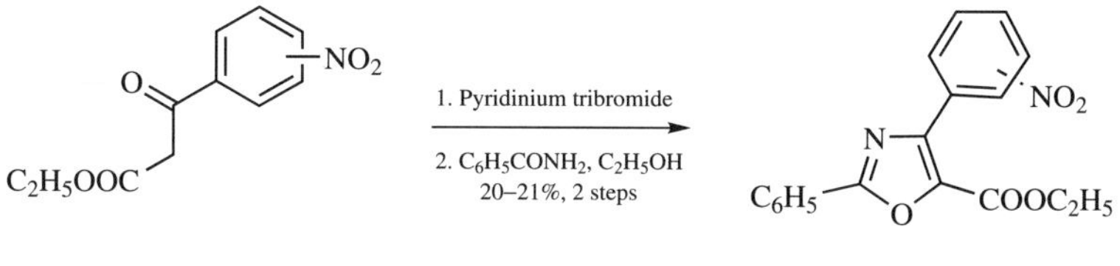

200a, 3-NO_2
200b, 4-NO_2

Scheme 1.54

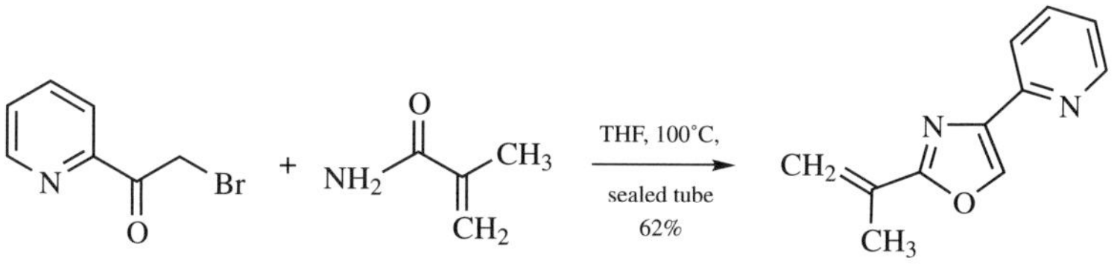

201

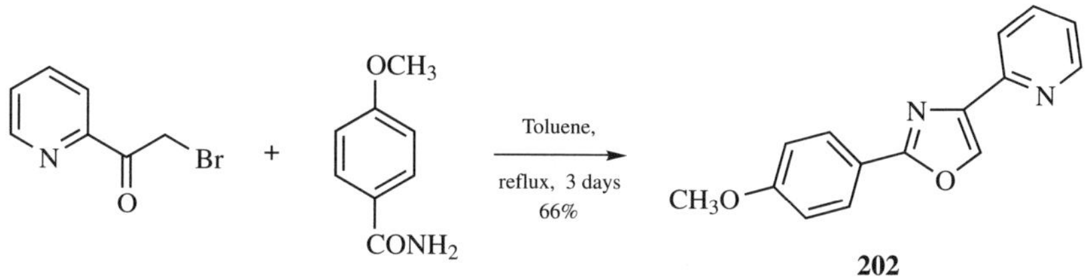

202

Scheme 1.55

were precursors to orally active inhibitors of lipoxygenases (Scheme 1.56). Panek and co-workers[106–110] also used this approach to prepare the tris-oxazole segment of the marine metabolite kabiramide C (see Section 1.5.6).

Lee and Hong[111] described the first direct synthesis of 2-alkyl-5-aryloxazoles **206** from aromatic α-methyl ketones (Scheme 1.57). They used thallium(III)triflate generated in situ to convert a ketone to the α-ketotriflate **204**. Reaction of **204** with a nitrile most likely produces a nitrilium salt **205**, as proposed by Meyers and Sicar,[112] which subsequently cyclizes to **206**. The reaction works well for aliphatic nitriles regardless of steric hindrance. Aromatic nitriles are not as useful,

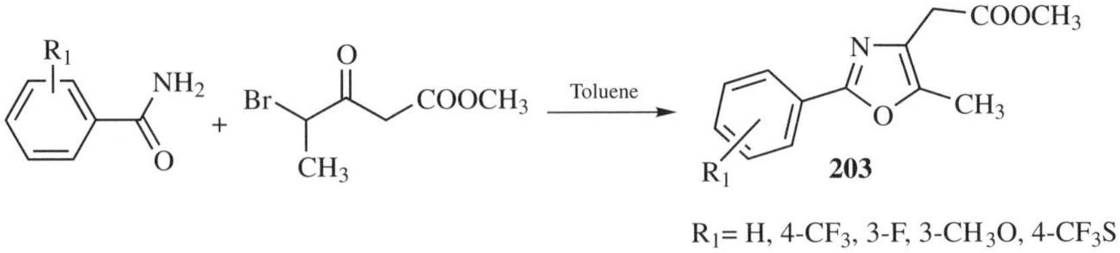

Scheme 1.56

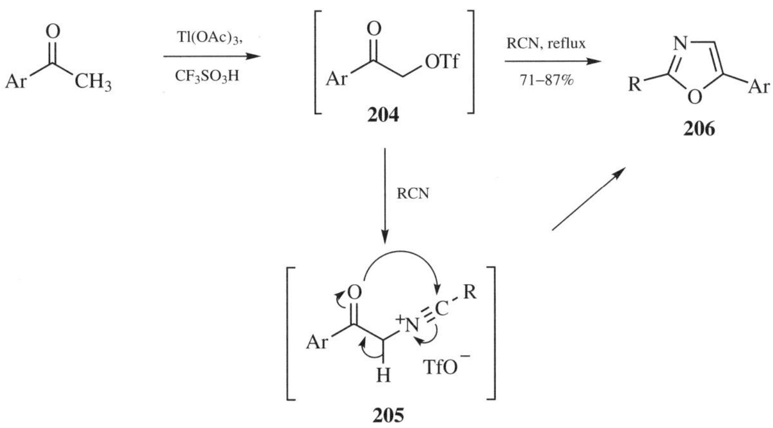

R_1 = H, 4-CF$_3$, 3-F, 3-CH$_3$O, 4-CF$_3$S

Scheme 1.57

e.g., benzonitrile gives 2,5-diphenyloxazole in 35% yield. This method is limited by the inherent toxicity of thallium reagents. Examples are shown in Table 1.15.

In a complementary approach for preparing α-ketotriflates Varma and Kumar[113] used the combination of iodobenzene diacetate and trifluoromethanesulfonic acid to prepare μ-oxobis[trifluoromethanesulfonato(phenyl)iodine] **207** in situ. This reagent cleanly converted aromatic α-methyl ketones to the corresponding α-ketotriflates **204**, which reacted with nitriles to afford 2-alkyl-5-aryloxazoles **206** in variable yield (Scheme 1.58). Mechanistically, the authors proposed the same

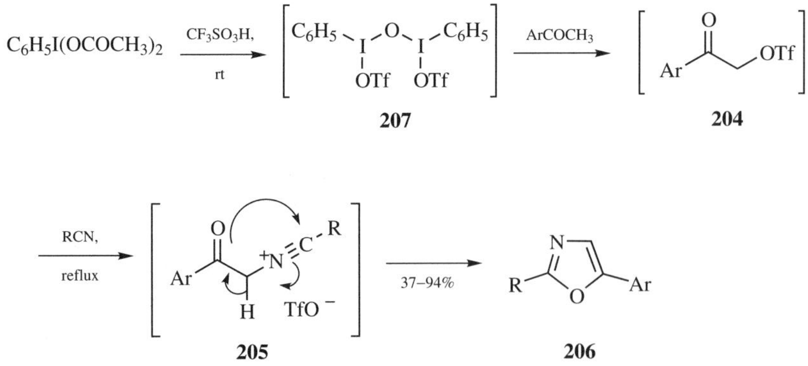

Scheme 1.58

TABLE 1.15. 2-ALKYL-5-ARYLOXAZOLES FROM AROMATIC α-METHYL KETONES, NITRILES, AND THALLIUM TRIFLATE[a]

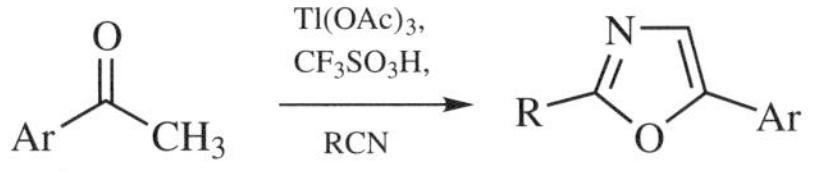

Figure 1.17

Entry	Ar	R	Oxazole	Yield (%)
1	C_6H_5	CH_3	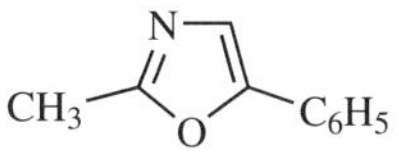	83
2	4-CH_3-C_6H_4	C_2H_5	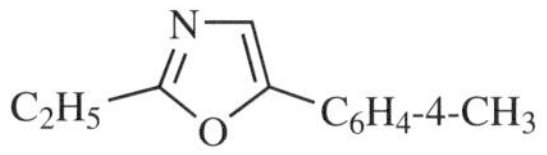	81
3	4-CH_3O-C_6H_4	C_2H_5	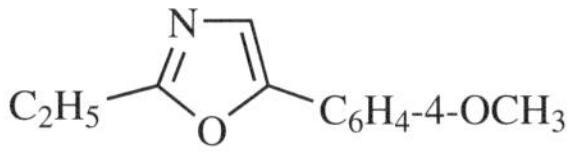	83
4	2-Thienyl	i-C_3H_7	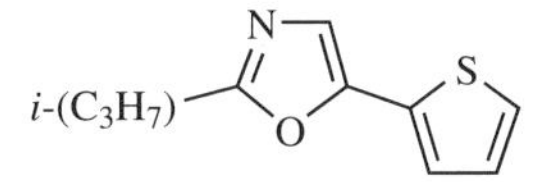	79

[a] Data from ref. 111.

rationale as Lee and Hong.[111] This method is advantageous in that it uses iodobenzene diacetate, a safe, relatively inexpensive, and nontoxic reagent. Representative examples are shown in Table 1.16.

Salgado-Zamora and co-workers[114] prepared a series of imidazo[5,1-b]oxazoles as potential radiosensitizers or antiprotozoal agents (Scheme 1.59). Alkylation of 4-bromo-2-methyl-5-nitroimidazole 208 with a series of phenacyl halides gave a 1:3 mixture of the regioisomeric imidazoles 209 and 210. These were separated, and 210 was cyclized to the desired 2-aryl-5-methyl-7-nitroimidazo[5,1-b]oxazoles 211. This reaction tolerates a wide variety of phenacyl halides. Selected examples are presented in Table 1.17.

Majo and Perumal[115] described an unprecedented synthesis of 5-aryl-4-oxazolecarboxaldehydes 215 (Scheme 1.60). They found that treatment of 2-azidoacetophenones 212 with Vilsmeier reagent at 80–90°C afforded 215. Alternatively, 215 could be prepared in two steps and one pot from a phenacyl halide and sodium azide in DMF, followed by cyclization with $POCl_3$. A mechanism involving an α-azido-imminium salt 213 was proposed. Thus the in situ generated Vilsmeier reagent reacts with a 2-azidoacetophenone 212 to generate 213. At elevated temperature, 213 cyclizes to 214, which aromatizes via loss of dimethylamine and nitrogen. Hydrolytic workup then affords the 5-aryl-4-oxazolecarboxaldehydes 215.

Synthesis and Reactions of Oxazoles

TABLE 1.16. 2-ALKYL-5-ARYLOXAZOLES FROM AROMATIC α-METHYL KETONES, NITRILES, AND IODOBENZENE DIACETATE[a]

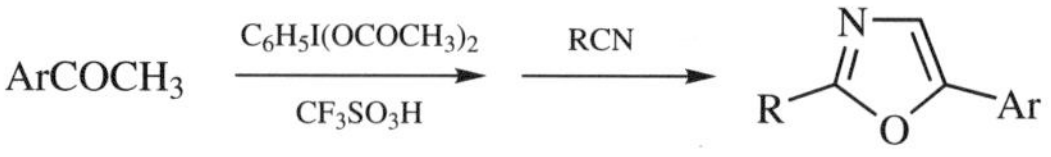

Figure 1.18

$$ArCOCH_3 \xrightarrow[CF_3SO_3H]{C_6H_5I(OCOCH_3)_2} \xrightarrow{RCN} \mathbf{206}$$

Entry	Ar	R	Oxazole	Yield (%)
1	C_6H_5	CH_3	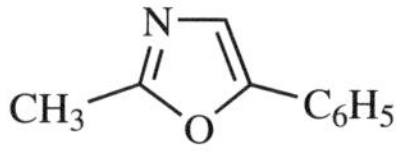 (2-CH₃, 5-C₆H₅)	94
2	4-CH_3-C_6H_4	CH_3	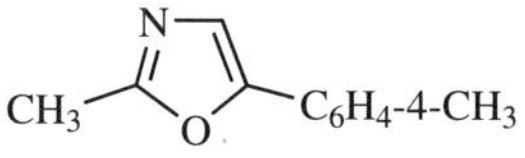 (2-CH₃, 5-C₆H₄-4-CH₃)	91
3	4-Cl-C_6H_4	CH_3	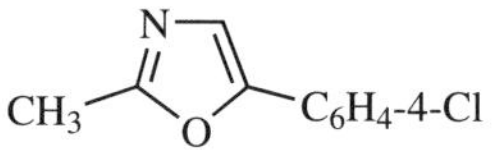 (2-CH₃, 5-C₆H₄-4-Cl)	90
4	C_6H_5	CH_2OCH_3	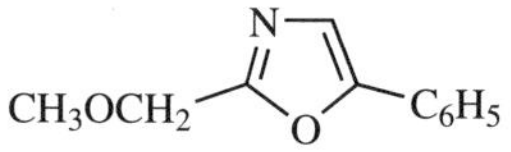 (2-CH₃OCH₂, 5-C₆H₅)	64
5	C_6H_5	C_6H_5	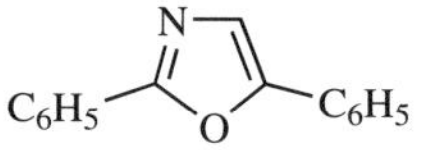 (2-C₆H₅, 5-C₆H₅)	37

[a] Data from ref. 113.

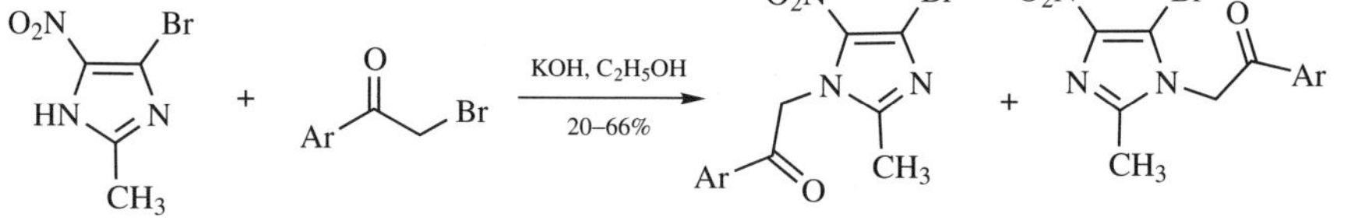

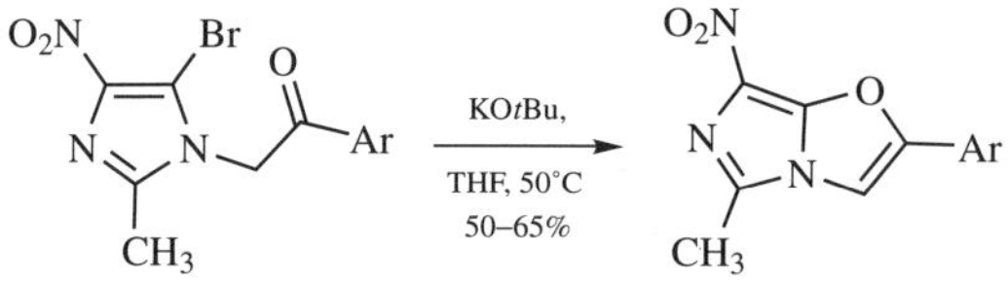

Scheme 1.59

TABLE 1.17. 2-ARYL-5-METHYL-7-NITROIMIDAZO[5,1-b]OXAZOLES[a]

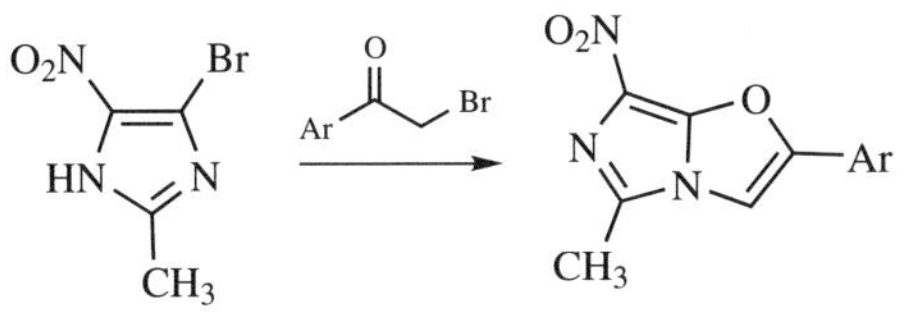

Figure 1.19

Entry	Ar	Oxazole	Yield (%)
1	C_6H_5	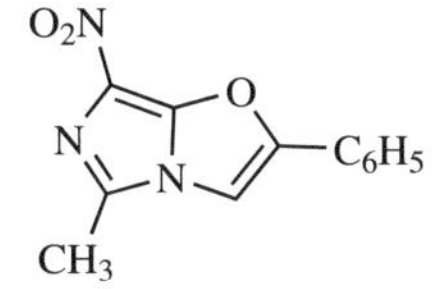	60
2	4-Cl-C_6H_4	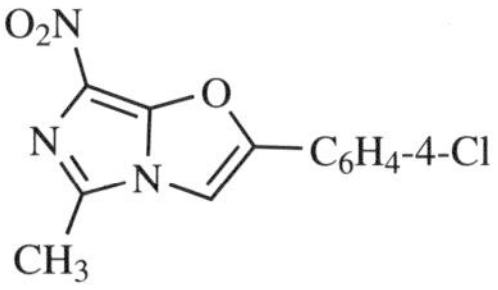	65
3	3-CF_3-C_6H_4	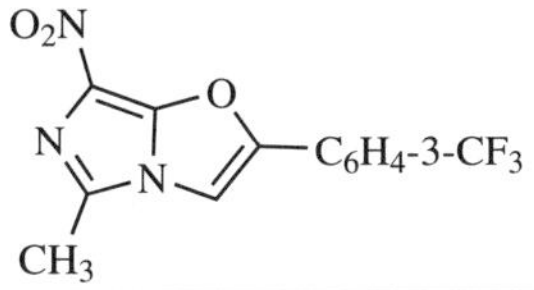	48

[a] Data from ref. 114.

Burger and co-workers[116] used a variation of an amide cyclization with an α-haloketone to prepare 2-alkyl(aryl)-5-fluoro-4-(trifluoromethyl)oxazoles **218** (Scheme 1.61). They condensed an amide with hexafluoroacetone to yield an N-(hexafluoroisopropylidene)carboxamide **217** after dehydration of the intermediate carbinolamine **216**. Reductive cyclization of **217** with $SnCl_2$ then gave a 2-alkyl(aryl)-5-fluoro-4-(trifluoromethyl)oxazole **218**. This $SnCl_2$ reductive cyclization is a useful method for introducing a trifluoromethyl group regioselectively. The yields are modest to good and the reaction is compatible with alkyl and aryl amides.

The authors adapted this methodology to prepare 3,3,3-trifluoroalanine **220**. Benzamide was first converted to **217** (Ar = C_6H_5) followed by reductive cyclization to 5-fluoro-2-phenyl-4-(trifluoromethyl)oxazole **218** (Ar = C_6H_5). Displacement of fluoride in **218** gave 5-ethoxy-2-phenyl-4-(trifluoromethyl)oxazole **219**, which was converted to **220** with iodotrimethylsilane (Scheme 1.61).

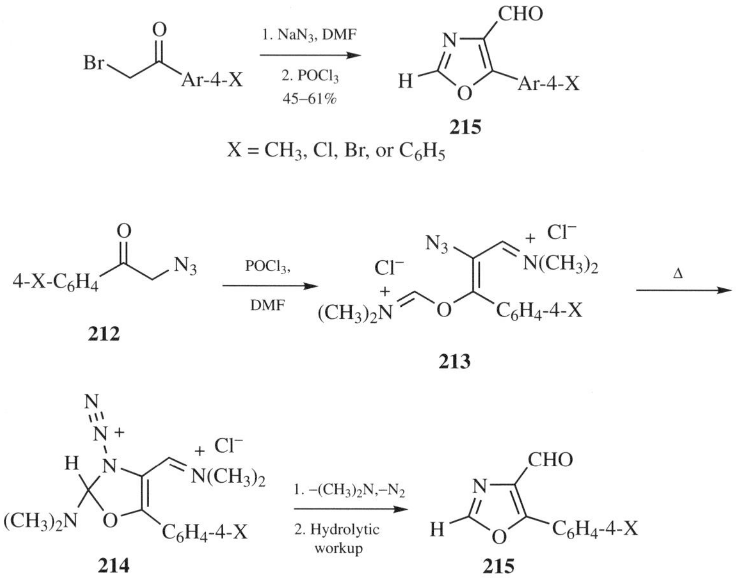

48   Synthesis and Reactions of Oxazoles

Scheme 1.60

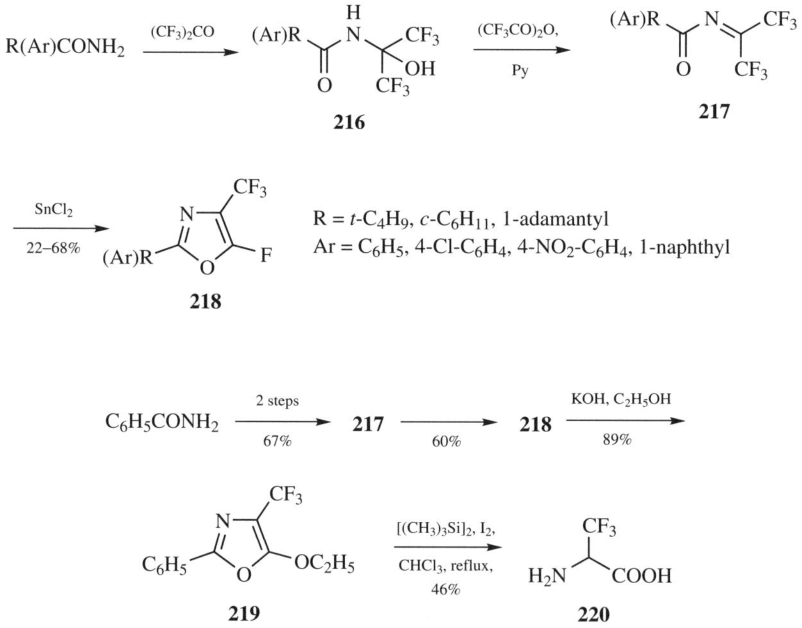

Scheme 1.61

Tobinaga and co-workers[117] prepared a series of 2,4,5-trisubstituted oxazoles **226** in one step from oxidation of ketones by the iron(III) solvates of nitriles (Scheme 1.62). Both $Fe(RCN)_6(ClO_4)_3$ and $Fe(RCN)_6(FeCl_4)_3$ were effective in this reaction. Mechanistically, the authors proposed that the starting ketone **221** enolizes, and it is this tautomer **222** that undegoes a one-electron oxidation by the iron(III) solvate to generate the radical **223**. Reaction of **223** with the nitrile solvent produces a new radical **224**, which is subsequently oxidized to the nitrilium cation **225**, followed by cyclization to the 2,4,5-trisubstituted oxazole **226**. Alternatively, **223** can undergo a one-electron oxidation to furnish **227**, which reacts with the nitrile solvent, yielding **225**. In addition, pathways involving an α-acetoxyketone **228** or an α-chloroketone **229** may also be operative. A number of these intermediates have been isolated, supporting the proposed mechanism. Representative examples are shown in Table 1.18.

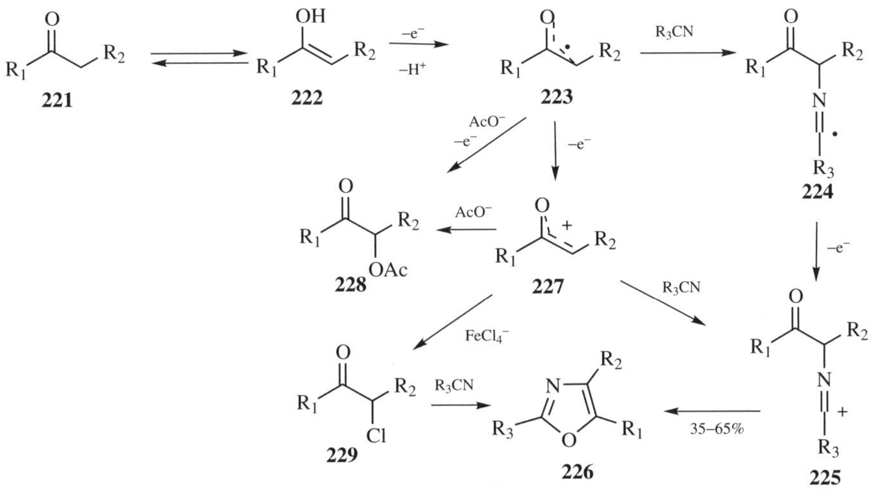

Scheme 1.62

Copper(II) triflate has also been used as an oxidant in a one-step preparation of 2,4,5-trisubstituted oxazoles from ketones and nitriles.[118] Kaufmann and co-workers[11] evaluated this strategy for preparation of a series of 2,4,5-trisubstituted oxazoles **231a–d** as flash-lamp pumped laser dyes (Scheme 1.63). They found

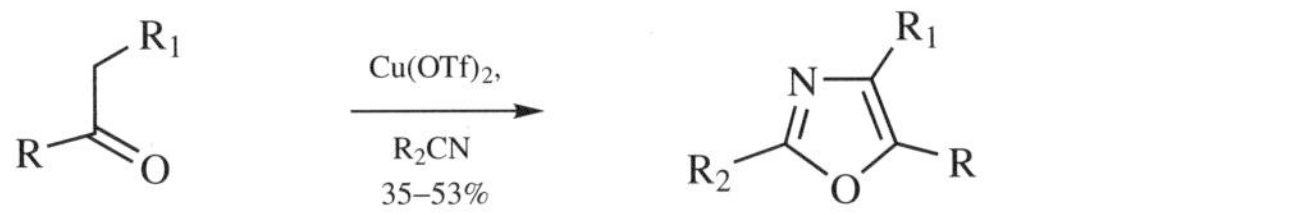

230a R = C_6H_5, R_1 = C_6H_5
230b R = CH_3, R_1 = C_6H_5
230c R = C_6H_5, R_1 = C_6H_5
230d R = C_6H_5, R_1 = C_6H_5

231a R = C_6H_5, R_1 = C_6H_5, R_2 = CH_3
231b R = CH_3, R_1 = C_6H_5, R_2 = CH_3
231c R = C_6H_5, R_1 = C_6H_5, R_2 = C_6H_5
231d R = C_6H_5, R_1 = C_6H_5, R_2 = 4-pyridyl

Scheme 1.63

TABLE 1.18. 2,4,5-TRISUBSTITUTED OXAZOLES FROM OXIDATION OF KETONES WITH IRON(III) SOLVATES OF NITRILES[a]

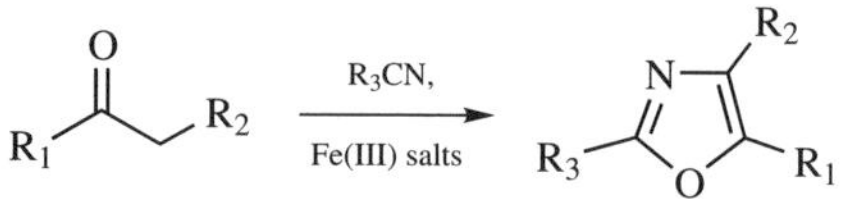

Figure 1.20   221   →   226

Entry	Ketone	Nitrile	Oxazole	Yield (%)[b]
1	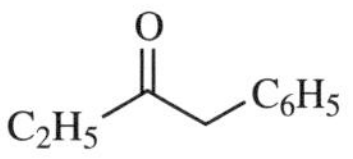 C_2H_5-CO-CH_2-C_6H_5	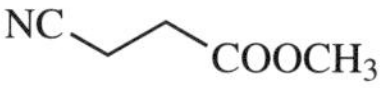 NC-CH_2CH_2-COOCH$_3$	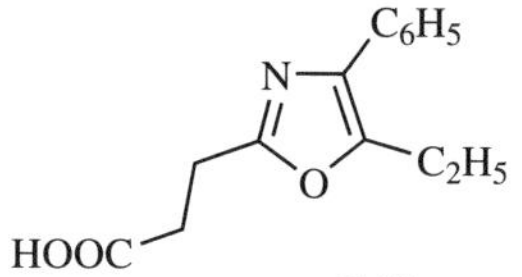 4-C_6H_5, 5-C_2H_5, 2-(CH_2CH_2COOH) oxazole	52
2	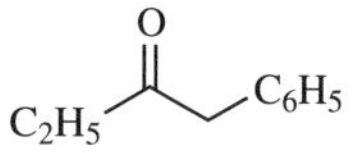 C_2H_5-CO-CH_2-C_6H_5	 NC-CH_2-COOCH$_3$	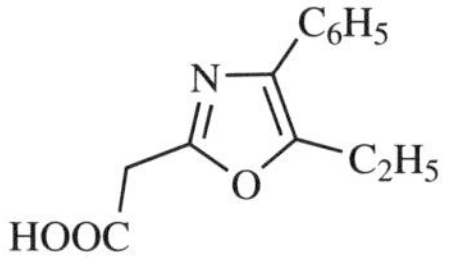 4-C_6H_5, 5-C_2H_5, 2-(CH_2COOH) oxazole	33
3	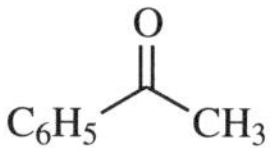 C_6H_5-CO-CH_3	CH_3CN	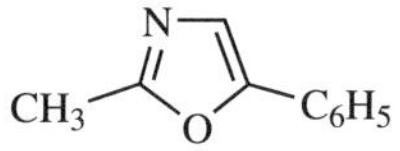 2-CH_3, 5-C_6H_5 oxazole	35
4	Cholestanone	CH_3CN	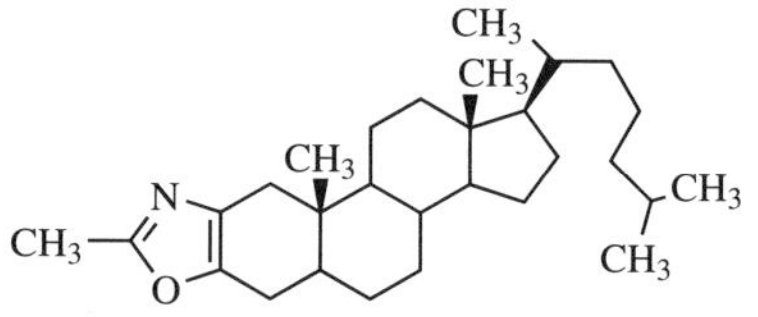 2-methyl oxazolo-fused cholestane	34

[a] Data from ref. 117.
[b] Fe(ClO$_4$)$_3$ or FeCl$_3$ was used as catalyst.

that reaction of a ketone **230a–c** with acetonitrile or benzonitrile in the presence of copper(II) triflate afforded 4,5-diphenyl-2-methyloxazole **231a**, 2,5-dimethyl-4-phenyloxazole **231b**, and 2,4,5-triphenyloxazole **231c** in 35–53% yield. Efforts to improve the yields of **231a–c** using 1:1 stoichiometry of **230**/R$_2$CN or solvents (e.g., ethyl acetate, glyme, and diglyme) were unsuccessful. There was no evidence for **231d** using 4-pyridylnitrile **230d** in this reaction. The authors concluded that this reaction was limited to only those examples wherein the nitrile was used as the solvent.

Kidwai and Kumar[120] reported a microwave-accelerated reaction of urea **232** with phenacylbromide to yield the iminooxazoline **233** (Scheme 1.64). In this case, the total reaction time required was only 1.0–1.5 min. Patel and Fernandes[121] prepared the 2-amino-4-substituted oxazoles **235** as precursors to novel oxazolo[5,4-c]pyrazoles and oxazolo[5,4-c]pyridazines, which were evaluated as antibacterial agents. Cyclization of α-bromoketones **234** with urea in refluxing DMF afforded **235** (Scheme 1.65).

Camoutsis[122] reviewed the syntheses and biological properties of heterocyclic-condensed steroids. Among the strategies used to prepare these analogs, reaction of

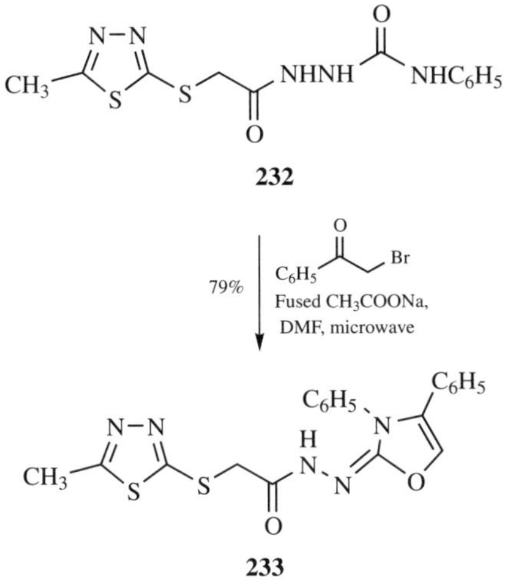

Scheme 1.64

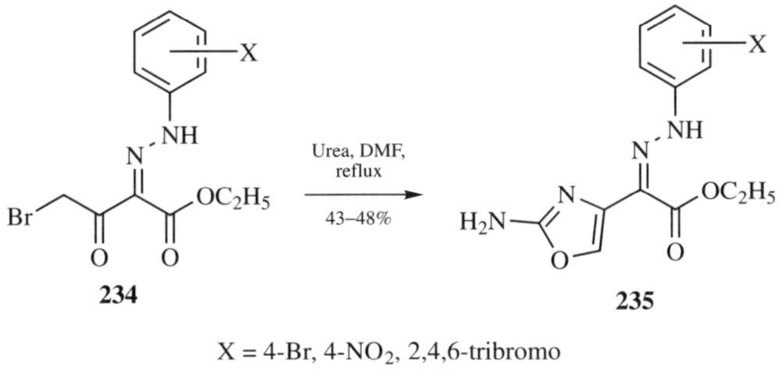

Scheme 1.65

X = 4-Br, 4-NO₂, 2,4,6-tribromo

an α-bromoketone with KOCN gave an intermediate 2(3H)-oxazolone, which was then converted to a fused oxazole. For example, reaction of α-bromoketones **236** with KOCN yields 2(3H)-oxazolones **237**, which were converted to a series of 2(alkylthio)steroidal oxazoles **238** (Scheme 1.66). The author also prepared the regioisomeric oxazoles **240** and **242** from reaction of the α-acetoxyketones **239** and **241** with ammonium acetate.

1.3.5. Oxazoles via Cyclizations

1.3.5.1. Propargylic Amides

Base-catalyzed cyclization of propargyl amides is an expeditious approach to a variety of substituted oxazoles. In addition, propargyl amides can serve as

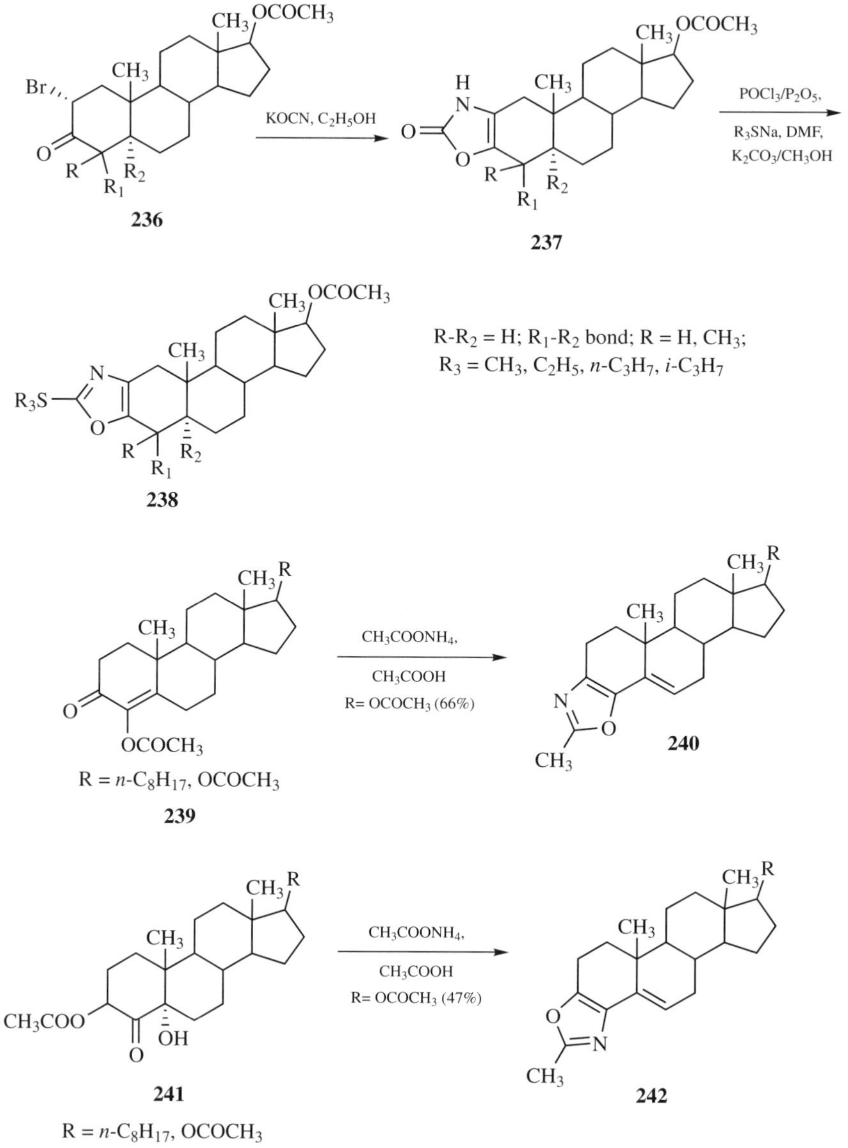

Scheme 1.66

precursors to (*E*)-β-iodo(vinyl)sulfones, useful intermediates to 2,5-disubstituted oxazoles. Several examples are highlighted in this section.

Nilsson and Hacksell[123] isolated 2,5-dimethyl-4-phenyloxazole **245** ($R_1 = CH_3$, $R_2 = C_6H_5$) as the only product from methylation of *N*-(1-phenyl-2-propynyl)acetamide **244** ($R_1 = CH_3$, $R_2 = C_6H_5$). The authors found that propargyl amides could be cyclized to 2,5-disubstituted and 2,4,5-trisubstituted oxazoles (Scheme 1.67). Thus a propargyl alcohol **243** was converted to the corresponding amide **244** in

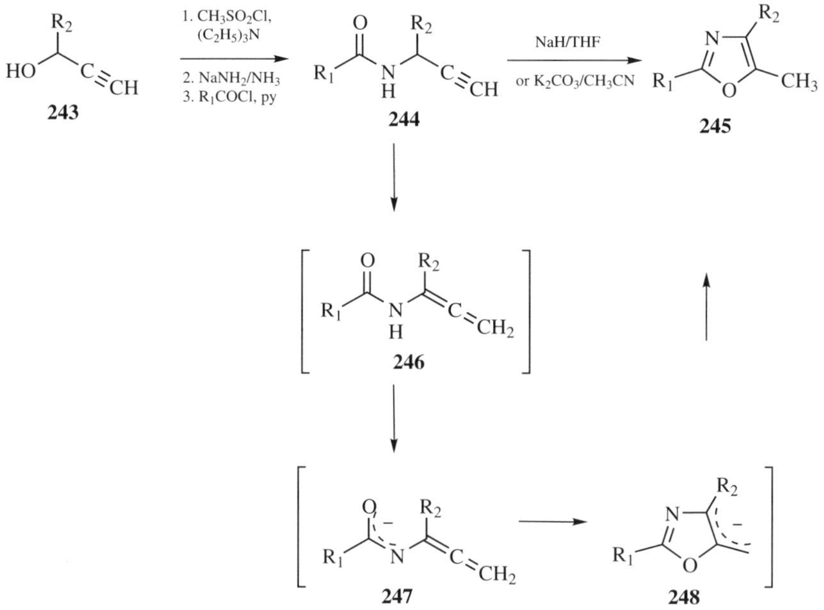

Scheme 1.67

three straightforward steps. Cyclization of **244** using NaH/THF or K$_2$CO$_3$/MeCN gave the desired oxazoles **245** in low to excellent yield. There are two possible mechanisms to account for the production of **245**, but the authors favored one in which **244** first isomerizes to the allene **246**. Deprotonation of **246** and attack at the central allenic carbon by the nucleophilic amide oxygen affects cyclization of **247**, producing **248** and ultimately **245**. The authors were able to obtain ^{1}H NMR evidence for allene intermediates. Selected examples of 2,5-disubstituted and 2,4,5-trisubstituted oxazoles prepared are shown in Table 1.19. The yields of this reaction vary widely.

Reisch and co-workers[124] prepared 5-methyleneoxazoles from the base-catalyzed cyclization of α,α-dialkylpropargyl amides (Scheme 1.68). In this case, the requisite α,α-dialkylpropargyl amides **250** were prepared by acylation of the corresponding propargyl amine **249** with isatoic anhydride. Refluxing **250** with alcoholic KOH then affected cyclization to **251** in good to excellent yields. Cyclization of **250** with triphosgene afforded both **251** and 2-methyleneoxazolo-quinazolines **252**.

Freedman and Huber[125] showed that the cyclization of **253** with NaH/DMSO did not afford 3-methyl-1,4-benzoxazepin-5(4H)-one **255**, as previously reported.[126] Instead, they isolated 2-(2-hydroxyphenyl)-5-methyloxazole **256** and confirmed the structure by independent synthesis from 2-hydroxy-N-(2-propynyl)benzamide **257** (Scheme 1.69). The authors considered an intermediate allene but were not able to unequivocally establish the mechanism for conversion of **253** to **256**.

Wipf and co-workers[127] developed a base-catalyzed cyclization of alkynyl ketones into a general synthesis of furans. The authors then extended the

TABLE 1.19. 2,5-DIALKYL-; 2,4,5-TRIALKYL-; AND 4-ARYL-2,5-DIALKYLOXAZOLES FROM N-PROPARGYLAMIDES[a]

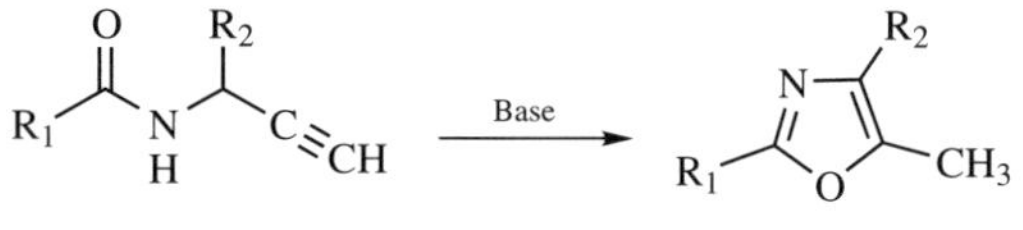

Figure 1.21 244 → 245 (Base)

Entry	R_1	R_2	Product	Yield (%)
1	CH_3	H	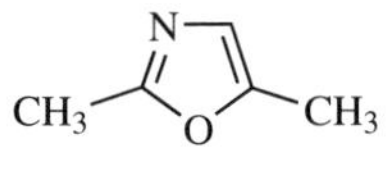 2,5-dimethyloxazole	5[b]
2	CH_3	CH_3	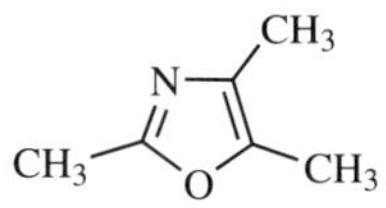 2,4,5-trimethyloxazole	16[b]
3	CH_3	4-NO_2-C_6H_4	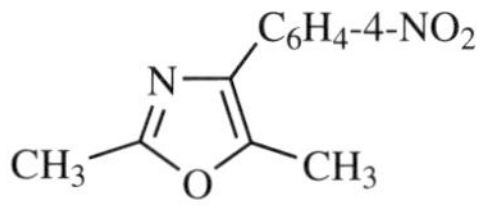 4-(4-NO_2-C_6H_4)-2,5-dimethyloxazole	95[c]
4	CF_3	C_6H_5	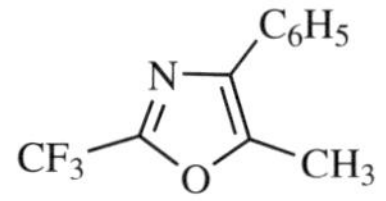 2-CF_3-4-C_6H_5-5-CH_3-oxazole	97[c]
5	CH_3	4-CH_3O-C_6H_4	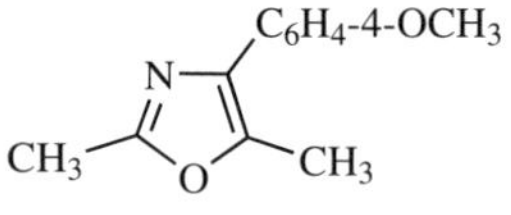 4-(4-CH_3O-C_6H_4)-2,5-dimethyloxazole	54[b]
6	CH_3	C_6H_5	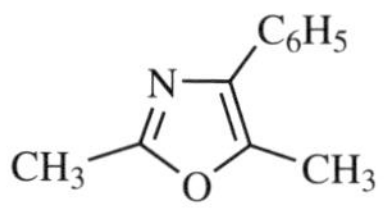 4-C_6H_5-2,5-dimethyloxazole	93[b]

[a] Data from ref. 123.
[b] NaH/THF.
[c] K_2CO_3 (10 equiv.)/MeCN.

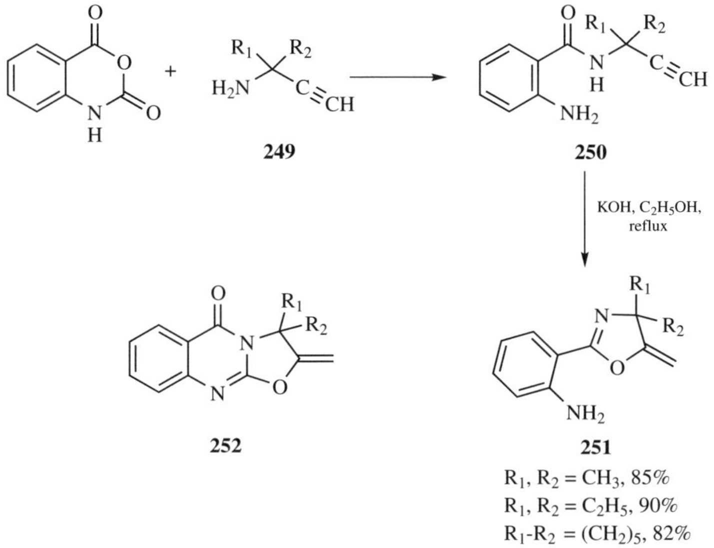

Scheme 1.68

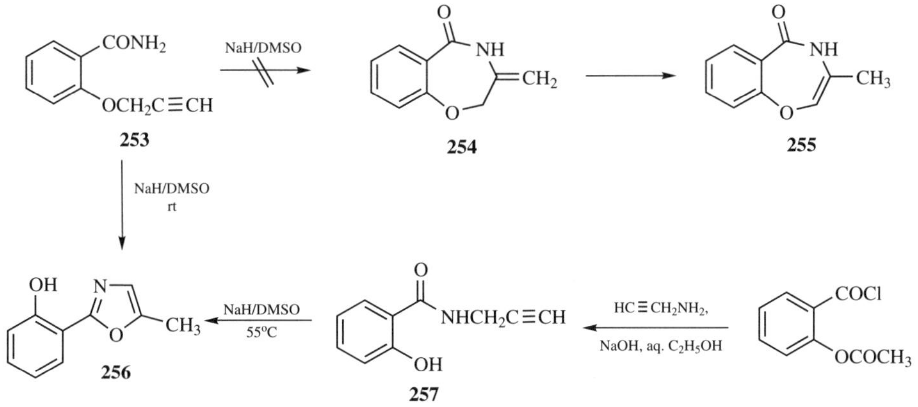

Scheme 1.69

methodology further to include propargyl amides and thioamides to prepare oxazoles and thiazoles. For example, benzoylation of propargyl amine **258** produced the propargyl amide **259**, which was treated with NaHMDS/THF at −78°C. Upon warming to 70°C the authors isolated 2-phenyl-5-vinyloxazole **262** in excellent yield (Scheme 1.70). They proposed a cycloelimination mechanism by which loss of OBn⁻ from **259** generates the cumulene **260**, which cyclizes to **261** and ultimately yields **262**. A more traditional intramolecular S_N2' O-alkylation of **259** was also considered to be mechanistically viable.

Other examples of more structurally complex oxazoles prepared using this methodology are shown in Scheme 1.71. It is noteworthy that cyclization of the

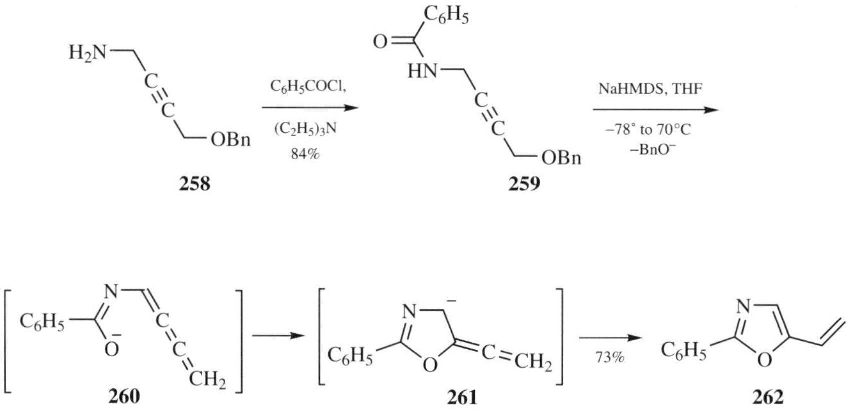

Scheme 1.70

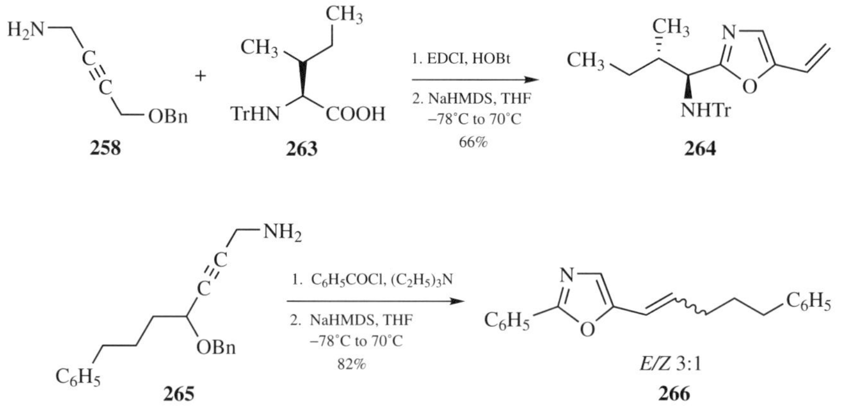

Scheme 1.71

propargyl amide derived from **258** and **263** furnished **264** as a single diastereomer, with no loss of stereochemical integrity. Benzoylation of **265** and treatment with NaHMDS afforded **266** in 82% yield isolated as a 3:1 mixture of *E/Z* isomers.

In Japan, workers[128] invoked conjugated allenyl esters as intermediates in their efforts to develop cysteine protease inhibitors. They envisioned that α-acetamido-α-alkynylmalonates (DAM) (e.g., **270**) would undergo selective hydrolysis to produce a chiral monoacid that could be decarboxylated to yield an allenyl ester intermediate. This reactive species could then capture an active-site nucleophile, thereby inactivating the enzyme. The α-acetamido-α-alkynylmalonates were readily prepared (Scheme 1.72). Staundinger reaction of diethyloxomalonate with iminophosphorane **267** gave the acetylimino malonate **268**, which was elaborated to **270** by reaction with an alkynllithium reagent **269**.

Treatment of **270** (R = 4-CH_3-C_6H_4) with porcine liver esterase (PLE) did not afford the expected chiral monoacid. Instead, PLE caused sequential hydrolysis with concomitant decarboxylation to produce an allene **271**, which cyclized to a 2,5-disubstituted 4-oxazolecarboxylic acid ester. The authors isolated a series of

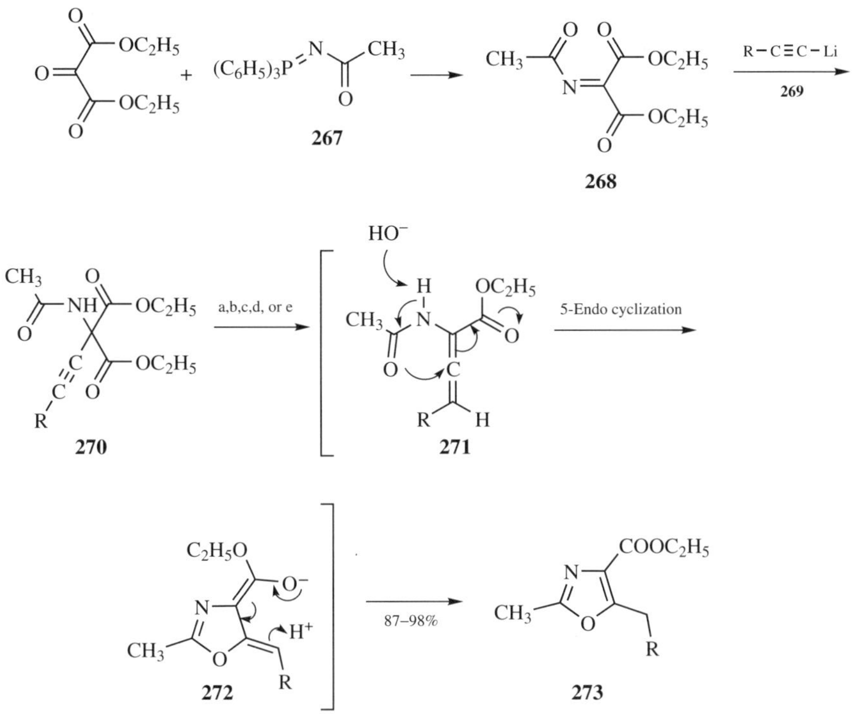

Scheme 1.72

(a) PLE 0.1M phosphate buffer (pH = 7.5) acetone/35–45°C; (b) 0.1M phosphate buffer (pH = 7.5)/acetone 35–45°C; (c) 1N KOH-C$_2$H$_5$OH/0°C; (d) C$_2$H$_5$NH$_2$, 1N KOH-C$_2$H$_5$OH/0°C; (e) imidazole, 1N KOH-C$_2$H$_5$OH/0°C

2,5-disubstituted 4-oxazolecarboxylic acid esters **273** in excellent yield from a variety of mildly basic conditions. They attributed the formation of **273** to a rare intramolecular 5-endo cyclization of **271** to **272**, which then gave rise to the observed products. Examples of this reaction are shown in Table 1.20. The yields are generally excellent and the reaction conditions are mild.

Short and Ziegler[129] employed propargylamides as precursors to (E)-β-iodo(vinyl)sulfones, which were key intermediates in their synthesis of 5-[(phenylsulfonyl)methyl]-2-substituted oxazoles (Scheme 1.73). Here, a propargylamide **274** was iodosulfonated photolytically to afford the key (E)-β-iodo(vinyl)sulfones **275**.

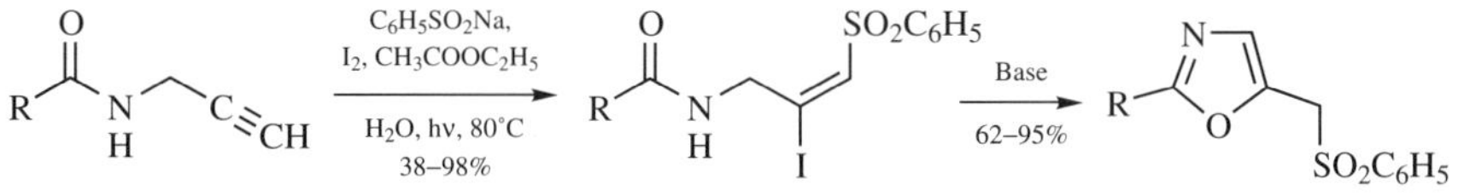

Base = LiHMDS, NaH, or DIPEA

Scheme 1.73

TABLE 1.20. 2,5-DISUBSTITUTED-4-OXAZOLECARBOXYLIC ACID ETHYL ESTERS FROM DIETHYL α-ALKYNYLMALONATES (DAM)[a]

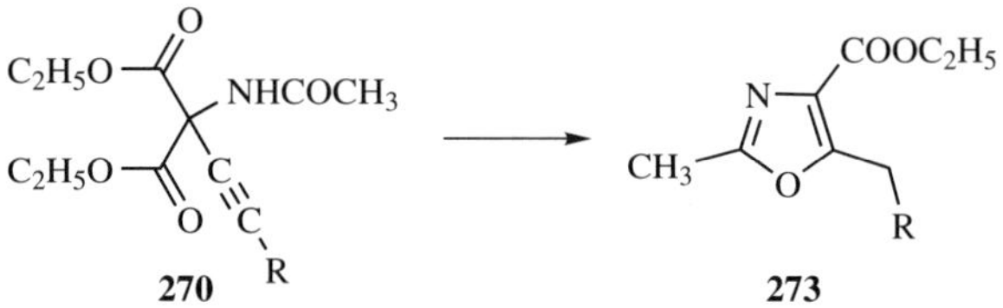

Figure 1.22

Entry	α-Alkynylmalonate	Product	Yield (%)
1	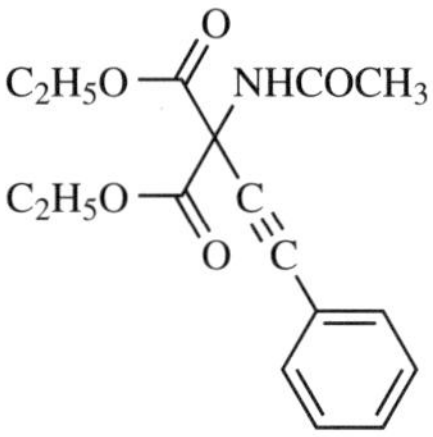 (R = CH₂C₆H₅)	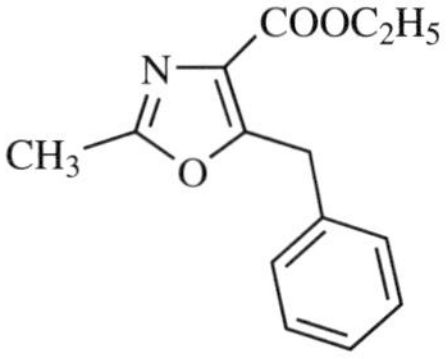	98
2	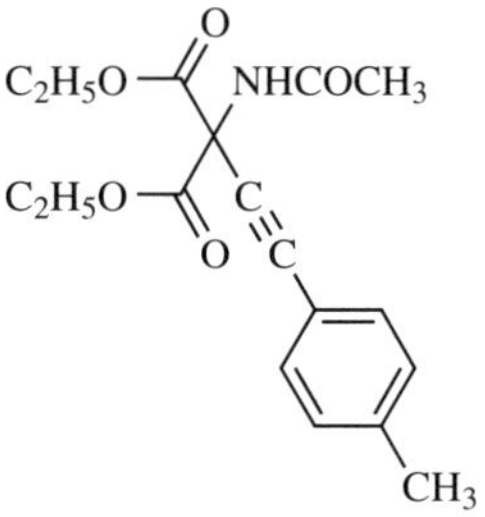 (R = CH₂-C₆H₄-CH₃)	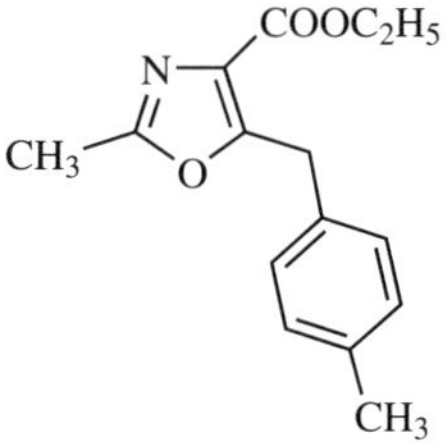	100 (87)[b]
3	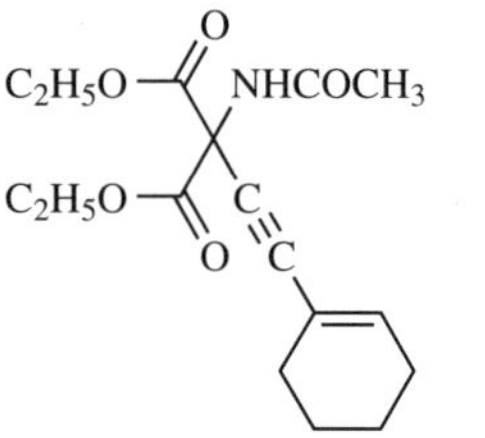 (R = cyclohexenyl)	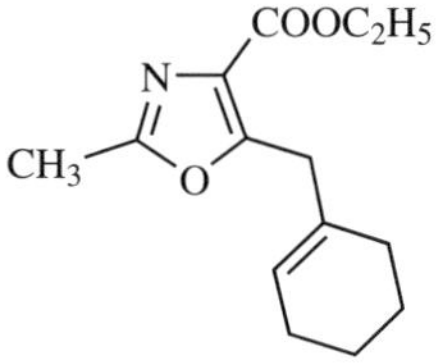	97
4	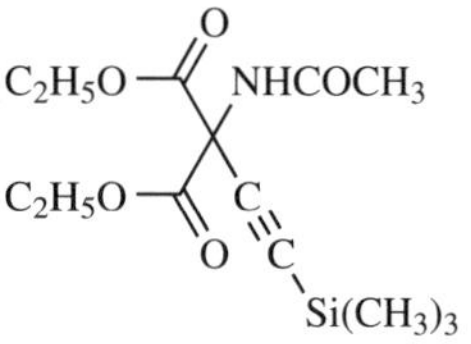 (R = Si(CH₃)₃)	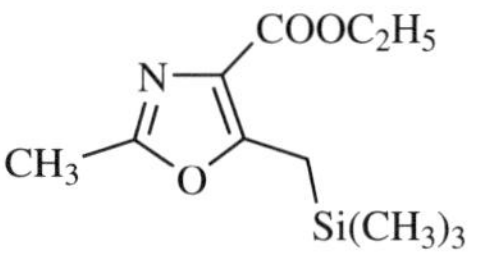	97

[a] Data from ref. 128.
[b] In 0.1 M phosphate buffer.

TABLE 1.21. 5-[(PHENYLSULFONYL)METHYL]-2-SUBSTITUTED OXAZOLES FROM PROPARGYL AMIDES VIA (E)-β-IODO(VINYL)SULFONES[a]

Entry	R	Conditions	Oxazole	Yield (%)
1	H	LiHMDS, 0°C, 2 h		95
2	4-CH$_3$O-C$_6$H$_4$	NaH, 65°C, 2 h		62
3	C$_2$H$_5$OOCCH=CH-	DIPEA, 65°C, 3 h		91
4	t-C$_4$H$_9$O	NaH, 0°C, 7.5 h		66

[a] Data from ref. 129.

Cyclization of **275** to a 5-[(phenylsulfonyl)methyl]oxazole **276** could be accomplished in good to excellent yield by treatment with base. Examples of oxazoles prepared using this methodology are shown in Table 1.21.

1.3.5.2. Azides

Azides have been used by a number of groups as precursors to oxazoles, and several examples that illustrate the diversity of available oxazole substitution patterns are described.

Eguchi and co-workers[130] developed a general synthesis of 2,5-disubstituted and 2,4,5-trisubstituted oxazoles from an intramolecular Aza-Wittig reaction of (Z)-β-(acyloxy)vinyl azides (Scheme 1.74). An α-bromoketone **277** was converted to an α-azidoketone **278**, which was O-acylated at −78°C to give, exclusively, a (Z)-β-(acyloxy)vinyl azide **279**. This was attributed to intramolecular chelation of lithium by the enol oxygen and the α-nitrogen atom of the azide.

Treatment of **279** with triethylphosphite produced an intermediate iminophosphorane that then cyclized to a 2,5-disubstituted or 2,4,5-trisubstituted oxazole **226** in good to excellent yield. The authors found triethylphosphite was more effective than tributylphosphine or triphenylphosphine in the Staudinger reaction.[131] This

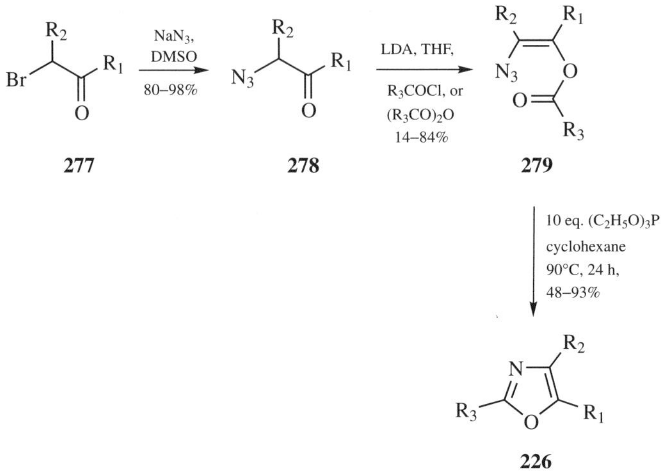

Scheme 1.74

method is advantageous, because it avoids strong dehydrating agents such as P_2O_5, H_2SO_4, $SOCl_2$, $POCl_3$, and Lewis acids. In addition, the mild, nonacidic reaction conditions afford a variety of oxazoles, as shown in Table 1.22.

Molina and co-workers[132] employed iminophosphoranes derived from α-azidoketones in their one-step syntheses of oxazole alkaloids (Scheme 1.75). Reaction of 4-methoxyphenacyl azide with nicotinoyl chloride in the presence of triphenylphosphine gave O-methylhalfordinol **280** directly in 56% yield. Similarly, reaction of 4-methoxyphenacyl azide with 3,4-dimethoxycinnamoyl chloride gave annuloline **281** in comparable yield after chromatographic purification. The authors proposed initial formation of an iminophosphorane, which was acylated with elimination of triphenylphosphine oxide to afford an imidoyl chloride, which then cyclized to the oxazole.

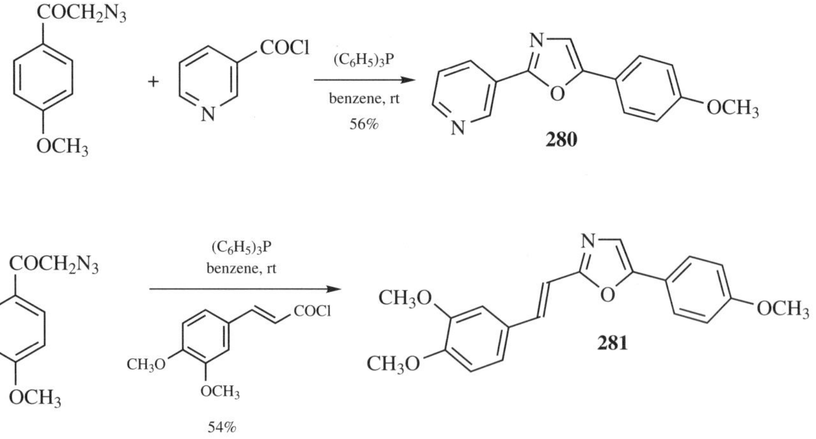

Scheme 1.75

TABLE 1.22. 2,5-DISUBSTITUTED- AND 2,4,5-TRISUBSTITUTED-OXAZOLES FROM INTRAMOLECULAR AZA-WITTIG REACTION OF (Z)-β-(ACYLOXY)VINYL AZIDES[a]

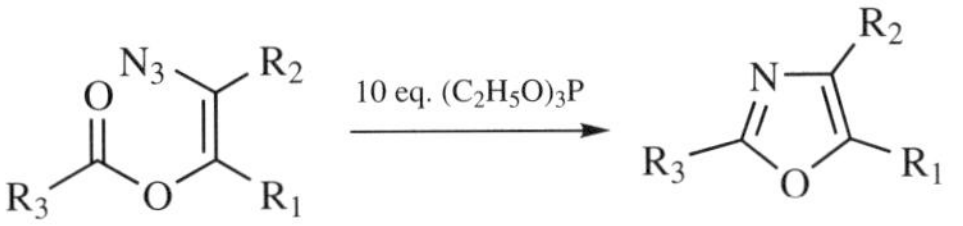

Figure 1.24

Entry	Reactant	Product	Yield (%)
1	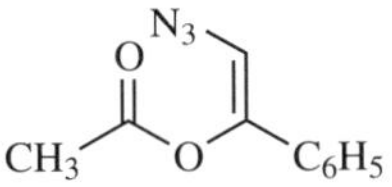 R_3=CH$_3$, R_1=C$_6$H$_5$, R_2=H	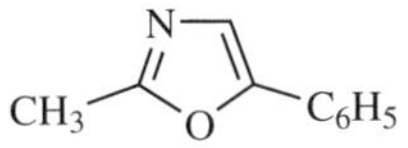 2-CH$_3$, 5-C$_6$H$_5$ oxazole	87
2	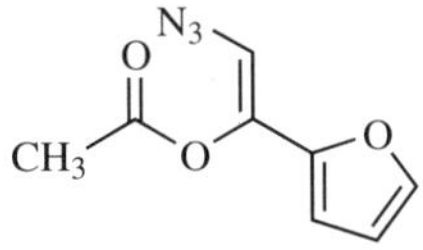 R_3=CH$_3$, R_1=2-furyl, R_2=H	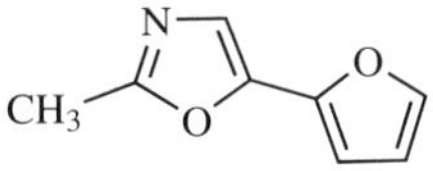 2-CH$_3$, 5-(2-furyl) oxazole	61
3	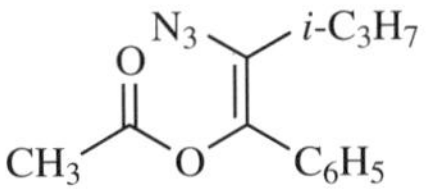 R_3=CH$_3$, R_1=C$_6$H$_5$, R_2=i-C$_3$H$_7$	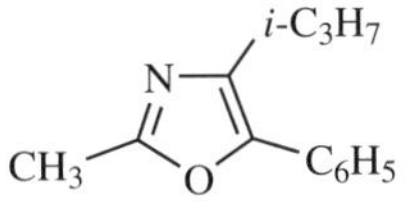 2-CH$_3$, 4-i-C$_3$H$_7$, 5-C$_6$H$_5$ oxazole	58
4	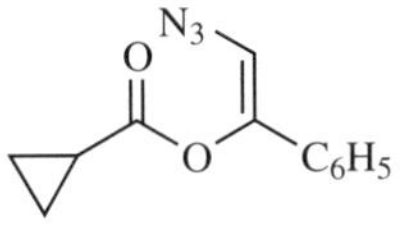 R_3=cyclopropyl, R_1=C$_6$H$_5$, R_2=H	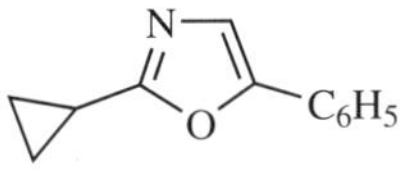 2-cyclopropyl, 5-C$_6$H$_5$ oxazole	61
5	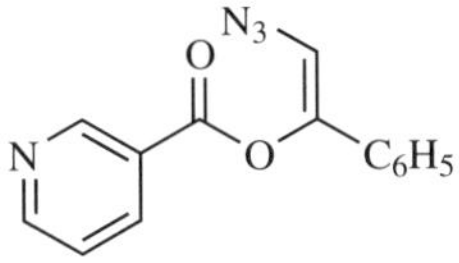 R_3=3-pyridyl, R_1=C$_6$H$_5$, R_2=H	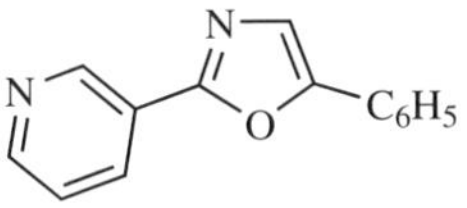 2-(3-pyridyl), 5-C$_6$H$_5$ oxazole	74
6	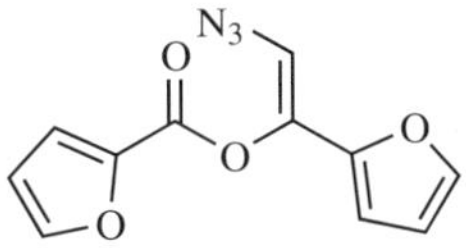 R_3=2-furyl, R_1=2-furyl, R_2=H	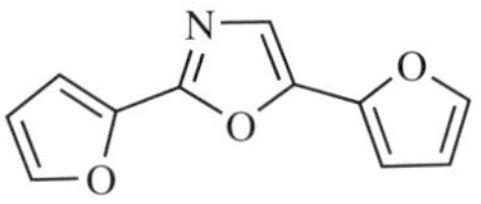 2,5-bis(2-furyl) oxazole	55

[a] Data from ref. 130.

Saalfrank and co-workers[133] prepared a series of 2-alkyl(aryl)amino-4-cyano-5-methoxyoxazoles **286** from methyl-3,3-diazidocyanoacrylate **282** (Scheme 1.76). Addition of a primary or secondary amine to **282** affords the crystalline 3-[alkyl(dialkyl)amino]-3-azido-2-cyanoacrylate esters **283** in good to excellent yield. A variety of alkyl, aryl, and heterocyclic primary amines and

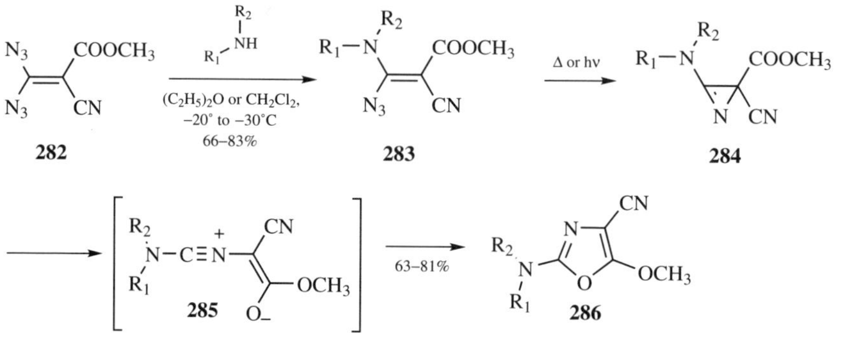

Scheme 1.76

pyrrolidine were used successfully. Thermolysis of **283** produced an azirine **284**, which ring opened to the nitrile ylide **285**. Cyclization of **285** then gave **286**. Representative examples are shown in Table 1.23.

Lautens and Roy[134] reported a synthesis of *N*-(2,4-dialkyloxazol-5-yl)oxazolidinones **288** from substituted *N*-acetoacetyl oxazolidinones **287** (Scheme 1.77). This reaction was initiated by acid-catalyzed addition of azide to the ketone in **287**. It seems that this method is limited in scope to the chiral oxazolidinone from (1*R*,2*S*)-(−)-norephedrine. Simple acyclic dialkylamide analogs of **287** were unreactive or decomposed. Simple oxazolidinone analogs of **287** produced isoxazoles or afforded complex mixtures that contained both isoxazoles and **288**. For aroylacetyl analogs of **287** (R = aryl), the authors recovered starting material but also encountered epimerization. In general, the yields were acceptable.

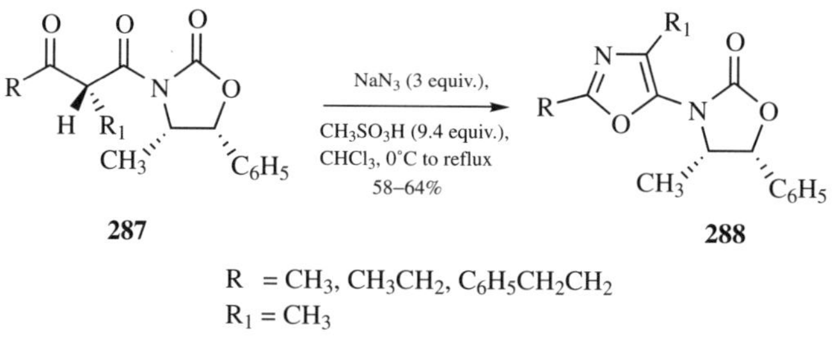

R = CH$_3$, CH$_3$CH$_2$, C$_6$H$_5$CH$_2$CH$_2$
R$_1$ = CH$_3$

Scheme 1.77

1.3.6. Oxazoles from Vinylogous Amides/Enamino Esters

During a study on the effects of ring size on the thermolysis of cyclic enamino esters **289** Gronsdemange-Pale and Chuche[135] prepared 2-alkyl-, 2,4-dialkyl-, and 2-alkyl-5-aryloxazoles **295** in 34–45% yield (Scheme 1.78). Pyrolysis of **289** ($n = 1$) at 420°C results in loss of ethanol to generate an iminoketene **290**, which rearranges to the vinylogous β-ketoamide **292**. This immediately cyclizes to **294**

TABLE 1.23. 2-ALKYL(ARYL)AMINO-4-CYANO-5-METHOXYOXAZOLES FROM METHYL-3,3-DIAZIDOCYANOACRYLATE[a]

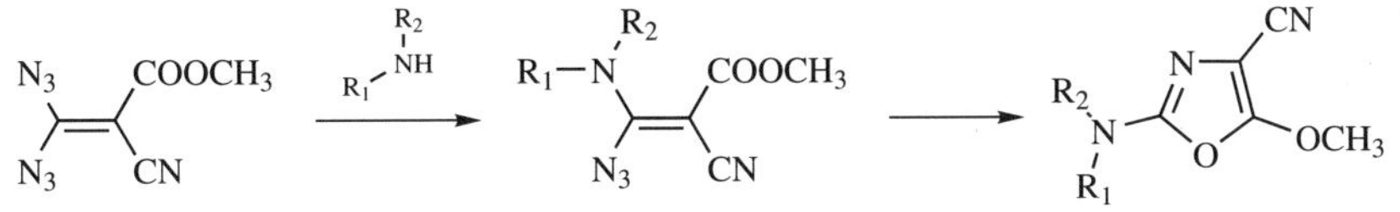

Figure 1.25

Entry	R_1	R_2	Product	Yield (%)
1	H	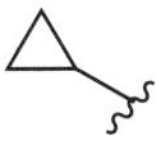 cyclopropyl	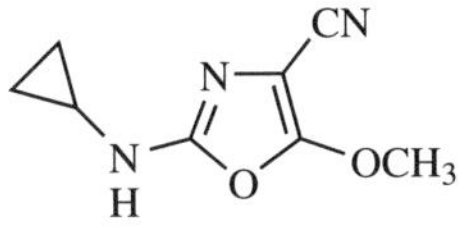 cyclopropyl-NH-oxazole(CN)(OCH₃)	75
2	H	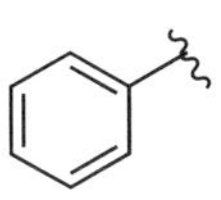 phenyl	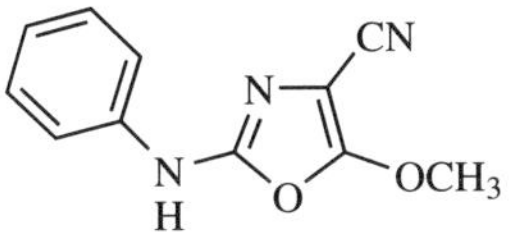 phenyl-NH-oxazole(CN)(OCH₃)	77
3	H	 norbornyl	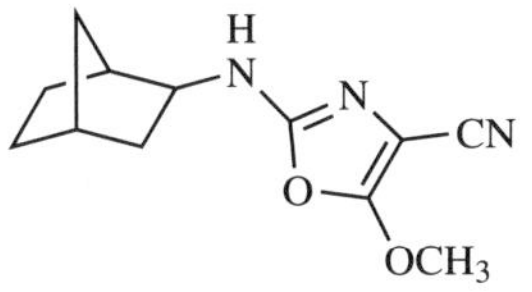 norbornyl-NH-oxazole(CN)(OCH₃)	70
4	H	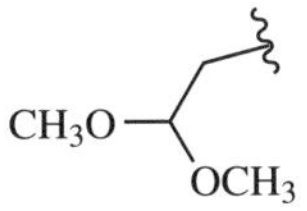 CH₃O-CH(OCH₃)-CH₂-	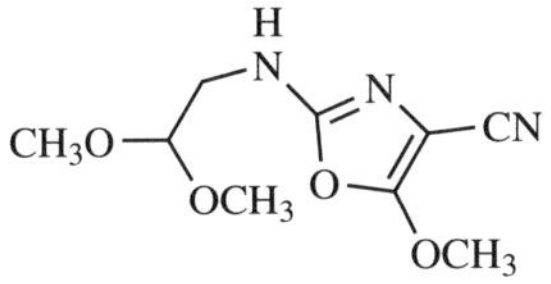 CH₃O-CH(OCH₃)-CH₂-NH-oxazole(CN)(OCH₃)	68

[a] Data from ref. 133.

from which **295** ($n = 3$) is produced via a retro ene reaction with rupture of the spriocyclopentane ring. Selected examples are shown in Table 1.24. This methodology has been applied to acyclic enaminoesters but produced 2-phenyloxazole (10%), 4-methyl-2-phenyloxazole (29%), 2-methyl-5-phenyloxazole (17%), and 2,5-diphenyloxazole (11%) in very low yields.[136] Attempts to extend this methodology to six- and seven-membered rings were not successful. For example, pyrolysis of the six-membered ring analog **289** afforded mixtures of acylated 4,5,6,7-tetrahydroindoles **296** and **297**, with no evidence for the expected oxazole (Scheme 1.79).

Threadgill and co-workers[137] described an interesting and unexpected synthesis of a 5-(trifluoromethyl)oxazole **303** as part of their investigations to prepare

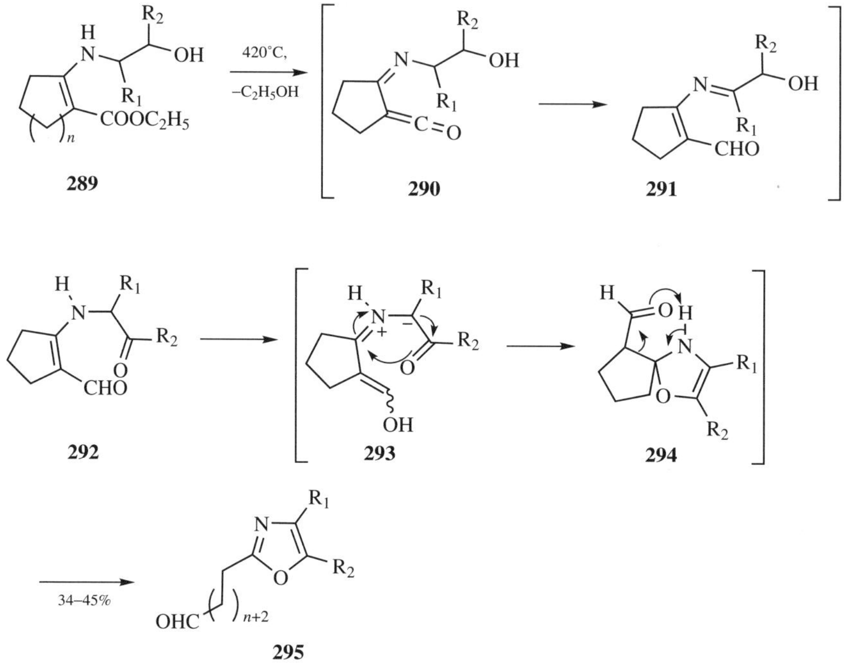

Scheme 1.78

TABLE 1.24. 2-ALKYL-, 2,4-DIALKYL-, AND 2-ALKYL-5-ARYL-OXAZOLES FROM PYROLYSIS OF CYCLIC ENAMINOESTERS[a]

Figure 1.26

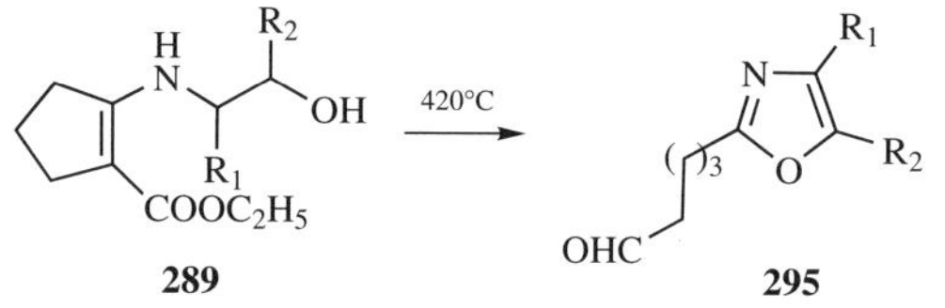

Entry	R_1	R_2	Product	Yield (%)
1	H	H	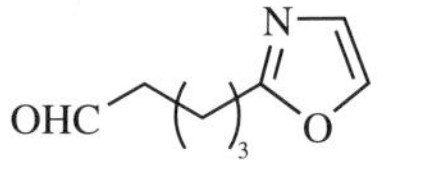	34
2	CH_3	H	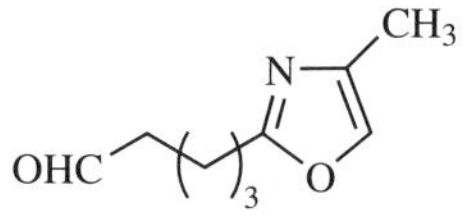	36
3	H	C_6H_5	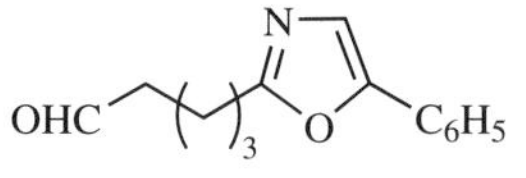	45

[a] Data from ref. 135.

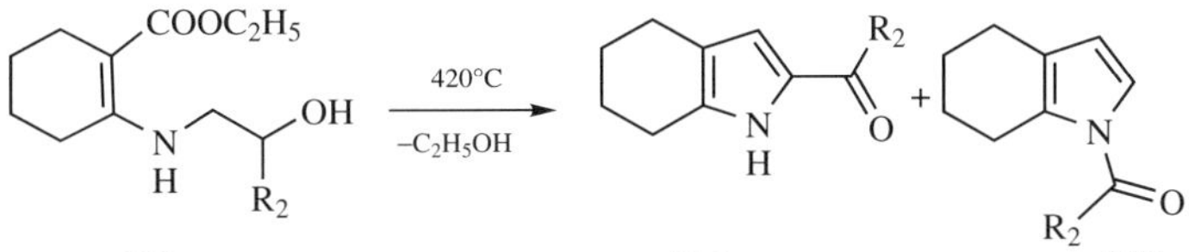

Scheme 1.79

2-(trifluoromethyl)histamine (Scheme 1.80). They isolated and identified **303** as (Z)-4-[1,3-bis(benzamido)prop-1-enyl]-2-phenyl-5-(trifluoromethyl)oxazole. It was produced as a significant by-product (18%) from cyclization of (Z)-1,2,4-tris(benzamido)butene **298** when treated with trifluoroacetic anhydride. Mechanistically, the authors proposed initial trifluoroacetylation of **298** to generate **299** that tautomerized to **300**. Cyclization of **300** followed by dehydration and double-bond isomerization then gave **303**.

1.3.7. Oxazoles from Oximes and Hydrazones

Derivatives of ketones—including oximes, imines, and hydrazones—have served as precursors to oxazoles, although many examples of such conversions are described for steroidal and other fused ring systems. A few examples of steroidal and fused heterocyclic oxazoles and monocyclic analogs are described in this section.

Nicolaides and co-workers[138–141] conducted extensive studies on preparing 2-aryl- and 2-heteroarylphenanthro[9,10-d]oxazoles. For instance, refluxing a solution of 10-(methoxyimino)phenanthrene-9-one **304** in an aromatic or heterocyclic methylated hydrocarbon affords a 2-aryl- or a 2-heteroarylphenanthro[9,10-d]oxazole **305** directly (Scheme 1.81). The yields are variable and poor to modest. However, using a substituted methyl analog, i.e., ArCH$_2$Y in this reaction markedly improved the yield of **305** for several cases. Examples of **305** prepared from aromatic and heterocyclic alkanes are shown in Table 1.25.

The authors advocated a redox reaction mechanism (Scheme 1.82), whereby **304** reacts with ArCH$_2$Y to produce the solvent cage radical pair **306**, which collapses to **307**. Elimination of methanol from **307** yielded **308**. Here, the loss of HY was believed to occur homolytically to yield **305**.[139]

In later studies, the authors[139,141] described examples of the somewhat capricious nature of this methodology. Although 3-methylthiophene and 2,5-dimethylthiophene did afford low yields of **305**, none of the expected product was isolated from N-methylpyrrole or 2-methylimidazole. In contrast, 2-methylpyrazine, 2-methylindole, and 2-methylfuran all produced the corresponding **305**

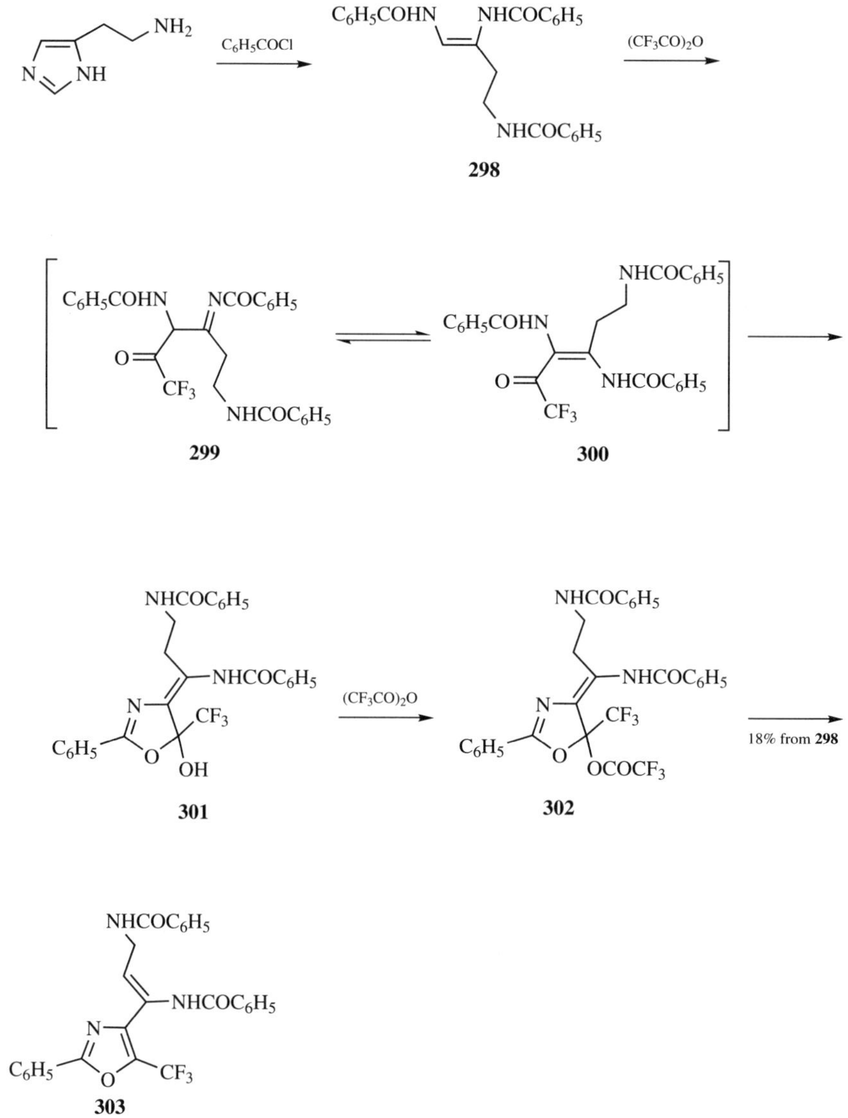

Scheme 1.80

albeit, again, in low yield. Finally, this methodology can be applied to 3,5-di-*t*-butyl-*O*-methyl orthoquinone monooxime **309** (Scheme 1.83). Refluxing **309** in benzyl alcohol affords 5,7-di-*t*-butyl-2-phenylbenzoxazole **310** in a respectable 61% yield. Overall, despite the low and variable yields, the authors noted this methodology is operationally simple, uses cheap starting materials, and affords the products in a single step.

Goodard[142] investigated a number of synthetic strategies to prepare 5-heteroaryl-2-thiophenecarboxylic acids as part of a broad pharmacological screening program.

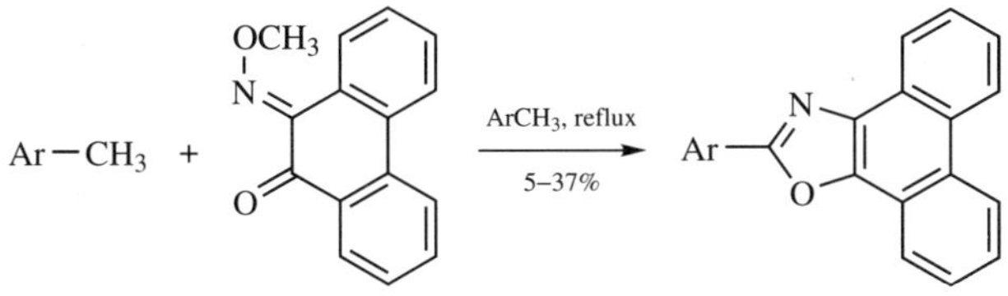

304 **305**

$Ar{-}CH_2Y$ + **304** $\xrightarrow[140\text{–}205^\circ C]{Ar-CH_2Y}$ **305**

Y = OH, 48%
Y = $OCOCH_3$, 63%
Y = SH, 24%
Y = COC_6H_5, 64%

Scheme 1.81

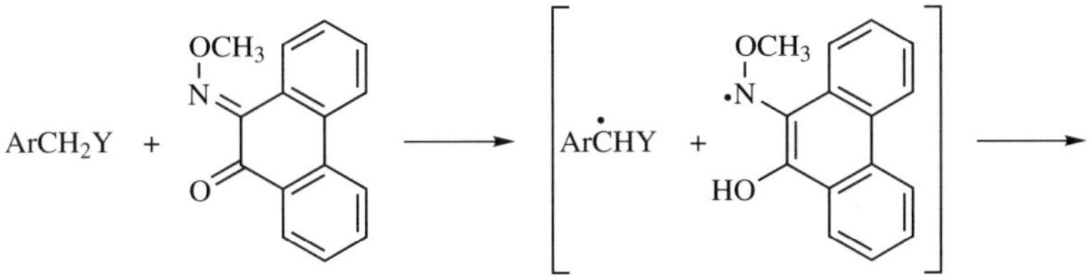

304 **306**

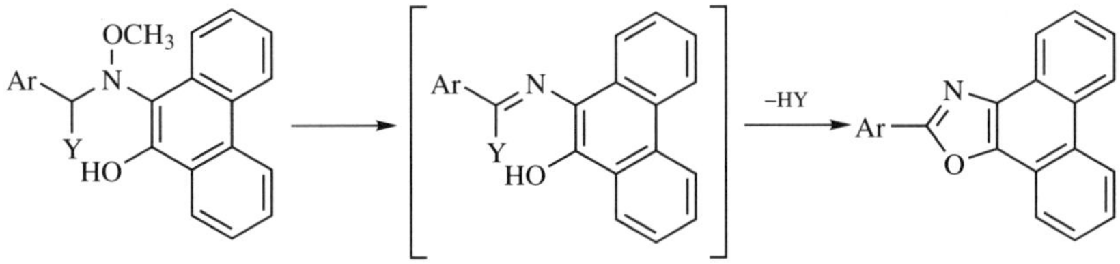

307 **308** **305**

Scheme 1.82

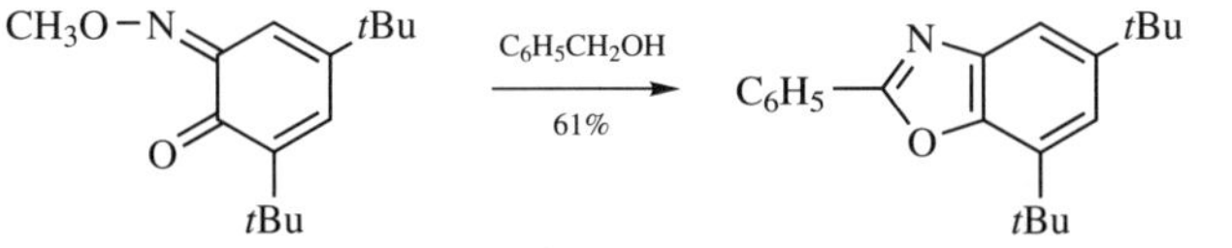

309 **310**

Scheme 1.83

TABLE 1.25. 2-ARYL- AND 2-HETEROARYLPHENANTHRO
[9,10-d]OXAZOLES FROM 10-(METHOXYIMINO)PHENANTHRENE-9-ONE[a]

Figure 1.27		304	305
Entry	Ar	Oxazole	Yield (%)
1	3,5-dimethylphenyl	2-(3,5-dimethylphenyl)phenanthro[9,10-d]oxazole	37
2	4-pyridyl	2-(4-pyridyl)phenanthro[9,10-d]oxazole	5
3	2-benzoxazolyl	2-(2-benzoxazolyl)phenanthro[9,10-d]oxazole	26

[a] Data from ref. 138.

When preparing 2,4,5-trisubstituted oxazoles, he found that acid-catalyzed condensation of methyl-5-formylthiophenecarboxylate **311** with 2,3-butanedione monooxime **312a** produced the oxazole N-oxide **313a**, which was reduced to **314a** with Zn/CH$_3$COOH (Scheme 1.84). Reaction of **311** with 1-phenyl-1,2-propanedione-2-oxime **312b** afforded **313b** and **314b** analogously.

Wasserman and Prowse[143] condensed **312b** with benzaldehyde to produce 2,5-diphenyl-4-methyloxazole N-oxide **315** (Scheme 1.85). Treatment of **315** with POCl$_3$ gave 4-(chloromethyl)-2,5-diphenyloxazole **316**, a key starting material in their synthesis of (±)-pyrenolide C **317**.

Jacobsen and Philippides[144] condensed α-hydroxyimino aromatic ketones with substituted benzaldehydes to prepare a series of metabolites of 2,5-diphenyloxazole, a noncarcinogenic, sensitive reagent used to assess aromatic hydrocarbon hydroxylase activity. In their work, acid-catalyzed cyclization of a series of

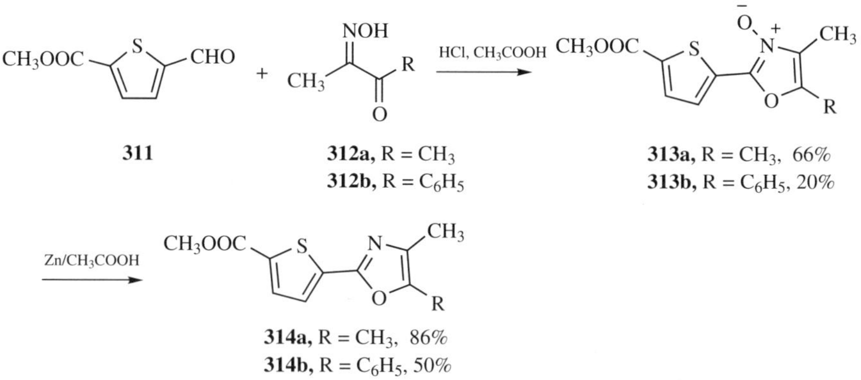

Scheme 1.84

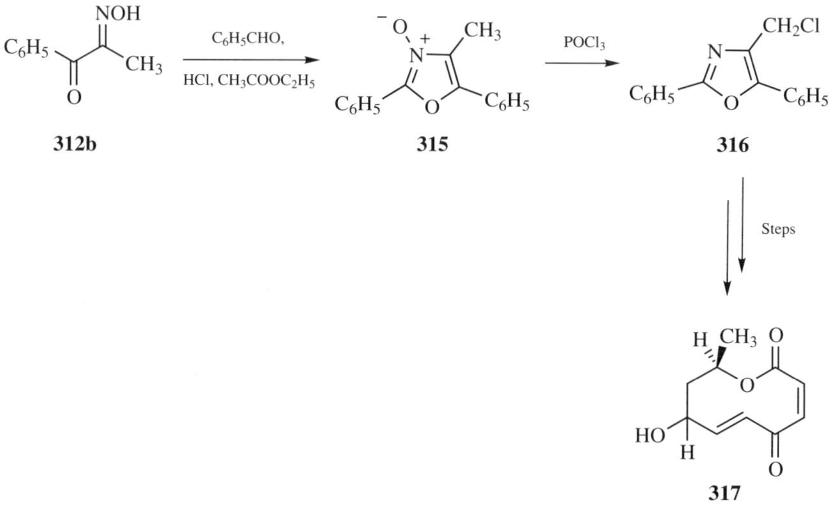

Scheme 1.85

α-hydroxyimino aromatic ketones **318** with substituted benzaldehydes produced the intermediate oxazole *N*-oxides, which were reduced to the target compounds **319** (Scheme 1.86).

Shafiullah and Ansari[145] found the steroidal oximes **320** to be useful starting materials for preparing the steroidal oxazoles **322** (Scheme 1.87). The authors also isolated small amounts of the corresponding ketones **323**.

Stanovnik and co-workers[146] described one of the first syntheses of the oxazolo[5,4-*c*]pyridazine ring system **325** by treatment of the *N*-heteroaryl formamidoxime **324** with *N*,*N*-dimethylformamide dimethyl acetal (Scheme 1.88). When this reaction was repeated at reflux, the authors isolated a 5:4 mixture of **325** and **326**.

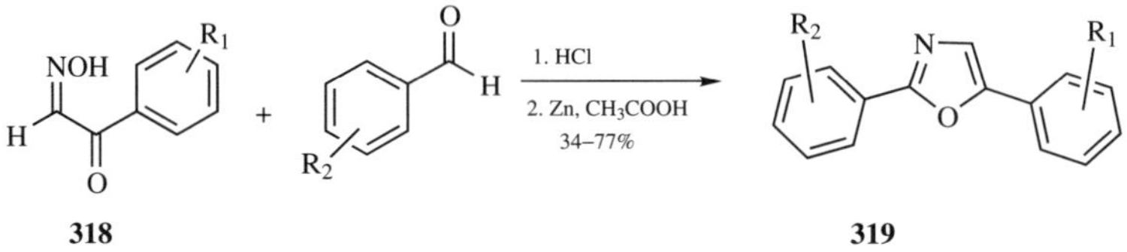

Synthesis and Reactions of Oxazoles

Scheme 1.86

R_1 = H, 2-OH, 3-OH, 4-OH, 2-OCH$_3$
R_2 = H, 2-OH, 3-OH, 4-OH, 2-OCH$_3$

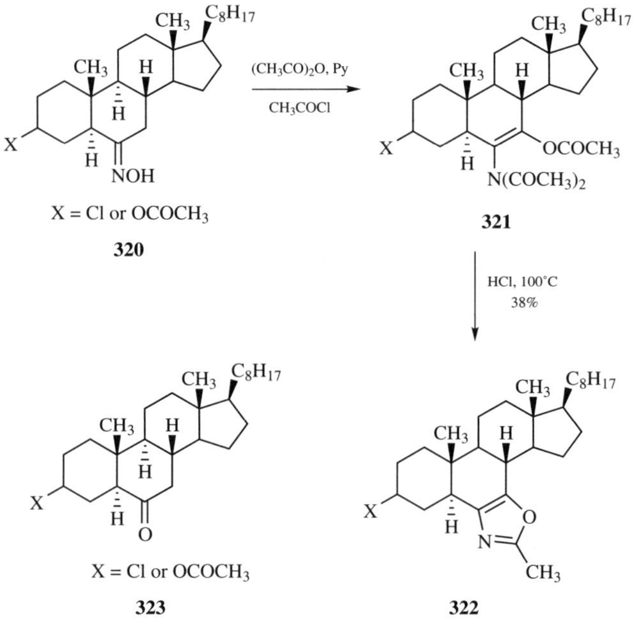

Scheme 1.87

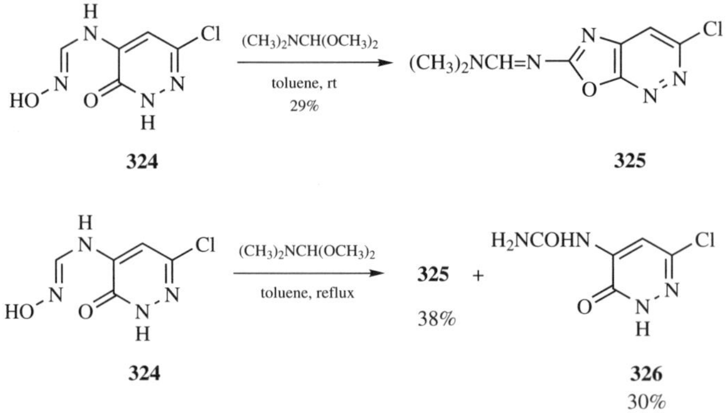

Scheme 1.88

Hojo and co-workers[147] described a novel synthesis of 4-aryl-5-(trifluoromethyl)oxazoles **331**, starting from aldehyde *tert*-butyl(methyl)hydrazones (Scheme 1.89). Trifluoroacetylation of a *tert*-butyl(methyl)hydrazone **327** yields **328** that cyclizes to a 4,5,5-trisubstituted oxazoline **329** upon refluxing in toluene in the presence of silica gel. Dehydration of **329** was then accomplished in two steps using $POCl_3$ and a base to generate the 4-aryl-5-(trifluoromethyl)oxazoles **331**.

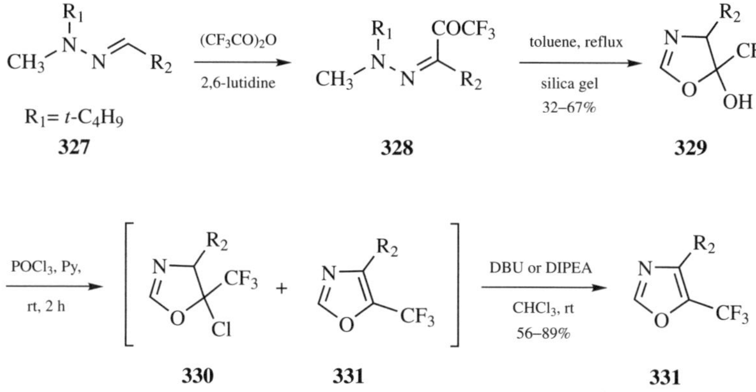

Scheme 1.89

Several observations regarding the cyclization of **328** to **329** are noteworthy. The authors reported that this reaction is exquisitely sensitive to the nature of the substituent R_1 and the reagents used to accomplish dehydration (Scheme 1.90). For example, reaction of a trifluoroacetylated dimethylhydrazone **332** in refluxing toluene or in refluxing toluene with silica gel affords regioisomeric imidazoles

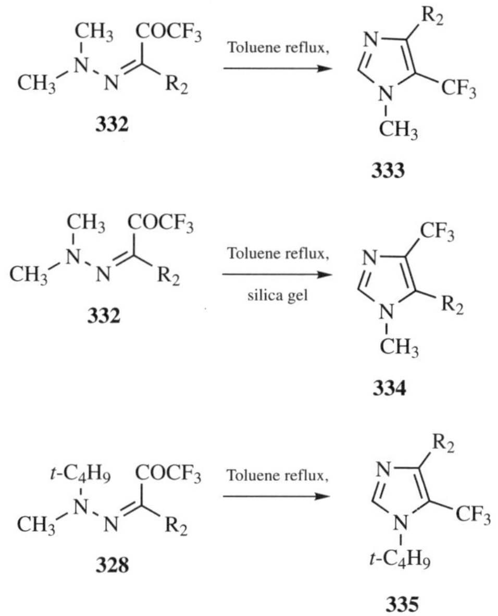

Scheme 1.90

333 and **334**. Similarly, the *N*-methyl-*N*-*tert*-butylhydrazone **328** also yields an imidazole **335** as the sole product on refluxing in toluene. However, refluxing a toluene solution of **328** with silica gel produces the oxazole **331**. The authors also noted that **329** was resistant to common methods of dehydration, requiring the two-step process shown. The mixture of **330** and **331** obtained from reaction of **329** with POCl$_3$/pyridine was cleanly converted, without isolation to the desired product.

This reaction is quite general and tolerates a variety of functional groups. In addition, this methodology was successfully used to prepare two simple 4-alkyl (alkenyl)-5-(trifluoromethyl)oxazoles. Selected examples are shown in Table 1.26.

TABLE 1.26. 4-ARYL-5-(TRIFLUOROMETHYL)OXAZOLES FROM ALDEHYDE *N*-METHYL-*N*-*tert*-BUTYLHYDRAZONES[a]

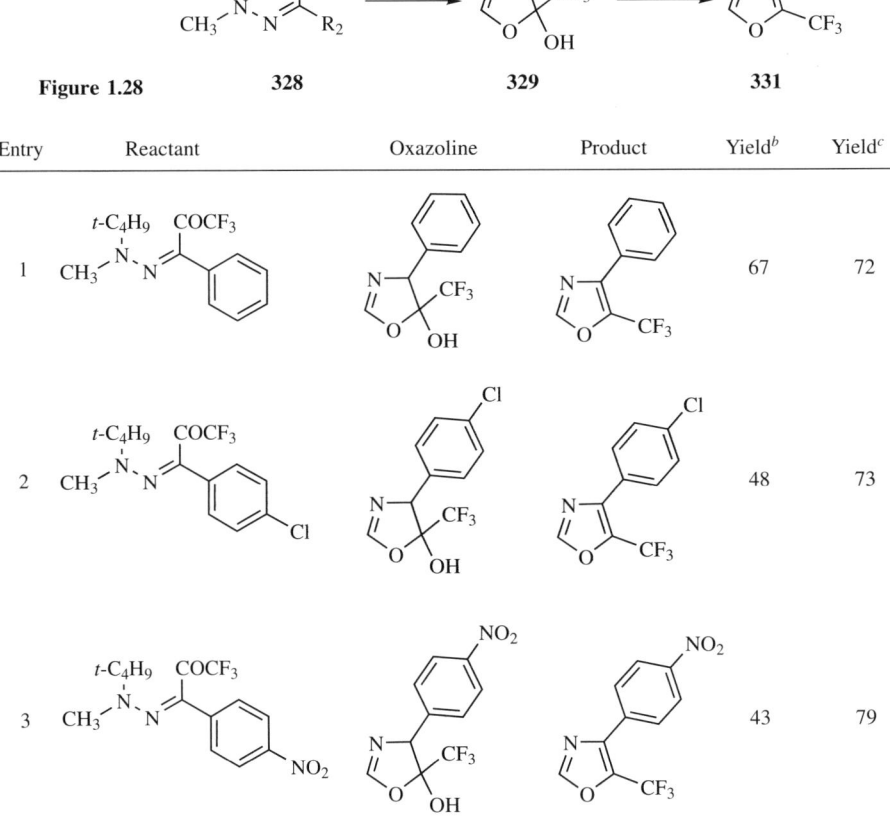

[a] Data from ref. 147.
[b] For oxazoline cyclization.
[c] For dehydration.

Venkataratnam and co-workers[148] condensed o-aminophenol or 2-amino-3-hydroxypyridine with trifluoroacetylketene diethyl acetal and isolated 2-(1,1,1-trifluoroacetonyl)benzoxazole or 2-(1,1,1-trifluoroacetonyl)pyridyloxazole in 70–80% yield.

1.3.8. Oxazoles from Imidates/Thioimidates—Cornforth Reaction

More than 50 years ago, Cornforth[149] described the synthesis of 4-oxazolecarboxylic acid esters via cyclization of the potassium salts of C-formylated imidates. This method continues to find further applications and has been used to prepare a variety of structurally diverse and complex oxazoles as shown in the following examples.

Yokoyama and co-workers[150] described a novel thermal cyclization of the D-ribose lactoxime O-vinyl ether **336** to the 2-substituted 4,5-oxazoledicarboxylic acid dimethyl ester **337** in excellent yield (Scheme 1.91). When the authors extended this methodology to other sugar lactoxime O-vinyl ethers, they found this rearrangement was accompanied by some epimerization at C-2 or suffered from low yields.

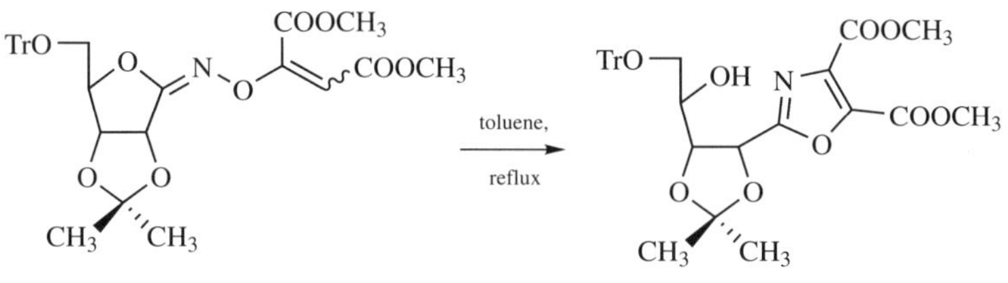

Scheme 1.91

From (Z)- **336**, 88%
From (E)- **336**, 70%

A structurally simpler analog, **338**, was studied mechanistically. The rearrangement is believed to proceed via initial N-O bond cleavage to produce two radical fragments **339** that then recombine to yield the more stable C-N-bonded species **340**. Intramolecular cyclization is accomplished through the enol tautotmer **341** via addition-elimination with loss of ethanol from the oxazoline **342** to afford 2-phenyl-4-oxazolecarboxylic acid ethyl ester **343** (Scheme 1.92).

Although the rearrangement was initially identified in sugar substrates it is very general. For example, the lactoxime O-vinyl ethers **345** were synthesized in two straightforward steps from thiolactones **344**. Thermolysis of **345** effects rearrangement to afford a series of 2-substituted-4,5-oxazoledicarboxylic acid dimethyl esters **346** from monocyclic precursors (Scheme 1.93). Additional examples of 2-substituted-4,5-oxazoledicarboxylic acid dimethyl esters available via this methodology are shown in Table 1.27.

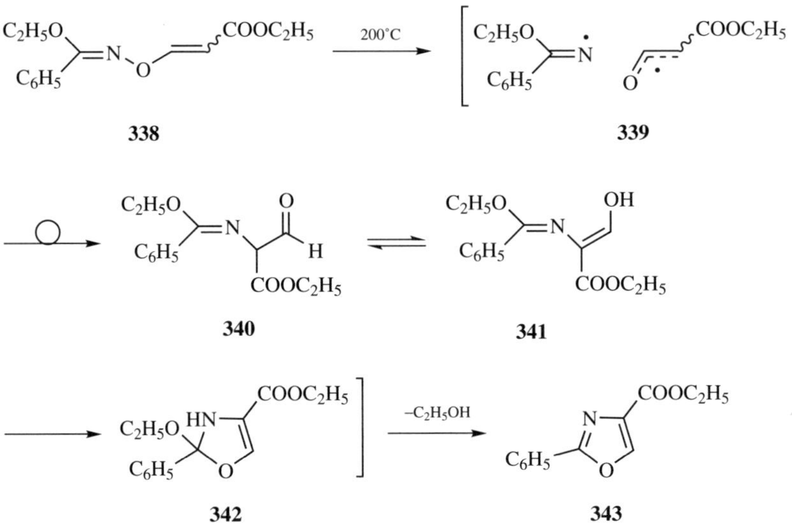

Scheme 1.92

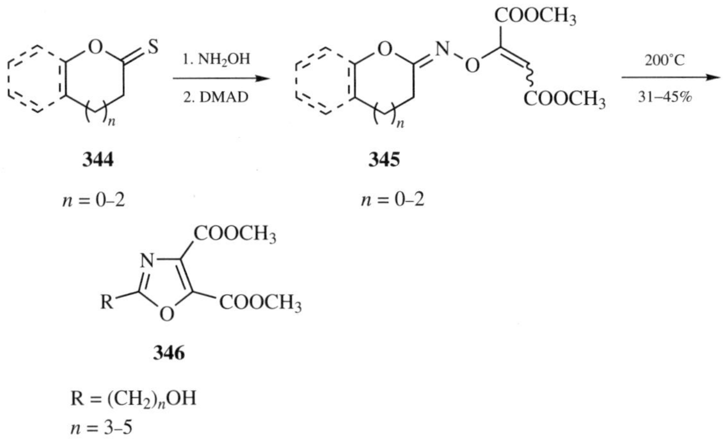

Scheme 1.93

Barrett and co-workers[151] prepared a key intermediate oxazole **350** in their synthesis of calyculin A using Cornforth methodology (Scheme 1.94). The benzyl ester of (R)-2-methyl-4-pentenoic acid **347** was converted to (R)-2-methyl-4-pentenenitrile **348** in two steps. Pinner reaction of **348**, followed by amine exchange with glycine methyl ester, gave the imidate **349** in 73% yield. Base-catalyzed formylation of **349** with in situ cyclization produced the 2-alkyl-4-oxazolecarboxylic acid methyl ester **350** in good yield. This entire sequence

Synthesis of Oxazoles

TABLE 1.27. 2-SUBSTITUTED-4,5-OXAZOLEDICARBOXYLIC ACID DIMETHYL ESTERS FROM CYCLIZATION OF LACTOXIME O-VINYL ETHERS[a]

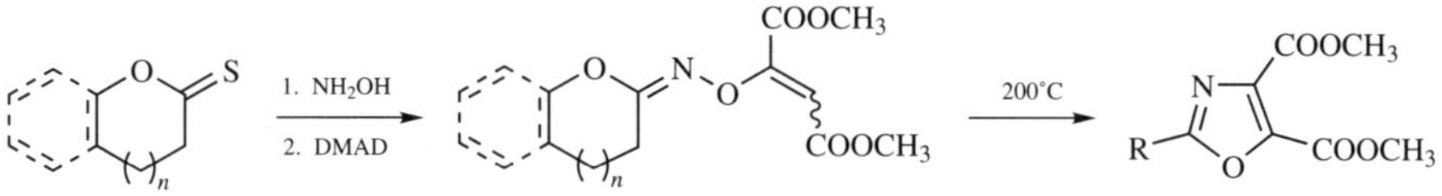

Figure 1.29

Entry	Reactant	R	Yield (%)
1	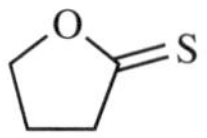 (γ-butyrolactone thione)	$(CH_2)_2CH_2OH$	31
2	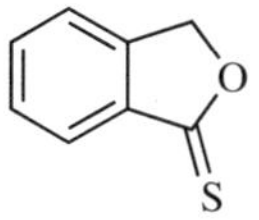 (phthalide thione)	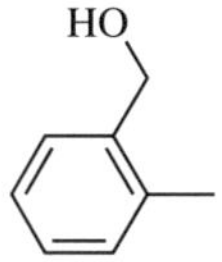 2-(hydroxymethyl)phenyl	51
3	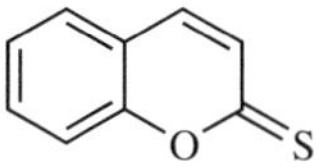 (coumarin thione)	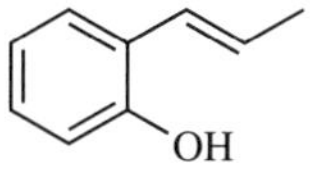 2-(1-propenyl)phenol	31

[a] Data from ref. 150.

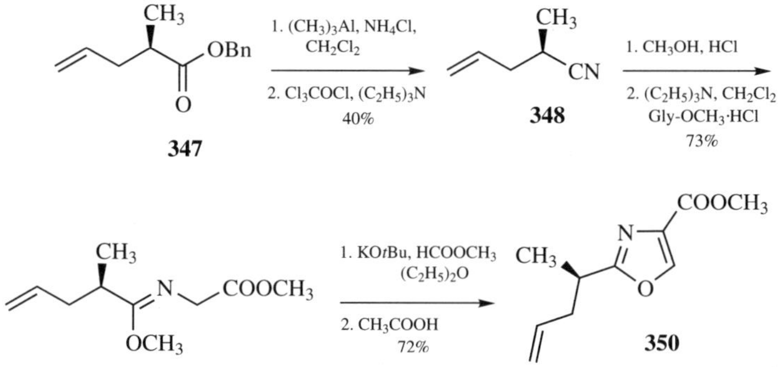

Scheme 1.94

was accomplished with no appreciable racemization. Elaboration of **350** and incorporation into calyculin A will be described in Section 1.5.3.

Hajjem and Baccar[152] used cyanomethylimidates as substrates for addition of organozinc reagents to prepare β-ketoimidates that cyclized to 2,5-disubstituted-oxazoles as shown in Scheme 1.95. For example, addition of an alkenylzinc bromide to a cyanomethylimidate **351** affords a β-ketoimidate **353** after hydrolysis that immediately cyclized to a 2,5-disubstituted-oxazole **354**. Organomagnesium and organolithium reagents were not useful in this sequence. Selected examples of 2,5-disubstituted-oxazoles are shown in Table 1.28.

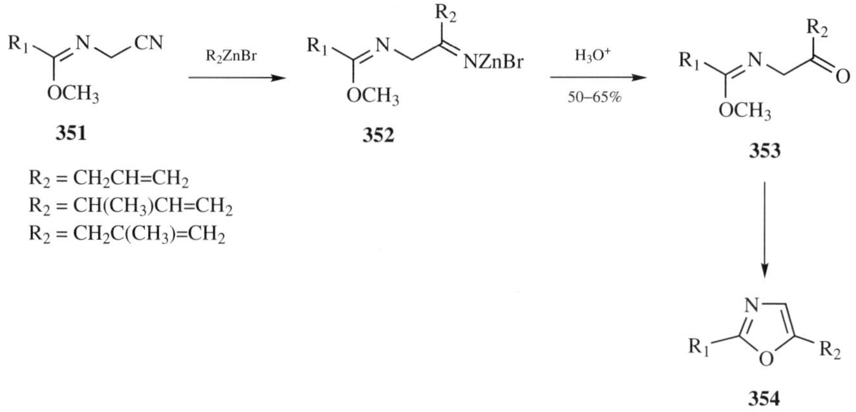

$R_2 = CH_2CH=CH_2$
$R_2 = CH(CH_3)CH=CH_2$
$R_2 = CH_2C(CH_3)=CH_2$

Scheme 1.95

Hermitage and co-workers at GlaxoWellcome[153] very recently described a novel, non-oxidative method to prepare 2-(chloromethyl)-4-oxazolecarboxylic acid methyl ester **357** shown in Scheme 1.96. This oxazole was a proposed intermediate in their scale up route to GW475151 **358**, a potent inhibitor of human neutrophil

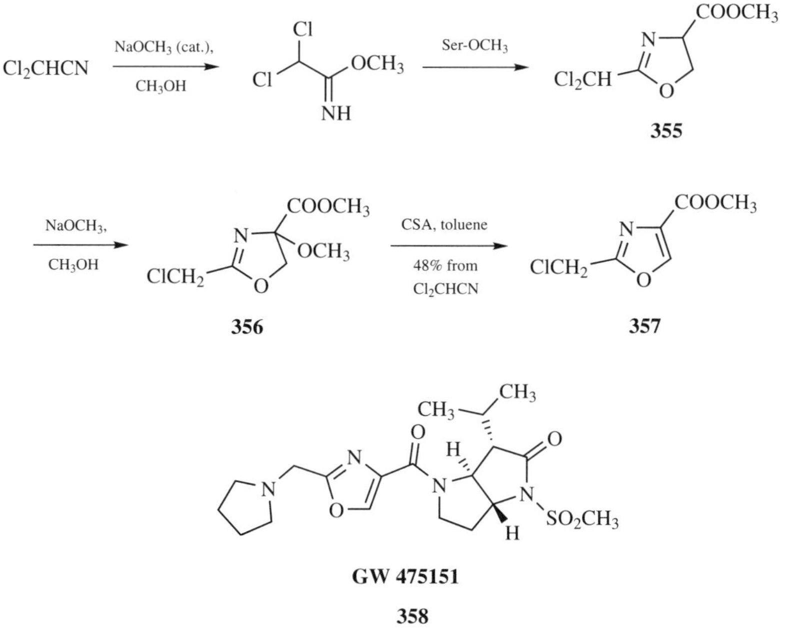

GW 475151

358

Scheme 1.96

TABLE 1.28. 2,5-DISUBSTITUTED OXAZOLES FROM CYANOMETHYLIMIDATES AND ORGANOZINC REAGENTS[a]

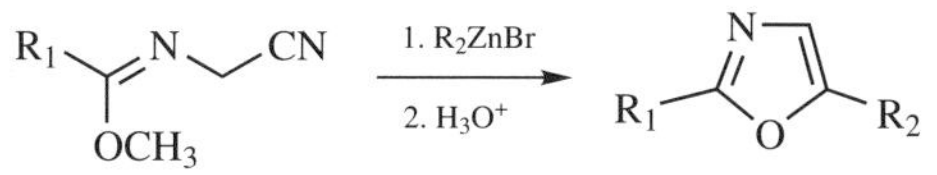

Figure 1.30

Entry	R_1	R_2	Product	Yield (%)
1	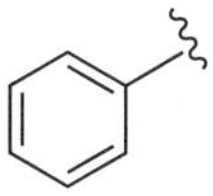 phenyl	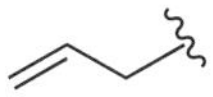 allyl	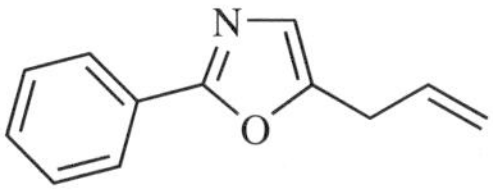 2-phenyl-5-allyloxazole	50
2	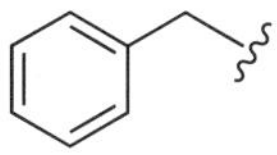 benzyl	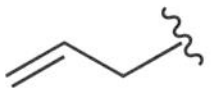 allyl	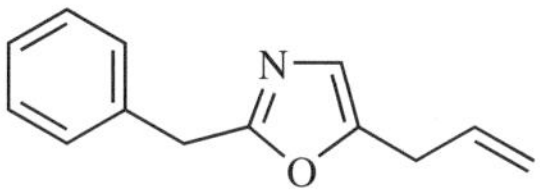 2-benzyl-5-allyloxazole	60
3	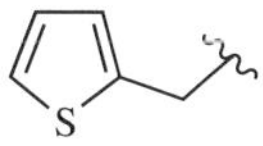 2-thienylmethyl	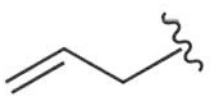 allyl	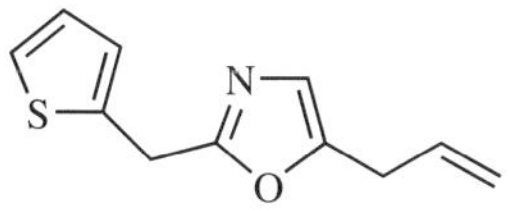 2-(2-thienylmethyl)-5-allyloxazole	65
4	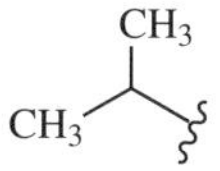 isopropyl	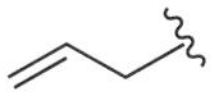 allyl	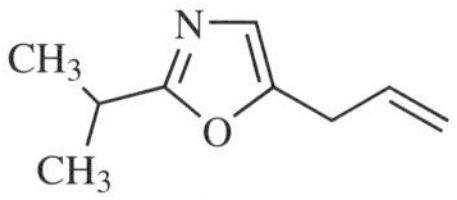 2-isopropyl-5-allyloxazole	65
5	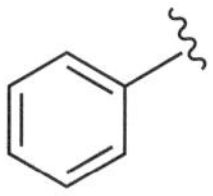 phenyl	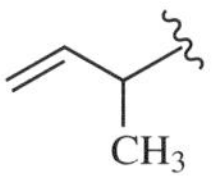 1-methylallyl	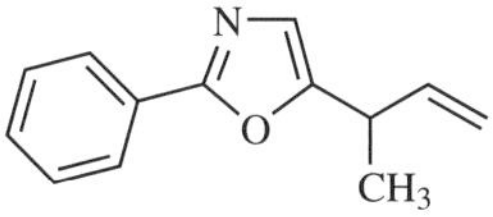 2-phenyl-5-(1-methylallyl)oxazole	55
6	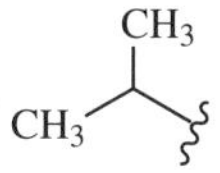 isopropyl	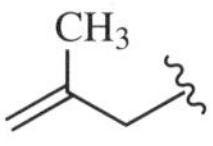 2-methylallyl	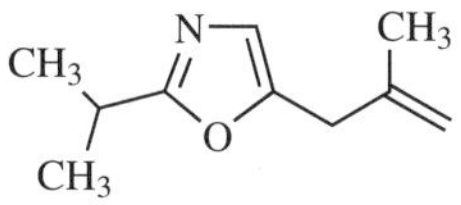 2-isopropyl-5-(2-methylallyl)oxazole	60

[a] Data from ref. 152.

elastase. The authors required a safe, scaleable and environmentally friendly route that would not require chromatographic purification.

To that end they converted dichloroacetonitrile to 2-(dichloromethyl)-4-oxazoline carboxylic acid methyl ester **355** in 88% yield. Treatment of **355** with one equivalent of sodium methoxide produced **356** via an "internal transfer of oxidation state through a molecular framework." The synthesis was completed by acid-catalyzed elimination of methanol to afford **357** in 48% overall yield from dichloroacetonitrile.

1.3.9. Oxazoles from Isocyanides

Schöllkopf and Gerhart[154] pioneered the use of metallated isocyanides in organic synthesis. This section discusses the versatility of such intermediates for preparing monosubstituted and disubstituted oxazoles as well as the parent **1**, itself. In addition, continuing advances in solid-phase synthesis have enhanced the use of tosylmethyl isocyanide (TosMIC) for oxazole synthesis; several examples are presented.

Yuan and co-workers[155] reported a general synthesis of 2-acyl-5-alkoxyoxazoles via cyclization of α-ketoimidoyl chlorides (Scheme 1.97). Acylation of ethyl isocyanoacetate with an acid chloride yielded an α-ketoimidoyl chloride **359**, which was not isolated but cyclized upon addition of triethylamine to afford a 2-acyl-5-alkoxyoxazole **360**.

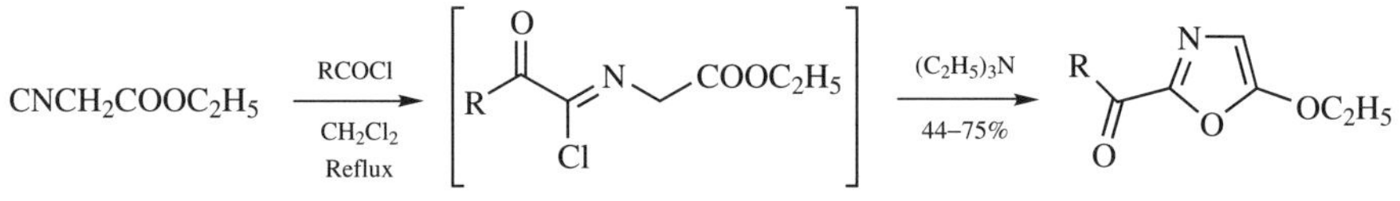

Scheme 1.97

Interestingly, only *p*-nitrobenzoyl chloride reacted with ethyl isocyanoacetate. Other aromatic acid chlorides, including *p*-chlorobenzoyl chloride, *p*-fluorobenzoyl chloride, and benzoyl chloride were unreactive even after 15 h at 110°C. This was attributed to marked differences in electrophilicity of the acid chlorides. The authors also considered an alternative mode of cyclization that would have produced 5-substituted 4-oxazolecarboxylic acid esters but found no evidence for these products. Selected examples are shown in Table 1.29.

Marcaccini's group[156] also used ethyl isocyanoacetate as a precursor to isothiocarbamoyl chlorides, which were cyclized to 5-alkoxy-2-(arylthio)oxazoles (Scheme 1.98). Thus low-temperature addition of an arylsulfenyl chloride to ethyl isocyanoacetate generated an isothiocarbamoyl chloride **361** quantitatively. These can be isolated if desired, but crude **361** cyclized readily with $(C_2H_5)_3N$ at $-20°C$ to produce 5-alkoxy-2-(arylthio)oxazoles **362**, also in near quantitative yield. The authors noted that these were the first examples of oxazoles containing both alkoxy and arylthio substituents.

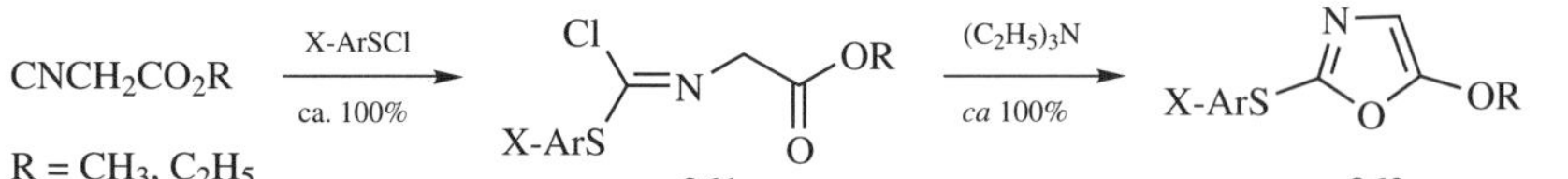

$Ar = C_6H_4$ or C_6H_3
$X = 2-NO_2, 4-Cl, 4-CH_3, H,$ and $2-NO_2-4-Cl$

Scheme 1.98

TABLE 1.29. 2-ACYL-5-ETHOXYOXAZOLES FROM CYCLIZATION OF α-KETOIMIDOYL CHLORIDES[a]

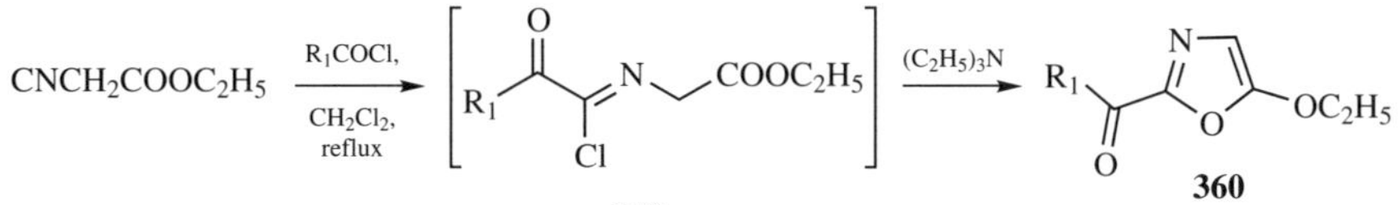

Figure 1.31

Entry	R₁	Product	Yield (%)
1	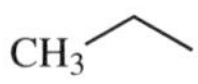 CH₃-CH₂-	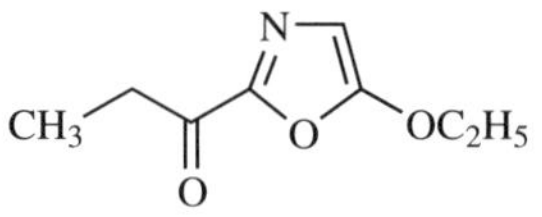 (ethyl-substituted oxazole)	53
2	cyclohexyl	(cyclohexyl-substituted oxazole)	62
3	(CH₃)₃C-	(tert-butyl-substituted oxazole)	75
4	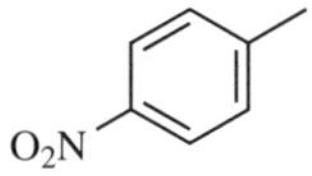 4-O₂N-C₆H₄-CH₂-	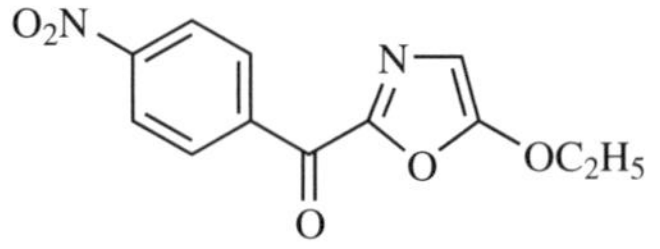 (4-nitrobenzyl-substituted oxazole)	46

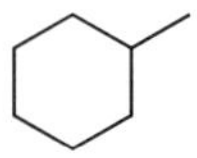

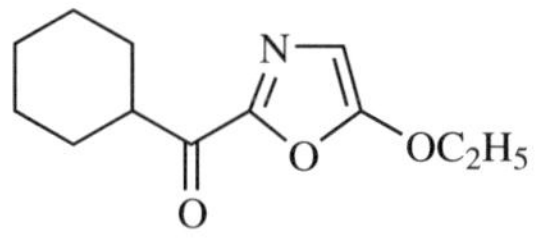

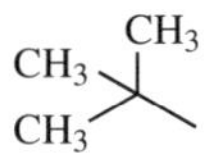

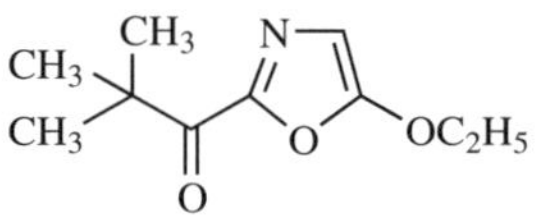

[a] Data from ref. 155.

These same workers[157] attempted to adapt this one-pot procedure for preparation of 5-(*N*-alkyl-*N*-phenylamino)-2-(arylthio)oxazoles **364** from arylsulfenyl chlorides and *N*-alkylisocyanoacetanilides. Instead, the authors isolated the novel 5-(*N*-alkyl-*N*-phenylamino)-2,4-(diarylthio)oxazoles **367** in 45–80% yield (Scheme 1.99). Mechanistically, they proposed that addition of the arylsulfenyl chloride to an *N*-alkylisocyanoacetanilide generated the expected isothiocarbamoyl chloride **363**. However, in this case, the increased tendency of **363** to enolize to **365** allowed for reaction with a second molecule of the arylsulfenyl chloride to produce **366**, which cyclized as expected to **367**. As expected, the best yields were obtained using a molar ratio of 1:2:2 for the *N*-alkylisocyanoacetanilide:ArSCl:(C₂H₅)₃N. A variety of electron-deficient arylsulfenyl chlorides could be used in this reaction sequence.

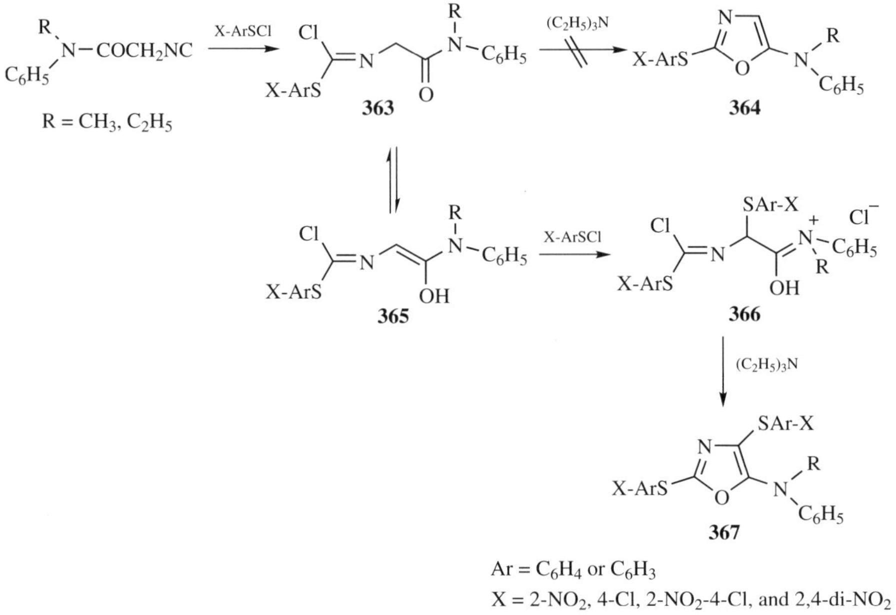

Scheme 1.99

Ar = C$_6$H$_4$ or C$_6$H$_3$
X = 2-NO$_2$, 4-Cl, 2-NO$_2$-4-Cl, and 2,4-di-NO$_2$

Ozaki and co-workers[158] prepared an extensive series of 5-aryl(heteroaryl)-4-oxazolecarboxylic acid methyl esters **368** (20 compounds) from acylation of methyl isocyanoacetate, followed by in situ cyclization (Scheme 1.100). These compounds were elaborated further and evaluated as blood platelet aggregation inhibitors.

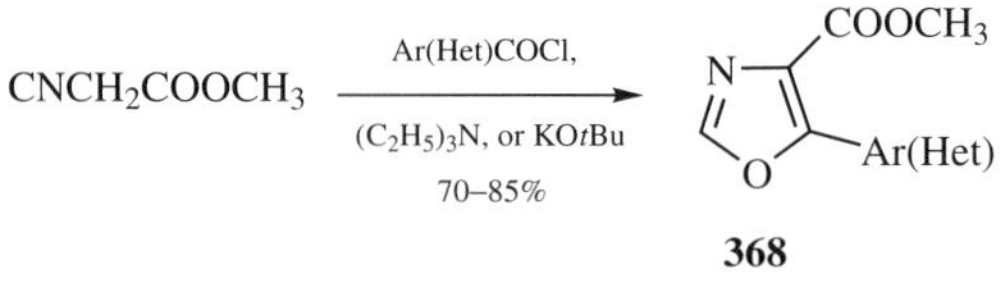

Scheme 1.100

Ohba and co-workers[159] developed a general synthesis of chiral N-protected 5-(aminomethyl)oxazoles from α-lithiated isocyanides and N-protected amino acid esters (Scheme 1.101). Metalation of an alkyl or benzyl isocyanide with n-BuLi or LDA and acylation with an N-Boc-α-amino acid methyl ester afforded an N-Boc-α-amino acid acyl isocyanide **369**, which cyclized to a chiral N-Boc-5-(aminomethyl) oxazole **370**. The products were obtained in >98% ee based on chiral HPLC analysis. The yields of **370** were somewhat variable, ranging from 45 to 91%. N-Boc-protected glycine, alanine, phenylalanine, proline, serine, and O-benzyl serine were all compatible with these reaction conditions.

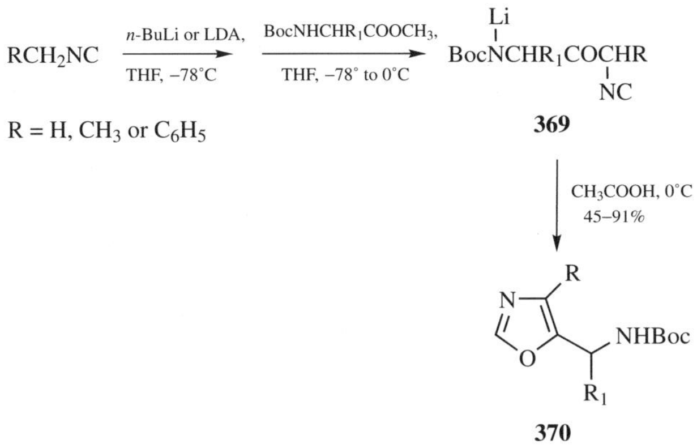

Synthesis of Oxazoles

Scheme 1.101

Ohba's group[160] employed this methodology in two different synthetic strategies for preparing **372**, a key intermediate in their synthesis of (−)-normalindine **374** (Scheme 1.102). Acylation of lithiomethyl isocyanide with *N*-Boc-L-alanine methyl ester gave **370** that was deprotected and alkylated with 2-(3-indolyl)ethyl bromide

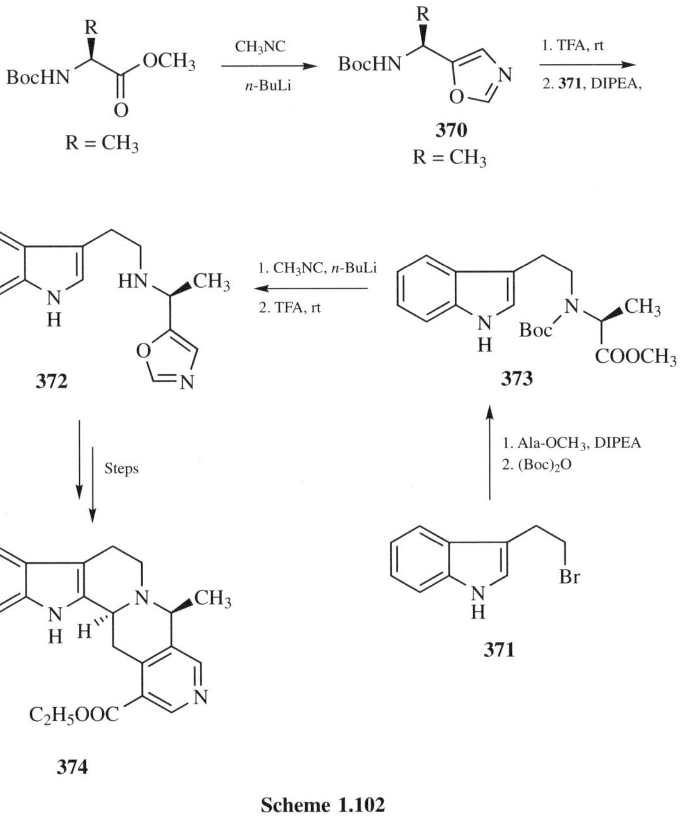

Scheme 1.102

371 to afford 372 in 41% overall yield. Although 372 was obtained in excellent enantiomeric purity, the alkylation step required 7 days in refluxing THF to provide 372 in 55% yield. Alternatively, sequential alkylation of L-alanine methyl ester with 371 and reaction with (Boc)$_2$O gave 373, which was converted to 372 with lithiomethyl isocyanide and deprotection with TFA. The overall yield of 372 was 46%, and the enantiomeric purity was also high. In a recent example, Ohba's group[161] again employed acylation of lithiomethyl isocyanide to prepare a chiral 5-substituted-oxazole of high enantiomeric purity. This intermediate was used in their total syntheses of (−)-plectrodorine and (+)-oxerine.

Tang and Verkade[162] found that the nonionic phosphazene superbase 375 was far superior to other common bases used to prepare oxazoles from isocyanides. For example, acylation of methyl isocyanoacetate with benzoyl chloride or 3,4,5-trimethoxybenzoyl chloride in the presence of one equivalent of 375 was complete in 30 min at room temperature (Scheme 1.103). The products, 5-phenyl-4-oxazolecarboxylic acid methyl ester 376 and 5-(3,4,5-trimethoxyphenyl)-4-oxazolecarboxylic acid methyl ester 377 were isolated in 99% and 98% yield, respectively. In contrast, acylation of methyl isocyanoacetate with benzoyl chloride or 3,4,5-trimethoxybenzoyl chloride, using one equivalent of DBU at room temperature, was incomplete after 2 h and showed 8% of 376. This remarkable difference is attributed to the differences in base strength, i.e., 375 is 10^{17} times stronger than DBU. Another advantage of the phosphazene base is that the salts are crystalline and readily separated from 376 and 377 by filtration. Regeneration of 375 from the salt is easily accomplished using potassium *tert*-butoxide. Similar results were obtained using acid anhydrides.

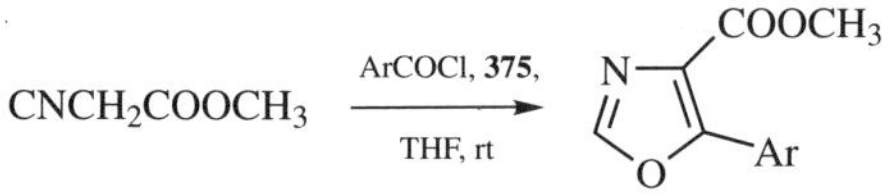

$$CNCH_2COOCH_3 \xrightarrow[\text{THF, rt}]{\text{ArCOCl, 375,}} \text{oxazole}$$

376 Ar = C$_6$H$_5$
377 Ar = 3,4,5-tri-(CH$_3$O)-C$_6$H$_2$

P(CH$_3$NCH$_2$CH$_2$)$_3$N
375

Scheme 1.103

Shiori and co-workers[163] used 5-substituted 4-oxazolecarboxylic acid esters 379 as β-hydroxy-α-amino acid synthons. They described a straightforward synthesis of 379 by acylation of an isocyanoacetic acid ester with an α-alkoxyacid 378 in the presence of diphenylphosphorylazide (DPPA) or diethylphosphoryl cyanide (DPPC) followed by base-catalyzed cyclization (Scheme 1.104). The reaction conditions do not epimerize optically active α-alkoxyacids. Dilute acid hydrolysis of 379 and reaction with (Boc)$_2$O affords the protected aminotetronic acids 380. Stereoselective hydrogenation of 380 then yields the 1,4-lactones 381, key intermediates in the synthesis of amino sugars. A variety of α-alkoxyacids were studied, and some examples are shown in Table 1.30.

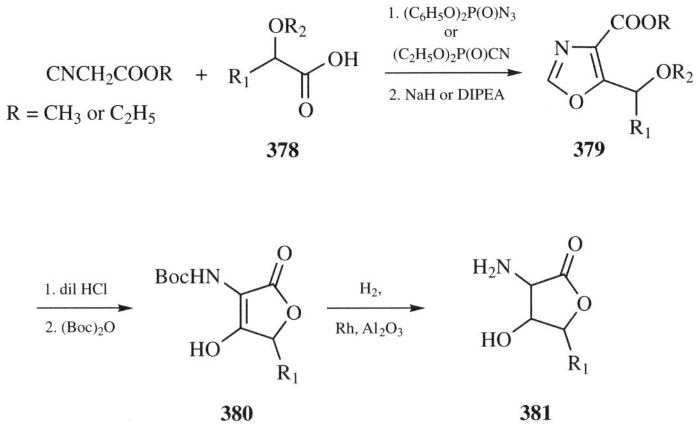

Synthesis of Oxazoles

Scheme 1.104

Marcaccini and co-workers[164] used the Passerini reaction[165] to prepare N-substituted α-acyloxy β-ketoamides **382**, valuable intermediates for their synthesis of 2,4-diaryl-5-oxazolecarboxamides **383** (Scheme 1.105). In their work, the authors used an arylglyoxal as the carbonyl component, whereby only the aldehyde was reactive. For example, condensation of phenylglyoxal (R = H), 2-chlorobenzoic acid (R_1 = 2-Cl) and cyclohexylisocyanide produced **382** in excellent yield.

TABLE 1.30. 5-SUBSTITUTED 4-OXAZOLECARBOXYLIC ACID ESTERS FROM α-ALKOXYACIDS AND ISOCYANOACETIC ACID ESTERS[a]

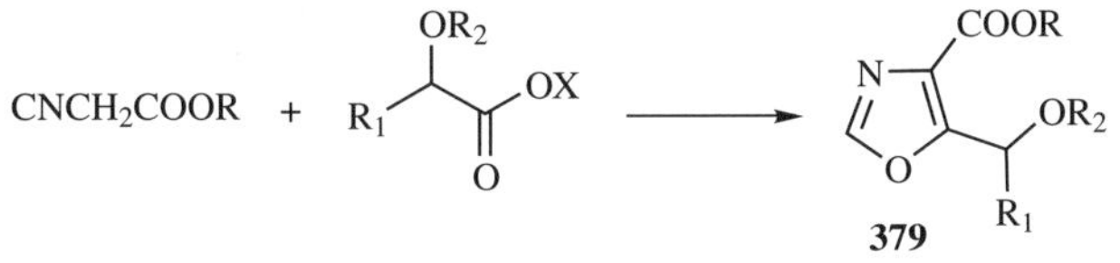

Entry	R_1, R_2	X	R	Yield (%)
1	CH_3, CH_3OCH_2	Li^+	CH_3	70
2	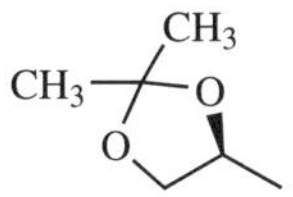 (dioxolane structure with CH_3)	K^+	C_2H_5	100
3	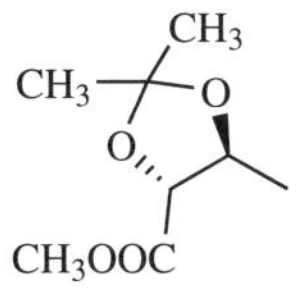 (dioxolane structure with CH_3OOC)	H	CH_3	79

[a] Data from ref. 163.

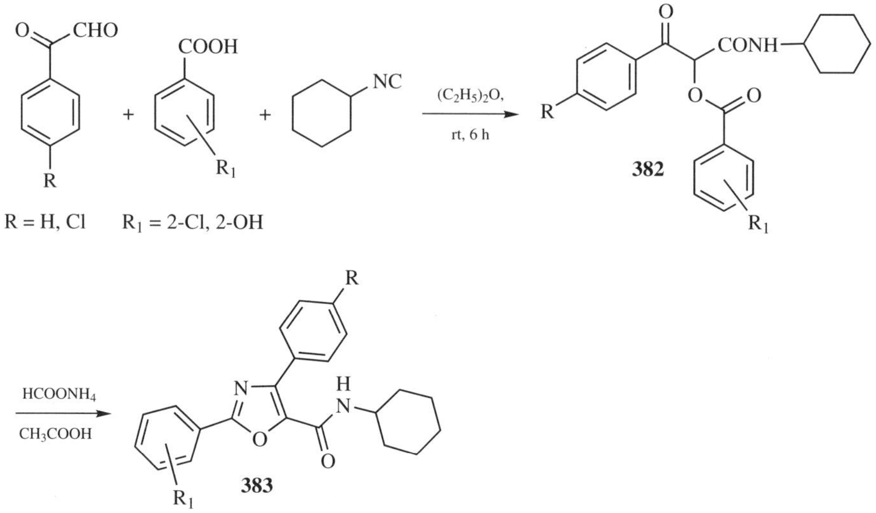

Scheme 1.105

Cyclization of **382** to *N*-cyclohexyl-2-(2′-chlorophenyl)-4-phenyl-5-oxazolecarboxamide **383** was accomplished using ammonium formate in refluxing acetic acid.[166] Selected examples of *N*-cyclohexyl-2,4-diaryl-5-oxazolecarboxamides **383** are shown in Table 1.31.

Lin and co-workers[167] condensed an *N,N*-dialkyl isocyanoacetamide **384** with two equivalents of an *N*-tosyl-benzaldimine **385** in the absence of an acid, base, or catalyst under mild conditions and isolated the singularly novel 5-(dialkylamino)-2,4-disubstituted oxazoles **390** (Scheme 1.106). This reaction was unique to **384** since methyl isocyanoacetate and TosMIC were unreactive under the same conditions. Polar solvents—including CH_3CN, DMF, CH_3OH, and CH_3OH/CH_2Cl_2 (1:1)—increased the yield of the reaction. In general, the yields of these unusually substituted oxazoles were good to excellent.

The authors initially considered two possible mechanisms. In one case, **384** tautomerized to a nitrile ylide (not shown) that added to **385** and then cyclized to a 5-(dialkylamino)-4-substituted oxazole. Because none of these products was observed, this mechanism was dismissed. Furthermore, according to this mechanism, reaction of an α-substituted isocyanoacetamide with **385** would give an intermediate incapable of cyclization. This experiment did give a 5-(dialkylamino)-2,4-disubstituted oxazole, which further discounted the viability of such a process.

Alternatively, initial nucleophilic addition of **384** to **385** gave rise to **386**, which underwent internal deprotonation to generate a nitrile ylide **387**. Conversion of **387** to **390** could occur via either of two pathways. In the first case, the nitrile ylide cyclized to a 5-(dialkylamino)-2-substituted oxazole **388**. Reaction of **388** with a second equivalent of **385** would then afford **390**. The authors showed that there was no reaction of an analog of **388** with a second equivalent of **385**, thereby discounting this proposal. These results led the authors to conclude that **387**

TABLE 1.31. *N*-CYCLOHEXYL-2,4-DIARYL-5-OXAZOLECARBOXAMIDES VIA THE PASSERINI REACTION

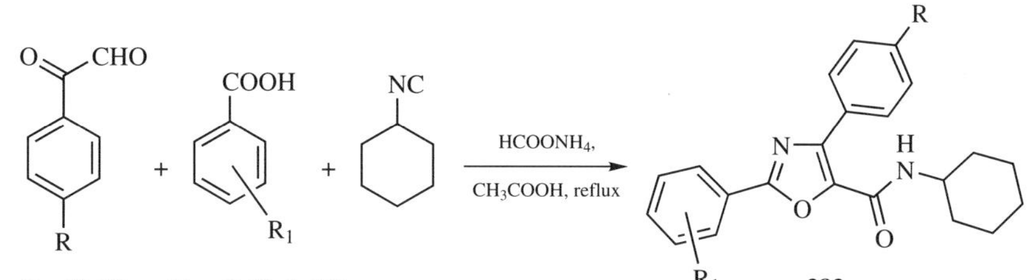

R = H, Cl R_1 = 2-Cl, 2-OH **383**

Figure 1.33

Entry	R	R_1	Product	Yield (%)
1	H	H	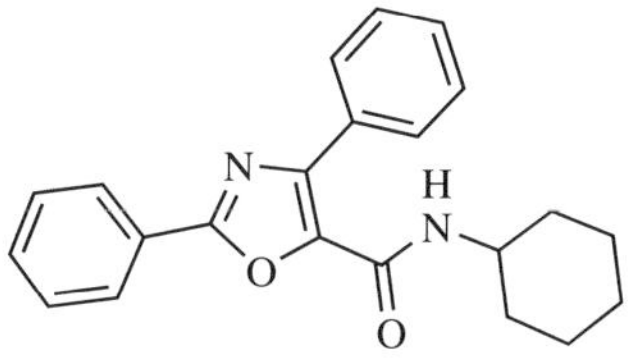	28
2	H	2-OH	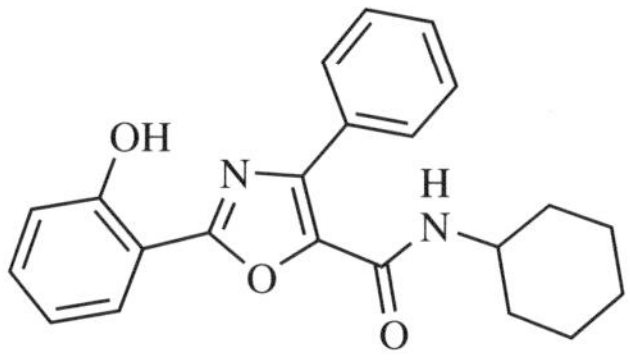	29
3	Cl	H	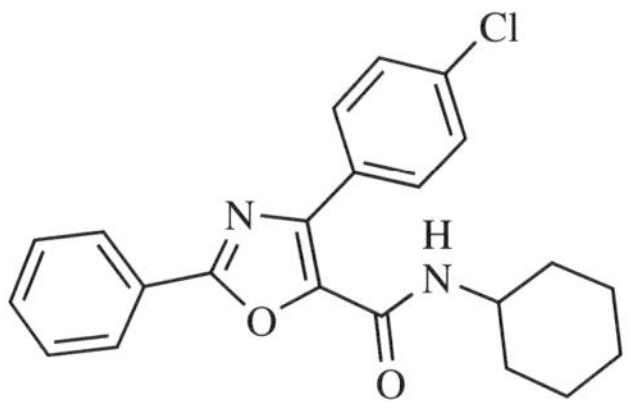	29
4	Cl	2-Cl	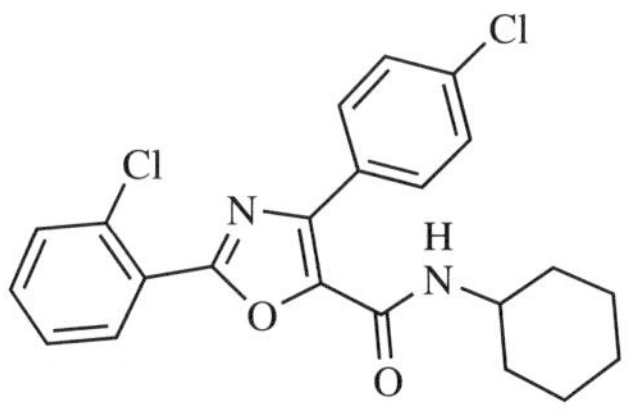	30

[a] Data from ref. 164.

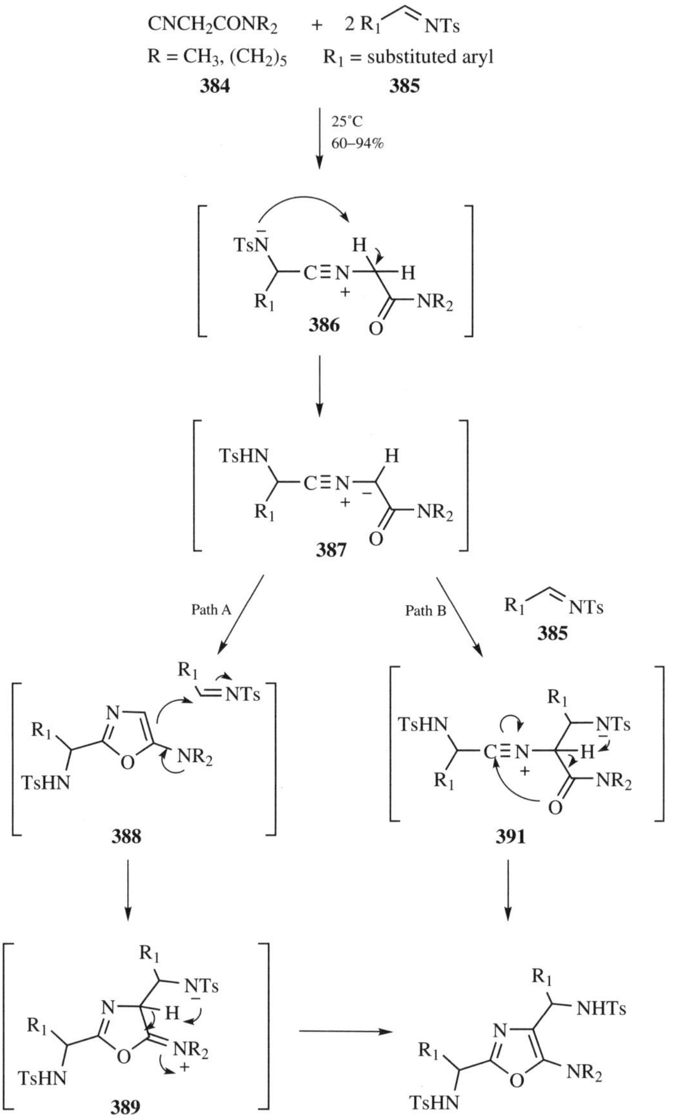

Scheme 1.106

reacted with a second equivalent of **385** to generate **391**, which cyclized to **390**. Some representative examples are shown in Table 1.32.

About 30 years ago van Leusen and co-workers[168] pioneered the condensation of TosMIC with aldehydes and ketones to prepare 4,5-disubstituted oxazolines that eliminate toluenesulfinic acid to yield 5-substituted oxazoles. This process continues today as an important method for constructing such analogues, as evidenced by the following examples.

TABLE 1.32. 5-(DIALKYLAMINO)-2,4-DISUBSTITUTED OXAZOLES FROM N-TOSYLIMINES AND N,N-DIALKYLISOCYANOACETAMIDES[a]

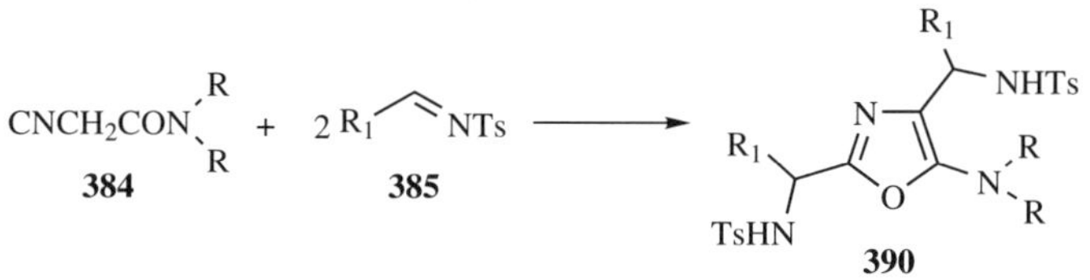

CNCH$_2$CON(R)(R) + 2 R$_1$CH=NTs → [oxazole structure 390]
     384   385

Figure 1.34

Entry	R$_1$	R	Yield (%)
1	4-Cl-C$_6$H$_4$	CH$_3$	78
2	4-NO$_2$-C$_6$H$_4$	CH$_3$	94
3	2,6-di-Cl-C$_6$H$_3$	(CH$_2$)$_5$	81
4	4-CH$_3$O-C$_6$H$_4$	(CH$_2$)$_5$	60

[a] Data from ref. 167.

Anderson and co-workers[169] were interested in preparing the oxazole ring analogs 392 and 393 of partial ergolines known to be potent 5-HT$_{1A}$ serotonin agonists. They found condensation of 394 with TosMIC afforded the oxazol-5-yl analog 393 directly and in excellent yield (Scheme 1.107). However, this methodology was not amenable to prepare the oxazol-2-yl analogue 392. In this case, the authors resorted to Pd(0) coupling of the iodide 396 with oxazol-2-ylzinc chloride. The target oxazol-2-yl analogue 392 was isolated in 54% yield. This compound was not accessible from conventional cyclodehydration protocols using PBr$_3$, PI$_3$, or TMSOTf.

Crozet and co-workers[170] converted the electron-rich aldehyde 397 to 5-(2,5-dimethoxy-3,4,6-trimethylphenyl)oxazole 399 (Scheme 1.108). Here, condensation of 397 with TosMIC was accomplished using K$_2$CO$_3$/CH$_3$OH to give the 5-aryl-4-methoxyoxazoline 398 from which elimination of methanol was achieved using a strongly acidic amberlyst resin. The authors converted the 5-aryloxazole 399 to 401 in four straightforward steps. This completed their synthesis of the first quinone-oxazole bioreductive alkylating agent.

Recently, Addie and Taylor[171] extended the synthetic utility of TosMIC by developing conditions to effect desulfonylation of a 5-substituted 4-tosyloxazole 405 (Scheme 1.109, p. 90). Their plan was to generate the dianion of 5-methyl-4-tosyloxazole 402 followed by quenching with an electrophile to produce a 5-substituted-4-tosyloxazole 404. Desulfonylation of 404 would provide a novel entry to 5-substituted oxazoles 405, thereby avoiding the use of toxic methyl isocyanide. In addition, desulfonylation of the 5-substituted 4-tosyloxazoles 406 normally obtained from reaction of TosMIC with an ester, acid chloride, or acid anhydride would further broaden the scope and use of this reaction.

The authors prepared 402 as usual by acylating TosMIC with acetyl chloride. Sequential lithiation of C(2) and the C(5) methyl group generated the dilithiated intermediate 403, which was quenched with one equivalent of an electrophile to selectively functionalize C(5). Alkyl halides and aldehydes reacted efficiently, but attempts to trap 403 with benzoyl chloride or ethyl chloroformate were unsuccessful.

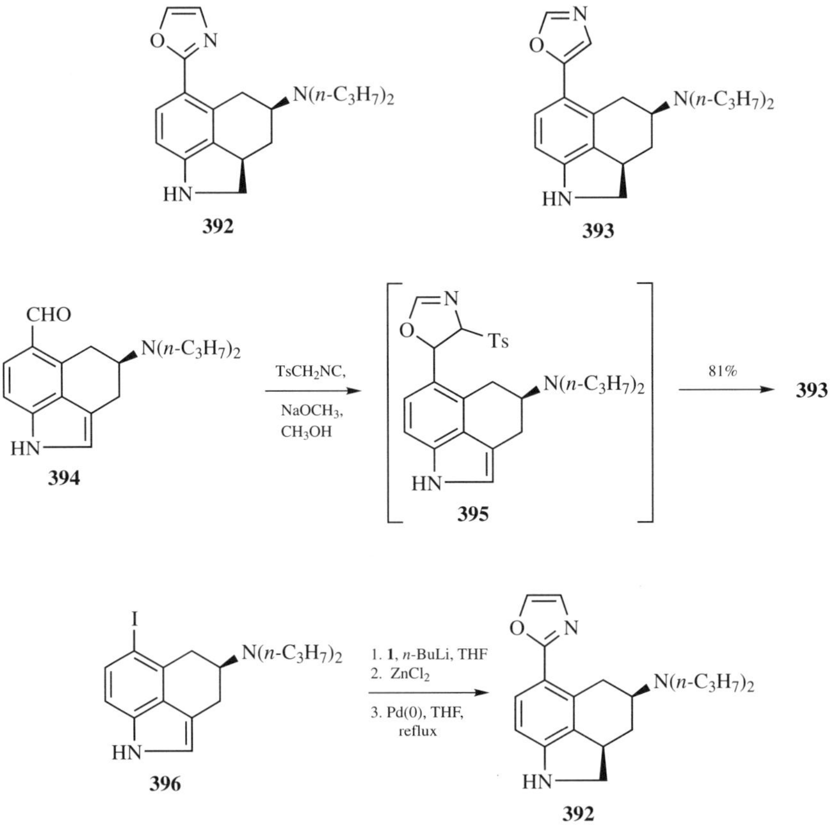

Scheme 1.107

Reductive desulfonylation of **404** proved to be nontrivial. Conventional methods gave no evidence of any reaction. Considerable experimentation was required to develop optimum conditions using sonication and 10% Na-Hg in THF/C_2H_5OH.

Shafer and Molinski[172] reported an efficient synthesis of the parent heterocycle **1**, which they required as a starting material to prepare substituted analogs as shown in Scheme 1.110. Condensation of ethyl isocyanoacetate with formic acid and 1,1′-carbonyldiimidazole gave 4-oxazolecarboxylic acid ethyl ester **408**. Hydrolysis and decarboxylation with CuO and quinoline afforded oxazole **1** in 24–33% yield from ethyl isocyanoacetate. The authors noted this route was amenable to scale up.

Kulkarni and Ganesan[173,174] adapted TosMIC for both solid-phase combinatorial synthesis and solution-phase parallel synthesis (Scheme 1.111). Initially, the authors developed a tentagel-SH supported TosMIC reagent to convert aromatic aldehydes to 5-aryloxazoles. However, this resin was unstable to the basic reaction conditions leading to unwanted impurities. They replaced tentagel with a polystyrene-based reagent and were gratified to find much less contamination in the

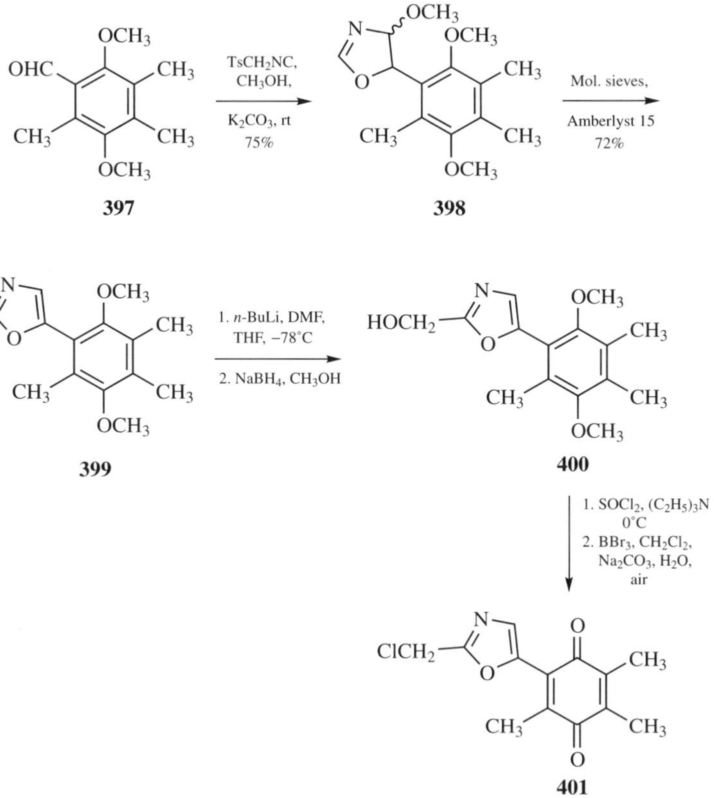

Scheme 1.108

products. Tetrabutylammonium hydroxide was a more efficient base than tetrabutylammonium fluoride or potassium carbonate for the preparation of **411**. Sodium ethoxide was completely unreactive. This methodology was compatible with a wide array of functional groups.

Further refinements from these workers[173] led to a general synthesis of 5-aryl(heteroaryl)oxazoles **411** from aromatic and heterocyclic aldehydes using TosMIC and a resin-bound quaternary ammonium hydroxide. This approach greatly simplifies purification. The base and toluenesulfinic acid are removed by simple filtration to afford **411** in quite acceptable purity for biological screening.

1.3.10. Oxazoles from Vinyl Bromides

Das and co-workers[175] at Bristol-Myers Squibb investigated several approaches for constructing the oxazole ring of the long-acting thromboxane receptor antagonist BMS 180,291 **413** (R = H). One particularly novel approach involved cyclization of a vinyl bromide **412**, readily prepared in five steps from N-pentyl-L-serine amide in a straightforward manner. The authors investigated several organic and

Synthesis and Reactions of Oxazoles

Scheme 1.109

Scheme 1.110

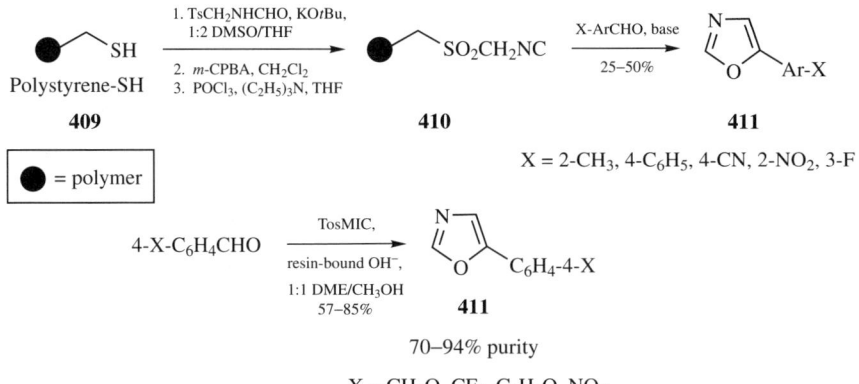

Scheme 1.111

inorganic bases and solvents to effect cyclization of **412**. They found Cs$_2$CO$_3$ to be the most effective. Indeed, prolonged reaction of **412** with Cs$_2$CO$_3$ in dioxane at room temperature produced **37** in 63% yield. The reaction time was shortened considerably with no loss in yield by raising the temperature to 50°C (Scheme 1.112). Previously, **37** had been prepared by CuBr$_2$/DBU/HMTA "oxdiation" of the oxazoline **36** (Scheme 1.9).

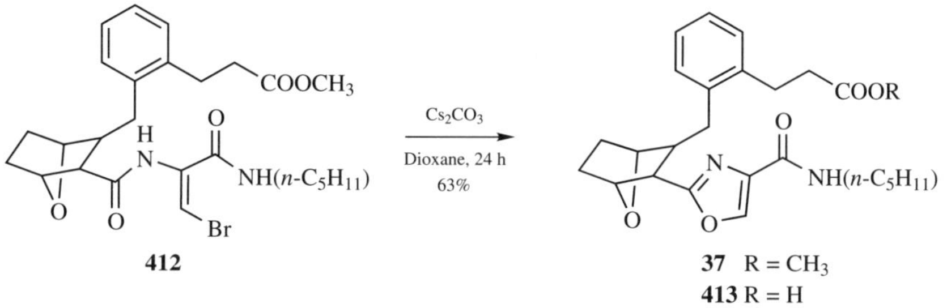

Scheme 1.112

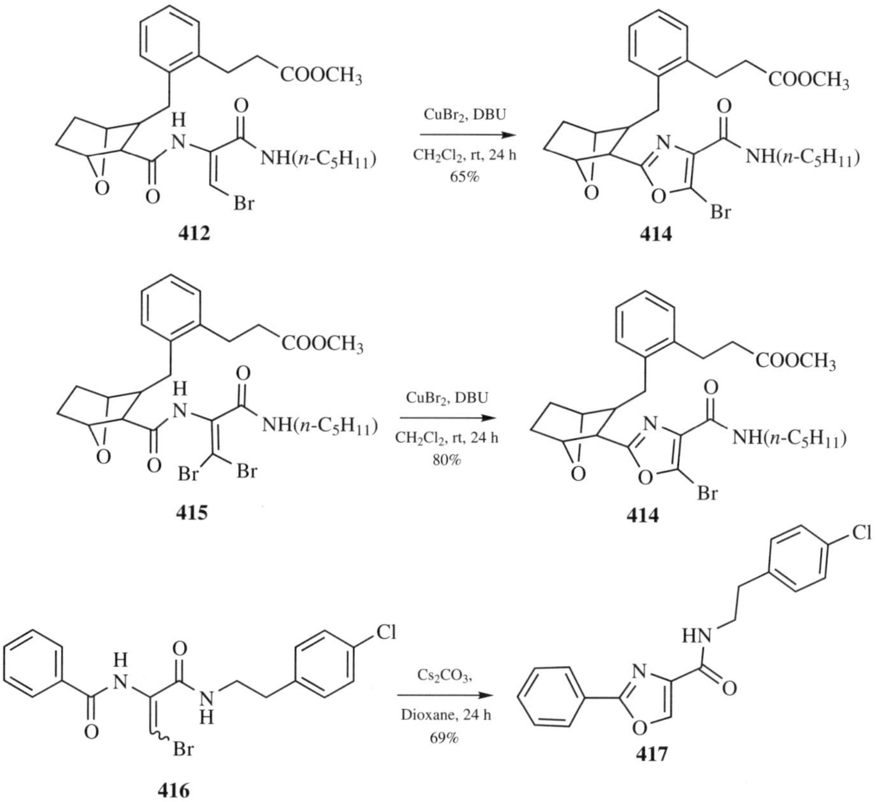

Scheme 1.113

92 Synthesis and Reactions of Oxazoles

Examples of other vinyl bromides that were cyclized to oxazoles are shown in Scheme 1.113. It should be noted that cyclization of **412** with CuBr$_2$/DBU produced **414**, which was also prepared from the dibromide **415**. Cyclization of **416** with Cs$_2$CO$_3$/dioxane gave the expected oxazole **417**.

1.3.11. Oxazoles from α-Acyloxyketones

The synthesis of oxazoles via cyclization of α-acyloxyketones with ammonium acetate or urea was described 65 years ago.[166] The yields of this reaction can vary significantly, and the yield can be complicated by formation of imidazoles. Nonetheless, it is still a useful method for preparing oxazoles given the ready availability of starting materials.

Shridhar and co-workers[176] used this method to prepare 2,4-diaryloxazoles that were intermediates for potential hypolipidemic agents (Scheme 1.114). The α-acyloxyketones **419** were readily prepared from **418** in modest to good yield. Cyclization of **419** to **420** was achieved using anhydrous ammonium acetate in refluxing acetic acid.

Workers in China[177] reported a novel and significant improvement in this methodology as part of an investigation to prepare naturally occurring 2,4-diaryloxazoles. They described a high-yielding two-step process in which cyclization was accomplished using acetamide-BF$_3$•OEt$_2$ (Scheme 1.115). The requisite α-acyloxyketones **422** were prepared from phenacyl bromide and a substituted benzoic acid **421** in excellent yield. Cyclization of **422** with acetamide-BF$_3$•OEt$_2$ in refluxing xylene then produced **423** in good to excellent yield with no evidence for the formation of any imidazole impurities. Selected examples are shown in Table 1.33.

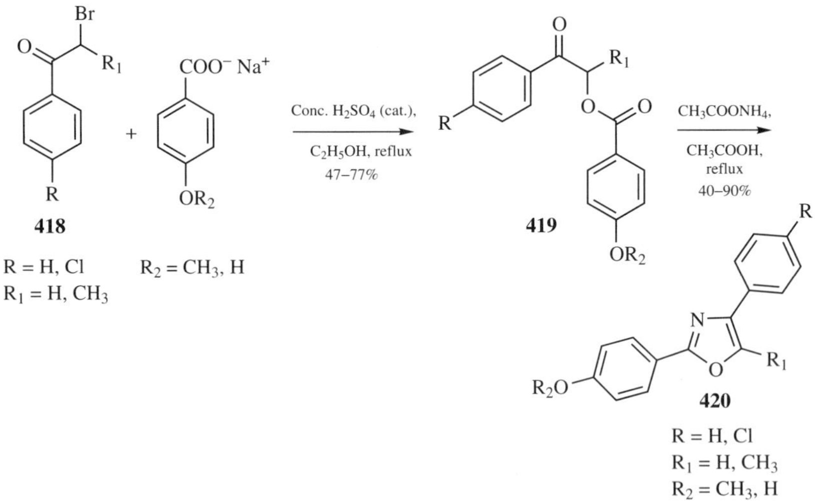

418

R = H, Cl
R$_1$ = H, CH$_3$

R$_2$ = CH$_3$, H

419

420

R = H, Cl
R$_1$ = H, CH$_3$
R$_2$ = CH$_3$, H

Scheme 1.114

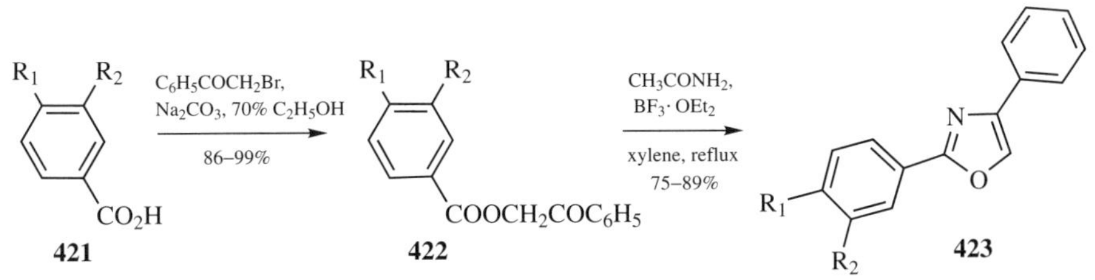

421 **422** **423**

Scheme 1.115

TABLE 1.33. 2,4-DIARYLOXAZOLES FROM CYCLIZATION OF α-ACYLOXYKETONES WITH ACETAMIDE-$BF_3 \cdot OEt_2$[a]

422 **423** **Figure 1.35**

Entry	R_1	R_2	Yield (%)
1	H	H	85
2	CH_3	H	81
3	CH_3O	H	87
4	CH_3O	CH_3O	89
5	OCH_2O		84
6	C_6H_5O	CH_3O	81
7	OH	CH_3O	81
8	Cl	H	78
9	Cl	Cl	75

[a] Data from ref. 177.

422 **424** **425**

426 **423**

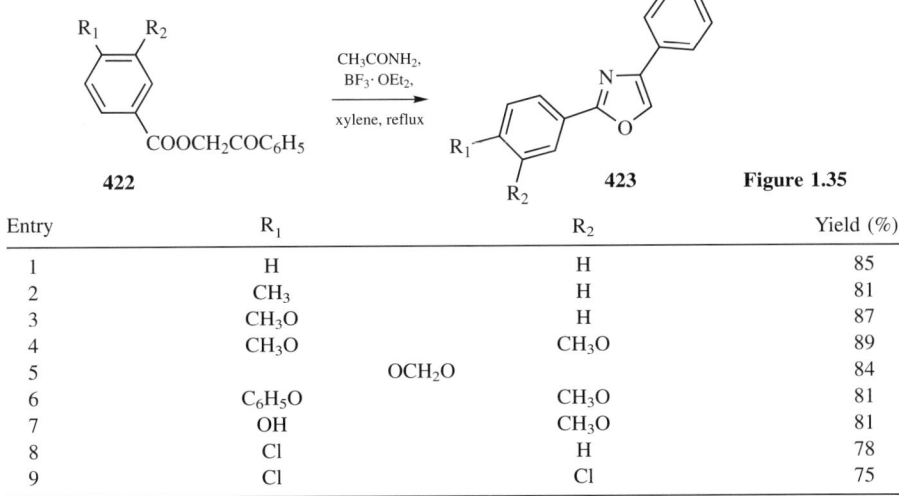

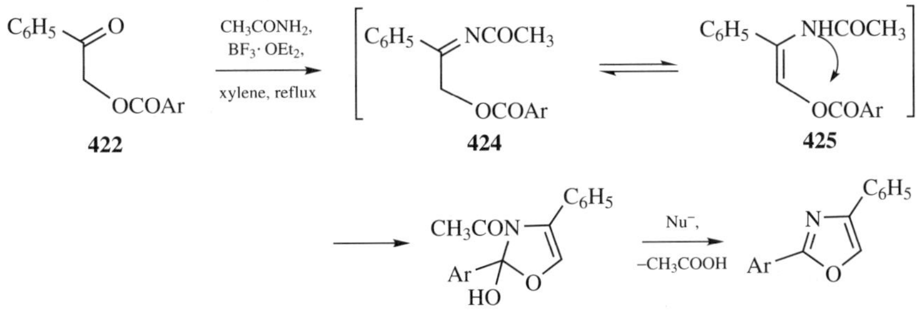

Scheme 1.116

The authors proposed a mechanism in which acetamide adds to the α-acyloxy-ketone **422** to produce an *N*-acylimine **424** that tautomerizes to an *N*-acylenamine **425**. Cyclization of **425** followed by deacetylation and dehydration furnishes **423** (Scheme 1.116). Other catalysts, including H_2SO_4, P_2O_5, $SOCl_2$, and polyphosphoric acid, were ineffective.

Maraccini and co-workers[164] cyclized α-acyloxyamides **382** with ammonium formate in refluxing acetic acid to prepare *N*-cyclohexyl-2,4-diaryl-5-oxazolecarboxamides **383**.

1.3.12. Robinson-Gabriel and Related Reactions

The classical dehydration of 2-acylaminoketones **427** is one of the oldest methods available for preparing a wide range of 2,5-disubstituted and 2,4,5-trisubstituted oxazoles **15** (Scheme 1.117). Robinson[178] and Gabriel[179] independently described this process early in the 20th century, and it continues to be modified and fine-tuned to meet the ever-increasing need for mild, versatile methods for constructing complex and sensitive oxazoles, particularly related to natural products. Much of the more recent work has been concerned with developing milder dehydrating agents that are compatible with a wide range of sensitive functional groups. This section is an overview of both classical dehydrating agents and newly developed reagents and reaction conditions. Examples of oxazoles prepared by cyclodehydration of *N*-acylamino ketones, *N*-acylamino acids/esters, *N*-acylamino nitriles, and *N*-acylpeptides are all included.

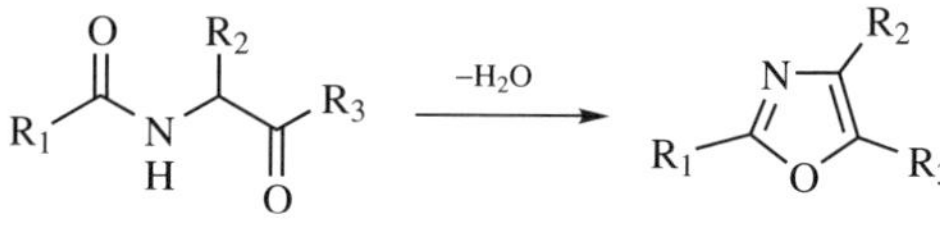

Scheme 1.117

Classical dehydrating agents, such as $POCl_3$, H_2SO_4, $SOCl_2$, P_2O_5, polyphosphoric acid (PPA), *p*-toluenesulfonic acid, and TFAA, are still widely used to prepare a variety of substituted oxazoles, as shown in the following examples.

Meguro and co-workers[37] explored a number of synthetic approaches for preparing oxazoles as hypoglycemic and hypolipemic agents. In some cases, they employed the Dakin-West reaction[180] of an *N*-acyl-aspartic acid 4-methyl ester **428** to prepare a series of 2-acylamino ketones **429**. Subsequent dehydration of **429** with $POCl_3$ then gave a 2-alkyl(aryl)- or a 2,5-dialkyl(aryl)-4-oxazoleacetic acid **430** (Scheme 1.118). Alternatively, Meguro and Fujita[181] prepared 2,5-disubstituted-4-[2-(4-nitrophenoxy)ethyl]oxazoles **432** by H_2SO_4-catalyzed cyclodehydration of the 2-acylamino ketones **431**.

White and co-workers[182] found $POCl_3$ cyclodehydration of the 2-acylamino ketones **433** to be a convenient method for preparing 1-[[[5-(substituted-phenyl)-2-oxazolyl]methylene]amino]-2,4-imidazolidinediones **434**, which were evaluated as skeletal-muscle relaxants. The reaction conditions tolerated both electron-donating

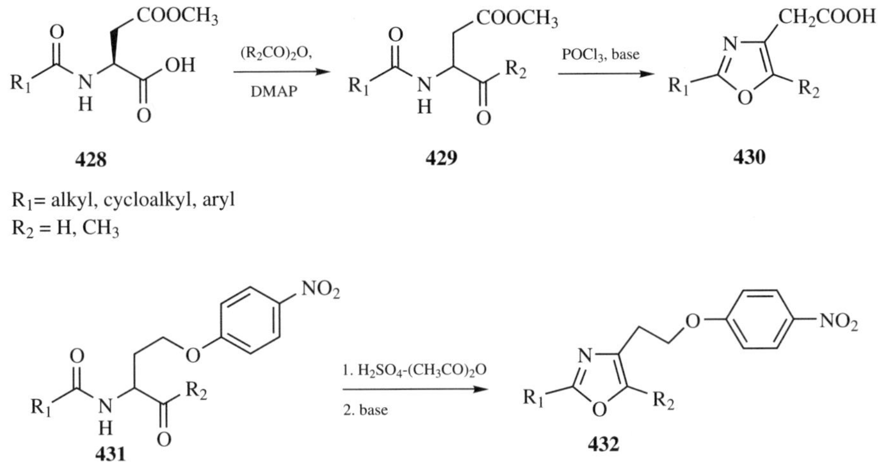

Scheme 1.118

R_1 = alkyl, cycloalkyl, aryl
R_2 = H, CH_3

and electron-withdrawing groups, and generally the yields are acceptable (Scheme 1.119). One exception noted was **434** (X = 4-CH_3), which was isolated in < 6% yield.

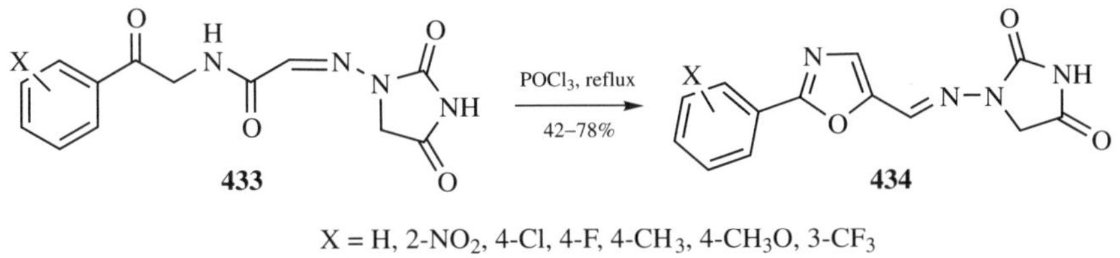

Scheme 1.119

X = H, 2-NO_2, 4-Cl, 4-F, 4-CH_3, 4-CH_3O, 3-CF_3

Loughlin and co-workers[183] prepared the pyrrole-oxazole skeleton of phorbazoles **437a–d**, novel marine alkaloids containing chlorinated pyrrole and oxazole rings, using $POCl_3$. In a model study, cyclodehydration of **435** in $POCl_3$ at 100°C gave **436** in excellent yield (Scheme 1.120). Attempts to prepare **436** in a similar manner using H_2SO_4 gave no reaction, whereas complex mixtures were obtained using $SOCl_2$.

Jacobi's group[184,185] cyclodehydrated N-carbobenzyloxy-D-prolyl-L-alanine methyl ester **438** using $POCl_3$/pyridine to prepare large quantities (50 g) of the 5-methoxy-4-methyl-2-substituted oxazole **439** (Scheme 1.121). Both **439** and ent-**439** were key starting points for their enantiospecific syntheses of (+)-norsecurinine and (−)-norsecurinine.

Kaufmann and co-workers[186] explored several complementary methods to prepare 2,5-disubstituted oxazoles and 2,4,5-trisubstituted oxazoles as flash-lamp pumped laser dyes (Scheme 1.122). Prolonged reflux of **440** with $POCl_3$ effected cyclodehydration to 5-(4-methoxyphenyl)-2-(4-pyridyl)oxazole **441** in modest

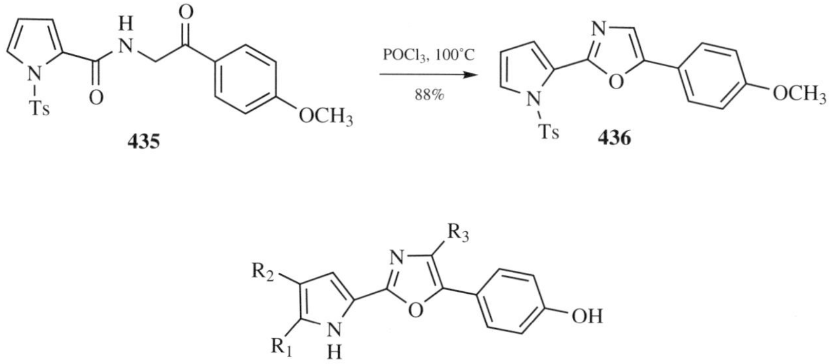

Scheme 1.120

437a R_1 = H, R_2 = R_3 = Cl   phorbazole A
437b R_1 = R_2 = Cl, R_3 = H   phorbazole B
437c R_1 = R_3 = H, R_2 = Cl   phorbazole C
437d R_1 = R_2 = R_3 = H   phorbazole D

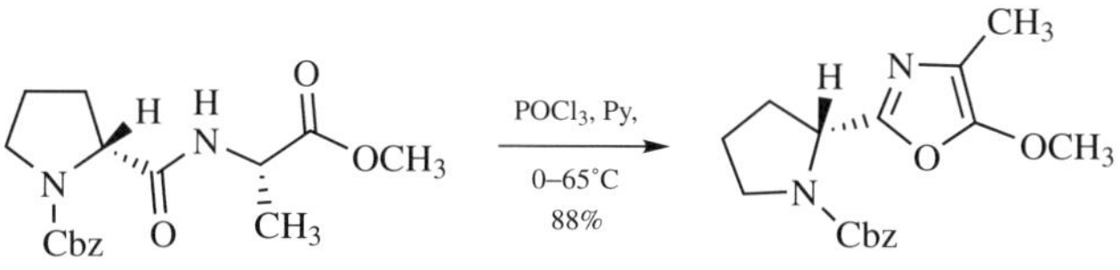

Scheme 1.121

yield. Interestingly, refluxing a mixture of isonicotinic acid and 4-methoxyphenacylammonium chloride in POCl$_3$ afforded **441** directly in 74% yield without isolation of **440**. The regioisomeric dye, 2-(4-methoxyphenyl)-5-(4-pyridyl) oxazole **443** was best prepared from **442** in refluxing SOCl$_2$. Finally, the 2,4,5-trisubstituted analogues **445** were available in excellent yield from cyclodehydration of the isonicotinamides of a 2-amino-6-substituted 1-tetralone **444**. Alternatively, refluxing a mixture of isonicotinic acid and the 6-substituted 2-amino-1-tetralone in POCl$_3$ was also effective (not shown).

Thionyl chloride is one of the most common and effective reagents to effect cyclodehydration of 2-acylamino ketones. For example, Litak and Kauffman[186] refluxed **446** in SOCl$_2$ and isolated 4-[2-(3,4-dihydro-2H-1-benzopyran-6-yl)oxazol-5-ylpyridine **447** in excellent yield (Scheme 1.123). Both **447** and **441** were starting materials for reactive fluorescent stains to be used in fluorescence microscopy. Reck and Friedrichsen[187] refluxed **448** in SOCl$_2$/chloroform and prepared 4-(methoxycarbonyl)-2-methyl-5-oxazoleacetic acid methyl ester **449** (Scheme 1.123). This material was a precursor of the novel diene 4-methoxy-2-methyl-furo[3,4-d]oxazole-6-carboxylic acid methyl ester **450**.

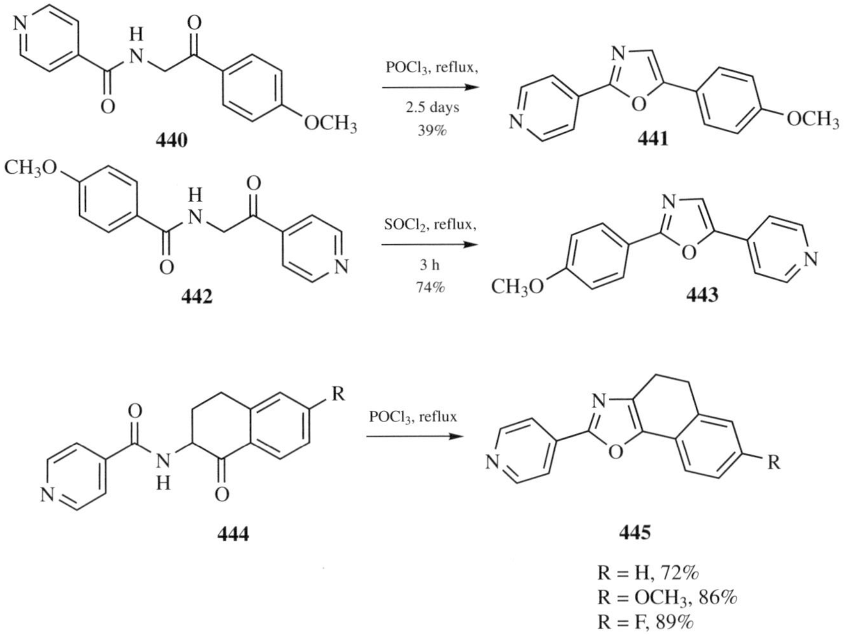

Synthesis of Oxazoles

Scheme 1.122

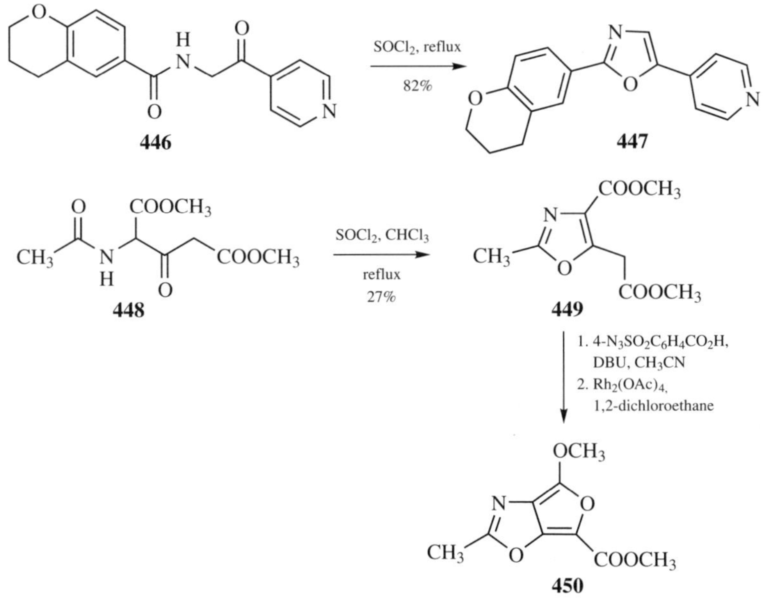

Scheme 1.123

Katritzky and co-workers[188] found that either **452** or **453** was cyclodehydrated in refluxing SOCl$_2$ to produce N,N-diethyl-2-phenyl-naphth[1,2-d]oxazole-5-sulfonamide **454** (Scheme 1.124).

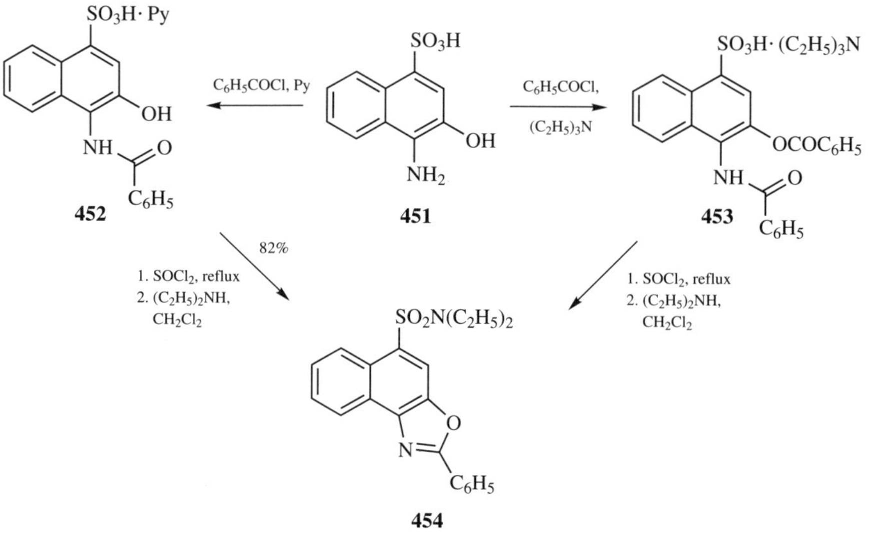

Scheme 1.124

Jacobi and co-workers[189,190] were quite successful using P$_2$O$_5$ or PPA to cyclodehydrate N-acylamino acid esters to prepare oxazole intermediates for natural product syntheses, as shown by the following examples. The authors treated N-acetyl-L-alanine ethyl ester **455** with P$_2$O$_5$ to give 2,4-dimethyl-5-ethoxyoxazole **456** to study regioselective metalation approaches in their synthesis of (±)-paniculide-A (Scheme 1.125).[189] In the same study, L-alanine ethyl ester was acylated with 3-methylglutaric anhydride to yield **457**. Surprisingly, **457** afforded none of the expected oxazole **461a** upon cyclodehydration with P$_2$O$_5$. Instead, the authors isolated only the glutarimide **458**. However, the diester **460** was smoothly converted to **461b** in excellent yield upon treatment with P$_2$O$_5$ in refluxing chloroform. Finally, the Weinreb amide[191] analogue **462** was also readily prepared in four straightforward steps from 3-methylglutaric anhydride.

Jacobi and Lee[190] found P$_2$O$_5$ to be effective for preparing larger quantities (20 g) of an oxazole required for their synthesis of (±)-stemoamide. In this case, 2-(3-chloropropyl)-5-methoxy-4-methyloxazole **464** was prepared from **463** (Scheme 1.126). The authors found that **464** was sufficiently pure as isolated to prepare the acetylenic oxazole **465**.

Brunner and co-workers[192] and Barni and co-workers[193] described examples of fused or heterocyclic oxazoles prepared using PPA (Scheme 1.127). Heating picolinic acid and 3-amino-2-naphthol in PPA afforded 2-(2-pyridyl)naphtho[2,3-d]oxazole **466** directly.[192] Similarly, 2-(nitrophenyl)oxazolo[5,4-b]pyridines **468**, precursors to heterocyclic azo dyes, were prepared from **467**.[193] These authors

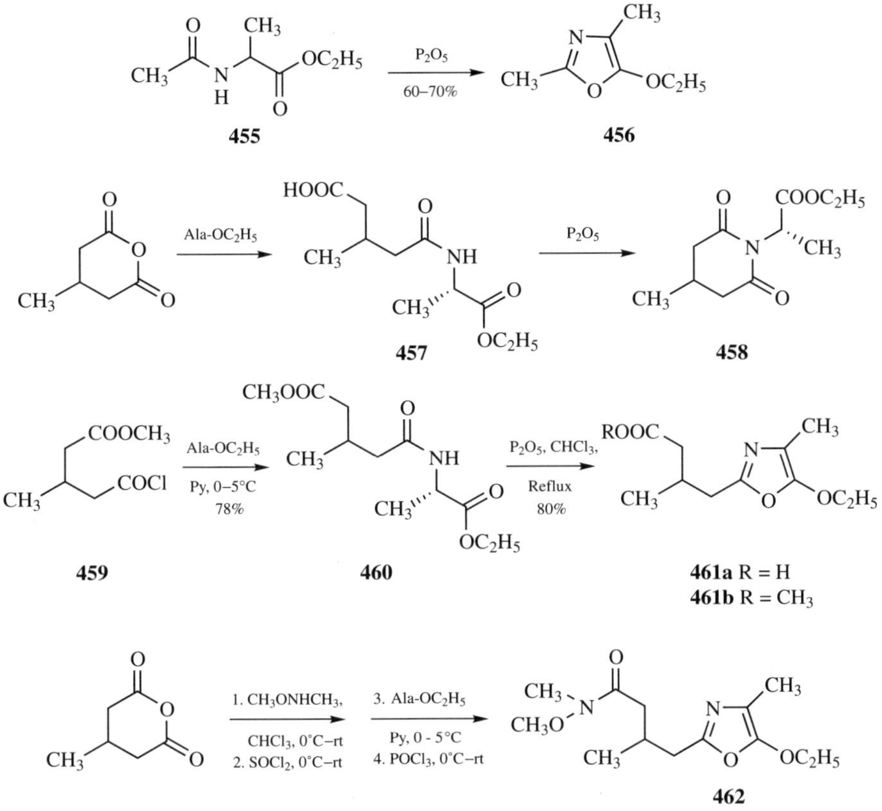

Synthesis of Oxazoles

Scheme 1.125

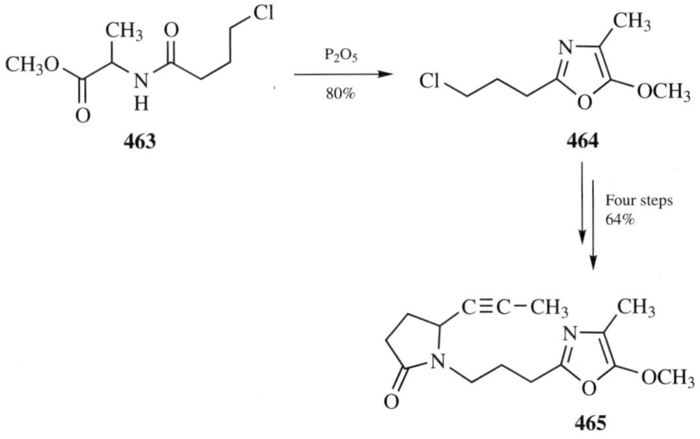

Scheme 1.126

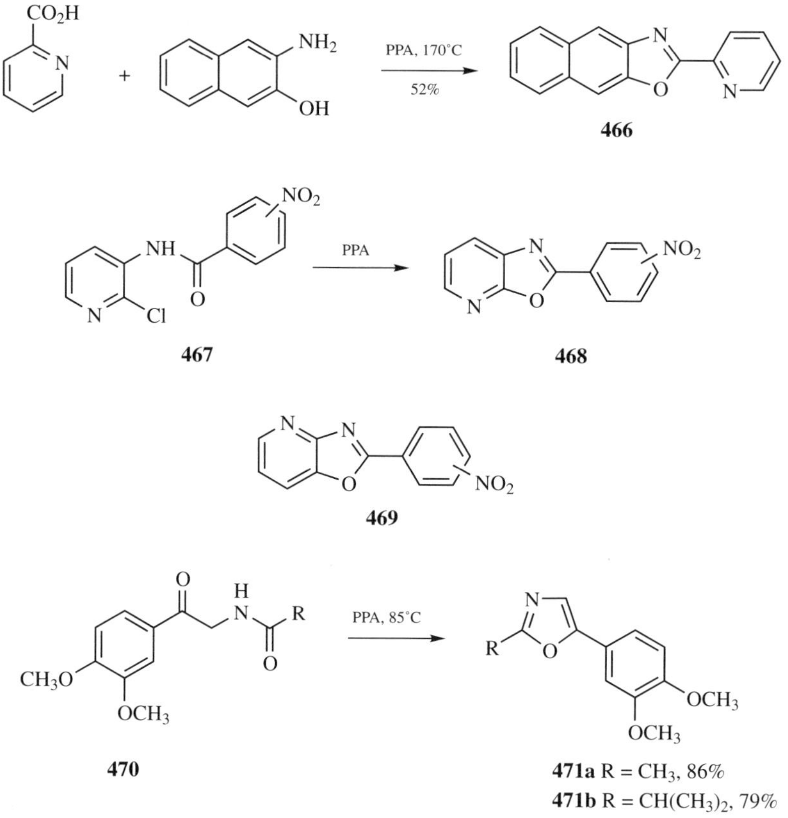

Scheme 1.127

also prepared 2-(nitrophenyl)oxazolo[4,5-*b*]pyridines **469**. Huth and co-workers[194] cyclized **470** (R = CH$_3$, CH(CH$_3$)$_2$) with PPA and isolated 5-(3,4-dimethoxyphenyl)-2-methyloxazole **471a** and 5-(3,4-dimethoxyphenyl)-2-isopropyloxazole **471b** in excellent yield.

Parsons and Heathcock[195] installed the oxazole ring as the penultimate step to complete their synthesis of (−)-tantazole B **473**. Prolonged reflux of **472** with anhydrous *p*-toluenesulfonic acid gave **473** in 34% yield (Scheme 1.128).

Nakahara and co-workers[196] also used *p*-toluenesulfonic acid to prepare the benzoxazole fragment of the antibiotic A23187 (calcimycin). Treatment of 2-amino-3-hydroxy-6-[(trifluoroacetyl)amino]benzoic acid methyl ester **474** with triethylorthoformate (TEOF) and a catalytic amount *p*-toluenesulfonic acid gave 5-[(trifluoroacetyl)amino]-4-benzoxazolecarboxylic acid methyl ester **475a** in excellent yield (Scheme 1.129). Methylation of **475a** gave **475b**, which was then incorporated to complete the synthesis of A23187.

Lipshutz and co-workers[197–200] investigated the interconversion of oxazoles and dipeptides in the context of total syntheses of cyclopeptide alkaloids, e.g., frangulanine **476** and pandamine **477**. In their early work, the authors[197,198]

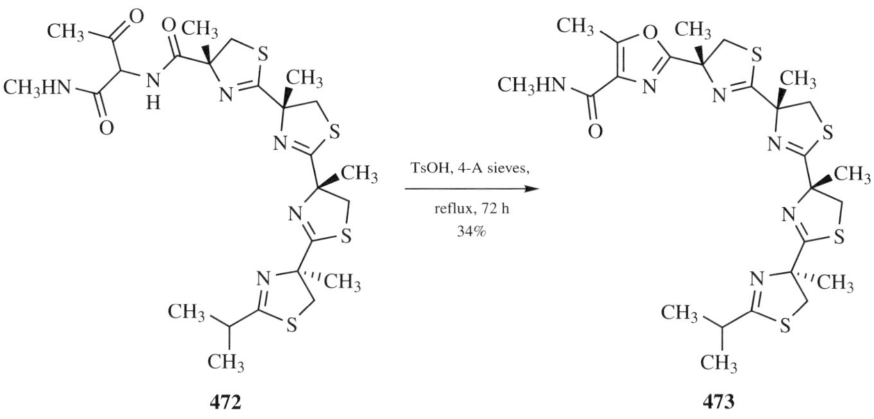

Synthesis of Oxazoles   101

Scheme 1.128

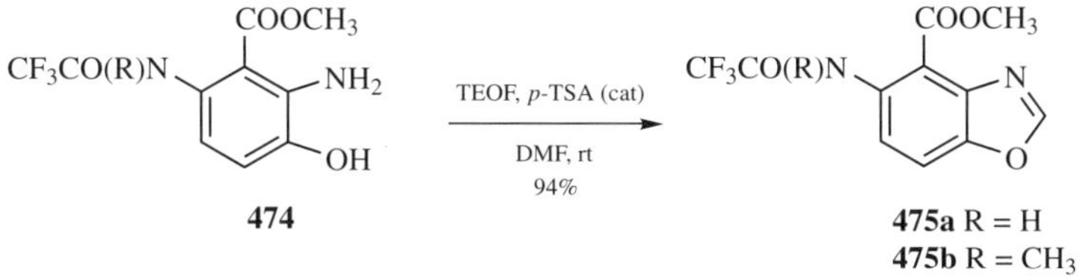

Scheme 1.129

investigated alternate processes to effect ring closure of the macrocyclic lactam of the cyclophane backbone. An intriguing approach is one in which the dipeptide was already incorporated into the cyclophane backbone, e.g., **478**. They anticipated that **478** would serve as a surrogate dipeptide **479** after mild hydrolysis. Retrosynthetically, **478** could be accessed via cyclization of **480** which, in turn, was expected to be available from **481** and **482** (Scheme 1.130).

The authors evaluated a number of routes to prepare the 5-(acylamino)-2,4-disubstituted oxazoles **484** required as the starting materials for their studies.[198] The most satisfactory approach involved acid-catalyzed cyclization of an in situ generated 2-acylamino nitrile **483** (Scheme 1.131). In this case the best yields of **484** were obtained using an acid bromide (acid chlorides were less effective) together with $BF_3 \cdot OEt_2$, although other Lewis acids were acceptable. This method was quite general and tolerated a wide range of functional groups. The authors also used TFAA/TFA (catalytic) extensively to prepare oxazoles via cyclodehydration of dipeptides.[199,200]

In a related study, the authors demonstrated the viability of their proposed synthetic strategy (Scheme 1.132).[197] Here, N-methoxycarbonyl phenylalanine was coupled with leucinonitrile to give the 2-acylamino nitrile **485**. Cyclization of **485**

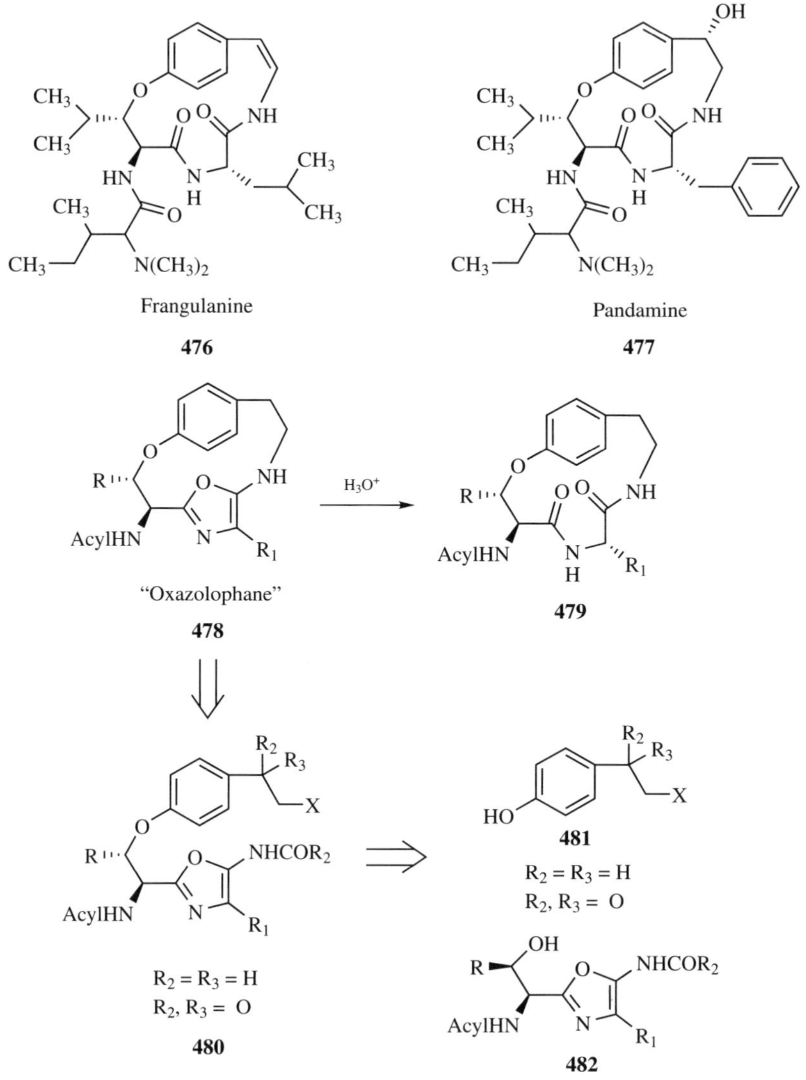

Scheme 1.130

was accomplished with TFAA/TFA (catalytic) and provided the key intermediate oxazole **486** in nearly quantitative yield. Methylation of **486** and acyl group exchange furnished **487**, which was alkylated with *p*-methoxyphenacyl bromide to give **488**. Catalytic hydrogenation of **488**, followed by acid hydrolysis, afforded the dipeptide **489**, the penultimate precursor to an oxazolophane.

At that time, the authors also made an interesting observation regarding cyclization and ring opening of oxazoles. Cyclization of a 1:1 mixture of the

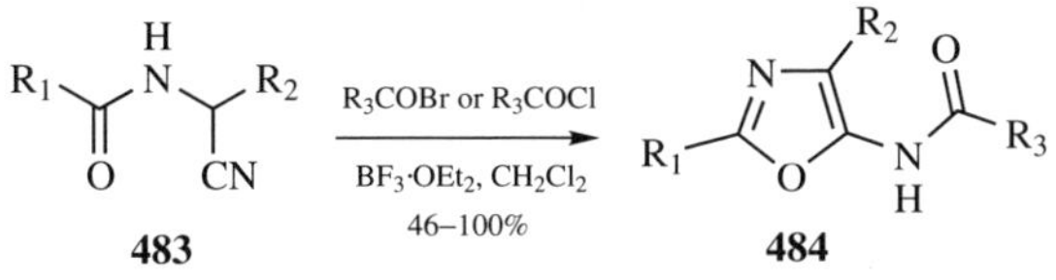

Scheme 1.131

R_1 = CH_3, $CH=CH-C_6H_5$, C_6H_5, OCH_3, OBn
R_2 = Bn, $CH_2CH(CH_3)_2$, H
R_3 = CH_3, t-C_4H_9, Cl_2CH, 4-AcO-$C_6H_4CH_2$

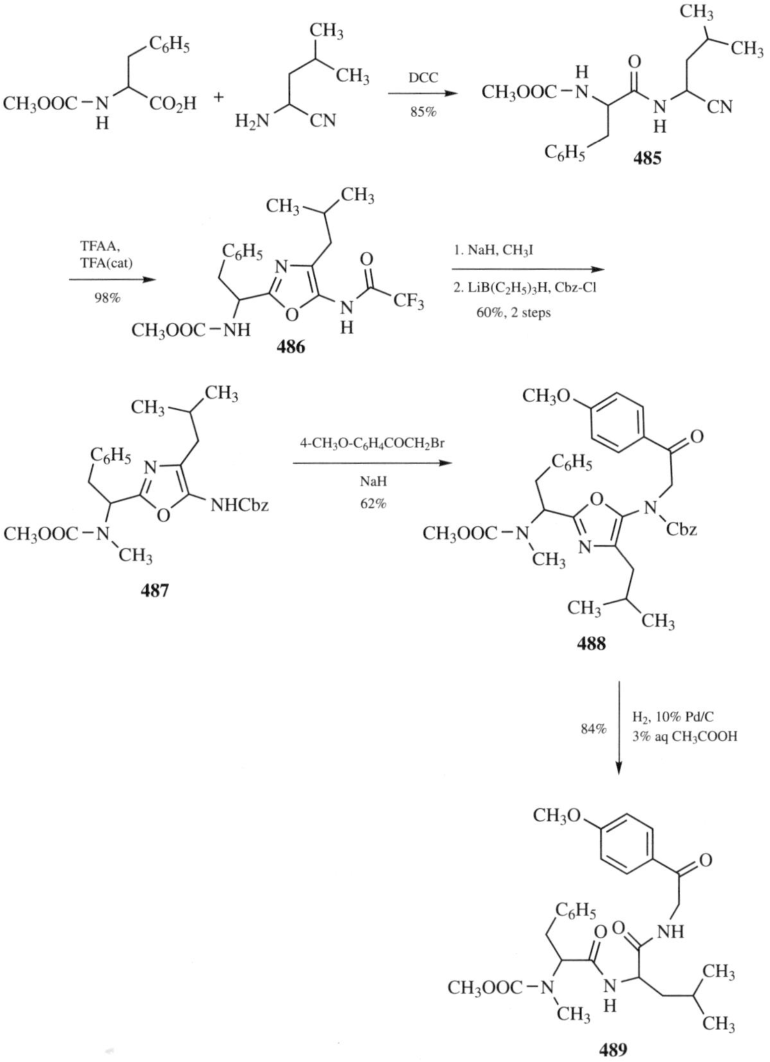

Scheme 1.132

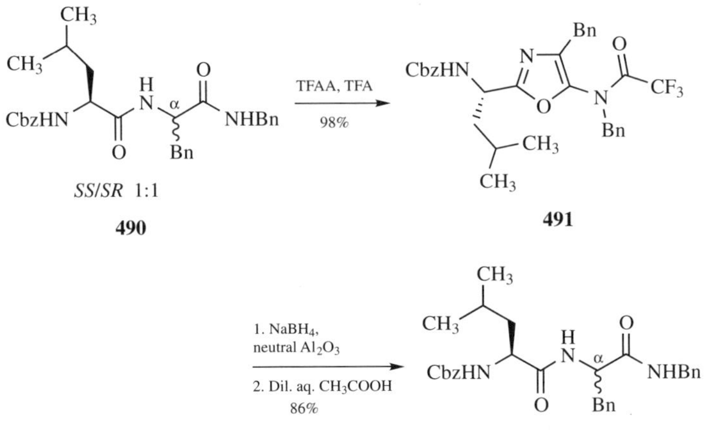

Scheme 1.133

SS/SR diastereomers of *N*-carbobenzyloxy-L-leucyl-D/L-phenylalanine benzylamide **490** with TFAA/TFA gave a single enantiomer (*S*) of the oxazole **491** (Scheme 1.133). Cleavage of the trifluoroacetyl group and subsequent hydrolysis regenerated **490**, but now as a 6.5:1 ratio of *SS/SR* diastereomers. Thus the sequence of cyclization and ring opening effected asymmetric induction at the α-carbon. The authors did not offer an explanation for these interesting results at the time, although later work from Lipshutz's group showed this chiral induction phenomenon to be a general, albeit modest reaction, characteristic of dipetides, diamides, and 2-acylamino nitriles.[199]

Lipshutz and co-workers[200] successfully implemented their strategy of "cyclic peptide chemistry without peptide couplings" during their efforts to prepare crenatine **492a** and integeressine **492b**. The oxazole **494** was prepared from **493** in four steps in 63% overall yield (Scheme 1.134). Here again, the authors found that the cyclization of **493** to a 5-(trifluoroacetamido)oxazole occurred in excellent yield (90%) using TFAA/TFA (cat.). This oxazole **494** then served as a template to induce a favorable macrocyclic conformation for cyclization to the oxazolophane **495**. Macrocyclization of **494** gave **495**, from which the conformationally constrained dipeptide **496** was obtained as a single diastereomer following deprotection. The unnatural configuration at C(8) and C(9) in **496** reflected only commercial availability of an early starting material in the synthesis.

Hulin and co-workers[201] at Pfizer prepared 5-ethyl-2-phenyl-4-(2-propynyl)oxazole **498** quantitatively by cyclization of *N*-(1-acetylbut-3-ynyl)benzamide **497** with a warm 1:2 mixture of TFAA/TFA (Scheme 1.135). This oxazole was a key precursor for new analogs of the hypoglycemic agent ciglitazone.

Kawase[202] described a direct and general synthesis of 2,4-disubstituted 5-(trifluoromethyl)oxazoles **500** from reaction of an *N*-acyl-*N*-benzyl amino acid **499** with trifluoroacetic anhydride in pyridine (Scheme 1.136). In general, the best

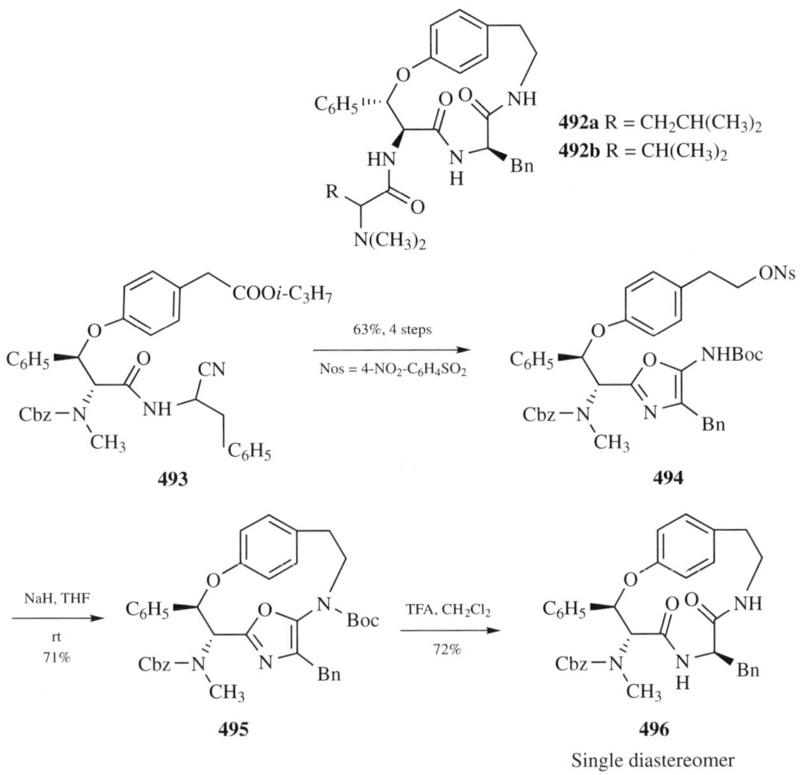

Synthesis of Oxazoles

Scheme 1.134

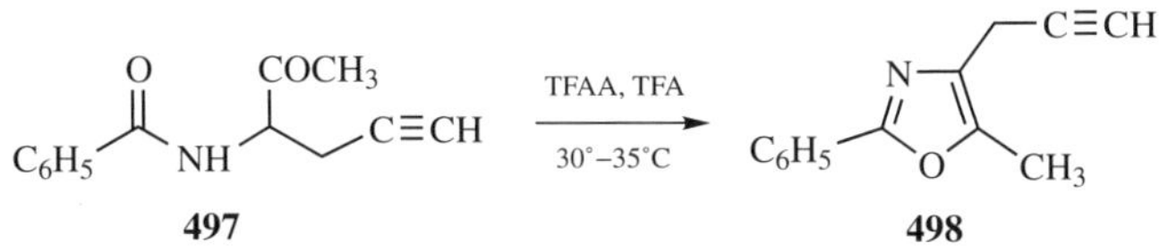

Scheme 1.135

yields were obtained from **499** ($R_2 = Bn$). Perfluoroacylation of **499** ($R_1 = R_3 = C_6H_5$) with $(C_2F_5CO)_2O$ and $(C_3F_7CO)_2O$ produced **501a** and **501b** in excellent yield. Similarly, reaction of an N-acylproline **502** afforded the 4-(3-hydroxypropyl)-2-substituted 5-(trifluoromethyl)oxazoles **503** together with varying amounts of a N-(6,6,6-trifluoro-5-hydroxy-4-oxohexyl)benzamide **504**.[203] In these cases, addition of a catalytic amount of DMAP improved the overall yields of **503**. Representative examples are shown in Table 1.34.

Mechanistically, the authors proposed that this reaction most probably proceeds by cyclodehydration of **499** with TFAA to afford a mesoionic 1,3-oxazolium-5-olate **505**.[203] Trifluoroacetylation of **505** yields **506**, which is ring opened with trifluoroacetate to produce **507** from which the enol trifluoroacetate **508** is produced by decarboxylation (Scheme 1.137). Cyclization of **508** gives the N-alkyloxazolium salt **509** that is dealkylated by trifluoroactate to furnish **500**. This dealkylation is consistent with the observation that **499** ($R_2 = Bn$) are better substrates than **499**

($R_2 = CH_3$ or C_2H_5). In addition, **511a** and **511b** have been isolated from reaction of **510a** and **510b** with TFAA. These intermediates are consistent with the mechanism of a Dakin-West reaction.[180]

Pattenden and co-workers[204] used an in situ Hantzch cyclization[205] to convert the aminoketone **512** to the benzofuranyloxazole **513**. Acetylation of **512** produced an intermediate 2-acylamino ketone (not shown) that cyclized to **513** under the reaction conditions (Scheme 1.138). This benzofuranyloxazole was a model for the quaternary center present in diazonamide A (see Section 1.5.5).

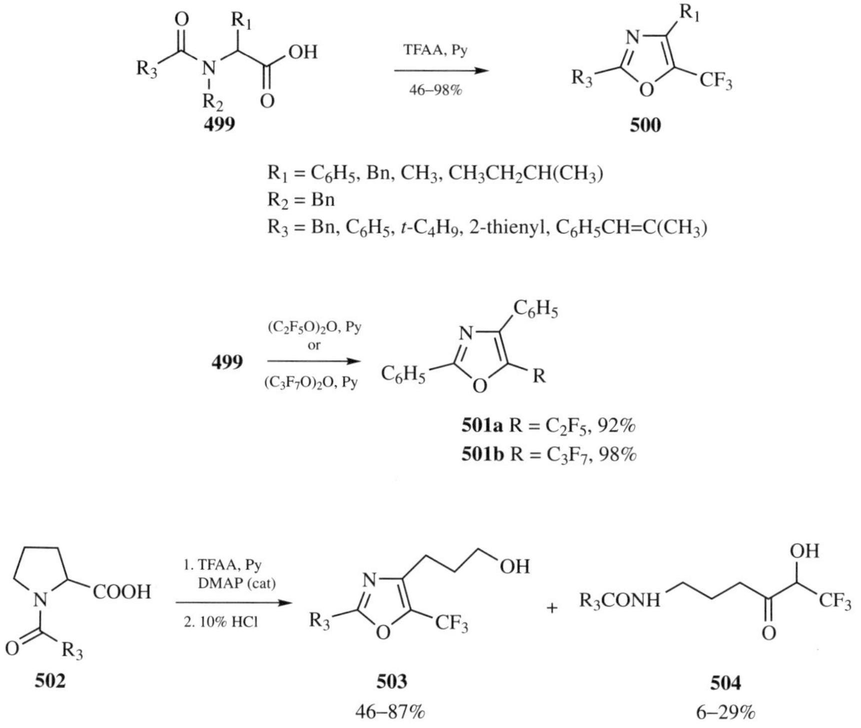
TFAA, Py
46–98%
499
500
$R_1 = C_6H_5$, Bn, CH_3, $CH_3CH_2CH(CH_3)$
R_2 = Bn
R_3 = Bn, C_6H_5, t-C_4H_9, 2-thienyl, $C_6H_5CH{=}C(CH_3)$
$(C_2F_5O)_2O$, Py
or
$(C_3F_7O)_2O$, Py
501a $R = C_2F_5$, 92%
501b $R = C_3F_7$, 98%
1. TFAA, Py
DMAP (cat)
2. 10% HCl
502
503
46–87%
504
6–29%
$R_3 = t$-C_4H_9, C_6H_5, 4-CH_3O-C_6H_4,
4-Cl-C_6H_4, $C_6H_5CH{=}C(CH_3)$

Scheme 1.136

Walker's group[206] described mild conditions suitable for cyclodehydration of a substituted *N*-phenacylamino ketone derived from monic acid **514**. Here, cyclodehydration of **514** ($R = C_2H_5$ or SO_2CH_3) with a mixture of Cl_3CCOCl and a weak base, such as pyridine, produced **515** in good yield (Scheme 1.139, p. 109). The authors had mixed success with other methods of cyclodehydration. For example, treatment of the tris-trimethylsilylether of **514** ($R = NO_2$) with PCl_5, $SOCl_2$, or thiophosgene gave no product. Reaction of the tris-trimethylsilylether of **514** ($R = NO_2$) with tosyl chloride/pyridine produced the enol tosylate, and reaction with $POCl_3$ gave **515** ($R = NO_2$) but in only 8% yield.

TABLE 1.34. 2,4-DISUBSTITUTED-5-(TRIFLUOROMETHYL)OXAZOLES FROM *N*-ACYLATED AMINO ACIDS AND TFAA

499 → (TFAA, Py) → 500   Figure 1.36

Entry	R_1	R_2	R_3	Oxazole	Yield (%)	Reference
1	Bn	Bn	C_6H_5		88	202
2	C_6H_5	Bn	C_6H_5		92	202
3	CH_3	Bn	C_6H_5		46	202
4	$CH_3CH_2CH(CH_3)$	Bn	C_6H_5		51	202
5	R_1-R_2 =		C_6H_5		61[a]	203
6	R_1-R_2 =		$t\text{-}C_4H_9$		87[a]	203

[a] After hydrolysis of CF_3COO group.

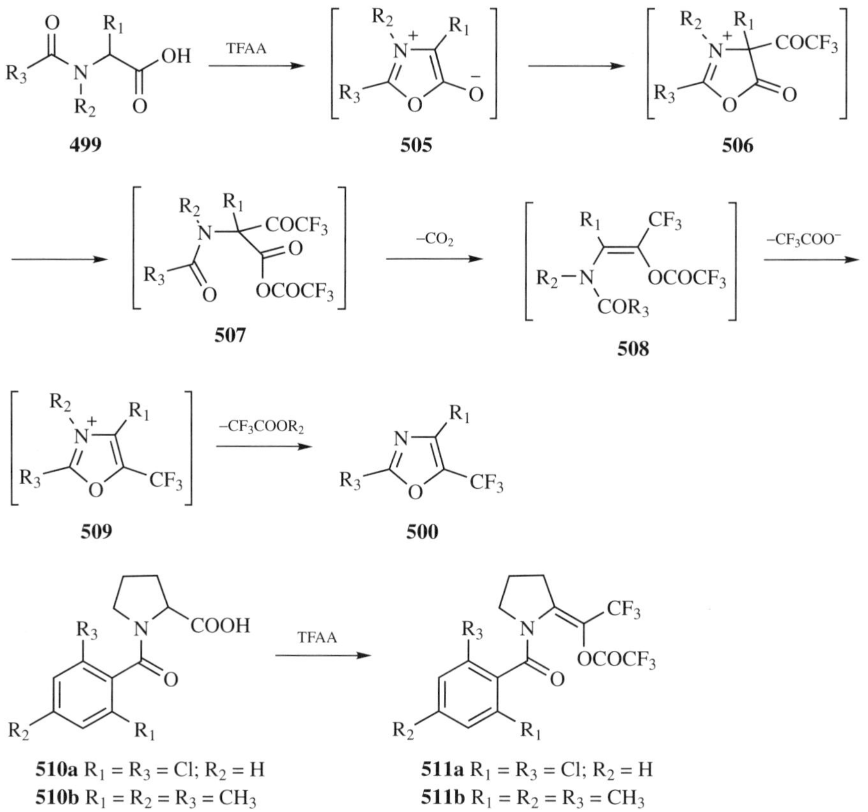
499
505
506
507
508
509
500
510a $R_1 = R_3 = Cl; R_2 = H$
510b $R_1 = R_2 = R_3 = CH_3$
511a $R_1 = R_3 = Cl; R_2 = H$
511b $R_1 = R_2 = R_3 = CH_3$

Scheme 1.137

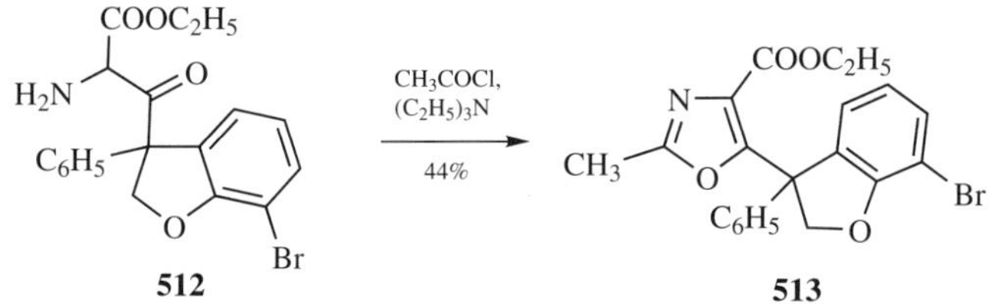
512
513

Scheme 1.138

Freeman's group[207] prepared a series of 5-amino-4-cyano-2-substituted oxazoles **517** from carboxylic acids and aminomalononitrile tosylate (AMNT) (Scheme 1.140). Presumably, AMNT is acylated by a carboxylic acid using DCC, which produces an intermediate *N*-acylamino nitrile **516** that is cyclized under the reaction conditions to yield **517**. The yields of **517** are fair to excellent, and the reaction tolerates a wide variety of functional groups. This further extended the scope of their earlier work.[208] Representative examples are shown in Table 1.35.

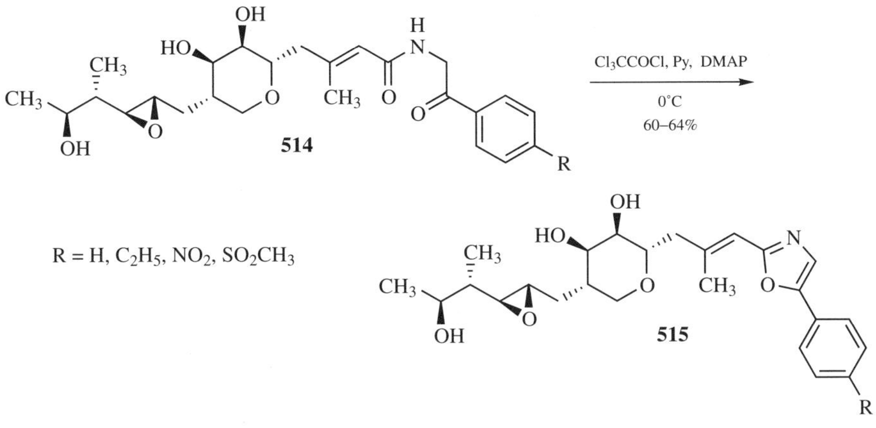

Scheme 1.139

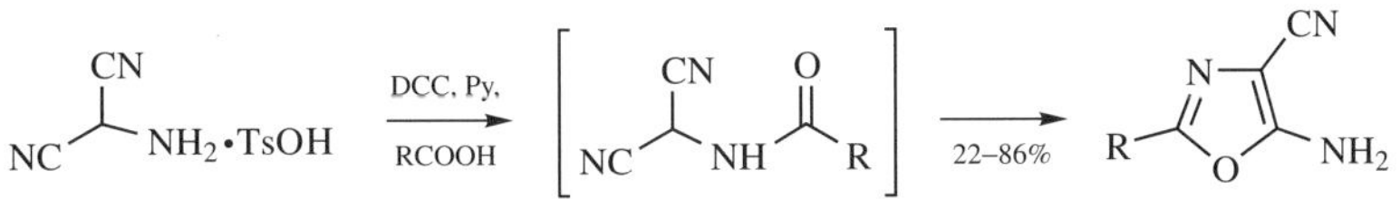

R = H, C$_2$H$_5$, NO$_2$, SO$_2$CH$_3$

Scheme 1.140

The foregoing examples clearly demonstrate the continued use of classical reagents to prepare a wide variety of structurally diverse oxazoles. Nonetheless, research efforts are ongoing to develop milder and more general reagents and reaction conditions that are compatible with the ever-increasing complexity of oxazole natural products. For example, cyclodehydration of optically active *N*-acylamino acids and amino acid esters to prepare peptide-derived oxazoles with minimal epimerization has become increasingly important in the context of total syntheses of natural products, such as thiangazole and tantazole (see Section 1.5.2). This has resulted in the introduction of several new and mild phosphorus-based reagent combinations, including (C$_6$H$_5$)$_3$P/I$_2$/(C$_2$H$_5$)$_3$N, (C$_6$H$_5$)$_3$P/CCl$_4$/DBU, and (C$_6$H$_5$)$_3$P/Br$_2$ for cyclodehydration of 2-acylamino ketones.

For example, Gordon and co-workers[209,210] were uniformly frustrated in their early attempts to prepare the azole peptide mimetic **519** from **518**. Traditional dehydrating agents, including concentrated H$_2$SO$_4$, PCl$_5$, P$_2$O$_5$, POCl$_3$, or SOCl$_2$, all failed to produce any **519**. However, reaction of **518** with three equivalents each of (C$_6$H$_5$)$_3$P and DBU in a 1:2.25:2.25 mixture of CCl$_4$/pyridine/CH$_3$CN afforded **519** in acceptable yield with low levels of epimerization (Scheme 1.141). Subsequently, **519** was incorporated into Pro-trpψ[oxzl]Phe-trp-LeuPheNH$_2$ and found to be a potent substance P antagonist.

Wipf and Miller[62] explored a number of reaction conditions for preparing highly functionalized oxazoles required for natural product synthesis. They introduced a mild two-step process (Scheme 1.142) wherein a β-hydroxy amide **520** is oxidized

TABLE 1.35. 5-AMINO-4-CYANO-2-SUBSTITUTED OXAZOLES FROM AMINOMALONONITRILE TOSYLATE AND CARBOXYLIC ACIDS[a]

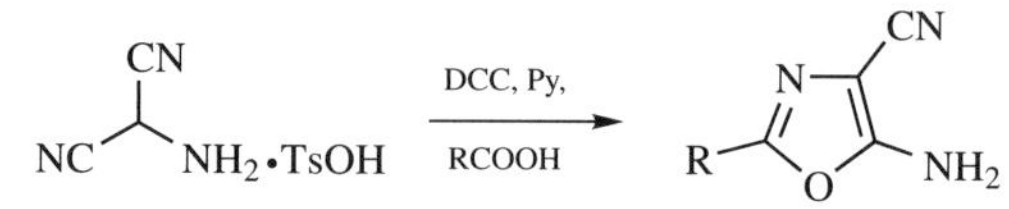

Figure 1.37   **517**

Entry	RCOOH	Product	Yield (%)
1	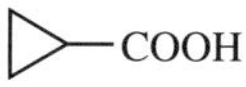 (cyclopropyl)—COOH	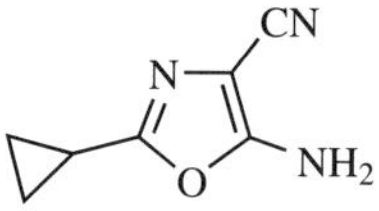 2-cyclopropyl-5-amino-4-cyano-oxazole	68
2	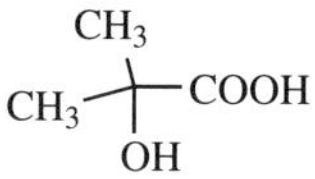 (CH$_3$)$_2$C(OH)—COOH	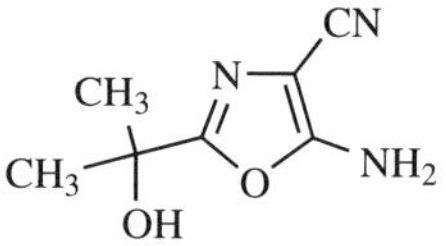 2-(2-hydroxypropan-2-yl)-5-amino-4-cyano-oxazole	22
3	C$_6$H$_5$CH$_2$OCH$_2$COOH	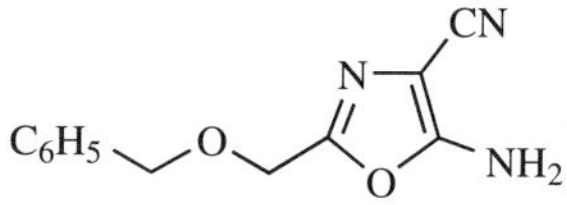 2-(benzyloxymethyl)-5-amino-4-cyano-oxazole	65
4	(C$_2$H$_5$O)$_2$P(O)CH$_2$COOH	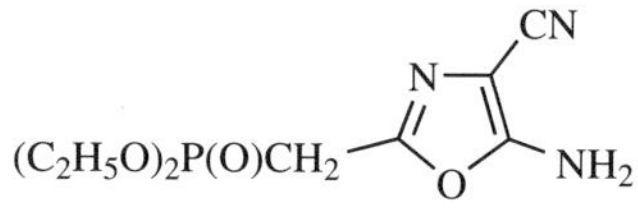 2-(diethoxyphosphorylmethyl)-5-amino-4-cyano-oxazole	86

[a] Data from ref. 207.

to a 2-acylamino ketone **521** via Dess-Martin periodinane[211] followed by in situ cyclization to **522** using (C$_6$H$_5$)$_3$P/I$_2$/(C$_2$H$_5$)$_3$N.

There are a number of noteworthy features in this approach. The starting β-hydroxy amides **520** are readily available and the Dess-Martin periodinane oxidation proceeds rapidly at room temperature. Cyclization of the intermediate N-acylamino ketones **521** also occurs readily under mild conditions, thus allowing for the synthesis of **522** with stereochemically sensitive side chains (Table 1.36, entry 1). The methodology is applicable to phenylserine analogs and can be used in tandem to afford 5-aryl-bis-oxazoles (Table 1.36, entry 5). This two-step process was not as efficient when applied to simple 2-acylamino primary alcohols (Table 1.36, entry 4), although the authors determined that, in this case, the oxidation was the problematic step. Nonetheless, this method was complementary to the SOCl$_2$ cyclodehydration of aldehydo amides.[212] Overall, this two-step protocol has been an extremely versatile extension of the traditional Robinson-Gabriel methodology. This work opens a new and completely nonoxidative pathway for cyclization of serine-derived aldehydes to oxazoles and circumvents cyclization using strong dehydrating agents.

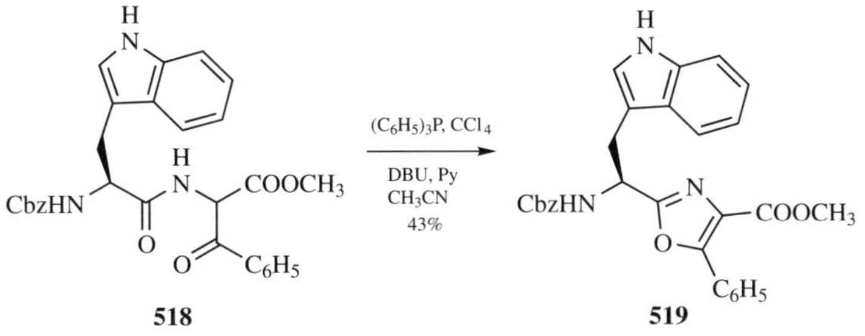

Synthesis of Oxazoles 111

Scheme 1.141

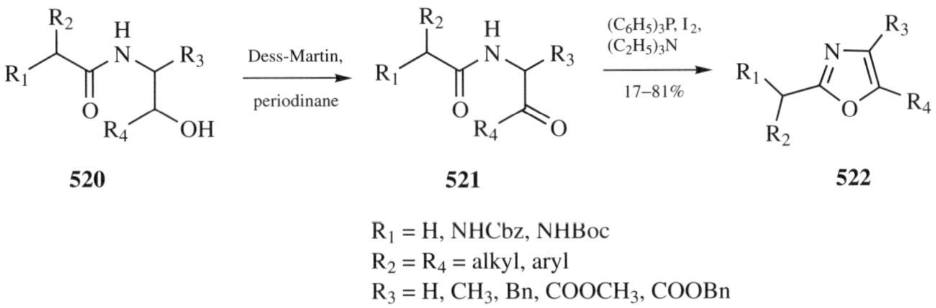

Scheme 1.142

R₁ = H, NHCbz, NHBoc
R₂ = R₄ = alkyl, aryl
R₃ = H, CH₃, Bn, COOCH₃, COOBn

The authors favored a mechanism in which **521** was trapped by (C₆H₅)₃P/I₂ to generate the enol phosphonium salt **523** (Scheme 1.143). Loss of triphenylphosphine oxide from **523** produced the acylimino carbene **524**, which cyclized directly to **522**. An alternative mechanistic rationale involving cyclization of **523** to an oxazoline **525** that then eliminated triphenylphosphine oxide to afford **522** was discounted.

Moody and co-workers[85,89–91] employed this reagent combination to prepare the oxazole fragments of natural products, including muscoride **156** and (+)-nostocyclamide **526**. In each case, they prepared the requisite 2-acylamino β-ketoesters, **162** and **165** (Scheme 1.45) and **527** (Scheme 1.144), using their rhodium-carbenoid N-H insertion methodology. Cyclization of **162**, **165**, and **527** with (C₆H₅)₃P/I₂, (C₂H₅)₃N produced the target oxazoles **163**, **166**, and **528** in good to excellent yield.

Hoekstra and co-workers[213] prepared the threonine-derived oxazole **530** in two steps from **529** using Wipf's (C₆H₅)₃P/I₂/(C₂H₅)₃N protocol[62] (Scheme 1.145). Interestingly, the serine-derived oxazole **532** was also prepared in two steps but was accessed via cyclization of **531** with Burgess reagent[49] followed by oxazoline oxidation using Meyers and Taveras's[35,36] methodology. Both **530** and **532** are key heterocyclic peptidomimetics that were incorporated into a series of small peptides evaluated as thrombin receptor (PAR-1) antagonists.

Falorni and co-workers[214] prepared a series of oxazole-containing amino acids **535a–d** as peptidomimetics via (C₆H₅)₃P/I₂/(C₂H₅)₃N cyclization of 2-acylamino ketones **534** (Scheme 1.146). The authors noted that **534** was prepared in high yield

TABLE 1.36. 2,5-DISUBSTITUTED 4-OXAZOLECARBOXYLIC ACID ESTERS, 2,4,5-TRISUBSTITUTED OXAZOLES, AND 2,4′-BIS-OXAZOLES FROM DESS-MARTIN PERIODINANE OXIDATION AND $(C_6H_5)_3P/I_2/(C_2H_5)_3N$ CYCLIZATION[a]

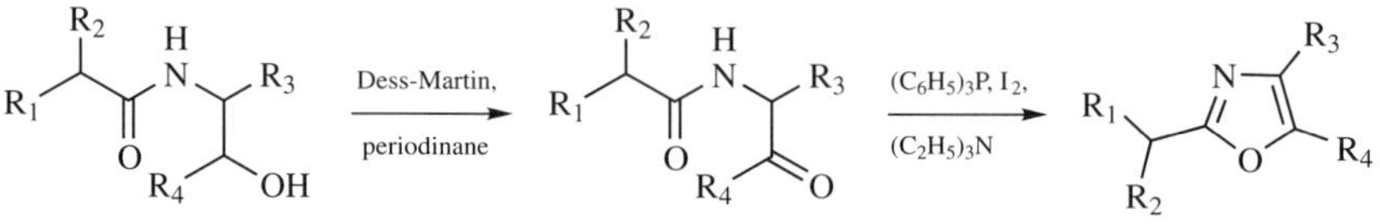

Figure 1.38

Entry	β-hydroxyamides	Oxazole	Yield (%)
1	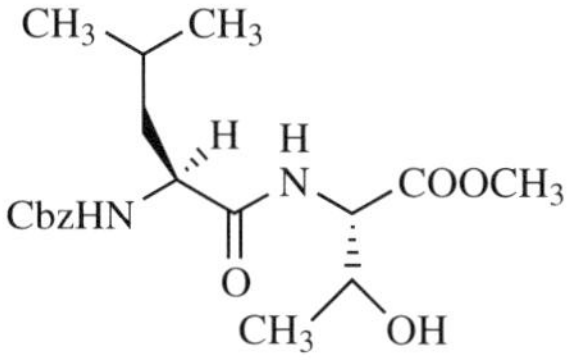 (CbzHN, CH(CH(CH₃)₂)CH₂-, C(O)NH-CH(COOCH₃)-CH(OH)CH₃)	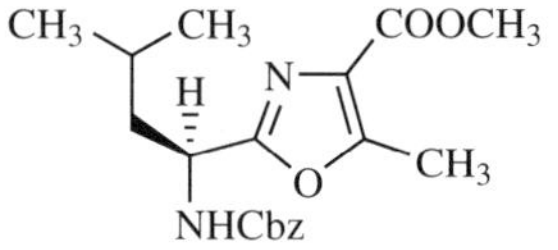 2-(CbzNH-CH(iBu))-4-COOCH₃-5-CH₃ oxazole	55
2	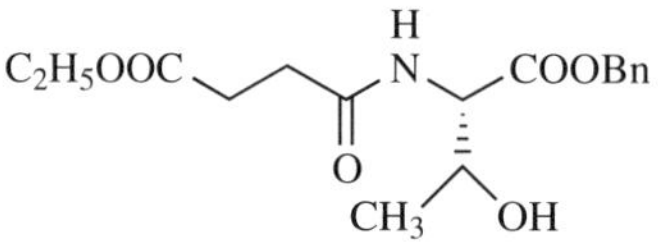 (C₂H₅OOC-CH₂CH₂-C(O)NH-CH(COOBn)-CH(OH)CH₃)	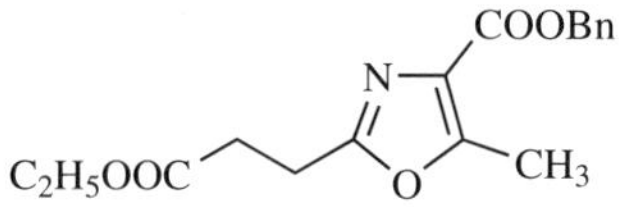 2-(C₂H₅OOC-CH₂CH₂-)-4-COOBn-5-CH₃ oxazole	61
3	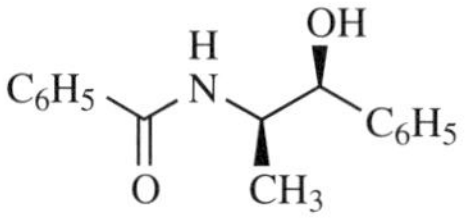 (C₆H₅-C(O)NH-CH(CH₃)-CH(OH)C₆H₅)	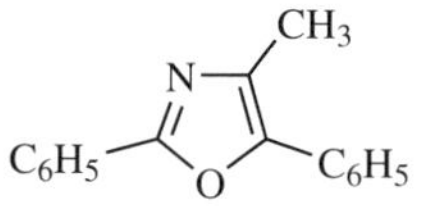 2-C₆H₅-4-CH₃-5-C₆H₅ oxazole	81
4	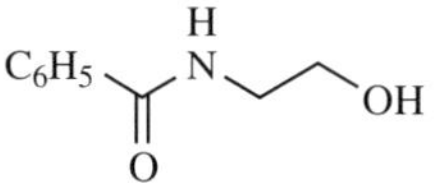 (C₆H₅-C(O)NH-CH₂-CH₂OH)	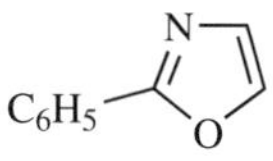 2-C₆H₅ oxazole	17
5	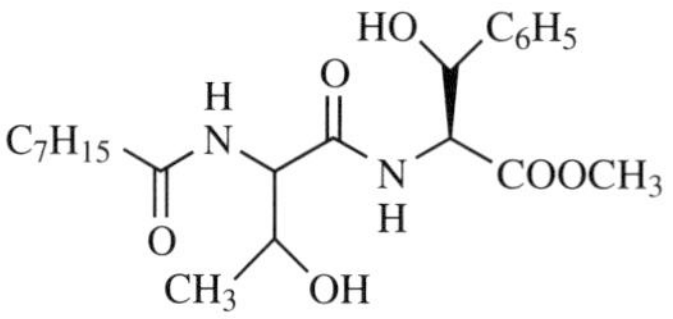 (C₇H₁₅-C(O)NH-CH(CH₃)-CH(OH)-C(O)-C(OH)(C₆H₅)-NH-CH(COOCH₃))	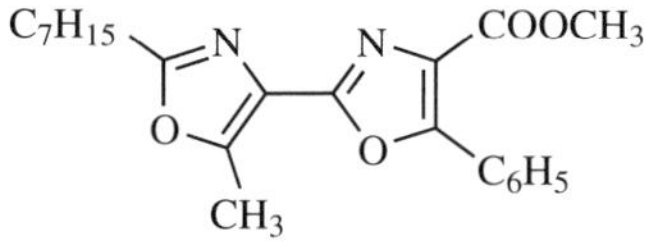 2-C₇H₁₅-5-CH₃-oxazole-bis-2-(4-COOCH₃-5-C₆H₅)oxazole	37

[a] Data from ref. 62.

only if they used dry, ethanol-free $CHCl_3$ as the solvent. The cyclization of **534** was straightforward, as expected, although the yields were quite variable.

Chemists at Bristol-Myers Squibb[45–47,175,215] explored several approaches to the serine-derived oxazole moiety present in the cardiovascular drug candidate ifetroban sodium **538**. Initially, the penultimate precursor **37** was prepared by $CuBr_2$/DBU/HMTA oxidation of the oxazoline **36** (Scheme 1.9).[45–47] In later work,

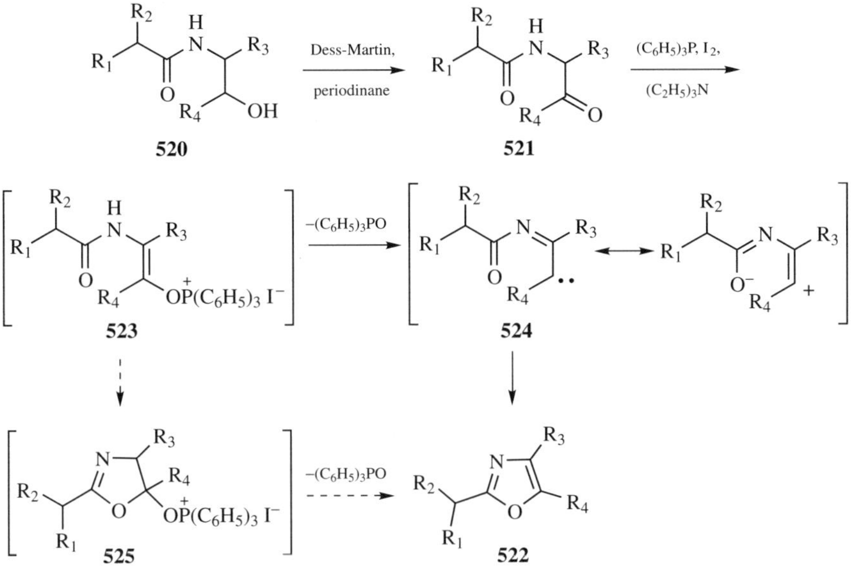

Synthesis of Oxazoles 113

Scheme 1.143

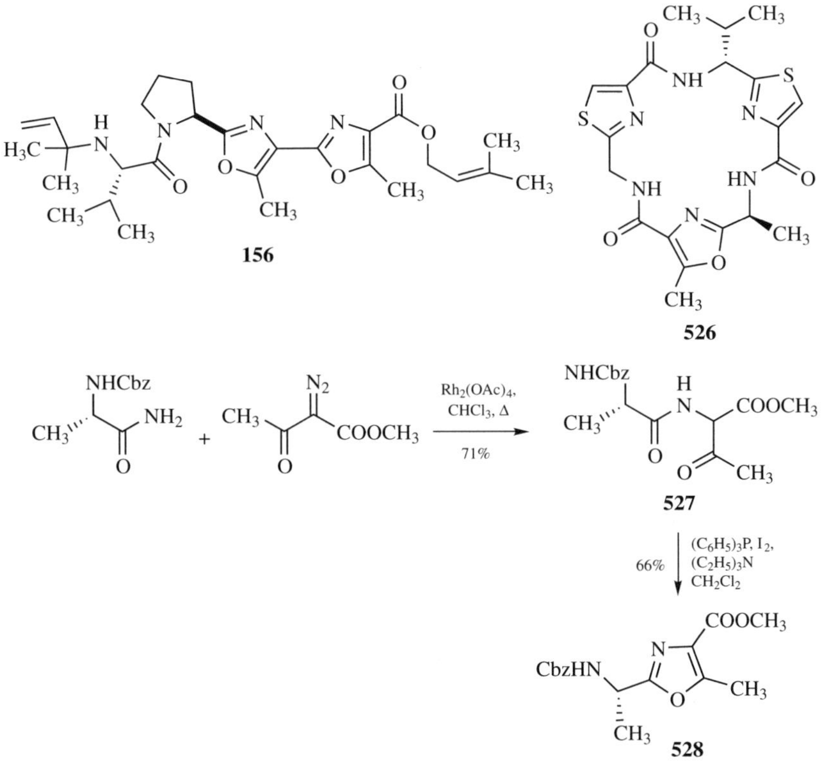

Scheme 1.144

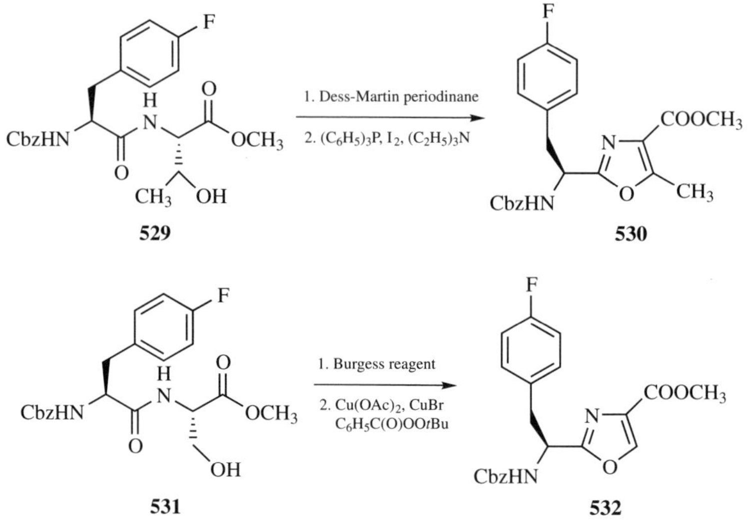

Scheme 1.145

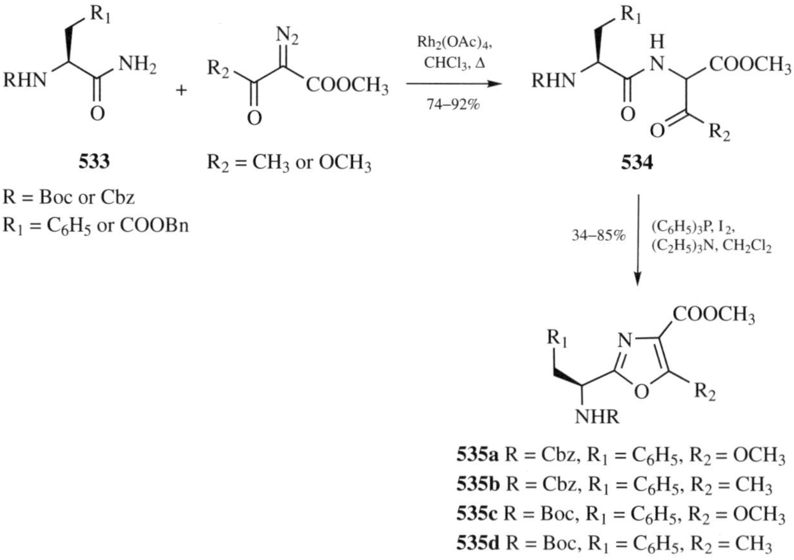

Scheme 1.146

535a R = Cbz, R_1 = C_6H_5, R_2 = OCH_3
535b R = Cbz, R_1 = C_6H_5, R_2 = CH_3
535c R = Boc, R_1 = C_6H_5, R_2 = OCH_3
535d R = Boc, R_1 = C_6H_5, R_2 = CH_3

the vinyl bromide **412** was cyclized with Cs_2CO_3 to afford **37** (Scheme 1.112).[175] The process group evaluated these approaches, as well as NiO_2 oxidation[27] and Wipf's protocol,[62] and determined that none of these routes was scalable. Concerns about environmental waste-disposal issues of copper bromide, large excesses of NiO_2, and poor to modest yields from NiO_2 or $(C_6H_5)_3P/I_2/(C_2H_5)_3N$ cyclization mandated a new approach to **37**.

Swaminathan and co-workers[215] discovered an unprecedented Lewis acid–mediated cyclization of the amide acetal **536** to afford **37** in excellent yield (Scheme 1.147). The crude oxazoline **537** is not isolated but immediately treated with NaOCH$_3$/CH$_3$OH to produce **37** in excellent yield and purity. The initial reaction conditions used an expensive Lewis acid TMSOTf together with the teratogenic solvent 1,2-dimethoxyethane. These were subsequently replaced with the inexpensive acid chlorosulfonic acid and with methyl acetate as the solvent. Ultimately, this process was refined to a one-pot procedure that afforded **37** in 81% overall yield. The amide acetal **540** was readily prepared in 45% overall yield from methyl 3-methoxyacrylate.

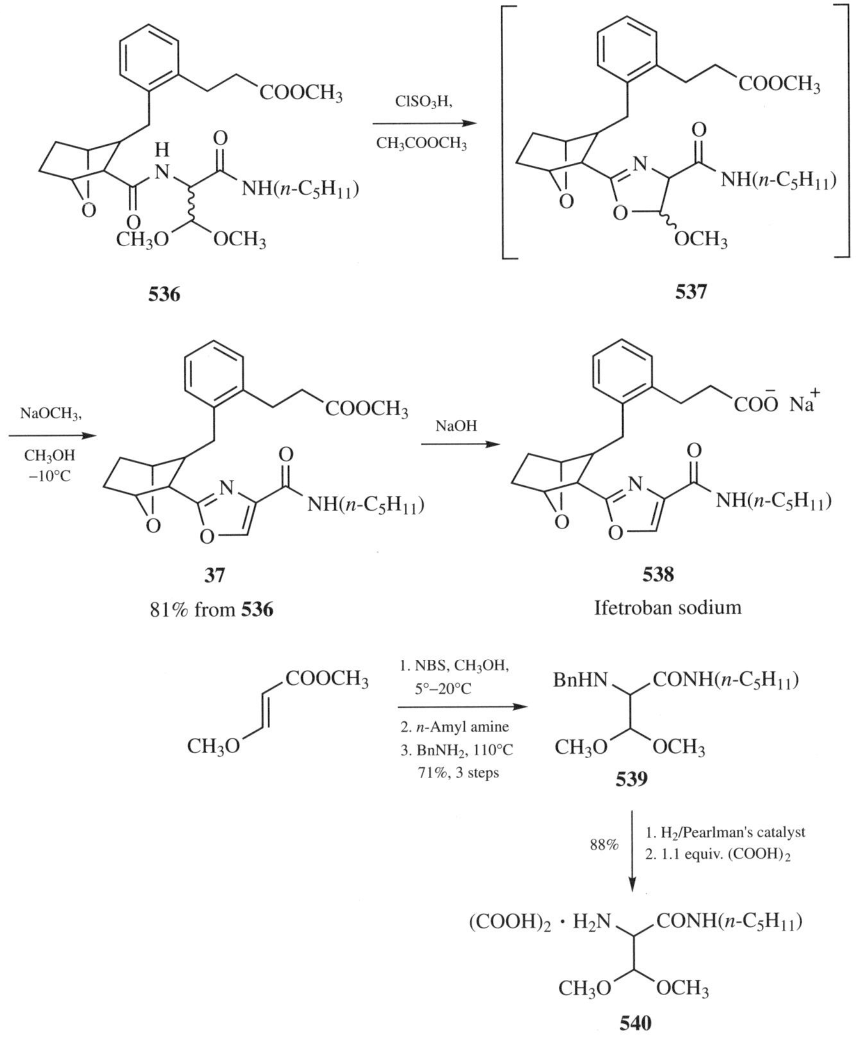

Scheme 1.147

Broom and co-workers[216,217] prepared and evaluated 5-heteroaryl-substituted oxazole analogs of pseudomonic acid **543** as antibacterial agents. Their synthetic strategy employed the addition of an electron-deficient aldehyde to a tosylmethyl amide **541** in the presence of $(C_6H_5)_3P$ and CCl_4 (Scheme 1.148). The resulting 4-tosyloxazoline **542** was treated with DBU to produce **543**. This methodology was not applicable to electron-rich aldehydes, since the intermediate 4-tosyloxazolines, **542** suffer ring opening. The yields vary widely in the examples studied.

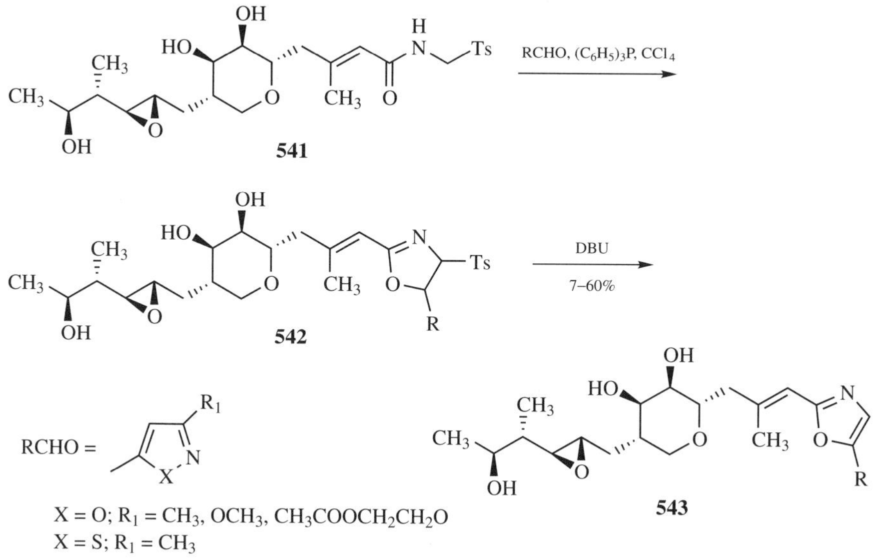

Scheme 1.148

Brain and Paul[218] described a significant improvement to the traditional Robinson-Gabriel methodology. They found that 2-acylamino carbonyl compounds **427** undergo a clean and rapid cyclodehydration by Burgess reagent[49] when flash heated by microwave irradiation to afford oxazoles **15** directly (Scheme 1.149). The reaction requires 1–8 min heating at 50–200 W and provides the desired oxazoles in 72–100% yield. Polyethylene glycol–supported Burgess reagent affords

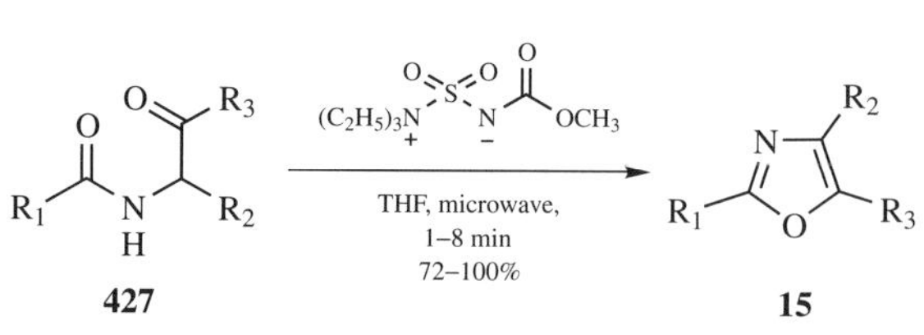

Scheme 1.149

oxazoles in yields comparable to those obtained using nonsupported Burgess reagent.

Notably, their approach is also applicable to the preparation of 2-substituted oxazoles, normally inaccessible from Robinson-Gabriel methodology given the labile nature of 2-acylamino aldehydes to the traditional reaction conditions.[62,212] This methodology broadens the scope of the Robinson-Gabriel synthesis while circumventing the troublesome removal of triphenylphosphine oxide. Further applications of this methodology to high-throughput syntheses are being investigated. Representative examples of oxazoles prepared via this methodology are shown in Table 1.37.

Mohapatra and Datta[219] isolated a series of 2,4-disubstituted-5-(*t*-butoxycarbonyloxy)oxazoles **546** by treating *N*-acyl amino acids **544** with di-*tert*-butyl pyrocabonate (Boc)$_2$O (Scheme 1.150). The authors proposed that (Boc)$_2$O effected cyclodehydration of **544**, producing a 2,4-disubstituted 5(4*H*)-oxazolone **545**. However, enolization of **545** under the reaction conditions and trapping with (Boc)$_2$O then led to **546**. Attempts to isolate **545** were unsuccessful. This one-pot reaction was quite general and afforded the oxazoles in good yield.

Interestingly, base-catalyzed removal of the Boc group in **547** did not afford the expected 5(4*H*)-oxazolone **548**. Instead, the authors isolated the novel quaternary amino acids **550a** and **550b**, which were believed to arise via an unprecedented *t*-butoxycarbonyl transposition in **547** to give **549**, followed by ring opening.

Some additional examples of the cyclodehydration of 2-acylamino ketones used to prepare oxazole intermediates in the context of total synthesis are described in Section 1.5.

TABLE 1.37. 2-SUBSTITUTED; 2,5-DISUBSTITUTED; AND 2,4,5-TRISUBSTITUTED OXAZOLES VIA MICROWAVE-ACCELERATED CYCLODEHYDRATION OF 2-ACYLAMINO KETONES OXAZOLES USING BURGESS REAGENT[a]

Entry	R_1	R_2	R_3	Yield (%)
1	CH_3	H	C_6H_5	78
2	2,4-di-CH_3-C_6H_3	H	H	90
3	C_6H_5	H	C_6H_5	84
4	3-C_6H_5O-C_6H_4	H	H	85
5	C_6H_5	CH_3	C_6H_5	82
6	(*S*)-CbzNHCH(*i*-C_3H_7)	$COOCH_3$	CH_3	95

[a] Data from ref. 218.

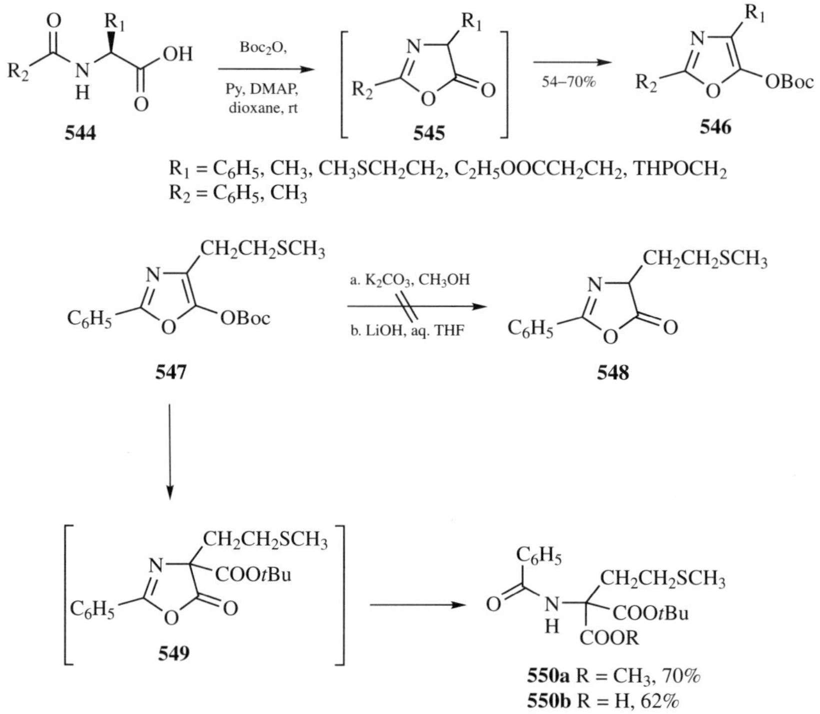

Scheme 1.150

1.3.13. Miscellaneous Reactions

There are a variety of other methods available for preparing substituted oxazoles. Some arbitrarily selected examples are discussed in this section. In some cases, these methods are specific for a particular substitution pattern, while others are more general in nature. As before, no attempt has been made to include all possible methods or oxazole-substitution patterns.

Lakhan and Singh[220] described the first direct synthesis of 4-amino-2,5-disubstituted oxazoles **552** from readily available starting materials. They condensed equimolar amounts of an aroyl cyanide **551** and an aldehyde with anhydrous ammonium acetate to afford **552** in 40–65% yield (Scheme 1.151). The reaction works better for aroyl cyanides containing electron-withdrawing groups. The

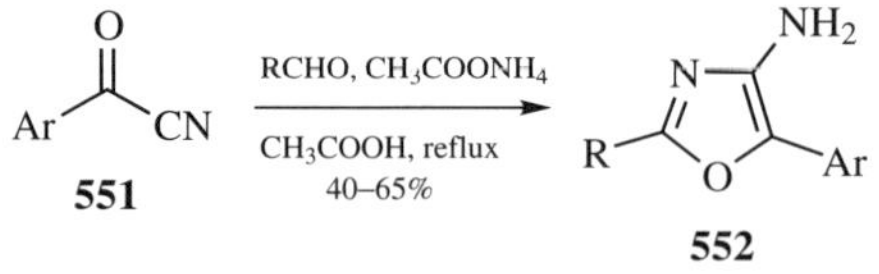

Scheme 1.151

TABLE 1.38. 4-AMINO-2,5-DIARYLOXAZOLES AND 2-ALKYL-4-AMINO-5-ARYLOXAZOLES FROM CONDENSATION OF AROYL CYANIDES, ALDEHYDES, AND AMMONIUM ACETATE[a]

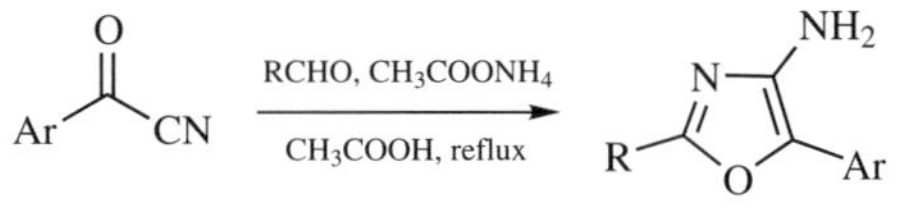

Ar-C(O)-CN (**551**) $\xrightarrow{\text{RCHO, CH}_3\text{COONH}_4}{\text{CH}_3\text{COOH, reflux}}$ 4-amino-2-R-5-Ar-oxazole (**552**)

Figure 1.40

Entry	Ar	R	Yield (%)
1	C_6H_5	C_6H_5	60
2	C_6H_5	2-NO_2-C_6H_4	50
3	4-CH_3-C_6H_4	4-Cl-C_6H_4	55
4	4-Cl-C_6H_4	4-CH_3O-C_6H_4	60
5	4-CH_3O-C_6H_4	4-NO_2-C_6H_4	50
6	C_6H_5	CH_3	65
7	C_6H_5	$CH_3CH_2CH_2$	50

[a] Data from ref. 220.

reaction fails if both **551** and the aldehyde contain electron-donating groups. Aliphatic and aromatic aldehydes are reactive. The authors noted the reaction proceeds without formation of by-products. In addition, the low yields are sometimes attributed to the recovery of starting materials. Representative examples are shown in Table 1.38.

Mechanistically, the authors proposed that ammonium acetate served as a source of ammonia, which condensed with **551** to generate an α-oxoamidine **553**. Condensation of **553** with an aldehyde gave **554**, which cyclized to the oxazoline **555**. Dehydration of **555** then produced **552** (Scheme 1.152). Efforts to isolate **553** or **555** were not successful.

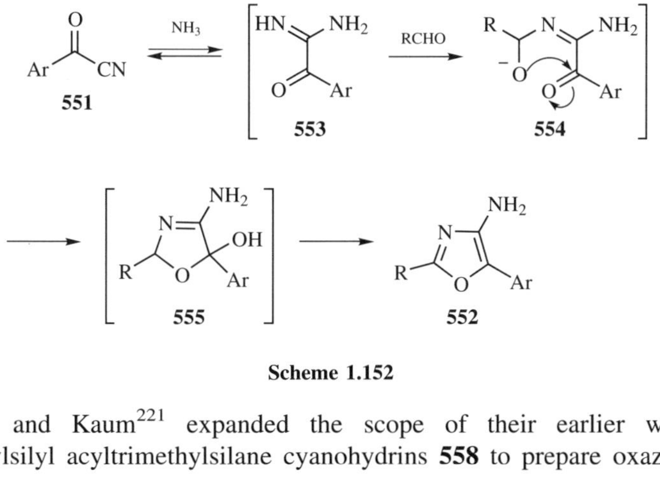

Scheme 1.152

Cunico and Kaum[221] expanded the scope of their earlier work using O-trimethylsilyl acyltrimethylsilane cyanohydrins **558** to prepare oxazoles.[222,223]

Previously, these authors reported a general synthesis of 2,4,5-trisubstituted oxazoles **561** from **558**[221,222] (Scheme 1.153). Addition of an alkyllithium reagent to **558** generated a lithium β-bis(trimethylsilyl)amino enolate **559** that was acylated in situ to produce a β-(acyloxy)-*N,N*-bis-(trimethylsilyl) enamine **560**. Cyclization of **560** to **561** was accomplished using FVP or TMSOTf.

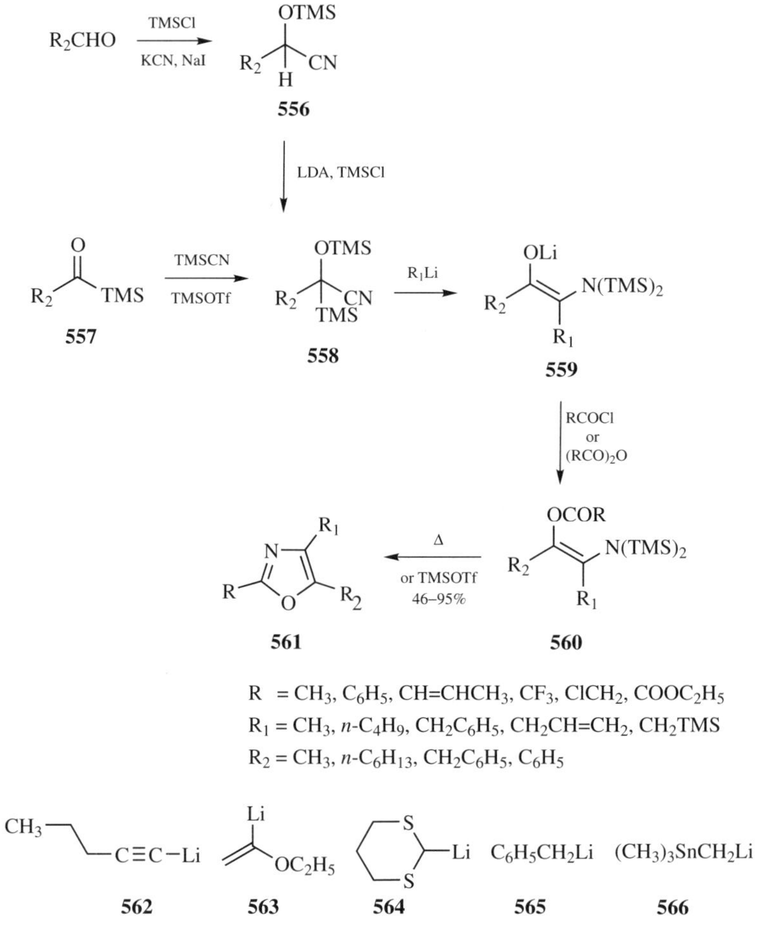

R = CH$_3$, C$_6$H$_5$, CH=CHCH$_3$, CF$_3$, ClCH$_2$, COOC$_2$H$_5$
R$_1$ = CH$_3$, *n*-C$_4$H$_9$, CH$_2$C$_6$H$_5$, CH$_2$CH=CH$_2$, CH$_2$TMS
R$_2$ = CH$_3$, *n*-C$_6$H$_{13}$, CH$_2$C$_6$H$_5$, C$_6$H$_5$

Scheme 1.153

Addition of a number of alkyllithium and alkenyllithium reagents and phenyllithium to **558** was successful. However, efforts to introduce functionality at C(4) in **561** via organolithiums **562–566** was unsuccessful, despite considerable experimentation. The authors reported that FVP of **560** usually afforded better yields of **561** and cleaner material. Cyclization of **560** with TMSOTf was favored for higher molecular weight substrates. A limitation of the methodology concerned preparation of **558**, which was not available by alkylation using LDA.

These authors[221] later reported they had discovered a solution to the alkylation dilemma for **556** (Scheme 1.54). The finding that O-trimethylsilylformyltrimethylsilane cyanohydrin **568** could now be alkylated using the less-hindered base lithium diethylamide (LDEA) and tetramethylethylene diamine (TMEDA) to produce **558** was the key to their new approach (Scheme 1.154). Addition of an alkyllithium reagent followed by in situ trapping of **559** with acetic anhydride gave the expected β-(acetoxy)-*N,N*-bis-(trimethylsilyl) enamine **560** (R= CH₃). Cyclization of **560** via FVP or TMSOTf produced 4,5-dialkyl-2-methyloxazoles **569** in good yield. A major advantage of this method is that one could independently vary the 4,5-oxazole substituents by choosing the appropriate organolithium or the alkylating agent. In addition, bulky substituents at C(5) of the oxazole should also be available. Representative examples are shown in Table 1.39.

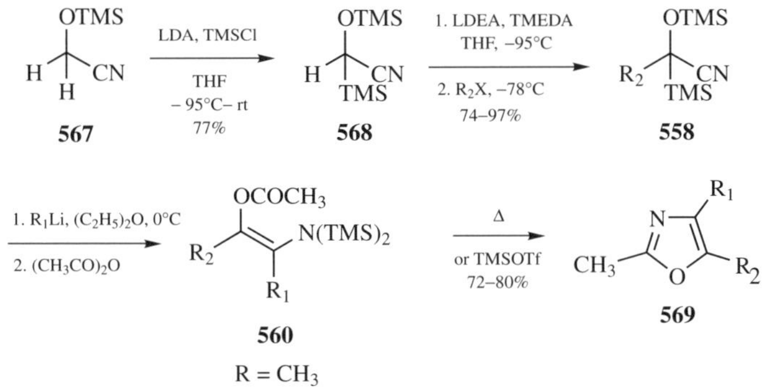

Scheme 1.154

Eissenstat and Weaver[224] developed a mild, neutral, retro-Diels-Alder approach to prepare some simple 2-substituted oxazoles (Scheme 1.155). The authors evaluated this approach in the context of preparing the potent antipicornaviral agent 5-[5-[2,6-dichloro-4-(2-oxazolyl)phenoxy]pentyl]-3-methylisoxazole **570**. Earlier attempts to prepare the oxazole moiety in **570** via classical condensation strategies or oxazoline oxidation failed, and they anticipated that it might be accessed via a retro-Diels-Alder reaction.

The requisite starting material, *endo*-3-aminobicyclo[2.2.1]hept-5-en-*endo*-2-ol **571** was readily prepared in gram quantities from Diels-Alder reaction of cyclopentadiene and 3-acetyl-2(3*H*)-oxazolone followed by base hydrolysis. Acylation of **571** gave the hydroxyamides **572**. The authors were gratified to find that thermal cyclization of **572** afforded the oxazoline **573**, since traditional cyclodehydration strategies using POCl₃ or SOCl₂ were ineffective. Furthermore, continued heating of the reaction mixture effected a retro-Diels-Alder reaction of **573** to afford 2-alkyl(aryl)oxazoles **82** in good to excellent yield. Representative examples are shown in Table 1.40. The yield for **82** (R= C₆H₅) is the only isolated yield. The authors noted the yields are lower for lower molecular weight analogues due to volatility and stability. In particular, these analogues were acid sensitive and

TABLE 1.39. 4,5-DISUBSTITUTED-2-METHYLOXAZOLES FROM
O-TRIMETHYLSILYLFORMYLTRIMETHYLSILANE CYANOHYDRIN[a]

Figure 1.41	568			569
Entry	R_2X	R_1	Product	Yield (%)
1	$n\text{-}C_4H_9I$	$n\text{-}C_4H_9$		72
2	$TMSCH_2I$	CH_3		53
3	$i\text{-}C_3H_7I$	CH_3		59
4	$C_6H_5CH_2Br$	CH_3		63
5	$CH_2=CHCH_2Br$	CH_3		71

Reagents: 1. LDEA, TMEDA, THF, −95°C; 2. R_2X, −78°C; then 1. R_1Li, $(C_2H_5)_2O$, 0°C; 2. $(CH_3CO)_2O$; Δ or TMSOTf.

[a] Data from ref. 221.

required base-washed glassware. This methodology was successfully used to prepare **570** from **571** in excellent yield (Scheme 1.156).[225]

Herdeis and Gebhard[226] prepared 2-amino-4-(5-benzyloxy-2-pyridyl)-5-hydroxyoxazole **576** as part of their on-going studies of heterocyclic amino acids (Scheme 1.157). They reported that the Bucherer-Bergs reaction of 5-benzyloxy-2-pyridinecarboxaldehyde with KCN, ammonium chloride, and ammonium carbonate in refluxing aqueous ethanol gave **576** as the sole product. The structure of **576** was confirmed by hydrolysis to 5-hydroxy-2-pyridylglycine **577**. Most likely, **576** exists as the 5(4H)-oxazolone tautomer (see Chapter 7).

Abraham and co-workers[227] investigated the photochemistry of acyl azides as a means of preparing acyl nitrenes (Scheme 1.158). They generated benzoyl nitrene from benzoyl azide in the presence of 2-butyne, 1,2-diphenylacetylene, and phenylacetylene and isolated 4,5-dimethyl-2-phenyloxazole **579**, 2,4,5-triphenyloxazole **231c**, and 2,5-diphenyloxazole **11**, respectively, in 18–30% yield. The

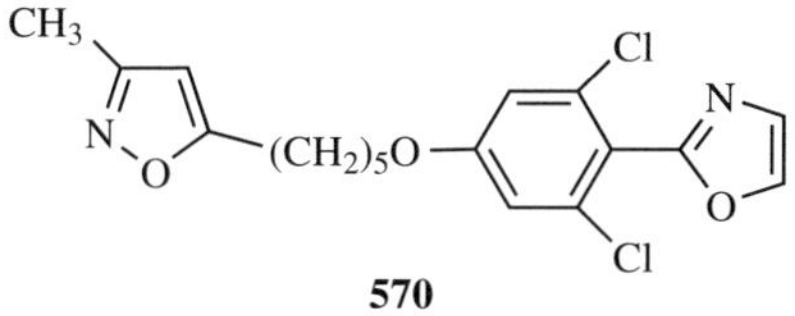

Synthesis of Oxazoles

570

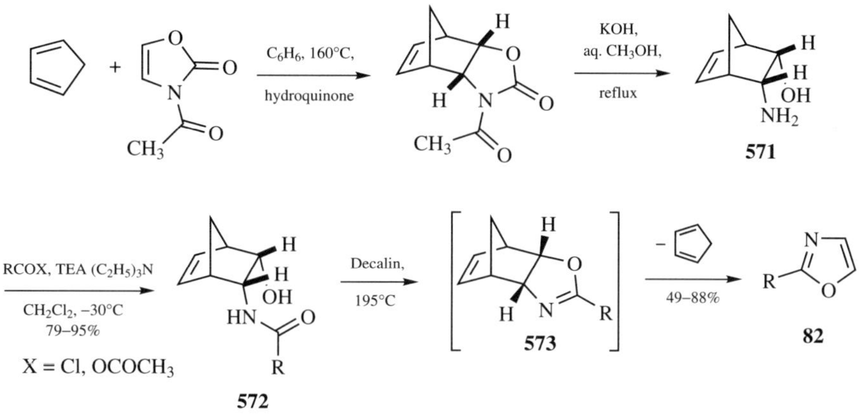

Scheme 1.155

TABLE 1.40. 2-ALKYL(ARYL)OXAZOLES VIA RETRO-DIELS-ALDER REACTION[a]

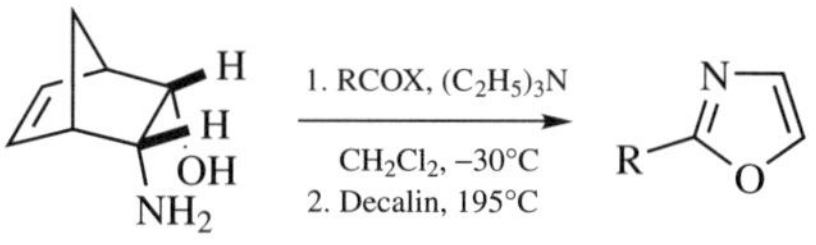

571 → **82** **Figure 1.42**

Entry	R	Product	Yield (%)
1	H	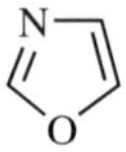 oxazole	49
2	CH$_3$	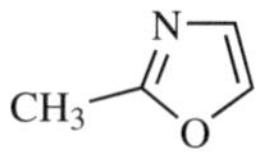 2-CH$_3$-oxazole	63
3	C$_2$H$_5$	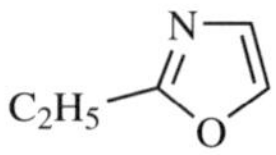 2-C$_2$H$_5$-oxazole	61
4	C$_6$H$_5$	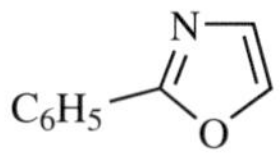 2-C$_6$H$_5$-oxazole	84

[a] Data from ref. 224.

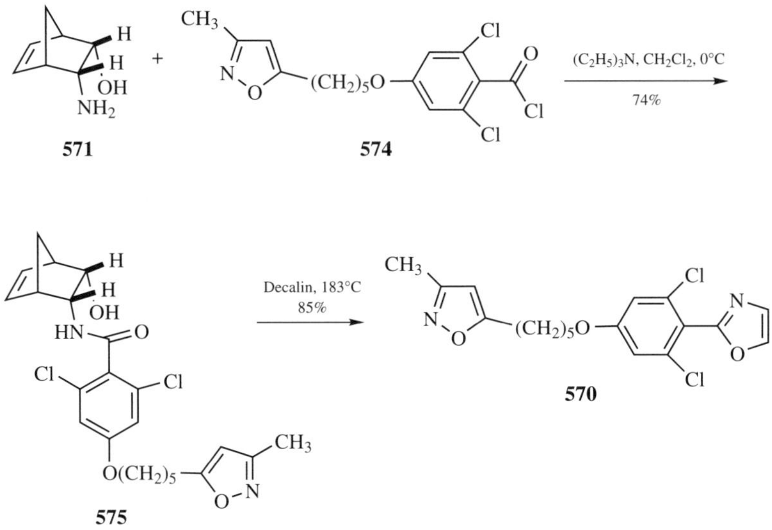

124 Synthesis and Reactions of Oxazoles

Scheme 1.156

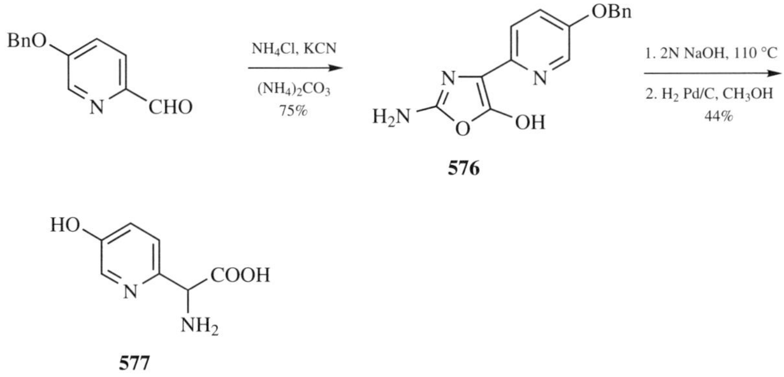

Scheme 1.157

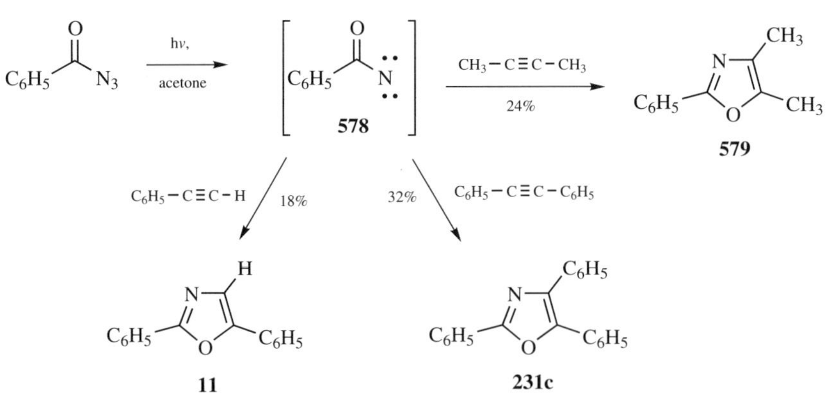

Scheme 1.158

oxazoles were isolated as components of mixtures that also contained 40–53% of phenyl isocyanate as well as 8–10% of 2-methyl-5-phenyl-1,3,4-oxadiazole.

Giori and co-workers[228] prepared pyrazolo[3,4-*d*]oxazoles **582** for evaluation as antitumor agents by photochemical cyclization of the internal zwitterionic diazonium salts **581** (Scheme 1.159).

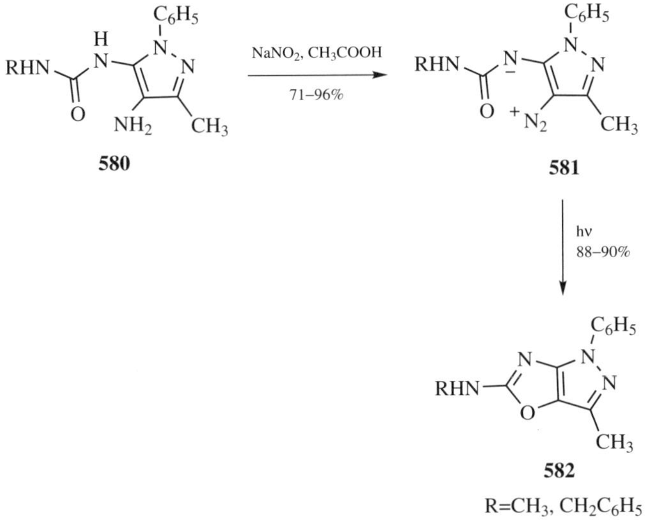

Scheme 1.159

Katritzky's group[229–234] has extensively explored the use of the benzotriazole (Bt) group in organic synthesis and synthetic methodology. This group[235] also adapted the methodology for a novel preparation of 5-(aroylamino)-2-aryloxazoles **587** (Scheme 1.160). Reaction of 1,2-bis(1*H*-benzotriazol-1-yl)-1,2-ethanediol **583** with two equivalents of an aryl amide affords a 1,2-(diaroylamino)-1,2-di(benzotriazol-1-yl)ethane, **584** in excellent yield. Heating **584** with sodium hydride in

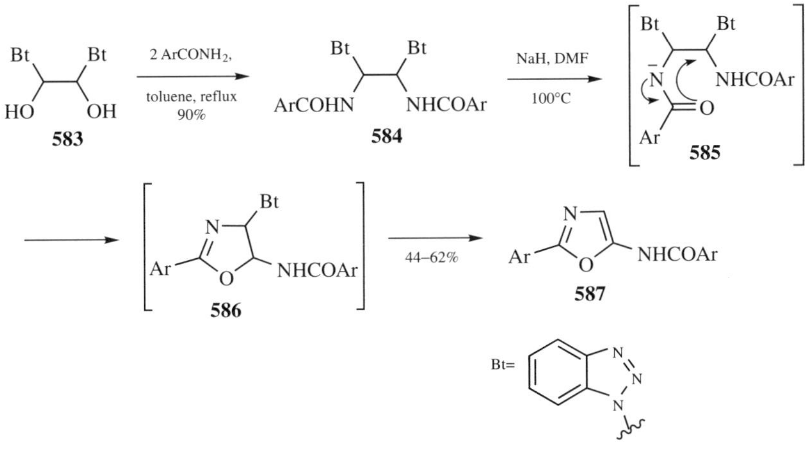

Scheme 1.160

TABLE 1.41. 5-(AROYLAMINO)-2-ARYLOXAZOLES FROM
1,2-BIS(1H-BENZOTRIAZOL-1-YL)-1,2-ETHANEDIOL[a]

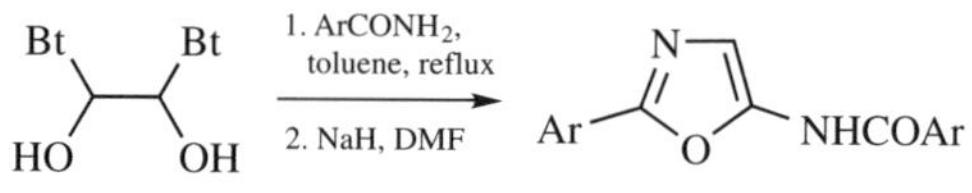

Figure 1.43   **583**   →   **587**

Reagents: 1. ArCONH$_2$, toluene, reflux; 2. NaH, DMF

Entry	Ar	Yield (%)
1	C$_6$H$_5$	62
2	4-CH$_3$–C$_6$H$_4$	61
3	4-CH$_3$O–C$_6$H$_4$	44
4	2-CH$_3$–C$_6$H$_4$	51

[a] Data from ref. 235.
[b] Bt = Benzotriazole

dimethylformamide effects sequential cyclization to **586** with subsequent elimination of Bt to produce a 5-(aroylamino)-2-aryloxazoles **587** in moderate to good yield. Representative examples are shown in Table 1.41.

Shaprio[236] described a nonoxidative method for preparing 2-substituted 4-oxazolecarboxylic acid esters **591** (Scheme 1.161). He prepared the key intermediate, dimethyl amino[(phenylthio)methyl]malonate **588**, in three straightforward steps from diethyl aminomalonate hydrochloride. Acylation of **588** gave the N-acyl derivative **589** in excellent yield, which was sequentially chlorinated and cyclized in one pot to afford the 2,4,4,5-tetrasubstituted oxazoline **590**. The author noted that anhydrous conditions were required to minimize sulfoxide formation. This was the only product isolated if the chlorination cyclization sequence was carried out in a hydroxylic solvent.

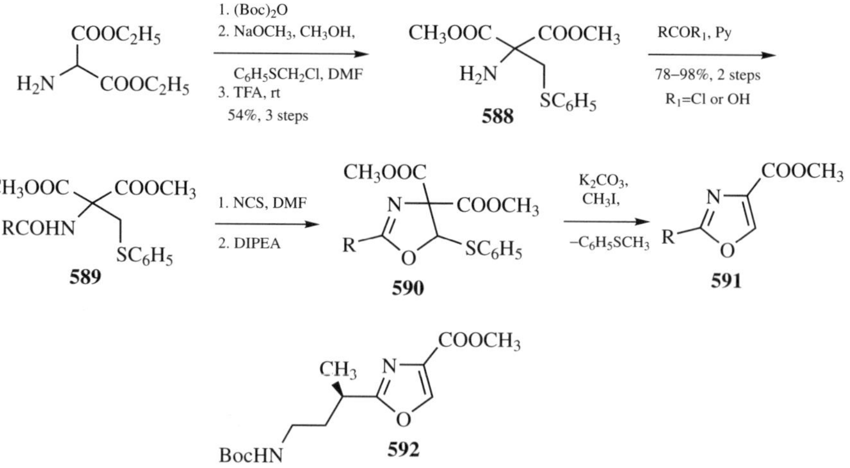

Scheme 1.161

TABLE 1.42. 2-SUBSTITUTED 4-OXAZOLECARBOXYLIC ACID METHYL ESTERS FROM DIMETHYL AMINO[PHENYLTHIO)METHYL]MALONATE[a]

$$CH_3OOC-C(H_2N)(SC_6H_5)-COOCH_3 \xrightarrow[R_1 = Cl\ or\ OH]{RCOR_1,\ Py} \text{oxazole with } R,\ COOCH_3$$

588   Figure 1.44   591

Entry	R	Yield (%)
1	CH_3	43
2	i-C_3H_7	46
3	C_6H_5	75
4	trans-C_6H_5–CH=CH	59
5	$C_6H_5CH_2OCH_2$	60
6	$BocNH(CH_2)_3$	40

[a] Data from ref. 236.

Decarbomethoxylation of **590** with K_2CO_3 in the presence of methyl iodide to trap thiophenoxide was the most efficient way of preparing **591**. Overall, this methodology appears to be general in scope although it was not an effective means of obtaining a tris-oxazole intermediate for ulapualide. In contrast, **592**, an oxazole fragment in calyculin, was prepared in 47% yield. Representative examples are shown in Table 1.42.

1.4. REACTIONS OF OXAZOLES

The oxazole ring exhibits rich and varied reactivity, which allows for functionalization at each ring atom other than oxygen. Oxazoles are weakly aromatic and, as such, display reactions characteristic of both aromatic substitutions and reactions of double bonds. Electrophilic aromatic substitutions—including bromination, nitration, and Friedel-Crafts reactions—preserve the aromatic character of the ring. On the other hand, additions across the C(4)-C(5) double bond that disrupt the aromatic nature of the ring are also known.

Nucleophilic aromatic substitution reactions of halooxazoles, alkoxyoxazoles, or oxazole sulfones afford both substituted oxazoles and ring-opening products, depending on the nucleophile. Hydrolytic ring cleavage of oxazoles has been used to prepare aminoketones, amino acids, and dipeptides, whereas reduction can produce amides or amino alcohols. Reaction of oxazoles with singlet oxygen affords triacylamines. This reaction continues to be exploited for natural product synthesis and has been extended to peptide synthesis. Oxazoles readily participate in Diels-Alder reactions either as a diene or as a dienophile.

The presence of 1,3-heteroatoms renders the C(2) position of oxazoles electron deficient. As such, this facilitates nucleophilic additions at C(2) and subsequent transformation of oxazoles into a variety of other heterocycles. Formal [3 + 2] cycloadditions of oxazoles with dipolarophiles also effects similar transformations.

The 1,3-disposition of the heteroatoms also imparts differential acidies to the ring positions, enabling sequential deprotonation. This rich acid–base chemistry permits selective functionalization of each ring-carbon atom. The synthesis of other organometallic oxazoles and transition-metal-catalyzed cross-coupling reactions are also described. Applications of new cross-coupling methods for preparing polyoxazoles in the context of the syntheses of natural products are described in Section 1.5.

The utility of 2-(trimethylsilyl)oxazoles, Wittig reactions, and the reactions of oxazolium salts are also included in this section.

1.4.1. Electrophilic Reactions

In general, the oxazole ring is electron deficient and as such does not readily undergo electrophilic substitution. However, experimental results and theoretical calculations indicate the order of reactivity to be $C(4) > C(5) > C(2)$,[2,17] and examples of deuteration, halogenation, nitration, and Friedel-Crafts acylation have been reported. Electrophilic substitution at the 2 position is usually achieved using a 2-(trimethylsilyl)oxazole, which is discussed in more detail in Section 1.4.9.

The reader is directed to several excellent reviews for further details. Hassner and Fischer's[2] general review of oxazoles covers both electrophilic aromatic substitution (EAS) reactions and addition reactions. Belen'kii and Chuvylkin[10] surveyed EAS reactions of oxazoles as part of a larger review for azoles. Larina and co-workers[237,238] published two extensive reviews of nitration of azoles, including oxazoles. The articles cover kinetics and the mechanism of nitrations as well as the synthesis of nitroazoles via heterocyclization and ring transformations and direct methods of nitration. In light of these reviews, only a few selected examples of EAS reactions of oxazoles are described in this section.

Venkatasubramanian and Krishnamachari[239] required deuterated oxazoles for a study of the flash photolysis of oxazole **1**. They obtained (2-D)-oxazole **593** exclusively from **1** using CF_3COOD and D_2O (Scheme 1.162). The authors did not speculate about this surprising result. The reaction is selective for monodeuteration, but **1** also suffers decomposition (up to 40%) under the reaction conditions leading to a mixture of **593** and deuterated pyrazines **594** n = 0–4. Trifluoroacetic acid-D was much more efficient than deuterosulfuric acid for this reaction.

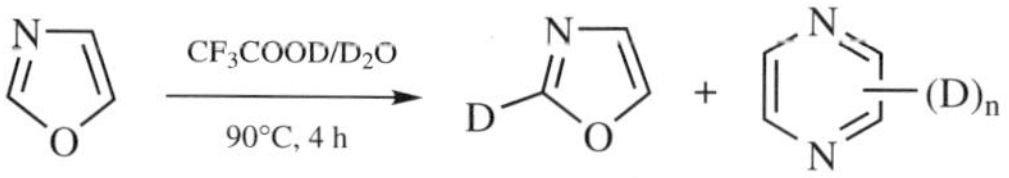

Scheme 1.162

Bromination of substituted oxazoles can yield normal aromatic substitution products or 4,5- and 2,5-addition products, depending on the reaction conditions. For example, Hassner and Fischer[240] brominated 2,5-diphenyloxazole **111** with bromine in acetic acid and sodium acetate to prepare 4-bromo-2,5-diphenyloxazole **595** (Scheme 1.163). Similarly, Belen'kii and co-workers[241,242] isolated a mixture of 5-bromo-2-phenyloxazole **596** and 4,5-dibromo-2-phenyloxazole **597** from treatment of 2-phenyloxazole **5** with bromine in refluxing benzene. Lawson and VanSant[243] isolated 2-amino-5-bromo-4-(trifluoromethyl)oxazole **599a** and 5-bromo-2-(methylamino)-4-(trifluoromethyl)oxazole **599b** from bromination of 2-amino-4-(trifluoromethyl)oxazole **598a** and 2-(methylamino)-4-(trifluoromethyl) oxazole **598b**, respectively, with bromine in acetic acid and sodium acetate.

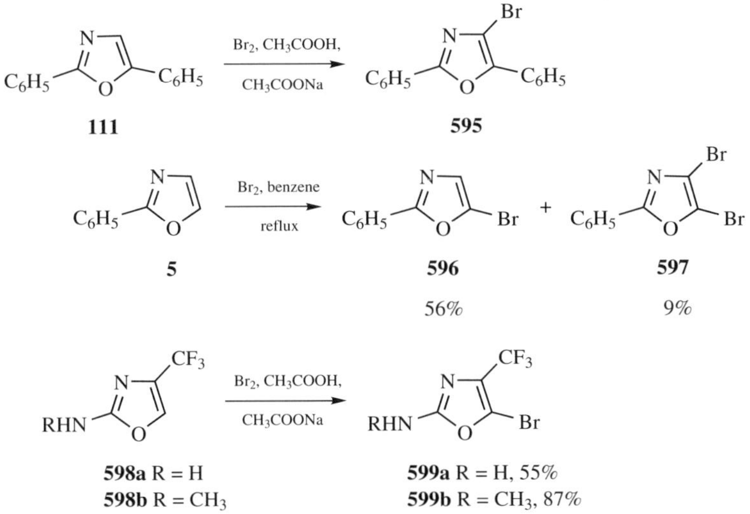

Scheme 1.163

In contrast, bromination of oxazoles with bromine in methanol in the presence of K_2CO_3 gave rise to nonaromatic addition products as well as ring-opened amides. Hassner and Fischer[240] conducted a thorough and systematic study of these reaction conditions and found that 4,5- and 2,5-addition products could be isolated. The product distribution depended highly on the nature and pattern of oxazole substitution and the reaction temperature, as shown by the following examples.

Treatment of a 2-aryloxazole **15** ($R_1 = C_6H_5$ or 4-NO_2-C_6H_5, $R_2 = R_3 = H$) with methanolic bromine in the presence of K_2CO_3 at 0°C produced a ring-opened trimethoxy amide **600** ($R_1 = C_6H_5$ or 4-NO_2-C_6H_5, $R_2 = R_3 = H$) in nearly quantitative yield (Scheme 1.164).

However, bromination of 5-methyl-2-phenyloxazole **15** ($R_1 = C_6H_5$, $R_2 = H$, $R_3 = CH_3$) at −78°C for 2 h afforded the 4,5-addition product, 4,5-dimethoxy-5-methyl-2-phenyl-2-oxazoline **603**, which was isolated in 43% yield.

In contrast, bromination of 2,4,5-trimethyloxazole **15** ($R_1 = R_2 = R_3 = CH_3$) at −78°C for 5 h gave the 2,5-addition product, 2,5-dimethoxy-2,4,5-trimethyl-3-oxazoline **602** ($R_1 = R_2 = R_3 = CH_3$) isolated as a mixture of *cis/trans* isomers.

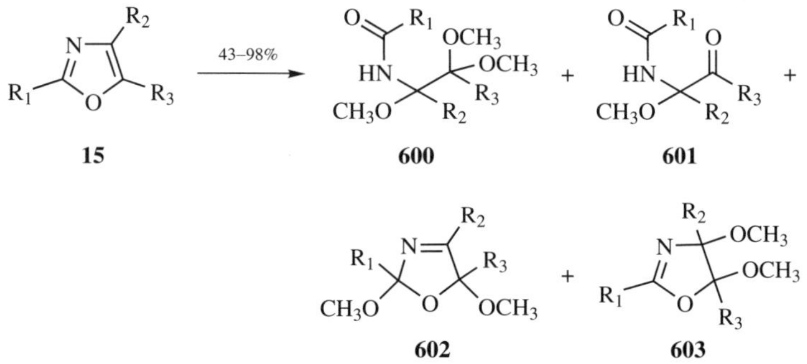

Scheme 1.164

Similarly, bromination of 4,5-diphenyl-2-methyloxazole **15** ($R_1 = CH_3$, $R_2 = R_3 = C_6H_5$) at $-78°$ to $-5°C$ afforded a mixture of *cis/trans* 2,5-dimethoxy-4,5-diphenyl-2-methyl-3-oxazoline **602** ($R_1 = CH_3$, $R_2 = R_3 = C_6H_5$) in 76% yield together with 13% of benzil, respectively.

In some cases, ring opening to produce benzil was a significant side reaction. For example, prolonged treatment of 4,5-diphenyloxazole **15** ($R_1 = H$, $R_2 = R_3 = C_6H_5$) with methanolic bromine at $-5°$ to $20°C$ generated a mixture of *cis/trans* 2,5-dimethoxy-4,5-diphenyl-3-oxazoline **602** ($R_1 = H$, $R_2 = R_3 = C_6H_5$) in 40% yield together with 38% of benzil, respectively.

Mechanistically, the authors studied these reactions using ^1H NMR, since the intermediates were unstable and difficult to isolate. They concluded that ring-opened products (e.g., **600** and **601**) predominated for 2-aryloxazoles. 3-Oxazolines **602** were the predominant product versus 2-oxazolines **603** for a 4-phenyl- or 2,4-dimethyl-substituted oxazole. Selected examples of oxazolines and ring-opened amides are shown in Table 1.43.

Nitration of oxazoles can afford a ring-substituted nitrooxazole or the reaction conditions can also effect nitration of a side chain aryl substituent (Scheme 1.165). For instance, Gompper and Christmann[244] nitrated 2-(dimethylamino)-4-phenyloxazole **604** using HNO_3 and H_2SO_4 and isolated 2-(dimethylamino)-5-nitro-4-(4-nitrophenyl)oxazole **605** (Scheme 1.165). Belen'kii and co-workers[241,242] reported that low temperature nitration of 2-phenyloxazole **5** under nonprotonating conditions was a better method of preparing 5-nitro-2-phenyloxazole **606** than traditional methods that use strong acids. Here, nitration was accomplished with *N*-nitropicolinium tetrafluoroborate prepared in situ from nitronium tetrafluoroborate and α-picoline. These conditions afforded a modest yield of a 90:5:5 mixture composed of 5-nitro-2-phenyloxazole **606**, 5-nitro-2-(3-nitrophenyl)oxazole **607**, and 5-nitro-2-(4-nitrophenyl)oxazole **608**.

Šindler-Kulyk and co-workers[245] investigated direct formylation of 4-methyloxazole **3** and 4,5-dimethyloxazole **611** as a means to prepare analogues that could be further elaborated in their synthesis of vitamin B_6 (Scheme 1.166). They isolated a 1:1 mixture of 4-methyl-5-oxazolecarboxaldehyde **609** and 4-methyl-2-oxazolecarboxaldehyde **610** in low yield from **3**. Similarly, **611** afforded 4,5-dimethyl-

TABLE 1.43. 2-OXAZOLINES AND 3-OXAZOLINES FROM SUBSTITUTED OXAZOLES AND BR_2/CH_3OH^a

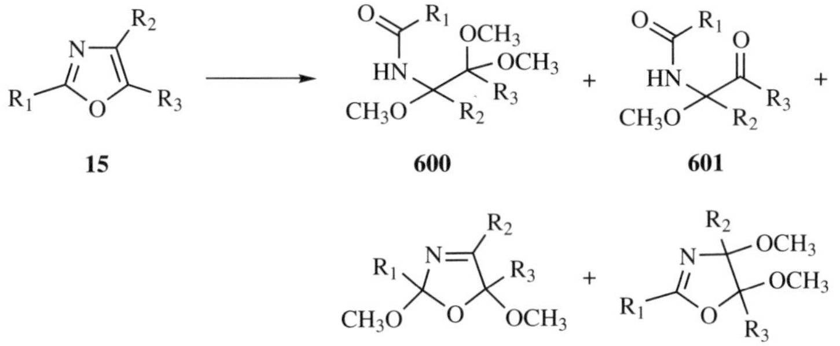

Oxazole **15** → **600** + **601** + **602** + **603**

Figure 1.45

Entry	Oxazole	Conditions	Product	Yield (%)
1	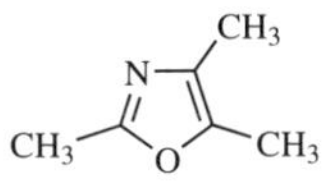 2,4,5-trimethyloxazole	−78°C, 5 h	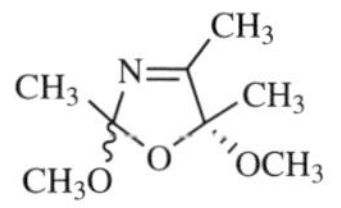 2-oxazoline (CH₃O, CH₃, CH₃, OCH₃)	60
2	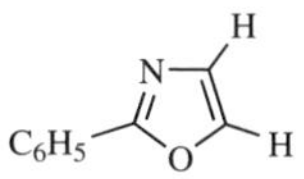 2-phenyloxazole (R₃=H)	0°C, 4 h	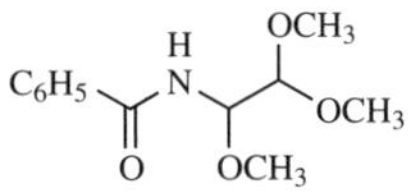 open-chain amide with OCH₃ groups	98
3	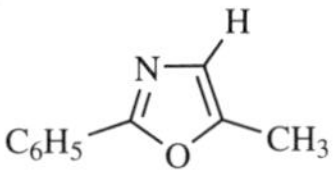 2-phenyl-5-methyloxazole	−78°C, 2 h	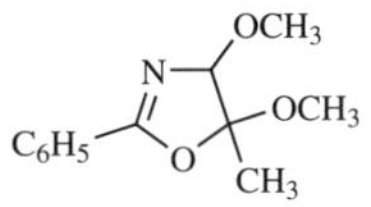 2-oxazoline (OCH₃, OCH₃, CH₃)	43
4	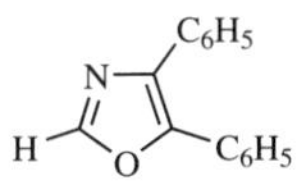 4,5-diphenyloxazole	−5° to 20°C, 12 h	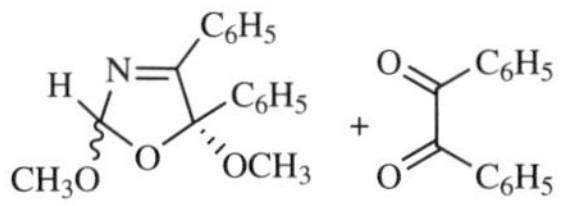 2-oxazoline + diketone	78[b]
5	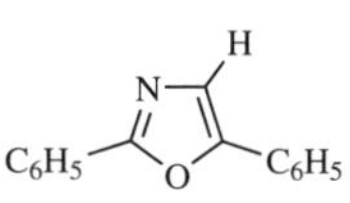 2,5-diphenyloxazole (R₂=H)	−15° to 5°C, 5 h	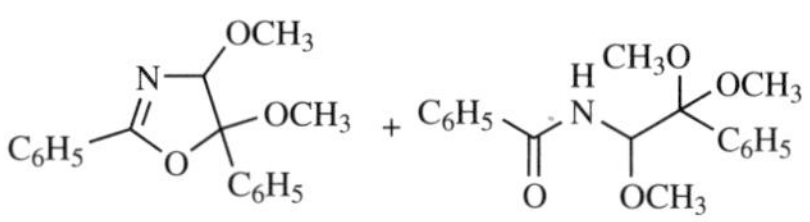 2-oxazoline + open-chain amide	78[c]

[a] Data from ref. 240.
[b] Product distribution 40% and 38%.
[c] Product distribution 23% and 55%.

2-oxazolecarboxaldehyde **612** in poor yield. The authors did prepare **610** and **612** in 48–50% yield via metalation chemistry. Formylation of **5** was reported to afford 2-phenyl-5-oxazolecarboxaldehyde (58%).[241,242]

Patsenker and co-workers[246] reported that 2-(2-furanyl)-5-phenyloxazole **613** is formylated or acetylated to produce **614a** and **614b** in good to excellent yield

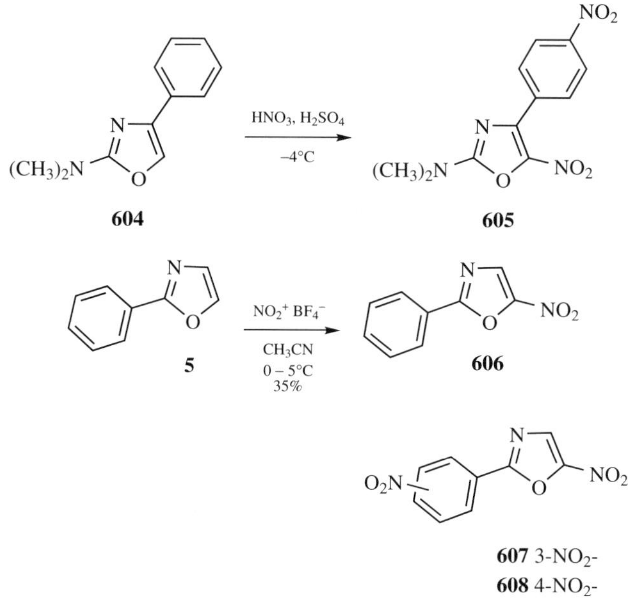

Scheme 1.165

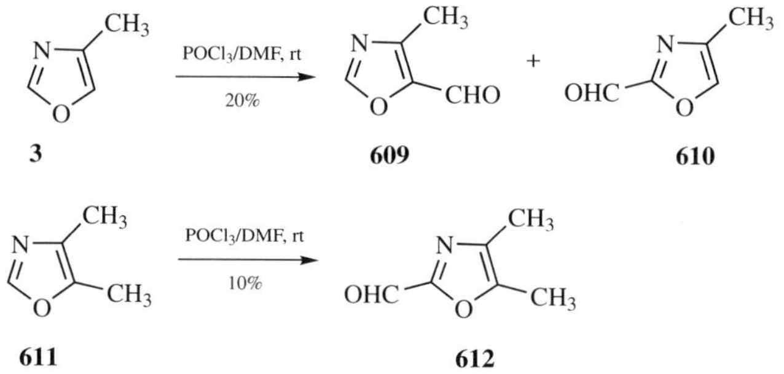

Scheme 1.166

(Scheme 1.167). There was no evidence for acylation of the oxazole ring. In contrast, formylation of 5-(4-dimethylaminophenyl)-2-(4-nitrophenyl)oxazole **615** effects Vilsmeier-Haack formylation of the oxazole ring as well as the electron-rich N,N-dimethylaniline. In this case, heterocyclization then leads to the observed quinazolinium salt **616**.[247]

Barbieri and co-workers[248] conducted a detailed study of the conformational behavior of 2-oxazolecarboxaldehyde using ab initio MO calculations. The authors found two ground states, with the O,O-cis conformer being the most stable.

Kim and Kong[249] isolated 2-aroyl-5-aryloxazoles **617** together with 2-aroyl-5-arylimidazoles **618** from reaction of arylglyoxals with tetrasulfur tetranitride (S_4N_4) (Scheme 1.168).

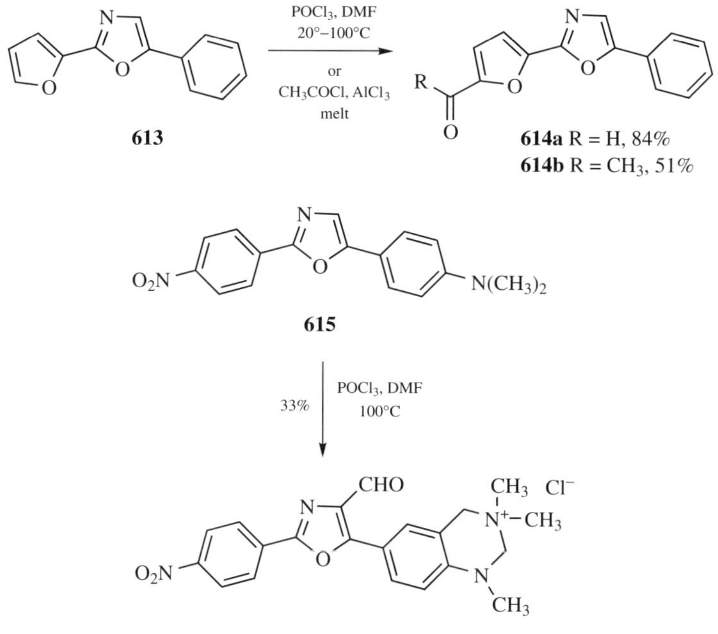

Scheme 1.167

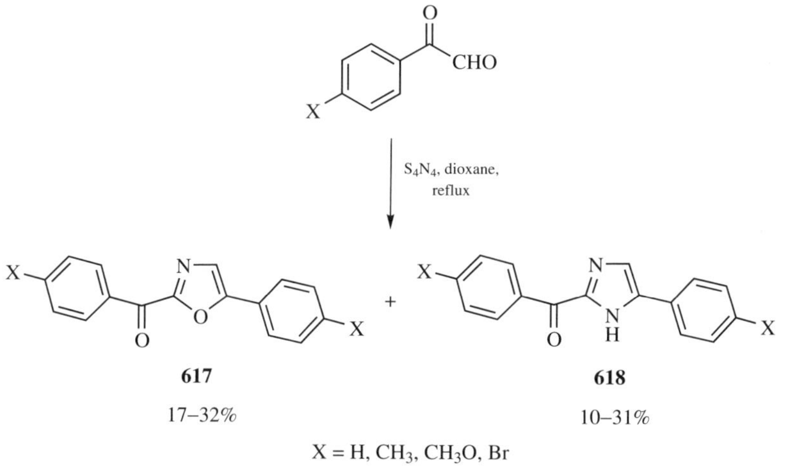

Scheme 1.168

BASF chemists[250] oxidized a variety of chloromethyl heterocycles to the corresponding aldehydes under mild conditions, using a trialkylamine N-oxide. For example, treatment of 4-(chloromethyl)-2-(4-fluorophenyl)oxazole **619** with N-methylmorpholine N-oxide monohydrate gave 2-(4-fluorophenyl)-4-oxazolecarboxaldehyde **620** in 48% yield (Scheme 1.169). This reaction was reported to be quite general for five- and six-membered ring heterocycles.

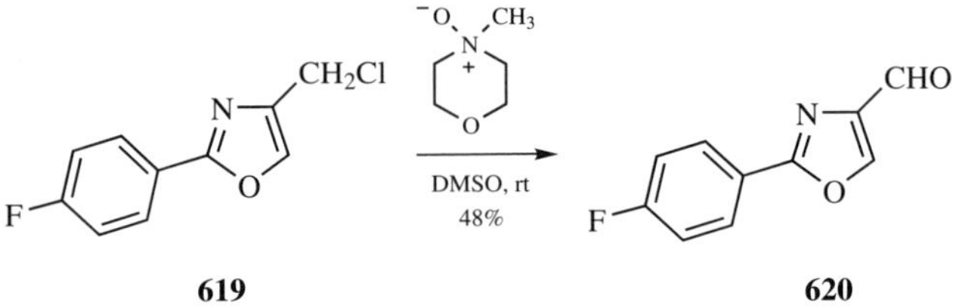

Scheme 1.169

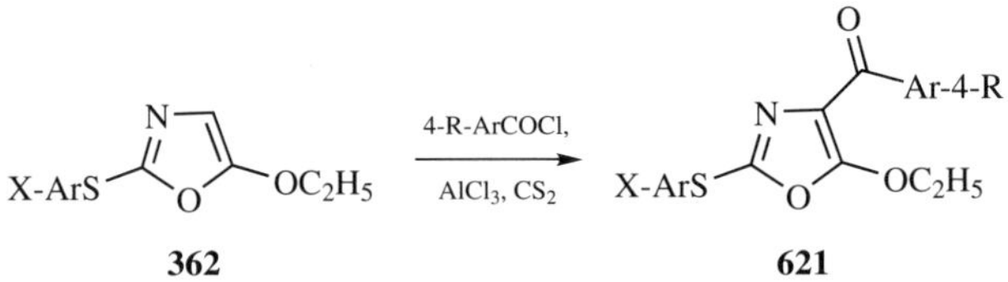

X = 2-NO_2, 2-NO_2-4-Cl

Scheme 1.170

Torroba and co-workers[251] described the first direct synthesis of a 4-acyloxazole (Scheme 1.170). The starting 2-(arylthio)-5-ethoxyoxazoles **362** were prepared from ethyl isocyanoacetate (Scheme 1.98). Treatment of **362** with an aroyl chloride and $AlCl_3$ produced 4-aroyl-2-(arylthio)-5-ethoxyoxazoles **621** in low to modest yields. Representative examples are shown in Table 1.44.

TABLE 1.44. 4-AROYL-2-(ARYLTHIO)-5-ETHOXYOXAZOLES FROM AROYLATION OF 2-(ARYLTHIO)-5-ETHOXYOXAZOLES[a]

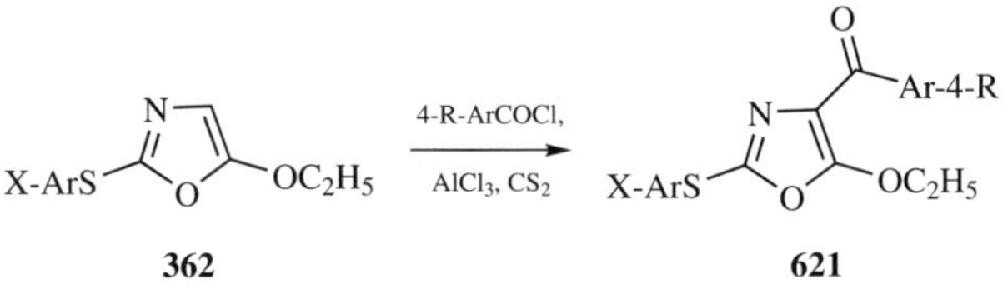

Figure 1.46

X = 2-NO_2, 2-NO_2-4-Cl

Entry	X	R	Yield (%)
1	2-NO_2	H	49
2	2-NO_2	4-CH_3	57
3	2-NO_2	4-CH_3O	60
4	2-NO_2	4-Cl	25
5	2-NO_2-4-Cl	H	42
6	2-NO_2-4-Cl	4-CH_3	47
7	2-NO_2-4-Cl	4-CH_3O	48

[a] Data from ref. 251.

Australian workers have extensively investigated EAS reactions of 4-alkyl(aryl)-2-aminoxazoles and described some interesting findings.[252–256] In an early study, Crank and co-workers[257] isolated the expected substituted thioureas from reaction of isothiocyanates and heterocyclic amines, including 2-aminooxazole. However, reaction of 4-alkyl(aryl)-2-aminooxazoles **622** (X = O) with alkyl- and arylisothiocyantes afforded both the expected thioureas **623** (X = O) and 4-alkyl(aryl)-2-amino-5-oxazolethiocarboxamides **624** (X = O) (Scheme 1.171).[252] The authors proposed that the 4-substitutent further activated C(5) for electrophilic reaction with the isothiocyanate.

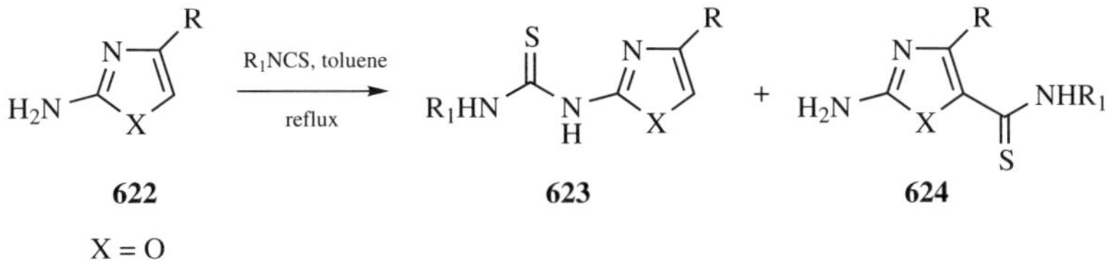

Scheme 1.171

However, they also noted that this reactivity was singularly unique. For example, reaction of 2-amino-4-methylthiazole **622** (X = S, R = CH$_3$) with CH$_3$NCS or C$_6$H$_5$NCS gave **623** (X = S, R = R$_1$ = CH$_3$) and **623** (X = S, R = CH$_3$, R$_1$ = C$_6$H$_5$) as expected, whereas 2-amino-4-methylimidazole **622** (X = NH, R = CH$_3$) gave only tar when treated with C$_6$H$_5$NCS. In addition, **622** (X = O) gave the expected ureas when treated with alkylisocyanates. Selected examples of **624** prepared via this methodology are shown in Table 1.45.

Khan and Crank[255] found this same pattern of reactivity with condensations of 2-amino-4-methyloxazole **622** (X = O, R = CH$_3$) with aromatic aldehydes. The authors did not see any evidence for the expected Schiff base. Instead, they isolated products of hydroxymethylation at C(5) **625** in good yield (Scheme 1.172). The structure of **625** (X = O, R = CH$_3$, Ar = 4-CH$_3$OC$_6$H$_4$) was confirmed by X-ray

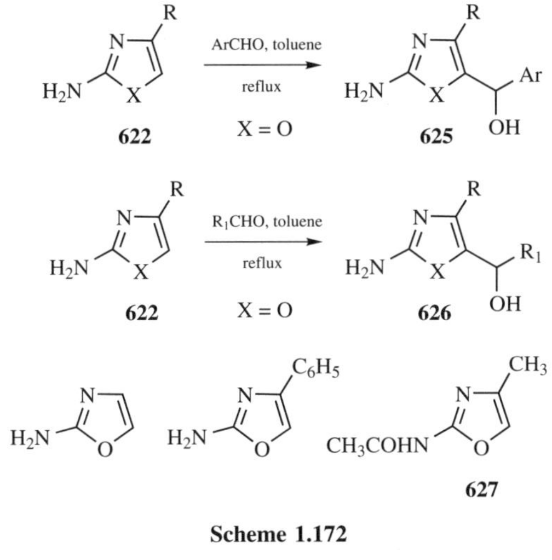

Scheme 1.172

TABLE 1.45. 4-ALKYL(ARYL)-2-AMINO-5-OXAZOLETHIOCARBOXAMIDES FROM 4-ALKYL(ARYL)-2-AMINOOXAZOLES AND ISOTHIOCYANATES[a]

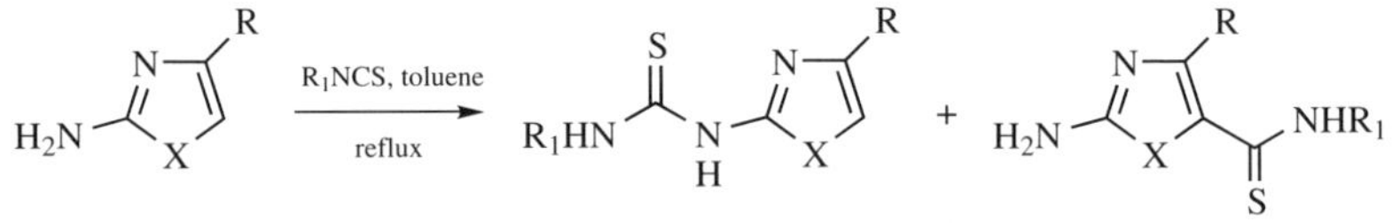

Figure 1.47

Entry	X	R	R_1	Yield 623 (%)	Yield 624 (%)
1	O	H	CH_3	60	—
2	O	CH_3	CH_3	15	63
3	O	CH_3	$n\text{-}C_4H_9$	30	40
4	O	CH_3	C_6H_5	—	85
5	O	CH_3	$CH_2=CHCH_2$	—	55
6	O	$n\text{-}C_3H_7$	CH_3	27	48
7	O	$t\text{-}C_4H_9$	CH_3	31	26
8	O	C_6H_5	CH_3	—	58

[a] Data from ref. 252.

crystallography. Later work from these authors[256] showed this hydroxymethylation reaction to be quite general for aromatic aldehydes and **622** (X = O, R = CH_3). The authors claimed that other 4-alkyl-2-aminooxazoles reacted similarly, but they gave no examples. Alkyl aldehydes (R_1 = CCl_3 and C_2H_5) reacted similarly to afford **626** (X = O, R = CH_3, R_1 = CCl_3 and C_2H_5). However, 2-aminooxazole **622** (X = O, R = H), 2-amino-4-phenyloxazole **622** (X = O, R = C_6H_5), and 2-acetamido-4-methyloxazole **627** were all unreactive. Representative examples are shown in Table 1.46.

Crank and Mekonnen[254] prepared a series of azo dyes **628** from **622** and aryl diazonium tetrafluoroborate salts (Scheme 1.173). In this case, both 2-amino-4-phenyloxazole **622** (X = O, R = C_6H_5) and 2-acetamido-4-methyloxazole **627** were also reactive and afforded the expected azo dyes uneventfully. The authors noted that isolated diazonium tetrafluoroborate salts gave much cleaner reactions than a diazonium salt generated in situ. 2-Amino-4-methyloxazole could not be

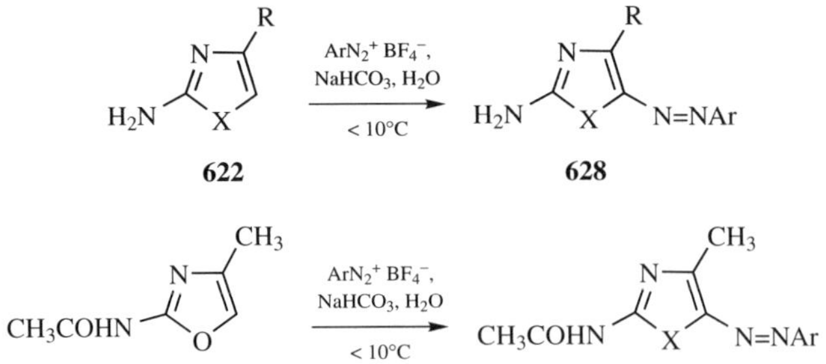

Scheme 1.173

TABLE 1.46. 2-AMINO-5-(HYDROXYMETHYL)-4-METHYLOXAZOLES FROM 2-AMINO-4-METHYLOXAZOLES AND ALDEHYDES[a]

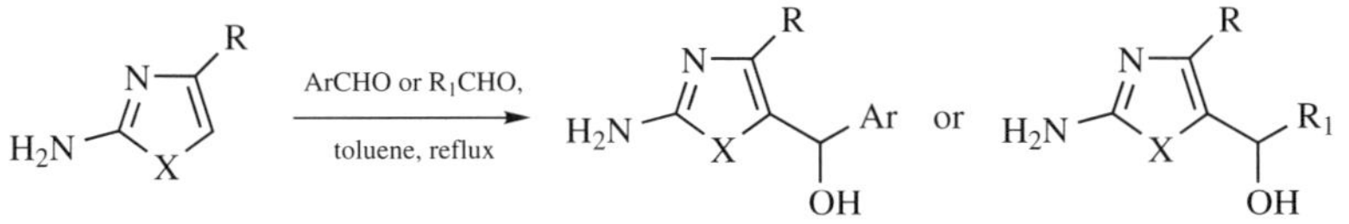

Figure 1.48

Entry	R	Ar or R_1	Yield 625 (%)	Yield 626 (%)
1	CH_3	4-CH_3O–C_6H_4	77	—
2	CH_3	4-NO_2–C_6H_4	87	—
3	CH_3	C_6H_5	69	—
4	CH_3	2-Cl–C_6H_4	20	—
5	CH_3	2-HO–C_6H_4	16	—
6	CH_3	CCl_3	—	45
7	CH_3	C_2H_5	—	62

[a] Data from ref. 256.

diazotized under a variety of reaction conditions. There was no product formed from 2-amino-4,5-dimethyloxazole and benzene diazonium tetrafluoroborate. Selected examples are shown in Table 1.47.

Mekonnen and Crank[253] also found the same reaction dichotomy in Friedel-Crafts reactions of **622** (X = O) with acid chlorides or acid anhydrides. In general, acylation of **622** (X = O, R = CH_3) with acetic anhydride or benzoyl chloride/pyridine afforded the corresponding 2-(acylamino)-4-methyloxazole **630a** and **630b** (Scheme 1.174). However, low-temperature acetylation or aroylation of **622** (X = O, R = CH_3) in the presence of two equivalents of $AlCl_3$ produced a 5-acyl-2-amino-4-methyloxazole **631** in low to modest yield. The authors reported that $AlCl_3$ was the most effective Lewis acid ($SnCl_4$ and $TiCl_4$) and that the reaction required two equivalents of Lewis acid. Selected examples are shown in Table 1.48.

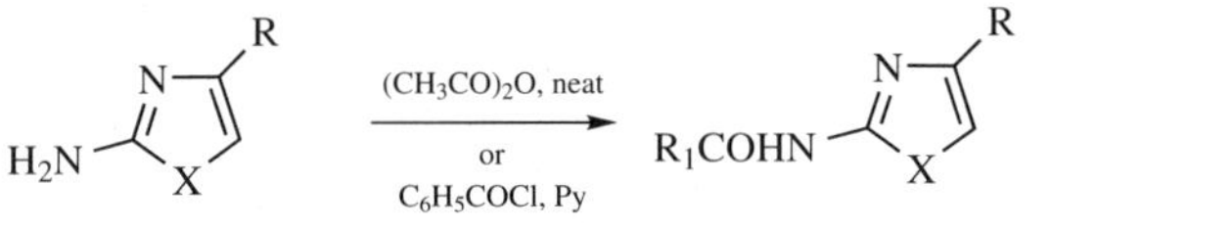

630a X = O, R = R_1 = CH_3
630b X = O, R = CH_3, R_1 = C_6H_5

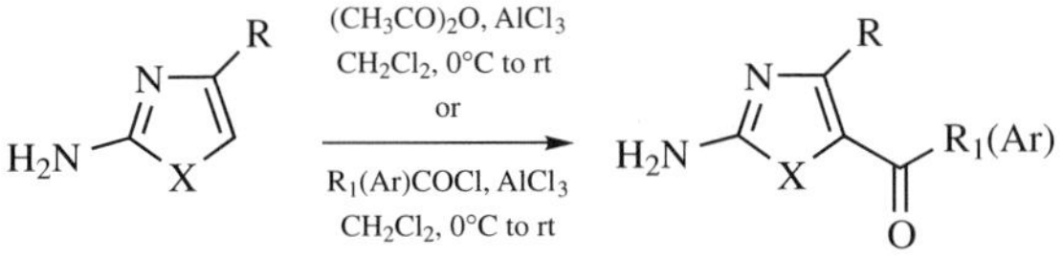

631 X = O

Scheme 1.174

TABLE 1.47. 4-ALKYL(ARYL)-2-AMINO(ACETAMIDO)-5-(ARYLAZO)OXAZOLES FROM 4-ALKYL(ARYL)-2-AMINO(ACETAMIDO)OXAZOLES AND ARYL DIAZONIUM TETRAFLUOROBORATE SALTS[a]

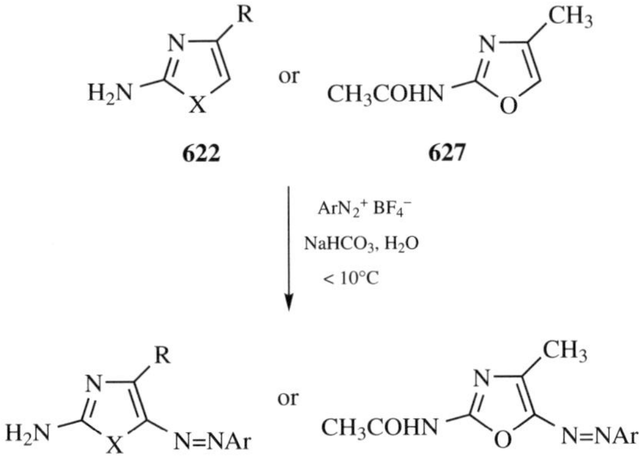

Figure 1.49

Entry	R	Ar	Yield **628** (%)	Yield **629** (%)	Color
1	CH_3	$4\text{-}CH_3O\text{-}C_6H_4$	57	—	light brown
2	CH_3	$4\text{-}NO_2\text{-}C_6H_4$	96	—	dark red
3	CH_3	C_6H_5	91	—	orange
4	CH_3	$2\text{-}Cl\text{-}C_6H_4$	72	—	yellow
5	C_6H_5	C_6H_5	73	—	yellow
6	CH_3	C_6H_5	—	98	yellow
7	CH_3	$4\text{-}CH_3O\text{-}C_6H_4$	—	44	yellow
8	CH_3	$3\text{-}Cl\text{-}C_6H_4$	—	90	red

[a] Data from ref. 254.

Kujundžić and Glunčić[258] isolated alkylated oxazoles from acylation of 5-(acylamino)-4-methyloxazoles **633**. Thus treatment of **633** with acetic- or propionic anhydride gave mixtures composed of 5-(diacylamino)-4-methyloxazoles **634**, 5-(acylamino)-2-alkyl-4-methyloxazoles **635** and 2-alkyl-5-(diacylamino)-4-methyloxazoles **636** (Scheme 1.175). The authors studied the effects of acid, base, stoichiometry, and temperature on the product distribution. They concluded that **634** was the major product when using the acid anhydride with pyridine or triethylamine. On the other hand, **636** was favored when using the neat acid anhydride or acetic acid as solvent. Succinic anhydride and phthalic anhydride gave products only analogous to **634**.

1.4.2. Nucleophilic Reactions and Hydrolysis

Nucleophilic reactions of oxazoles can be characterized by several different outcomes. In many cases nucleophilic addition and elimination of a suitable leaving

TABLE 1.48. 5-ACYL-4-ALKYL-2-AMINOOXAZOLES FROM FRIEDEL-CRAFTS REACTIONS OF 4-ALKYL-2-AMINOOXAZOLES[a]

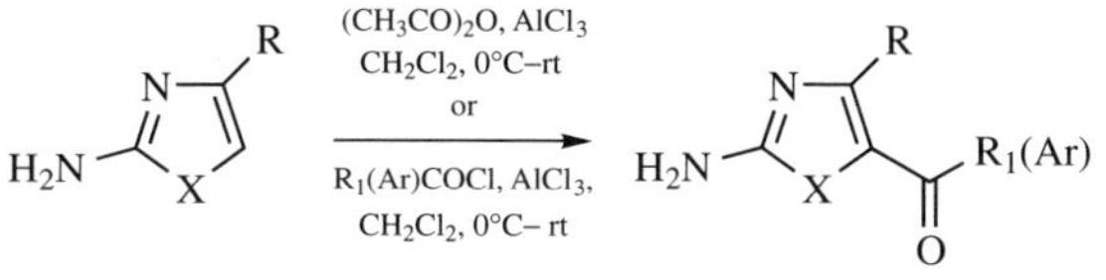

622 **631** X = O **Figure 1.50**

Entry	R	R_1 or Ar	Yield **631** (%)
1	CH_3	4-CH_3O–C_6H_4	36
2	CH_3	4-Cl–C_6H_4	49
3	CH_3	C_6H_5	47
4	CH_3	2-Cl–C_6H_4	40
5	$HOCH_2$	3-CH_3O–C_6H_4	19
6	CH_3	CH_3	35
7	CH_3	3-CH_3O–C_6H_4	46
8	CH_3	2-naphthyl	61

[a] Data from ref. 253.

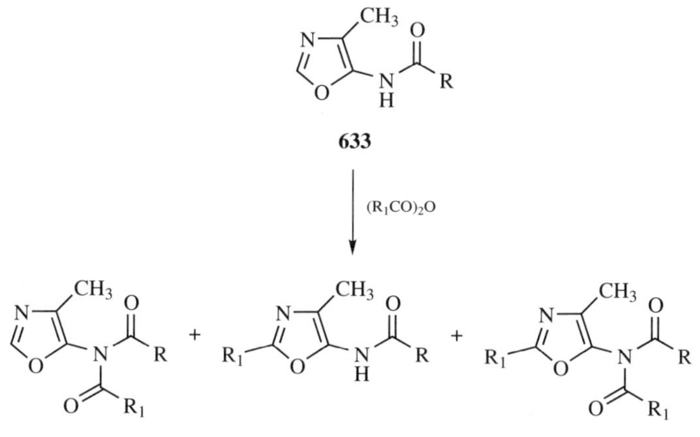

634 10–80% **635** 2–8% **636** 8–72%

Scheme 1.175

group, e.g., halide or sulfinate affords an aromatic substitution product. Alternatively, the initial substitution product can react further, e.g., in a Claisen rearrangement to afford ring-opened products after hydrolytic workup. Finally, nucleophilic addition can effect ring opening and recyclization to generate a new heterocyclic ring system. Selected examples of substitution reactions are discussed in this section, and ring opening and recyclization are discussed in Section 1.4.6.

140   Synthesis and Reactions of Oxazoles

There were only two reports[259,260] of the displacement of halides in 1,3-azoles by carbanions of active methylene compounds before Yamanaka and co-workers[261] detailed studies. Yamanaka's group investigated the reactivity of halo- or methanesulfonyl-substituted 1,3-azoles in aromatic S_N2-substitution reactions and demonstrated the reactivity order of oxazole ≫ thiazole > N-methylimidazole for displacement of methanesulfinate or halide at C(2) by carbanions of active methylene compounds (Scheme 1.176). For oxazoles, nucleophilic substitution follows the general order C(2) > C(5) > C(4).

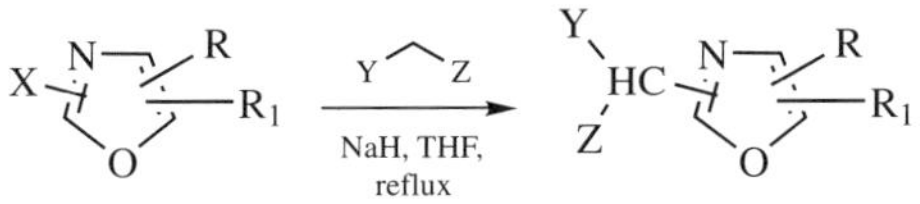

637 → 638

NaH, THF, reflux

X = Cl, Br, SO_2CH_3 Y, Z = CN, $COOC_2H_5$, $COCH_3$, C_6H_5
R, R_1 = H, C_6H_5

Scheme 1.176

For example, reaction of 2-chloro-4,5-diphenyloxazole **639** (X = Cl) with the anion of phenylacetonitrile afforded α,4,5-triphenyl-2-oxazoleacetonitrile **640** in very good yield (Scheme 1.177). 5-Chloro-2,4-diphenyloxazole **641** (X = Cl) reacted similarly to yield α,2,4-triphenyl-5-oxazoleacetonitrile **642**. 4-Chloro-2,5-diphenyloxazole **643** (X = Cl) was unreactive under these reaction conditions. However, 4-bromo-2,5-diphenyloxazole **643** (X = Br) did yield α,2,5-triphenyl-4-oxazoleacetonitrile **644**. The SO_2CH_3 group of 4,5-diphenyl-2-methanesulfonyloxazole **639** (X = SO_2CH_3) was readily substituted by carbanions of active methylene compounds to afford the corresponding analogues **645**. The order of nucleophilicity for these carbanions was $^-CH(CN)C_6H_5$ ≫ $^-CH(CN)_2$ > $^-CH(CN)COOC_2H_5$ > $^-CH(COOC_2H_5)_2$ ≫ $^-CH(COCH_3)_2$.

Burger and co-workers[116,262,263] extensively investigated the syntheses and reactions of 2-alkyl(aryl)-5-fluoro-4-(trifluoromethyl)oxazoles **218**. They found that **218** react with a variety of oxygen, nitrogen, sulfur, and carbon nucleophiles to afford substitution products **646** in modest to excellent yield (Scheme 1.178).[262,264,265] Representative examples are shown in Table 1.49.

Displacement of fluoride in **218** (Ar = C_6H_5) using bis-nucleophiles afforded 1,3-bis-[2-phenyl-4-(trifluoromethyl)-5-oxazolyloxy]benzene **647**, 1,4-bis-[2-phenyl-4-(trifluoromethyl)-5-oxazolyloxy]benzene **648**, and 2,5-bis-[2-phenyl-4-(trifluoromethyl)5-oxazolylthio]-1,3,4-thiadiazole **649**, respectively.[262]

These same authors[263] observed a similar reactivity pattern with 5-fluoro-2-heteroaryl-4-(trifluoromethyl)oxazoles **650**. The starting materials were readily prepared from a heterocyclic amide according to their earlier protocol.[116] Here again, displacement of fluoride in **650** was readily accomplished to afford **651**, **652** and **653** (Scheme 1.179).

This same group[116,266] treated 5-fluoro-2-phenyl-4-(trifluoromethyl)oxazole **218** (Ar = C_6H_5) with KOH/C_2H_5OH or KOtBu/tBuOH to prepare **219** and **654**, respectively. Reaction of **219** with iodotrimethylsilane gave 3,3,3-trifluoroalanine **220**

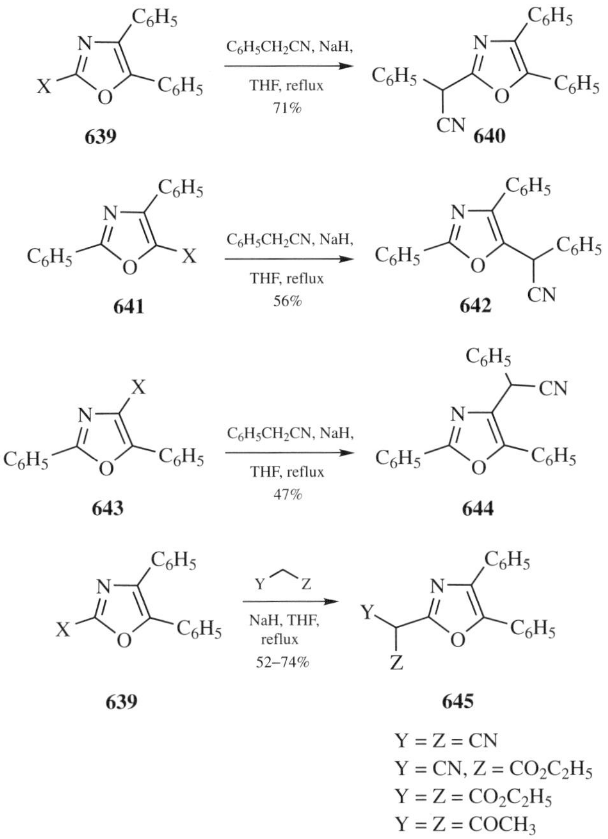

Scheme 1.177

(Scheme 1.61), whereas treatment of **654** with TFA afforded *N*-benzoyl-3,3,3-trifluoroalanine **655** (Scheme 1.180).

Burger and co-workers[267–269] adapted this methodology for an elegant synthesis of α-substituted α-trifluoromethyl amino acids. For example, reaction of **218** (Ar = C_6H_5) with an allylic alcohol gives **656** that undergoes a hetero-Cope rearrangement to produce a 4-alkenyl-5(4*H*)-oxazolone **657** (Scheme 1.181). Addition of water effects ring opening of **657** to **658**, which is then deprotected with HCl to yield **659**. Substitution of a propargyl alcohol for the allylic alcohol in this reaction sequence leads to the previously unknown 2-allenyl-α-trifluoromethyl amino acid analogues **660**.[267]

Burger and co-workers[268,269] extended this methodology further and prepared α-substituted α-trifluoromethyl aromatic and heterocyclic α-amino acids (Scheme 1.182). Here, displacement of fluoride in **218** (Ar = C_6H_5) was accomplished by using a benzylic alcohol to produce **661** in good to excellent yield. The presumed Claisen rearrangement of **661** to a 4-substituted 5(4*H*)-oxazolone **662** was effected in $CHCl_3$, toluene, or xylene from room temperature to 120°C. Ring opening of **662** with 1N HCl gave the *N*-benzoyl amino acid **663**, from which

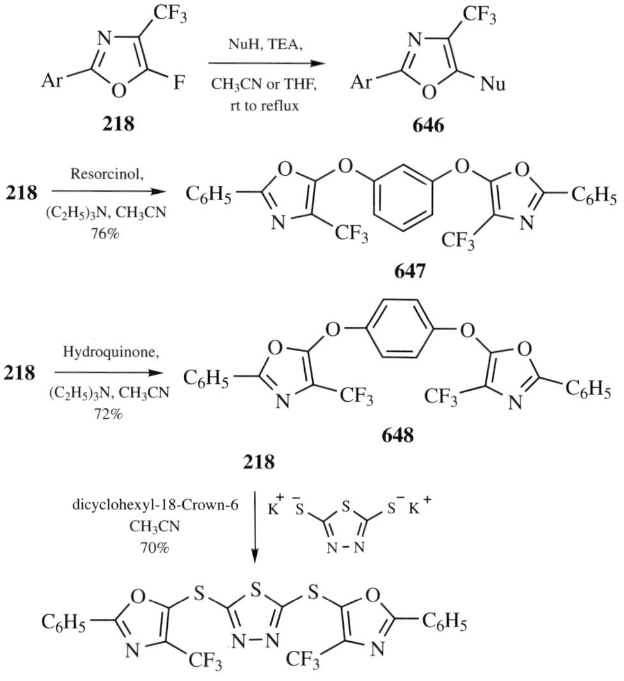

649

Scheme 1.178

TABLE 1.49. 2-ARYL-5-SUBSTITUTED 4-(TRIFLUOROMETHYL)OXAZOLES FROM NUCLEOPHILIC SUBSTITUTION OF 2-ARYL-5-FLUORO-4-(TRIFLUOROMETHYL)OXAZOLES

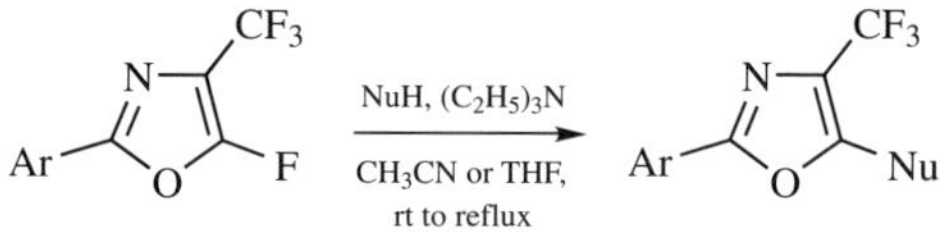

Figure 1.51 **218** **646**

Entry	Ar	NuH	Nu	Yield (%)	Reference
1	4-CH_3O–C_6H_5	CH_3CH_2OH	CH_3CH_2O	79	262
2	4-F–C_6H_4	CF_3CH_2OH	CF_3CH_2O	89	262
3	C_6H_5	CH_3NHCH_2COOH	CH_3NCH_2COOH	77	262
4	C_6H_5	$HSCH_2CO_2CH_3$	$SCH_2CO_2CH_3$	73	262
5	4-Cl–C_6H_4	$(CH_3)_2NNH_2$	$(CH_3)_2NNH$	66	262
6	C_6H_5	LiCl	Cl	53	262
7	C_6H_5	LiBr	Br	62	262
8	C_6H_5	C_6H_5Li	C_6H_5	47	262
9	C_6H_5	$(CH_3)_3SiN_3$	N_3	63	264
10	4-CH_3O–C_6H_4	$(CH_3)_3SiN_3$	N_3	51	264
11	4-F–C_6H_4	$(CH_3)_3SiN_3$	N_3	37	264
12	4-CH_3–C_6H_4	$(CH_3)_3SiN_3$	N_3	57	264
13	C_6H_5	KCN	CN	74	265

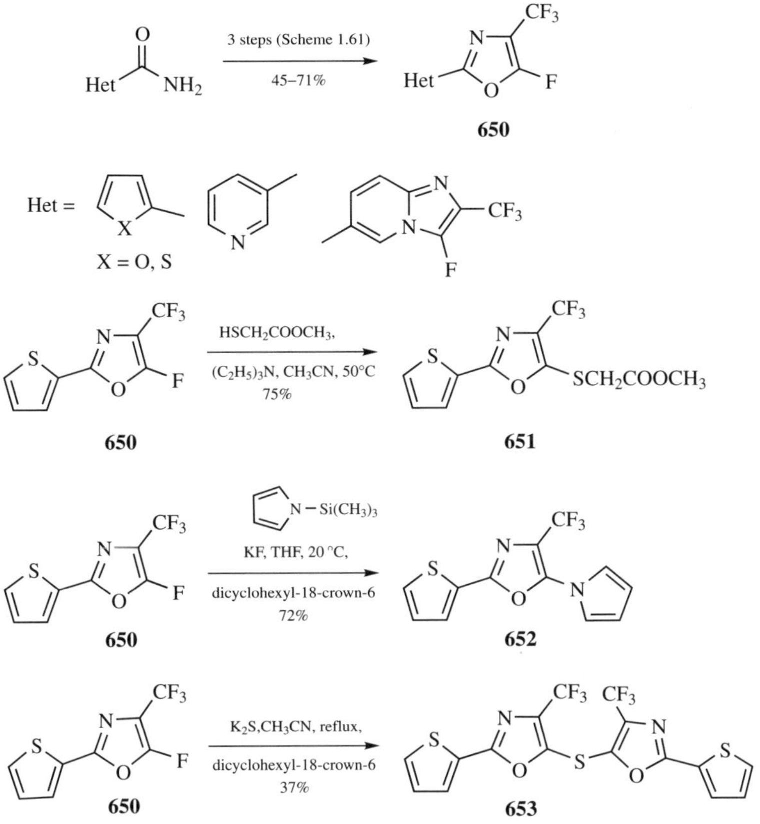

Scheme 1.179

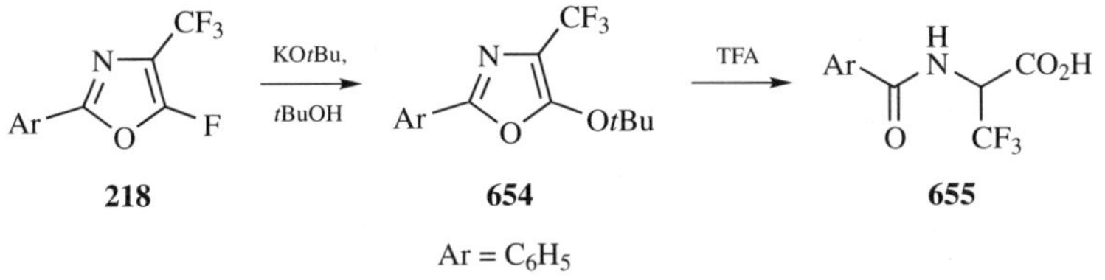

Scheme 1.180

Ar = C₆H₅

hydroylsis with 12M HCl gave **664**. The heterocyclic analogues **665** and **666** were prepared analogously. The authors determined the mechanism of the rearrangement of **661** to **662** to be a 1,3-benzyl shift from oxygen to carbon and not a concerted 1,3-sigmatropic rearrangement. This was supported by isolation of mixed products from crossover experiments. Some representative examples of **663** are shown in Table 1.50.

Burger and Helmreich[270] reported improved conditions for preparing **218** analogs via ultrasonic mediated cyclization of *N*-(hexafluoroisopropylidene)

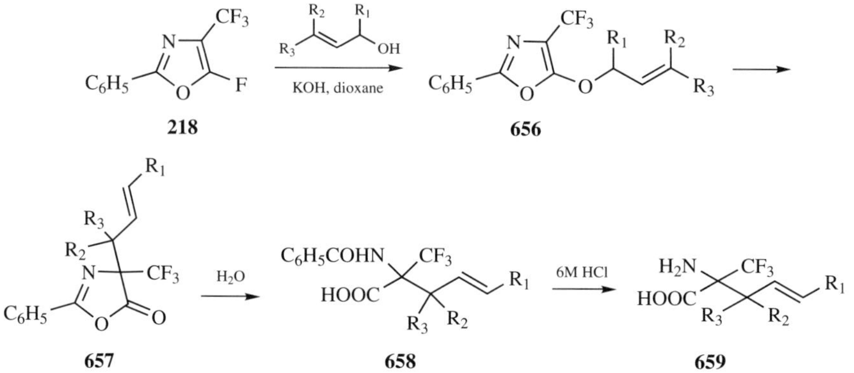

Scheme 1.181

$R_1 = H, CH_3, C_2H_5, n\text{-}C_3H_7, CH=CH_2$
$R_2 = H, CH_3, n\text{-}C_3H_7$
$R_3 = H, CH_3$

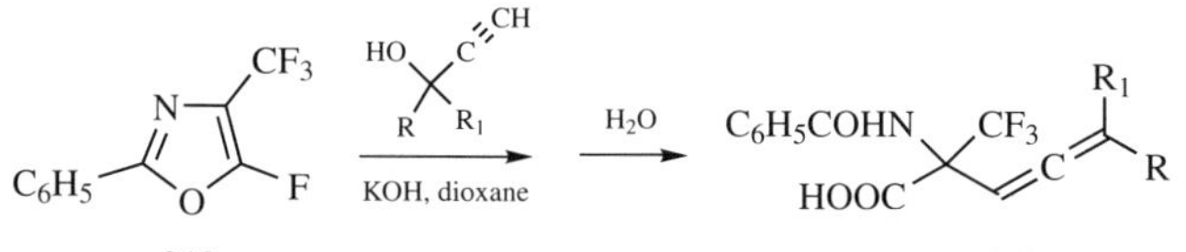

$R = H, CH_3, C_2H_5$
$R_1 = H, C_2H_5$

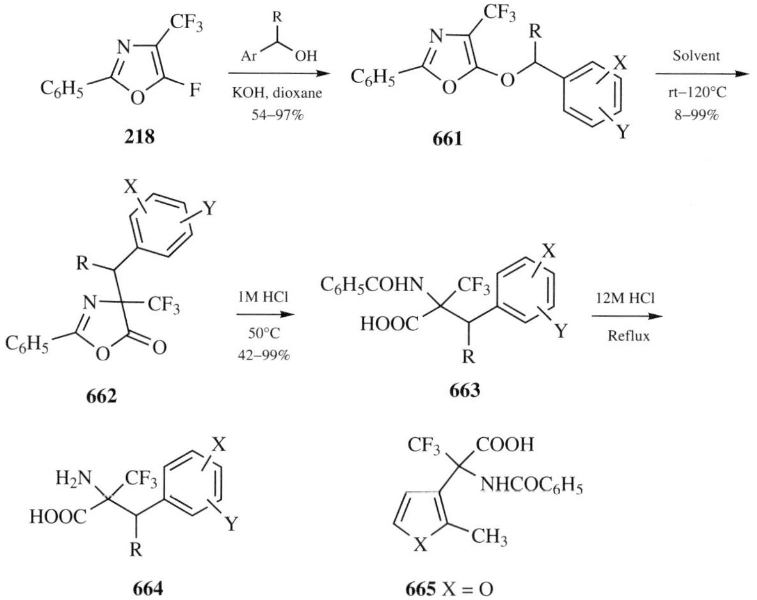

Scheme 1.182

665 X = O
666 X = S

TABLE 1.50. N-BENZOYL-α-(TRIFLUOROMETHYL) AROMATIC α-AMINO ACIDS FROM 5-FLUORO-2-PHENYL-4-(TRIFLUOROMETHYL)OXAZOLE[a]

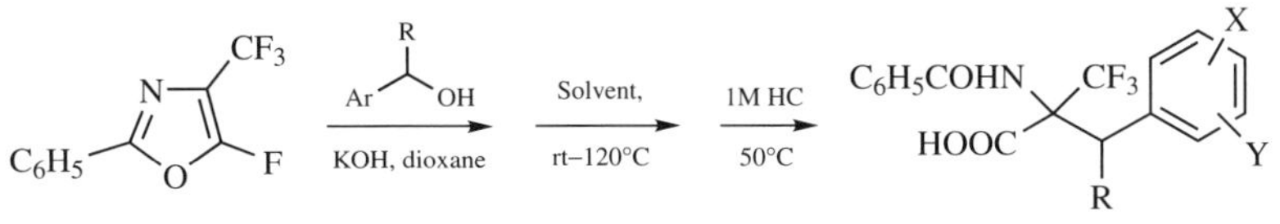

Figure 1.52

Entry	X	Y	Yield (%)[b]
1	H	H	87
2	3-OCH$_3$	4-OCH$_3$	73
3	H	4-F	99
4	H	4-Cl	99
5	H	4-NO$_2$	99

[a] Data from ref. 268.
[b] Refers to the hydrolytic ring opening of the 5(4H)-oxazolone

carboxamides **217** in the presence of zinc (Scheme 1.183). The authors evaluated a variety of metals, including zinc, lead, tin, cadmium, thallium, germanuim, copper, aluminum, magnesium, zirconium, mercury, and gallium/indium and concluded that zinc dust gave superior results.

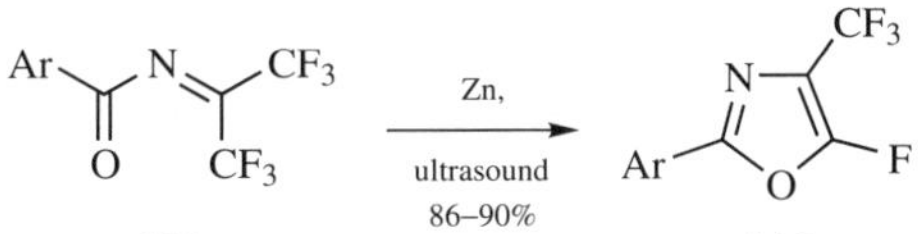

Ar = C$_6$H$_5$, 2-CH$_3$-C$_6$H$_4$-, 4-CH$_3$-C$_6$H$_4$-, 4-CH$_3$O-C$_6$H$_4$-, 2-naphthyl

Scheme 1.183

Padwa and Cohen[271] prepared a series of 2-alkoxy-4,5-diphenyloxazoles **668** from 2-chloro-4,5-diphenyloxazole **667** and the corresponding sodium alkoxide (Scheme 1.184). These 2-alkoxy-4,5-diphenyloxazoles **668** were substrates for the authors' systematic investigations of the thermal and photochemical [3,3]-sigmatropic rearrangements of suitably substituted oxazoles (see Section 1.4.5).

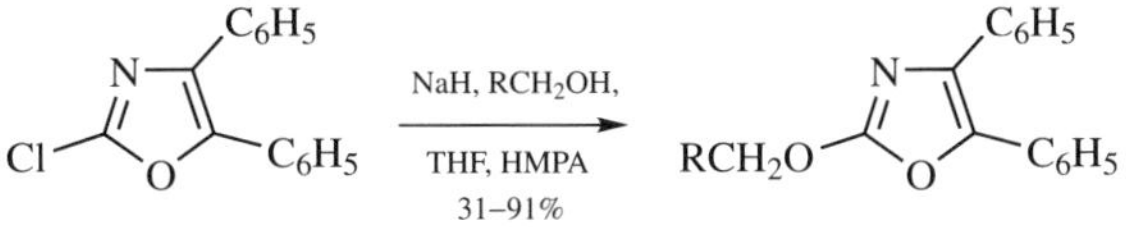

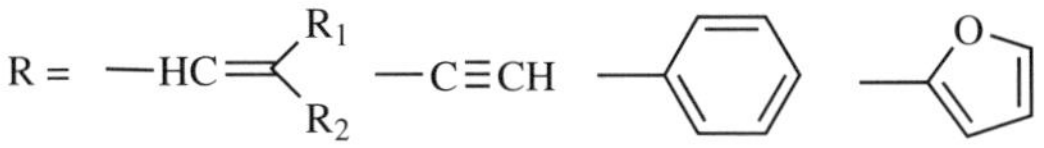

R$_1$, R$_2$ = H, CH$_3$

Scheme 1.184

Schnell and co-workers[272,273] reported that 4-diethylphosphono-5-substituted oxazoles are useful precursors to other oxazoles or to phosphonoglycine amides. For example, reaction of 5-chloro-4-diethylphosphonooxazole **669** with C_2H_5SH/KO*t*Bu gave 4-diethylphosphono-5-ethylthiooxazole **670** (Scheme 1.185).[272]

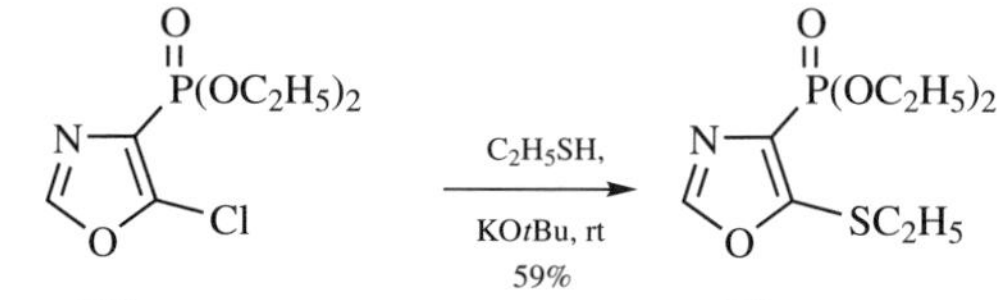

669 **670**

Scheme 1.185

Alternatively, hydrolysis of a 5-amino-4-diethylphosphonooxazole **672**, prepared from 1-formylamino-2,2,2-trichloroethanephosphonate **671**,[274,275] afforded different products, depending on the reaction conditions (Scheme 1.186).[273] Mild solvolysis produced diethoxyphosphonoglycine amide HCl salts **673** in excellent yield, whereas hydrolysis of **672** with dilute aqueous HCl generated *N*-formyl-diethoxyphosphonoglycine amides **674**. Finally, methylation of **672** followed by basic hydrolysis yielded *N*-formyl-*N*-methyl-diethoxyphosphonoglycine amides **675**. Representative examples of each analog are shown in Table 1.51.

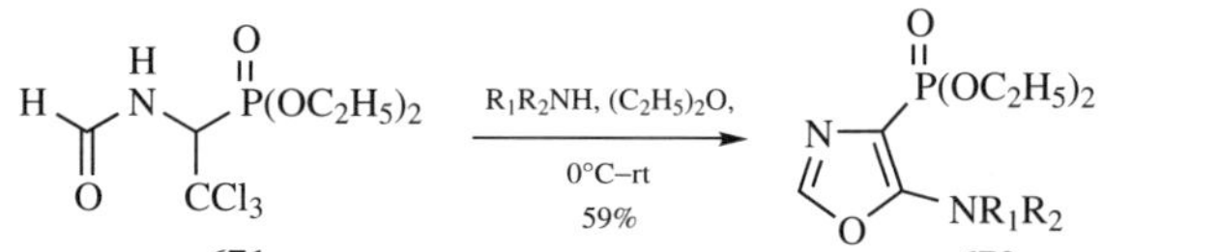

671 **672**

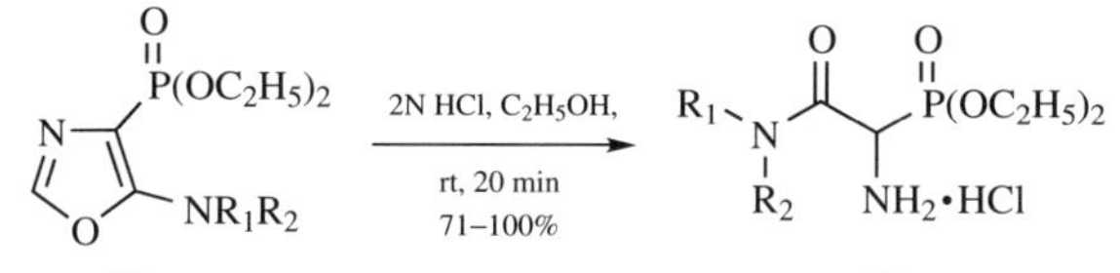

672 **673**

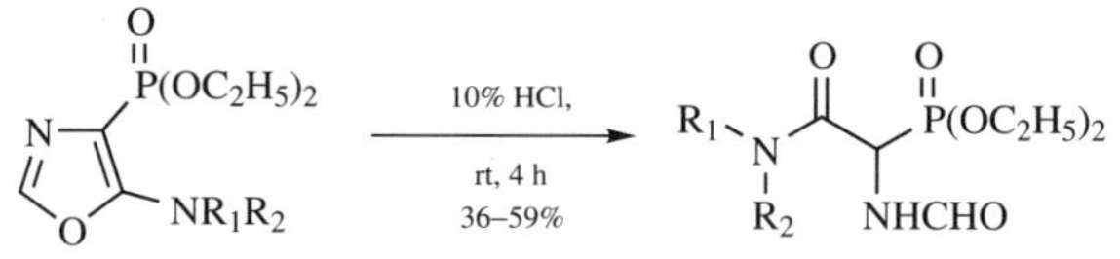

672 **674**

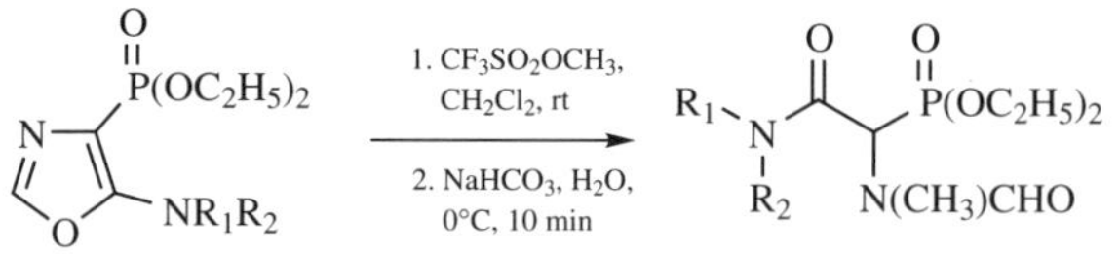

672 **675**

Scheme 1.186

Ueda and co-workers[276] at Bristol-Myers used mild acid hydrolysis to cleave the oxazole moiety in oxazolocoumarins to produce 3-amino-4-hydroxycoumarins.

TABLE 1.51. DIETHYLPHOSPHONOGLYCINE AMIDES FROM HYDROLYSIS OF 5-AMINO-4-DIETHYLPHOSPHONOOXAZOLES[a]

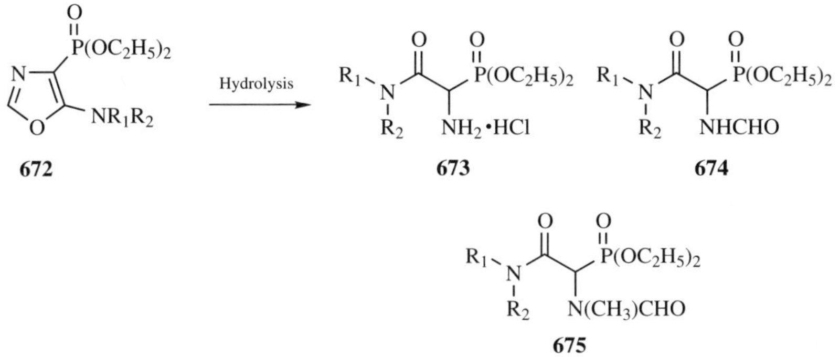

Figure 1.53

Entry	R_1	R_2	Conditions[b]	Yield 673 (%)	Yield 674 (%)	Yield 675 (%)
1	H	CH_3	A	71	—	—
2	H	CH_3	B	—	36	—
3	H	CH_3	C	—	—	93
4	CH_3	CH_3	A	100	—	—
5	CH_3	CH_3	B	—	53	—
6	CH_3	CH_3	C	—	—	90
7	$-(CH_2)_5-$		A	92	—	—
8	$-(CH_2)_5-$		B	—	59	—
9	$-(CH_2)_5-$		C	—	—	93

[a] Data from ref. 273.
[b] A, 2N HCl, C_2H_5OH, rt, 20 min; B, 10% HCl, rt, 4 h; C, (1) $CF_3SO_2OCH_3$, CH_2Cl_2, (2) $NaHCO_3$, 0°C, 10 min.

This was a key step in their synthesis of coumermycin analogues that were prepared as potential antibacterial agents against methicillin-resistant strains of staphylococci. For example, treatment of **676** with exactly one equivalent of HCl generated in situ from acetyl chloride gave **677** in excellent yield (Scheme 1.187). It is noteworthy that the oxazole ring was cleaved in the presence of a pyrrole carboxylic acid ester, the coumarin lactone, and a glycosidic linkage. This transformation could not be effected cleanly using alkaline conditions or mineral acid. The authors also successfully adapted their mild cleavage conditions to the novobiocin series.

Hamada and Shioiri[277] described the first synthesis of mugineic acid **680**, an iron-chelating amino acid isolated from barley roots. Their synthetic strategy included an elegant application of an oxazole ring as a masked α-amino ketone surrogate. To implement their strategy, the authors hydrolyzed the intermediate oxazole **678** with methanesulfonic acid in aqueous benzyl alcohol to give the bis-methanesulfonate salt, **679** (Scheme 1.188). The synthesis of **680** was completed in six steps from **679**.

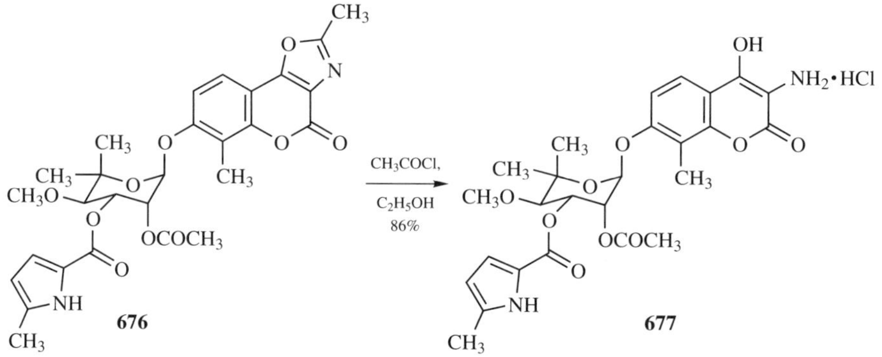

Scheme 1.187

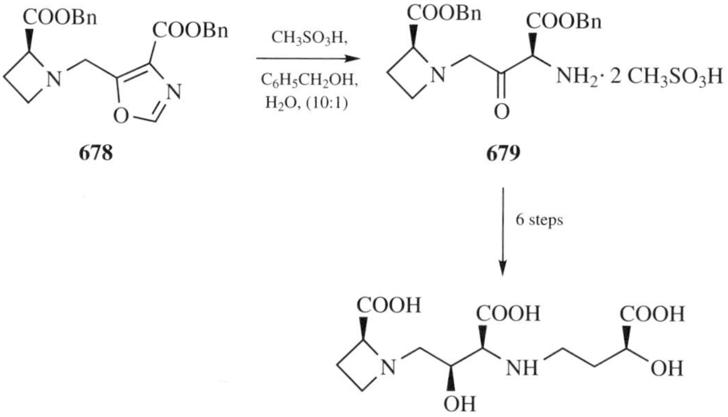

Scheme 1.188

Lipshutz and co-workers[197–199] effected mild hydrolysis of an oxazole ring as a key step in their use of oxazoles as masked dipeptides. The authors examined NaOCH$_3$, NaOH, Triton B, and K$_2$CO$_3$/CH$_3$OH as well as Li(C$_2$H$_5$)$_3$BH, DIBAL-H, NaBH$_3$CN, and 9-BBN to cleave the trifluoroacetamide present in **491** and **681**. They found that a reduction/hydrolysis reaction sequence was a particularly effective protocol to unmask the dipeptide with a minimum of side reactions. The authors favored NaBH$_4$ supported on neutral alumina to minimize epimerization before hydrolysis, although NaBH$_4$ at −78°C was also effective in achiral systems, e.g., **681** → **682**. Hydrogenolysis was used to cleave a Cbz group, e.g., **488**. In all cases, hydrolysis of the oxazole ring was accomplished in excellent yield (Scheme 1.189).

Loupy and co-workers[278] prepared 2-methyl-5-palmityl-4-oxazolecarboxylic acid ethyl ester **683** and investigated a number of reactions including alcoholysis (Scheme 1.190). There was no reaction after prolonged reflux of **683** in CH$_3$OH or C$_2$H$_5$OH. However, refluxing **683** in CH$_3$OH in the presence of HCl or KOCH$_3$

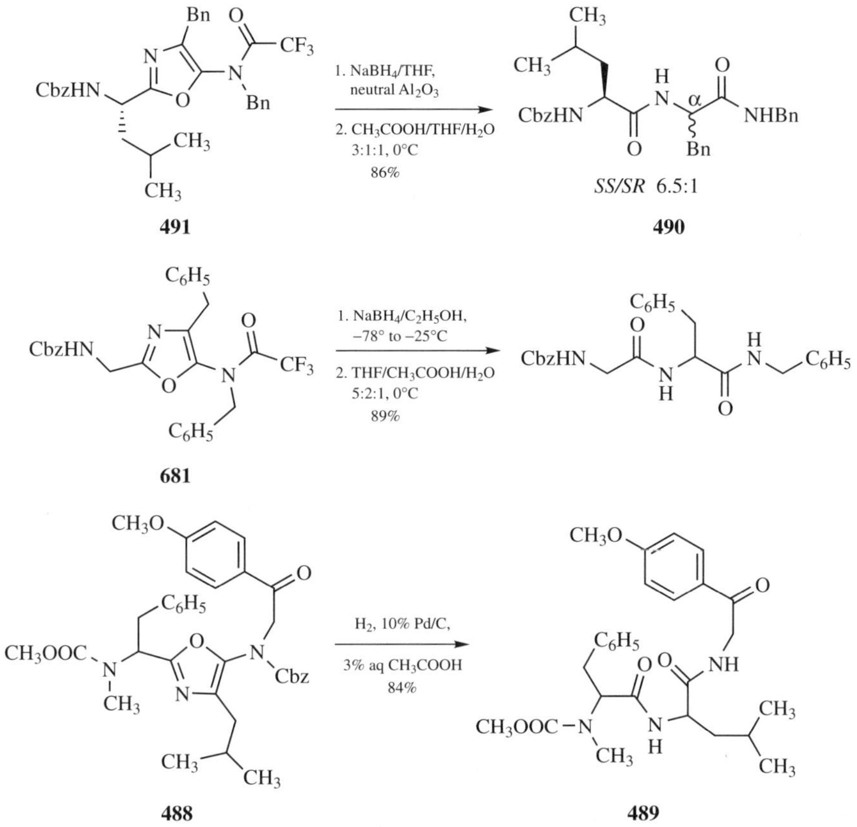

Scheme 1.189

effected both transesterification and ring cleavage to yield a 1:3-4 mixture of 2-methyl-5-palmityl-4-oxazolecarboxylic acid methyl ester **684** and methyl palmitate **685a**. Heating **683** with KOCH$_3$ in DMF cleanly generated palmitoic acid dimethylamide **685b**, whereas heating **683** with KCN in DMF produced a 2:1 mixture of 2-methyl-5-palmityl-4-oxazolecarboxylic acid dimethylamide **686** and **685b**.

Examples of oxazole ring cleavage effected by nucleophilic addition and recyclization to other heterocyclic ring systems are discussed in Section 1.4.6. Section 1.4.10 contains examples of nucleophilic addition to N-alkyl- and N-acyloxazolium salts.

1.4.3. Reduction

Oxazoles can be reduced to afford ring-opened products or oxazolines, depending on the reaction conditions. For example, Lunn[279] investigated the reduction of heterocycles with a nickel-aluminum alloy in aqueous potassium hydroxide.

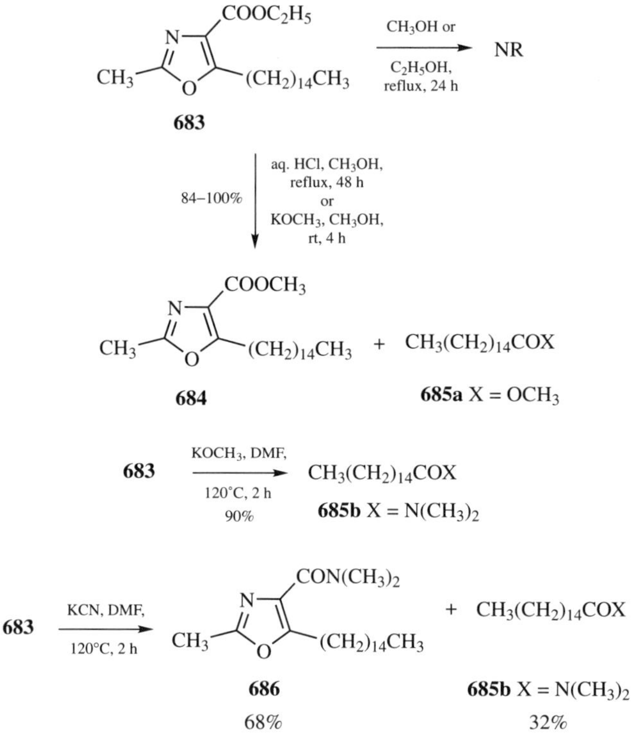

150   Synthesis and Reactions of Oxazoles

Scheme 1.190

Oxazole **1** gave a low yield of a highly water-soluble product, 2-(methylamino)ethanol (Scheme 1.191). This was the only product isolated and resulted from an unexpected C-O bond cleavage. More highly substituted oxazoles, e.g., 2,4,5-trimethyloxazole were inert.

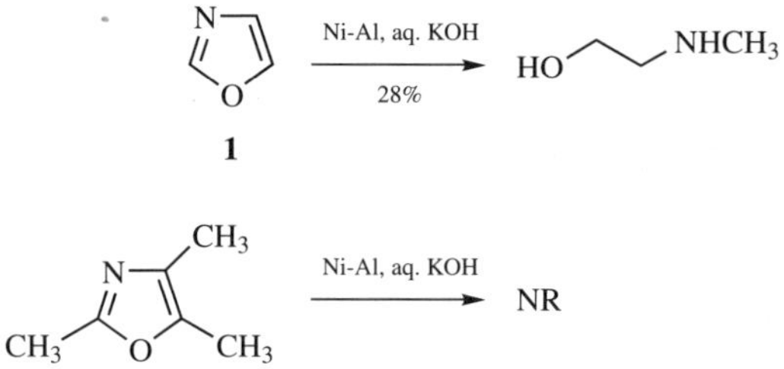

Scheme 1.191

Loupy and co-workers[278] found that prolonged treatment of **683** with NaBH$_4$ in THF or CH$_3$OH at reflux afforded hexadecanol (Scheme 1.192). The authors proposed that initial 1,4-addition of hydride to **683** produced the 2-oxazoline **687**. This oxazoline was then cleaved under the reaction conditions to generate pentadecanal, which was reduced to the final product. This same transformation was effected in 93% yield using Red-Al® in toluene.

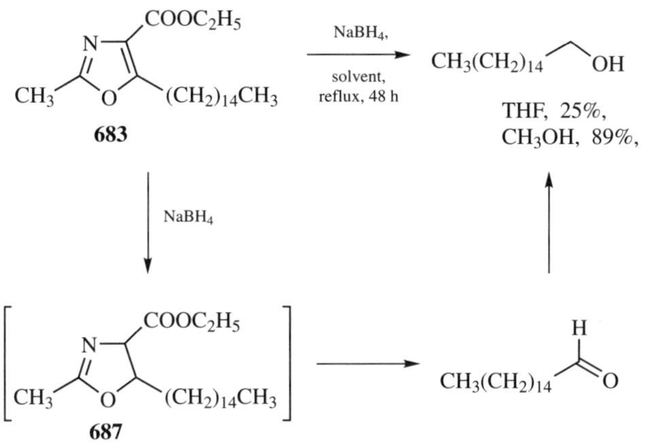

Scheme 1.192

Kamat and co-workers[280] investigated dissolving metal reduction of oxazoles using potassium metal. The authors isolated only ring cleavage products that depended on the substitution pattern of the oxazole (Scheme 1.193). For example, reduction of 2,4,5-triphenyloxazole **231c** with potassium in THF gave *N*-(1,2-diphenylethyl)benzamide **688** together with a small amount of benzoic acid. Alternatively, an alkyl-diaryloxazole was only partially reduced to afford an enamide. Thus 4,5-diphenyl-2-methyloxazole **231a** yielded (*E*)-*N*-(1,2-diphenylvinyl)acetamide **689**, whereas 2,4-diphenyl-5-methyloxazole **196** produced

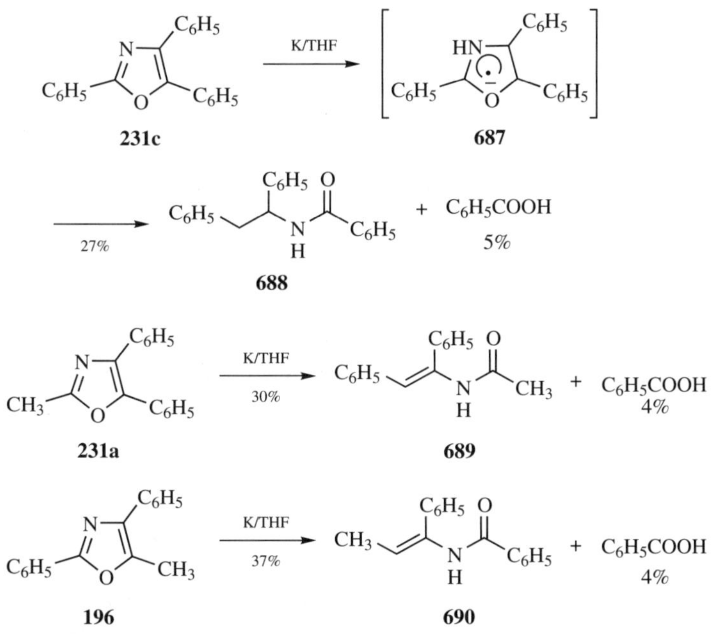

Scheme 1.193

(*E*)-*N*-(1-phenyl-1-propenyl)benzamide **690**. In all three cases, the reactions were incomplete, with up to 59% recovered starting material (**231a**). These authors determined the reversible reduction half-peak potentials of **231c**, **231a**, and **196** to be −2.21 V, −2.36 V, and −2.46 V, respectively, versus saturated sodium chloride-calomel electrode (SSCE).

Tsveniashvili and co-workers[281,282] acquired polarographic data for series of 2,4-disubstituted, 2,5-disubstituted, and 2,4,5-trisubstituted oxazoles and investigated the effects of substituents on the electrochemical reduction potential. They explained their results by two one-electron reversible steps to produce a dianion, which was then protonated. Rogers and co-workers[283] described anion-pi radicals formed during electrochemical reduction of fused oxazoles, e.g., 2-phenylbenzoxazole.

Tanaka and Kuriyama[284] prepared *cis*-4,5-disubstituted-oxazolines by catalytic hydrogenation of 2-amino-4,5-disubstituted oxazoles using PtO_2 or Pd/C. The authors isolated ring cleavage products from reactions employing a high catalyst load.

1.4.4. Oxidation

Oxazoles are normally resistant to oxidation yet they are readily converted to triamides via a dye-sensitized photooxidation with singlet oxygen. This reaction has been the subject of extensive mechanistic investigations and synthetic applications in natural product synthesis by Wasserman's group.[6,143,285–289] Gollnick and Koegler[290,291] obtained the first definitive evidence for the formation of a 2,5-oxazole endoperoxide from low-temperature (−50°C) ^1H and ^{13}C NMR spectra. In addition, Kashima and co-workers[12,292–298] conducted detailed studies on ozonolysis of oxazoles to produce a variety of synthetically useful anhydrides, including some applicable for peptide synthesis. Ito and co-workers[299–302] elegantly applied the dye-sensitized photooxidation of polymeric oxazoles for the preparation of photoresists and holographic media. Selected examples of these reactions and applications are discussed in this section.

In principle, reaction of an oxazole, e.g., **15** with $^1O_2$ could yield triamides **693** and/or **696** via two reasonable pathways (Scheme 1.194). In the first case, path A, [4 + 2] cycloaddition of $^1O_2$ to **15** would yield an oxazole-2,5-endoperoxide **691**. Rearrangement of **691** to an imino-anhydride **692** by a Bayer-Villiger-like reaction and subsequent *O*-acyl to *N*-acyl transfer would produce triamide **693**.[303] Alternatively, [2 + 2] cycloaddition of $^1O_2$ to **15** would produce the dioxetane **694**, cleavage of which would then yield a second imino-anhydride **695**, which could also undergo an *O*-acyl to *N*-acyl transfer reaction to produce triamide **696**.

Wasserman's[304] early work involved labeling experiments using both oxygen-18-enriched singlet oxygen and ^{18}O-labeled oxazoles to distinguish between these routes. The authors designed a masterful set of labeling experiments that, coupled with high-resolution mass spectrometry, allowed them to identify the position of the ^{18}O label in the triamides by mass spectral fragmentation patterns. These studies

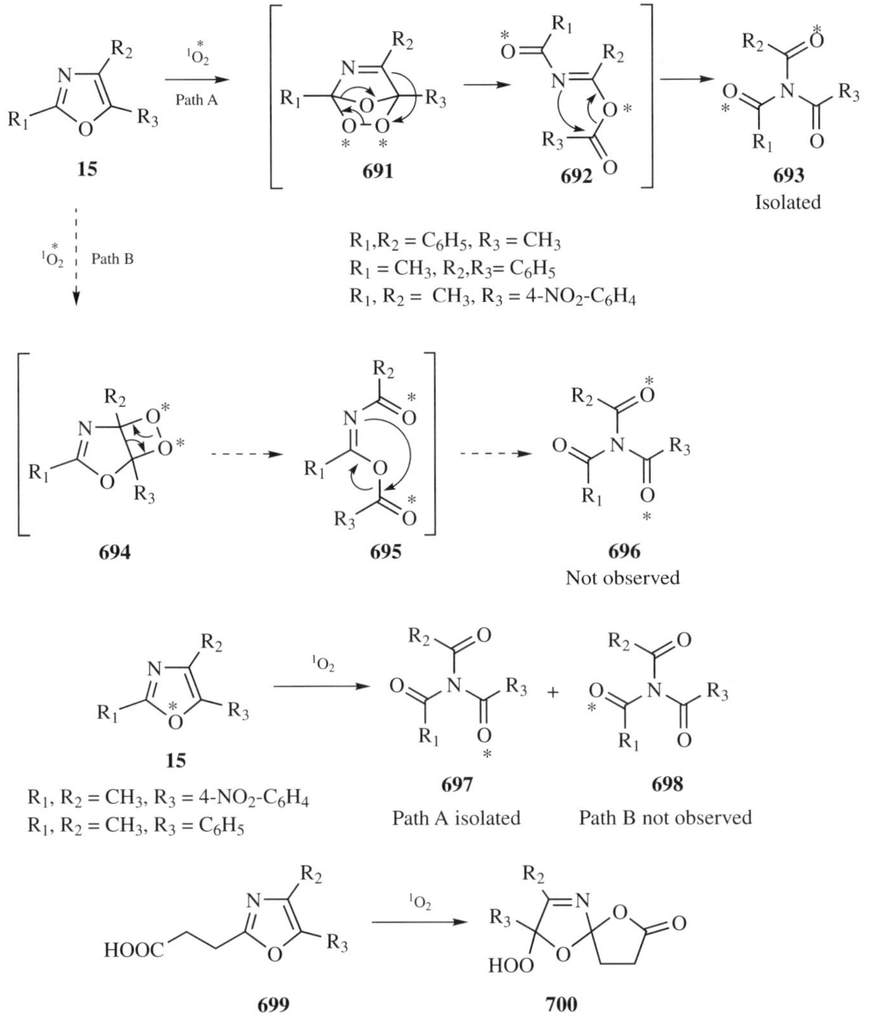

Reactions of Oxazoles 153

Scheme 1.194

enabled them to identify **693** as the product and, therefore, conclude that path A was operable for the 2,4,5-trisubstituted oxazoles investigated. A similar mechanistic rationale and interpretation applied to reaction of ^{18}O-labeled **15** with $^1O_2$ led to the triamide **697**.

Additional evidence to support an oxazole 2,5-endoperoxide intermediate was obtained by isolation of the spirolactone **700** (Scheme 1.194).[305] The authors proposed that **700** was generated from **699** by intramolecular capture of an intermediate carbocation with the suitably positioned carboxylic acid side chain. They also concluded that addition of $^1O_2$ to oxazoles was a concerted reaction based on a lack of solvent or substituent effects.

Gollnick and Koegler[290] also conducted detailed mechanistic studies of the [4 + 2] cycloaddition of $^1O_2$ with oxazoles. The authors characterized the first examples of disubstituted oxazole 2,5-endoperoxides **702** (Scheme 1.195). These endoperoxides were particularly interesting since the presence of an olefinic proton permitted structure assignment based on low-temperature ($-50°C$) ^1H and ^{13}C NMR data. The authors were able to rule out **694** and **703–705** as possible intermediates in this reaction, and they unequivocally identified the intermediates as the oxazole 2,5-endoperoxides **702**.

701

$R_1 = H, R_2 = R_3 = CH_3$
$R_1 = CH_3, R_2 = H, R_3 = CH_3$
$R_1 = R_2 = CH_3, R_3 = H$
$R_1 = CH_3, R_2 = H, R_3 = C_6H_5$
$R_1 = C_6H_5, R_2 = H, R_3 = C_6H_5$

$^1O_2$, CH$_3$CN, TPP or RB, hv, CDCl$_3$/CFCl$_3$, < $-50°C$

TPP = tetraphenylporphin
RB = rose bengal

702

694   **703**   **704**   **705**

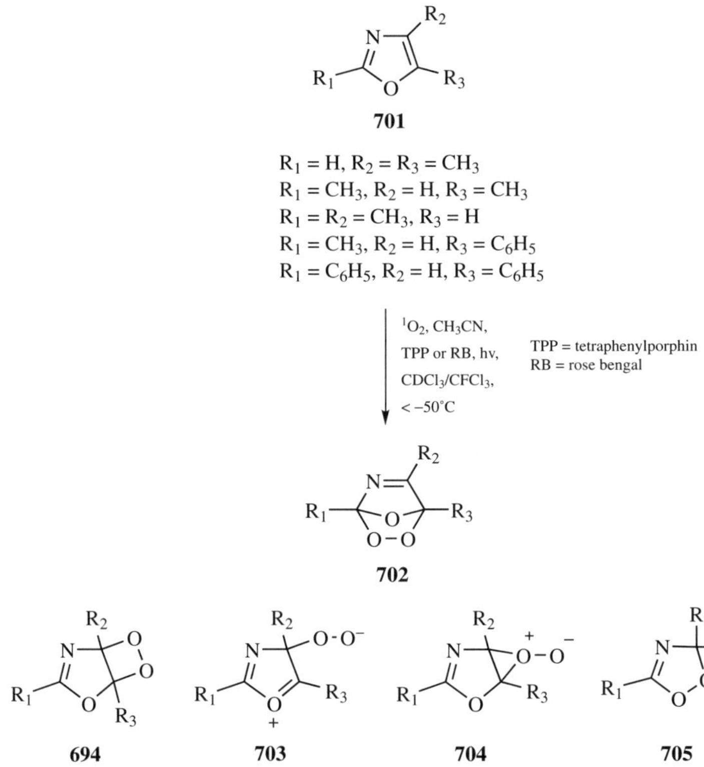

Scheme 1.195

Gollnick and Koegler[291] also investigated rearrangements of **702** in protic and aprotic solvents. Examples of **702** derived from monosubstituted oxazoles and oxazole itself were prepared and included in this work. The authors found that **702** was stable in CDCl$_3$ at temperatures < $-30°C$, regardless of the oxazole substitution pattern. However, **702** rearranged at temperatures above $-30°C$ and the product distribution depended on the oxazole substitution pattern and the solvent. For example, a 2-phenyloxazole **701** ($R_1 = C_6H_5$) produced the corresponding triamide **706** quantitatively (Scheme 1.196, Path A). On the other hand, a 2-unsubstituted oxazole **701** ($R_1 = H$) fragmented to yield a 1:1 mixture of a nitrile (or HCN) **707** and an anhydride **708** (Scheme 1.196, Path B), although the authors

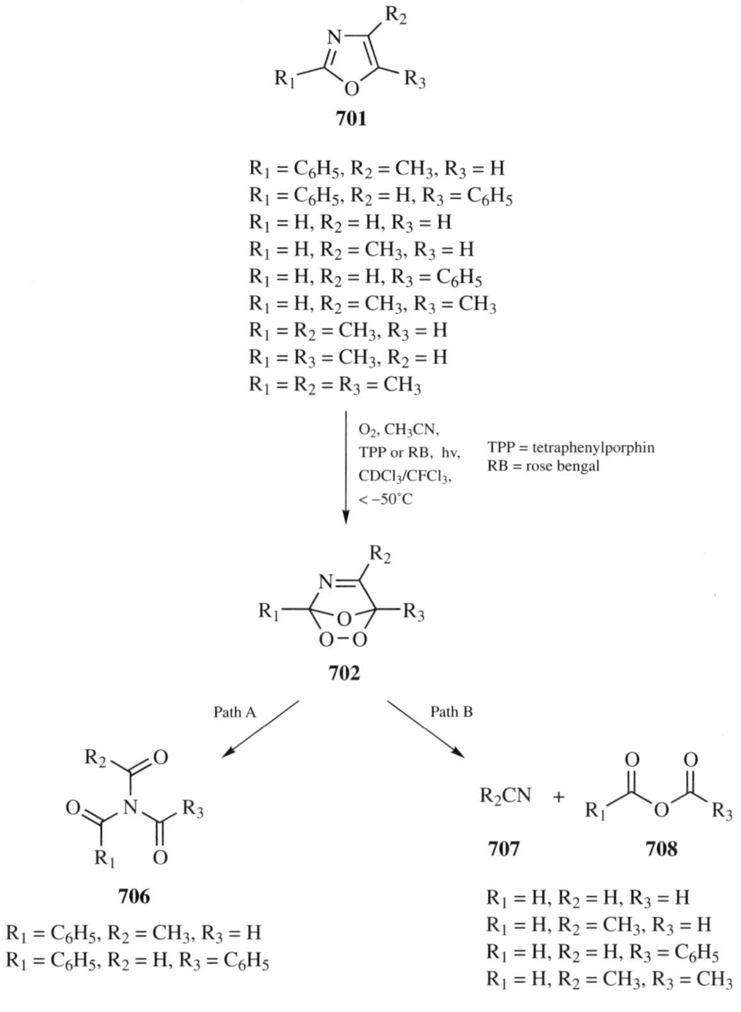

701

$R_1 = C_6H_5, R_2 = CH_3, R_3 = H$
$R_1 = C_6H_5, R_2 = H, R_3 = C_6H_5$
$R_1 = H, R_2 = H, R_3 = H$
$R_1 = H, R_2 = CH_3, R_3 = H$
$R_1 = H, R_2 = H, R_3 = C_6H_5$
$R_1 = H, R_2 = CH_3, R_3 = CH_3$
$R_1 = R_2 = CH_3, R_3 = H$
$R_1 = R_3 = CH_3, R_2 = H$
$R_1 = R_2 = R_3 = CH_3$

O_2, CH_3CN, TPP or RB, hv, $CDCl_3/CFCl_3$, < −50°C

TPP = tetraphenylporphin
RB = rose bengal

702

Path A   Path B

706

$R_1 = C_6H_5, R_2 = CH_3, R_3 = H$
$R_1 = C_6H_5, R_2 = H, R_3 = C_6H_5$

R_2CN + **707** **708**

$R_1 = H, R_2 = H, R_3 = H$
$R_1 = H, R_2 = CH_3, R_3 = H$
$R_1 = H, R_2 = H, R_3 = C_6H_5$
$R_1 = H, R_2 = CH_3, R_3 = CH_3$

Scheme 1.196

could not spectroscopically detect any intermediates leading to **707** and **708**. Finally, a 2-alkyloxazole **701** ($R_1 = CH_3$) generated a mixture of **706**, **707** and **708**. The ratio of **706** to (**707** + **708**) also depended on the substitution at C(4) and C(5).

Irradiation of a methanolic solution of an oxazole at −78°C in the presence of $^1O_2$ and rose bengal produced a 2,5-disubstituted 5-hydroperoxy-4-methoxy-4,5-dihydrooxazole **709** (Scheme 1.197).[291] Here again, the oxazole substitution pattern determined the product distribution. For example, 4,5-diphenyl-2-methyloxazole **231a** produced only N, N-dibenzoyl acetamide **706** ($R_1 = CH_3, R_2 = R_3 = C_6H_5$). On the other hand, 4-methyl-2-phenyloxazole **701a** gave **709** ($R_1 = C_6H_5$, $R_2 = CH_3, R_3 = H$), which decomposed on warming to room temperature to yield

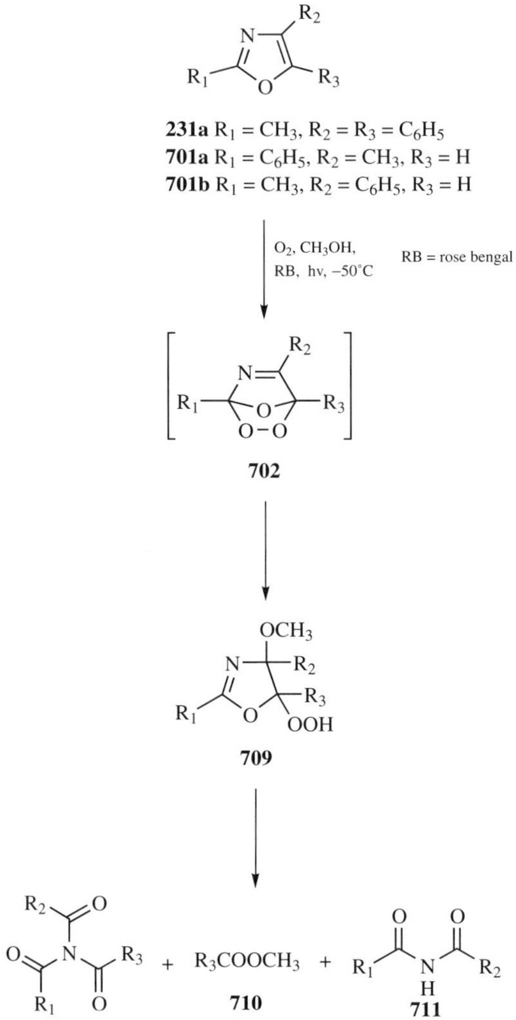

231a $R_1 = CH_3$, $R_2 = R_3 = C_6H_5$
701a $R_1 = C_6H_5$, $R_2 = CH_3$, $R_3 = H$
701b $R_1 = CH_3$, $R_2 = C_6H_5$, $R_3 = H$

O_2, CH_3OH, RB, hv, $-50°C$ RB = rose bengal

702

709

706 + R_3COOCH_3 **710** + **711**

Scheme 1.197

methyl formate **710** ($R_3 = H$) and N-acetyl benzamide **711** ($R_1 = C_6H_5$, $R_2 = CH_3$). However, 2-methyl-4-phenyloxazole **701b** ($R_1 = CH_3$, $R_2 = C_6H_5$, $R_3 = H$) yielded ca. 30% **706** ($R_1 = CH_3$, $R_2 = C_6H_5$, $R_3 = H$) together with ca. 70% of a mixture of methyl formate **710** ($R_3 = H$) and N-acetyl benzamide **711** ($R_1 = C_6H_5$, $R_2 = CH_3$).

Wasserman's group showcased the synthetic utility of singlet oxygen cleavage of oxazoles, whereby the oxazole effectively functions as a carbonyl 1,1-dipole. The researchers[285,288] prepared small and medium rings and applied this methodology to natural product syntheses, including antimycin A_3 **722**[286] and ($\pm$)-pyrenolide C **726**.[6,287]

Reactions of Oxazoles 157

For example, 4,5-diphenyl-2-methyloxazole **231a** was converted to **712** in five steps and 52% yield (Scheme 1.198).[285] Here, the oxazole ring serves as a protected acid surrogate and is unmasked by oxidation with singlet oxygen to produce an "activated acid" in the form of the triamide **713**. Lactonization of **713** afforded the α-alkylidene-γ-butyrolactones **714** as a 1.1:1 mixture of Z and E isomers. This reaction sequence is noteworthy in that singlet oxygen unmasks the oxazole and simultaneously oxidizes the phenylselenide. Similarly, conversion of 2,5-diphenyloxazole **715** to **716**, followed by oxidation, gave **717** in 28% yield from **715**. Macrolactonization of **717** using CPTS then produced 7-heptanolide **718** in good yield.[288]

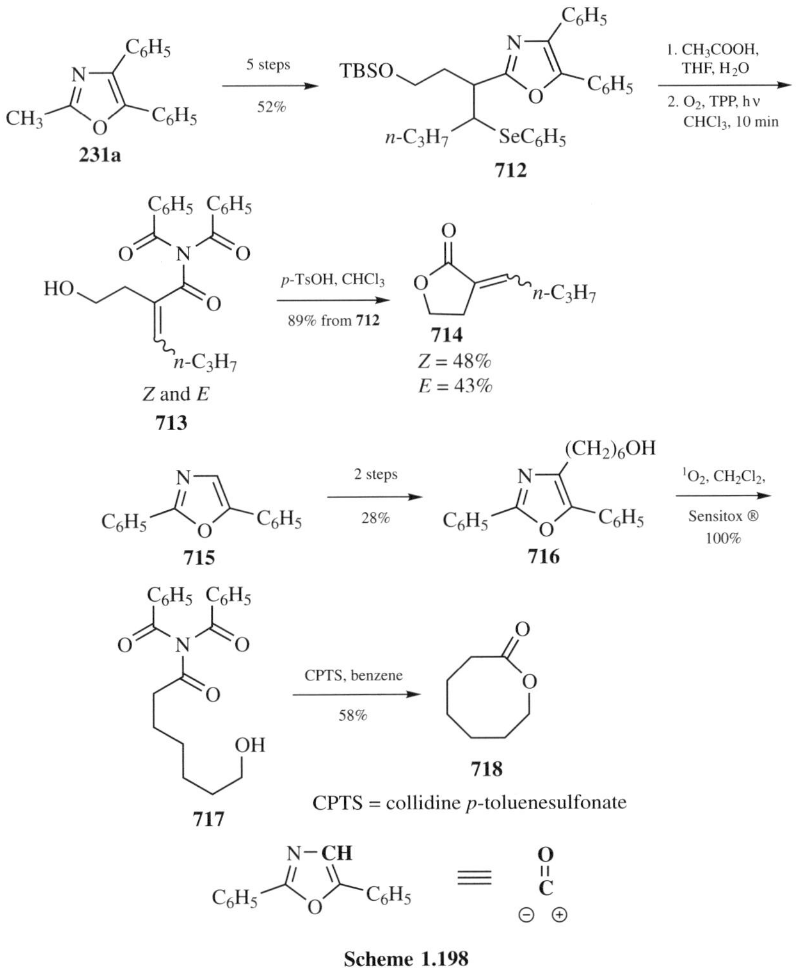

Scheme 1.198

Examples of highly substituted oxazoles that have been converted to key intermediates for syntheses of (+)-antimycin A$_3$ **722**[286] and (±)-pyrenolide C **726**[12,287] are shown in Schemes 1.199 and 1.200.

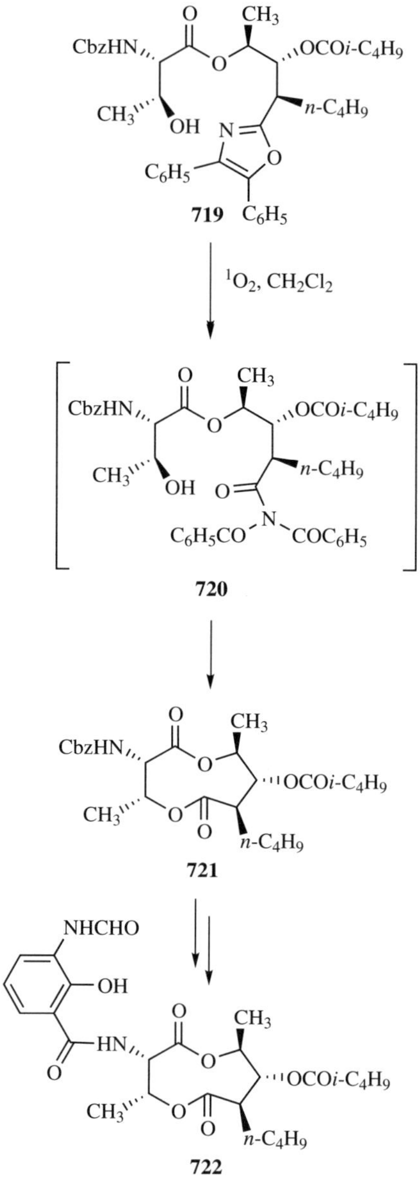

(+)-antimycin A_3

Scheme 1.199

SmithKline chemists[306] adapted singlet oxygen oxidation methodology to prepare (−)-methyl (5*S*), (6S)-oxido-7-hydroxyheptanoate **730** ($n = 1$), a key fragment of leukotrienes A–E (Scheme 1.201). 4,5-Diphenyl-2-methyloxazole **231a** was converted to the allylic alcohol **727** in five or six straightforward steps. Katsuki and Sharpless's[307] asymmetric epoxidation of **727** gave **728** only if the

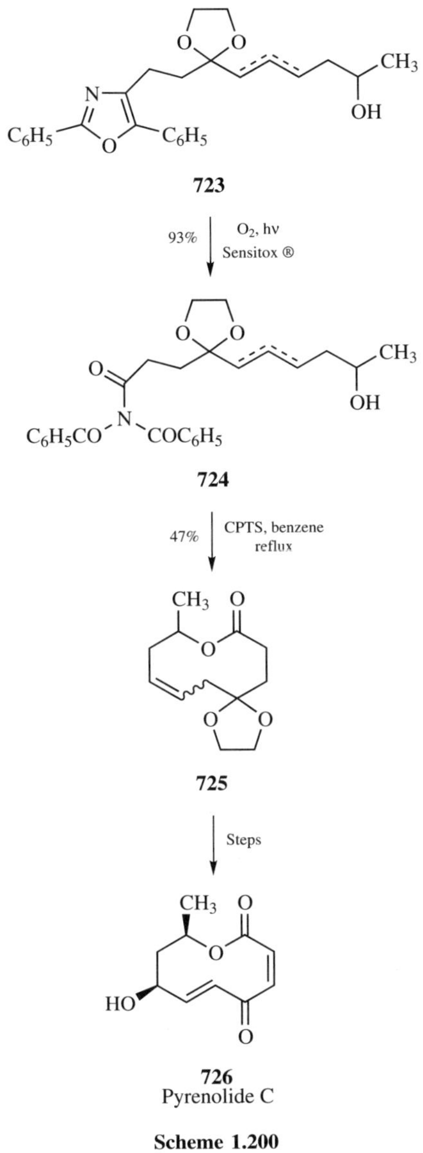

Scheme 1.200

ω-ester was masked as an oxazole. Both ω-methyl and ω-t-butyl esters interfered with this reaction. Acetylation of **728** and reaction with singlet oxygen gave **729**, which was converted to **730** ($n = 1$). This methodology was also successful for **730** ($n = 2$) but failed for **730** ($n = 0$).

Oxazoles also react with dilute solutions of ozone to produce products similar to those obtained from singlet oxygen. In 1997 Kashima and co-workers[12] reviewed in detail their investigations on the ozonolysis of oxazoles. This group[292] initially treated 2,5-diphenyloxazole **731a**, 2-methyl-5-phenyloxazole **731b**, and 5-methyl-2-phenyloxazole **731c** with a dilute solution of ozone in methylene chloride followed by NaBH$_4$/C$_2$H$_5$OH and isolated benzoic acid and acetic acid (Scheme 1.202). Subsequently, they modified their reaction conditions and added methylamine directly to the reaction mixture after ozonolysis to isolate the corresponding *N*-methylamide.[293] The authors proposed that ozonolysis of **731a** and **732a–c** initially gave a mixed anhydride, which was then converted to the observed *N*-methylamides **733** and **734**.

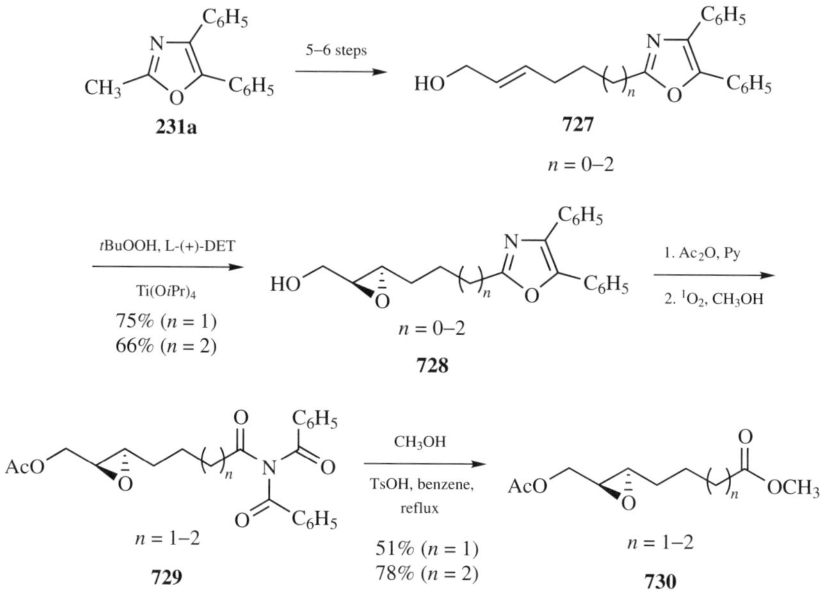

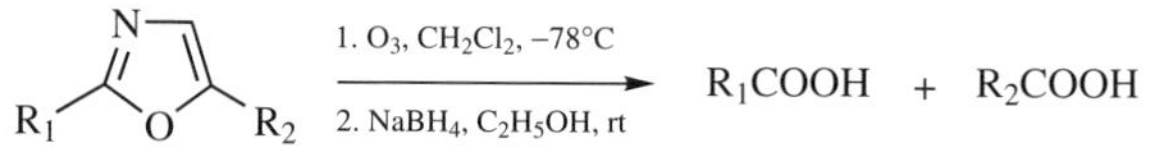

Scheme 1.201

731a R$_1$ = R$_2$ = C$_6$H$_5$
731b R$_1$ = CH$_3$, R$_2$ = C$_6$H$_5$
731c R$_1$ = C$_6$H$_5$, R$_2$ = CH$_3$

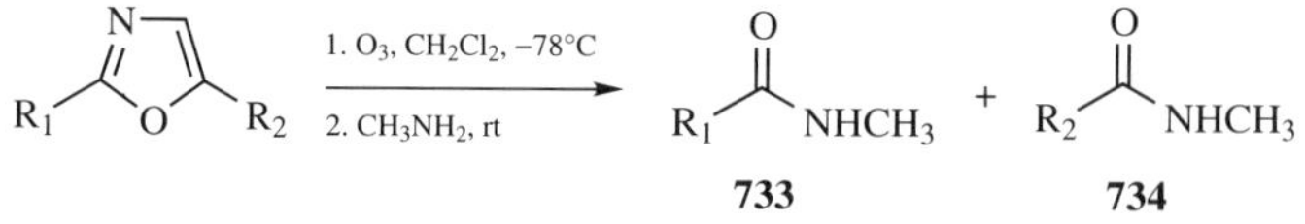

731a R$_1$ = R$_2$ = C$_6$H$_5$ — 96%
732a R$_1$ = 4-Cl-C$_6$H$_4$; R$_2$ = C$_6$H$_5$ — 39% / 34%
732b R$_1$ = 4-CH$_3$-C$_6$H$_4$; R$_2$ = C$_6$H$_5$ — 19% / 25%
732c R$_1$ = 4-CH$_3$O-C$_6$H$_4$; R$_2$ = C$_6$H$_5$ — — / 89%

Scheme 1.202

Kashima's group[294] described a convenient formylation reaction using the ozonolysate of oxazole **1** iteslf. IR and mass spectral evidence confirmed the presence of carbon dioxide in the ozonolysate. This suggested that formic anhydride (unstable above $-40°C$) was not the active formylating agent. Instead, the authors proposed the formic anhydride equivalent, N-formylformamide, **736**. Mechanistically, **736** could arise as shown in Scheme 1.203. This formylation methodology has significant advantages, because formyl chloride and formic anhydride are not readily available. In addition, **1** is commercially available and the by-product formamide is easily removed from the products during workup. The reaction is quite general and convenient via a simple workup procedure. Formylation of alcohols can be achieved similarly. Examples of formamides and formate esters are shown in Table 1.52. Kashima and Arao[295] extended this methodology to include deuterioformylation.

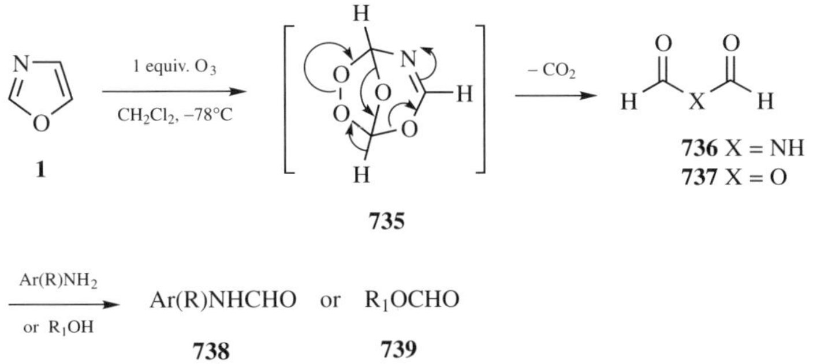

Scheme 1.203

TABLE 1.52. FORMYLATION OF AMINES AND ALCOHOLS VIA N-FORMYLFORMAMIDE[a]

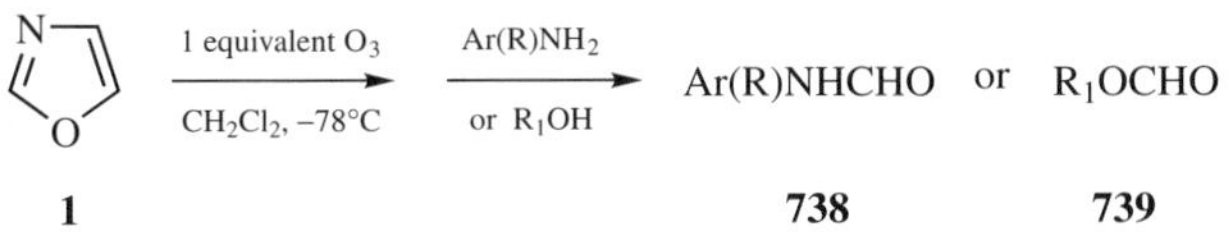

Entry	Ar(R)NH$_2$	R$_1$OH	Yield **738** (%)	Yield **739** (%)
1	C$_6$H$_5$NH$_2$	—	87	—
2	4-CH$_3$O–C$_6$H$_4$NH$_2$	—	88	—
3	4-CH$_3$–C$_6$H$_4$NH$_2$	—	87	—
4	4-Cl–C$_6$H$_4$NH$_2$	—	73	—
5	C$_6$H$_5$NHCH$_3$	—	74	—
6	C$_6$H$_5$CH$_2$CH$_2$NH$_2$	—	99	—
7	pyrrolidine	—	79	—
8	H$_2$NCH(Bn)COOCH$_3$	—	99	—
9	—	C$_6$H$_5$CH$_2$OH	—	55
10	—	1-menthol	—	60
11	—	cholesterol	—	59
12	—	n-C$_4$H$_9$OH	—	76

[a] Data from ref. 294.

A mechanistic rationale that explains the formation of acyl isocyanates, mixed anhydrides, and N-acylamides from ozonolysis of substituted oxazoles was presented by Kashima's group.[296,298] In addition, Kashima and co-workers[297] adapted their methodology for peptide synthesis (Scheme 1.204). Here, the starting amino acid was readily converted to a 5-phenyloxazole analogue **740** in several straightforward steps. Ozonolysis of **740** followed by reaction with glycine methyl ester

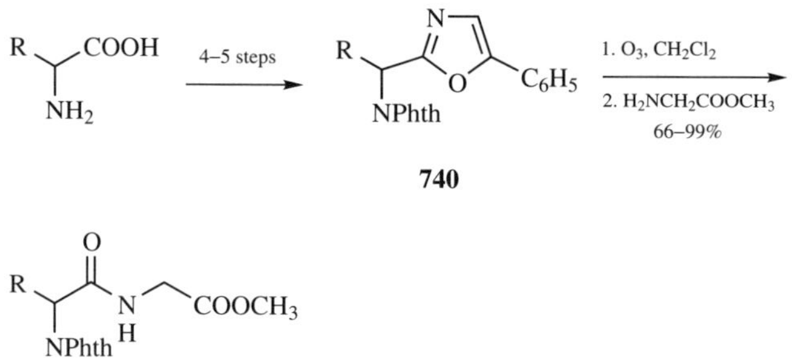

Scheme 1.204

gave the protected dipeptides **741** in good to excellent yield. Thus the oxazole ring served as a carboxylic acid surrogate and as an activating group. This methodology was nicely complementary to the pioneering work from Wasserman's group[308] nearly 10 years earlier.

Wasserman and co-workers[289] also investigated ozonolysis of oxazoles. They found that ozonolysis of 2,4,5-triphenyloxazole **231c** afforded only 31% of the expected tribenzamide **743**, together with an unstable by-product **744**, which regenerated **231c** on treatment with triphenylphosphine (Scheme 1.205). The unknown was not 2,4,5-triphenyloxazole N-oxide **745**, as shown by comparison with an authentic sample. The authors proposed the oxazole 4,5-epoxide **744** for the unknown, although they could not rule out the oxaziridine **746**.

Ito and co-workers[299–302] applied dye-sensitized photooxidation of oxazoles to prepare positive-working photoresists. The authors prepared the key oxazole monomer, 4,5-diphenyl-2-(3-methacryloyloxypropyl)oxazole **747**, from **231a** in three straightforward steps (Scheme 1.206). Co-polymerization of **747** with methyl methacrylate, N-isopropylacrylamide, or N,N-dimethylacrylamide produced **748a**, **748b**, and **748c**, respectively. Oxygen saturated solutions of the individual co-polymers containing meso-tetraphenylporphin (TPP) were then exposed to visible light. This effected the photooxidation of the oxazole moiety to the corresponding triamide. Subsequent reaction with ethanolamine produced **749a**, **749b**, and **749c**. These modified co-polymers were then soluble in aqueous media, allowing the creation of a positive-working photoresist image by dissolution of the exposed areas of the film. Inclusion of different sensitizer dyes led to panchromatic

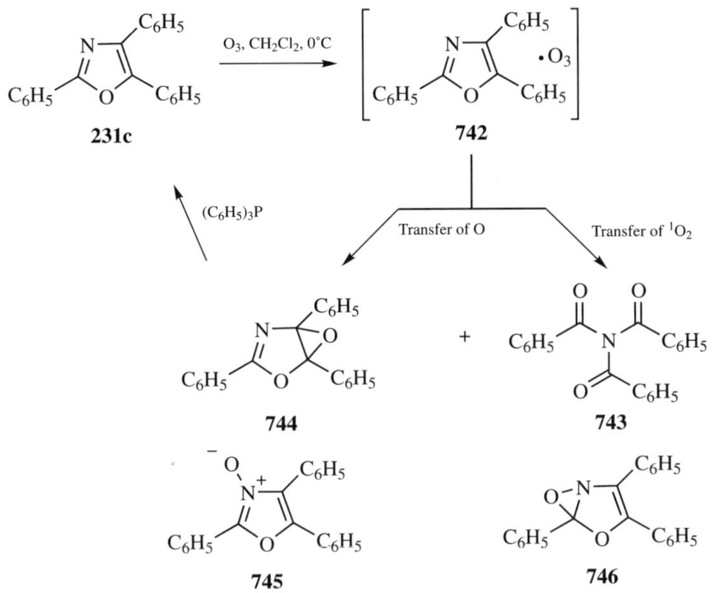

Scheme 1.205

photoresists. Further refinements in this process include incorporation of ionizable amino acids in place of ethanolamine.[302]

1.4.5. Cycloaddition Reactions and Sigmatropic Rearrangements

Oxazoles readily participate in cycloaddition reactions as dienophiles and as dienes in Diels-Alder reactions, and suitably substituted oxazoles participate in sigmatropic rearrangements (e.g., aza-Claisen rearrangements). In particular, the Diels-Alder reaction of oxazoles is one of the most widely explored and synthetically useful reactions, and as such, it has been used extensively both in natural product syntheses and to convert oxazoles to other heterocyclic ring systems. For example, a partial list of heterocyclic systems readily accessible from oxazoles via Diels-Alder reactions or other cycloadditions include pyridines; hydroxypyridines; isoindoles; pyridazines; tetrahydronaphthyridines; benzo[h]-1,6-naphthyridines; benzopyrano[3,4-b]pyridines; 2-substituted, 2,4-disubstituted, 3,4-disubstituted, and 2,3,4-trisubstituted furans; isobenzofurans; 1,2,4-triazolines; 1,2,4-oxadiazoles; oxazolines; and thiazolines. See Chapter 3 for a comprehensive discussion of the diene and dienophilic character of oxazoles in Diels-Alder reactions. In addition, Chapter 4 is devoted to the synthesis and reactions of mesoionic oxazoles, a rich source of new and interesting heterocycles.

This section focuses on cycloaddition reactions wherein an oxazole reacts as an exo-diene. In addition, selected examples of formal [3 + 2] cycloadditions of

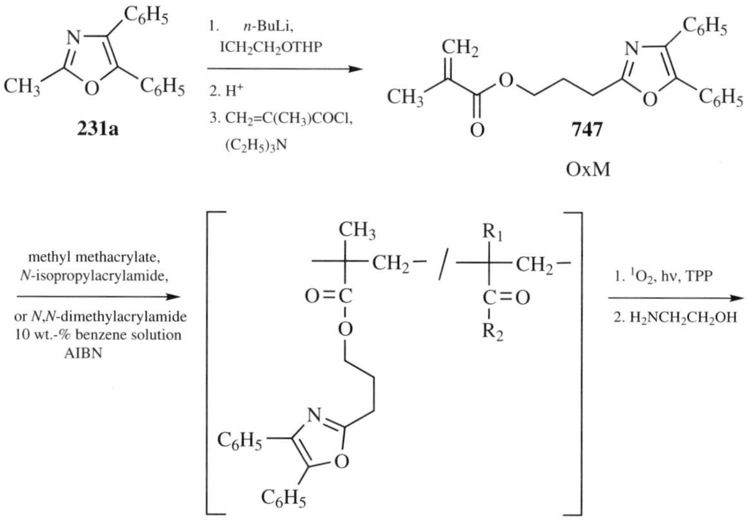

748a pOxM-co-methyl methacrylate; $R_1 = CH_3$, $R_2 = OCH_3$
748b pOxM-co-N-isopropylacrylamide; $R_1 = H$, $R_2 = NHCH(CH_3)_2$
748c pOxM-co-N,N-dimethylacrylamide; $R_1 = H$, $R_2 = N(CH_3)_2$

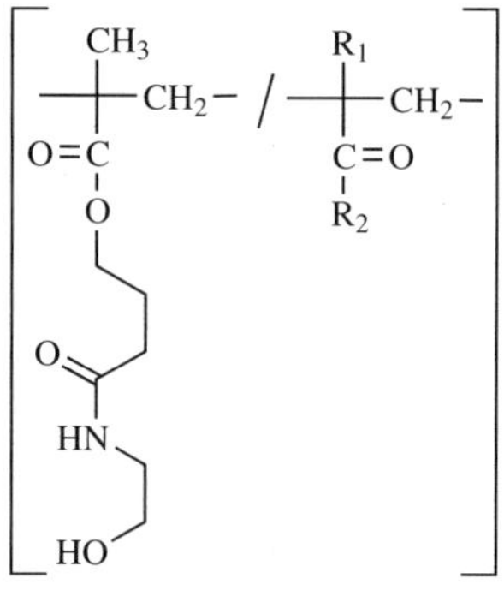

749a pOxM-co-methyl methacrylate; $R_1 = CH_3$, $R_2 = OCH_3$
749b pOxM-co-N-isopropylacrylamide; $R_1 = H$, $R_2 = NHCH(CH_3)_2$
749c pOxM-co-N,N-dimethylacrylamide; $R_1 = H$, $R_2 = N(CH_3)_2$

Scheme 1.206

oxazoles and dienophiles/heterodienophiles involving zwitterionic intermediates and sigmatropic rearrangements are also discussed.

Storr and co-workers[309] at ICI reported the first example of an oxazole-o-quinodimethane in 1990 (Scheme 1.207). FVP of 5-[[(p-chlorobenzoyl)oxy]-methyl]-4-methyloxazole **750** generated oxazole-o-quinodimethane **751**. Trapping **751** with thiophenol gave rise to a mixture of 4-methyl-5-[(phenylthio)methyl]oxazole **752** and 5-methyl-4-[(phenylthio)methyl]oxazole **753**, whereas reaction of **751** with sulfur dioxide produced 4,6-dihydro-thieno[3,4-d]oxazole 5,5-dioxide

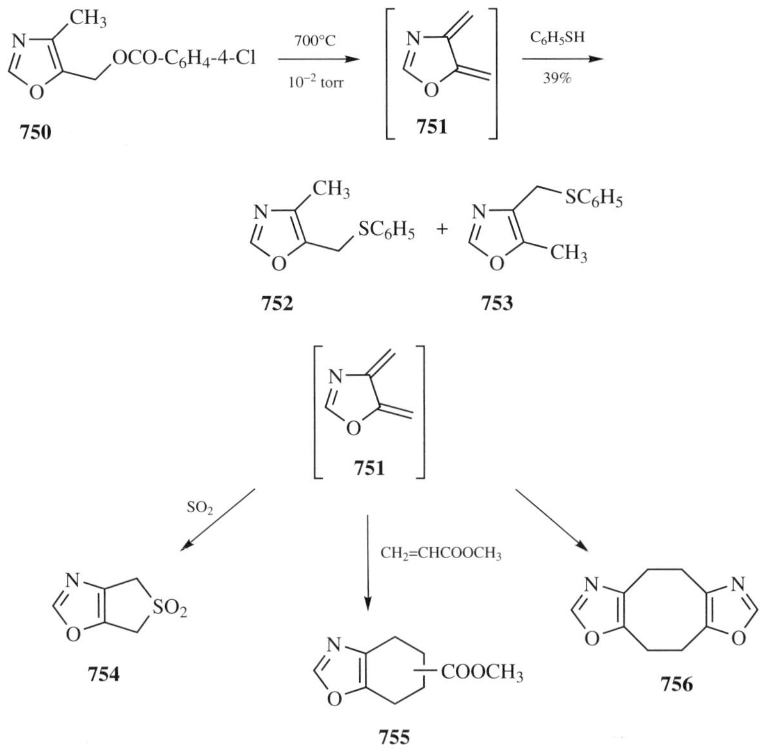

Scheme 1.207

754 in low yield. Generation of **751** in the presence of methyl acrylate afforded the Diels-Alder adduct **755** as a mixture of regioisomers, whereas the head-to-head [4 + 4] dimer **756** was isolated upon warming **751** to room temperature in the absence of a trapping agent.

Chou and co-workers[310,311] investigated 4,6-dihydro-2-substituted-thieno[3,4-d] oxazole 5,5-dioxides **758** both experimentally and theoretically. The authors were initially disappointed to find that 4-bromo-3-sulfolanone **757** did not react with formamide or acetamide to yield **754** or **758** (R = CH$_3$) (Scheme 1.208). This synthetic approach had been successful for preparing 2-substituted thiazolo-3-sulfolenes **759**.[312]

The authors next prepared a series of amidosulfones **760** but were frustrated by the complex mixtures isolated from classical cyclodehydration using Ac$_2$O, H$_2$SO$_4$, SOCl$_2$, P$_2$O$_5$, or polyphosphoric acid. However, heating **760** with PCl$_5$ gave a mixture of a 4,5-bis(chloromethyl)-2-substituted oxazole **761** together with **758**. The product distribution depended on the molar ratio of reactants, reaction temperature, and time. For example, the desired **758** analog (e.g., R = C$_6$H$_5$) was isolated in good yield using a slight excess of PCl$_5$ in CHCl$_3$ at room temperature for 1 h.

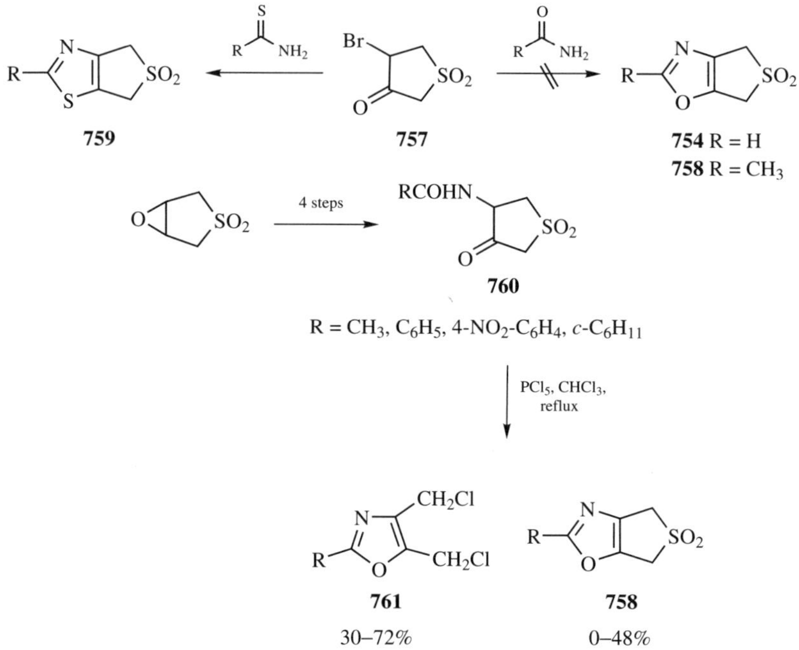

Scheme 1.208

Both **758** and **761** were useful precursors to oxazole-*o*-quinodimethane **762**, although consistently higher yields of cycloadducts were obtained from **758** (Scheme 1.209).

Chao and co-workers[311] also studied **758** (R = C_6H_5) theoretically in an effort to explain the remarkable reactivity they observed experimentally. Their results from ab initio HF/6-31G*//HF/6-31G* calculations indicated that oxazole **1** dearomatizes more readily than furan, pyrazole, isoxazole, or thiazole. This dearomatization is further enhanced by the presence of the C(2)-phenyl substituent in **758**. Moreover, extrusion of SO_2 is associated with significant strain relief for the oxazole-fused sulfolene **758**.

Geerlings and co-workers[313] examined Diels-Alder reactions of ethylene with quinodimethanes, including **751**, using ab initio molecular orbital calculations and density functional theory (DFT).

Eguchi and co-workers[314] extruded SO_2 from **758** (R = C_6H_5) in the presence of fullerene and isolated the 1:1 cycloadduct **767** (Scheme 1.210). The authors found **767** to be photosensitive, which was not unexpected given that C_{60} is known to be a good sensitizer to produce singlet oxygen.[315,316] Indeed, irradiation of a toluene solution of **767** in the presence of oxygen followed by addition of methanol produced the diester **768**. This was attributed to a self-sensitized photooxygenation of the oxazole ring followed by rearrangement and methanolysis.

Ibata and co-workers[317,318] reported that electron-rich oxazoles react with electron-deficient olefins, e.g., tetracyanoethylene (TCNE) to afford cycloadducts

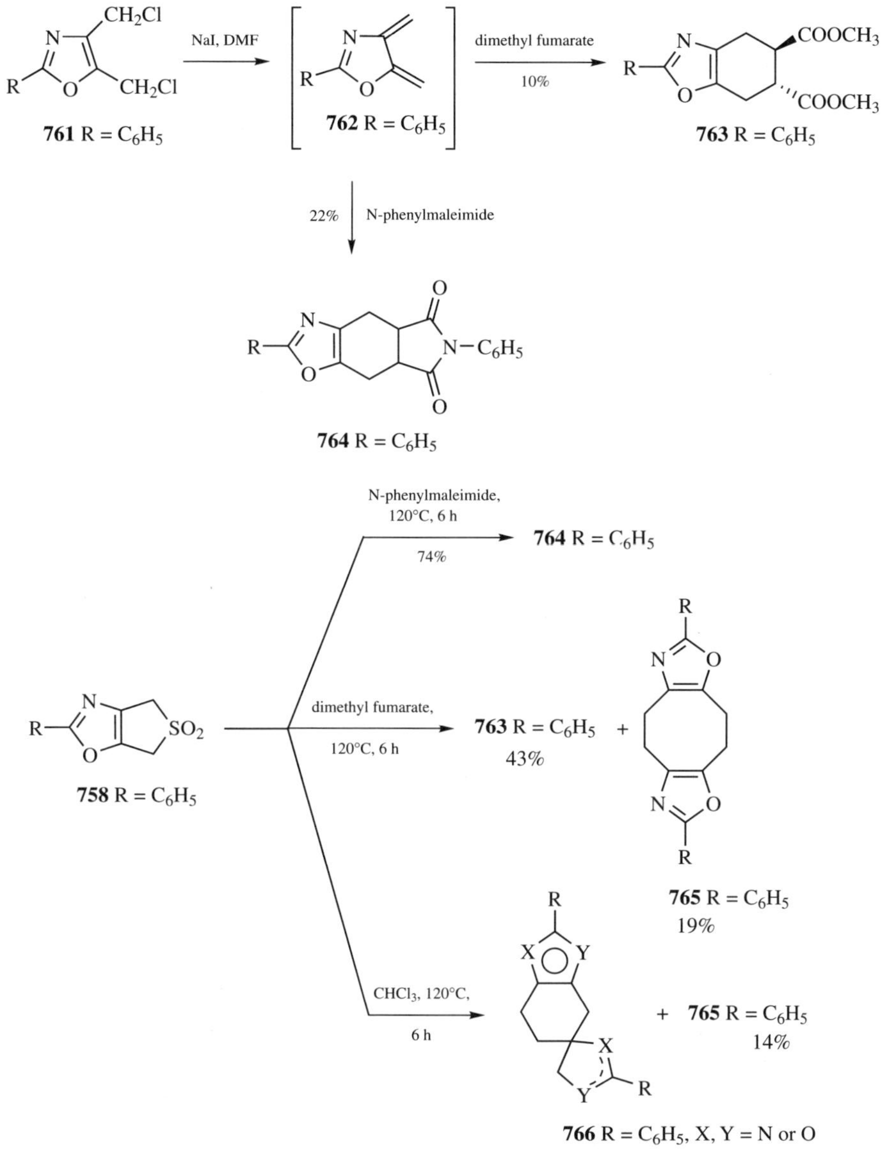

Scheme 1.209

that are not the usual Diels-Alder products.[319] Thus refluxing a 2-alkyl-5-methoxy-4-(4-nitrophenyl)oxazole **769** with TCNE in acetonitrile produced a 2-alkyl-3,3,4,4-tetracyano-4,5-dihydro-5-(4-nitrophenyl)pyrrole-5-carboxylic acid methyl ester **770** in good to excellent yield (Scheme 1.211).

The authors observed that a mixture of **769** (R = CH$_3$) and TCNE afforded an olive solution at room temperature. Heating this solution to reflux discharged the

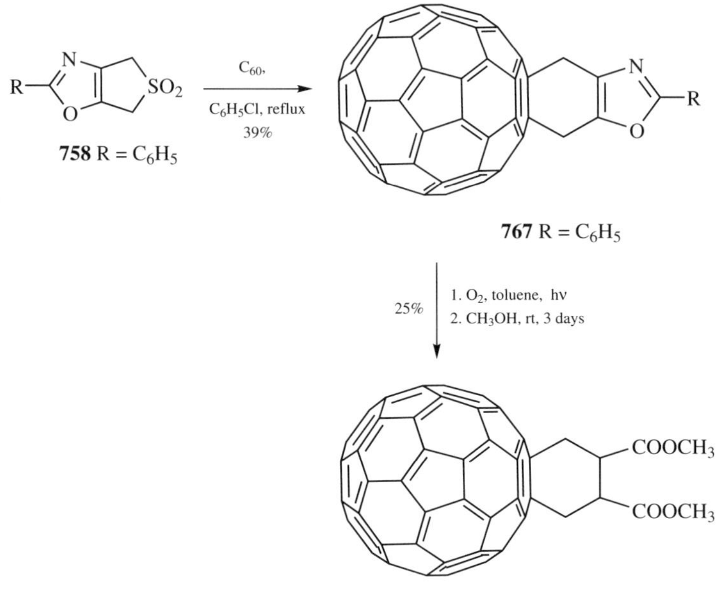

168   Synthesis and Reactions of Oxazoles

Scheme 1.210

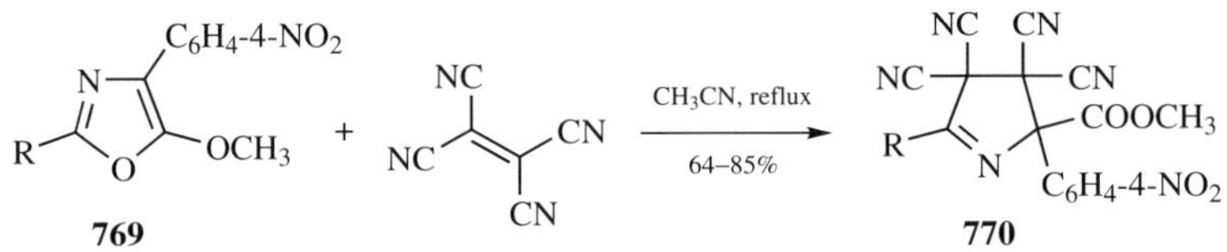

Scheme 1.211

R = CH$_3$, C$_2$H$_5$, i-C$_3$H$_7$, t-C$_4$H$_9$

color and effected formation of **770** (R = CH$_3$). They proposed a mechanism wherein **769** and TCNE first reacted to generate a charge transfer (CT) complex **771** (Scheme 1.212). Once formed, **771** collapsed by electrophilic attack of TCNE at C(2) or C(4) of the oxazole to generate a zwitterionic intermediate, **772** or **773**. Ring opening of **772** or **773** would afford **774** or **775**, either of which could cyclize to afford the isolated product. The reaction was accelerated using polar solvents that also supported the proposed mechanistic rationale. The authors were not able to distinguish between these routes, since efforts to trap **772** and/or **773** were unsuccessful.

Oxazoles typically react with heterodieneophiles including C=O,[320–326] C=S,[327,328] N=O,[328–330] and N=N[194,328–332] in a [3 + 2] cycloaddition reaction to afford "abnormal" products. A few selected examples follow, although additional discussion can be found in Chapter 3.

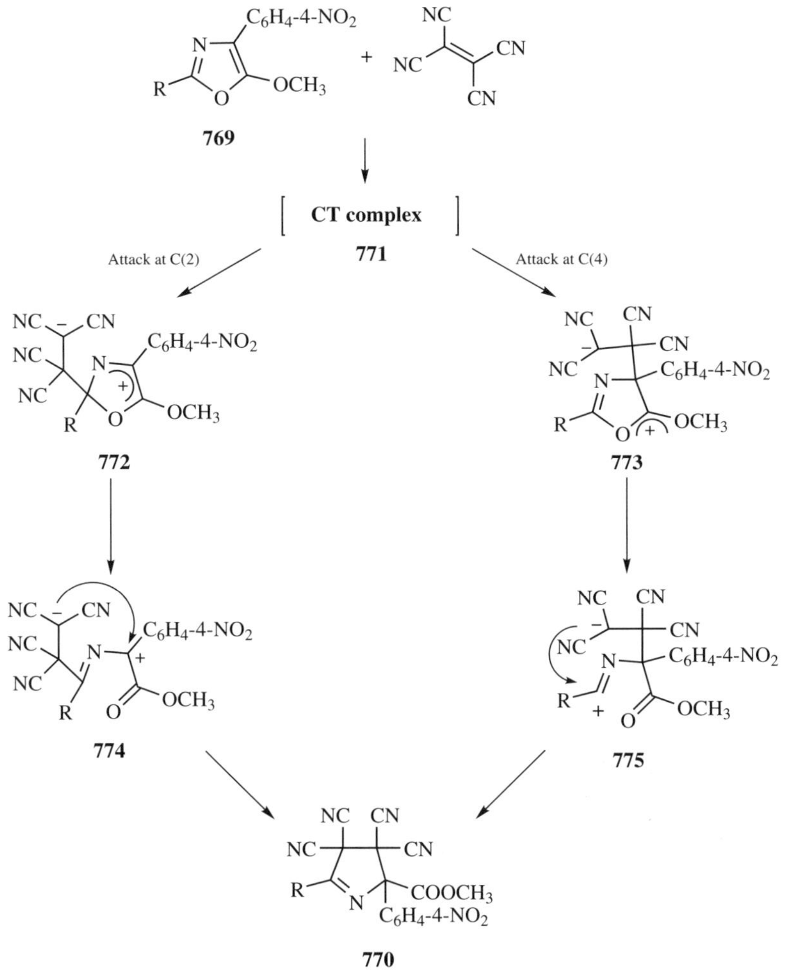

Scheme 1.212

Huth and co-workers[194] described the first examples of [3 + 2] cycloadducts of oxazoles and N=N dienophiles. The starting 5-aryloxazoles **776** (R = R$_1$ = H, R$_2$ = Ar-X, X = H, 4-Cl, 4-OCH$_3$, 4-CF$_3$, 4-CN, 4-NO$_2$, 3,4-di-OCH$_3$) were readily prepared from a substituted benzaldehyde and TosMIC. Cycloaddition of **776** with 4-phenyl-1,2,4-triazoline-3,5-dione (PTAD) **777** was accomplished in dioxane at room temperature to afford the corresponding Δ^3-1,2,4-triazoline-1,2-dicarboximides **778** in low to good yield (Scheme 1.213). The authors did not isolate an analog **778** from **776** (R = R$_1$ = H, R$_2$ = Ar-X, X = 4-CN or 4-NO$_2$) and PTAD even under forcing conditions, e.g., dioxane/reflux. PTAD also reacted with 2-alkyl(aryl)-5-aryloxazoles **776** (R = CH$_3$, i-C$_3$H$_7$, C$_6$H$_5$, R$_1$ = H, R$_2$ = Ar-X, X = H or 3,4-di-OCH$_3$) under similar conditions to produce **778** derivatives in comparable yield.

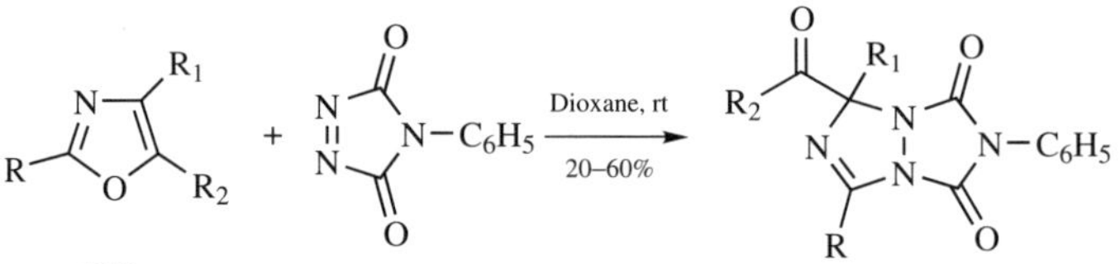

776 777 778

R = R_1 = H
R_2 = Ar-X

↑ TosMIC, K_2CO_3, CH_3OH

X-ArCHO

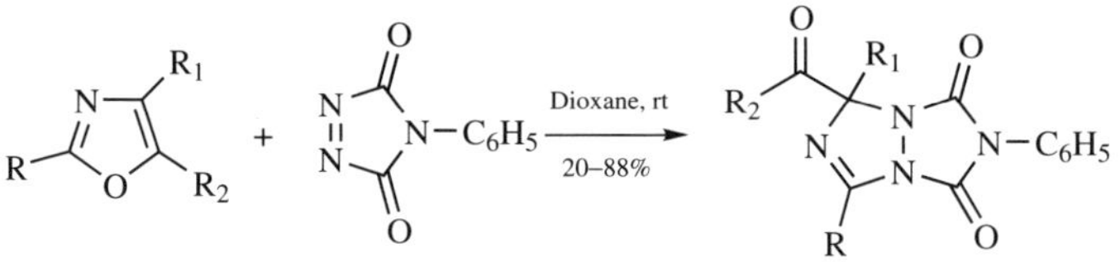

776 777 778

R = CH_3, n-C_3H_7, C_6H_5; R_1 = H; R_2 = Ar-X

Scheme 1.213

Ibata and co-workers[331,332] extended Huth's work to include 5-alkoxyoxazoles **776** ($R_2 = OCH_3$, OC_2H_5) and developed a general high-yield synthesis of *N*-phenyl-5-acyl-Δ^3-1,2,4-triazoline-1,2-dicarboximides **779** for evaluation as potential fungicides (Scheme 1.214). The structure of **779** (R = 4-CH_3-C_6H_4, $R_1 = CH_3$, $R_2 = OCH_3$) was confirmed by single-crystal X-ray crystallography. Representative examples of **779** are shown in Table 1.53.

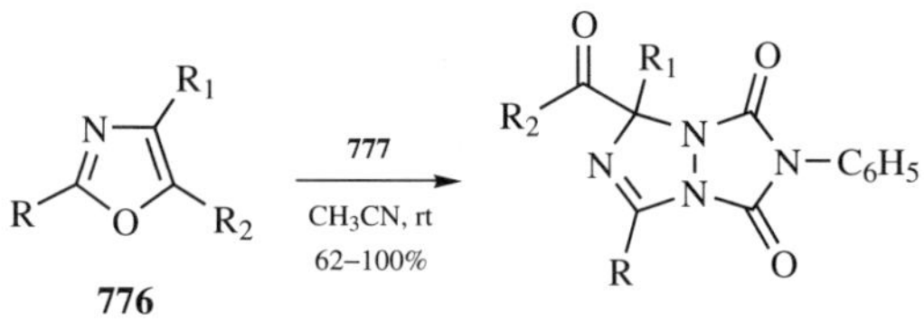

776 779

R = alkyl, aryl
R_1 = $COOC_2H_5$, CH_3, C_6H_5, 4-NO_2-C_6H_4
R_2 = OCH_3, OC_2H_5, CH_3, C_6H_5, 4-CH_3O-C_6H_4

Scheme 1.214

Similarly, both Hassner and Fischer[320] and Ibata's group[332] reported that **776** reacted with diethyl azodicarboxylate (DEAD) to afford 1,2-dicarbethoxy-3,5-disubstituted- and 3,5,5-trisubstituted Δ^3-1,2,4-triazolines **780** (Scheme 1.215, Table 1.54). Mechanistically, Ibata's group[332] favored a stepwise addition process analogous to that for TCNE (vide supra). Hassner and Fischer[320] also proposed

TABLE 1.53. N-PHENYL-5-ACYL-Δ^3-1,2,4-TRIAZOLINE-1,2-DICARBOXIMIDES FROM SUBSTITUTED OXAZOLES AND PTAD

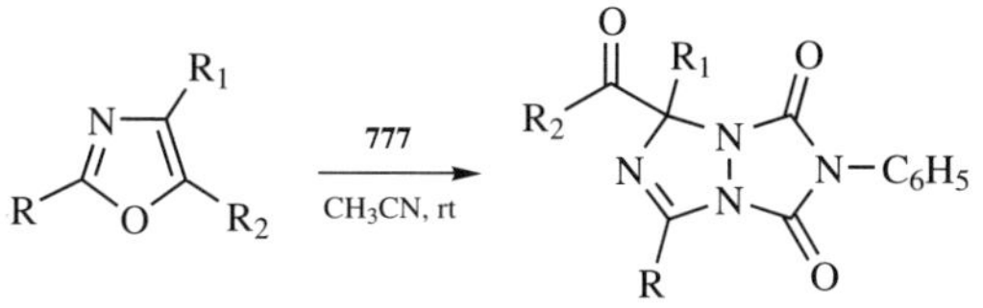

776 →(777, CH$_3$CN, rt)→ **779**

Figure 1.55

Entry	R	R$_1$	R$_2$	Yield (%)	Reference
1	CH$_3$	COOC$_2$H$_5$	OC$_2$H$_5$	99	331
2	C$_2$H$_5$	4-NO$_2$–C$_6$H$_4$	OCH$_3$	100	331
3	t-C$_4$H$_9$	4-NO$_2$–C$_6$H$_4$	OCH$_3$	98	331
4	C$_6$H$_5$	4-NO$_2$–C$_6$H$_4$	OCH$_3$	99	331
5	CH$_3$	H	4-CH$_3$O–C$_6$H$_4$	89	332
6	CH$_3$	CH$_3$	CH$_3$	62	332
7	CH$_3$	C$_6$H$_5$	C$_6$H$_5$	78	332[a]

[a] Reaction conducted at 70°C for 24 h.

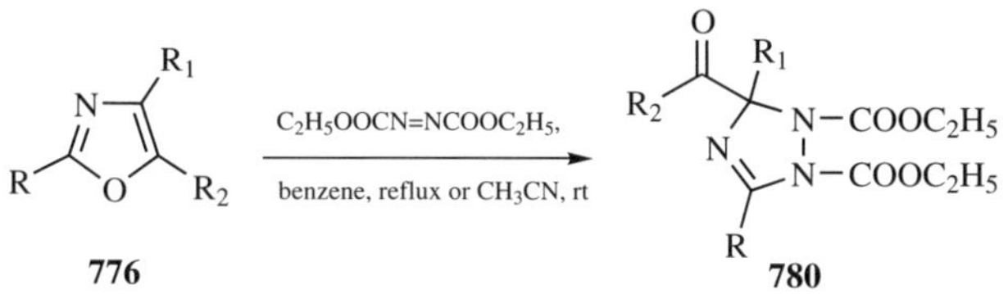

776 →(C$_2$H$_5$OOCN=NCOOC$_2$H$_5$, benzene, reflux or CH$_3$CN, rt)→ **780**

Scheme 1.215

such a process but could not rule out alternative mechanistic pathways with certainty, since DEAD is a symmetrical reagent.

Ibata and co-workers[329,330] described the reaction of nitroso compounds with oxazoles as a new method of preparing 2,5-dihydro-1,2,4-oxadiazoles. Here, [3 + 2] cycloaddition of a substituted 5-alkoxyoxazole **776** (R$_2$ = OCH$_3$ or OC$_2$H$_5$) with an arylnitroso compound yields a 2,5-dihydro-1,2,4-oxadiazole **781** regioselectively in fair to excellent yield (Scheme 1.216). The authors favored a stepwise mechanism whereby nucleophilic attack by C(2) of the oxazole on the arylnitroso compound produces a zwitterionic intermediate **782**, which ring opens to an imino nitrone **783**. Cyclization of **783** then yields **781**. An alternative mechanism involving nucleophilic attack by C(4) of the oxazole on the nitrosobenzene to generate **784** followed by ring opening to **785** was discounted based on the relative stabilities of **783** and **785**. Representative examples are shown in Table 1.55.

Vedejs and Fields[327] prepared a series of 3-thiazolines **787** from cycloaddition of thioaldehydes and oxazoles **776** (Scheme 1.217). The authors generated the

TABLE 1.54. 1,2-DICARBETHOXY-3,5-DISUBSTITUTED- AND 3,5,5-TRISUBSTITUTED Δ^3-1,2,4-TRIAZOLINES FROM SUBSTITUTED OXAZOLES AND DEAD

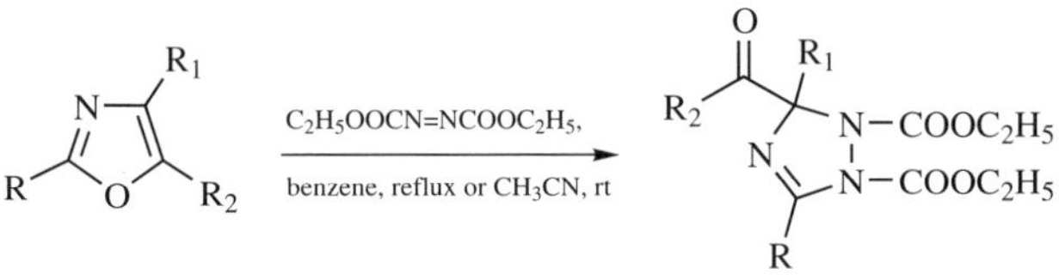

Figure 1.56   776   780

Entry	R	R_1	R_2	Yield (%)	Reference
1	C_6H_5	H	OC_2H_5	99	320[a]
2	C_6H_5	H	$OSi(CH_3)_3$	100	320[a]
3	CH_3	H	OC_2H_5	41	320[a]
4	OC_2H_5	H	C_6H_5	35	320[a]
5	CH_3	C_6H_5	C_6H_5	0	320[a]
6	CH_3	$COOC_2H_5$	OC_2H_5	84	332[b]
7	t-C_4H_9	4-NO_2–C_6H_4	OCH_3	75	332[b]
8	CH_3	CH_3	CH_3	80	332[b]
9	CH_3	C_6H_5	C_6H_5	25	332[b]
10	4-Cl–C_6H_4	CH_3	CH_3	52	332[b]

[a] Reaction conditions: benzene, reflux.
[b] Reaction conditions: CH_3CN, room temperature, or reflux.

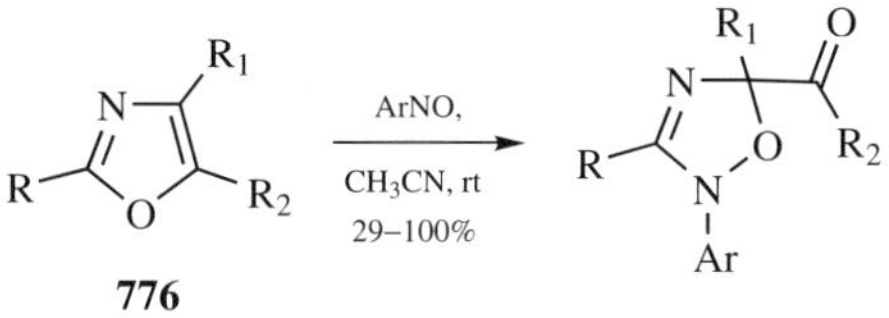

776   781

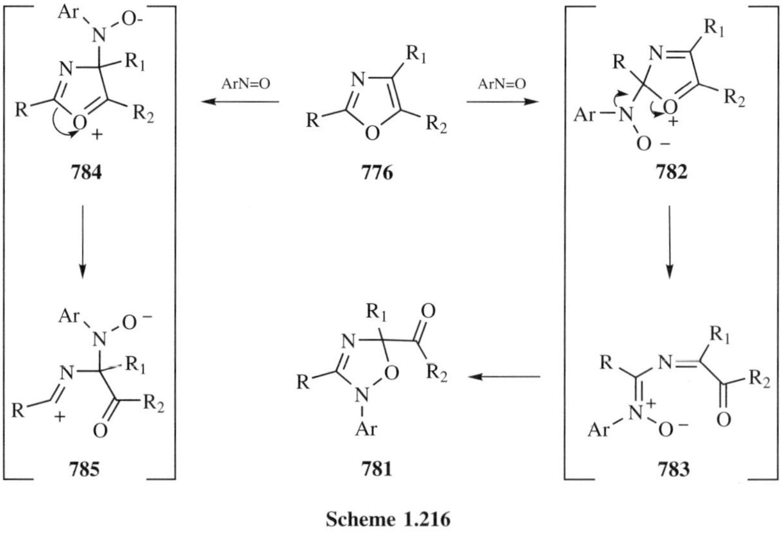

784   776   782

785   781   783

Scheme 1.216

TABLE 1.55. 2,5-DIHYDRO-1,2,4-OXADIAZOLES FROM OXAZOLES AND ARYLNITROSO COMPOUNDS

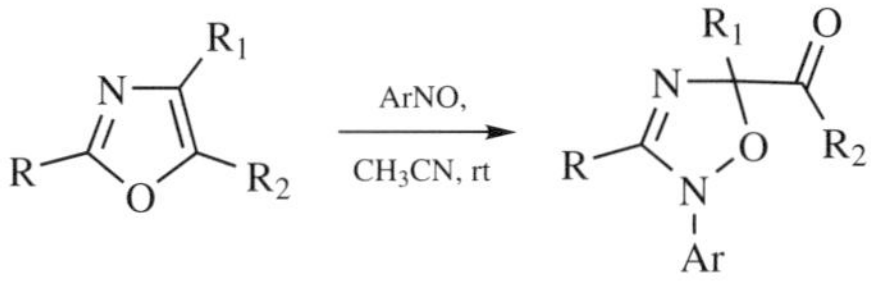

Figure 1.57

Entry	R	R_1	R_2	Ar	Yield (%)	Reference
1	CH_3	4-NO_2–C_6H_4	OCH_3	C_6H_5	98	329[a]
2	4-CH_3–C_6H_4	4-NO_2–C_6H_4	OCH_3	C_6H_5	70	329[a]
3	CH_3	CH_3	OCH_3	C_6H_5	65	329[a]
4	CH_3	H	OC_2H_5	C_6H_5	80	329[a]
5	CH_3	CH_3	OCH_3	4-Cl–C_6H_4	38	330[b]
6	4-CH_3O–C_6H_4	C_6H_5	OCH_3	4-Cl–C_6H_4	100	330[b]
7	4-CH_3O–C_6H_4	CH_3	OCH_3	4-Cl–C_6H_4	99	330[b]
8	4-CH_3O–C_6H_4	C_6H_5	OCH_3	4-CH_3–C_6H_4	98	330[b]
9	CH_3	H	OC_2H_5	4-CH_3–C_6H_4	87	330[b]
10	CH_3	CH_3	CH_3	4-CH_3–C_6H_4	29	330[b]

[a] Reaction conditions: CH_3CN, room temperature.
[b] Reaction conditions: CH_3CN, room temperature.

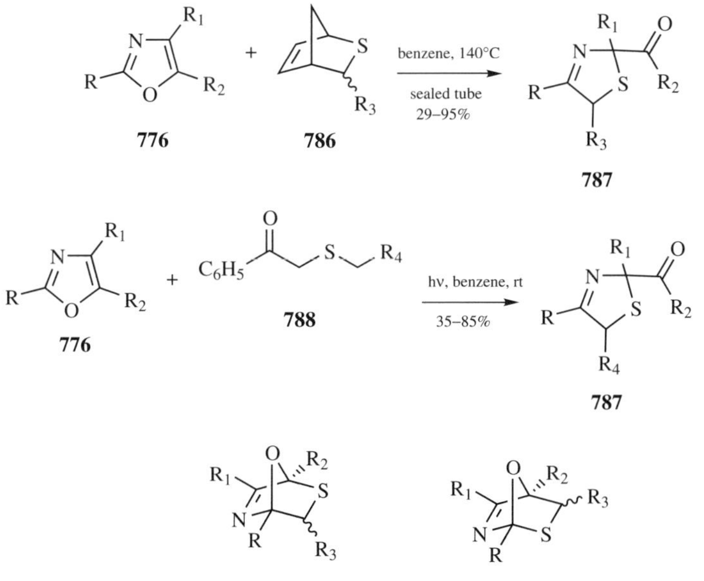

Scheme 1.217

TABLE 1.56. 3-THIAZOLINES FROM OXAZOLES AND THIOALDEHYDES[a]

Entry	R	R_1	R_2	R_3	R_4	Conditions[b]	Yield (%)
1	CH_3	H	OCH_3	H	—	A	81
2	CH_3	H	OCH_3	CH_3	—	A	59[c]
3	CH_3	H	OCH_3	$CH_2CH_2C_6H_5$	—	A	95[c]
4	CH_3	H	OCH_3	C_6H_5	—	A	70[c]
5	CH_3	H	OCH_3	$COOC_2H_5$	—	A	29[c]
6	C_6H_5	CH_3	OCH_3	H	—	A	25
7	C_6H_5	CH_3	OCH_3	—	H	B	55
8	C_6H_5	CH_3	OCH_3	$CH_2CH_2C_6H_5$	—	A	92[c]
9	C_6H_5	CH_3	OCH_3	—	$CH_2CH_2C_6H_5$	B	67[c]
10	C_6H_5	CH_3	OCH_3	C_6H_5	—	A	93[c]
11	C_6H_5	CH_3	OCH_3	—	C_6H_5	B	50[c]

[a] Data from ref. 327.
[b] A, 135–140°C or 105–115°C; B, hv, benzene, room temperature.
[c] Product is a mixture of diastereomers.

requisite thioaldehydes by thermolysis of a 3-substituted 2-thiabicyclo[2.2.1]hept-5-ene **786** or by photolysis of a phenacyl sulfide **788**. In general, they found thermolysis of **786** to be the method of choice for preparing the thioaldehydes. The authors could not unequivocally establish the mechanism of this reaction, although they favored formation of the Diels-Alder adducts **789** or **790** in the first step. Selected examples of **787** prepared via this methodology are shown in Table 1.56.

Hassner and Fischer[328] described an intramolecular cycloaddition of heterodienophiles and oxazoles. Here, an oxazole tethered with a suitably positioned aldehyde, thioaldehyde, or N-carbethoxyurethane afforded the expected heterocyclic product. Examples of each type of reaction are shown in Scheme 1.218.

Refluxing a toluene solution of 5-(5-ethoxy-4-methyloxazol-2-yl)pentanal **791b** gave 2-methyl-2-carbethoxycyclohexa[d]-2,7a-dihydrooxazole **792b** in 78% yield. In contrast, **791a** was unreactive after prolonged reflux in toluene and decomposed in refluxing xylene.

Reaction of **791a** or **791b** with bis-(trimethylsilyl)disulfide gave the thioaldehydes **793a** and **793b**, which spontaneously cyclized at room temperature to yield 2-methyl-2-carbethoxycyclopenta[d]-2,6-dihydrothiazole **794a** and 2-methyl-2-carbethoxycyclohexa[d]-2,7a-dihydrothiazole **794b**.

Conversion of **791a** to **795a** with ethyl urethane followed by refluxing in the presence of p-toluenesulfonic acid gave 2-methyl-2-carbethoxycyclopenta[d]-2,6a-dihydroimidazole **796a** in 43% yield. The analogous reaction with **795b** gave a poor yield of 2-methyl-2-carbethoxycyclohexa[d]-2,7a-dihydroimidazole **796b**,

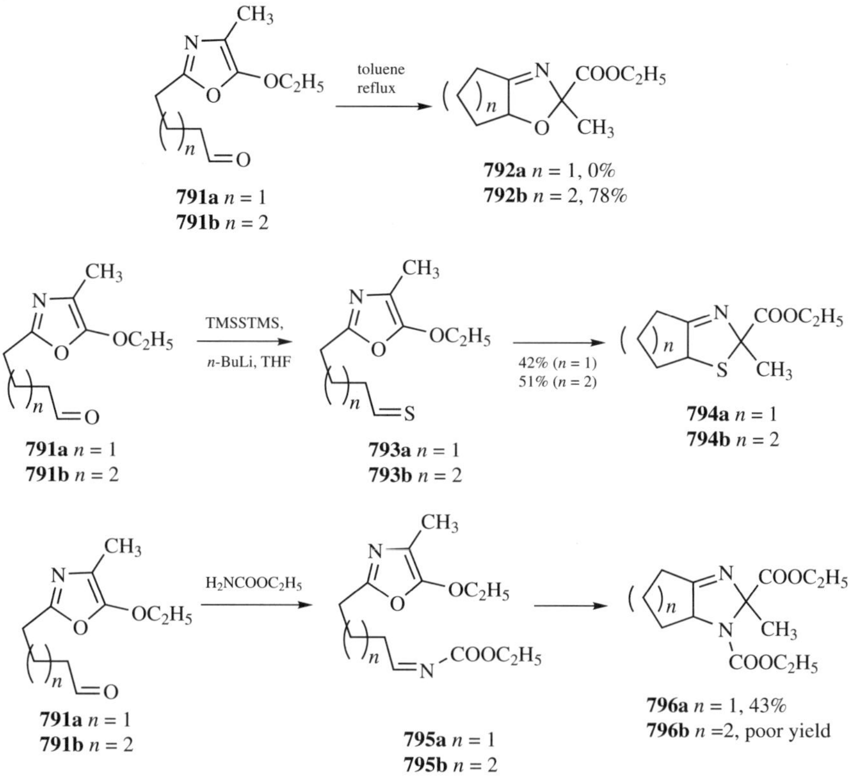

Scheme 1.218

owing to a competitive cycloaddition during the synthesis of **795b**. Among four possible mechanisms, the authors favored an initial Diels-Alder reaction to afford the unstable adduct **797** followed by fragmentation and ring opening to a 1,3-dipole **799**, which recyclized to the observed products (Scheme 1.219).

Hassner and Fischer[320] also found that 5-ethoxy-2-phenyloxazole **15** ($R_1 = C_6H_5$, $R_2 = H$, $R_3 = OC_2H_5$) reacted with diethyloxomalonate to give several products, **800–802**, depending on the temperature (Scheme 1.220). In contrast, 2-ethoxy-5-phenyloxazole **15** ($R_1 = OC_2H_5$, $R_2 = H$, $R_3 = C_6H_5$) reacted with diethyloxomalonate only in the presence of $BF_3 \cdot OEt_2$ to afford **803**.

Ibata and co-workers[321] investigated the [3 + 2] cycloaddition of oxazoles with diethyloxomalonate and reported that the regiochemistry could be controlled using a Lewis acid (Scheme 1.221). Thus a 5-alkoxy-2-aryloxazole **15** ($R_1 = $ aryl, $R_2 = $ H, alkyl, or aryl, $R_3 = $ alkoxy) reacted readily and regioselectively with diethyloxomalonate in the presence of tin(IV) chloride to afford exclusively a 2-oxazoline-4,5,5-tricarboxylate **804** in high yield. In contrast, a 5-alkoxy-2-methyl-oxazole **15** ($R_1 = $ methyl, $R_2 = $ alkyl or aryl, $R_3 = $ alkoxy) reacted regioselectively with diethyloxomalonate under high pressure or elevated temperature to afford exclusively a 3-oxazoline-2,5,5-tricarboxylate **805** in widely varying yield.

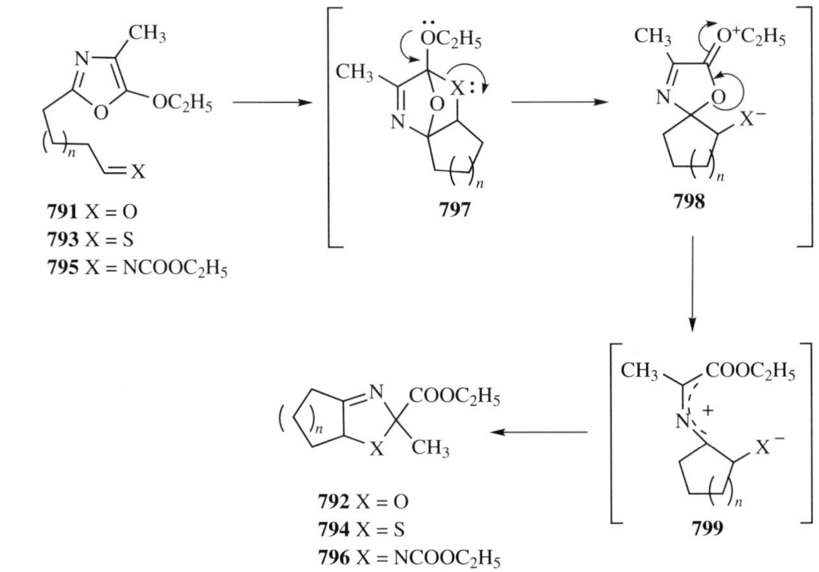

791 X = O
793 X = S
795 X = NCOOC$_2$H$_5$

797

798

792 X = O
794 X = S
796 X = NCOOC$_2$H$_5$

799

Scheme 1.219

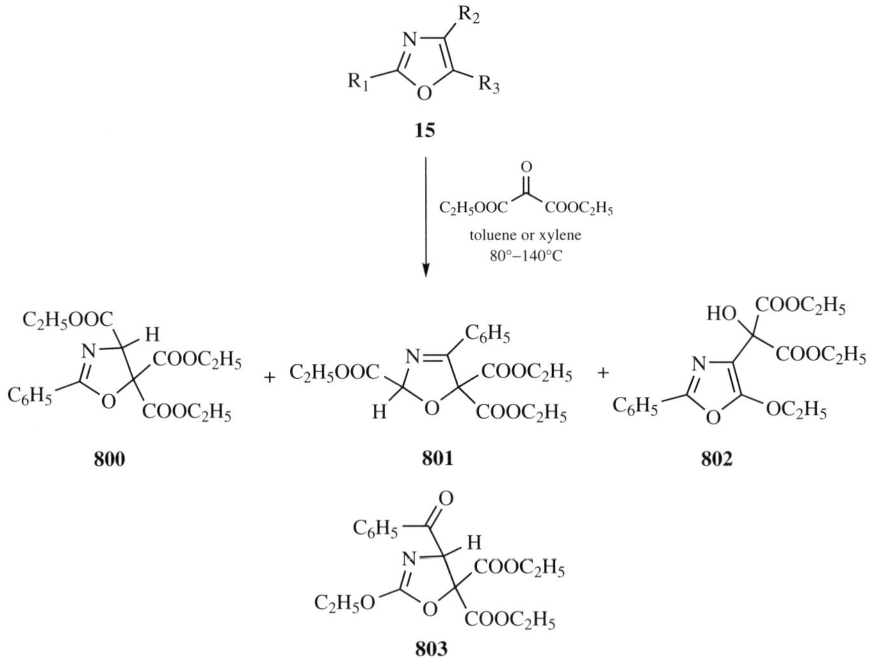

15

800

801

802

803

Scheme 1.220

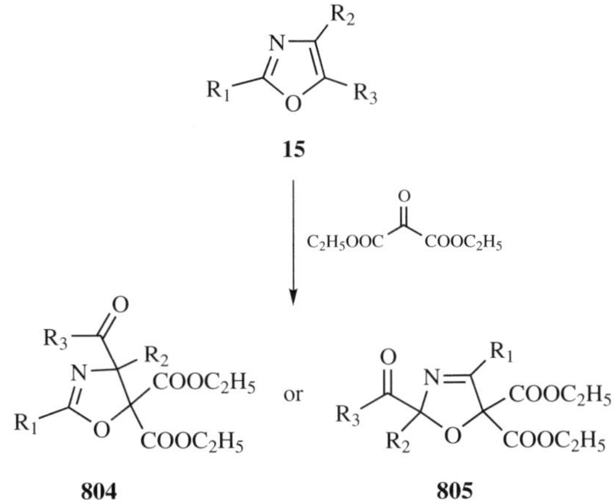

Reactions of Oxazoles 177

Scheme 1.221

TABLE 1.57. 2-OXAZOLINE-4,5,5-TRICARBOXYLATES AND
3-OXAZOLINE-2,5,5-TRICARBOXYLATES FROM [3 + 2] CYCLOADDITION
OF OXAZOLES AND DIETHYLOXOMALONATE[a]

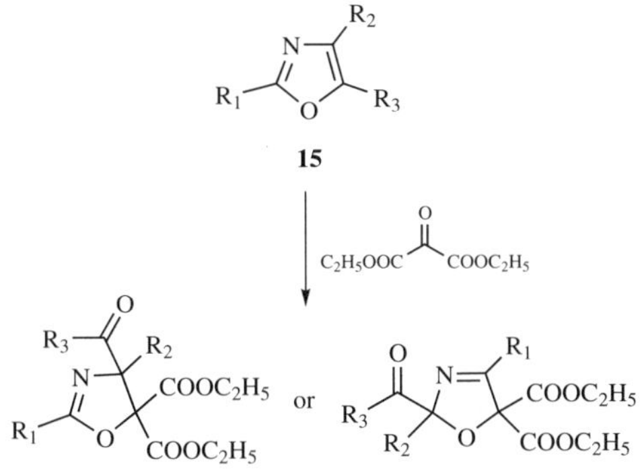

Figure 1.59

Entry	R_1	R_2	R_3	Yield (%)	**804:805**	Conditions[b]
1	4-CH$_3$O–C$_6$H$_4$	H	OCH$_3$	79	100:0	A
2	C$_6$H$_5$	H	OC$_2$H$_5$	79	100:0	A
3	4-CH$_3$O–C$_6$H$_4$	C$_6$H$_5$	OCH$_3$	95	100:0	A
4	4-CH$_3$O–C$_6$H$_4$	4-NO$_2$–C$_6$H$_4$	OCH$_3$	60	100:0	A
5	4-CH$_3$O–C$_6$H$_4$	CH$_3$	OCH$_3$	97	100:0	A
6	CH$_3$	H	OC$_2$H$_5$	10	0:100	C
7	CH$_3$	4-NO$_2$–C$_6$H$_4$	OCH$_3$	47	0:100	B
8	CH$_3$	CH$_3$	OCH$_3$	68	0:100	B
9	CH$_3$	i-C$_3$H$_7$	OCH$_3$	78	0:100	B

[a] Data from ref. 321.
[b] A, One equivalent SnCl$_4$, CH$_3$CN, room temperature; B, High pressure (0.85 GPa), CH$_3$CN, 40°C;
C, xylene, reflux.

Selected examples of 2- and 3-oxazolines prepared by this methodology are shown in Table 1.57.

Ibata's group[322–326] extensively investigated [3 + 2] cycloaddition reactions of 5-alkoxyoxazoles with aldehydes as a means of preparing precursors of β-hydroxy amino acids or 2-amino-1,3-diols. In particular, acid hydrolysis of cis-2-oxazoline-4-carboxylates produces the relatively inaccessible *erythro*-β-hydroxy amino acids. The authors studied the effects of solvent, temperature, Lewis acid, and aldehyde substituent on the product distribution and developed a synthesis of 2-oxazoline-4-carboxylates that is both *cis*-selective and enantioselective.

Initially, Suga and co-workers[322] investigated the Lewis acid–catalyzed reaction of aromatic and heterocyclic aldehydes with 5-methoxy-2-(4-methoxyphenyl) oxazole **806** to afford *cis*- and *trans*-oxazolines **807** and **808**, respectively (Scheme 1.222). The desired *cis* isomer **807** was the major product isolated using a number of Lewis acids. In all cases, the *cis:trans* ratio (**807**:**808**) depended on the

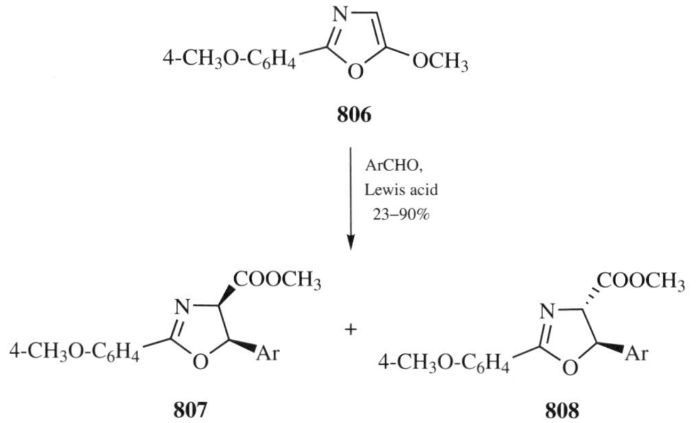

Scheme 1.222

Lewis acid and the nature of the aldehyde. A systematic study of the effects of the Lewis acid showed that a bulky Lewis acid maximized the *cis* selectivity, and the best catalyst was found to be methylaluminum β-binaphthoxide **809** (Fig. 1.60). In contrast, the *cis* selectivity was completely reversed using a titanium catalyst **810** prepared from 2,2′-dihydroxy-1,1′-binaphthyl and titanium tetrachloride.

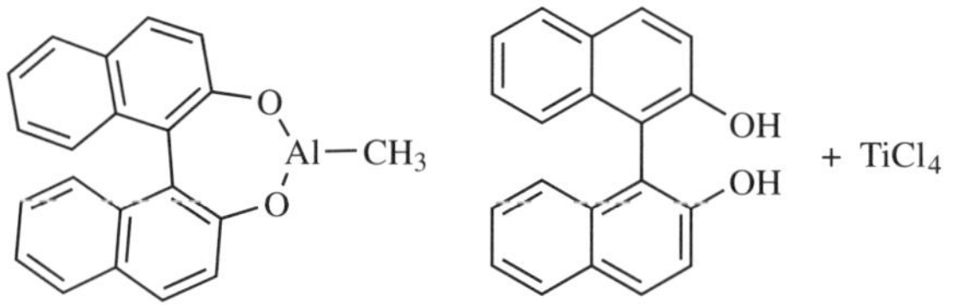

Figure 1.60. Lewis acid catalysts.

TABLE 1.58. *CIS*-2-OXAZOLINE-4-CARBOXYLATES AND *TRANS*-2-OXAZOLINE-4-CARBOXYLATES FROM LEWIS ACID–CATALYZED [3 + 2] CYCLOADDITION OF OXAZOLES AND ALDEHYDES[a]

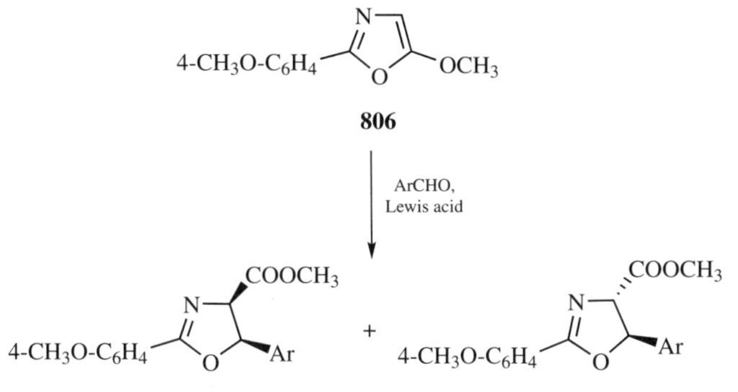

Figure 1.61

Entry	Ar	Lewis Acid	Yield (%)	**807:808**
1	C_6H_5	$SnCl_4$	84	36:64
2	C_6H_5	$TiCl_4$	77	14:86
3	C_6H_5	$EtAlCl_2$	68	26:74
4	C_6H_5	**810**	69	13:87
5	C_6H_5	**809**	78	98:2
6	4-CH_3O–C_6H_4	**809**	49	76:24
7	4-NO_2–C_6H_4	**809**	90	93:7
8	3-NO_2–C_6H_4	**809**	87	85:15
9	2-pyridyl	**809**	28	79:21
10	4-Cl–C_6H_4	**809**	82	98:2

[a] Data from ref. 322.

Similarly, hard Lewis acids, including $SnCl_4$, $TiCl_4$, and $(C_2H_5)_2AlCl$, also afforded the *trans*-oxazoline **808** as the major product. Some selected examples of *cis*-2-oxazoline-4-carboxylates prepared by this method are shown in Table 1.58.

Reaction of **806** with ethyl glyoxylate and chloral in the presence of $SnCl_4$, $TiCl_4$, or $(C_2H_5)_2AlCl$ also yielded *trans*-2-oxazolines **812a** and **812b** as the major products (Scheme 1.223). However, in some cases, **812a** and **812b** resulted from isomerization under the reaction conditions of the initially formed *cis*-2-oxazolines **811a** and **811b**. In contrast, good *cis* selectivity was obtained when **806** reacted with ethyl glyoxylate and chloral using a 1:1 mixed catalyst of $TiCl_4$ and Ti(IV) alkoxide.

Ibata's group[323] extended this methodology to include α-alkoxyaldehydes and developed an elegant asymmetric synthesis of chiral *erythro*-2-amino polyol derivatives, such as those found in D-AB1 **813**, 1-deoxynorjirimycin **814**, sphingosine **815**, and phytosphingosine **816** (Fig. 1.62). For example, reaction of **806** with

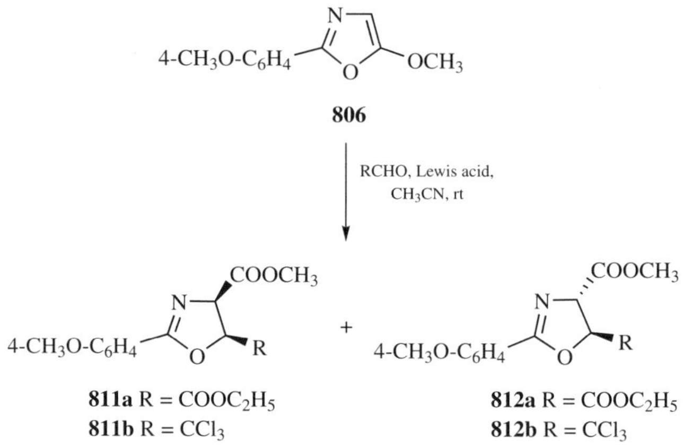

Scheme 1.223

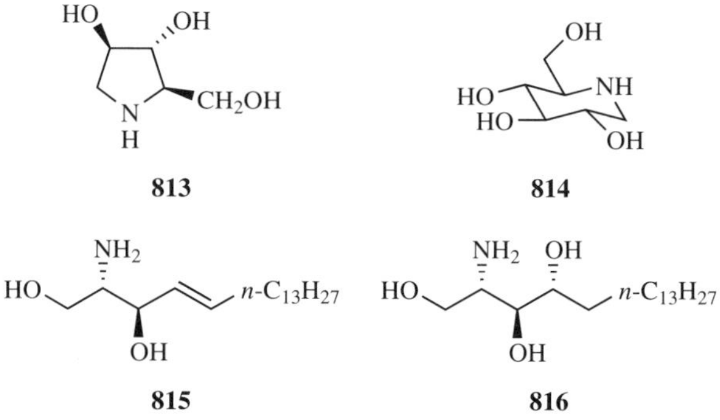

Figure 1.62. Chiral *erythro*-2-aminopolyols.

2,3-di-*O*-benzyl-D-glyceraldehyde **817** in the presence of SnCl₄ gave the *cis*-2-oxazoline-4-carboxylate **818** in 48% yield and >95% diastereoselectivity (Scheme 1.224). Scale-up of this reaction to 10 mmol afforded **818** in 61% yield together with 26% recovered **806**. Conversion of **818** to **825** was accomplished in four straightforward steps with 78% overall yield. This amino polyol (**825**) is an important intermediate, since it has the same absolute configuration as the amino polyol moieties present in **813** and **814**.

Finally, Ibata and co-workers[324–326] described further refinements and improvements in the *cis* selectivity and enantioselectivity of the [3 + 2] cycloaddition of 5-alkoxyoxazoles and α-alkoxyaldehydes.

Padwa and Cohen[271] systematically investigated thermal and photochemical [3,3]-sigmatropic rearrangements of suitably substituted oxazoles. The authors hoped to exploit this methodology as a means of generating reactive 1,3-dipoles via extrusion of carbon dioxide under mild conditions. The requisite 2-alkoxy 4,5-diphenyloxazoles **826** were readily prepared, as described previously (Scheme 1.184).

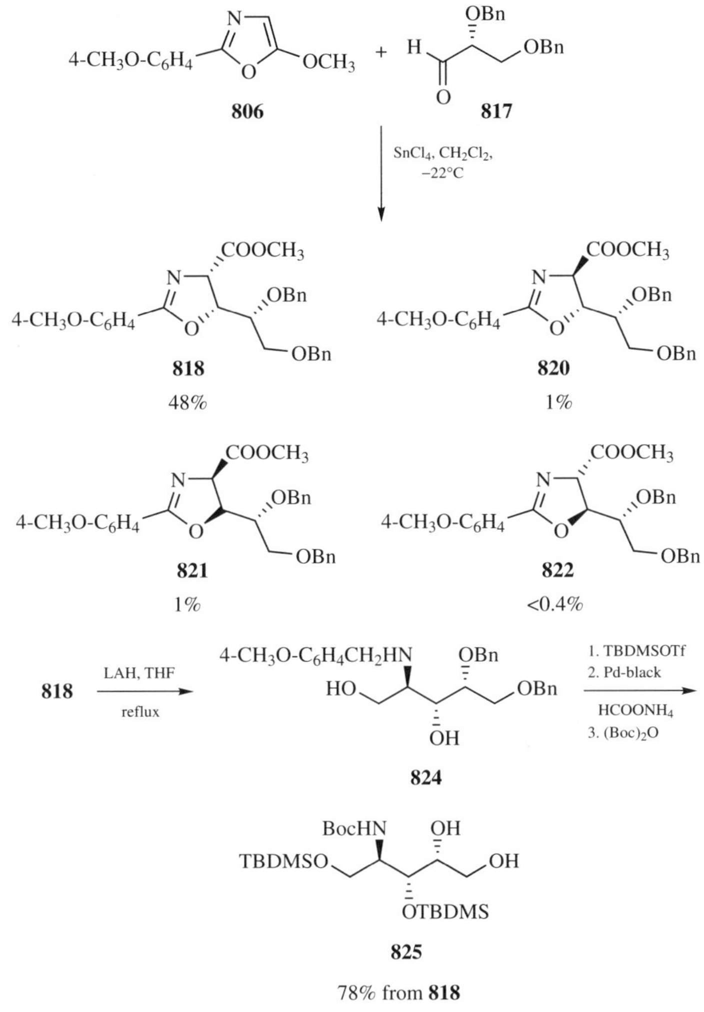

Scheme 1.224

Thermally, 2-allyloxy-4,5-diphenyloxazole **826a** undergoes a facile aza-Claisen rearrangement to produce 4,5-diphenyl-3-(2-propenyl)-2(3H)-oxazolone **827a** in excellent yield (Scheme 1.225). Similarly, 4,5-diphenyl-2-[(2-butenyl)oxy]oxazole **826b** gave **827b** in 97% yield. Rearrangement of 4,5-diphenyl-2-[(3-methyl-2-butenyl)oxy]oxazole **826c** was exceptionally facile and occurred readily to afford **827c** exclusively. Finally, 4,5-diphenyl-2-(propynyloxy)oxazole **828** rearranged to 4,5-diphenyl-3-(1,2-propadienyl)-2(3H)-oxazolone **829** in 90% yield. In all cases, the reactions occurred via a [3,3] sigmatropic process.

In contrast, **826a** rearranged photolytically to give 4,5-diphenyl-5-(2-propenyl)-2(5H)-oxazolone **830**. Unfortunately, no conditions for extruding carbon dioxide from **827a** or **830** could be identified.

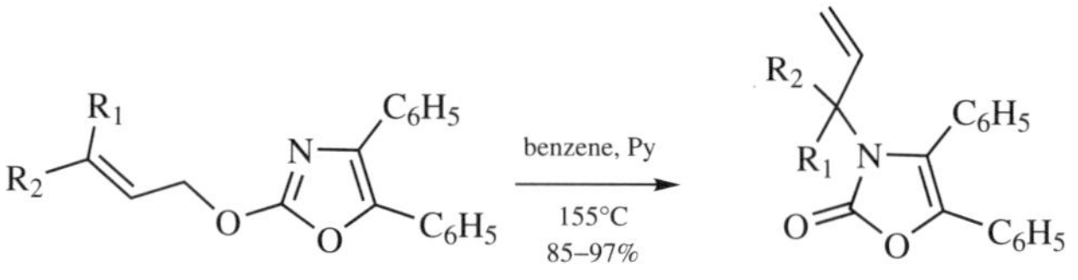

826a $R_1 = R_2 = H$
826b $R_1 = H, R_2 = CH_3$
826c $R_1 = R_2 = CH_3$

827a $R_1 = R_2 = H$
827b $R_1 = H, R_2 = CH_3$
827c $R_1 = R_2 = CH_3$

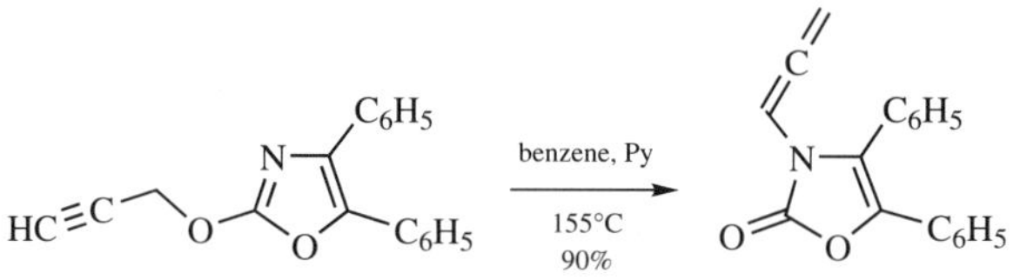

828 **829**

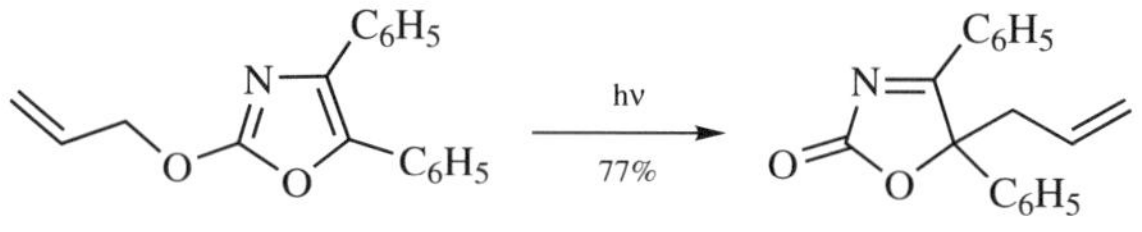

826a **830**

Scheme 1.225

Thermal rearrangement of 2-benzyloxy-4,5-diphenyloxazole **831** produced 3-benzyl-4,5-diphenyl-2(3*H*)-oxazolone **833** (Scheme 1.226). However, **833** was not the product of a simple [3,3]-sigmatropic rearrangement. Rather, the authors suggested a mechanism involving an aza-Claisen rearrangement to **832** followed by a rapid 1,3 shift of the amido group. This mechanism was favored over a radical scission-recombination, although the authors could not distinguish between the two mechanisms based on the existing experimental data. Photolysis of **831** generated a mixture of **833**, **834** and bibenzyl.

Diels-Alder reactions of 2-alkenyloxy-4,5-diphenyloxazoles **835** ($n = 3, 4$) were not successful. There was no reaction up to 230°C, whereas FVP of **835** ($n = 3, 4$) produced 4,5-diphenyl-2(3*H*)-oxazolone **837** nearly quantitatively via an intramolecular ene reaction. Photolysis of 5-[(5-hexenyl)oxy]-2-phenyloxazole **838** ($n = 3$) afforded the azirine **840** in 54% yield as the sole photoproduct. This was postulated to arise via ring opening to the nitrile ylide **839** and recyclization.

Steglich and co-workers[333] prepared optically active 4-allenyl-4-phenyl-2-substituted 5(4*H*)-oxazolones **843** from chiral *N*-acyl-*O*-propargyl phenylglycine derivatives **841** (Scheme 1.227). Cyclization of **841** produced intermediate 4-phenyl-5-(propargyloxy)-2-substituted oxazoles **842**, which could not be isolated. Instead, **842** immediately undergoes Claisen rearrangement, yielding **843**. The reaction was quite general and afforded **843** analogues in modest to good yield. A facile aza-Cope rearrangement of **843** generated optically active 4-phenyl-2-propargyl-2-substituted 5(2*H*)-oxazolones **844**, which were then converted to optically active ketones in two steps.

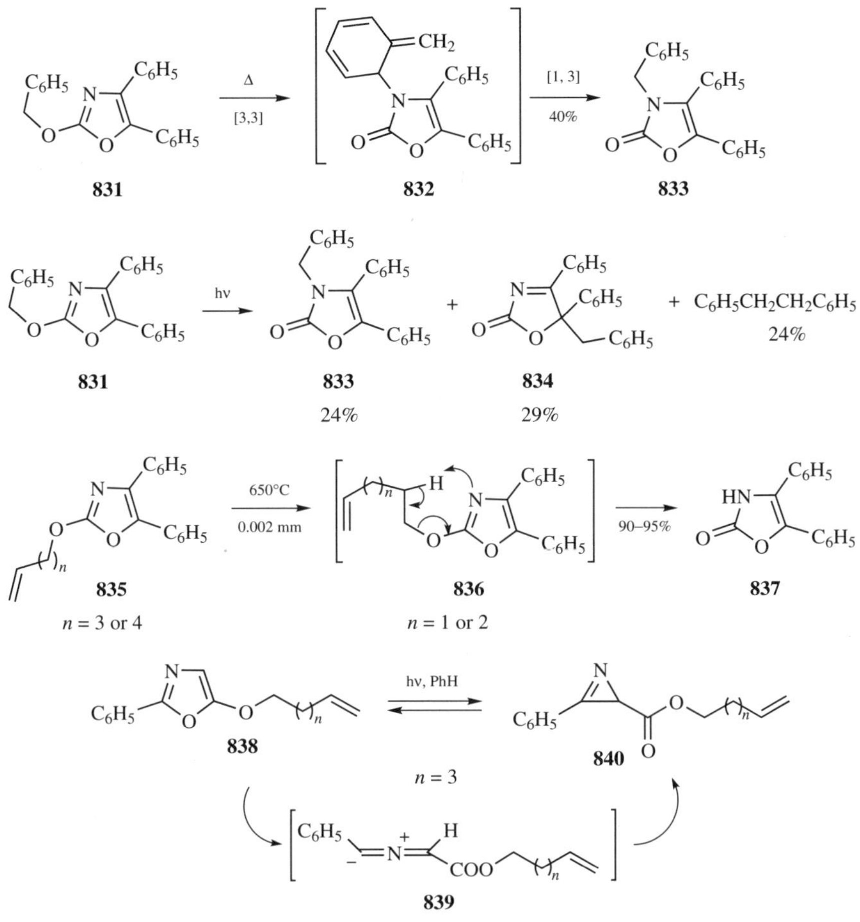

Scheme 1.226

Colombo and co-workers[334] elegantly applied a Claisen rearrangement of an allyloxyoxazole in their stereoselective synthesis of C-glycosyl α-amino acids, precursors to glycopeptide-based immunomodulating agents. Here, the authors envisioned that a suitably protected glycal would function as the allyl moiety. Thus the protected glucal **846** was coupled with *N*-benzoylalanine to furnish **847** in excellent yield (Scheme 1.228). Cyclization of **847** generated the intermediate 5-(allyloxy)oxazole **848**, which spontaneously rearranged to afford the β-glycosidic 5(4*H*)-oxazolone **849** isolated as a 3:1 mixture of diastereomers. Acid hydrolysis of **849** then afforded the target amino acid **850**.

1.4.6. Conversion of Oxazoles to Other Heterocycles

Oxazoles are readily converted to a variety of other heterocyclic rings, including pyrroles, pyrazoles, pyrimidines, imidazoles, and thiazoles by nucleophilic addition

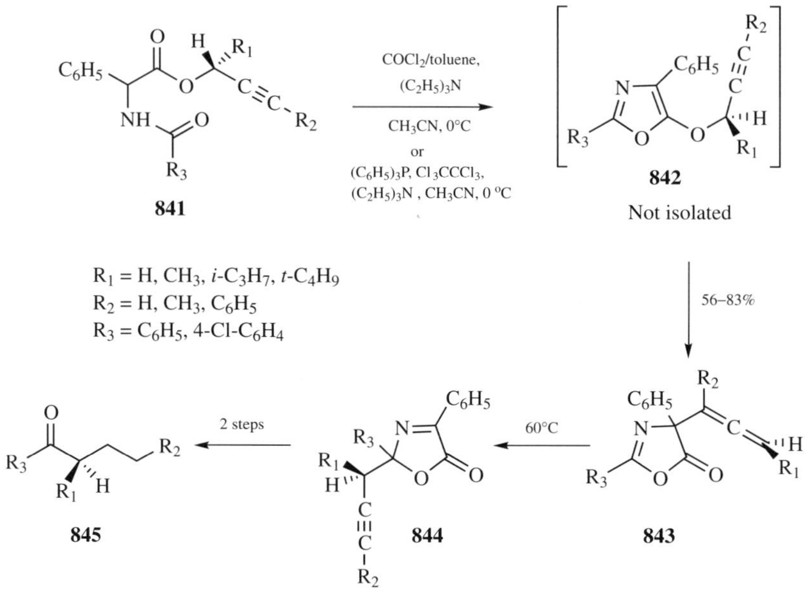

184 Synthesis and Reactions of Oxazoles

$R_1 = H, CH_3, i\text{-}C_3H_7, t\text{-}C_4H_9$
$R_2 = H, CH_3, C_6H_5$
$R_3 = C_6H_5, 4\text{-}Cl\text{-}C_6H_4$

Scheme 1.227

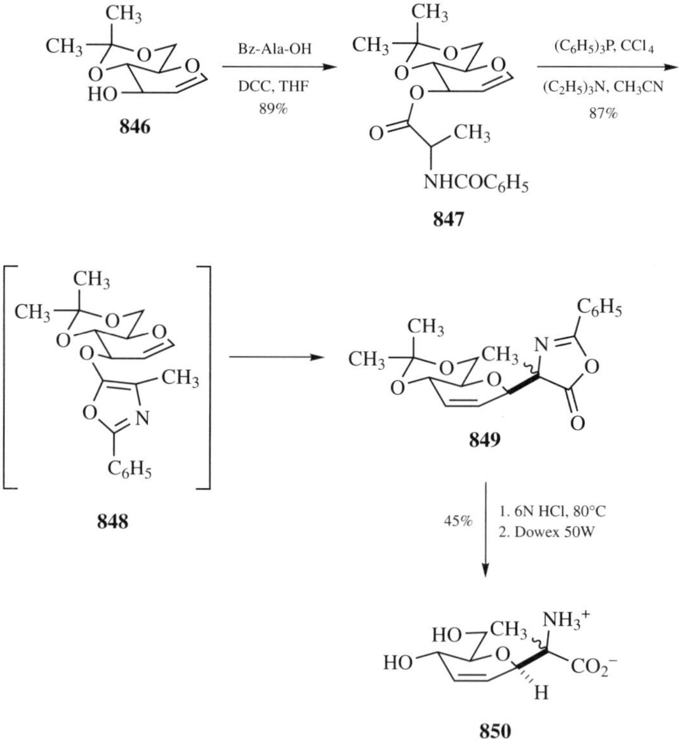

Scheme 1.228

to the oxazole ring followed by ring opening and recyclization or by a Cornforth rearrangement process.[335–345] Some selected examples are detailed in this section.

Moriya and co-workers[335] prepared a series of potential hypolipidemic β-aminopyrroles from 4-oxazoleacetic acid esters. Thus polyphosphate ester (PPE) cyclodehydration of an aspartic acid diamide β-ester **851** produced a 2-aryl-5-(dialkylamino)-4-oxazoleacetic acid ethyl ester **852** (Scheme 1.229). Treatment of **852** under Vilsmeier-Haack reaction conditions then gave, unexpectedly, a 5-aryl-2-(dialkylaminocarbonyl)-4-(dimethylamino)-3-pyrrolecarboxylic acid ethyl ester **853** in good to excellent yield. The reaction is quite general for the synthesis of these tetrasubstituted pyrroles (Table 1.59). Several analogs were reported to show high hypolipidemic activity.

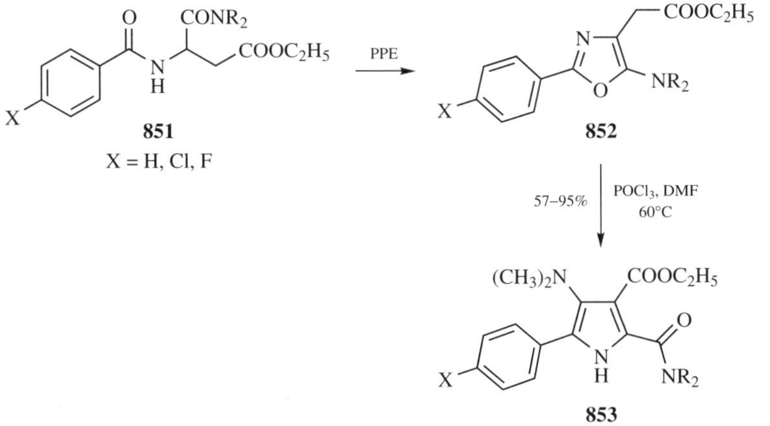

Scheme 1.229

A one-pot synthesis of **853** was developed by cyclodehydration of **851** with three equivalents of POCl$_3$/DMF at 60°C.

Mechanistically, the authors proposed that **852** was first converted to an enaminoester **854** by the in situ generated Vilsmeier reagent (Scheme 1.230). Nucleophilic addition of this enamine to C(2) of the oxazole would generate the oxazabicyclo[2.2.1]heptane **855**, which ring opened to a 3*H*-pyrrole **856**. Subsequent proton transfer in **856** then yielded **853**. The authors also noted that direct attack of an enamine on a heterocycle was most interesting.

Turchi and Cullen[336] described the first examples of pyrazoles prepared from oxazoles (Scheme 1.231). Treatment of 5-ethoxy-β-oxo-2-phenyl-4-oxazolepropanoic acid ethyl ester **857** with hydrazine did not afford the expected pyrazolone **858**. Instead, the authors isolated ethyl-4-(benzoylamino)-5-ethoxy-1*H*-pyrazole-3-acetate **860** (R = H) in good yield. Similarly, reaction of **857** with methylhydrazine produced **860** (R = CH$_3$) in comparable yield.

The authors proposed that **857** reacts with a hydrazine to generate a hydrazone (not shown) that cyclizes to a zwitterionic intermediate **859** by nucleophilic attack at C(5). Ring opening **859** then yields the isolated products. Treatment of **857** with

TABLE 1.59. 5-ARYL-2-(DIALKYLAMINOCARBONYL)-4-(DIMETHYLAMINO)-3-PYRROLECARBOXYLIC ACID ETHYL ESTERS FROM 2-ARYL-5-(DIALKYLAMINO)-4-OXAZOLEACETIC ACID ETHYL ESTERS UNDER VILSMEIER-HAACK CONDITIONS[a]

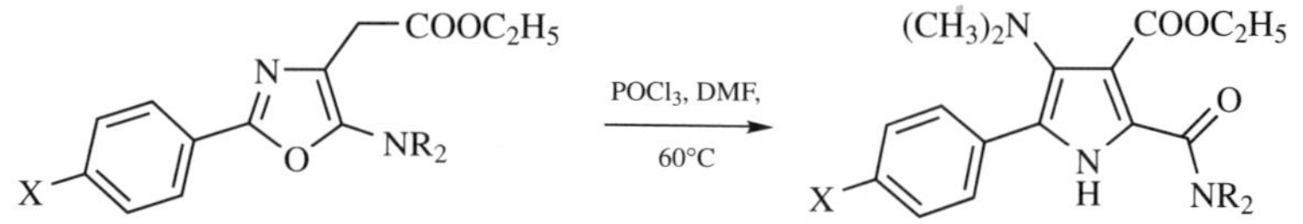

Figure 1.63

Entry	X	R_2	Yield (%)
1	Cl	$(CH_2)_4$	80
2	Cl	$(CH_2)_2O(CH_2)_2$	82
3	F	$(CH_2)_2O(CH_2)_2$	57
4	Cl	$(CH_2)_2NCH_3(CH_2)_2$	95
5	H	$(CH_2COOCH_3)_2$	78

[a] Data from ref. 335.

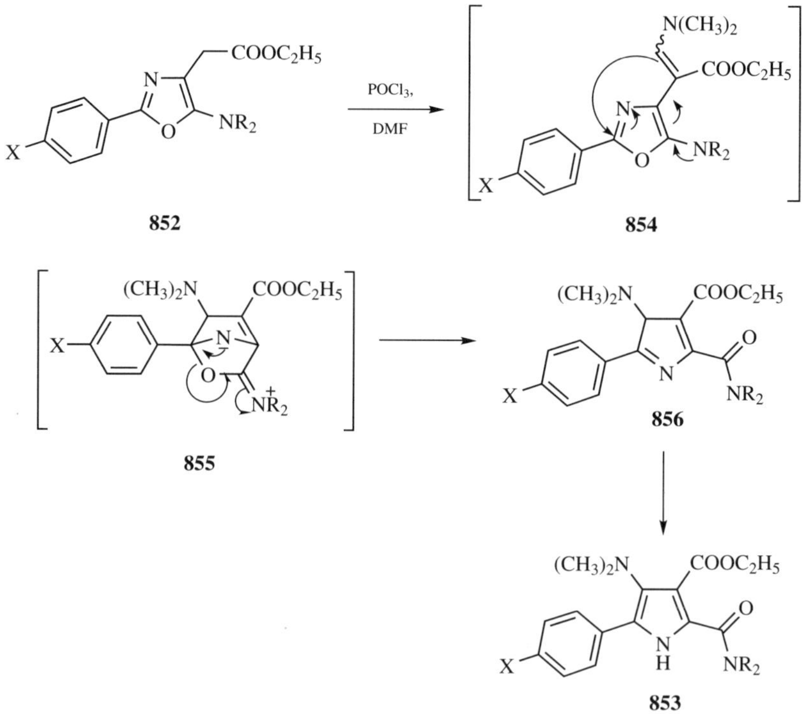

Scheme 1.230

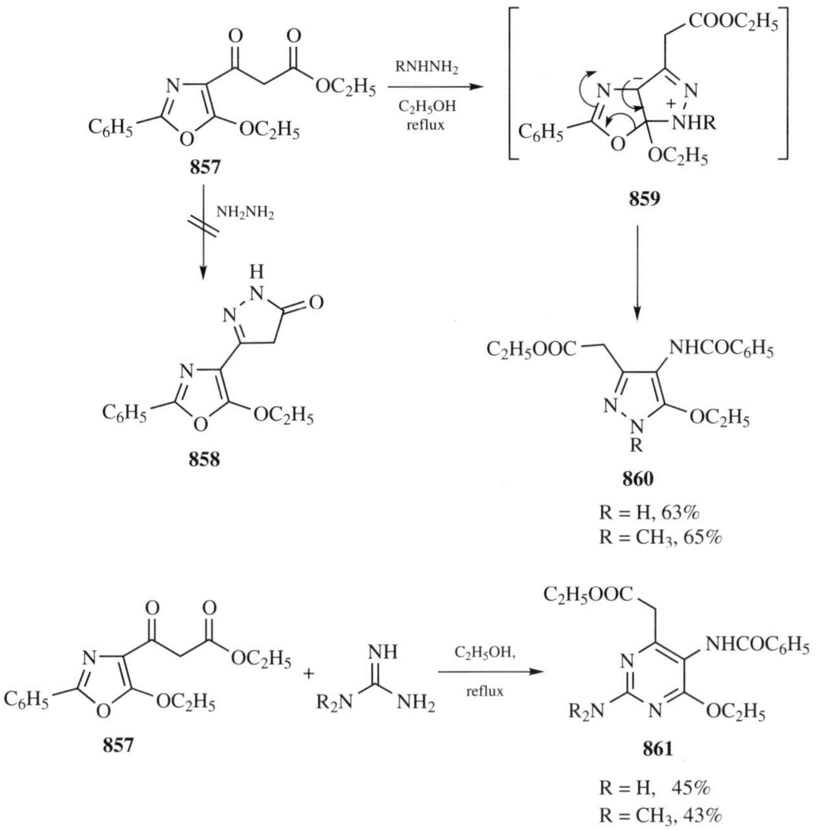

Reactions of Oxazoles

[Scheme 1.231 with structures 857, 858, 859, 860, 861]

Scheme 1.231

guanidine or 1,1-dimethylguanidine produced ethyl-2-amino-5-(benzoylamino)-6-ethoxy-4-pyrimidineacetate **861** (R = H) and ethyl-5-(benzoylamino)-2-(dimethylamino)-6-ethoxy-4-pyrimidineacetate **861** (R = CH$_3$), respectively, presumably via the same mechanism. The authors noted that reaction of 4-formyl- and 4-acyloxazoles with hydrazines may be a general approach to prepare 4-aminopyrazoles.

Amines react readily with 5-acyloxazoles to afford mixtures of imidazoles and pyrimidines. Several groups have investigated the mechanism of this rearrangement and the product distribution.[337–340]

Pfizer chemists[337] reacted 5-acetyl-2-aminooxazole **862** with a primary or secondary amine and isolated mixtures of 1(H)-5-acetyl-2-aminoimidazoles **863** and 2-amino-4-methyl-5-pyrimidinols **864** (Scheme 1.232). The reaction was independent of the amine substituent, since both primary and secondary amines yield **863** as the major product. The ratio of **863:864** was ca. 3:1 with primary amines but nearly 1:1 with secondary amines. It is noteworthy that water was requisite for this reaction sequence. Otherwise, the product is the Schiff base derived from **862**.

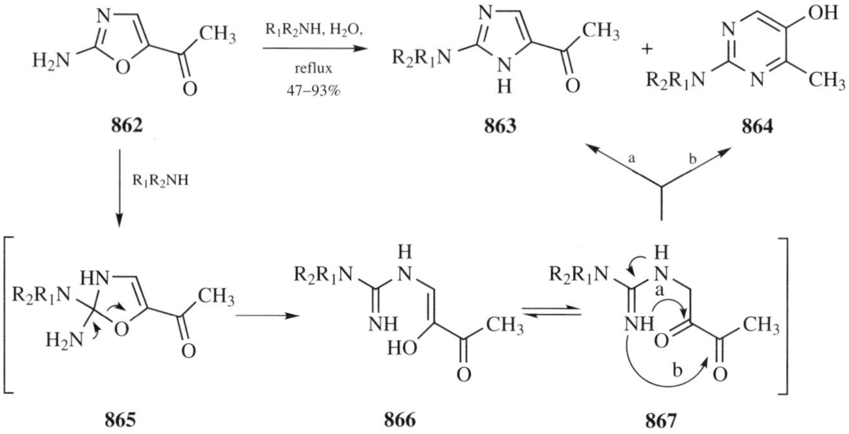

Scheme 1.232

Mechanistically, the authors proposed initial nucleophilic addition of the amine at C(2) to give **865**, followed by ring opening to the enol-keto tautomers **866** and **867**. Recyclization of the α-diketo amidine tautomer **867** via pathway a or pathway b then affords **863** or **864**, respectively. The reaction is general for a variety of primary and secondary amines; selected examples are shown in Table 1.60.

In contrast, a second Pfizer group[338] found that the pyrimidinol **870** was the major, if not exclusive, product from reaction of 2-amino-5-aroyloxazoles **868** and dimethylamine (Scheme 1.233). In this case, the authors found that electron-withdrawing substituents favored cyclization to yield **870**. A similar mechanistic rationale (vide supra) was proposed to explain the product distribution. Selected examples are shown in Table 1.61.

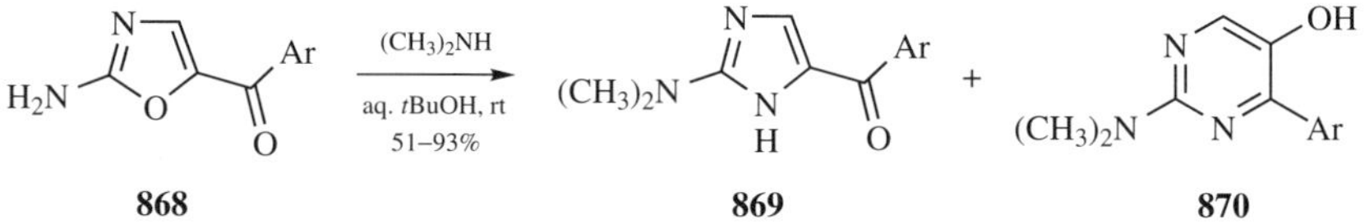

Scheme 1.233

In India, workers[339,340] converted 5-acyl-4-methyloxazoles **871a–c** to 4-(hydroxymethyl)-6-methyl-5-pyrimidinol **873** for evaluation as a pyridoxine antagonist for chemotherapy. Sen and Sengupta[339] reacted **871a** and **871b** with methanolic ammonia in a sealed tube to prepare the protected 4-(hydroxymethyl)-6-methyl-5-pyrimidinols **872a** and **872b** (Scheme 1.234). Acid hydrolysis of the tetrahydropyranyl (THP) ether in **872b** ultimately yielded **873**.

Ray and co-workers[340] prepared **872c** from **871c**, also using methanolic ammonia in a sealed tube. Interestingly, there was no reaction when **871c** was heated with a mixture of NH$_4$Cl and ammonia in C$_2$H$_5$OH at 170°C. Hydrogenolysis of the benzyl ether then gave **873**.

TABLE 1.60. 1(H)-5-ACETYL-2-AMINOIMIDAZOLES AND 2-AMINO-4-METHYL-5-PYRIMIDINOLS FROM 5-ACETYL-2-AMINOOXAZOLE AND AMINES[a]

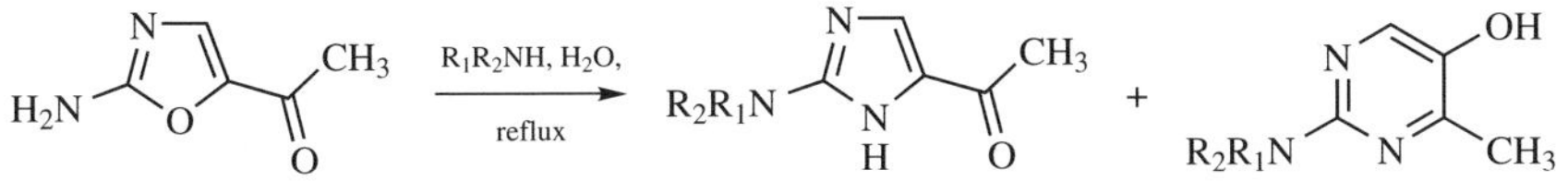

H_2N—[862] —CH_3 $\xrightarrow{R_1R_2NH, H_2O, reflux}$ R_2R_1N—[863]—CH_3 + R_2R_1N—[864]—OH, CH_3

Figure 1.64

Entry	R_1	R_2	Yield 863 (%)	Yield 864 (%)
1	H	C_2H_5	52	17
2	H	n-C_6H_{13}	62	19
3	H	c-C_6H_{11}	44	24
4	H	H	32	15
5	H	$CH_2C_6H_5$	43	15
6	CH_3	CH_3	46	44
7	$(CH_2)_4$		48	35

[a] Data from ref. 337.

Sasaki and Kitagawa[341] prepared 1,3-bis-(4-methyloxazol-5-yl)benzene **875** and 1,4-bis-(4-methyloxazol-5-yl)benzene **878** from the bis-isonitriles **874** and **877** and acetaldehyde. Upon heating **875** or **878** in formamide at 170°C, the authors isolated 1,3-bis-(4-methylimidazol-5-yl)benzene **876** or 1,4-bis-(4-methylimidazol-5-yl)benzene **879** (Scheme 1.235). Here, formamide acts as a source of ammonia at elevated temperatures.

TABLE 1.61. 1(H)-5-AROYL-2-(DIMETHYLAMINO)IMIDAZOLES AND 4-ARYL-2-(DIMETHYLAMINO)-5-PYRIMIDINOLS FROM 2-AMINO-5-AROYLOXAZOLES AND DIMETHYLAMINE[a]

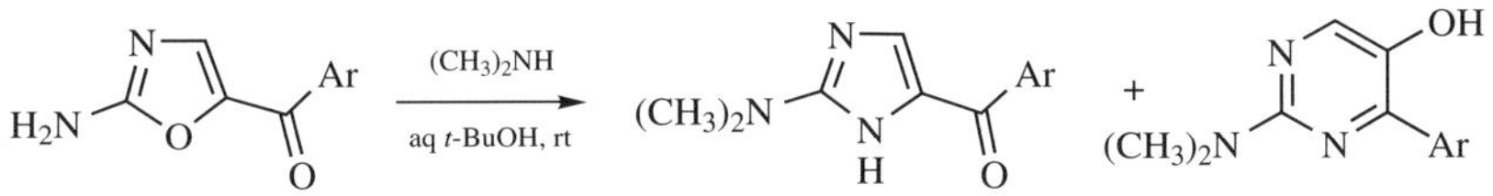

H_2N—[868]—Ar $\xrightarrow{(CH_3)_2NH, aq\ t\text{-BuOH, rt}}$ $(CH_3)_2N$—[869]—Ar + $(CH_3)_2N$—[870]—OH, Ar

Figure 1.65

Entry	Ar	Yield 869 (%)	Yield 870 (%)	870:869
1	C_6H_5	6	67	11:1
2	4-CH_3O–C_6H_4	22	51	2.3:1
3	4-Cl–C_6H_4	20	64	3.2:1
4	3-CH_3–C_6H_4	0	91	>20:1
5	3-F–C_6H_4	0	93	>20:1
6	3,4-Cl–C_6H_3	0	68	>20:1
7	3-CF_3–C_6H_4	0	65	>20:1
8	2,4-Cl–C_6H_3	0	51	>20:1

[a] Data from ref. 338.

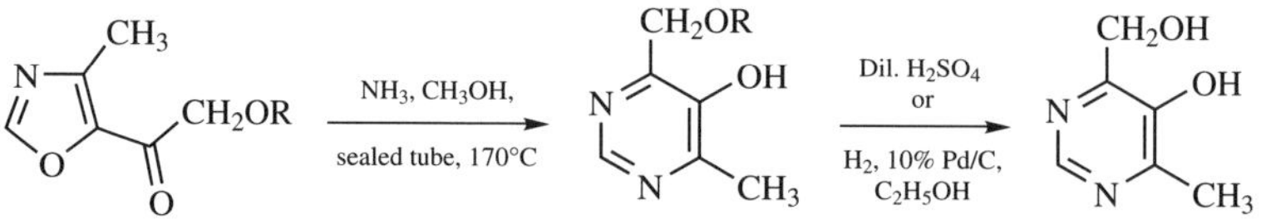

871a R = CH_3
871b R = THP
871c R = Bn

872a R = CH_3, 36%
872b R = THP, 59%
872c R = Bn, 61%

873
68% from **872b**
57% from **872c**

Scheme 1.234

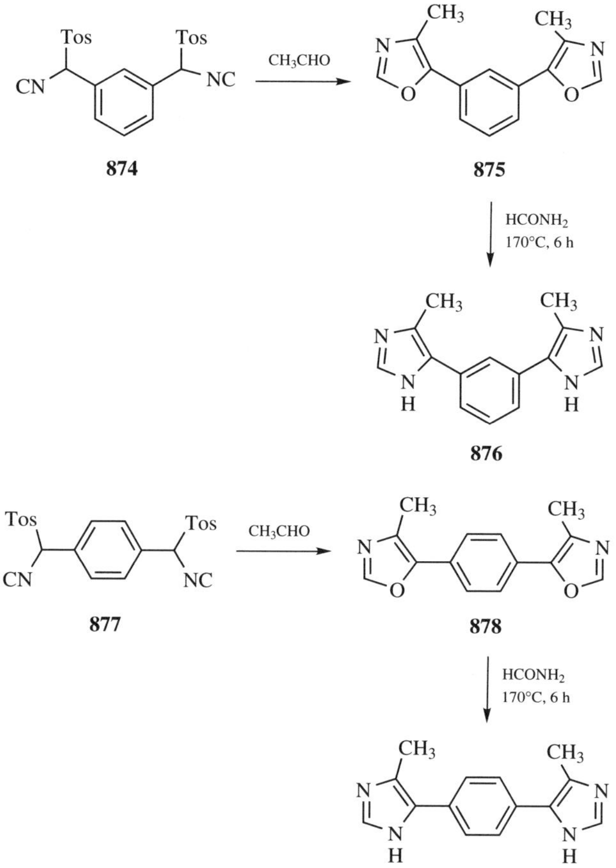

874 **875** **876** **877** **878** **879**

Scheme 1.235

The authors then applied this methodology to prepare the diimidazolo-[3^2]metacyclophanes **881, 883**, and **885** in good to excellent yield from the corresponding dioxazolo[3^2]metacyclophanes **880**, **882**, and **884** (Scheme 1.236). Sasaki's group[342] developed a general synthesis of metacyclophanes that incorporates three oxazole rings as well as analogs of **880**, wherein isophthalaldehyde was replaced by 2,5-furandicarboxaldehyde[343] and 2,5-thiophenedicarboxaldehyde.[344]

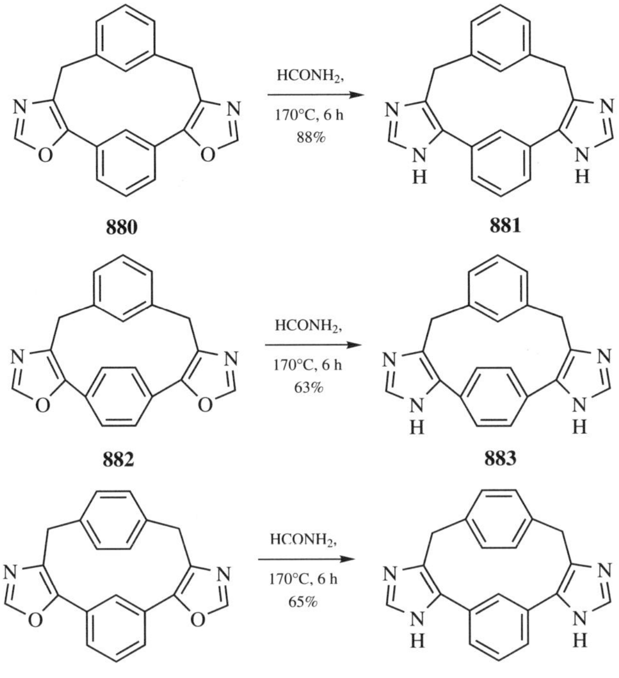

Reactions of Oxazoles 191

Scheme 1.236

Lawson and VanSant[243] observed the hydantoin by-products **886a** and **886b** during bromination of **598a** and **598b** with one equivalent of Br_2. However, bromination of **598a** and **598b** with two equivalents of Br_2 afforded **886a** and **886b** directly in 20% and 54% yield, respectively, (Scheme 1.237). The authors were not able to explore the generality of this rearrangement since **598** (R = C_2H_5, i-C_3H_7, t-C_4H_9) could not be prepared satisfactorily from the corresponding urea.

Turchi and co-workers[345] employed a Cornforth rearrangement to convert oxazoles to thiazoles (Scheme 1.238). In their work, a 2-alkyl(aryl)-5-ethoxy-4-oxazolecarboxamide **887** was converted to the thioamide **888** using Lawesson's reagent.[346] Rearrangement of **888** in refluxing toluene afforded a 2-alkyl(aryl)-5-amino-4-carboethoxythiazole **890** in good to excellent yield. The authors proposed an electrocyclic ring opening of **888** to the nitrile ylide **889**, which recyclized via a 1,5-dipolar cycloaddition reaction. MINDO/3 MO calculations supported this mechanistic rationale. A limitation of this methodology was encountered with 4-[(N-arylamino)carbonyl]-5-ethoxyoxazoles that fail to react with Lawesson's reagent. Examples of **890** prepared by this method are shown in Table 1.62.

1.4.7. Organometallic Reactions

Oxazoles have a rich organometallic chemistry with diverse reactivity. As such, metalation reactions have been used to funtionalize each position of the

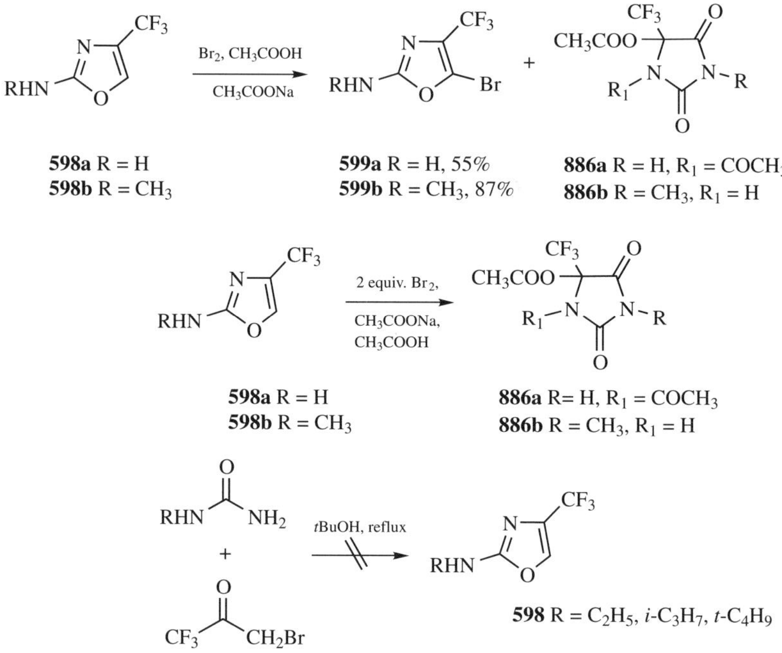

Scheme 1.237

598a R = H
598b R = CH_3

599a R = H, 55%
599b R = CH_3, 87%

886a R = H, R_1 = $COCH_3$
886b R = CH_3, R_1 = H

598a R = H
598b R = CH_3

886a R = H, R_1 = $COCH_3$
886b R = CH_3, R_1 = H

598 R = C_2H_5, i-C_3H_7, t-C_4H_9

Scheme 1.238

887 → **888** (Lawessons reagent, THF, reflux, 55–69%) → (110°C, Toluene) → [**889**] → **890** (78–92%)

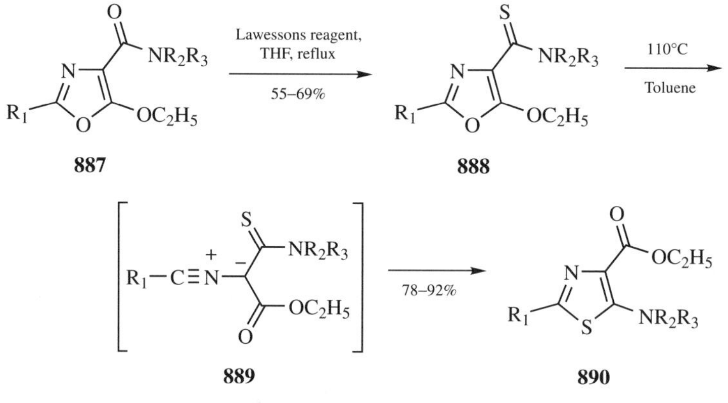

heterocyclic ring as well as of the side chain alkyl groups. Lithiation of oxazoles has been an extremely fruitful area of research. This is a challenging arena and the synthetic use of such derivatives is predicated on both a predictable and a regioselective pattern of reaction. For example, the well-documented equilibrium of a 2-lithiooxazole with the acyclic isocyanovinyllithium alkoxide has produced several clever and novel approaches to effect regioselective functionalization.

TABLE 1.62. 2-ALKYL(ARYL)-5-AMINO-4-CARBOETHOXYTHIAZOLES VIA CORNFORTH REARRANGEMENT OF 2-ALKYL(ARYL)-5-ETHOXY-4-OXAZOLETHIOCARBOXAMIDES[a]

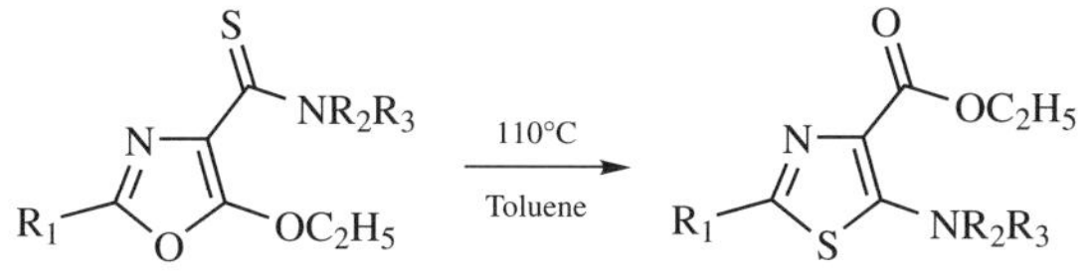

Figure 1.66

Entry	R_1	R_2	R_3	Yield (%)
1	CH_3	H	H	53
2	CH_3	CH_3	H	78
3	CH_3	$(CH_2)_4$		85
4	C_6H_5	H	H	87
5	C_6H_5	CH_3	H	88
6	C_6H_5	$C_6H_5CH_2$	H	85
7	C_6H_5	$(CH_2)_5$		85
8	C_6H_5	$(CH_2)_2O(CH_2)_2$		93

[a] Data from ref. 345.

The reader should consult the excellent and comprehensive review of lithiooxazoles published in 1994 by Iddon.[14] This compilation of the literature from 1968 through 1993 thoroughly covers ring metalation, halogen-lithium exchange reactions, lateral metalation, polylithiated derivatives, and other organometallic analogues. As such, the information contained therein is not further discussed in this chapter.

This section details some recent examples of regioselective ring metalation and regioselective lateral metalation and subsequent functionalization. The examples were arbitrarily selected to demonstrate the versatility of such analogues. There is no attempt to include every metalated oxazole and the reader should consult the primary literature for additional examples. The synthesis and cross-coupling reactions of stannyloxazoles, silyloxazoles, and other organometallic oxazole analogues are discussed in Sections 1.4.8 and 1.4.9, respectively.

Hughes and co-workers[347] studied the lithiation of 5-(3-methyl-5-isoxazolyl)oxazole 891 as part of their program to prepare suitable species for cross-coupling reactions (Scheme 1.239). They investigated the equilibrium of the 2-lithiooxazole 892 and the acyclic isocyanovinyllithium alkoxide 893 using ^{1}H-NMR and ^{13}C-NMR spectroscopy. The authors concluded that 893 was the predominant species under the reaction conditions investigated based on comparison of the ^{1}H-NMR and ^{13}C-NMR spectra with the spectra of 894 together with an extensive series of deuterium quenches. However, they also cautioned that the results of "deuterium quenches of systems where dynamic equilibria are possible or expected must be interpreted with care."

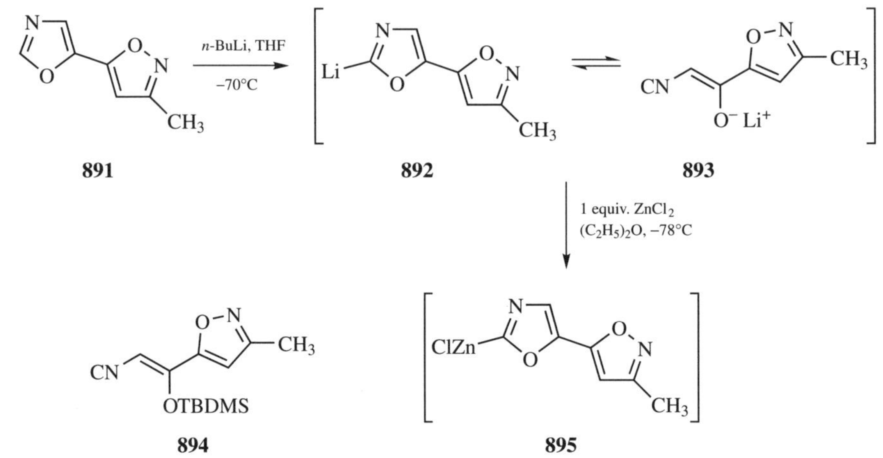

Scheme 1.239

In contrast, if the mixture of **892** ⇌ **893** was transmetalated at −78°C with ZnCl$_2$ and the resulting organozinc species quenched with D$_4$-acetic acid, > 85% of the deuterium incorporation was found at C(2). The authors concluded that this organozinc species was best represented as the oxazole **895**, which was subsequently used in palladium-catalyzed cross-coupling reactions (see Section 1.4.8).

Boche and co-workers[348,349] extensively investigated the equilibrium of a 2-lithiooxazole **897** (M = Li) with the acyclic isocyanovinyllithium alkoxide **898** (M = Li) (Scheme 1.240), using ^{13}C-NMR, IR, single-crystal X-ray, and molecular orbital calculations. The authors observed no change in the ^{13}C-NMR spectrum from −105° to 20°C when **896** (R = R$_1$ = H) was treated with n-BuLi. There was no evidence of line broadening in the spectrum, and the resonances observed were consistent with **898** (M = Li, R = R$_1$ = H). The isocyanide in THF solution absorbed at ν = 2120 cm^{-1} in the IR spectrum. The authors concluded that **898** (M = Li, R = R$_1$ = H) was the predominant species in solution up to 95 ± 5%. Similar behavior was observed for 4-phenyloxazole **896** (R = C$_6$H$_5$, R$_1$ = H) and 4-t-butyloxazole **896** (R = t-C$_4$H$_9$, R$_1$ = H). In addition, they also crystallized a

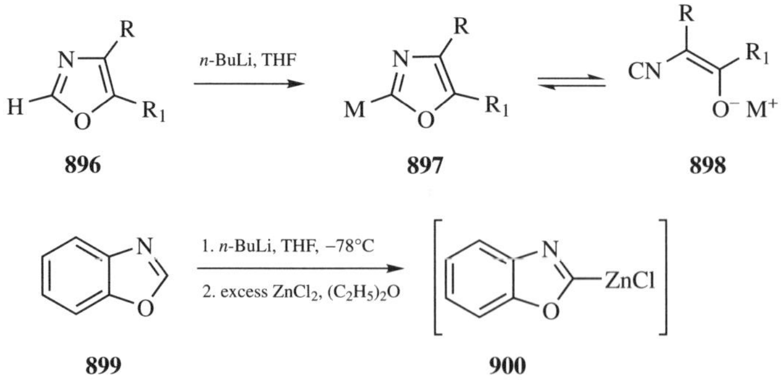

Scheme 1.240

Reactions of Oxazoles 195

digylme-complexed dimer of **898** (M = Li, R = R$_1$ = H) and confirmed this by an X-ray crystal structure. However, when benzoxazole **899** was sequentially lithiated at −78°C followed by treatment with excess ZnCl$_2$, the authors observed only the 2-zincated benzoxazole **900** in the ^{13}C-NMR.

Subsequently, Boche's group[349] crystallized a dimer of **900** from THF and reported the results of their detailed solid-state structure elucidation. In the same study, the authors examined the structures of **896** (R = R$_1$ = H), **897** (M = Li or ZnCl, R = R$_1$ = H), and **898** (M = Li or ZnCl, R = R$_1$ = H) using ab initio B3LYP/ 6-31G + G(d) molecular orbital calculations. Their calculations were in total agreement with experimental observations, i.e., **897** (M = Li, R = R$_1$ = H) was significantly less stable than **898** (M = Li, R = R$_1$ = H), owing to the oxophilic nature of Li$^+$. In contrast, the more covalent C-Zn bond contributes to the enhanced stability of **897** (M = ZnCl, R = R$_1$ = H) vs **898** (M = ZnCl, R = R$_1$ = H).

Crews and co-workers[350,351] employed a 2-lithiooxazole in a key step of their synthetic strategy to prepare bengazole A **904**, an antihelminthic agent isolated from marine sponges. Lithiation of **901** at −50°C followed by addition of 5-oxazolecarboxaldehyde **902** gave the bis-oxazolyl methanol **903** (Scheme 1.241). Acylation of **903** with myristoyl chloride and deprotection completed the synthesis of **904**.

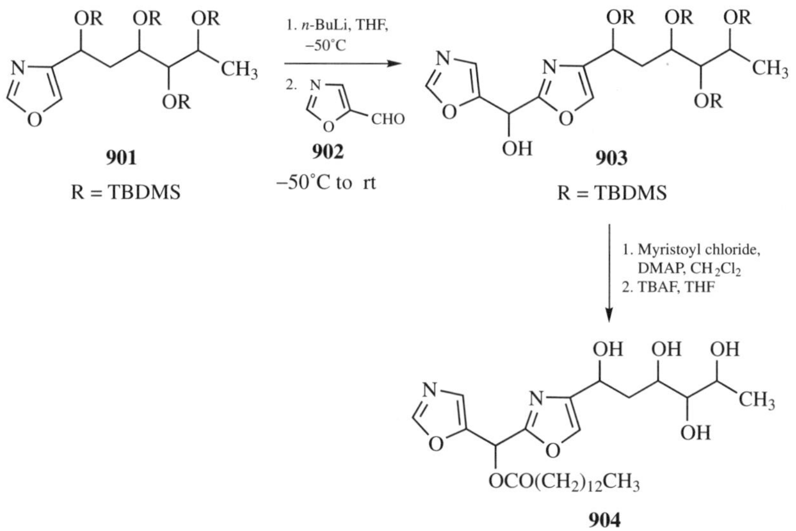

Scheme 1.241

Shioiri's group[352] also employed a 2-lithiooxazole to construct the bis-oxazolyl methanol core of **904**. Here, lithiation of a silyl-protected 4-oxazolemethanol **905a/b** followed by condensation with **902** afforded a racemic bis-oxazolyl methanol **906a/b** in modest yield (Scheme 1.242). Attempts to improve the yield of **906** using a Lewis acid to coordinate the oxazole nitrogen, alternate bases, or addition of co-solvents were not successful. Barton-McCombie[353] radical deoxygenation was used

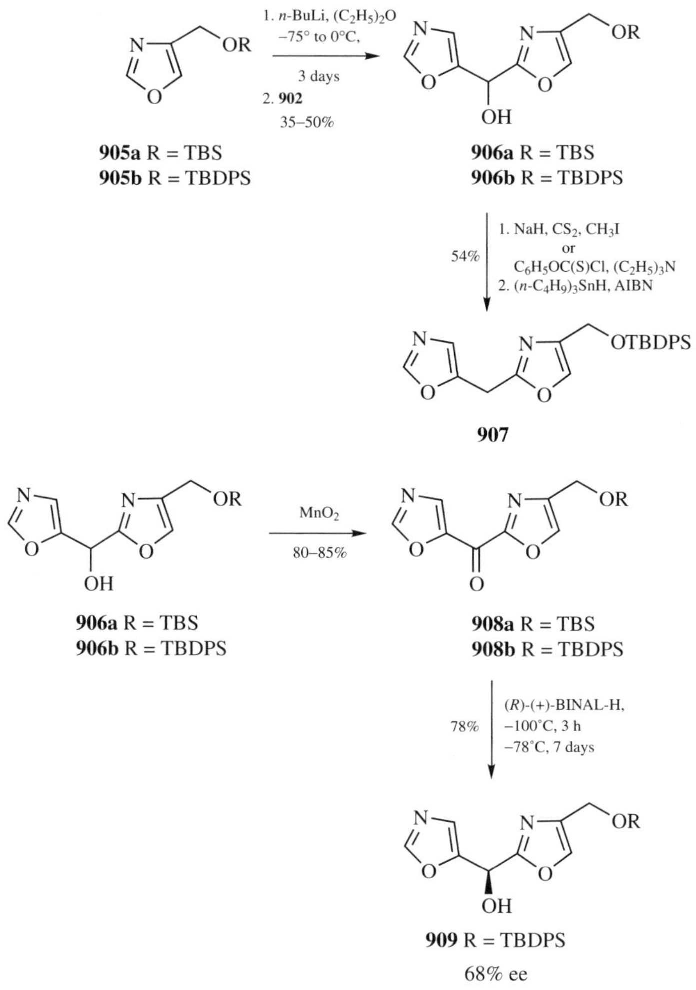

Scheme 1.242

to convert **906b** to **907**, a key intermediate leading to bengazoles C and D. Alternatively, oxidation of **906a/b** with MnO_2 gave the bis-oxazolyl ketones **908a/b** in very good yield. Asymmetric reduction of **908b** with (R)-(+)-BINAL-H afforded the (S)-bis-oxazolyl methanol **909**, a key intermediate for bengazole A.

Shafer and Molinski[354] exploited the unusual ambident reactivity of **897** (M = Li, R = R$_1$ = H) with aldehydes in the first report of a C(4) directed oxazole coupling in natural product synthesis. Condensation of a large excess of **897** (M = Li, R = R$_1$ = H) with **910** generated the desired targets **911a/911b** as a 1:1 mixture of epimers (Scheme 1.243). The relatively low yield of **911a/911b** was attributed to low reactivity of 2-lithiooxazole (relative to phenyllithium), competing ß-elimination, and enolization of **910**. The authors did not observe any 2-substituted-oxazole by-products and noted that **897** failed to react with the unprotected

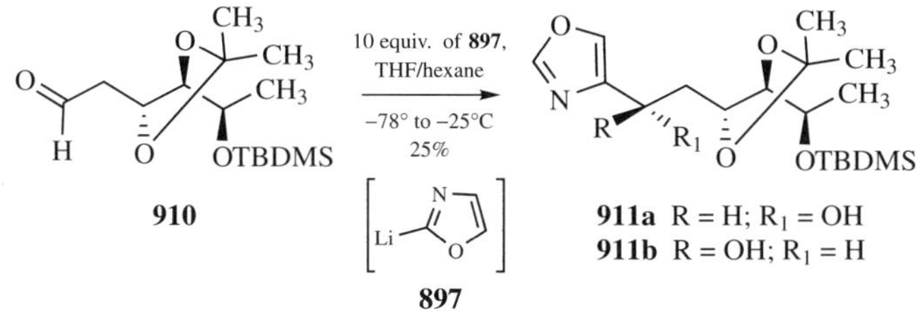

Scheme 1.243

(TBS) precursor to **910**. Nonetheless, **911a** is an advanced intermediate leading to bengazole A, and this methodology does provide rapid late-stage assembly of a core fragment that incorporates an oxazole ring.

Merck chemists[355,356] prepared novel quinuclidine-based ligands for the muscarinic cholinergic receptor (Scheme 1.244). These ligands were evaluated in a program directed to the treatment of Alzheimer disease or senile dementia of the Alzheimer type. For example, lithiation of 4-methyloxazole **896** (R = CH_3, R_1 = H) afforded **912** (R = CH_3, R_1 = H) in modest yield. Interestingly, **912** proved refractory to dehydration. However, conversion of the tertiary alcohol to a chloride followed by treatment with $(n\text{-}C_4H_9)_3SnH$ provided an entry to **913** (R = CH_3, R_1 = H). Additional examples of **913** (R = H, C_2H_5, R_1 = H) were prepared from the corresponding **896** precursor.

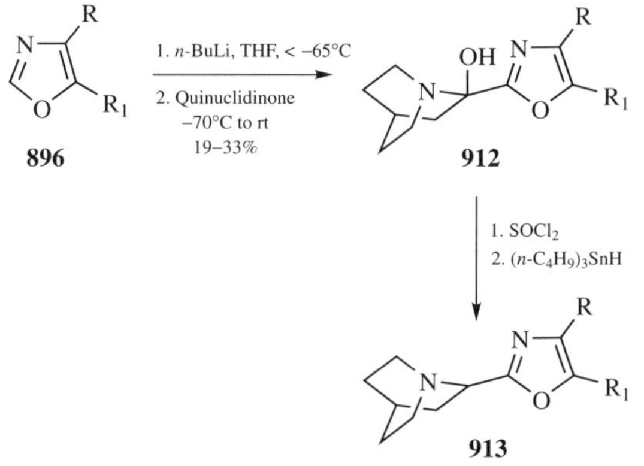

Scheme 1.244

As noted earlier in this section lithiation of oxazoles with *n*-butyllithium can be plagued by the formation of products derived from the tautomeric isocyanovinyllithium alkoxide species. Vedejs and Monahan[357] conceived an elegant solution to avoid this side reaction (Scheme 1.245). The authors precomplexed an oxazole **896** with THF-borane to form the stable, isolable Lewis acid–base complex **914**. This complex effectively coordinated the lone pair of electrons on nitrogen that is essential for the undesired electocyclic ring opening reaction. In addition, the authors anticipated that the coordination would enhance the acidity of the C(2)-H. Deprotonation of the complex with LiTMP (*n*-BuLi and *s*-BuLi are also effective)

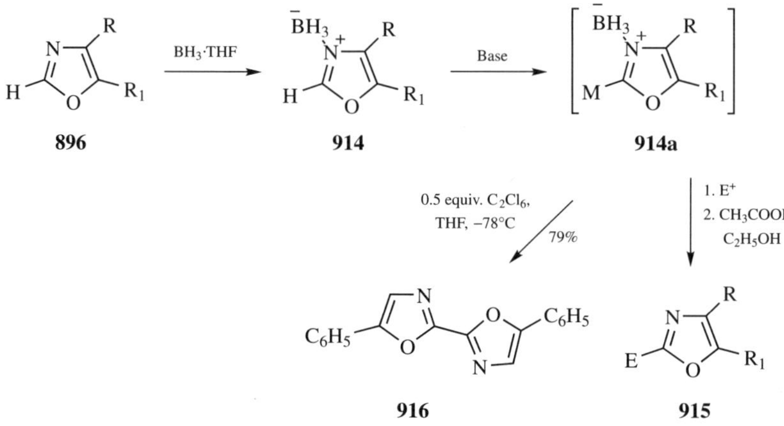

Scheme 1.245

followed by addition of an electrophile provided the C(2)-substituted oxazole **915** in good to excellent yield. The oxazole-borane complex was then decomposed with acetic acid/ethanol workup in most cases. The reaction is quite general, and a variety of electrophiles were investigated. Selected examples are shown in Table 1.63.

Quenching **914a** (M = Li, R = H, R_1 = C_6H_5) with one half equivalent of C_2Cl_6 gave rise to a 2,2′-bis-oxazole **916** in excellent yield. This product was presumed to arise via reaction of **914a** with the initially produced intermediate **914a** (M = Cl, R = H, R_1 = C_6H_5).

TABLE 1.63. 2-SUBSTITUTED AND 2,5-DISUBSTITUTED OXAZOLES FROM 2-LITHIOOXAZOLE-BORANE COMPLEXES AND ELECTROPHILES[a,b]

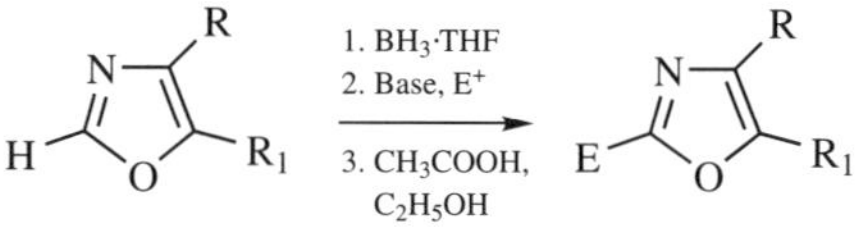

Figure 1.67

Entry	R	R_1	E^+	E	Yield (%)
1	H	C_6H_5	C_6H_5CHO	C_6H_5CHOH	94
2	H	C_6H_5	$(CH_3)_3$SiCl	$(CH_3)_3$Si	78
3	H	C_6H_5	CH_3I	CH_3	74
4	H	C_6H_5	$C_6H_5CH_2CH_2$OTf	$C_6H_5CH_2CH_2$	65
5	H	C_6H_5	C_2Cl_6 (2 equiv.)	Cl	86
6	H	H	C_6H_5CHO	C_6H_5CHOH	70
7	H	H	$C_6H_5CH_2CH_2$OTf	$C_6H_5CH_2CH_2$	76
8	H	$CH_2CH_2CH(OTBS)CH_3$	C_6H_5CHO	C_6H_5CHOH	84

[a] Data from ref. 357.
[b] Reaction conditions: base LiTMP, n-BuLi, or s-BuLi; $-78°C$ or $-20°C$.

Vedejs and Luchetta[358] exploited the ambident reactivity of 2-lithiooxazoles to prepare a series of 4-iodo-5-substituted-oxazoles **917** (Scheme 1.246). Initially, they found that **896** (R = H, $R_1 \neq$ H) reacted with LiHMDS and I_2 to produce a mixture containing the desired 4-iodo-5-substituted oxazole **917** together with the 2-iodo-5-substituted oxazole **918** and the 2,4-diiodo-5-substituted oxazole **919**. The ratio of **917**:**918** depended on the nature of the electrophile, the solvent, and the base.

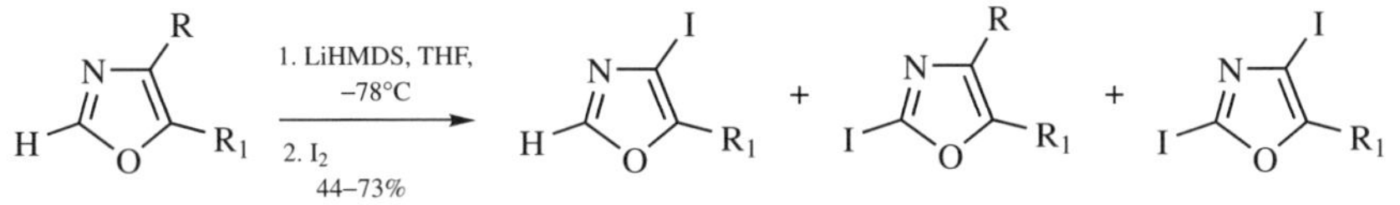

Scheme 1.246

Addition of DMPU as a co-solvent favored formation of **917** as the major product. Interestingly, lithiation of **896** (R = H, R_1 = 4-$CH_3C_6H_4$) with *n*-BuLi in THF in the absence of DMPU and quenching with 1,2-diiodoethane yielded **918** exclusively. The authors found that the overall yield of (**917** + **918**) and the ratio of (**917** + **918**):**919** could be dramatically improved for **896** (R = H, R_1 = alkyl). In these cases, after addition of iodine, the crude reaction mixture was treated with 2 to 2.3 equivalents of *n*-BuLi to facilitate halogen metal exchange at C(2) and thus effect overall net reduction. Examples of 4-iodooxazoles prepared are shown in Table 1.64.

TABLE 1.64. 4-IODO-5-SUBSTITUTED AND 2-IODO-5-SUBSTITUTED OXAZOLES FROM 2-LITHIOOXAZOLES AND IODINE[a]

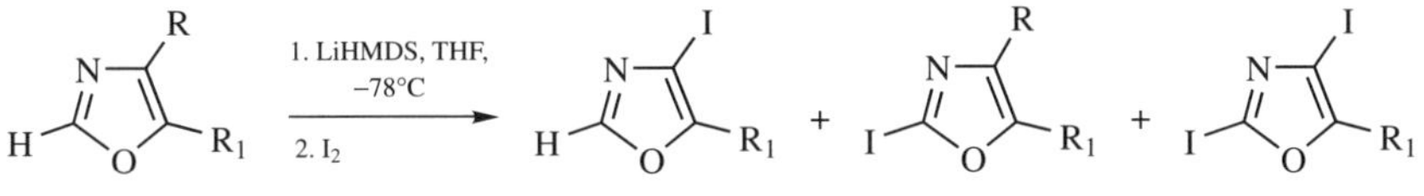

Figure 1.68

Entry	R	R_1	Yield **917** + **918** (%)	**917**:**918**	(**917** + **918**):**919**
1	H	4-CH_3–C_6H_4	73	32:1	15:1
2	H	C_6H_5	67	32:1	52:1
3	H	$COOC_2H_5$	44	5.3:1	6.3:1
4	H	$C_6H_5CH_2CH_2$	43	>49:1	4:1
5[b]	H	$C_6H_5CH_2CH_2$	64	>49:1	99:1
6	H	$(CH_2)_3$OTBS	34	>49:1	1:1
7[b]	H	$(CH_2)_3$OTBS	63	>49:1	99:1

[a] Data from ref. 358.
[b] After the addition of iodine, the crude product was treated with *n*-BuLi at $-78°C$ to a phenanthroline end point.

200   Synthesis and Reactions of Oxazoles

Shafer and Molinski[359] described a two-step, general method of preparing 5-substituted oxazoles. This method is complementary to the work of Vedejs and Monahan (vide supra). The authors found that with introduction of a methylthio group at C(2), lithiation occurs smoothly at C(5) and subsequent reaction with an electrophile affords **922** readily (Scheme 1.247). Reductive desulfurization of **922** then yields **923**.

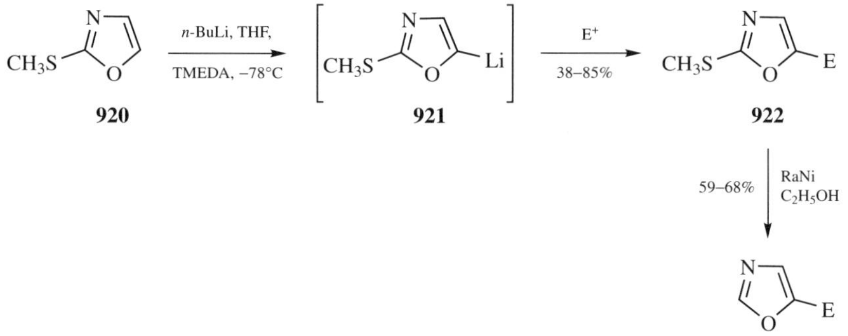

Scheme 1.247

The starting 2-(methylthio)oxazole **920** was best prepared from the thione precursor using KH/CH_3I. Metalation of **920** occured smoothly using 10 equivalents of TMEDA, while inverse addition of **921** to the electrophile at $-78°C$ provided the best results for **922**. Reactions with aldehydes were rapid and gave high yields of the corresponding carbinols. Acid chlorides also reacted rapidly and completely with a large excess of the acid chloride. Ketones reacted more slowly and were prone to complications from enolization. Deuterium incorporation occurred almost exclusively at C(5) (>90%) when the authors used their standard metalation conditions; < 5% deuterium incorporation was observed in the methylthio group. Selected examples of 5-substituted oxazoles prepared via this methodology are shown in Table 1.65.

Williams and co-workers[360] described a regioselective metalation of 2,4'-bis-oxazoles in the context of natural product synthesis. Regioselective lithiation of 4-(methoxymethyl)-2'-methyl-2,4'-bis-oxazole **924** followed by quenching with a variety of electrophiles produced the 4-(methoxymethyl)-2'-methyl-5-substituted-2,4'-bis-oxazoles **925**, respectively, in modest to excellent yield (Scheme 1.248).

The authors attributed these results to heteroatom complexation and an internally directed deprotonation. Semiempirical calculations using AM1 and PM3 Hamiltonians supported this rationale. Conformer **924A** was predicted to be slightly more stable than conformer **924B**. However, internal stabilization of the 5'-lithiooxazole by complexation with the nitrogen atom (as shown in **926**) was considerably more stable than any of the alternative structures, **927–929**. Examples of **925** are shown in Table 1.66.

TABLE 1.65. 5-SUBSTITUTED OXAZOLES VIA LITHIATION OF 2-(METHYLTHIO)OXAZOLE AND REDUCTIVE DESULFURIZATION[a]

CH$_3$S-[oxazole] →(1. n-BuLi, THF, TMEDA, −78°C; 2. Electrophile)→ CH$_3$S-[oxazole]-E →(RaNi, C$_2$H$_5$OH)→ [oxazole]-E

920 → 922 → 923   Figure 1.69

Entry	Electrophile	E	Yield 922 (%)	Yield 923 (%)
1	C$_6$H$_5$CHO	C$_6$H$_5$CHOH	84	68
2	2-naphthylCHO	2-naphthylCHOH	77	60
3	C$_9$H$_{19}$CHO	C$_9$H$_{19}$CHOH	73	60
4	t-C$_4$H$_9$CHO	t-C$_4$H$_9$CHOH	69	59
5	cyclohexanone	c-C$_6$H$_{10}$COH	53	60
6	C$_6$H$_5$COCl	C$_6$H$_5$CO	38	—
7	t-C$_4$H$_9$COCl	t-C$_4$H$_9$CO	71	—

[a] Data from ref. 359.

Williams and McClymont[361] described detailed studies of the alkylation and acylation of 5-(1,3-dithian-2-yl)oxazole 932. The authors showed from deuterium incorporation studies that sequential deprotonation of 932 occurred first at C(2) and then at the dithiane carbon atom. Thus treatment of 932 with excess LiHMDS followed by an alkylating agent afforded exclusively the side chain alkyl analogues 933 (Scheme 1.249, Table 1.67). Reactive electrophiles, e.g., CH$_3$I and TMSCl afforded complex mixtures of C- and O-alkylated products derived from the acyclic isocyanovinyllithium alkoxides 938a and 938b (discussed below), whereas secondary alkyl halides were unreactive.

On the other hand, acylation of 932 did not produce either of the expected products 933 or 934. Instead, the authors isolated good to excellent yields of 4,5-disubstituted oxazoles 935 and/or 936, wherein a carbonyl group had been inserted between the oxazole ring and the dithiane ring (Scheme 1.250, Table 1.67).

Mechanistically, these interesting results were explained in the following manner. Initial deprotonation of 932 afforded dianion 937, which was alkylated to produce 933 (Scheme 1.251, p. 205). However, 937 equilibrates with the isocyanovinyllithium alkoxides 938a and 938b. A C-selective acylation of 938b would yield 939, which can cyclize through either carbonyl group to generate the regioisomeric 4,5-disubstituted oxazoles 935 and 936. This acylation represents the first example of a base-induced, low-temperature Cornforth rearrangement.

Liu and Panek[362] also investigated lithiation of a dithiane containing oxazole during their synthesis of kabiramide C. Lithiation of 940 with excess LDA followed by D$_2$O produced the 5,5′,5″-trideutero analog (not shown). Similarly, treatment of 940 with one or two equivalents of tBuLi followed by two equivalents of 941 afforded 943 or 944 (Scheme 1.252, p. 206). However, the authors were pleased to find that lithiation of 940 with excess tBuLi and alkylation with one equivalent of 941 afforded 942, a key intermediate in their synthetic approach.

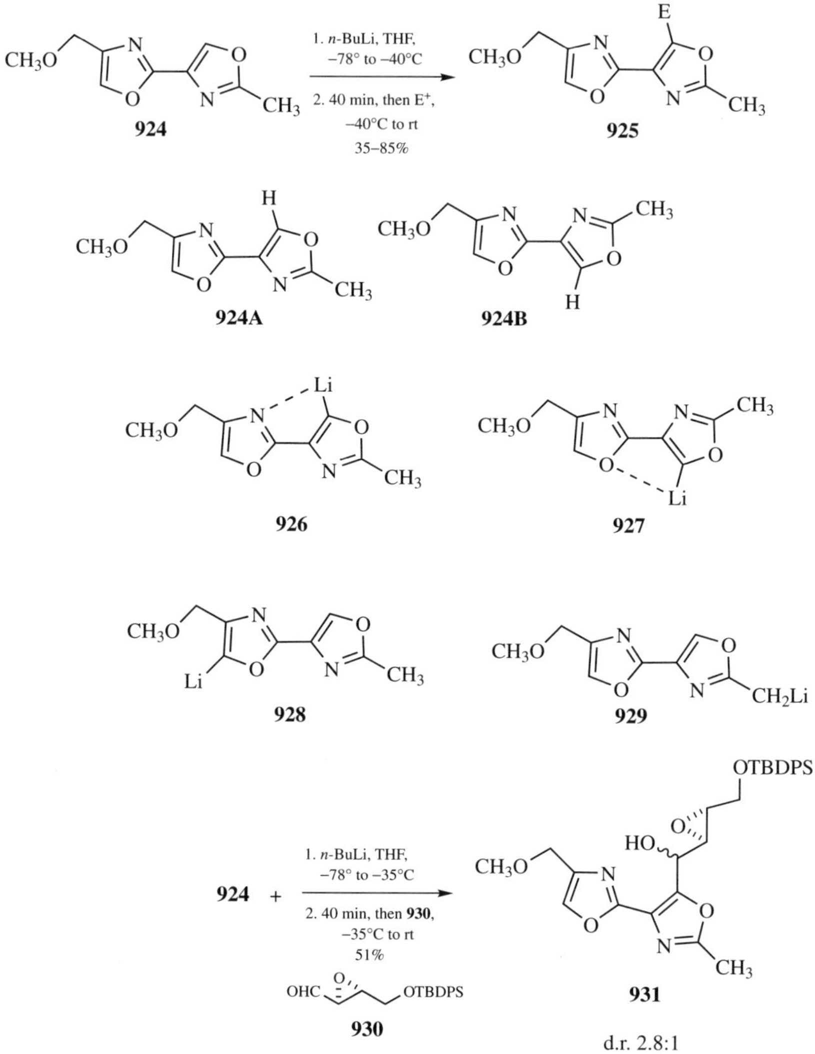

Scheme 1.248

Evans and co-workers[363] developed a regioselective lithiation protocol of 2-methyl-4-substituted oxazoles **945** during their synthesis of phorboxazoles. In particular, the authors required a general method to generate regioselectively a 2-(lithiomethyl)oxazole and to functionalize the intermediate without competitive lithiation and reaction at C(5). Among the bases investigated, lithium diethylamide was particularly effective and selective for the required transformation (Scheme 1.253, p. 206).

The authors investigated the factors controlling this remarkable selectivity and concluded that the "unique ability of diethylamine to effect the rapid equilibration

TABLE 1.66. 4-(METHOXYMETHYL)-2′-METHYL-5-SUBSTITUTED-2,4′-BIS-OXAZOLES VIA REGIOSELECTIVE LITHIATION OF 4-(METHOXYMETHYL)-2′-METHYL-2,4′-BIS-OXAZOLE[a]

Entry	Electrophile	E	Yield **925** (%)
1	CH$_3$I	CH$_3$	63
2	(CH$_3$)$_3$SiCl	(CH$_3$)$_3$Si	83
3	NCS	Cl	50
4	C$_6$H$_5$CHO	C$_6$H$_5$CHOH	85
5	(CH$_3$)$_2$CHCHO	(CH$_3$)$_2$CHCHOH	84

[a] Data from ref. 360.

of lithiated regioisomers at low temperatures" was the single most important factor. Examples relevant to their phorboxazole synthesis are shown in Scheme 1.254, p. 207.

Several examples of other metallooxazoles prepared from lithiated oxazoles by transmetalation reactions are discussed in the remaining portion of this section.

TABLE 1.67. ALKYLATION AND ACYLATION OF 5-(1,3-DITHIAN-2-YL)OXAZOLE)[a]

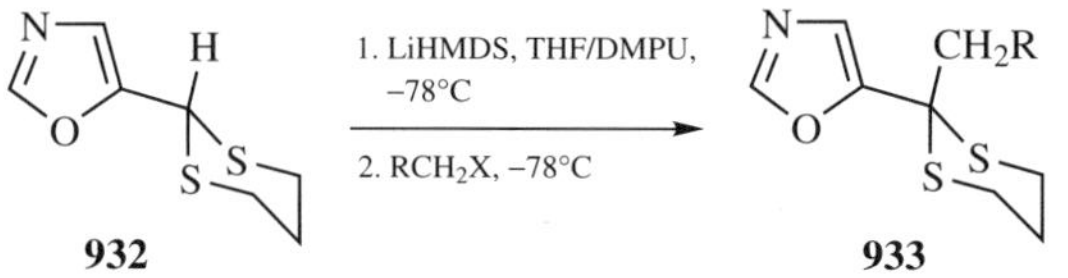

Figure 1.71

Entry	Electrophile	Product	Yield (%)
1	n-C$_4$H$_9$I	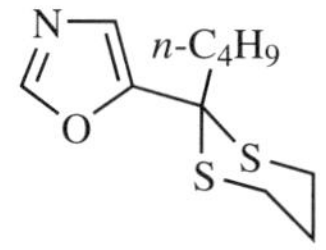	82
2	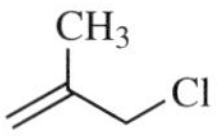 methallyl chloride (CH$_2$=C(CH$_3$)CH$_2$Cl)	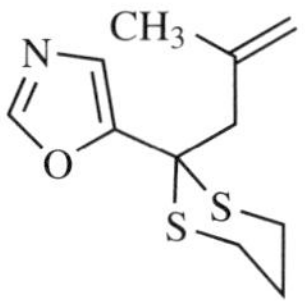	88
3	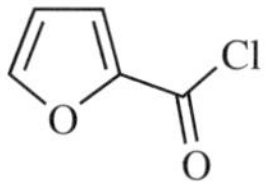 2-furoyl chloride	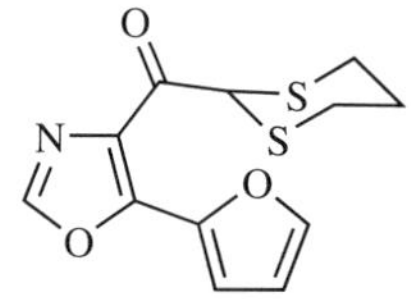	55
4	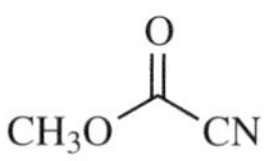 CH$_3$OC(O)CN	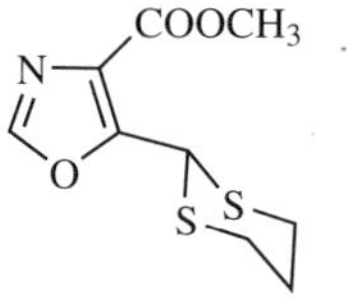	84
5	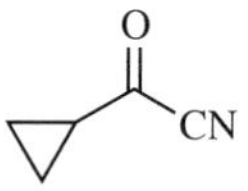 cyclopropyl-C(O)CN	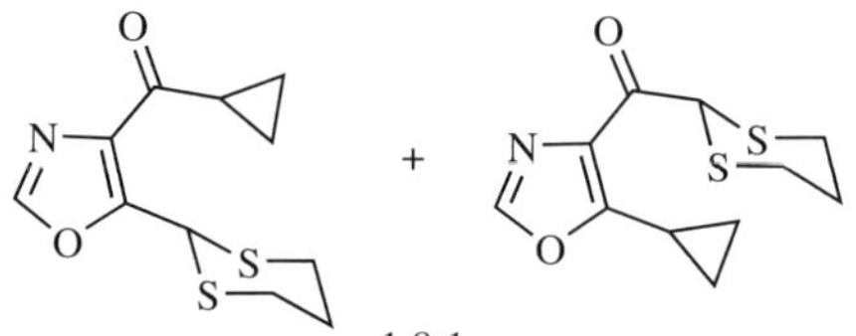 1.8:1 mixture	98

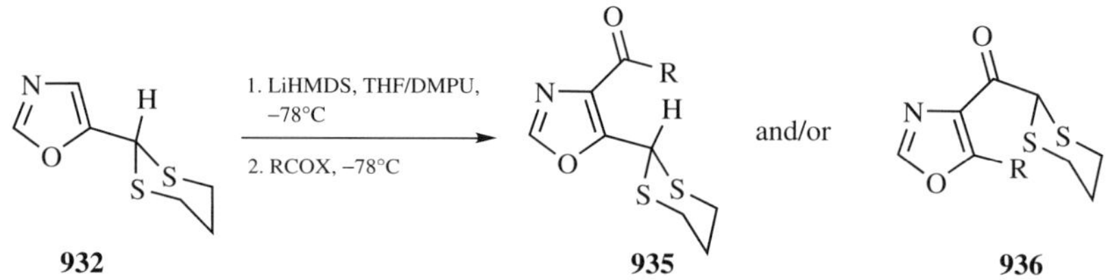

[a] Data from ref. 361.

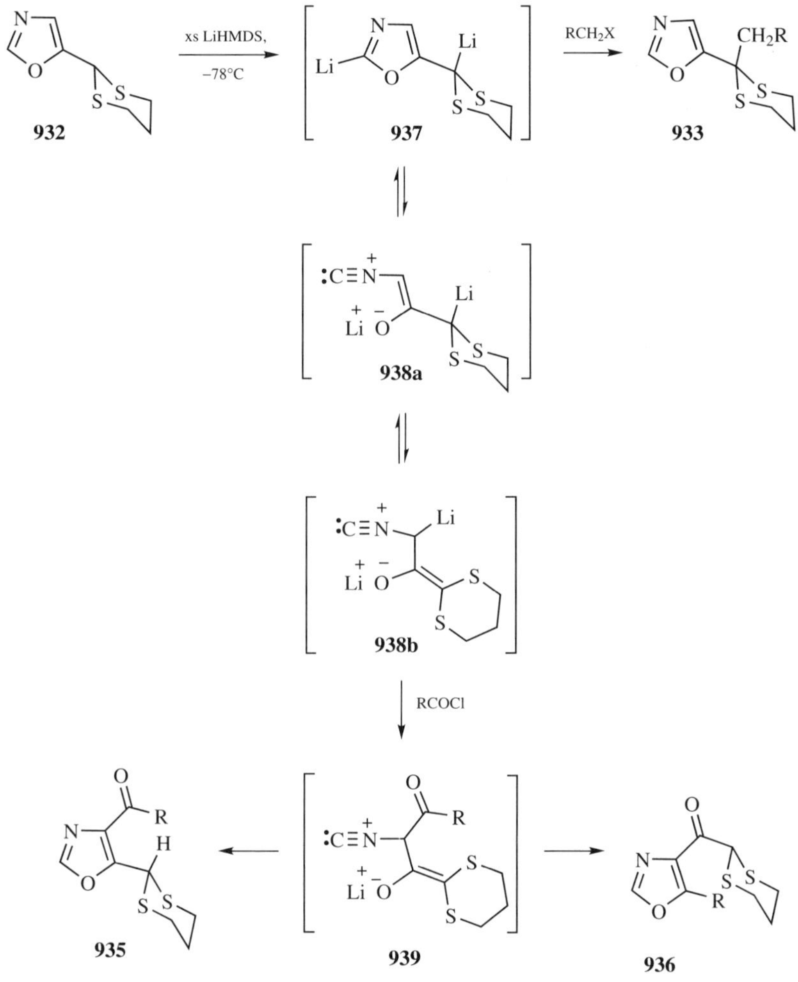

Scheme 1.251

Anderson and co-workers[364] reported a general synthesis of 2-acyl-5-phenyloxazoles **956**, which are generally inaccessible by traditional methods. The key intermediate, 5-phenyloxazol-2-ylzinc chloride **954** was prepared via transmetalation of 5-phenyloxazol-2-yllithium with $ZnCl_2$ (Scheme 1.255). However, this reagent was unreactive and required conversion to the bimetallic species **955** using CuI. Acylation of **955** then proceeded smoothly to afford **956** in good to excellent yields. The reaction worked best in THF with a stoichiometric amount of CuI and excess acyl chloride. Selected examples of 2-acyl-5-phenyloxazoles prepared are shown in Table 1.68. General applications of oxazolylzinc halide reagents in coupling reactions are discussed in Section 1.4.8.

Helquist and co-workers[365] developed an oxazole organozinc Reformatsky-type reagent **958** as a solution to the problem of introducing a 2-methyloxazole moiety in

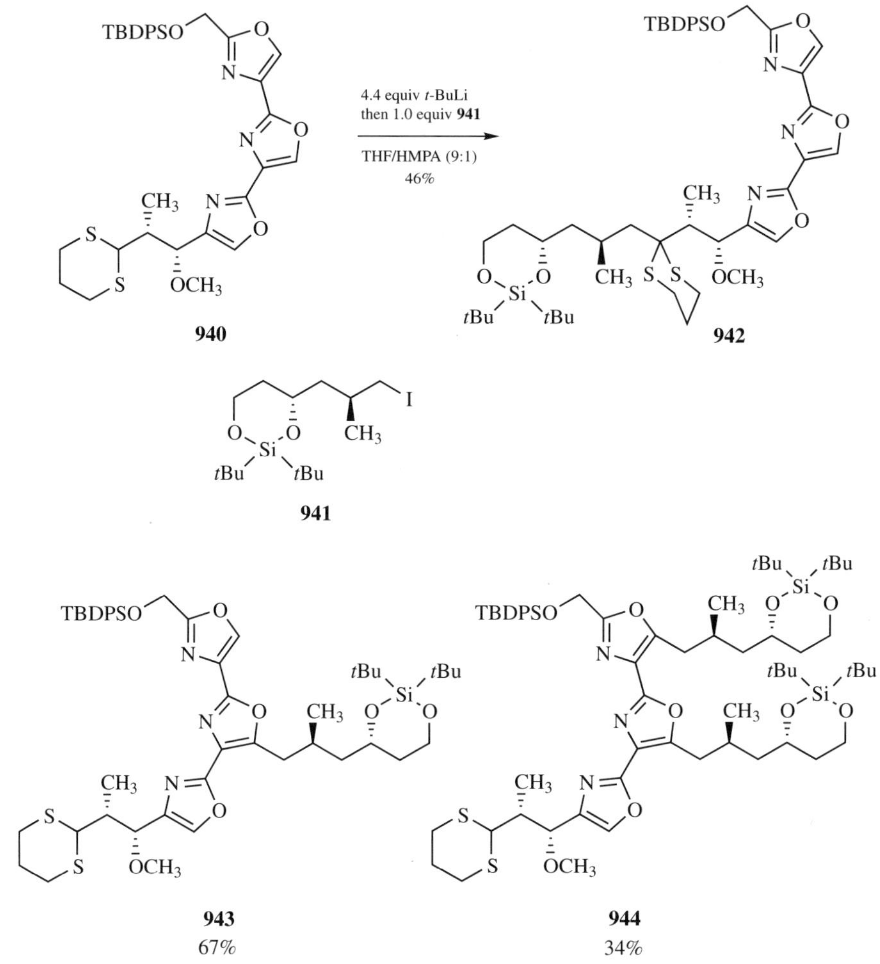

Scheme 1.252

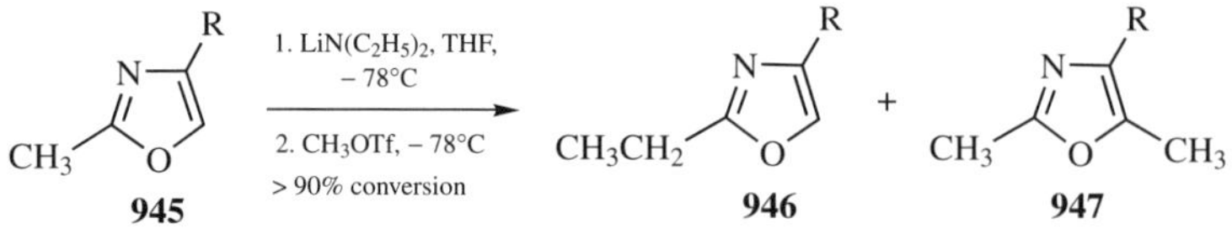

946:947
> 95:5 up to 99:1

R = C_6H_5, CH_2OH, $CH{=}C(CH_3)CH_2OTIPS$, $CH(OTES)CH_2CH_2OTBS$

Scheme 1.253

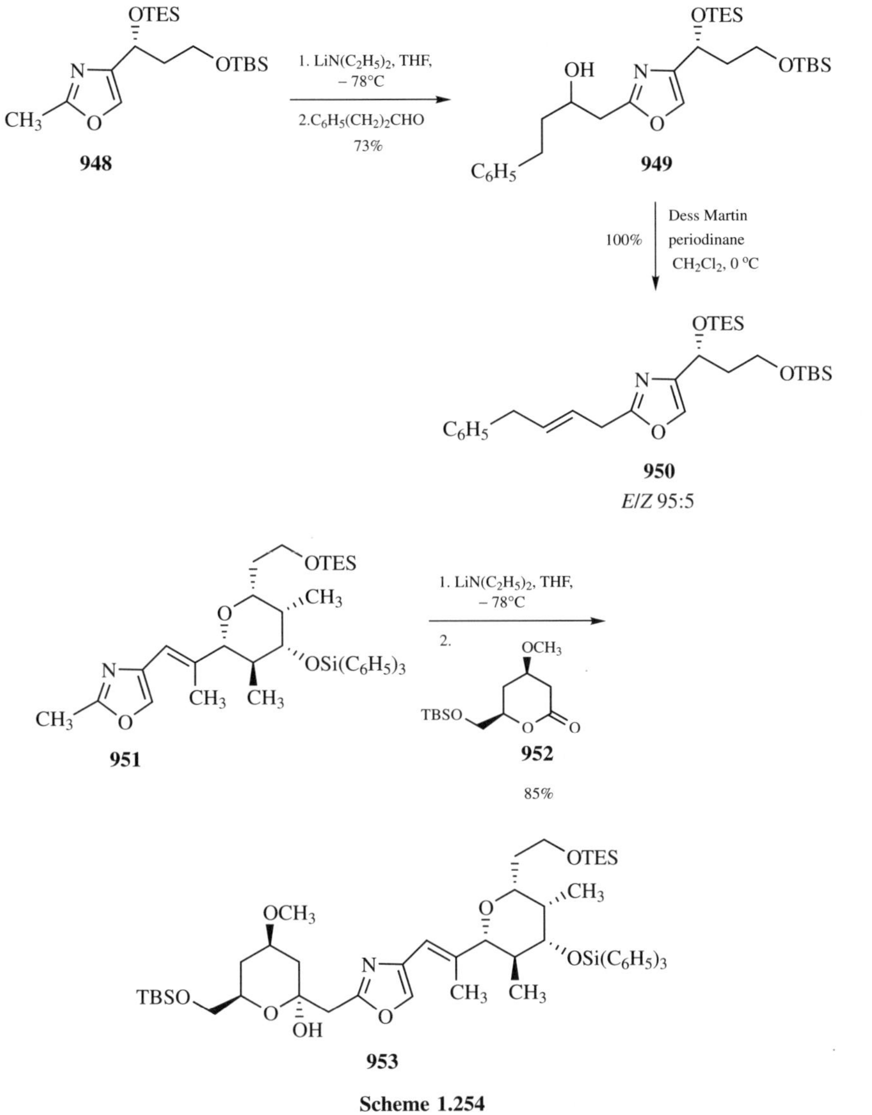

Scheme 1.254

their syntheses of virginiamycin and madumycin (see Section 1.5). In addition, the authors developed this type of reagent into a general method for preparing 2-(2-hydroxyethyl)-4-oxazolecarboxylic acid esters **959** starting from 2-(bromomethyl)-4-oxazolecarboxylic acid ethyl ester **957** (Scheme 1.256). Originally, the authors envisioned using **957** as an electrophile but found instead that it reacted readily with zinc dust to generate the Reformatsky-type reagent **958**. Condensation of **958** with ketones and aldehydes yielded the target alcohols **959**.

Advantages of this method are that it does not require the preparation of a mixed Zn/Cu derivative. In addition, Lewis acids are not required as catalysts. The authors did not observe any Michael addition products from α,β-unsaturated

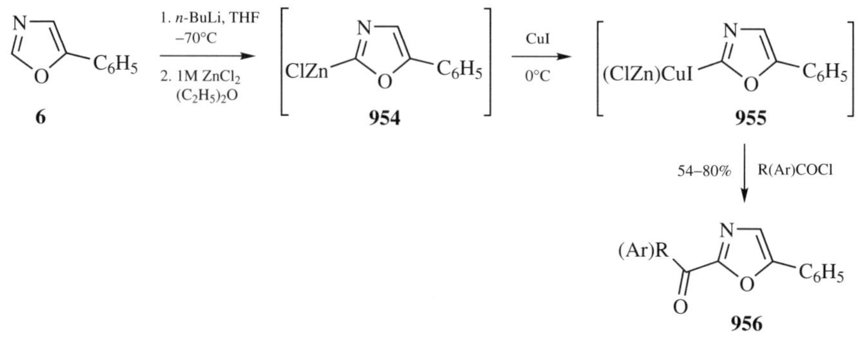

Synthesis and Reactions of Oxazoles

Scheme 1.255

TABLE 1.68. 2-ACYL-5-PHENYLOXAZOLES FROM ACYLATION OF 5-PHENYLOXAZOL-2-YLZINC CHLORIDE[a]

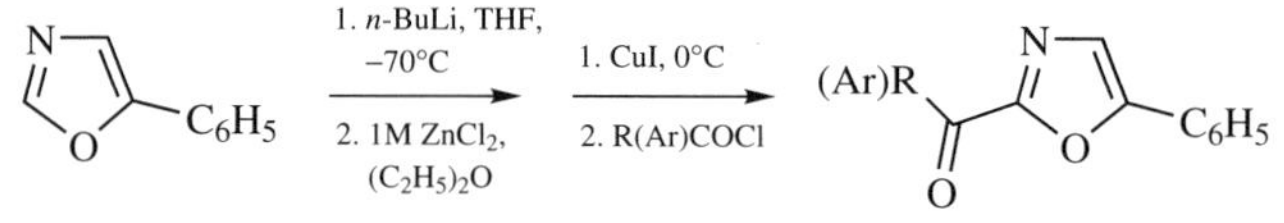

Figure 1.72

Entry	R(Ar)COCl	R(Ar)	Yield (%)
1	C_6H_5COCl	C_6H_5	70
2	4-CH_3O-C_6H_4COCl	4-CH_3O-C_6H_4	65
3	4-NO_2-C_6H_4COCl	4-NO_2-C_6H_4	80
4	$C_6H_5CH=CHCOCl$	$C_6H_5CH=CH$	58
5	$CH_3(CH_2)_2COCl$	$CH_3(CH_2)_2$	67
6	$(CH_3)_2CHCOCl$	$(CH_3)_2CH$	67
7	$(CH_3)_3CCOCl$	$(CH_3)_3C$	64

[a] Data from ref. 364.

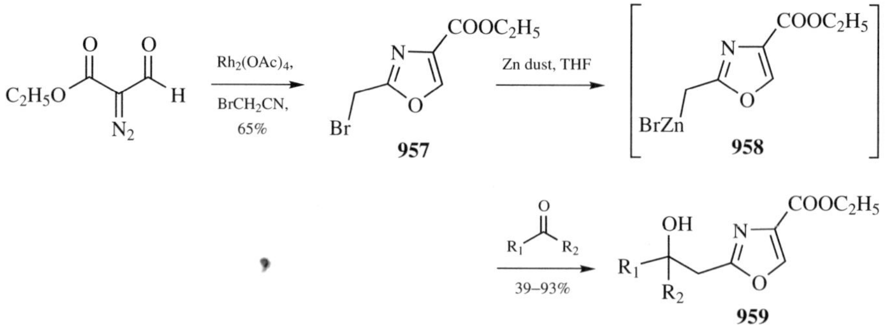

Scheme 1.256

TABLE 1.69. 2-(2-HYDROXYETHYL)-4-OXAZOLECARBOXYLIC ACID ETHYL ESTERS FROM 2-(BROMOMETHYL)-4-OXAZOLECARBOXYLIC ACID ETHYL ESTER REFORMATSKY REAGENT AND ALDEHYDES OR KETONES[a]

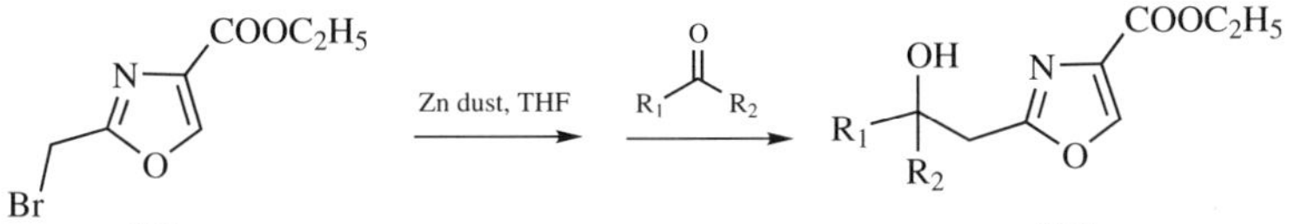

Figure 1.73

Entry	R_1	R_2	Product	Yield %
1	$n\text{-}C_5H_{11}$	H	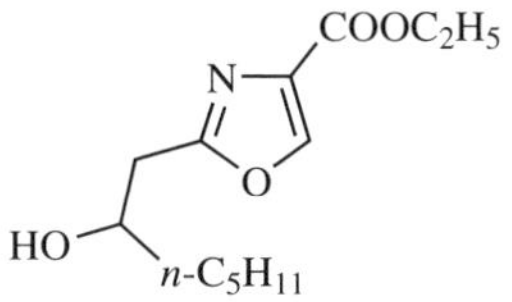	92
2	C_6H_5	H	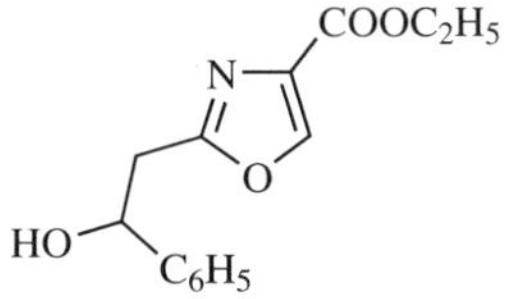	96
3	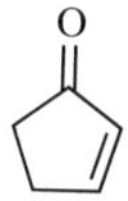 cyclopentenone		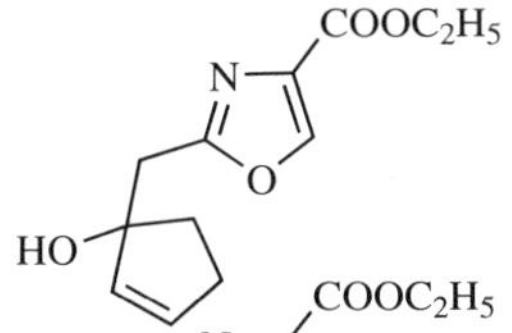	90
4	$CH_2=CH$	CH_3	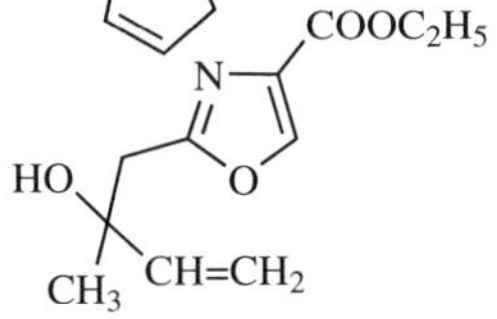	52

[a] Data from ref. 365.

aldehydes and ketones. However, acid chlorides do not react cleanly, although oxidation of a β-hydroxyoxazole can provide the same analogue. Representative examples are shown in Table 1.69.

In a later study, Helquist and co-workers[366] adapted this methodology for preparing an advanced optically active intermediate in a synthetic approach to type A streptogramin antibiotics. Treatment of **960** and "activated" zinc[367] with **957** afforded the alcohol (not shown) that was subjected to a Swern oxidation[367a] to produce the desired *E,E*-diene **961** (Scheme 1.257). The authors noted that the addition reaction yielded 50% of the intermediate alcohol together with 30–40% recovered **960** regardless of the amount of **957** used.

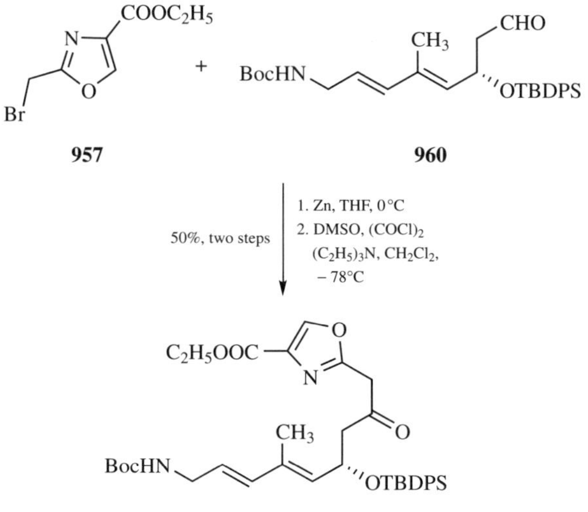

Scheme 1.257

Baker[368] described the transmetalation of 4,5-diphenyl-2-[2-(trimethylstannyl)ethyl]oxazole **962** with titanium tetrachloride in anticipation that it could function as a homoenolate equivalent. Reaction of **231a** with LDA followed by alkylation with (iodomethyl)trimethyltin gave **962** in very good yield. Treatment of **962** with TiCl$_4$ in the presence of benzaldehyde afforded α,4,5-triphenyl-2-oxazolepropanol **963** in 80% yield (Scheme 1.258).

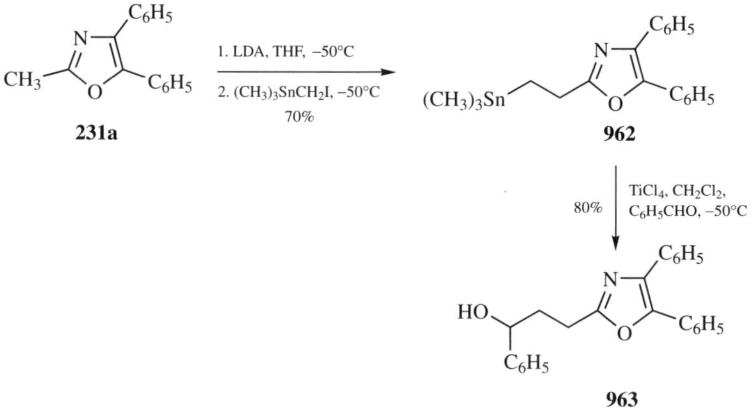

Scheme 1.258

Baker's mechanistic investigation of the transmetalation revealed the central role of the oxazole nitrogen atom in this reaction. He proposed a model in which this nitrogen coordinates to titanium tetrachloride to generate a chelate together with the titanium coordinated carbon-tin σ bond that facilitates the transmetalation. His studies also showed that compounds lacking a coordinating nitrogen atom fail to transmetalate. A representation of the proposed transition-state for transmetalation **964** is shown in Figure 1.74

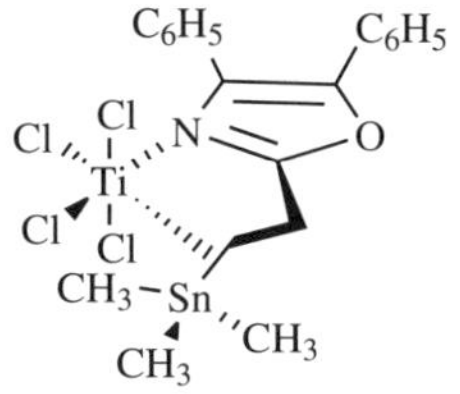

Figure 1.74. Proposed transition state for transmetalation.

1.4.8. Transition Metal-Catalyzed Cross-Coupling Reactions

Transition metal-catalyzed cross-coupling reactions are among the most useful carbon–carbon bond-forming reactions available to synthetic chemists. Construction of substituted oxazoles, particularly bis-oxazole and tris-oxazole containing natural products, continues to fuel significant research efforts in this arena. In addition, a number of remarkably novel and ingenious methodologies have resulted from efforts to prepare intricate and complex oxazole containing pharmaceutical agents.

Appropriately substituted oxazoles can participate in transition metal-catalyzed cross-coupling reactions either as the organometallic reagent or as the coupling partner. Halide-, OTf-, or SCH_3-substituted oxazoles have been employed quite successfully as coupling partners to prepare 2-, 4-, and 5-substituted analogs. Examples of each type of reaction from the recent literature are discussed. Additional examples are described in Section 1.5.

Pridgen[369] adapted his $Ni(dppe)Cl_2$ catalyzed cross-coupling methodology of Grignard reagents to include 4,5-diphenyl-2-(methylthio)oxazole **965** (Scheme 1.259). Thus coupling **965** with a variety of Grignard reagents afforded a 2-alkyl(aryl)-4,5-diphenyloxazole **966** in 76–95% yield. This methodology is particularly useful for preparing 2-alkyl(aryl)oxazoles in which the other functional groups are stable to Grignard reagents. Representative examples are shown in Table 1.70.

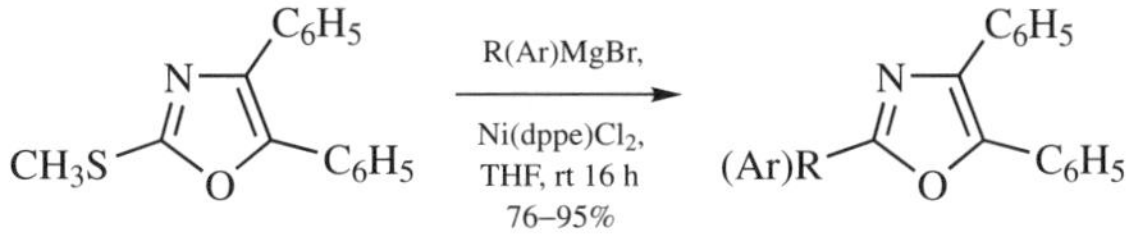

Scheme 1.259

Vedejs and Luchetta[358] coupled 5-(4-methylphenyl)oxazol-2-ylzinc chloride **968** with 4-iodo-5-(2-phenylethyl)oxazole **969** and isolated 5-(4-methylphenyl)-

TABLE 1.70. 2-ALKYL(ARYL)-4,5-DIPHENYLOXAZOLES FROM Ni(dppe)Cl$_2$ CATALYZED CROSS-COUPLING OF 4,5-DIPHENYL-2-(METHYLTHIO)OXAZOLE AND GRIGNARD REAGENTS[a]

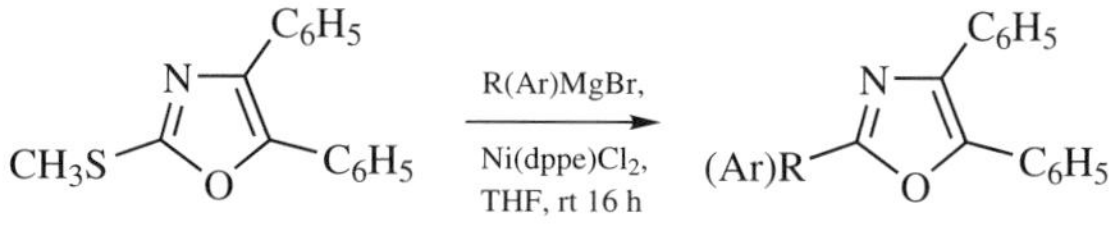

Figure 1.75

Entry	R(Ar)MgBr	Product	Yield (%)
1	C$_6$H$_5$MgBr	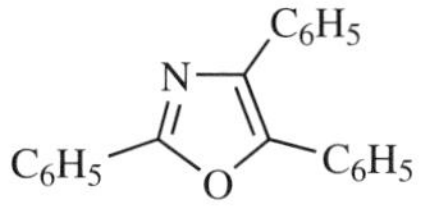 2-C$_6$H$_5$-4,5-diphenyloxazole	90
2	C$_2$H$_5$MgBr	2-C$_2$H$_5$-4,5-diphenyloxazole	95
3	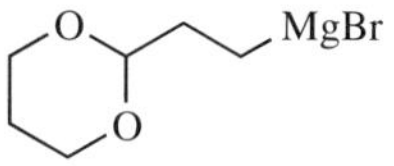 (1,3-dioxan-2-yl)CH$_2$CH$_2$MgBr	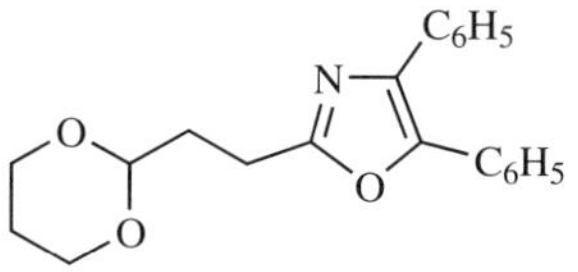 2-[2-(1,3-dioxan-2-yl)ethyl]-4,5-diphenyloxazole	76
4	4-CH$_3$-C$_6$H$_4$MgBr	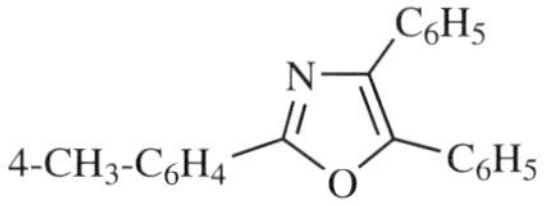 2-(4-CH$_3$-C$_6$H$_4$)-4,5-diphenyloxazole	82

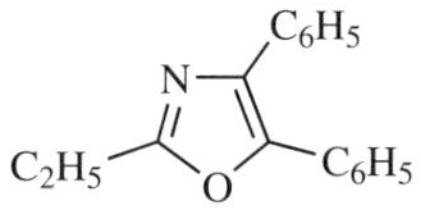

[a] Data from ref. 369.

5′-(2-phenylethyl)-2,4′-bis-oxazole **970** in a synthetically acceptable yield (Scheme 1.260). The authors observed no homocoupling products and recovered 38% of **969**.

Yamanaka and co-workers[370] prepared 2-methyl-5-phenyl-4-(phenylethynyl)oxazole **972** and 2-methyl-4-phenyl-5-(phenylethynyl)oxazole **974** in excellent yield via palladium-catalyzed coupling of the bromooxazoles **971** and **973** with phenylacetylene (Scheme 1.261). In contrast, the products derived from coupling with trimethylsilylacetylene decomposed during purification.

Coupling **971** and **973** with ethyl acrylate, acrylonitrile, and styrene produced **975a–c** and **976a–c**, respectively, albeit in lower and more variable yields. Finally, the authors were unsuccessful in their efforts to couple a 2-halooxazole with a terminal acetylene or an olefin. In these cases, they encountered significant decomposition that was attributed to ring cleavage of the 2-palladated oxazole.

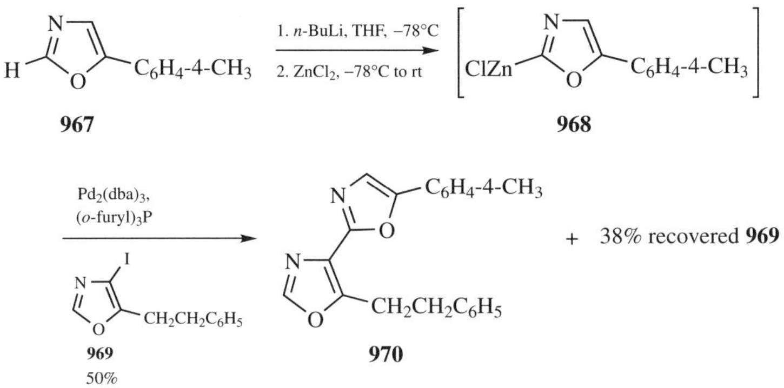

Scheme 1.260

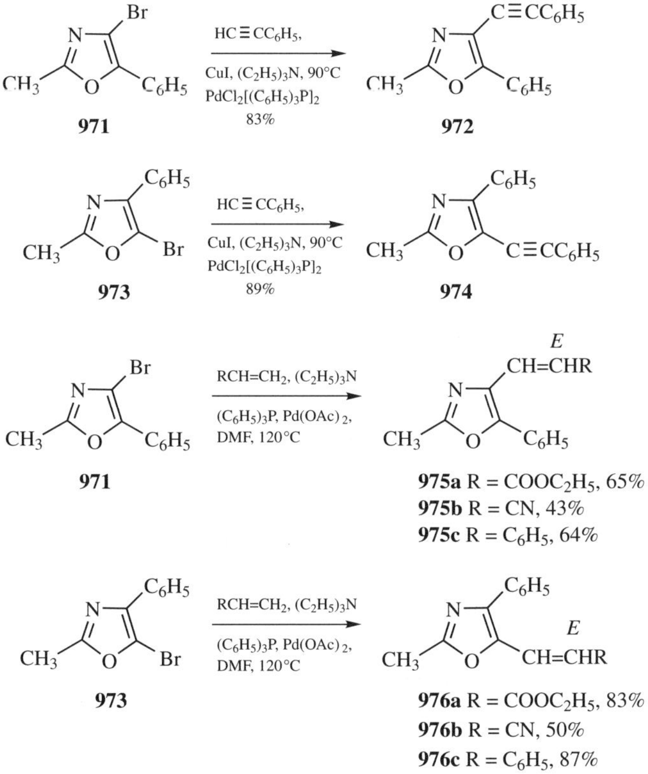

Scheme 1.261

Kelly and Lang[104,371a] reported the first examples of oxazole triflates as coupling partners in Stille couplings during their studies directed to the synthesis of dimethylsulfomycinamate. In a model study, *p*-anisamide was readily acylated, cyclized, and converted to 2-(4-methoxyphenyl)-4-oxazole triflate **977**

(Scheme 1.262). Palladium-catalyzed cross-coupling of **977** with a variety of organostannanes then afforded the 4-substituted oxazoles **202** and **978a–c** in excellent yield.

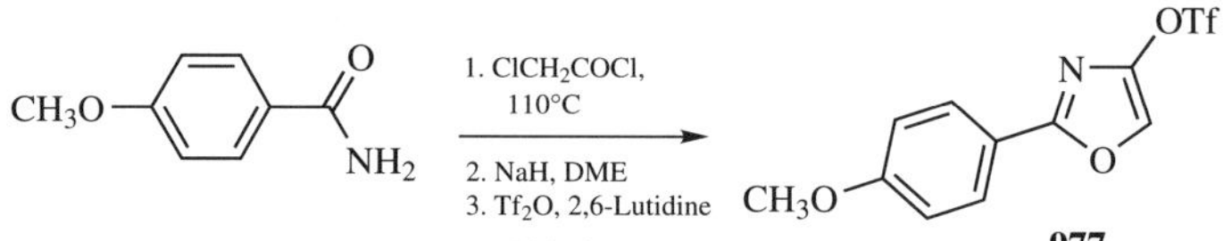

977

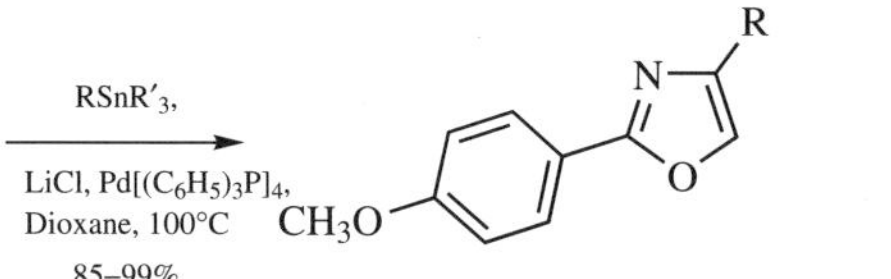

202 R = 2-pyridyl
978a R = C_6H_5
978b R = $CH{=}CH_2$
978c R = $C{\equiv}CC_6H_5$

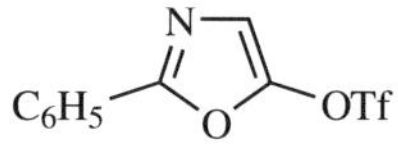

979

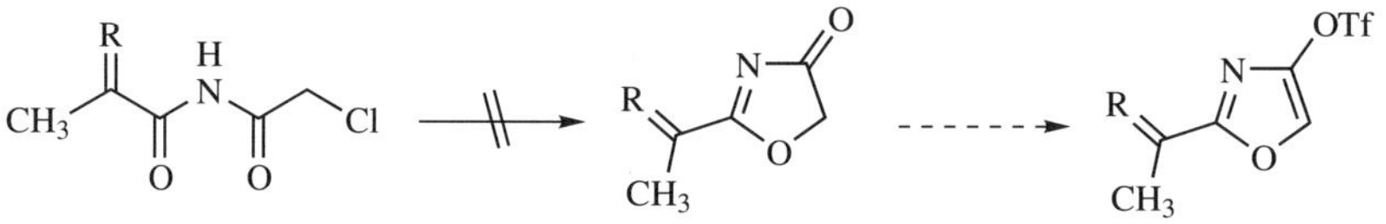

980a R = CH_2 **981a** R = CH_2 **982a** R = CH_2
980b R = O **981b** R = O **982b** R = O

Scheme 1.262

Unfortunately, the authors were frustrated in their efforts to adapt this methodology for preparing the core structure of dimethylsulfomycinamate. 2-Phenyl-5-oxazole triflate **979** was stable only at low temperature and decomposed under the coupling conditions. In contrast, the authors found that neither chloroamide **980a** or **980b** could be cyclized to the corresponding 4(5*H*)-oxazolone, thereby precluding preparation of the requisite triflates.

Schaus and Panek[370] also employed oxazole triflates as coupling partners in their palladium-catalyzed synthesis of vinyloxazoles for application to the C(26)-C(31) subunit of phorboxazole. They reported an improved procedure of preparing 2-phenyl-4-oxazole triflate **984** from 2-phenyl-4(5*H*)oxazolone **983** (Scheme 1.263). With **984** in hand, they developed a one-pot Cp_2ZrCl_2 catalyzed carboalumination of a terminal alkyne to produce an intermediate vinyl alane (not shown), which was then coupled with **984** to generate a 4-(*E*)-alkenyl-2-phenyloxazole, e.g., **985** or **987**, respectively.

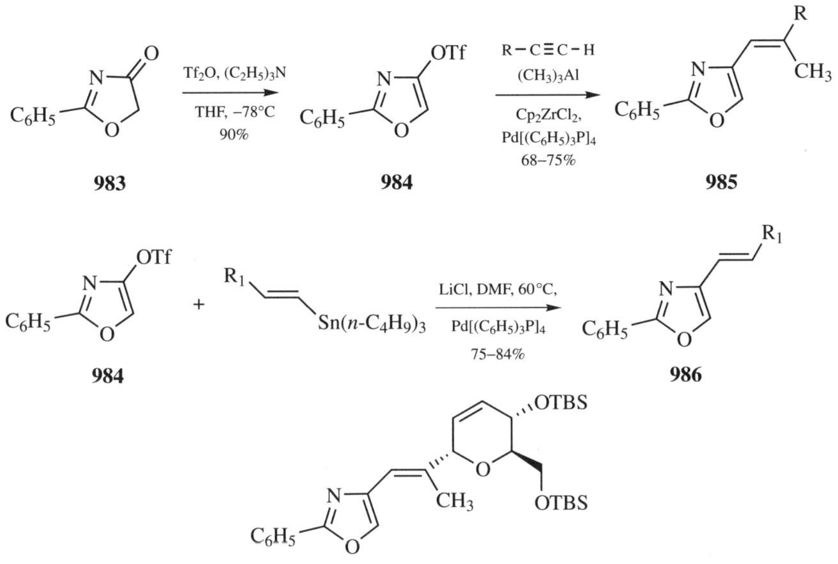

Scheme 1.263

In addition, the authors also prepared **986** via a Stille coupling of **984** with (E)-alkenylstannanes. After considerable experimentation, they also prepared **987**, a structural analog of the C(26)-C(31) phorboxazole subunit. These examples further broaden the scope of transition metal catalyzed couplings of oxazole triflates. Examples of 4-(E)-alkenyl-2-phenyloxazoles **985** and **986** are shown in Table 1.71.

Harran and co-workers[371] investigated the $PdCl_2(CH_3CN)_2$-catalyzed Stille coupling of stannyl styrenes **989a/b** with complex 5-bromo-4-cyano-2-substituted oxazoles **988b** and **991** (Scheme 1.264). This methodology was developed as a means to prepare key intermediates for the construction of the diazonamide ring skeleton (see Section 1.5). Initially, Stille coupling of **991** with bis-stannyl styrene **989b** afforded the undesired α-styryl stannane **992**. However, the authors found that **989b** was converted to **989a** in excellent yield by treatment with oxalic acid/silica gel. The completely functionalized A-ring coupling partner, **988b**, was prepared in two steps from N-t-butyloxycarbonyl-L-valine and aminomalononitrile tosylate (see Section 1.3.9) followed by bromination. Then, Stille coupling of **988b** with **989a** produced **990** in excellent yield. Further elaboration of **990** to a macrocyclic lactam containing the A, E, and F rings of diazonamide is discussed in Section 1.5.

Dondoni's group[372] also studied the synthesis and coupling reactions of 2-(stannyl)oxazoles. This work was part of their extensive investigations of the synthetic utility of stannylated and silylated azoles. Lithiation of 4-methyloxazole **3** followed by quenching with trimethyltin chloride gave the air- and moisture-sensitive 4-methyl-2-(trimethylstannyl)oxazole **993** in 60% yield after distillation

TABLE 1.71. 4-(E)-ALKENYL-2-PHENYLOXAZOLES FROM PALLADIUM-CATALYZED CROSS-COUPLING REACTIONS OF 2-PHENYL-4-OXAZOLE TRIFLATE[a]

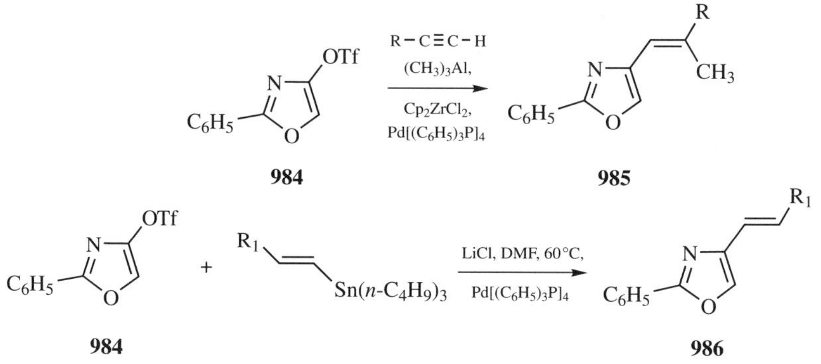

Figure 1.76

Entry	R	R_1	Conditions[b]	Yield **985** (%)	Yield **986** (%)
1	n-C_5H_{11}	—	A	75	—
2	C_6H_5	—	A	70	—
3	CH_2CH_2OTIPS	—	A	72	—
4	—	C_6H_5	B	—	75
5	—	n-C_5H_{11}	B	—	80
6	—	$CH_2CH_2COCH_3$	B	—	84

[a] Data from ref. 370.
[b] A, Cp_2ZrCl_2, $(CH_3)_3Al$, $Pd[(C_6H_5)_3P]_4$; B, LiCl, $Pd[(C_6H_5)_3P]_4$, DMF, 60°C.

(Scheme 1.265). The authors did not observe any evidence (IR) for the isocyanovinyltrimethylstannyl enol ether **994**. They attributed this to the lower affinity of tin compared to silicon for oxygen (see Section 1.4.9).

Treatment of **993** with an aryl- or heteroaryl halide in the presence of 5 mol % $Pd[(C_6H_5)_3P]_4$ gave the 2-aryl(heteroaryl)-4-methyloxazoles **995** in excellent yield. This reaction provides a facile entry to 2-aryloxazoles and 2-(heteroaryl)oxazoles and is a complementary approach to Pridgen's nickel(II)-catalyzed coupling of 2-(methylthio)oxazoles with Grignard reagents (vide supra).[369] Representative examples are shown in Table 1.72. The authors also prepared 4,4-dimethyl-2-tributylstannyl-2-oxazoline using tributyltin chloride, which avoids using highly toxic trimethyltin chloride.

Lilly chemists[373] prepared **998b** as part of a series of 4-amino-6-heteroaryl-1,3,4,5-tetrahydrobenz[c,d]indoles evaluated as anti-emetic agents via coupling of 2-(tributylstannyl)oxazole **996** with (+)-(2aR, 4S)-1-benzoyl-6-iodo-4-(di-n-propylamino)-1,2,2a,3,4,5-hexahydrobenz[c,d]indole **997**. They isolated **998b** after deprotection as shown in Scheme 1.266.

Hughes and co-workers[347] transmetalated **892** with $ZnCl_2$ and prepared **895** as described previously (Scheme 1.239). The authors found that **895** readily cross-coupled with vinyliodides and aryliodides using $PdCl_2[(C_6H_5)_3P]_2$ prereduced with DIBAL-H to afford 2-substituted-5-(3-methyl-5-isoxazolyl)oxazoles **999** in modest

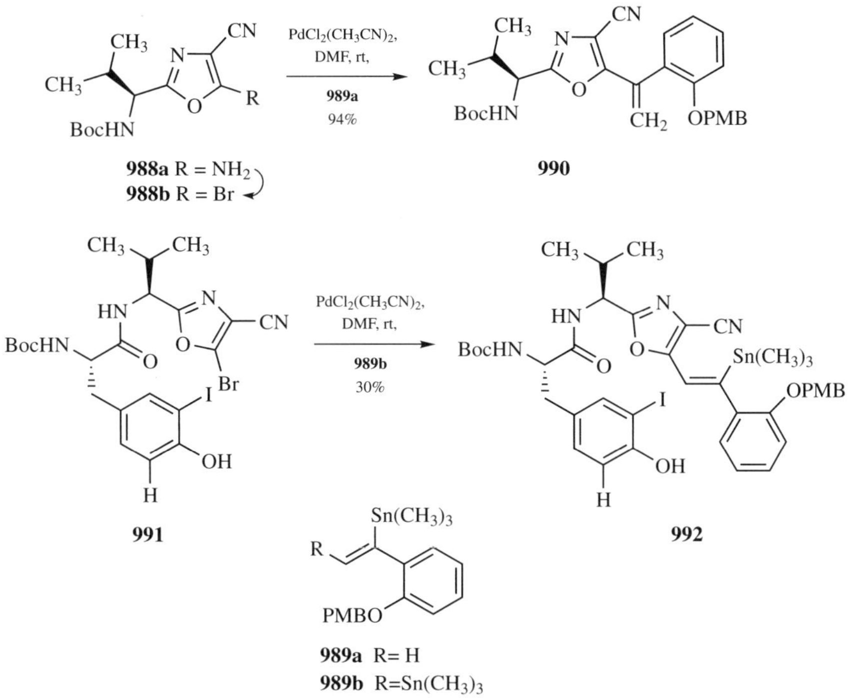

Scheme 1.264

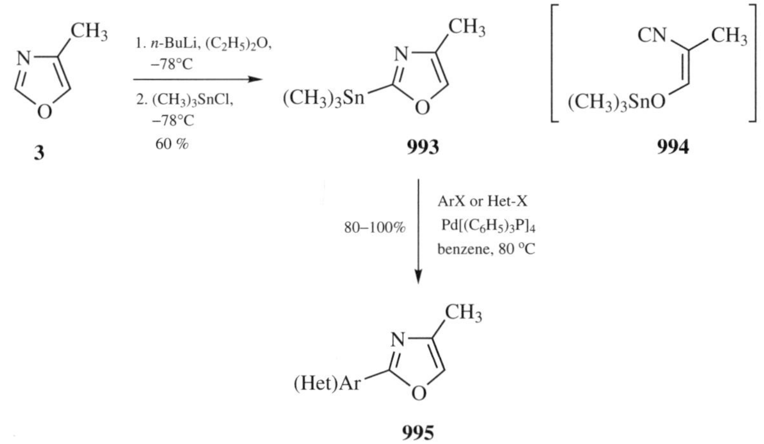

Scheme 1.265

yield (Scheme 1.267). Vinyl bromides were much less effective due to a competitive decomposition of the intermediate organozinc complex. Selected examples of 2-substituted oxazoles prepared via this methodology are shown in Table 1.73.

TABLE 1.72. 2-ARYL(HETEROARYL)-4-METHYLOXAZOLES FROM PALLADIUM-CATALYZED CROSS-COUPLING OF 4-METHYL-2-(TRIMETHYLSTANNYL)OXAZOLE[a]

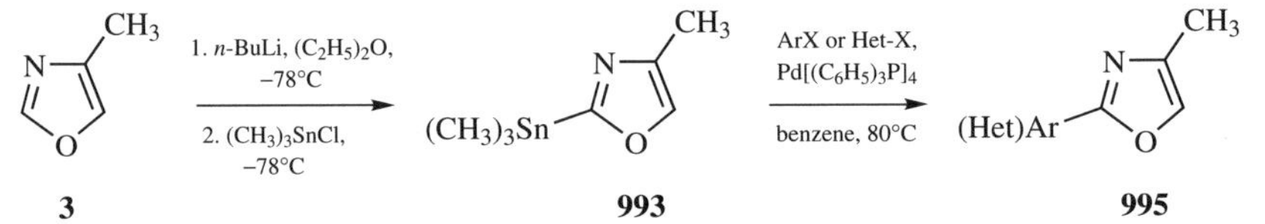

Figure 1.77

Entry	ArX or Het-X	Ar(Het)	Yield (%)
1	C_6H_5I	C_6H_5	80
2	4-CH_3CO-C_6H_4Br	4-CH_3CO-C_6H_4	100
3	3-Br-pyridine	3-pyridyl	100
4	2-Br-$C_{10}H_7$	2-naphthyl	92

[a] Data from ref. 372.

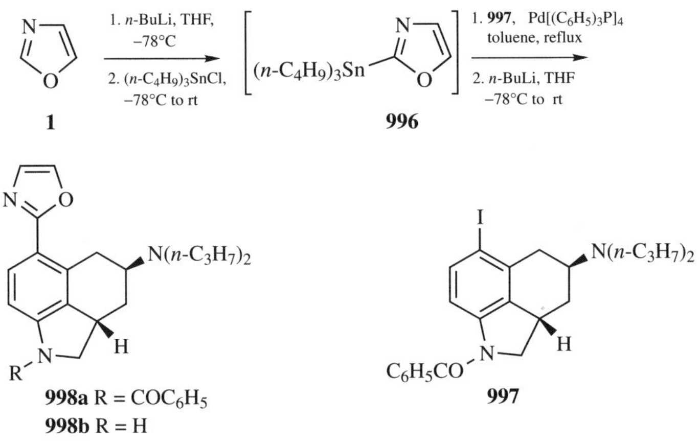

Scheme 1.266

Anderson and Harn[374] described a general synthesis of 2-substituted- and 2,5-disubstituted oxazoles that also employed a palladium-catalyzed cross-coupling reaction of oxazol-2ylzinc chlorides (Scheme 1.268). Transmetalation was best accomplished using a threefold excess of $ZnCl_2$. The authors investigated a number of cross-coupling catalysts, including Pd[$(C_6H_5)_3$P]$_4$, Pd(OAc)$_2$/$(C_6H_5)_3$P (two equiv), PdCl$_2$[(o-Tol)$_3$P]$_2$/n-BuLi (two equivalents), PdCl$_2$[(o-Tol)$_3$P]$_2$/DIBAL-H, and PdCl$_2$[$(C_6H_5)_3$P]$_2$/n-BuLi (two equivalents). The most effective reaction conditions used 5 mol % PdCl$_2$[$(C_6H_5)_3$P]$_2$ and 10 mol % n-BuLi in refluxing THF. Triflates were the best coupling partners, although aryl iodides were also synthetically useful; aryl bromides were not particularly reactive.

The authors also prepared **998a** in much improved yield (52%) relative to their preparation of **998a** from **996**. Finally, they extended the utility of the methodology

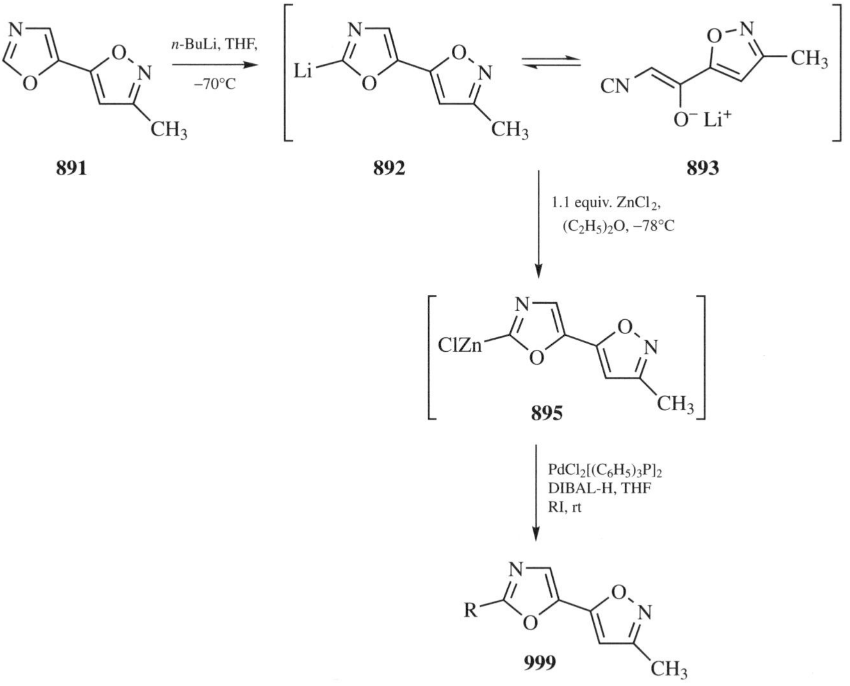

Scheme 1.267

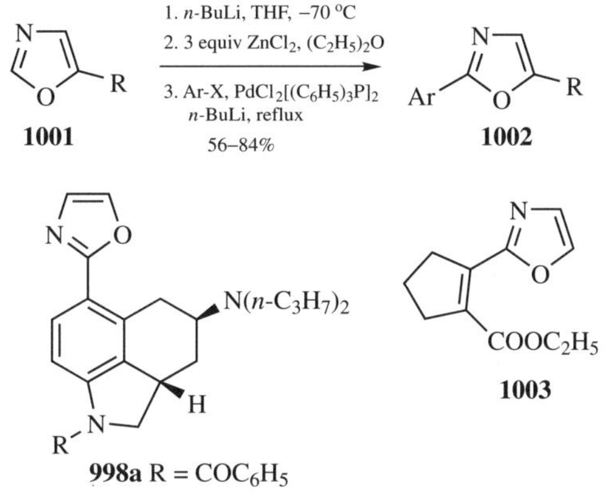

Scheme 1.268

to include synthesis of the functionalized cyclopentene ester **1003**. Overall, synthetically, oxazol-2-ylzinc reagents are comparable to, and in many cases superior to, oxazol-2-yltin reagents in terms of yield, product isolation, reagent toxicity, and waste disposal. Selected examples are shown in Table 1.74.

Miller and co-workers[375] prepared 2-(4-methoxyphenyl)oxazole (Table 1.74, entry 2) for evaluation as a nonlinear optical chromophore. These authors coupled

TABLE 1.73. 2-ALKENYL(ARYL)OXAZOLES OR 2-ALKENYL(ARYL)-5-SUBSTITUTED OXAZOLES FROM PALLADIUM-CATALYZED CROSS-COUPLING OF OXAZOL-2YLZINC CHLORIDES[a]

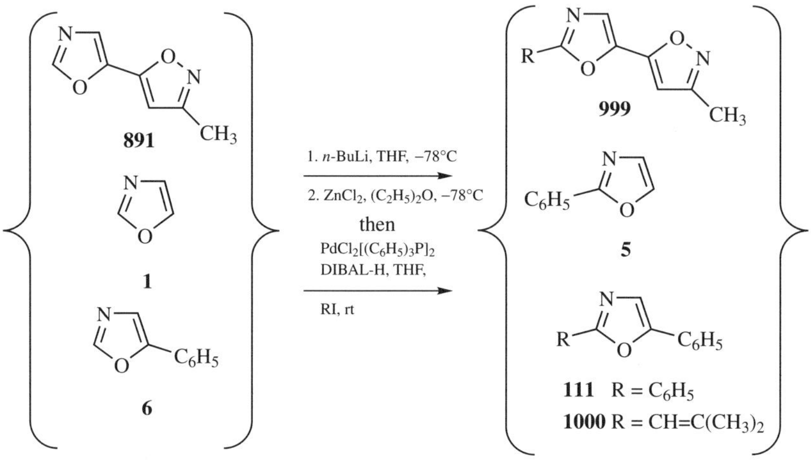

Figure 1.78

Entry	Oxazole	RX	R	Product	Yield (%)
1	895	C_6H_5I	C_6H_5	999a	68
2	895	$(CH_3)_2C=CHI$	$(CH_3)_2C=CH$	999b	78
3	1	C_6H_5I	C_6H_5	5	53
4	6	C_6H_5I	C_6H_5	111	69
5	6	$(CH_3)_2C=CHI$	$(CH_3)_2C=CH$	1000	78

[a] Data from ref. 347.

TABLE 1.74. 2-ARYLOXAZOLES AND 2-ARYL-5-PHENYLOXAZOLES FROM PALLADIUM CATALYZED CROSS-COUPLING OF OXAZOL-2YLZINC CHLORIDES[a]

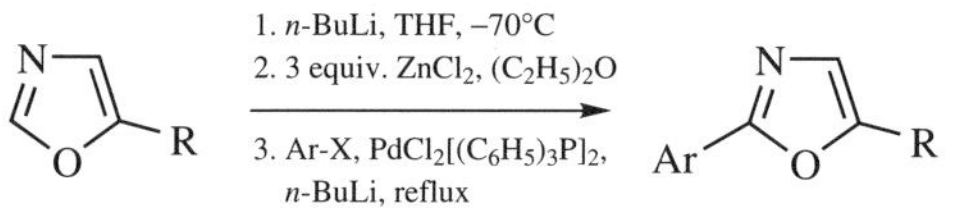

Figure 1.79   1001   1002

Entry	R	Ar	X	Yield (%)
1	H	1-Naphthyl	OTf	67
2	H	4-CH_3CO–C_6H_4	OTf	67
3	C_6H_5	1-Naphthyl	OTf	83
4	C_6H_5	1-Naphthyl	I	65
5	C_6H_5	2-CH_3–C_6H_4	I	68
6	C_6H_5	4-CH_3O–C_6H_4	I	64
7	C_6H_5	4-NO_2–C_6H_4	I	64

[a] Data from ref. 374.

(4-methoxyphenyl)zinc chloride with 2-bromooxazole in the presence of Pd[$(C_6H_5)_3P$]$_4$ and isolated 2-(4-methoxyphenyl)oxazole in 75% yield.

1.4.9. Trimethylsilyloxazoles

Dondoni and co-workers[7,372,372a,376–378] have a long-standing program to explore the synthesis and reactions of 2-silylated and 2-stannylated oxazoles as stable oxazol-2-yl anion equivalents (Fig. 1.80). The reactivity of such derivatives parallels that of the corresponding 2-silylated and 2-stannylated thiazoles in many cases. Some representative examples highlight the versatility of these important intermediates in organic synthesis.

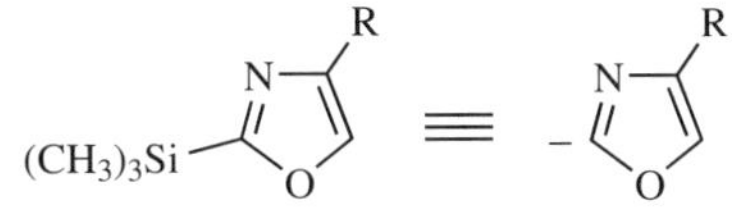

Figure 1.80. 2-(Trimethylsilyl)oxazole: an oxazol-2-yl anion equivalent.

Dondoni's group[376] reported the first synthesis of a 2-(trimethylsilyl)oxazole (Scheme 1.269). Lithiation of 4-methyloxazole **3** and trapping with

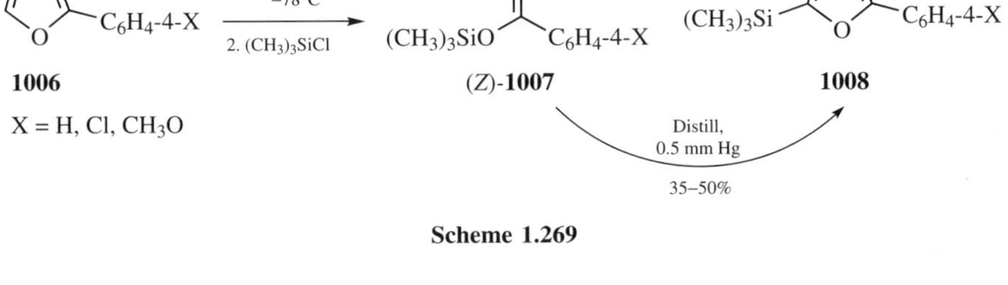

Scheme 1.269

chlorotrimethylsilane gave predominately (95% by ^{1}H NMR) (Z)-2-isocyano-O-(trimethylsilyl)-1-propen-1-ol **1004** in 85% yield. Distillation of **1004** in the presence of a KOH pellet then gave the desired 4-methyl-2-(trimethylsilyl)oxazole **1005**, together with a small amount of **1004**.

5-Aryl-2-(trimethylsilyl)oxazoles can be prepared similarly.[7,372a] For example, lithiation of 5-phenyloxazole **1006** (X = H) and reaction with chlorotrimethylsilane generates a ~1:1 mixture of (Z)-2-isocyano-1-phenyl-O-(trimethylsilyl)-1-propen-1-ol **1007** (X = H) and 5-phenyl-2-(trimethylsilyl)oxazole **1008** (X = H). As before, distillation effected rearrangement of **1007** to **1008**. In both cases, the rearrangement was unexpected in that it results in the formation of a thermodynamically less stable C-Si bond from a stable O-Si bond. However, the rearrangement is energetically favored due to the formation of the aromatic oxazole ring.[372a]

2-(Trimethylsilyl)oxazoles behave as soft carbanion equivalents and readily react with a wide variety of electrophiles to afford 2-substituted oxazoles.[372a,376] For example, 4-methyl-2-(trimethylsilyl)oxazole **1005** readily reacts with aldehydes to yield 4-methyl-2-oxazolemethanol derivatives **1009** after desilylation. Aliphatic and aromatic aldehydes are useful in this reaction. In addition, modest diastereoselection was observed with chiral aldehydes, such as glyceraldehyde acetonide and 2-phenylpropanal. Examples are shown in Table 1.75.

Acylation of 2-(trimethylsilyl)oxazoles **1005** and **1008** (X = H) can be effected using acid halides, chloroformates, or oxalates and affords 2-acyloxazoles in poor to good yield (Scheme 1.270; Table 1.76).[372a,376] Only benzoyl chloride was sufficiently reactive among aromatic acid chlorides to acylate **1005** or **1008** (X = H), albeit in low to modest yield. In contrast, benzoylation of 4-methyl-2-(trimethylstannyl)oxazole **993** produced 2-benzoyl-4-methyloxazole in 90–95%

TABLE 1.75. 4-METHYL-2-OXAZOLEMETHANOL DERIVATIVES FROM 4-METHYL-2-(TRIMETHYLSILYL)OXAZOLE AND ALDEHYDES[a]

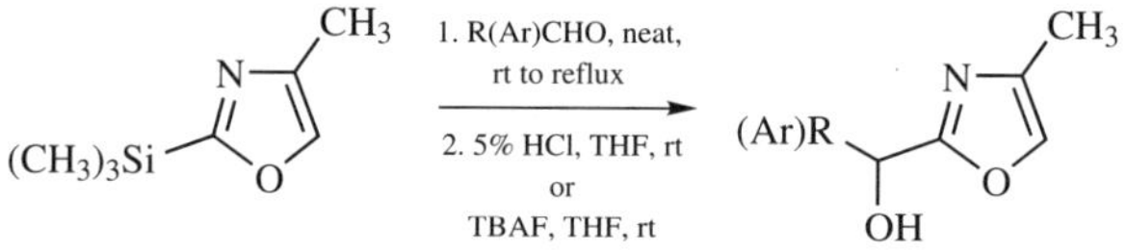

Figure 1.81

Entry	Aldehyde	R(Ar)	Yield (%)
1	C_6H_5CHO	C_6H_5	41
2	i-C_3H_7CHO	i-C_3H_7	74
3	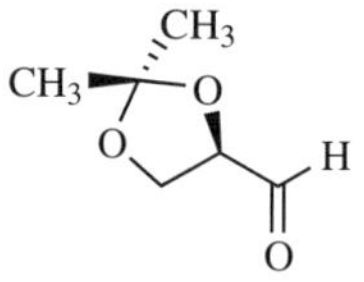 (glyceraldehyde acetonide)	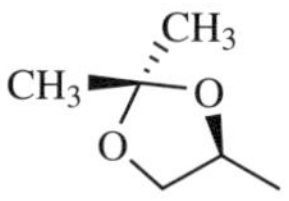 (acetonide product), d.r. (R:S) = 79:21	65
4	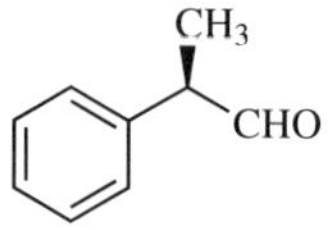 2-phenylpropanal	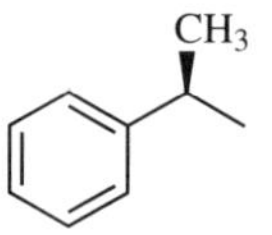 (1-phenylethyl), d.r. (R:S) = 31:69	60

[a] Data from refs. 372a and 376.

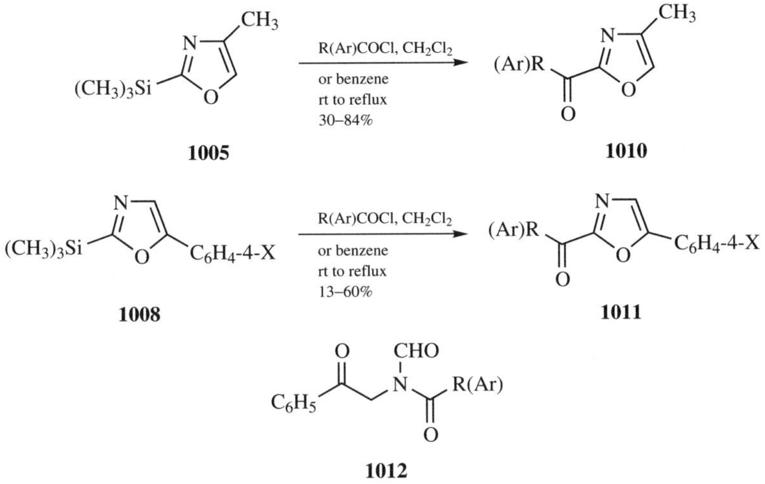

Reactions of Oxazoles 223

Scheme 1.270

TABLE 1.76. 2-ACYL-4-METHYLOXAZOLES AND 2-ACYL-5-ARYLOXAZOLES FROM 4-METHYL-2-(TRIMETHYLSILYL)OXAZOLE OR 5-ARYL-2-(TRIMETHYLSILYL)OXAZOLES AND ELECTROPHILES

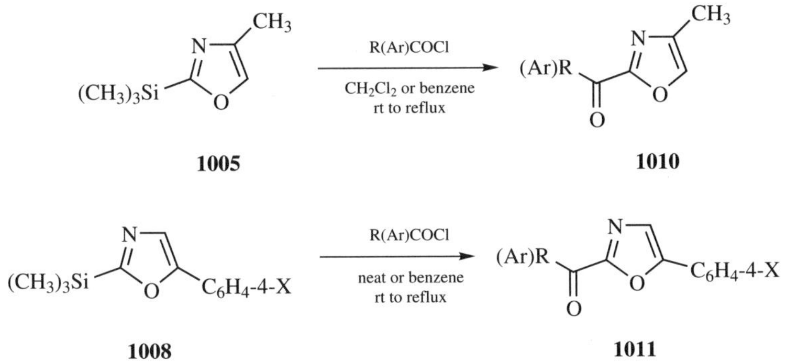

Figure 1.82

Entry	Oxazole	R(Ar)COCl	Yield **1010** (%)	Yield **1011** (%)	References
1	**1005**	C_6H_5COCl	40	—	372a, 376
2	**1005**	CH_3COCl	85	—	372a, 376
3	**1005**	C_2H_5OCOCl	84	—	372a, 376
4	**1005**	$CH_3OOCCOCl$	32	—	372a, 376
5	**1008**	C_6H_5COCl	—	29	372a
6	**1008**	CH_3COCl	—	32	372a
7	**1008**	C_2H_5OCOCl	—	13[a]	372a
8	**1008**	$CH_3OOCCOCl$	—	23	372a
9	**1008**	$CH_3OOCCH_2CH_2COCl$	—	30	372a
10	**1008**	$CH_3(CH_2)_4COCl$	—	25[b]	372a
11	**1008**	9-octadecenyl	—	60	372a

[a] Isolated 17% **1012** R = OC_2H_5.
[b] Isolated 37% **1012** R = $CH_3(CH_2)_4$.

yield (Table 1.76, entry 1; **1010** Ar = C_6H_5). Acid anhydrides were unreactive, even under forcing conditions or using fluoride ion catalysis. Overall, this methodology does complement the synthesis of 2-acyloxazoles via bimetallic 2-organozinc/ cuprate oxazole derivatives (see Scheme 1.255) although the yields are variable.

Both **1005** and **1008** (X = H) are readily acylated by an activated ketene, e.g., dichloroketene or *t*-butylcyanoketene (Scheme 1.271).[372a] The authors isolated excellent yields of the α,α-disubstituted 2-acyl-5-phenyloxazoles **1015a/b** (X = H). In contrast, the α,α-disubstituted 2-acyl-4-methyloxazoles **1014a/b** were isolated in modest to good yield. Nonethless, the yield of **1014a** was an improvement over direct acylation of **3** using dichloroketene.[379]

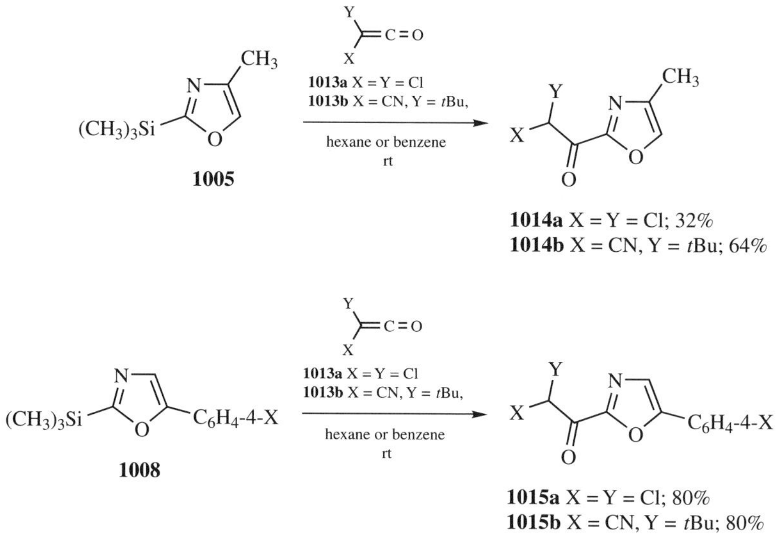

Scheme 1.271

Nucleophilic addition of **1005** to an *N*-acyliminium ion, e.g., a Reissert salt, and oxidation of the resulting adduct affords a 2-(heteroaryl)oxazole **1019**.[372a] Acylation of a nitrogen heterocycle **1016** with ethyl chloroformate generated the intermediate Reissert salts **1017** in situ (Scheme 1.272). Addition of **1005** to **1017** gave adducts **1018**, which produced **1019** after oxidative deacylation with *o*-chloranil. Examples of 2-(heteroaryl)oxazoles prepared from thiazole, benzothiazole, pyridine, quinoline, and isoquinoline are shown in Table 1.77. Donodni and co-workers[5] summarized their early work on preparation and reaction of 2-(trimethylstannyl)oxazoles and 2-(trimethylsilyl)oxazoles.

Dondoni and co-workers[378] conducted a detailed kinetic and spectroscopic study of the reaction of 2-(trimethylsilyl)thiazole **1020** with aldehydes to elucidate the reaction mechanism. They proposed a sequence involving a thiazolium-2-ylide (Scheme 1.273). Thus **1020** was reversibly quaternized with an aldehyde to produce the betaine **1021**, which undergoes a C → O silyl transfer to generate the

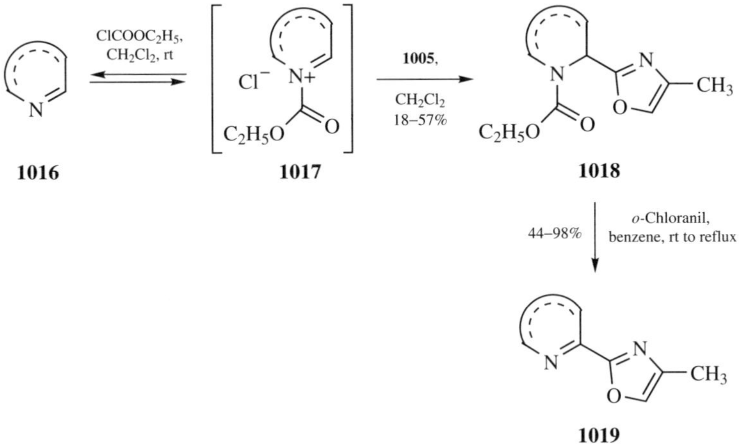

Reactions of Oxazoles   225

Scheme 1.272

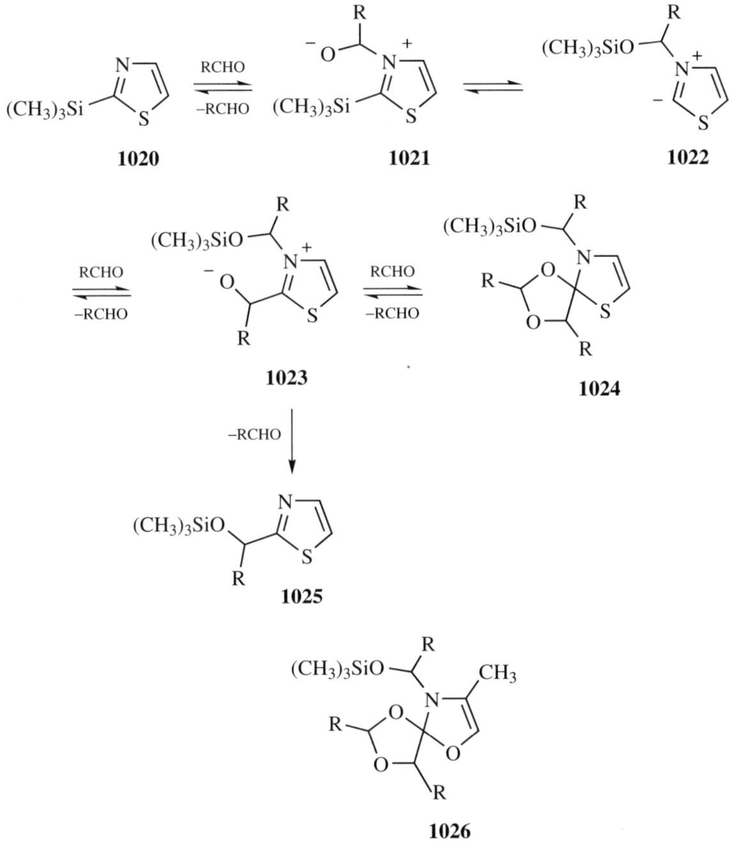

Scheme 1.273

TABLE 1.77. 2-(HETEROARYL)OXAZOLES FROM 4-METHYL-2-(TRIMETHYLSILYL)OXAZOLE AND REISSERT SALTS[a]

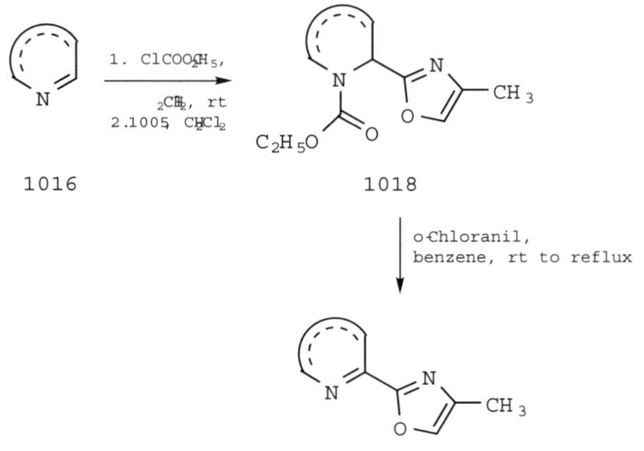

Figure 1.83

Entry	Heterocycle	1019	Yield 1018 (%)	Yield 1019 (%)
1	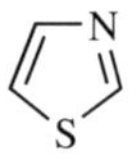 thiazole	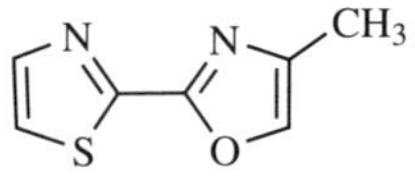	35	77
2	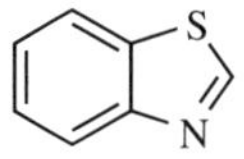 benzothiazole	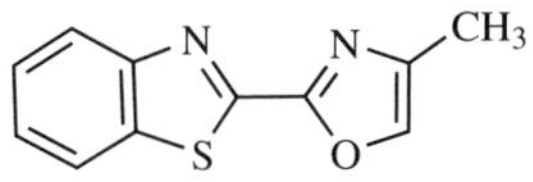	35	66
3	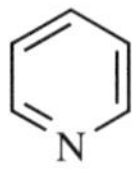 pyridine	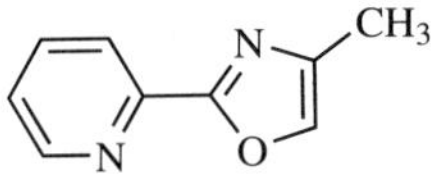	18	72
4	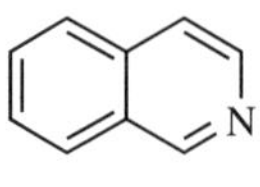 isoquinoline	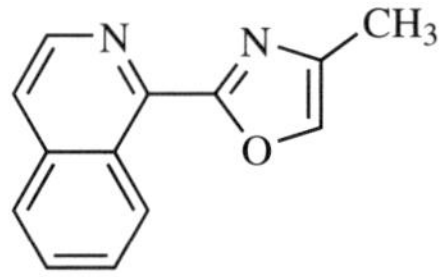	57	98
5	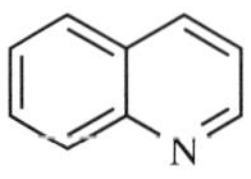 quinoline	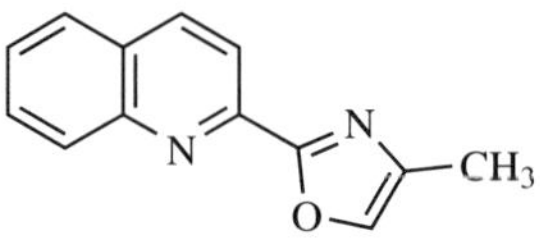	43	44

[a] Data from refs. 372a and 377.

thiazolium-2-ylide **1022**. The authors noted that the facile silyl group transfer from carbon to oxygen was probably facilitated by formation of the O-Si bond.

A second equivalent of the aldehyde reacts reversibly with **1022** to afford a new betaine **1023** that can further condense with a third equivalent of the aldehyde to generate the spiro[thiazoline-2,4′]dioxolane **1024** (R = H). Conversion of **1024** to a 2-(silyloxymethyl)thiazole **1025** is then accomplished by successive loss of 2 mol of the aldehyde. The authors obtained detailed ^{1}H, ^{13}C and NOE data to support the existence of **1024** (R = H and R = CH(CH$_3$)$_2$).

The authors were also able to monitor the reaction of 4-methyl-2-(trimethylsilyl)oxazole **1005** with formaldehyde, using the same spectroscopic techniques. At −70°C they characterized completely the spiro[oxazoline-2,4′]dioxolane **1026** (R = H) using ^{1}H and ^{13}C NMR. Reaction of acetaldehyde with **1005** at 25°C proved to be ∼1/100 of the reaction rate of acetaldehyde with **1020**. The authors could not unequivocally establish the structure of **1026** (R = CH$_3$) due to the complexity of the spectra. Nonetheless, they speculated that both **1005** and **1020** reacted with aldehydes via a similar mechanism.

Ito and co-workers[380,381] described aldol condensations of 5-methoxy-2-(*t*-butyldimethylsilyl)oxazoles with aldehydes to produce *cis*- and *trans*-oxazolines, important precursors to β-hydroxy-α-amino acids. The starting oxazoles **1027a–c** were readily assembled in one step from an α-isocyanoester (Scheme 1.274). Lewis acid–catalyzed aldol condensation of **1027a** with an aldehyde gave a mixture of the *trans*-oxazoline **1028** and the *cis*-oxazoline **1029** in highly variable yields. The

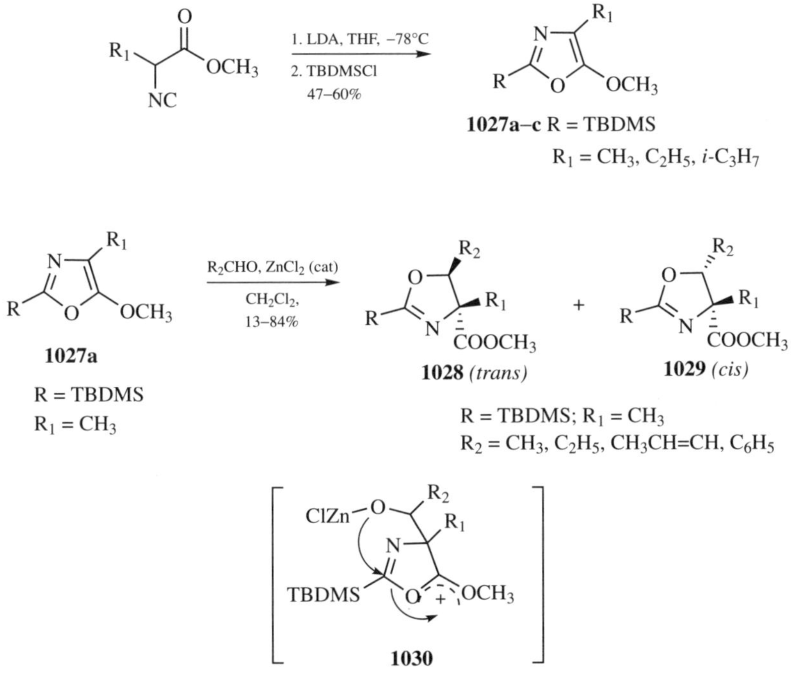

Scheme 1.274

authors found that $ZnCl_2$ was the best catalyst among the Lewis acids examined ($TiCl_4$, $BF_3 \cdot OEt_2$, $(C_2H_5)_2AlCl$, $CuCl$). However, the diastereoselectivities were uniformly unsatisfactory regardless of the nature of the catalyst or the stoichiometry, i.e., **1028:1029** = 65:35 to 25:75.

Mechanistically, the authors proposed that the aldehyde suffered nucleophilic attack by the C(4)-C(5) double bond in **1027a** to produce **1030**, which then rearranged to the observed product mixture. The 5-methoxy group was presumed to facilitate this process, which is reasonable given Dondoni's results for reaction of **1005** with aldehydes to afford **1009** (see Table 1.75).

In a later study, Ito's group[381] found that TBAF also effected aldol reactions of **1027** and aldehydes to produce oxazolines. Here, the authors proposed fluoride-catalyzed desilylation of **1027** generated an equilibrium mixture of the oxazol-2-yl carbanion **1031** and the isocyanovinyl enolate **1032** (Scheme 1.275). Condensation of **1032** with an aldehyde and cyclization then gave the oxazolines **1028** and **1029**. Aromatic aldehydes gave predominately the *cis*-oxazoline **1029** in good yield. Interestingly, aliphatic aldehydes were unreactive. The authors also prepared **1028** and **1029** directly from an α-isocyanoacetate ester and an aldeyhde, but used only a catalytic amount of TBAF.

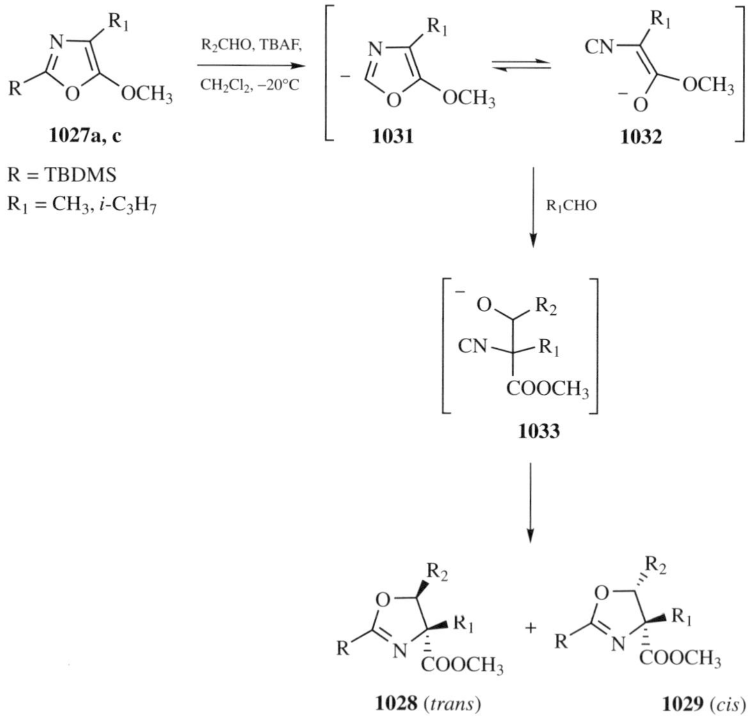

Scheme 1.275

1.4.10. Oxazolium Salts

N-Alkyl- and *N*-acyloxazolium salts are readily prepared from intramolecular and intermolecular reactions of oxazoles and reactive reagents, including Meerwein's salts, methyltriflate, and chloroformates. These salts, in turn, react with nucleophiles, including organolithium and Grignard reagents, alkyltin reagents, cyanide, amines, and hydride. In addition, oxazolium salts are precursors to azomethine ylides, which undergo 1,3-dipolar cycloaddition reactions to generate nitrogen heterocycles. Selected examples of these reactions are discussed in this section.

Workers in Spain[382] described the addition of organolithium reagents, Grignard reagents, and lithium dialkylcuprates to *N*-ethyloxazolium salts (Scheme 1.276). Alkylation of an oxazole with triethyloxonium tetrafluoroborate gave the expected *N*-ethyloxazolium salts **1035a–d** in excellent yield. Addition of methylmagnesium iodide, methyllithium, or lithium dimethylcuprate furnished the oxazolines **1036a–d** in fair to excellent yield. Of particular note, *N*-ethyl-4-cyano-2,5-dimethyloxazolium tetrafluoroborate **1035d** produced **1036d** in up to 85% yield, whereas these same

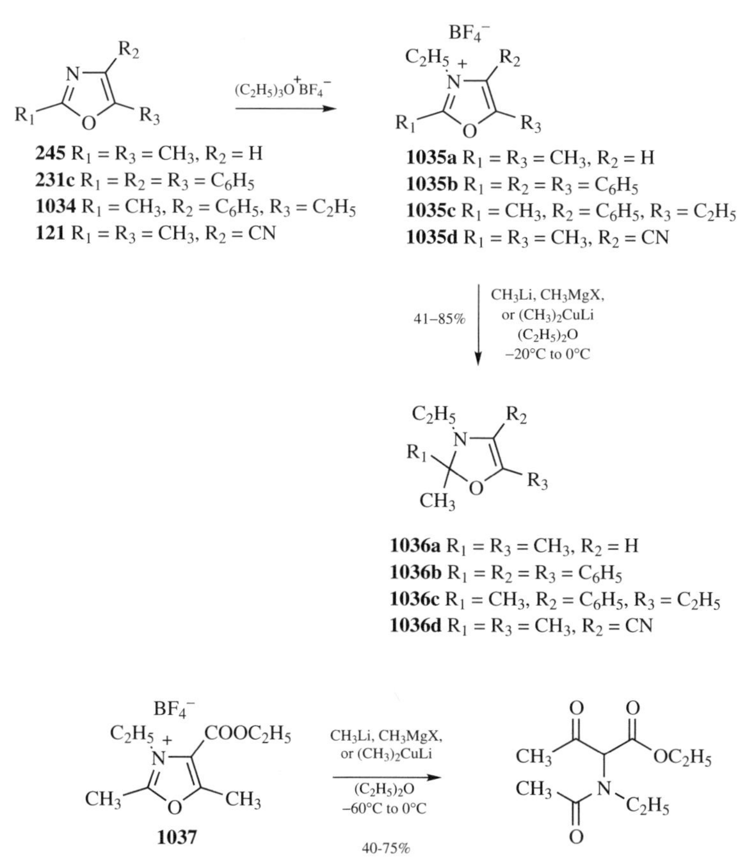

245 $R_1 = R_3 = CH_3$, $R_2 = H$
231c $R_1 = R_2 = R_3 = C_6H_5$
1034 $R_1 = CH_3$, $R_2 = C_6H_5$, $R_3 = C_2H_5$
121 $R_1 = R_3 = CH_3$, $R_2 = CN$

1035a $R_1 = R_3 = CH_3$, $R_2 = H$
1035b $R_1 = R_2 = R_3 = C_6H_5$
1035c $R_1 = CH_3$, $R_2 = C_6H_5$, $R_3 = C_2H_5$
1035d $R_1 = R_3 = CH_3$, $R_2 = CN$

41–85% | CH_3Li, CH_3MgX, or $(CH_3)_2CuLi$
$(C_2H_5)_2O$
−20°C to 0°C

1036a $R_1 = R_3 = CH_3$, $R_2 = H$
1036b $R_1 = R_2 = R_3 = C_6H_5$
1036c $R_1 = CH_3$, $R_2 = C_6H_5$, $R_3 = C_2H_5$
1036d $R_1 = R_3 = CH_3$, $R_2 = CN$

1037

CH_3Li, CH_3MgX, or $(CH_3)_2CuLi$
$(C_2H_5)_2O$
−60°C to 0°C
40-75%

Scheme 1.276

reagents cleanly effected ring opening of N-ethyl-4-carboethoxy-2,5-dimethyl-oxazolium tetrafluoroborate **1037** to generate ethyl 2-(N-ethylacetamido)acetoacetate.

The authors also reported two distinct pathways for reduction of oxazolium salts with a metal hydride (Scheme 1.277). In both cases, only ring-opened products were observed. Thus treatment of an oxazolium salt with excess sodium borohydride or lithium aluminum hydride gave **1039a–c** in good yield. Mechanistically, it was proposed that hydride addition at C(2) produced **1040**, which was then cleaved in the presence of excess hydride to afford **1039a–c** upon workup. In contrast, treatment of an oxazolium salt with one equivalent of lithium tri-*tert*-butoxyaluminum hydride afforded **1042a,b** in modest yield. In this case, the authors proposed that, in the absence of excess hydride, **1040** was hydrolyzed with concomitant ring

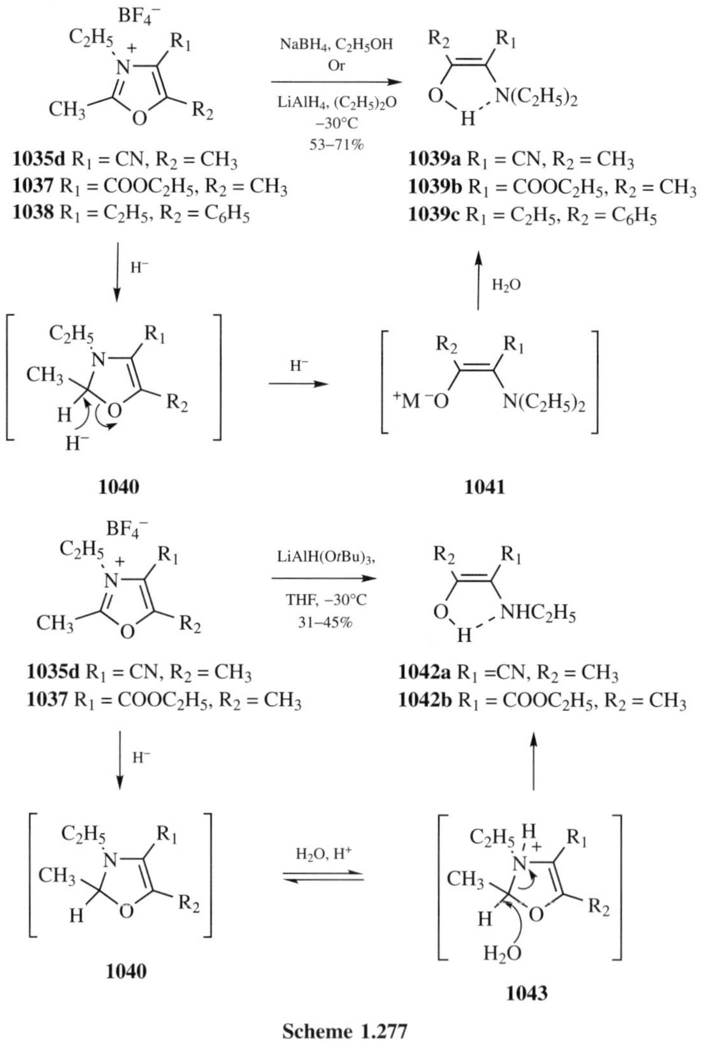

Scheme 1.277

opening to yield an intermediate hemiacetal (not shown), which then lost acetaldehyde to yield **1042a,b**.

Oshawa and co-workers[383,384] investigated the allylation of azoles and benzo-fused analogs. In the course of this work, they prepared 2-allyl-3-[(1-chloroethoxy)carbonyl]-4-oxazoline **1046** (X = Cl) in good yield (Scheme 1.278). Acylation of **1** with 1-chloroethyl chloroformate **1044** (X = Cl) gave **1045** (X = Cl) that reacted with allyltributyltin to produce **1046** as expected. There was no reaction of oxazole with ethyl chloroformate **1044** (X = H). Unfortunately, attempts to aromatize **1046** by oxidation with potassium ferricyanide in aqueous base resulted in decomposition and produced none of 2-allyloxazole **1047**. The authors also prepared 3-[(1-chloroethoxy)carbonyl]-2-ethynyl-4-oxazoline **1048** (X = Cl) in modest yield.

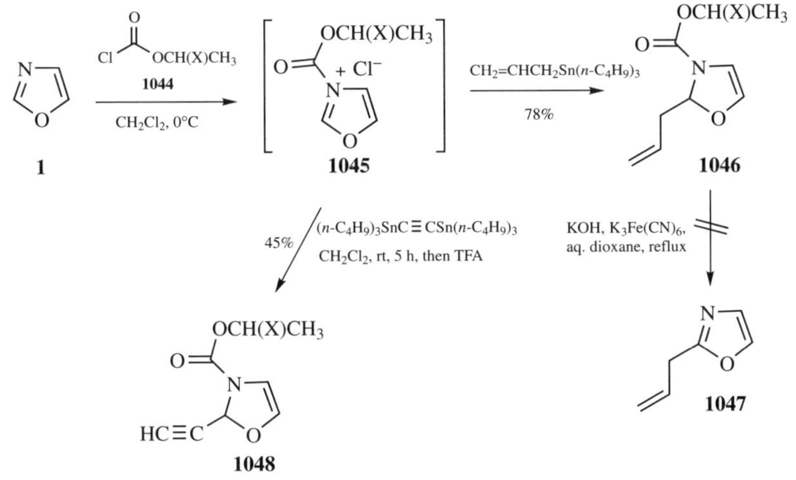

Scheme 1.278

This allylation/ethynlyation methodology was far more synthetically useful for imidazoles and thiazoles. In addition, it could be applied to benzoxazoles rather than the monocyclic derivatives.

Iyer and co-workers[385,386] invoked the presence of an open-chain bis-Reissert salt to explain the facile conversion of **1049** to **1051** (Scheme 1.279). Thus cyclization of **1049** with perchloric acid generated the bis-oxazolium perchlorate **1050**, which was isolated and completely characterized spectroscopically. The predominant tautomer in solution was **1050**, based on ^{1}H NMR. Hydrolysis of **1050** then produced the bis amino acid **1051**, a known metabolite of the antitubercular drug ethambutol, **1052**.[387]

The authors also investigated Diels-Alder reactions of **1050** (Scheme 1.280). For example, reaction of **1050** with methyl acrylate first afforded the bis-oxazabicyclo[2.2.1]heptene **1053**, which fragmented to the diketone **1054**. Recyclization of **1054** gave the bis-pyrrole **1055**, from which 5-phenyl-2-propanoyl-3-pyrrolecarboxylic acid methyl ester **1056** was isolated after hydrolysis. Ethylene diamine

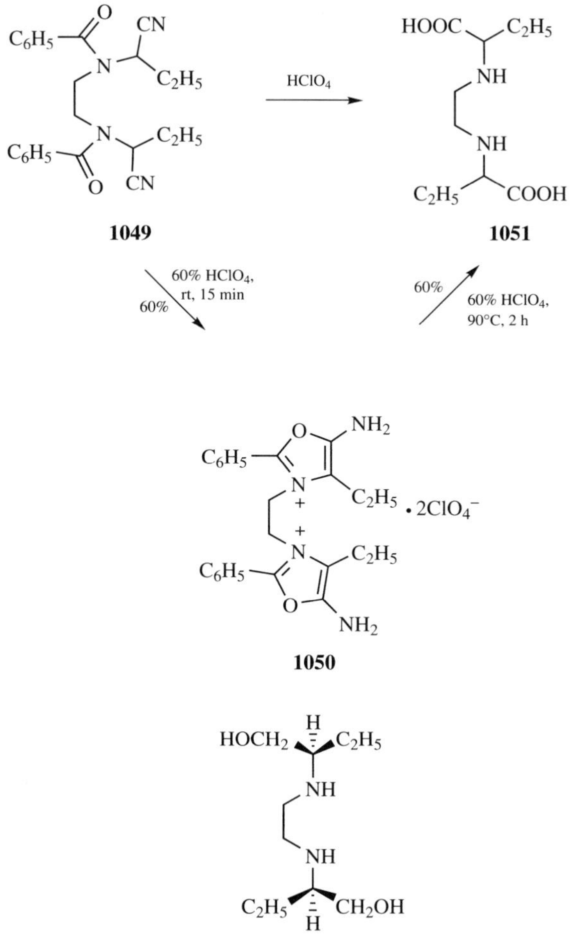

1049 1051 1050 1052

Scheme 1.279

was also isolated from this reaction mixture, which provided further support for their mechanistic rationale.

The authors favored the azafulvene structure **1057** for the product, since they observed a chelated enolic hydroxyl group in the IR spectrum. In addition, the product gave a positive ferric chloride test but did not afford an oxime or a hydrazone. Finally, the ^{13}C NMR spectrum was more consistent with **1057** than with **1056**.

Cycloaddition of **1050** with DMAD gave the tetrasubstituted bispyrrole **1058**. In this case, loss of 2 mol $HClO_4$ from the bis-oxazolium salt was presumed to generate a mesoionic species (not shown) that underwent cycloaddition with DMAD, followed by a retro Diels-Alder reaction and loss of 2 mol of HNCO to yield **1058**.

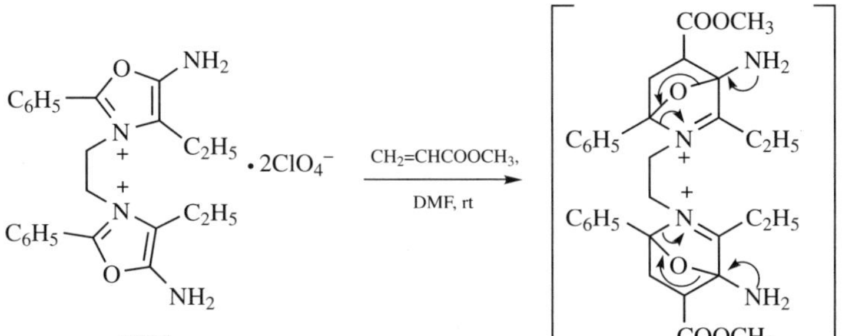

Reactions of Oxazoles 233

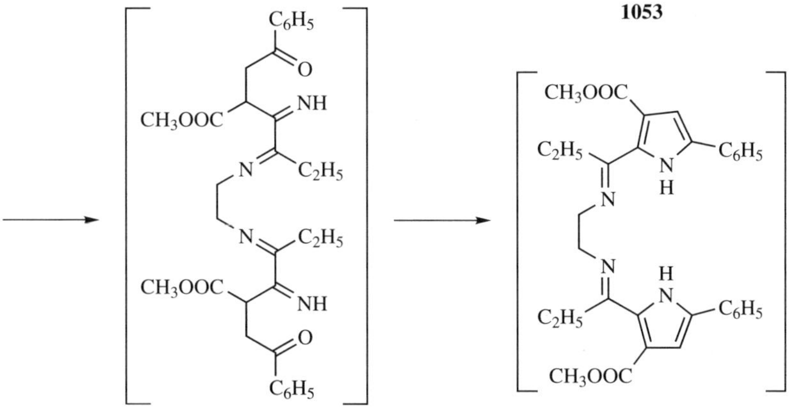

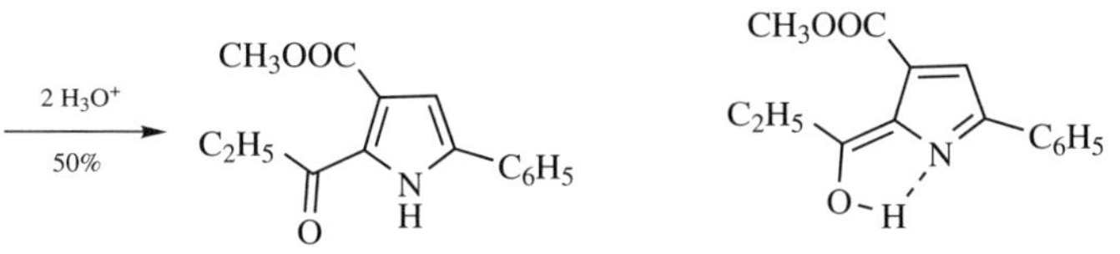

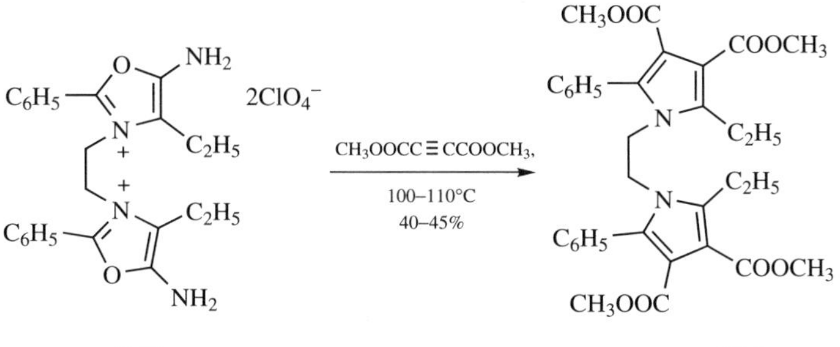

Scheme 1.280

McEwen and co-workers[388] also investigated the regioselectivity and the mechanism of the Diels-Alder reaction of oxazolium salts using ethylpropiolate and ethyl phenylpropiolate. These authors favored a general mechanism wherein an oxazolium tetrafluororborate salt first dissociated in part via loss of fluoroboric acid to generate a mesoionic intermediate. They also proposed the formation of spin-paired diradicals as unstable intermediates resulting from the 1,3-dipolar addition reactions. Their studies were undertaken to assess the relative effectiveness of merostabilization vs. Linnett stabilization to control the regioselectivity in these [4 + 2] cycloadditions. They concluded that the product distribution was best explained in that merostabilization of the diradicals was more effective in controlling the regiochemical outcome.

Hassner and Fischer[389,390] described the first examples of intramolecular oxazolium salts and subsequent ring-opening reactions (Scheme 1.281). Their early NMR studies showed that 2-(4-bromobutyl)-5-ethoxyoxazole **1059** ($n = 1$) slowly equilibrated with the oxazolium salt **1061** ($n = 1$). When **1059** ($n = 1$) was treated

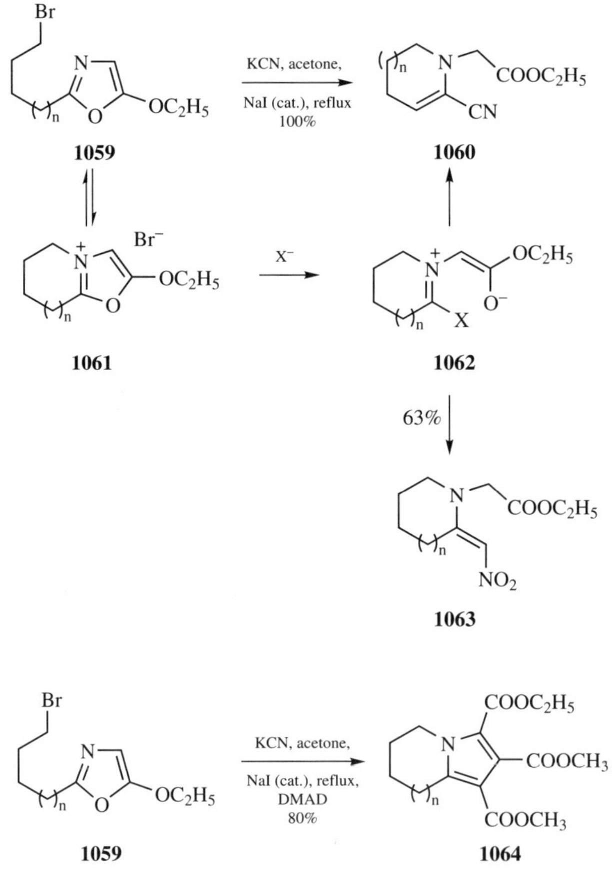

Scheme 1.281

with KCN in the presence of a catalytic amount of NaI, 1-(carbethoxymethyl)-2-cyano-1,4,5,6-tetrahydropyridine **1060** ($n = 1$) was isolated quantitatively. Similarly, reaction of **1059** ($n = 1$) with nitromethane and triethylamine gave 1-(carbethoxymethyl)-2-(nitromethylene)piperidine **1063** ($n = 1$). These products were proposed to arise via an intermediate azomethine ylide **1062** (X = CN or CH_2NO_2, $n = 1$) followed by a proton shift. The azomethine ylide **1062** (X = CN, $n = 1$) was trapped with DMAD to yield 1,2-dicarbomethoxy-3-carbethoxy-5,6,7,8-tetrahydroindolizine **1064**. Additional studies by the authors, as well as FMO calculations, supported their mechanistic interpretation.[390]

These same authors found that, in the absence of a dipolarophile, generation of an azomethine ylide intermediate from an oxazolium salt followed distinctly different pathways. Azomethine ylides that contain an acidic proton undergo an internal proton shift to yield a stable product (vide supra). However, an internal alkyl shift was observed in cases where such a proton shift was not possible.

For example, sodium cyanide reacted with 5-ethoxy-3-ethyl-2-phenyloxazolium tetrafluoroborate **1065** in the absence of a dipolarophile and unexpectedly produced ethyl hippurate (Scheme 1.282). Mechanistically, they rationalized this result by initial addition of cyanide at C(2) to produce **1066**. Ring opening of **1066** then generated the azomethine ylide **1067** that, in the *S*-conformation, undergoes intramolecular *N*-deethylation to yield the ketene acetal **1068** (not isolated). Chromatographic purification of the reaction mixture effected hydrolysis of **1068** with concomitant loss of HCN and C_2H_5OH to produce ethyl hippurate.

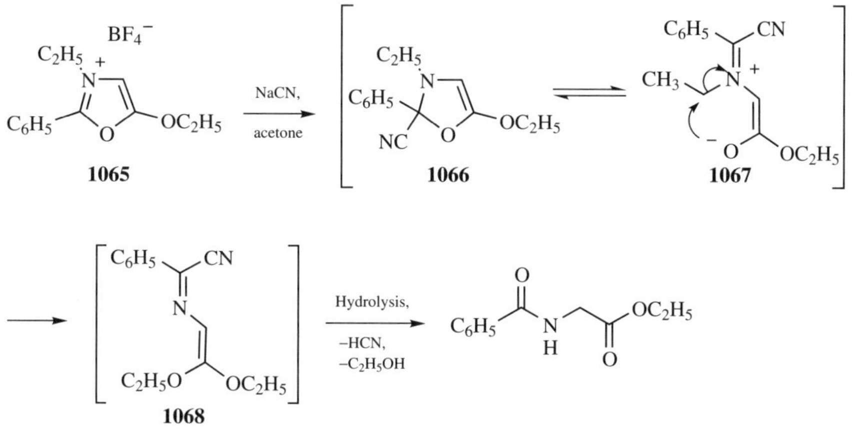

Scheme 1.282

Laude and co-workers[391] prepared fused oxazolium tetrafluoroborates **1070** and **1072** via intramolecular cyclization of the Reissert compounds **1069** and **1071** (Scheme 1.283). The authors studied the Diels-Alder reaction of **1070** and **1072** with a variety of dieneophiles. Cycloaddition of **1070** with a dienophile initially gave rise to **1073** via a reaction sequence very similar to that described by Iyer and co-workers (see Scheme 1.280).

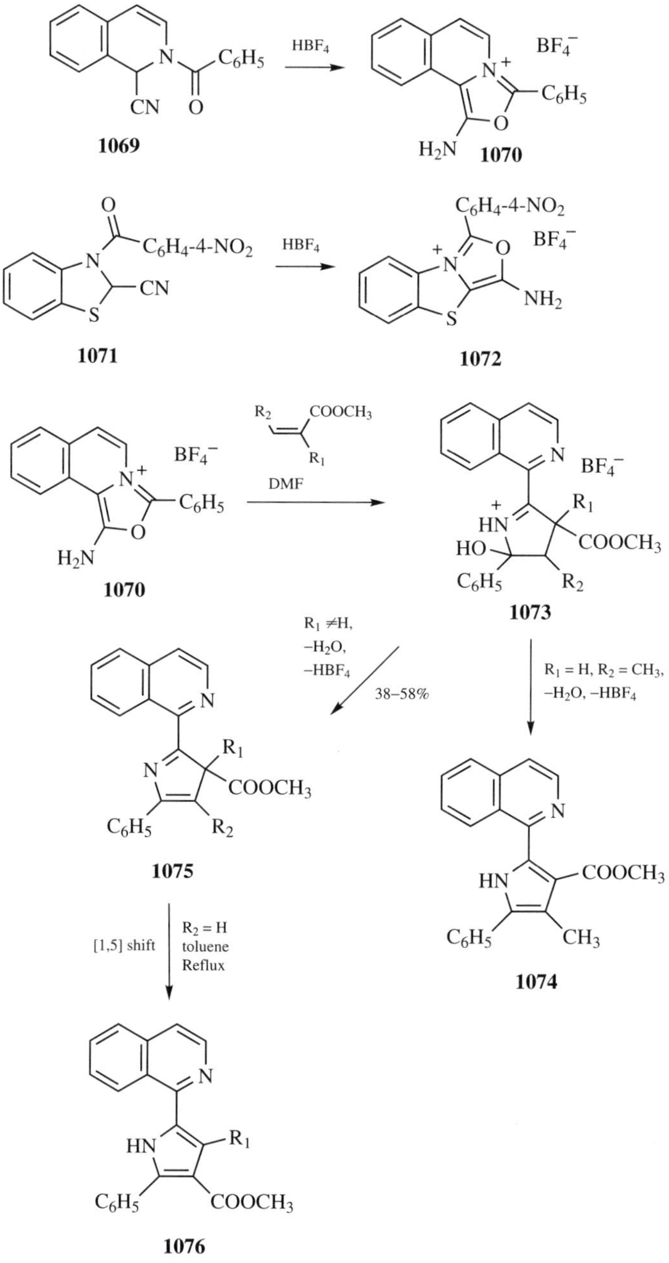

Synthesis and Reactions of Oxazoles

Scheme 1.283

$R_1 = CH_3, CH_2COOCH_3, COOCH_3$

Elimination of water and fluoroboric acid from **1073** ($R_1 = H$) afforded the 1*H*-pyrrole **1074**, directly. However, for **1073** ($R_1 \neq H$), elimination of water and fluoroboric acid produced the 3*H*-pyrrole-3-carboxylic acid methyl ester **1075** in modest to good yield. In these cases, **1075** ($R_2 = H$) was readily converted to the 1*H*-pyrrole **1076** in refluxing toluene via a [1,5] sigmatropic rearrangement of the carbomethoxy group.

Babaev and co-workers[392] prepared and isolated several oxazolo[3,2-*a*]pyridinium perchlorates (Scheme 1.284). For example, using Bradsher and Zinn's protocol,[393] the authors treated 1-phenacyl-2-pyridone **1077** ($R = X = H$) with concentrated H_2SO_4 to produce an intermediate sulfo betaine (not isolated), which was heated in 70% $HClO_4$ to afford 2-phenyloxazolo[3,2-*a*]pyridinium perchlorate **1078** ($R = X = H$) in good yield.

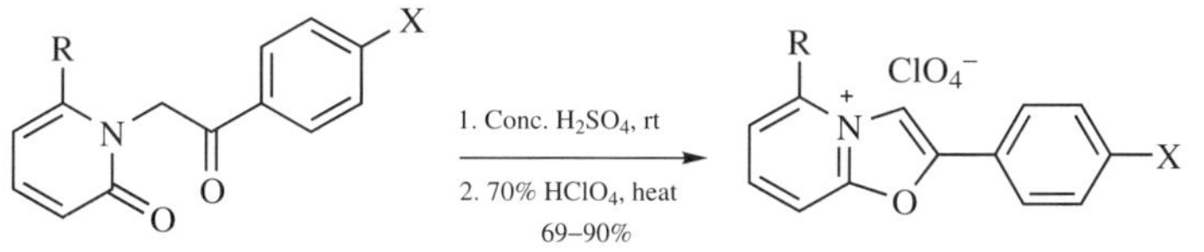

1. Conc. H_2SO_4, rt
2. 70% $HClO_4$, heat
69–90%

1077

$R = H, CH_3$
$X = H, NO_2$

1078

Scheme 1.284

Babaev's group[392,394–402] extensively investigated the reactions of **1078** analogs with nucleophiles and described examples of interesting reactions and rearrangements. Reaction of **1078** ($R = X = H$) with gaseous ammonia produced 2-phenylimidazo[1,2-*a*]pyridine **1079** ($R = X = H$) in good yield (Scheme 1.285). However, treatment of **1078** ($R = X = H$) with liquid ammonia gave 40% of **1079** ($R = X = H$) together with 5% of 1-amino-4-(5-phenyloxazol-2-yl)-1,3-butadiene **1080** ($R = X = H$). In contrast, reaction of **1078** ($R = X = H$) with piperidine gave rise to 1-[(1*E*, 3*E*)-4-(5-phenyloxazol-2-yl)-1,3-butadienylpiperidine **1081** ($R = X = H$).[398]

Nitrophenyl analogs of **1078** ($R = H$ or CH_3; $X = NO_2$) reacted with amines to afford products that highly depended on the structure of **1078**.[392,394–396,400,402] Piperidine effected ring opening of 2-(4-nitrophenyl)oxazolo[3,2,-*a*]pyridinium perchlorate **1078** ($R = H$, $X = NO_2$) to afford isomeric (1*E*, 3*E*)- **1083** or (1*E*, 3*Z*)-butadienes **1082**, depending on the temperature (Scheme 1.286).[392] This ring-opening process was analogous to that reported for reaction of 2-chloro-*N*-(4-nitrophenacyl)pyridinium bromide **1084** with secondary amines to produce a mixture composed primarily of a (1*E*, 3*E*)-1-dialkylamino-4-[5-(4-nitrophenyl)oxazol-2-yl]-1,3-butadiene **1085** together with the (1*E*, 3*Z*)-isomer **1086**. Here, the authors favored a mechanism whereby **1084** cyclized to the oxazolo[3,2-*a*]pyridinium bromide **1087**, followed by ring opening to the observed product mixture.[394,402]

Heating 5-methyl-2-(4-nitrophenyl)oxazolo[3,2-*a*]pyridinium perchlorate **1078** ($R = CH_3$, $X = NO_2$) with ammonia gave the expected 5-methyl-2-(4-nitrophenyl)

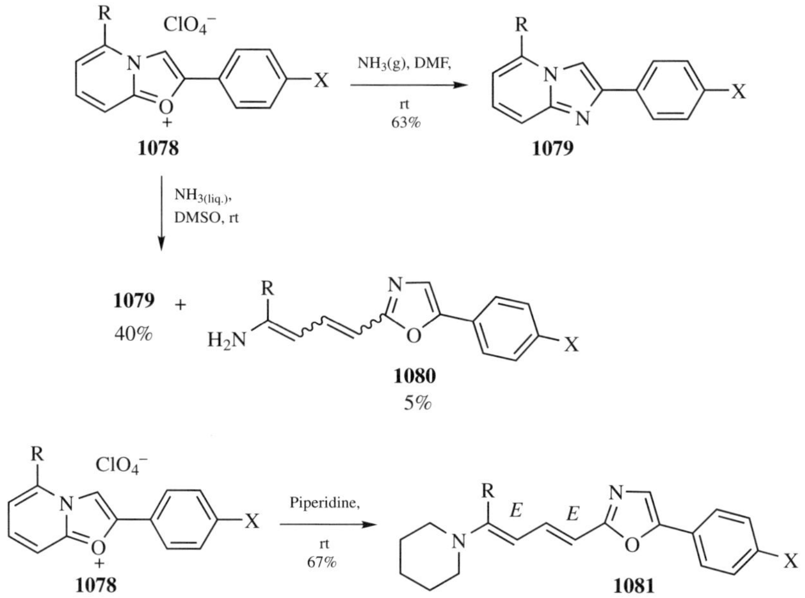

Scheme 1.285

imidazo[1,2-a]pyridine **1088** (R = CH$_3$, X = NO$_2$) quantitatively (Scheme 1.287).[400] The authors observed considerable decomposition when **1078** (R = CH$_3$, X = NO$_2$) was heated with primary amines (e.g., butylamine, sec-butylamine, or benzylamine). There was no reaction observed using aniline or p-anisidine. Hydrazine reacted with **1078** (R = CH$_3$, X = NO$_2$) to furnish 6-methyl-3-(4-nitrophenyl)-4H-pyrido[2,1-c][1,2,4]triazine perchlorate **1089** (R = CH$_3$, X = NO$_2$) in good yield. The structure of **1089** (base:HClO$_4$ = 2:1) was confirmed by X-ray crystallography.

However, heating **1078** (R = CH$_3$, X = NO$_2$) with a secondary amine afforded unexpected rearrangement products, 5-(dialkylamino)-2-(4-nitrophenyl)imidazo[1,2-a]pyridines **1090a–c** (R = NR$_1$R$_2$, X = NO$_2$) in good yield (Scheme 1.287).[392,395,396] The authors proposed that initial addition of the amine was accompanied by ring opening to generate the betaine **1091**, which tautomerized to the enamine **1092**. Cyclodehydration of **1092** then gave **1090a–c**. The structure of **1090b** (R$_1$, R$_2$ = (CH$_2$)$_6$, X = NO$_2$) was confirmed by X-ray crystallography. A limitation to the scope of this reaction is related to the instability of 5-aminoimidazo[1,2-a]pyridines. Reaction of **1078** (R = CH$_3$, X = H or Br) also yielded the corresponding 5-(dialkylamino)-2-(4-X-phenyl)imidazo[1,2-a]pyridine but these derivatives could not be isolated cleanly, owing to facile oxidation.

Hetzheim[403] summarized her investigations of ring transformations of 2-amino-3-phenacyloxazolium salts with oxygen and nitrogen nucleophiles to prepare substituted imidazolinones, imidazoles, imidazo[1,2-a]imidazoles, and 1,4-dihydro-imidazo[2,1-b][1,2,4]triazines.

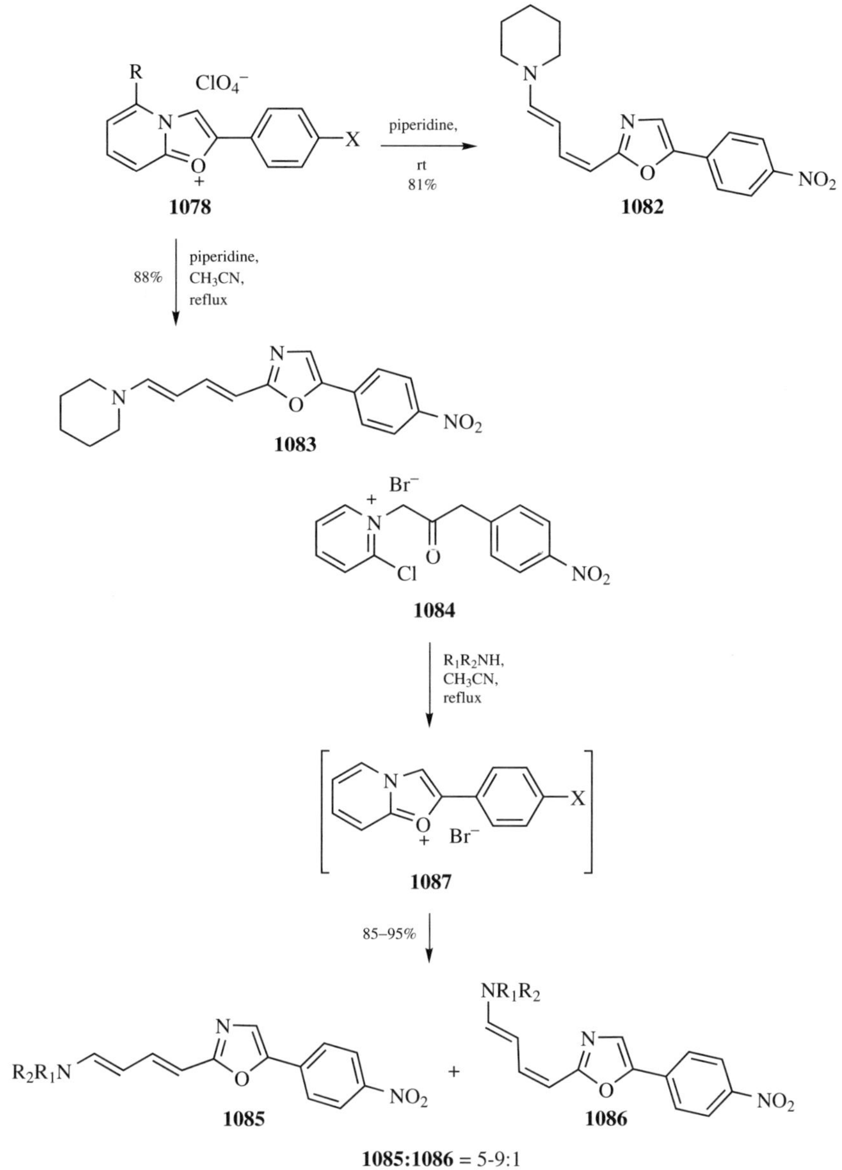

Scheme 1.286

R_1R_2NH = dimethylamine, piperidine, morpholine, 4-methylpiperazine, hexamethyleneimine

1085:1086 = 5-9:1

Padwa and co-workers[404] converted 4-phenyloxazole **7** and 5-phenyloxazole **6** to the corresponding trimethylsilylmethyl oxazolium salts **1093** and **1094** (Scheme 1.288). Desilylation of **1093** and **1094** with anhydrous CsF generated the nonstabilized *N*-oxazolium methylides **1095** and **1096**, which were immediately

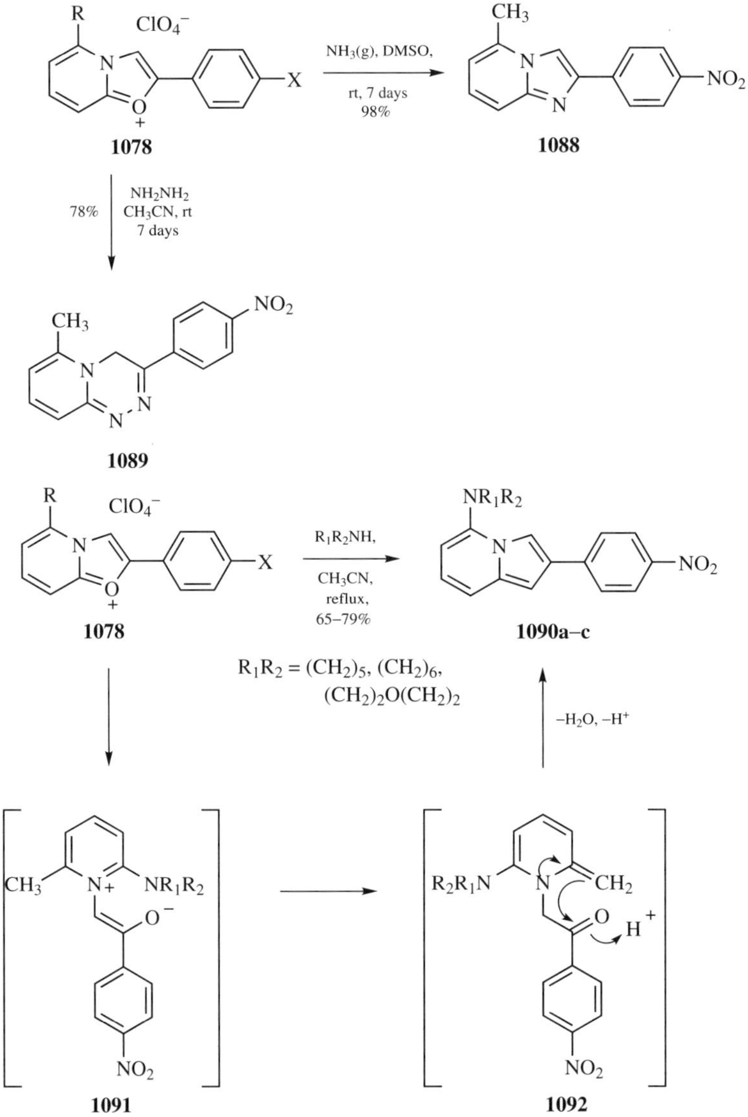

Scheme 1.287

trapped with an acetylenic dipolarophile. The resulting cycloadducts spontaneously fragmented to afford the substituted pyrroles **1099a–d** in excellent yield. The authors noted that a one-pot process gave the best overall results. This methodology was also successfully applied to 4-methylthiazole and benzothiazole.

Vedejs and Grissom[405–407] described the synthesis and reactions of azomethine ylides generated from 4-oxazolines. They found that alkylation of a 2,5-disubstituted or a 2,4,5-trisubstituted oxazole **15** gave an *N*-alkyloxazolium salt, which was

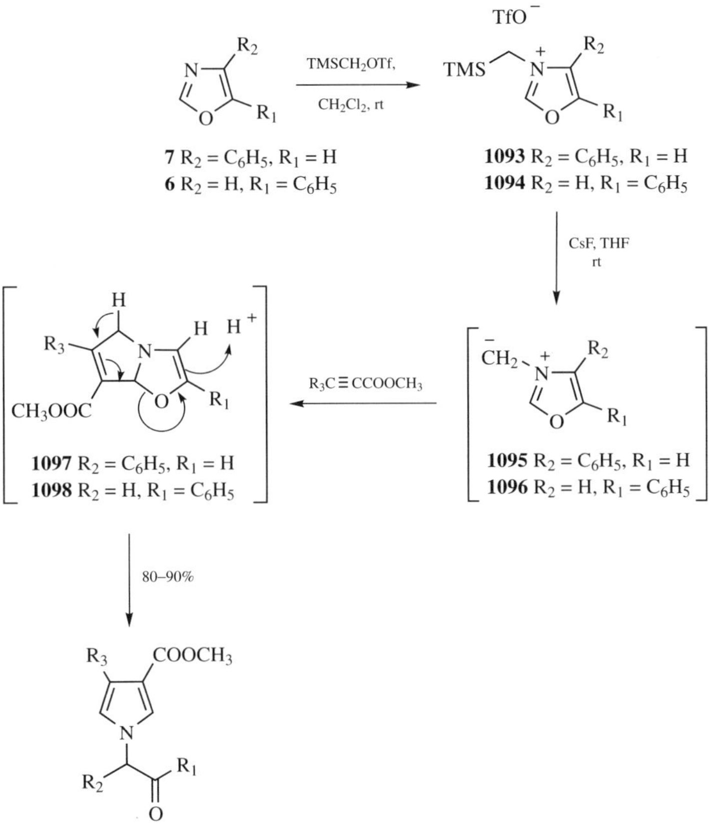

1099a $R_1 = C_6H_5$, $R_2 = R_3 = H$
1099b $R_1 = C_6H_5$, $R_2 = H$, $R_3 = COOCH_3$
1099c $R_1 = R_3 = H$, $R_2 = C_6H_5$
1099d $R_1 = H$, $R_2 = C_6H_5$, $R_3 = COOCH_3$

Scheme 1.288

reduced, in the presence of a dipolarophile, with phenylsilane/cesium fluoride to an intermediate 4-oxazoline **1100**. This oxazoline then reacted with the dipolarophile in two distinct manners, depending on the substitution pattern. If the oxazole was substituted at C(4) then **1100** ($R_1 = R_2 = R_3 = CH_3$ or C_6H_5) reacted with DMAD to generate a [2 + 2] cycloadduct **1101** ($R_1 = R_2 = R_3 = CH_3$ or C_6H_5) (Scheme 1.289).

Alternatively, in the absence of a 4-substituent, **1100** ($R_1 = CH_3$ or C_6H_5; $R_2 = H$; $R_3 = CH_3$, C_6H_5, OC_2H_5) spontaneously ring opened, yielding an azomethine ylide **1102** that reacted with DMAD to give [2 + 3] cycloadducts **1103**. The intermediate dihydropyrroles **1103** were oxidized to the pyrroles **1104** in good to excellent yield. Cycloaddition of **1102** with ethyl acrylate afforded **1105** in 40–87% yield together with 9–20% of the regiosiomer **1106**. Other dipolarophiles, including *N*-phenylmaleimide, methyl or ethyl propiolate, and phenylvinyl sulfone also gave good to excellent yields of the corresponding cycloadducts.[405,407]

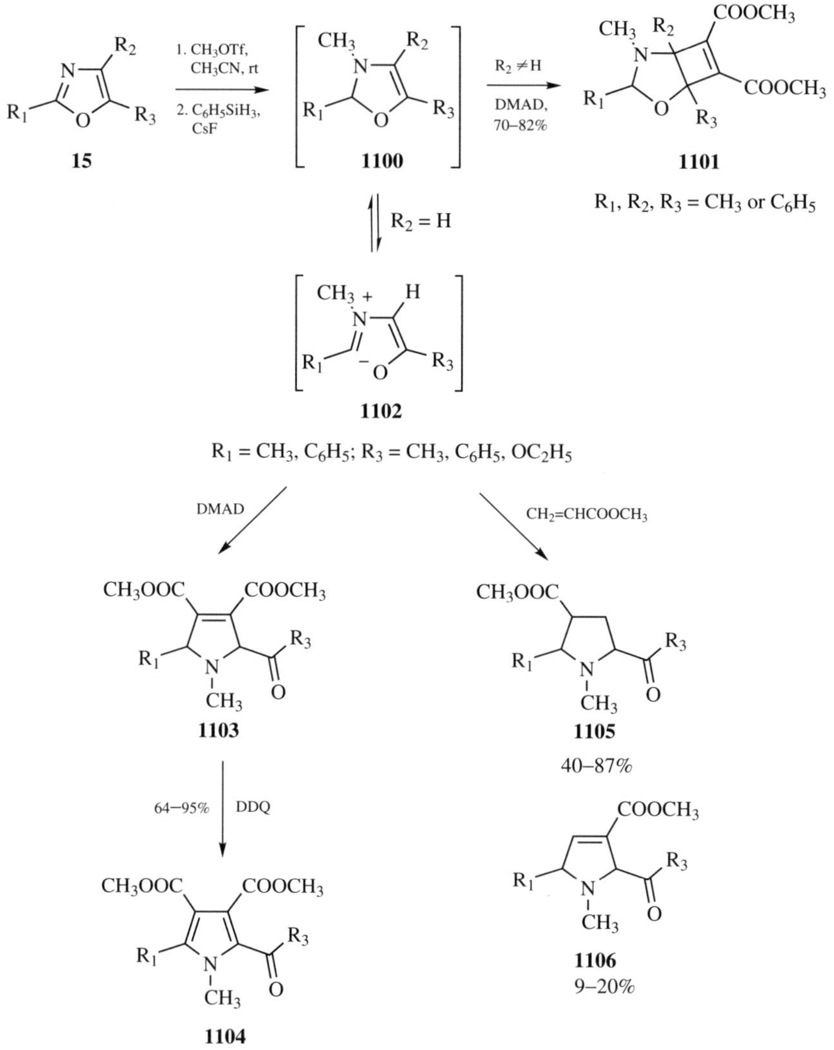

Scheme 1.289

The authors found that phenylsilane/anhydrous cesium fluoride was the preferred reagent combination for this reaction sequence. Other silanes were effective, whereas tributyltin hydride afforded low yields (< 20%) and only in the presence of a fluoride source. Tris(dimethylamino)sulfonium difluorotrimethylsilicate (TASF) and tetrabutylammonium fluoride (TBAF) destroyed the silane. Acetonitrile was the best solvent among the ethers and halogenated solvents evaluated.[407]

In a subsequent study[406] the authors converted a 2,5-disubstituted oxazole **15** ($R_1 = R_3 = CH_3$, C_6H_5, or OC_2H_5, $R_2 = H$) to the oxazolium salt followed by treatment with trimethylsilylcyanide (Scheme 1.290). The resulting intermediate 2-cyano-4-oxazoline **1107** spontaneously ring opened to the stabilized azomethine

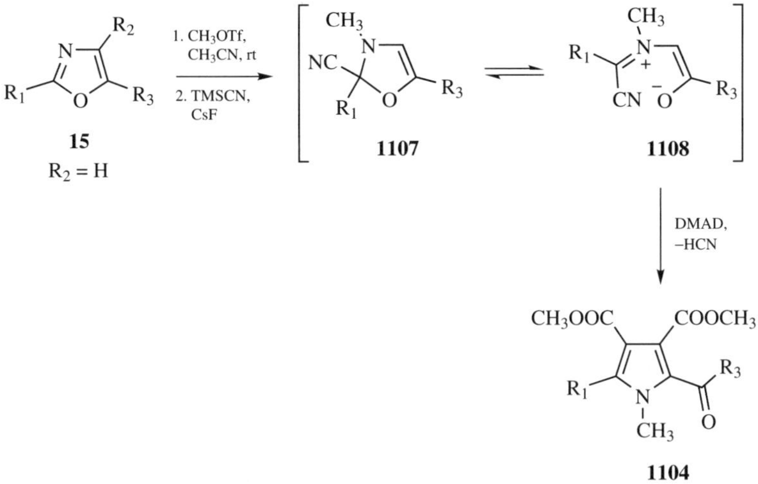

Reactions of Oxazoles

Scheme 1.290

ylide **1108**. In this case, [2 + 3] cycloaddition with DMAD afforded **1104** directly via loss of HCN, thereby obviating the need for the DDQ oxidation. The authors found that tetraethylammonium cyanide and thiolates were also effective in this reaction sequence, although the yields were consistently lower. Examples of **1104**

TABLE 1.78. 2-ACYL-1-METHYL-5-SUBSTITUTED PYRROLE-3,4-DICARBOXYLIC ACID DIMETHYL ESTERS FROM 2,5-DISUBSTITUTED OXAZOLES, METHYLTRIFLATE, AND DMAD

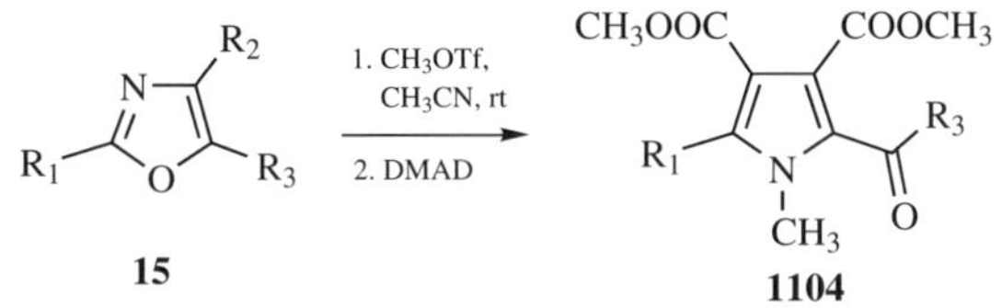

Figure 1.84

Entry	R_1	R_2	R_3	Conditions[a]	Yield (%)	Reference
1	C_6H_5	H	C_6H_5	A	95	405
2	C_6H_5	H	C_6H_5	B	80	406
3	C_6H_5	H	CH_3	A	90	405
4	C_6H_5	H	CH_3	B	57	406
5	C_6H_5	H	OC_2H_5	A	93	405
6	C_6H_5	H	OC_2H_5	B	95	406
7	C_6H_5	H	H	A	57	405
8	H	H	C_6H_5	B	79	406
9	CH_3	H	C_6H_5	A	85	405
10	CH_3	H	C_6H_5	B	16	406
11	CH_3	H	OC_2H_5	A	64	405
12	CH_3	H	OC_2H_5	B	74	406

[a] A, $C_6H_5SiH_3$/CsF, then DDQ oxidation; B, TMSCN/CsF.

prepared via $C_6H_5SiH_3$/CsF and TMSCN/CsF methodologies are shown in Table 1.78.

Vedejs and co-workers[407–410] applied $C_6H_5SiH_3$/CsF reduction methodology of oxazolium salts as a means to access the indoloquinone ring present in aziridino-mitosene **1109**. In a model study,[408] the oxazolium salt **1111** was readily prepared

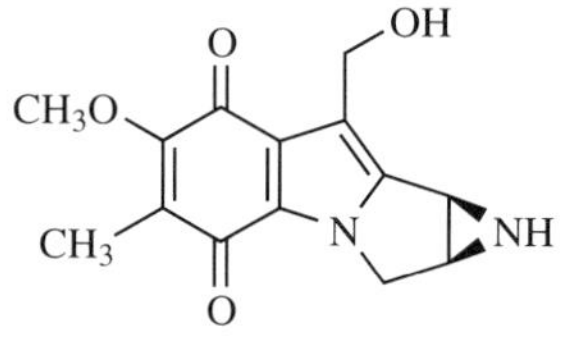

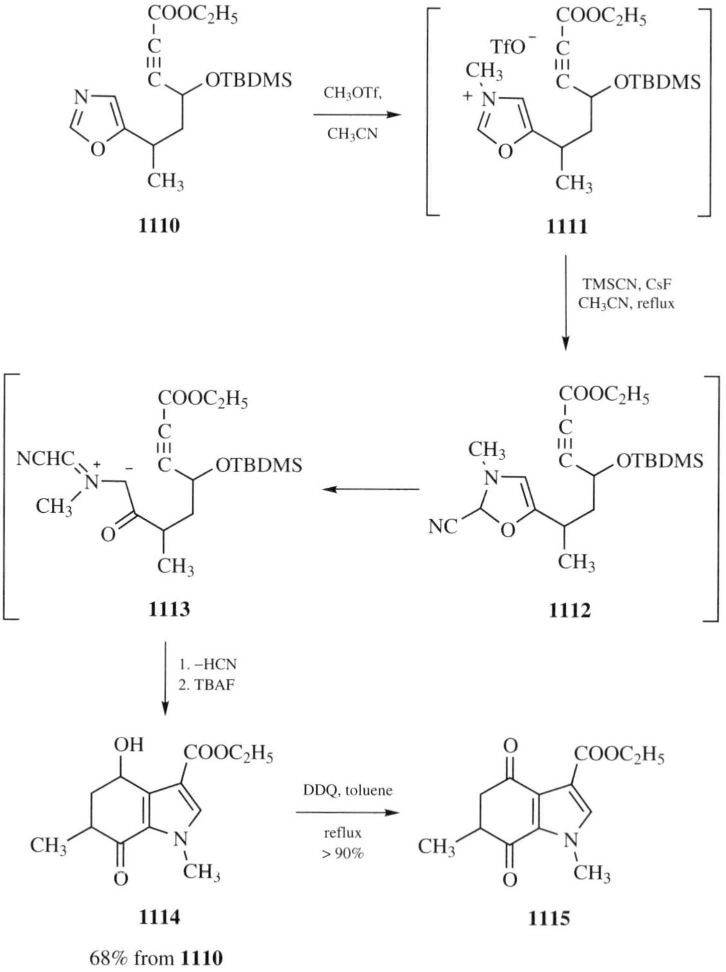

Scheme 1.291

Reactions of Oxazoles 245

without incident and was immediately reduced with TMSCN/CsF (Scheme 1.291). The resulting oxazoline spontaneously ring opened and cyclized to produce **1114**, after cleavage of the silyl ether. Oxidation of **1114** with DDQ in refluxing toluene then gave the target indoloquinone **1115**. It is significant that oxidation of **1114** was not accompanied by any aromatization.

In later work, Vedejs and Piotrowski[409] expanded the model study to prepare an advanced tricyclic intermediate containing the complete carbon skeleton of **1109**. In this case, the authors generated an intramolecular oxazolium salt to achieve their goal. Thus a mixture of 4-chlorobutyronitrile and diazo ketone **1116** was added to 4-chlorobutyronitrile and boron trifluoride etherate to afford 2-(3-chloropropyl)-5-(5-pentynyloxy)oxazole **1117** (Scheme 1.292). Oxidative carbonylation of **1117** gave the acetylenic ester **1118** requisite for further elaboration. Refluxing a mixture of **1118** and anhydrous sodium iodide yielded the oxazolium salt **1119**, which was immediately subjected to the $C_6H_5SiH_3$/CsF protocol. The authors isolated the advanced pyrrolo[1,2-a]indolone intermediate, **1120** in 48% yield from **1118**.

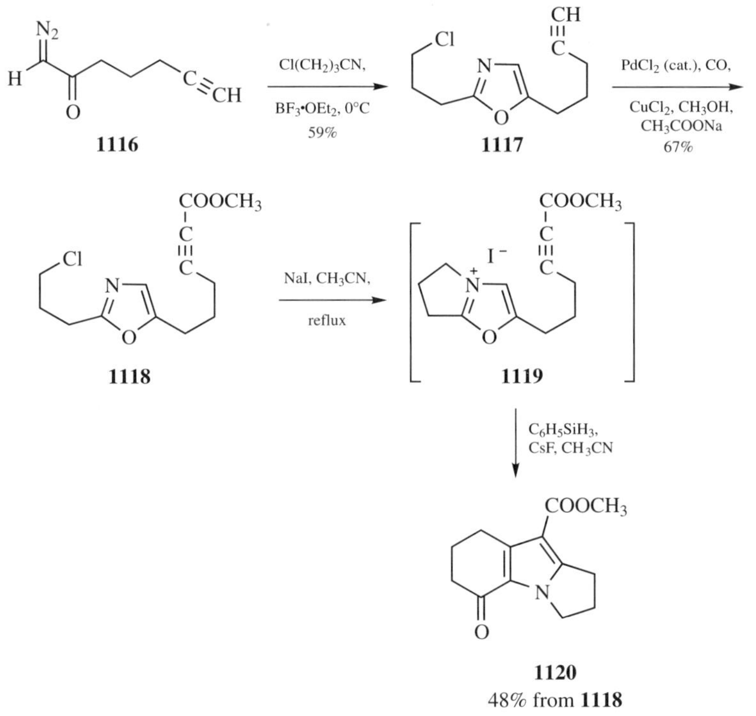

Scheme 1.292

The reduction of a fully elaborated oxazolium salt and subsequent trapping of the azomethine ylide allowed Vedejs and co-workers[410] to complete their synthesis of aziridinomitosene. Alkylation of **1121** with 2-(phenylsulfonyl)ethyl triflate produced the oxazolium salt **1122**, which was converted to the key bicyclic pyrrole **1125**, using $C_6H_5CH_2N(CH_3)_3^+ CN^-$ as an organic soluble cyanide source (Scheme 1.293). The authors isolated **1125** in 66% overall yield from **1121**. The structure of **1125** was confirmed by a two-step conversion to the indoloquinone **1126**.

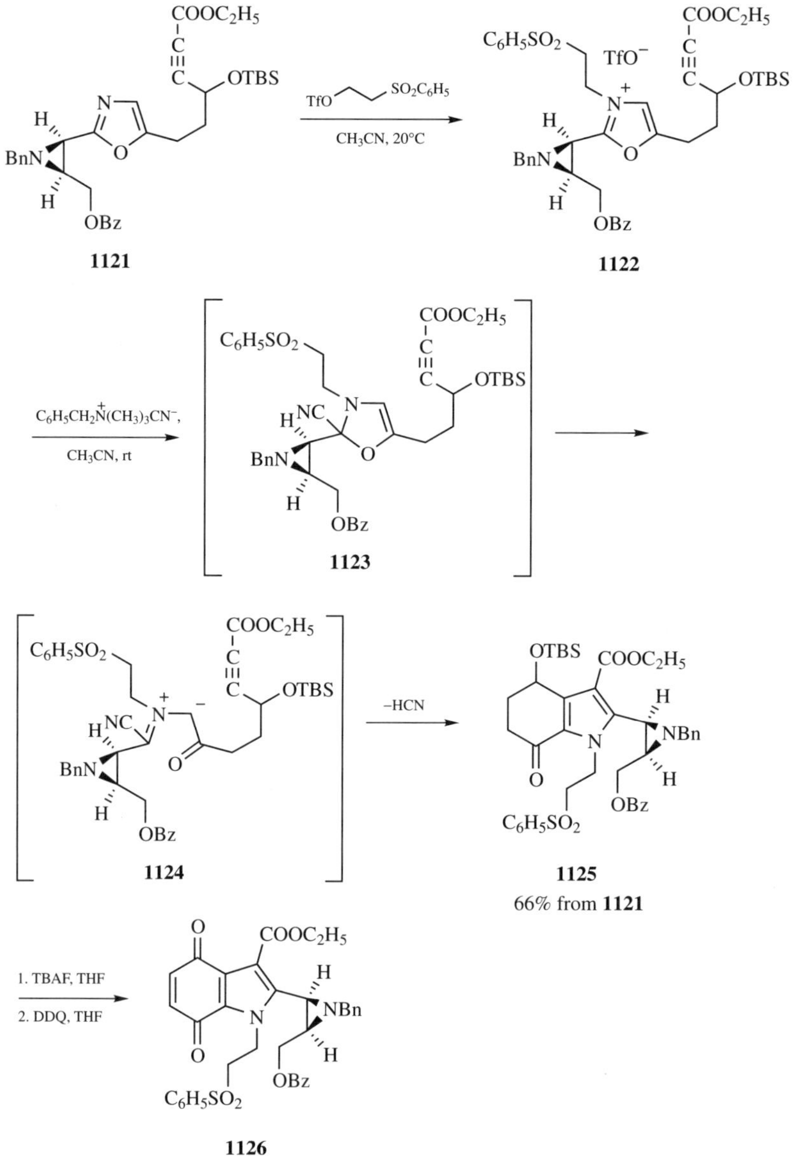

Scheme 1.293

This reaction sequence employing $C_6H_5CH_2N(CH_3)_3^+ CN^-$ was far superior to one in which **1122** was reduced with sodium borohydride (Scheme 1.294). Here,

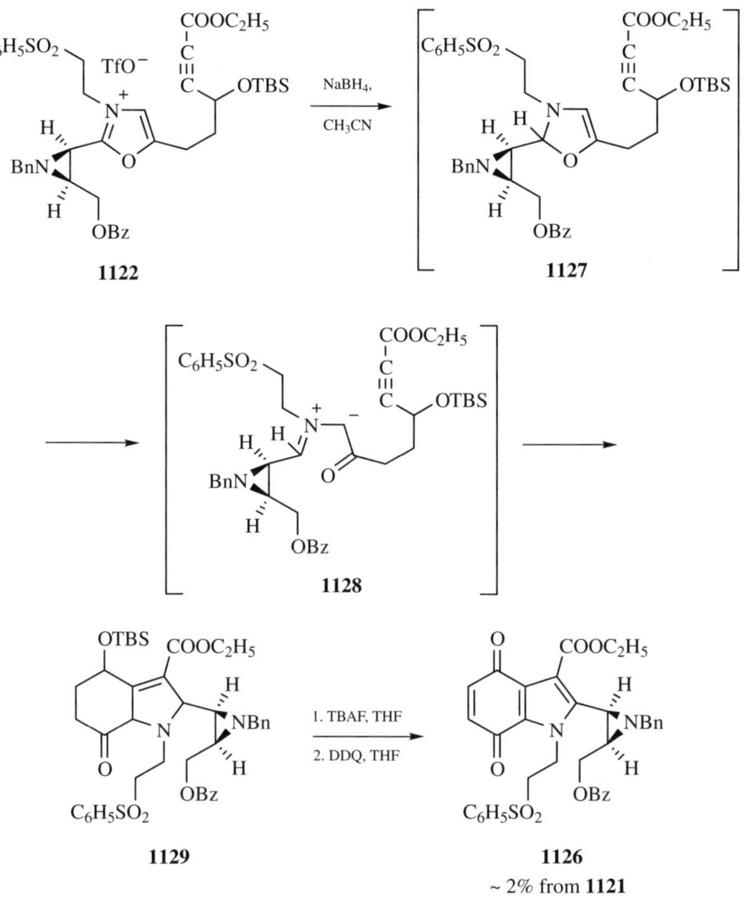

Scheme 1.294

the complex mixture containing the unstable intermediate **1129** was oxidized with DDQ to afford a mixture containing only ca. 2% of the indoloquinone **1126**. Reduction of **1122** using the $C_6H_5SiH_3$/CsF protocol provided a complex product mixture that was also converted in two steps to a mixture containing only 3–5% of **1126**.

Dondoni and co-workers[379] investigated reactions of oxazoles with reactive ketenes as a means to prepare [2 + 2] cycloadducts related to clavam or isoclavam antibacterial agents. Although the authors did not observe any [2 + 2] cycloaddition products, they did discover a new method of carbon–carbon bond formation in oxazoles that was complementary to acylation of 2-(trimethylsilyl)oxazoles (see Scheme 1.271).

Acylation of an oxazole with dichloroketene **1013a**, generated in situ, furnished low yields of the 2-(dichloroacetyl)oxazoles **1134a** and **1134b** (Scheme 1.295). The authors proposed a rationale analogous to the reaction of thiazoles with ketenes. Thus acylation of **1** or **3** produced an N-acyloxazolium ylide **1132a,b**, which was

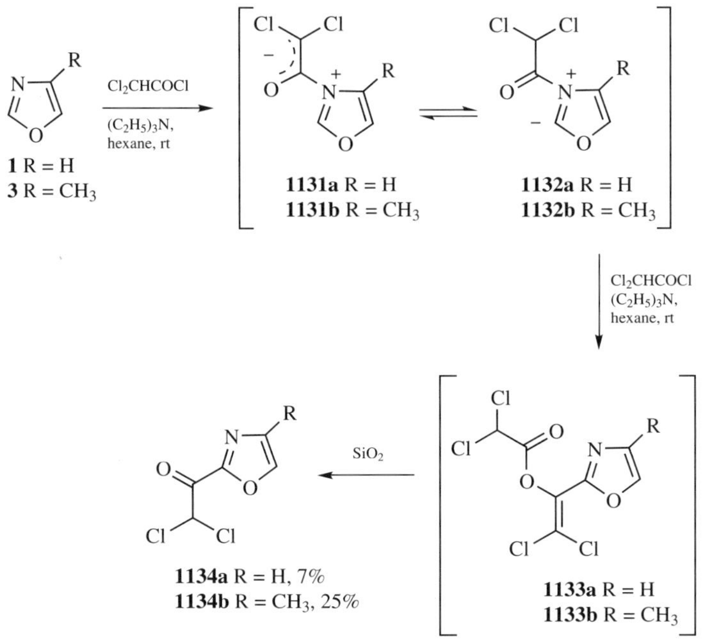

Scheme 1.295

acylated by a second equivalent of **1013a** to produce **1133a,b**. Purification of **1133a,b** then gave the 2-(dichloroacetyl)oxazoles **1134a,b**. The authors also examined reactions of 2-(*N*-benzyl-*N*-methylamino)oxazole, 2-(*N*-benzyl-*N*-methylamino)-4-methyloxazole, and 2-(*N*-benzyl-*N*-methylamino)-4,5-dimethyloxazole with reactive ketenes. In these examples the (C5)-acylated products were derived from Michael addition.

1.4.11. Oxazole Wittig Reagents

Oxazole-derived Wittig reagents and Horner-Emmons reagents have been investigated in the context of natural product syntheses. In particular, these types of reagents are key intermediates in several synthetic approaches leading to the mono-oxazole containing calyculins and the tris-oxazole containing ulapualides, kabiramides, and halichondramides. See Sections 1.5.3 and 1.5.6 for a more detailed discussion. Examples of model oxazole Wittig reagents and Horner-Emmons reagents are discussed in this section.

Armstrong and co-workers[411] described a general synthesis of *trans*-4-alkenyloxazoles as a model reaction for their synthetic approach to calyculins. The authors prepared a 4-(iodomethyl)-2-substituted-oxazole **1135**, starting from the corresponding amide (Scheme 1.296). Arbuzov reaction of **1135** with triethylphosphite gave the phosphonate **1136**. Heating **1135** with tri-*n*-butylphosphine using

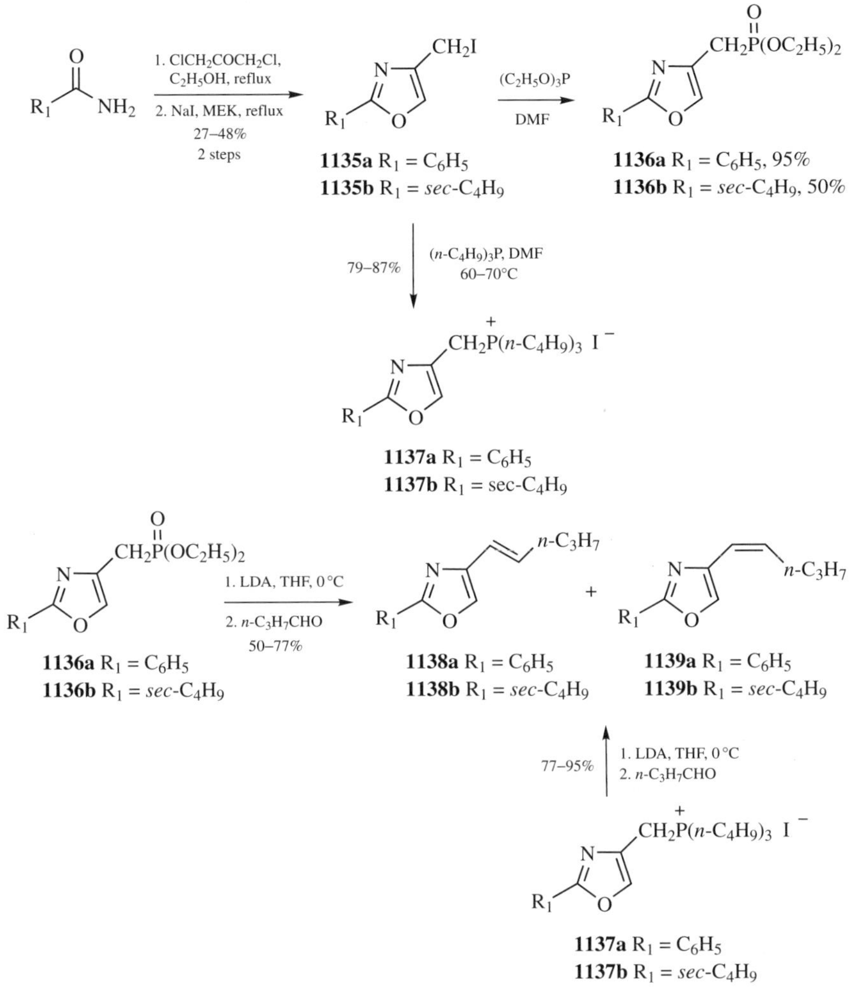

Scheme 1.296

Schlosser's modification[412] of the Wittig reaction produced the phosphonium salts **1137** in excellent yield.

Aldehydes were converted to predominantly *trans* olefins with **1136** and **1137** although *trans:cis* ratios of up to 99:1 were obtained using **1137b**. Both LDA and KH were effective bases in the reaction. As expected, the phosphonium salts were more reactive than the stabilized phosphonates. Examples of *trans*-4-alkenyloxazoles **1141–1143** prepared as advanced intermediates for natural product synthesis are shown in Scheme 1.297.

Armstrong's success in preparing *trans*-4-alkenyloxazoles (vide supra) was a good precedent for Panek and co-workers'[106] general approach to *trans*-2-alkenyloxazoles. These workers explored synthetic routes to *trans*-2-alkenyloxazoles as a means of introducing the C(25)-C(26) *trans* double bond of the tris-oxazole containing

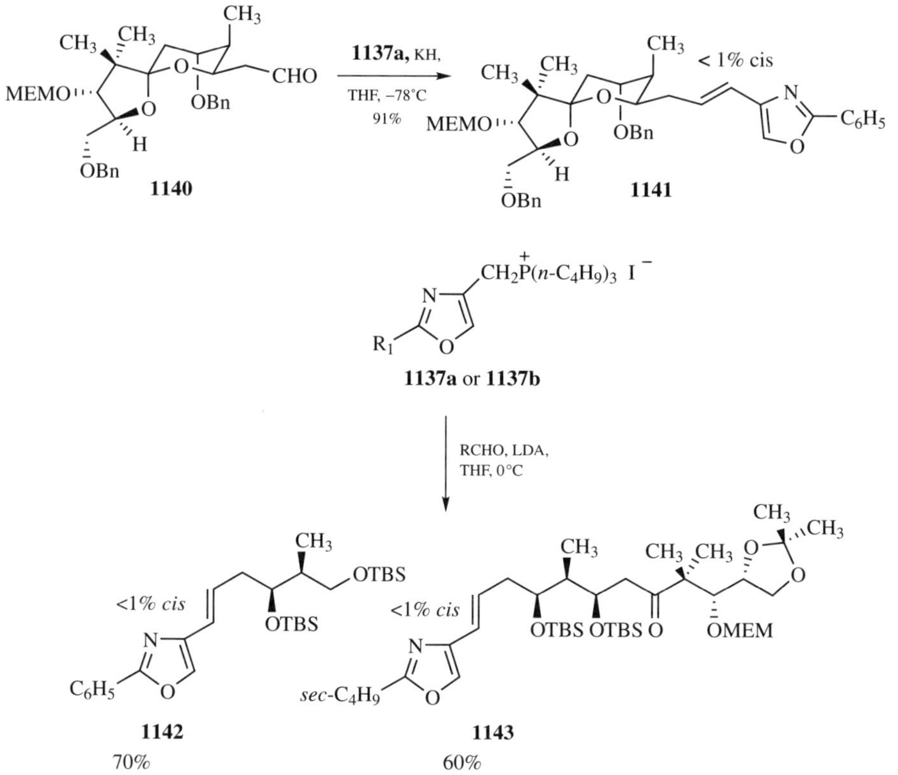

Scheme 1.297

ulapualides. They prepared a series of phosphonium salts from 2-(iodomethyl)-4-oxazolecarboxylic acid ethyl ester **1144** and evaluated these Wittig reagents with tiglic aldehyde using Schlosser-Wittig olefination conditions.[413]

The authors found that branched alkyl phosphines were not useful at all, whereas aryl phosphines afforded **1145** in acceptable yield but modest *trans/cis* selectivity (Scheme 1.298). However, both triethylphosphine and tri-*n*-butylphosphine produced **1145** in nearly quantitative yield with 100:0 *trans/cis* selectivity. Operationally, the phosphonium salts were generated in situ, triethylphosphine was the reagent of choice. Similarly, the *trans*-2-alkenyl bis- and tris-oxazoles **1147–1150** were readily prepared. A further application of this methodology for the synthesis of ulapualide A is discussed in Section 1.5.6.

1.4.12. Miscellaneous Reactions

Naito and co-workers[414] employed an oxazole ring as a template to construct the piperidine ring of the novel antineoplastic alkaloids, pseudodistomin A and B, **1151** and **1152**, respectively. This work was the first synthesis of pseudodistomin

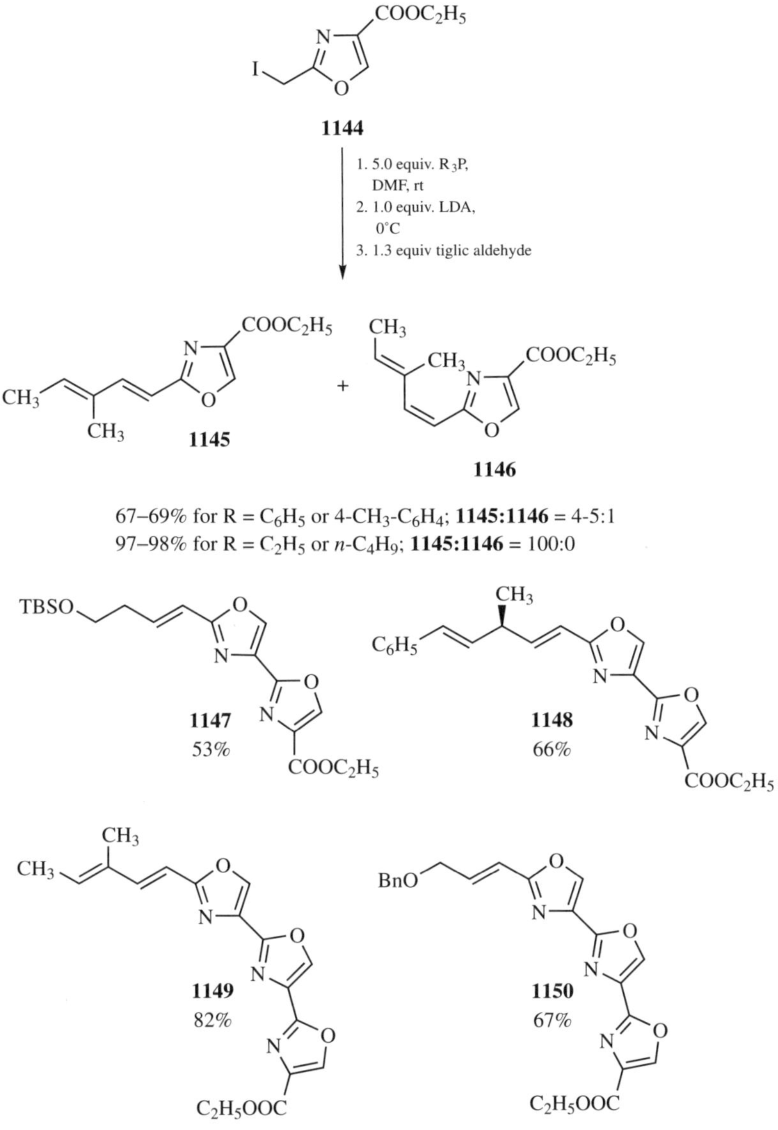

Reactions of Oxazoles 251

Scheme 1.298

tetrahydroacetate **1159** and led to a revision of the diene regiochemisty first proposed for pseudodistomin B acetate **1152** (R = Ac).

Acylation of methyl-*N*-benzylthioacetimidate with 2-phenyl-4-oxazolecarbonyl chloride **1153** gave the (methylthio)enamide **1154** (Scheme 1.299). Reductive cyclization was effected by irradiation of **1154** in the presence of NaBH$_4$ to generate a 9:1 mixture of the diastereomeric lactams **1155** and **1156**. The diastereomers were separated and **1155** was converted to **1157** that was then

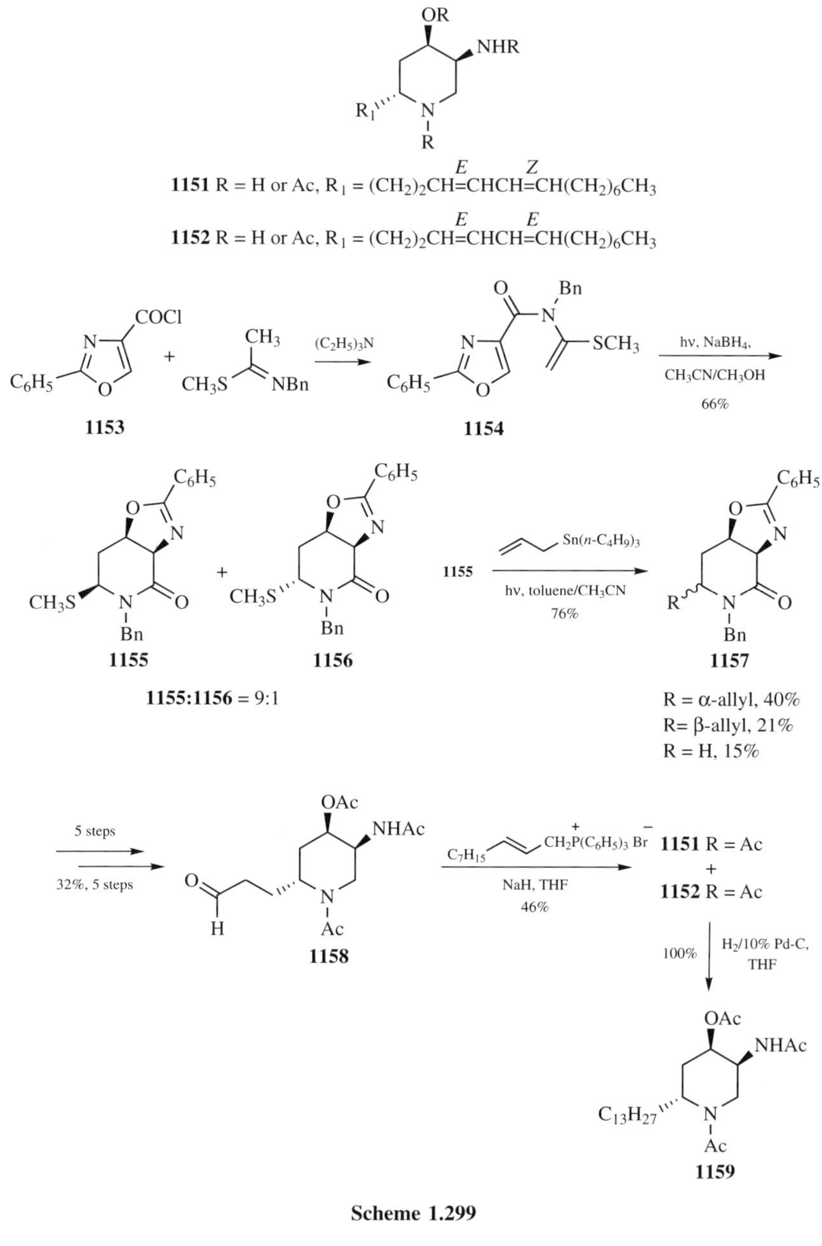

1151 R = H or Ac, R$_1$ = (CH$_2$)$_2$CH=CHCH=CH(CH$_2$)$_6$CH$_3$ (E,Z)

1152 R = H or Ac, R$_1$ = (CH$_2$)$_2$CH=CHCH=CH(CH$_2$)$_6$CH$_3$ (E,E)

1155:1156 = 9:1

1157 R = α-allyl, 40%; R = β-allyl, 21%; R = H, 15%

Scheme 1.299

elaborated in five steps to generate the aldehyde **1158**. Wittig olefination of **1158** gave a mixture of **1151** and **1152**, which was reduced to **1159** and was identical with an authentic sample.

Griesbeck and co-workers[415] described the first examples of oxazoles reacting as the alkene component in a Paternò-Büchi photocycloaddition reaction with aldehydes and α-ketoesters. The authors envisioned that such a process would generate masked α-amino-β-hydroxycarbonyl compounds wherein the oxazole would serve as a latent α-amino ketone or α-amino aldehyde (Fig. 1.85).

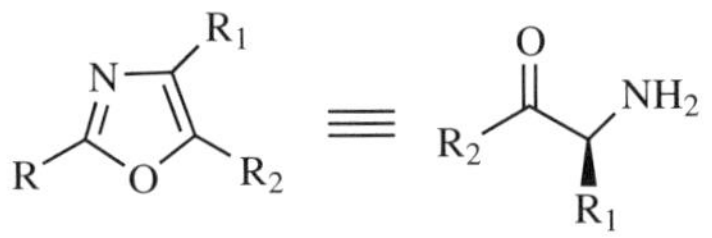

Figure 1.85. Oxazole as a latent α-amino ketone or α-amino aldehyde.

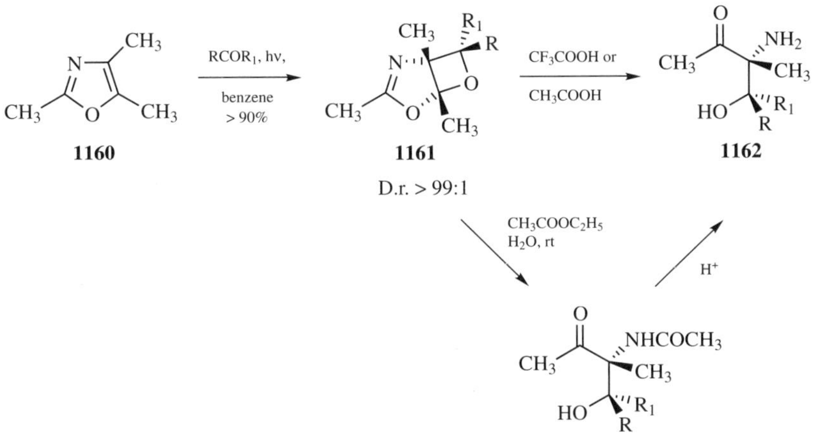

Scheme 1.300

Photolysis of 2,4,5-trimethyloxazole **1160** and an aliphatic or aromatic aldehyde generated a single regioisomer of the bicyclic oxetane **1161** with excellent *syn* diastereoselection (Scheme 1.300). These oxetanes were thermally and hydrolytically labile and readily cleaved to produce an α-N-acetamido β-hydroxy methyl ketone **1163**. Mild acid treatment then converted **1163** to an *erythro* α-amino β-hydroxy methyl ketone **1162**. Cleavage of **1161** with CF$_3$COOH or CH$_3$COOH afforded **1162** directly. Mechanistically, the authors proposed that the diastereoselectivity obtained in this photocycloaddition resulted from "spin-orbit coupling controlled ISC-geometries favorable for spin inversion and transition to closed-shell products and the methyl-group effect." The authors have described these factors previously for other photocycloadditions. Examples of **1161** and **1162** prepared via this methodology are shown in Table 1.79.

Liebscher and co-workers[416] reported a general method of preparing optically active 2-(α-aminoalkyl)oxazoles **1167** via an asymmetric α-alkylation of 2-(aminomethyl)-4,5-diphenyloxazole **1164** (Scheme 1.301). Their work complements the known methodology for asymmetric α-alkylation of other aminomethyl heterocycles.[417,418] Thus Lewis acid–catalyzed condensation of **1164** with commercially available (1S, 2S, 5S)-2-hydroxypinanone generated (1S, 2S, 5S)-3-(4,5-diphenyloxazol-2-ylmethylimino)-2,6,6-trimethylbicyclo[3.1.1]heptan-2-ol **1165** in 65% yield.

TABLE 1.79. *ERYTHRO* α-AMINO-β-HYDROXY METHYL KETONES FROM PATERNÒ-BÜCHI PHOTOCYCLOADDITION OF 2,4,5-TRIMETHYLOXAZOLE WITH ALDEHYDES OR α-KETOESTERS[a]

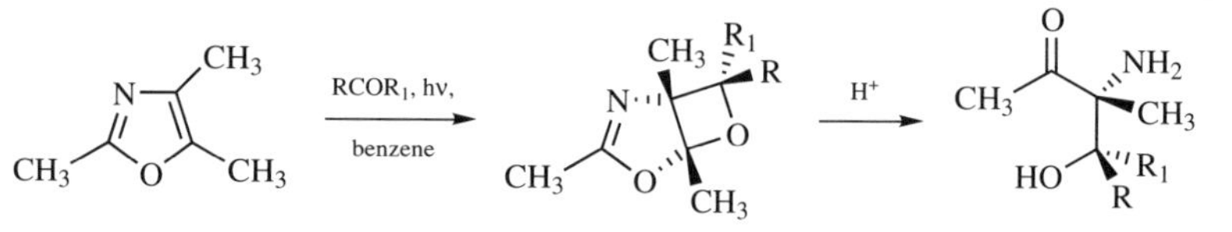

Figure 1.86

Entry	R	R_1	d.r. 1161	Yield 1161 (%)
1	C_2H_5	H	> 99:1	> 90%
2	t-C_4H_9	H	> 99:1	> 90%
3	C_6H_5	H	> 99:1	> 90%
4	C_6H_5	$COOCH_3$	74:26	> 90%

[a] Data from ref. 415.

Asymmetric alkylation of **1165** gave **1166** in good yield with d.r. ratios typically > 95:5. LDA was the base of choice since *n*-BuLi afforded inseparable product mixtures. Mild acid hydrolysis of **1166** with citric acid then gave the desired targets and permitted recovery of the chiral auxillary. Imine exchange using hydroxylamine in acetic acid [419] was much less effective for this transformation. Representative examples of **1166** and **1167** are shown in Table 1.80.

The authors combined their methodology with the known ability of oxazoles to function as masked carboxylic acids. This resulted in the first synthesis of (2*S*, 4*S*, 5*S*)-4,5-dihydroxypipecolinic acid **1172**, a naturally occurring seed component of the legume species *Derris eliptica*[420] using a chiral auxiliary. In this case, their synthetic strategy began with *ent*-**1165** prepared from **1164** and commercially available (1*R*, 2*R* 5*R*)-2-hydroxypinanone (Scheme 1.302). Asymmetric α-alkylation of *ent*-**1165** gave **1168**, which was hydrolyzed uneventfully to **1169**. Cyclization of **1169** to the oxazol-2yl piperidine **1170a** was accomplished with NaHCO$_3$/C$_2$H$_5$OH. Singlet oxygen oxidation of the Cbz-protected analog **1170b** unmasked the carboxylic acid to yield **1171**. Finally, straightforward deprotection of **1171** yielded **1172**.

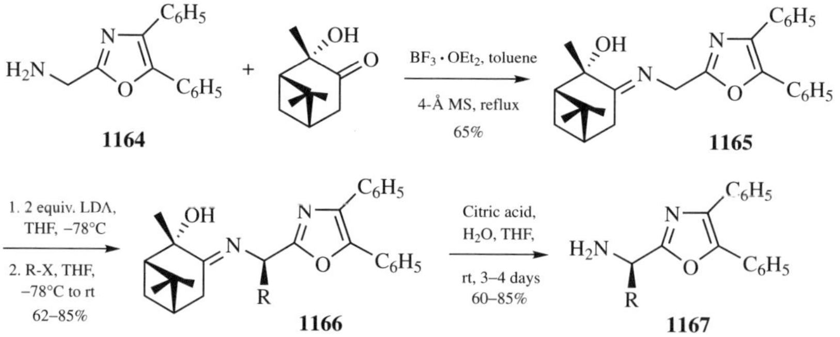

Scheme 1.301

This is the shortest and most straightforward synthesis of **1172**. In addition, both enantiomers of 2-hydroxypinanone are commercially available, thereby allowing access to unnatural analogs of **1172**, relatively inaccessible from other synthetic strategies.

TABLE 1.80. OPTICALLY ACTIVE 2-(α-AMINOMETHYL)OXAZOLES FROM ASYMMETRIC α-ALKYLATION OF 2-(AMINOMETHYL)-4,5-DIPHENYLOXAZOLE[a]

Entry	RX	R	Yield 1166 (%)	d.r. 1166	Yield 1167 (%)	ee 1167
1	CH_3I	CH_3	75	>95:5	63	>99%
2	$i\text{-}C_3H_7I$	$i\text{-}C_3H_7$	82	>95:5	80	NR
3	$c\text{-}C_6H_{11}I$	$c\text{-}C_6H_{11}$	62	>95:5	60	NR
4	$CH_2CH=CH_2Br$	$CH_2CH=CH_2$	72	90:10	85	NR
5	$CH_2=CHCH_2CH_2Br$	$CH_2=CHCH_2CH_2$	72	>95:5	66	NR
6	$ClCH_2CH_2CH_2I$	$ClCH_2CH_2CH_2$	85	>95:5	60	NR
7	$ClCH_2CH_2CH_2CH_2I$	$ClCH_2CH_2CH_2CH_2$	67	>95:5	61	>99%

[a] Data from ref. 416.

1.5. OXAZOLE NATURAL PRODUCTS

A variety of natural products, particularly those of marine origin, incorporate at least one oxazole ring into their molecular architecture. These compounds possess a broad array of exciting and potentially therapeutically useful biological activities and have shown potency as antibacterials, antialgicidals, peripheral analgesics, herpes simplex virus type 1 (HSV-1) inhibitors, antitumor agents, antileukemic agents, antifungal agents, antiviral agents, ichthyotoxic agents, and highly selective serine-theronine phosphatase inhibitors. Selected representative examples of naturally occuring oxazoles are shown in the figures and are arranged in order of increasing number of oxazole rings.

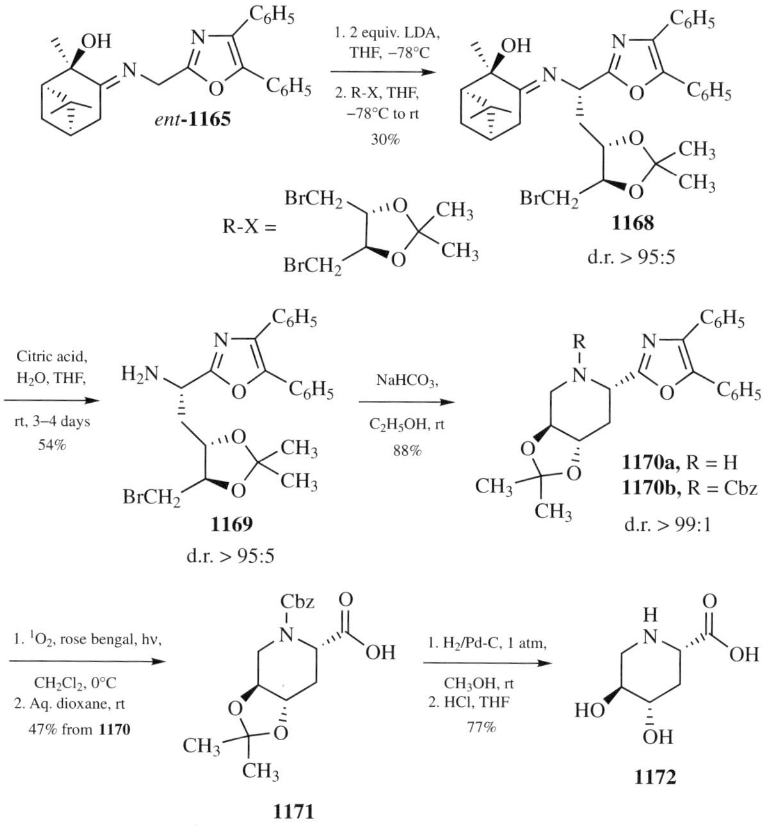

Scheme 1.302

Mono-oxazoles. Oxazole alkaloids can be relatively simple (e.g., the 2,5-diaryloxazoles texamine **1173** and texaline **1174** isolated from the roots of *Amyris texana* P. Wilson) (Fig. 1.88).[421,422] Pimprinine **154** (R = CH$_3$, R$_1$ = H) and pimprinethine **154** (R = C$_2$H$_5$, R$_1$ = H) are examples of 3-(2-alkyloxazol-5-yl) indoles isolated from *Streptoverticillum olivoreticuli*.[423] Melanoxazal **1175** and melanoxadin **1176** are insect melanin biosynthesis inhibitors. These 4-alkenyloxazoles are isolated from fungal fermentation broths.[424] Melanoxazal inhibited melanin production in silkworm larvae (*Bombyx mori*) and is an inhibitor of mushroom tyrosinase. The deceptively simple oxazole-γ-pyrone phenoxan **28** possess potent cytotoxic activity and is an NADH ubiquinone oxidoreductase inhibitor.[41,61,90,425,426] Phthaloxazolins **1177** and **1178** are oxazole-triene antibiotics isolated from *Streptomyces* sp. KO-7888 and *Streptomyces* sp. OM-5714.[427,428] The selective in vitro antifungal activity of these oxazoles against *Phytophthora parasitica* is related to the configuration of the triene side chain. Rhizoxin **1179** was isolated from *Rhizopus chinensis*[429] and possesses antifungal and antibiotic activity. In addition, **1179** has potent antitumor activity, including activity against drug-resistant cells.[51,430]

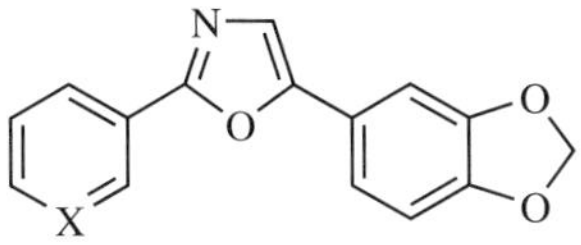

1173 X = CH
1174 X = N

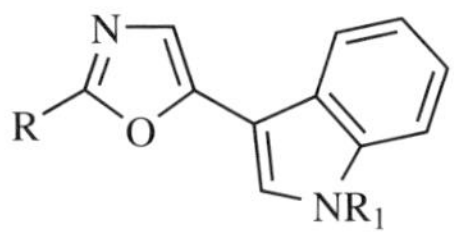

154 R = CH_3, R_1 = H
154 R = C_2H_5, R_1 = H

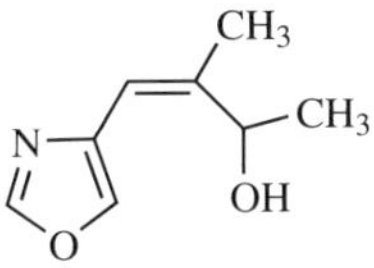

1175

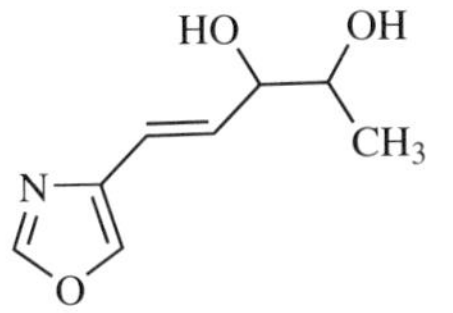

1176

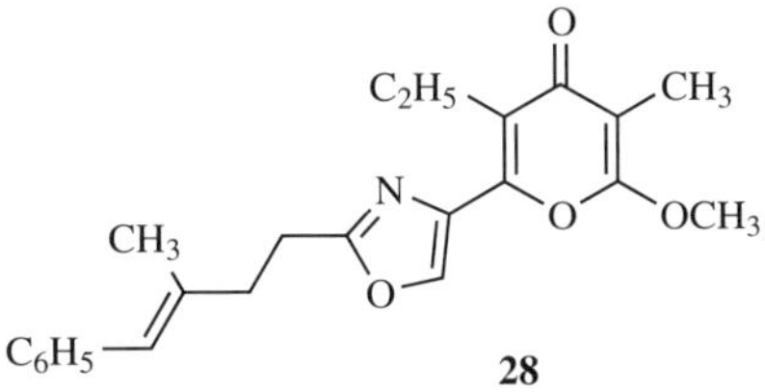

28

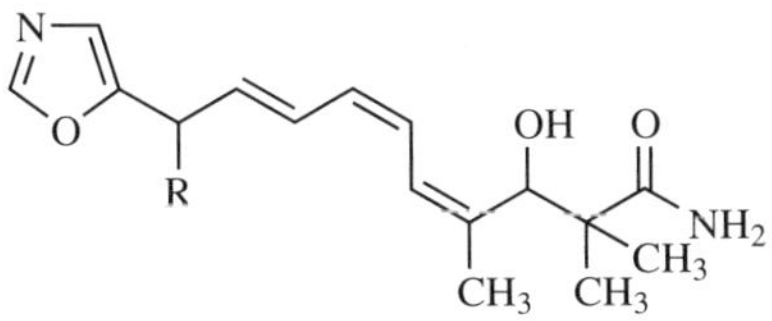

1177 R = H, OH

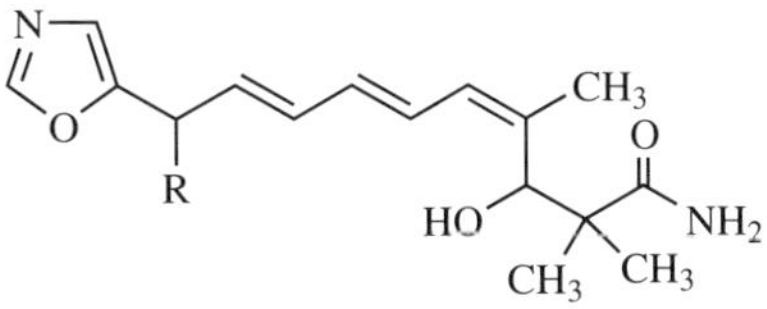

1178 R = H, OH

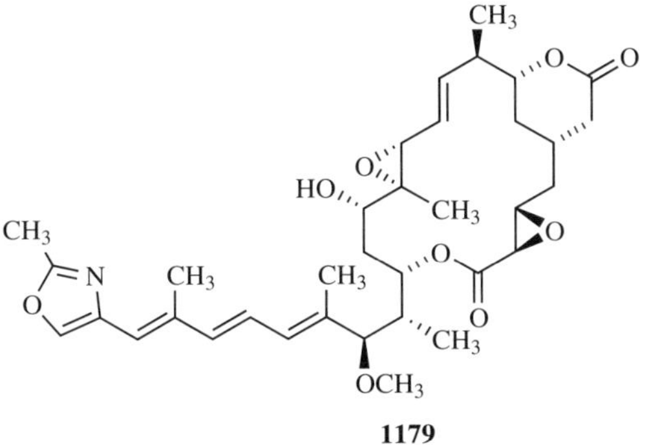

1179

Figure 1.88. Mono-oxazole natural products.

Incorporation of an oxazole ring into a macrocycle gives rise to a number of synthetically novel and challenging structures (Fig. 1.89). Nostocyclamide **526** and dendroamide A **1180** are cyclic hexapeptides isolated from the cyanobacterium *Nostoc* sp.[85,89,431,432] These hexapeptides are active antialgicidal and anticyanobacterial agents. Keramamides **1181a–d** are a family of oxazole containing cyclic peptides isolated from an Okinawan *Theonella* sponge.[433]

Virginiamycin M_2 **1182**, madumycin II **1183**, and griseoviridin **1184** are cyclic peptide oxazole containing streptogramin A antibiotics isolated from the *Streptomyces* family of soil microorganisms (Fig. 1.90).[434–436] Theonezolide A **1185**, the first member of a novel oxazole containing 37-membered ring polyketide was also isolated from the Okinawan *Theonella* sponge.[437]

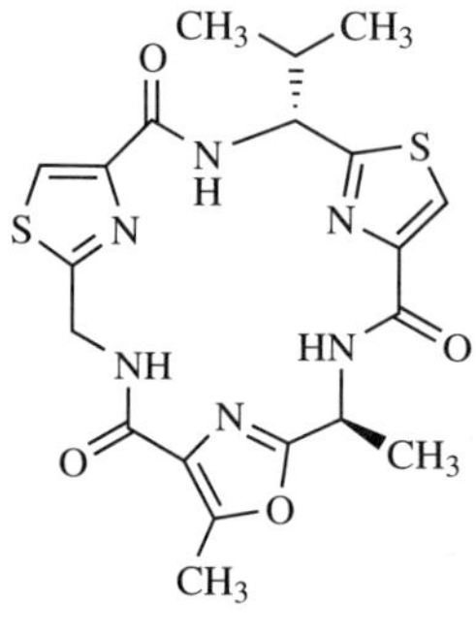

526

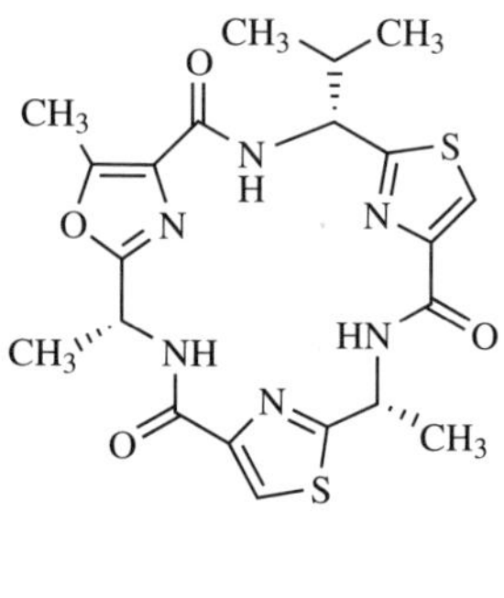

1180

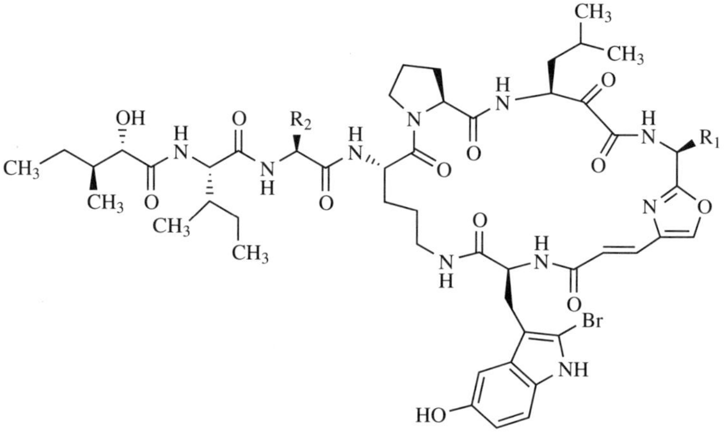

1181a R₁ = CH₃, R₂ = CH₂CH₂CH₃
1181b R₁ = CH₂CH₃, R₂ = CH₂CH₂CH₃
1181c R₁ = CH₂CH₃, R₂ = CH₂CH₃
1181d R₁ = CH₃, R₂ = CH₂CH₃

Figure 1.89. Mono-oxazole natural products.

Thiangazole **1187** and the tantazoles **1188a–c** are cytotoxic alkaloids that possess a unique array of three or four linearly fused 2,4-disubstituted thiazoline rings together with a terminal oxazole ring (Fig. 1.91).[438] Thiangazole is isolated from a metabolite of the gliding bacterium *Polyangium* sp. strain P13007 and is reported to be a potent HIV-1 inhibitor. Tantazoles, on the other hand, are isolated from the blue-green cyanophyte *Scytonema mirabile* (strain BY-8-1) and possess cytotoxic activity, particularly against murine solid tumors.[439] Calyculins, **1189a–d** are isolated from the marine sponge *Discodermia calyx*.[440,441] These structurally intriguing compounds are cytotoxic agents but also highly selective serine-threonine phosphatase inhibitors.

Bis-oxazoles. Two oxazole rings can be incorporated into the structure of natural products in one of several patterns (Fig. 1.92). Bengazoles **1190a–m**, a family of unusual methylene bis-oxazoles, are isolated from sponges of the Astrophorida

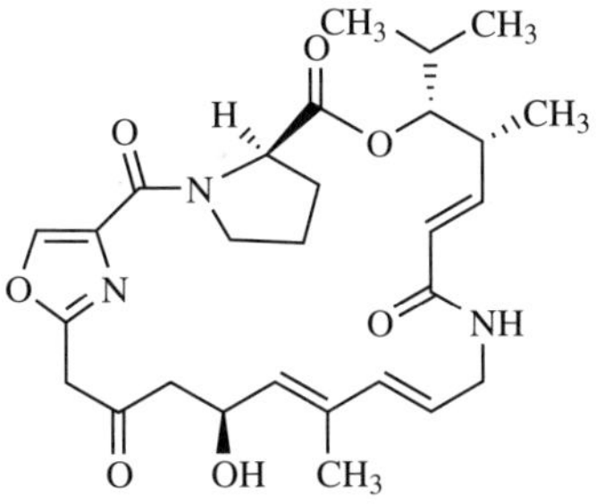

1182

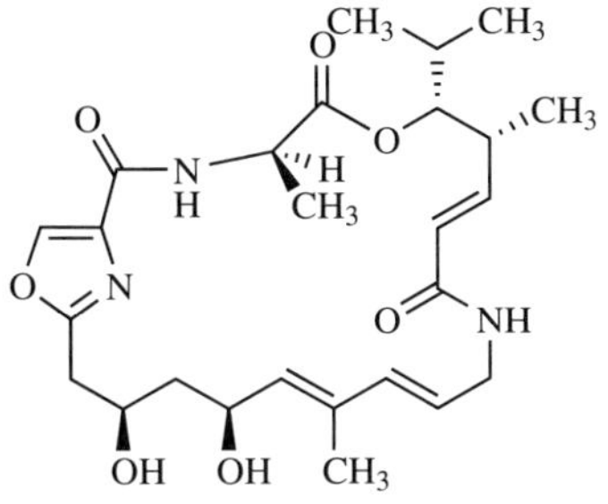

1183

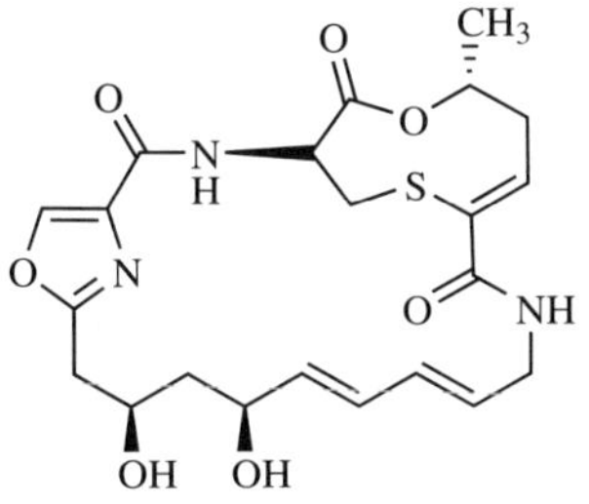

1184

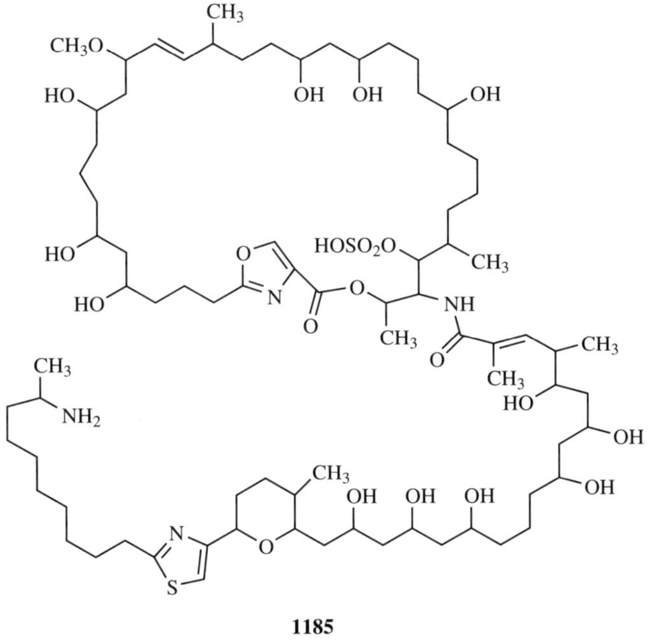

1185

Figure 1.90. Mono-oxazole natural products.

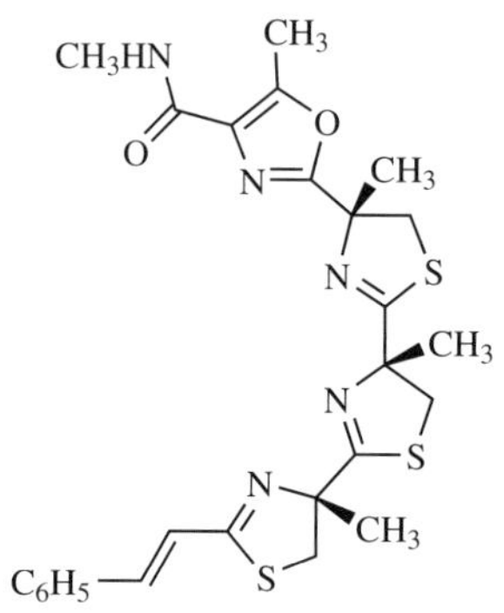

1187

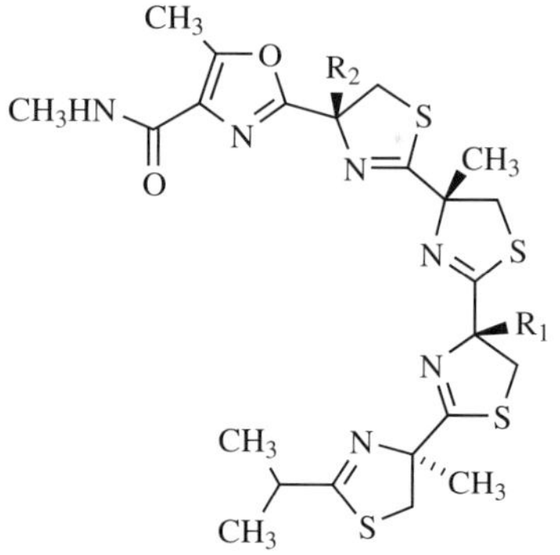

1188a $R_1 = H, R_2 = CH_3$
1188b $R_1 = R_2 = CH_3$
1188c $R_1 = CH_3, R_2 = H$

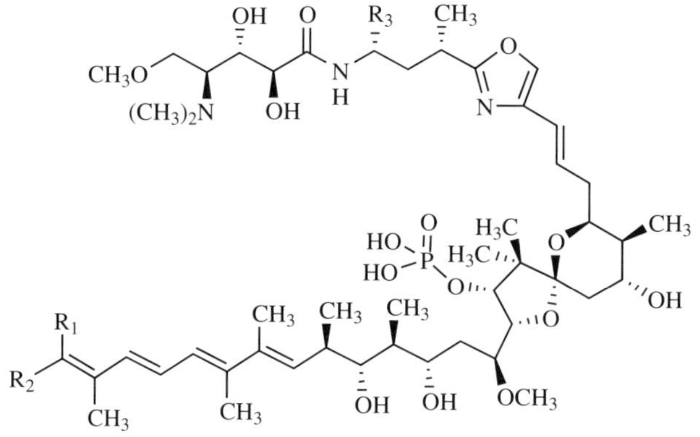

1189a $R_1 = CN, R_2 = R_3 = H$
1189b $R_1 = R_3 = H, R_2 = CN$
1189c $R_1 = CN, R_2 = H, R_3 = CH_3$
1189d $R_1 = H, R_2 = CN, R_3 = CH_3$

Figure 1.91. Mono-oxazole natural products.

order.[350,442–444] Bengazoles contain a carbohydrate like polyol side chain together with a fatty acid ester. Both **1190a** and **1190b** are potent antihelminthic agents.

A linearly linked 2,4′-bis-oxazole is a common structural feature of several complex natural products. The freshwater cyanobacterium *Nostoc muscorum* is the source of muscoride A **156**, a novel peptide alkaloid isolated in 1995.[90,91,445] Muscoride A is the first example of a compound that contains an *N*-(2-methyl-3-buten-2-yl)valine residue and two linearly connected methyloxazoles. Hennoxazoles A-D **1191a–d** are isolated from the marine sponge *Polyfibrospongia* sp.[446,447] Hennoxazole A **1191a** is an exceptionally active inhibitor of herpes simplex virus (HSV-1) and also posseses peripherial analgesic activity. Diazonamides **1192a,b** are secondary metabolites of the colonial ascidian *Diazonia*

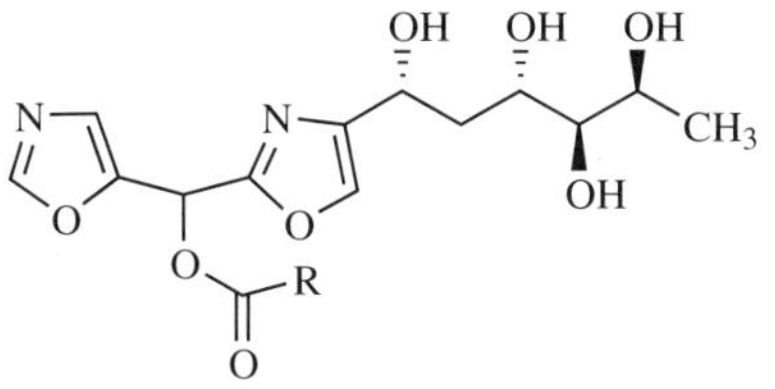

1190a R = $C_{13}H_{27}$
1190b R = $C_{14}H_{29}$

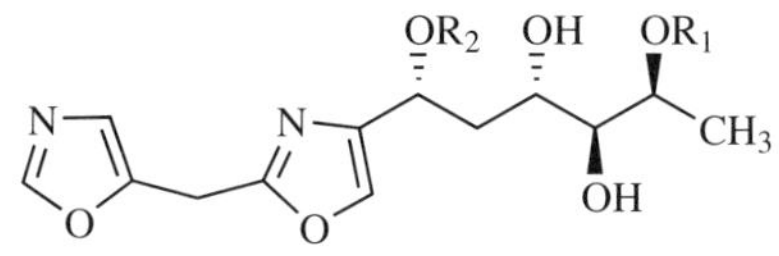

1190c–m R_1, R_2 = H, $CO(CH_2)_{12}CH_3$
$CO(CH_2)_{11}CH(CH_3)_2$
$CO(CH_2)_{13}CH_3$
$CO(CH_2)_{12}CH(CH_3)_2$
$CO(CH_2)_{14}CH_3$

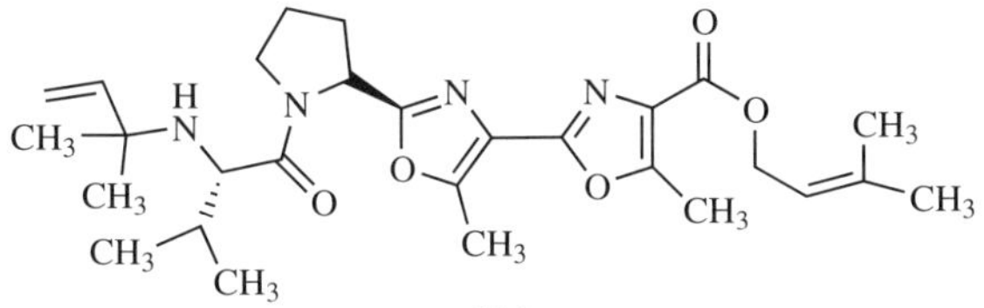

156

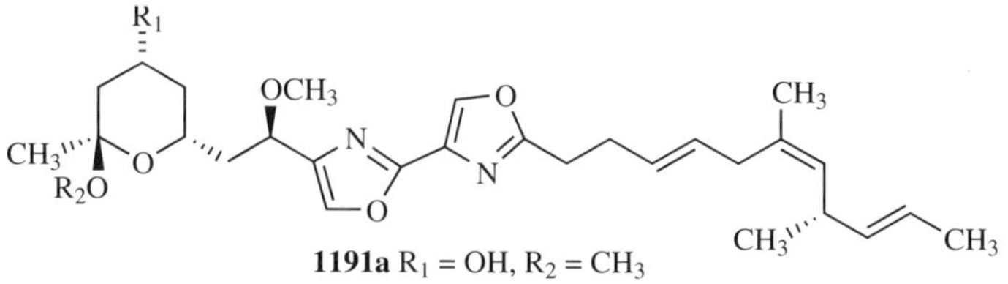

1191a R_1 = OH, R_2 = CH_3
1191b R_1 = OH, R_2 =C_2H_5
1191c R_1 = OH, R_2 = n-C_4H_9
1191d R_1 = H, R_2 = CH_3

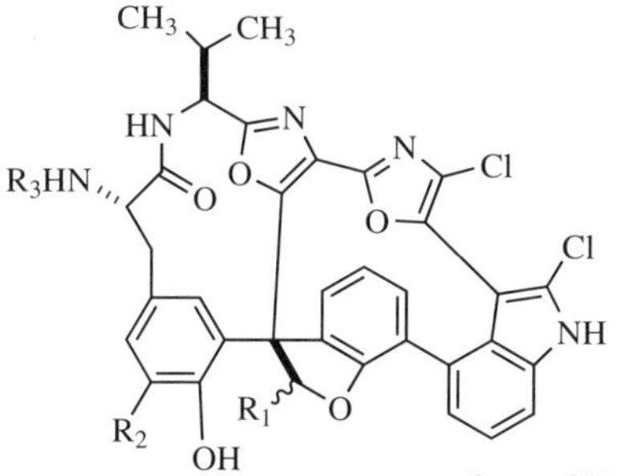

1192a R_1 = OH, R_2 = H, R_3 = 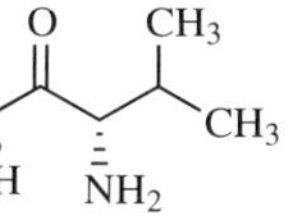
1192b R_1 = OH, R_2 = Br, R_3 = H

Figure 1.92. Bis-oxazole natural products.

chimensis.[448] Diazonamide A **1192a** has pronounced in vitro cytotoxicity against human coleorectal carcinoma (HCT-116) and B-16 murine melanoma cell lines.

Phorboxazoles A and B, **1193a,b** are extremely cytotoxic agents, isolated in 1995 from the marine sponge *Phorbas* sp. (Fig. 1.93).[449] One 2,4-disubstituted

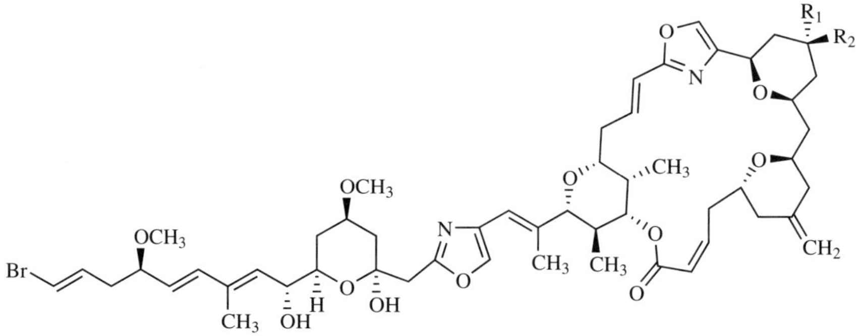

1193a $R_1 = OH, R_2 = H$
1193b $R_1 = H, R_2 = OH$

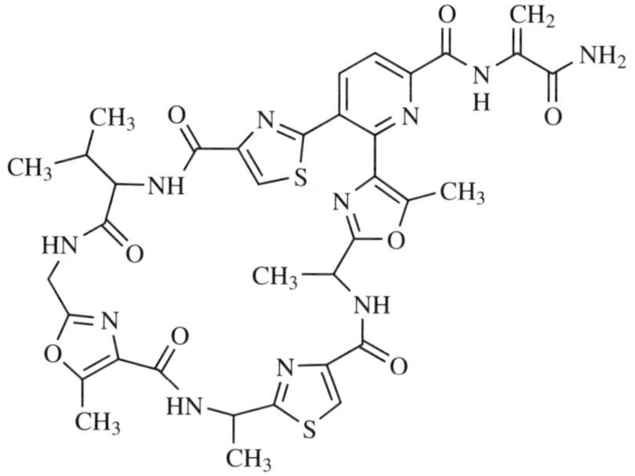

1194

Figure 1.93. Bis-oxazole natural products.

oxazole is incoporated into the complex macrolide ring, whereas the second 2,4-disubstituted oxazole is a critical component of the side chain. Phorboxazoles are among the most toxic natural products and arrest cell growth in the S phase of the cell cycle. Thiopeptide antibiotics isolated from *Streptomyces* sp. SF2741 include promothiocin A **1194** in which both 2,4-disubstituted oxazoles are incorporated into the macrocyle.[450]

Tris-oxazoles and tetra-oxazoles. The ulapualides are a group of secondary metabolites isolated from nudibranches and sponges.[451–455] These metabolites include halichondramides, kabiramides, and mycalolides, all of which possess a unique arrangement of three contiguous oxazoles arrayed in a macrolide structure (Fig. 1.94). Structural differences among the ulapualides are usually reflected in the oxygenation pattern and the alkyl substitutents. Ulapualide A **1195** is a representative example of these compounds. Microcin B17 (MccB17) **1196** is an *E. coli* peptide antibiotic that contains four oxazole and five thiazole rings.[456] In this case, the 2,4-disubstituted oxazole rings are incorporated independently and as part of a linear oxazole-thiazole array. Microcin B17 is a DNA gyrase inhibitor. For

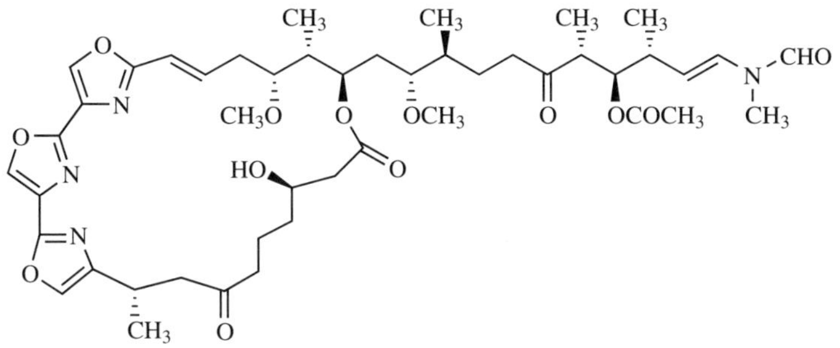

Figure 1.94. Tris- and tetra-oxazole natural products.

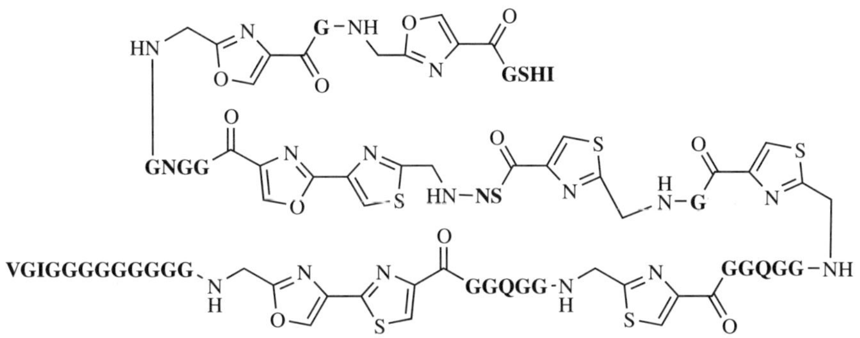

additional examples of other oxazole natural products and more detailed discussions, see more specialized reviews.[19–22]

It is beyond the scope or intent of this section to describe the synthesis of every oxazole containing natural product. Instead, a number of representative examples—including group A stretogramin antibiotics **1182–1184**, thiangazole **1187**, calyculin A **1189a**, calyculin C **1189c**, hennoxazole A **1191a**, diazonamide A **1192a**, and ulapulaide A **1195**—were arbitrarily selected to illustrate synthetic challenges and the diversity of solutions formulated to address these challenges. This section discusses a number of synthetic strategies for each class of natural product and specifically focuses on the construction of the intermediate mono-, bis-, and tris-oxazole moieties critical for completion of the total syntheses. See the cited references for the complete details of any specific synthesis. In addition, earlier sections of this chapter described synthetic methodology specifically targeted to intermediates of natural products, which will not be discussed further.

For example, approaches to phenoxan **28** by Yamamura and co-workers,[41] Peña and co-workers,[61] and Moody and co-workers[90] are discussed in Schemes 1.05, 1.18, and 1.43, respectively. In addition, Peña and co-workers[426] refined their earlier work to prepare a phenoxan analog. Moody and co-workers[90] completed the synthesis of pimprinine **154** (R = CH$_3$, R$_1$ = H) and pimprinethine **154** (R = C$_2$H$_5$, R$_1$ = H) during their early work on diazonamide (Scheme 1.43). Sengupta and

Mondal[422] described a one-step synthesis of texamine **1173** from benzonitrile and α-diazo-(3′,4′-methylenedioxy)acetophenone using $InCl_3$. This approach is similar to Fukushima and Ibata's[83] earlier synthesis of 2-aryl-5-phenyloxazoles (Scheme 1.37).

Leahy and co-workers[51] prepared 2-methyl-4-oxazolecarboxylic acid ethyl ester **34** (R = C_2H_5), a starting material in their synthetic approach to rhizoxin **1179** in excellent yield (Scheme 1.08). Keck and co-workers[430] described an asymmetric synthesis of rhizoxin D in which the oxazole-containing moiety was incorporated using samarium diiodide–mediated Julia-Lythgoe olefination methodology.[457] Approaches to the oxazole amino acid of nostocyclamide **526** by Moody and co-workers[85,89] are shown in Scheme 1.144. Recently, Bertram and Pattenden [432] reported a one-pot synthesis of **526** from the constituent heterocyclic amino acids via a metal-catalyzed assembly in the presence of pentafluorophenyldiphenylphosphinate.

Several synthetic strategies to bengazoles **1190a–m**, encompassing the work of Crews and co-workers,[351] Shioiri and co-workers,[352] and Shafer and Molinski,[354] are described in Schemes 1.241, 1.242, and 1.243, respectively. Moody and co-workers'[90] synthesis of an advanced intermediate for muscoride **156** is detailed in Scheme 1.45. The authors did not complete their synthesis, however, in view of the synthetic efforts of Pattenden and co-workers[458] and Wipf and Venkatraman.[459] Early work from Leahy's group to prepare intermediates leading to phorboxazole **1193a,b** has been described.[51] Evans and co-workers'[363] and Schaus and Panek's[370a] syntheses of key intermediates requisite for **1193a,b** are described in Schemes 1.253, 1.254, and 1.263, respectively. The key oxidation reaction of Meyers and co-workers'[55] approach to the oxazole fragment of bistratamide D is shown in Schemes 1.14 and 1.16. Loughlin and co-workers'[183] one-step synthesis of intermediates for the pyrrole-oxazole skeleton of the halogenated marine alkaloids phorbazoles **437a–d** is presented in Scheme 1.120. The chiral oxazole requisite for the oxazole-thiazole-pyridine fragment of promothiocin A **1194** was prepared by Moody's group,[86] using their metallocarbenoid insertion methodology.

The complemetary work of Warner and co-workers[48] and Jung and co-workers[58,59] for preparing the oxazole-thiazole fragment present in microcin B17 **1196** is shown in Schemes 1.07 and 1.17, respectively.

1.5.1. Group A Streptogramin Antibiotics

Virginiamycin M_2 **1182**, madumycin II **1183**, and griseoviridin **1184** are representative examples of the group A streptogramin antibiotics that incorporate a 2,4-disubstituted oxazole as part of the 23-membered ring macrolactone (Fig. 1.95). Additional charactcristic structural features include an (*E*, *E*)-dienyl amine and a 1,3 dioxygenation pattern. The numbering shown is accepted by Chemical Abstracts Service and reported by Meyers and co-workers.[459a] The streptogramin antibiotics are widely used compounds and have been the subject of intense synthetic investigations in a number of research groups.[459a–464,466,467,469,471,472,475]

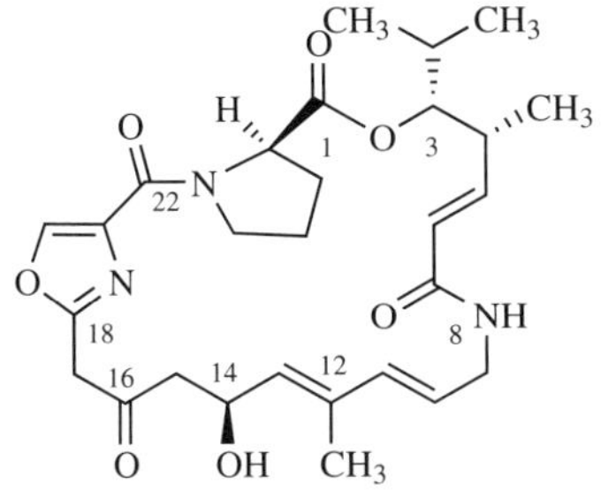

Oxazole Natural Products 265

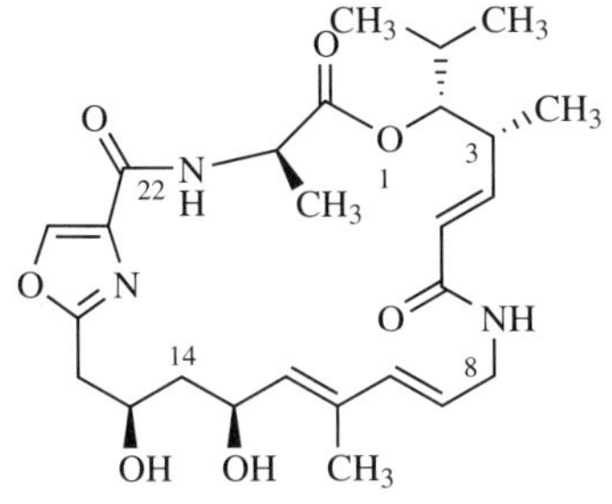

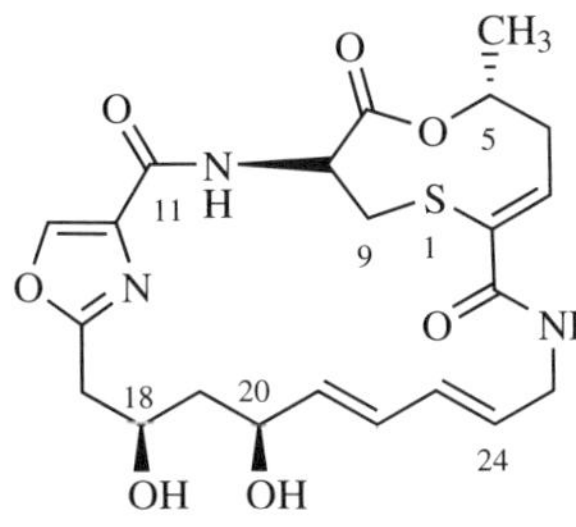

Figure 1.95. Group A streptogramin antibiotics.

Early synthetic work was directed to address the problem of regioselectivity of deprotonation and functionalization of 2-methyloxazole analogues. Fujita and co-workers[460] were not able to selectively acylate the methyl group of 2-methyl-4-oxazolecarboxylic acid ethyl ester or 2-methyl-5-(trimethylsilyl)-4-oxazolecarboxylic acid *t*-butyl ester in their initial model studies. They increased the acidity of the methyl group by preparing 2-(phenylsulfonylmethyl)-4-oxazole carboxylic acid *t*-butyl ester **1197** in two steps (Scheme 1.303). Acylation of **1197** then proceeded smoothly to genereate **1198**, from which the target oxazoles **1199** were prepared by desulfonylation. The authors noted that **1197** effectively functions as the anion equivalent **1200** and should be useful in further approaches to streptogramin A antibiotics.

Wood and Ganem[461] described a more direct approach for deprotonation of a 2-methyl group (Scheme 1.304). They prepared 2-methyl-5-(trimethylsilyl)-4-oxazole carboxylic acid **1201** in which the 5-position was blocked. Dianion formation, followed by quenching with an electrophile, afforded **1202** in excellent yield. In one example investigated, desilylation with cesium fluoride was quantitative. This methodology was developed to complement their convergent strategy to **1182**.

Meyers and co-workers[459a] also investigated lateral lithiation of a 2-methyloxazole derivative during their synthesis of the oxazole dienyl amine moiety **1203**, common to **1183** and **1184**. The authors proposed a retrosynthetic path, wherein selective functionalization of a 2-methyl-4-oxazolecarboxylic acid ester would be the starting point for **1203** (Scheme 1.305).

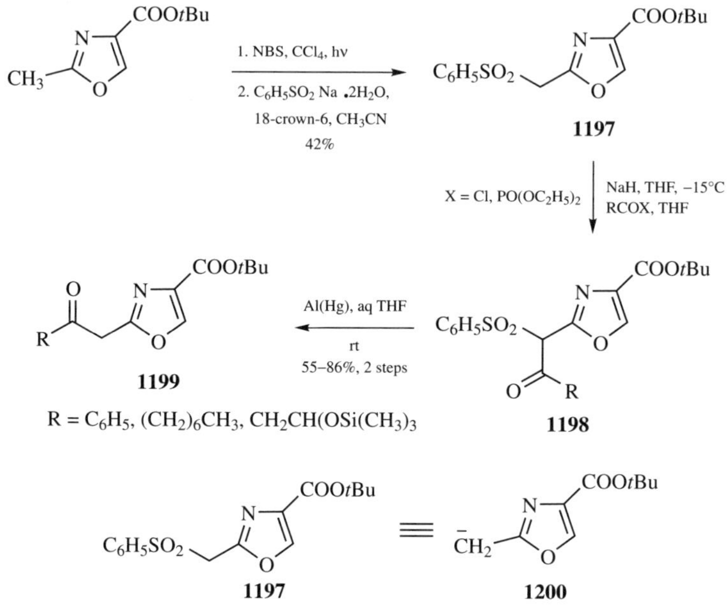

Scheme 1.303

The Cornforth synthesis of the starting 2-methyl-4-oxazolecarboxylic acid methyl ester was straightforward. Unfortunately, this material was not useful in their studies. The ester was hydroylzed to the acid, but attempts to lithiate the acid laterally were also frustrating (Scheme 1.306). The authors isolated only 2-methyl-4-oxazole-5-D-carboxylic acid-D **1207b** (92% D incorporation). Similar results were obtained from metalation of 4-oxazolecarboxylic acid and quenching with CH_3OD, only 4-oxazole-5-D-carboxylic acid-D **1207a** was isolated. Furthermore, lithiation of **1207b** with two equivalents of tBuLi followed by quenching with

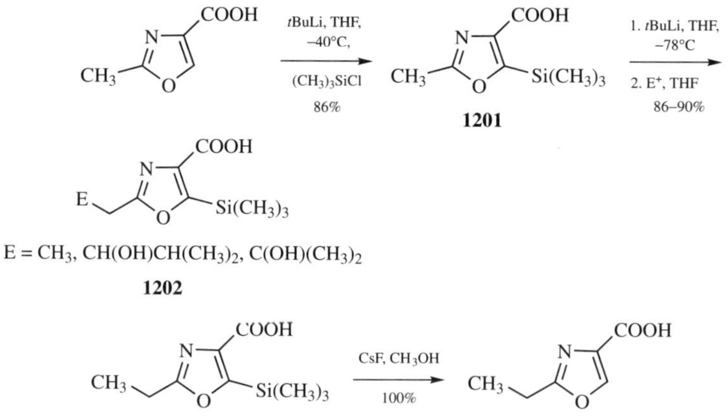

Scheme 1.304

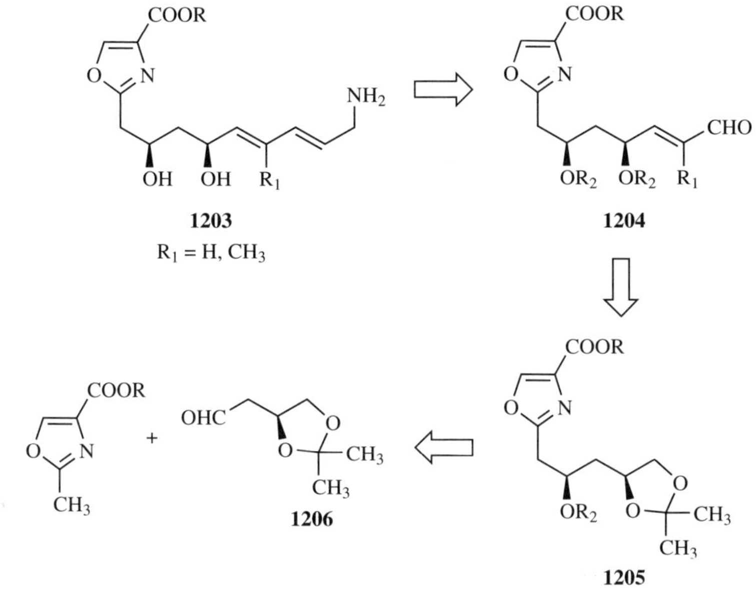

Scheme 1.305

methyl iodide and aqueous workup returned 90% of the starting 2-methyl-4-oxazolecarboxylic acid together with 10% of 2,5-dimethyl-4-oxazolecarboxylic acid **1208**. There was no evidence for the expected 2-ethyl-4-oxazolecarboxylic acid **1209**.

These results are in sharp contrast to the general reactivity for metalation of oxazoles. The authors proposed that the C(4)-carboxylic acid significantly enhanced the acidity of the 5-H and resulted in exclusive deprotonation at C(5) under kinetic and thermodynamic conditions.

The solution to these unexpected results was realized when the authors envisioned intercepting the traditional Cornforth intermediate **1210**. Low-temperature

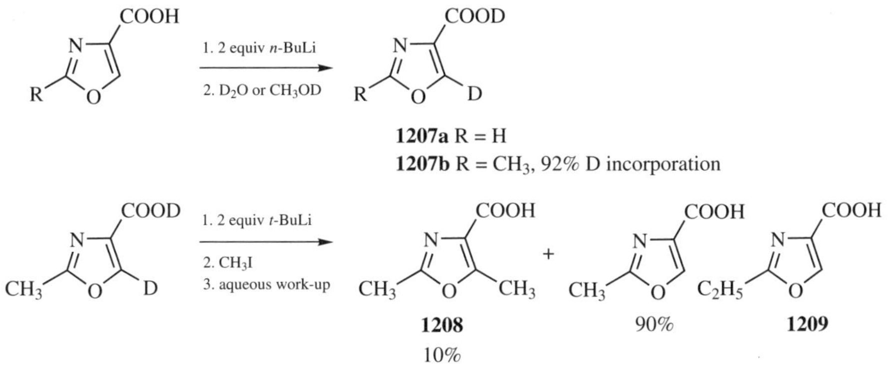

1207a R = H
1207b R = CH$_3$, 92% D incorporation

1208
10%

90%   **1209**

Scheme 1.306

lithiation of **1210** gave a dianion **1211**. Quenching **1211** in model studies with simple aldehydes or benzyl bromide followed by cyclization with boron trifluoride afforded the expected oxazoles **1212** (Scheme 1.307).

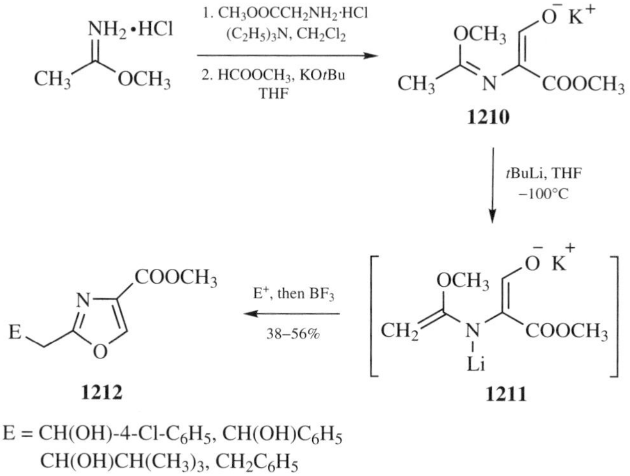

Scheme 1.307

The successful application of this methodology for preparing **1203** analogs is shown in Scheme 1.308. A notable feature of this synthesis is the isolation of the requisite 1,3-dioxane protected triol side chain with simultaneous resolution of the diastereomeric acetonides **1213** through an acetal/acetonide exchange reaction with mesitaldehyde dimethylacetal. Enantiomerically pure **1214** was isolated in 90% yield after chromatographic purification. Swern oxidation[367a] of **1214** to the aldehyde, followed by olefination, produced **1215a** or **1215b** in excellent yield. Final elaboration of **1215a,b** to **1216a,b** was accomplished in modest yield.

Yokoyama and co-workers[462] extended their novel rearrangement of lactoxime O-vinylethers (Section 1.3.8, Schemes 1.90–1.92) to prepare 2-(phenylsulfonylmethyl)-4-oxazolecarboxylic acid t-butyl ester **1197**, a starting material for their approach to virginiamycin M2. Ethyl bromoacetate was readily converted to ethyl (phenylthio)acetohydroximate **1217** in three steps (Scheme 1.309). The key lactoxime O-vinylether **1218** was prepared in excellent yield from **1217** and t-butylpropiolate. Thermal rearrangement of **1218** in refluxing o-dichlorobenzene followed by oxidation with m-CPBA then yielded **1197**. The authors modified their original mechanistic explanation and proposed a charge-separated 1,2-oxazetidine intermediate based on the results of cross-over experiments and other kinetic parameters. This synthesis is longer than Fujita's[460] approach to **1197**, but it does offer an alternate process from readily available starting materials.

Pattenden and co-workers[463,464] reported the total synthesis of 14,15-anhydropristinamycin II$_B$ **1219**. Retrosynthetically, the authors proposed a late-stage construction of the macrocycle via an sp^2-sp^2 Stille coupling of the vinyl stannane **1220**

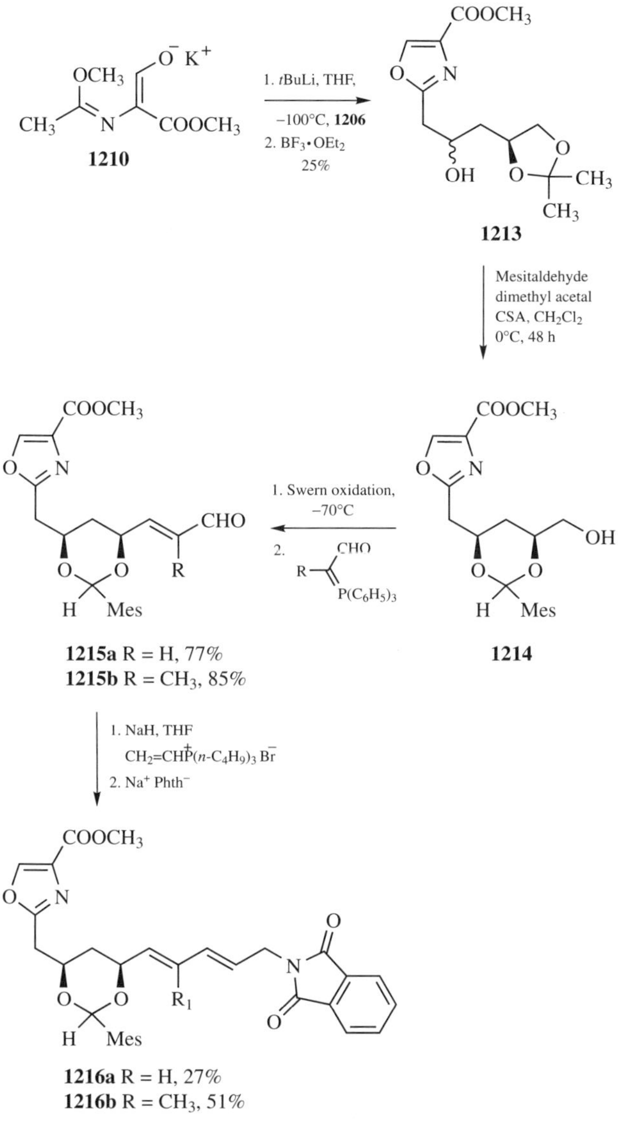

Scheme 1.308

(Scheme 1.310). Further disconnection of **1220** would give rise to the oxazole-diene moiety **1221** and the proline-based vinyl stannane **1222**.

The authors realized their synthetic goals as shown (Scheme 1.311). Careful oxidation of **1223a** with activated MnO_2 generated (2E)-3-bromobut-2-enal **1223b** in excellent yield. Conversion of **1223b** to (2E, 4E)-5-bromo-hexadienal **1224** was accomplished in two steps without incident. The starting 4-(hydroxymethyl)-2-methyloxazole **1225**, prepared via Cornforth methodology,[465] was laterally lithiated and condensed with **1224** to produce **1221a** as an epimeric mixture.

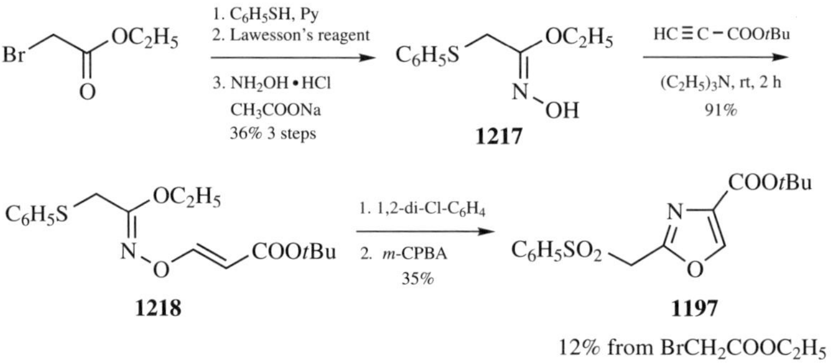

Scheme 1.309

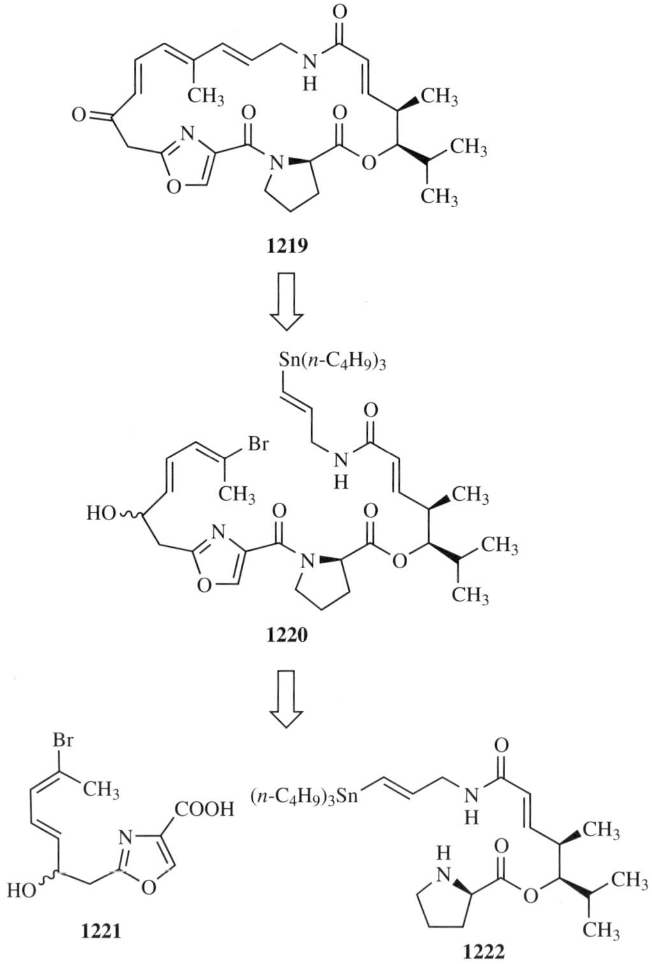

Scheme 1.310

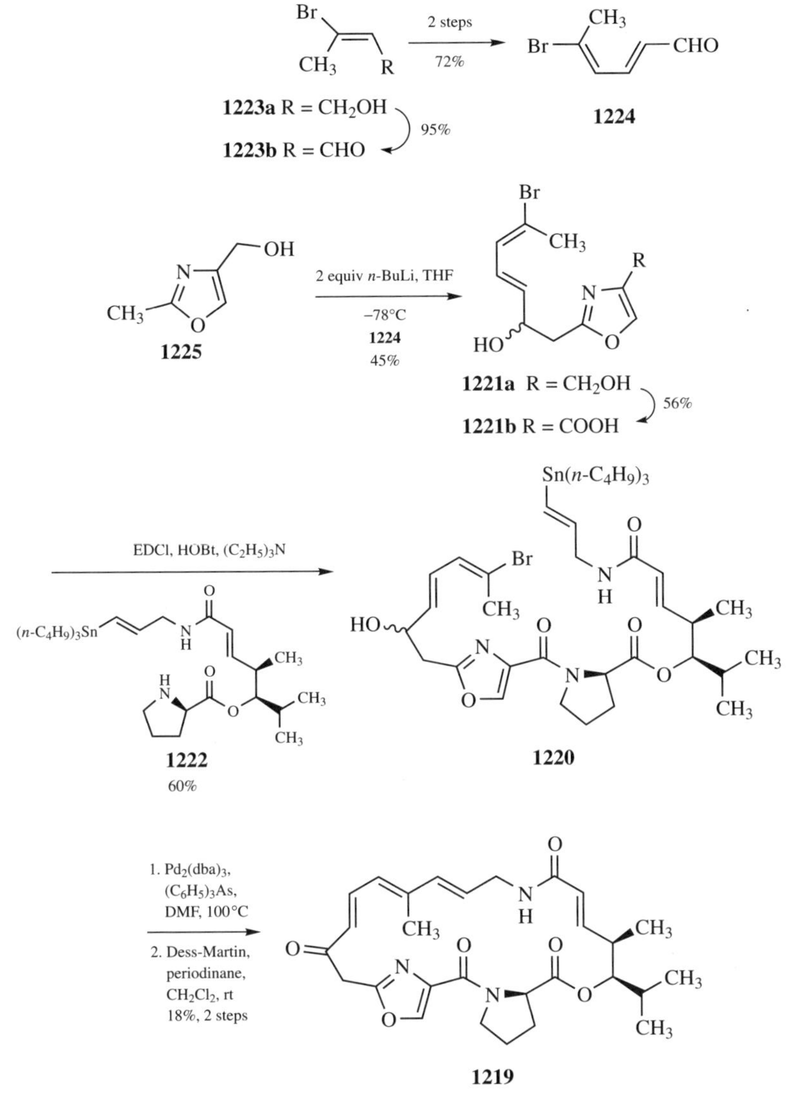

Scheme 1.311

Oxidation of **1221a** to the acid **1221b** was not trivial and was complicated by competitive oxidation of the secondary alcohol. The solution was a stepwise oxidation of **1221a** first, with manganese dioxide to afford the aldehyde (not shown), followed by buffered sodium chlorite to produce **1221b** in 56% yield for the two steps. Standard peptide coupling of **1221b** with the fully elaborated proline-based vinyl stannane **1222** generated **1220** in an acceptable yield. Macrocyclization of **1220** was then accomplished under Farina conditions followed by Dess-Martin periodinane oxidation[211] to complete the synthesis of **1219**.

272   Synthesis and Reactions of Oxazoles

Helquist and co-workers[365,466] targeted a late-stage macrolactamization of **1227** to assemble the entire carbon skeleton of madumycin I **1226**. Further retrosynthetic analysis focused on the final assembly of **1227**, involving a 5-hydroxy-2-alkenoic acid fragment **1228**, D-alanine, and an oxazole-pentadiene unit **1229** (Scheme 1.312). These authors had adapted their oxazol-2-ylzinc Reformatsky-type reagent[365,366] to prepare **961**, a protected oxazole-diene C(9)–C(23) fragment common to both **1184** and **1226** (see Scheme 1.257).

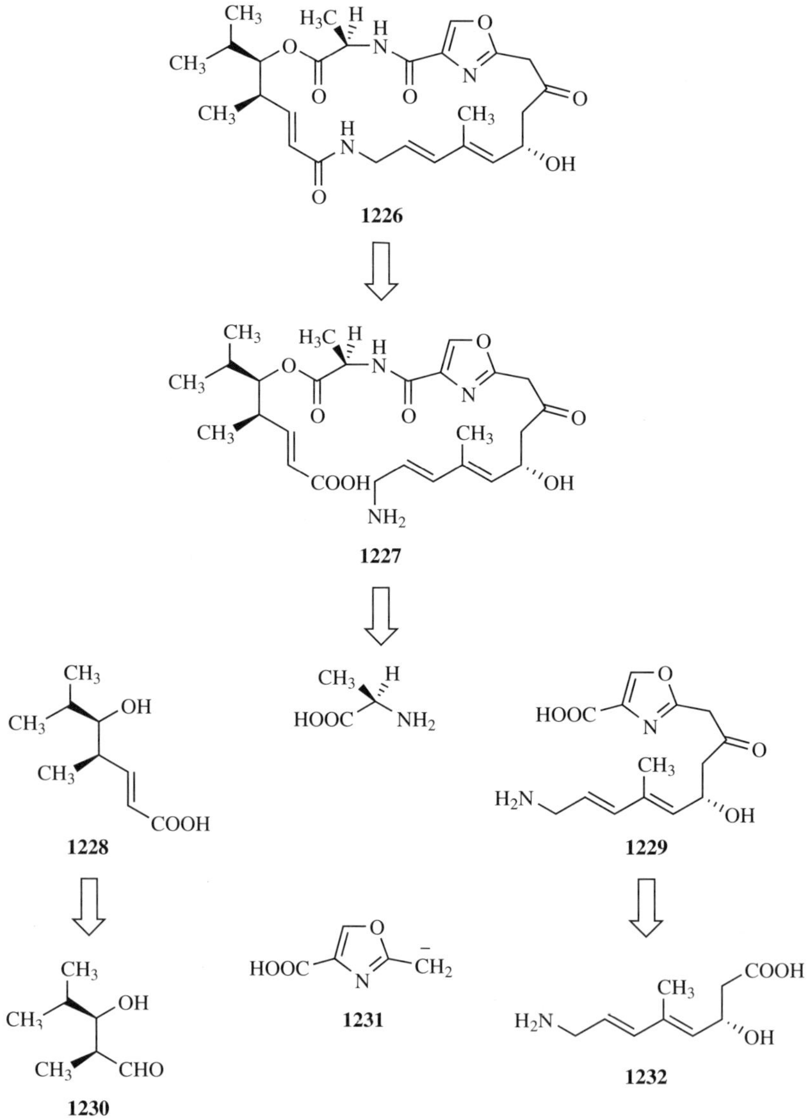

Scheme 1.312

Their synthesis began with conversion of ethyl 3,3-dimethylacrylate to (2E, 4E)-6-t-butoxycarbonylamino-3-methyl-hexadienal **1233** in six steps (Scheme 1.313). This unsaturated aldehyde was relatively labile and was immediately converted to **960** in four additional steps. The starting oxazole **957** was prepared in one step from ethyl α-formyldiazoacetate and bromoacetonitrile. Coupling **960** with **957** in the presence of activated zinc afforded the target (E, E)-oxazole-diene fragment **961** after Swern oxidation.[367a]

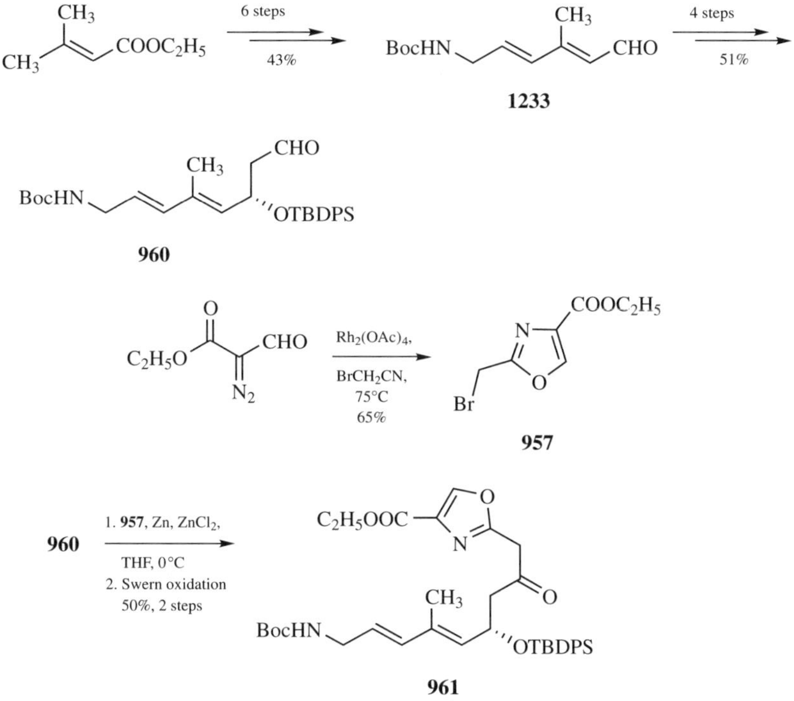

Scheme 1.313

In later work, Helquist and co-workers[466] evaluated much of their overall synthetic strategy to **1226** in a model study. Cleavage of the silyl ether in **1234** and esterification of the resulting alcohol with a protected D-alanine furnished **1235** in excellent yield (Scheme 1.314). The requisite 4-oxazolecarboxylic acid, **1237** was readily prepared in two steps from **1236** and **957**. After cleavage of the Boc group, **1235** was coupled with **1237**, followed by Swern oxidation,[367a] to produce the advanced model target **1238** in 45% overall yield. The authors noted that the experimental conditions to prepare the appropriate protected precursor of **1226** and subsequent macrolactamization were in progress.

Brennan and Campagne[467] also targeted a protected form of **1229** in their approach to the oxazole-dienyl amine fragment of the group A streptogramin antibiotics (Scheme 1.315). However, these authors installed the oxazole ring as the

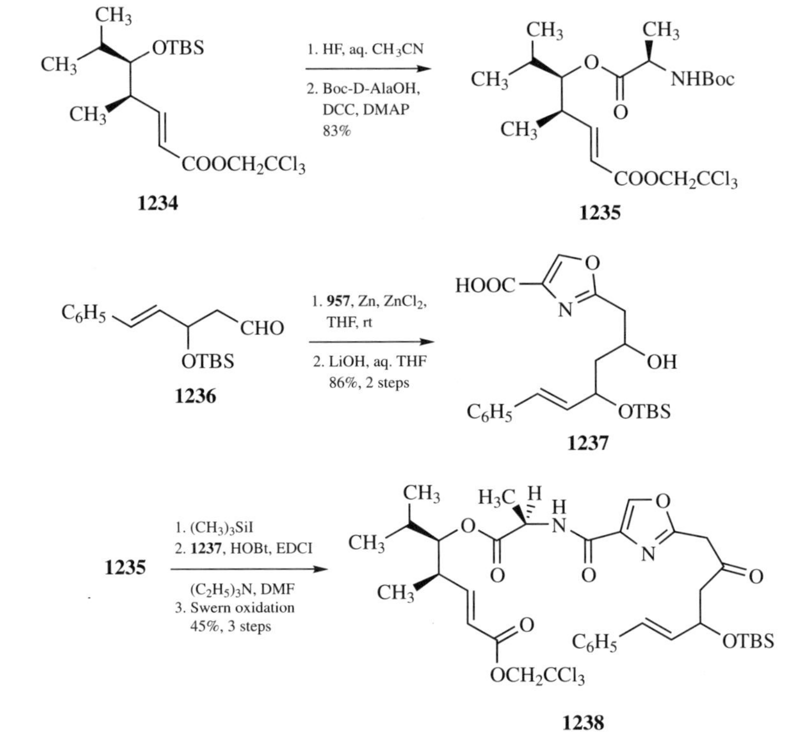

274 Synthesis and Reactions of Oxazoles

Scheme 1.314

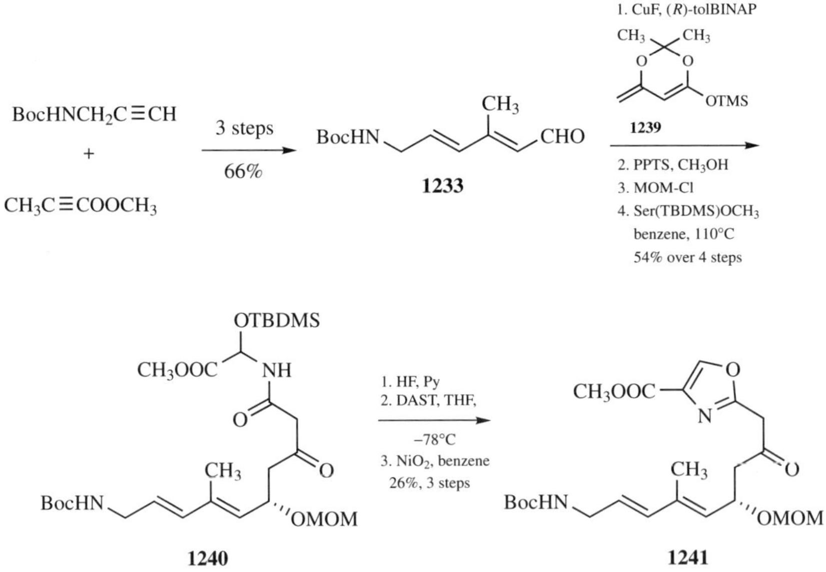

Scheme 1.315

last step in their synthetic approach. They prepared **1233** from *N*-Boc propargylamine and methyl 2-butynoate in three steps. It took considerable experimentation before the authors discovered that LAH effected reduction of the ester and alkyne simultaneously, generating the desired (*E, E*) allylic alcohol (not shown).

An asymmetric acetoacetate vinylogous Mukaiyama aldol condensation of **1233** with **1239** using the precatalyst CuF, (*R*)-tolBINAP[468] was followed by deprotection and ring opening of the dioxinone with Ser(TBS)OCH$_3$ to give **1240** in excellent overall yield. The TBS group was cleaved and DAST cyclization of the resulting *N*-acylamino acid ester generated an intermediate oxazoline (not shown). Oxidation of this oxazoline was problematic. Standard methods afforded complex mixtures that included products derived from elimination of the OMOM group. Ultimately, NiO$_2$ was the reagent of choice to prepare **1241** albeit in 34% yield.

Schlessinger and Li[469] reported the first enantioselective synthesis of (−)-virginiamycin M$_2$ **1182** in 1996. The authors proposed initial disconnection

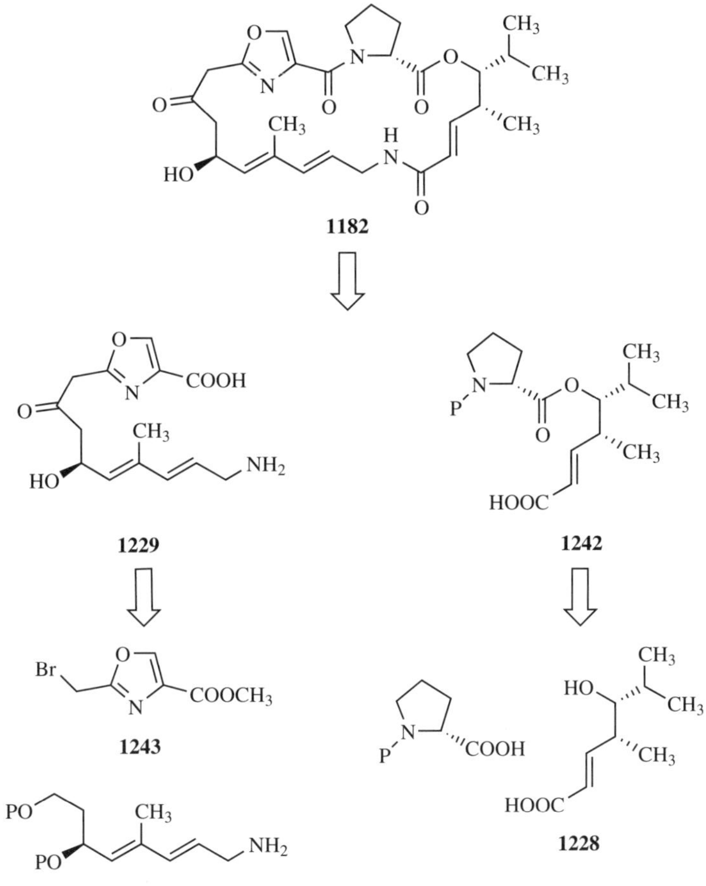

Scheme 1.316

of **1182** at both amides and targeted the oxazole-dienyl amine fragement **1229** and the proline-ester-based 5-hydroxy-2-alkenoic acid **1242** as their late-stage intermediates (Scheme 1.316). Further disconnection of **1229** and **1242** generated **1243**, **1244**, **1228**, and proline as the initial synthetic targets.

The authors assembled the key proline-ester-based 5-hydroxy-2-alkenoic acid **1246** in five steps from **1245** (Scheme 1.317). The octadienylamine containing a differentially protected diol **1248** was prepared from **1247** in eight steps. The authors had noted earlier that the timing of the fragment assembly together with the nature of the protecting groups would be critical for the success of their approach. They apparently altered their original synthetic plan, because **1246** and **1248** were now coupled to produce **1249**, which is the fully elaborated skeleton, lacking only the oxazole and ring closure to complete the synthesis.

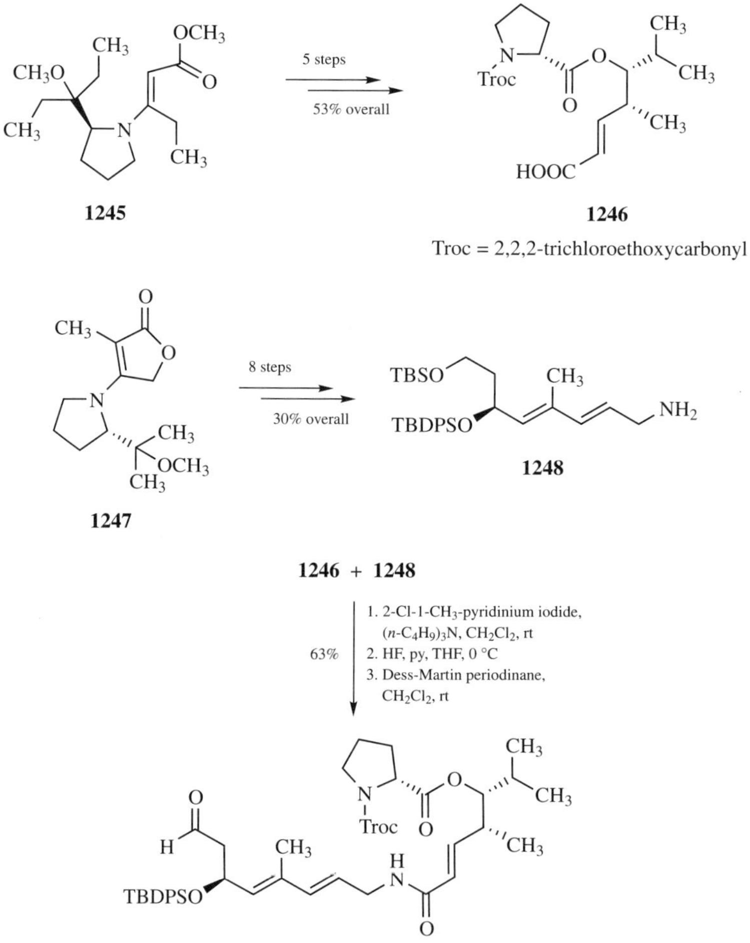

Troc = 2,2,2-trichloroethoxycarbonyl

Scheme 1.317

The oxazole was installed via an oxazol-2-yl Reformatsky-type reagent, prepared from a 2-(bromomethyl)-4-oxazolecarboxylic acid ester **1243** (Scheme 1.318). Thus the organozinc reagent, generated from **1243** and zinc in the presence of $(C_2H_5)_2AlCl$, added to **1249** to afford **1250** as a diastereomeric mixture of silylated secondary alcohols. After cleavage of the Troc group and saponification, macrolactamization to **1251** was accomplished using a Mukaiyama amide coupling under dilute conditions.[470] The authors attributed the facility of this coupling to the ability of **1246** to function as a β-turn mimic. Conversion of **1251** to **1182** was straightforward.

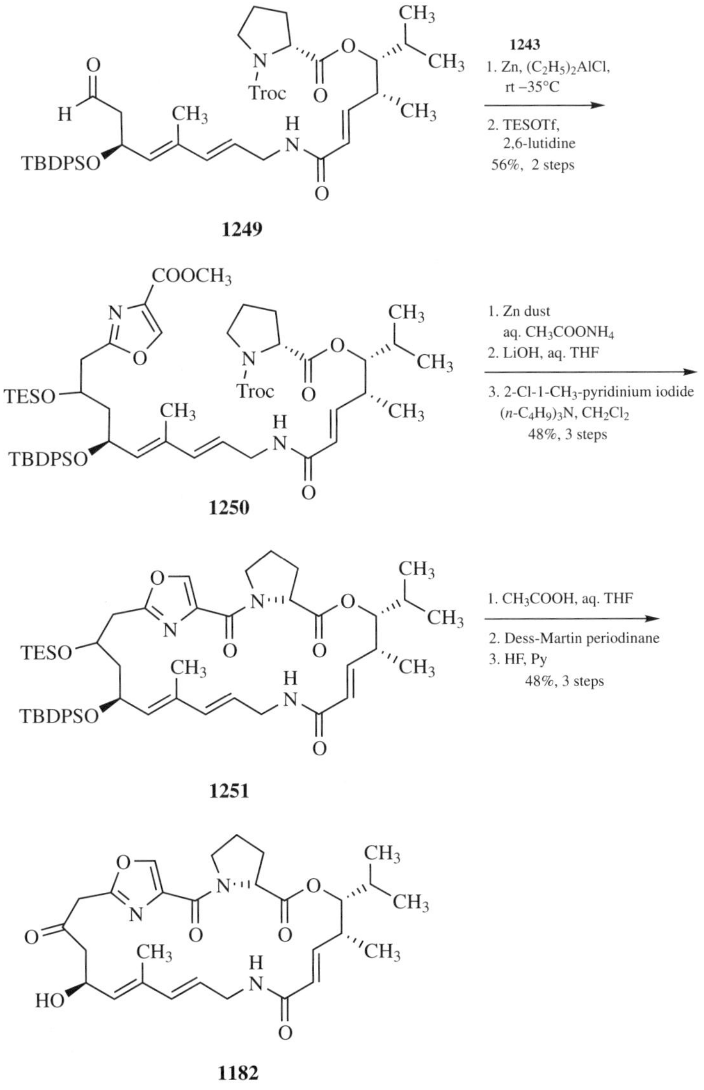

Scheme 1.318

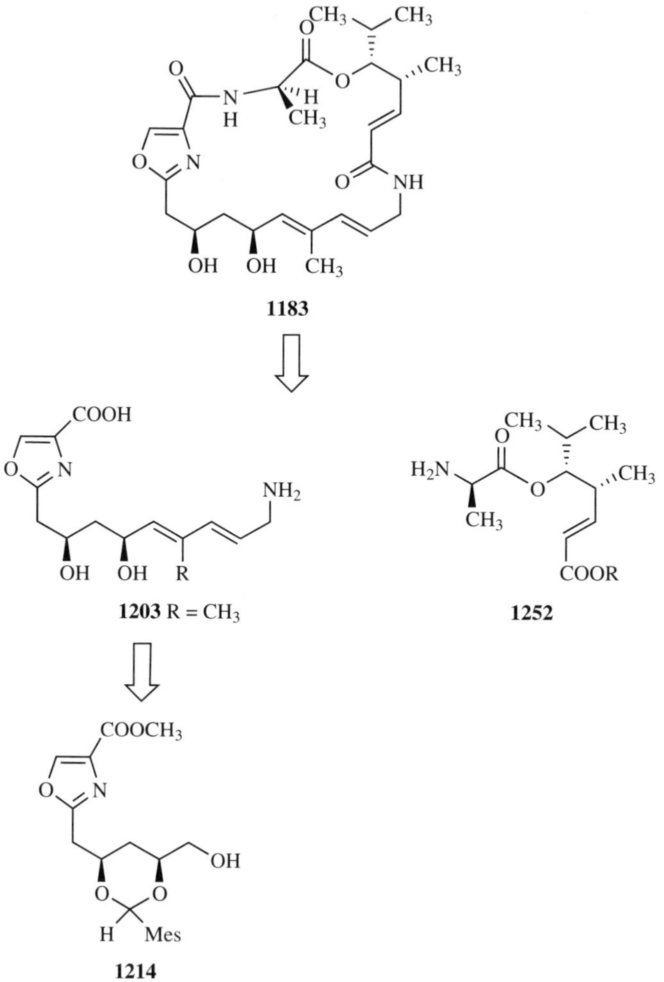

Scheme 1.319

Simultaneously, Meyers and co-workers[471] described the first total synthesis of (−)-madumycin II **1183**. They envisioned construction of **1183** from the oxazole-dienylamine **1203** and the D-alanyl ester **1252** (Scheme 1.319).

Meyer's group[459a] previously prepared the protected oxazole-dienylamine fragment **1216b** (see Scheme 1.308), but they pursued an alternate approach to the key protected triol **1214** in this latest synthesis (Scheme 1.320). The Weinreb amide[191] **1253** was transformed to the acid **1254** in four straightforward steps. A chelation-controlled reduction established the required *syn* 1,3-configuration of the protected triol. Coupling **1254** with serine methyl ester and cyclization with Burgess reagent[49] gave an intermediate oxazoline, which was oxidized to the oxazole acetonide **1255a** using *t*-butylperbenzoate and CuBr/Cu(OAc)$_2$.[35,36] After cleavage of the TBS ether, **1255b** was elaborated to **1216b**, exactly as described previously. The D-alanyl ester

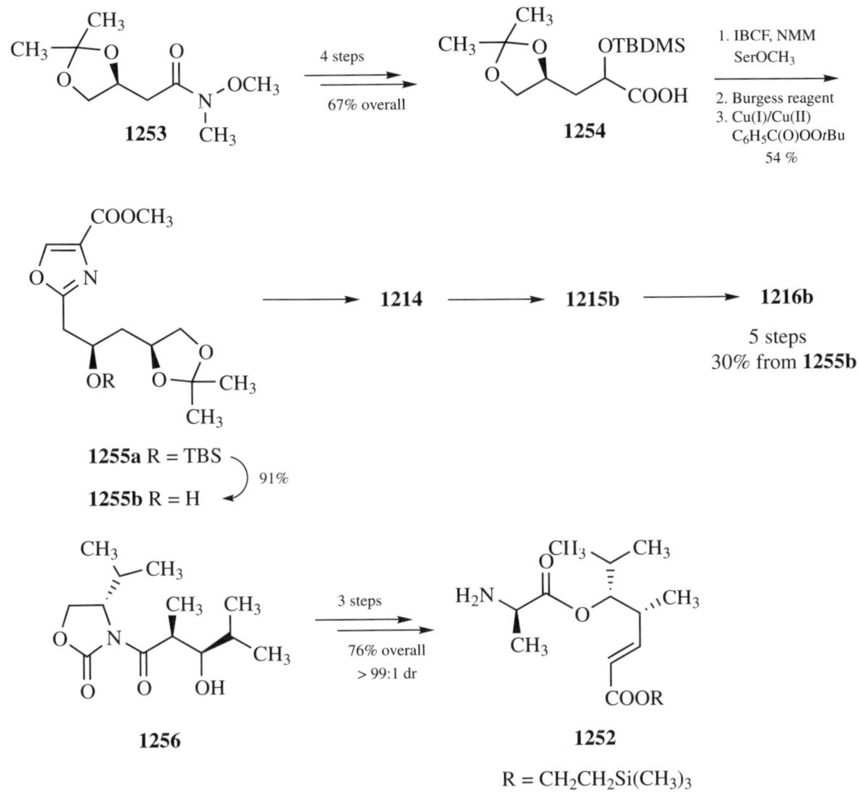

Scheme 1.320

fragment **1252** (R = CH₂CH₂Si(CH₃)₃) was readily prepared in three steps in excellent yield and diastereomeric purity.

Amino acid coupling of **1257** with **1252** (R = CH₂CH₂Si(CH₃)₃) was accomplished using DCC/DMAP to afford **1258** (Scheme 1.321). Sequential deprotection of the phthalimide and β-trimethylsilylethyl ester was straightforward. Macrolactamization of the resulting amino acid and hydrolysis of the dioxane moiety completed the synthesis of **1183**. The authors observed a small amount (8–10%) of double bond isomer, but showed that this resulted from hydrolysis of the mesitylene acetal.

Shortly thereafter, Ghosh and Liu[472] reported a convergent and enantioselective synthesis of **1183**. In their strategy, macrolactonization of **1259** is the key step for assembling the core structure. Further disconnection of **1259** then revealed the oxazole dienylamine **1203** (R = CH₃), D-alanine, and the 5-hydroxy-2-alkenoic acid **1228** as intermediate synthetic targets (Scheme 1.322).

Enzymatic desymmetrization of *cis*-1,4-diacetoxy-2-cyclopentene **1260** followed by a series of straightforward functional group manipulations and

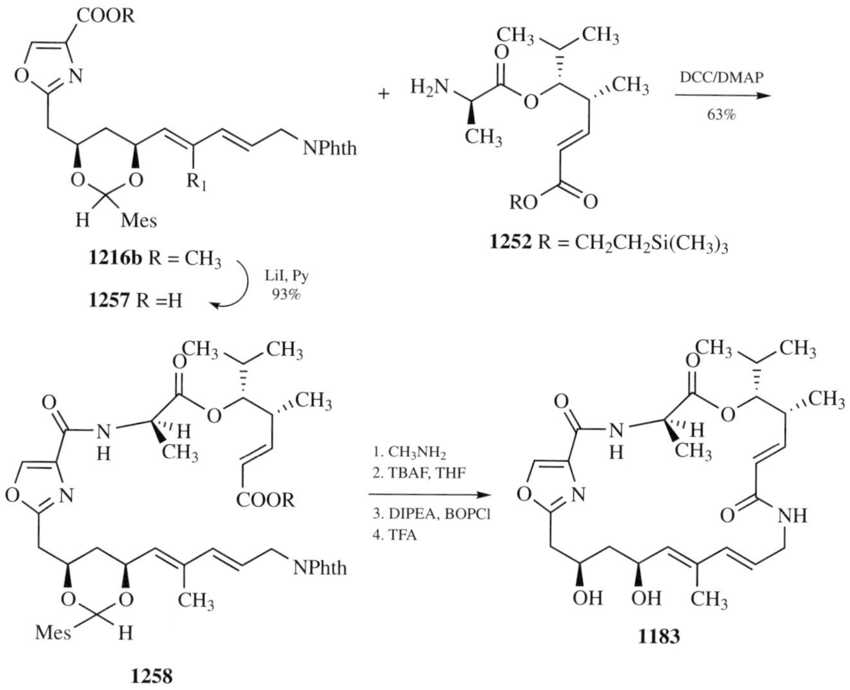

Scheme 1.321

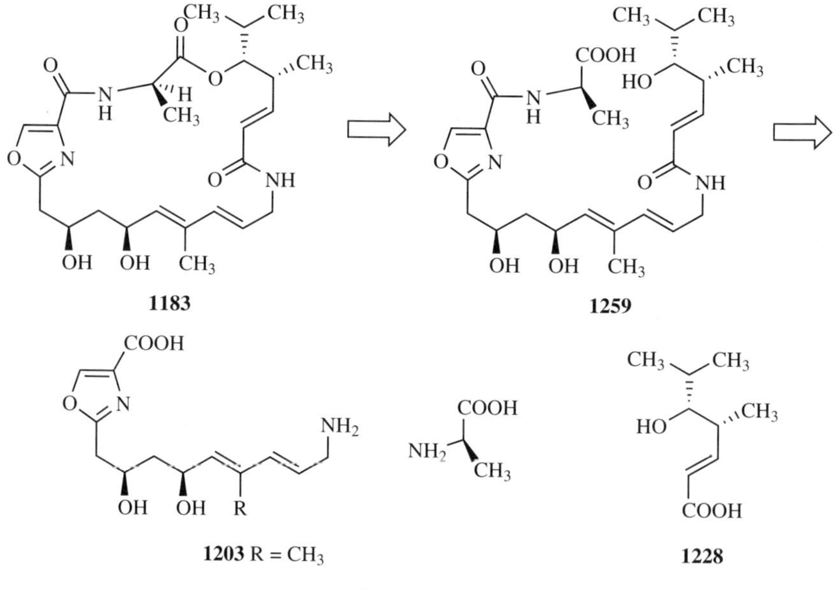

Scheme 1.322

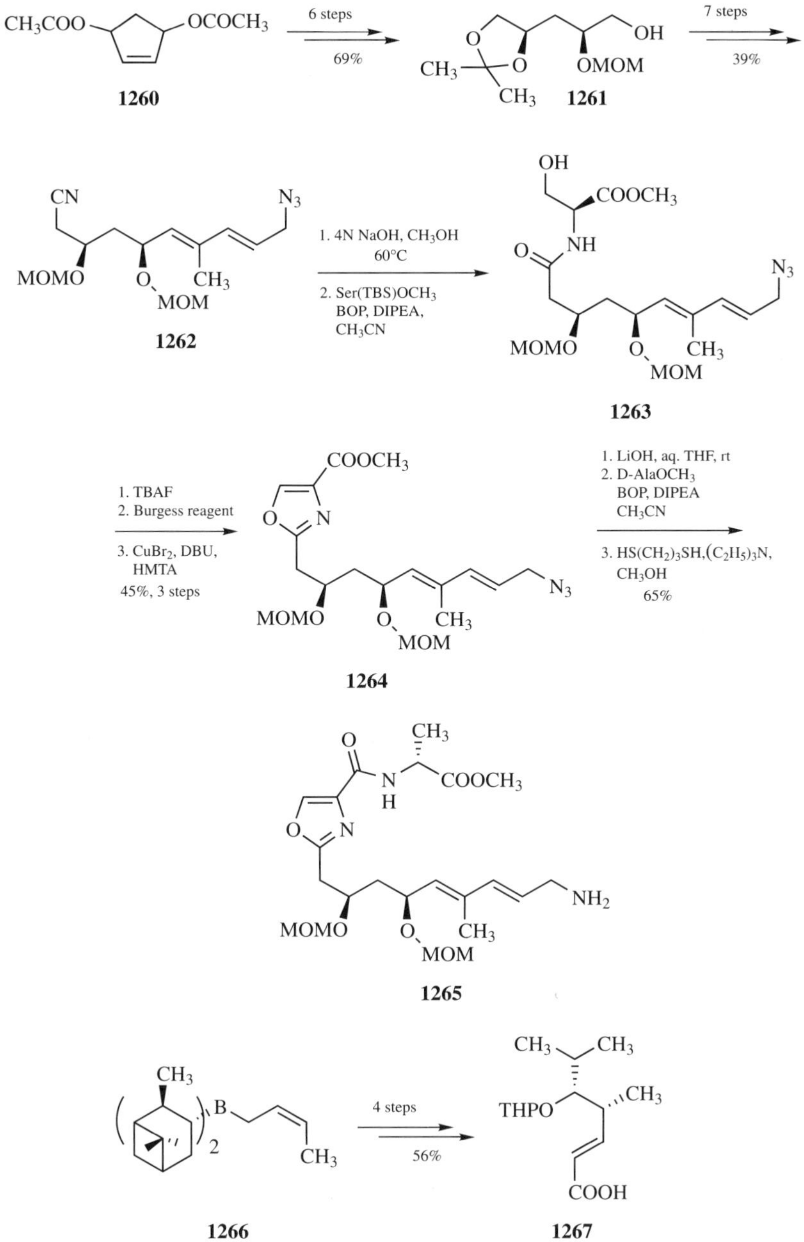

Scheme 1.323

elaborations gave rise to **1261** in excellent overall yield (Scheme 1.323). The bis-MOM-protected nitrile **1262** was then prepared from **1261** in seven steps. The oxazole ring was then constructed. Hydrolysis of **1262** gave the acid (not shown) that was coupled with Ser(TBS)OCH$_3$ to give **1263** after deprotection of the silyl group. The resulting *N*-acylamino alcohol was cyclized to the oxazoline with Burgess reagent[49] followed by oxidation with $CuBr_2$/DBU/HMTA (see Section 1.3.1.3; Schemes 1.06–1.08)[43–47] to produce **1264** in 45% overall yield from **1263**. Conversion of **1264** to **1265** was accomplished in three straightforward steps. The protected 5-hydroxy-2-alkenoic acid fragment **1267** was readily avaialable from **1266**.

The final fragment assembly is shown in Scheme 1.324. Coupling **1265** with **1267** was accomplished with BOP reagent[473] to afford **1268**. The THP group was removed and macrolactonization was achieved using Yamaguchi[474] conditions followed by removal of the MOM groups to produce **1183**.

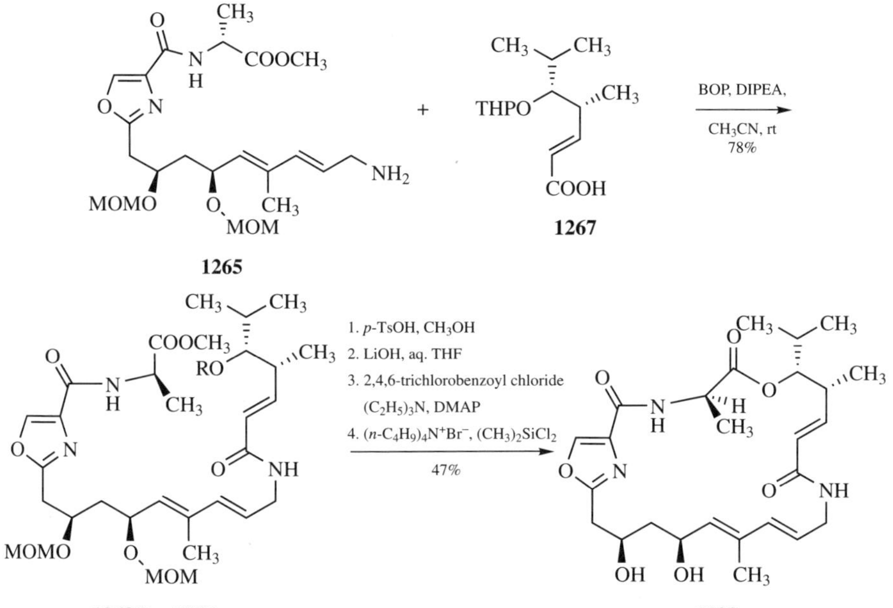

Scheme 1.324

Meyers and co-workers[475] completed the first synthesis of (−)-griseoviridin **1184**. In this case, the authors envisioned an unprecedented olefin metathesis reaction of **1269** would construct the macrocycle (Scheme 1.325). Amide bond disconnection of **1269** then led to the key intermediate fragments, an oxazole diene moiety **1270**, and the sulfur-containing nine-membered ring lactone **1271**.

Synthesis of the oxazole diene fragment **1270** began with the aldehyde **1272**, which had been prepared earlier via Swern oxidation[367a] of **1214**.[459a,471] Thus **1272** was converted to a 5:1 (*Z/E*) mixture of stereoisomeric dienes **1273**, which was photoisomerized to **1274** (Scheme 1.326). Saponification of **1274** then produced **1270**. The sulfur-containing nine-membered ring lactone **1271** was prepared from

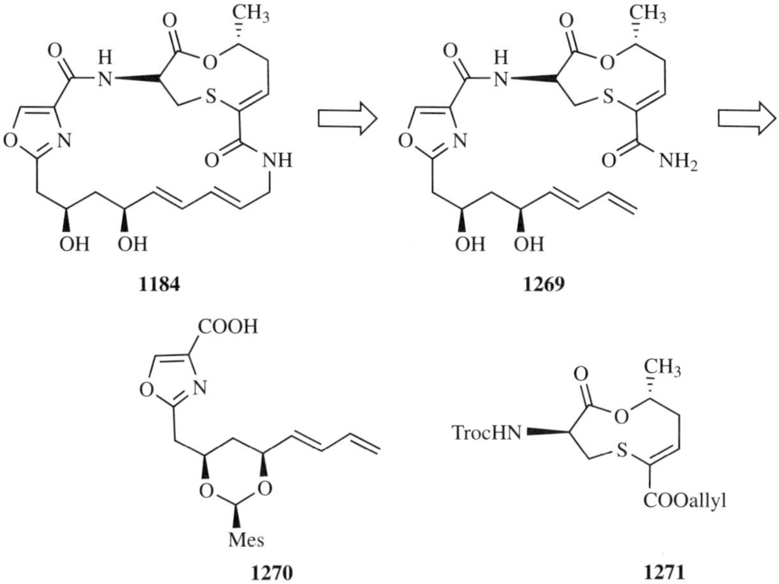

Scheme 1.325

1275 as shown. A key step in this synthesis was the introduction of the protected D-cysteine residue using **1276**.[476]

Peptide coupling of **1270** and **1279** furnished **1280** in good yield (Scheme 1.327). The allyl ester **1280** was cleaved with simultaneous formation of the allyl amide (not shown), which was subjected to the ring-closing metathesis reaction using Grubbs's catalyst at high dilution.[477] The authors isolated **1281** as a single product in 37–42% yield, despite their efforts to improve the yield. Nonetheless, **1281** afforded **1184** as a single diastereomer after removal of the protecting group. The authors also prepared C-8 *epi*-griseoviridin using the (*R*)-analog of **1276**. This synthesis is the first example of **1184** being prepared by a "novel ring-closing metathesis which involved a highly diastereoselective triene to diene macrocyclic ring formation."

1.5.2. Thiangazole and Tantazole

Thiangazole **1187** and tantazole **1188a–c** are cytotoxic natural products containing an array of three or four linearly fused thiazolines, together with a terminal oxazole ring (Fig. 1.96). These challenging targets have interesting biological activity. For example, **1187** is an extremely potent HIV-1 inhibitor that was reported to have no cell toxicity at millimolar concentrations.[438] A number of research groups have completed total syntheses of **1187**, which are described in this section.[30,31,478–483]

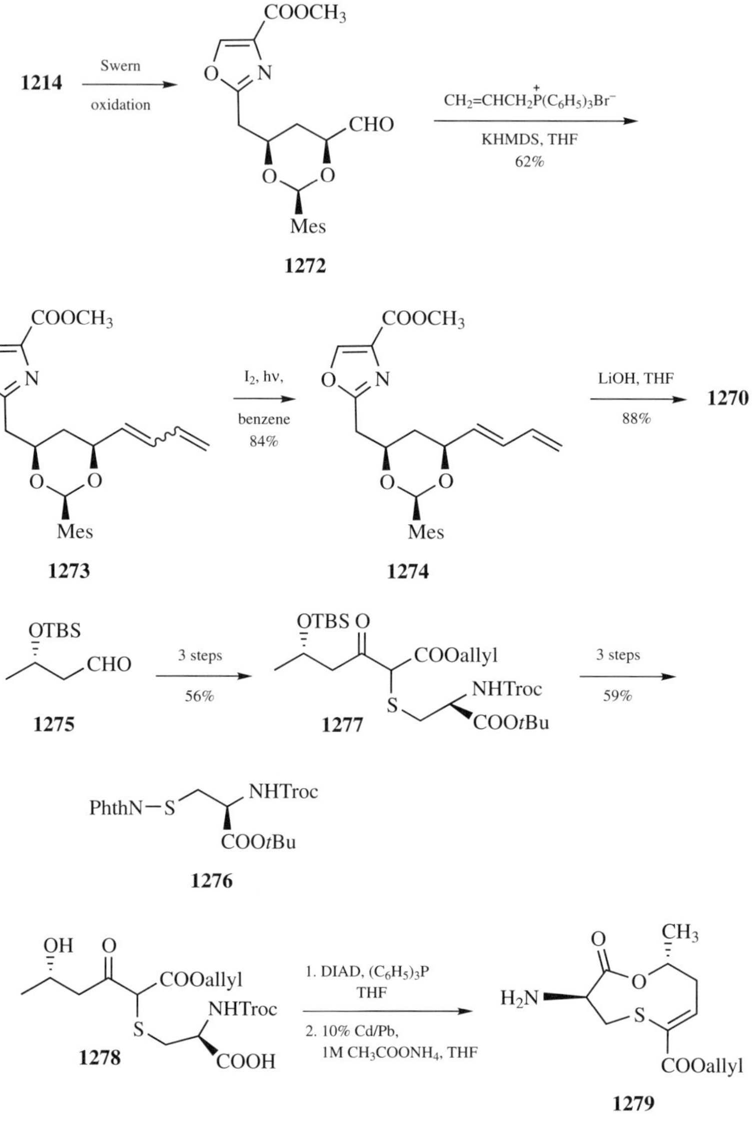

Scheme 1.326

In 1994, three total syntheses of **1187** were described by Pattenden and co-workers,[30,31] Ehrler and Farooq,[478] and Heathcock and Parsons.[479] Pattenden and co-workers[30,31] envisioned that the key step in their synthesis of **1187** would be cyclocondensation of the bis-thiazoline nitrile **1283** with the oxazole **1282** (Scheme 1.328). The authors had previously described a useful method of preparing a variety of chiral 4-methylthiazolines derived from (R)-2-methylcysteine methyl ester.[484]

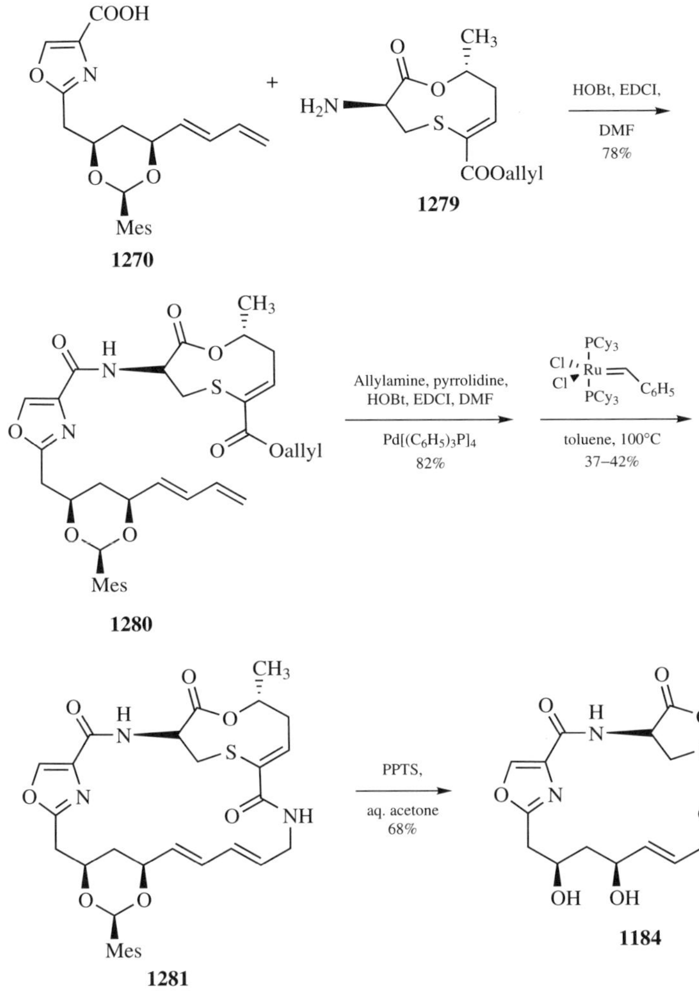

Scheme 1.327

The starting 2-cinnamyl-4-cyano-4-methylthiazoline **1285** was prepared in three steps from (R)-2-methylcysteine methyl ester hydrochloride **1284** (Scheme 1.329). The synthesis of **1283** was accomplished by condensation of **1284** with **1285**, followed by reption of steps two and three (Scheme 1.329). For **1289**·HCl ($R_1 = CH_3$), bis-Boc-(R)-2-methylcysteine, **1286** was coupled with *rac*-threonine methyl ester hydrochloride followed by cyclization with Burgess reagent[49] to afford a mixture of diastereomeric oxazolines **1288**. Oxidation of **1288** with Cu(I)Br/*t*-butylperbenzoate (Section 1.3.1.4) proved to be more efficient than NiO_2. Deprotection then yielded **1289**·HCl ($R_1 = CH_3$).

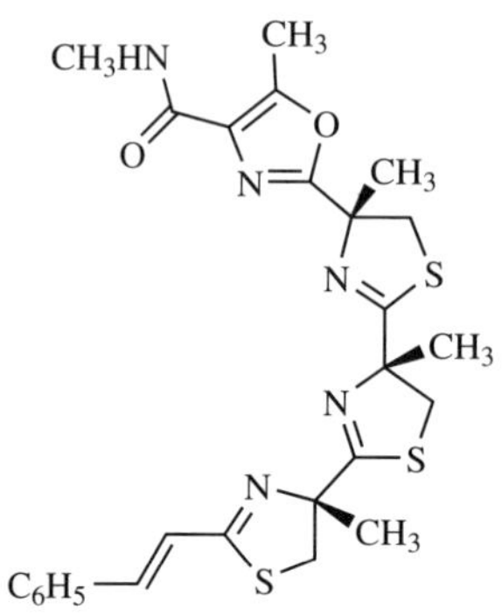

1187

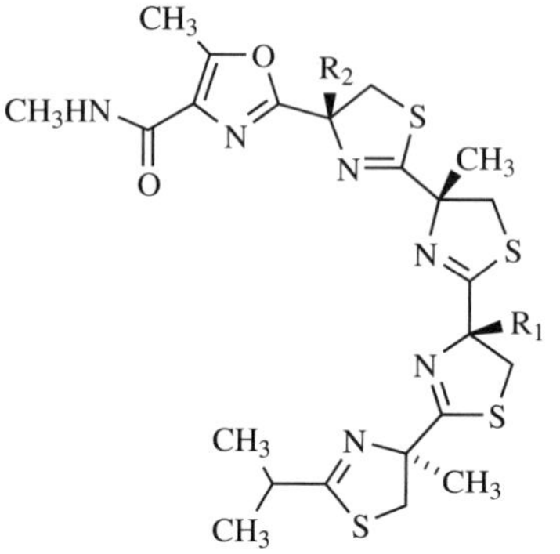

1188a R_1= H, R_2= CH_3
1188b R_1= R_2= CH_3
1188c R_1= CH_3, R_2= H

Figure 1.96. Thiangazole and tantazoles.

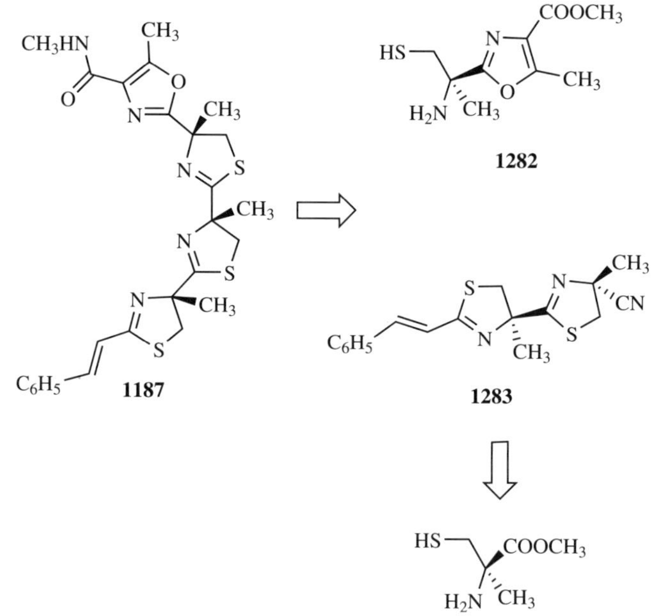

1187 **1282** **1283** **1284**

Scheme 1.328

Cyclocondensation of **1283** and **1289**·HCl (R_1 = CH_3) installed the last thiazoline ring to yield **1290** (R_1 = CH_3), from which **1187** was obtained by reaction with methylamine (Scheme 1.330).

Heathcock and Parsons[479] pursued a synthetic strategy to **1187** wherein the terminal oxazole ring was constructed in the presence of the tris-thiazoline array.

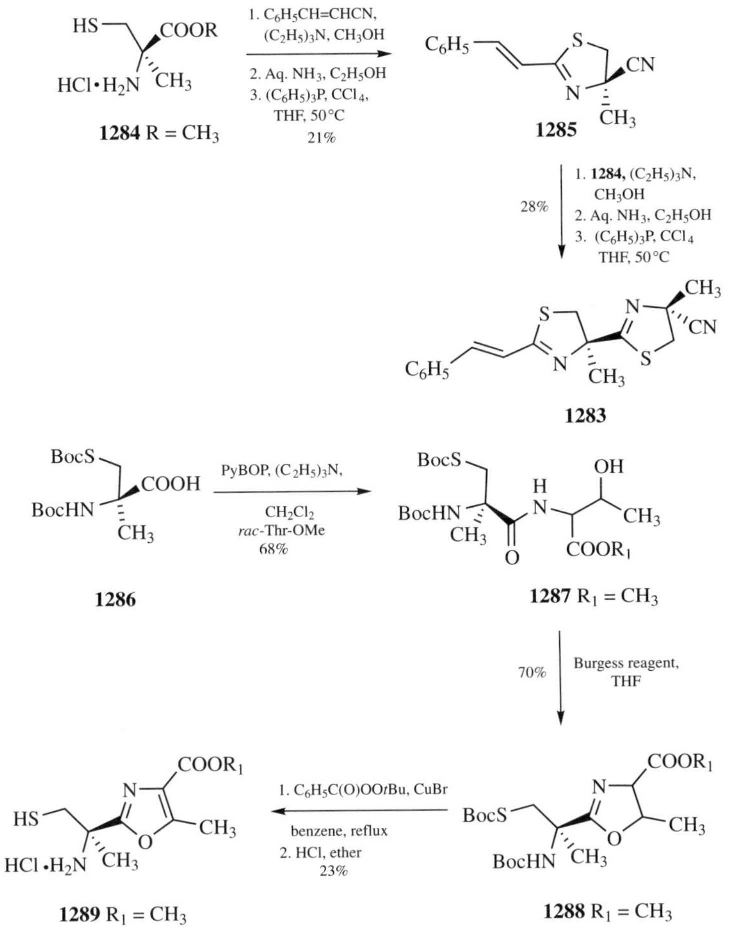

Scheme 1.329

Heathcock and Walker[485] previously developed an extremely efficient iterative coupling sequence to similar peptides as part of their synthetic efforts leading to mirbazoles. In addition, these same authors described an elegant $TiCl_4$-mediated cyclization of such peptides for preparing the polythiazoline rings of mirbazoles simultaneously.

Heathcock and Parsons[479] first converted N-Cbz-D-alanine to both key protected (R)-α-methylcysteine derivatives **1292** and **1293** via the cis-oxazolidinone **1291** (Scheme 1.331).[485,486] Coupling **1292** with **1293** produced the dipeptide **1294**. An iterative reaction sequence furnished the expected tripeptide (not shown), which was acylated with dihydrocinnamoyl chloride to produce **1295**. Saponification of **1295** followed by coupling with O-benzylthreonine N-methylamide **1296** afforded the key tetrapeptide **1297** in excellent yield. This tetrapeptide then contained the entire backbone requisite for construction of the three-thiazoline rings, the terminal oxazole, and introduction of the cinnamoyl side chain.

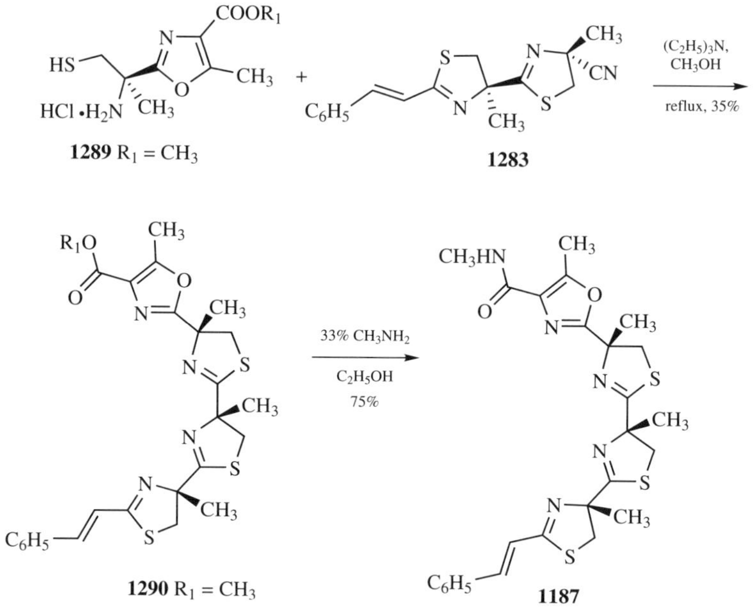

288   Synthesis and Reactions of Oxazoles

Scheme 1.330

Deprotection of **1297** followed by treatment with TiCl$_4$ effected cyclization and installed the three thiazoline rings of **1298** in good yield (Scheme 1.332). Dess-Martin periodinane oxidation[211] of **1298** followed by cyclodehydration of the resulting β-ketoamide with TsOH gave **1299**. The styryl side chain of **1187** was then installed using DDQ. These authors have used TsOH in a similar synthetic strategy to install the oxazole ring of tantazole B **1188b** (R$_1$ = R$_2$ = CH$_3$) (see Scheme 1.128).[195]

The key to this approach was the formation of the oxazole ring in the presence of the thiazoline rings. Cyclization of **1298** with Burgess reagent[49] gave the expected intermediate oxazoline (not shown), but this could not be oxidized to **1299** using NiO$_2$,[27] NBS/benzoyl peroxide, or CuBr/*tert*-butyl perbenzoate.[35] Thionyl chloride, phosphorus oxychloride, titanium tetrachloride, or triphenylphosphine/iodine did not effect cyclodehydration of the β-ketoamide produced from the Dess-Martin periodinane oxidation.

Ehrler and Farooq[478] prepared **1187** via a stepwise construction of the three thiazoline rings, introduction of the styryl group, and preparation of the oxazole ring (Scheme 1.333). Condensation of L-cysteine ethyl ester hydrochloride **1284** (R = C$_2$H$_5$) with an imino ether derived from cinnamamide was unsuccessful, owing to complications from Michael addition. This necessitated a strategy in which the styryl double bond was introduced after assembly of thiazoline rings.

Initial efforts to introduce the styryl side chain via a 2-formylthiazoline or the equivalent were unsuccessful. However, the 2-(diphenylphosphinylmethyl)thiazoline **1300**, readily prepared from **1284** (R = C$_2$H$_5$), was elaborated to the tris-thiazoline **1301** by an iterative process using the TiCl$_4$ protocol[479] to close the second and third thiazoline rings. Wittig olefination of **1301** with benzaldehyde yielded

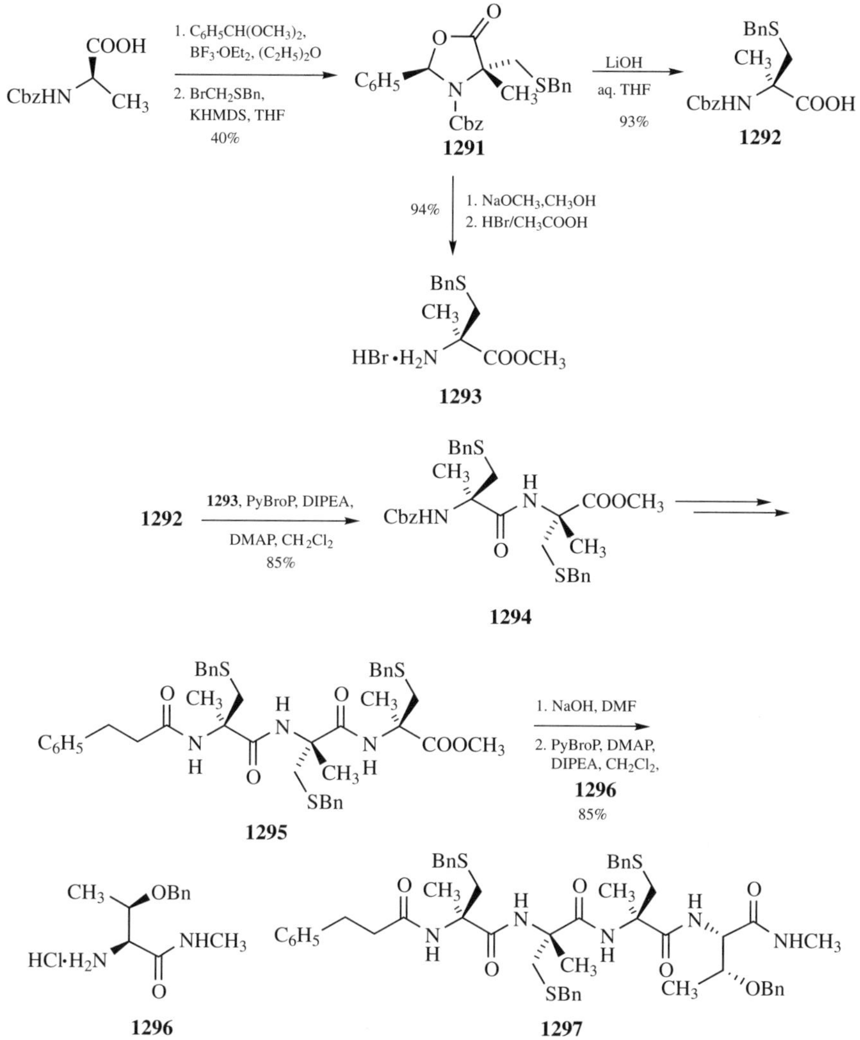

Scheme 1.331

1302a, which was converted to the amide 1302b in three steps using Ghosez's reagent[487] to prepare the acid chloride.

The oxazole ring was then introduced by a process using Masamune's protocol[488] for calyculin C. Thus reaction of 1302b with ethyl 3-bromo-2-oxobutyrate followed by dehydration of the intermediate 4-hydroxyoxazoline (not shown) with TFAA in pyridine produced 1290 ($R_1 = C_2H_5$), which was easily converted to 1187.

Wipf and Venkatraman[480,481] employed a novel conceptual approach in their synthesis of 1187 (Scheme 1.334). They targeted a suitably functionalized trisoxazoline 1303 as the precursor to the tris-thiazoline array. Such a strategy

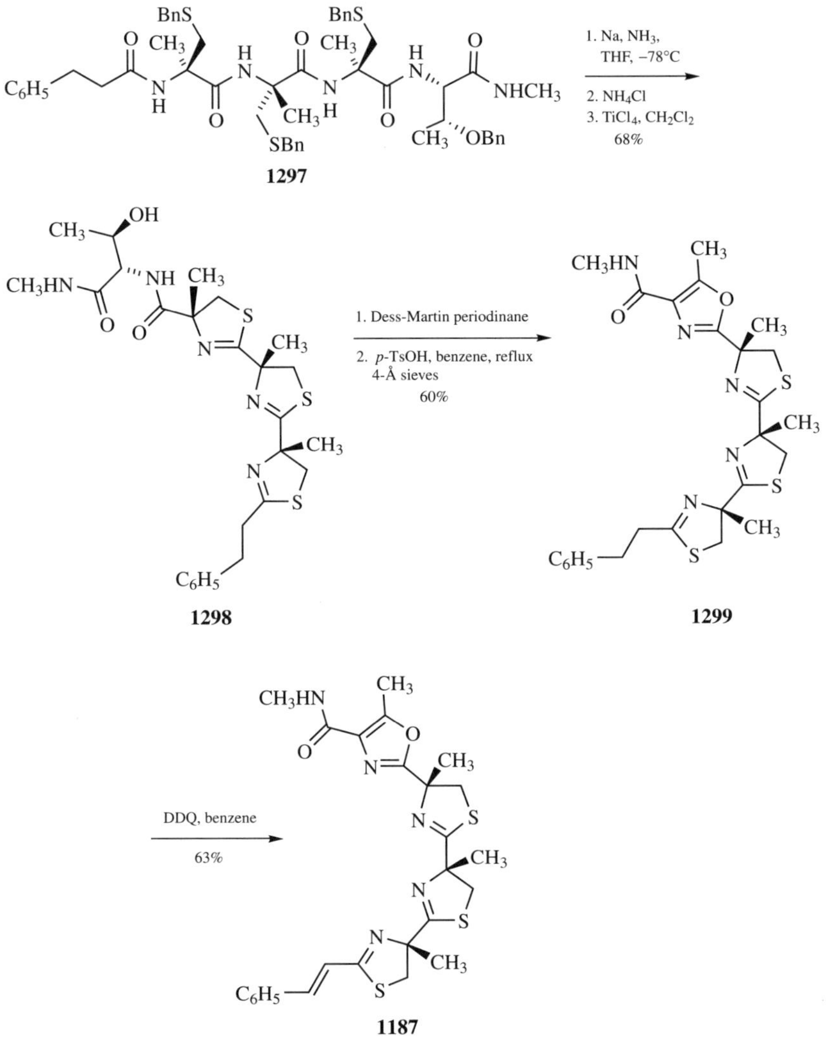

290   Synthesis and Reactions of Oxazoles

Scheme 1.332

minimized the synthetic manipulations involving sensitive sulfur-containing intermediates. Furthermore, **1303** would be derived from the oxazole-based tripeptide **1304** prepared from **1305** using an iterative assembly with a suitably protected α-methylserine derivative (e.g., **1306**). Finally, **1305** itself, would be derived from **1306**.

The synthesis began with acylation of D-threonine methyl ester with O-benzyl-N-[(trimethylsilyl)ethyl]sulfonyl-α-methylserine **1306** (P = Ses, R = Bn)[489] to give the dipeptide **1307** (Scheme 1.335). The authors found that D-threonine methyl ester gave markedly improved yields and the reaction rate was comparable to the L-enantiomer. This was attributed to unfavorable steric interactions in the acylation

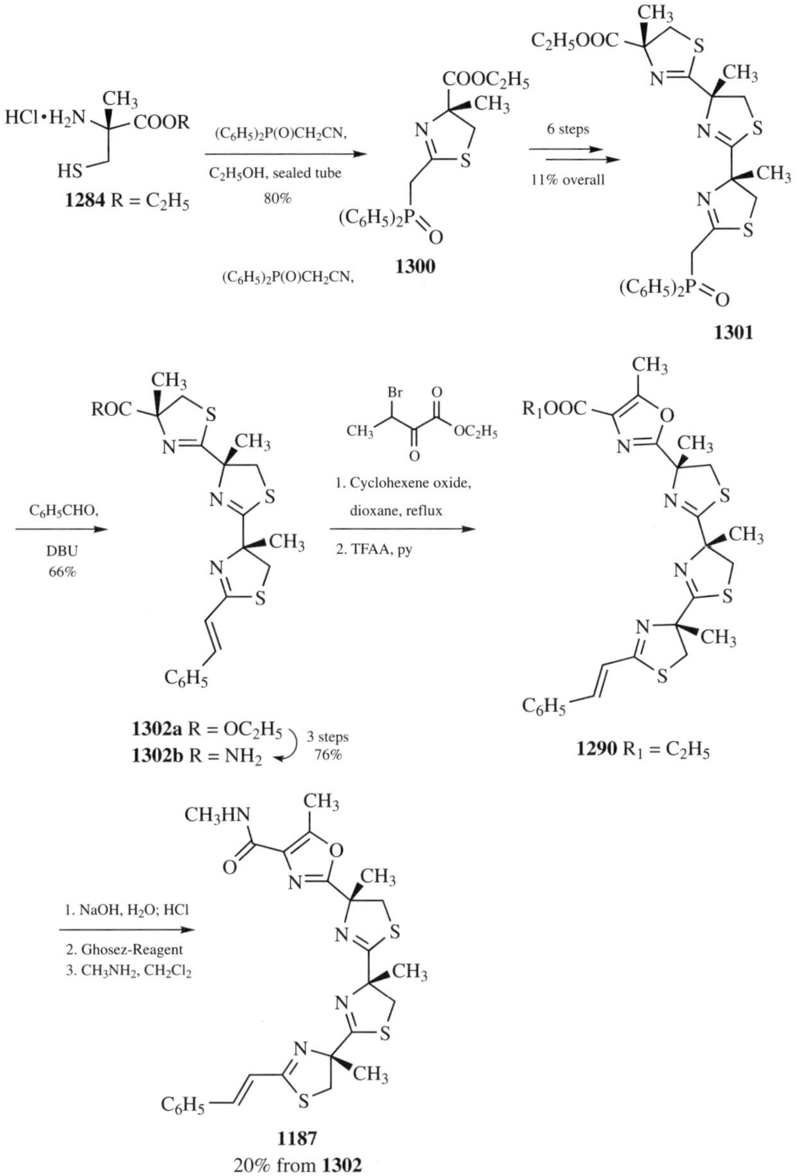

Scheme 1.333

transition state. Conversion of **1307** to **1305** (P = Ses, R = Bn) was accomplished in good yield. After preparation of the N-methyl amide and removal of the Ses group, the resulting amine was coupled a second time with **1306** (P = Ses, R = Bn). This process was iterated a third time, whereupon coupling with 3-phenylpropionic acid produced the oxazole tripeptide **1304**, isolated in 20% overall yield from **1305**.

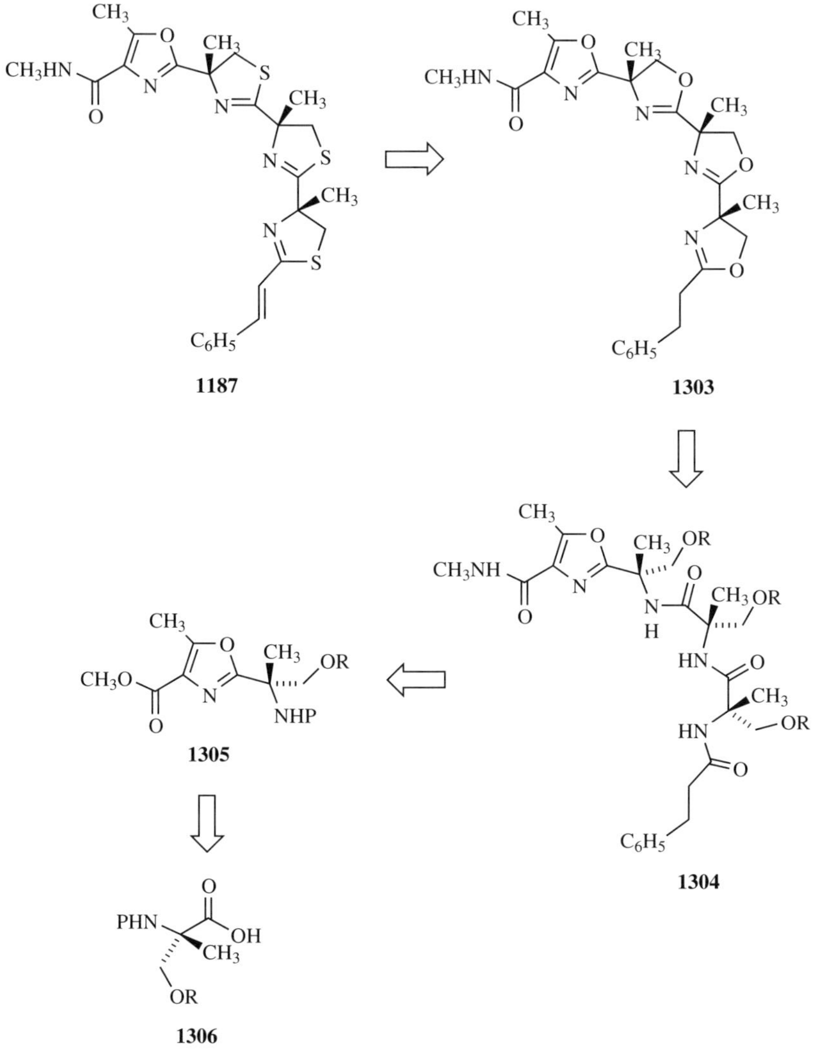

P = protecting group

Scheme 1.334

Cleavage of the benzyl groups and cyclization of the resulting *N*-acylamino alcohols with Burgess reagent[49] afforded the tris-oxazoline **1303** in good yield. The key oxazoline to thiazoline interconversion was now effected by ring opening **1303** with thiolacetic acid, cleavage of the acetyl groups, and $TiCl_4$-mediated cyclization[479] to furnish **1299**. Finally, the styryl side chain of **1187** was installed using benzeneseleninic acid.

The authors readily adapted this synthetic strategy to prepare a number of analogs. Interestingly, they found **1187** was extremely cytotoxic in contrast to earlier reports.[438] Wipf and Venkatraman[481] published an excellent review and comparison of synthetic strategies leading to **1187**.

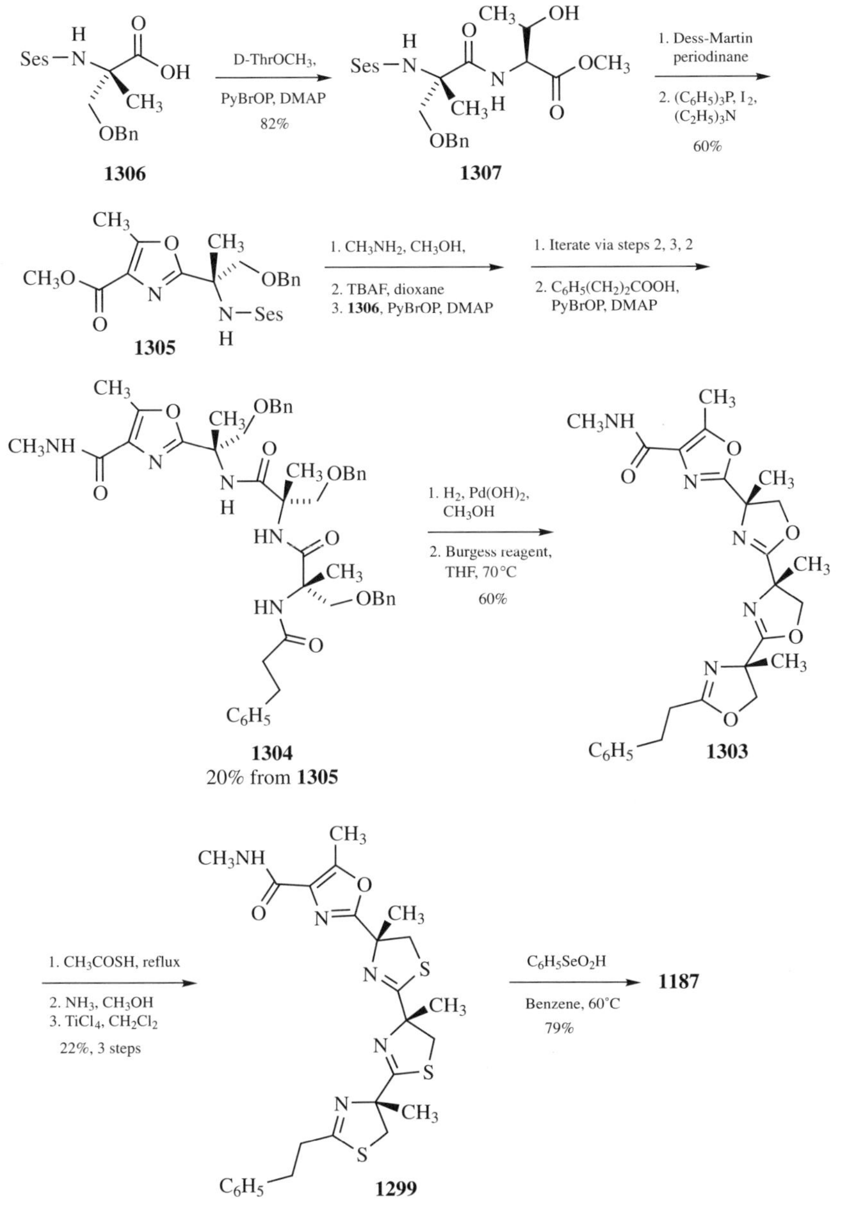

Scheme 1.335

Akaji and Kiso[482] reported an efficient synthesis of **1297** (Scheme 1.331) from Boc-(R)-MeCys(Bn)-OH, Boc-Thr(Bn)-NHCH$_3$ and 3-phenylpropionyl chloride. They improved the peptide-coupling conditions for the sterically hindered α-methylcysteine and threonine analogs normally used to prepare **1187** and **1188**.

These authors employed the reagent combination of 2-chloro-1,3-dimethyl-imidazolium hexafluorophosphate (CIP)/HOAt that had been developed previously

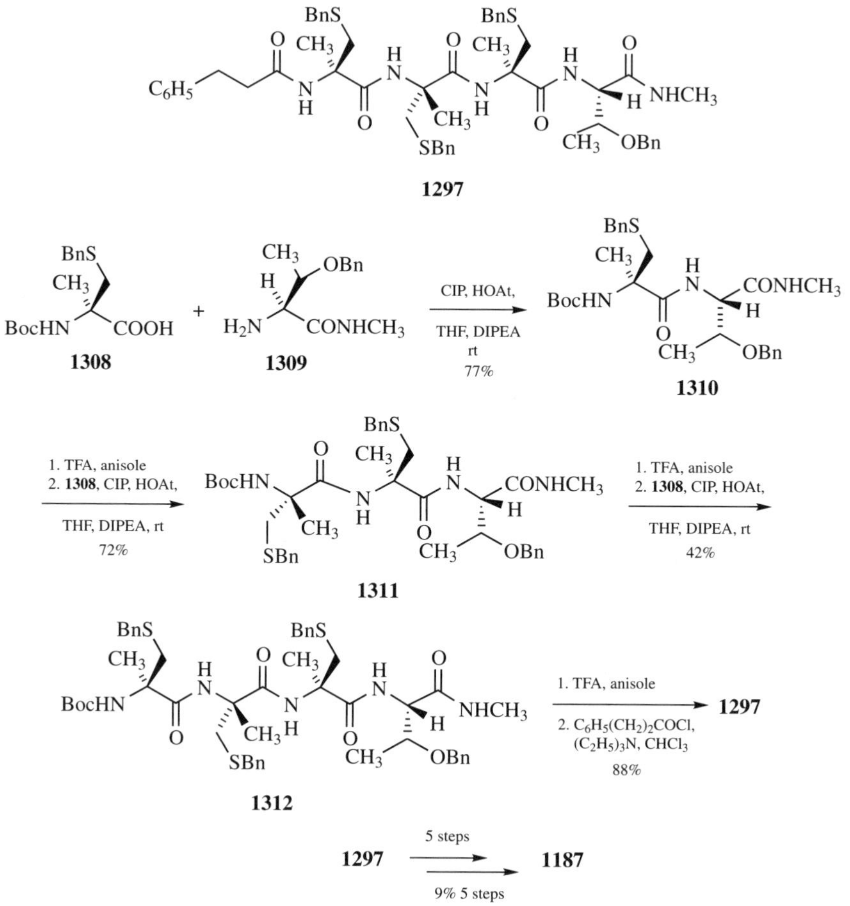

Scheme 1.336

in their laboratories to incorporate α,α-dialkylamino acids and N-methylamino acids.[490–492] Their coupling strategy and approach to **1297** are shown in Scheme 1.336. Synthesis of **1187** from **1297** was accomplished employing Heathcock's protocol.[479] Here, the authors noted that the reductive debenzylation of **1297** using sodium and liquid ammonia must be carefully controlled to avoid significant reductions in yield.

Fukuyama and Xu[483] reported the first total synthesis of (−)-tantazole B **1188b** ($R_1 = R_2 = CH_3$) in 1993. This synthesis also corrected the originally assigned stereochemistry of the methyl group in ring E. The authors discovered a novel and versatile synthesis of thiazolines **1314** via cyclodehydration of ammonium thiol esters **1313** and adapted this methodology to prepare **1188b** (Scheme 1.337).

The enantiomerically pure β-lactones **1315** and **1316** were each prepared in three steps from the malonate. Ring opening of the (+)-β-lactone **1315** with thioisobutyric acid followed by deprotection with TFA and heating gave the thiazoline,

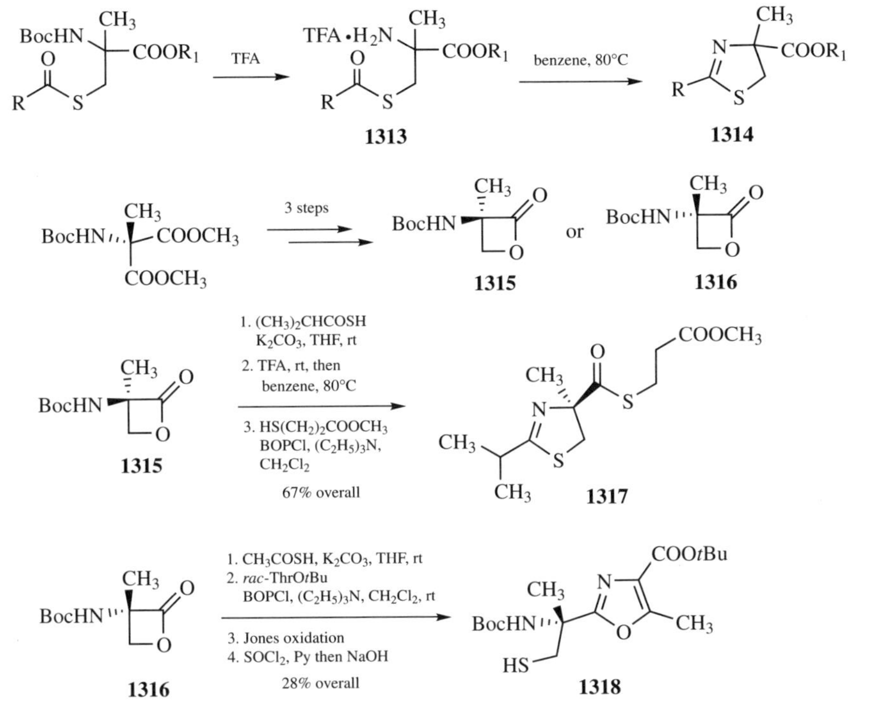

Scheme 1.337

which was immediately converted to the protected thiocarboxylic acid **1317** in 67% overall yield.

Similarly, the (−)-β-lactone **1316** was ring opened with thiolacetic acid to yield (S)-N-t-Boc-(S-acetyl)-α-methylcysteine (not shown). In this case, Jones oxidation followed by cyclodehydration generated the key oxazole **1318**.

Retro-Michael reaction of **1317** gave the thiol acid that was condensed with the (−)-β-lactone **1316** to yield **1319** (Scheme 1.338). Deprotection of **1319** followed by cyclization as above gave a bis-thiazoline analog of **1319** (not shown), which was converted to the tris-thiazoline **1320**. Coupling **1318** and **1320** was accomplished in high yield using BOPCl to afford **1321**. The remaining thiazoline ring was installed at this point in the usual manner. The synthesis of **1188b** was completed by conversion of **1321** to the N-methyl amide.

1.5.3. Calyculins

The calyculins make up a family of potent, selective serine-threonine phosphatase inhibitors, which have been isolated from the Japanese sponge *Discodermia calyx*.[440,441] There are at least 13 different members of this family, and the most abundant are calyculin A **1189a** and calyculin C **1189c**.[44] As nanomolar inhibitors of protein phosphatases 1 and 2A (PP1 and PP2A), **1189a** and analogs have

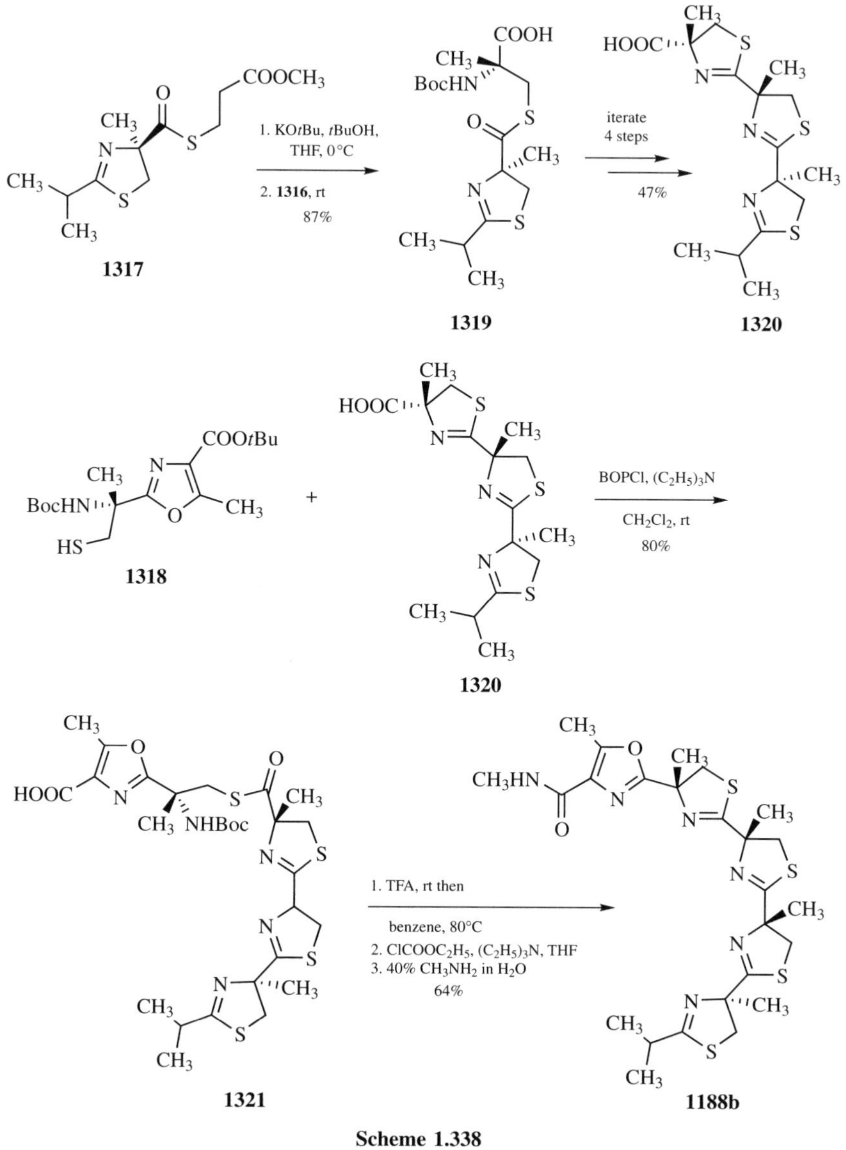

Scheme 1.338

attracted the interest of a number of synthetic groups (Fig. 1.97). Among the synthetic challenges these structures contain are the [5,6]-spiroketal, the light-sensitive C(1)–C(9) cyanotetraene, and the 2,4-disubstituted oxazole containing a *trans* alkene.

Calyculin A **1189a** and calyculin C **1189c** have attracted the most attention from synthetic groups.[32,33,44,60,151,236,411,488,493–505] All groups have targeted the C(26)-C(37) oxazole–containing fragment **1322a–c** as a key intermediate in their retrosynthetic strategies (Scheme 1.339). However, the different approaches to the

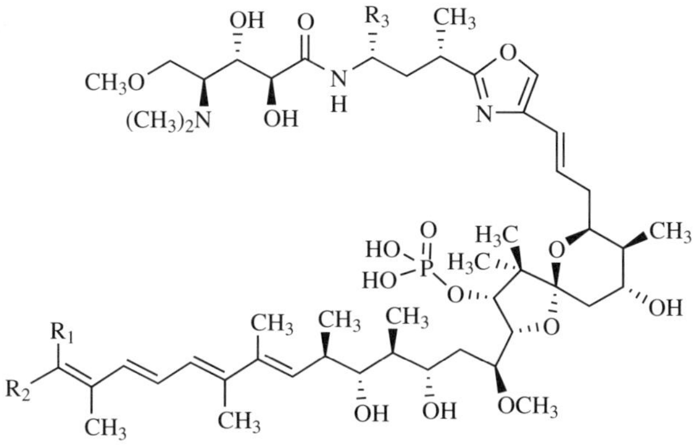

1189a $R_1 = CN, R_2 = R_3 = H$
1189b $R_1 = R_3 = H, R_2 = CN$
1189c $R_1 = CN, R_2 = H, R_3 = CH_3$
1189d $R_1 = H, R_2 = CN, R_3 = CH_3$

Figure 1.97. Calyculins.

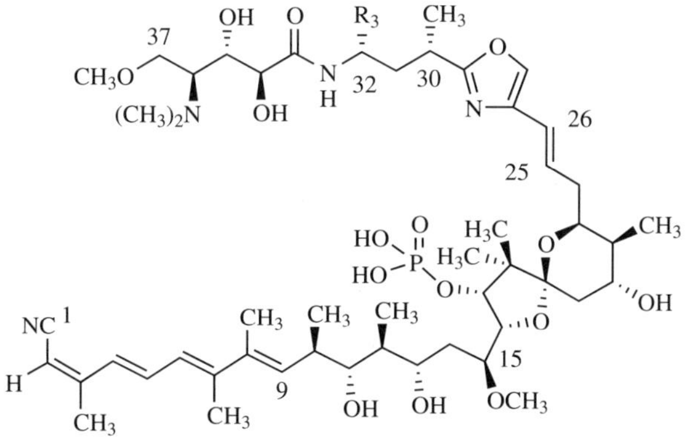

1189a $R_3 = H$
1189c $R_3 = CH_3$

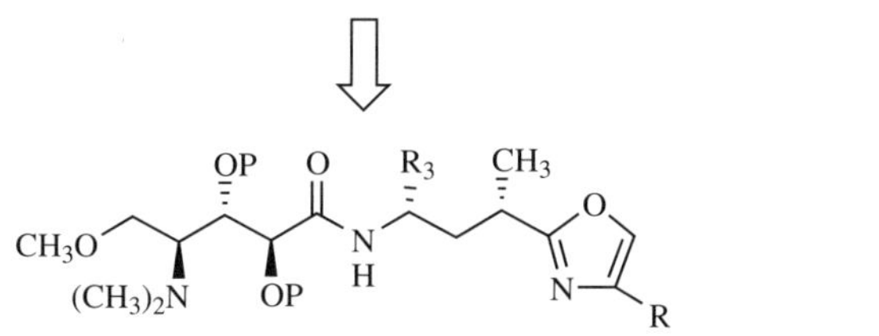

1322a-c

a. $R = CH_2Cl, CH_2\overset{+}{P}(C_4H_9)_3\ X^-, CH_2PO(OC_2H_5)_2, COOCH_3, CH_2OP, CHO$
b. P = protecting group or H
c. $R_3 = H$ or CH_3

Scheme 1.339

assembly of **1322a–c** and introduction of the C(25)-C(26) *trans* double bond are the focus of this section.

Early in their approach to calyculins, Armstrong and co-workers[411] described an efficient, general synthesis of oxazole phosphonium salts and phosphonates (Scheme 1.296). Olefination of these salts produced 2,4-disubstituted oxazoles containing a trans alkene with up to 99:1 *trans:cis* ratios using **1137b** (R_1 = sec-C_4H_9, $R_2 = n$-C_4H_9).

Armstrongs's group[493] extended this methodology to prepare the fully elaborated phosphonium salt **1322** required for calyculin C **1189c** (Scheme 1.340). The authors identified **1326** and **1327b** as key intermediates. In particular, earlier work from this group[411] provided an excellent precedent for direct prepartion of **1326** (Scheme 1.296). Thus **1323** was converted to (2S, 4R)-4-[(*tert*-butyloxycarbonyl) amino]-2,4-dimethylbutanamide **1325** in five steps. Condensation of **1325** with 1,3-dichloroacetone afforded a 10:1 ratio of the 4-(chloromethyl)-2-substituted oxazole **1326** together with recovered **1325** in 61% combined yield. After cleavage of the Boc group, attempts to couple the resulting amine (not shown) with **1327a** failed to yield any of the desired amide. However, cleavage of the Boc group followed by reaction of the amine (not shown) with trimethylaluminum generated an aluminum-amide complex that reacted with **1327b** to furnish **1328** and **1329** as a 2.7:1

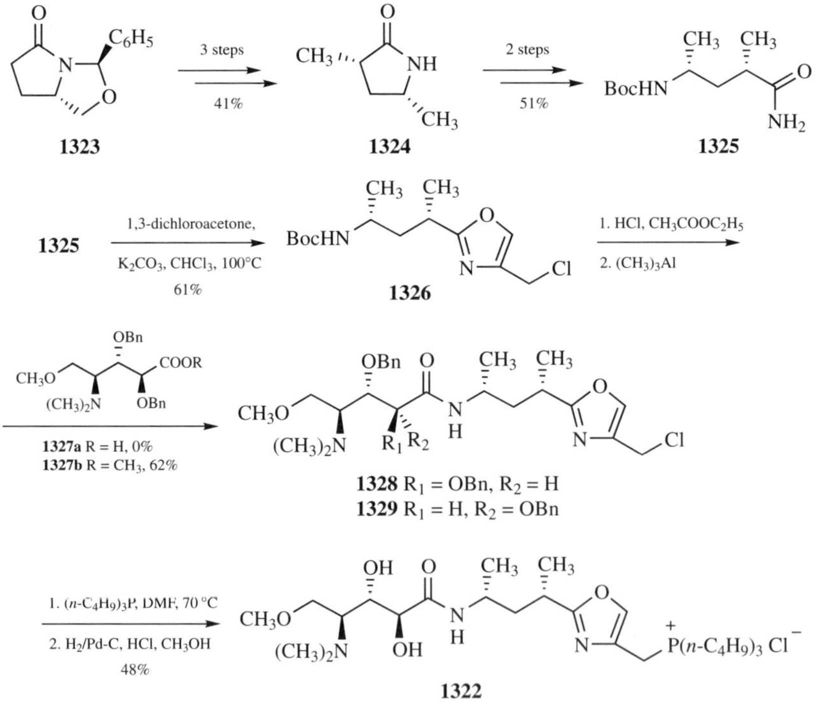

Scheme 1.340

separable mixture of diastereomers in good yield. The authors showed that epimerization had occurred before amide formation. Finally, reaction of **1328** with (n-C$_4$H$_9$)$_3$P followed by hydrogenolysis afforded the fully elaborated oxazole fragment **1322** (R = CH$_2$P(n-C$_4$H$_9$)$_3^+$ Cl$^-$, P = H, R$_3$ = CH$_3$).

Wittig condensation of **1322** (R = CH$_2$P(n-C$_4$H$_9$)$_3^+$ Cl$^-$, P = H, R$_3$ = CH$_3$) with the model aldehyde **1330** generated **1331** (Scheme 1.341), thereby confirming the viability of this approach to **1189c**.

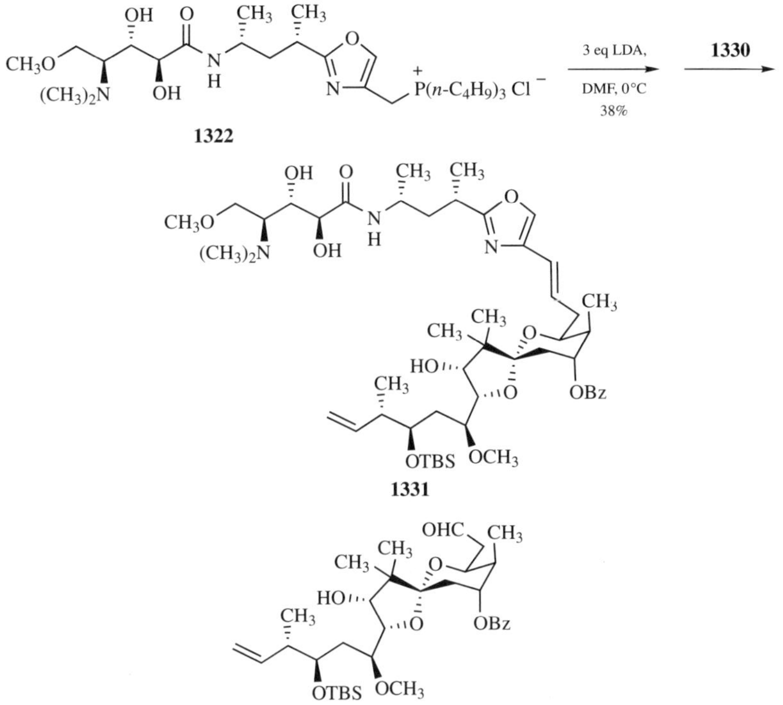

Scheme 1.341

Ogawa and Armstrong[494] successfully implemented this strategy to prepare calyculin C **1189c**. In the process, the authors revised their earlier stereochemical assignments of **1328** and **1329**. They found that the minor diastereomer **1329** produced **1189c**, whereas the major diastereomer **1328** afforded C(34)-*epi*-calyculin C.

Shioiri and co-workers[32] developed a general synthesis of 2,4-disubstituted oxazoles as intermediates leading to calyculins. In this early work directed to calyculin A **1189a**, the authors prepared the antipode of the C(26)-C(32) oxazole fragment, starting from methyl (*S*)-3-hydroxy-2-methylpropionate **1332**. Homologation of **1332** to (*R*)-*tert*-butyl-5-hydroxy-4-methyl-pentanoate, **1333** was accomplished in four steps with excellent overall yield (Scheme 1.342). Oxidation of **1333** to the acid and coupling with L-serine methyl ester was problematic owing to epimerization, which occurred during the oxidation. The best results were obtained using the RuCl$_3$-NaIO$_4$ protocol of Sharpless.[506] The diastereomers (92:8) were separated, and the major diastereomer **1334** was converted to the acid chloride and cyclized to the oxazoline **1335** with AgOTf/CaCO$_3$.[507] Oxidation of **1335** with NiO$_2$,[27] cleavage of the Boc group, Curtius rearrangement, and LAH reduction then afforded the C(26)-C(32) oxazole fragment **1337**. Alternatively, **1334** was converted to **1335** in a single step, using a general methodology developed by these same authors.[33]

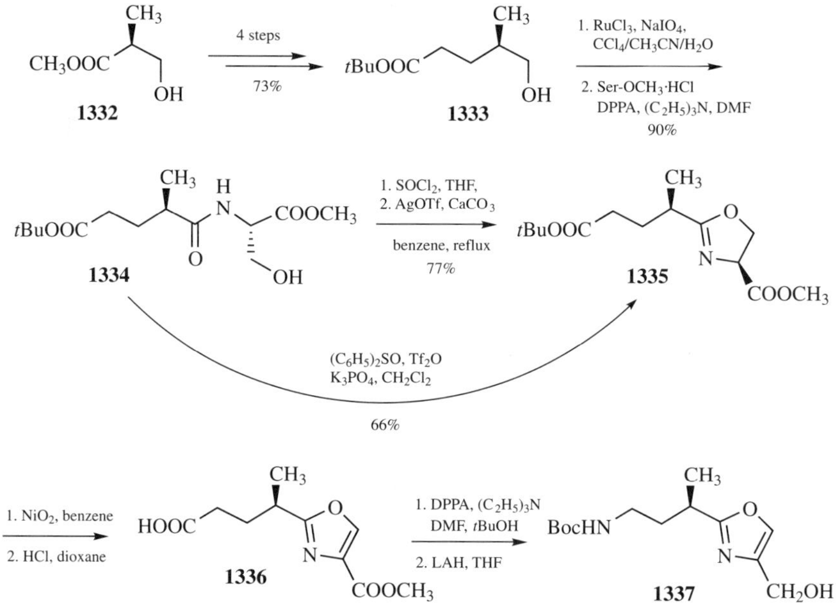

Scheme 1.342

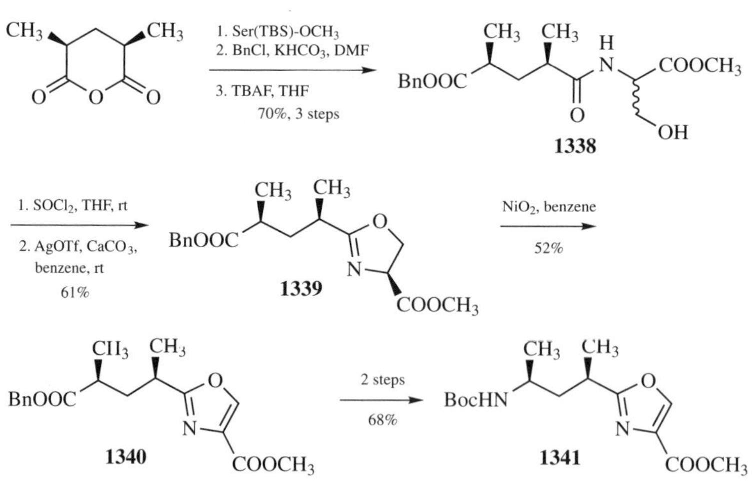

Scheme 1.343

Shioiri and co-workers[495] also applied their general synthesis of oxazolines[33] for preparing the C(26)-C(32) oxazole fragment of calyculin C, **1189c** (Scheme 1.343). In this case, the authors cyclized a serine dipeptide to an oxazoline without epimerization, using triflic anhydride and diphenyl sulfoxide (see Scheme 1.03). For **1189c**, *meso*-2,4-dimethylglutaric anhydride was converted to an 85:15 ratio of diastereomeric serine dipeptides **1338** in acceptable yield. The diastereomers were separated, and the major one was converted to **1339**. Oxidation of **1339** to the oxazole **1340** was straightforward, as was the conversion of **1340** to **1341**, the antipode of the C(26)-C(32) fragment of calyculin C.

In later work Shioiri's group[496] reported a formal total synthesis of **1189a**. In this approach, **1342** was coupled with **1343** to yield a fully elaborated oxazole **1344a** (Scheme 1.344). After a straightforward functional group interchange, **1344b** was

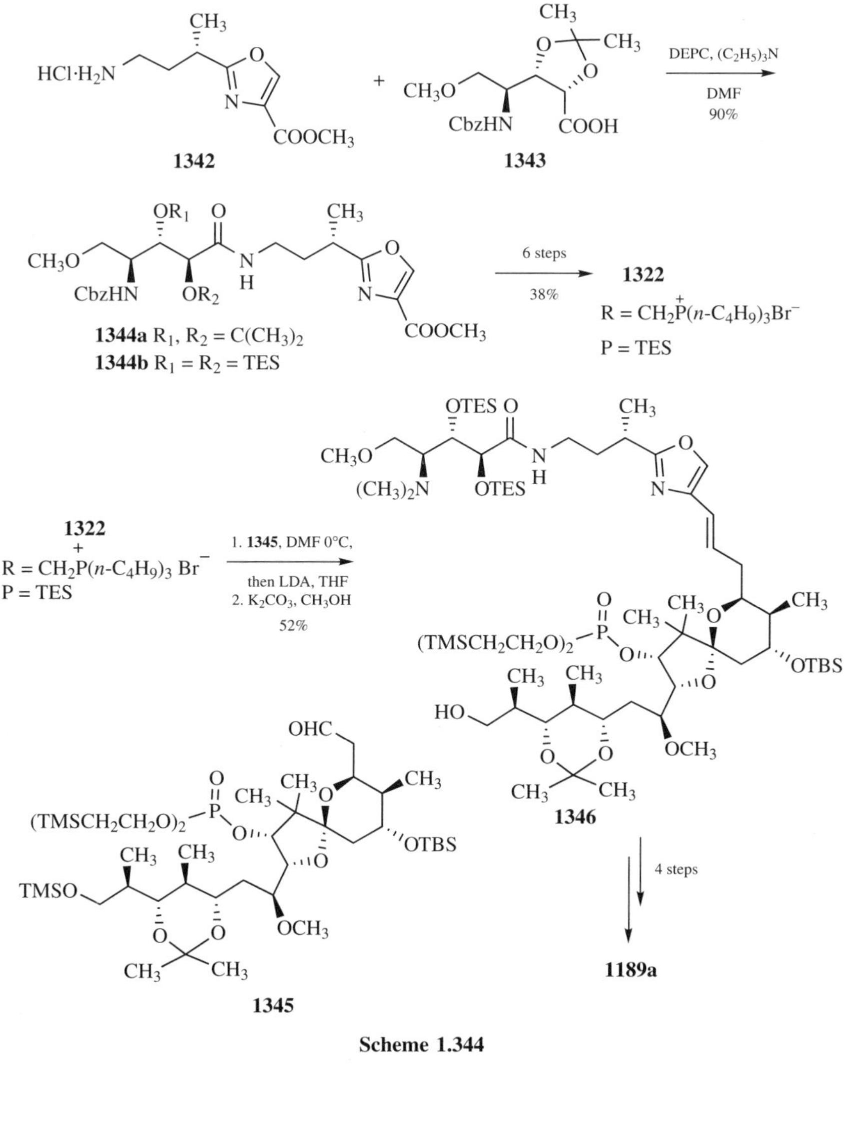

Scheme 1.344

converted to the key phosphonium bromide **1322** (R = CH$_2$P(n-C$_4$H$_9$)$_3^+$ Br$^-$, P = TES, R$_3$ = H) in six steps, following the reaction sequence described by Evans and co-workers[497] for the antipode. Wittig reaction of **1322** (R = CH$_2$P(n-C$_4$H$_9$)$_3^+$ Br$^-$, P = TES, R$_3$ = H) with the fully elaborated, protected aldehyde **1345** afforded **1346** after deprotection of the C(9) silyl group. Masamune and co-workers[500] previously converted this advanced intermediate to **1189a**.

Evans and co-workers described the first synthesis of *ent*-calyculin A, i.e. (+)-calyculin A, the antipode of **1189a**.[497–499] This group also targeted a C(26)-C(37) oxazole fragment as a key intermediate in their initial strategy, particularly the phosphonium salt **1347a** and the phosphonate **1347b**. Further disconnection of **1347** gave rise to the initial targets **1348** and **1349** (Fig. 1.98).

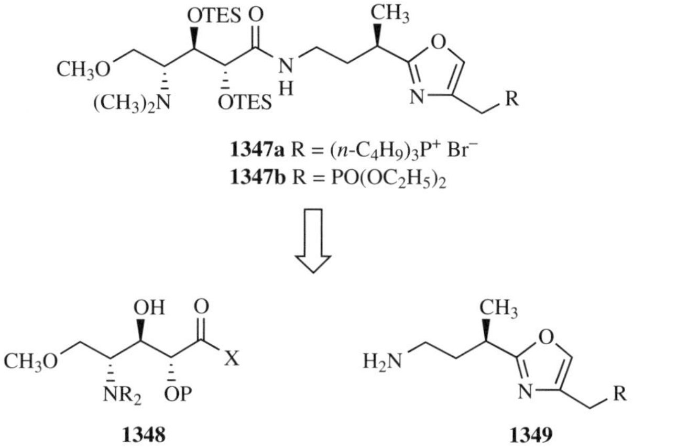

1347a R = (n-C$_4$H$_9$)$_3$P$^+$ Br$^-$
1347b R = PO(OC$_2$H$_5$)$_2$

Figure 1.98. C(26)-C(37) fragment of *ent*-calyculin A.

The starting point for **1349** was the *N*-propionyloxazolidinone **1350**, which was converted to **1351** as a single diastereomer in excellent yield (Scheme 1.345).[497] Curtius rearrangement of **1351**, cleavage of the chiral auxillary and coupling of the resulting acid with serine methyl ester gave **1352**. Cyclization of **1352** generated an intermediate oxazoline (not shown), which was oxidized to **1353** with NiO$_2$[27] but in erratic yield. Instead, **1352** was converted to **1354** via Boc protection, KHMDS/C$_6$H$_5$SeCl, and 30% H$_2$O$_2$. This three-step process was much more reliable on a larger scale. The model phosphorus reagents **1356a,b** were readily prepared in excellent yield from **1353**. Both reagents were effective for preparation of *trans*-olefins from simple aldehydes.

The fully elaborated C(26)-C(37) oxazole fragments **1347a,b** were prepared from **1354** (Scheme 1.346). Cleavage of the Boc groups followed by reaction of the amine (not shown) with trimethylaluminum generated an aluminum-amide complex that was acylated with **1357** to produce **1358a**, together with some of the deprotected analog **1358b**. The authors found that excess trimethylaluminum effected clean conversion to **1358b**. Protection of **1358b** as the bis-TES derivative **1359** was followed by a straightforward four-step sequence to yield **1347a**. The target phosphonate analog **1347b** was prepared by a more efficient four-step process, starting with **1356b**.

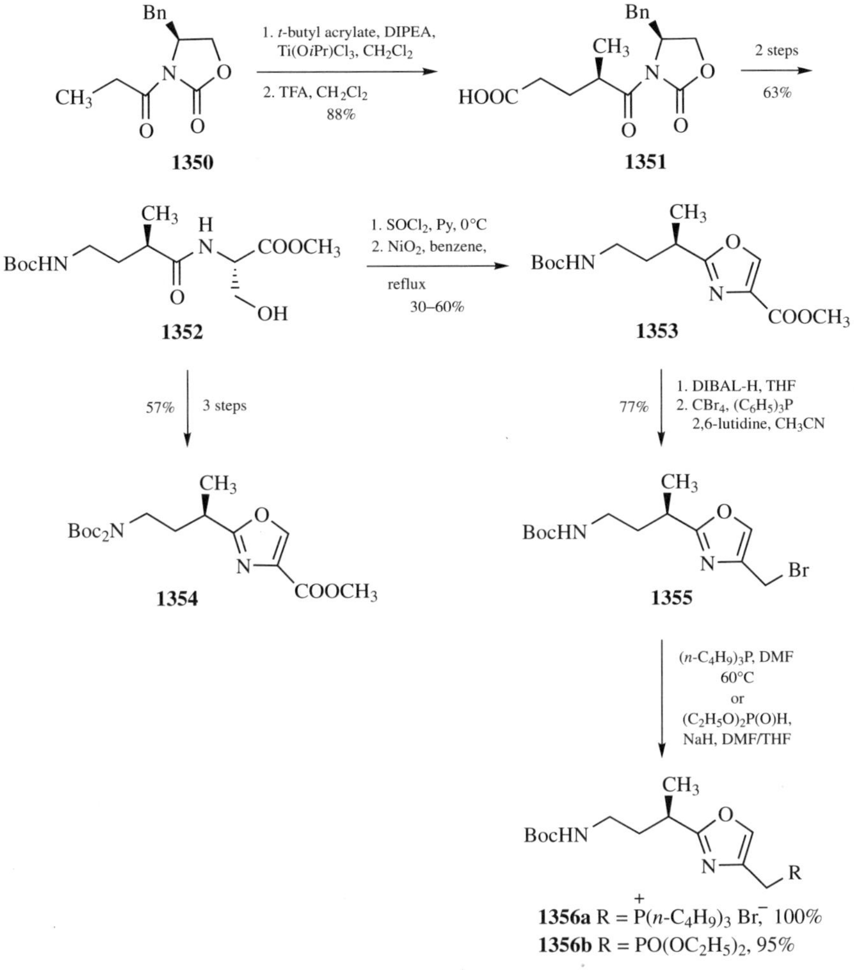

Scheme 1.345

Attempts to olefinate **1360** with the sodium dianion of **1356a** or **1356b** were completely unsuccessful.[498] In both cases β-elimination of the labile spiroketal oxygen was observed. However, treatment of a precooled mixture of **1356a** and **1360** with one equivalent of KHMDS gave **1361** with high (E/Z) selectivity (Scheme 1.347). The authors noted that these reaction conditions were previously described by Kishi and co-workers during their synthesis of palytoxin.[508] The success of this protocol was attributed to a "higher apparent kinetic acidity of the stabilized phosphonium salt relative to the aldehyde or carbamate." Similarly, treatment of a precooled mixture of **1347a** and **1360** with one equivalent of KHMDS afforded the fully elaborated skeleton of *ent*-calyculin A **1362** (Scheme 1.348). Only minor modifications of this methodology were required to complete the first total synthesis of *ent*-calyculin A.[499]

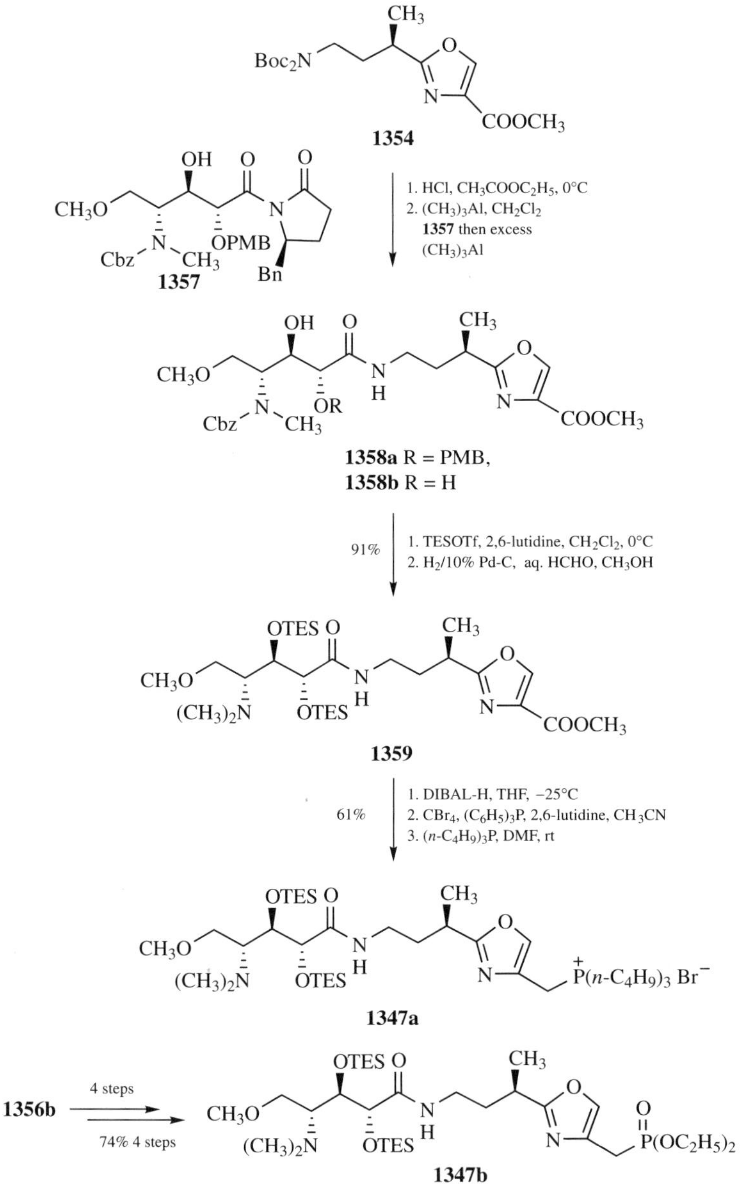

Scheme 1.346

Masamune and co-workers[500] reported the first synthesis of naturally occurring (−)-calyculin A, **1189a**. An earlier report[488] focused specifically on construction of a fully elaborated oxazole containing a C(26)-C(37) fragment **1322** and the intermediates **1363** and **1364** (Scheme 1.349). Note that the configurations of **1363**, **1364**, and **1369** correspond to those present in *ent*-**1189a** (i.e., (+)-

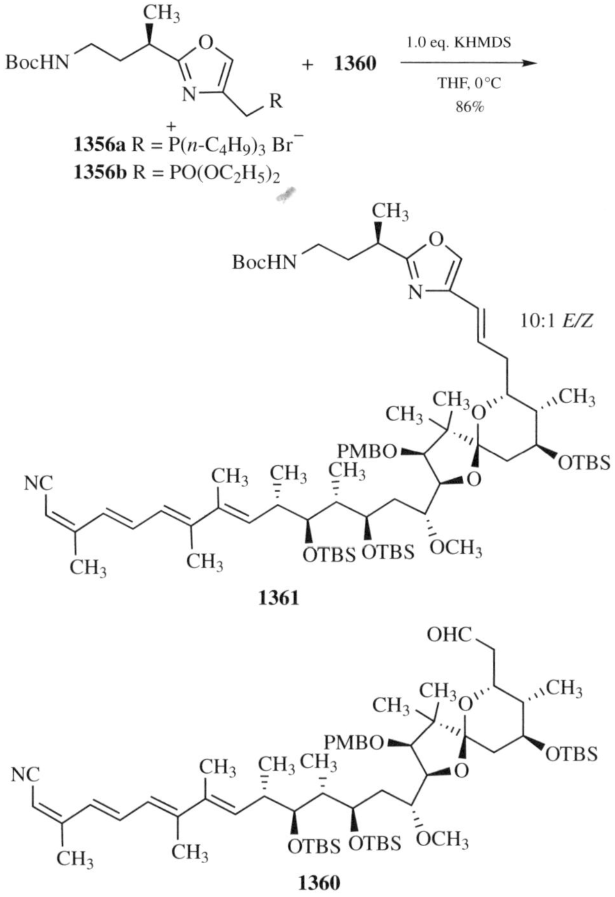

Scheme 1.347

calyculin A). This was not apparent until the absolute stereochemistry of **1189a** was revised in 1991.[509,510]

Gulonolactone served as the starting material for **1365**, which was converted to **1363** in eight steps and 55% overall yield (Scheme 1.350). For the oxazole intermediate **1364**, the authors began by converting the epoxide **1366** to the amide **1367** in five steps and 88% overall yield. At this point, the authors encountered epimerization at the α-carbon when **1367** was treated with ethylbromopyruvate using classical reaction conditions. However, after considerable experimentation, this epimerization was circumvented by addition of excess 3,4-epoxycyclopentene as an acid scavenger and conducting the reaction at room temperature. The intermediate oxazoline (not shown) was immediately dehydrated with TFAA to afford the 2-substituted 4-oxazolecarboxylic acid ethyl ester **1368**. Conversion of **1368** to **1364** was straightforward and accomplished in excellent overall yield. Saponification of **1363** and coupling the resulting acid with **1364** then furnished the desired C(26)-C(37) fragment **1369**.[488]

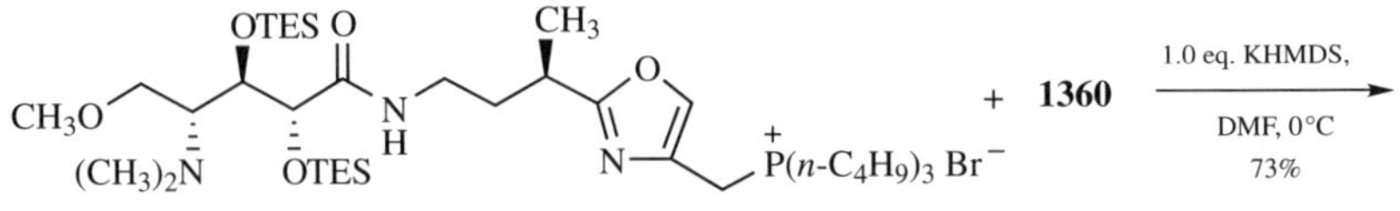

1347a

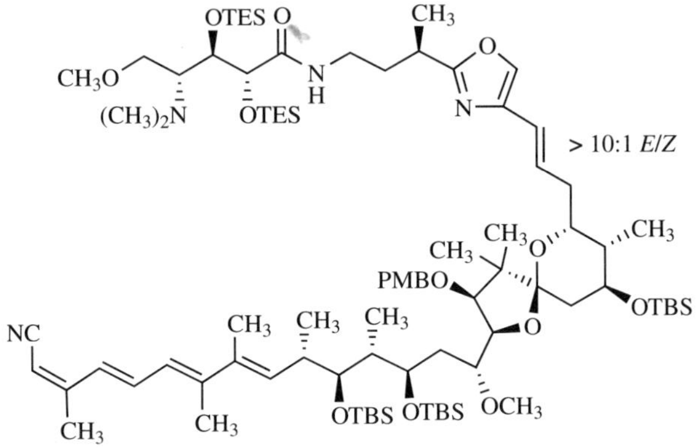

1362

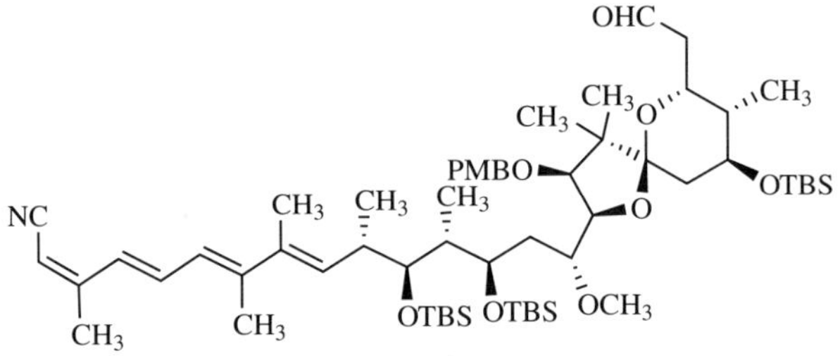

1360

Scheme 1.348

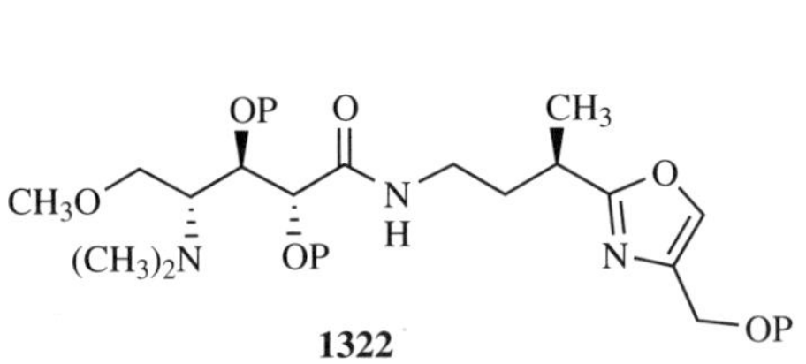

1322

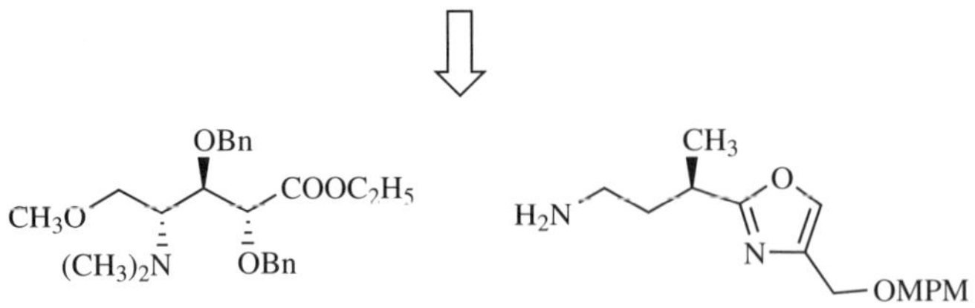

1363 **1364**

Scheme 1.349

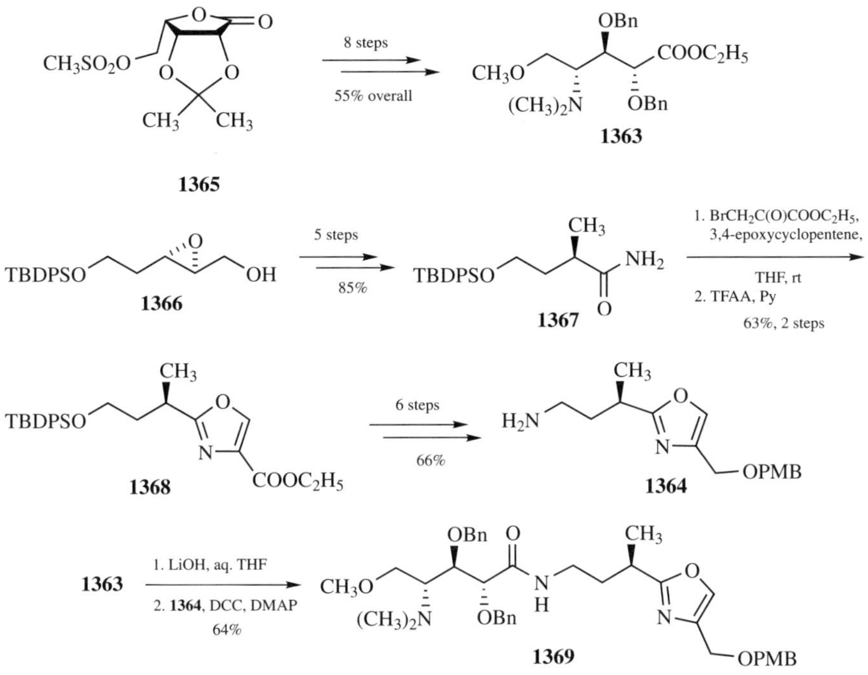

Scheme 1.350

Once the absolute stereochemistry of **1189a** had been established, Masamune's group[500] revised the synthetic strategy and focused on completing the first synthesis of (−)-calyculin A, **1189a**. D-Galactose served as the starting material for preparation of **1370**, which was converted to the bis-TES protected fragment **1371** (Scheme 1.351). The authors noted that cleavage of the C(34) and C(35) benzyl ethers in the presence of the C(26) OPMB group required specific reaction conditions for success. The conversion of **1371** to the aldehyde containing C(26)-C(37) fragment **1372** was uneventful.

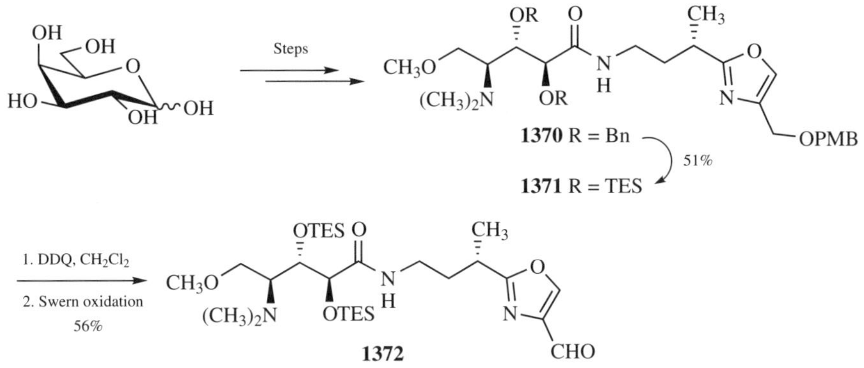

Scheme 1.351

The synthesis of **1189a** was completed as shown in Scheme 1.352. The anion of the sulfone **1373** was generated with phenyllithium, followed by addition of aldehyde **1372**. The resulting adduct (not shown) was first benzolyated and then subjected to reductive-elimination to introduce the *trans*-C(25)-C(26) double bond in **1374**. After preparation of the aldehyde **1375**, installation of the cyanotetraene side chain was accomplished in two stages to afford synthetic calyculin A, identical in all respects to the natural product **1189a**.

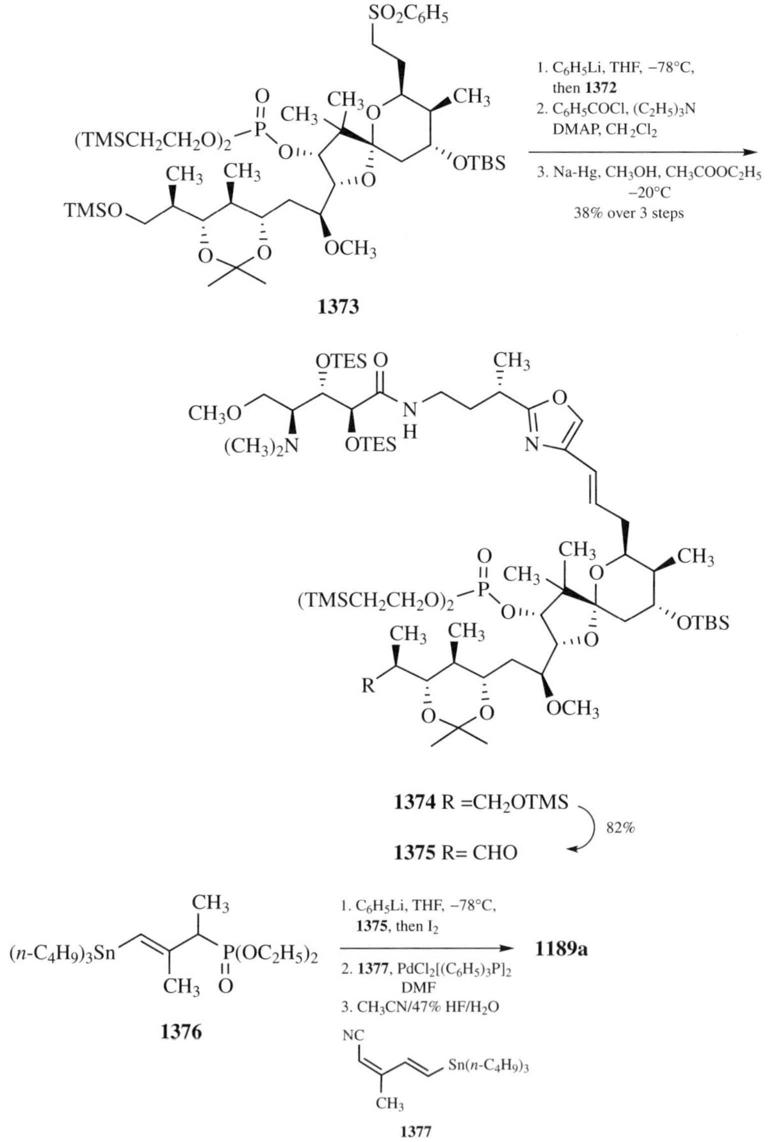

Scheme 1.352

Barrett and co-workers'[151] synthetic strategy to their key C(26)-C(37) oxazole fragment **1382** incorporated the classical Cornforth methodology, as modified by Meyers.[459a] The starting 2-substituted-4-oxazolecarboxylic acid methyl ester **350** was readily prepared from benzyl (R)-2-methyl-4-pentenoate **347** (see Scheme 1.94). Elaboration of **350** to **1382** was accomplished as shown in Scheme 1.353. The terminal olefin was oxidized to the corresponding aldehyde **1378**, from which **1379** was prepared in four steps and 72% overall yield by straightforward functional group manipulation. The authors prepared a Mosher ester of the intermediate alcohol, leading to **1379**, and confirmed there was no significant racemization in the process. The most efficient coupling conditions for preparing **1381a** were DCC/HOBt. Sequential treatment of **1381a** with TFA and then methyl iodide produced **1382**.

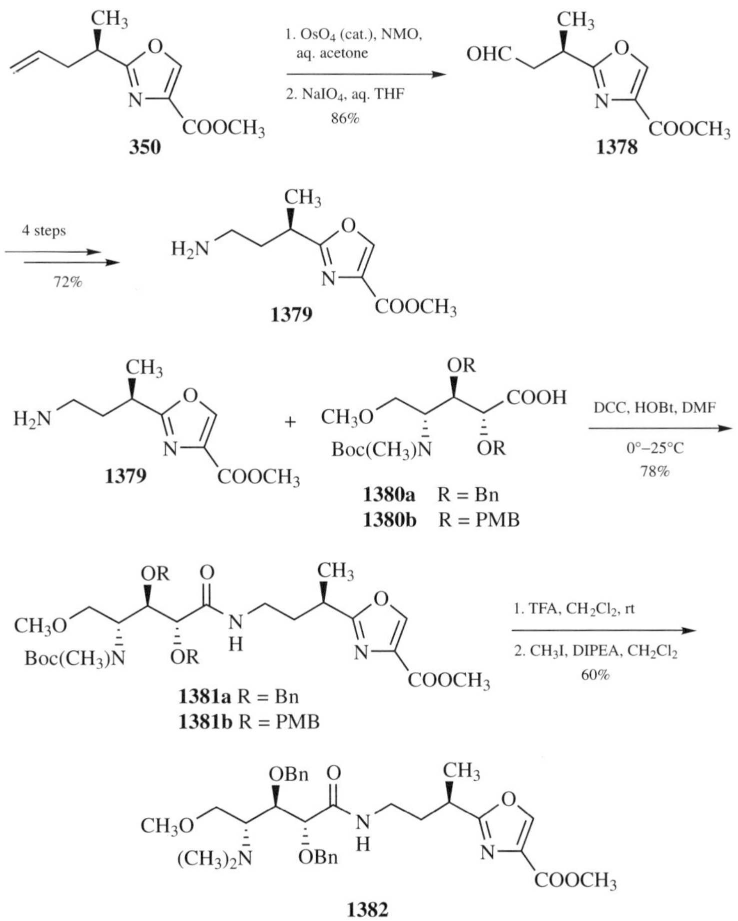

Scheme 1.353

Barrett's group[501] completed a formal total synthesis of (+)-calyculin A. The authors' synthetic target for the C(26)-C(37) oxazole fragment **1322** (R = CH$_2$OH, P = TES, R$_3$ = H) was modified somewhat for this work. Their approach to **1379** was the same (vide supra). Here, low yields and difficulties in removing the dicyclohexylurea by-product complicated attempts to couple **1379** with **1380b** using DCC/HOBt (Scheme 1.354). However, the authors found that 1-ethyl-3-(3-dimethylaminopropyl)carbodiimide hydrochloride (EDCI), anhydrous HOBt, and 4-Å molecular sieves dramatically improved the yield of **1383** from **1379** and **1380b**. At that point, **1383** was converted to **1359**, which was identical in all respects to the material prepared by Evans and co-workers.[497] The authors also prepared the fully elaborated cyanotetraene **1360**,[497] thereby completing their formal total synthesis of (+)-calyculin A.

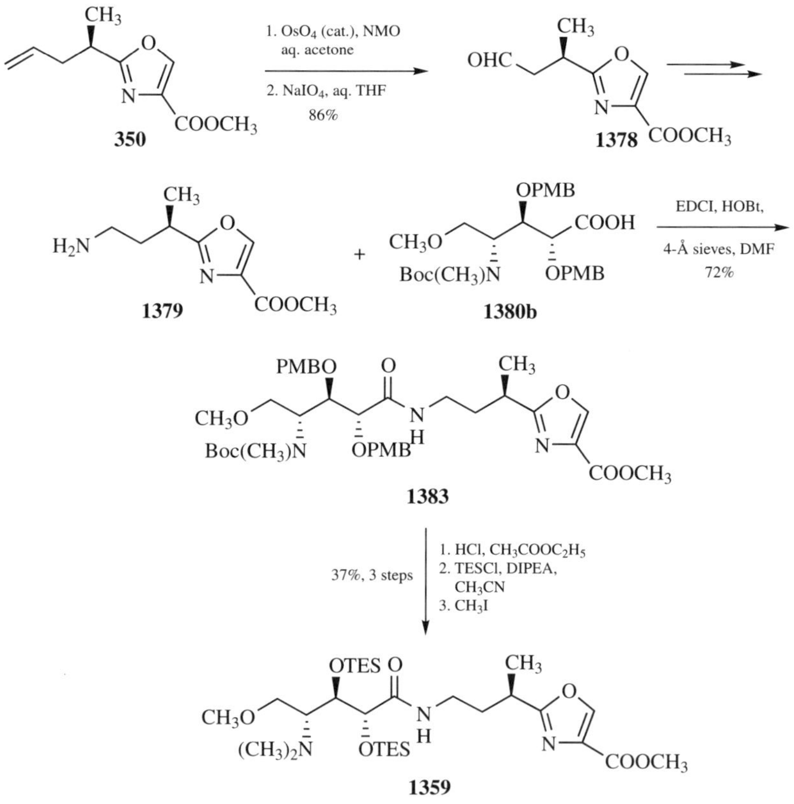

Scheme 1.354

Koskinen and co-workers[44,60] evaluated several methodologies for converting oxazolines to oxazoles as part of their synthetic studies leading to the C(26)-C(32) oxazole fragment **58** of calyculin C **1189c**. D-Alaninal was converted to (5*R*)-*N*-(*tert*-butoxycarbonyl)-3,5-dimethylpyrrolin-2-one **1384** in excellent yield

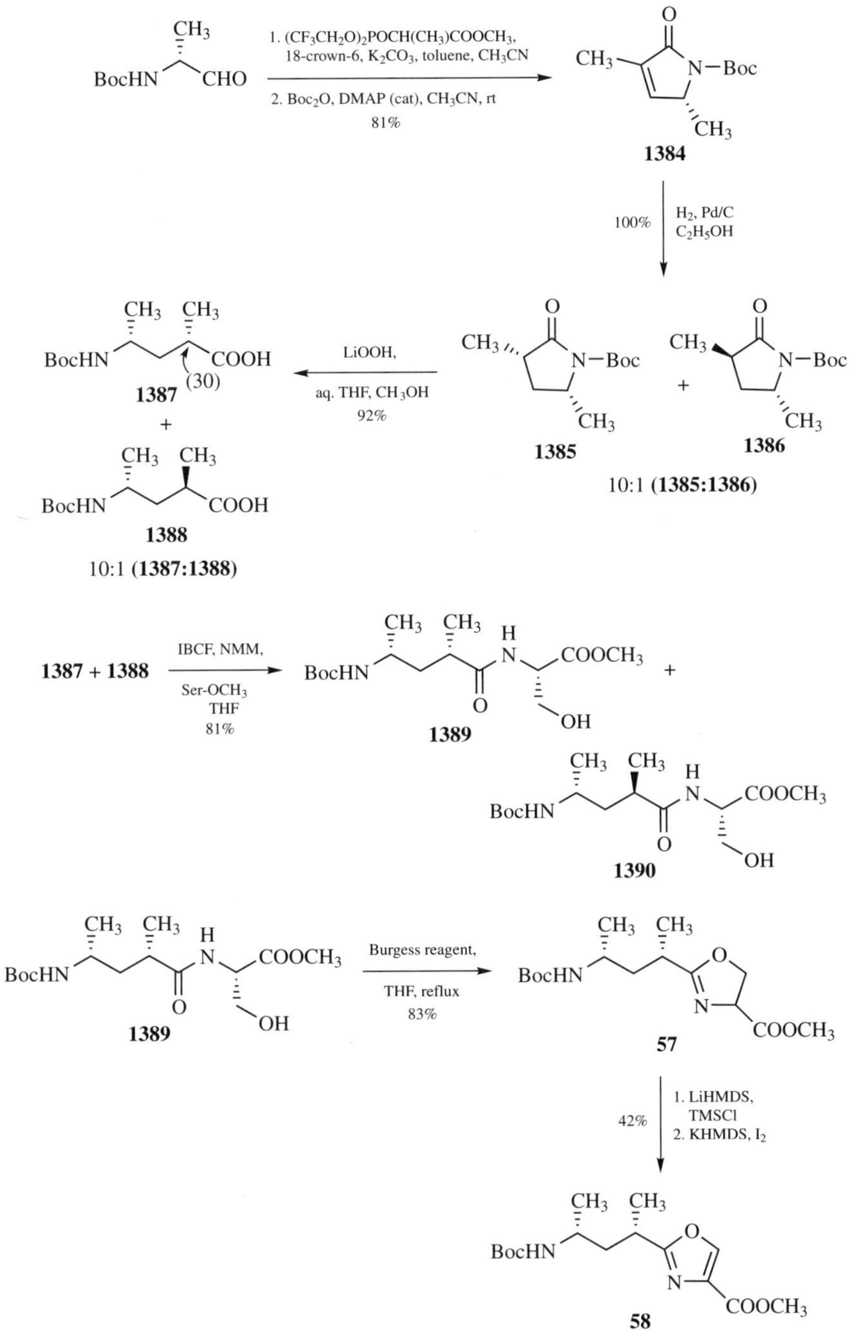
1. (CF3CH2O)2POCH(CH3)COOCH3, 18-crown-6, K2CO3, toluene, CH3CN
2. Boc2O, DMAP (cat), CH3CN, rt
81%
1384
100%
H2, Pd/C
C2H5OH
1385
1386
10:1 (1385:1386)
LiOOH,
aq. THF, CH3OH
92%
1387
(30)
1388
10:1 (1387:1388)
1387 + 1388
IBCF, NMM,
Ser-OCH3
THF
81%
1389
1390
Burgess reagent,
THF, reflux
83%
57
42%
1. LiHMDS, TMSCl
2. KHMDS, I2
58

Scheme 1.355

(Scheme 1.355). Catalytic reduction of **1384** produced a 10:1 mixture of diastereomers **1385** and **1386** quantitatively. Hydrolysis of this diastereomeric mixture with lithium hydroperoxide[511] gave the acids **1387** and **1388**, which were immediately coupled with serine methyl ester to afford the diastereomeric dipeptides **1389** and **1390**. Hydrolysis of **1385** and **1386** with lithium hydroxide also gave **1387** and **1388** but effected significant epimerization at C(30). The diastereomers were separated, and **1389** was cyclized to **57** with Burgess reagent.[49]

The best conditions for converting **57** to the target oxazole **58** involved temporary TMS protection of the urethane followed by KHMDS and iodine (see Scheme 1.18). Temporary protection of the urethane as a silyl carbamate was critical to the success of this process. Attempts to employ a second Boc-protecting group or bulkier silyl-protecting groups failed.

Comparable yields of **58** were also obtained using $CuBr_2$/HMTA/DBU.[43] Other methods—including activated MnO_2,[37] NiO_2,[27] and Cu(II)OAc/*tert*-butyl perbenzoate[36]—resulted in poor yields and extensive fragmentation. Koshkinen and co-workers[60] also isolated diastereomeric spiro[4,4]orthoester aminals **59** as byproducts from the conversion of **57** to **58**. The structure of one diastereomer was confirmed by X-ray crystallography.

Smith and co-workers[502–505] evaluated several routes for preparing the C(26)-C(37) segment of (+)-calyculin A. In their early studies,[502] 4-chlorobutyryl chloride was converted to (*R*)-4-azido-2-methylbutanoic acid **1391** in four steps and 72% overall yield (Scheme 1.356). Coupling **1391** with L-serine methyl ester was accomplished using diethyl phosphoryl cyanide (DEPC) to give **1392a**. Attempts to prepare **1392b** using $SOCl_2$/THF[507] resulted in 15–20% levels of epimerization at the carbon α to the amide carbonyl group. However, the authors overcame this problem by using a two-step process that afforded **1392b** with <5% epimerization.

Cyclization of **1392b** then produced the oxazoline **1393** in 65% yield from **1392a**. Oxidation of **1393** with NiO_2[27] furnished the oxazole **1394**, which was converted to **1395** in three straighforward steps. The fully elaborated, protected C(26)-C(37) oxazole fragment **1397** for (+)-calyculin A was then prepared by coupling the aluminum amide of **1395** with **1396** followed by deprotection.

Salvatore and Smith[503] found that the epimerization encountered during preparation of **1392b** was effectively circumvented by cyclization of **1392a** directly to **1393** with Burgess reagent[49] (Scheme 1.357). The authors isolated **1393** in excellent yield but were unable to determine the extent of epimerization reliably. Conversion of **1393** to the oxazolidine-2-selone **1399** was achieved using the Silks-Odom protocol.[512,513] The ^{77}Se NMR spectrum showed complete resolution of the Se singlets and allowed for an accurate determination of any epimerization. There was <2% epimerization at C(30) in **1393** prepared using Burgess reagent vs. 14% epimerization using $SOCl_2$/THF.

Smith and co-workers[504,505] described the total syntheses of (+)-calyculin A and (−)-calyculin B in later studies. At that time, the most efficient synthesis of **1394** employed the Burgess reagent[49] for cyclization of **1392a** followed by $CuBr_2$/HMTA/DBU[43] oxidation of **1393**. In this manner, **1394** was prepared in 67% yield with little epimerization.

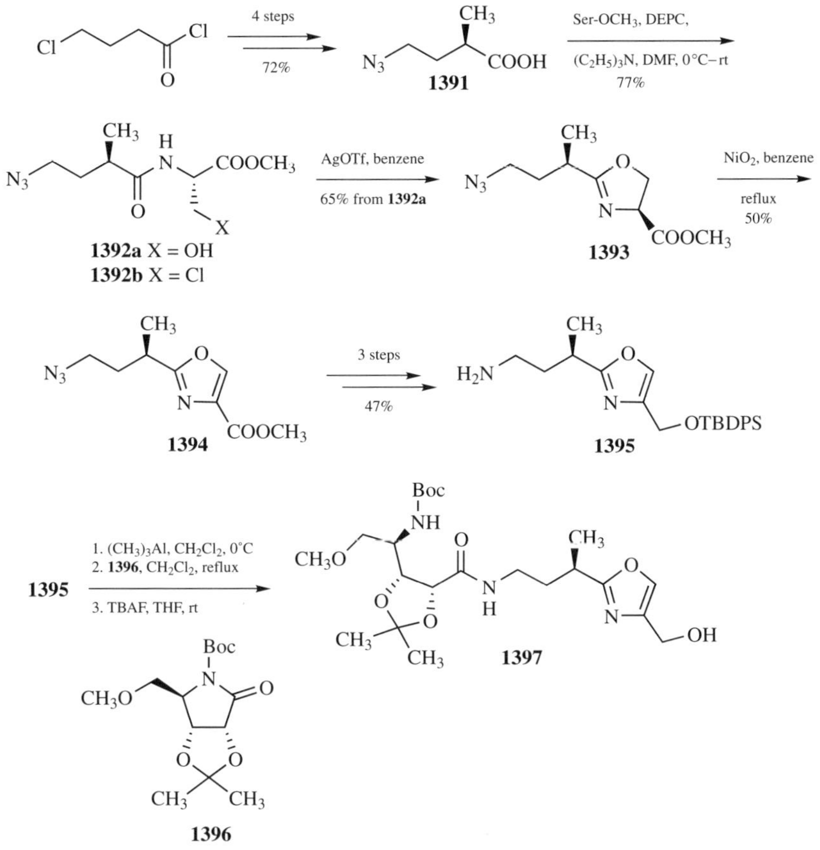

Oxazole Natural Products

Scheme 1.356

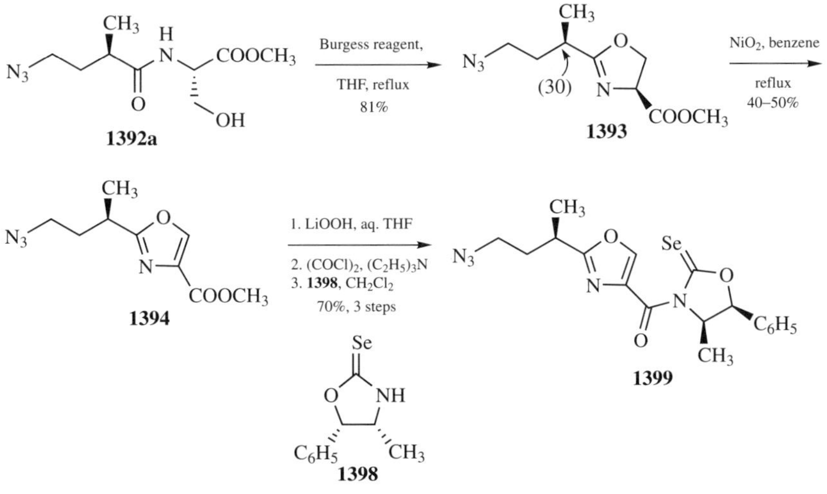

Scheme 1.357

After considerable experimentation the authors determined that the best approach to the key C(26)-C(37) fragment **1404** was via an alternate coupling strategy (Scheme 1.358). The unstable oxazole amino ester **1400** was prepared from **1394** uneventfully, and hydrolysis of **1396** was also straightforward and provided **1401** in excellent yield. Coupling **1400** and **1401** using DEPC was an efficient method of preparing **1402**. The authors found that the acetonide could not be removed from a fully protected calyculin A analog, necessitating an alternate protection strategy. In addition, it was more practical to introduce the dimethylamino group at this time. Therefore, **1402** was converted to **1403** in four steps in 50% overall yield. The desired phosphonium salt **1404** was then prepared in three additional steps. The 4-(chloromethyl)oxazole (not shown) proved to be the precursor of choice to **1404** and is stable to low-temperature storage conditions.

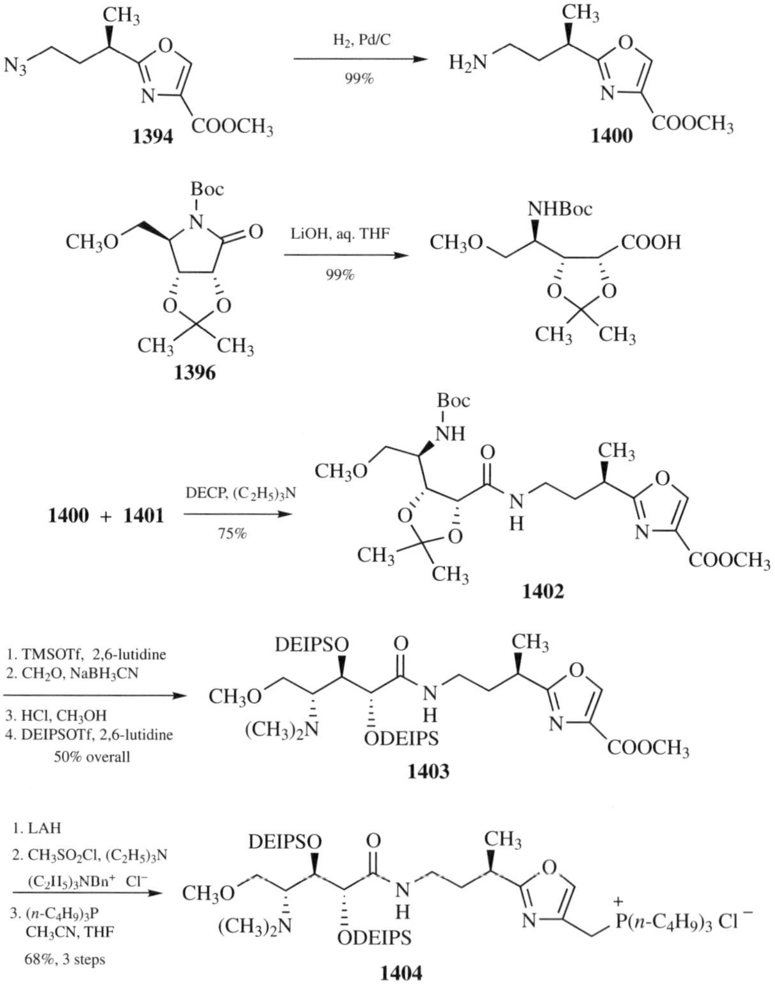

Scheme 1.358

Model studies of the Wittig olefination of hydrocinnamaldehyde were undertaken with phosphonium iodides derived from **1397**. These results indicated the best *E/Z* ratios were obtained with a tributylphosphonium salt,[411] using an excess of *n*-BuLi. These conditions were modified slightly and then applied to **1404** (Scheme 1.359). Thus **1404** and **1405** were treated with LiHMDS in DMF to produce **1406a** in excellent yield with 9:1 *E/Z* selectivity. A two-step sequence converted **1406a** to the aldehyde **1406b**.

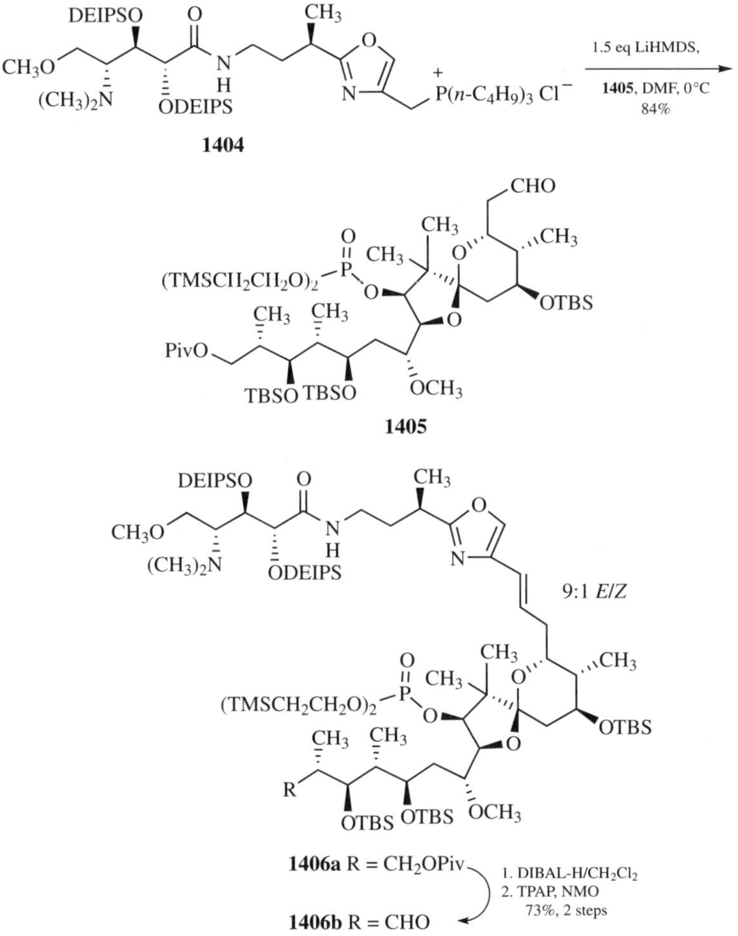

1404

1405

1406a R = CH_2OPiv

1406b R = CHO

Scheme 1.359

The synthesis was completed as shown in Scheme 1.360. Horner-Emmons reaction of the triene phosphonate **1407** with **1406b** and workup with 0.5 N HCl furnished the trienone **1408** in 92% yield with 15:1 *E/Z* selectivity. Peterson olefination of **1408** with $(CH_3)_3SiCH_2CN$ afforded a 1.7:1 *E/Z* mixture of protected calyculins in 94% yield. The regioisomers were separated and subjected to global

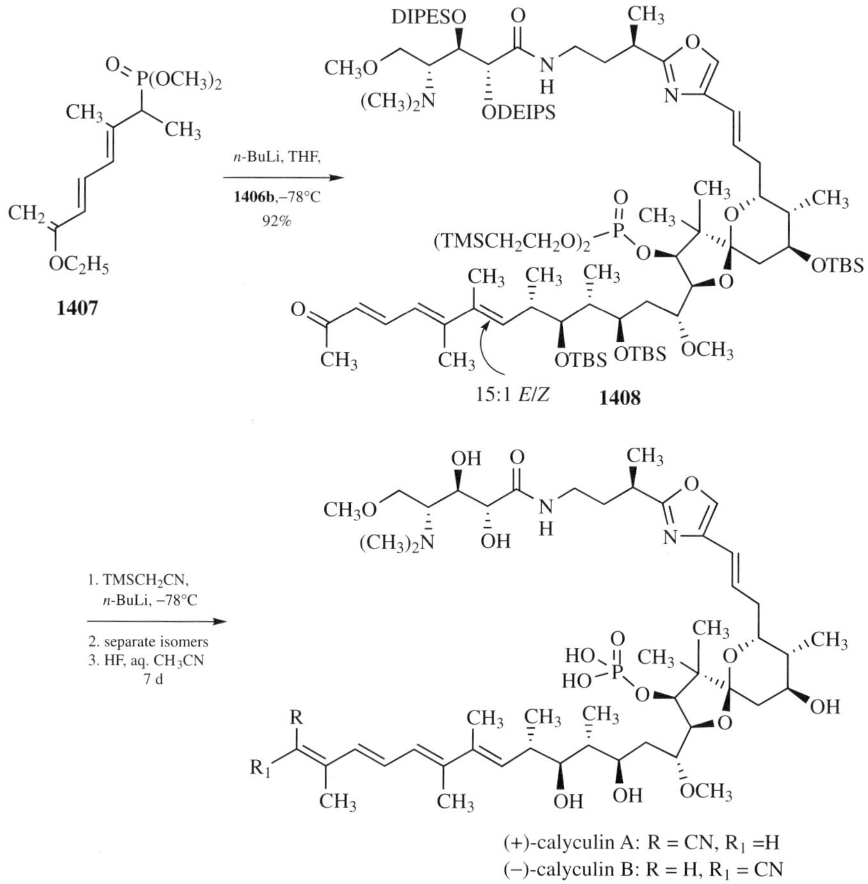

Scheme 1.360

deprotection to complete the syntheses of *ent*-**1189a** and *ent*-**1189b**, respectively. In both cases, the products were identical to **1189a** and **1189b**, except for optical rotation and CD spectra.

1.5.4. Hennoxazole

Hennoxazoles **1191a–d** are 2,4′-bis-oxazole-containing natural products isolated from the sponge *Polyfibrospongia* sp. (Fig. 1.99).[446,447] (−)-Hennoxazole A **1191a** is highly active against herpes simplex virus 1 (HSV-1) and possesses peripheral analgesic activity comparable to indomethacin. Total syntheses of these intriguing natural products have been pursued by several research groups.[514–519]

Efficient syntheses of 2,4′-bis-oxazoles were described in earlier sections of this chapter and are further discussed here. For example, the reader should consult the work of Prager and co-workers[73] (Scheme 1.29), Moody and co-workers[87,88] (Schemes 1.41 and 1.45), Williams and co-workers[360] (Scheme 1.248), and Vedejs

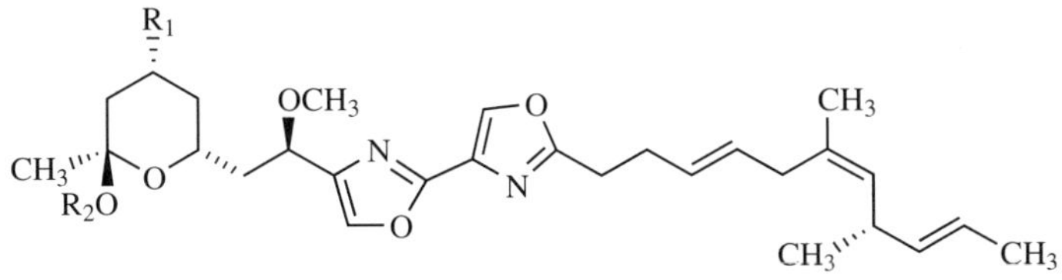

Figure 1.99. Hennoxazoles.

1191a R_1 = OH, R_2 = CH_3
1191b R_1 = OH, R_2 = C_2H_5
1191c R_1 = OH, R_2 = n-C_4H_9
1191d R_1 = H, R_2 = CH_3

and Luchetta[358] (Scheme 1.260) for further details. In addition, relevant references are also found in the introduction to Section 1.5.

In early model studies leading to **1191a**, Barrett and Kohrt[517] prepared a 2,4′-bis-oxazole via a Pd(0) catalyzed Stille coupling[520] (Scheme 1.361). The authors

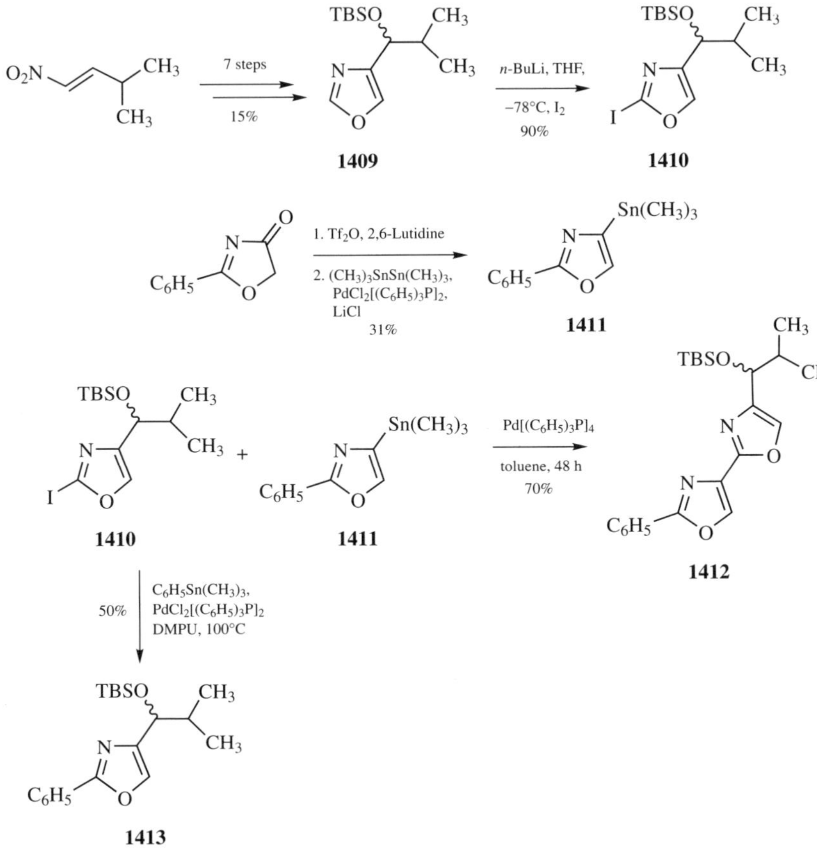

Scheme 1.361

318   Synthesis and Reactions of Oxazoles

converted 3-methyl-1-nitro-1-butene to the racemic 4-substituted-oxazole **1409**, using classical Robinson-Gabriel methodology. Iodination of **1409** was uneventful and proceeded in excellent yield to generate **1410**. The authors were not able to prepare a 2-(stannyl)oxazole from the intermediate 2-lithiooxazole generated during the iodination reaction.

The coupling partner, 2-phenyl-4-(trimethylstannyl)oxazole **1411** was readily prepared, in two steps, from 2-phenyl-4(5H)-oxazolone. Stille coupling of **1410** with **1411** using tetrakis-(triphenylphosphine)-palladium[0] then afforded the model 2,4'-bis-oxazole **1412** in good yield. Similarly, **1410** was coupled with phenyltrimethyltin to generate **1413**.

Wipf and Lim[514] reported the first synthesis of a hennoxazole, **1420**, although it was found to be the enantiomer of **1191a**. The authors adopted a convergent strategy wherein the first oxazole ring was prepared in the triene-containing fragment. After coupling the triene fragment with a suitably protected pyran derivative, the second oxazole was elaborated on the fully intact carbon skeleton.

The synthesis of the skipped triene-oxazole fragment **1417** began with (Z)-2-butene-1,4-diol, which was converted to the unsaturated ester **1414** in three

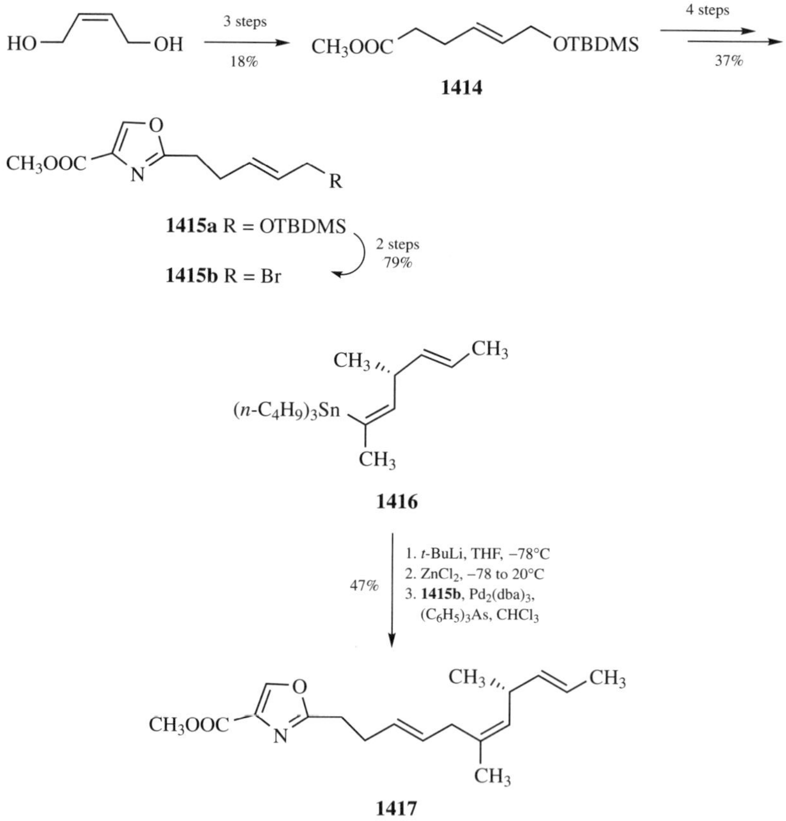

Scheme 1.362

Oxazole Natural Products 319

straightforward steps (Scheme 1.362). Saponification of **1414**, coupling the resulting acid with serine methyl ester followed by cyclization with Burgess reagent[49] and oxidation using $CuBr_2$/DBU/HMTA,[43] gave **1415a**, which was converted to 2-(5-bromo-3-penten-1-yl)-4-oxazolecarboxylic acid methyl ester **1415b**. An attempt to prepare **1417** directly from **1415b** and **1416** via palladium-catalyzed coupling failed. Therefore, the authors prepared an intermediate vinylzinc reagent from **1416**, which was successfully coupled with **1415b** to give **1417** in acceptable yield.

At this point, **1417** was saponified and coupled with the fully elaborated pyran fragment **1418** to afford **1419** (Scheme 1.363). The second oxazole ring was then installed by sequential Dess-Martin oxidation[211] of **1419** followed by cyclization with $(C_6H_5)_3P$, $BrCl_2CCCl_2Br$ in the presence of 2,6-di-*tert*-butyl-4-methylpyridine. These conditions were critical given the acid-sensitivity of the pyran moiety.

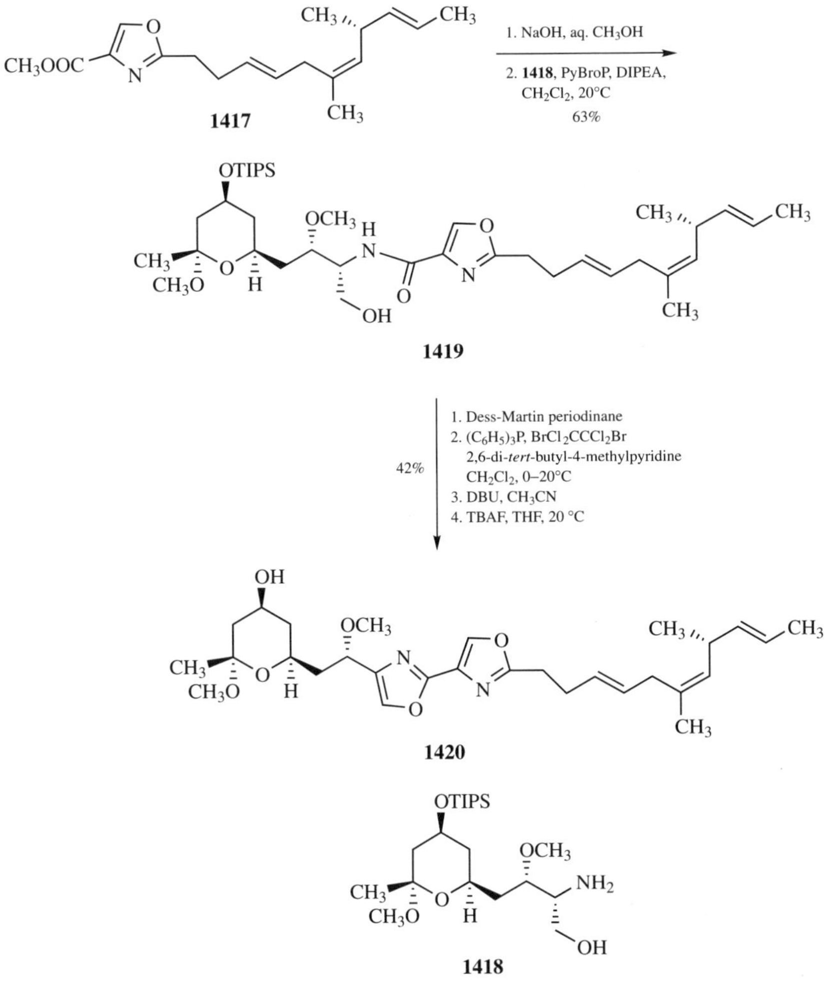

Scheme 1.363

320 Synthesis and Reactions of Oxazoles

Dehydrobromination of the intermediate bromooxazoline (not shown) was best accomplished with DBU, after which cleavage of the TIPS protecting group afforded **1420** in 42% yield from **1419**. Comparison of the optical rotation and CD spectra confirmed that **1420** was the enantiomer of **1191a**.

In their early work on **1191a**, Shiori and co-workers[516] also proposed a strategy for constructing the second oxazole ring at a late stage in the synthesis. However, these authors elected to prepare the fully elaborated pyran fragment containing the oxazole first, followed by coupling and then elaboration of the second oxazole ring. Thus **1421** and **1422** were key intermediates in their approach. Further retrosynthetic analysis led to the dithiane substituted oxazole **1423** (Scheme 1.364).

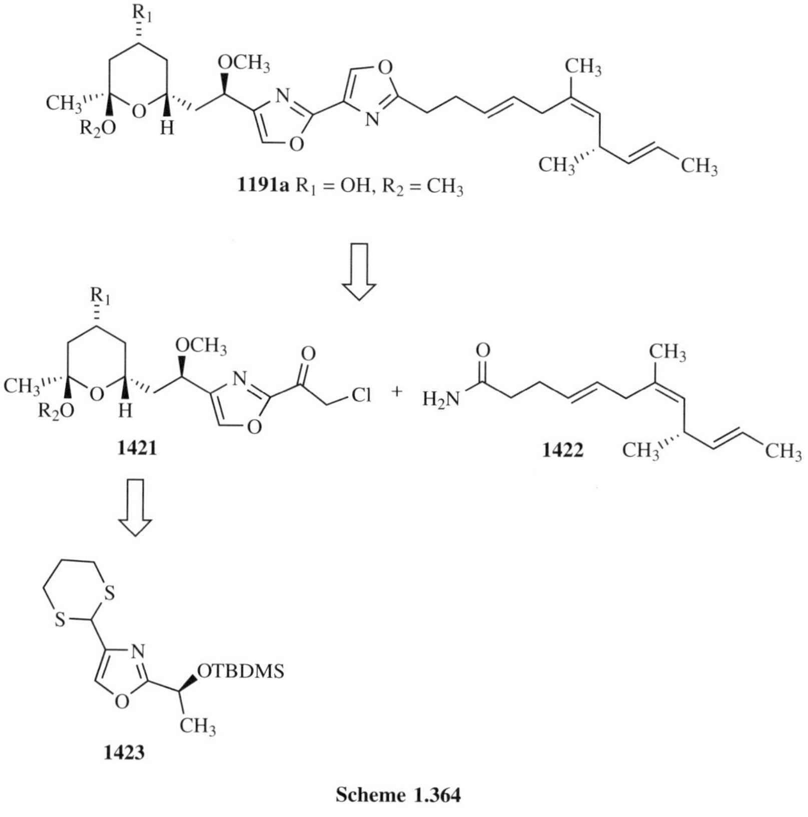

Scheme 1.364

L-Lactic acid was converted to the chloride **1424** in three steps (Scheme 1.365). Cyclization of **1424** to the oxazoline (not shown) was accomplished using AgOTf/CaCO$_3$[507] followed by oxidation[43] to yield **1425**. A straightforward three-step sequence then gave the target **1423** in very good yield.

Further work from Shiori's group[518,519] described the total synthesis of **1191a**, but not via their original strategy. Originally, the authors focused on an oxazole-containing linear precursor to the pyran fragment, which was to be prepared from 2-styryl-4-oxazolecarboxylic acid methyl ester **1426** (Scheme 1.366). Preparation of

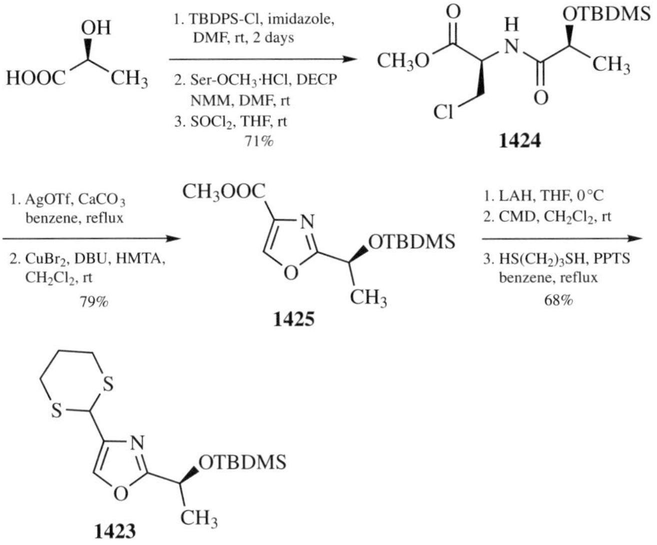

Oxazole Natural Products

Scheme 1.365

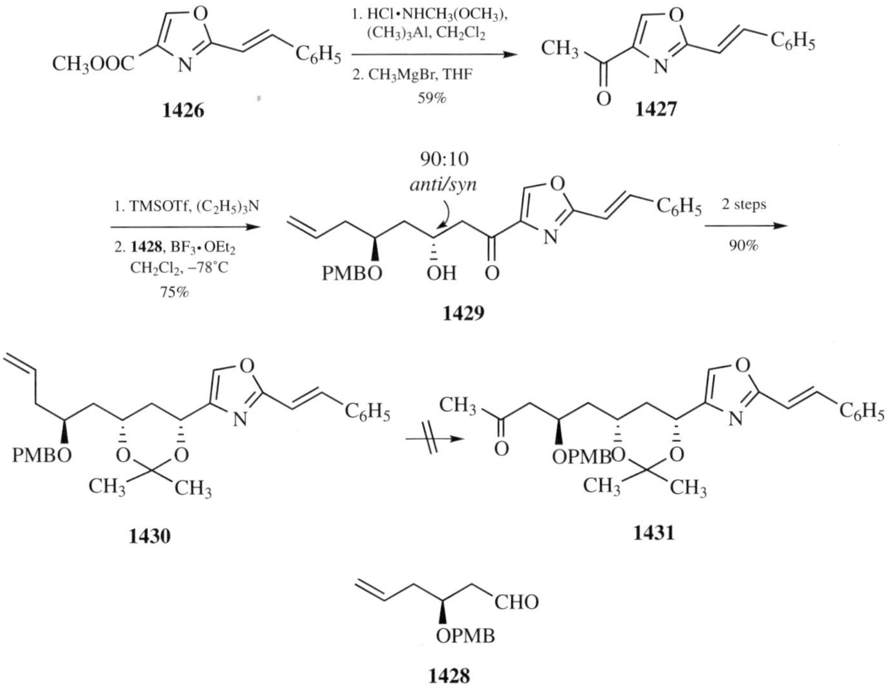

Scheme 1.366

4-acetyl-2-styryloxazole **1427** was straightforward. Diastereoselective Mukaiyama aldol methodology coupled with chelation controlled reduction converted **1427** to the acetonide **1430**. Unfortunately, the authors were unable to convert the terminal olefin of **1430** to the methyl ketone **1431**. Therefore, they modified their strategy to parallel that employed by Wipf and Kim (vide supra).[514]

The C(14)-C(25)-skipped trienoic acid **1432** was prepared from (S)-3-hydroxy-2-methylpropionate in 12 steps and 8% overall yield (Scheme 1.367). Coupling **1432** with L-serine methyl ester yielded the β-hydroxyamide **1433**. Sequential cyclization of **1433** with Deoxo-fluor[553] to an oxazoline (not shown) and oxidation with CBrCl$_3$/DBU[57] then produced the desired skipped triene-oxazole fragment **1417** in 73% yield.

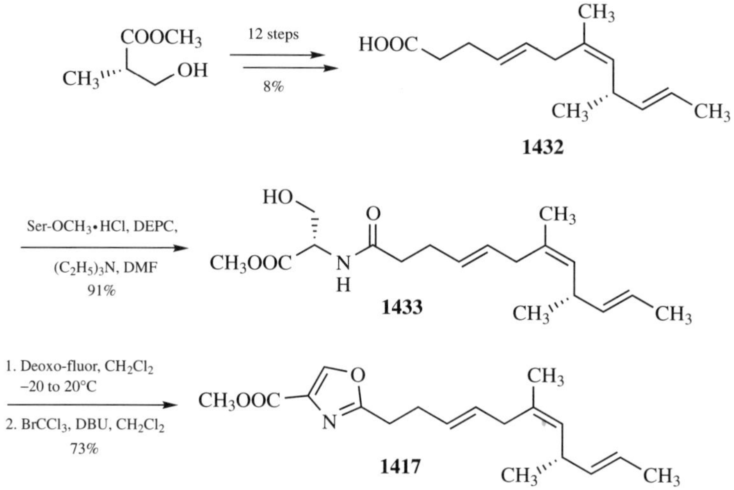

Scheme 1.367

The last steps of the synthesis were accomplished as shown in Scheme 1.368. Saponification of **1417** followed by coupling with the fully elaborated pyran fragment **1434** gave **1435** in excellent yield. The second oxazole ring was installed using the Wipf and Lim protocol,[514] followed by oxidative cleavage of the PMB-protecting group. Synthetic hennaxazole A was identical in all respects to **1191a**.

Williams and co-workers[515] also reported an enantio-controlled convergent synthesis of **1191a**. These authors constructed a 2,4′-difunctionalized-2,4′-bis-oxazole early in their synthesis and then elaborated both functional groups to incorporate the skipped triene and pyran moieties. The key 2,4′-bis-oxazole **1440** was prepared in good yield (Scheme 1.369). Mixed anhydride coupling of **1436** with (rac)-serine methyl ester produced the β-hydroxyamide **1437**, which was cyclized to the oxazole **1438**. The second oxazole ring was introduced in a three-step sequence using standard methodology to furnish **1439**, which was reduced to the key 2,4′-bis-oxazole **1440**.

The plan was to elaborate the pyran moiety through the aldehyde while the skipped triene would be elaborated through the protected hydroxypropyl side chain.

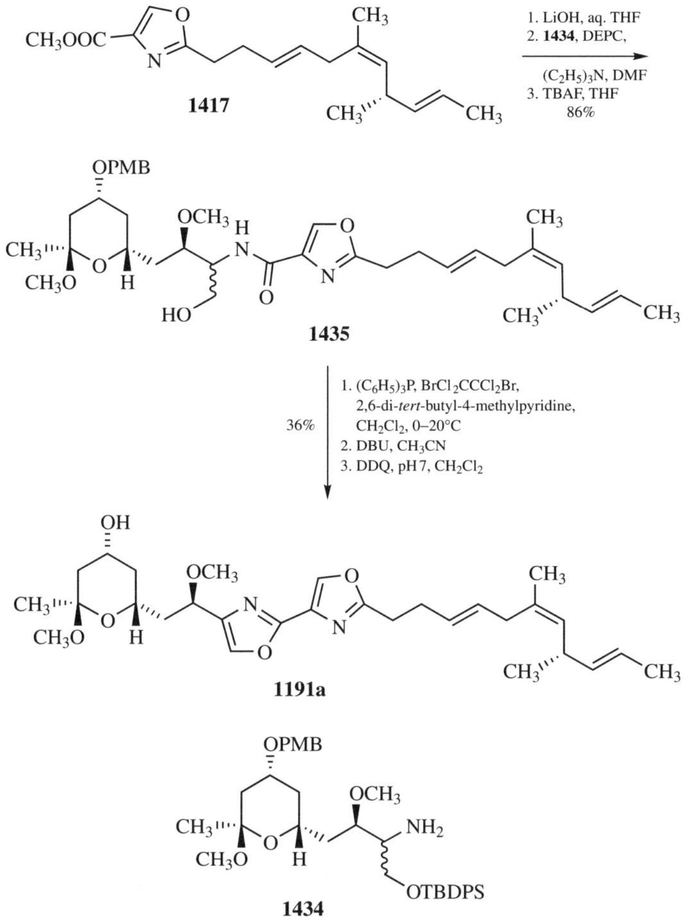

Scheme 1.368

The successful implementation of this strategy is shown in Scheme 1.370. The optically pure stannane **1441** was transmetalated in situ with **1442**, preparing a borane reagent that condensed with **1440**. After methylation, **1443** was isolated in excellent yield with 10.5:1 diastereoselection. At this point, **1443** was elaborated to incorporate the pyran moiety (five steps), and the side chain was converted to an aldehyde (two steps). Overall, this reaction sequence to **1444** was accomplished in 46% yield. The skipped triene fragment was then incorporated via the anion of sulfone **1445**. Saponification of the pivalate ester then completed the synthesis of **1191a**.

1.5.5. Diazonamides

The diazonamides A and B **1192a** and **1192b** are singularly unique natural products isolated from the colonial ascidian *Dizona chinensis*.[448] Diazonamide A

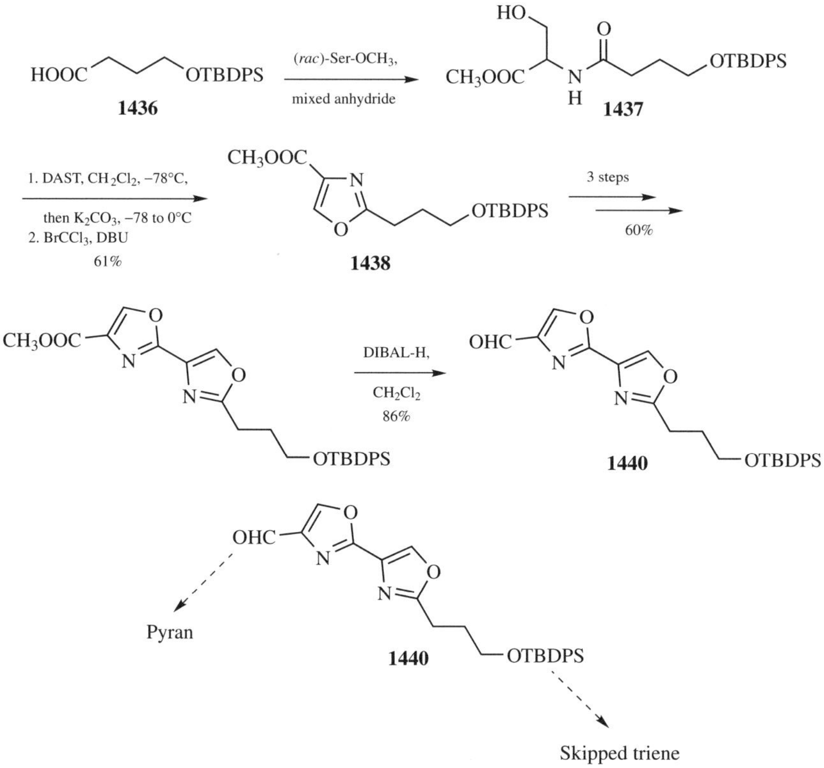

Scheme 1.369

has shown potent in vitro cytotoxicity against HCT-116 human colon carcinoma and B-16 murine cancer cell lines. These densely substituted heterocycles present formidable synthetic challenges (Fig. 1.100). They possess a complex and strained heterocyclic core made up of two directly linked halogenated trisubstituted-oxazoles and indole and biphenyl rings linked via a chiral quaternary carbon atom. The macrocylic core is further linked to a tyrosine-valine cyclopeptide residue, and the entire molecular ensemble exists as a single atropisomer. A total synthesis of these natural products has not yet been described. This section focuses on synthetic efforts of a number of research groups[87,88,91–94,204,371,521–531] to prepare the halogenated trisubstituted bis-oxazole fragment present in both **1192a** and **1192b**.

Early on, Moody and co-workers[87,88] adapted their rhodium(II) carbenoid methodology for reaction of diazocarbonyl compounds with nitriles to prepare 3-(oxazol-5-yl)indoles **154** as models for the C-D-E rings of a diazonamide (see Scheme 1.43).

Later studies[92] that employed the more complex model systems **1446** and **1448** were frustrated by surprising and unexpected results (Scheme 1.371). Initially, the authors planned to prepare the oxazole ring from **1446** followed by

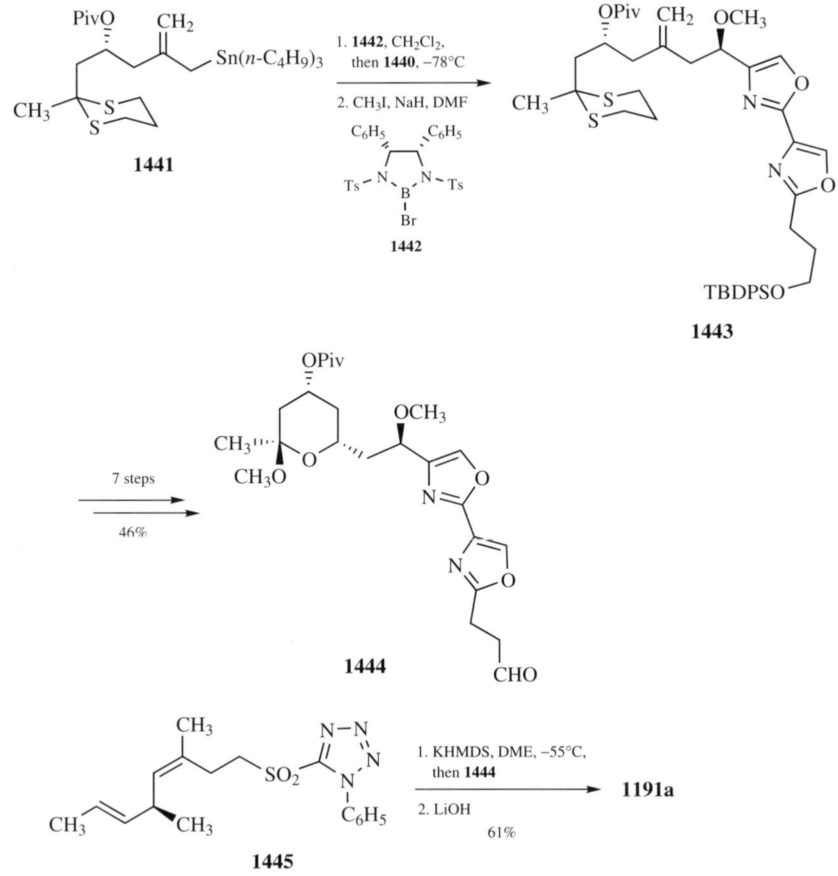

Oxazole Natural Products 325

Scheme 1.370

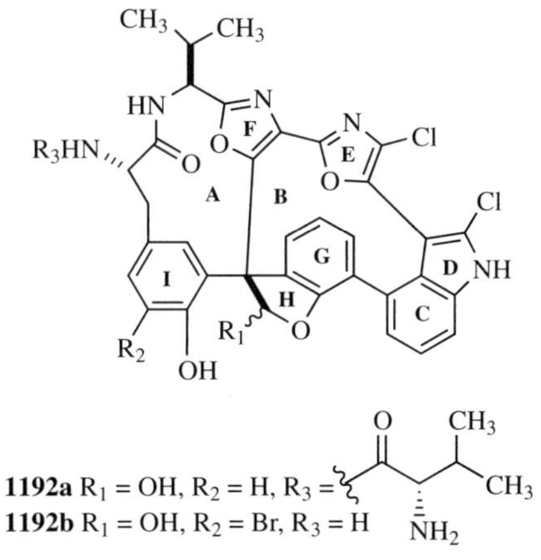

Figure 1.100. Diazonamide A and B.

1192a R$_1$ = OH, R$_2$ = H, R$_3$ = (isobutyryl group)
1192b R$_1$ = OH, R$_2$ = Br, R$_3$ = H

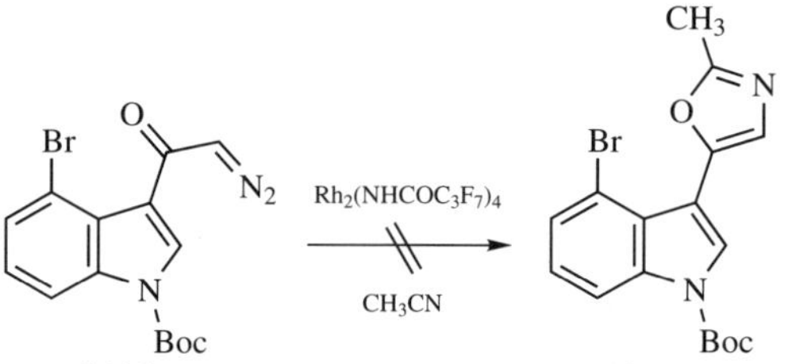

Scheme 1.371

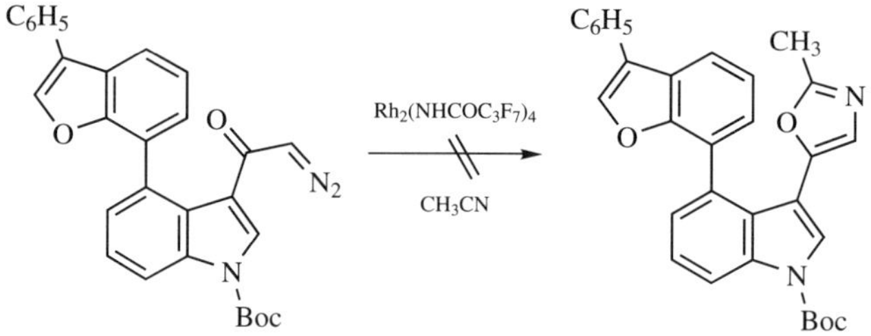

a palladium-catalyzed cross-coupling reaction to introduce the benzofuran ring. However, rhodium (II)-catalyzed decomposition of **1446** in acetonitrile gave none of the expected C-D-E fragment **1447**. This was attributed to a possible competitive formation of a cyclic bromonium ylide. Therefore, the oxazole ring was to be constructed after the biaryl coupling. Surprisingly, in this case, using **1448** as a model afforded none of the expected C-D-E-G-H fragment **1449**. Instead, the authors isolated a low yield of the 3,4-bridged indole **1450**. This unexpected product was attributed to an intramolecular C-H insertion reaction.

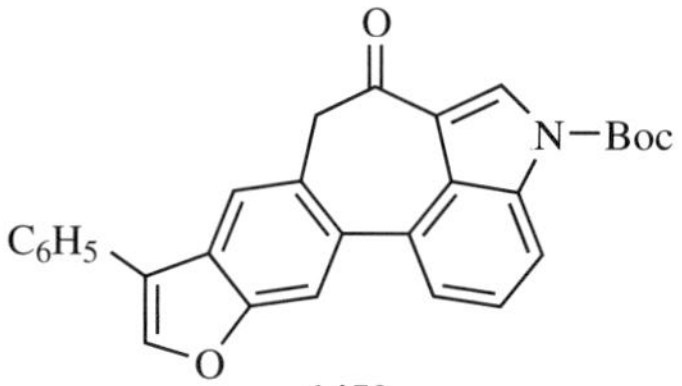

More recently, Moody and co-workers[93] applied their rhodium carbenoid N-H insertion methodology to construct **1452a,b** (Scheme 1.372). Here, (S)-benzyl-oxycarbonylvalinamide first reacted with **1451a** to produce a 1,4-dicarbonyl intermediate (not shown), which was converted to **1452a** in 38% overall yield. Mixed anhydride coupling of **1452a** with tryptophan methyl ester gave rise to **1453**, which was converted to **1454** via DDQ oxidation followed by cyclodehydration. This clearly demonstrated the viability of this approach to construct an indole bis-oxazole (C-D-E-F) fragment.

The authors then extended this methodology to prepare **1457**, the requisite precursor for testing the critical intramolecular biaryl coupling. The expected

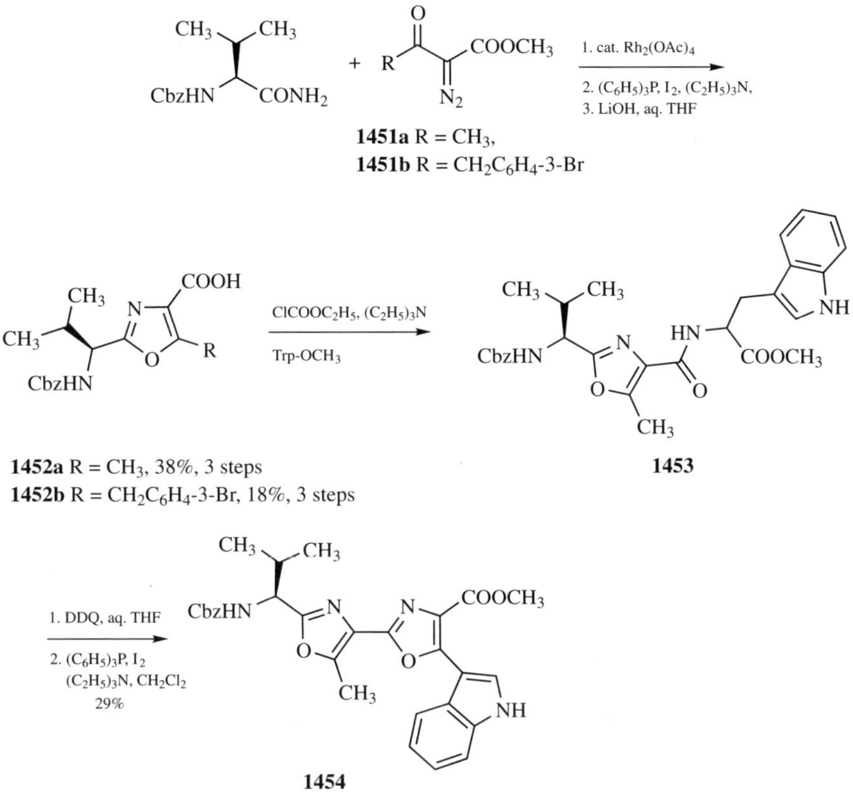

Scheme 1.372

product, a fragment containing the B-C-D-F-G rings, **1458** would then be elaborated to install the second oxazole (E) ring as for **1453** → **1454** (Scheme 1.373). Conversion of (S)-benzyloxycarbonylvalinamide to **1452b** was accomplished in 19% yield using the α-diazocarbonyl compound **1451b**. The other starting material, 4-bromotryptophan methyl ester **1456** was prepared in three steps from **1455**. Mixed anhydride coupling of **1452b** and **1456** then produced the desired intermediate **1457**. Unfortunately, their synthetic strategy was thwarted at this point. The authors evaluated $NiCl_2$, $(C_6H_5)_3P$, NaI, Zn, $(CH_3)_6Sn_2$, and $Pd[(C_6H_5)_3P]_4$ to effect the biaryl coupling, but none of these generated the desired fragment, **1458**. Pattenden and co-workers[204] observed a similar problem in their approach to a B-C-D-F-G-H model fragment.

In an alternative strategy, Moody's group[94] prepared a highly substituted 4-aryltryptamine precursor **1463**, which contained the C-D-F-G rings and was suitably functionalized to prepare the B ring via macrolactamization. In addition, a methoxy group was incorporated into the G ring for later functionalization. The authors effected an early stage intermolecular biaryl coupling to prepare a C-D-G fragment, which was a scaffold for the remaining synthetic steps.

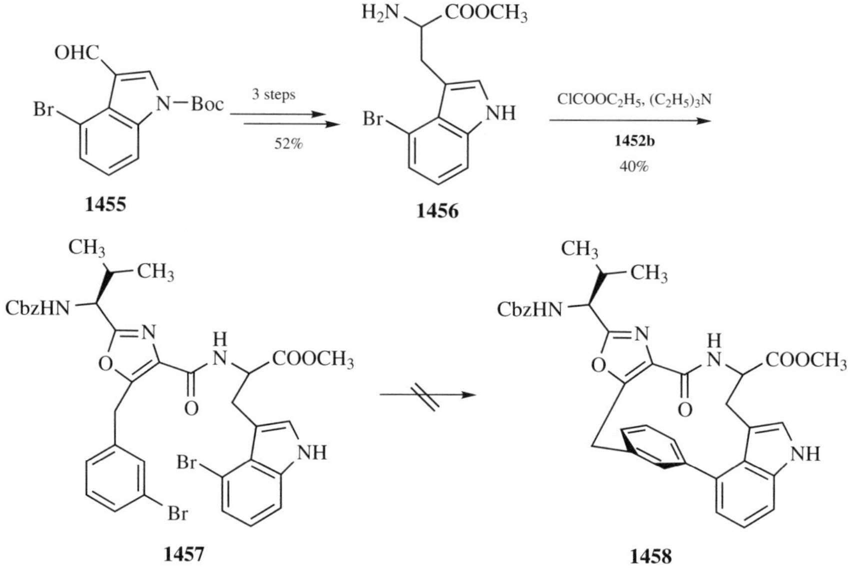

Scheme 1.373

The requisite bis-Boc-protected 4-bromotryptamine **1459** was available in four steps from 4-bromoindole (Scheme 1.374). The other component of the biaryl coupling, the boronic acid **1460**, was prepared from 2-(allyloxy)-bromobenzene. Suzuki coupling of **1459** and **1460** proceeded in excellent yield to afford **1461**, which was elaborated to the α-diazo-β-ketoester **1462**. It required some experimentation to prepare **1462**. Dirhodium(II) acetate–catalyzed N-H insertion of **1462** with (S)-benzyloxycarbonylvalinamide, followed by cyclodehydration, gave **1463**, albeit in low overall yield (13%, two steps). Nonetheless, the authors had prepared a highly functionalized 4-aryltryptamine analog that could serve as a precursor to construct the B, E, and H rings.

Konopelski and co-workers[521] also used an α-diazo-β-ketoester to prepare a 2-chloroindole C-D-E fragment **1467** in a model study for **1192** (Scheme 1.375). They converted commercially available isatin to N-Boc-2-chloro-3-indolecarboxaldehyde **1464** in three steps. The β-ketoester **1465** was prepared in high yield and then converted to the α-diazo-β-ketoester **1466**. The authors found that standard reaction conditions failed to convert **1465** to **1466**. In this case, substitution of DBU for TEA was critical for success. The oxazole ring was then constructed using Doyle's protocol.[532]

Pattenden and co-workers[204] described a variety of synthetic strategies, including Suzuki, Heck, and Stille sp^2-sp^2 coupling protocols to prepare key heterocyclic fragments leading to **1192a** (Scheme 1.376). For example, the F-G-H benzofuranyl oxazole fragment **513** was prepared via an in situ Hantzch cyclization[205] as a model for the quaternary center of **1192a** (see Scheme 1.138). The authors described a facile Stille coupling of N-tosyl-3-(tributylstannyl)indole **1468** with 5-bromo-2-phenyloxazole **63** to afford the C-D-E model fragment **1469**. Acylation of

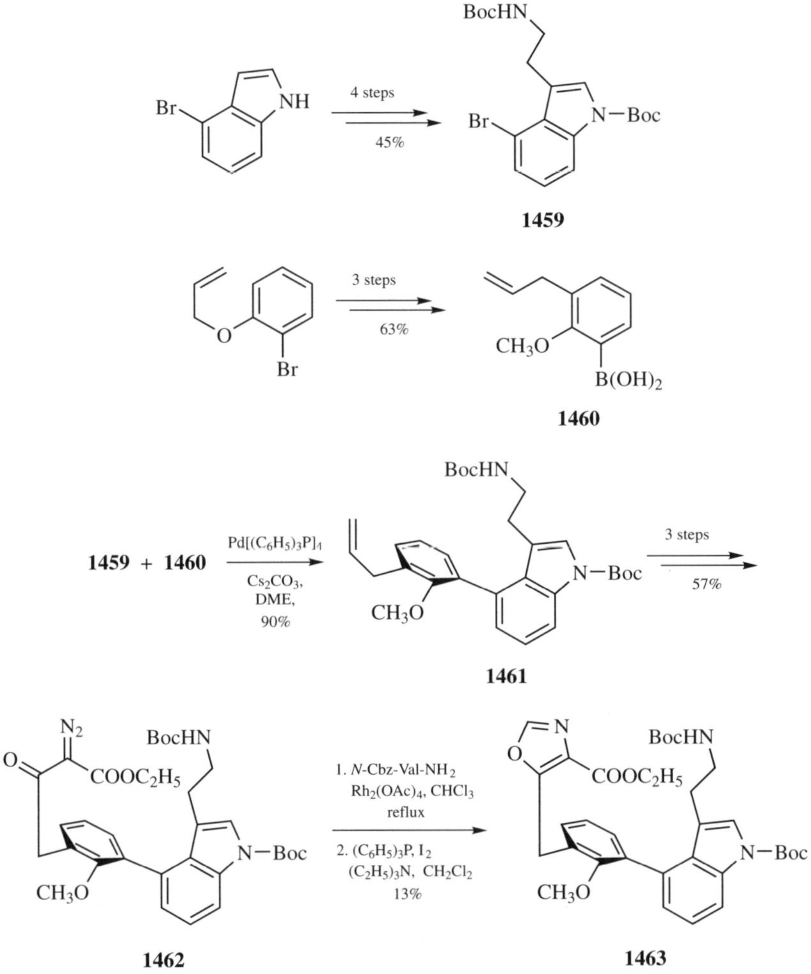

Scheme 1.374

N-tosyltryptamine with 2-methyl-5-oxazolecarbonyl chloride generated **1470a**, which was cyclized to the C-D-E-F indole bis-oxazole model fragment **1471** via the *N*-acylamino ketone **1470b**. Finally, the authors noted they were unable to effect an intramolecular Ullmann coupling of **1472** to prepare the B-C-D-F-G-H model fragment **1473**.

Wipf and Yokokawa[522] also employed both Stille and intramolecular Heck couplings to generate advanced intermediates leading to **1192a**. In particular, these authors prepared model C-D-E, **1476** and C-D-E-G-H, **1480** fragments (Schemes 1.377 and 1.378).

For **1476**, tryptamine was readily converted to the iodoindole **1474** in four steps. Conversion of **1474** to the requisite iodoindolyloxazole precursor **1475** was

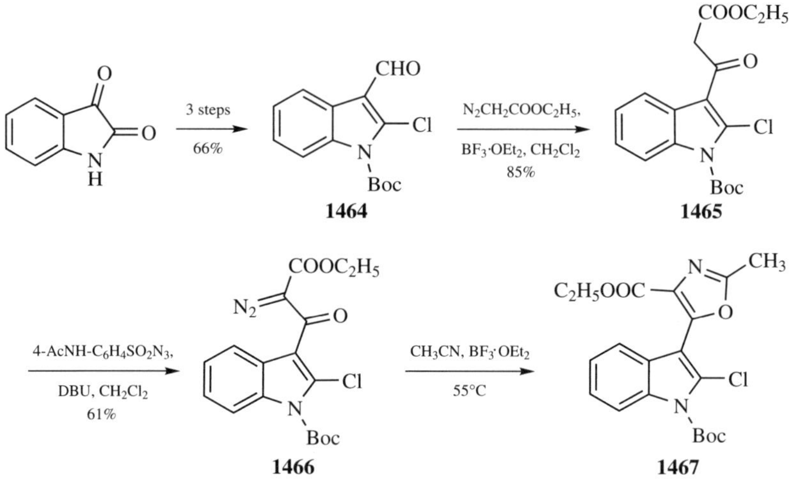

Scheme 1.375

achieved in two steps. Cyclodehydration of **1475** and protection of the indole nitrogen then completed the synthesis of **1476**.

The authors then incorporated **1476** into a more advanced C-D-E-G-H fragment **1480** (Scheme 1.378). Stille coupling of **1476** with **1477** required considerable experimentation, including both CuI and $(C_6H_5)_3As$ as co-catalysts. After chromatographic purification, **1478** and recovered **1476** were isolated in a 1:1 ratio. There was no improvement in yield using longer reaction times or the trimethyltin analog of **1477**. Nonetheless, **1478** was elaborated in two steps to the C-D-E-G-H fragment, **1480**. A modest asymmetric induction at the quaternary center of **1480** was noted using (R)-BINAP. The authors also simultaneously introduced chlorine atoms in the indole and oxazole rings of **1476**.

Later studies from Wipf's group[523] described a direct and efficient synthesis of polyoxazoles based on a Chan-type rearrangement of imides. Application of this rearrangement to prepare a C-D-E-F indole bis-oxazole fragment is shown in Scheme 1.379, p. 333. The starting amino ketone **1482**, prepared from indole, was converted to the C-D-E fragment **1483** in a manner analogous to that used to prepare **1476**. Cleavage of the Cbz group, coupling the resulting amine with 2,2-diphenylpropionic acid and imide formation with Boc_2O, generated **1484**, the precursor for the expected rearrangement. Treatment of **1484** with LDA cleanly produced **1485** in 78% yield. Elaboration of **1485** to the target C-D-E-F indole bis-oxazole fragment **1486** was uneventful. The enantiomeric purity of **1486** was determined to be > 90% based on analysis of the corresponding Mosher amides.[533] The authors have extended this methodology iteratively to prepare polyoxazoles such as **1487**.

Magnus and McIver[524] recently described a synthesis of a C-D-E-F indole bis-oxazole fragment **1490** that also incorporates both chlorine atoms. Their approach was based on Wipf and Yokokawa's[522] earlier observation regarding chlorination of

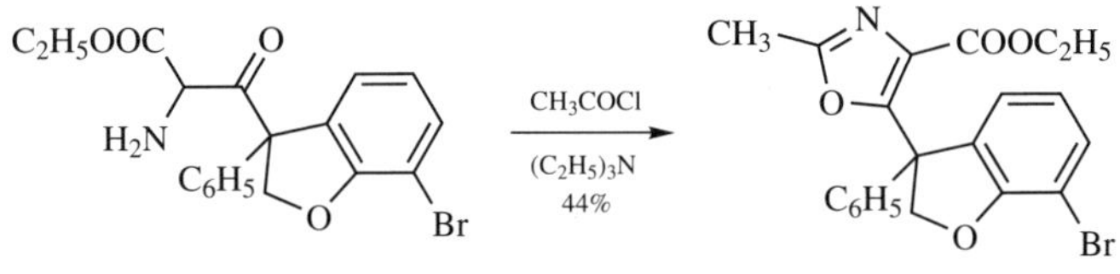

512 **513**

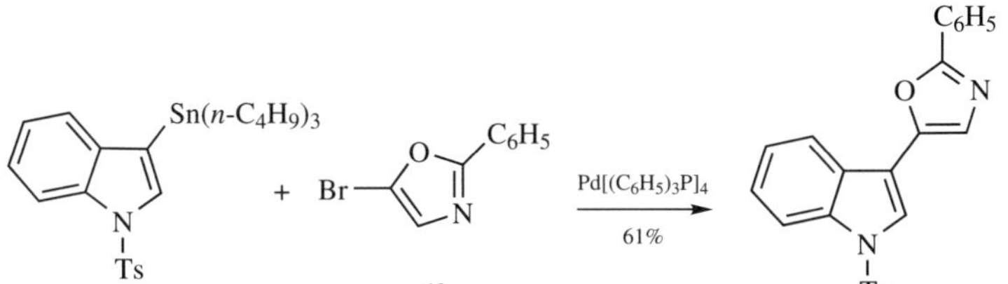

1468 **63** **1469**

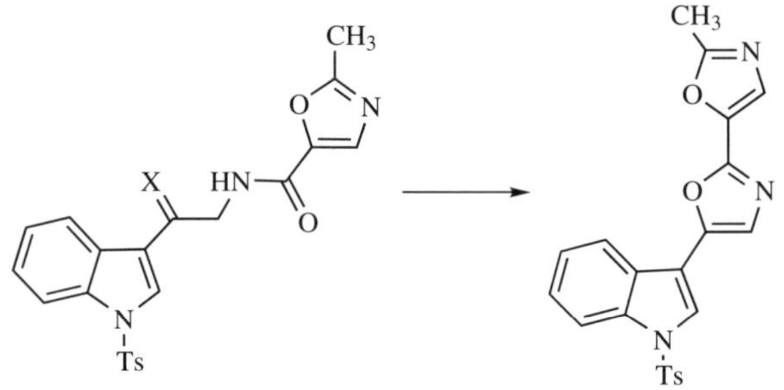

1470a X = H, H
1470b X = O
DDQ

1471

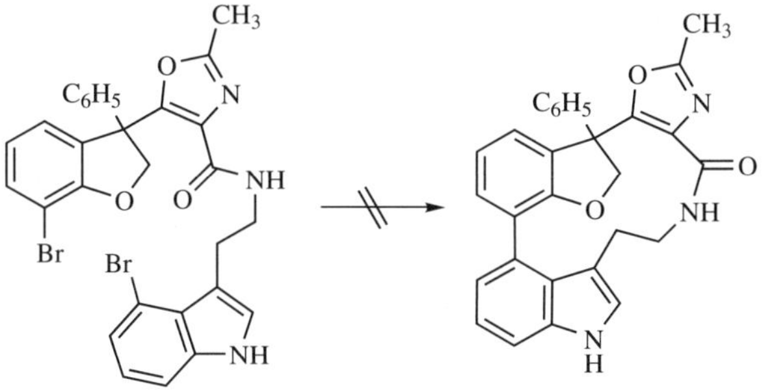

1472 **1473**

Scheme 1.376

1476. The authors envisioned that chlorine could be introduced into an appropriately substituted indolyl bis-oxazole by a chlorodecarboxylation reaction (Scheme 1.380, p. 334). Model systems to test this hypothesis were then evaluated.

The valine-derived-4-oxazolecarboxylic acid **1491** was coupled with tryptophan methyl ester to generate an intermediate that underwent oxidation and cyclodehydration with DDQ to afford the C-D-E-F indole bis-oxazole fragment

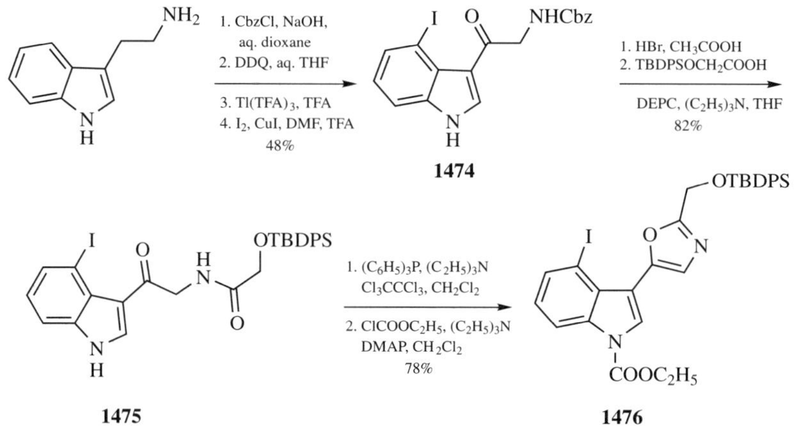

Scheme 1.377

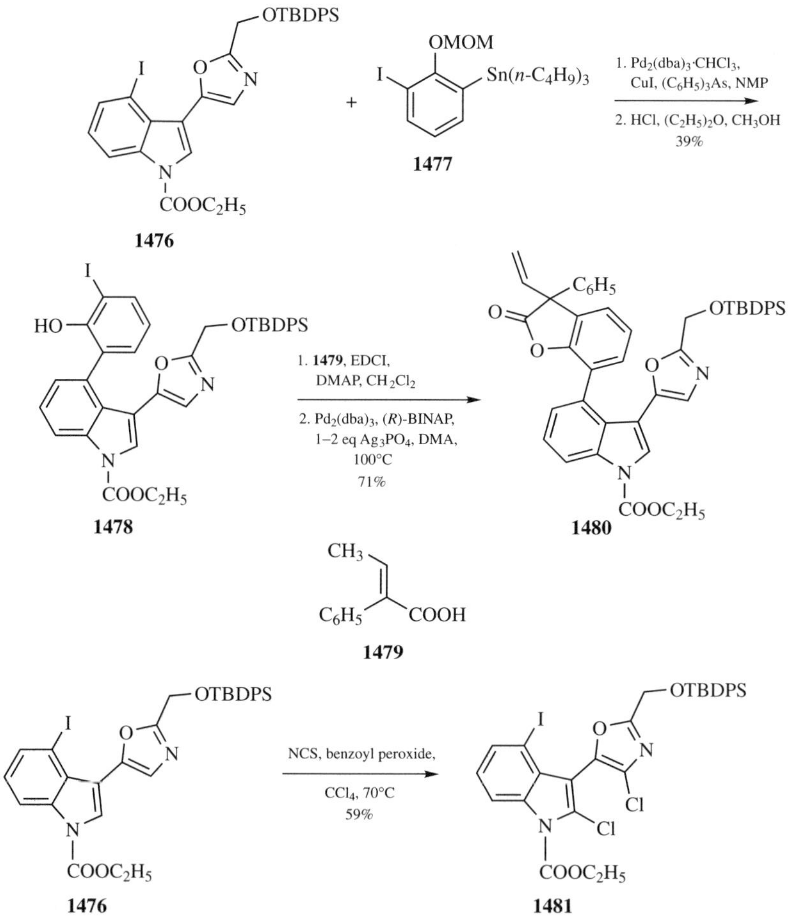

Scheme 1.378

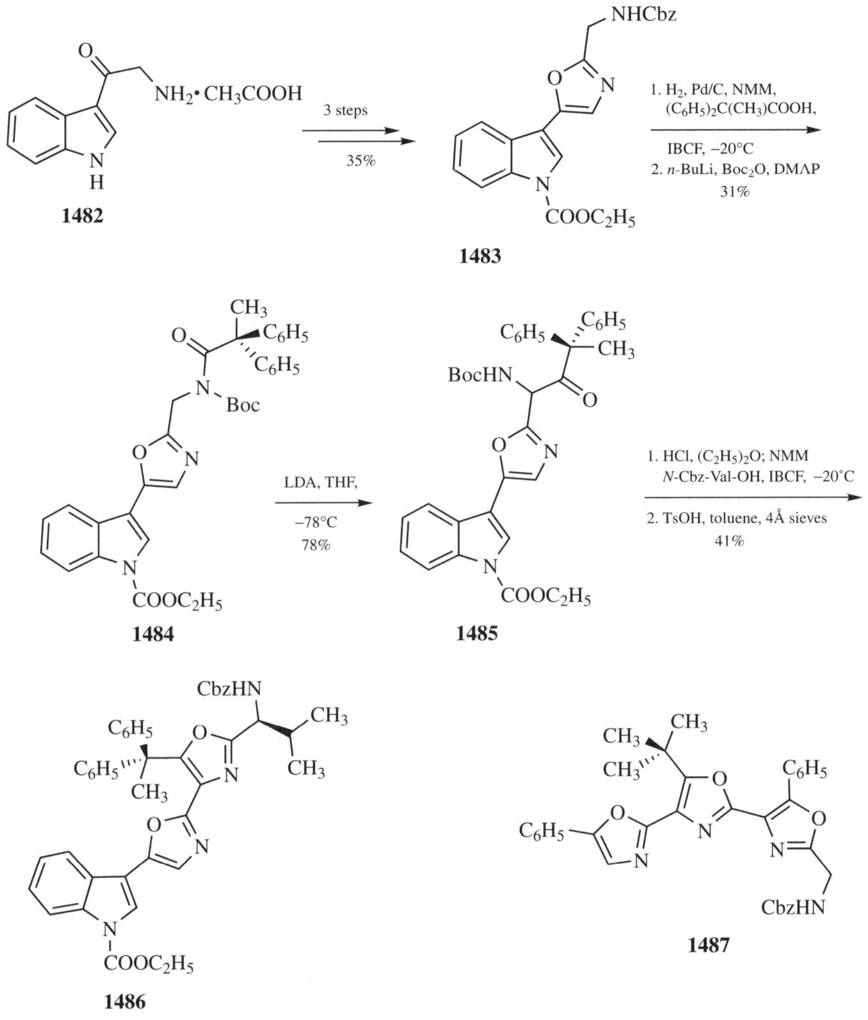

Oxazole Natural Products 333

Scheme 1.379

1492a (Scheme 1.381). The authors noted that this DDQ dehydrogenation/cyclodehydration sequence does not work well on tryptamine derived substrates. Chlorination of **1492a** with NCS/CCl$_4$ afforded **1493a** in excellent yield. However, all attempts to saponify **1493a** and effect decarboxylation resulted in extensive decomposition.

At this point, the authors reversed the sequence of reactions. Typical saponification conditions for **1492a** led to extensive decomposition. Ultimately, prolonged exposure of **1492a** to KOSi(CH$_3$)$_3$ in aqueous THF yielded **1492b**. This crude material was chlorinated as before to afford the trichloride **1493b**, albeit in only 5% yield. Reductive dechlorination of **1493b** then produced the target **1494** in excellent

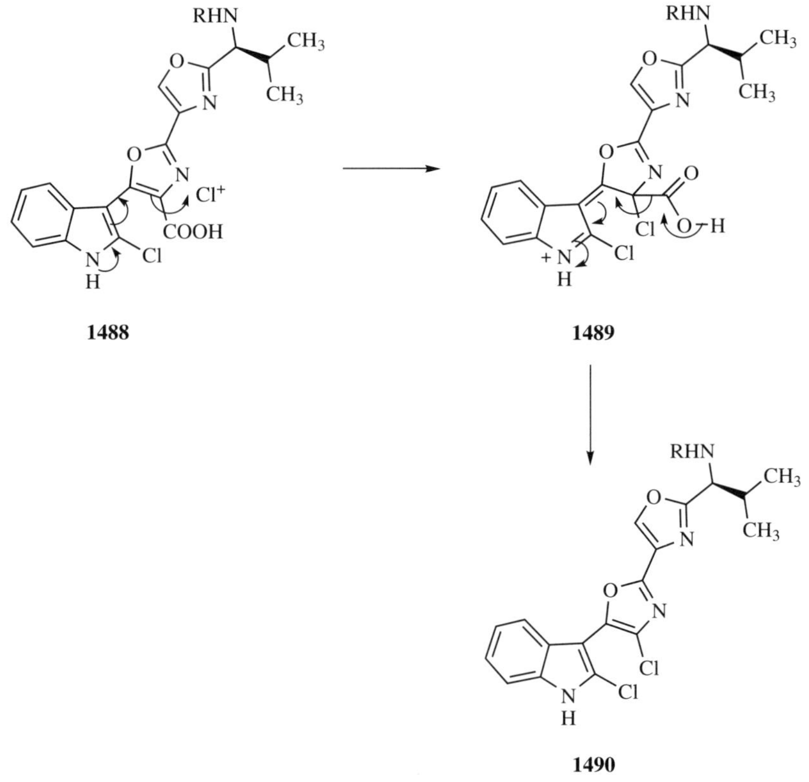

334 Synthesis and Reactions of Oxazoles

Scheme 1.380

yield, but the poor yield for the chlorodecarboxylation step mitigated against further development of this approach.

They repeated the reaction sequence with tryptamine to circumvent the problems associated with saponification and decarboxylation. Thus **1495a** was prepared from **1491** (Scheme 1.382). Chlorination of **1495a** then generated a 17:1 mixture of **1494** and **1493b** in 91% yield. Deprotection of **1494** produced the C-D-E-F dichloro containing indole bis-oxazole fragment **1490** in excellent yield.

Protection of the indole nitrogen after DDQ oxidation followed by cyclization as before gave **1495b** in 69% yield from which **1495a** was isolated in 89% yield after deprotection. This three-step sequence was a substantial improvement and eliminated the variability encountered previously with the dehydration step.

Vedejs and Barda[525] adapted Schollköpf's methodology[154] and prepared an indole-oxazole C-D-E model fragment, which was subsequently incorporated into a more advanced C-D-E-G-I fragment. 4-Benzyloxyindole was converted to **1496** in four steps and excellent overall yield (Scheme 1.383, p. 337). Treatment of **1496** with lithiomethyl isocyanide produced a 2:1 mixture of the acylisocyanide (not shown) and **1497**. Reaction of this mixture with PPTS cleanly provided the C-D-E model fragment, **1497**. The triflate **1497a** was then prepared in three straightforward steps.

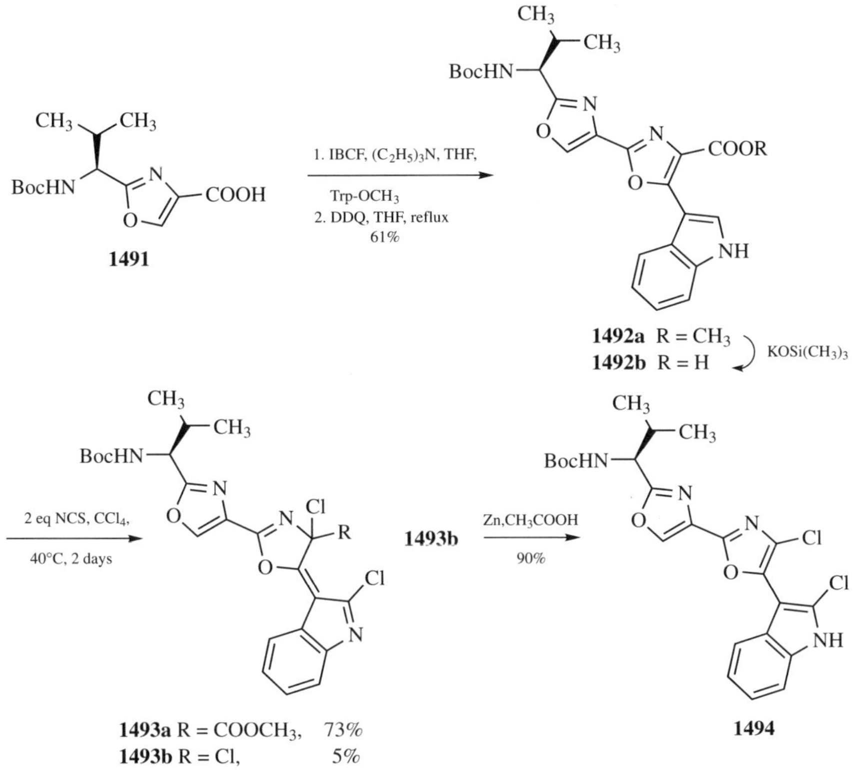

Scheme 1.381

A model Suzuki coupling of **1497a** with **1498** furnished the C-D-E-G fragment **1499** in excellent yield, thereby demonstrating the utility of this synthetic strategy. This was the first example of a Suzuki coupling with an indole-4-triflate.

The authors extended the cross-coupling strategy to **1500** and produced the target C-D-E-G-I fragment **1501**, which then included a model for the benzofuran moiety of **1192a**. The authors noted that **1501** existed as a 1:1 mixture of atropisomers, as shown by NMR experiments. Finally, a method for preparing a 4-chlorooxazole that did not involve an electrophilic reaction was described.

Nicolaou's group[526] described the first synthesis of a B-C-D-E-G-H-I fragment **1510** in atropisomerically pure form, which represents the basic heterocyclic core of **1192a**. These authors focused on assembly of this fragment by olefination methodology and olefin metathesis reactions in conjuction with Suzuki and/or Stille biaryl couplings. Their initial approach to **1510** is shown in Scheme 1.384.

They first prepared the indole-oxazole C-D-E fragment **1506**. The O-TBDPS protected glycolamide **1504** was readily prepared from 4-bromoindole in three steps and 69% overall yield. Oxidation of **1504** to the intermediate N-acylamino ketone (not shown) was best accomplished in two steps using DDQ followed by IBX, prepared in situ from oxone and 2-iodobenzoic acid. Following cyclization

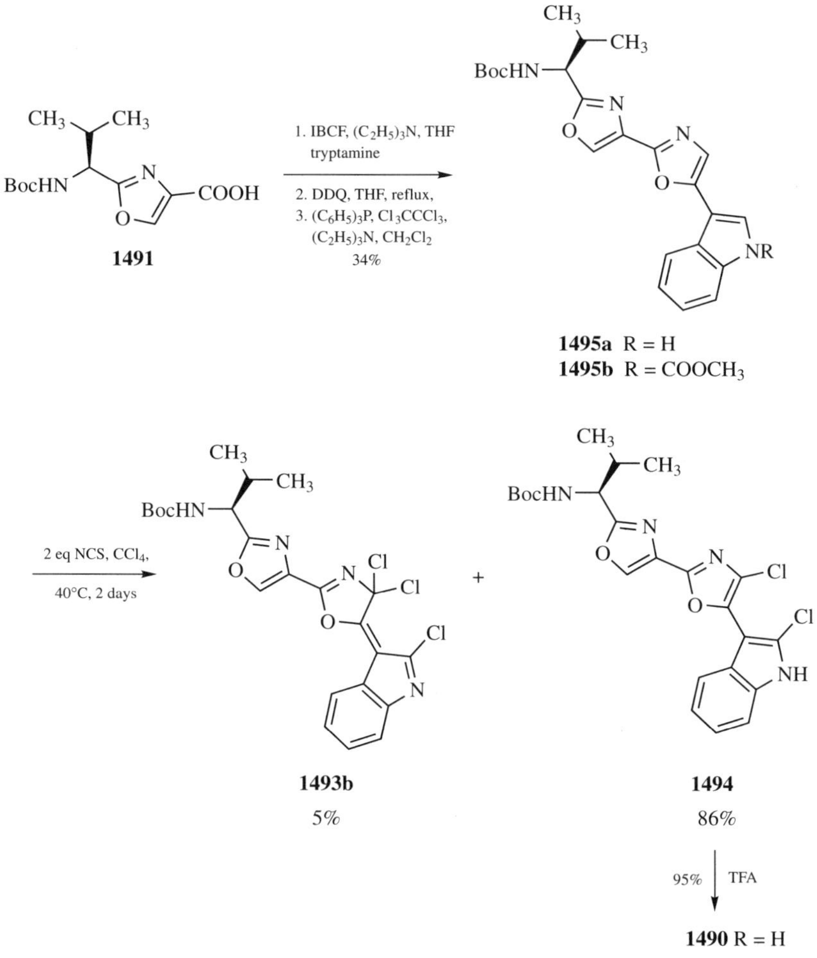

336   Synthesis and Reactions of Oxazoles

Scheme 1.382

and methylation, **1505** was isolated in 79% yield from **1504**. Conversion of **1505** to the 2-vinyloxazole **1506** was straightforward.

Biaryl coupling of **1506** with **1507** was not trivial. Standard conditions for preparing a boronic acid of **1506** or **1507** failed. However, a boronate ester was prepared from **1507** and bis(pinocolato)diboron **1508** and then coupled with **1506**. This afforded the penultimate precursor **1509**, a C-D-E-G-H-I fragment. Olefin metathesis of **1509** was expected to yield **1510**, but the authors were unable to define conditions to effect this transformation. Instead, they encountered either decomposition or recovered **1509** intact. This was attributed to steric hindrance about the double bonds and necessitated that they modify their synthetic plan accordingly.

In their second approach to **1510**, the authors proposed an intramolecular Horner-Wadsworth-Emmons reaction[527] to construct the critical carbon–carbon

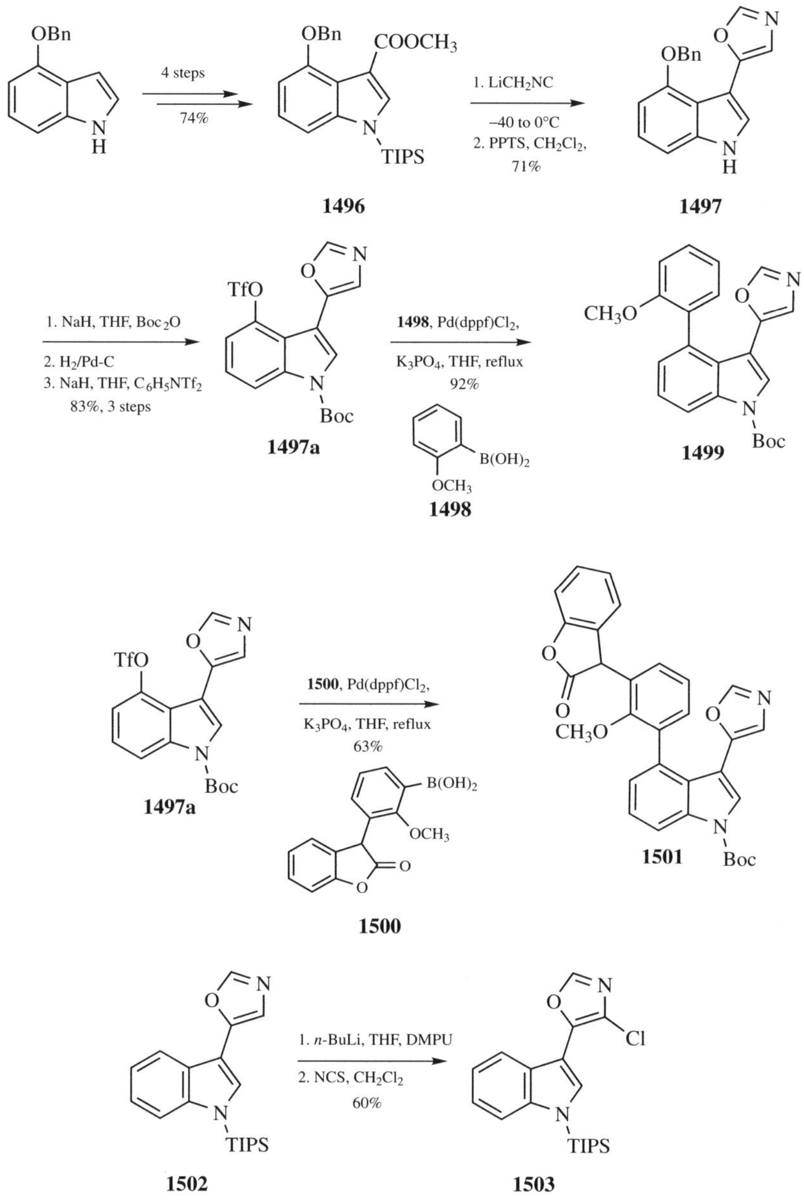

Scheme 1.383

double bond. To that end, **1505** was converted to the phosphonate **1511** in three steps and 74% overall yield (Scheme 1.385). The other coupling partner, **1513**, was prepared from **1512** in two steps. Treatment of **1513** with **1508** afforded an in situ boronate ester (not shown) which coupled with **1511** to afford **1514** in acceptable yield. The authors noted that this was the first example of a Suzuki coupling of a

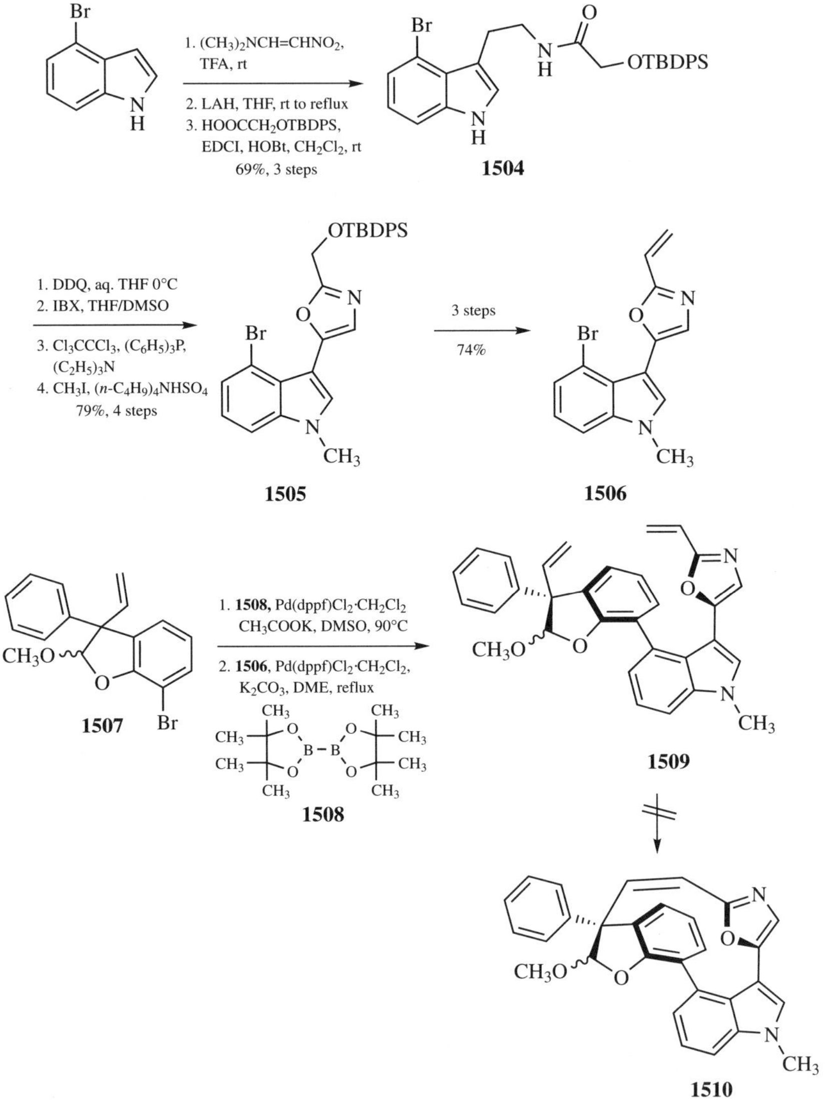

Scheme 1.384

phosphonate. Cleavage of the TBS-protecting group followed by oxidation gave the intermediate aldehyde, which was subjected to intramolecular Horner-Wadsworth-Emmons reaction. Unfortunately, only one diastereomer had reacted, and this afforded a product **1515** with the undesired stereochemistry. The stereochemistry of **1515** was confirmed by multiple NOE interactions. Nonetheless, encouraged by the results, the authors adapted this methodology to their original approach.

Now, the C-D-E phosphonate fragment **1511** was coupled with an in situ-generated boronate ester derived from **1512** and **1508** to afford **1516** as a mixture of four diastereomers (Scheme 1.386). This mixture of diastereomers was then

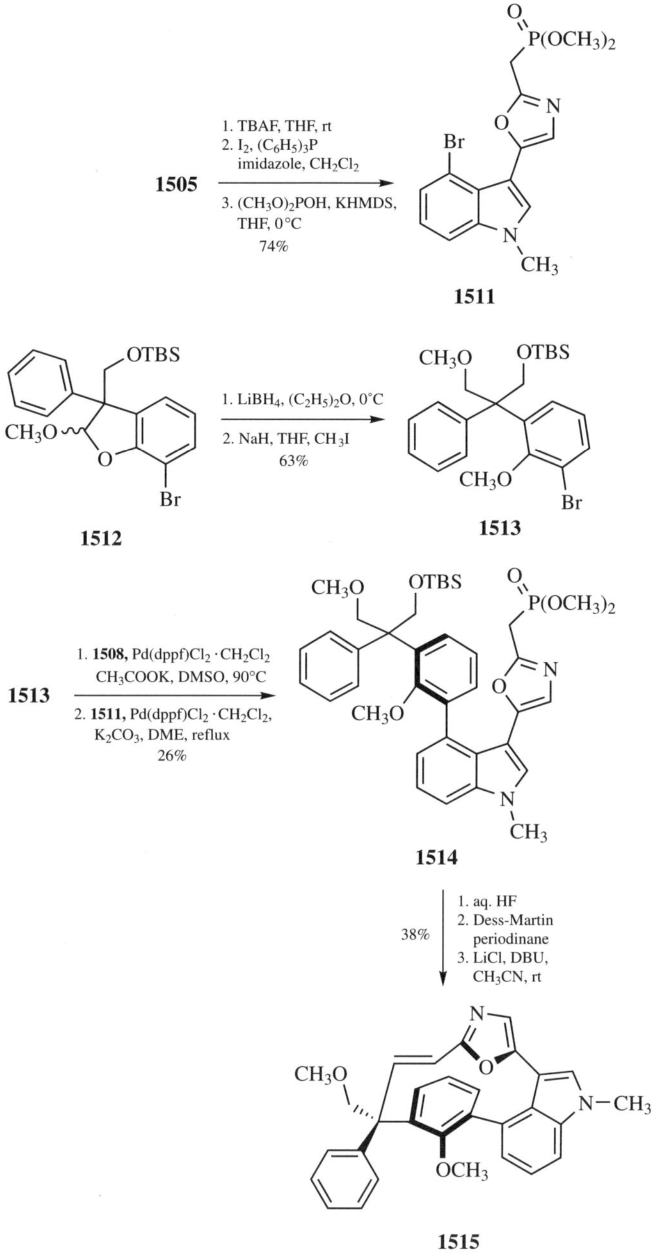

Scheme 1.385

converted to the aldehyde **1517**, requisite to test the viability of the intramolecular Horner-Wadsworth-Emmons reaction. Macrocyclization of **1517** with sodium hydride indeed afforded **1518** as a single diastereomer with the correct stereochemistry needed for **1192a**.

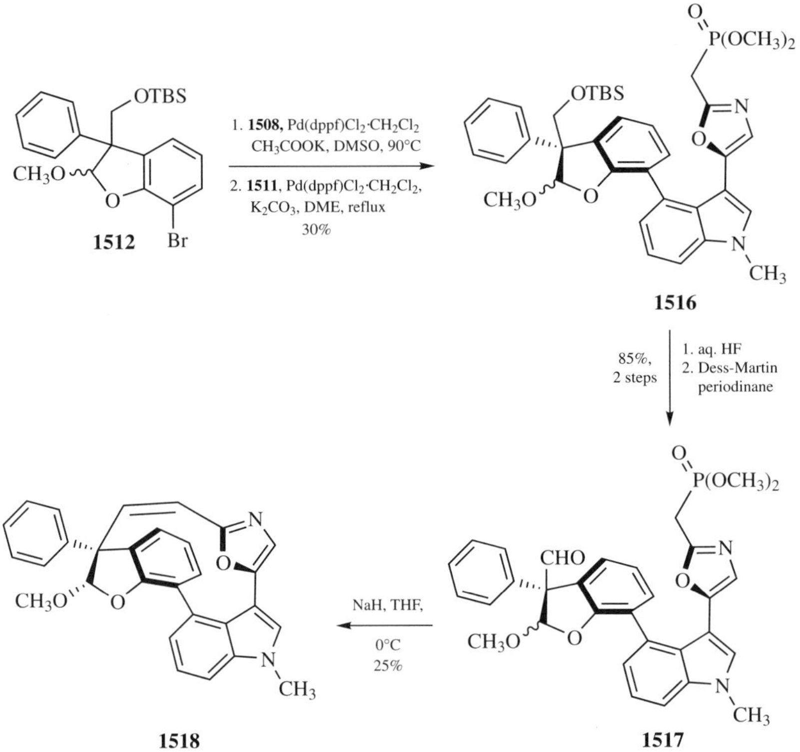

Scheme 1.386

Radspieler and Liebscher[528] prepared several 4-chloro-2,5-disubstituted oxazoles **1520** from aromatic acyl cyanides **1519**. These oxazoles are the first examples in which the 2-substitutent was not derived from an aromatic aldehyde.[529] They extended this methodology to prepare suitably functionalized dichloro indole bis-oxazole C-D-E fragments. Their approach is noteworthy as the first synthesis of such fragments that does not involve chlorination of an intact oxazole ring.

The authors converted oxindole to 2-chloro-3-(cyanocarbonyl)-1-(2,2,2-trichloro-ethoxycarbonyl)indole, **1522** via the acid **1521** (Scheme 1.387). Condensation of **1522** with acrolein or bromoacetaldehyde in the presence of boron trifluoride etherate and HCl gave **1523** and **1524**, in good and modest yields, respectively.

Harran and co-workers[371,530,531] pursued a conceptually novel approach to **1192a**. The authors reasoned that issues of atropisomerism could be avoided if the indole-dihydrobenzofuran biaryl coupling was deferred to the end of the synthesis. Furthermore, they proposed installing the chlorine atoms at a late stage in the synthesis. Finally, they considered that the 12-membered macrocyclic ring of **1526** could be derived from a pinacol ring contraction of a vicinal diol derived from **1527**. Therefore, they targeted the oxazole containing tyrosine-valine cyclopeptide, i.e., an A-F-G-I fragment **1527** as a scaffold from which to construct the remaining heterocyclic core (Scheme 1.388).

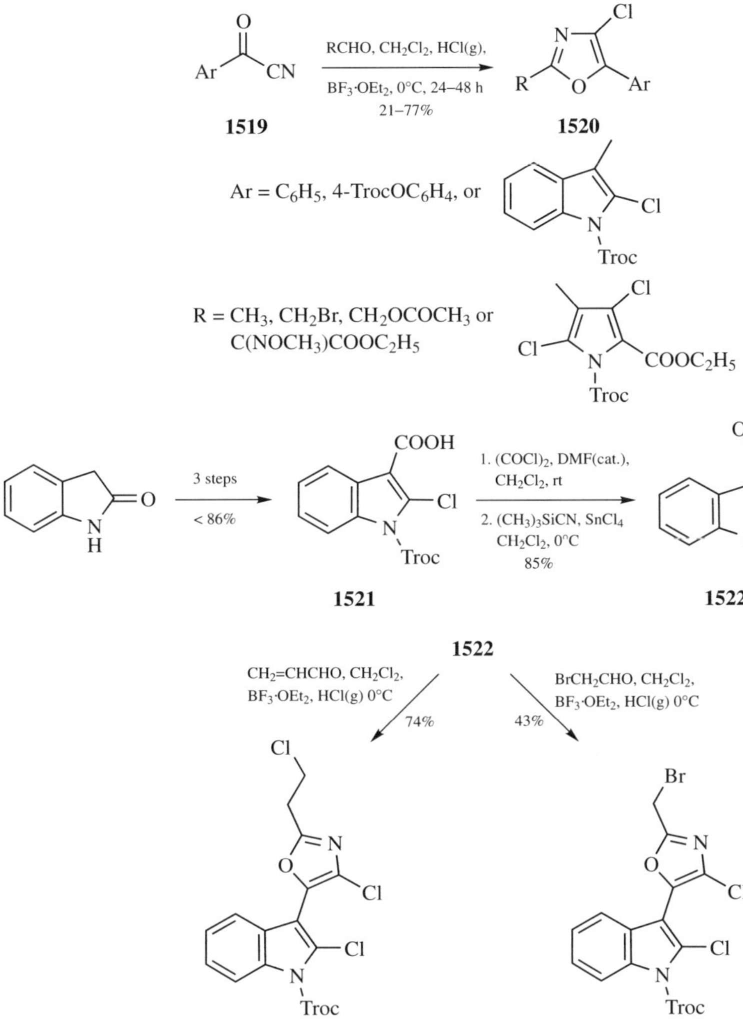

Scheme 1.387

In their early work, the authors[371] prepared the 4-cyanooxazole **988b** by coupling N-Boc-L-valine and aminomalononitrile tosylate (AMNT) to generate the o-aminonitrile **988a**, which was diazotized to furnish **988b** in 35% yield (Scheme 1.389). Coupling **988b** with 3,5-dibromo-N-Boc-L-tyrosine or 3-iodo-N-Boc-L-tyrosine gave **1528** and **991**, respectively. Stille coupling of **1528** or **991** with the bis-stannyl styrene **989b** produced the synthetically undesirable α-styryl stannanes **1529** and **992** in modest yields. At this point, the authors discovered that **989b** could be selectively converted to **989a**, and they modified their strategy accordingly. Then **1528** or **991** was coupled with **989a** to afford **1530a** or **1530b** in good yield. Unfortunately, all attempts to effect Heck cyclization of **1530a** or **1530b** to

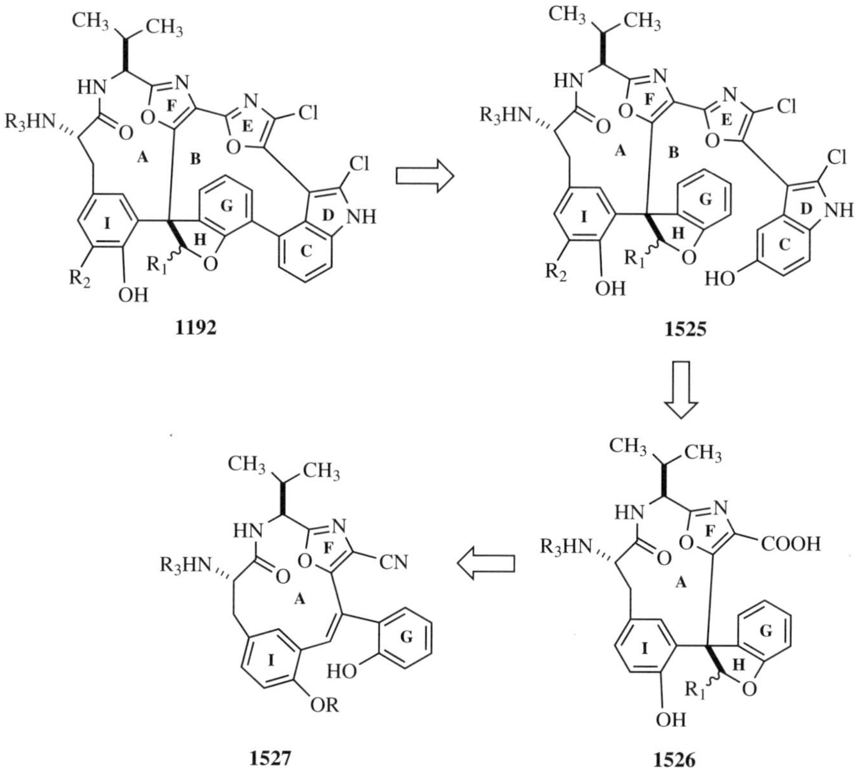

Scheme 1.388

produce an analog of **1527** were unsuccessful. A contributing factor to these results was thought to be the steric crowding present in the G-ring. Therefore, the authors changed the order of the reaction sequence and modified their protecting group strategy.

Cross-coupling **988b** with **989a** and deprotection with BBr$_3$ produced the amino phenol **1531** in excellent yield (Scheme 1.390). Standard peptide coupling of **1531** with a benzyl protected ether of 3,5-dibromo- or 3-iodo-*N*-Boc-L-tyrosine furnished the expected dipeptides **1532a** and **1532b** in 89% and 86% yields, respectively. After considerable experimentation the authors developed effective conditions for Heck cyclization to yield the **1527** derivatives **1533a** and **1533b**. They noted that the anisole analog of **1532b** afforded only 6% of the expected macrocyclic olefin. They interpreted these results to indicate the possible importance of preorganization rather than steric affects on this reaction.

Harran's group[530] later refined and improved the coupling conditions used to prepare **1533b** as part of their continuing efforts leading to **1192a**. Their previous method (vide supra) suffered from limitations derived from incomplete reaction as well as by-product formation at high conversions. These problems were precluded by addition of 2-(di-*tert*-butylphosphanyl)biphenyl[534] to the reaction mixture that

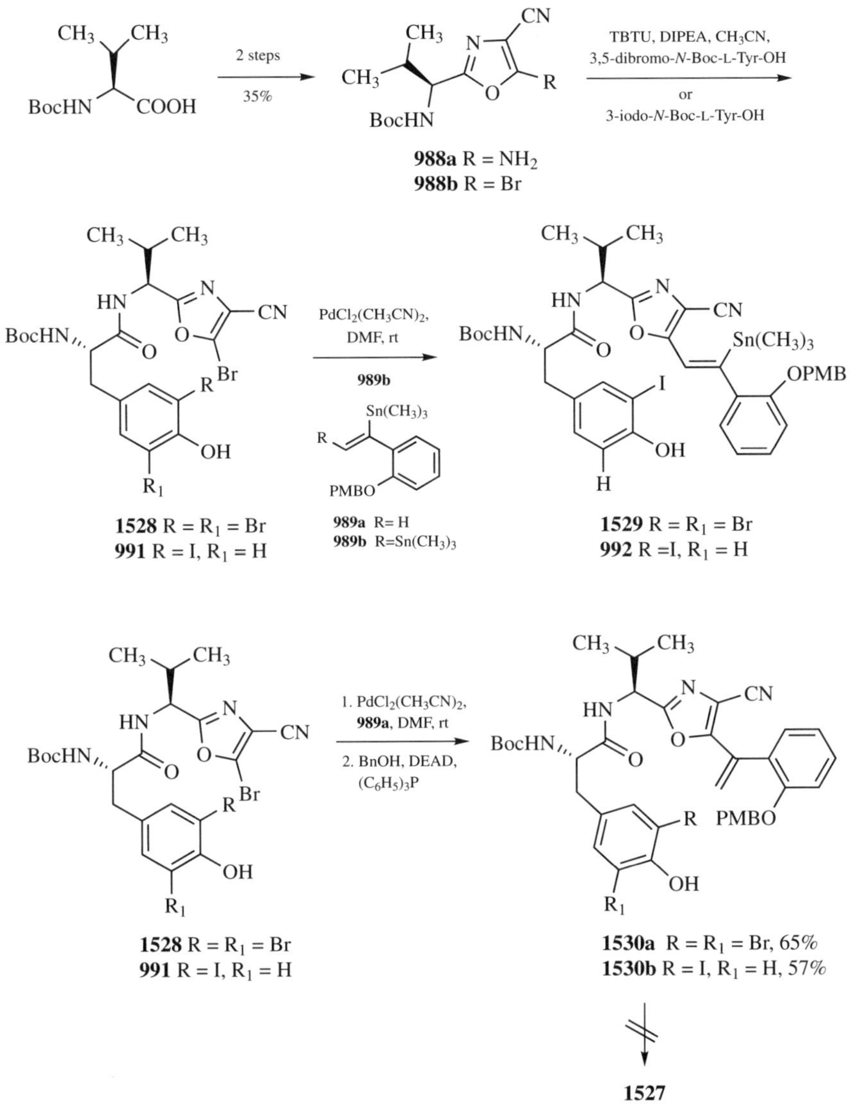

Scheme 1.389

afforded **1533b** in improved yield and consumed less metal (Scheme 1.391). The authors found that all other phosphane additives were inhibitors. The A-F-G-I fragment **1534**, representing the first example of an intact diazonamide triarylacetaldehyde, was then prepared from **1533b** in five steps. Further elaboration of **1534** produced the valerolactone **1535**, a key precursor to more advanced intermediates.

Harran's group[531,535] used **1536** as a scaffold from which to construct the remainder of the heterocyclic core of **1192a** (Scheme 1.392). The triarylacetaldehyde **1537** was prepared from **1536** in three steps. Reductive cleavage of the

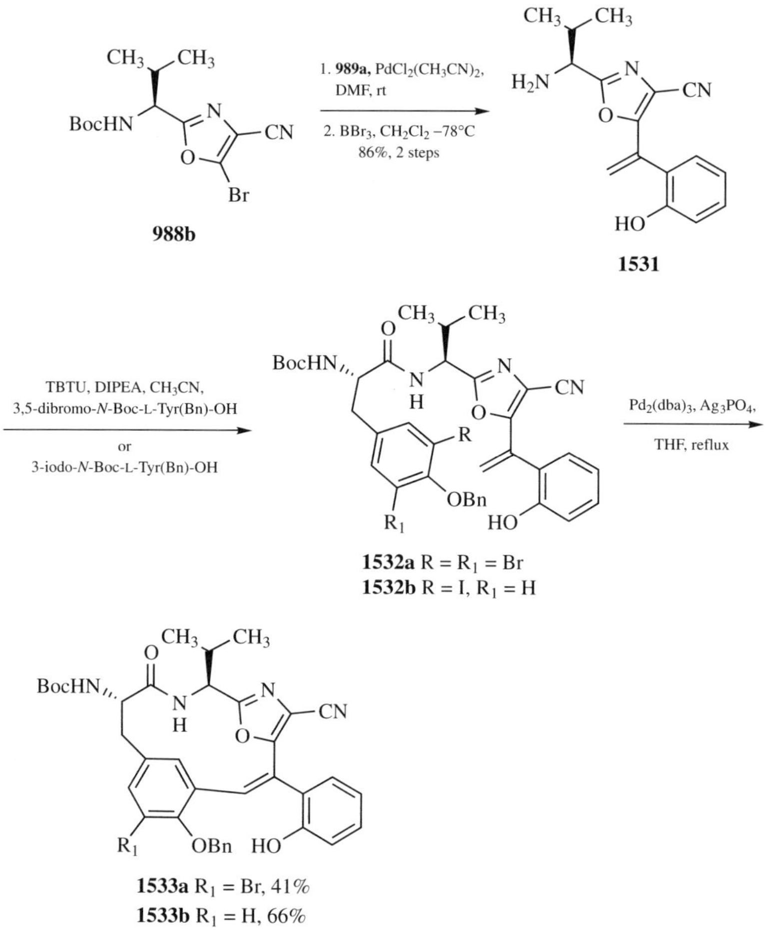

Scheme 1.390

bromoethyl ether of **1537** followed by benzylic oxidation with DDQ gave an intermediate *N*-acylamino ketone (not shown), which was cyclized to the A-C-D-E-F-G-H-I fragment **1538a** in 47% yield. Introduction of the two chlorine atoms was then accomplished using NCS to furnish **1538b** in excellent yield, although the authors noted that care must be taken with the stoichiometry to avoid a trichloroindolenine impurity. This material has been further elaborated to an even more advanced intermediate, **1539**.

1.5.6. Ulapualides

There is a family of marine natural products, isolated recently, that incorporates an unprecedented three contiguous oxazoles in a 25-membered macrocyclic

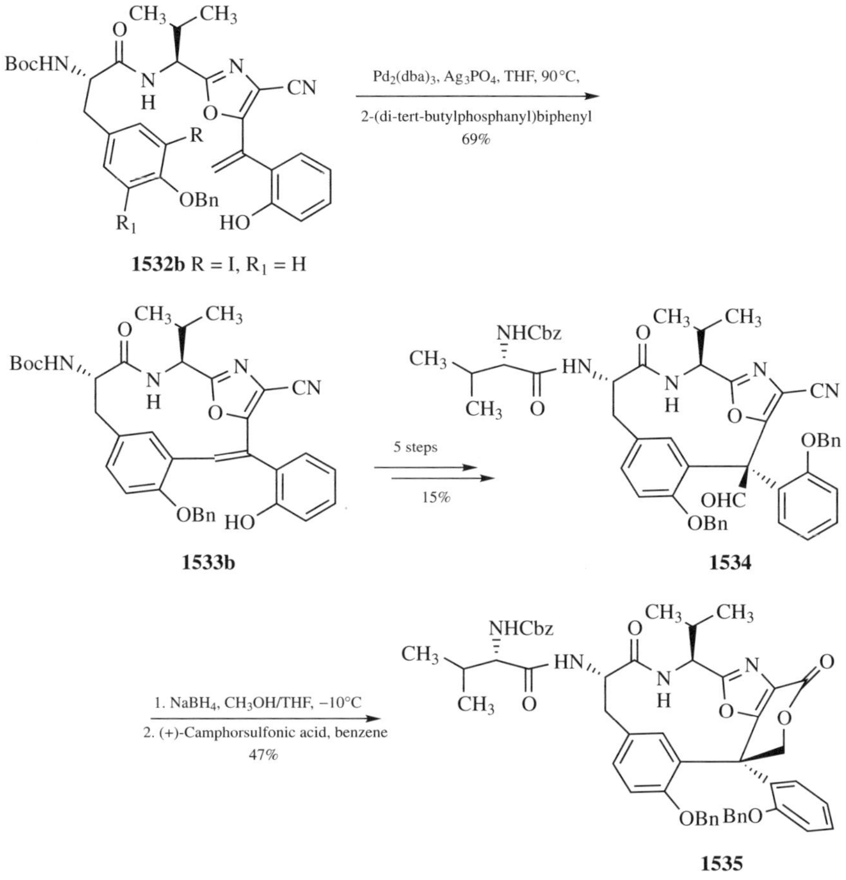

Scheme 1.391

lactone, which is attached to an 11 carbon side chain terminated by an *N*-methyl-*N*-formylenamine. These secondary metabolites are isolated from nudibranch (sea slugs) egg masses and sponges and include ulapualide A,[451] halichondramides,[452,453] kabiramides,[454] mycalolides,[455] and halishigamides.[536] Members of this family possess interesting biological properties, including antileukemic, antifungal, and ichthyotoxic activites, and these unique structures have attracted considerable interest in the synthetic community. Selected examples of syntheses of the tris-oxazole core of ulapualide A **1195** are described in this section (Fig. 1.101).

Pattenden and co-workers[537,538] described the first synthesis of a 2,4-disubstituted tris-oxazole **1542** (Scheme 1.393). Their approach was patterned on a probable biogentic pathway derived from three serine residues. Condensation of L-serine ethyl ester hydrochloride with ethylacetimidate hydrochloride gave ethyl-4,5-dihydro-2-methyl-4-oxazolecarboxylic acid **33**, which was oxidized to 2-methyl-4-oxazolecarboxylic acid ethyl ester **34**. Saponification, conversion to the

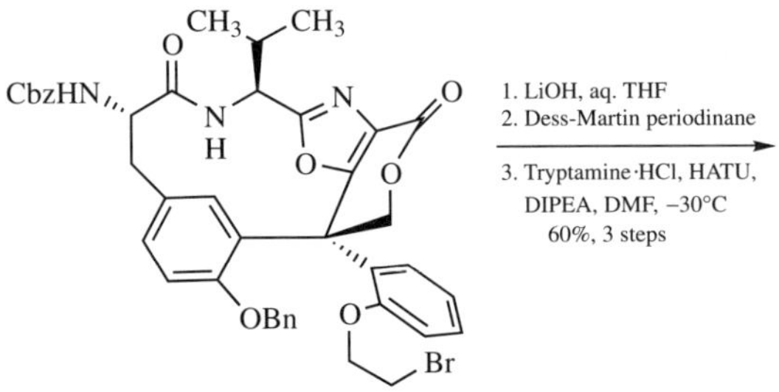

1536

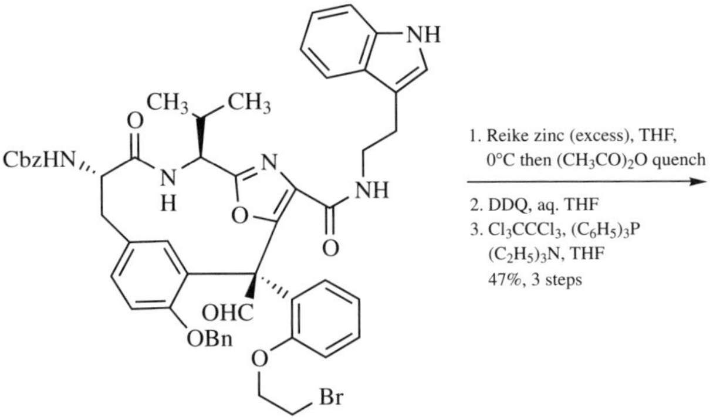

1537

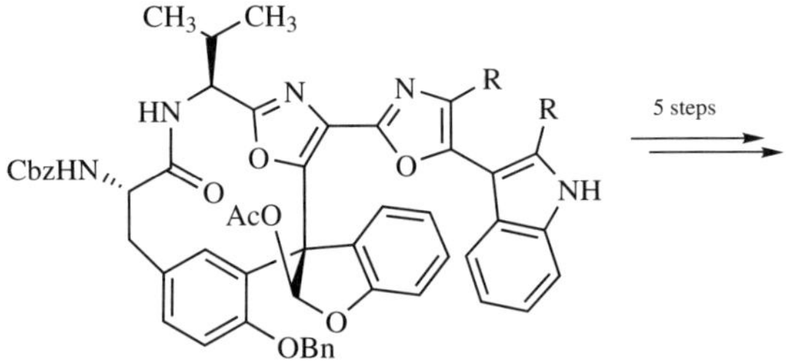

1538a R = H
1538b R = Cl

NCS, CCl_4/THF 85%

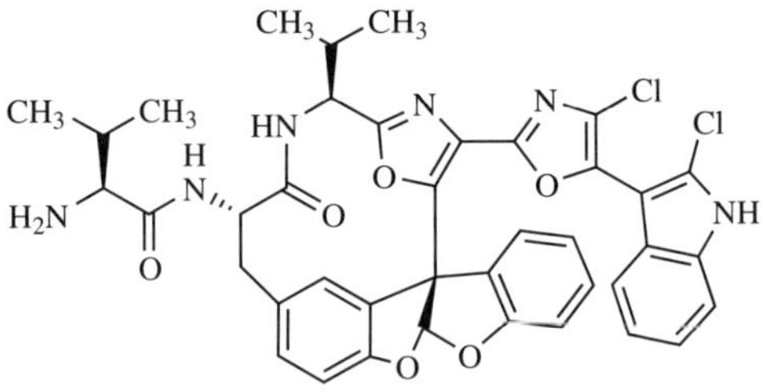

1539

Scheme 1.392

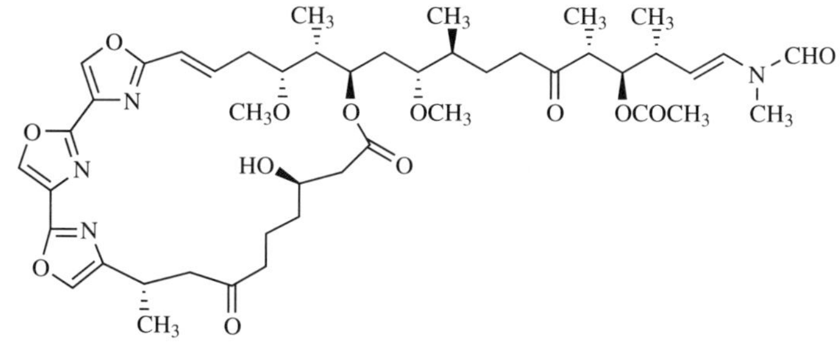

Figure 1.101. Ulapualide A.

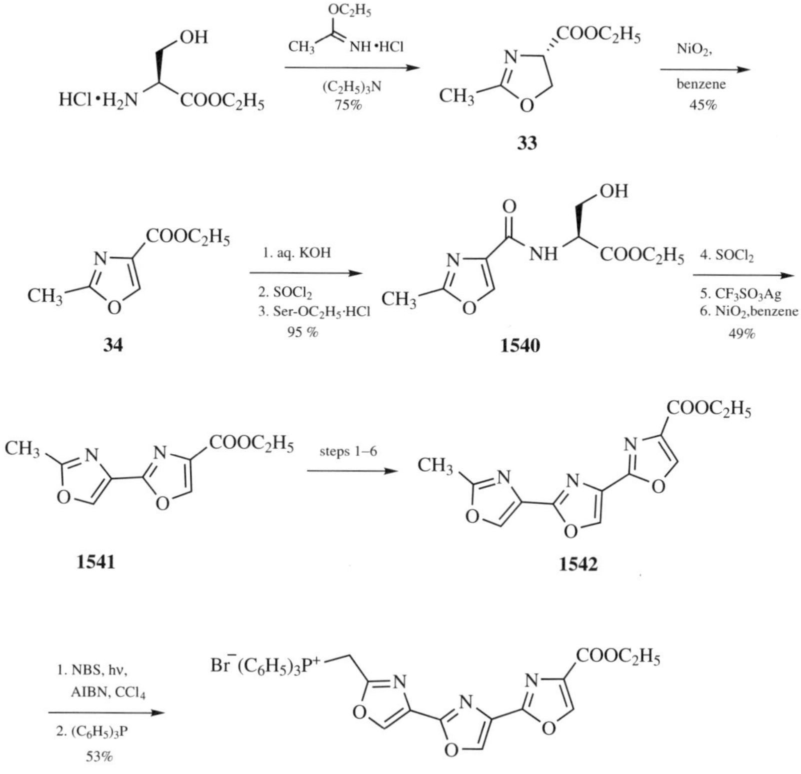

Scheme 1.393

acid chloride, and condensation with a second molecule of L-serine ethyl ester hydrochloride gave **1540**, which was cyclized and oxidized to the 2,4′-bis-oxazole **1541**. Iteration of this reaction sequence a third time then gave the desired trisoxazole **1542**, which was elaborated to the Wittig reagent **1543** uneventfully. Pattenden[538] reviewed and summarized his group's early work in this area.

Yoo[539] assembled the tris-oxazole **1549** via a series of sequential rhodium-catalyzed [3 + 2] cycloaddition reactions of diazomalonates with nitriles (Scheme 1.394). In his approach, dimethyl diazomalonate was added slowly to the *O*-TBS protected cyanohydrin of pivaldehyde in the presence of Rh$_2$(OAc)$_4$ to generate the 5-methoxy-2-substituted 4-oxazolecarboxylic acid methyl ester **1544**. Reduction of **1544** with LiAlH$_4$ gave the 4-(hydroxymethyl)oxazole **1545**, presumably via a simultaneous reduction of both the ester and methoxy group. Doyle and Moody[80] also cleaved 5-methoxyoxazoles using LAH; Helquist and co-workers[540] successfully cleaved 5-methoxyoxazoles (e.g., **129**) using Li(C$_2$H$_5$)$_3$BH.

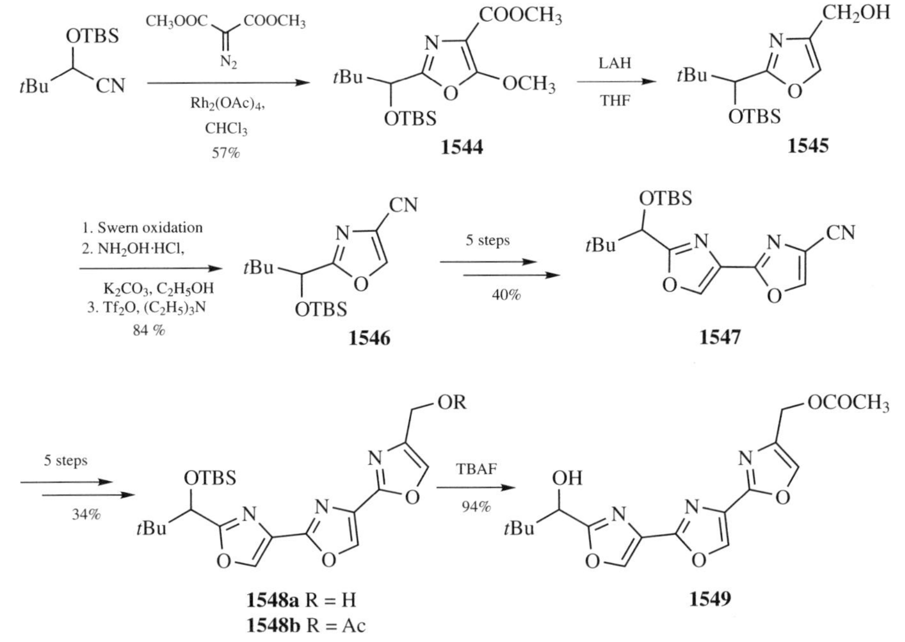

Scheme 1.394

Conversion of **1545** to the 4-cyanooxazole **1546** occurred uneventfully via a three-step sequence. Repetition of the entire five-step reaction sequence with **1546** gave the bis-oxazole, **1547** from which the tris-oxazole, **1548a** was prepared by a third five-step iteration. Finally, acetylation of **1548a** and cleavage of the *O*-TBS group afforded the desired target tris-oxazole **1549**, which was then suitably functionalized for further elaboration to **1195**.

Panek and co-workers[106–110] described syntheses of suitably bis-functionalized tris-oxazoles as intermediates leading to kabiramide C, halichondramide, and ulapualide A based on sequential application of a modified Hantzsch condensation methodology (Scheme 1.395). For example, condensation of cinnamide with ethyl bromopyruvate followed by cyclodehydation gave the starting material, 2-styryl-4-oxazolecarboxylic acid ethyl ester **1550** in excellent yield.[107] The authors noted that alkyl- and α-alkoxyamides were not as effective in this reaction sequence. Ammonolysis of **1550** gave an intermediate amide (not shown), which was elaborated to the bis-oxazole **1551** by repetition of the Hantzsch reaction sequence.

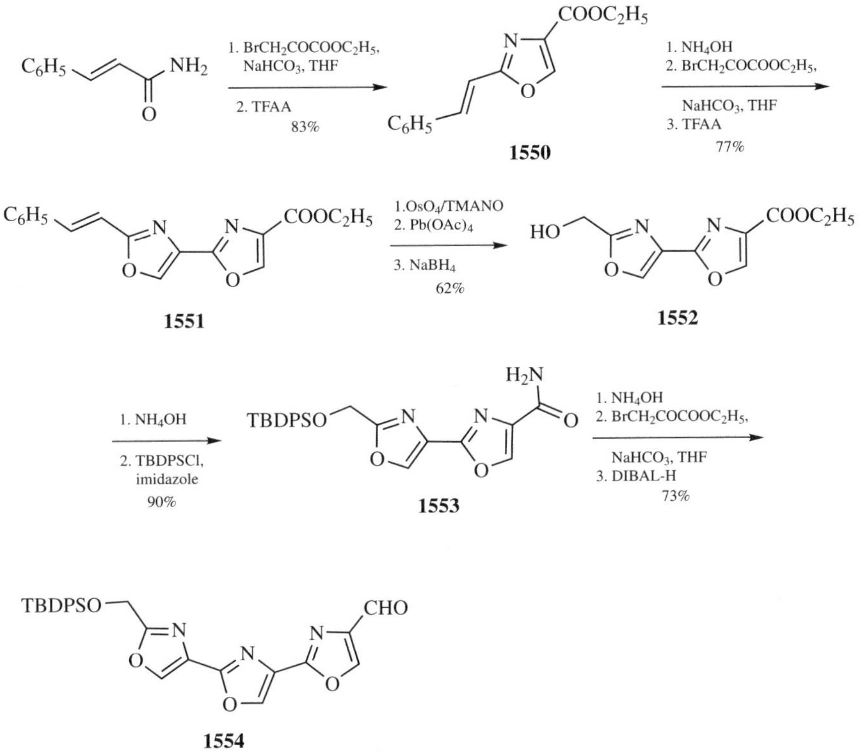

Scheme 1.395

At this point, the 2-styryl group was converted to the corresponding aldehyde via a two-step hydroxylation-oxidation sequence followed by reduction to the alcohol **1552**. The timing for unmasking the primary alcohol was critical to the success of their strategy. The overall synthesis of the tris-oxazole was less efficient when the primary alcohol was unmasked at the monooxazole stage, whereas solubility issues were encountered when this reaction sequence was attempted on the fully elaborated tris-oxazole.

Amidation of the ester and protection of the alcohol gave **1553**, which was elaborated to the tris-oxazole **1554** via a third modified Hantzsch condensation-cyclodehydration sequence followed by DIBAL-H reduction of the ester.

Panek and co-workers[109] modified this reaction sequence to complete the first asymmetric synthesis of a tris-oxazole **1559**, which incorporated the C(9) stereocenter of **1195** (Scheme 1.396). The bis-oxazole **1556** was readily assembled as described in Scheme 1.395. However, in this case, the styryl group of **1550** was converted to a MOM ether **1555** before elaboration of the second oxazole ring. Coupling **1556** with **1557** and cleavage of the benzyl group gave the bis-oxazole peptide **1558**, which was oxidized and cyclized to **1559** in excellent yield. Liu and Panek[110] described further applications of this methodology in their studies leading to kabiramide C.

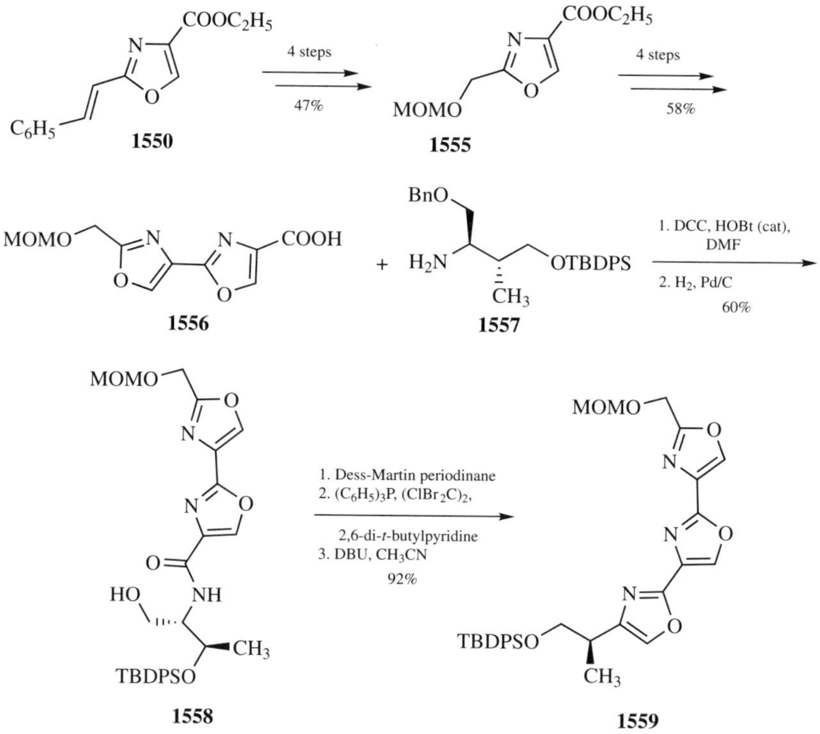

Scheme 1.396

Panek's group[106] employed the Schlosser-Wittig[412,413] olefination reaction of an in situ-generated tris-oxazole phosphonium salt to incorporate the *trans* double bond of an advanced intermediate leading to ulapualide A (Scheme 1.397). Model studies that precedent this work were described in Scheme 1.298. Thus reaction of **1560** with **1561** in the presence of excess triethylphosphine and one equivalent of LDA generated **1562** in excellent yield as a single double-bond isomer. This result was noteworthy in that it demonstrated the viability of this type of bond construction in the presence of a potentially labile β-alkoxy substituent.

Pattenden's group[56,541–545] completed and described the total synthesis of **1195**. In their early work,[538] the phosphonium salt **1543** was elaborated sequentially to the (*E,E*)-diene **1566** (Scheme 1.398).

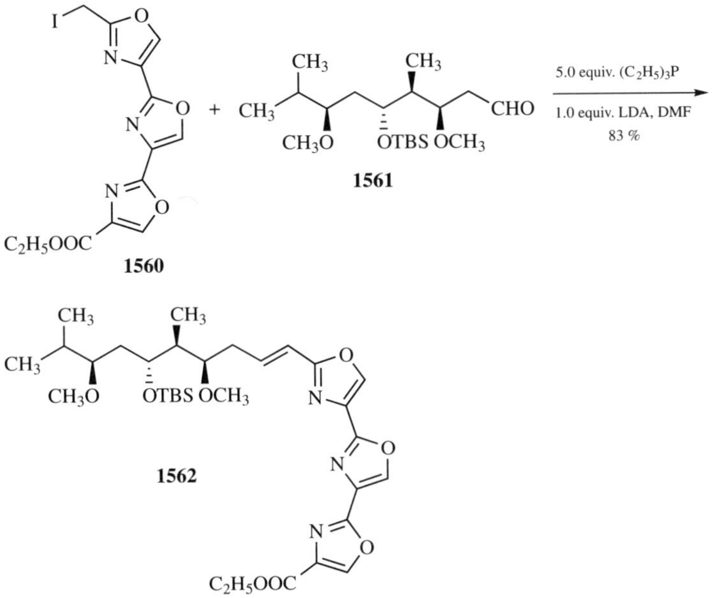

1560 1561 1562

Scheme 1.397

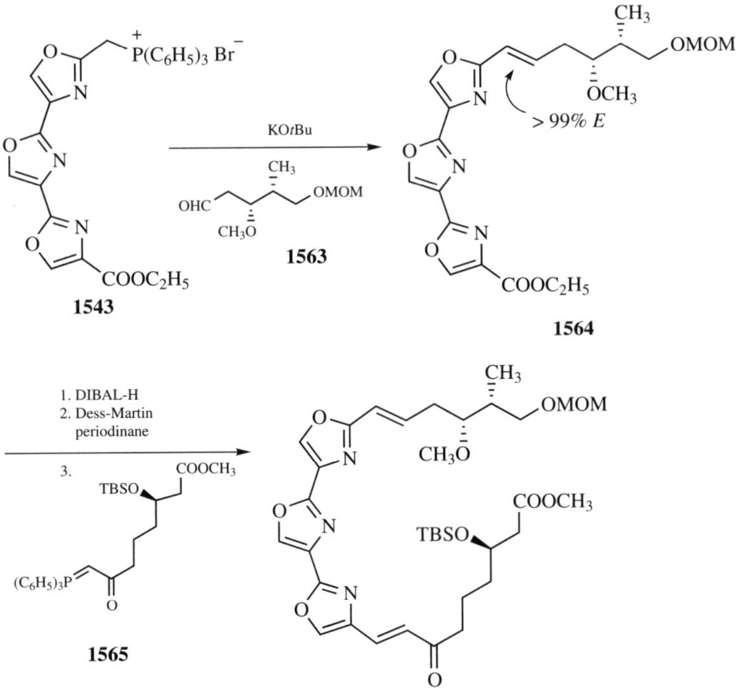

1543 1563 1564 1565 1566

Scheme 1.398

Chattopadhay and Pattenden[541] adopted a similar strategy in an attempt to assemble the entire carbon skeleton of **1195** (Scheme 1.399). Here, the authors prepared **1567** and **1568** independently from **1543**. It was anticipated that a late-stage macrolactonization of **1567** or **1568** would then give rise to entire carbon skeleton. Unfortunately, this strategy was frustrated by the disappointingly low yields (5–10%) encountered during this crucial step.

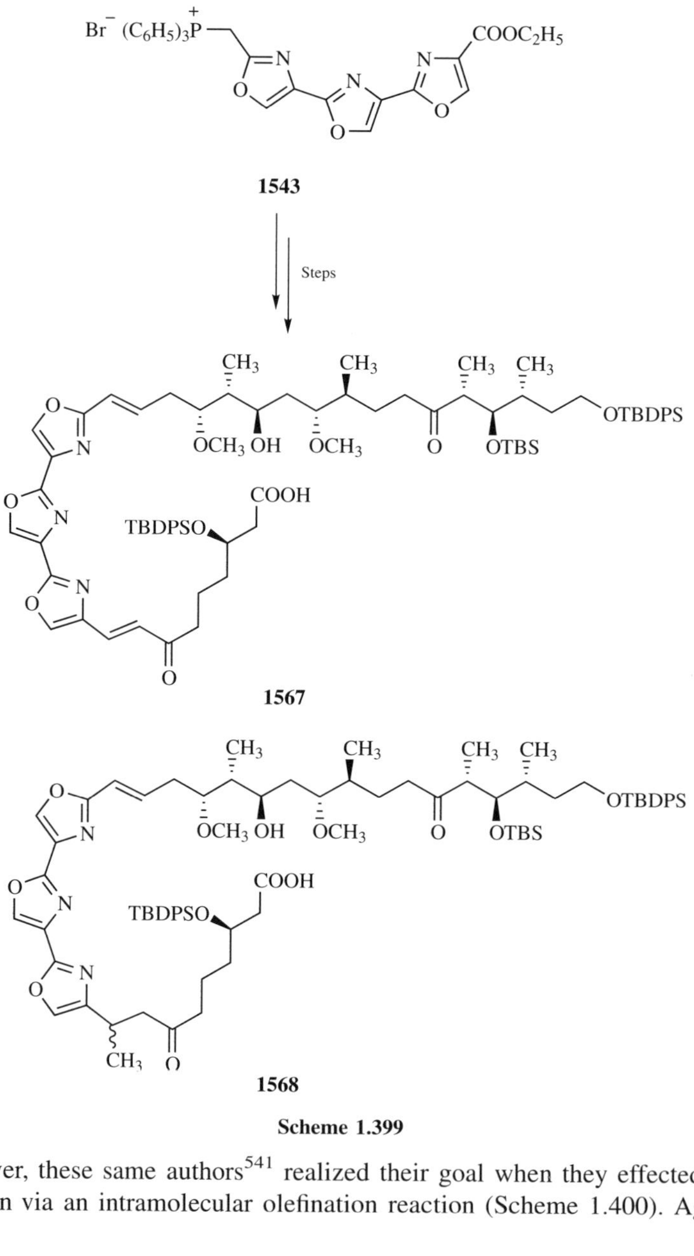

Scheme 1.399

However, these same authors[541] realized their goal when they effected macrocyclization via an intramolecular olefination reaction (Scheme 1.400). Again, the

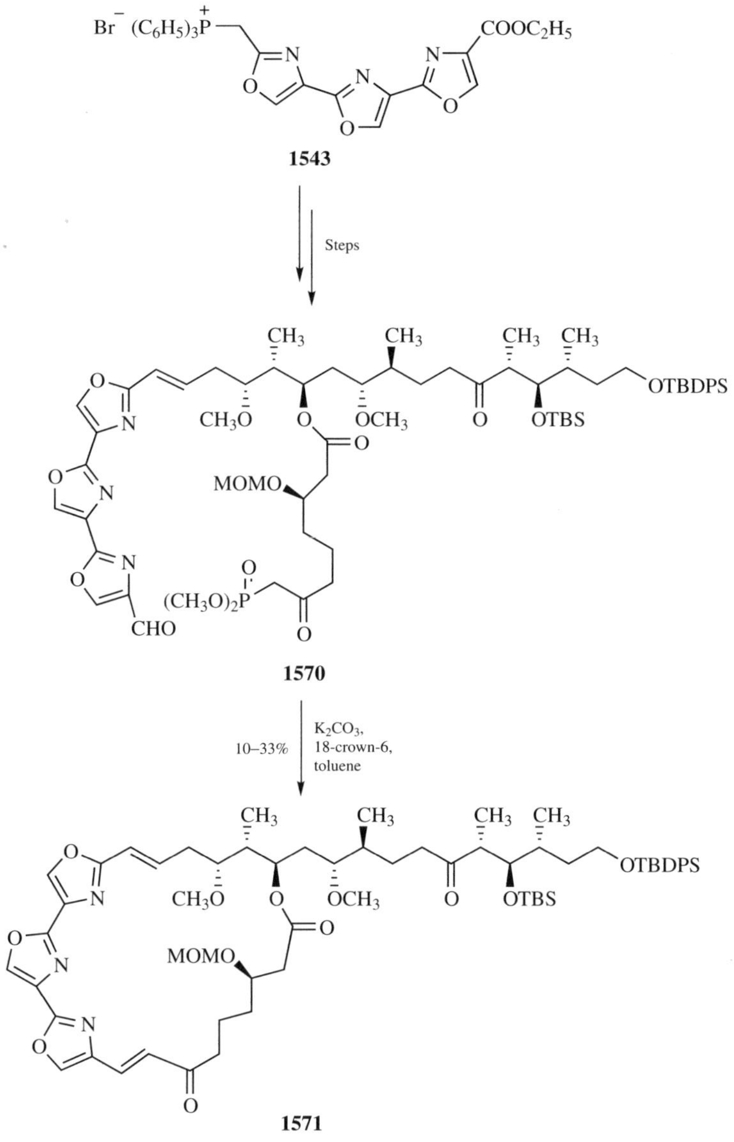

Scheme 1.400

phosphonium salt **1543** was elaborated sequentially to the penultimate precursor **1570**. Treatment of **1570** in a Horner-Wadsworth-Emmons reaction gave the enone **1571** after macrocyclization. This enone then possessed the complete macrolide tris-oxazole core as well as the entire carbon skeleton of **1195**.

The authors then extended this strategy to provide the first total synthesis of **1195**.[542] They noted a very slight difference in the ^{13}C spectrum of the C(33) methyl group in synthetic vs. naturally occurring ulapualide A.

Pattenden and co-workers[56,543] proposed a convergent alternative strategy to the ulapualide core, wherein suitably functionalized monooxazoles would be coupled first. Construction of the final, central oxazole ring would then complete the macrocyclization process.

Chattopadhyay and Pattenden[56] first demonstrated the viability of such a synthetic strategy in a model system designed to construct the basic tris-oxazole core. The oxazole amino alcohol **1573** was prepared from Garner's acid[546] **1572** in four steps (Scheme 1.401). Serine benzyl ester was the starting material for 2-(acetoxymethyl)-4-oxazolecarbonyl chloride **1574**. Acylation of **1573** with **1574** produced the bis-oxazole amide **1575**. The differentially functionalized model tris-oxazole **1576** was then prepared from **1575** in two straightforward steps.

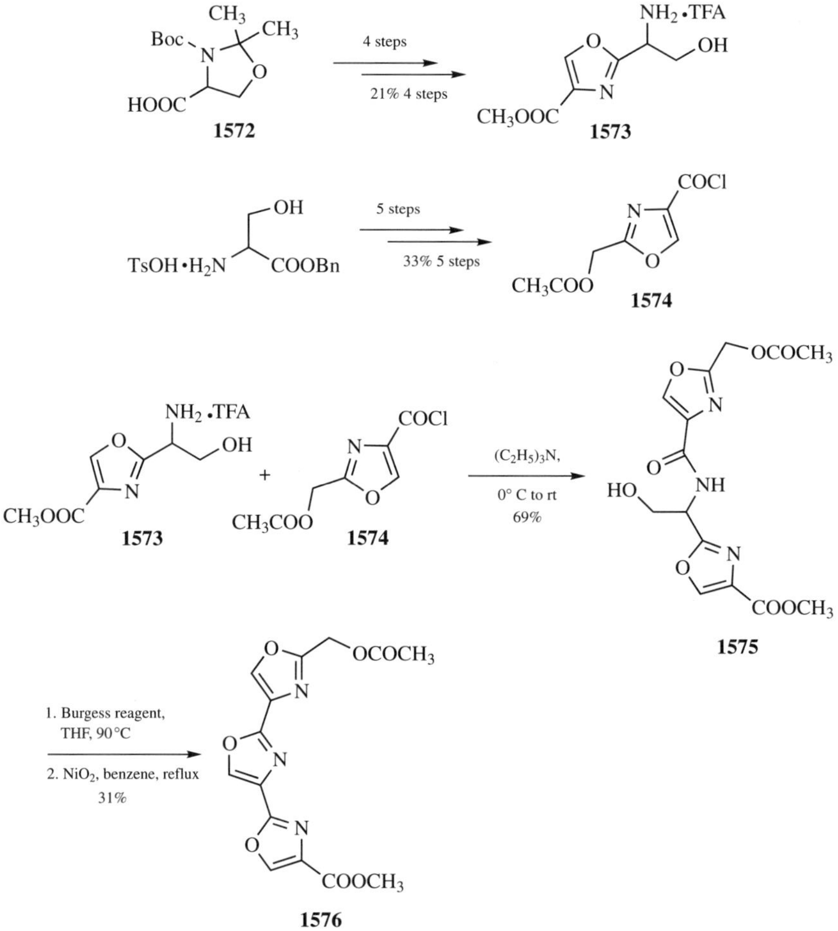

Scheme 1.401

Subsequently, Kempson and Pattenden[543] applied this strategy to prepare a model for the complete tris-oxazole macrolide core of ulapualides (Scheme 1.402).

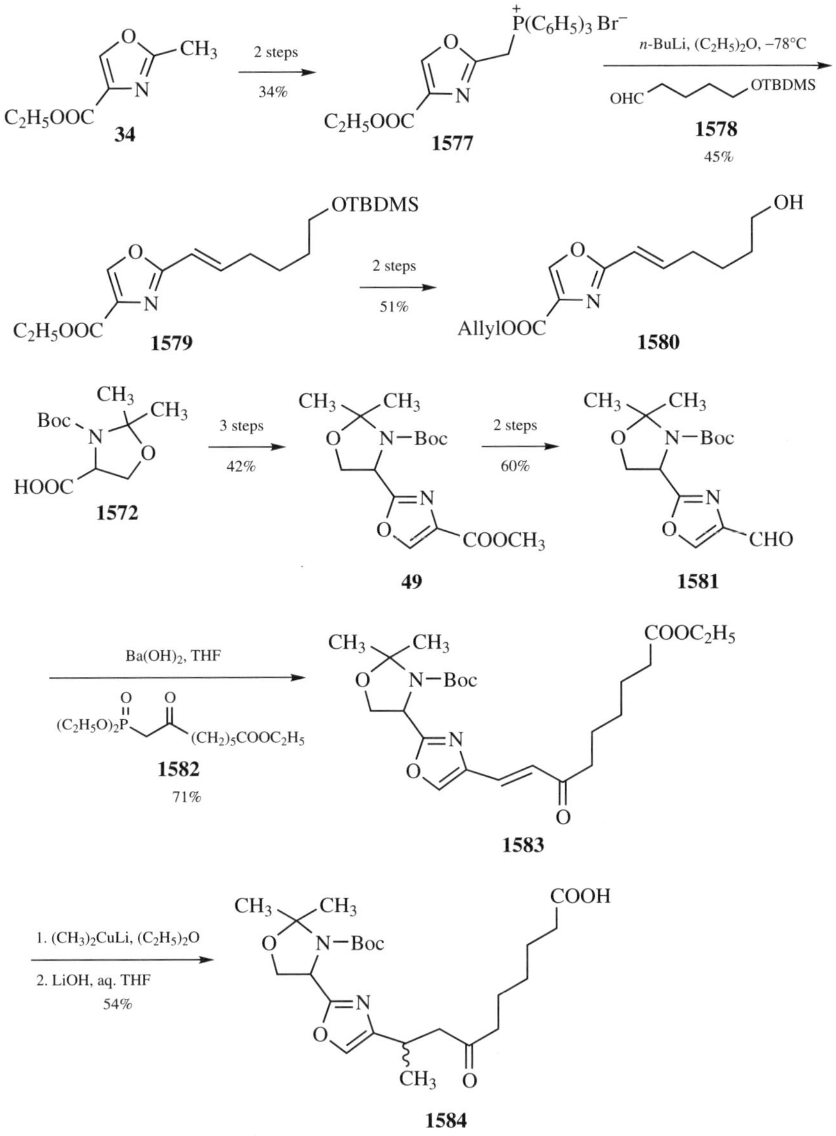

Scheme 1.402

The requisite monooxazoles **1580** and **1584** were assembled as follows. 2-Methyl-4-oxazolecarboxylic acid ethyl ester **34** was converted to the phosphonium salt **1577** in two steps. Wittig reaction of **1577** with **1578** furnished **1579**, which was converted to the allyl ester **1580** uneventfully.

The synthesis of the other monooxazole, **1584** also began with Garner's acid[546] **1572**. Chattopadhyay and Pattenden[56] described the conversion of **1572** to **49** (Scheme 1.14). A straightforward reduction–oxidation sequence afforded **1581** from **49**. Horner-Wadsworth-Emmons olefination of **1581** with the keto-phosphonate **1582** produced the (*E*)-enone **1583**, from which **1584** was obtained in two steps.

Esterification of **1584** with **1580** gave **1585** in excellent yield (Scheme 1.403). Sequential deprotection of **1585** unmasked the amino alcohol (not shown), which was subjected to macrolactamization using DPPA to afford the β-hydroxyamide **1586**. The synthesis was then completed by cyclization/oxidation of **1586** to install the central oxazole ring of **1587**.

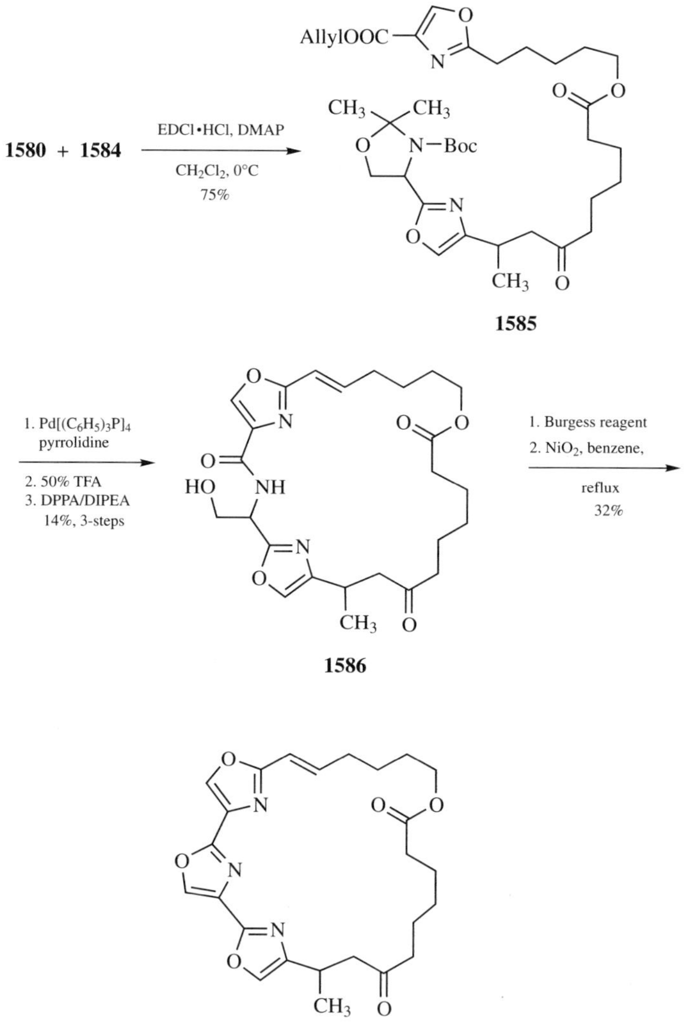

Scheme 1.403

Pattenden and co-workers[544,545] summarized all of the synthetic strategies they evaluated to prepare **1195**. In addition, the complete details of the total synthesis of

1195 were described. As noted earlier, the authors concluded that there are one or more stereogenic centers, which differ in the C(28)-C(33) region of the side chain.

1.6. ADDENDUM

Several reviews on various aspects of oxazole chemistry have been published. Cicchi and co-workers[547] and Gilchrist[548] reviewed synthesis and reactions of oxazoles. Walsh and co-workers[549] reviewed the biosynthesis of thizazole and oxazole peptides, including microcin B17 **1196**. Kawase and co-workers[550] reviewed the synthesis of 5-(trifluoromethyl)oxazoles via Dakin-West chemistry. Suga and Ibata[551] reviewed much of their work on [3 + 2] cycloadditions of 5-alkoxyoxazoles. In addition, Mrozek and co-workers[552] applied the harmonic oscillator model of aromaticity (HOMA) index to five-membered ring heterocycles, including oxazoles.

Wipf and co-workers[553] developed an efficient one-pot process to prepare a variety of highly functionalized 2-substituted 4-oxazolecarboxylic acid esters **39**. The starting β-hydroxyamides **1588** were cyclodehydrated using Deoxo-fluor (bis(2-methoxyethyl)aminosulfur trifluoride) or DAST (diethylaminosulfur trifluoride) to afford intermediate oxazolines **38**, which were oxidized in situ with BrCCl$_3$/DBU[57] to afford **39** (Scheme 1.404). DAST was found to be preferable with serine-derived β-hydroxyamides, whereas Deoxo-fluor was more useful for threonine derived β-hydroxyamides.

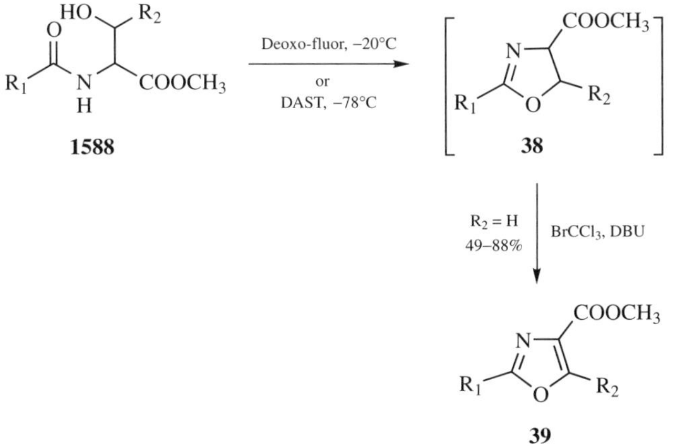

Scheme 1.404

Advantages of this methodology are mild reaction conditions that tolerate a variety of sensitive functional groups and little observed epimerization with optically active side chains. It is also noteworthy that 5-unsubstituted oxazoles are available, which complements the Dess-Martin periodinane-I$_2$/(C$_6$H$_5$)$_3$P protocol[62] for preparation of 5-substituted oxazoles.

Merck chemists[554] found BrCCl$_3$/DBU[57] to be the preferred method for preparing 2″-oxazole analogs of nodulisporic acid **1591**, which were evaluated as potential vetinary antiflea and antitick agents (Scheme 1.405).

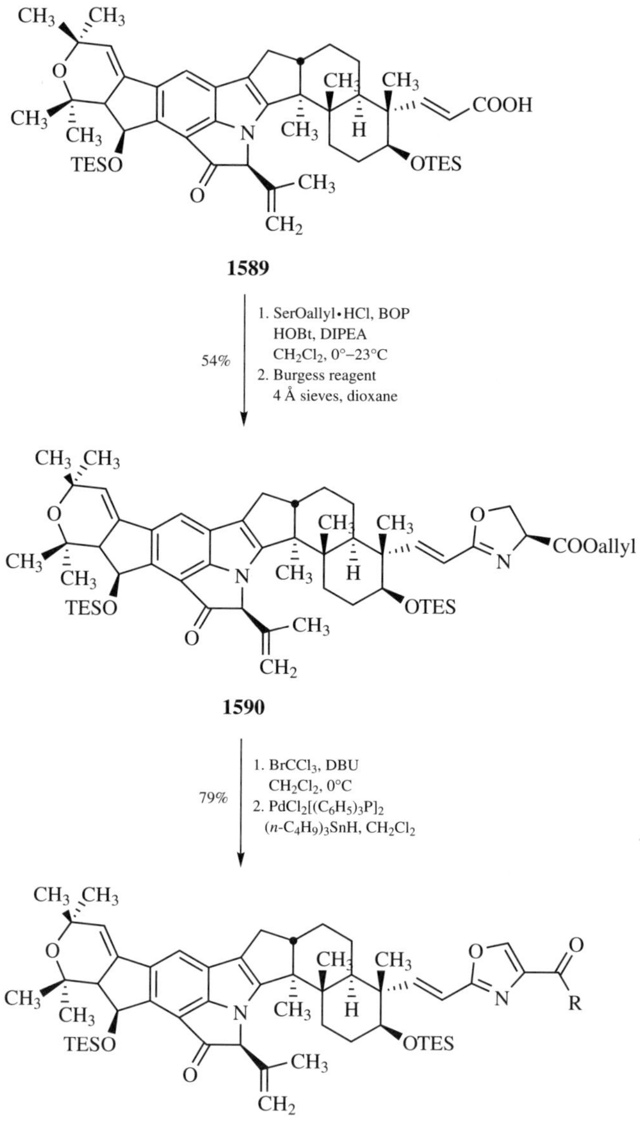

1589

1. SerOallyl·HCl, BOP
HOBt, DIPEA
CH$_2$Cl$_2$, 0°–23°C
2. Burgess reagent
4 Å sieves, dioxane

54%

1590

1. BrCCl$_3$, DBU
CH$_2$Cl$_2$, 0°C
2. PdCl$_2$[(C$_6$H$_5$)$_3$P]$_2$
(n-C$_4$H$_9$)$_3$SnH, CH$_2$Cl$_2$

79%

1591a R = OH
1591b R = N(CH$_3$)$_2$

Scheme 1.405

Stanchev and co-workers[555] prepared nonproteinogenic oxazole containing amino acid esters **1593** via oxidation of the intermediate oxazolines **1592**, using

Gordon's protocol (Scheme 1.406).[209] The oxazole containing amino acid esters **1593** were converted to **1594a–c**, which possessed significant in vitro activity against Gram-positive and Gram-negative organisms.

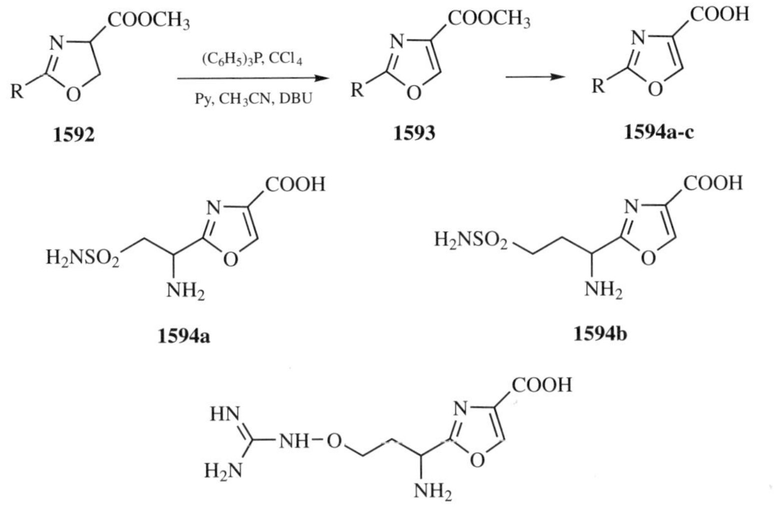

Scheme 1.406

Prager and co-workers[556] described a general synthesis of 2-(1-aminoalkyl)-4-oxazolecarboxylic acid esters **1596** and 2-(1-aminoalkyl)-5-oxazolecarboxylic acid esters **1598** from photolysis of *N*-acylisoxazol-5-ones **1595** and **1597** (Scheme 1.407). A key finding was the use of the phthalamido protecting group in the photolysis reaction since *N*-acylisoxazolones are not usually efficiently converted to oxazoles photolytically. The optical integrity of **1596** and **1597** was maintained throughout the photolysis reaction and the hydrazinolysis of the phthalamido group.

In later work, Pragers group[557] applied this methodology for preparing almazole A and almazole B, **1599a** and **1599b**, respectively.

Wang and Zhu[558] reported a general synthesis of 5-(perfluoroalkyl)-2-substituted 4-oxazolecarboxylic acid ethyl esters **1601** from decomposition of ethyl 2-diazoperfluoroalkylacetoacetates **1600** in the presence of a nitrile as solvent (Scheme 1.408). Rhodium(II) acetate was the preferred catalyst for this reaction. The yields of **1601** varied considerably.

Searle chemists[559] described an improved synthesis of 2-(bromomethyl)-4-oxazolecarboxylic acid ethyl ester **957**. Thus slow addition of ethyl 2-diazo-3-oxo-propanoate in bromoacetonitrile to $Rh_2(OAc)_4$ in bromoacetonitrile at 70°C gave **957** in 78% yield (Scheme 1.409). Incorporation of **957** into a series of PGE_2 antagonists **1604** was straightforward. Here, the 2,4-disubstituted oxazole was an isostere for a diacylhydrazine moiety.

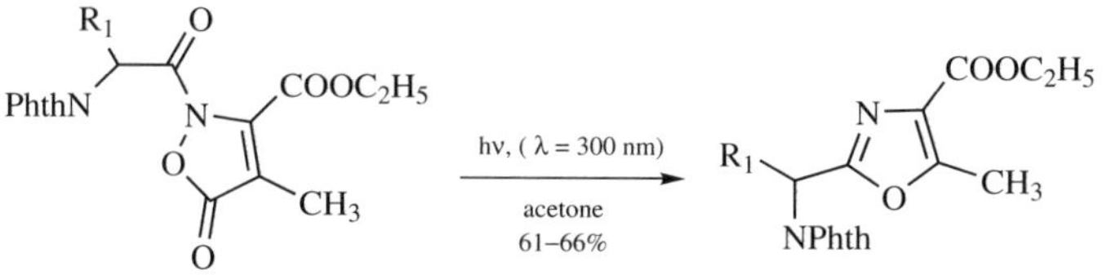

1595 **1596**

R_1 = H,CH_3

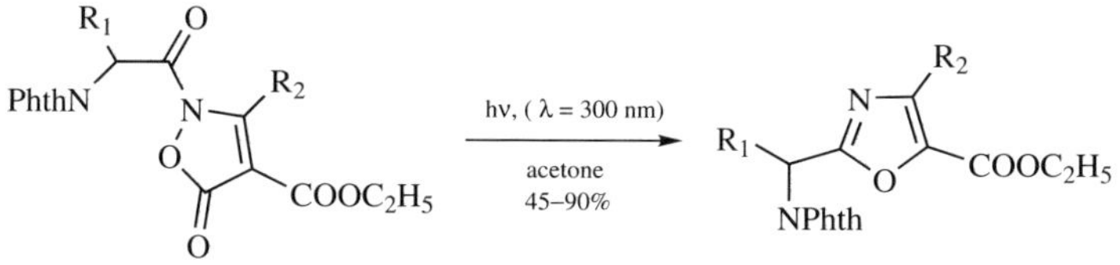

1597 **1598**

R_1 = H, CH_3
R_2 = CH_3, H, COC_6H_5

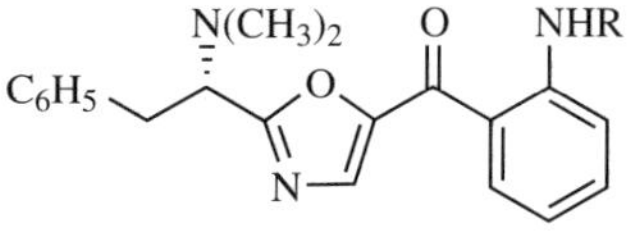

1599a R = CHO
1599b R = H

Scheme 1.407

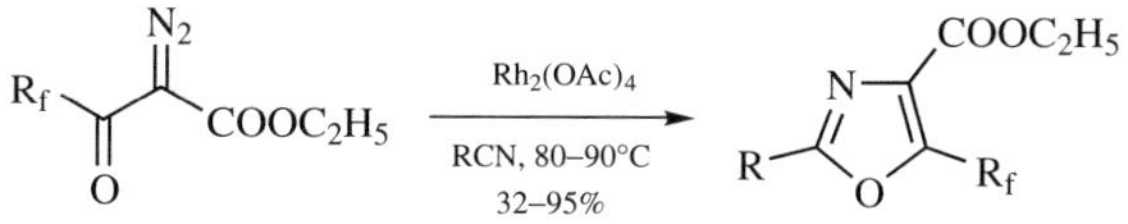

1600 **1601**

R = C_6H_5, 3-CH_3-C_6H_4, $CH_2C_6H_5$, *E*- or *Z*-CH=CHCH$_3$, ClCH$_2$
R_f = CF_3, $CF_2CF_2CF_2Cl$, C_5F_{11}

Scheme 1.408

Janda and co-workers[560] developed a general, solid-phase synthesis of 2-aryl-4-substituted oxazoles from polymer-bound α-diazo-β-ketoesters **1605**. In this case, hydroxybutyl *J*anda*J*el™ was reacted with diketene, followed by diazo transfer, to prepare **1605** (Scheme 1.410). Rhodium(II)octanoate-catalyzed insertion of **1605** with an amide gave the polymer-bound *N*-acylamino acid ester **1606**, which was cyclodehydrated to **1607**, using commercially available $(C_6H_5)_3PCl_2$. Cleavage of **1607** was then accomplished reductively, via transesterification or by amidation. The reaction sequence afforded oxazoles in modest yields but of good to excellent purities. The authors extended the methodology to prepare other 5-substituted

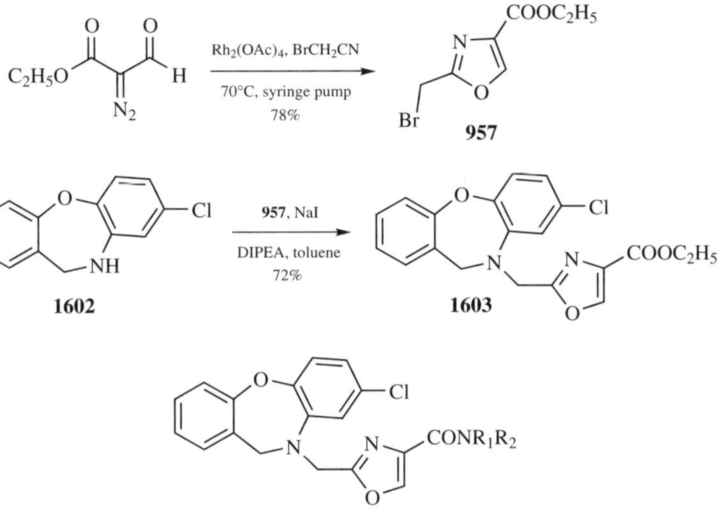

Scheme 1.409

analogs **1611** and **1612**, but these required Burgess reagent[49,218] to effect cyclodehydration of the analogous polymer-bound N-acylamino acid ester.

Kim and co-workers[561] prepared a series of 2,4,5-trisubstituted oxazoles **1614** in low to good yields from desoxybenzoins **1613** and nitriles (Scheme 1.411). This Ritter-like reaction works best in the presence of a 1:1 mixture of trifluoromethanesulfonic acid–sulfuric acid. Trifluoromethanesulfonic acid or sulfuric acid alone is ineffective. Fuming sulfuric acid gave < 10% of an oxazole. The authors proposed that sulfuric acid acts as an oxidant, analogous to $Tl(OTf)_3$[111,113] or $Cu(OTf)_2$.[118,119]

Cacchi's group[562] reported an elegant synthesis of 2,5-diaryloxazoles **1616** using a palladium-catalyzed reaction of N-propargyl benzamides **1615** and aryl iodides (Scheme 1.412). The authors had evidence to support two different mechanistic pathways, depending on the substitution pattern of **1615** and the aryl iodide. The optimal reaction conditions were found to be $Pd_2(dba)_3$, $P(2-furyl)_3$ as the ligand, NaOtBu as the base, and acetonitrile as the solvent. These conditions significantly reduce the reaction time and favor oxazole formation.

Dehaen and co-workers[563] thermolyzed 5-azidothiazoles **1617** and prepared several 4-cyanooxazoles **1618** in modest yields (Scheme 1.413). The authors ruled out a Cornforth-type rearrangement. Instead, they favored ring opening of **1617** to a thioamide followed by cis/trans isomerization and a 1,6-electrocyclization to produce an unstable oxathiazine **1619**. Extrusion of sulfur from **1619** then affords **1618**.

Barrett and co-workers[564] adapted ring-opening metathesis polymerization (ROMP) to prepare the ROMPgel TosMIC reagent **1621** (Scheme 1.414, p. 364). Thus copolymerization of the norbornene-derived formamide **1620** with norbornene in the presence of Grubbs's catalyst[477] gave **1621** in quantitative yield after dehydration. Aromatic aldehydes react with **1621** in the presence of a strong

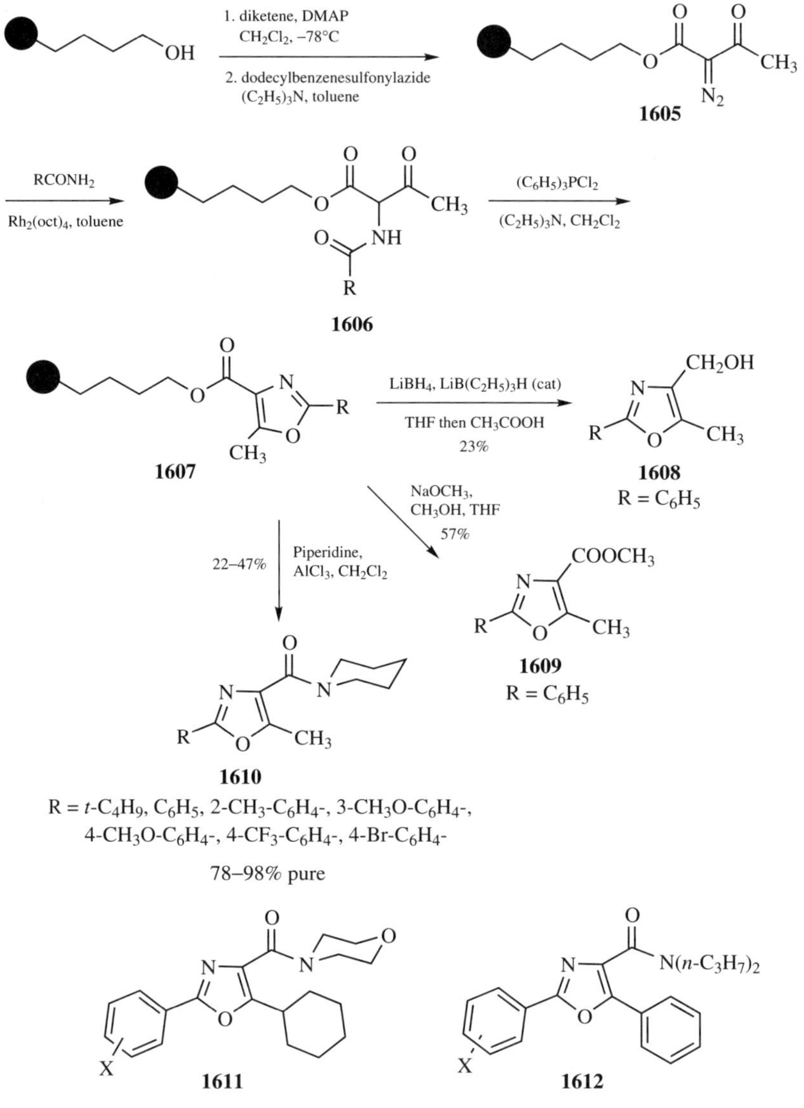

Synthesis and Reactions of Oxazoles

Scheme 1.410

amine base (e.g., *tert*-butyltetramethylguanidine) to afford 5-substituted oxazoles **411** in good to excellent yield. The desired oxazoles were routinely isolated in >95% purity, thereby precluding the need for chromatographic purification. This method complements that of Kulkarni and Ganesan.[173,174]

Hénaff and Whiting[565,566] prepared the 5-(3-tri-*n*-butylstannylallyl)oxazole **1624** from 3-butyn-1-ol and TosMIC (Scheme 1.415). This was a key intermediate in their synthetic approach to phthaloxazolin A **1177** (R = H).

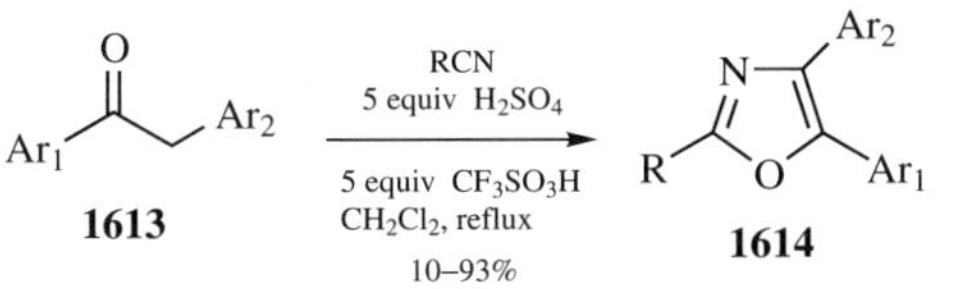

1613 1614

R = CH_3, C_2H_5, $ClCH_2$, $NCCH_2$, C_6H_5
Ar_1 = C_6H_5, 4-CH_3O-C_6H_4, 2,5-di-CH_3-C_6H_3, 2,4,6-tri-CH_3-C_6H_2
Ar_2 = C_6H_5, 4-CH_3O-C_6H_4

Scheme 1.411

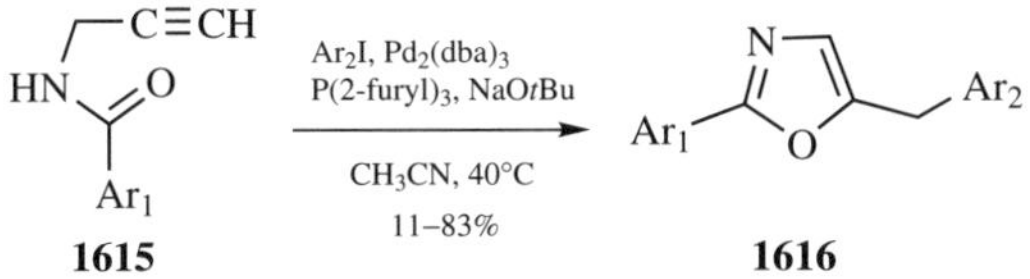

1615 1616

Ar_1 = C_6H_5, 4-CH_3O-C_6H_4, 4-CH_3-C_6H_4, 3-CF_3-C_6H_4,
Ar_2 = C_6H_5, 4-CH_3CO-C_6H_4, 4-CH_3-C_6H_4, 3-CF_3-C_6H_4, 4-Cl-C_6H_4, 3-F-C_6H_4, 4-F-C_6H_4, 3,5-di-CH_3-C_6H_3

Scheme 1.412

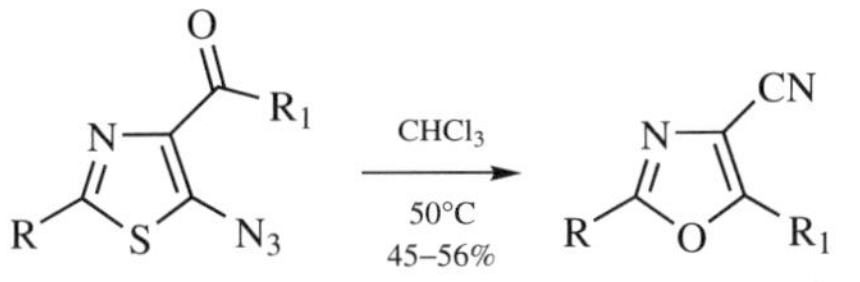

1617 1618

R = CH_3, C_6H_5
R_1 = H, CH_3

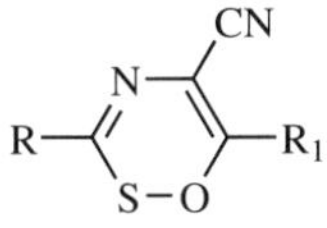

1619

Scheme 1.413

Bristol-Myers Squibb chemists[567] converted substituted 4-nitrobenzaldehydes to 5-(substituted 4-nitrophenyl)oxazoles **1625** using TosMIC (Scheme 1.416). As an example, 5-(2-methoxy 4-nitrophenyl)oxazole **1625** (R = CH_3O), evaluated as an IMPDH antagonist, was prepared from 2-methoxy-4-nitrobenzaldehyde in 84% yield.

Duarte and co-workers[568] correctly identified the product from the Michael addition of benzohydroxamic acid to a propiolate ester to be a 1,4,2-dioxazole **1626**, not an oxaziridine **1627** (Scheme 1.417). The authors converted **1626** to a mixture of (*E/Z*) lactoxime *O*-vinylether sodium salts **1628**, from which 2-phenyl-4-oxazolecarboxylic acid methyl ester was isolated upon thermolysis.

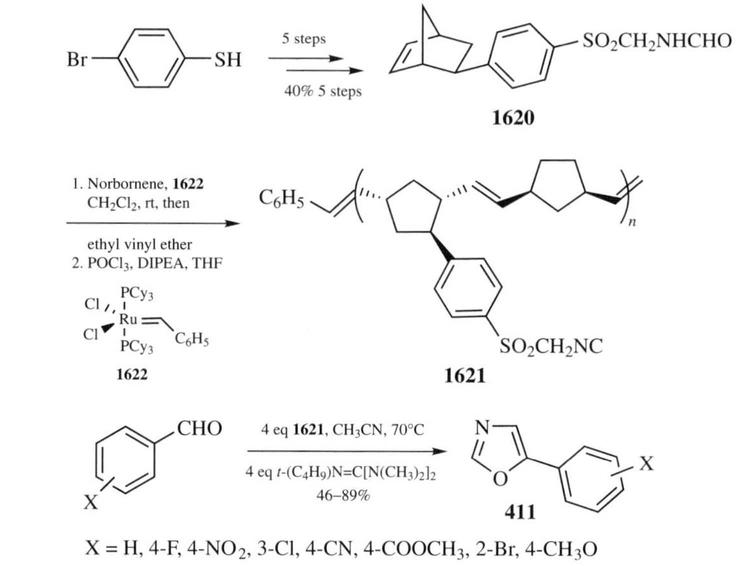

Scheme 1.414

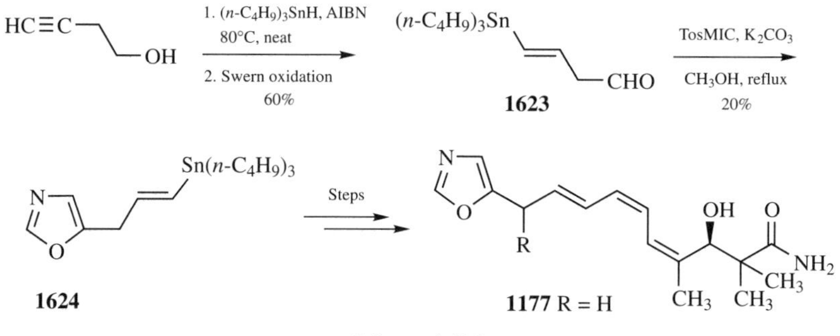

X = H, 4-F, 4-NO₂, 3-Cl, 4-CN, 4-COOCH₃, 2-Br, 4-CH₃O

Scheme 1.415

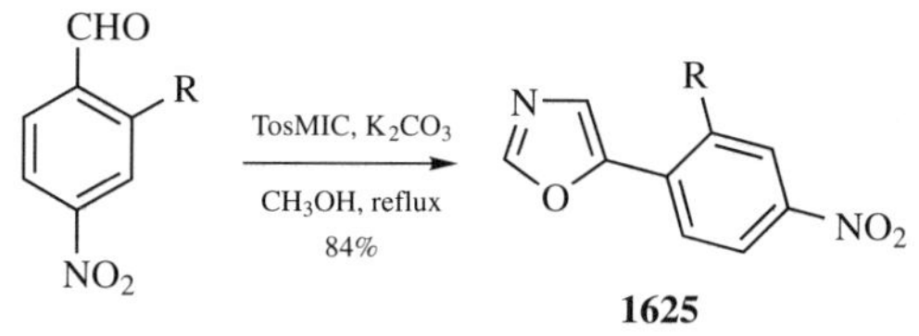

Scheme 1.416

Hermitage and co-workers[569] refined and improved their process of preparing 2-(chloromethyl)-4-oxazolecarboxylic acid ethyl ester **1630** (Scheme 1.418). Several improvements are noteworthy. The authors converted dichloroacetonitrile to **1630** in three steps, using DIPEA to effect oxidation of **1629**. Quaternization of other tertiary amine bases (e.g., triethylamine) by **1630** was a problem. The use of serine

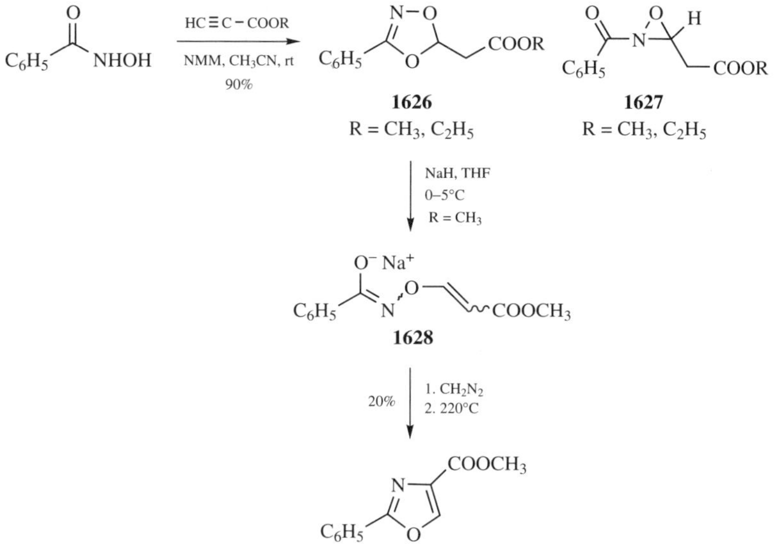

Scheme 1.417

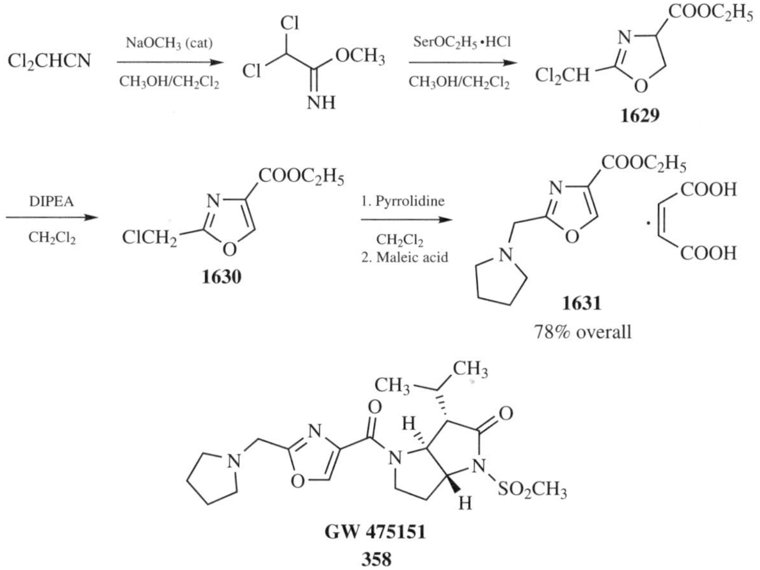

Scheme 1.418

ethyl ester rather than the methyl ester in this sequence completely precluded amidation during conversion of **1630** to **1631**. Overall, the authors telescoped the reaction sequence from dichloroacetonitrile and isolated **1631** in excellent yield.

In Japan, workers[570] prepared 4-cyclohexyl-5-(4-methanesulfonylphenyl)-2-methyloxazole **1633**, a potential cyclooxygenase 2 inhibitor, from **1632** using ammonium acetate (Scheme 1.419). In France, workers[571] described an improved process of preparing 2,4,5-trisubstituted oxazoles **1635** in which an α-acyloxyketone **1634** was refluxed with two equivalents of thiourea in DMF. For example, **1634** was converted to 2,4,5-triphenyloxazole **231c** in 68% yield. The authors reported this to be a general process readily amenable to scale up, although the yields varied considerably (20–68%).

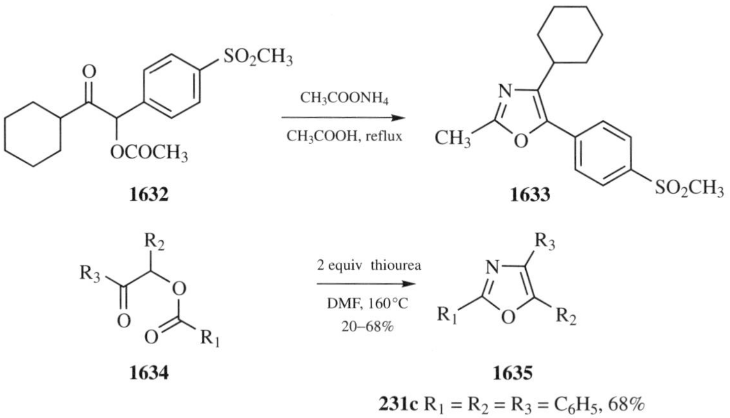

Scheme 1.419

Drach and co-workers[572] prepared 4,5-(diarylthio)-2-phenyloxazoles **1638** from β,β-dichloro-α-(toluenesulfonyl)benzamide **1636** (Scheme 1.420). The intermediate N-[1,2,2-tri(arylthio)ethenyl]benzamides **1637a** and **1637b** were isolated in excellent yield and were characterized spectroscopically. Cyclization of **1637** to **1638** was accomplished in modest yield.

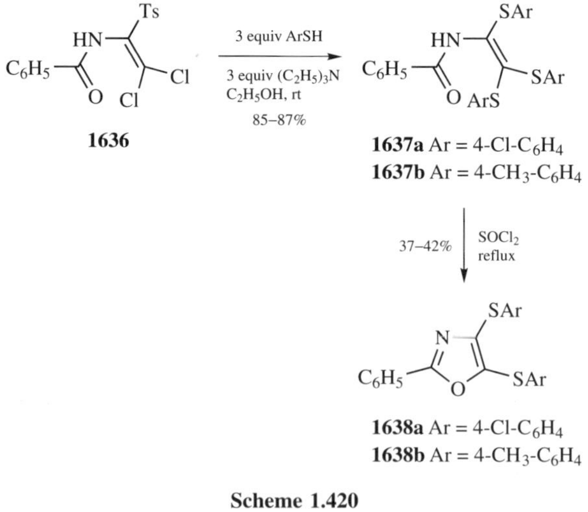

Scheme 1.420

Drach and co-workers[573] described a general synthesis of 2-alkyl(aryl)-5-hydrazino-4-phosphorylated-oxazoles **1640** from **1639** or **1641** (Scheme 1.421). Reactions of **1640** analogs with *p*-tolualdehyde, *p*-toluic chloride, and phenyl isothiocyanate gave the expected products.

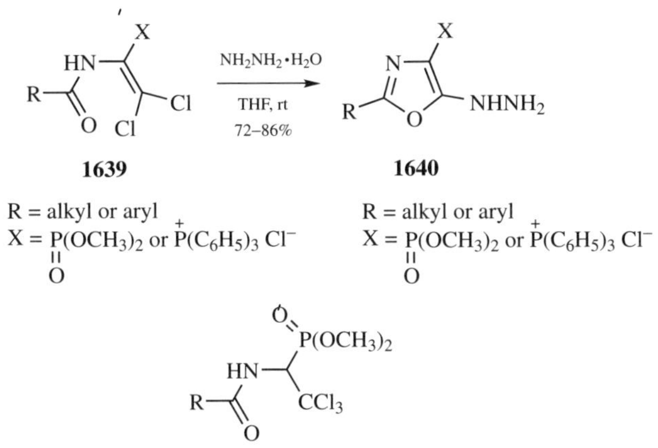

Scheme 1.421

Yamane[574] prepared 4-(chloromethyl)-5-methyl-2-phenyloxazole **1643** in good yield from 4,5-dimethyl-2-phenyloxazole **1642** and NCS (Scheme 1.422).

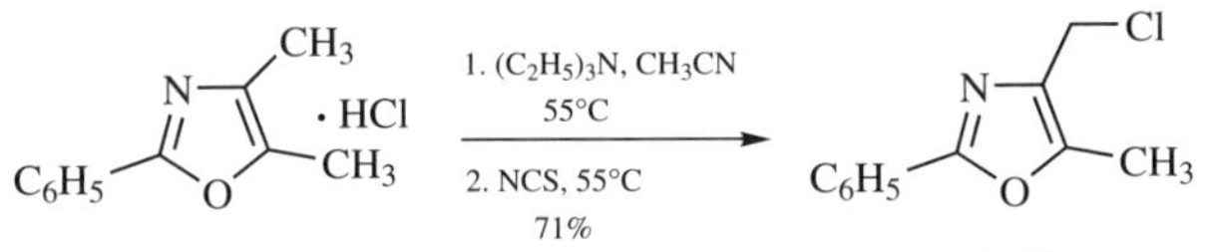

Scheme 1.422

Drach and co-workers[575] acylated **1640** (X = P(O)(OCH$_3$)$_2$) to afford **1644** in excellent yield. Refluxing **1644** in aqueous ethanol effected hydrolysis and recyclization to the 2,5-disubstituted 1,3,4-oxadiazoles **1646**, presumably via the *N,N'*-diacylhydrazide **1645** (Scheme 1.423).

Later work from Drach's group[576] resulted in a general synthesis of the previously unknown 2-aryl-4-cyano-5-hydrazinooxazoles **1648** from a 2-(acylamino)-3,3-dichloroacrylonitrile **1647** (Scheme 1.424). Refluxing **1648** in acetic acid afforded the 2-methyl-5-substituted 1,3,4-oxadiazoles **1650**. The structure of **1650** (X = CH$_3$O) was confirmed by X-ray crystallography. All attempts to cyclize **1648** to an oxazolopyrazole **1651** failed.

Iwase and Aoki[577] reported a very efficient synthesis of 3- or 4-substituted pyridines via high-pressure Diels-Alder reaction of oxazole **1**. For example, **1** reacted with acrylonitrile at 1.1 Mpa to produce 3-cyanopyridine **1652** in 95% yield (Scheme 1.425). Kondrat'eva and co-workers[578] described cycloaddition reactions of 2-aminooxazoles **1653** with maleimide (Scheme 1.425). The authors isolated several different

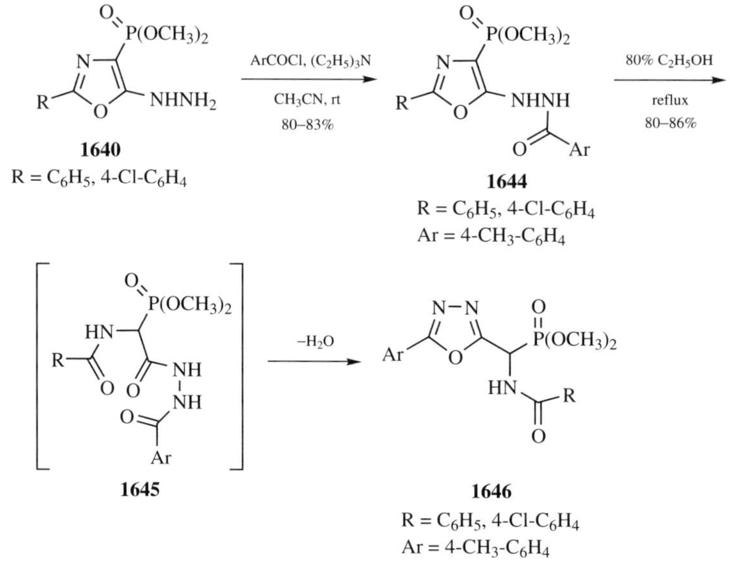

Scheme 1.423

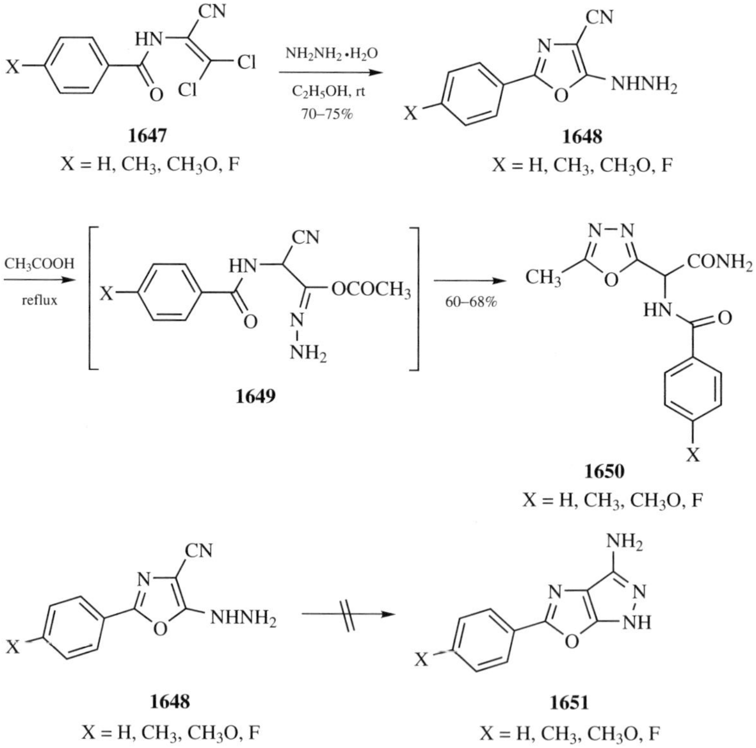

Scheme 1.424

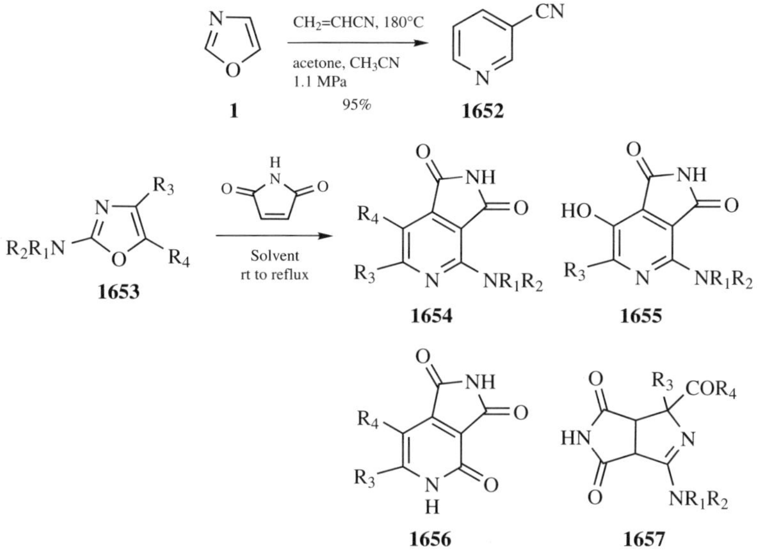

Scheme 1.425

products **1654–1657**, depending on the solvent—ether, benzene, or glacial acetic acid—the temperature, and the substitution pattern. Note that 2-pyridones **1656** were the major or exclusive products isolated from refluxing acetic acid. The 1,3-cycloaddition products **1657** were isolated only from refluxing benzene.

Clapham and Sutherland[579] developed a general synthesis of 2,5-diphenyl-4-substituted oxazoles **1659** in the context of preparing monomers for scintillating polymers. Stille coupling of 4-bromo-2,5-diphenyloxazole **1658** with a variety of stannanes produced the target oxazoles in modest to excellent yield (Scheme 1.426). Alternatively, the authors reversed the Stille-coupling approach

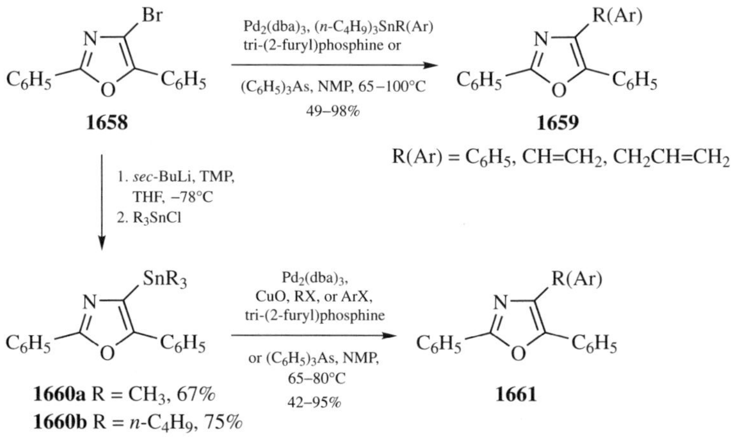

Scheme 1.426

and prepared the 2,5-diphenyl-4-(trialkylstannyl)oxazoles **1660a** and **1660b** from **1658**. Standard Stille-coupling conditions using **1660a** or **1660b** were not successful. However, addition of a stoichiometric amount of CuO dramatically improved the reaction rate and yield. The reaction was successful using aryl- and heteroaryl halides, benzyl halides, and acid chlorides.

Wenkert and co-workers[580] prepared a series of 5-alkenyl-substituted oxazolium salts and investigated their intramolecular Diels-Alder reactions as a means to construct alkaloids (Scheme 1.427). The starting oxazoles **1662a–c**, **1663**, and **1664** were prepared from an acyclic methyl ester and lithiomethyl isocyanide.[154] Methylation of **1662a–c**, **1663**, and **1664** with methyl triflate then afforded the corresponding *N*-methyloxazolium salt quantitatively. The oxazolium salts **1665**, **1666**, and **1667** were stable for up to 3 h at 90°C.

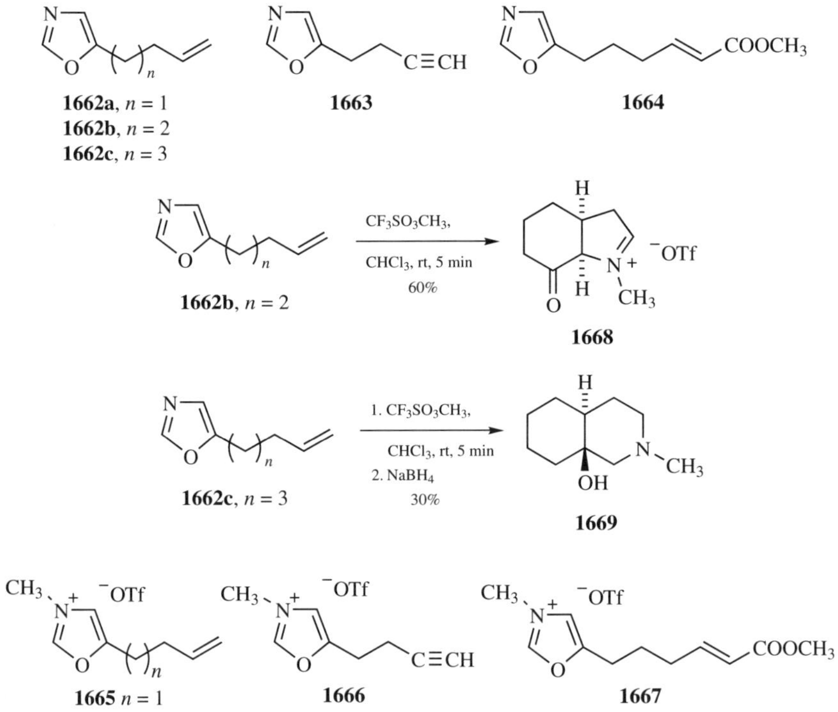

Scheme 1.427

On the other hand, 5-(4-pentenyl)oxazole **1662b** reacted with methyl triflate at room temperature to produce (±)-1-methyl-7(6*H*)-oxo-*cis*-hexahydroindolenium triflate **1668** as the only product in 60% yield. Under similar conditions, 5-(5-hexenyl)oxazole **1662c** reacted with methyl triflate to give an intermediate adduct, which was reduced with NaBH₄ to afford (±)-8a-hydroxy-2-methyl-*trans*-decahydroisoquinoline **1669** as the major product. The formation of **1668** and **1669** were consistent with an intramolecular Diels-Alder reaction.

Pour and co-workers[581] reported an unusual fragmentation of an oxazole ring duirng their studies to prepare and evaluate 3-heteroaryl-2,5-dihydrofuran-2-ones as potential antifungal agents (Scheme 1.428). The authors prepared **1670** in six steps, uneventfully from benzamide. However, oxidation of **1670** to the selenoxide **1671** did not install the carbon–carbon double bond leading to **1672** as anticipated. Instead, 3-(1-benzamido-2-oxo-ethylidene)-5-methyltetrahydrofuran-2-one **1673** was isolated in low yield. Mechanistically, this result was rationalized by a [2,3]-sigmatropic rearrangement of **1671**, followed by ring opening and hydrolysis during workup.

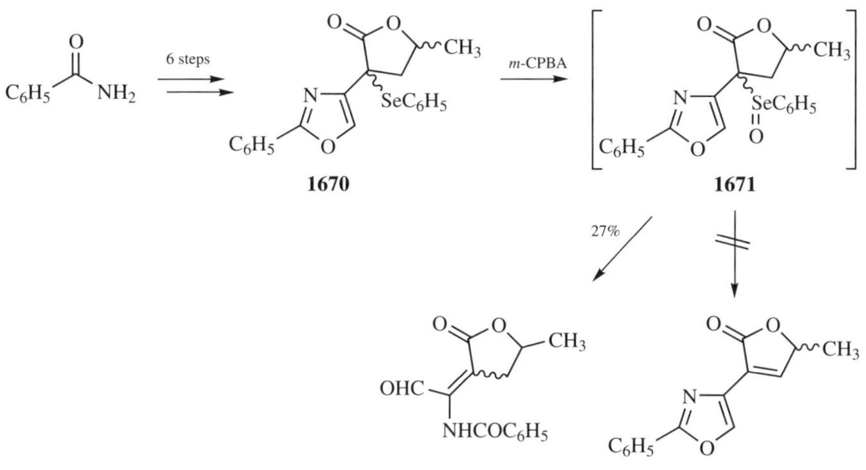

Scheme 1.428

Zhu and co-workers[582] described novel examples of an Ugi three-component condensation[583] for preparing 2,4,5-trisubstituted oxazoles. They elegantly combined this reaction with an acylation and an intramolecular Diels-Alder reaction to prepare pyrrolo[3,4-b]pyridines **1675** as cyclic analogs of nicotinamide. Thus reaction of an aldehyde, an amine, and a substituted isocyanoacetamide produced a series of 2-(aminomethyl)-4-benzyl-5-morpholinooxazoles **1674** in good yield (Scheme 1.429). Acylation of **1674** with an α,β-unsaturated acid chloride followed by heating in toluene then gave the novel pyrrolo[3,4-b]pyridine scaffolds **1675** in good yield. Examples of highly functionalized **1675** available via this methodology are structures **1676–1678**. The authors described this "acylation/IMDA/retro-Michael cycloreversion" methodology as a triple domino sequence that should be eminently useful for preparing combinatorial libraries via solid-phase synthesis or automated techniques.

Zhu's group[584] applied their synthetic methodology for **1675** analogs to the synthesis of macrocyclic depsipeptides wherein an oxazole was the activating group

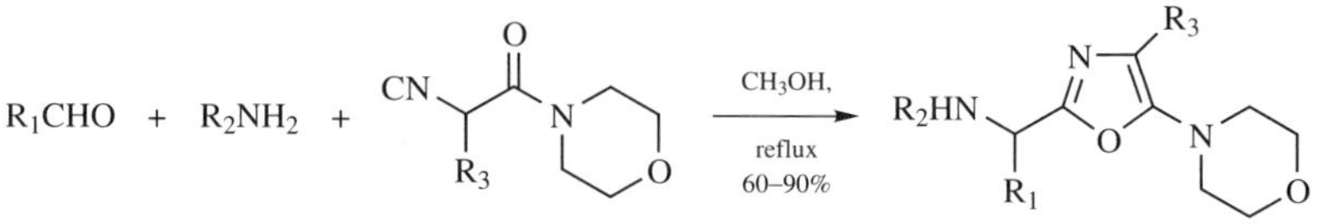

1674

R_1 = alkyl, cycloalkyl
R_2 = substituted benzyl, phenethyl
R_3 = phenyl, benzyl

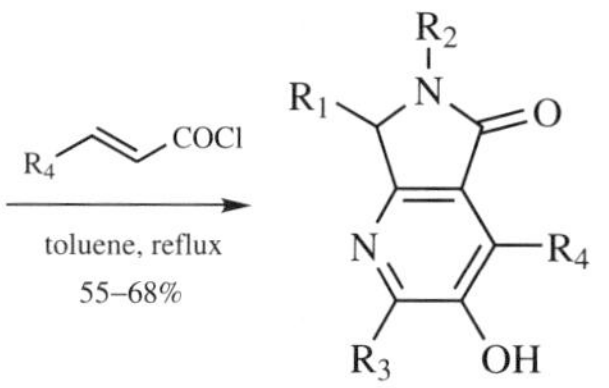

1675

R_1 = alkyl, cycloalkyl
R_2 = substituted benzyl, phenethyl
R_3 = phenyl, benzyl
R_4 = C_2H_5OOC, 4-X-C_6H_4

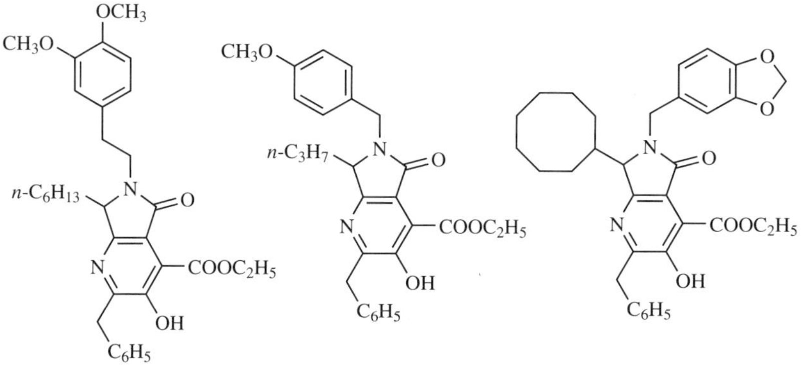

1676 **1677** **1678**

Scheme 1.429

to effect macrolactonization. The key oxazole intermediates **1679** were readily assembled from a three-component condensation reaction of an aldehyde, an amino alcohol, and a dipeptide isocyanide (Scheme 1.430). Saponification of the methyl ester followed by macrocyclization with TFA produced the macrocylic depsipeptides **1680** in good to excellent yields. Examples of 13-, 14-, 15-, and 16-membered ring lactones prepared using this methodology are structures **1681a**, **1681b**, **1681c**, and **1682–1683**, respectively.

This synthetic approach represents a new concept for preparing complex marcocylodepsipetides. The authors noted that the process is atom economic,[585] is ecologically friendly, and should be readily amenable to combinatorial synthesis.

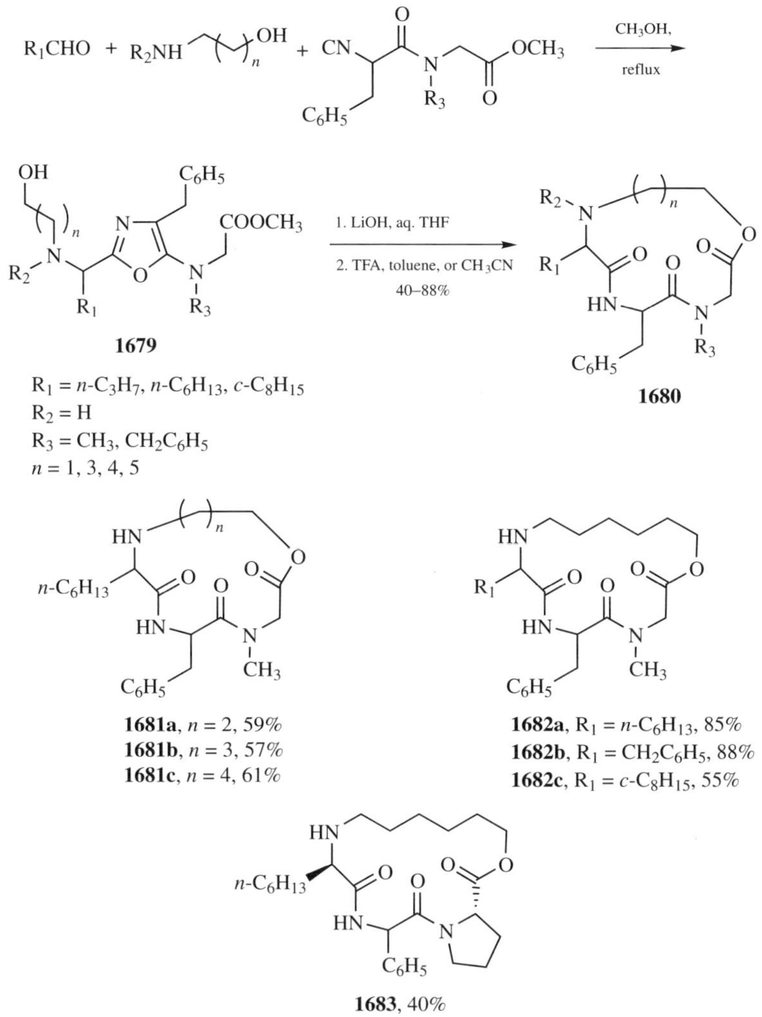

$R_1CHO + R_2NH\text{-}(\,)_n\text{-}OH + CN\text{-}...\text{-}OCH_3 \xrightarrow{CH_3OH,\ reflux}$

1679

R₁ = n-C₃H₇, n-C₆H₁₃, c-C₈H₁₅
R₂ = H
R₃ = CH₃, CH₂C₆H₅
n = 1, 3, 4, 5

1. LiOH, aq. THF
2. TFA, toluene, or CH₃CN
40–88%

1680

1681a, n = 2, 59%
1681b, n = 3, 57%
1681c, n = 4, 61%

1682a, R₁ = n-C₆H₁₃, 85%
1682b, R₁ = CH₂C₆H₅, 88%
1682c, R₁ = c-C₈H₁₅, 55%

1683, 40%

Scheme 1.430

1.7. SUMMARY

The chemistry of oxazoles continues to be an important focus of academic and industrial laboratories around the world. Indeed, this small-ring heterocycle has elicited extraordinary creativity from medicinal and process chemists, polymer chemists, materials scientists, photographic dye chemists, and natural products chemists engaged in basic and applied research.

Synthetic strategies for almost any oxazole-substitution pattern are available from classical methods or have been developed as new methodologies. Oxazoles are readily prepared from cyclic precursors such as oxazolines, *N*-acyltriazoles, *N*-acylaziridines, isoxazoles, *N*-acylisoxazolones, and imidazoles. On the other hand, virtually any type of functionalized acyclic moiety has been converted to an oxazole. For example, α-diazoketones, -esters, -nitriles, -sulfones, and -phosphonates have been converted to oxazoles using rhodium-carbene methodology. Acetylenes, acyl cyanides, aldehydes, amides, amino acids, amino nitriles, azides, enamines, hydrazones, imidates, imines, isocyanides, ketones, oximes, nitriles, thioimidates, and vinyl halides are all useful precursors to oxazoles. Oxazoles are now routinely prepared very efficiently via solid-phase methodologies that have been adapted for combinatorial libraries and parallel syntheses.

Oxazoles serve as versatile precursors to a remarkable variety of heterocyclic ring systems via nucleophilic addition, ring opening, and recyclization as well as [2 + 2], [3 + 2], and [4 + 2] cycloaddition reactions. The oxazole ring has found significant use as a precursor to amino ketones, amino acids, dipeptides, and triacylamines. The predictable and regioselective metalation chemistry of oxazoles has been combined with transition metal–catalyzed cross-coupling reactions to provide synthetic approaches to a vast array of increasingly complex natural products. In addition, the complexity and sensitivity of oxazole natural products, particularly those of marine origin, has fostered several novel, efficient, and mild syntheses of mono-, bis-, and tris-oxazoles. Oxazoles have served as scaffolds from which macrocyclization reactions have produced a variety of interesting and novel depsipeptides and macrolactones. Despite these significant advances, challenges remain. For example, a mild, efficient, and general method to oxidize 4-alkyloxazolines to 4-alkyloxazoles is still needed.

The future of oxazole syntheses and methodology is limited only by the structures that await discovery to provide a stimulus for the creativity of synthetic chemists around the world.

Acknowledgments

D. C. P. thanks Dr. Fuqiang Liu for his comments, suggestions, and review of this chapter.

REFERENCES

1. D'Auria, M. *Targets Heterocycl. Syst.* **1998**, *2*, 233–279.
2. Hassner, A.; Fischer, B. *Heterocycles* **1993**, *35*, 1441–1465.
3. Boyd, G. V. *Prog. Heterocycl. Chem.* **1992**, *4*, 150–167.
4. Boyd, G. V. *Prog. Heterocycl. Chem.* **1991**, *3*, 166–185.
5. Dondoni, A.; Fantin, G.; Fogagnolo, M.; Mastellari, A.; Medici, A.; Negrini, E.; Pedrini, P. *Gazz. Chim. Ital.* **1988**, *118*, 211–231.
6. Wasserman, H. H.; McCarthy, K. E.; Prowse, K. S. *Chem. Rev.* **1986**, *86*, 845–856.
7. Dondoni, A. *Lect. Heterocycl. Chem.* **1985**, *8*, 13–20.
8. Turchi, I. J. In Turchi, I. J., ed. *Oxazoles*, Vol. 45, *The Chemistry of Heterocyclic Compounds*, John Wiley, New York, 1986, pp. 1–341.

References

9. Shvekhgeimer, G. A.; Zvolinskii, V. I.; Kobrakov, K. I. *Khim. Geterotsikl. Soedin.* **1986**, 435–452 [*Chem. Abstr.* **1987**, *106l*, 50078].
10. Belen'kii, L. I.; Chuvylkin, N. D. *Khim. Geterotsikl. Soedin.* **1996**, 1535–1563 [*Chem. Abstr.* **1997**, *126l*, 224892].
11. Moody, C. J.; Doyle, K. J. *Prog. Heterocycl. Chem.* **1997**, *9*, 1–16.
12. Kashima, C.; Maruyama, T.; Arao, H. *Rev. Heteroat. Chem.* **1997**, *16*, 197–212.
13. Boyd, G. V. *Prog. Heterocycl. Chem.* **1996**, *8*, 192–208.
14. Iddon, B. *Heterocycles* **1994**, *37*, 1321–1346.
15. Asaoka, M.; Takei, H. *Yuki Gosei Kagaku Kyokaishi* **1983**,, *41*, 718–727 [*Chem. Abstr.* **1984**, *100*, 6577].
16. Lipshutz, B. *Chem. Rev*, **1986**, *86*, 795–820.
17. Hartner, F. W. In Shinkai, I., Ed. *Comprehensive Heterocyclic Chemistry II*, Vol. 3, Pergamon, Oxford, UK, 1996, pp. 262–318.
18. Boyd, G. V. In Potts, K. T., Ed. *Comprehensive Heterocyclic Chemistry II*, Vol. 6, Pergamon, Oxford, UK, 1984, pp. 177–233.
19. Lewis, J. R. *Nat. Prod. Rep.* **1992**, *9*, 81–101.
20. Lewis, J. R. *Nat. Prod. Rep.* **1996**, *13*, 435–467.
21. Lewis, J. R. *Nat. Prod. Rep.* **1998**, *15*, 371–395.
22. Lewis, J. R. *Nat. Prod. Rep.* **2000**, *17*, 57–84.
23. Mehicic, M.; Pesa, F. A.; Grasselli, J. G. *J. Phys. Chem.* **1984**, *88*, 581–586.
24. Chen, B. C.; Philipsborn, W. V.; Nagarajan, K. *Helv. Chim. Acta.* **1983**, *66*, 1537–1555.
25. Deady, L. W. *Aust. J. Chem.* **1973**, *26*, 1949–1953.
26. Witanowski, M.; Stefaniak, L.; Januszewaski, H.; Grabowski, Z. *Tetrahedron* **1972**, *28*, 637–653.
27. Evans, D. L.; Minster, D. K.; Jordis, U.; Hecht, S. M.; Mazzu, A. Jr.; Meyers, A. I. *J. Org. Chem.* **1979**, *44*, 497–501.
28. Misra, R. N.; Sher, P. M.; Stein, P. D.; Hall, S. E.; Floyd, D.; Barrish, J. C. Eur. Pat. Appl., EP 476994, 1992 [*Chem. Abstr.* **1992**, *117*, 111371].
29. Levin, J. I.; Weinreb, S. M. *J. Am. Chem. Soc.* **1983**, *105*, 1397–1398.
30. Boyce, R. J.; Mulqueen, G. C.; Pattenden, G. *Tetrahedron Lett.* **1994**, *35*, 5705–5708.
31. Boyce, R. J.; Mulqueen, G. C.; Pattenden, G. *Tetrahedron* **1995**, *51*, 7321–7330.
32. Yokokawa, F.; Hamada, Y.; Shioiri, T. *Synlett* **1992**, 149–151.
33. Yokokawa, F.; Hamada, Y.; Shioiri, T. *Synlett* **1992**, 153–155.
34. Klein, R. F. X.; Horak, V.; Baker, G. A. S. *Collect. Czech. Chem. Commun.* **1993**, *58*, 1631–1635.
35. Meyers, A. I.; Tavares, F. X. *Tetrahedron Lett.* **1994**, *35*, 2481–2484.
36. Meyers, A. I.; Tavares, F. X. *J. Org. Chem.* **1996**, *61*, 8207–8215.
37. Meguro, K.; Tawada, H.; Sugiyama, Y.; Fujita, T.; Kawamatsu, Y. *Chem. Pharm. Bull.* **1986**, *34*, 2840–2851.
38. Shin, C. G.; Nakamura, Y.; Okumura, K. *Chem. Lett.* **1993**, 1405–1408.
39. Shin, C. G.; Okumura, K.; Ito, A.; Nakamura, Y. *Chem. Lett.* **1994**, 1305–1306 (1994).
40. Okumura, K.; Ito, A.; Saito, H.; Nakamura, Y.; Shin, C. G. *Bull. Chem. Soc. Jpn.* **1996**, *69*, 2309–2316.
41. Ishibashi, Y.; Ohba, S.; Nishiyama, S.; Yamamura, S. *Tetrahedron Lett.* **1996**, *37*, 2997–3000.
42. Mitsunobu, O. *Synthesis* **1981**, 1–28.
43. Barrish, J. C.; Singh, J.; Spergel, S. H.; Han, W-C.; Kissick, T. P.; Kronenthal, D. R.; Muellar, R. H. *J. Org. Chem.* **1993**, *58*, 4494–4496.
44. Pihko, P. M.; Koskinen, A. M. P. *J. Org. Chem.* **1998**, *63*, 92–98.
45. Misra, R. N.; Harris, D. N.; Michel, I. M.; Goldenberg, H. J.; Webb, M. L.; Brown, B. R. *Bioorg. Med. Chem.* **1992**, *2*, 937–940.

46. Misra, R. N.; Brown, B. R.; Sher, P. M.; Patel, M. M.; Hall, S. E.; Han, W.-C.; Barrish, J. C.; Kocy, O.; Harris, D. N.; Goldenberg, H. J.; Michel, I. M.; Schumacher, W. A.; Webb, M. L.; Monshizadegam, H.; Ogletree, M. L. *J. Med. Chem.* **1993**, *36*, 1401–1417.

47. Misra, R. N.; Brown, B. R.; Sher, P. M.; Patel, M. M.; Hall, S. E.; Han, W.-C.; Barrish, J. C.; Floyd, D. M.; Sprague, P. W.; Morison, R. A.; Ridgewell, R. E.; White, R. E.; DiDonato, G. C.; Harris, D. N.; Hedberg, A.; Schumacher, W. A.; Webb, M. L.; Ogletree, M. L. *Bioorg. Med. Chem. Lett.* **1992**, *2*, 73–76.

48. Li, G.; Warner, P. M.; Jebaratnam, D. J. *J. Org. Chem.* **1996**, *61*, 778–780.

49. Burgess, E. M.; Penton, H. R. Jr.; Taylor, E. A. *J. Org. Chem.* **1973**, *38*, 26–31.

50. Wipf, P.; Miller, C. P. *Tetrahedron Lett.* **1992**, *33*, 907–910.

51. Provencal, D. P.; Gardelli, C.; Lafontaine, J. A.; Leahy, J. W. *Tetrahedron Lett.* **1995**, *36*, 6033–6036.

52. Wolbers, P.; Hoffmann, H. M. R. *Synthesis* **1999**, 797–802.

53. Rawlinson, D. J.; Sosnovsky, G. *Synthesis* **1972**, 1–28.

54. Rawlinson, D. J.; Sosnovsky, G. *Synthesis* **1973**, 567–603.

55. Downing, S. V.; Aguilar, E.; Meyers, A. I. *J. Org. Chem.* **1999**, *64*, 826–831.

56. Chattopadhyay, S. K.; Pattenden, *Synlett.* **1997**, 1342–1344.

57. Williams, D. R.; Lowder, P. D.; Gu, Y-G.; Brooks, D. A. *Tetrahedron Lett.* **1997**, *38*, 331–334.

58. Videnov, G.; Kaiser, D.; Kempter, C.; Jung, G. *Angew. Chem. Int. Ed. Engl.* **1996**, *35*, 1503–1506.

59. Videnov, G.; Kaiser, D.; Brooks, M.; Jung, G. *Angew. Chem. Int. Ed. Engl.* **1996**, *35*, 1506–1508.

60. Pihko, P. M.; Koskinen, A. M. P.; Nissinen, M. J.; Rissanen, K. *J. Org. Chem.* **1999**, *64*, 652–654.

61. Garey, D.; Ramirez, M.; Gonzales, S.; Wertsching, A.; Tith, S.; Keefe, K.; Peña, M. R. *J. Org. Chem.* **1996**, *61*, 4853–4856.

62. Wipf, P.; Miller, C. P. *J. Org. Chem.* **1993**, *58*, 3604–3606.

63. McGarvey, G. J.; Wilson, K. J.; Shanholtz, C. E. *Tetrahedron Lett.* **1992**, *33*, 2641–2644.

64. Balaban, M.-C.; Schiketanz, I.; Balaban, A. T.; Ana, G.; Mircea, D. *Rev. Roum. Chim.* **1987**, *32*, 975–978.

65. Kashima, C.; Arao, H. *Synthesis* **1989**, 873–874.

66. Eastwood, F. W.; Perlmutter, P.; Yang, Q. *Tetrahedron Lett.* **1994**, *35*, 2039–2042.

67. Williams, E. L. *Tetrahedron Lett.* **1992**, *33*, 1033–1036.

68. Maquestiau, A.; Puk, E.; Flammang, R. *Tetrahedron Lett.* **1986**, *27*, 4023–4024.

69. Pérez, J. D.; Wunderlin, D. A. *J. Org. Chem.* **1987**, *52*, 3637–3640.

70. Pascual, A. *Helv. Chim. Acta* **1989**, *72*, 556–569.

71. Shibanuma, T.; Shiono, M.; Mukaiyama, T. *Chem. Lett.* **1977**, 575–576.

72. Ang, K. H.; Prager, R. H.; Smith, J. A.; Weber, B.; Williams, C. M. *Tetrahedron Lett.* **1996**, *37*, 675–678.

73. Prager, R. H.; Smith, J. A.; Weber, B.; Williams, C. M. *J. Chem. Soc. Perkin Trans 1* **1997**, 2665–2672.

74. Clark, A. D.; Janowski, W. K.; Prager, R. H. *Tetrahedron* **1999**, *55*, 3637–3648.

75. Doleschall, G.; Seres, P. *J. Chem. Soc. Perkin Trans 1* **1988**, 1875–1879.

76. Buscemi, S.; Frenna, V.; Vivona, N. *Heterocycles* **1991**, *32*, 1765–1772.

77. Liu, K. C.; Howe, R. K. *Org. Prep. Proced. Int.* **1983**, *15*, 265–268.

78. Tanaka, H.; Osamura, Y.; Matsushita, T.; Nishimoto, K. *Bull. Chem. Soc. Jpn.* **1981**, *54*, 1293–1298.

79. Tanaka, H.; Matsushita, T.; Nishimoto, K. *J. Am. Chem. Soc.* **1983**, *105*, 1753–1760.

80. Doyle, K. J.; Moody, C. J. *Tetrahedron* **1994**, *50*, 3761–3772.

81. Connell, R. D.; Tebbe, M.; Gangloff, A. R.; Helquist, P.; Aakermark, B. *Tetrahedron* **1993**, *49*, 5445–5459.

82. Tullis, J. S.; Helquist, P. *Org. Synth.* **1997**, *74*, 229–240.
83. Fukushima, K.; Ibata, T. *Bull. Chem. Soc. Jpn.* **1995**, *68*, 3469–3481.
84. Shi, G.; Xu, Y.; Xu, M. *J. Fluorine Chem.* **1991**, *52*, 149–157 (1991).
85. Moody, C. J.; Bagley, M. C. *J. Chem. Soc. Perkin Trans. 1* **1998**, 601–607.
86. Moody, C. J.; Bagley, M. C. *Synlett* **1998**, 361–362.
87. Doyle, K. J.; Moody, C. J. *Tetrahedron Lett.* **1992**, *33*, 7769–7770.
88. Moody, C. J.; Doyle, K. J.; Elliott, M. C.; Mowlem, T. J. *Pure Appl. Chem.* **1994**, *66*, 2107–2110.
89. Moody, C. J.; Bagley, M. C. *Synlett* **1996**, 1171–1172.
90. Bagley, M. C.; Buck, R. T.; Hind, S. L.; Moody, C. J. *J. Chem. Soc. Perkin Trans. 1* **1998**, 591–600.
91. Bagley, M. C.; Buck, R. T.; Hind, S. L.; Moody, C. J.; Slawin, A. M. Z. *Synlett* **1996**, 825–826.
92. Moody, C. J.; Doyle, K. J.; Elliott, M. C.; Mowlem, T. J. *J. Chem. Soc. Perkin Trans. 1* **1997**, 2413–2419.
93. Bagley, M. C.; Hind, S. L.; Moody, C. J. *Tetrahedron Lett.* **2000**, *41*, 6897–6900.
94. Bagley, M. C.; Moody, C. J.; Pepper, A. C. *Tetrahedron Lett.* **2000**, *41*, 6901–6904.
95. Gogonas, E. P.; Hadjiarapoglou, L. P. *Tetrahedron Lett.* **2000**, *41*, 9299–9303.
96. Iso, Y.; Shindo, H.; Hamana, H. *Tetrahedron* **2000**, *56*, 5353–5361.
96. (a) Miyaura, N.; Suzuki, A. *Chem. Rev.* **1995**, *95*, 2457–2483.
97. Ducept, P. C.; Marsden, S. P. *Synlett* **2000**, 692–694.
98. Sonogashira, K.; Tohda, Y.; Hagihara, N. *Tetrahedron Lett.* **1975**, 4467–4470.
99. Ohno, M.; Itoh, M.; Ohashi, T.; Eguchi, S. *Synthesis* **1993**, 793–796.
100. Ibata, T.; Isogami, Y. *Bull. Chem. Soc. Jpn.* **1989**, *62*, 618–620.
101. Fukumoto, T.; Aso, Y.; Otsubo, T.; Ogura, F. *J. Chem. Soc. Chem. Commun.* **1992**, 1070–1072.
102. Lee, J. C.; Song, I.-G. *Tetrahedron Lett.* **2000**, *41*, 5891–5894.
103. Ohkubo, M.; Kuno, A.; Sakai, H.; Takasugi, H. *Chem. Pharm. Bull.* **1995**, *43*, 947–954.
104. Kelly, T. R.; Lang, F. *Tetrahedron Lett.* **1995**, *36*, 5319–5322.
105. Malamas, M. S.; Carlson, R. P.; Grimes, D.; Howell, R.; Glaser, K.; Gunawan, I.; Nelson, J. A.; Kanzelberger, M.; Shah, U.; Hartman, D. A. *J. Med. Chem.* **1996**, *39*, 237–245.
106. Celatka, C. A.; Liu, P.; Panek, J. S. *Tetrahedron Lett.* **1997**, *38*, 5449–5452.
107. Liu, P.; Celatka, C. A.; Panek, J. S. *Tetrahedron Lett.* **1997**, *38*, 5445–5448.
108. Panek, J. S.; Beresis, R. T.; Celatka, C. A. *J. Org. Chem.* **1996**, *61*, 6494–6495.
109. Panek, J. S.; Beresis, R. T. *J. Org. Chem.* **1996**, *61*, 6496–6497.
110. Liu, P.; Panek, J. S. *Tetrahedron Lett.* **1998**, *39*, 6143–6146.
111. Lee, J. C.; Hong, T. *Tetrahedron Lett.* **1997**, *38*, 8959–8960.
112. Meyers, A. I.; Sircar, J. C. In Rappoport, Z., Patai, S., Eds. *The Chemistry of the Cyano Group*, Interscience, New York, 1970, pp. 341–421.
113. Varma, R. S.; Kumar, D. *J. Heterocycl. Chem.* **1998**, *35*, 1533–1534.
114. Salgado-Zamora, H.; Campos, E.; Jiménez, R.; Sánchez-Pavon, E.; Cervantes, H. *Heterocycles* **1999**, *50*, 1081–1090.
115. Majo, V. J.; Perumal, P. T. *Tetrahedron Lett.* **1997**, *38*, 6889–6892.
116. Burger, K.; Huebl, D.; Gertitschke, P. *J. Fluorine Chem.* **1985**, *27*, 327–332.
117. Kotani, E.; Kobayashi, S.; Adachi, M.; Tsujioka, T.; Nakamura, K.; Tobinaga, S. *Chem. Pharm. Bull.* **1989**, *37*, 606–609.
118. Nagayoshi, K.; Sato, Y. *Chem. Lett.* **1983**, 1355–1356.
119. Hall, J. H.; Chien, J. Y.; Kauffman, J. M.; Litak, P. T.; Adams, J. K.; Henry, R. A.; Hollins, R. A. *J. Heterocycl. Chem.* **1992**, *29*, 1245–1273.
120. Kidwai, M.; Kumar, R. *Gazz. Chim. Ital.* **1997**, *127*, 263–267.
121. Patel, H. V.; Fernandes, P. S. *Indian J. Chem. Sect. B* **1989**, *28*, 782–785.

122. Camoutsis, C. C. *J. Heterocycl. Chem.* **1996**, *33*, 539–557.
123. Nilsson, B. M.; Hacksell, U. *J. Heterocycl. Chem.* **1989**, *26*, 269–275.
124. Reisch, J.; Usifoh, C. O.; Oluwadiya, J. O. *J. Heterocycl. Chem.* **1989**, *26*, 1495–1498.
125. Freedman, J.; Huber, E. W. *J. Heterocycl. Chem.* **1990**, *27*, 343–346.
126. Scherrer, V.; Jackson-Mülly, M.; Zsindely, J.; Schmid, H. *Helv. Chim. Acta* **1978**, *61*, 716–731.
127. Wipf, P.; Rahman, L. T.; Rector, S. R. *J. Org. Chem.* **1998**, *63*, 7132–7133.
128. Nagao, Y.; Kim, K.; Sano, S.; Kakegawa, H.; Lee, W. S.; Shimizu, H.; Shiro, M.; Katunuma, N. *Tetrahedron Lett.* **1996**, *37*, 861–864.
129. Short, K. M.; Ziegler, C. B. Jr., *Tetrahedron Lett.* **1993**, *34*, 71–74.
130. Takeuchi, H.; Yanagida, S.; Ozaki, T.; Hagiwara, S.; Eguchi, S. *J. Org. Chem.* **1989**, *54*, 431–434.
131. For a review, see Gololobov, Y. G.; Zhmurova, I. N.; Kasukhin, L. F. *Tetrahedron* **1981**, *37*, 437–472.
132. Molina, P.; Fresneda, P. M.; Almendros, P. *Heterocycles* **1993**, *36*, 2255–2258.
133. Saalfrank, R. W.; Ackermann, E.; Fischer, M.; Wirth, U.; Zimmermann, H. *Chem. Ber.* **1990**, *123*, 115–120.
134. Lautens, M.; Roy, A. *Org. Lett.* **2000**, *2*, 555–557.
135. Gronsdemange-Pale, C.; Chuche, J. *Bull. Soc. Chim. Fr.* **1989**, 644–649.
136. Gronsdemange-Pale, C.; Chuche, J. *Tetrahedron* **1989**, *45*, 3397–3414.
137. Jones, B. G.; Branch, S. K.; Threadgill, M. D.; Wilman, D. E. V. *J. Fluorine Chem.* **1995**, *74*, 221–222.
138. Nicolaides, D. N.; Varella, E. A.; Awad, R. W. *Tetrahedron* **1993**, *49*, 7779–7786.
139. Nicolaides, D. N.; Papageorgiou, G. K.; Stephanidou-Stephanatou, J. *Tetrahedron* **1989**, *45*, 4585–4592.
140. Nicolaides, D. N.; Awad, R. W.; Varella, E. A. *J. Heterocycl. Chem.* **1996**, *33*, 633–637.
141. Nicolaides, D. N.; Awad, R. W.; Papageorgiou, G. K.; Kojanni, E.; Tsoleridis, C. A. *J. Heterocycl. Chem.* **1997**, *34*, 1651–1656.
142. Goddard, C. J. *J. Heterocycl. Chem.* **1991**, *28*, 17–28.
143. Wasserman, H. H.; Prowse, K. S. *Tetrahedron* **1992**, *48*, 8199–8212.
144. Jacobsen, N. W.; Philippides, A. *Aust. J. Chem.* **1985**, *38*, 1335–1338.
145. Shafiullah, A. J. A.; Ansari, A. J. *Synth. Commun.* **1983**, *13*, 419–425.
146. Merslavic, M.; Stanovnik, B.; Tišer, M. *Monatsh. Chem.* **1986**, *117*, 221–230.
147. Kamitori, Y.; Hojo, M.; Masuda, R.; Takahashi, T.; Wada, M. *Heterocycles* **1992**, *34*, 1047–1054.
148. Narsaiah, B.; Sivaprasad, A.; Venkataratnam, R. V. *J. Fluorine Chem.* **1994**, *66*, 47–50.
149. Cornforth, J. W. *The Chemistry of Penicillin*, Princeton University Press: Princeton, NJ, 1949.
150. Yokoyama, M.; Irie, M.; Sujino, K.; Kagemoto, T.; Togo, H.; Funabashi, M. *J. Chem. Soc. Perkin Trans. 1* **1992**, 2127–2134.
151. Barrett, A. G. M.; Edmunds, J. J.; Hendrix, J. A.; Malecha, J. W.; Parkinson, C. J. *J. Chem. Soc. Chem. Commun.* **1992**, 1240–1242.
152. Hajjem, B.; Baccar, B. *Synth. Commun.* **1991**, *21*, 1501–1509.
153. Cardwell, K. S.; Hermitage, S. A.; Sjolin, A. *Tetrahedron Lett.* **2000**, *41*, 4239–4242.
154. Schöllkopf, U.; Gerhart, F. *Angew. Chem. Int. Ed. Engl.* **1968**, *7*, 805–806.
155. Huang, W.-S.; Zhang, Y. X.; Yuan, C-Y. *Synth. Commun.* **1996**, *26*, 1149–1154.
156. Bossio, R.; Marcaccini, S.; Pepino, R. *Heterocycles* **1986**, *24*, 2003–2005.
157. Bossio, R.; Marcaccini, S.; Pepino, R.; Polo, C.; Torroba, T. *Heterocycles* **1989**, *29*, 1829–1833.
158. Ozaki, Y.; Maeda, S.; Iwasaki, T.; Matsumoto, K.; Odawara, A.; Sasaki, Y.; Morita, T. *Chem. Pharm. Bull.* **1983**, *12*, 4417–4424.
159. Ohba, M.; Kubo, H.; Seto, S.; Fujii, T.; Ishibashi, H. *Chem. Pharm. Bull.* **1998**, *46*, 860–862.

160. Ohba, M.; Kubo, H.; Fujii, T.; Ishibashi, H.; Sargent, M. V.; Arbain, D. *Tetrahedron Lett.* **1997**, *38*, 6697–6700.
161. Ohba, M.; Izuta, R.; Shimizu, E. *Tetrahedron Lett.* **2000**, *41*, 10251–10255.
162. Tang, J.; Verkade, J. G. *J. Org. Chem.* **1994**, *59*, 7793–7802.
163. Hamada, Y.; Kawai, A.; Matsui, T.; Hara, O.; Shioiri, T. *Tetrahedron* **1990**, *46*, 4823–4846.
164. Bossio, R.; Marcaccini, S.; Pepino, R. *Liebigs, Ann. Chem.* **1991**, 1107–1108.
165. Passerini, M. *Gazz. Chem. Ital.* **1921**, *51*, 126–129.
166. Davidson, D.; Weiss, M.; Jelling, M. *J. Org. Chem.* **1937**, *2*, 328–334.
167. Zhou, X.-T.; Lin, Y.-R.; Dai, L.-X.; Sun, J. *Tetrahedron* **1998**, *54*, 12445–12456.
168. van Leusen, A. M.; Hoogenboom, B. E.; Siderius, H. *Tetrahedron Lett.* **1972**, 2369–2373.
169. Anderson, B. A.; Becke, L. M.; Booher, R. N.; Flaugh, M. E.; Harn, N. K.; Kress, T. J.; Varie, D. L.; Wepsiec, J. P. *J. Org. Chem.* **1997**, *62*, 8634–8639.
170. Crozet, M. P.; Sabuco, J. F.; Tamburlin, I.; Barreau, M.; Giraud, L.; Vanelle, P. *Heterocycles* **1993**, *36*, 45–54.
171. Addie, M. S.; Taylor, R. J. K. *J. Chem. Soc. Perkin Trans 1* **2000**, 527–531.
172. Shafer, C. M.; Molinski, T. F. *Heterocycles* **2000**, *53*, 1167–1170.
173. Kulkarni, B. A.; Ganesan, A. *Tetrahedron Lett.* **1999**, *40*, 5637–5638.
174. Kulkarni, B. A.; Ganesan, A. *Tetrahedron Lett.* **1999**, *40*, 5633–5636.
175. Das, J.; Reid, J. A.; Kronenthal, D. R.; Singh, J.; Pansegrau, P. D.; Mueller, R. H. *Tetrahedron Lett.* **1992**, *33*, 7835–7838.
176. Shridhar, D. R.; Ram, B.; Sarma, C. R.; Thapar, G. S.; Krishnamurthy, A. *Indian J. Chem. Sect. B* **1984**, *23*, 183–185.
177. Huang, W.; Pei, J.; Chen, B.; Pei, W.; Ye, X. *Tetrahedron* **1996**, *52*, 10131–10136.
178. Robinson, R. *J. Chem. Soc.* **1909**, *95*, 2167–2174.
179. Gabriel, S. *Chem. Ber.* **1910**, *43*, 1283–1287.
180. For a review, see Buchanan, G. L. *Chem. Soc. Rev.* **1988**, *17*, 91–109.
181. Meguro, K.; Fujita, T. Jpn. Kokai Tokkyo Koho JP 61,282,369, 1986; *Chem. Abstr.* **1987**, *106*, 176372.
182. White, R. L. Jr.; Wessels, F. L.; Schwan, T. J.; Ellis, K. O. *J. Med. Chem.* **1987**, *30*, 263–266.
183. Loughlin, W. A.; Muderawan, I. W.; McCleary, M. A.; Volter, K. E.; King, M. D. *Aust. J. Chem.* **1999**, *52*, 231–234.
184. Jacobi, P. A.; Blum, C. A.; DeSimone, R. W.; Udodong, U. E. S. *Tetrahedron Lett.* **1989**, *30*, 7173–7176.
185. Jacobi, P. A.; Blum, C. A.; DeSimone, R. W.; Udodong, U. E. S. *J. Am. Chem. Soc.* **1991**, *113*, 5384–5392.
186. Litak, P. T.; Kauffman, J. M. *J. Heterocycl. Chem.* **1994**, *31*, 457–479.
187. Reck, S.; Friedrichsen, W. *J. Org. Chem.* **1998**, *63*, 7680–7686.
188. Katritzky, A. R.; Rachwal, B.; Rachwal, S.; Macomber, D.; Smith, T. P. *J. Heterocycl. Chem.* **1993**, *30*, 135–139.
189. Jacobi, P. A.; Kaczmarek, C. S. R.; Udodong, U. E. S. *Tetrahedron* **1987**, *43*, 5475–5488.
190. Jacobi, P. A.; Lee, K. *J. Am. Chem. Soc.* **1997**, *119*, 3409–3410.
191. Nahm, S.; Weinreb, S. M. *Tetrahedron Lett.* **1981**, *22*, 3815–3818.
192. Brunner, H.; Olschewski, G.; Nuber, B. *Synthesis* **1999**, 429–434.
193. Savarino, P.; Viscardi, G.; Carpignano, R.; Borda, A.; Barni, E. *J. Heterocycl. Chem.* **1989**, *26*, 289–292.
194. Huth, A.; Rosenberg, D.; Schumann, I.; Thielert, K. *Liebigs Ann. Chem.* **1984**, 641–648.
195. Parsons, R. L.; Heathcock, C. H. *Synlett* **1996**, 1168–1170.

196. Nakahara, Y.; Fujita, A.; Beppu, K.; Ogawa, T. *Tetrahedron* **1986**, *42*, 6465–6476.
197. Lipshutz, B. H.; Hungate, R. W.; McCarthy, K. E. *Tetrahedron Lett.* **1983**, *24*, 5155–5158.
198. Lipshutz, B. H.; Hungate, R. W.; McCarthy, K. E. *J. Am. Chem. Soc.* **1983**, *105*, 7703–7713.
199. Lipshutz, B. H.; Hungate, R. W.; McCarthy, K. E. *Isr. J. Chem.* **1986**, *27*, 49–55.
200. Lipshutz, B. H.; Huff, B. E.; McCarthy, K. E.; Mukkarram, S. M. J.; Siahann, T. J.; Vaccaro, W. D.; Webb, H.; Falick, A. M.; Miller, T. A. *J. Am. Chem. Soc.* **1990**, *112*, 7032–7041.
201. Hulin, B.; Newton, L. S.; Lewis, D. M.; Genereux, P. E.; Gibbs, E. M.; Clark, D. A. *J. Med. Chem.* **1996**, *39*, 3897–3907.
202. Kawase, M. *Heterocycles* **1993**, *36*, 2441–2444.
203. Kawase, M.; Miyamae, H.; Kurihara, T. *Chem. Pharm. Bull.* **1998**, *46*, 749–756.
204. Boto, A.; Ling, M.; Meek, G.; Pattenden, G. *Tetrahedron Lett.* **1998**, *39*, 8167–8170.
205. Theilig, G. *Chem. Ber.* **1953**, *86*, 96–109.
206. Crimmin, M. J.; O'Hanlon, P. J.; Rogers, N. H.; Sime, F. M.; Walker, G. *J. Chem. Soc. Perkin Trans. 1* **1989**, 2059–2063.
207. Freeman, F.; Chen, T.; Van der Linden, J. B. *Synthesis* **1997**, 861–862.
208. Freeman, F.; Kim, D. S. H. L. *Tetrahedron Lett.* **1989**, *30*, 2631–2632.
209. Gordon, T. D.; Singh, J.; Hansen, P. E.; Morgan, B. A. *Tetrahedron Lett.* **1993**, *34*, 1901–1904.
210. Gordon, T.; Hansen, P.; Morgan, B.; Singh, J.; Baizman, E.; Ward, S. *Bioorg. Med. Chem. Lett.* **1993**, *3*, 915–920.
211. Dess, D. B.; Martin, J. C. *J. Am. Chem. Soc.* **1991**, *113*, 7277–7287.
212. Sen, P. K.; Veal, C. J.; Young, D. W. *J. Chem. Soc. Perkin Trans. 1* **1981**, 3053–3058.
213. Hoekstra, W. J.; Hulshizer, B. L.; Mccomsey, D. F.; Andrade-Gorden, P.; Kauffman, J. A.; Addo, M. F.; Oksenberg, D.; Scarborough, R. M.; Maryanoff, B. E. *Bioorg. Med. Chem. Lett.* **1998**, *8*, 1649–1654.
214. Falorni, M.; Dettori, G.; Giacomelli, G. *Tetrahedron: Asymmetry* **1998**, *9*, 1419–1426.
215. Swaminathan, S.; Singh, A. K.; Li, W.-S.; Venit, J. J.; Natalie, K. J. Jr.; Simpson, J. H.; Weaver, R. E.; Silverberg, L. J. *Tetrahedron Lett.* **1998**, *39*, 4769–4772.
216. Broom, N. J. P.; Elder, J. S.; Hannan, P. C. T.; Pons, J. E.; O'Hanlon, P. J.; Walker, G.; Wilson, J.; Woodall, P. *J. Antibiot.* **1995**, *48*, 1336–1344.
217. Broom, N. J. P.; Cassels, R.; Cheng, H.-Y.; Elder, J. S.; Hannan, P. C. T.; Masson, N.; O'Hanlon, P. J.; Pope, A.; Wilson, J. M. *J. Med. Chem.* **1996**, *39*, 3596–3600.
218. Brain, C. T.; Paul, J. M. *Synlett* **1999**, 1642–1644.
219. Mohapatra, D. K.; Datta, A. *Synlett* **1996**, 1129–1130.
220. Lakhan, R.; Singh, R. L. *J. Heterocycl. Chem.* **1988**, *25*, 1413–1417.
221. Cunico, R. F.; Kaun, C. P. *J. Org. Chem.* **1992**, *57*, 6999–7000.
222. Cunico, R. F.; Kaun, C. P. *Tetrahedron Lett.* **1990**, *31*, 1945–1948.
223. Cunico, R. F.; Kaun, C. P. *J. Org. Chem.* **1992**, *57*, 3331–3336.
224. Eissenstat, M. A.; Weaver, J. D. III. *J. Org. Chem.* **1993**, *58*, 3387–3390.
225. Bailey, T. R.; Diana, G. D.; Kowalczyk, P. J.; Akullian, V.; Eissenstat, M. A.; Cutcliffe, D.; Mallamo, J. P.; Carabateas, P. M.; Pevear, D. C. *J. Med. Chem.* **1992**, *35*, 4628–4633.
226. Herdeis, C.; Gebhard, R. *Heterocycles* **1986**, *24*, 1019–1024.
227. Clauss, K.-U.; Buck, K.; Abraham, W. *Tetrahedron* **1995**, *51*, 7181–7192.
228. Vicentini, C. B.; Veronese, A. C.; Poli, T.; Guarneri, M.; Giori, P. *Heterocycles* **1991**, *32*, 727–734.
229. Katritzky, A. R.; Yang, Z.; Cundy, D. J. *Aldrichim. Acta* **1994**, *27*, 31–38.
230. Katritzky, A. R.; Lan, X. *Chem. Soc. Rev.* **1994**, 363–373.

231. Katritzky, A. R.; Lan, X.; Fan, W.-Q. *Synthesis* **1994**, 445–456.
232. Katritzky, A. R.; Lan, X.; Yang, J. Z.; Denisko, O. V. *Chem. Rev. (Washington, D.C.)* **1998**, *98*, 409–548.
233. Katritzky, A. R.; Qi, M. *Tetrahedron* **1998**, *54*, 2647–2668.
234. Katritzky, A. R. *J. Heterocycl. Chem.* **1999**, *36*, 1501–1522.
235. Katritzky, A. R.; Wu, H.; Xie, L. *J. Heterocycl. Chem.* **1995**, *32*, 1651–1652.
236. Shapiro, R. *J. Org. Chem.* **1993**, *58*, 5759–5764.
237. Larina, L. I.; Lopyrev, V. A.; Larina, L. I.; Voronkov, M. G. *Rev. Heteroat. Chem.* **1994**, *11*, 27–67.
238. Larina, L. I.; Lopyrev, V. A.; Voronkov, M. G. *Zh. Org. Khim.* **1994**, *30*, 1081–1118 [*Chem. Abstr.* **1995**, *122*, 314472].
239. Venkatasubramanian, R.; Krishnamachari, S. L. N. G. *Indian J. Chem. Sect. B* **1990**, *29*, 562–563.
240. Hassner, A.; Fischer, B. *Tetrahedron* **1989**, *45*, 6249–6262.
241. Belen'kii, L. I.; Cheskis, M. A. *Khim. Gesterotsikl. Soedin.* **1984**, 881–884 [*Chem. Abstr.* **1984**, *101*, 191754].
242. Belen'kii, L. I.; Gromova, G. P.; Cheskis, M. A.; Gol'dfarb, Y. L. *Chem. Scr.* **1985**, *25*, 295–299 [*Chem. Abstr.* **1987**, *106*, 17638].
243. Lawson, J. P.; VanSant, K. A. *J. Heterocycl. Chem.* **1999**, *36*, 283–285.
244. Gompper, R.; Christmann, O. *Chem. Ber.* **1959**, *92*, 1944–1949.
245. Šindler-Kulyk, M.; Vojnović, D.; Defterdarović, N.; Marinić, Ž.; Srzić, D. *Heterocycles* **1994**, *38*, 1791–1796.
246. Patsenker, L. D.; Lokshin, A. I.; Drushlyak, T. G.; Baumer, V. N. *Chem. Heterocyl. Compd. (N.Y.)* **1997**, *33*, 1266–1271.
247. Patsenker, L. D.; Ermolenko, I. G.; Fedunyaeva, I. A.; Popova, N. A.; Krasovitskii, B. M. *Chem. Heterocyl. Compd. (N.Y.)* **2000**, *36*, 623–625.
248. Barbieri, G.; Benassi, R.; Taddei, F. *Theochem* **1989**, *53*, 269–276.
249. Kong, Y. C.; Kim, K. *J. Heterocycl. Chem.* **1999**, *36*, 911–915.
250. Harreus, A.; Kirstgen, R.; Oberdorf, K.; Wingert, H.; Koehler, U. Eur. Pat. Appl. EP0557834, 1993 [*Chem. Abstr.* **1994**, *120*, 134481].
251. Bossio, R.; Marcaccini, S.; Pepino, R.; Polo, C.; Torroba, T. *Org. Prep. Proced. Int.* **1991**, *23*, 670–672.
252. Crank, G.; Kahn, H. R. *Aust. J. Chem.* **1985**, *38*, 447–458.
253. Mekonnen, B.; Crank, G. *J. Heterocycl. Chem.* **1997**, *34*, 567–572.
254. Crank, G.; Mekonnen, B. *J. Heterocycl. Chem.* **1992**, *29*, 1469–1472.
255. Khan, H. R.; Crank, G. *Tetrahedron Lett.* **1987**, *28*, 3381–3382.
256. Khan, H. R.; Crank, G.; Jesdapaulpaan, S. *J. Heterocycl. Chem.* **1988**, *25*, 815–817.
257. Crank, G.; Neville, M.; Ryden, R. *J. Med. Chem.* **1973**, *16*, 1402.
258. Kujundžić, N.; Glunčić, B. *Croat. Chem. Acta* **1990**, *63*, 215–224.
259. Gompper, R.; Effenberger, F. *Chem. Ber.* **1959**, *92*, 1928–1934.
260. Mizuno, Y.; Adachi, K.; Ikeda, K. *Chem. Pharm. Bull.* **1954**, *2*, 225–234.
261. Yamanaka, H.; Ohba, S. Sakamoto, T. *Heterocycles* **1990**, *31*, 1115–1127.
262. Burger, K.; Hűbl, D.; Geith, K. *Synthesis* **1988**, 194–198.
263. Burger, K.; Geith, K.; Hűbl, D. *Synthesis* **1988**, 199–203.
264. Burger, K.; Geith, K.; Hoess, E. *Synthesis* **1990**, 352–356.
265. Burger, K.; Hoess, E.; Geith, K. *Synthesis* **1990**, 360–365.
266. Burger, K.; Hoess, E.; Gaa, K.; Sewald, N.; Schierlinger, C. *Z. Naturforsch. B Chem. Sci.* **1991**, *46*, 361–384.

267. Burger, K.; Geith, K.; Gaa, K. *Angew. Chem.* **1988**, *100*, 860–861.
268. Burger, K.; Gaa, K.; Geith, K.; Schierlinger, C. *Synthesis* **1989**, 850–855.
269. Burger, K.; Geith, K.; Gaa, K. Eur. Pat. Appl. EP 336305, 1989 [*Chem. Abstr.* **1990**, *112*, 199122].
270. Burger, K.; Helmreich, B. *Chem. Ztg.* **1991**, *115*, 253–255.
271. Padwa, A.; Cohen, L. A. *J. Org. Chem.* **1984**, *49*, 399–406.
272. Röhr, G.; Köckritz, A.; Schnell, M. *Phosphorus Sulfur Silicon Relat. Elem.* **1992**, *71*, 157–163.
273. Röhr, G.; Schnell, M.; Köckritz, A. *Synthesis* **1992**, 1031–1034.
274. Scheidecker, S.; Köckritz, A.; Schnell, M. *J. Prakt. Chem.* **1990**, *332*, 968–976.
275. Köckritz, A.; Schnell, M. *Phosphorus Sulfur Silicon Relat. Elem.* **1993**, *83*, 125–133.
276. Ueda, Y.; Chuang, J. M.; Crast, L. B.; Partyka, R. A. *J. Org. Chem.* **1988**, *53*, 5107–5113.
277. Hamada, Y.; Shioiri, T. *J. Org. Chem.* **1986**, *51*, 5489–5490.
278. Loupy, A.; Petit, A.; Zaparaucha, A.; Mahieu, C.; Semeria, D. *Bull. Soc. Chim. Fr.* **1994**, 642–647.
279. Lunn, G. *J. Org. Chem.* **1987**, *52*, 1043–1046.
280. Muneer, M.; Kamat, P. V.; George, M. V. *Can. J. Chem.* **1990**, *68*, 969–975.
281. Tsveniashvili, V. S.; Malashkhiya, M. V.; Shvaika, O. P.; Khavtasi, N. S. *Zh. Obshch. Khim.* **1984**, *54*, 624–630 [*Chem. Abstr.* **1984**, *101*, 72019].
282. Tsveniashvili, V. S.; Shvaika, O. P.; Malashkhiya, M. V.; Korzhenevskaya, N. G.; Snagoshchenko, L. P. *Zh. Obshch. Khim.* **1985**, *55*, 1774–1777 [*Chem. Abstr.* **1986**, *104*, 5736].
283. Rogers, J. W.; Sund, E. H.; Hughlett, R. K. *Tex. J. Sci.* **1991**, *43*, 149–156.
284. Tanaka, C.; Kuriyama, S. *Yakugaku Zasshi* **1979**, *99*, 78–82 [*Chem. Abstr.* **1979**, *90*, 203922].
285. Wasserman, H. H.; Lu, T. J. *Recl. Trav. Chim. Pays.-Bas.* **1986**, *105*, 345–346.
286. Wasserman, H. H.; Gambale, R. J. *Tetrahedron* **1992**, *48*, 7059–7070.
287. Wasserman, H. H.; Prowse, K. S. *Tetrahedron Lett.* **1992**, *33*, 5423–5426.
288. Wasserman, H. H.; DeSimone, R. W.; Ho, W.-B.; McCarthy, K. E.; Prowse, K. S.; Spada, A. P. *Tetrahedron Lett.* **1992**, *33*, 7207–7210.
289. Wasserman, H. H.; Yoo, J. U.; DeSimone, R. W. *J. Am. Chem. Soc.* **1995**, *117*, 9772–9773.
290. Gollnick, K.; Koegler, S. *Tetrahedron Lett.* **1988**, *29*, 1003–1006.
291. Gollnick, K.; Koegler, S. *Tetrahedron Lett.* **1988**, *29*, 1007–1010.
292. Kashima, C.; Hibi, S.; Maruyama, T.; Harada, K.; Omote, Y. *J. Heterocycl. Chem.* **1987**, *24*, 637–639.
293. Kashima, C.; Hibi, S.; Harada, K.; Omote, Y. *J. Chem. Soc. Perkin Trans. 1* **1988**, 529–533.
294. Kashima, C.; Arao, H.; Hibi, S.; Omote, Y. *Tetrahedron Lett.* **1989**, *30*, 1561–1562.
295. Kashima, C.; Arao, H. *Heterocycles* **1990**, *31*, 1513–1516.
296. Kashima, C.; Arao, H. *J. Heterocycl. Chem.* **1991**, *28*, 805–806.
297. Kashima, C.; Okada, R.; Arao, H. *J. Heterocycl. Chem.* **1991**, *28*, 1241–1244.
298. Kashima, C.; Arao, H.; Hibi, S. *J. Chem. Research (S)* **1991**, 34–35.
299. Ichimura, K; Ikeda, T.; Ito, H. *Macromol. Chem. Rapid Commun.* **1992**, *13*, 415–420.
300. Ito, H.; Ikeda, T.; Ichimura, K. *Macromolecules* **1993**, *26*, 4533–4538.
301. Ito, H.; Ichimura, K. *Macromol. Chem. Phys.* **1995**, *196*, 995–1003.
302. Ito, H.; Ichimura, K. *Macromol. Chem. Phys.* **1998**, *199*, 2547–2551.
303. Curtin, D. Y.; Miller, L. *Tetrahedron Lett.* **1965**, 1869–1876.
304. Wasserman, H. H.; Vinick, F. J.; Chang, Y. C. *J. Am. Chem. Soc.* **1972**, *94*, 7180–7182.
305. Wasserman, H. H.; Pickett, J. A.; Vinick, F. J. *Heterocycles* **1981**, *15*, 1069–1073.
306. Pridgen, L. N.; Shilcrat, S. C.; Lantos, I. *Tetrahedron Lett.* **1984**, *25*, 2835–2838.
307. Katsuki, T.; Sharpless, K. B. *J. Am. Chem. Soc.* **1980**, *102*, 5974–5976.
308. Wasserman, H. H.; Lu, T.-J. *Tetrahedron Lett.* **1982**, *23*, 3831–3834.

References

309. Chauhan, P. M. S.; Crew, A. P. A.; Jenkins, G.; Storr, C.; Walker, S. M.; Yelland, M. *Tetrahedron Lett.* **1990**, *31*, 1487–1490.
310. Chou, T.-S.; Chen, H.-C.; Tsai, C.-Y. *J. Org. Chem.* **1994**, *59*, 2241–2245.
311. Chao, I.; Lu, H.-F.; Chou, T.-S. *J. Org. Chem.* **1997**, *62*, 7882–7884.
312. Chou, T. S.; Tsai, C. Y. *Tetrahedron Lett.* **1992**, *33*, 4201–4204.
313. Manoharan, M.; De Proft, F.; Geerlings, P. *J. Org. Chem.* **2000**, *65*, 7971–7976.
314. Ohno, M.; Koide, N.; Sato, H.; Eguchi, S. *Tetrahedron* **1997**, *53*, 9075–9086.
315. Arbogast, J. W.; Darmanyan, A. P.; Foote, C. S.; Rubin, Y.; Diederich, F.; Alvarez, M. M.; Anz, S. J.; Whetten, R. J. *J. Phys. Chem.* **1991**, *95*, 11–12.
316. Wasielewsky, M. R.; O'Neil, M. P.; Lykke, K. R.; Pellin, M. J.; Gruen, D. M.; *J. Am. Chem. Soc.* **1991**, *113*, 2774–2776.
317. Ibata, T.; Isogami, Y.; Nakano, S.; Nakawa, H.; Tamura, H.; *J. Chem. Soc., Chem. Commun.* **1986**, 1692–1693.
318. Ibata, T.; Isogami, Y.; Nakawa, H.; Tamura, H.; Suga, H.; Shi, X.; Fujieda, H. *Bull. Chem. Soc. Jpn.* **1992**, *65*, 1771–1778.
319. Ibata, T.; Nakano, S.; Nakawa, H.; Toyoda, J.; Isogami, Y. *Bull. Chem. Soc. Jpn.* **1986**, *59*, 433–437.
320. Hassner, A.; Fischer, B. *Tetrahedron* **1989**, *45*, 3535–3546.
321. Suga, H.; Shi, X.; Ibata, T. *Chem. Lett.* **1994**, 1673–1676.
322. Suga, H.; Shi, X.; Ibata, T. *J. Org. Chem.* **1993**, *58*, 7397–7405.
323. Suga, H.; Fujieda, H.; Hirotsu, Y.; Ibata, T. *J. Org. Chem.* **1994**, *59*, 3359–3364.
324. Suga, H.; Shi, X.; Fujieda, H.; Ibata, T. *Tetrahedron Lett.* **1991**, *32*, 6911–6914.
325. Suga, H.; Ikai, K.; Ibata, T. *Tetrahedron Lett.* **1998**, *39*, 869–872.
326. Suga, H.; Ikai, K.; Ibata, T. *J. Org. Chem.* **1999**, *64*, 7040–7047.
327. Vedejs, E.; Fields, S. *J. Org. Chem.* **1988**, *53*, 4663–4667.
328. Hassner, A.; Fischer, B. *J. Org.Chem.* **1991**, *56*, 3419–3425.
329. Suga, H.; Ibata, T. *Chem. Lett.* **1991**, 1221–1224.
330. Suga, H.; Shi, X.; Ibata, T. *Bull. Chem. Soc. Jpn.* **1998**, *71*, 1231–1236.
331. Ibata, T.; Isogami, Y.; Tamura, H. *Chem. Lett.* **1988**, 1551–1554.
332. Ibata, T.; Suga, H.; Isogami, Y.; Tamura, H.; Shi, X. *Bull. Chem. Soc. Jpn.* **1992**, *65*, 2998–3007.
333. Fischer, J.; Kilpert, C.; Klein, U.; Steglich, W. *Tetrahedron* **1986**, *42*, 2063–2074.
334. Colombo, L.; Casiraghi, G.; Pittalis, A. *J. Org. Chem.* **1991**, *56*, 3897–3900.
335. Kubota, H.; Moriya, T.; Matsumoto, K. *Chem. Pharm. Bull.* **1990**, *38*, 570–572.
336. Turchi, I. J.; Cullen, T. G. *Heterocycles* **1984**, *22*, 2463–2466.
337. LaMattina, J. L.; Mularski, C. J. *Tetrahedron Lett.* **1984**, *25*, 2957–2960.
338. Walker, F. J.; Kraus, K. G. *J. Heterocycl. Chem.* **1987**, *24*, 1485–1486.
339. Sen, A. K.; Sengupta, D. K. *Indian J. Chem. Sect. B* **1985**, *24*, 535–538.
340. Ray, S.; Pal, S. K.; Saha, C. K. *Indian J. Chem. Sect. B.* **1995**, *34*, 112–115.
341. Sasaki, H.; Kitagawa, T. *Chem. Pharm. Bull.* **1988**, *36*, 1593–1596.
342. Sasaki, H.; Nakagawa, H.; Khuhara, M.; Kitagawa, T. *Chem. Lett.* **1988**, 1531–1534.
343. Sasaki, H.; Egi, R.; Kawanishi, K.; Kitagawa, T.; Shingu, T. *Chem. Pharm. Bull.* **1989**, *37*, 1176–1178.
344. Sasaki, H.; Kawanishi, K.; Kitagawa, T.; Shingu, T. *Chem. Pharm. Bull.* **1989**, *37*, 2303–2306.
345. Corrao, S.; Macielag, M. J.; Turchi, I. J. *J. Org. Chem.* **1990**, *55*, 4484–4487.
346. Scheibye, S.; Pederson, B. S.; Lawesson, S.-O. *Bull. Soc. Chim. Belg.* **1978**, *87*, 229–238.
347. Crowe, E.; Hossner, F.; Hughes, M. J. *Tetrahedron* **1995**, *51*, 8889–8900.
348. Hilf, C.; Bosold, F.; Harms, K.; Marsch, M.; Boche. G. *Chem. Ber. Recl.* **1997**, *130*, 1213–1221.

349. Boche, G.; Bosold, F.; Hermann, H.; Marsch, M.; Harms, K.; Lohrenz, J. C. W. *Chem. Eur. J.* **1998,** *4,* 814–817.
350. Adamczeski, M.; Quiñoà, E.; Crews, P. *J. Am. Chem. Soc.* **1988,** *110,* 1598–1602.
351. Crews, P.; Matthews, T. R.; Cabana, E. Q.; Adamczeski, M. U. S. Patent, US 4,785,012, 1988 [*Chem. Abstr.* **1989** *110,* 231615].
352. Chittari, P.; Hamada, Y.; Shioiri, T. *Synlett* **1998,** 1022–1023.
353. Barton, D. H. R.; McCombie, S. W. *J. Chem. Soc., Perkin Trans. 1* **1975,** 1574–1585.
354. Shafer, C. M.; Molinski, T. F. *Tetrahedron Lett.* **1998,** *39,* 2903–2906.
355. Saunders, J.; Cassidy, M.; Freedman, S. B.; Harley, E. A.; Iversen, L. L.; Kneen, C.; MacLeod, A. M.; Merchant, K. J.; Snow, R. J.; Baker, R. *J. Med. Chem.* **1990,** *33,* 1128–1138.
356. Baker, R.; Snow, R. J.; Saunders, G. A.; Showell, G. A. Eur. Pat. Appl. EU0307141, 1988 [*Chem. Abstr.* **1989,** *111,* 153786].
357. Vedejs, E.; Monahan, S. D. *J. Org. Chem.* **1996,** *61,* 5192–5193.
358. Vedejs, E.; Luchetta, L. M. *J. Org. Chem.* **1999,** *64,* 1011–1014.
359. Shafer, C. M.; Molinski, T. F. *J. Org. Chem.* **1998,** *63,* 551–555.
360. Williams, D. R.; Brooks, D. A.; Meyer, K. G.; *Tetrahedron Lett.* **1998,** *39,* 8023–8026.
361. Williams, D. R.; McClymont, E. L. *Tetrahedron Lett.* **1993,** *34,* 7705–7708.
362. Liu, P.; Panek, J. S. *Tetrahedron Lett.* **1998,** *39,* 6147–6150.
363. Evans, D. A.; Cee, V. J.; Smith, T. E.; Santiago, K. J. *Org. Lett.* **1999,** *1,* 87–90.
364. Harn, N. K.; Gramer, C. J.; Anderson, B. A. *Tetrahedron Lett.* **1995,** *36,* 9453–9456.
365. Gangloff, A. R.; Akermark, B.; Helquist, P. *J. Org. Chem.* **1992,** *57,* 4797–4799.
366. Bergdahl, M.; Hett, R.; Friebe, T. L.; Gangloff, A. R.; Iqbal, J.; Wu, Y.; Helquist, P. *Tetrahedron Lett.* **1993,** *34,* 7371–7374.
367. Fieser, L. F.; Fieser, M. *Reagents for Organic Synthesis*, Vol. 1, Wiley, New York, 1967.
367. (a) Mancuso, A. J.; Swern, D. *Synthesis*, **1981,** 165–185.
368. Baker, W. R. *J. Org. Chem.* **1985,** *50,* 3943–3945.
369. Pridgen, L. N. *Synthesis* **1984,** 1047–1048.
370. Sakamoto, T.; Nagata, H.; Kondo, Y.; Shiraiwa, M.; Yamanaka, H. *Chem. Pharm. Bull.* **1987,** *35,* 823–828.
370. (a) Schaus, J. V.; Panek, J. S. *Org. Lett.* **2000,** *2,* 469–471.
371. Jeong, S.; Chen, X.; Harran, P. G. *J. Org. Chem.* **1998,** *63,* 8640–8641.
371. (a) Kelly, T. R.; Lang, F. *J.Org. Chem.* **1996,** *61,* 4623–4633.
372. Dondoni, A.; Fantin, G.; Fogagnolo, M.; Medica, A.; Pedrini, P. *Synthesis* **1987,** 693–696.
372. (a) Dondoni, A.; Fantin, G.; Fogagnolo, M.; Medici, A.; Pedrini, P. *J. Org. Chem.* **1987,** *52,* 3413–3420.
373. Booher, R. N.; Flaugh, M. E.; Lawhorn, D. E.; Paget, C. J. Jr.; Schaus, J. M. Eur. Pat. Appl. EU0590971, 1994 [*Chem. Abstr.* **1994,** *121,* 134151].
374. Anderson, B. A.; Harn, N. K. *Synthesis* **1996,** 583–585.
375. Miller, R. D.; Lee, V. Y.; Moylan, C. R. *Chem. Mat.* **1994,** *6,* 1023–1032.
376. Dondoni, A.; Dall'Occo, T.; Fantin, G.; Fogagnolo, M.; Medici, A.; Pedrini, P. *J. Chem. Soc. Chem. Commun.* **1984,** 258–260.
377. Dondoni, A.; Dall'Occo, T.; Galliani, G.; Mastellari, A.; Medici, A. *Tetrahedron Lett.* **1984,** *25,* 3637–3640.
378. Dondoni, A.; Douglas, A. W.; Shinkai, I. *J. Org. Chem.* **1993,** *58,* 3196–3200.
379. Dondoni, A.; Fantin, G.; Fogagnolo, M.; Mastellari, A.; Medici, A.; Pedrini, P. *J. Org. Chem.* **1984,** *49,* 3478–3483.
380. Murakami, M.; Higuchi, N.; Ito, Y. *Chem. Express* **1990,** *5,* 411–414.

381. Ito, Y.; Higuchi, N.; Murakami, M. *Heterocycles* **2000**, *52*, 91–93.
382. Alberola, A.; Cuadrado, P.; Gonzalez, A. M.; Laguna, M. A.; Pulido, F. J. *An. Quim. Ser. C* **1988**, *84*, 49–52.
383. Itoh, T.; Hasegawa, H.; Nagata, K.; Okada, M.; Ohsawa, A. *Tetrahedron* **1994**, *50*, 13089–13100.
384. Itoh, T.; Hasegawa, H.; Nagata, K.; Ohsawa, A. *J. Org. Chem.* **1994**, *59*, 1319–1325.
385. Iyer, R. P.; Sonaseth, M. S.; Kulkarni, S. P.; Gopalan, R.; Ratnam, K. R.; Prabhu, A. V. *Indian J. Chem. Sect. B.* **1984**, *23*, 289–290.
386. Iyer, R. P.; Ratnam, R. K.; Kulkarni, S. P.; Sonaseth, M. S. *J. Heterocycl. Chem.* **1986**, *23*, 991–997.
387. Wilkinson, R. C.; Shephard, R. G.; Thomas, J. P.; Baughn, C. *J. Am. Chem. Soc.* **1961**, *83*, 2212–2213.
388. Langridge, D. C.; Hixson, S. S.; McEwen, W. E. *J. Org. Chem.* **1985**, *50*, 5503–5507.
389. Hassner, A.; Fischer, B. *Tetrahedron Lett.* **1990**, *31*, 7213–7214.
390. Hassner, A.; Fischer, B. *J. Org. Chem.* **1992**, *57*, 3070–3075.
391. Perrin, S.; Monnier, K.; Laude, B. *Bull. Soc. Chim. Belg.* **1996**, *105*, 777–783.
392. Babaev, E. V.; Efimov, A. V.; Maiboroda, D. A.; Jug, K. *Eur. J. Org. Chem.* **1998**, *1*, 193–196.
393. Bradsher, C. K.; Zinn, M. F. *J. Heterocycl. Chem.* **1967**, *4*, 66–70.
394. Babaev, E. V.; Tsisevich, A. A. *Chem. Heterocycl. Compd. (N.Y.)* **1998**, *34*, 254–255.
395. Babaev, E. V.; Efimov, A. V.; Zhukov, S. G.; Rybakov, V. B. *Chem. Heterocycl. Compd. (N.Y.)* **1998**, *34*, 852–854.
396. Babaev, E. V.; Tsisevich, A. A. *Chem. Heterocycl. Compd. (N.Y.)* **1997**, *33*, 875–876.
397. Babaev, E. V.; Bozhenko, S. V. *Chem. Heterocycl. Compd. (N.Y.)* **1997**, *33*, 125–126.
398. Babaev, E. V.; Pasichnichenko, K. Yu.; Maiboroda, D. A. *Chem. Heterocycl. Compd. (N.Y.)* **1997**, *33*, 338–342.
399. Babaev, E. V.; Orlova, I. A. *Chem. Heterocycl. Compd. (N.Y.)* **1997**, *33*, 489–490.
400. Babaev, E. V.; Efimov, A. V.; Rybakov, V. B.; Zhukov, S. G. *Chem. Heterocycl. Compd. (N.Y.)* **1999**, *35*, 486–491.
401. Babaev, E. V.; Rybakov, V. B.; Zhukov, S. G.; Orlova, I. A. *Chem. Heterocycl. Compd. (N.Y.)* **1999**, *35*, 479–485.
402. Babaev, E. V.; Tsisevich, A. A. *J. Chem. Soc. Perkin Trans. 1* **1999**, 399–401.
403. Hetzheim, A. *Greifswald, Math.-Nat. Wiss. Reihe* **1988**, *37*, 7–10.
404. Padwa, A.; Chiaccio, U.; Venkatramanan, M. K. *J. Chem. Soc. Chem. Commun.* **1985**, 1108–1109.
405. Vedejs, E.; Grissom, J. W. *J. Am. Chem. Soc.* **1986**, *108*, 6433–6434.
406. Vedejs, E.; Grissom, J. W. *J. Org. Chem.* **1988**, *53*, 1876–1882.
407. Vedejs, E.; Grissom, J. W. *J. Am. Chem. Soc.* **1988**, *110*, 3238–3246.
408. Vedejs, E.; Dax, S. L. *Tetrahedron Lett.* **1989**, *30*, 2627–2630.
409. Vedejs, E.; Piotrowski, D. W. *J. Org. Chem.* **1993**, *58*, 1341–1348.
410. Vedejs, E.; Piotrowski, D. W.; Tucci, F. C. *J. Org. Chem.* **2000**, *65*, 5498–5505.
411. Zhao, Z.; Scarlato, G. R.; Armstrong, R. W. *Tetrahedron Lett.* **1991**, *32*, 1609–1612.
412. Schlosser, M. In McEwen, W. E.; Berlin, K. D., Eds. *Organophosphorous Stereochemistry*, Part II, Halsted Press, New York, 1977, pp. 5–35.
413. Schlosser, M.; Schaub, B. *J. Am. Chem. Soc.* **1982**, *104*, 5821–5823.
414. Naito, T.; Yuumoto, Y.; Ninomiya, I.; Kiguchi, T. *Tetrahedron Lett.* **1992**, *33*, 4033–4036.
415. Griesbeck, A. G.; Fiege, M.; Lex, J. *J. Chem. Soc. Chem. Commun.* **2000**, 589–590.
416. Thieme, M.; Vieira, E.; Liebscher, J. *Synthesis* **2000**, 2051–2059.

417. Roth, G. P.; Landi, J. J.; Salvagno, A. M.; Muller-Bottischer, H. *Org. Process Res. Dev.* **1997**, *1*, 331–338.
418. Pohl, M.; Thieme, M.; Jones, P. G.; Liebscher, J. *Liebigs Ann. Chem.* **1995**, 1539–1545.
419. Oguri, T.; Kawai, N.; Shioiri, T.; Yamada, S. I. *Chem. Pharm. Bull.* **1978**, *26*, 803–808.
420. Marlier, M.; Dardenne, G.; Casimir, J. *Phytochemistry* **1976**, *15*, 183–185.
421. Dominguez, X. A.; de la Fuente, G.; Gonzalez, A. G.; Reina, M.; Timón, I. *Heterocycles*, **1988**, *27*, 35–38.
422. Sengupta, S.; Mondal, S. *Tetrahedron Lett.* **1999**, *40*, 8685–8688.
423. Koyama, Y.; Yokose, K.; Dolby, L. J. *Agric. Biol. Chem.* **1981**, *45*, 1285–1287.
424. Takahashi, S.; Hashimoto, R.; Hamano, K.; Suzuki, T.; Nakagawa, A. *J. Antibiot.* **1996**, *49*, 513–518.
425. Kunze, B.; Jansen, R.; Pridzun, L.; Jurkiewicz, E.; Hunsmann, G.; Höfle, G.; Reichenbach, H. *J. Antibiot.* **1992**, *45*, 1549–1552.
426. Zhang, X.; Hinkle, B.; Ballantyne, L.; Gonzales, S.; Peña, M. R. *J. Heterocycl. Chem.* **1997**, *34*, 1061–1064.
427. Henkel, T.; Zeeck, A. *Liebigs Ann. Chem.* **1991**, 367–373.
428. Shiomi, K.; Arai, N.; Shinose, M.; Takahashi, Y.; Yoshida, H.; Iwabuchi, J.; Tanaka, Y.; Omura, S. *J. Antibiot.* **1995**, *48*, 714–719.
429. Iwasaki, S.; Namikoshi, M.; Kobayashi, H.; Furukawa, J.; Okuda, S. *Chem. Pharm. Bull.* **1986**, *34*, 1387–1390.
430. Keck, G. E.; Wager, C. A.; Wager, T. T.; Savin, K. A.; Covel, J. A.; Mclaws, M. D.; Krishnamurthy, D.; Cee, V. J. *Angew. Chem. Int. Ed.* **2001**, *40*, 231–234, and references therein.
431. Todorova, A. K.; Jüttner, F.; Linden, A.; Plüss, T.; von Philipsborn, W. *J. Org. Chem.* **1995**, *60*, 7891–7895.
432. Bertram, A.; Pattenden, G. *Synlett* **2001**, 1873–1874.
433. Kobayashi, J.; Itagaki, F.; Shigemori, H.; Takao, T.; Shimonishi, Y. *Tetrahedron* **1995**, *51*, 2525–2532.
434. Brazhnikova, M. G.; Kudinova, M. K.; Potapova, N. P.; Filippova, T. M.; Borowski, E.; Zelinski, Y.; Golik, J. *Bioorgan. Khim.* **1976**, *2*, 149–157.
435. Chamberlin, J. W.; Chen, S. *J. Antibiot.* **1977**, *30*, 197–201.
436. Delpierre, G. R.; Eastwood, F. W.; Gream, G. E.; Kingston, D. G. I.; Sarin, P. S.; Todd, Lord, A. R.; Williams, D. H. *J. Chem. Soc. C* **1966**, 1653–1669.
437. Kobayashi, J.; Kondo, K.; Ishibashi, M.; Wälchli M. R.; Nakamura, T. *J. Am. Chem. Soc.* **1993**, *115*, 6661–6665.
438. Jansen, R.; Kunze, B.; Reichenbach, H.; Jurkiewicz, E.; Hunsmann, G.; Höfle, G. *Liebigs Ann. Chem.* **1992**, 357–359.
439. Carmeli, S.; Moore, R. E.; Patterson, G. M. L.; Corbett, T. H.; Valeriote, F. A. *J. Am. Chem. Soc.* **1990**, *112*, 8195–8197.
440. Kato, Y.; Fusetani, N.; Matsunaga, S.; Hashimoto, K.; Fujita, S.; Furuya, T. *J. Am. Chem. Soc.* **1986**, *108*, 2780–2781.
441. Kato, Y.; Fusetani, N.; Matsunaga, S.; Hashimoto, K.; Koseki, K. *J. Org. Chem.* **1988**, *53*, 3930–3932.
442. Rodriguez, J.; Nieto, R. M.; Crews, P. *J. Natr. Prod.* **1993**, *56*, 2034–2040.
443. Rudi, A.; Kashman, Y.; Benayahu, Y.; Schleyer, M. *J. Natr. Prod.* **1994**, *57*, 829–836.
444. Fernández, R.; Dherbomez, M.; Letourneux, Y.; Nabil, M.; Verbist, J. F.; Biard, J. F. *J. Nat. Prod.* **1999**, *62*, 678–680.
445. Nagatsu, A.; Kajitani, H.; Sakakibara, J. *Tetrahedron Lett.* **1995**, *36*, 4097–4100.
446. Ichiba, T.; Yoshida, W. Y.; Scheuer, P. J.; Higa, T.; Gravalos, D. G. *J. Am. Chem. Soc.* **1991**, *113*, 3173–3174.

447. Higa, T.; Tanaka, J.; Kitamura, A.; Koyama, T.; Takahashi, M.; Uchida, T. *Pure Appl. Chem.* **1994**, *66*, 2227–2230.
448. Lindquist, N.; Fenical, W.; Van Duyne, G. D.; Clardy, J. *J. Am. Chem. Soc.* **1991**, *113*, 2303–2304.
449. Searle, P. A.; Molinski, T. F. *J. Am. Chem. Soc.* **1995**, *117*, 8126–8131.
450. Yun, B.-S.; Hidaka, T.; Furihata, K.; Seto, H. *J. Antibiot.* **1994**, *47*, 510–514.
451. Roesener, J. A.; Scheuer, P. J. *J. Am. Chem. Soc.* **1986**, *108*, 846–847.
452. Kernan, M. R.; Molinski, T. F.; Faulkner, D. J. *J. Org. Chem.* **1988**, *53*, 5014–5020.
453. Matsunaga, S.; Fusetani, N.; Hashimoto, K.; Koseki, K.; Noma, M.; Noguchi, H.; Sankawa, U. *J. Org. Chem.* **1989**, *54*, 1360–1363.
454. Matsunaga, S.; Fusetani, N.; Hashimoto, K.; Koseki, K.; Noma, M. *J. Am. Chem. Soc.* **1986**, *108*, 847–849.
455. Fusetani, N.; Yasumuro, K.; Matsunaga, S.; Hashimoto, K. *Tetrahedron Lett.* **1989**, *30*, 2809–2813.
456. Li, Y.-M.; Milne, J. C.; Madison, L. L.; Kolter, R.; Walsh, C. T. *Science*, **1996**, *274*, 1188–1193.
457. Keck, G. E.; Savin, K. A.; Weglarz, M. A. *J. Org. Chem.* **1995**, *60*, 3194–3204.
458. Muir, J. C.; Pattenden, G.; Thomas, R. M. *Synthesis* **1998**, 613–618.
459. Wipf, P.; Venkatraman, S. *J. Org. Chem.* **1996**, *61*, 6517–6522.
459. (a) Meyers, A. I.; Lawson, J. P.; Walker, D. G.; Linderman, R. J. *J. Org. Chem.* **1986**, *51*, 5111–5123.
460. Nagao, Y.; Yamada, S.; Fujita, E. *Tetrahedron Lett.* **1983**, *24*, 2287–2290.
461. Wood, R. D.; Ganem, B. *Tetrahedron Lett.* **1983**, *24*, 4391–4392.
462. Yokoyama, M.; Menjo, Y.; Ubukata, M.; Irie, M.; Watanabe, M.; Togo, H. *Bull. Chem. Soc. Jpn.* **1994**, *67*, 2219–2226.
463. Entwhistle, D. A.; Jordan, S. I.; Montgomery, J.; Pattenden, G. *J. Chem. Soc. Perkin Trans. 1* **1996**, 1315–1317.
464. Entwhistle, D. A.; Jordan, S. I.; Montgomery, J.; Pattenden, G. *Synthesis* **1998**, 603–612.
465. Cornforth, J. W.; Cornforth, R. H. *J. Chem. Soc.* **1947**, 96–102.
466. Helquist, P.; Bergdahl, M.; Hett, R.; Gangloff, A. R.; Demillequand, M.; Cottard, M.; Mader, M. M.; Friebe, T.; Iqbal, J.; Wu, Y.; Åkermark, B.; Rein, T.; Kann, N. *Pure Appl. Chem.* **1994**, *66*, 2063–2066.
467. Brennan, C. J.; Campagne, J.-M. *Tetrahedron Lett.* **2001**, *42*, 5195–5197.
468. Kruger, J.; Carreira, E. M. *J. Am. Chem. Soc.* **1998**, *120*, 837–838.
469. Schlessinger, R. H.; Li, Y.-J. *J. Am. Chem. Soc.* **1996**, *118*, 3301–3302.
470. Mukaiyama, T.; Bald, E.; Saigo, K. *Chem. Lett.* **1975**, 1163–1166.
471. Tavares, F.; Lawson, J. P.; Meyers, A. I. *J. Am. Chem. Soc.* **1996**, *118*, 3303–3304.
472. Ghosh, A. K.; Liu, W. *J. Org. Chem.* **1997**, *62*, 7908–7909.
473. Castro, B.; Dormoy, J. R.; Evin, G.; Selve, C. *Tetrahedron Lett.* **1975**, 1219–1222.
474. Inanaga, J.; Hirata, K.; Saeki, H.; Katsuki, T.; Yamaguchi, M. *Bull. Chem. Soc. Jpn.* **1979**, *52*, 1989–1993.
475. Dvorak, C. A.; Schmitz, W. D.; Poon, D. J.; Pryde, D. C.; Lawson, J. P.; Amos, R. A.; Meyers, A. I. *Angew. Chem. Int. Ed. Engl.* **2000**, *39*, 1664–1666.
476. Liu, L.; Tanke, R. S.; Miller, M. M. *J. Org. Chem.* **1986**, *51*, 5332–5337.
477. Grubbs, R. H.; Chang, S. *Tetrahedron* **1998**, *54*, 4413–4450.
478. Ehrler, J.; Farooq, S. *Synlett* **1994**, 702–704.
479. Parsons, R. L. Jr.; Heathcock, C. H. *J. Org. Chem.* **1994**, *59*, 4733–4734.
480. Wipf, P.; Venkatraman, S. *J. Org. Chem.* **1995**, 60, 7224–7229.
481. Wipf, P.; Venkatraman, S. *Synlett* **1997**, 1–10.

482. Akaji, K.; Kiso, Y. *Tetrahedron* **1999**, *55*, 10685–10694.
483. Fukuyama, T.; Xu, L. *J. Am. Chem. Soc.* **1993**, *115*, 8449–8450.
484. Thom, S. M.; Pattenden, G.; Jones, M. F. *Tetrahedron* **1993**, *49*, 2131–2138.
485. Walker, M. A.; Heathcock, C. H. *J. Org. Chem.* **1992**, *57*, 5566–5568.
486. Karady, S.; Amato, J. S.; Weinstock, L. M. *Tetrahedron Lett.* **1984**, *25*, 4337–4340.
487. Devos, A.; Remion, J.; Frisque-Hesbain, A.-M.; Colens, A.; Ghosez, L. *J. Chem. Soc. Chem. Commun.* **1979**, 1180–1181.
488. Vaccaro, H. A.; Levy, D. E.; Sawabe, A.; Jaetsch, T.; Masamune, S. *Tetrahedron Lett.* **1992**, *33*, 1937–1940.
489. Wipf, P.; Venkatraman, S.; Miller, C. P. *Tetrahedron Lett.* **1995**, *36*, 3639–3642.
490. Akaji, K.; Kuriyama, N.; Kiso, Y. *Tetrahedron Lett.* **1994**, *35*, 3315–3319.
491. Akaji, K.; Tamai, Y.; Kiso, Y. *Tetrahedron* **1997**, *53*, 567–584.
492. Akaji, K.; Hayashi, Y.; Kiso, Y.; Kuriyama, N.; *J. Org. Chem.* **1999**, *64*, 405–411.
493. Ogawa, A. K.; DeMattei, J. A.; Scarlato, G. R.; Tellew, J. E.; Chong, L. S.; Armstrong, R. W. *J. Org. Chem.* **1996**, *61*, 6153–6161.
494. Ogawa, A. K.; Armstrong, R. W. *J. Am. Chem. Soc.* **1998**, *120*, 12435–12442.
495. Yokokawa, F.; Hamada, Y.; Shioiri, T. *Synlett* **1992**, 151–152.
496. Yokokawa, F.; Hamada, Y.; Shioiri, *Chem. Commun.* **1996**, 871–872.
497. Evans, D. A.; Gage, J. R.; Leighton, J. L.; Kim, A. S. *J. Org. Chem.* **1992**, *57*, 1961–1963.
498. Evans, D. A.; Gage, J. R.; Leighton, J. L. *J. Org. Chem.* **1992**, *57*, 1964–1966.
499. Evans, D. A.; Gage, J. R.; Leighton, J. L. *J. Am. Chem. Soc.* **1992**, *114*, 9434–9453.
500. Tanimoto, N.; Gerritz, S. W.; Sawabe, A.; Noda, T.; Filla, S. A.; Masamune, S. *Angew. Chem. Int. Ed. Engl.* **1994**, *33*, 673–675.
501. Anderson, O. P.; Barrett, A. G. M.; Edmunds, J. J.; Hachiya, S.-I.; Hendrix, J. A.; Horita, K.; Malecha, J. W.; Parkinson, C. J.; VanSickle, A. *Can. J. Chem.* **2001**, *79*, 1562–1592.
502. Smith, A. B. III; Salvatore, B. A. Hull, K. G.; Duan, J. J. W. *Tetrahedron Lett.* **1991**, *32*, 4859–4862.
503. Salvatore, B. A.; Smith, A. B. III *Tetrahedron Lett.* **1994**, *35*, 1329–1330.
504. Smith, A. B. III; Friestad, G. K.; Duan, J. J.-W.; Barbosa, J.; Hull, K. G.; Iwashima, M.; Qui, Y.; Spoors, P.; Bertounesque, E.; Salvatore, B. A. *J. Org. Chem.* **1998**, *63*, 7596–7597.
505. Smith, A. B. III; Friestad, G. K.; Barbosa, J.; Bertounesque, E.; Duan, J. J.-W.; Hull, K. G.; Iwashima, M.; Qiu, Y.; Spoors, P. G.; Salvatore, B. A. *J. Am. Chem. Soc.* **1999**, *121*, 10478–10486.
506. Carlsen, P. H. J.; Katsuki, T.; Martin, V. S.; Sharpless, K. B. *J. Org. Chem.* **1981**, *46*, 3936–3938.
507. Hamada, Y.; Shibata, M.; Shioiri, T. *Tetrahedron Lett.* **1985**, *26*, 6501–6504.
508. Armstrong, R. W.; Beau, J.-M.; Cheon, S. H.; Christ, W. J.; Fujioka, H.; Ham, W.-H.; Hawkins, L. D.; Jin, H.; Kang, S. H.; Kishi, Y.; Martinelli, M. J.; McWhorter, W. J. Jr.; Mizuno, M.; Nakata, M.; Stutz, A. E.; Talamas, F. X.; Taniguchi, M.; Tino, J. A.; Ueda, K.; Uenishi, J.; White, J. B.; Yonaga, M. *J. Am. Chem. Soc.* **1989**, *111*, 7525–7530.
509. Matsunaga, S.; Fusetani, N. *Tetrahedron Lett.* **1991**, *32*, 5605–5606.
510. Hamada, H.; Tanada, Y.; Yokokawa, F.; Shioiri, T. *Tetrahedron Lett.* **1991**, *32*, 5983–5986.
511. Evans, D. A.; Britton, T. C.; Ellman, J. A. *Tetrahedron Lett.* **1987**, *28*, 6141–6144.
512. Silks, L. A. III; Dunlop, R. B.; Odom, J. D. *J. Am. Chem. Soc.* **1990**, *112*, 4979–4982.
513. Silks, L. A. III; Peng, J.; Odom, J. D.; Dunlap, R. B. *J. Org. Chem.* **1991**, *56*, 6733–6736.
514. Wipf, P.; Lim, S. *J. Am. Chem. Soc.* **1995**, *117*, 558–559.
515. Williams, D. R.; Brooks, D. A.; Berliner, M. A. *J. Am. Chem. Soc.* **1999**, *121*, 4924–4925.
516. Cheng, Z.; Hamada, Y.; Shioiri, T. *Synlett* **1997**, 109–110.
517. Barrett, A. G. M.; Kohrt, J. T. *Synlett* **1995**, 415–416.

518. Yokokawa, F.; Asano, T.; Shioiri, T. *Org. Lett.* **2000**, *2*, 4169–4172.
519. Yokokawa, F.; Asano, T.; Shioiri, T. *Tetrahedron* **2001**, *57*, 6311–6327.
520. Milstein, D.; Stille, J. K. *J. Am. Chem. Soc.* **1979**, *101*, 4992–4998.
521. Konopelski, J. P.; Hottenroth, J. M.; Oltra, H. M.; Veliz, E. A.; Yang, Z. C. *Synlett* **1996**, 609–611.
522. Wipf, P.; Yokokawa, F. *Tetrahedron Lett.* **1998**, *39*, 2223–2226.
523. Wipf, P.; Methot, J.-L. *Org. Lett.* **2001**, *3*, 1261–1264.
524. Magnus, P.; McIver, E. G. *Tetrahedron Lett.* **2000**, *41*, 831–834.
525. Vedejs, E.; Barda, D. A. *Org. Lett.* **2000**, *2*, 1033–1035.
526. Nicolaou, K. C.; Snyder, S. A.; Simonsen, K. B.; Koumbis, A. E. *Angew. Chem. Int. Ed. Engl.* **2000**, *39*, 3473–3478.
527. See Nicolaou, K. C.; Härtner, M. W.; Gunzner, J. L.; Nadin, A. *Liebigs Ann. Chem.* **1997**, 1283–1301, for a review of the Wittig and Horner-Wadsworth-Emmons reactions.
528. Radspieler, A.; Liebscher, J. *Synthesis* **2001**, 745–750.
529. Davis, M.; Lakhan, R.; Ternai, B. *J. Heterocycl. Chem.* **1977**, *14*, 317–318.
530. Chen, X.; Esser, L.; Harran, P. G. *Angew. Chem. Int. Ed. Engl.* **2000**, *39*, 937–940.
531. Li, J.; Chen, X.; Burgett, A. W. G.; Harran, P. G. *Angew. Chem. Int. Ed. Engl.* **2001**, *40*, 2682–2685.
532. Doyle, M. P.; Buhro, W. E.; Davidson, J. G.; Elliott, R. C.; Hoekstra, J. W.; Oppenhuizen, M. *J. Org. Chem.* **1980**, *45*, 3657–3664.
533. Dale, J. A.; Dull, D. L.; Mosher, H. S. *J. Org. Chem.* **1969**, *34*, 2543–2549.
534. Aranyos, A.; Old, D. W.; Kiyomori, A.; Wolfe, J. P.; Sadighi, J. P.; Buchwald, S. L. *J. Am. Chem. Soc.* **1999**, *121*, 4369–4378.
535. Chen, X.; Esser, L.; Harran, P. G. *Angew. Chem. Int. Ed. Engl.* **2000**, *39*, 937–940.
536. Kobayashi, J.; Tsuda, M.; Fuse, H.; Sasaki, T.; Mikami, Y. *J. Nat. Prod.* **1997**, *60*, 150–154.
537. Knight, D. W.; Pattenden, G.; Rippon, D. E. *Synlett* **1990**, 36–37.
538. Pattenden, G. *J. Heterocycl. Chem.* **1992**, *29*, 607–618.
539. Yoo, S.-K. *Tetrahedron Lett.* **1992**, *33*, 2159–2162.
540. Connell, R.; Scavo, F.; Helquist, P.; Akermark, B. *Tetrahedron Lett.* **1986**, *27*, 5559–5562.
541. Chattopadhyay, S. K.; Pattenden, G. *Synlett* **1997**, 1345–1347.
542. Chattopadhyay, S. K.; Pattenden, G. *Tetrahedron Lett.* **1998**, *39*, 6095–6098.
543. Kempson, J.; Pattenden, G. *Synlett* **1999**, 533–536.
544. Chattopadhyay, S. K.; Kempson, J.; McNeil, A.; Pattenden, G.; Reader, M.; Rippon, D. E.; Waite, D. *J. Chem. Soc. Perkin Trans. 1* **2000**, 2415–2428.
545. Chattopadhyay, S. K.; Pattenden, G. *J. Chem. Soc. Perkin Trans. 1* **2000**, 2429–2454.
546. Garner, P.; Park, J. M. *J. Org. Chem.* **1987**, *52*, 2361–2364.
547. Cicchi, S.; Cordero, F. M.; Giomi, D. *Prog. Heterocycl. Chem.* **2001**, *13*, 217–237.
548. Gilchrist, T. L. *J. Chem. Soc. Perkin Trans. 1* **2001**, 2491–2515.
549. Roy, R. S.; Gehring, A. M.; Milne, J. C.; Belshaw, P. J.; Walsh, C. T. *Nat. Prod. Rep.* **1999**, *16*, 249–263.
550. Kawase, M.; Hirabayashi, M.; Saito, S. *Recent Res. Devel. Organic Chem.* **2000**, *4*, 283–293.
551. Suga, H.; Ibata, T. *Rev. Heteroat. Chem.* **1999**, *21*, 195–221.
552. Mrozek, A.; Karolak-Wojciechowska, J.; Amiel, P.; Barbe, J. *J. Mol. Struct.* **2000**, *524*, 151–157.
553. Phillips, A. J.; Uto, Y.; Wipf, P.; Reno, M. J.; Williams, D. R. *Org. Lett.* **2000**, *2*, 1165–1168.
554. Berger, R.; Shoop, W. L.; Pivnichny, J. V.; Warmke, L. M.; Zakson-Aiken, M.; Owens, K. A.; deMontigny, P.; Schmatz, D. M.; Wyvratt, M. J.; Fisher, M. H.; Meinke, P. T.; Colletti, S. L. *Org. Lett.* **2001**, *3*, 3715–3718.

555. Stanchev, M. S.; Grueva, E. S.; Pajpanova, T. I.; Golovinsky, E. V. *Bulg. Chem. Commun.* **2001**, *33*, 73–78.
556. Khalafy, J.; Svensson, C. E.; Prager, R. H.; Williams, C. M. *Tetrahedron Lett.* **1998**, *39*, 5405–5408.
557. Khalafy, J.; Prager, R. H.; Williams, C. M. *Aust. J. Chem.* **1999**, *52*, 31–36.
558. Wang, Y.; Zhu, S. *J. Fluorine Chem.* **2000**, *103*, 139–144.
559. Hallinan, E. A.; Hagen, T. J.; Tsymbalov, S.; Stapelfeld, A.; Savage, M. A. *Bioorg. Med. Chem.* **2001**, *9*, 1–6.
560. Clapham, B.; Spanka, C.; Janda, K. D. *Org. Lett.* **2001**, *3*, 2173–2176.
561. Kim, T. Y.; Kim, H. S.; Chung, Y. M; Kim, J. N. *Bull. Korean Chem. Soc.* **2000**, *21*, 673–674.
562. Arcadi, A.; Cacchi, S.; Cascia, L.; Fabrizi, G.; Marinelli, F. *Org. Lett.* **2001**, *3*, 2501–2504.
563. Ceulemans, E.; Dyall, L. K.; Dehaen, W. *Tetrahedron* **1999**, *55*, 1977–1988.
564. Barrett, A. G.; Cramp, S. M.; Hennessy, A. J.; Procopiou, A.; Roberts, R. S. *Org. Lett.* **2001**, *3*, 271–273.
565. Hénaff, N.; Whiting, A. *Tetrahedron* **2000**, *56*, 5193–5204.
566. Hénaff, N.; Whiting, A. *Org. Lett.* **2000**, *2*, 1137–1139.
567. Iwanowicz, E. J.; Watterson, S. H.; Dhar, T. G. M.; Pitts, W. J.; Gu, H. H. PCT Intl. Appl. WO 0181340, 2001 [*Chem. Abstr.* **2001**, *135*, 344472].
568. Duarte, M. P.; Lobo, A. M.; Prabhakar, S. *Tetrahedron Lett.* **2000**, *41*, 7433–7435.
569. Hermitage, S. A.; Cardwell, K. S.; Chapman, T.; Cooke, J. W. B.; Newton, R. *Org. Process Res. Dev.* **2001**, *5*, 37–44.
570. Haruta, J.; Hashimoto, H.; Matsushita, M. PCT Int. Appl. WO 9619462, 1996 [*Chem. Abstr.* **1996**, *125*, 167967].
571. Dalko, M.; Dumats, J. Eur. Patent Appl. EP 1095937, 2001 [*Chem. Abstr.* **2001**, *134*, 326522].
572. Babii, S. B.; Zyabrev, V. S.; Drach, B. S. *Russ. J. Gen. Chem.* [Engl. Transl.] **2001**, *71*, 644–645.
573. Brovarets, V. S.; Vydzhak, R. N.; Pil'o, S. G.; Zyuz', K. V.; Drach, B. S. *Russ. J. Gen. Chem.* [Engl. Transl.] **2001**, *71*, 1726–1728.
574. Yamane, T. PCT Int. Appl. WO 0119805, 2001 [*Chem. Abstr.* **2001**, *134*, 237473].
575. Brovarets, V. S.; Pil'o, S. G.; Popovich, T. P.; Vydzhak, R. N.; Drach, B. S. *Russ. J. Gen. Chem.* [Engl. Transl.] **2001**, *71*, 1825–1826.
576. Pil'o, S. G.; Brovarets, V. S.; Vinogradova, T. K.; Chernega, A. N.; Drach, B. S. *Russ. J. Gen. Chem.* [Engl. Transl.] **2001**, *71*, 280–285.
577. Iwase, K.; Aoki, T. Jpn. Kokai Tokkyo Koho JP 20011270866, 2001 [*Chem. Abstr.* **2001**, *135*, 256166].
578. Kondrt'eva, G. Y.; Aitzhanova, M. A.; Bogdanov, V. S.; Stashina, G. A.; Sedishev, I. P. *Chem. Heterocycl. Compd.* [N.Y.] **2000**, *36*, 584–592.
579. Clapham, B.; Sutherland, A. J. *J. Org. Chem.* **2001**, *66*, 9033–9037.
580. Wenkert, D.; Chen, T.-F.; Ramachandran, K.; Valasinas, L.; Weng, L.-l.; McPhail, A. T. *Org. Lett.* **2001**, *3*, 2301–2503.
581. Kuneš, J.; Balšánek, V.; Pour, M.; Buchta, V. *Collect. Czech. Chem. Commun.* **2001**, *66*, 1809–1830.
582. Zhao, G.; Sun, X.; Bienayme, H.; Zhu, J. *J. Am. Chem. Soc.* **2001**, *123*, 6700–6701.
583. Dömling, A.; Ugi, I. *Angew. Chem. Int. Ed. Engl.* **2000**, *39*, 3168–3210.
584. Sun, X.; Janvier, P.; Zhao, G.; Bienayme, H.; Zhu, J. *Org. Lett.* **2001**, *3*, 877–880.
585. Trost, B. M. *Angew. Chem. Int. Ed. Engl.* **1995**, *34*, 259–281.

CHAPTER 2

Spectroscopic Properties of Oxazoles

Derek Lowe

Department of Chemistry Research
Bayer Corporation
West Haven, Connecticut

2.1. Introduction
2.2. Nuclear Magnetic Resonance
    2.2.1. Proton NMR
    2.2.2. Carbon NMR
    2.2.3. Nitrogen NMR
    2.2.4. Oxygen NMR
    2.2.5. Fluorine NMR
2.3. Mass Spectrometry
2.4. Optical Spectroscopy
    2.4.1. Infrared
    2.4.2. Ultraviolet and Visible
2.5. Microwave Spectroscopy
2.6. Miscellaneous Spectroscopic Methods
2.7. Summary

2.1. INTRODUCTION

Most physical properties of oxazoles have now been extensively explored. This chapter serves as an overview of the most important areas and updates the previous edition, in which the spectroscopic chapter remains relevant in all details.[1] NMR (surely now the single most important technique to the practicing organic chemist) is covered first and in the most detail, followed by a review of mass spectrometry, infrared and ultraviolet/visible spectroscopy, microwave spectroscopy, and other techniques. This order parallels that used in the previous edition, with some changes: the proton and carbon NMR tables have been expanded, oxygen and fluorine NMR are now covered, as are microwave spectroscopy and other methods, such as photoelectron spectra.

The Chemistry of Heterocyclic Compounds, Volume 60: Oxazoles: Synthesis, Reactions, and Spectroscopy, Part A, edited by David C. Palmer
ISBN 0-471-39494-7 Copyright © 2003 John Wiley & Sons, Inc.

2.2. NUCLEAR MAGNETIC RESONANCE

2.2.1. Proton NMR

The general features of oxazole ^{1}H NMR are well established: The ring protons are deshielded, and $H_2 > H_5 > H_4$. The chemical shift of H_4 is generally least affected by substitution in the ring, but is affected more strongly by groups at C_5 rather than C_2 (Fig. 2.1). Protonation of oxazoles is known to deshield the ring protons, with H_2 being most deshielded, followed by H_4, and with minimal effects on H_5. Similarly, more polar solvents tend to deshield H_2 slightly. Coupling constants between protons on the oxazole ring are often noted. J_{4-5} is generally the largest, ranging from 0.75 to 1.5 Hz. J_{2-5} is typically ≤ 1 Hz, whereas J_{2-4}, when seen, is <0.5 Hz. These values can vary with different solvents. Tables 2.1–2.3 summarize typical resonances for monosubstituted, disubstituted, and trisubstituted oxazoles. As for the other NMR tables in this chapter, the reference cited refers to a recent characterization and is not necessarily the only literature reference to the compound.

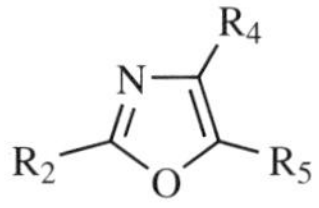

Figure 2.1

In specialized ^{1}H NMR topics, variable-temperature experiments have been used to study the conformational behavior of oxazole-containing cyclophanes, contrasting them to the phenyl or thiazolyl analogs.[67–69]

2.2.2. Carbon NMR

A large body of oxazole ^{13}C NMR data exists and is readily accessible, thus only representative functional group combinations are shown here to illustrate the general principles involved. In addition to the previous volume of this series,[1] Bogdanov and co-workers,[2] Hiemstra and co-workers,[70] and Schiketanz and co-workers[71] are the more detailed discussions of the field.

The ring carbons are deshielded in the order $C_2 > C_5 > C_4$. The chemical shifts for the various carbon atoms are typically 150–165 ppm for C_2, 100–150 ppm for C_4, and 130–160 ppm for C_5. Substitution at C_2 is known to shield or deshield C_5 by approximately ≤ 2 ppm, with lesser effects on C_4. An anionic carbon α to C_2 deshielded it by 5.8 ppm and shielded C_5 by 14.4 ppm; C_4 was hardly affected.[14] Substituents at C_4 generally shield an unsubstituted C_5 carbon (β effect) and vice versa. An unsubstituted C_4 generally resonates in the range of 119–127 ppm with an R group on C_5; an unsubstituted C_5 resonates in the range of 132–140 ppm with

TABLE 2.1. ^{1}H NMR OF MONOSUBSTITUTED OXAZOLES^a

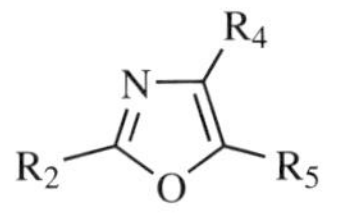

Figure 2.2

Structure	R_2	R_4	R_5	δR_2	δR_4	δR_5	Solvent	Reference
1	H	H	H	7.84	7.12	7.66	CCl$_4$	2
2	H	CH$_3$	H	7.7	(2.17)	7.32	CCl$_4$	3
3	H	Ph	H	7.94	—	7.94	CDCl$_3$	4
4	H	(4-NO$_2$)Ph	H	8.12	—	8.01	CDCl$_3$	4
5	H	(4-OCH$_3$)Ph	H	7.92	—	7.86	CDCl$_3$	4
6	H	CO$_2$Et	H	8.20	—	7.91	CDCl$_3$	5
7	H	STol	H	7.86	—	7.67	CDCl$_3$	6
8	H	SO$_2$Tol	H	8.26	—	7.96	CDCl$_3$	7
9	H	H	CH$_3$	7.75	6.77	(2.28)	CDCl$_3$	8
10	H	H	Ph	7.85	7.30	—	CDCl$_3$	9
11	H	H	(3-CH$_3$-5-isoxazoyl)	7.98	7.55	—	CDCl$_3$	10
12	H	H	CHO	7.93	7.48	—	CDCl$_3$	11
13	H	H	CO$_2$Et	8.03	7.78	—	CDCl$_3$	12
14	H	H	OEt	7.34	6.05	—	CDCl$_3$	13
15	CH$_3$	H	H	(2.40)	6.86	7.43	CCl$_4$	3
16	Bn	H	H	(4.10)	7.04	7.56	CDCl$_3$	14
17	Ph	H	H	—	7.26	7.71	CDCl$_3$	15
18	CO$_2$Et	H	H	—	7.30	7.76	CDCl$_3$	16
19	NH$_2$	H	H	—	6.63	7.12	CDCl$_3$	17
20	NHAc	H	H	—	6.96	7.42	CDCl$_3$	17
21	SCH$_3$	H	H	(2.65)	7.10	7.66	CDCl$_3$	18

^{a} Numbers in parentheses refer to the chemical shift of the substituent.
—, No relevent measure.

TABLE 2.2. ^{1}H NMR OF DISUBSTITUTED OXAZOLES^a

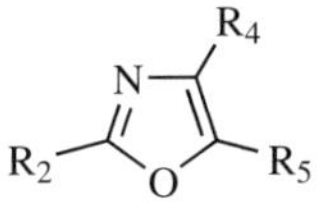

Figure 2.3

Structure	R_2	R_4	R_5	δR_2	δR_4	δR_5	Solvent	Reference
22	H	CH$_3$	CH$_3$	7.63	(2.01)	(2.16)	CDCF$_3$/CDCl$_3$^{b}	19
23	H	CH$_3$	COCH$_3$	7.65	nd	nd	CCl$_4$	2
24	H	CH$_3$	NCO	7.60	(2.12)	—	CDCl$_3$	20
25	H	CH$_3$	O(n-Pr)	7.14	nd	nd	CCl$_4$	2
26	H	Ph	Ph	7.94	—	—	CDCl$_3$	21
27	H	Ph	CF$_3$	7.73	—	—	CDCl$_3$	22
28	H	Ph	OCH$_3$	7.45	—	(4.02)	CDCl$_3$	23
29	H	CN	SCH$_3$	7.90	—	(2.65)	CDCl$_3$	24
30	H	CN	SPh	7.90	—	—	CDCl$_3$	24
31	H	CO$_2$CH$_3$	CH$_3$	7.84	—	(2.65)	CDCl$_3$	25
32	H	CO$_2$Et	CO$_2$Et	7.98	—	—	CDCl$_3$	26
33	H	CF$_3$	OEt	7.25	—	(4.30)	CDCl$_3$	27

TABLE 2.2 (Continued)

Structure	R_2	R_4	R_5	δR_2	δR_4	δR_5	Solvent	Reference
34	H	OBt	Ph	8.04	—	—	CDCl$_3$	28
35	H	SCH$_3$	Ph	7.98	(2.61)	—	CDCl$_3$	29
36	H	SCH$_3$	(4-Cl)Ph	8.07	(2.67)	—	CDCl$_3$	29
37	H	I	Ph	7.89	—	—	CDCl$_3$	12
38	H	I	CO$_2$Et	7.96	—	—	CDCl$_3$	12
39	CH$_3$	H	CH$_3$	(2.33)	6.51	(2.22)	CDCF$_3$/CDCl$_3$[b]	19
40	CH$_3$	H	Ph	(2.48)	nd	—	CDCF$_3$/CDCl$_3$[b]	19
41	CH$_3$	H	(4-NO$_2$)Ph	(2.58)	7.42	—	CDCl$_3$	30
42	CH$_3$	H	CO$_2$Et	(2.56)	7.65	—	CDCl$_3$	15
43	Ph	H	CH$_3$	—	6.82	(2.37)	CDCl$_3$	31
44	Ph	H	Ph	—	~7.45	—	CDCl$_3$	32
45	Ph	H	(4-OCH$_3$)Ph	—	7.31	—	CDCl$_3$	30
46	Ph	H	(4-Cl)Ph	—	7.42	—	CDCl$_3$	30
47	Ph	H	OEt	—	6.20	(4.16)	CDCl$_3$	30
48	Ph	H	Br	—	7.09	—	CDCl$_3$	31
49	COCH$_3$	H	Ph	(2.67)	7.5	—	CDCl$_3$	33
50	COPh	H	Ph	—	7.6	—	CDCl$_3$	33
51	CO$_2$Et	H	n-pentyl	—	6.97	(2.75)	CDCl$_3$	34
52	CO$_2$Et	H	Ph	—	(7.3–7.8)	—	CDCl$_3$	33
53	NH$_2$	H	CH$_3$	(6.2)	6.2	(2.1)	d-6 DMSO	35
54	N(CH$_3$)$_2$	H	Ph	(3.06)	7.02	—	CDCl$_3$	36
55	N(CH$_3$)$_2$	H	(4-NO$_2$)Ph	—	7.31	—	CDCl$_3$	30
56	N(CH$_3$)$_2$	H	(4-OCH$_3$)Ph	(3.05)	6.90	—	CDCl$_3$	36
57	N(CH$_3$)$_2$	H	(4-Cl)Ph	(3.11)	7.16	—	CDCl$_3$	36
58	N(i-Pr)$_2$	H	OEt	—	5.82	(4.03)	CDCl$_3$	37
59	OPh	H	(4-NO$_2$)Ph	—	7.40	—	CDCl$_3$	30
60	SH	H	Ph	(~13)	8.0	—	d-6 DMSO	38
61	SH	H	(4-Cl)Ph	(~13)	7.83	—	d-6 DMSO	38
62	SCH$_3$	H	Ph	(2.67)	7.27	—	CDCl$_3$	36
63	SCH$_3$	H	(4-NO$_2$)Ph	(2.69)	7.42	—	CDCl$_3$	36
64	SCH$_3$	H	(4-OCH$_3$)Ph	(2.67)	7.14	—	CDCl$_3$	36
65	SCH$_3$	H	(4-Br)Ph	(2.67)	7.28	—	CDCl$_3$	36
66	SCH$_3$	H	COPh	(2.72)	7.73	—	CDCl$_3$	18
67	SPh	H	OEt	—	6.14	(~4)	CDCl$_3$	39
68	S(4-Cl)Ph	H	OEt	—	6.17	(~4.1)	CDCl$_3$	39
69	CH$_3$	CH$_3$	H	(2.29)	(1.99)	7.13	CDCl$_3$	15
70	CH$_3$	Ph	H	(2.52)	—	7.81	CDCl$_3$	21
71	CH$_3$	CN	H	—	—	8.07	CDCl$_3$	40
72	cyclo-C$_6$H$_{11}$	CH$_2$CO$_2$H	H	—	(3.63)	7.5	CDCl$_3$	41
73	Ph	CH$_3$	H	—	(2.22)	~8.0	CDCl$_3$	15
74	Ph	Ph	H	—	—	7.97	CDCl$_3$	15
75	Ph	CN	H	—	—	8.21	CDCl$_3$	40
76	Ph	NO$_2$	H	—	—	8.50	CDCl$_3$	42
77	(4-NO$_2$)Ph	Ph	H	—	—	8.07	CDCl$_3$	21
78	(4-Cl)PhCH$_2$CO$_2$	H	H	—	(3.66)	7.75	CDCl$_3$	43
79	(4-Cl)PhCH$_2$CO$_2$	Et	H	—	(3.66)	7.68	CDCl$_3$	43
80	(3,5-di-NO$_2$)Ph	Ph	H	—	—	8.88	d-6 DMSO	21
81	COCH$_3$	CH$_3$	H	(2.67)	(2.3)	7.65	CDCl$_3$	33
82	COEt	CO$_2$CH$_3$	H	—	—	8.37	CDCl$_3$	44
83	COPh	CH$_3$	H	—	(2.38)	7.7	CDCl$_3$	33
84	CO$_2$Et	CH$_3$	H	—	(2.3)	7.65	CDCl$_3$	33

TABLE 2.2 (Continued)

Structure	R_2	R_4	R_5	δR_2	δR_4	δR_5	Solvent	Reference
85	CF_3	Ph	H	—	—	8.03	$CDCl_3$	15
86	$N(Et)_2$	CH_3	H	nd	nd	6.75	CCl_4	2
87	t-BuNH	CN	H	(5.0, NH)	—	7.60	$CDCl_3$	45
88	TMS	CH_3	H	(0.36)	(2.21)	7.47	$CDCl_3$	33

[a] Numbers in parentheses refer to the chemical shift of the substituent.
[b] 3:1 at $-50°C$.
nd, not determined; —, no relevant measure; Bt, benzotriazolyl; cyclo-C_6H_{11}, cyclohexyl.

TABLE 2.3. ^{1}H NMR OF TRISUBSTITUTED OXAZOLES[a]

Figure 2.4

Structure	R_2	R_4	R_5	δR_2	δR_4	δR_5	Solvent	Reference
89	CH_3	CH_3	CH_3	(2.28)	(1.96)	(2.15)	CCl_4	5
90	CH_3	CH_3	Ph	(2.46)	(2.36)	—	$CDCl_3$	46
91	CH_3	CH_3	CHO	(2.6)	(2.5)	—	$CDCl_3$	47
92	CH_3	CH_3	CO_2Et	(2.5)	(2.4)	—	$CDCl_3$	47
93	CH_3	t-Bu	CH_3	(2.31)	—	(2.28)	$CDCl_3$	46
94	CH_3	vinyl	CH_3	(2.38)	—	(2.26)	$CDCl_3$	46
95	CH_3	Ph	CH_3	(2.49)	—	(2.45)	$CDCl_3$	46
96	CH_3	Ph	$CH_2CO_2CH_3$	(2.47)	—	(3.75)	$CDCl_3$	48
97	CH_3	Ph	Ph	(2.51)	—	—	$CDCl_3$	15
98	CH_3	Ph	OEt	(2.03)	—	(4.45)	$CDCl_3$	36
99	CH_3	$(4-NO_2)$Ph	OCH_3	(2.43)	—	(4.13)	$CDCl_3$	36
100	CH_3	$(4-NO_2)$Ph	$N(CH_3)_2$	(2.41)	—	(2.82)	$CDCl_3$	36
101	CH_3	4-pyridyl	CH_3	(2.6)	—	(2.5)	$CDCl_3$	49
102	CH_3	CO_2Et	CH_3	(2.60)	—	(2.46)	$CDCl_3$	50
103	CH_3	CO_2Et	CO_2Et	(2.57)	—	—	$CDCl_3$	26
104	CH_3	$CONH_2$	SCH_3	(2.36)	—	—	$CDCl_3$	24
105	CH_3	CN	CH_3	(2.40)	—	(2.35)	$CDCl_3$	51
106	CH_3	CN	Ph	(2.45)	—	—	$CDCl_3$	24
107	CH_3	CN	OCH_3	(2.38)	—	(4.22)	$CDCl_3$	52
108	CH_3	CN	SCH_3	(2.46)	—	(2.58)	$CDCl_3$	24
109	CH_3	CN	SPh	(2.43)	—	—	$CDCl_3$	24
110	CH_3	SCH_3	Ph	(~2.5)	(~2.5)	—	$CDCl_3$	29
111	CH_3	SO_2Bn	Ph	(2.54)	(4.48)	—	$CDCl_3$	53
112	CH_3	SO_2Tol	Ph	(2.46)	—	—	$CDCl_3$	53
113	Et	SO_2Ph	OEt	(2.65)	—	(4.45)	$CDCl_3$	54
114	cyclo-C_3H_5	CN	NH_2	(1.87, 1H)	—	(4.89)	$CDCl_3$	55
115	Ph	CH_3	CH_3	—	(2.14)	(2.30)	$CDCl_3$	46
116	Ph	CH_3	Ph	—	(2.52)	—	$CDCl_3$	46
117	Ph	CH_3	CO_2Et	—	(2.56)	—	$CDCl_3$	15
118	Ph	CH_3	OCH_3	—	(2.12)	(3.98)	$CDCl_3$	56
119	Ph	CH_3	S-allyl	—	(2.28)	—	$CDCl_3$	57
120	Ph	CH_3	Br	—	(2.57)	—	$CDCl_3$	15
121	Ph	Ph	CH_3	—	—	(2.42)	$CDCl_3$	58
122	Ph	4-pyridyl	CH_3	—	—	(2.5)	$CDCl_3$	49
123	Ph	CHO	CH_3	—	(10.0)	(2.72)	$CDCl_3$	59

396 Spectroscopic Properties of Oxazoles

TABLE 2.3 (Continued)

Structure	R_2	R_4	R_5	δR_2	δR_4	δR_5	Solvent	Reference
124	Ph	CO_2CH_3	CH_3	—	—	(2.33)	$CDCl_3$	60
125	Ph	CN	CH_3	—	—	(2.59)	$CDCl_3$	40
126	Ph	CN	OCH_3	—	—	(4.31)	$CDCl_3$	54
127	Ph	SO_2Ph	OEt	—	—	(4.59)	$CDCl_3$	54
128	Ph	$P(O)(OEt)_2$	OEt	—	—	(4.54)	$CDCl_3$	54
129	(4-OCH_3)Ph	$CH_2CO_2CH_3$	CH_3	—	(3.56)	(2.34)	$CDCl_3$	61
130	(4-F)Ph	CH_2CH_2OH	CH_3	—	(2.71)	(2.32)	$CDCl_3$	61
131	(4-F)Ph	$CH_2CO_2CH_3$	CH_3	—	(3.56)	(2.56)	$CDCl_3$	61
132	2-pyridyl	CH_2CH_2OH	CH_3	—	(2.72)	(2.32)	$CDCl_3$	61
133	2-pyridyl	CH_2CO_2Me	CH_3	—	(3.53)	(2.33)	$CDCl_3$	61
134	CHO	CH_3	CH_3	(9.61)	(2.21)	(2.37)	$CDCl_3$	62
135	CO_2Et	CH_3	CH_3	—	(2.14)	(2.30)	$CDCl_3$	63
136	CO_2Et	CH_3	CO_2Et	—	(2.51)	—	$CDCl_3$	15
137	CO_2Et	CO_2Et	CH_3	—	—	(2.74)	$CDCl_3$	15
138	CF_3	CH_3	CH_3	—	(2.12)	(2.30)	$CDCl_3$	63
139	CF_3	CH_3	Ph	—	(2.38)	—	$CDCl_3$	46
140	CF_3	Ph	CH_3	—	—	(2.61)	$CDCl_3$	64
141	t-BuNH	CN	CH_3	(4.95, NH)	—	(2.40)	$CDCl_3$	45
142	PhNH	CN	CH_3	(10.3, NH)	—	(2.40)	d-6 DMSO	45
143	(4-OCH_3)PhNH	CN	CH_3	(10.15, NH)	—	(2.35)	d-6 DMSO	45
144	(4-Cl)PhNH	CN	CH_3	(10.0, NH)	—	(2.5)	d-6 DMSO	45
145	OCH_3	CO_2Et	CH_3	(4.11)	—	(2.53)	$CDCl_3$	15
146	OEt	CH_3	CH_3	—	(1.98)	(2.12)	$CDCl_3$	65
147	SCH_3	CO_2Et	CH_3	(2.65)	—	(1.37)	$CDCl_3$	66

[a] Numbers in parentheses refer to the chemical shift of the substituent.
—, no relevant resonance, $cyclo$-C_3H_5, cyclopropyl.

TABLE 2.4. ^{13}C NMR OF MONOSUBSTITUTED OXAZOLES

Figure 2.5

Structure	R_2	R_4	R_5	δR_2	δR_4	δR_5	Solvent	Reference
1	H	H	H	151.2	126.7	138.9	$CDCl_3$	73
3	H	Ph	H	150.9	nd	139.8	$CDCl_3$	4
4	H	(4-NO_2)Ph	H	151.8	nd	138.7	$CDCl_3$	4
5	H	(4-OCH_3)Ph	H	151.1	nd	140.1	$CDCl_3$	4
10	H	H	Ph	150.4	121.5	151.5	$CDCl_3$	9
148	H	H	2-furyl	149.8	121.3	143.9	$CDCl_3$	74
149	H	H	3-pyridyl	151.0	122.5	149.2	$CDCl_3$	74
13	H	H	CO_2Et	157.5[a]	133.2	142.7	$CDCl_3$	12
16	Bn	H	H	162.6	127	139.7	$CDCl_3$	14
17	Ph	H	H	162	nd	138.6	$CDCl_3$	15
21	SCH_3	H	H	161.1	128.1	139.8	$CDCl_3$	18

[a] Chemical shift assignment not certain.
nd, not determined.

TABLE 2.5. ^{13}C NMR OF DISUBSTITUTED OXAZOLES

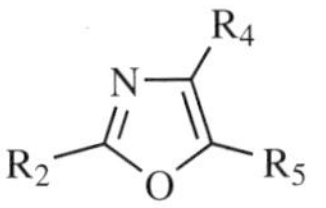

Figure 2.6

Structure	R_2	R_4	R_5	δR_2	δR_4	δR_5	Solvent	Reference
22	H	CH$_3$	CH$_3$	148.7	130.1	143.4	CDCF$_3$/CDCl$_3$[a]	19
23	H	CH$_3$	COCH$_3$	142.7	112.2	155.1	CCl$_4$	2
24	H	CH$_3$	NCO	149.6	133.8	144.6	CDCl$_3$	20
25	H	CH$_3$	O(n-Pr)	142.7	112.2	155.1	CCl$_4$	2
150	H	Ph	CH$_3$	149.5	134.2	144.2	CCl$_4$	2
37	H	I	Ph	151.3	77.6	150.4	CDCl$_3$	12
38	H	I	CO$_2$Et	157.1	61.7	143.4	CDCl$_3$	12
39	CH$_3$	H	CH$_3$	160.1	122.7	148.2	CDCF$_3$/CDCl$_3$[a]	19
40	CH$_3$	H	Ph	160.9	121.4	151.1	CDCF$_3$/CDCl$_3$[a]	19
41	CH$_3$	H	(4-NO$_2$)Ph	163.0	125.5	149.1	CDCl$_3$	30
42	CH$_3$	H	CO$_2$Et	164.7	134.0	142.5	CDCl$_3$	15
151	t-Bu	H	OCH$_3$	161.7	97.2	160.3	CDCl$_3$	75
44	Ph	H	Ph	161.0	123.5	151.2	CDCl$_3$	71
45	Ph	H	(4-OCH$_3$)Ph	160.5	121.9	151.3	CDCl$_3$	30
46	Ph	H	(4-Cl)Ph	161.4	123.8	150.3	CDCl$_3$	30
152	Ph	H	OCH$_3$	152.5	99.7	160.7	CDCl$_3$	75
47	Ph	H	OEt	152.6	100.8	159.8	CDCl$_3$	30
153	CN	H	CH$_3$	135.6	126.0	154.5	CDCl$_3$	76
54	N(CH$_3$)$_2$	H	(4-NO$_2$)Ph	162.1	127.9	142.8	CDCl$_3$	30
59	OPh	H	(4-NO$_2$)Ph	161.1	126.3	145	CDCl$_3$	30
60	SH	H	Ph	177.4	nd	147.3	d-6 DMSO	38
65	SCH$_3$	H	COPh	167.2	nd	150.6	CDCl$_3$	18
69	CH$_3$	CH$_3$	H	161.0	133.7	136.4	CDCl$_3$	15
71	CH$_3$	CN	H	163.1	114.9	146.1	CDCl$_3$	40
154	i-Pr	COCl	H	171.1	136.9	147.7	CDCl$_3$	75
73	Ph	CH$_3$	H	161.3	nd	137.5	CDCl$_3$	15
74	Ph	Ph	H	161.9	nd	142.0	CDCl$_3$	15
75	Ph	CN	H	163.1	116.2	145.8	CDCl$_3$	40
155	Ph	COCl	H	163.5	136.3	147.8	CDCl$_3$	75
156	(4-CH$_3$)Ph	CF$_3$	H	163.7	133.7	137.8	CDCl$_3$	77
76	Ph	NO$_2$	H	161.5	149.4	137.7	CDCl$_3$	42
85	CF$_3$	Ph	H	150.8	nd	142.0	CDCl$_3$	15
86	N(Et)$_2$	CH$_3$	H	162.4	137.0	128.2	CCl$_4$	2
87	t-BuNH	CN	H	159.5	107.0	158.2	CDCl$_3$	45
88	TMS	CH$_3$	H	170.4	136.2	136.6	CDCl$_3$	33

[a] 3:1 at −50°C.
nd, not determined.

TABLE 2.6. ^{13}C NMR OF TRISUBSTITUTED OXAZOLES

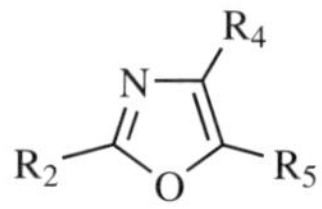

Figure 2.7

[Oxazole structure with R$_2$ at 2-position, R$_4$ at 4-position, R$_5$ at 5-position]

Structure	R$_2$	R$_4$	R$_5$	δR$_2$	δR$_4$	δR$_5$	Solvent	Reference
89	CH$_3$	CH$_3$	CH$_3$	158.6	130.0	142.5	CDCl$_3$	73
90	CH$_3$	CH$_3$	Ph	159.4	nd	145.1	CDCl$_3$	46
157	CH$_3$	CH$_3$	CN	164.5	149.4	120.9	CDCl$_3$	76
158	CH$_3$	CH$_3$	N(Ph)$_2$	156.9	128.3	144.7	CCl$_4$	2
159	CH$_3$	CH$_3$	OEt	152.0	112.9	154.3	CCl$_4$	2
93	CH$_3$	t-Bu	CH$_3$	157.5	140.5	141.8	CDCl$_3$	46
94	CH$_3$	vinyl	CH$_3$	159.5	133.4	144.4	CDCl$_3$	46
95	CH$_3$	Ph	CH$_3$	159.1	nd	143.3	CDCl$_3$	46
96	CH$_3$	Ph	CH$_2$CO$_2$CH$_3$	167.9	nd	138.9	CDCl$_3$	48
97	CH$_3$	Ph	Ph	160.1	135.1	145.3	CDCF$_3$/CDCl$_3$[a]	19
101	CH$_3$	4-pyridyl	Ph	159.4	125.7	139.4	CDCl$_3$	49
105	CH$_3$	CN	CH$_3$	160.8	110.9	158.0	CDCl$_3$	76
106	CH$_3$	CN	OCH$_3$	152.3	87.4	164.7	CDCl$_3$	52
160	CH$_3$	N(i-Pr)$_2$	iPr	156.6	135.6	151.5	CDCl$_3$	46
113	Et	SO$_2$Ph	OEt	157.9	141.5	155.7	CDCl$_3$	54
161	cyclo-C$_3$H$_5$	CN	NH$_2$	160.0	86.5	156.2	CDCl$_3$	55
162	t-Bu	CHO	N$_3$	166.1	124.0	148.8	CDCl$_3$	75
163	t-Bu	CHO	Cl	170.9	132.4	142.3	CDCl$_3$	75
164	1-adamantyl	CF$_3$	F	161.8	103.3	154.4	CDCl$_3$	78
115	Ph	CH$_3$	CH$_3$	159.1	nd	143.4	CDCl$_3$	63
165	Ph	CH$_3$	CF$_3$	161.8	139.7	134.3	CDCl$_3$	79
117	Ph	CH$_3$	CO$_2$Et	162.2	nd	147.1	CDCl$_3$	15
118	Ph	CH$_3$	S-allyl	162.6	127.1	144.5	CDCl$_3$	57
166	Ph	Bn	CF$_3$	162.2	142.3	134.2	CDCl$_3$	80
167	Ph	Ph	CF$_3$	157.9	132.9	142.6	CDCl$_3$	79
168	Ph	Ph	TMS	164.1	149.8	151.3	CDCl$_3$	65
122	Ph	4-pyridyl	CH$_3$	159.2	125.7	139.4	CDCl$_3$	49
169	Ph	CHO	Cl	160.8	133.7	142.5	CDCl$_3$	75
170	Ph	CHO	N$_3$	156.5	125.5	149.1	CDCl$_3$	75
124	Ph	CO$_2$CH$_3$	CH$_3$	167.2	nd	nd	CDCl$_3$	60
125	Ph	CN	CH$_3$	161.2	112.4	158.4	CDCl$_3$	40
126	Ph	CN	OCH$_3$	162.1	nd	152.5	CDCl$_3$	54
171	Ph	CO$_2$Et	Ph	159.8	121.6	155.1	CDCl$_3$	30
172	Ph	CF$_3$	Ph	160.6	127.0	150.1	CDCl$_3$	81
173	Ph	CF$_3$	CN	165.4	141.3	122.8	CDCl$_3$	82
174	Ph	CF$_3$	1-pyrrolidinyl	149.9	102.2	152.7	CDCl$_3$	81
175	Ph	CF$_3$	N$_3$	156.7	115.1	141.8	CDCl$_3$	83
176	Ph	CF$_3$	F	152.6	105.4	154.7	CDCl$_3$	78
177	Ph	CF$_3$	Cl	161.0	129.5	136.0	CDCl$_3$	81
178	Ph	NO$_2$	COPh	160.6	147.3	142.5	CDCl$_3$	84
179	Ph	NO$_2$	CO$_2$CH$_3$	160.9	148.5	134.9	CDCl$_3$	84
180	Ph	TMS	Ph	161.1	135.6	157.02	CDCl$_3$	65
127	Ph	SO$_2$Ph	OEt	158.0	141.5	151.5	CDCl$_3$	54
128	Ph	P(O)(OEt)$_2$	OEt	152.8	163.3	131.8	CDCl$_3$	54
181	(4-NO$_2$)Ph	CF$_3$	F	150.6	106.8	155.4	CDCl$_3$	78

TABLE 2.6 (Continued)

Structure	R$_2$	R$_4$	R$_5$	δR$_2$	δR$_4$	δR$_5$	Solvent	Reference
182	2-thiazolyl	CF$_3$	1-imidazolyl	146.0	120.2	151.2	CDCl$_3$	85
183	(4-F)Ph	CF$_3$	1-triazolyl	159.2	123.5	173.7	CDCl$_3$	81
134	CHO	CH$_3$	CH$_3$	155.9	134.8	148.6	CDCl$_3$	62
135	CO$_2$Et	CH$_3$	CH$_3$	155.9	133.8	147.7	CDCl$_3$	63
137	CO$_2$Et	CO$_2$Et	CH$_3$	~155.0–160.0	129.5	150.1	CDCl$_3$	15
138	CF$_3$	CH$_3$	CH$_3$	146.6	132.3	148.2	CDCl$_3$	63
139	CF$_3$	CH$_3$	Ph	~147.0–150.0	nd	nd	CDCl$_3$	46
140	CF$_3$	Ph	CH$_3$	146.7	nd	136.0	CDCl$_3$	64
141	t-BuNH	CN	CH$_3$	158.5	109.5	151.4	CDCl$_3$	45
142	PhNH	CN	CH$_3$	155.5	108.4	152.4	d-6 DMSO	45
143	(4-OCH$_3$)PhNH	CN	CH$_3$	155.8	108.1	152.1	d-6 DMSO	45
144	(4-Cl)PhNH	CN	CH$_3$	155.2	108.3	152.7	d-6 DMSO	45
145	OCH$_3$	CO$_2$Et	CH$_3$	162.4	126.4	151.4	CDCl$_3$	15
146	OEt	CH$_3$	CH$_3$	159.9	128.9	137.3	CDCl$_3$	46
184	NHAc	CH$_3$	CH$_3$	151.9	129.0	139.6	CDCl$_3$	2
185	N(CH$_3$)$_2$	Ph	CH$_3$	160.4	134.4	137.2	melt	2
147	SCH$_3$	CO$_2$Et	CH$_3$	159.1	128.4	156.8	CDCl$_3$	66
186	SCH$_3$	CO$_2$Et	Ph	160.0	126.5	156.2	CDCl$_3$	66
187	SCH$_3$	CO$_2$Et	CF$_3$	nd	152.9	156.6	CDCl$_3$	86
188	I	I	Ph	101.3	79.6	nd	CDCl$_3$	12
189	I	I	CO$_2$Et	92.3	62.2	149.4	CDCl$_3$	12

^{a} 3:1 at −50°C.
nd, not determined.

an R group on C$_4$. These relationships have proven helpful in assignment of oxazole-containing natural products.[72] Little variation is seen with changes in solvent. Tables 2.4–2.6 parallel those for the proton NMR data, with some additional compounds.

In more specialized carbon NMR topics, a summary of ^{13}C assignments for 2,5-diaryloxazoles **190** has been published (Fig. 2.8).[71] Girault and Perronnet[29] give extensive spectral data for this class of compounds. NMR-based investigations of 2-metalated oxazoles **191** and their related anionic species have appeared (Fig. 2.9).[10,87]

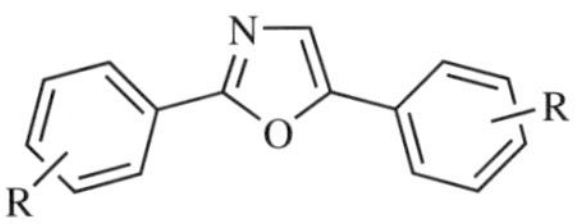

190

Figure 2.8

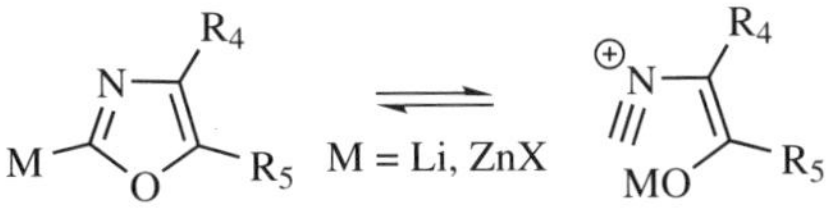

191

M = Li, ZnX

Figure 2.9

The reader is referred to Chapter 2 in the previous edition,[1] Bogdanor and co-workers,[2] and Hiemstra and co-workers[70] for tables and discussions of carbon–proton coupling constants. Briefly, $^1J_{C-H}$ values vary around 230 Hz at C_2, around 195 Hz at C_4, and from 205 to 210 Hz at C_5. $^2J_{C-H}$ values range from 14 to 20 Hz. These values are largely insensitive to substituent effects but depend on position and the coupling path involved.[70] Carbon–carbon spin-spin couplings have been determined in five-membered heterocycles.[88] $^1J_{C-C}$ is 70.7 Hz between C_4 and C_5 of oxazole, the largest value of the 11 nitrogen, oxygen, and sulfur-containing heterocycles studied. The long-range couplings are 4.2 Hz between C_2 and C_4, and 4.4 Hz between C_2 and C_5.

2.2.3. Nitrogen NMR

^{14}N NMR provides mainly chemical shift information due to linewidth difficulties. The oxazole nitrogen is found at +100 to +177 ppm (note the convention for reporting nitrogen NMR shieldings upfield of nitromethane reference).[2] At the lower end of this range are compounds substituted with electron-withdrawing groups at C_5, whereas compounds substituted with electron-donating heteroatoms (or α-anions[14]) at C_2 are found at the upper end of this range. The nitrogen atom of oxazole itself is found at +124 ppm (in CCl_4). These values, and other heterocyclic nitrogen chemical shifts, are sensitive to both the polarity and hydrogen-bonding ability of the solvent environment: The nitrogen shift of oxazole itself can vary from +119.5 ppm in cyclohexane to +138 ppm in trifluoroethanol.[89]

^{15}N spectra can provide coupling information as well as chemical shift data and can be determined using either with specifically labeled compounds or (with greater effort) at natural abundance. For recent work in using ^{15}N techniques in assigning oxazole-containing natural products, see Martin and co-workers.[90] ^{15}N spectra of azoles in general have been studied[91,92] and ^{15}N (as well as ^{13}C) shifts have been used to map charge distribution in azoles and azole anions.[14] ^{15}N-^1H couplings in oxazole (d_6-DMSO solution) are $^2J_{(3,2)} = 13.4$ Hz, $^2J_{(3,4)} = 10.4$ Hz, and $^3J_{(3-5)} = 1.2$ Hz.[91] 4-cyano-5-methoxy-2-methyloxazole **107** has been prepared in ^{15}N-labeled form (Fig. 2.10).[52] The $^2J_{^{12}C-^{15}N}$ couplings are 13.3 Hz with the methyl carbon (C2) and 7.8 Hz with the nitrile carbon (C4).

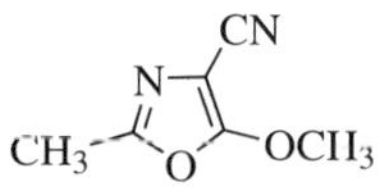

107

Figure 2.10

2.2.4. Oxygen NMR

Relative to the other nuclei discussed here, ^{17}O spectroscopy remains relatively uncommon.[93,94] A study of ^{17}O spectra of various azoles and sydnones has appeared,[95] and 2,5-diphenyloxazole **44** is reported therein to have a shift of 247 ppm relative to water (Fig. 2.11). This is reasonably similar to the ^{17}O shift of furan (228 ppm), indicating that the oxazole nitrogen (and presumably much ring substitution) has only a small influence on the oxygen chemical shift.

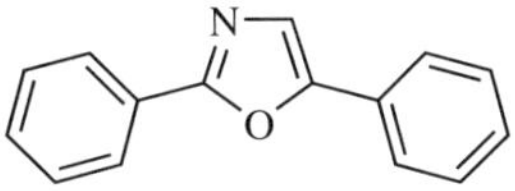

44

Figure 2.11

2.2.5. Fluorine NMR

Relatively few directly fluorinated oxazoles have been reported—e.g., compounds **164**, **176**, and **187**—but a number of fluoroalkyl-substituted compounds have appeared in the literature. Tables 2.7–2.9 provide a summary of chemical shift and coupling information for selected classes of compounds. Note that the chemical shifts have all been reset to a CFCl$_3$ standard, with the previously used trifluoroacetic acid standard taken as δ −76.55.

TABLE 2.7. ^{19}F NMR OF 2-(TRIFLUOROMETHYL)OXAZOLES

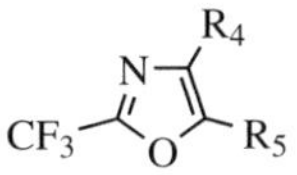

Figure 2.12

Structure	R$_4$	R$_5$	^{19}F δCF$_3$	$^1J_{C-F}$	$^2J_{C-CF_3}$	$^4J_{C-CF_3}$	Reference
85	Ph	H	nd	177	nd	—	15
140	Ph	CH$_3$	−65.4	270	—	—	64
138	CH$_3$	CH$_3$	nd	270	44	—	63
139	CH$_3$	Ph	nd	270	44	—	46

nd, not determined.

TABLE 2.8. ^{19}F NMR OF 4-(TRIFLUOROMETHYL)OXAZOLES^a

$$\underset{\text{Figure 2.13}}{\underset{R_2}{\overset{CF_3}{\underset{N}{\bigcirc}}}\!\!\!-\!\!R_5}$$

Structure	R$_2$	R$_5$	^{19}F δF	^{19}F δCF$_3$	$^1J_{C-F}$	C$_2$ $^2J_{C-O-CF}$	C$_4$ $^2J_{C-CF_3}$	C$_4$ $^2J_{C-CF}$	C$_5$ $^1J_{C-F}$	C$_5$ $^3J_{C-C-CF_3}$	Reference
7	H	OEt	—	−61.2	—	—	—	—	—	—	27
—	Ph	ArCH$_2$O	—	−59.8 to −60.8	121–122	—	40–41	—	—	3–4	96
192	(4-CH$_3$)Ph	H	—	−62.6	265	—	40	—	—	4	77
193	(4-CH$_3$)Ph	Cl	—	−61.3	nd	—	—	—	—	—	77
—	Alk/Ar	F	−114.5 to −123.5	−60.6 to −61.3	nd	3–7	42–43	7	294–298	3–5	78
—	CO$_2$R	OEt	—	−59.9 to −61.2	—	—	—	—	—	—	27
194	CF$_3$	F	−113.5	−62.5, −65.8	nd	nd	nd	nd	nd	nd	97
—	Ar	XR (X = O, N, S)	—	−51.5 to −62.5	nd	—	39–41	—	—	2–4	81
173	Ph	CN	—	−62.4	270	—	41	—	—	small	82
175	Ph	N$_3$	—	−60.1	267	—	41	—	—	3	83
195	Ph	CH$_2$NH$_2$	—	−60.4	269	—	39	—	—	3	82
—	Het	F	−118.6 to −121.4	−61.2 to −63.4	nd	6–7	42–50	6–7	292–297	3	85
—	Subst. biphenyl	F	−116.6 to −119.6	−61 to −61.2	nd	nd	42	7	295	3	85
182	2-thiazolyl	1-imidazoyl	—	−59.8	nd	—	40	—	—	3	85

^{a} In 5-fluoro-4-trifluoromethyl oxazoles, the ^{19}F $^4J_{CF_3,F}$ couplings are generally 10–15 Hz (vide supra). nd, not determined.

TABLE 2.9. ^{19}F NMR OF 5-(PERFLUOROALKYL)OXAZOLES

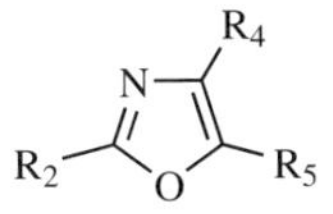

Figure 2.14

Structure	R_2	R_4	R_5	^{19}F δCF_3	$^1J_{CF_3}$	$C_4\ ^3J_{C\text{-}C\text{-}CF_3}$	$C_5\ ^2J_{C\text{-}CF_3}$	Reference
—	Alk/Ar	Alk/Ar	CF_3	nd	267–268	2–2.5	41–42.5	79
196	Ph	a	CF_3	−58.9	268	—	44	98
197	Ph	i-Pr	CF_2CF_3	−85.1, −114.7	nd	—	nd	99
166	Ph	Bn	CF_3	nd	267.9	2	42	80
187	SCH_3	CO_2Et	CF_3	nd	288.5	1.22	2.77	86

^{a} 1,3-bis(benzamido)prop-1-enyl.
nd, not determined.

2.3. MASS SPECTROMETRY

The only review of this subject continues to be Traldi and co-workers.[100] The majority of oxazoles follow the main ring-cleavage fragmentation pattern of unsubstituted oxazole: radical cation formation, cleavage of the O-C_2 bond, and loss of CO, followed by loss of HCN or nitrile (Fig. 2.15).[101] In the case of oxazole itself, an initial loss of HCN is believed to lead to oxirene radical cation.[102]

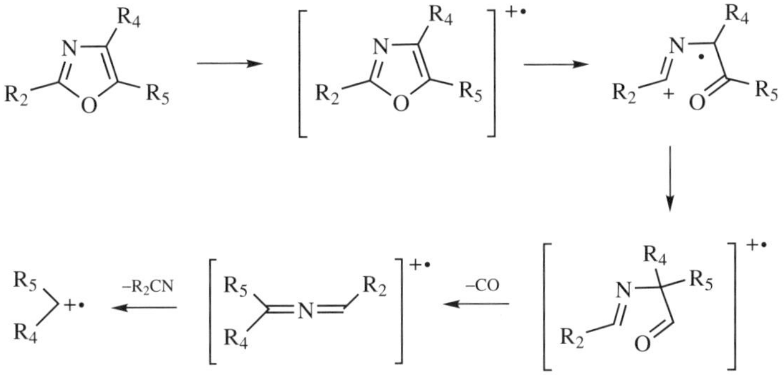

Figure 2.15

As expected, substituents can significantly modify these fragmentation pathways. For example, 2-methyloxazole **15** and 4-methyloxazole **2** have fragmentations reminiscent of the parent (CO loss), whereas 5-methyloxazole **9** shows unexpected loss of the methyl group (Fig. 2.16).[101] Nonetheless, this cleavage appears to take place after radical cation formation and O-C_2 bond cleavage.

5-Alkyl-substituted oxazoles can also give benzylic cleavage products from the initial radical cation (Fig. 2.17). With substituents at C_2, cleavage with McLafferty rearrangement is often noted (Fig. 2.18).

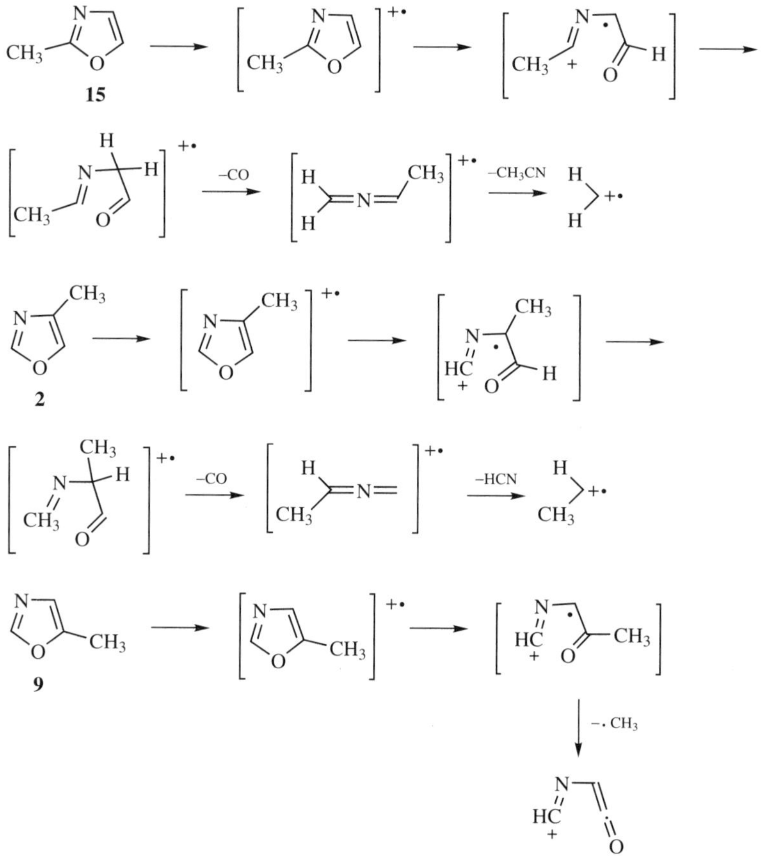

Figure 2.16

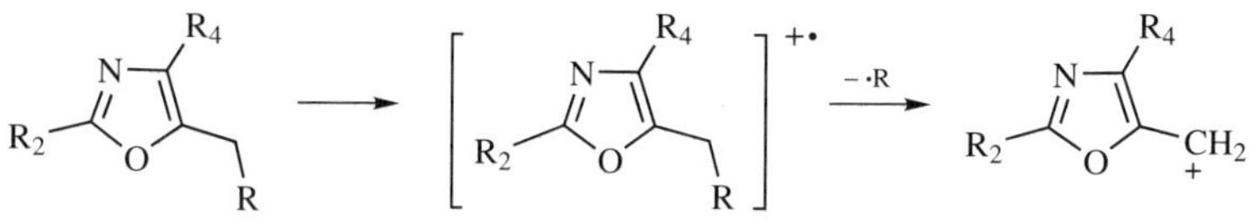

Figure 2.17

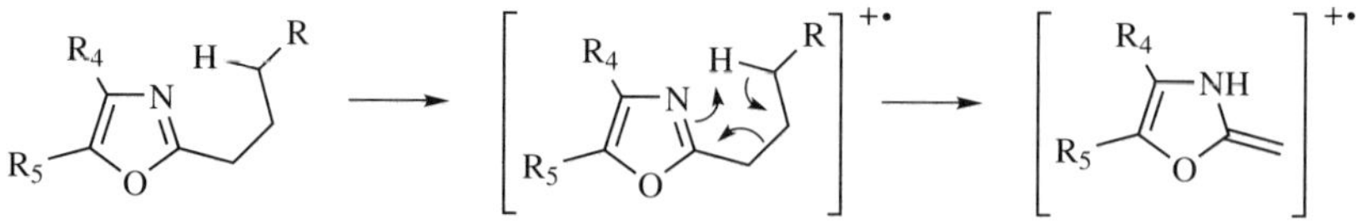

Figure 2.18

It should be emphasized that the ease of these processes is relative. In a study of microcin B17, a polypeptide antibiotic containing numerous oxazole and thiazole moieties, the presence of such biogenic heterocycles in the peptide chain was apparent from the electrospray MS/MS spectra owing to the suppression of bond cleavage compared to the comparatively labile peptide linkages (Fig. 2.19).[103]

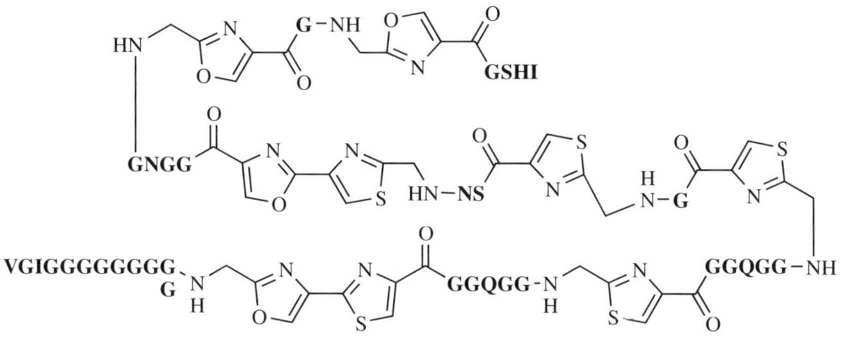

Figure 2.19

The mass spectra of 2-aryl-5-phenyloxazoles **198** have also been studied,[104] as have linked bis-oxazoles of types **199** and **200**, which fragment through ring cleavage and with facile formation of multiply charged molecular ions.[105] A study of cleavage patterns and phenyl migration in 2,4,5-triphenyloxazole **201** has appeared (Fig. 2.20).[106]

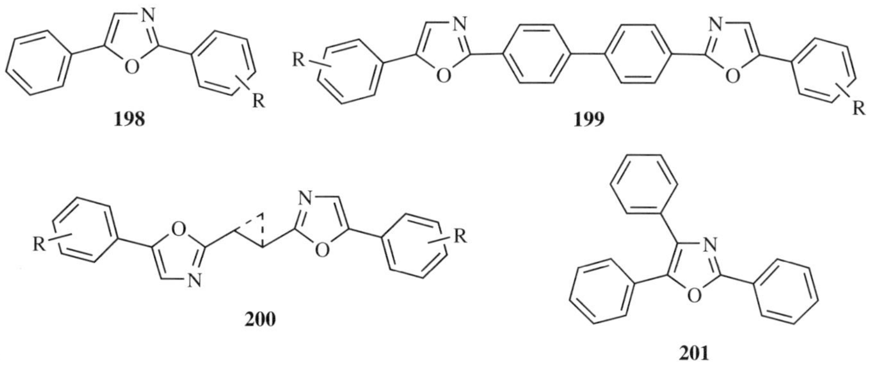

Figure 2.20

Negative ion fragmentations have been studied in various azole systems.[107] Oxazole anions have a somewhat more complex fragmentation pattern than the cationic species, but the principles involved are similar.

2.4. OPTICAL SPECTROSCOPY

2.4.1. Infrared

The IR and Raman properties of oxazoles have not been comprehensively reviewed, and no such treatment now seems likely. The most characteristic IR feature of the heterocycle is a strong absorbance, which generally falls in the range of 1555–1590 cm^{-1}, but can be shifted down to 1500 or above 1600 in some cases. This is assigned as the NCO ring stretch. Oxazole itself has been investigated and the spectrum interpreted.[1,108] Calculated geometries and spectra of oxazole and thiazole have been compared to experimental results.[109] The symmetrical CH vibrations of 2,4,5-trimethyloxazole **89** (and related methyl-substituted azoles) have been investigated,[110] and a report has appeared on the spectra and vibrational modes of 4,5-diaryl-2-substituted oxazoles **202** (Fig. 2.21).[111]

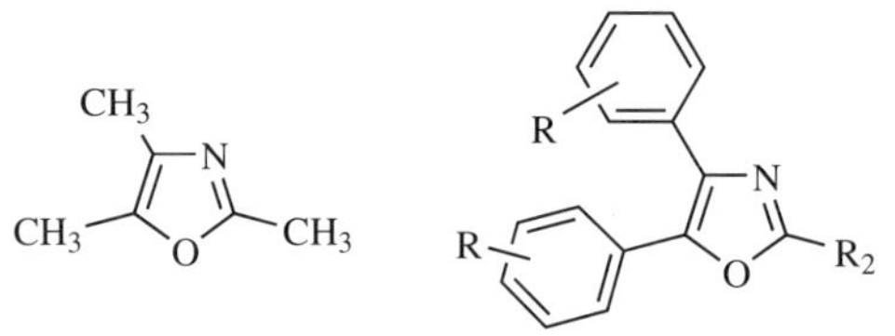

Figure 2.21

Structural analyses (including X-ray, IR, and Raman spectra) of bis-arylazoles **203** related to the scintillant POPOP (1,4-bis-(5-phenyloxazol-2-yl)benzene **204**) have been published (Fig. 2.22).[112] No useful information on intermolecular hydrogen bonding in such compounds could be obtained from the vibrational spectra. The ring-stretch IR bands mentioned above confirmed that the nonplanar and asymmetric conformations seen in the X-ray crystal structures were present in solution as well.

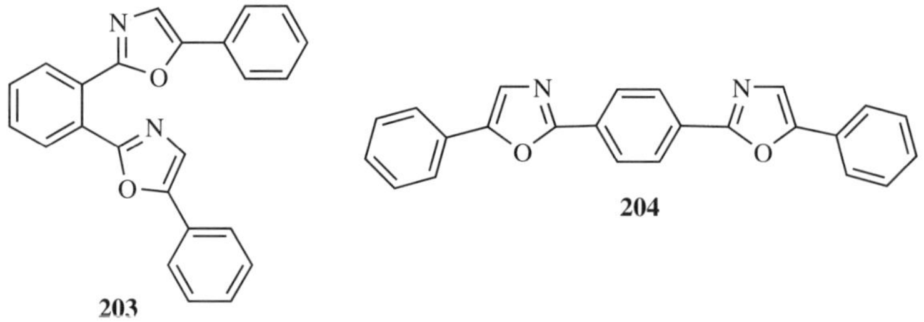

Figure 2.22

Complexes of phenol with 5-aryl-2-methyloxazoles **205** and related azoles[113] as well as 2-aryl-5-phenyloxazole **198**, 2-(2-furyl)-5-phenyloxazole **206** and 5-phenyl-2-(2-thienyl)oxazole **207** have been investigated in different solvents by both NMR and IR (Fig. 2.23).[114] Finally surface-enhanced Raman scattering (SERS) was investigated with oxazole and related heterocycles in their adsorption on colloidal silver particles.[115] Oxazole (and isoxazole) showed weak interactions in this system, with no charge-transfer character.

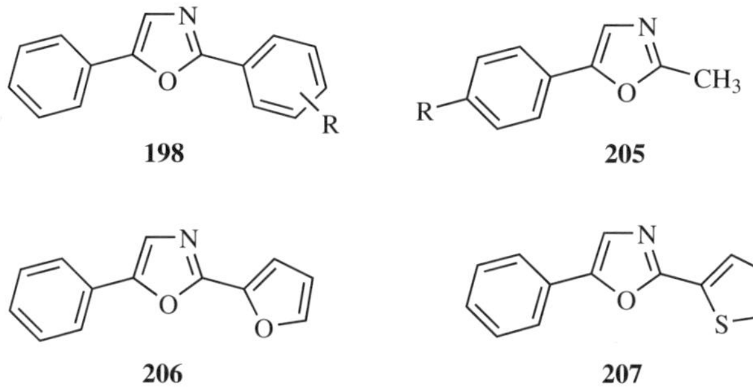

Figure 2.23

2.4.2. Ultraviolet and Visible

Since the last edition, the field of oxazole UV and visible spectroscopy has seen a great deal of work, as the fluorescent and luminescent properties of aryloxazoles have become apparent.[116,117]

Oxazole itself has a UV absorption maximum at 205 nm, and monoaryl oxazoles generally show a strong absorbance between 240 and 270 nm.[118] The ground and excited states of oxazoles and 2-aryloxazoles have been studied.[119,120] The photophysical properties of phenolic 2-aryloxazoles **208** and the related thiazoles were determined in various solvents and media (Fig. 2.24).[121,122]

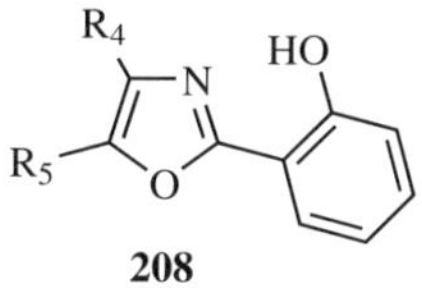

Figure 2.24

2,5-Diaryloxazoles show a much different spectrum with strong absorbances noted at 314–350 nm and 220–240 nm, with a less intense band observed at 260–300 nm. Early work in this field was summarized by Balaban and co-workers[123] and Pavlopoulos and Hammond.[124] More recent studies of the spectral, fluorescent, and luminescent properties of 2,5-diaryloxazoles include their dipole moments,[125] UV-amplified spontaneous emission (ASE) laser spike spectroscopy,[126,127] and the effects of magnetic fields on heavy-ion-induced luminescence.[128] Among the molecules in which general spectral behavior have been investigated are 2-cyanoaryl analogs,[129] solvent effects on the spectra of 2-aryl-5-(aryloxadiazolyl)phenyloxazoles **209**,[130] 2-(2-furyl)-5-phenyloxazole **206**,[131] 2,5-diphenyl-[132] and 5-aryl-2-(4-pyridyl)oxazoles **210** at selected pHs,[133] the water-soluble 2-(4-benzenesulfonic acid)-5-(4-dimethylaminophenyl)oxazole sodium salt **211**,[134] 2,5-bis(biphenylyl)oxazoles **212**,[135] POPOP (**204**) crystals,[136] ortho-analogs of POPOP such as **203**,[137–141] ortho-phenolic 2,5-diaryloxazoles such as **208** in the condensed phase[142] and in jet-cooled molecular beams,[143] aryl-2,5′-bis-oxazoles **213**,[144] and thiophene analogs such as **214** (Fig. 2.25).[144] 2,4,5-Triaryloxazoles were also investigated for absorbance and emission properties.[145]

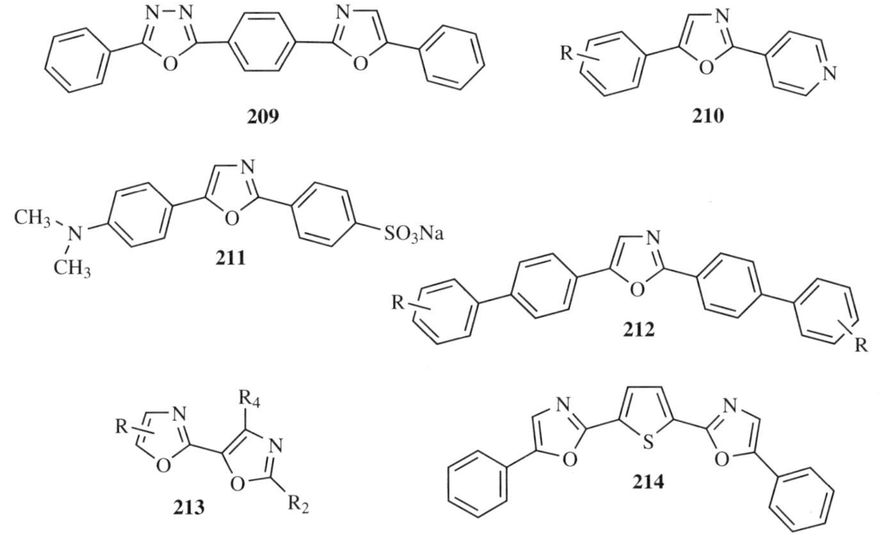

Figure 2.25

In general, combinations of electron-rich aryl groups at one position with electron-deficient or neutral groups in the other have provided compounds with a wide range of photophysical properties, and research in this area continues apace.

Polymeric derivatives of aryloxazoles have been investigated for use in optical and electronic applications. These have been prepared by doping existing polyimides[146] and have been studied in dinonyl bisoxazole systems.[147,148] A red shift in the maximum absorbance is noted as the degree of oligomerization increases in these latter compounds.

It should be noted that there is also a large body of literature on the photophysics of benzoxazole derivatives, which is outside of the scope of this review but clearly of relevance to those in the field. Leading references are provided for benzoxazole itself[149,150] and the widely studied "oxazole yellow" fluorescent dyes and their interactions with DNA strands.[151–153]

2.5. MICROWAVE SPECTROSCOPY

The fundamental papers on oxazole microwave spectroscopy appeared in 1978.[154,155] These studies confirmed a planar structure for oxazole with an NCO bond angle of 115°, a dipole moment of 1.50 D, and measured the nitrogen quadrupole coupling constants. A reanalysis of the microwave rotational spectra of 2-, 4-, and 5-methyloxazoles **15**, **2**, and **9** has been reported, which corrects the previously obtained rotational constants.[156] The dipole moments are 1.37 D, 1.08 D, and 2.16 D for **15**, **2**, and **10**, respectively. The Van der Waals complex between argon and oxazole has been investigated through microwave spectroscopy, which allows for determination of rotational constants, centrifugal distortion constants, and nitrogen quadrupole coupling constants.[157] The argon atom lies above the ring, displaced toward the ring oxygen, and there is no charge transfer between the partners.

2.6. MISCELLANEOUS SPECTROSCOPIC METHODS

No spectral determination of a true oxazole radical has been reported. A radical produced from irradiation of oxazole and sodium in an argon matrix was studied.[158] The ESR spectrum indicated rupture of the CO bond, with a radical at what had been C_2. Species formed from attack of the hydroxyl, sulfate, and phosphate radicals on oxazole, 4-methyloxazole **2**, 5-methyloxazole **9**, and other heterocycles have been analyzed by ESR (Fig. 2.26).[159] In all cases, a dihydrooxazole radical at C_4 is produced by addition of the radical to the double bond at C_5; g for these species varies between 2.00251 and 2.00258.

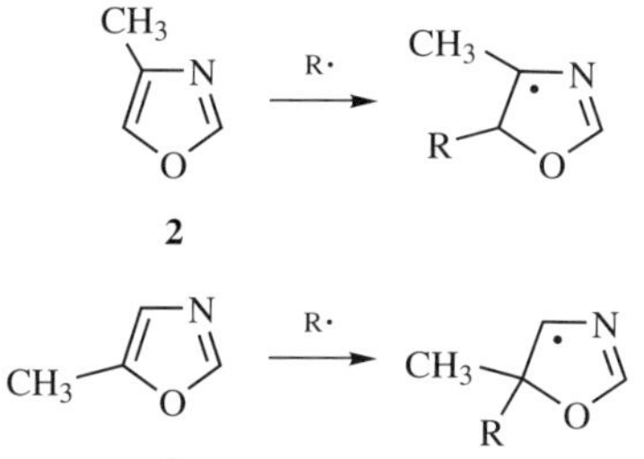

Figure 2.26

Photoelectron spectra of oxazole and related heterocycles have been investigated in the He(I) and He(II) regions.[160] Eleven ionization potentials were noted, from 9.83 eV to approximately 26.5 eV. The photoelectron spectrum of 2-methyl-5-phenyloxazole **40** has been determined as well.[161]

The X-ray absorption near-edge structure technique (N K XANES) has been used to probe the bonding of atoms in various heterocycles.[162] Compared to pyrrole, oxazole shows a much lower electron density at nitrogen, as expected, with corresponding effects on the π^* resonances.[163]

2.7. SUMMARY

The well-characterized nature of the oxazole system can be attested to by the number of references cited—and indeed, by the rest of this volume. This chapter provides sufficient data to assist the practicing chemist, but the reader is referred to the primary literature for further details and new advances. The outstanding problems in oxazole spectroscopy appear to lie in studies of highly substituted derivatives whose electronic character has been usefully altered by delocalization and/or heteroatom groups. Investigations of heterocyclic physical phenomena (such as surface interactions) by modern spectroscopic techniques also seem to be fruitful areas of research.

REFERENCES

1. Maryanoff, C. A. In Turchi, I. J., ed. *Oxazoles*, Vol. 45, *The Chemistry of Heterocyclic Compounds*, John Wiley, New York, 1986, pp. 343–360.
2. Bogdanov, V. S.; Aitzhanova, M. A.; Abronin, I. A.; Medvedskaya, L. B. *Izv. Akad. Nauk. SSSR Ser. Khim.* **1980**, 305–316 [*Chem. Abstr.* **1980**, *92*, 214449].
3. Bowie, J. H.; Donaghue, P. F.; Rodda, H. J. *J. Chem. Soc. B* **1969**, 1122–1125.
4. Whitney, S. E.; Winters, M.; Rickborn, B. *J. Org. Chem.* **1990**, *55*, 929–935.
5. Brown, D. J.; Ghosh, P. B. *J. Chem. Soc. B* **1969**, 270–276.
6. Schöllkopf, U.; Blume, E. *Tetrahedron Lett.* **1973**, *14*, 629–632.
7. Bull, J. R.; Tuinman, A. *Tetrahedron* **1975**, *31*, 2151–2155.
8. Maquestiau, A.; Puk, E.; Flammang, R. *Tetrahedron Lett.* **1986**, *27*, 4023–4024.
9. Nyce, P.; Puar, M. S.; Gala, D.; Jaret, R. S. *J. Heterocycl. Chem.* **1987**, *24*, 505–508.
10. Crowe, E.; Hossner, F.; Hughes, M. J. *Tetrahedron* **1995**, *51*, 8889–8900.
11. Hodges, J. C.; Patt, W. C.; Connolly, C. J. *J. Org. Chem.* **1991**, *56*, 449–452.
12. Vedejs, E.; Luchetta, L. M. *J. Org. Chem.* **1999**, *64*, 1011–1014.
13. Grigg, R.; Jackson, J. L. *J. Chem. Soc. C* **1970**, 552–556.
14. Abbotto, A.; Bradamante, S.; Pagani, G. A. *J. Org. Chem.* **1996**, *61*, 1761–1769.
15. Prager, R. H.; Smith, J. A.; Weber, B.; Williams, C. M. *J. Chem. Soc. Perkin Trans. 1* **1997**, 2665–2672.
16. Jones, G.; Good, R. H. *J. Chem. Soc. C* **1971**, 1196–1198.
17. Bödeker, J.; Burmester, K. *Z. Chem.* **1987**, *27*, 258–259.
18. Shafer, C. M.; Molinski, T. F. *J. Org. Chem.* **1998**, *63*, 551–555.
19. Gollnick, K.; Koegler, S. *Tetrahedron Lett.* **1988**, *29*, 1003–1006.
20. Ray, S.; Ghosh, S. *Indian J. Chem. Sect. B* 1999, *38*, 986–988.
21. Pei, W.; Li, S.; Nie, X.; Li, Y.; Pei, J.; Chen, B.; Wu, J.; Ye, X. *Synthesis* **1998**, 1298–1304.
22. Kamitori, Y.; Hojo, M.; Matsuda, R.; Takahashi, T.; Wada, M. *Heterocycles* **1992**, *34*, 1047–1054.
23. Kakehi, A.; Ito, S.; Funahashi, T.; Ota, Y. *J. Org. Chem.* **1976**, *41*, 1570–1574.

24. Matsumura, K.; Miyashita, O.; Shimazu, H.; Hashimoto, N. *Chem. Pharm. Bull.* **1976**, *24*, 948–959.
25. Suzuki, M.; Iwasaki, T.; Miyoshi, M.; Okumura, K.; Matsumoto, K. *J. Org. Chem.* **1973**, *38*, 3571–3575.
26. Yokoyama, M.; Menjo, Y.; Watanabe, M.; Togo, H. *Synthesis* **1994**, 1467–1470.
27. Shi, G.; Xu, Y.; Xu, M. *J. Fluorine Chem.* **1991**, *52*, 149–157.
28. Sasaki, H. *Chem. Pharm. Bull.* **1997**, *45*, 1369–1371.
29. Girault, P.; Perronnet, J. *J. Heterocycl. Chem.* **1981**, *18*, 419–421.
30. Fukushima, K.; Ibata, T. *Bull. Chem. Soc. Jpn.* **1995**, *68*, 3469–3481.
31. Kashima, C.; Arao, H. *Synthesis*, **1989**, 873–874.
32. Pivsa-Art, S.; Satoh, T.; Kawamura, Y.; Miura, M.; Nomura, M. *Bull. Chem. Soc. Jpn.* **1998**, *71*, 467–473.
33. Dondoni, A.; Fantin, G.; Fogagnolo, M.; Medici, A.; Pedrini, P. *J. Org. Chem.* **1987**, *52*, 3413–3420.
34. Yamanaka, H.; Mizugaki, M.; Sakamoto, T.; Sagi, M.; Nakagawa, Y.; Takayama, H.; Ishibashi, M.; Miyazaki, H. *Chem. Pharm. Bull.* **1983**, *31*, 4549–4553.
35. Pérez, J. D.; de Díaz, R. G.; Yranzo, G. I. *J. Org. Chem.* **1981**, *46*, 3505–3508.
36. Ibata, T.; Yamashita, T.; Kashiuchi, M.; Nakano, S.; Nakawa, H. *Bull. Chem. Soc. Jpn.* **1984**, *57*, 2450–2455.
37. Fukushima, K.; Lu, Y.-Q; Ibata, T. *Bull. Chem. Soc. Jpn.* **1996**, *69*, 3289–3295.
38. Yamamoto, I.; Okuda, K.; Nagai, S.; Motoyoshiya, J.; Gotoh, H.; Matsuzaki, K. *J. Chem. Soc. Perkin Trans. 1* **1984**, 435–438.
39. Bossio, R.; Marcaccini, S.; Pepino, R. *Heterocycles* **1986**, *24*, 2003–2005.
40. Ceulemans, E.; Dyall, L. K.; Dehaen, W. *Tetrahedron* **1999**, *55*, 1977–1988.
41. Meguro, K.; Tawada, H.; Sugiyama, Y.; Fujita, T.; Kawamatsu, Y. *Chem. Pharm. Bull.* **1986**, *34*, 2840–2851.
42. Nesi, R.; Turchi, S.; Giomi, D. *J. Org. Chem.* **1996**, *61*, 7933–7936.
43. Seki, M.; Moriya, T.; Matsumoto, K.; Takashima, K.; Mori, T.; Odawara, A.; Takeyama, S. *Chem. Pharm. Bull.* **1988**, *36*, 4435–4440.
44. Okumura, K.; Ito, A.; Saito, H.; Nakamura, Y.; Shin, C.-G *Bull. Chem. Soc. Jpn.* **1996**, *69*, 2309–2316.
45. Pascual, A. *Helv. Chim. Acta* **1989**, *72*, 556–569.
46. Cunico, R. F.; Kuan, C. P. *J. Org. Chem.* **1992**, *57*, 3331–3336.
47. Boulos, J.; Schulman, J. *J. Heterocycl. Chem.* **1998**, *35*, 859–863.
48. Sá, M. C. M.; Kascheres, A. *J. Org. Chem.* **1996**, *61*, 3749–3752.
49. Braña, M. F.; Castellano, J. M.; de Miguel, P.; Posada, P.; Sanz, C. R.; Migallón, A. S. *Tetrahedron* **1994**, *50*, 10061–10072.
50. Effenberger, F.; Beisswenger, T. *Chem. Ber.* **1984**, *117*, 1497–1512.
51. Doleschall, G.; Seres, P. *J. Chem. Soc. Perkin Trans. 1* **1988**, 1875–1879.
52. Perrocheau, J.; Carrié, R.; Fleury, J.-P. *Can. J. Chem.* **1994**, *72*, 2458–2467.
53. Kuo, Y.-C; Aoyama, T.; Shioiri, T. *Chem. Pharm. Bull.* **1982**, *30*, 526–533.
54. Doyle, K. J.; Moody, C. J. *Tetrahedron* **1994**, *50*, 3761–3772.
55. Freeman, F.; Chen, T.; van der Linden, J. B. *Synthesis* **1997**, 861–862.
56. Iesce, M. R.; Cermola, F.; Graziano, M. L.; Scarpati, R. *Synthesis* **1994**, 944–948.
57. Jenny, C.; Heimgartner, H. *Helv. Chim. Acta* **1989**, *72*, 1639–1646.
58. Hoppe, I.; Schöllkopf, U. *Liebigs Ann. Chem.* **1979**, 219–226.
59. Hulin, B.; Clark, D. A.; Goldstein, S. W.; McDermott, R. E.; Dambek, P. J.; Kappeler, W. H.; Lamphere, C. H.; Lewis, D. M.; Rizzi, J. P. *J. Med. Chem.* **1992**, *35*, 1853–1864.

60. Wipf, P.; Cunningham, A.; Rice, R. L.; Lazo, J. S. *Bioorg. Med. Chem.* **1997**, *5*, 165–177.
61. Collins, J. L.; Blanchard, S. G.; Boswell, G. E.; Charifson, P. S.; Cobb, J. E.; Henke, B. R.; Hull-Ryde, E. A.; Kazmierski, W. M.; Lake, D. H.; Leesnitzer, L. M.; Lehmann, J.; Lenhard, J. M.; Orband-Miller, L. A.; Gray-Nunez, Y.; Parks, D. J.; Plunkett, K. D.; Tong, W.-Q *J. Med. Chem.* **1998**, *41*, 5037–5054.
62. Šindler-Kulyk, M.; Vojnović, D.; Defterdarović, N.; Marinić, Ž.; Srzić, D. *Heterocycles* **1994**, *38*, 1791–1796.
63. Cunico, R. F.; Kuan, C. P. *Tetrahedron Lett.* **1990**, *31*, 1945–1948.
64. Nilsson, B. M.; Hacksell, U. *J. Heterocycl. Chem.* **1989**, *26*, 269–275.
65. Whitney, S. E.; Rickborn, B. *J. Org. Chem.* **1991**, *56*, 3058–3063.
66. Alvarez-Ibarra, C.; Mendoza, M.; Orellana, G.; Quiroga, M. L. *Synthesis* **1989**, 560–562.
67. Mashraqui, S. H.; Keehn, P. M. *J. Am. Chem. Soc.* **1982**, *104*, 4461–4465.
68. Sasaki, H.; Kitagawa, T. *Chem. Pharm. Bull.* **1987**, *35*, 4747–4756.
69. Sasaki, H.; Kawanishi, K.; Kitagawa, T.; Shingu, T. *Chem. Pharm. Bull.* **1989**, *37*, 2303–2306.
70. Hiemstra, H.; Houwing, H. A.; Possel, O.; van Leusen, A. M. *Can. J. Chem.* **1979**, *57*, 3168–3170.
71. Schiketanz, I.; Racoveanu-Schiketanz, A.; Gheorghiu, M. D.; Balaban, A. T. *Rev. Roum. Chim.* **1992**, *37*, 1315–1323 [*Chem. Abstr.* **1994**, *120*, 54093].
72. Adamczeski, M.; Quiñoà, E.; Crews, P. *J. Am. Chem. Soc.* **1988**, *110*, 1598–1602.
73. Pouchert, C. J.; Behnke, J., eds. *The Aldrich Library of ^{13}C and ^{1}H FT-NMR Spectra*, Vol. 3, Aldrich Chemical Co.: Milwaukee, 1993, p. 101.
74. Katritzky, A. R.; Chen, Y.-X.; Yannakopoulou, K.; Lue, P. *Tetrahedron Lett.* **1989**, *30*, 6657–6660.
75. L'abbé, G.; Ilisiu, A.-M.; Dehaen, W.; Toppet, S. *J. Chem. Soc. Perkin Trans. 1* **1993**, 2259–2261.
76. Terui, Y.; Yamakawa, M.; Honma, T.; Tada, Y.; Tori, K. *Heterocycles* **1982**, *19*, 221–228.
77. Burger, K.; Ottlinger, R.; Goth, H.; Firl, J. *Chem. Ber.* **1982**, *115*, 2494–2507.
78. Burger, K.; Geith, K.; Hübl, D. *Synthesis* **1988**, 189–194.
79. Kawase, M.; Miyamae, H.; Kurihara, T. *Chem. Pharm. Bull.* **1998**, *46*, 749–756.
80. Kawase, M. *Heterocycles* **1993**, *36*, 2441–2444.
81. Burger, K.; Hübl, D.; Geith, K. *Synthesis* **1988**, 194–198.
82. Burger, K.; Höß, E.; Geith, K. *Synthesis* **1990**, 360–365.
83. Burger, K.; Geith, K.; Höß, E. *Synthesis* **1990**, 352–356.
84. Nesi, R.; Giomi, D.; Turchi, S. *J. Org. Chem.* **1998**, *63*, 6050–6052.
85. Burger, K.; Geith, K.; Hübl, D. *Synthesis* **1988**, 199–203.
86. Alvarez-Ibarra, C.; López-Ranz, M. M.; López-Sánchez, M. I.; Orellana, G.; Ortiz, P.; Quiroga, M. L. *J. Chem. Soc., Perkin Trans. 1* **1989**, 1577–1584.
87. Hilf, C.; Bosold, F.; Harms, K.; Marsch, M.; Boche, G. *Chem. Ber./Recl.* **1997**, *130*, 1213–1221.
88. Witanowski, M.; Biedrzycka, Z. *Magn. Reson.* **1994**, *32*, 62–66.
89. Witanowski, M.; Beidrzycka, Z.; Sicinska, W.; Grabowski, Z.; Webb, G. A. *J. Magn. Reson., Ser. A* **1996**, *120*, 148–154.
90. Martin, G. E.; Crow, F. W.; Kaluzny, B. D.; Marr, J. G.; Fate, G. D.; Gilbertson, T. J. *Magn. Reson.* **1998**, *36*, 635–644, and references therein.
91. Chen, B. C.; Von Philipsborn, W.; Nagarajan, K. *Helv. Chim. Acta* **1983**, *66*, 1537–1555.
92. Stefaniak, L.; Roberts, J. D.; Witanowski, M.; Webb, G. A. *Org. Magn. Res.* **1984**, *22*, 215–220.
93. Boykin, D. W., ed. *^{17}O NMR Spectroscopy in Organic Chemistry*, CRC Press, Boca Raton, FL, 1991.
94. Gerothanassis, I. P. *Prog. Nucl. Magn. Reson. Spectrosc.* **1994**, *26*, 239–292.
95. Dahn, H.; Ung-Truong, M-N. *Helv. Chim. Acta* **1988**, *71*, 241–248.

References

96. Burger, K.; Gaa, K.; Geith, K.; Schierlinger, C. *Synthesis* **1989**, 850–855.
97. Koshelev, V. M.; Barsukov, I. N.; Vasil'ev, N. V.; Gontar, A. F. *Khim. Geterotsikl. Soedin.* **1989**, 1699–1700 [*Chem. Abstr.* **1990**, *113*, 78215].
98. Jones, B. G.; Branch, S. K.; Threadgill, M. D.; Wilman, D. E. V. *J. Fluorine Chem.* **1995**, *74*, 221–222.
99. Curran, T. T. *J. Fluorine Chem.* **1995**, *74*, 107–112.
100. Traldi, P.; Vettori, U.; Clerici, A. *Heterocycles* **1980**, *14*, 847–865.
101. Flammang, R.; Plisnier, M.; Bouchoux, G.; Hoppilliard, Y.; Humbert, S.; Wentrup, C. *Org. Mass. Spectrom.* **1992**, *27*, 317–325.
102. Hop, C. E. C. A.; Holmes, J. L.; Terlouw, J. K. *J. Am. Chem. Soc.* **1989**, *111*, 441–445.
103. Kelleher, N. L.; Belshaw, P. J.; Walsh, C. T. *J. Am. Chem. Soc.* **1998**, *120*, 9716–9717.
104. Yu, Z.; Lin, X.; He, X.; Wu, J. *Beijing Daxue Xuebao, Ziran Kexueban* **1996**, *32*, 773–776 [*Chem. Abstr.* **1997**, *127*, 121665].
105. Zhang, W.; You, D.; Zhu, H.; Zhang, X. *Tianjin Daxue Xuebao* **1996**, *29*, 874–882 [*Chem. Abstr.* **1997**, *126*, 343251].
106. Guesten, H.; Klasinc, L.; Marić, D.; Srzić, D. *Int. J. Mass Spectrom. Ion Phys.* **1983**, *47*, 423–426.
107. Adams, G. W.; Bowie, J. H.; Hayes, R. N. *Int. J. Mass. Spectrom. Ion Processes* **1992**, *114*, 163–182.
108. Sbrana, G.; Castellucci, E.; Ginanneschi, M. *Spectrochim. Acta Part A* **1967**, *23*, 751–758.
109. El-Azhary, A. A. *J. Chem. Res., Synop.* **1995**, 174–175.
110. Zatsepina, N. N.; Tupitsyn, I. F.; Belyashova, A. I.; Kane, A. A.; Kolodina, N. S.; Sudakova, G. N. *Khim. Geterotsikl. Soedin.* **1977**, 1110–1119 [*Chem. Abstr.* **1977**, *87*, 183818].
111. Shi, X. *Fenxi Huaxue* **1992**, *20*, 1135–1139 [*Chem. Abstr.* **1993**, *118*, 101852].
112. Doroshenko, A. O.; Patsenker, L. D.; Baumer, V. N.; Chepeleva, L. V.; Vankevich, A. V.; Shilo, O. P.; Yarmolenko, S. N.; Shershukov, V. M.; Mitina, V. G.; Ponomarev, O. A. *Zh. Obshch. Khim.* **1994**, *64*, 646–652 [*Chem. Abstr.* **1995**, *122*, 55936].
113. Panomarev, O. A.; Surov, Y. N.; Pivnenko, N. S.; Popova, N. A.; Fedyunyaeva, I. A. *Chem. Heterocycl. Comp. (N.Y.)* **1997**, *33*, 707–711.
114. Patsenker, L. D.; Surov, Y. N.; Lokshin, A. I.; Shkumat, A. P. *Russ. J. Gen. Chem.* **1999**, *69*, 1810–1816 [*Chem. Abstr.* **2000**, *133*, 73666].
115. Muniz-Miranda, M. *Vibrational Spec.* **1999**, *19*, 227–232.
116. Obukhov, A. E. *Zh. Fiz. Khim.* **1995**, *69*, 1015–1024 [*Chem. Abstr.* **1995**, *123*, 111259].
117. Obukhov, A. E. *Laser Phys.* **1999**, *9*, 699–722.
118. Bredereck, H.; Gompper, R.; Reich, F. *Chem. Ber.* **1960**, *93*, 1389–1397.
119. Obukhov, A. E.; Belen'kii, L. I.; *Chem. Heterocycl. Comp. (N.Y.)* **1999**, *35*, 832–854.
120. Obukhov, A. E.; Belen'kii, L. I. *Chem. Heterocycl. Comp. (N.Y.)* **1998**, *34*, 1011–1022.
121. Guallar, V.; Moreno, M.; Lluch, J. M.; Amat-Guerri, F.; Douhal, A. *J. Phys. Chem.* **1996**, *100*, 19789–19794.
122. LeGourrierec, D.; Kharlanov, V. A.; Brown, R. G.; Rettig, W. *J. Photochem. Photobiol., A* **2000**, *130*, 101–111.
123. Balaban, A. T.; Bally, I.; Frangopol, P. T.; Bacescu, M.; Cioranescu, E.; Birladeanu, L. *Tetrahedron* **1963**, *19*, 169–176.
124. Pavlopoulos, T. G.; Hammond, P. R. *J. Am. Chem. Soc.* **1974**, *96*, 6568–6579.
125. Doroshenko, A. O.; Artyukhov, A. N.; Shershukov, V. M.; Fedyunyaeva, I. A. *Functional Materials* **1996**, *3*, 431–436.
126. del Valle, J. C.; Kasha, M.; Catalan, J. *Chem. Phys. Lett.* **1996**, *263*, 154–160.
127. del Valle, J. C.; Kasha, M.; Catalan, J. *J. Phys. Chem. A* **1997**, *101*, 3260–3272.
128. LaVerne, J. A.; Brocklehurst, B. *J. Phys. Chem.* **1996**, *100*, 1682–1688.

129. Krasovitskii, B. M.; Shershukov, V. M.; Volkov, V. L. *Khim. Geterotsikl. Soedin.* **1986**, *22*, 1265–1266 [*Chem. Abstr.* **1988**, *108*, 21758].
130. Chen, J.; Pan, J.; Kao, C.; *Gaodeng Xuexiao Huaxue Xuebao* **1991**, *12*, 56–58 [*Chem. Abstr.* **1991**, *115*, 48694].
131. Patsenker, L. D.; Lokshin, A. I. *Chem. Heterocycl. Comp. (N.Y.)* **1999**, *35*, 275–280.
132. Trifonov, R. E.; Rtishchev, N. I.; Ostrovskii, V. A. *Spectrochim. Acta, Part A* **1996**, *52*, 1875–1882.
133. Breusova, E. G.; Kuznetsova, R. T.; Maier, G. V. *Khim. Vys. Energ.* **1994**, *28*, 154–158 [*Chem. Abstr.* **1994**, *121*, 34687].
134. Diwu, Z.; Zhang, C.; Klaubert, D. H.; Haugland, R. P. *J. Photochem. Photobiol., A* **2000**, *131*, 95–100.
135. Gryczynski, I.; Malak, H.; Hell, S. W.; Lakowicz, J. R. *J. Biomed. Opt.* **1996**, *1*, 473–480.
136. Agal'tsov, A. M.; Gorelik, V. S.; Rakhmatullaev, I. A. *Zh. Prikl. Spektrosk.* **1996**, *63*, 998–1002 [*Chem. Abstr.* **1997**, *126*, 178366].
137. Doroshenko, A. O.; Kirichenko, A. V.; Mitina, V. G.; Ponomarev, O. A. *J. Photochem. Photobiol., A* **1996**, *94*, 15–26.
138. Kirichenko, A. V.; Doroshenko, A. O.; Shershukov, V. M. *Khim. Fiz.* **1998**, *17*, 41–48 [*Chem. Abstr.* **1998**, *130*, 88022].
139. Doroshenko, A. O. *Khim. Fiz.* **1999**, *18*, 40–44 [*Chem. Abstr.* **1999**, *132*, 3177].
140. Doroshenko, A. O.; Kyrychenko, A. V.; Waluk, J. *J. Fluoresc.* **2000**, *10*, 41–48.
141. Doroshenko, A. O.; Kyrychenko, A. V.; Baumer, V. N.; Verezubova, A. A.; Ptyagina, L. M. *J. Mol. Struct.* **2000**, *524*, 289–296.
142. Doroshenko, A. O.; Posokhov, E. A.; Shershukov, V. M.; Mitina, V. G.; Ponomarev, O. A. *High Energy Chem.* **1997**, *31*, 388–394.
143. Douhal, A.; Lahmani, F.; Zehnacker-Rentien, A.; Amat-Guerri, F. *J. Phys. Chem.* **1994**, *98*, 12198–12205.
144. Belen'kii, L. I.; Vasil'eva, I. A.; Galanin, M. D.; Nikitina, A. N.; Chizhikova, Z. A. *Opt. Spektrosk.* **1990**, *68*, 801–806 [*Chem. Abstr.* **1990**, *113*, 122904].
145. Buttke, K.; Baumgaertel, H.; Niclas, H. J.; Schneider, M. *J. Prakt. Chem. Chem. Ztg.* **1997**, *339*, 721–728.
146. Cahill, P. A.; Seager, C. H.; Meinhardt, M. B.; Beuhler, A. J.; Wargowski, D. A.; Singer, K. D.; Kowalczyk, T. C.; Kosc, T. Z. In Moehlmann, G. R., ed. *Nonlinear Optical Properties of Organic Materials VI*, SPIE Press, Bellingham, WA, 1993, pp. 48–55.
147. Politis, J. K.; Curtis, M. D. *Polym. Prepr.* **1998**, *39*, 181–182.
148. Politis, J. K.; Somoza, F. B.; Kampf, J. W.; Curtis, M. D.; Ronda, L. G.; Martin, D. C. *Chem. Mater.* **1999**, *11*, 2274–2284.
149. Catalan, J.; Mena, E.; Fabero, F.; Amat-Guerri, F. *J. Chem. Phys.* **1992**, *96*, 2005–2016.
150. Dey, J. K.; Dogra, S. K. *Indian J. Chem. Sect. A* **1990**, *29*, 1153–1164.
151. Larsson, A.; Carlsson, C.; Jonsson, M.; Albinsson, B. *J. Am. Chem. Soc.* **1994**, *116*, 8459–8465.
152. Larsson, A.; Carlsson, C.; Jonsson, M. *Biopolymers* **1995**, *36*, 153–167.
153. Abramo, K. H.; Pitner, J. B.; Mcgown, L. B. *Biospectroscopy* **1998**, *4*, 27–35.
154. Kumar, A.; Sheridan, J.; Stiefvater, O. L. *Z. Naturforsch., A* **1978**, *33*, 145–152.
155. Kumar, A.; Sheridan, J.; Stiefvater, O. L. *Z. Naturforsch., A* **1978**, *33*, 549–558.
156. Fliege, E. R. L. *Z. Naturforsch., A: Phys. Sci.* **1990**, *45*, 911–922.
157. Krafka, E.; Cremer, D.; Spoerel, U.; Merke, I.; Stahl, W.; Dreizler, H. *J. Phys. Chem.* **1995**, *99*, 12466–12477.
158. Kasai, P. H.; McLeod, D. Jr. *J. Am. Chem. Soc.* **1973**, *95*, 4801–4805.
159. Dogan, I.; Steenken, S.; Schulte-Frohlinde, D.; Içli, S. *J. Phys. Chem.* **1990**, *94*, 1887–1894.

160. Palmer, M. H.; Findlay, R. H.; Egdell, R. G. *J. Mol. Struct.* **1977**, *40*, 191–210.
161. Bychkov, N. N.; Vinogradov, N. N.; Marshakova, S. A.; Redchenko, V. V.; Rodin, O. G. *Zh. Obshch. Khim.* **1993**, *63*, 205–211 [*Chem. Abstr.* **1993**, *119*, 72128].
162. Bender, S.; Franke, R.; Hormes, J.; Pavlychev, A. A.; Fominykh, N. G.; Shemelev, V. V. In Goulon, J.; Goulon-Ginet, C.; Brookes, N. B., eds. *Journal of Physics IV: Proceedings of the 9th International Conference on X-Ray Absorption Fine Structure*, Vol. 1, Editions de Physique, 1997, pp. 527–528.
163. Hennig, C.; Hallmeier, K. H.; Bach, A.; Bender, S.; Franke, R.; Hormes, J.; Szargan, R. *Spectrochim. Acta, Part A* **1996**, *52*, 1079–1083.

CHAPTER 3

Oxazole Diels-Alder Reactions

Jeremy I. Levin

Wyeth Research
Pearl River, New York

Leif M. Laakso

Wyeth Research
Pearl River, New York

3.1　Introduction
3.2　Oxazole-Olefin Cycloadditions
　　　3.2.1　Intermolecular Reactions
　　　　　　3.2.1.1　Pyridines and Hydroxypyridines
　　　　　　3.2.1.2　Aminooxazole cycloadditions
　　　　　　3.2.1.3　Isoindoles
　　　　　　3.2.1.4　Oxazoles as Dienophiles
　　　3.2.2　Intramolecular Reactions
　　　　　　3.2.2.1　Eupolauramine
　　　　　　3.2.2.2　Benzopyranopyridines and Benzonaphthyridines
　　　　　　3.2.2.3　Bromoleptoclinidinone
　　　　　　3.2.2.4　Amphimedine
　　　　　　3.2.2.5　Synthesis of Normalindine
　　　　　　3.2.2.6　Amidooxazole Cycloadditions
3.3　Oxazole-Alkyne Cycloadditions
　　　3.3.1　Intermolecular Reactions
　　　　　　3.3.1.1　Furans
　　　　　　3.3.1.2　Reactions with Aminooxazoles
　　　　　　3.3.1.3　Reactions with Benzyne
　　　3.3.2　Intramolecular Reactions
　　　　　　3.3.2.1　Synthesis of Evodone
　　　　　　3.3.2.2　Synthesis of Ligularone and Petasalbine
　　　　　　3.3.2.3　Synthesis of Paniculide-A
　　　　　　3.3.2.4　Synthesis of Gnididione
　　　　　　3.3.2.5　Synthesis of Norsecurinine
　　　　　　3.3.2.6　Synthetic Approach to Geigerin
　　　　　　3.3.2.7　Synthesis of Stemoamide
　　　　　　3.3.2.8　Thiofurans and Amidofurans
3.4　Oxazole-Heterodienophile Reactions
　　　3.4.1　Intermolecular Reactions
　　　　　　3.4.1.1　Nitrogen-Containing Heterodienophiles

The Chemistry of Heterocyclic Compounds, Volume 60: Oxazoles: Synthesis, Reactions, and Spectroscopy, Part A, edited by David C. Palmer
ISBN 0-471-39494-7　Copyright © 2003 John Wiley & Sons, Inc.

3.4.1.2 Carbonyl Heterodienophiles
3.4.1.3 Thioaldehyde Heterodienophiles
3.4.1.4 Singlet Oxygen
3.4.2 Intramolecular Oxazole-Heterodienophile Reactions
3.5 Summary
3.6 Addendum

3.1. INTRODUCTION

The Diels-Alder reaction of oxazoles with alkenes, alkynes, and heterodienophiles has become a valuable tool for the construction of highly substituted pyridines, furans, and other heterocycles and has now been exploited for the synthesis of diverse compounds from pharmaceuticals to complex natural products. These reactions have been extensively reviewed.[1-12] The purpose of this chapter is to provide an introduction to the use of oxazoles in Diels-Alder cycloadditions and an update on these reactions since 1985.

3.2. OXAZOLE-OLEFIN CYCLOADDITIONS

The first use of an oxazole as an azadiene in a Diels-Alder cycloaddition was reported by Kondrat'eva in 1957.[13,14] In these seminal studies a variety of alkyl-substituted oxazoles **1** (R_1, R_2, R_3 = H, alkyl) reacted with maleic anhydride in either benzene or ether to provide the cinchomeronic anhydrides **2**, rather than the expected bicyclic ethers analogous to the Diels-Alder adducts of furans (Fig. 3.1).

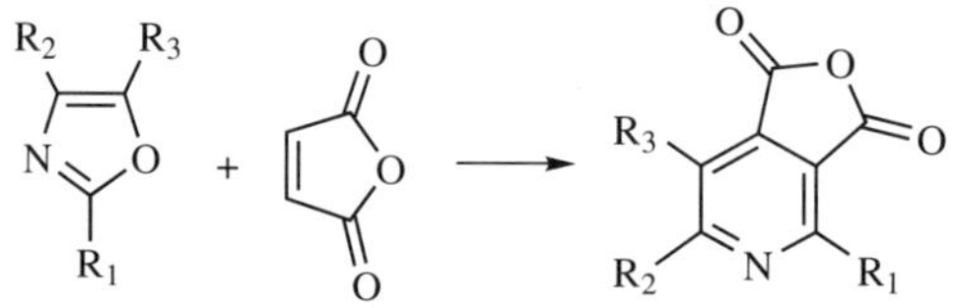

Figure 3.1

In addition, the Diels-Alder reactions of oxazoles **3** bearing alkoxy substituents at the 5 position were disclosed by Kondrat'eva and Huang in 1961 (Fig. 3.2).[15] In contrast to the alkyloxazoles, the cycloadditions of derivatives such as maleimide with **3** resulted in the formation of pyridinols **4** rather than the expected 3-alkoxypyridines.

In 1965, Ishikawa and co-workers[16-19] investigated the oxazole cycloaddition reactions of 4-methyloxazole, **5**, with a variety of dienophiles (Fig. 3.3). The product distribution for these reactions was found to depend on the substituents on the dienophile as well as the reaction conditions. For example, the reaction of 4-methyloxazole with diethyl fumarate gave a mixture of pyridinols **6** and **7**.

Furthermore, the addition of hydrogen peroxide to the reaction of **5** and diethyl fumarate dramatically increased the yield of **6** at the expense of **7**.[17]

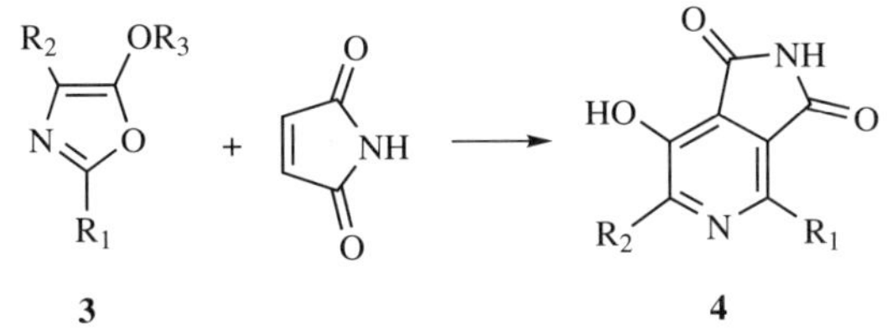

Figure 3.2

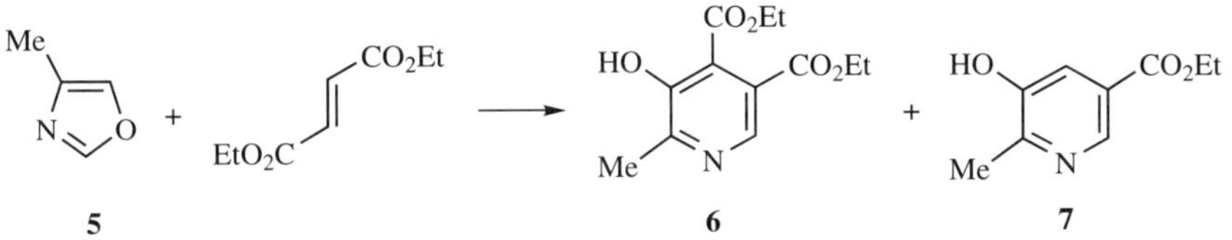

Figure 3.3

That these reactions proceed via the intermediacy of a Diels-Alder cycloaddition adduct has been affirmed by the isolation of a variety of the 1:1 Diels-Alder adducts.[20–24] For example, the reaction of 5-ethoxy-4-methyloxazole **8** with *cis*-2,5-dimethoxy-2,5-dihydrofuran **9** provided the isolable *endo* and *exo* adducts **10** and **11** respectively, in a 2:1 ratio (Fig. 3.4).[25] Similarly, 5-ethoxy-4-oxazoleacetic acid ethyl ester **12** reacted with maleic anhydride to provide the stable endo and exo adducts **13** and **14**, in which the olefin has moved into conjugation with the ester moiety.[26] In this case, compound **13** was the sole product when the reaction proceeded at 10°C, but only the *exo*-adduct **14** was isolated if the cycloaddition was conducted at 80°C. Heating at 50°C for 3 h converted **13** into **14**. The 2-carboethoxy analog of oxazole **12** behaved similarly.[27]

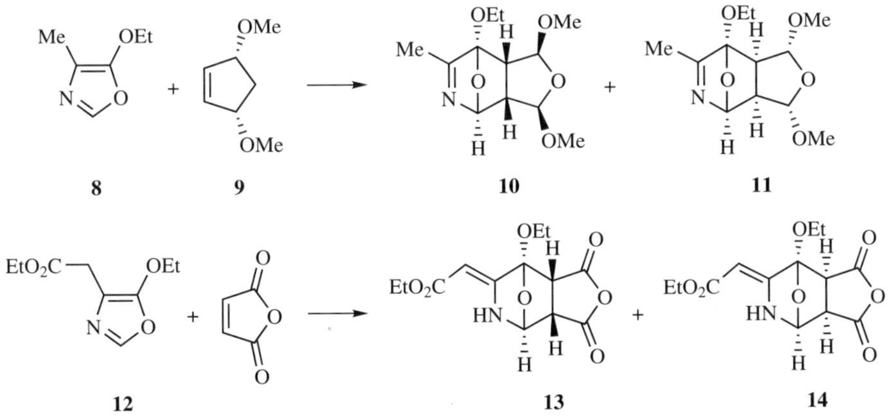

Figure 3.4

420   Oxazole Diels-Alder Reactions

Based on these and other examples, the general outcome of the reactions of oxazoles with olefins may be summarized as shown in Figure 3.5. Oxazoles and olefins first undergo a [4 + 2] cycloaddition to form a bridged adduct **15**. This reaction is predicted to be concerted with an asynchronous transition-state geometry, similar to that for the reaction of 1-azabutadiene with ethylene, based on computational studies of the Diels-Alder cycloaddition of oxazole with ethylene and other simple dienophiles.[28–30] The forming bond lengths in the oxazole-olefin Diels-Alder reaction are slightly shorter than for the analogous azabutadiene-olefin cycloaddition. Both the azabutadiene and oxazole cycloaddition transition states display a small charge transfer from the dienophile to the heterodiene. The reaction is calculated to be exothermic, with a heat of reaction of −19.7 kcal/mol and an activation energy of approximately 28 kcal/mol.

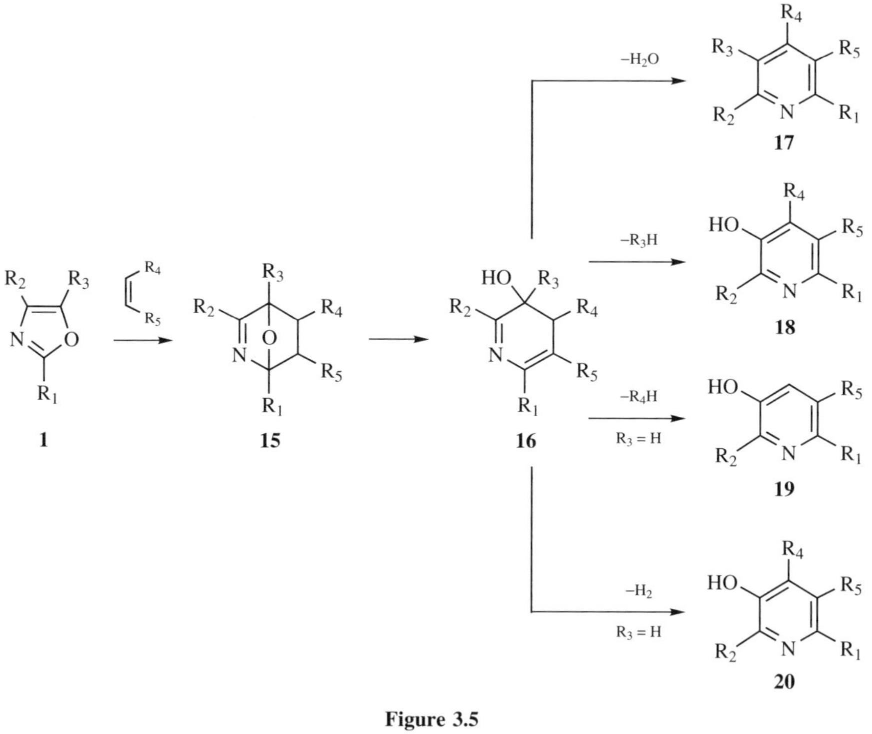

Figure 3.5

Cycloadduct **15** next undergoes cleavage of the ether linkage to give azadiene **16**. This intermediate can then decompose via four distinct pathways to provide pyridine or hydroxypyridine products. Elimination of water from **16** gives the simple pyridine products **17**. This is the major fragmentation pathway when neither R_3 nor R_4 is an electron-withdrawing group. The loss of R_3H from **16** yields the 3-hydroxypyridine **18**. This is the predominant pathway when R_3 is an alkoxy, carbonate, cyano moiety, or another suitable leaving group.[15,20,31–34] Oxazoles lacking a 5-substituent can eliminate R_4H, where R_4 is a leaving group (such as a cyano group) to give hydroxypyridines **19**.[17] In the reaction of 4-methyloxazole with various dienophiles, the amount of hydroxypyridine formed via elimination of R_4H has been shown to be directly proportional to the ability of R_4 to stabilize a negative charge.[17] Oxazoles unsubstituted at the 5 position also undergo the loss of hydrogen from **16** to provide hydroxypyridines **20**. This oxidative decomposition of **16** is rarely seen, unless a hydride acceptor such as nitrobenzene or hydrogen peroxide is present in the reaction.

Although the regiochemistry of oxazole-olefin cycloadditions has not been extensively studied, some generalizations can be made. The major product is the 4-substituted pyridine in those instances when the alkene dienophile bears an electron-withdrawing substituent.[35–40] This regioselectivity is exemplified by the Diels-Alder reaction used in a synthesis of the natural product ellipticine by Kozikowski and Hasan,[41] starting from the reaction of oxazole **21** with acrylonitrile to give 4-cyanopyridine **22** (Fig. 3.6).

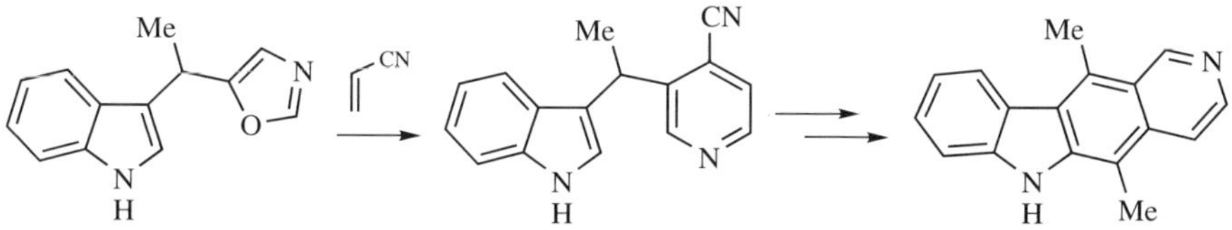

Figure 3.6

This regioselectivity is reversed for olefins substituted with electron-releasing groups, and the 3-substituted pyridines are the predominant products. For example, reaction of **8** with allyl *t*-butyl ether gives **23** as the major product in 61% yield (Fig. 3.7).[42] The 4-substituted pyridine **24** was isolated in 13% yield. The less-encumbered diallyl ether provided essentially the same ratio of 3- to 4-substituted pyridine products, indicating that this regiochemistry is in fact driven by electronic factors.

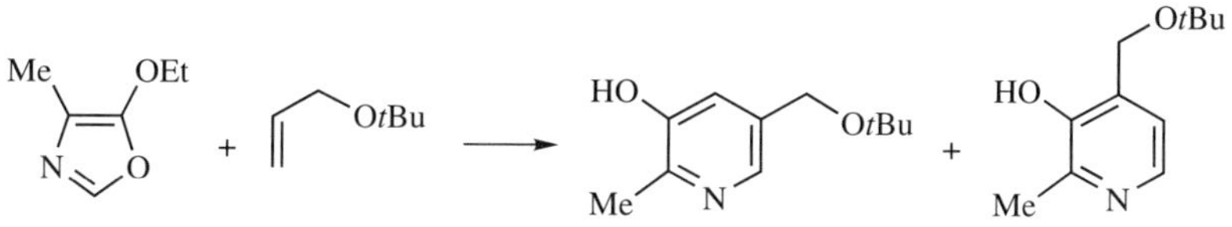

Figure 3.7

Steric interactions can also effect the regiochemistry of oxazole-olefin cycloadditions.[42] Thus the bulky alkene **25a** provided a 2:1 mixture of the 3- and 4-isopropylpyridines **26a** and **27a**, respectively (Fig. 3.8). However, a 1:1 ratio of pyridines **26b** and **27b** was obtained from the reaction of **8** with the less sterically demanding 1-hexene, **25b**.

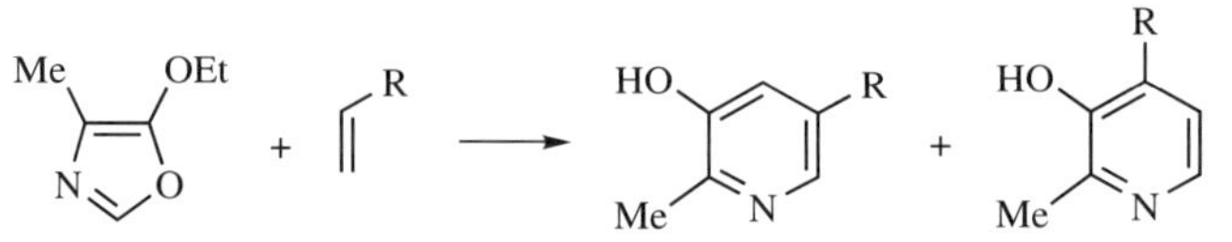

8 25a R = CH(CH₃)₂   26a R = CH(CH₃)₂   27a R = CH(CH₃)₂
   25b R = (CH₂)₃CH₃ 26b R = (CH₂)₃CH₃ 27b R = (CH₂)₃CH₃

Figure 3.8

The rate of Diels-Alder reactions of oxazoles with electron-deficient olefins and acetylenes depends on the electron-donating ability of the oxazole substituent. Rates decrease in the order $OR > NR_1R_2 > $ alkyl > 4-phenyl $> COCH_3 > CO_2CH_3 \gg 2$-,5-phenyl; 5-alkoxy oxazoles have reactivity roughly equivalent to an all-carbon diene.

On occasion, oxazole-alkene Diels-Alder adducts such as **29**, and more generally **15**, fragment to provide products other than pyridines, depending on the reaction conditions used (Fig. 3.9). The reaction of 5-isopropoxy-4-methyloxazole **28** with nitroethylene therefore gave **31**, rather than the expected nitropyridine.[43,44] Compound **31** was presumed to form via acid-catalyzed decomposition of the initial Diels-Alder adduct **29** to provide amino-diketone **30**, which then cyclized to give 2-acetyl-3-nitropyrrole **31**.

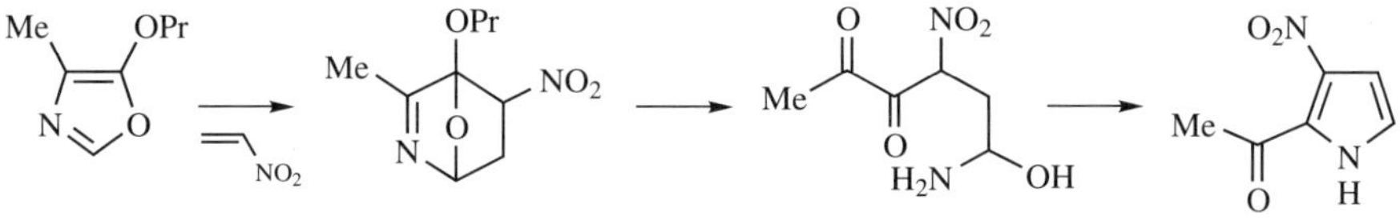

Figure 3.9

Another example of a Diels-Alder adduct forming products other than pyridines is the intramolecular Diels-Alder reaction of oxazole carboxamide **32** (Fig. 3.10).[45] After refluxing in xylene for 24 h, compound **32** provided a 38% yield of Diels-Alder adduct **33** together with 32% of recovered starting material. Refluxing adduct **33** in 1:1 water:dioxane for 1 h then gave a quantitative yield of tetrahydrofuran **34**. However, if compound **33** was resubjected to refluxing xylene for 5 days, a 32% yield of pyridine **35** was isolated. A third product, phenol **36**, could be obtained on heating **33** in acetic acid at 110°C for 30 min.

3.2.1. Intermolecular Reactions

Examples of intermolecular cycloadditions of oxazoles with alkenes since 1985 include approaches to the ubiquitous analogs of the 3-hydroxypyridine pyridoxol, several applications in medicinal chemistry, and the first demonstrations of oxazole acting as a dienophile in [4 + 2] cycloadditions.

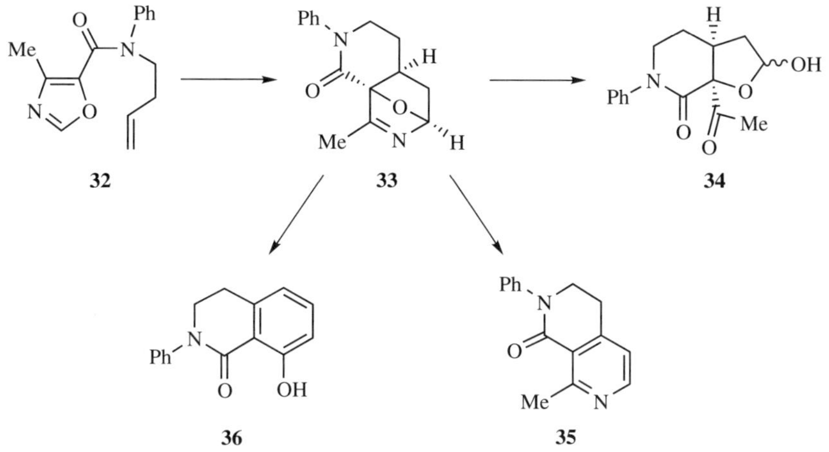

Oxazole-Olefin Cycloadditions   423

Figure 3.10

3.2.1.1. Pyridines and Hydroxypyridines

A major driving force in the early study of Diels-Alder cycloadditions of oxazoles was the enormous annual commercial demand for pyridoxol, a form of vitamin B_6. The numerous studies, beginning in the 1960s, describing the utility of oxazoles for the construction of pyridoxol and its analogs have been detailed (Fig. 3.11).[3]

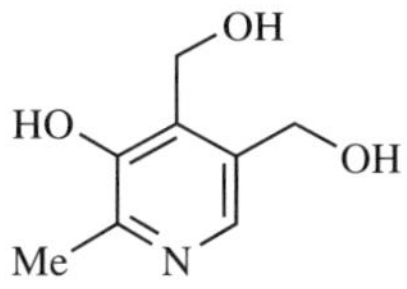

Pyridoxol

Figure 3.11

This methodology is exemplified by two relatively recent syntheses of pyridoxol analogs. For the synthesis of [CD_3]-pyridoxal **39b**, the requisite deuterated oxazole **37**, readily available from deutero-L-alanine, underwent Diels-Alder cycloaddition when heated to 135°C with a 20-fold excess of alkoxy dihydrofuran **38** to provide a 1:1 mixture of the regioisomeric pyridines **39a** and **40** in 50–60% yield (Fig. 3.12).[46] Interestingly, the undesired regioisomer **40** can be quantitatively converted into **39a** by treatment with ethanolic HCl. Treatment of **39a** with dilute acid then provided hemiacetal **39b** in 35–42% overall yield from the deutero-alanine. This route to **39a** is clearly superior to the three-step conversion of pyridoxol to [CD_3]-pyridoxal, which proceeds in 3% overall yield.[47]

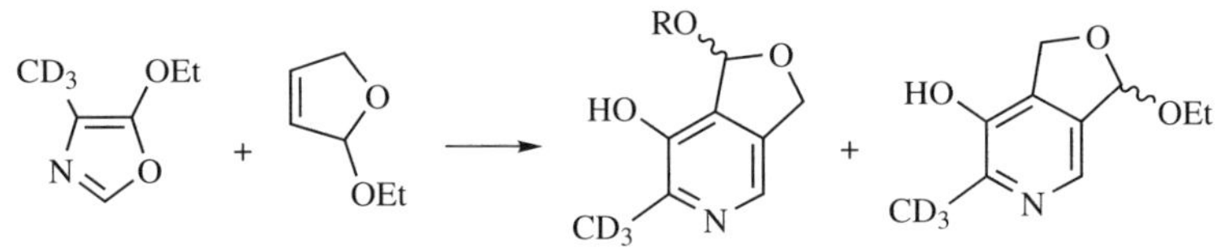

37 **38** **39a** R = Et **39b** R = H **40**

Figure 3.12

An oxazole-alkene Diels-Alder reaction has also been used to synthesize a trifluoromethylated pyridine related to pyridoxol. 5-Ethoxy-4-(trifluoromethyl)-2-oxazolecarboxylic acid **41** decarboxylated and then reacted with acrylic acid to give the expected 3-hydroxy-2-(trifluoromethyl)pyridine-4-carboxylic acid **42** in 63% yield (Fig. 3.13).[48] The electron-withdrawing trifluoromethyl group of **41** strongly modulates its reactivity, however, and it does not react with ethyl acrylate or *N*-phenyl maleimide.

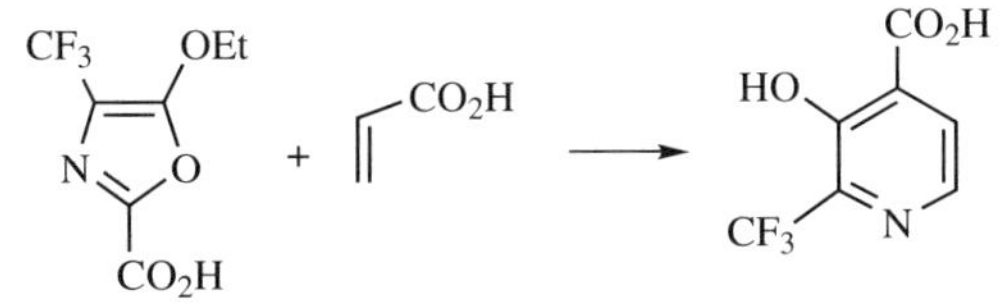

41 **42**

Figure 3.13

Another example of an approach to pyridoxol began with the cycloaddition of 4-methyloxazole **5** with the sulfonyl-alkene **43** (Fig. 3.14).[49] The bridged cycloadduct **44** was formed with the expected regiochemistry, directed by the electron-withdrawing sulfonyl group. Elimination of methanesulfinic acid then provided hydroxypyridine **45**, a precursor of pyridoxol. In a related synthesis reaction of the nitro-olefin **46** with **5** afforded the pyridoxol precursor **47** in low yield.[50]

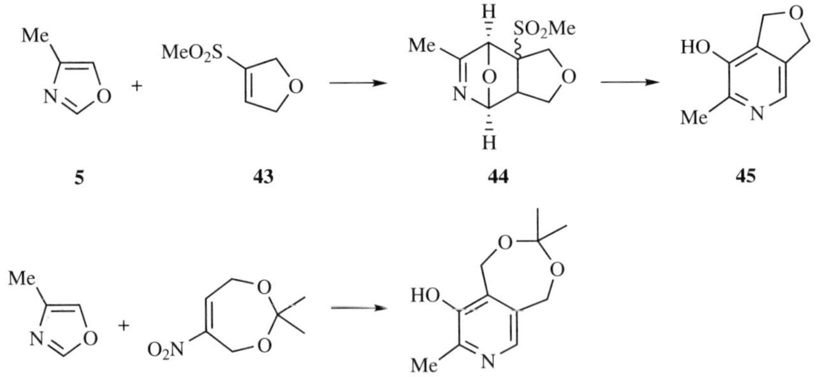

5 **43** **44** **45**

5 **46** **47**

Figure 3.14

In contrast, vinyl sulfonyl fluoride **48** reacted with 5-ethoxy-4-methyloxazole **8** to form adduct **49**, which subsequently underwent acid-catalyzed aromatization via elimination of ethanol to give sulfonate ester **50** (Fig. 3.15).[12] A more recent example of a sulfonyl-alkene dienophile is the isothiazolone **51**, which combined with 5-ethoxy-4-methyloxazole at room temperature to yield the *exo*-adduct **52** in 55% isolated yield, along with a minor amount of the *endo*-isomer.[51] In this case, the more electronegative carbonyl group of the isothiazolone controls the cycloaddition regiochemistry. Treatment of **52** with ethanolic HCl produced an 81% yield of hydroxypyridine **53**. Compound **53** was then converted in several steps to pyridine **54**, an analog of the oxicam nonsteroidal anti-inflammatory drugs (NSAIDs).

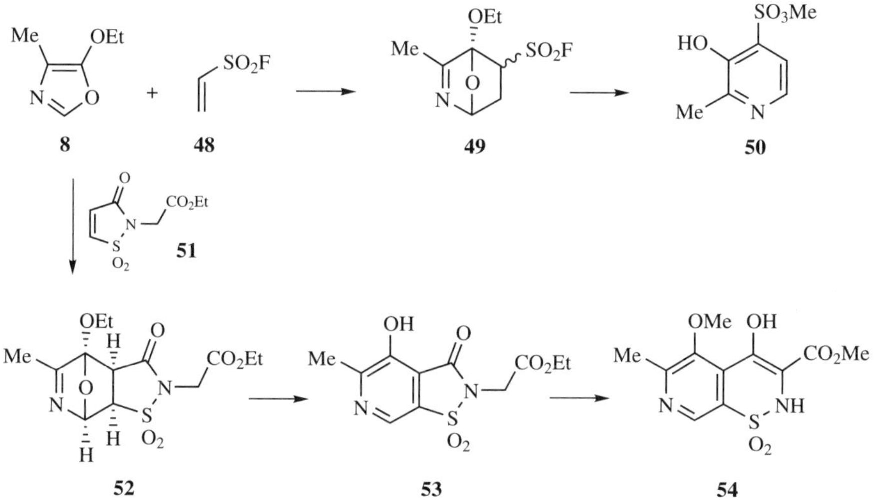

Figure 3.15

An oxazole-olefin Diels-Alder reaction was also used in an approach to nonpeptide neurokinin-3 receptor antagonists.[52] Thus when 4-phenyloxazole **55** is melted together with maleic acid at 110°C for 15 min, 2-phenyl-4-pyridinecarboxylic acid **56** is obtained in 18% yield after decarboxylation (Fig. 3.16). This compound is then coupled with methyl phenylglycinate to give the desired final product **57**.

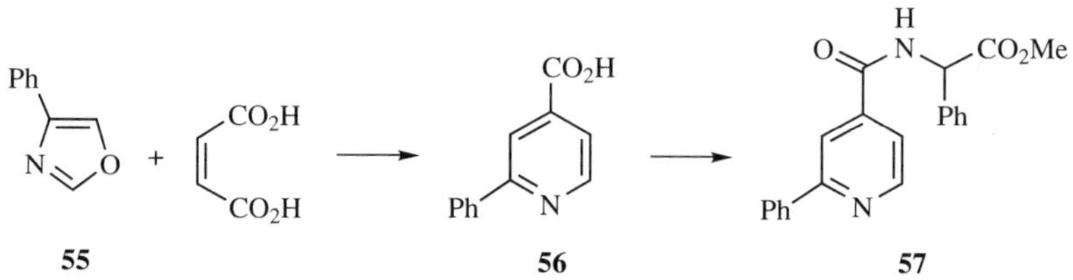

Figure 3.16

Quinones have been shown to react with 5-alkoxyoxazoles to yield hydroxypyridines. The isoquinolinequinone **60** was prepared in 60% yield from 4-ethyl-5-methoxy-2-methyloxazole **58** and the quinone **59** in benzene at 60°C for 30 h (Fig. 3.17).[53]

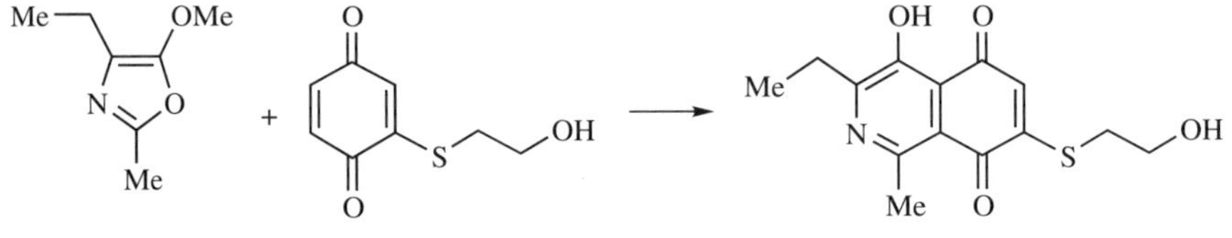

Figure 3.17

The effect of pressure on the rate and product distribution of oxazole Diels-Alder reactions has been studied for 5-methoxy-2-methyl-4-(4-nitrophenyl)oxazole **61**, a 5-alkoxyoxazole (Fig. 3.18).[24,54,55] At atmospheric pressure, **61** gave low yields of cycloaddition products with N-phenyl maleimide and did not react at all with dimethyl maleate or dimethyl fumarate, even when heated at 110°C for 100 h.[22] However, after 10 min at 10 kbar, a benzene solution of **61** and N-phenyl maleimide provided a 4:1 mixture of endo (**62**) and exo (**63**) adducts in quantitative yield. Longer reaction times result in equilibration of **62** and **63**, which is postulated to occur via a retro-Diels-Alder reaction, producing a 65:35 ratio after 25 h at 10 kbar. If the reaction is carried out at 10 kbar for 25 h using methanol as the solvent, hydroxypyridine **64** is isolated in quantitative yield. Dimethyl maleate could be induced to react with **61** only at elevated temperature and pressure (10 kbar, 60°C, 100 h, C_6H_6) to give a 71% yield of hydroxypyridine **65**. However, even these

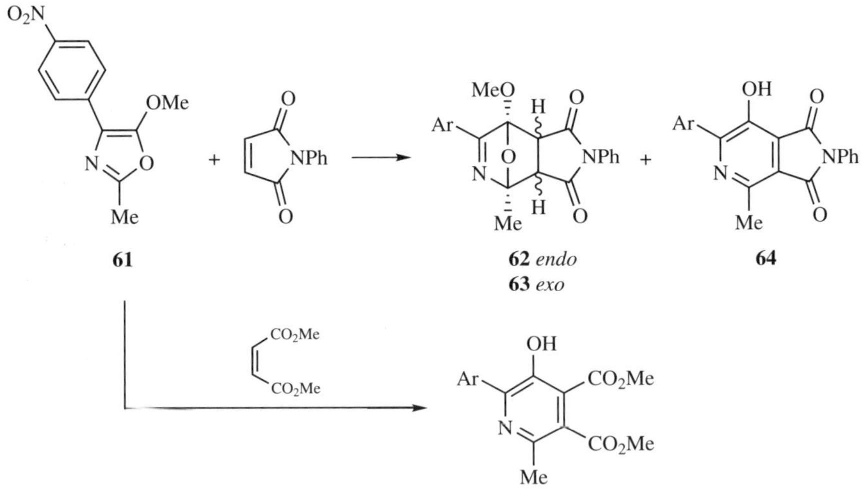

Figure 3.18

forcing conditions were not sufficient to provide synthetically useful yields of cycloaddition products from the reaction of **61** and dimethyl fumarate.

3.2.1.2. Aminooxazole Cycloadditions

Compared to the Diels-Alder reactions of 5-alkoxyoxazoles, relatively few examples exist for intermolecular cycloadditions of aminooxazoles and their derivatives. Since the initial studies of 5-aminooxazoles by Kondrat'eva[21] only four others have been reported. 2-Amino-4-methyloxazole **66a** reacted with diethyl maleate at room temperature in ethanol to provide a 35% yield of **67** and 11% of **68** (Fig. 3.19).[56] The acetamide derivative **66b** reacted with maleic anhydride at 50°C in toluene to give only the decarboxylated 3-hydroxypyridine **69**.

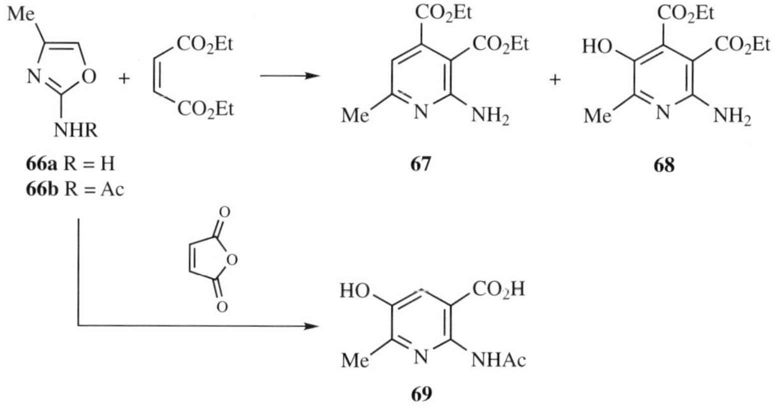

Figure 3.19

A systematic study of the reactions of 5-aminooxazole derivatives has also been disclosed. Thus the isolable cycloadducts **71a–c**, obtained from the reaction of oxazoles **70a–c** and *N*-phenylmaleimide were treated with acetic acid to produce the expected pyridine products (Fig. 3.20).[57] The acetamide **71a** gave a 6:1 ratio of

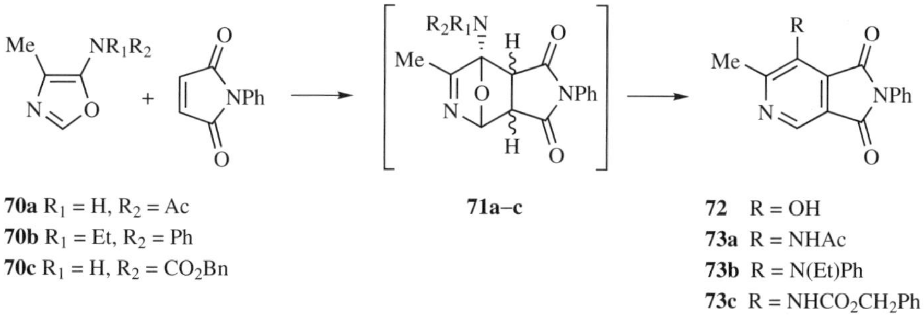

70a $R_1 = H$, $R_2 = Ac$
70b $R_1 = Et$, $R_2 = Ph$
70c $R_1 = H$, $R_2 = CO_2Bn$

71a–c

72 $R = OH$
73a $R = NHAc$
73b $R = N(Et)Ph$
73c $R = NHCO_2CH_2Ph$

Figure 3.20

3-hydroxypyridine **72** to aminopyridine **73a**. The ethyl-anilino derivative **71b** gave only the anilino-pyridine **73b** in 64% yield. Benzyl carbamate **71c** produced the carbamoyl-pyridine **73c** as the major product in a 2:1 ratio. The ratio of 3-aminopyridine derived products to 3-hydroxypyridine products was therefore proportional to the basicity of the 5-substituent of the starting oxazole.

An expeditious route to 1,2,3,4-tetrahydropyrido[3,4-d]-pyrimidine-2,4-diones (potential analogs of the biologically active pteridines) starting from oxazole-ureas has been reported.[58] For example, N-(4-methyloxazol-5-yl)urea **74** and acrylonitrile afforded a 58% yield of pyridopyrimidine **76** after 6 h at reflux in ethanol (Fig. 3.21). Thus cycloaddition of the electron-deficient acrylonitrile provided the expected intermediate Diels-Alder adduct **75** regioselectively. Cyclization of the urea onto the nitrile moiety to form the pyrimidine ring and dehydration to form the pyridine ring occurred concomitantly to give **76**. Hydrolysis of **76** in refluxing dilute hydrochloric acid then led to the desired pyrimidine-2,4-dione **77** in 90% yield.

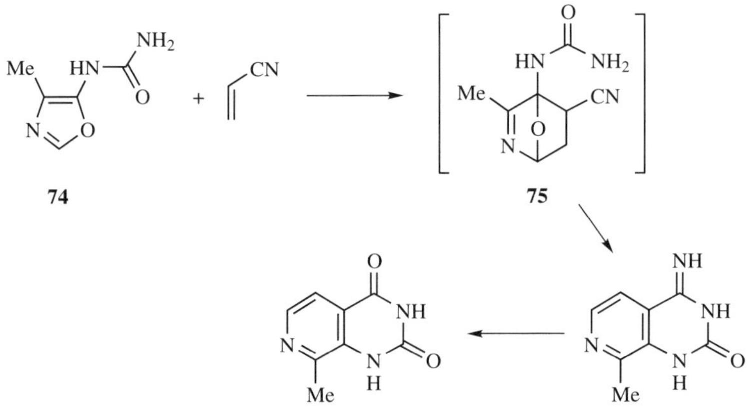

Figure 3.21

3.2.1.3. Isoindoles

The use of oxazole-alkene Diels-Alder cycloadditions to form biologically relevant molecules has recently been applied to the synthesis of isoindoles, useful intermediates for the preparation of substance P antagonists.[59] Thus 5-ethoxy-4-methyloxazole **8** reacted with 4,4-dimethyl-2-cyclopentenone **78** in refluxing benzene in the presence of catalytic zinc bromide to give the (1H)-cyclopenta(c)-pyrrole **80** as a separable 3:1 mixture of cis and trans isomers in 53% yield (Fig. 3.22). The reaction is presumed to proceed via the intermediacy of cyclo-adduct **79**. When cyclohexenone **81** was used as the dienophile, the product was the hydro-(1H)-isoindole **82**, obtained via dehydroformylation of the cycloadduct, in 85% yield after only 30 min in refluxing benzene.

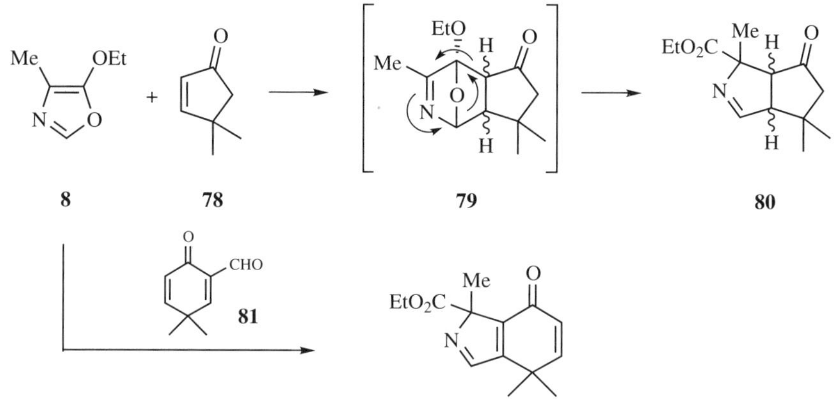

Figure 3.22

3.2.1.4. Oxazoles as Dienophiles

The first example of an oxazole participating as a dienophile in a [4 + 2] cycloaddition was the inverse electron demand cycloaddition of the 2-aminooxazole **83** with o-chloranil **84** to afford the oxazoline **85** in 87% yield after 1 h at room temperature (Fig. 3.23).[60] Hexachlorocyclopentadiene and 1,1-dimethoxytetrachlorocyclopentadiene also reacted with **83** to give 86 and 85% yields of endo cycloadducts (not shown), respectively.

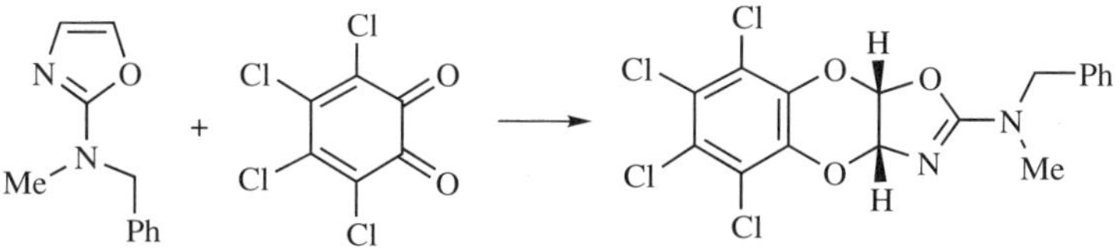

Figure 3.23

4-Nitro-2-phenyloxazole **86** functions as a dienophile with a variety of unactivated dienes. Thus **86** reacted with a fivefold excess of 2,3-dimethyl-1,3-butadiene, **87**, in a sealed tube at 110°C to give a 71% yield of the bicyclic oxazole **89** and 13% of 5,6-dimethyl-2-phenylbenzoxazole **90** (Fig. 3.24).[61,62] Although the intermediate cycloadduct **88** could not be isolated in this case, cyclohexadiene and **86** gave a separable mixture of the *endo*- and *exo*-2-oxazolines in 33 and 30% yield, respectively. In addition to all-carbon dienes, **86** reacted with 1-(dimethylamino)-3-methyl-1-azabuta-1,3-diene **91** at 55°C in chloroform to provide, via loss of nitrous acid and dimethylamine from intermediate adduct **92**, a 30% yield of the oxazolo[4,5-*b*]pyridine **93**.[62]

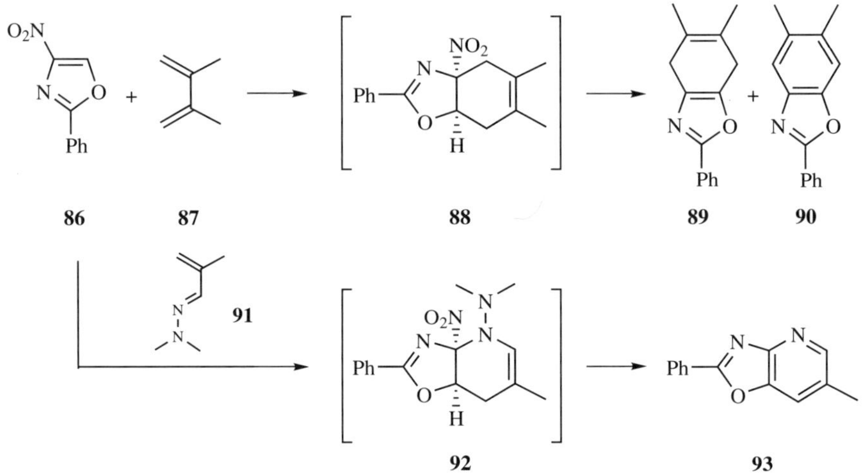

Figure 3.24

Oxazole itself participates as a dienophile in a Diels-Alder reaction with the electron-deficient 3,6-bis(trifluoromethyl)-1,2,4,5-tetrazine **94** to give *N*-[3,6-bis-(trifluoromethyl)-pyridazin-4-yl]formamide **96** in 80% yield after 55 h in refluxing toluene (Fig. 3.25).[63] The reaction is postulated to proceed via initial cycloaddition followed by loss of nitrogen to give **95**. Ring-opening aromatization of **95** then gave **96**.

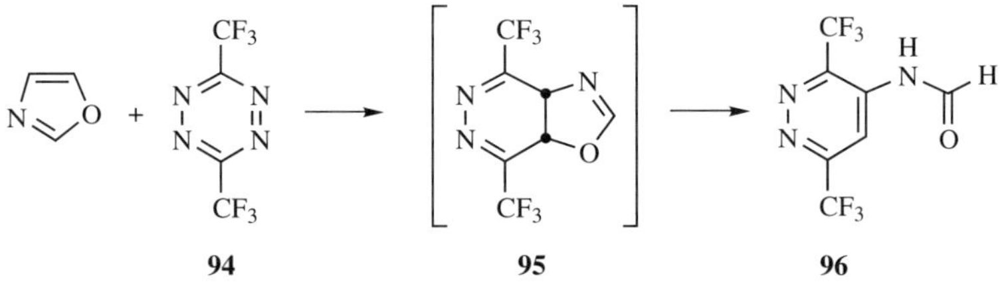

Figure 3.25

3.2.2. Intramolecular Reactions

The first examples of intramolecular Diels-Alder reactions between an oxazole and an alkene were reported independently by Levin and Weinreb[64,65] and Shimada and Tojo[45] in 1983. Shortly thereafter Turchi[12] investigated the intramolecular reactions of 5-ethoxyoxazole-olefins.

Since 1985, the major application of intramolecular oxazole-alkene Diels-Alder reactions has been in the field of natural product synthesis, in which readily available oxazole-olefins can provide access to highly substituted pyridines. All of the intramolecular reactions described here have four-atom linkers between the oxazole and olefin, found by Turchi[12] to be optimal for these cycloadditions.

TABLE 3.1. INTERMOLECULAR OXAZOLE-OLEFIN DIELS-ALDER REACTIONS

Oxazole	Dienophile	Product	Yield (%)	References
37	38	39a	50	46
41	(CO$_2$H olefin)	42	63	48
8	51	52	55	51
70a	(maleimide NPh)	71a	54	57
8	78	80	53	59
86	87	89	71	61, 62

3.2.2.1. Eupolauramine

The azaphenanthrene alkaloid eupolauramine appeared to be an ideal target for exploring the applicability of an intramolecular Kondrat'eva pyridine synthesis in the construction of natural products. Despite the fact that 2-phenyloxazoles are unreactive in intermolecular cycloadditions, the Diels-Alder reaction of oxazole-olefin **97** was expected to provide the tricyclic framework of eupolauramine

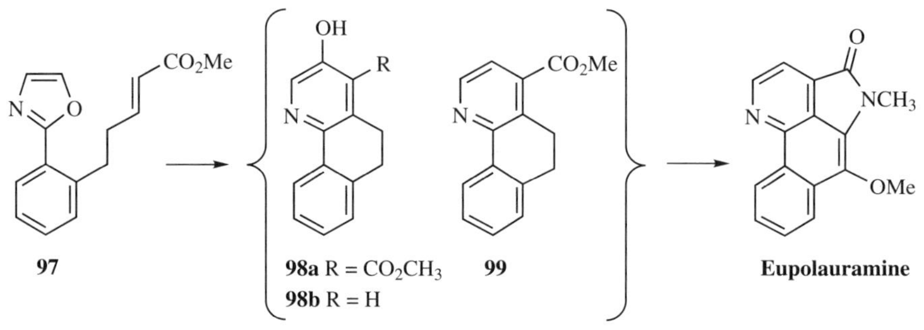

Figure 3.26

(Fig. 3.26).[64,65] Instead, upon refluxing **97** for 3 h in *o*-dichlorobenzene, a 1:1.7 ratio of 3-hydroxypyridine **98a** and the decarboxylation product **98b** were obtained in 76% overall yield. This is one of the first examples of the efficient formation of 3-hydroxypyridines from an oxazole Diels-Alder reaction without the addition of a hydride acceptor. Thorough deoxygenation of the reaction mixture and the use of lower boiling solvents, such as bromobenzene, altered the ratio of the hydroxypyridine products to 4.4:1 in favor of ester **98a** but failed to provide the desired pyridyl-ester **99**. However, the addition of 0.75 equivalent of the strong base DBN to the reaction mixture resulted in the formation of **99** as the sole product of the reaction in 76% yield. Whether the base catalyzes the dehydration of the intermediate cycloadduct or acts as an acid scavenger is unclear. Compound **99** was then converted into eupolauramine in six or seven steps, depending on which of two routes was used.

3.2.2.2. *Benzopyranopyridines and Benzonaphthyridines*

A further series of examples of intramolecular Diels-Alder reactions of 2-phenyloxazoles was applied to the synthesis of the tricyclic benzo[*h*]-1,6-naphthyridines and benzopyrano[4,3-*b*]pyridines.[66] The latter class of molecules are known to possess inotropic, analgesic, and anti-allergy activity. These reactions were also the first oxazole-olefin cycloadditions to proceed with Lewis acid catalysis. Thus, although the oxazole-olefin **100a** returned only starting material when refluxed for 16 h in *o*-dichlorobenzene, the same reaction with the addition of 7 mol % of europium(fod)$_3$ provided a 46% yield of the desired benzopyrano[4,3-*b*]pyridine **101a** (Fig. 3.27). The analogous amide **100b** gave a 75% yield of the benzo[*h*]-1,6-naphthyridine **101b** under identical conditions.

In contrast, the europium(fod)$_3$ catalyzed Diels-Alder reactions of the monoactivated olefins **102a** and **102b** led to mixtures of products (Fig. 3.28). After refluxing for 2 h in *o*-dichlorobenzene ether **102a** gave a 6:1 ratio of the pyridine **103a** and the 3-hydroxypyridine **104a**, respectively, in 35% overall yield. The

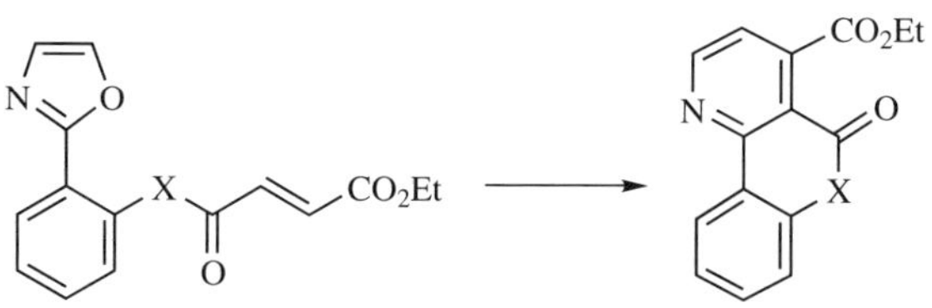

100a X = O
100b X = NH

101a X = O
101b X = NH

Figure 3.27

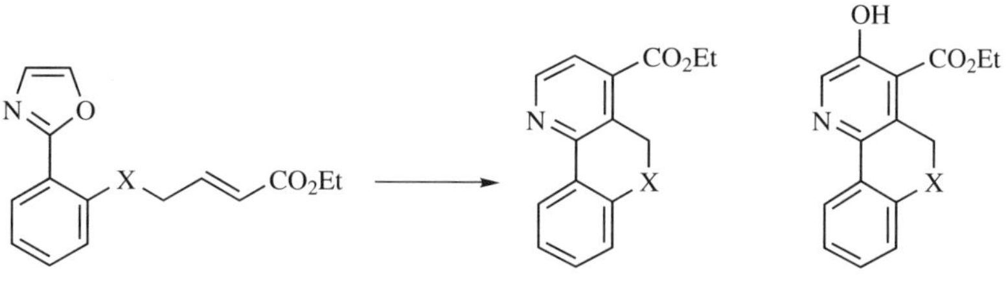

102a X = O
102b X = NCOCF$_3$

103a X = O
103b X = NCOCF$_3$

104a X = O
104b X = NCOCF$_3$

Figure 3.28

3-hydroxypyridine **104a** could be obtained as the sole product of the cycloaddition of **102a** in 46% yield when the reaction was carried out in nitrobenzene. The trifluoroacetamide **102b** was less reactive than **102a**, requiring 18 h at reflux in *o*-dichlorobenzene to provide a 4.3:1 mixture of **103b** and **104b** in 53% overall yield. It was also noted that no Diels-Alder reaction took place with oxazole-olefins lacking an electron-withdrawing group on the olefin carbon, which must form a bond with C-5 of the oxazole, even in the presence of a Lewis acid catalyst.[66]

3.2.2.3. Bromoleptoclinidinone

The initially reported structure of the cyctotoxic ascidian alkaloid 2-bromoleptoclinidinone **105** was amenable to a synthesis approach using an intramolecular oxazole-alkene Diels-Alder reaction, as shown in the retro-synthetic analysis in Figure 3.29.[67] In a model system, the *N*-benzyl-substituted amide **106** afforded a 50% yield of pyridine **107** after refluxing in benzene for 18 h with 0.75 equivalent of DMAP. The analogous NH-carboxamide failed to provide any of the desired tricyclic pyridine. This was attributed to a conformational preference that allows an internal hydrogen bond between the amide-NH and the oxazole, rather than the conformation that allows efficient overlap of the oxazole and olefin. The yield of **107** could be increased to 87% if the reaction was performed in the presence of the Lewis acid europium(hfc)$_3$. This was not further elaborated since the structure of 2-bromoleptoclinidinone was subsequently revised in 1989.

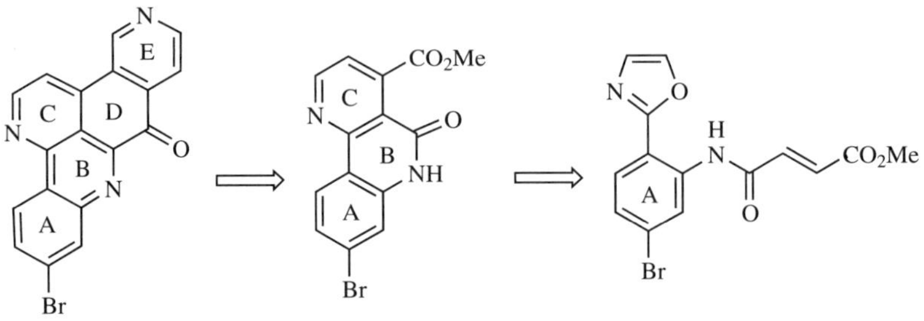

434 Oxazole Diels-Alder Reactions

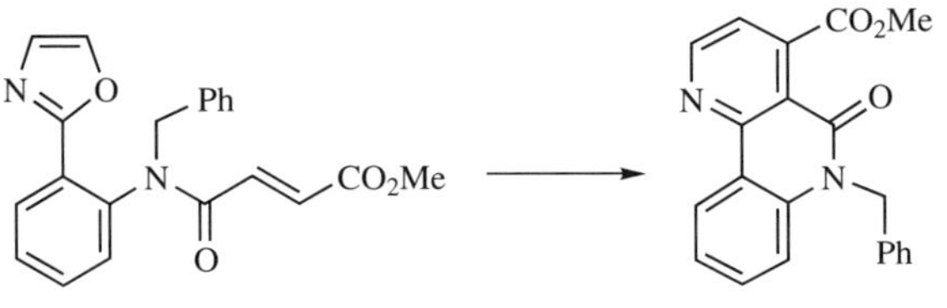

Figure 3.29

3.2.2.4. Amphimedine

Weinreb and co-workers[68] also used an intramolecular Kondrat'eva pyridine synthesis in their approach to the antineoplastic marine alkaloid amphimedine. Initial studies on model system **108a**, again using DBN in refluxing *o*-dichlorobenzene, resulted in the formation of only the decarboxylated pyridine **109a** (Fig. 3.30). Because oxazoles bearing electron-withdrawing groups at the 4 position are generally unreactive dienophiles, the ester functionality of compound **108a** was

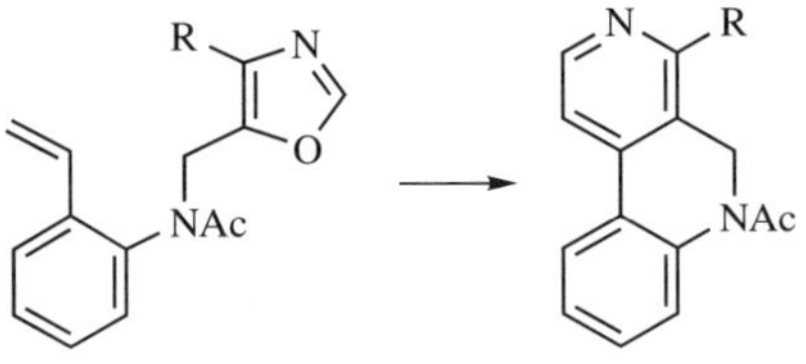

108a R = CO$_2$Me
108b R = CH$_2$OTBS

109a R = H
109b R = CH$_2$OTBS

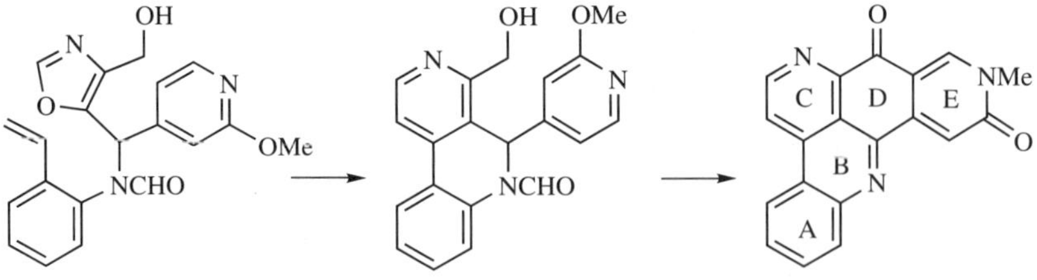

110   111   Amphimedine

Figure 3.30

reduced and the resulting alcohol was protected as the TBS-ether before the cycloaddition. On heating in *o*-dichlorobenzene, oxazole **108b** cyclized to provide pyridine **109b**. Similarly, when the more highly functionalized oxazole **110** was heated to reflux in *o*-dichlorobenzene with one equivalent of DBN, a 71% yield of the desired pyridine **111** was obtained. Unfortunately, the subsequent formation of the D-ring was unsuccessful.

3.2.2.5. Synthesis of Normalindine

The *Strychnos* alkaloid (−)-normalindine, first isolated in 1987, was efficiently constructed by Ohba's group[69] using an intramolecular Kondrat'eva pyridine synthesis. In the key step, a 3:1 mixture of oxazole-alkenes **112a** and **112b** was refluxed in toluene for 24 h to give a separable 10:1 mixture of the 1α-H cycloadducts **113** and **114** in 58% overall yield (Fig. 3.31). After separation,

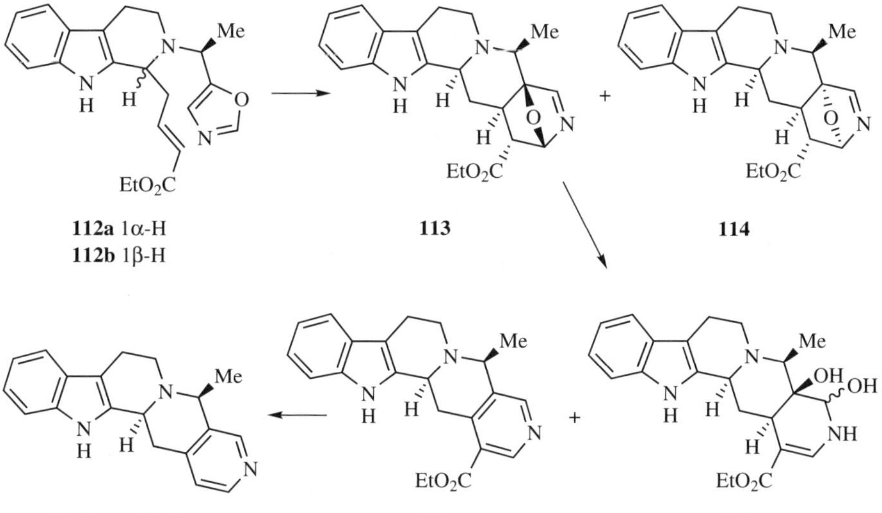

Figure 3.31

compound **113** was then refluxed in acetic acid-xylene for 5 h to afford a mixture of diol **115** and pyridine **116** in 18% and 64% yields, respectively. Interestingly, Diels-Alder adduct **114** did not afford either diol **115** or pyridine **116** after treatment with acetic acid. When diol **115** was resubjected to acetic acid–xylene an additional 13% yield of **116** was obtained along with 62% of recovered starting material. Pyridyl-ester **116** was then converted into (−)-normalindine in four steps.

3.2.2.6. Amidooxazole Cycloadditions

A single example of an intramolecular amidooxazole-olefin Diels-Alder cycloaddition has been reported.[70] A toluene solution of acetamidooxazole **117** and 0.8 equivalent of DBU when heated to 180°C in a sealed tube provided a 39% yield of the tetrahydronaphthyridine **118** (Fig. 3.32). The intermediate Diels-Alder adduct was not observed.

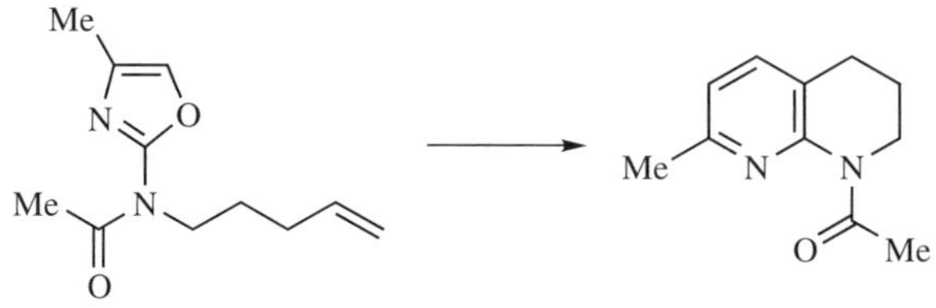

Figure 3.32

The intermolecular oxazole-olefin Diels-Alder reactions are summarized in Table 3.1, p. 431.

3.3. OXAZOLE-ALKYNE CYCLOADDITIONS

The synthetic utility of oxazoles as azadienes was further advanced when Grigg and co-workers[71,72] reported that the Diels-Alder reactions of oxazoles with alkynes provided furans via a tandem Diels-Alder retro-Diels-Alder sequence. Thus 5-ethoxy-4-substituted oxazoles **119** reacted with dimethyl acetylenedicarboxylate in cold ether to yield 2-ethoxy-3,4-furandicarboxylic acid dimethyl ester **121** in >50% yield (Fig. 3.33). In this case, the intermediate cycloaddition adduct **120** extrudes a molecule of hydrogen cyanide or a nitrile derived from the C-4 substituent of the oxazole, via a retro-Diels-Alder reaction to provide a substituted

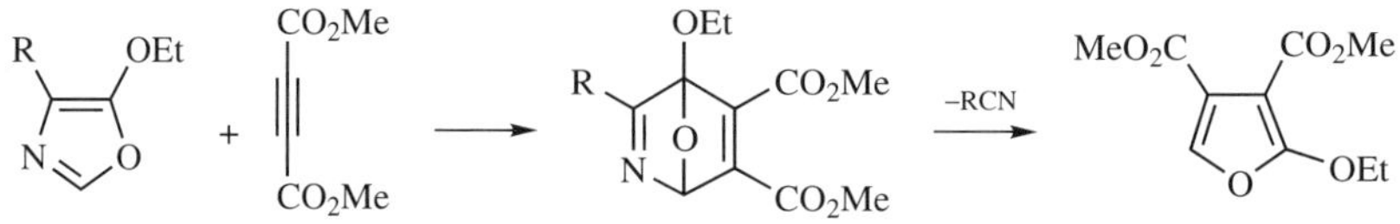

Figure 3.33

furan. A series of papers by Kondrat'eva's group[73–75] in 1971 expanded on this new route to substituted furans.

A variety of substituents—including alkyl, alkenyl, cyano, acetyl, and alkoxy— is tolerated at the 2 and 5 positions of the oxazole ring for these cycloadditions.[76] Acetylenic dienophiles with alkyl, trialkylsilyl, phenyl, ester, ketone, and acetal substituents, as well as terminal alkynes, are precedented.[77,78] Ab initio calculations predict a slightly higher activation energy for the cycloaddition of oxazole with acetylene compared to the oxazole-ethylene reaction.[29]

The regiochemistry of oxazole-alkyne cycloadditions is ill-defined except for the reactions of electron-rich 5-alkoxyoxazoles with electron-deficient alkynes. In these cases, as for the analogous oxazole-olefin cycloadditions, the major product results from bond formation between the alkyne carbon bearing the electron-withdrawing moiety and the 5 position of the oxazole to give the 2-substituted furan. A computational study of the regiochemistry of oxazole-alkyne Diels-Alder reactions has been reported.[79] Steric effects have also been noted in oxazole-alkyne cycloadditions. Thus 2-benzyloxazole **122** reacted with propargyl acetate to give predominantly the 2,4-disubstituted furan **123a** rather than the 2,3-disubstituted regioisomer **123b** (Fig. 3.34).[76]

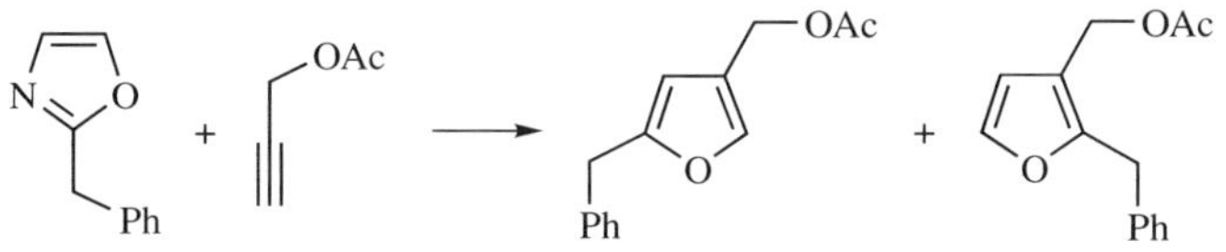

Figure 3.34

Similar to oxazole-olefin cycloadditions, the rate of Diels-Alder reactions of oxazoles with acetylenes depends on the electron-donating ability of the oxazole substituents. Rates decrease with decreasing electron-donating capacity of the substituents, in the order $OR > NR_1R_2 >$ alkyl $>$ 4-phenyl $> COCH_3 > CO_2CH_3 \gg$ 2-,5-phenyl. 5-Alkoxyoxazoles are approximately equivalent in reactivity to an all-carbon diene.

3.3.1. Intermolecular Reactions

3.3.1.1. Furans

The Diels-Alder reaction of oxazoles with alkynes has become a preferred method for the synthesis of substituted furans with diverse applications.[80–82] A large number of early examples of this reaction have been tabulated.[3] Activated dienophiles such as acetylenic ketones and esters can be used,[83–85] although unactivated alkyl, aryl, and silyl alkynes have been used as well.[78] In particular, the reaction of 4-phenyloxazole with substituted acetylenes is frequently used for preparing 3,4-disubstituted furans. Cycloadducts derived from 4-phenyloxazole typically decompose under milder conditions than 4-alkyloxazoles, allowing the synthesis of a wider range of functionalized furans. Several examples are shown below.

Diethylacetylene reacted with 4-phenyloxazole **55** in an autoclave for 3 days at 250°C to afford a 70% yield of 3,4-diethylfuran **124** (Fig. 3.35). This furan was then converted into the interesting tetraoxaporphyrin **125** in two steps.[86] The novel diimidazole copper ligand **127** is formed in 84% yield by the cycloaddition of 4-phenyloxazole with the acetylenic diimidazole **126** at 185°C for 18 h (Fig. 3.36).[87]

An application of the oxazole-alkyne Diels-Alder reaction in natural product synthesis is exemplified by the construction of the maleic anhydride portion of the

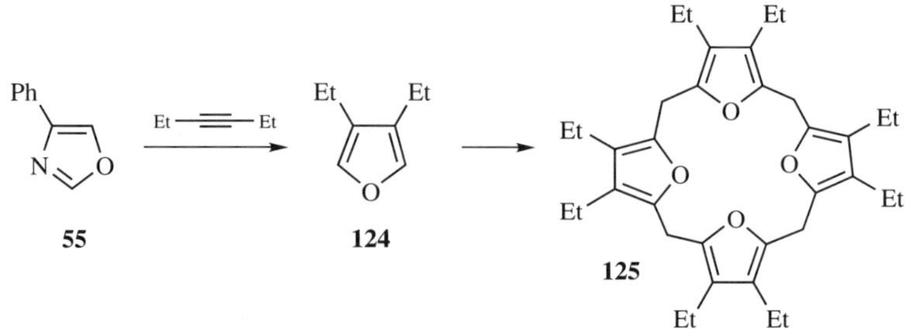

438 Oxazole Diels-Alder Reactions

Figure 3.35

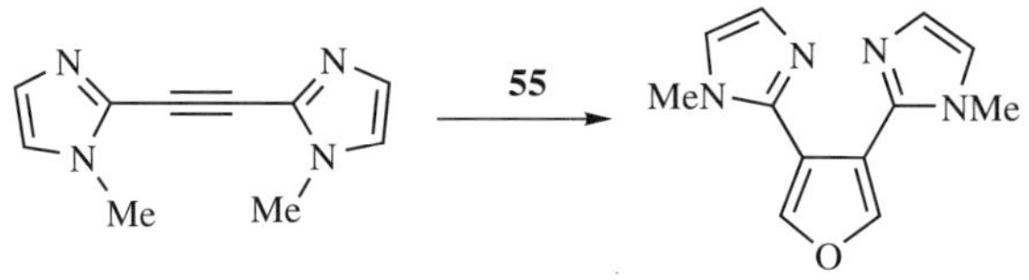

Figure 3.36

antiboitic tautomycin.[88] Reaction of a benzene solution of 4-phenyloxazole and ethyl 2-butynoate at 190°C in a sealed tube for 24 h provided an 86% yield of 4-methyl-3-furancarboxylic acid ethyl ester **128** (Fig. 3.37). Manipulation of the ester side chain and oxidation of the furan with singlet oxygen, followed by PCC oxidation of the resulting butenolide and removal of protecting groups, gave the desired maleic anhydride **129** in eight steps.

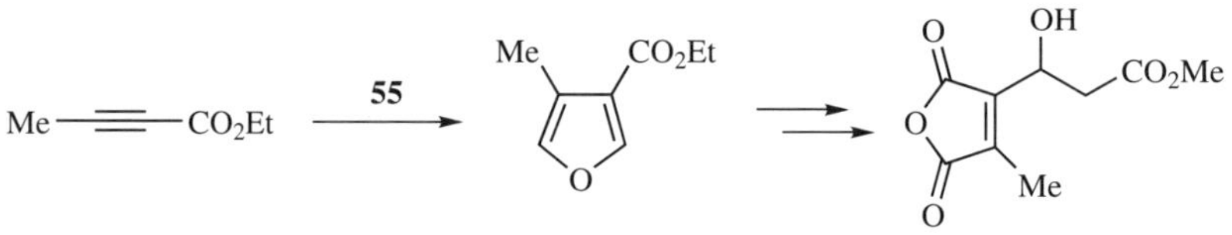

Figure 3.37

In a series of papers, Wong and co-workers[89–92] detailed the preparation and use of 3,4-bis(trimethylsilyl)furans. Thus, after heating an equimolar mixture of 4-phenyloxazole and bis(trimethylsilyl)acetylene with 5 mol % of triethylamine in a sealed tube at 250°C for 2 days, an 80% yield of the desired furan **130** was obtained (Fig. 3.38).[91,92] If the same reaction is performed in the presence of a catalytic amount of formic or trifluoroacetic acid instead of triethylamine, the analogous 2,4-bis-trimethylsilylfuran is obtained in good yield. Furan **130** itself reacts with alkynes in a tandem Diels-Alder retro-Diels-Alder sequence to provide other furans, e.g., 3,4-furandicarboxylic acid dimethyl ester **131**. Friedel-Crafts acylations can also be performed using **130**.[89] In addition, **130** reacted with boron trichloride followed by mild acid or base workup to give the tris[(4-trimethylsilyl)furan-3-yl]boroxine **132**.[91,92] Boroxine **132** was found to undergo palladium(0)-catalyzed Suzuki couplings in high yield to afford the mono-silylated furans **133**. A second iteration of this two-step sequence then gave the 3,4-disubstituted furans **134**.

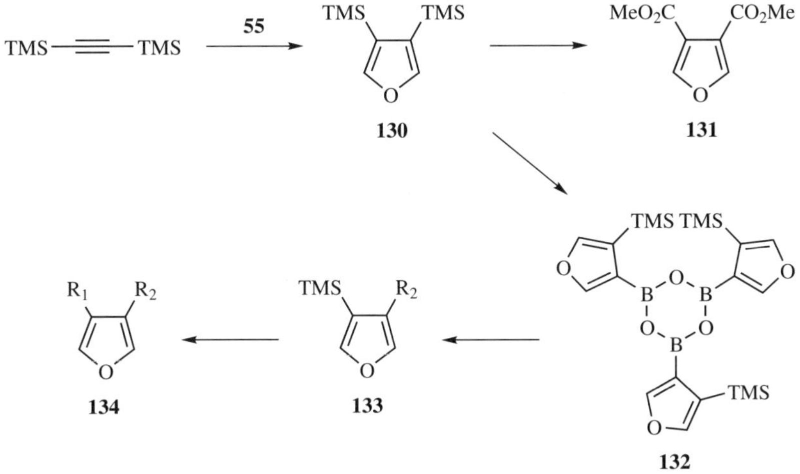

Figure 3.38

Wong and co-workers[93] reported a concise synthesis of rosefuran. Heating 4-phenyloxazole **55** and trimethylsilylpropyne in the presence of DBU for 5 days at 270°C in a sealed tube gave 3-methyl-4-(trimethylsilyl)furan **135** in 75% yield (Fig. 3.39). Metalation of **135** with t-butyllithium and quenching with prenyl bromide followed by desilylation with trifluoroacetic acid then gave rosefuran.

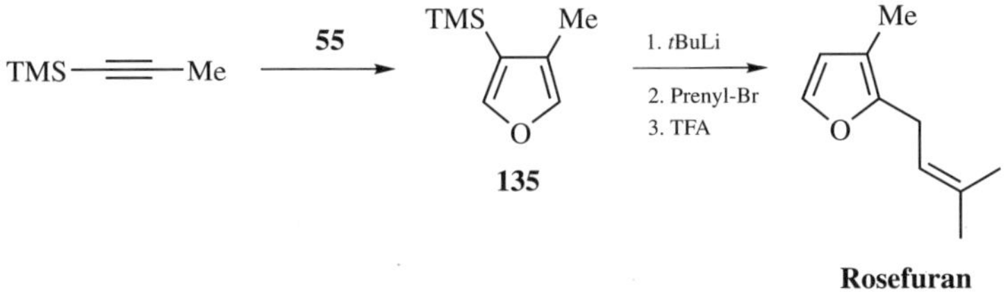

Figure 3.39

Access to 2,4-disubstituted furans is exemplified by the reaction of 4-phenyloxazole and trimethylsilylacetylene to give 3-(trimethylsilyl)furan **136** in 70% yield (Fig. 3.40). As before, metalation with t-butyllithium and quenching with benzyl bromide gave 2-benzyl-4-(trimethylsilyl)furan **137**. Conversion of **137** to the corresponding boroxine and subsequent Suzuki coupling afforded a 4-substituted 2-benzylfuran **138**.[92]

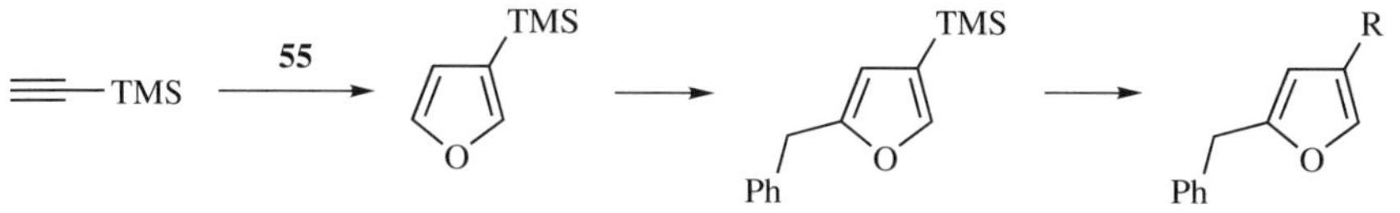

Figure 3.40

The same methodology has also been applied to the synthesis of stannylated furans.[94,95] Heating bis(tributylstannyl)acetylene and **55** in a sealed tube at 185°C for 10 days provided 3,4-bis(tributylstannyl)furan **139** and 3-(tributylstannyl)furan **140** in 22 and 10% isolated yield, respectively (Fig. 3.41). These stannanes underwent palladium-catalyzed couplings with aryl, vinyl, and benzylic halides as well as acid chlorides.

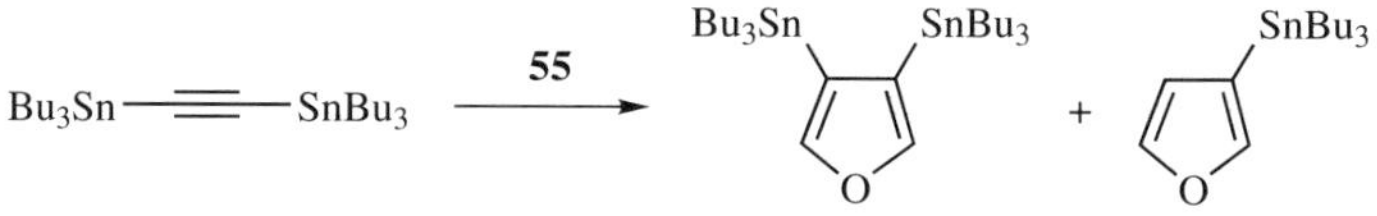

Figure 3.41

A series of anti-inflammatory hydroxy butenolides was synthesized starting from **55**. Thus heating **55** with ethyl phenyl-prop-1-ynoate **141** for 16 h at 210°C afforded 3-phenyl-4-furancarboxylic acid ethyl ester **142** (Fig. 3.42).[96] Functionalization of the ester followed by reaction of the furan ring with singlet oxygen gave the biologically active butenolides, **143**.

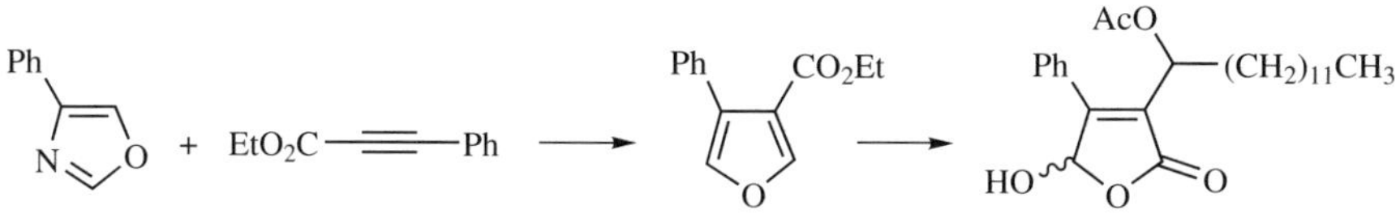

Figure 3.42

The 2-aryl-substituted 4-phenyloxazole **144** has been used in the synthesis of the pharmacologically active tetrahydrofuran lignan natural products (Fig. 3.43). Refluxing a mixture of **144** with 2-butyn-1,4-diol diacetate in the presence of sodium carbonate and hydroquinone for 22 h provided an 89% yield of the 2,3,4-trisubstituted furan **145**. Elaboration of **145** by hydrolysis of the diacetate and selective oxidation of the 4-carbinol then afforded the lignan **146** as a mixture of diastereomers after three additional steps.[97]

3-(Trifluoromethyl)furan **147** was prepared in 67% yield by heating 4-methyloxazole **5** and 3,3,3-trifluoropropyne in toluene at 180°C for 13 h (Fig. 3.44). Metalation of **147** at the 2 position with *n*-butyllithium and subsequent reaction with aldehydes gave 3-(trifluoromethyl)-2-furyl carbinols **148**.[98]

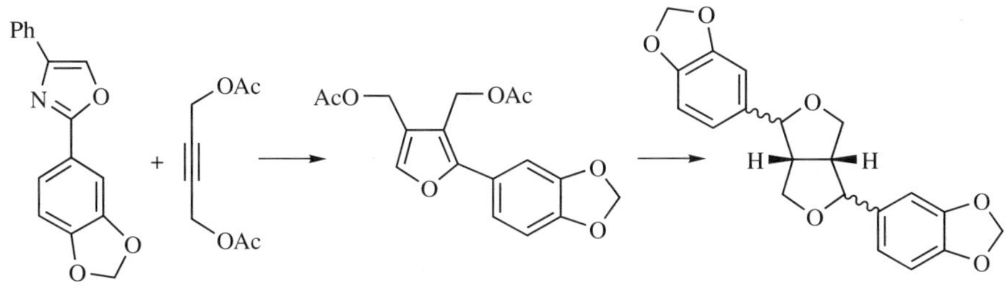

Figure 3.43

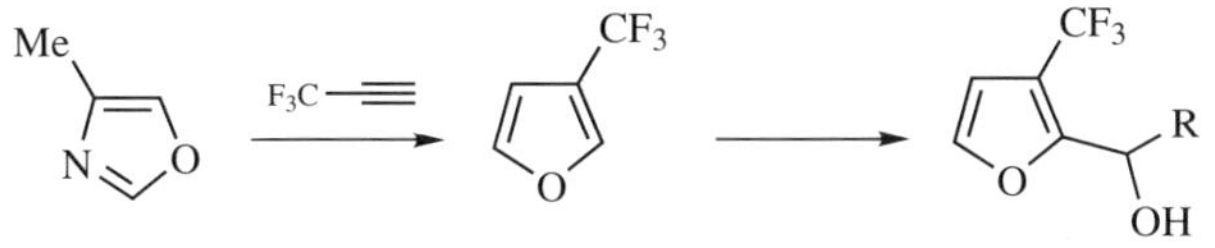

Figure 3.44

5-Alkoxyoxazoles have also proven to be valuable heterodienes in the synthesis of highly substituted furans and biologically active natural products via oxazole-alkyne Diels-Alder reactions.[99,100] Thus the synthesis of a synthon for the DEF ring system of fredericamycin began with the cycloaddition of 5-ethoxy-4-methyl-oxazole **8** with enyne ester **149** (Fig. 3.45). This reaction proceeded in 24 h in refluxing toluene to afford a 65% yield of the 2-ethoxyfuran **150**. Furan **150** was then converted to the isoquinoline **151** in six steps.[101]

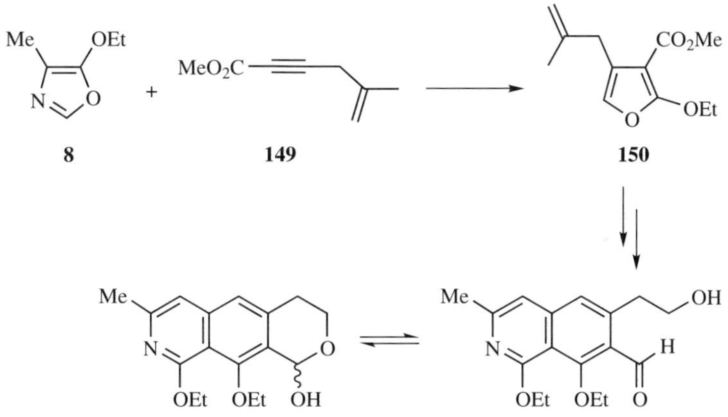

Figure 3.45

An expeditious route to hydroxy butenolides, found in several natural products, also uses a 5-alkoxyoxazole-acetylene cycloaddition. Reaction of **8** with ethyl phenyl-prop-1-ynoate **141** in toluene for 3 h gave 2-ethoxy-4-phenyl-3-furan-carboxylic acid ethyl ester **152** as the only regioisomer in 90% yield (Fig. 3.46).[102] Oxidation of **152** with manganese dioxide and treatment with HCl provided the desired hydroxy butenolide **153** in 78% yield. This methodology was then applied to the synthesis of A-factor, an inducer of streptomycin biosynthesis. Cycloaddition of alkyne **154** and **8** gave the furan **155** in 80% yield. This 2,3,4-trisubstituted furan **155** was converted to the hydroxy butenolide, which was subsequently transformed to racemic A-factor in three additional steps.

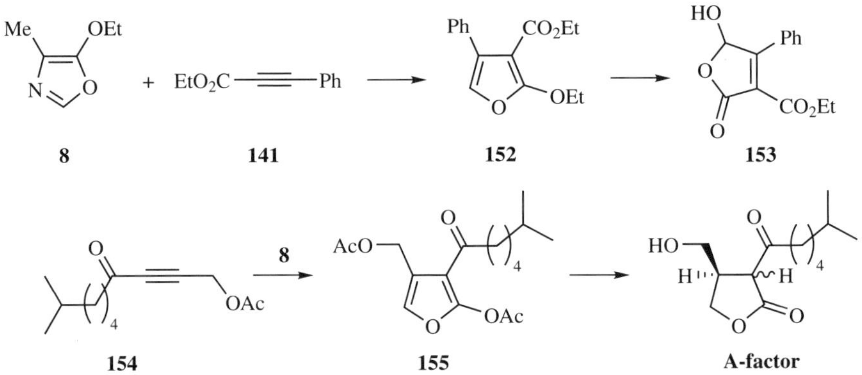

Figure 3.46

Rao and co-workers[103] effected cycloaddition of **8** with the propargylic aldehyde **156** to afford the 2,3,4-trisubstituted furan **157** in 90% yield in their regioselective synthesis of the antitumor agent camptothecin (Fig. 3.47). Several steps were required to convert **157** to the butenolide **158**, a key synthon for the D- and E-rings of camptothecin.

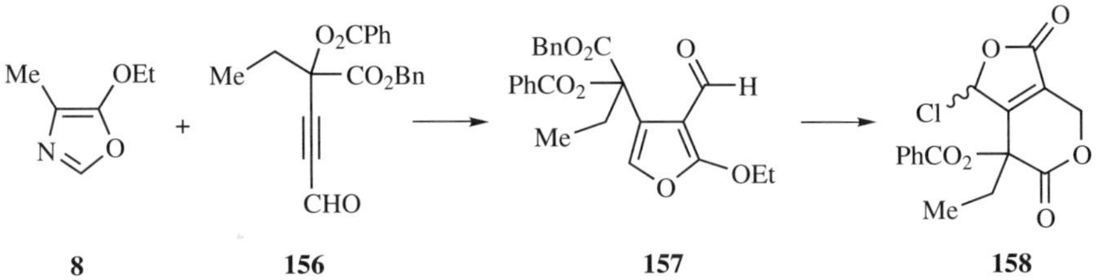

Figure 3.47

3.3.1.2. Reactions with Aminooxazoles

Crank and Kahn[56] reported a rare example of an isolable Diels-Alder adduct from the reaction of an oxazole with an acetylene in 1985. Thus 2-aminooxazole reacted with dimethylacetylene dicarboxylate to give a mixture of the cycloadduct **159** in 44% yield together with the bicyclic Michael addition product **160** in 16% yield

(Fig. 3.48). Several other 2-amino-4-alkyloxazoles also afford the bridged Diels-Alder adducts as the major products in 20–50% yields.

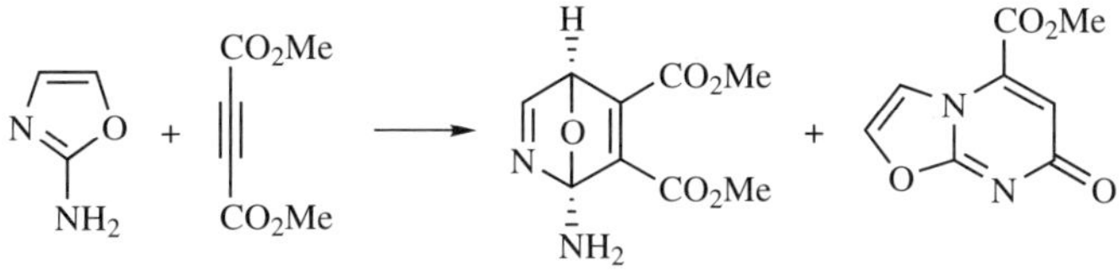

159   **160**

Figure 3.48

3.3.1.3. Reactions with Benzyne

The first report of a [4 + 2] cycloaddition of an oxazole with benzyne was the reaction of a trisubstituted oxazole with benzyne generated from anthranilic acid and isoamyl nitrite in refluxing dioxane.[104] The product of this reaction was not the oxazole Diels-Alder adduct or a furan, but rather the bis(benzyne) adduct **163** resulting from initial cycloaddition to give **161**, subsequent retro-Diels-Alder reaction to give the isobenzofuran **162** and then a second benzyne Diels-Alder reaction to afford **163** (Fig. 3.49).

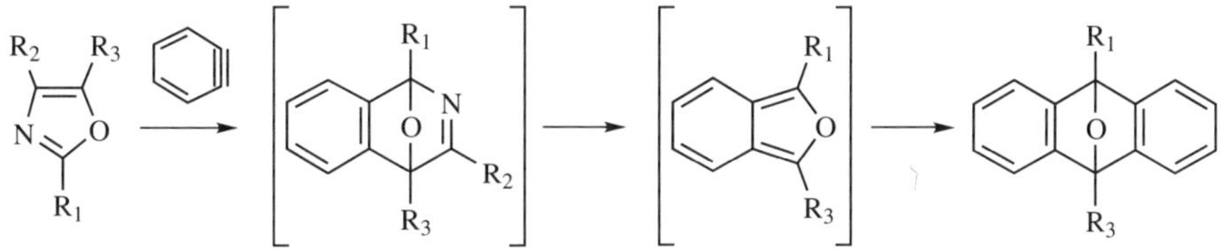

161   **162**   **163**

Figure 3.49

Rickborn and co-workers[105,106] isolated the cycloadduct **164** from 4-phenyloxazole through careful manipulation of the experimental conditions (Fig. 3.50). They generated benzyne at 0°C from 1-aminobenzotriazole and lead tetraacetate. Compound **164** was stable at room temperature but, on heating, eliminated benzonitrile to give isobenzofuran **162**, which could be trapped with N-methylmaleimide to afford a quantitative yield of the tetracyclic derivative **166** as an 88:12 mixture of endo and exo isomers. The cyclic aminal **164** was also sensitive to acid and rearranged to 4-hydroxy-3-phenyl-isoquinoline **165** on exposure to silica gel or a catalytic amount of trifluoroacetic acid. The benzyne cycloadditions were also carried out on 4-(4-nitrophenyl)oxazole and 4-(4-methoxyphenyl)oxazole. A four-fold rate increase was seen for the cycloaddition of the nitrophenyl-substituted oxazole relative to the methoxyphenyl analog, indicating a concerted process with little contribution from a polar intermediate.

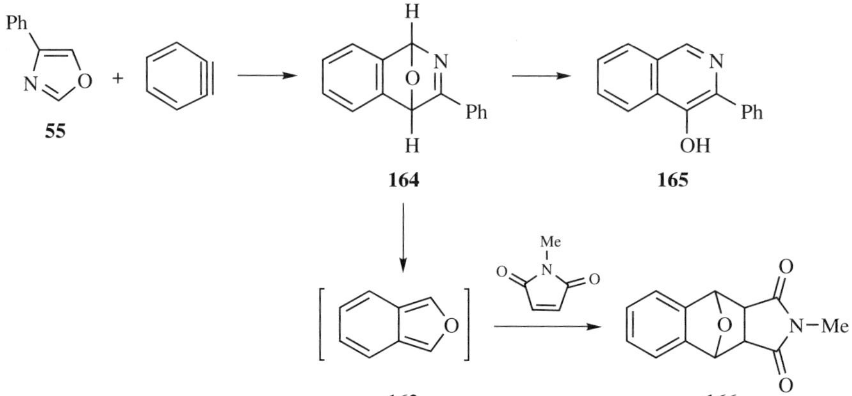

Figure 3.50

Lead tetraacetate has also used to generate 4,5-dehydrotropone **168** from the *N*-aminotriazole **167** in the presence of several 4-phenyloxazoles (Fig. 3.51).[107] In this case, the 1:1 oxazole-tropone cycloaddition adducts were not observed and only the furo[3,4-*d*]tropones **169** were isolated in 20–57% yields.

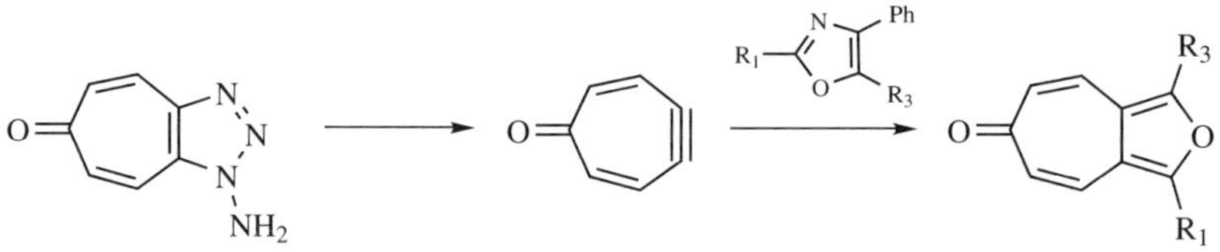

Figure 3.51

The intermolecular reactions are summarized in Table 3.2.

3.3.2. Intramolecular Reactions

Jacobi has used intramolecular oxazole-alkyne Diels-Alder reactions, elegantly and with great success, to prepare a number of natural products containing furan, lactone, and butenolide rings. These syntheses and the methodology based on the oxazole Diels-Alder reaction, termed bis-heteroannulation, have been reviewed in detail.[108]

3.3.2.1. Synthesis of Evodone

The first natural product synthesized using the bis-heteroannulation strategy was evodone, a structurally simple member of the naturally occurring furanoterpenes.[110] The synthesis began with commercially available 4-methyl-δ-valerolactone **170**, which was converted to the cycloaddition precursor **171** in four steps (Fig. 3.52).

Oxazole-Alkyne Cycloadditions

TABLE 3.2. INTERMOLECULAR OXAZOLE-ALKYNE DIELS-ALDER REACTIONS

Oxazole	Dienophile	Product	Yield (%)	References
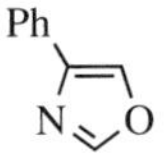 Ph-oxazole **55**	Et—≡—Et	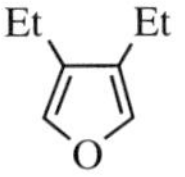 3,4-diethylfuran **124**	70	86
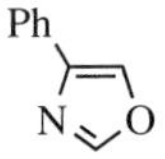 Ph-oxazole **55**	Me—≡—CO$_2$Et	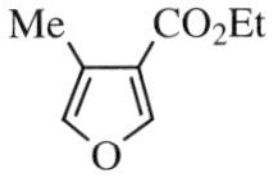 Me, CO$_2$Et furan **128**	86	88
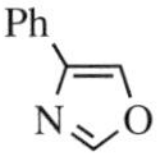 Ph-oxazole **55**	TMS—≡—Me	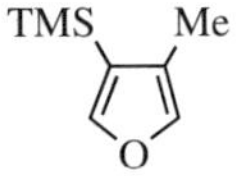 TMS, Me furan **135**	75	92, 93
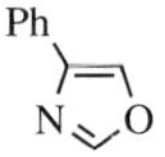 Ph-oxazole **55**	Bu$_3$Sn—≡—SnBu$_3$	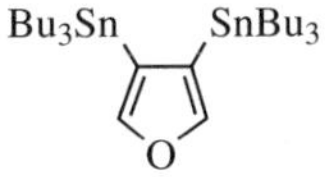 Bu$_3$Sn, SnBu$_3$ furan **139**	22	94, 95, 109
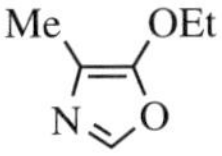 Me, OEt oxazole **8**	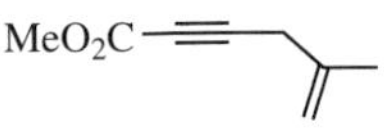 MeO$_2$C—≡—C(Me)=CH$_2$ **149**	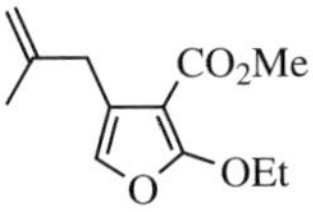 furan **150**	65	101
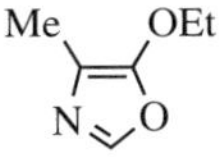 Me, OEt oxazole **8**	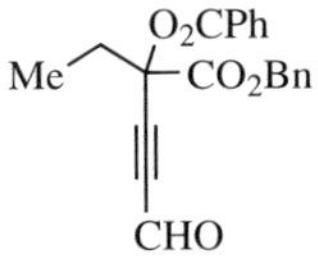 Me-C(O$_2$CPh)(CO$_2$Bn)-C≡C-CHO **156**	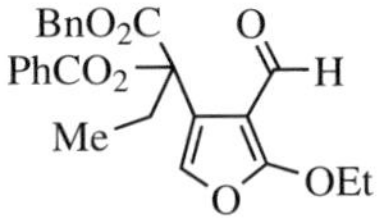 furan **157**	90	103
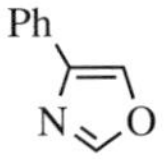 Ph-oxazole **55**	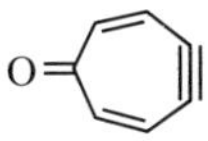 tropone **168**	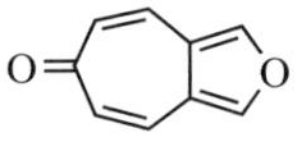 **169**	57	107

Figure 3.52: **170** (methyl δ-valerolactone) → (4 steps) → **171** (oxazole-alkyne intermediate) → (−HCN) → **Evodone**

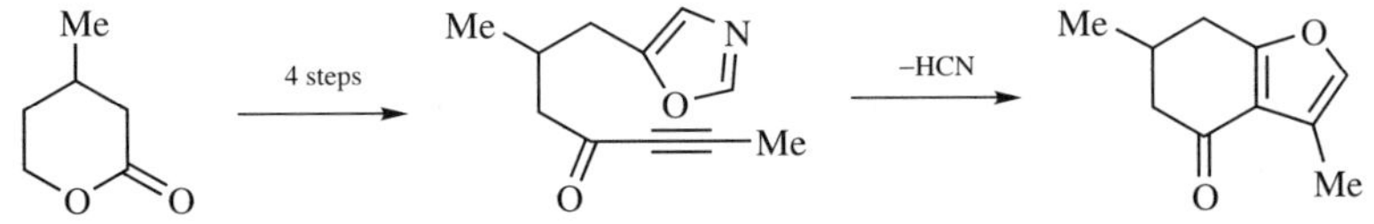

Upon refluxing in ethylbenzene for 96 h, **171** underwent the requisite Diels-Alder retro-Diels-Alder sequence to furnish evodone in 76% yield.

3.3.2.2. Synthesis of Ligularone and Petasalbine

Jacobi[111] also completed syntheses of the sesquiterpenes ligularone and petasalbine using this intramolecular oxazole-acetylene Diels-Alder reaction. A 55:45 mixture of the diastereomeric alcohols **173a** and **173b**, prepared in three steps from lactone **172**, was quantitatively oxidized to acetylenic ketone **174** under Swern conditions (Fig. 3.53). Refluxing **174** in ethylbenzene for 26 h effected clean cycloaddition to give racemic ligularone in 92% yield. Since ligularone was previously converted to petasalbine, this synthesis also represents a formal total synthesis of petasalbine. In addition, a total synthesis of petasalbine was completed by the thermolysis of acetylenic alcohol **173a** under the same conditions used for the cycloaddition of **174**, thereby furnishing racemic petasalbine directly in 84% yield.

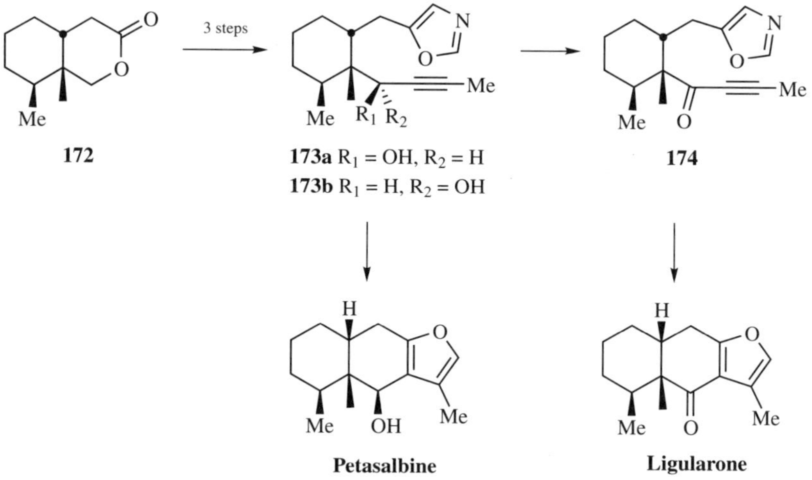

Figure 3.53

3.3.2.3. Synthesis of Paniculide-A

Paniculide-A is structurally the least complicated of the paniculides, a family of highly oxygenated sesquiterpenes from *Andrographis paniculata*. Jacobi's[112,113] synthesis of paniculide-A began with a four-step conversion of 3-methyl glutaric anhydride **175** to the oxazole acetylenic ketone **176** (Fig. 3.54). Cycloaddition of **176** was accomplished in refluxing ethylbenzene for 11 h to provide methoxyfuran **177** in 94% yield. Unfortunately, the direct conversion of **177** to the corresponding keto-butenolide was unsuccessful. Therefore, **177** was deprotonated, and the anion was quenched with phenylselenenyl chloride, affording a 1:1 mixture of epimeric phenyl selenides. This ratio was improved to 96:4 by kinetic deprotonation of **177** with LDA and subsequent protonation with acetic acid. Reduction of the ketone moiety with diisobutylaluminum hydride and hydrolysis of the methoxyfuran moiety at pH 5 then gave alcohol **178**. The formal total synthesis of the natural product was completed via a five-step conversion of phenyl selenide **178** to the allylic alcohol **179**, a known intermediate in the synthesis of racemic paniculide-A.

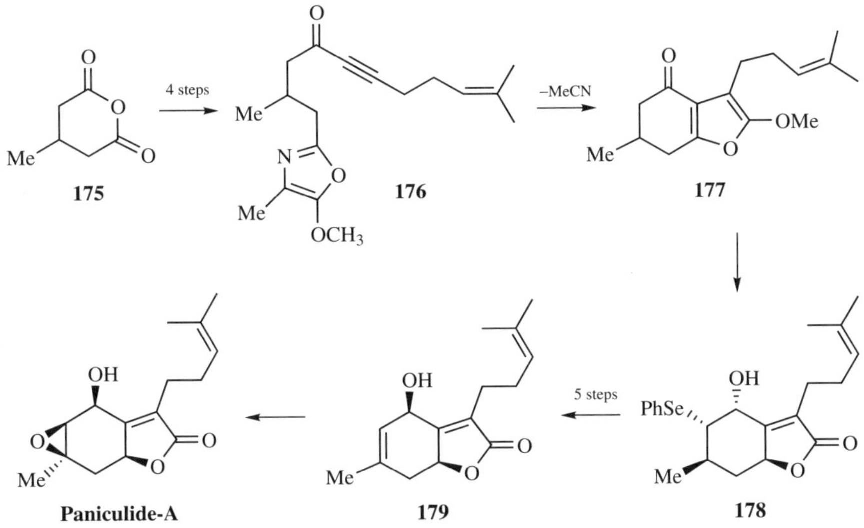

Figure 3.54

3.3.2.4. Synthesis of Gnididione

The furanosesquiterpene gnididione was isolated from *Gnidia latifolia* in the course of searches for plant-based tumor inhibitors. Racemic gnididione and isognididione were synthesized by employing a chemoselective sequential intramolecular oxy-Cope Diels-Alder approach.[114] Oxazole-aldehyde **181** was available in six steps from the known diester **180** (Fig. 3.55). This aldehyde, **181** was then converted in three steps to **182a**, the precursor to gnididione and to **182b**, the precursor to isognididione.

Thermolysis of the Z-olefin **182a** in refluxing toluene for 4 h furnished the desired oxy-Cope product **183a** (β-Me) in 87% yield. The product of the acetylene-olefin oxy-Cope reaction of **182a** was not observed. Similarly, the *E*-olefin **182b** afforded a 75% yield of **183b**. Moreover, in refluxing mesitylene, **182a** was directly converted to gnididione ketal **184a** in 48% yield via a tandem oxy-Cope Diels-Alder retro-Diels-Alder sequence. Finally, hydrolysis of **184a** afforded racemic gnididione in 98% yield. In the same manner, **182b** was directly converted to **184b** in 60% yield, followed by hydrolysis to afford racemic isognididione in 96% yield.

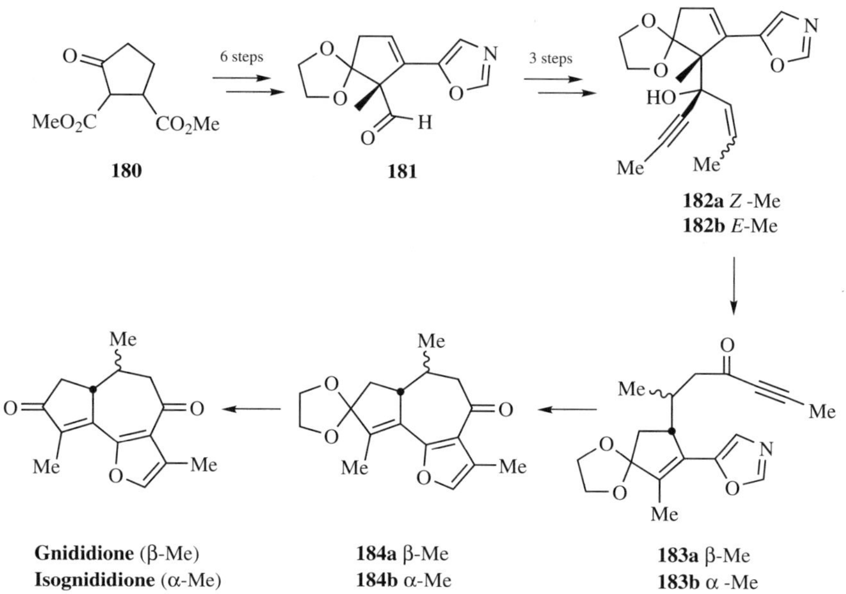

448   Oxazole Diels-Alder Reactions

Gnididione (β-Me)   **184a** β-Me   **183a** β-Me
Isognididione (α-Me) **184b** α-Me   **183b** α-Me

180   **181**   **182a** Z-Me
                    182b E-Me

Figure 3.55

3.3.2.5. Synthesis of Norsecurinine

Norsecurinine, a member of the *Securinega* class of alkaloids, occurs naturally in both enantiomeric forms. Jacobi and co-wrokers[115,116] fashioned enantiospecific syntheses of both antipodes starting from either D- or L-proline.

The feasibility of an oxazole Diels-Alder approach for the synthesis of norsecurinine was initially investigated with a model system. Oxazole-ynone **185a**, available from proline in six steps, underwent thermolysis in ethylbenzene for 7 h to furnish the trimethylsilylfuranoketone **187a** via extrusion of acetonitrile from the Diels-Alder adduct **186** (Fig. 3.56). A considerably lower yield of the unsubstituted-furano ketone **187b** was obtained upon thermolysis of acetylenic ketone **185b**. The attenuation of the reaction rate in the unsubstituted analog **185b** relative to **185a** was expected, since trimethylsilyl groups have previously been noted to enhance the reactivity of acetylenic dienophiles.[117]

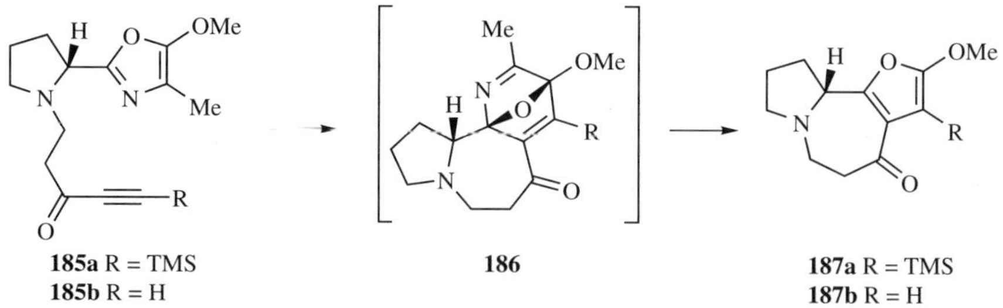

185a R = TMS   **186**   **187a** R = TMS
185b R = H   **187b** R = H

Figure 3.56

The application of this cycloaddition strategy to the synthesis of norsecurinine was then undertaken. Thus the oxazole-pyrrolidine derivative **188** was prepared in four steps from D-proline (Fig. 3.57). Coupling **188** with enynone **189**, prepared in four steps from maleic anhydride, afforded a near-quantitative yield of the acetylenic ketone **190** as a 2:1 mixture of C-7 epimers. After refluxing for 30 min in mesitylene, **190** underwent cycloaddition to produce a 2:1 mixture of the tricyclic furano-ketone **191** and the C-7 epimer in 50% overall yield from **189**. The undesired epimer could be recycled by epimerizing with sodium carbonate to give a 1:1 mixture of epimers. The tricyclic furano-ketone **191** was then converted to (−)-norsecurinine in six steps. (+)-Norsecurinine was synthesized using an identical route starting with L-proline.

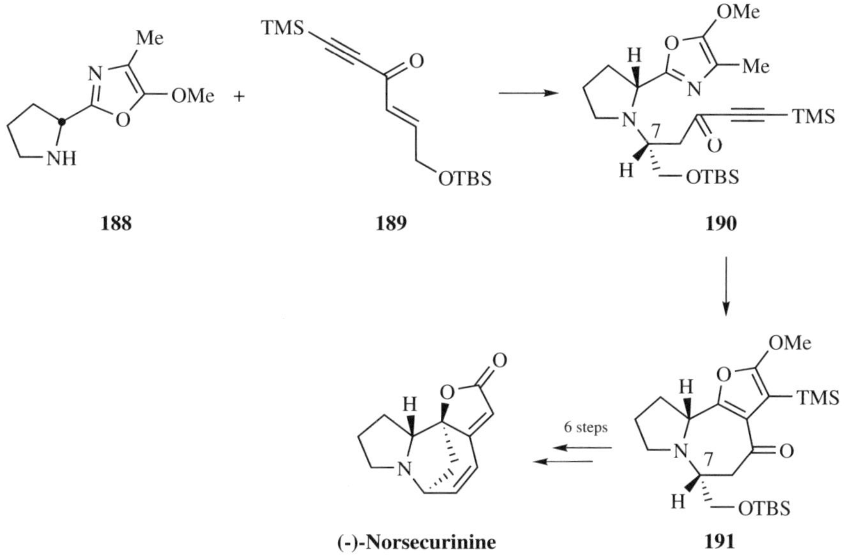

Figure 3.57

3.3.2.6. Synthetic Approach to Geigerin

The sesquiterpene geigerin, first isolated from *G. aspera Harv.* in 1936 was also synthesized by Jacobi's group[118,119] using their bis-heteroannulation strategy. Synthetically, the lactone ring of geigerin was postulated to be accessible from a 2-alkoxyfuran derived from an intramolecular oxazole-alkyne cycloaddition. The viability of this strategy was initially tested with a model system. Thus oxazole **192** was subjected to thermolysis in ethylbenzene at 134°C for 72 h in the presence of hydroquinone to furnish the bicyclic ethoxyfuran **193** in 94% yield (Fig. 3.58). Conversion of **193** to the lactone **194** with the requisite stereochemistry for geigerin was accomplished in two steps and 70% overall yield.[118]

Their attempted total synthesis of geigerin started from the acetylenic ketone **197**, which was generated in 76% yield from the oxy-Cope rearrangement of

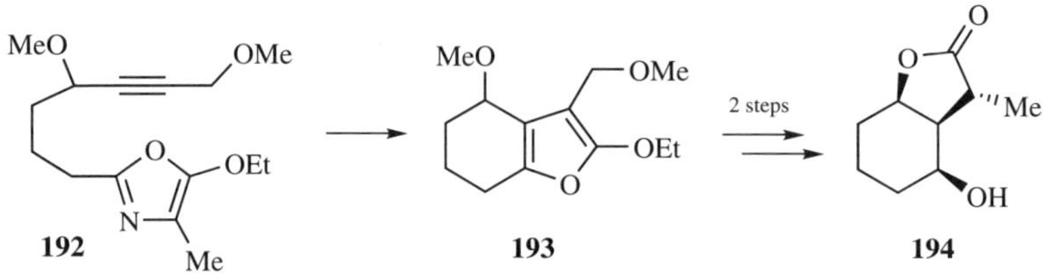

450   Oxazole Diels-Alder Reactions

Figure 3.58

oxazole **195** followed by desilylation of the resulting enol ether **196** (Fig. 3.59).[118] Refluxing **197** for 4.5 h in toluene effected the oxazole-acetylene Diels-Alder reaction to give **198a**. Subsequent reduction of the ketone with sodium borohydride furnished the desired furan-alcohol **198b** as a 3:1 mixture of epimers in 80% overall yield. In stark contrast to the successful conversion of **193** to **194** in the model system, **198** could not be converted to geigerin.

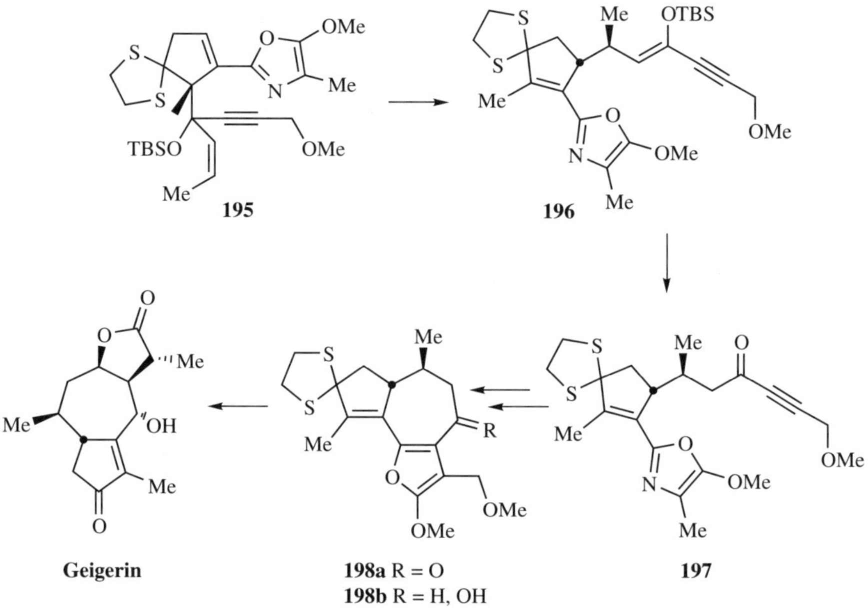

Figure 3.59

3.3.2.7. Synthesis of Stemoamide

Jacobi and co-workers[120,121] described the total synthesis of racemic stemoamide, a member of the stemona alkaloids isolated in 1992 from *Stemona tuberosa*, and an enantioselective route to (−)-stemoamide. The starting alkyne oxazoles **200** and **203** were prepared from γ-chlorobutyryl chloride **199** in five steps (Fig. 3.60).

The activated ester alkyne-oxazole **200** underwent facile intramolecular Diels-Alder reaction in refluxing toluene with concomitant loss of acetonitrile to give an 89% yield of the desired tricyclic methyl ester **201**. However, **201** could not be converted to the methyl butenolide **202** required to complete the synthesis.

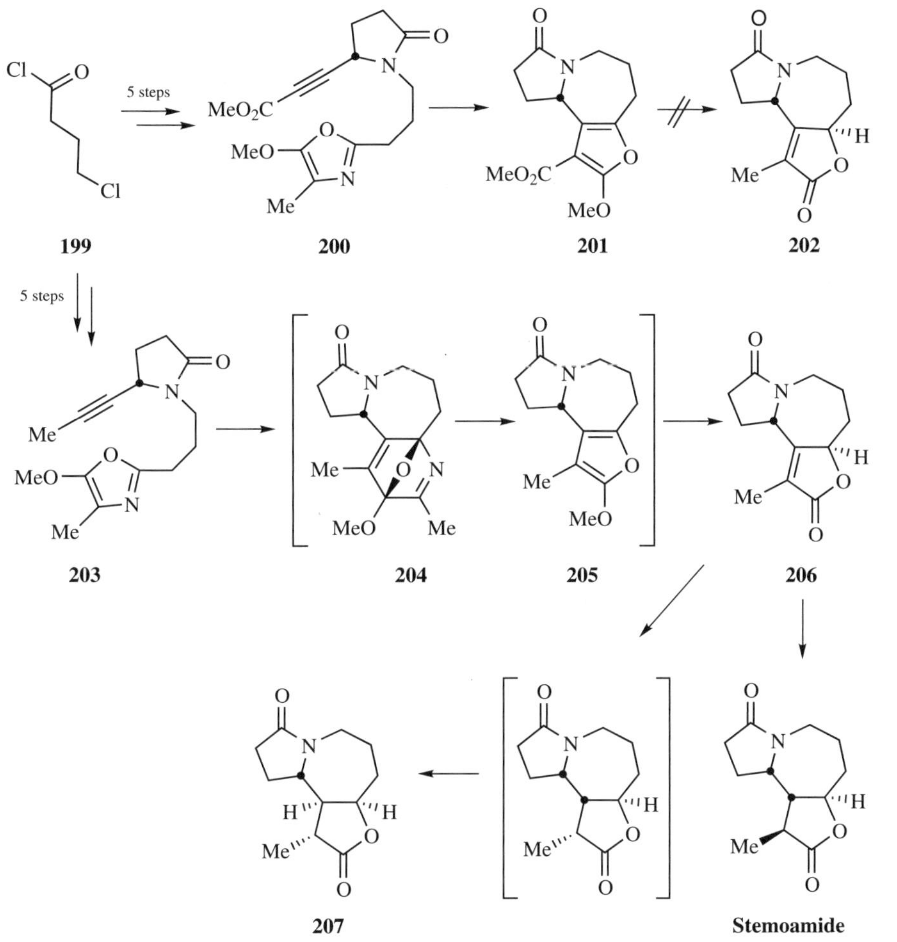

Figure 3.60

The unactivated alkyne-oxazole **203** was considerably less reactive in the intramolecular Diels-Alder retro-Diels-Alder sequence. In ethyl benzene at 135°C, only trace amounts of the tricyclic furan **205**, formed via the Diels-Alder adduct **204**, were produced. Attempts to catalyze the reaction with Lewis acids were unsuccessful.

Fortunately, when **203** was refluxed in diethyl benzene at 182°C, a 50–55% yield of butenolide **206** (the proposed precursor to stemoamide) was obtained directly.

When this reaction was followed by GCMS, the tricyclic furan **205** was the major product observed; the facile hydrolysis to butenolide **206** occurred during workup. The lability of **205** to hydrolysis is in marked contrast to the inert nature of the methyl ester **201** to acid-catalyzed conversion to the corresponding butenolide.

TABLE 3.3. INTRAMOLECULAR OXAZOLE DIELS-ALDER REACTIONS

Oxazole-Dienophile	Product	Yield (%)	Reference
100a	**101a**	46	66
110	**111**	71	68
112a	**113**	53	69
197	**198a**	90	118
200	**201**	89	121

The synthesis was completed by reduction of **206** with the nickel boride catalyst derived from nickel(II) chloride and sodium borohydride directly producing (±)-stemoamide in 73% yield. In addition, a 15% yield of the *cis*-lactone **207**, derived from α-face reduction of **206** followed by epimerization at C-10, was also obtained.

An enantioselective synthesis of natural (−)-stemoamide was carried out precisely as described for the synthesis of racemic stemoamide. In this case, enantiomerically pure **203** was prepared from L-pyroglutamic acid in eight steps.

An electron-transfer mechanism involving radical cation **208** was proposed to account for the presence of several by-products obtained in low yield from the cycloaddition of **203** (Fig. 3.61). Electrochemical studies indicated that formation of **208** from **203** is possible in the presence of mild oxidants. Radical cation **208** could potentially undergo an inverse electron demand Diels-Alder reaction, giving radical cation **209**. Pathway A, the major pathway, generates **205** via a single electron reduction of **209**, which subsequently leads to the major product **206** upon workup. Pathway B leads to **210** and **211** through methyl radical abstraction by C-8 and C-10, respectively, followed by a single electron reduction. Deprotonation of **209** at C-9α leads to the radical **212**, which in turn may be oxidized and then hydrolyzed to generate the oxidation products **213** and **214**, as shown in pathway C.

The precise mechanism of this transformation was not deduced. An electron-transfer mechanism may represent the predominant pathway or it may operate in conjunction with a thermal Diels-Alder mechanism. It is important to note that cyclization of **203** to **205**, when carried out on large scale, is facilitated by benzoquinone and other electron acceptors.

3.3.2.8. *Thiofurans and Amidofurans*

Selnick and Brookes[122] prepared a series of thiopyrans as carbonic anhydrase II inhibitors via intramolecular oxazole-alkyne cycloadditions. The starting 4-methyl-2-thiooxazole-alkyne **215** was readily constructed from 2-mercapto-4-methyloxazole (Fig. 3.62). Refluxing **215** in toluene afforded an 88% yield of the furanothiopyran **216**. Varying the length of the oxazole-acetylene linker provided furans fused to 5- and 7-membered rings as well.

Padwa and co-workers[70,123] reported an approach to the pyrrolophenanthridone alkaloids (e.g., hippadine) that uses an intramolecular oxazole-alkyne Diels-Alder reaction. Thus the 2-acetamidooxazole model system **217** was thermolyzed at 200°C to produce the tricyclic furan **218** in 93% yield (Fig. 3.63).

In contrast, palladium(0)-catalyzed coupling of the requisite starting oxazole for the synthesis of hippadine **219** with trimethylsilylacetylene at 80°C did not afford the expected oxazole-alkyne **220** (Fig. 3.64). Instead, they isolated the tricyclic furan **221** derived from a Diels-Alder retro-Diels-Alder reaction in 77% yield. Thermolysis of **221** at 320°C effected an intramolecular Diels-Alder reaction with concomitant desilylation. Subsequent DDQ oxidation of this product (not shown) then provided hippadine.

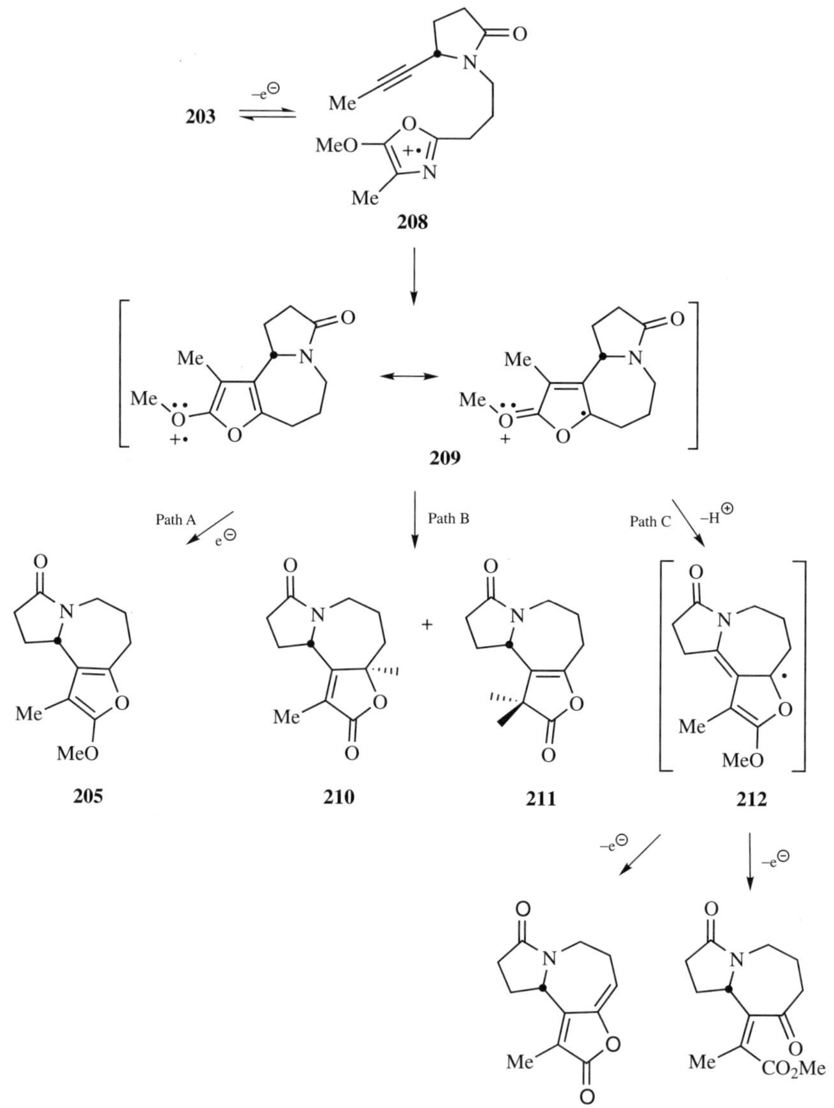

454 Oxazole Diels-Alder Reactions

Figure 3.61

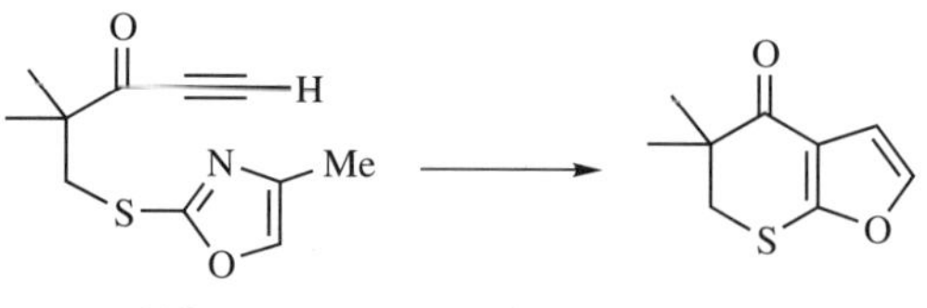

Figure 3.62

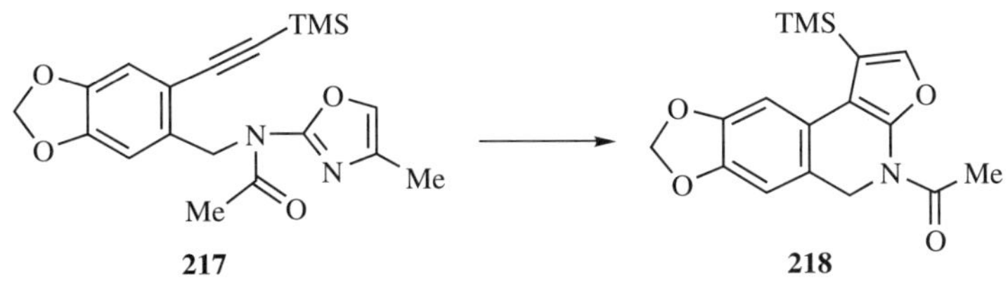

Figure 3.63

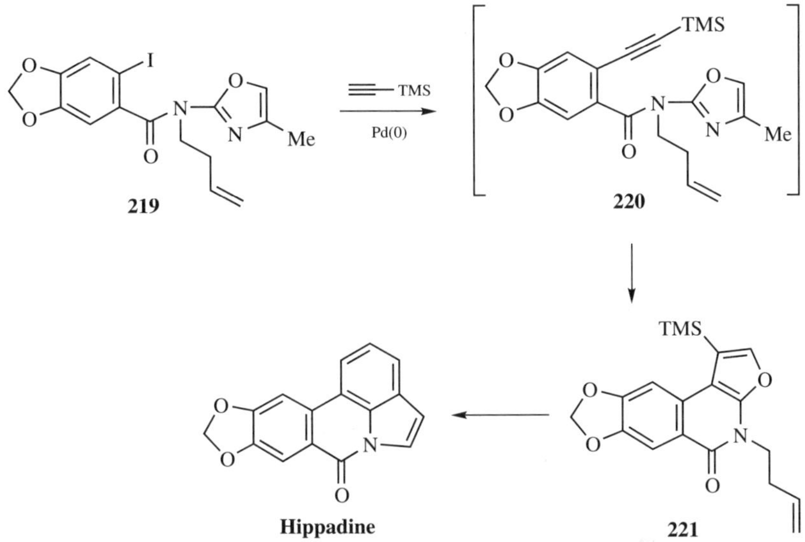

Figure 3.64

For intramolecular oxazole Diels-Alder reactions see Table 3.3, p. 452.

3.4. OXAZOLE-HETERODIENOPHILE REACTIONS

Reactions of oxazoles with heterodienophiles have been reported starting with work by Grigg and co-workers in 1969.[71] The mechanism of these reactions has not been determined. It has been suggested that, for at least some cases, they proceed via an initial Diels-Alder cycloaddition, although an isolable 1:1 oxazole-heterodienophile Diels-Alder adduct has never been obtained.[124,125] Thus the initial hetero-Diels-Alder adduct **222** can fragment to generate **223** (Fig. 3.65). The zwitterionic intermediate **223** then ring opens to produce **224**, which subsequently cyclizes to provide **225**. It is not possible to distinguish between reactions that proceed through the Diels-Alder adduct **222** and those in which **223** is formed directly from the oxazole and heterodienophile. Stepwise mechanisms proceeding through diradical or dipolar intermediates have also been proposed.[21,126] A computational study of oxazole-heterodienophile Diels-Alder reactions has been reported.[127]

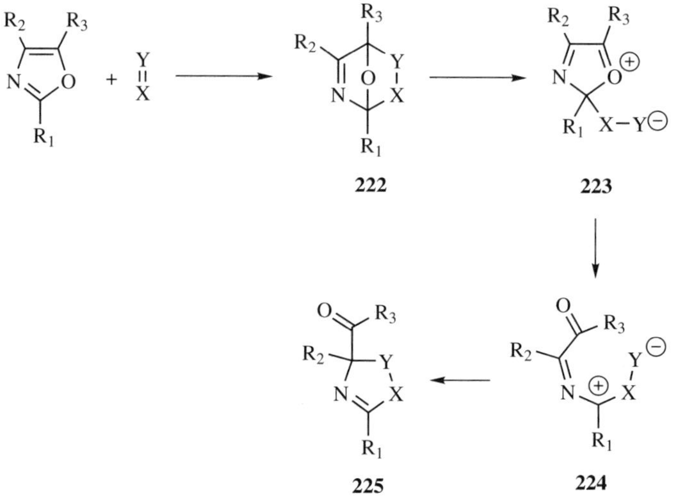

Figure 3.65

3.4.1. Intermolecular Reactions

3.4.1.1. Nitrogen-Containing Heterodienophiles

5-Ethoxyoxazoles **1** (R_3 = OEt) react readily with diethyl azodicarboxylate, frequently at room temperature, to afford 1,2,4-triazol-3-ine products **226** (Fig. 3.66).[22–24,124,128] Early studies misidentified the reaction products as the 1:1 Diels-Alder adducts **227**.[22,71] The corresponding 2-ethoxyoxazoles (**1**, R_1 = OEt) are less reactive than the 5-alkoxy analogs but still produced triazolines **226**.[124] Several examples of trisubstituted oxazoles lacking an alkoxy moiety also provide

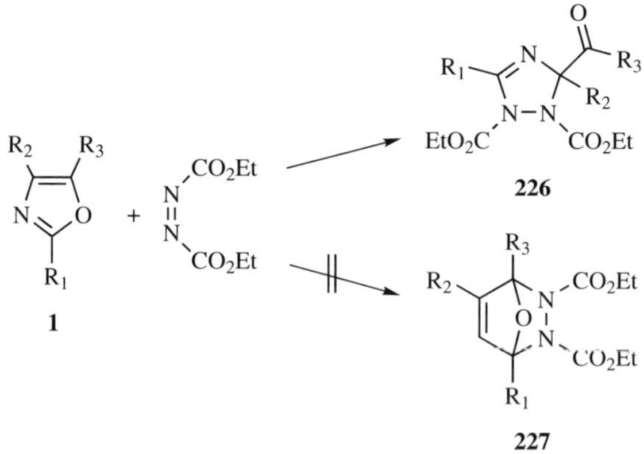

Figure 3.66

triazolines upon reaction with diethyl azodicarboxylate, although these reactions required increased temperature or pressure.[128]

Similarly, oxazoles and 4-phenyl-1,2,4-triazoline-3,5-dione (PTAD) **229** reacted to give the bicyclic triazolines **230** (Fig. 3.67).[22,24,124,128] An X-ray crystal structure has been obtained for one example of the triazoline series **230** (R_1 = p-tolyl, R_2 = Me, R_3 = OMe).[128] Ibata and co-workers[128] proposed direct formation of **223** from reaction of an oxazole with PTAD or DEAD without the prior formation of a Diels-Alder adduct. A reaction mechanism involving dipolar cycloaddition of the nitrile ylide **228** with PTAD has been suggested to explain the increase in chemical yield with more electron-withdrawing R_2 groups.[124]

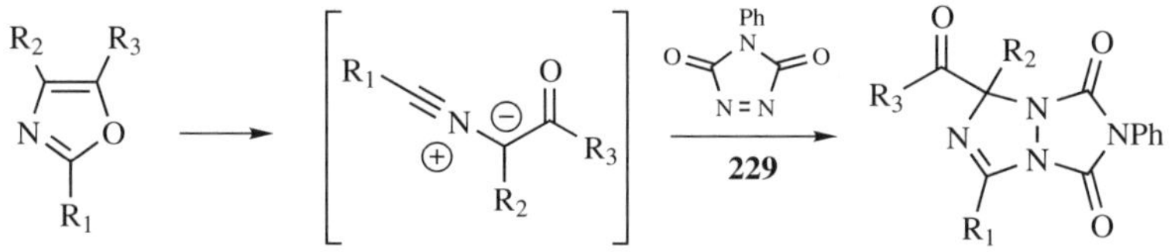

Figure 3.67

There is a single example of a cycloaddition involving an oxazole and a hydantoin. Thus 5-ethoxy-2-phenyloxazole **231** and hydantoin **232** reacted in refluxing xylene for 22 h to provide a 3:2 mixture of cis- and trans-**233** in 50% overall yield (Fig. 3.68).[124] Only one regioisomer was obtained. The reaction could be catalyzed by boron trifluoride etherate to give **233** in 90% yield after only 4 h at 110°C.

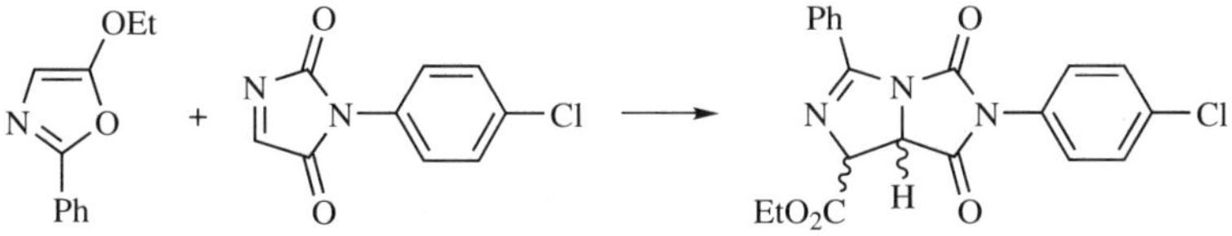

Figure 3.68

Ibata and co-workers[129,130] also investigated the room temperature reaction of nitrosobenzenes **234** with oxazoles in acetonitrile to prepare 2,5-dihydro-1,2,4-oxadiazoles **235** (Fig. 3.69). As in the case of **232**, these authors obtained a single regioisomer independent of the nature of the substituents on the oxazole or nitrosobenzene. A simple trialkyl oxazole (**1**, $R_1 = R_2 = R_3 =$ Me) as well as 5-alkoxyoxazoles afforded oxadiazoles in 29–100% yield.

3.4.1.2. Carbonyl Heterodienophiles

Reaction of 5-ethoxy-2-phenyloxazole **231** with diethyl ketomalonate was reported by Hassner in 1989 (Fig. 3.70).[124] At 80°C, the 2,4,5-trisubstituted oxazole

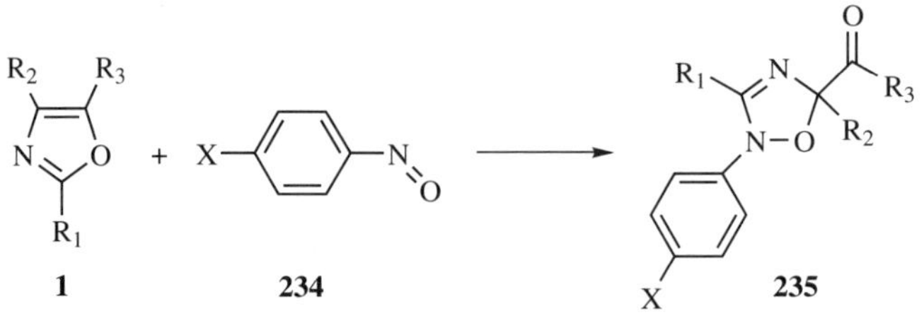

458   Oxazole Diels-Alder Reactions

Figure 3.69

236 was obtained in 24% yield together with recovered starting material. However, at 140°C, only oxazolines **237** and **238** were obtained in a 1.2:1 ratio in 96% overall yield. Ibata[131] determined that the 2-oxazoline **238** was the only product if the reaction of **231**, or other 2-aryloxazoles, and diethyl ketomalonate was catalyzed by tin tetrachloride at room temperature in acetonitrile. Changes in pressure or the Lewis acid alter the product distribution to favor a 3-oxazoline **237**, particularly for 2-alkyloxazoles. It has been proposed that these reactions proceed through initial nucleophilic attack by C-4 of the oxazole at the carbonyl carbon of diethyl ketomalonate.

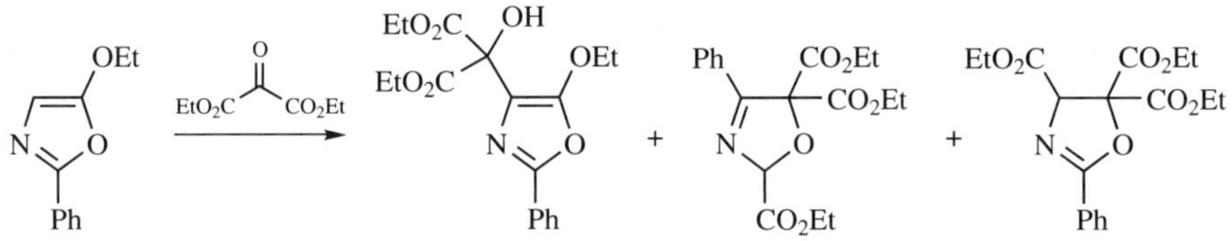

Figure 3.70

Lewis acid catalysis was required for the reaction of 2-ethoxy-5-phenyloxazole **239** and diethyl ketomalonate, which proceeded at 80°C in 5.5 h to give oxazoline **240** as the sole product in 43% yield (Fig. 3.71).[124]

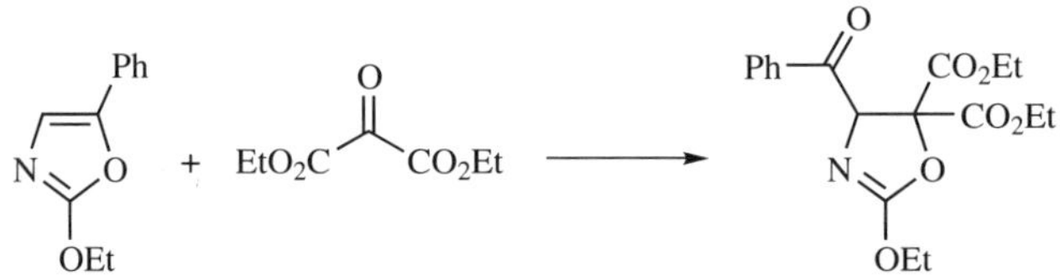

Figure 3.71

The reaction of 5-alkoxyoxazoles with carbonyl heterodienophiles were extended to simple aldehydes in a series of papers by Ibata and co-workers.[132–135] For example, 5-methoxy-2-(4-methoxyphenyl)oxazole **241** and benzaldehyde reacted in the presence of one equivalent of Lewis acid **242** (0°C, 93 h) to provide a 76% yield of the *cis*-2-oxazoline **243** and only 1.7% of the analogous *trans*-2-oxazoline

(Fig. 3.72).[132,133] Furthermore, if 30 mol % of chiral (R)-**242** is used as the catalyst for this transformation, **243** is obtained in 75% yield with an 88% enantiomeric excess.[135] Substitution of electron-donating and electron-withdrawing groups on the benzaldehyde did not affect enantioselectivity. The nonaromatic aldehydes cinnamaldehyde and propanal also reacted with **241** to give the corresponding cis-2-oxazolines with good selectivity, but in poor yield.[132,133]

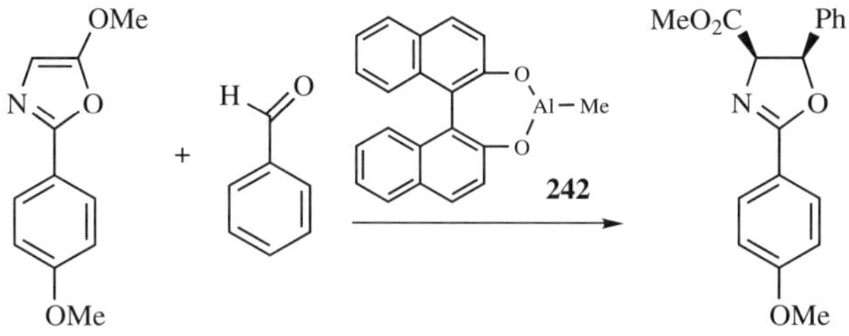

Figure 3.72

Tin(IV) chloride effectively catalyzed the reaction of α-benzyloxy aldehydes with 5-alkoxyoxazoles to produce cis-5-substituted 2-aryl-2-oxazoline-4-carboxylates.[134] Thus in the presence of one equivalent of tin(IV) chloride at −20°C for 2 h, 2-(S)-(benzyloxy)propanal **244** and **241** provided a 94:3:2:1 mixture of the oxazolines **245**, **246a**, **247**, and **248a**, respectively, in approximately 85% overall yield (Fig. 3.73). Interestingly, the analogous t-butyldimethylsilyl-protected α-hydroxy aldehyde gave only the two trans substituted-diastereomeric 2-oxazolines, **246b** and **248b**, in a 92:8 ratio and 53% yield. Presumably, these arise via epimerization of the cis-oxazolines to the thermodynamically more stable products under the reaction conditions (0°C, 30 h).

3.4.1.3. Thioaldehyde Heterodienophiles

In 1988, Vedejs and Fields[125] prepared 3-thiazolines from 5-methoxy-2-methyloxazole **249** and a series of thioaldehydes, generated in situ from their cyclopentadiene adducts. Thus heating **249** and thioaldehyde **250** in a sealed tube at 140°C for 48 h afforded a 95% yield of **251** as a 1:1 mixture of diastereomers (Fig. 3.74). The unactivated oxazole, 5-methyl-2-phenyloxazole was unreactive under the same conditions. A Diels-Alder mechanism was proposed for this reaction but no intermediate Diels-Alder adduct was observed. A nitrile ylide pathway was discounted, since these reactions proceed at room temperature if the thioaldehyde is generated photolytically and oxazoles do not form nitrile ylides under those conditions. Thioformaldehyde, thioacetaldehyde, thiobenzaldehyde, and thioacetone all successfully underwent similar reactions to provide thiazolines.

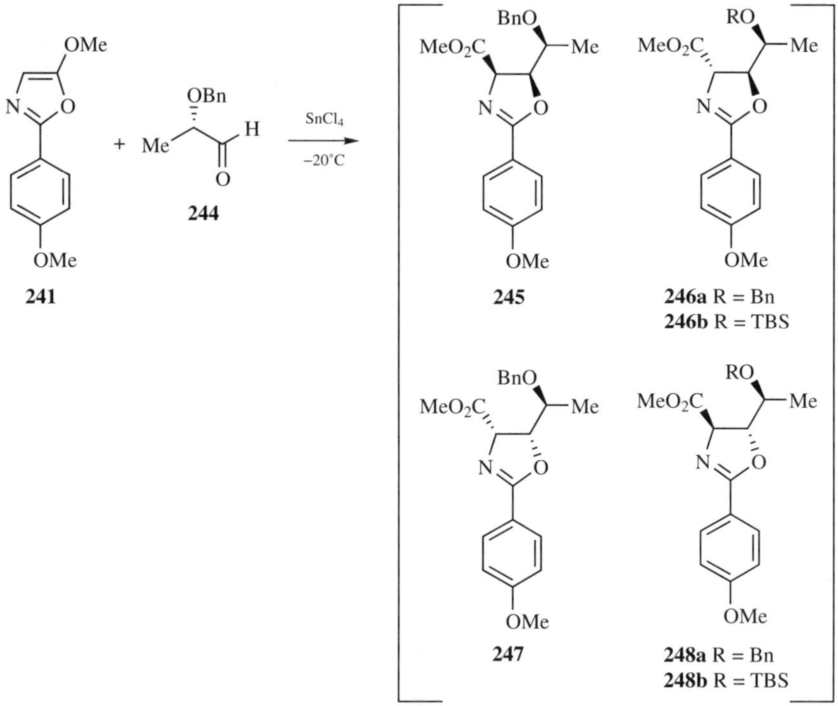

Figure 3.73

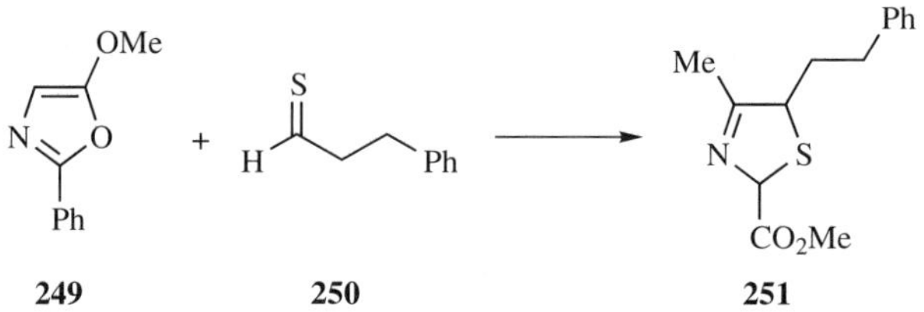

Figure 3.74

3.4.1.4. Singlet Oxygen

Wasserman and Floyd[136] first demonstrated the dye-sensitized photooxidation of oxazoles to prepare triamides **255** in 1966 (Fig. 3.75). This reaction has since become a valuable synthetic tool.[137] Several reports have now appeared that establish the mechanism of this transformation to proceed through an initial [4 + 2]-cycloaddition to provide an unstable 2,5-endoperoxide, **252**. Oxygen-18 studies were consistent with the formation of **252** and excluded the formation of an

TABLE 3.4. INTERMOLECULAR OXAZOLE-HETERODIENOPHILE DIELS-ALDER REACTIONS

Oxazole	Dienophile	Product	Yield (%)	Reference
231		226	99	124
231	229	230	100	124
231	232	233	50	124
1	234	235	50	129
241		243	76	135
249	250	251	95	125

initial [2 + 2]-cycloaddition, resulting in a dioxetane intermediate.[138] However, dioxetanes have been observed in the photooxygenation of 2-alkoxyoxazoles.[139]

More recently, the use of low-temperature NMR has provided convincing evidence for the 2,5-endoperoxide as well as each of the subsequent intermediates

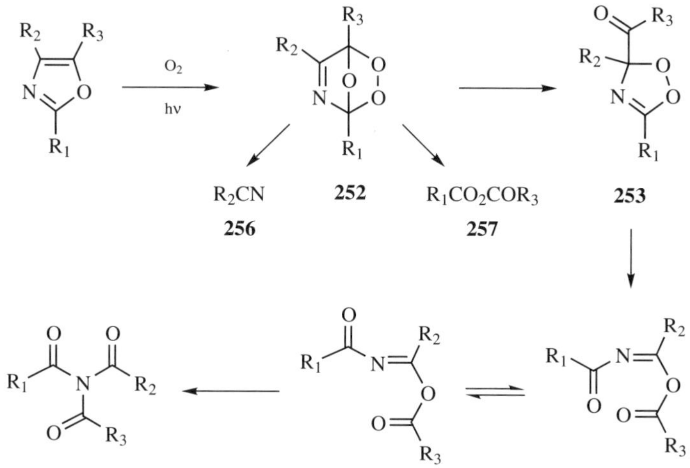

Figure 3.75

leading to triamides.[140–142] Thus an endoperoxide **252** fragments to the dioxazole **253**, possibly via a concerted rearrangement. A formal Baeyer-Villiger rearrangement converts **253** to the imino anhydrides **254a** and **254b**. Subsequently, **254a** and **254b** undergo a 1,3-acyl transfer to afford the triamide **255**. Endoperoxides **252** in which R_1 is hydrogen decompose to form nitriles **256** and anhydrides **257**.[141]

3.4.2. Intramolecular Oxazole-Heterodienophile Reactions

Hassner and Fischer[143] also studied the intramolecular bis-heteroannulation of oxazoles tethered to a variety of heterodienophiles. The azo-compounds **258** ($n = 3$–4), prepared in situ via (diacetoxyiodo)benzene oxidation of the corresponding hydrazide in refluxing benzene, gave a 40% yield of **259** ($n = 1$–2) (Fig. 3.76).

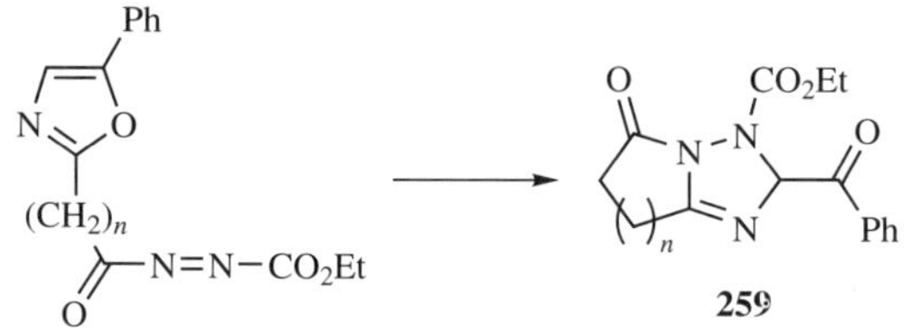

Figure 3.76

In a similar fashion the imine, aldehyde, and thioaldehyde precursors **260a–c** afforded the bicyclic adducts **261a–c** (Fig. 3.77). Imine **260a** ($n = 3$) gave a 43%

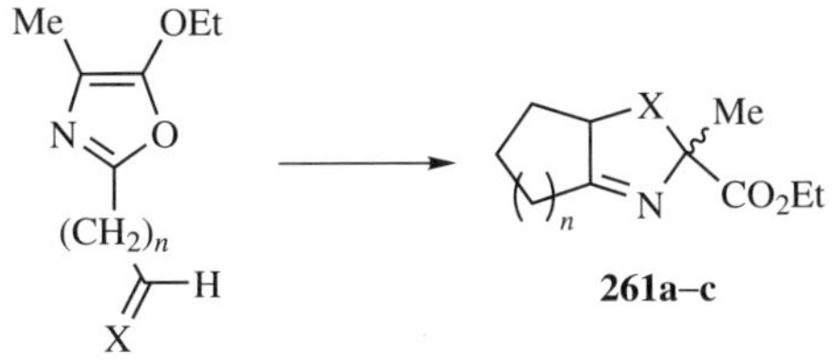

260a X = NCO$_2$Et
260b X = O
260c X = S

Figure 3.77

yield of the imidazoline **261a** ($n = 2$) after refluxing in benzene. The aldehyde **260b** ($n = 4$) reacted at 110°C for 5 h, in the absence of a Lewis acid catalyst, to give a 78% yield of **261b** ($n = 3$) as a 2:1 mixture of diastereomers. The analogous aldehyde with a 3-carbon tether, **260b** ($n = 3$) is unreactive. Finally, thioaldehydes **260c** ($n = 3$–4) reacted at room temperature to give thiazolines **261c** ($n = 1$–2) in 42 and 51% yield, respectively, as 1:1 mixtures of diastereomers.

Intermolecular oxazole-heterodienophile reactions are summarized in Table 3.4, p. 461.

3.5. SUMMARY

Diels-Alder reactions of oxazoles with alkenes, alkynes, and heterodienophiles have shown great utility in several areas of organic synthesis. The intermolecular [4 + 2] cycloadditions have been a valuable tool for the synthesis of highly substituted pyridines and furans, which are accessible only with difficulty through other routes. These compounds have found uses in the synthesis of more complex molecules, pharmaceuticals, and natural products. The intramolecular variant of this methodology has been particularly important in the synthesis of a wide variety of complex natural products containing pyridine, furan, and butenolide ring systems. Finally, although debate continues as to whether the reactions of oxazoles and heterodienophiles proceed via a Diels-Alder mechanism, these reactions provide an efficient entry into several interesting and useful heterocyclic ring systems.

3.6. ADDENDUM

Several additional examples of oxazole Diels-Alder reactions applied to the synthesis of natural products have appeared. Using the same synthetic strategy that resulted in the total synthesis of normalindine,[69,144] Ohba and co-workers[145] adapted the oxazole-olefin Diels-Alder cycloaddition for the synthesis of the monoterpene alkaloids plectrodorine and oxerine. The synthesis of (−)-plectrodorine began with the requisite Diels-Alder precursor **262a**, prepared from (S)-(−)-malic acid (Fig. 3.78). The intramolecular Diels-Alder reaction of this 2-alkyloxazole with a doubly activated olefin surprisingly required forcing conditions, o-dichlorobenzene at 150°C for 48 h, to provide both recovered starting material and a 37% yield of the cyclopenta[*c*]pyridine **263a**. Reduction of the ketone with sodium borohydride occurs predominantly from the side opposite to the bulky *t*-butyldimethylsilyl ether to give a 75% yield of alcohol **264a**. Deprotection of the *t*-butyldimethylsilyl ether in **264a** with tetrabutylammonium fluoride affords (–)-plectrodorine.

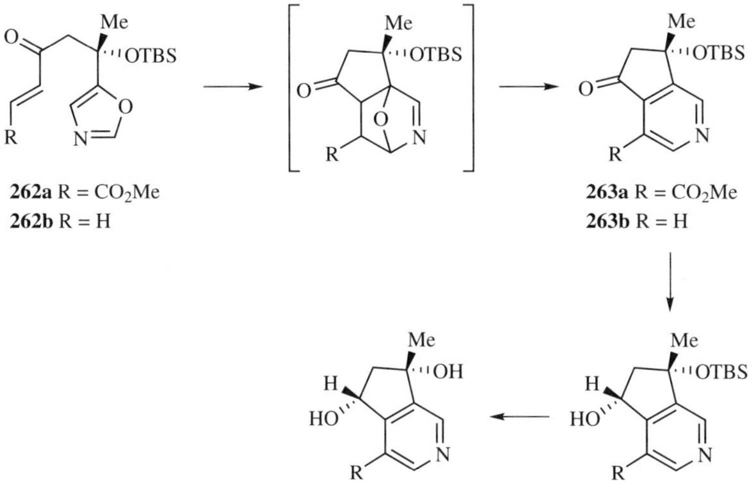

262a R = CO_2Me
262b R = H

263a R = CO_2Me
263b R = H

(-)-Plectrodorine R = CO_2Me
(+)-Oxerine R = H

264a R = CO_2Me
264b R = H

Figure 3.78

Oxerine was prepared similarly from oxazole-olefin **262b**. After heating at 150°C for 9 h, the starting olefin was completely consumed, and a 23% yield of the desired ketone **263b** was obtained. Borohydride reduction and subsequent desilylation of **264b** then produced (+)-oxerine.

The low yields of the key oxazole-olefin cycloaddition steps in the above syntheses of plectrodorine and oxerine prompted Ohba to investigate the use of Lewis acid catalysis in the formation of the parent cyclopenta[*c*]pyridine ring system.[146] The uncatalyzed reaction of oxazole **265a** in *o*-dichlorobenzene at 150°C for 24 h furnishes a 21% yield of cyclopenta[*c*]pyridine **266a** (Fig. 3.79). Addition of 10 mol % of copper triflate reduces the reaction time to 3 h and increases the yield to 48%. At 180°C, a 55% yield of **266a** is obtained after only 1 h using 2 mol % of copper triflate. Scandium and ytterbium triflates were slightly less effective. An even more striking improvement is seen upon addition of copper triflate in the case of the unsaturated ester **265b**. Here, addition of 10 mol % of copper triflate to **265b** in refluxing toluene reduced the reaction time from > 24 h to 3 h and increased the yield from 3% to 93%. Enhanced reaction rates and yields are also seen with **265c** and **265d**. In these compounds, which lack the geminal dimethyl group on the tether linking the oxazole and olefin, the effects are much less pronounced.

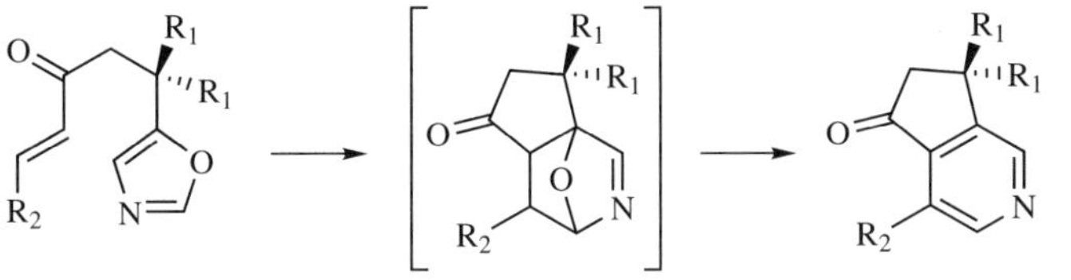

265a R_1 = Me, R_2 = H
265b R_1 = Me, R_2 = CO_2Me
265c R_1 = H, R_2 = H
265d R_1 = H, R_2 = CO_2Me

266a R_1 = Me, R_2 = H
266b R_1 = Me, R_2 = CO_2Me
266c R_1 = H, R_2 = H
266d R_1 = H, R_2 = CO_2Me

Figure 3.79

The total synthesis of (−)-colchicine was accomplished via an intermediate furan derived from an intramolecular oxazole-acetylene cycloaddition.[147,148] Thus refluxing the regioisomeric oxazole-acetylenes **267a** and **267b** in *o*-dichlorobenzene produced the furan **268** in 60–70% yield (Fig. 3.80). The 5-substituted oxazole isomer **267b** underwent cycloaddition substantially faster than 2-substituted oxazole isomer **267a** (40 h vs. 66 h) although the yields were comparable. Conversion of **268** to (−)-colchicine was accomplished in six steps, which included a novel [4 + 3]-oxyallyl cycloaddition reaction using an α-heteroatom-substituted oxyallyl system to construct the 7-membered ring.

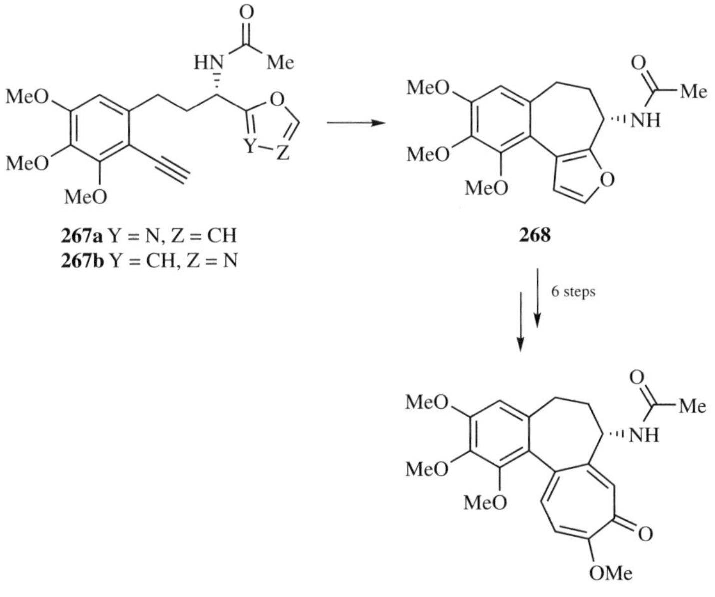

267a Y = N, Z = CH
267b Y = CH, Z = N

268

6 steps

(-)-**Colchicine**

Figure 3.80

466   Oxazole Diels-Alder Reactions

Imerubrine, a tropolone natural product related to colchicine was also synthesized via an oxazole-acetylene Diels-Alder reaction followed by a [4 + 3]-oxyallyl cycloaddition.[149] Here, 8-iodo-5,6,7-trimethoxyisoquinoline **269** was converted to 5-substituted oxazole **270** in four steps and 42% overall yield (Fig. 3.81). Thermolysis of **270** in refluxing *o*-dichlorobenzene effected the desired intramolecular Diels-Alder cycloaddition with concomitant loss of the Boc-protecting group to afford the tetracyclic furan **271** in 90% yield. At this point, **271** was subjected to the [4 + 3] cycloaddition in the presence of 1,3,3-trichloro-2-propanone and 2,2,2-trifluoroethanol. Subsequent dechlorination of the intermediate (not shown) with zinc provided the oxabicyclic **272** as a single regioisomer in 73% yield. The synthesis of imerubrine was completed in three steps from **272**.

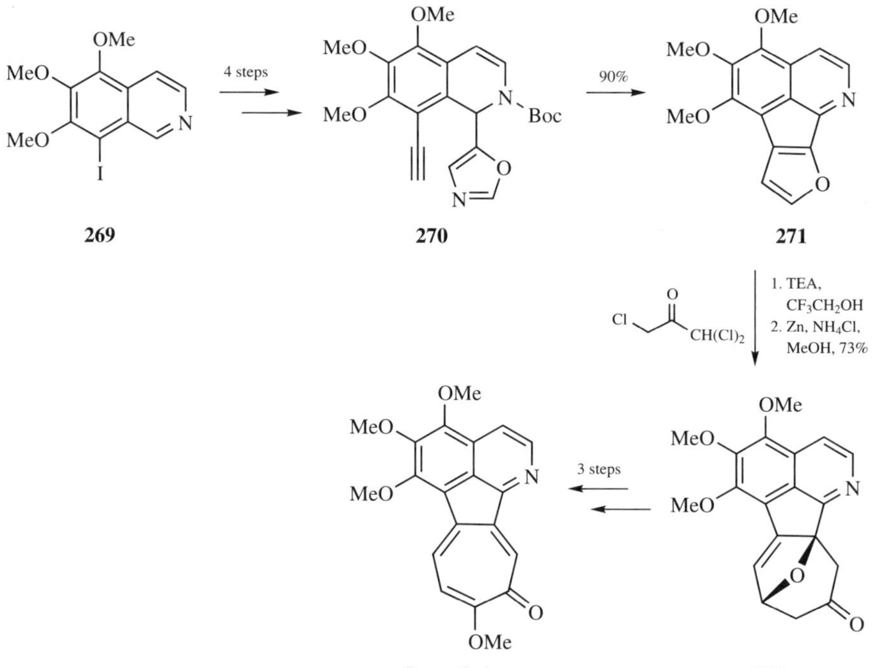

Figure 3.81

A recent and succinct approach to 2-aryltetrahydrofuran precursors to lignan natural products involves the cycloaddition of 2-aryl-4-phenyloxazoles with dimethylacetylene dicarboxylate.[150] For example, 2,4-diphenyloxazole **273** (R = H) reacts with dimethylacetylene dicarboxylate in refluxing xylenes in the presence of sodium carbonate and hydroquinone to provide 2-phenyl-3,4-furandicarboxylic acid dimethyl ester **274** in 98% yield (Fig. 3.82). Not surprisingly, 4,5-diphenyloxazole and 2,4,5-triphenyloxazole gave lower yields of the corresponding furans

(65 and 23%, respectively). Selective hydrogenation of the furan ring in **274** then gave a 91% yield of the 2,3,4-trisubstituted tetrahydrofuran **275** as a single diastereomer. Hydrogenation of analogs of **274** in which the phenyl ring bears alkoxy substituents afford 2:1 mixtures of 2,3-*cis*-3,4-*cis* tetrahydrofurans and the corresponding 2,3-*cis*-3,4-*trans* isomers.

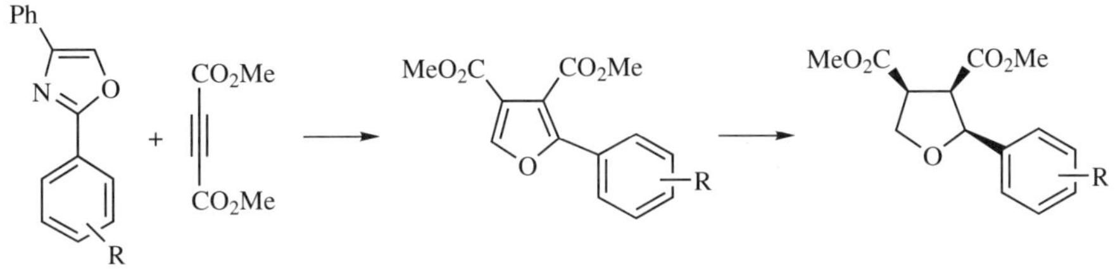

273 **274** **275**

Figure 3.82

The first examples of what appear to be Diels-Alder reactions of oxazolium salts with olefins have been reported.[151] Methylation of the 5-substituted oxazole **276a** with methyl triflate at room temperature for 36 h produced the cis-fused iminium salt **278** in 60% yield (Fig. 3.83). The structure of **278** was confirmed by an X-ray crystal structure. Presumably, the initially formed oxazolium salt undergoes an intramolecular Diels-Alder cycloaddition to give iminium salt **277a**, as evidenced by proton NMR. Rearrangement via cleavage of the ether bridge of **277a** then provides **278**. Similarly, the addition of one equivalent of methyl triflate converts oxazole **276b** into an iminium salt, likely **277b**, after 6 days at room temperature or 1 h at 74°C. Reduction of the iminium salt with sodium borohydride affords the trans-fused decahydroisoquinoline **279** as a crystalline solid in 30% yield. This promising methodology was not generally applicable, however. Shorter tethers, unsaturated esters, and alkynes gave only the *N*-methylated oxazolium salts, even after heating at 90°C for 3 h.

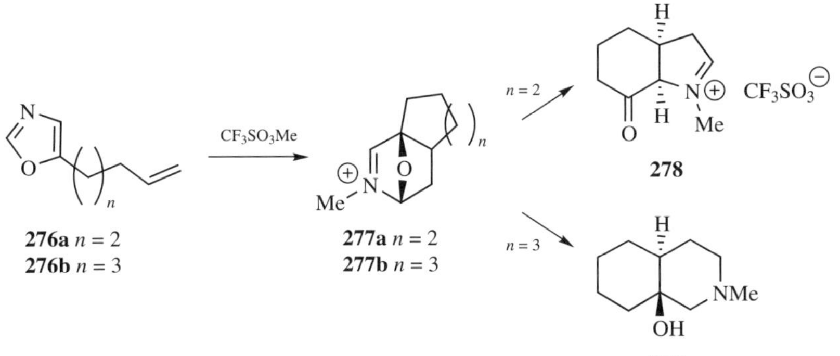

276a $n = 2$
276b $n = 3$

277a $n = 2$
277b $n = 3$

278

279

Figure 3.83

Acknowledgments

J. I. L. would like to dedicate this chapter to the memory of Marvin H. Levin. The authors would like to thank Dr. Jerauld Skotnicki and Dr. Tarek Mansour for proofreading this manuscript.

REFERENCES

1. Boger, D. L. *Tetrahedron* **1983**, *39*, 2869–2939.
2. Boger, D. L. *Chem Rev.* **1986**, *86*, 781–794.
3. Boger, D. L.; Weinreb, S. M. *Hetero Diels-Alder Methodology in Organic Synthesis*, Academic Press: San Diego, 1987.
4. Lipshutz, B. H. *Chem. Rev.* **1986**, *86*, 795–820.
5. Karpeiskii, M. Y.; Florent'ev, V. L. *Usp. Khim.* **1969**, *38*, 1244–1256 [*Chem. Abstr.* **1969**, *71*, 9134].
6. Lakhan, R.; Ternai, B. *Adv. Heterocycl. Chem.* **1974**, *17*, 99–211.
7. Turchi, I. J.; Dewar, M. J. S. *Chem. Rev.* **1975**, *75*, 389–437.
8. Turchi, I. J. *Ind. Eng. Chem. Prod. Res. Dev.* **1981**, *20*, 32–76.
9. Boyd, G. V. In Potts, K. T., ed. *Comprehensive Heterocyclic Chemistry*, Vol. 6, Pergamon: Oxford, UK, 1984, pp. 195–197.
10. Boyd, G. V. *Prog. Heterocycl. Chem.* **1992**, *4*, 150–167.
11. Hassner, A.; Fischer, B. *Heterocycles* **1993**, *35*, 1441–1465.
12. Turchi, I. J. In Turchi, I. J., ed. *Oxazoles, The Chemistry of Heterocyclic Compounds*, Vol. 45, Wiley: New York, 1986, pp. 1–341.
13. Kondrat'eva, G. Y. *Khim. Nauka Prom.* **1957**, *2*, 666–667 [*Chem. Abstr.* **1958**, *52*, 35255].
14. Kondrat'eva, G. Y. *Izvest. Akad. Nauk SSSR Otdel. Khim. Nauk* **1959**, 484–490 [*Chem. Abstr.* **1959**, *53*, 122148].
15. Kondrat'eva, G. Y.; Huang, C.-H. *Dokl. Akad. Nauk SSSR* **1962**, *142*, 593–595 [*Chem. Abstr.* **1962**, *57*, 10824].
16. Naito, T.; Yoshikawa, T.; Ishikawa, F.; Isoda, S.; Omura, Y.; Takamura, I. *Chem. Pharm. Bull.* **1965**, *13*, 869–872.
17. Yoshikawa, T.; Ishikawa, F.; Omura, Y.; Naito, T. *Chem. Pharm. Bull.* **1965**, *13*, 873–878.
18. Yoshikawa, T.; Ishikawa, F.; Naito, T. *Chem. Pharm. Bull* **1965**, *13*, 878–881.
19. Naito, T.; Yoshikawa, T. *Chem. Pharm. Bull.* **1966**, *14*, 918–921.
20. Kimel, W.; Leimgruber, W. U. S. Pat. 3,250,778 [*Chem. Abstr.* **1965**, *63*, 24058].
21. Kondrat'eva, G. Y.; Aitzhanova, M. A.; Bogdanov, V. S.; Chizhov, O. S. *Izv. Akad. Nauk SSSR Ser. Khim.* **1979**, 1313–1322 [*Chem. Abstr.* **1979**, *91*, 140753].
22. Ibata, T.; Nakano, S.; Nakawa, H.; Toyoda, J.; Isogami, Y. *Bull. Chem. Soc. Jpn.* **1986**, *59*, 433–437.
23. Ibata, T.; Nakawa, H.; Isogami, Y.; Matsumoto, K. *Bull. Chem. Soc. Jpn.* **1986**, *59*, 3197–3200.
24. Ibata, T.; Isogami, Y.; Tamura, H. *Chem. Lett.* **1988**, 1551–1554.
25. Naito, T.; Ueno, K.; Sano, M.; Omura, Y.; Itoh, I.; Ishikawa, F. *Tetrahedron Lett.* **1968**, 5767–5770.
26. Matsuo, T.; Miki, T. *Chem. Pharm. Bull.* **1972**, *20*, 669–676.
27. Kondrat'eva, G. Y.; Bogdanov, V. S.; Ostval'd, G. V.; Zhulin, V. M. *Izv. Akad. Nauk SSSR Ser. Khim.* **1987**, 2535–2540 [*Chem. Abstr.* **1988**, *109*, 92677].
28. Gonzalez, J.; Taylor, E. C.; Houk, K. N. *J. Org. Chem.* **1992**, *57*, 3753–3755.
29. Jursic, B. S. *J. Chem. Soc. Perkin Trans.* **1996**, *2*, 1021–1026.

30. Jursic, B. S.; Zdravkovski, Z. *Glas. Hem. Tehnol. Maked.* **1994**, *13*, 55–59 [*Chem. Abstr.* **1995**, *122*, 238986].
31. Firestone, R. A.; Harris, E. E.; Reuter, W. *Tetrahedron* **1967**, *23*, 943–955.
32. Florent'ev, V. L.; Drobinskaya, N. A.; Ionova, L. V.; Karpieskii, M. Y. *Tetrahedron Lett.* **1967**, 1747–1751.
33. Murakami, M.; Iwanami, M. *Bull. Chem. Soc. Jap.* **1968**, *41*, 726–727.
34. Kondrat'eva, G. Y.; Huang, C.-H. *Dokl. Akad. Nauk SSSR* **1961**, *141*, 628–631 [*Chem. Abstr.* **1962**, *56*, 73409].
35. Kondrat'eva, G. Y.; Huang, C.-H. *Dokl. Akad. Nauk SSSR* **1965**, *164*, 816–819 [*Chem. Abstr.* **1966**, *64*, 11467].
36. Maruyama, T.; Yasumatu, M.; Araki, E.; Kurizono K. U. S. Pat. 3,257,408 [*Chem. Abstr.* **1965**, *63*, 98210].
37. Pfister, K. III; Harris, E.; Firestone, R. A. U. S. Pat. 3,227,721 [*Chem. Abstr.* **1966**, *64*, 51946].
38. Matsuo, T.; Miki, T. *Chem. Pharm. Bull.* **1972**, *20*, 806–814.
39. Kondrat'eva, G. Y.; Medvedskaya, L. B.; Ivanova, Z. N. *Izv. Akad. Nauk SSSR Ser. Khim.* **1972**, 2125–2126 [*Chem. Abstr.* **1972**, *77*, 164570].
40. Medvedskaya, L. B.; Kondrat'eva, G. Ya. *Izv. Akad. Nauk SSSR, Ser. Khim.* **1980**, 2164–2165; *Chem. Abstr.* **1980**, *93*, 185242.
41. Kozikowski, Λ. P.; Hasan, N. M. *J. Org. Chem.* **1977**, *42*, 2039–2040.
42. Morisawa, Y.; Kataoka, M.; Watanabe, T. *Chem. Pharm. Bull.* **1976**, *24*, 1089–1093.
43. Stepanova, S. V.; L'vova, S. D.; Belikov, A. B.; Gunar, V. I. *Zh. Org. Khim.* **1977**, *13*, 889–892 [*Chem. Abstr.* **1977**, *87*, 53135].
44. Stepanova, S. V.; L'vova, S. D.; El'yanov, B. S.; Gunar, V. I. *Khim.-Farm. Zh.* **1977**, *11*, 92–94 [*Chem. Abstr.* **1977**, *87*, 117803b].
45. Shimada, S.; Tojo, T. *Chem. Pharm. Bull.* **1983**, *31*, 4247–4258.
46. Bringmann, G.; Schneider, S. *Tetrahedron Lett.* **1986**, *27*, 175–178.
47. Coburn, S. P.; Lin, C. C.; Schaltenbrand, W. E.; Mahuren, J. D. *J. Labelled Compd. Radiopharm.* **1982**, *19*, 703–716.
48. Shi, G.; Xu, Y.; Xu, M. *J. Fluorine Chem.* **1991**, *52*, 149–157.
49. Boell, W.; Koenig, H. *Leibigs Ann. Chem.* **1979**, 1657–1664.
50. Jadrijevic-Mladar Takac, M.; Butula, I.; Vinkovic, M.; Dumic, M. *Croat. Chem. Acta* **1997**, *70*, 649–666 [*Chem. Abstr.* **1997**, *127*, 176411].
51. Burri, K. *Helv. Chim. Acta* **1990**, *73*, 69–80.
52. Giardina, G. A. M.; Sarau, H. M.; Farina, C.; Medhurst, A. D.; Grugni, M.; Raveglia, L. F.; Schmidt, D. B.; Rigolio, R.; Luttmann, M.; Vecchietti, V.; Hay, D. W. P. *J. Med. Chem.* **1997**, *40*, 1794–1807.
53. Tolstikov, G. A.; Shul'ts, E. E.; Vafina, G. F. *Zh. Org. Khim.* **1989**, *25*, 2249–2250 [*Chem. Abstr.* **1990**, *113*, 6126].
54. Zhulin, V. M.; Bogdanov, V. S.; Ostvald, G. V.; Kabotianskaya, E. B.; Koreshkov, Y. D. *Dokl. Akad. Nauk SSSR* **1990**, *310*, 362–365 [*Chem. Abstr.* **1990**, *113*, 39646].
55. Zhulin, V. M.; Ostval'd, G. V.; Bogdanov, V. S.; Kabotianskaya, E. B.; Koreshkov, Y. D.; Kondrat'eva, G. Ya. *Dokl. Akad. Nauk SSSR* **1988**, *302*, 1408–1412 [*Chem. Abstr.* **1990**, *112*, 76085].
56. Crank G.; Khan, H. R. *J. Heterocycl. Chem.* **1985**, *22*, 1281–1284.
57. Shimada, S. *J. Heterocyclic Chem.* **1987**, *24*, 1237–1241.
58. Kujundzic, N.; Gluncic, B. *Croat. Chem. Acta* **1991**, *64*, 599–606 [*Chem. Abstr.* **1992**, *116*, 214456].
59. Reddy, P. V.; Bhat, S. V. *Tetrahedron Lett.* **1997**, *38*, 9039–9042.
60. Dondoni, A.; Fogagnolo, M.; Mastellari, A.; Pedrini, P.; Ugozzoli, F. *Tetrahedron Lett.* **1986**, *27*, 3915–3918.

61. Nesi, R.; Turchi, S.; Giomi, D.; Papaleo, S. *J. Chem. Soc. Chem. Commun.* **1993**, 978–979.
62. Nesi, R.; Turchi, S.; Giomi, D. *J. Org. Chem.* **1996**, *61*, 7933–7936.
63. Seitz, G.; Hoferichter, R.; Mohr, R. *Arch. Pharm.* [Weinheim, Ger.] **1989**, *322*, 415–417.
64. Levin J. I.; Weinreb, S. M. *J. Am. Chem. Soc.* **1983**, *105*, 1397–1398.
65. Levin, J. I.; Weinreb, S. M. *J. Org. Chem.* **1984**, *49*, 4325–4332.
66. Levin, J. I. *Tetrahedron Lett.* **1989**, *30*, 2355–2358.
67. Jung M. E.; Dansereau, S. M. K. *Heterocycles* **1994**, *39*, 767–778.
68. Subramanyam, C.; Noguchi, M.; Weinreb, S. M. *J. Org. Chem.* **1989**, *54*, 5580–5585.
69. Ohba, M.; Kubo, H.; Fujii, T.; Ishibashi, H.; Sargent, M. V.; Arbain, D. *Tetrahedron Lett.* **1997**, *38*, 6697–6700.
70. Padwa, A.; Brodney, M. A.; Liu, B.; Satake, K.; Wu, T. *J. Org. Chem.* **1999**, *64*, 3595–3607.
71. Grigg, R.; Hayes R.; Jackson, J. L. *J. Chem. Soc. D* **1969**, 1167–1168.
72. Grigg R.; Jackson, J. L. *J. Chem. Soc. C* **1970**, 552–556.
73. Kondrat'eva, G. Y.; Medvedskaya, L. B.; Ivanova, Z. N.; Shmelev, L. V. *Dokl. Akad. Nauk SSSR* **1971**, *200*, 1358–1360 [*Chem. Abstr.* **1972**, *76*, 153650].
74. Kondrat'eva, G. Y.; Medvedskaya, L. B.; Ivanova, Z. N. *Izv. Akad. Nauk SSSR, Ser. Khim.* **1971**, 2276–2279 [*Chem. Abstr.* **1972**, *76*, 59332].
75. Kondrat'eva, G. Y.; Medvedskaya, L. B; Ivanova, Z. N.; Shmelev, L. V. *Izv. Akad. Nauk SSSR Ser. Khim.* **1971**, *6*, 1363–1364.
76. Koenig, H.; Graf, F.; Weberndoerfer, V. *Liebigs Ann. Chem.* **1981**, 668–682.
77. Hutton, J.; Potts B.; Southern, P. F. *Synth. Commun.* **1979**, *9*, 789–797.
78. Liotta, D.; Saindane, M.; Ott, W. *Tetrahedron Lett.* **1983**, *24*, 2473–2476.
79. Medvedskaya, L. B.; Makarov, M. G.; Kondrat'eva, G. Y. *Izv. Akad. Nauk SSSR, Ser. Khim.* **1973**, 1311–1315 [*Chem. Abstr.* **1973**, *79*, 91452].
80. Turner, S.; Ohlsen, S. R. *J. Chem. Soc. C* **1971**, 1632–1633.
81. Jurasek, A.; Zvak, V.; Kovac, J.; Rajaniakova, O.; Stetinova, J. *Collect. Czech. Chem. Commun.* **1985**, *50*, 2077–2083.
82. Ansell, M. F.; Caton, M. P. L.; North, P. C. *Tetrahedron Lett.* **1981**, *22*, 1727–1728.
83. Stetinova, J.; Lesko, J.; Dandarova, M.; Kada, R.; Koren, R. *Collect. Czech. Chem. Commun.* **1994**, *59*, 2721–2726 [*Chem. Abstr.* **1995**, *122*, 187296].
84. Bengston, G.; Keyaniyan, S.; De Meijere, A. *Chem. Ber.* **1986**, *119*, 3607–3630.
85. Iesce, M. R.; Cermola, F. M.; Graziano, L.; Scarpati, R. *Synthesis* **1994**, 944–948.
86. Vogel, E.; Doerr, J.; Herrmann, A.; Lex, J.; Schmickler, H.; Walgenbach, P.; Gisselbrecht, J. P.; Gross, M. *Angew. Chem. Int. Ed. Engl.* **1993**, *32*, 1597–1600.
87. Traylor, T. G.; Hill, K. W.; Tian, Z. Q.; Rheingold, A. L.; Peisach, J.; McCracken, J. *J. Am. Chem. Soc.* **1988**, *110*, 5571–5573.
88. Ichikawa, Y.; Naganawa, A.; Isobe, M. *Synlett* **1993**, 737–738.
89. Ho, M. S.; Wong, H. N. C. *J. Chem. Soc. Chem. Commun.* **1989**, 1238–1240.
90. Song, Z. Z.; Ho, S M. S.; Wong, H. N. C. *J. Org. Chem.* **1994**, *59*, 3917–3926.
91. Song, Z. Z.; Zhou, Z. Y.; Mak, T. C.; Wong, H. N. C. *Angew. Chem., Int. Ed. Engl.* **1993**, *32*, 432–434.
92. Song, Z. Z.; Wong, H. N. C. *J. Chin. Chem. Soc. (Taipei)* **1995**, *42*, 673–679; *Chem. Abstr.* **1995**, *123*, 340299.
93. Wong, M. K.; Leung, Y.; Wong, H. N. C. *Tetrahedron* **1997**, *53*, 3497–3512.
94. Yang Y.; Wong, H. N. C. *J. Chem. Soc. Chem. Commun.* **1992**, 656–658.
95. Ferkous, F.; Rahm, A. *J. Soc. Alger. Chim.* **1995**, *5*, 49–60 [*Chem. Abstr.* **1996**, *124*, 261203].
96. Lee, G. C. M.; Garst, M. E.; Sachs, G. U.S. pat. 5,183,906, 1993.

97. Chen, B.; Ye, X.; Chen, Q. *Synth. Commun.* **1998**, *28*, 2831–2841.
98. Kawada, K.; Kitagawa, O.; Kobayashi, Y. *Chem. Pharm. Bull.* **1985**, *33*, 3670–3674.
99. Caesar, J. C.; Griffiths, D. V.; Griffiths, P. A.; Tebby, J. C. *J. Chem. Soc. Perkin Trans. 1* **1990**, 2329–2334.
100. Medvedskaya, L. B.; Kondrat'eva, G. Y.; Bykanova, N. V. *Izv. Akad. Nauk SSSR Ser. Khim.* **1979**, 1613–1615 [*Chem. Abstr.* **1979**, *91*, 175251].
101. Rao, A. V. R.; Reddy, D. R. *J. Chem. Soc. Chem. Commun.* **1987**, 574–575.
102. Yadav, J. S.; Valluri, M.; Rama Rao, A. V. *Tetrahedron Lett.* **1994**, *35*, 3609–3612.
103. Rama Rao, A. V.; Yadav, J. S.; Valluri, M. *Tetrahedron Lett.* **1994**, *35*, 3613–3616.
104. Reddy G. S.; Bhatt, M. V. *Tetrahedron Lett.* **1980**, *21*, 3627–3628.
105. Whitney, S. E.; Rickborn, B. *J. Org. Chem.* **1988**, *53*, 5595–5596.
106. Whitney, S. E.; Winters, M.; Rickborn, B. *J. Org. Chem.* **1990**, *55*, 929–935.
107. Nakazawa, T.; Ishihara, M.; Jinguji, M.; Miyatake, R.; Sugihara, Y.; Murata, I. *Tetrahedron Lett.* **1994**, *35*, 8421–8424.
108. Jacobi, P. A. In Pearson, W. H. ed. *Advances in Heterocyclic Natural Product Synthesis*, Vol. 2, JAI Press: Greenwich, 1992, pp. 251–298.
109. Yang Y.; Wong, H. N. C. *Tetrahedron* **1994**, *50*, 9583–9608.
110. Jacobi, P. A.; Walker, D. G.; Odeh, I. M. A. *J. Org. Chem.* **1981**, *46*, 2065–2069.
111. Jacobi, P. A.; Walker, D. G. *J. Am. Chem Soc.* **1981**, *103*, 4611–4613.
112. Jacobi, P. A.; Kaczmarek, C. S. R.; Udodong, U. E. *Tetrahedron Lett.* **1984**, *25*, 4859–4862.
113. Jacobi, P. A.; Kaczmarek, C. S. R.; Udodong, U. E. *Tetrahedron* **1987**, *43*, 5475–5488.
114. Jacobi, P. A.; Selnick, H. G. *J. Org. Chem.* **1990**, *55*, 202–209.
115. Jacobi, P. A.; Blum, C. A.; DeSimone, R. W.; Udodong, U. E. S. *Tetrahedron Lett.* **1989**, *30*, 7173–7176.
116. Jacobi, P. A.; Blum, C. A.; DeSimone, R. W.; Udodong, U. E. S. *J. Am. Chem. Soc.* **1991**, *113*, 5384–5392.
117. Nicolaou, K. C.; Li, W. S. *J. Chem. Soc. Chem. Commun.* **1985**, 421.
118. Jacobi, P. A.; Touchette, K. M.; Selnick, H. G. *J. Org Chem.* **1992**, *57*, 6305–6313.
119. Jacobi, P. A.; Craig, T. *J. Am. Chem. Soc.* **1978**, *100*, 7748–7750.
120. Jacobi, P. A.; Lee, K. *J. Am. Chem. Soc.* **1997**, *119*, 3409–3410.
121. Jacobi, P. A.; Lee, K. *J. Am. Chem. Soc.* **2000**, *122*, 4295–4303.
122. Selnick, H. G.; Brookes, L. M. *Tetrahedron Lett.* **1989**, *30*, 6607–6610.
123. Liu, B.; Padwa, A. *Tetrahedron Lett.* **1999**, *40*, 1645–1648.
124. Hassner, A.; Fischer, B. *Tetrahedron* **1989**, *45*, 3535–3546.
125. Vedejs E.; Fields, S. *J. Org. Chem.* **1988**, *53*, 4663–4667.
126. Padyukova, N. S.; Florent'ev, V. L. *Khim. Geterotsikl. Soedin* **1973**, 600–607 [*Chem. Abstr.* **1973**, *79*, 52483].
127. Jursic, B. S.; Zdravkovski, Z. *J. Chem. Soc. Perkin Trans.* **1994**, *2*, 1877–1881.
128. Ibata, T.; Suga, H.; Isogami, Y.; Tamura, H.; Shi, X. *Bull. Chem. Soc. Jpn.* **1992**, *65*, 2998–3007.
129. Suga H.; Ibata, T. *Chem. Lett.* **1991**, 1221–1224.
130. Suga, H.; Shi, X.; Ibata, T. *Bull. Chem. Soc. Jpn.* **1998**, *71*, 1231–1236.
131. Suga, H.; Shi, X.; Ibata, T. *Chem. Lett.* **1994**, 1673–1676.
132. Suga, H.; Shi, X.; Fujieda, H.; Ibata, T. *Tetrahedron Lett.* **1991**, *47*, 6911–6914.
133. Suga, H.; Shi, X.; Ibata, T. *J. Org. Chem.* **1993**, *58*, 7397–7405.
134. Suga, H.; Fujieda, H.; Hirotsu, Y.; Ibata, T. *J. Org. Chem.* **1994**, *59*, 3359–3364.
135. Suga, H.; Ikai, K.; Ibata, T. *Tetrahedron Lett.* **1998**, *39*, 869–872.

136. Wasserman, H. H.; Floyd, M. B. *Tetrahedron* (Suppl. 7) **1966**, 441–448.
137. Wasserman, H. H.; McCarthy, K. E.; Prowse, K. S. *Chem. Rev.* **1986**, *86*, 845–856.
138. Wasserman, H. H.; Vinick, F. J.; Chang, Y. C. *J. Am. Chem. Soc.* **1972**, *94*, 7180–7182.
139. Graziano, M. L.; Iesce, M. R.; Cimminiello, G.; Scarpati, R.; Parrilli, M. *J. Chem. Soc. Perkin Trans. 1* **1990**, 1011–1017.
140. Gollnick, K.; Koegler, S. *Tetrahedron Lett.* **1988**, *29*, 1003–1006.
141. Gollnick, K.; Koegler, S. *Tetrahedron Lett.* **1988**, *29*, 1007–1010.
142. Iesce, M. R.; Graziano, M. L.; Cimminiello, G.; Cermola, F.; Parrilli, M.; Scarpati, R. *J. Chem. Soc. Perkin Trans. II* **1991**, 1085–1089.
143. Hassner, A.; Fischer, B. *J. Org. Chem.* **1991**, *56*, 3419–3425.
144. Ohba, M.; Kubo, H.; Ishibashi, H. *Tetrahedron* **2000**, *56*, 7751–7761.
145. Ohba, M.; Izuta, R.; Shimizu, E. *Tetrahedron Lett.* **2000**, *41*, 10251–10255.
146. Ohba, M.; Izuta, R. *Heterocycles* **2001**, *55*, 823–826.
147. Lee, J. C.; Cha, J. K. *Tetrahedron* **2000**, *56*, 10175–10184.
148. Lee, J. C.; Jin, S.; Cha, J. K. *J. Org. Chem.* **1998**, *63*, 2804–2805.
149. Lee, J. C.; Cha, J. K. *J. Am. Chem. Soc.* **2001**, *123*, 3243–3246.
150. Pei, W.; Pei, J.; Li.; S.; Ye, X. *Synthesis* **2000**, 2069–2077.
151. Wenkert, D.; Chen, T.-F.; Ramachandran, K.; Valasinas, L.; Weng, L.; McPhail, A. T. *Org. Lett.* **2001**, *3*, 2301–2303.

CHAPTER 4

Mesoionic Oxazoles

Gordon W. Gribble

Department of Chemistry
Dartmouth College
Hanover, New Hampshire

4.1. Introduction
4.2. 1,3-Oxazolium 5-Oxides (Munchnones)
    4.2.1. Synthesis
    4.2.2. Structure and Spectral Properties
        4.2.2.1. Structure
        4.2.2.2. Spectra Properties
    4.2.3. Reactions
        4.2.3.1. Electrophilic Reactions
        4.2.3.2. Nucleophilic Additions
        4.2.3.3. Cycloaddition Reactions
            4.2.3.3.1. Acetylenic Dipolarophiles
            4.2.3.3.2. Aryne Dipolarophiles
            4.2.3.3.3. Olefinic Dipolarophiles
            4.2.3.3.4. Carbonyl Compounds
            4.2.3.3.5. Thiocarbonyl Compounds
            4.2.3.3.6. Oxygen
            4.2.3.3.7. Nitrogen-Containing Dipolarophiles
            4.2.3.3.8. Heterocumulenes
            4.2.3.3.9. Phosphorus-Containing Dipolarophiles
        4.2.3.4. Ring-Chain Valence Tautomerism
4.3. 1,3-Oxazolium 5-Imines (Munchnone Imines)
4.4. 1,3-Oxazolium 4-Oxides (Isomunchnones)
    4.4.1. Synthesis
    4.4.2. Structure and Spectra Properties
        4.4.2.1. Infrared and Mass Spectra
        4.4.2.2. NMR Spectra
        4.4.2.3. Molecular Orbital Calculations
    4.4.3. Reactions
        4.4.3.1. Nucleophiles
        4.4.3.2. Cycloaddition
4.5. 1,3-Oxazolium 4-Imines (Isomunchnone Imines)
4.6. Summary

The Chemistry of Heterocyclic Compounds, Volume 60: Oxazoles: Synthesis, Reactions, and Spectroscopy, Part A, edited by David C. Palmer
ISBN 0-471-39494-7 Copyright © 2003 John Wiley & Sons, Inc.

4.1. INTRODUCTION

In the nearly 20 years since the publication of the outstanding exhaustive review of mesoionic oxazoles by Gingrich and Baum,[1] the mystique of munchnones **1** and isomunchnones **2** has vanished, and these remarkable heterocycles have found a permanent place in the arsenal of the synthetic organic chemist (Fig. 4.1). Indeed, the major focus of this chapter is the diverse and efficient applications of these mesoionic oxazoles in synthesis.

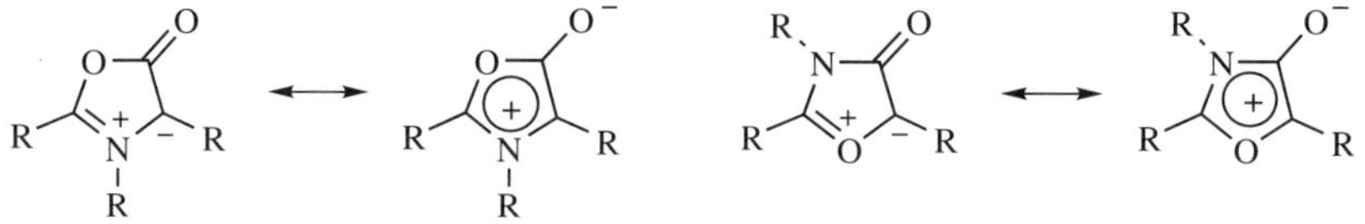

Figure 4.1

The coverage herein begins where Gingrich and Baum left off, in mid-1983, and duplication of that excellent review is avoided. The reader is urged to consult Gingrich and Baum for the rich history and early synthetic applications of mesoionic oxazoles. Several reviews published in the interim have also discussed these heterocycles.[2-9]

Although several resonance structures can be imagined for munchnones (1,3-oxazolium 5-oxides or 1,3-oxazolium 5-olates) and isomunchnones (1,3-oxazolium 4-oxides or 1,3-oxazolium 4-olates), I will draw those forms depicted in **1** and **2**, respectively, which capture the flavor of the 1,3-dipolar reactivity of these heterocycles. In addition, this chapter covers the relatively few new developments involving munchnone imines and isomunchnone imines. The format follows that used by Gingrich and Baum.

4.2. 1,3-OXAZOLIUM 5-OXIDES (MUNCHNONES)

4.2.1. Synthesis

The traditional synthesis of munchnones involves the cyclodehydration of *N*-acylamino acids usually with acetic anhydride or another acid anhydride. Potts and Yao[10] were apparently the first to employ dicyclohexylcarbodiimide (DCC) to generate mesoionic heterocycles, including munchnones; e.g., **3** → **4** → **5** (Fig. 4.2).

Anderson and Heider[11] discovered that munchnones can be constructed by the cyclodehydration of *N*-acylamino acids using *N*-ethyl-*N'*-dimethylaminopropylcar-

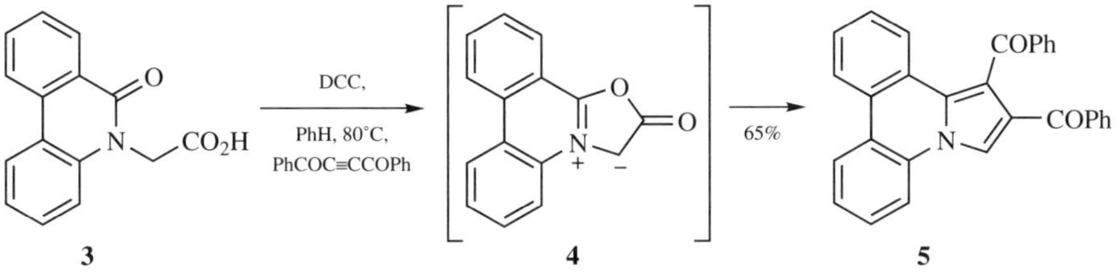

1,3-Oxazolium 5-Oxides (Munchnones) 475

Figure 4.2

bodiimide (EDC) or silicon tetrachloride. The advantage of EDC over DCC is that the urea by-product is water soluble and easily removed, in contrast to dicyclohexylurea formed from DCC, which is difficult to remove completely from the reaction product. Although the authors conclude that the traditional Huisgen method of acetic anhydride is "still the method of choice," the present two new methods are important alternatives. Some examples from the work of Anderson and Heider are listed in Scheme 4.1. The in situ–generated munchnones (not shown) were trapped with either dimethyl acetylenedicarboxylate (DMAD) or ethyl propiolate.

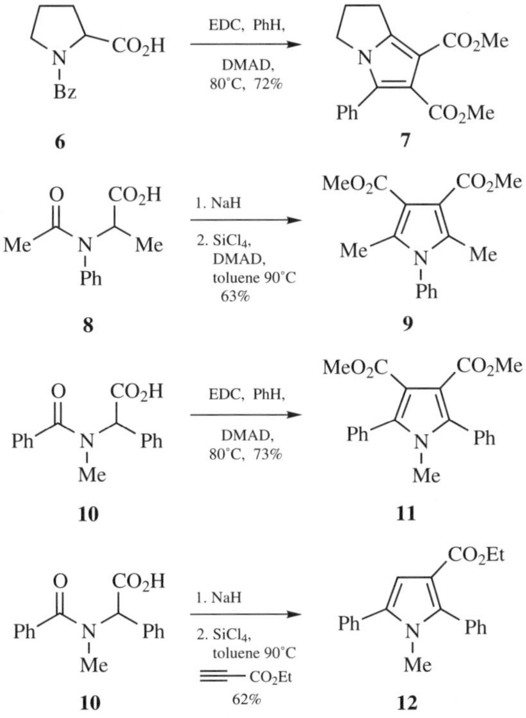

Scheme 4.1

During a peptide synthesis study, Slebioda[12] apparently independently observed the formation of munchnone **13** from *N*-benzoylphenylalanine and DCC (Fig. 4.3).

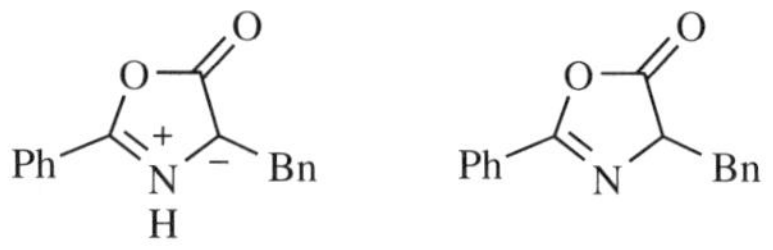

13 **14**

Figure 4.3

Although this method of munchnone generation is precedented,[10,11] the author was able to isolate and fully characterize this crystalline material (mp 132°–133°C; C=O at 1709 cm^{-1}). Moreover, Slebioda examined the tautomerization of oxazolone **14** to **13** as a function of base and solvent and found that this conversion was best effected with triethylamine in DMF (68% yield). Triethylamine in carbon tetrachloride affords **13** in only 3% yield.

Another important route to munchnones is that described by Boyd and Wright[13,14] involving the cyclodehydration of N-acylamino acids with acetic anhydride in the presence of perchloric acid to give 1,3-oxazolonium salts (N-alkylated azlactones). Upon exposure to mild base (triethylamine or sodium carbonate) munchnones are formed. Hershenson and Pavia[15] reported a variation involving the in situ alkylation of azlactones **15**, deprotonation with 2,6-di-*tert*-butylpyridine, and trapping of the resulting munchnones **16** with DMAD to give pyrroles **17** in good yield (Fig. 4.4). Highly reactive alkylating agents are necessary for successful reaction as is 2,6-di-*tert*-butylpyridine. The azlactone derived from leucine yielded dimethyl 1-ethyl-2-methyl-5-isobutyl-1*H*-pyrrole-3,4-dicarboxylate in similar fashion (39% yield).

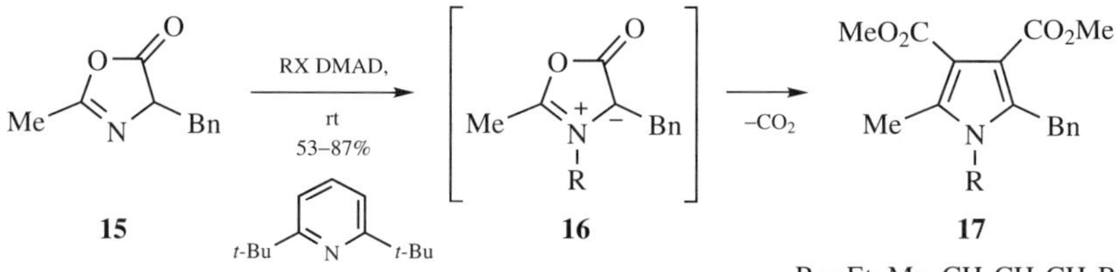

15 **16** **17**

R = Et, Me, CH$_2$CH$_2$CH$_2$Br

RX = Et$_3$OBF$_4$, MeOSO$_2$CF$_3$, Br(CH$_2$)$_3$OSO$_2$CF$_3$

Figure 4.4

Wilde[16] generated N-acyl munchnones for the first time via the acylation and desilylation of 5-siloxyoxazoles.[17] Thus, for example, exposure of **18** to acetyl chloride in the presence of DMAD affords N-acetylpyrrole **20** via N-acetyl munchnone **19** (Fig. 4.5). Other trapping agents (thiocarbonyls, N-phenylmaleimide) failed, and the analogous trimethylsiloxyoxazoles were unusually labile and gave lower yields of pyrroles (13–15%). Ethyl chloroformate was also used in this sequence and gave, for example, pyrroles **21**–**22**.

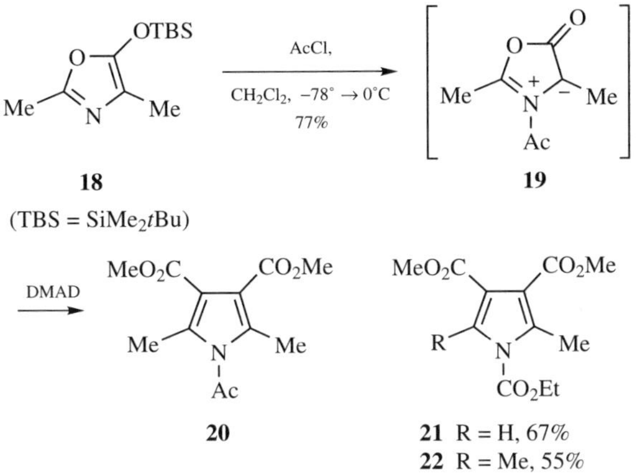

1,3-Oxazolium 5-Oxides (Munchnones)

Figure 4.5

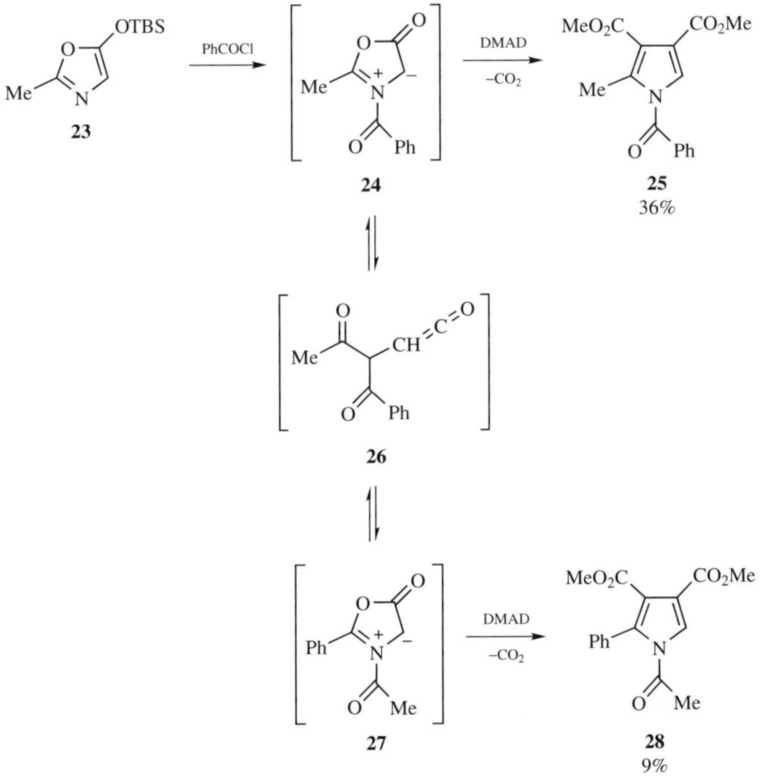

Scheme 4.2

Moreover, Wilde[16] also found that N-acyl munchnones are prone to equilibrate through ring-chain tautomerism, a process well known for N-alkyl munchnones.[1,18–20] Thus oxazole **23** reacts with benzoyl chloride in the presence of DMAD to give a mixture of pyrroles **25** and **28**, presumably via ketene **26**, which (slowly) equilibrates munchnones **24** and **27** (Scheme 4.2). Capture of munchnone **24** by DMAD is faster than equilibration to munchnone **27**. Among the other examples reported, oxazole **29** affords a mixture of pyrroles **30** and **31** (Fig. 4.6).

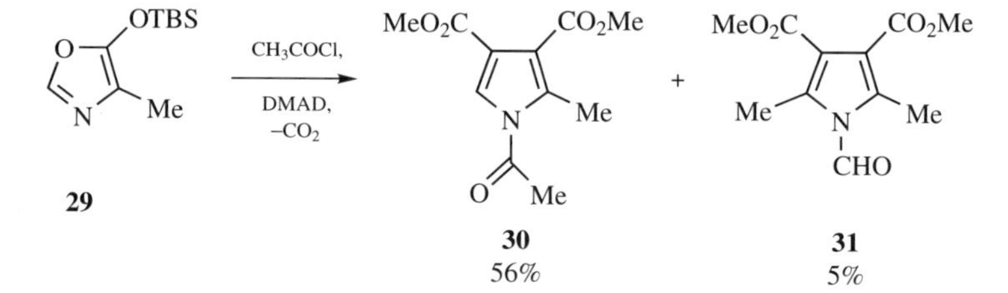

Figure 4.6

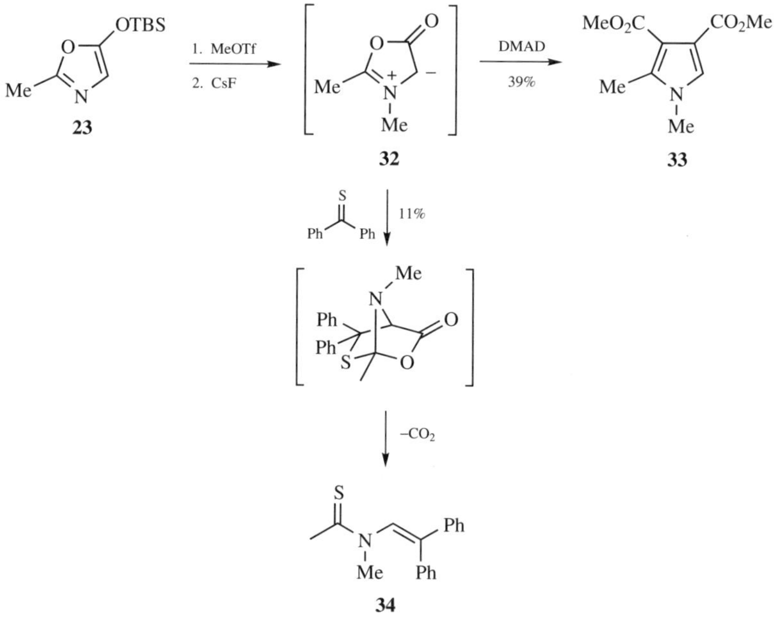

Scheme 4.3

Wilde[16] also found that N-methyl munchnone **32** could be generated using methyl triflate and trapped with DMAD and thiobenzophenone to give **33** and **34**, respectively (Scheme 4.3). This protocol is similar to that developed by Hershenson and Pavia[15] (vide supra). Other trapping agents, such as trimethylsilyl chloride and p-toluenesulfonyl chloride, afforded lower yields of cycloadducts with DMAD.

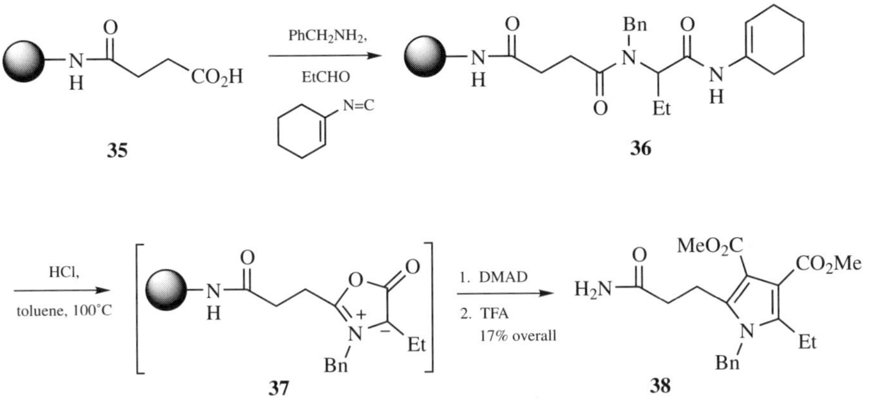

1,3-Oxazolium 5-Oxides (Munchnones)

Scheme 4.4

In a series of papers, Armstrong and co-workers[21–24] employed the elegant Ugi four-component condensation[25,26] to construct munchnone precursors and, following deprotonation, munchnones. The overall sequence, which has been adapted to the solid-phase synthesis of pyrroles by Armstrong and co-workers[24] and, independently, by Mjalli and co-workers,[27] is illustrated in Scheme 4.4 for the synthesis of pyrrole **38**.[24] A Rink or Wang resin can be used as a precursor to **35**. The Ugi condensation leads directly to **36** and munchnone **37** formation is induced with acid. Trapping of **37** with DMAD and cleavage from the resin yields **38**.

Mjalli and colleagues[27] extended this concept and generated several solid-supported munchnones **39**, which were converted into pyrroles by trapping with DMAD and other alkynes (Fig. 4.7). Merlic and co-workers[28] reported a novel

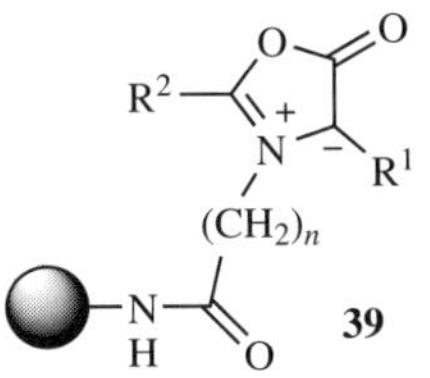

$n = 1, 2$
$R^1 = $ Et, i-Pr, n-Pr, n-Bu, i-Bu
$R^2 = $ Ph, Bn, 4-Br-Ph, 4-MeO-Ph, 4-Me-Ph, 4-CF$_3$-Ph

Figure 4.7

generation of munchnones from acylamino chromium carbene complexes. Thus exposure of complex **40** to carbon monoxide affords the chromium ketene complex **41**, which cyclizes to munchnone **42**, which can be isolated in 27% yield. Performing this reaction in the presence of DMAD furnishes pyrrole **11** in 78% yield (Fig. 4.8).

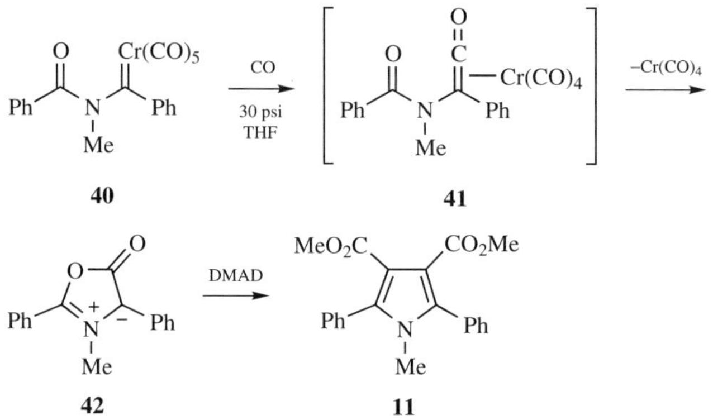

Figure 4.8

The related but more stable chelate carbene complex **43** also reacts with pressurized carbon monoxide to give munchnones that can be trapped with alkynes to afford the expected pyrroles (Fig. 4.9). The relatively sensitive acylamino chromium complexes (e.g., **45**) can be prepared in situ from stable amino carbene complexes (e.g., **44**), as shown for the generation of munchnone **46** and conversion to pyrrole **47** with DMAD (Fig. 4.10).

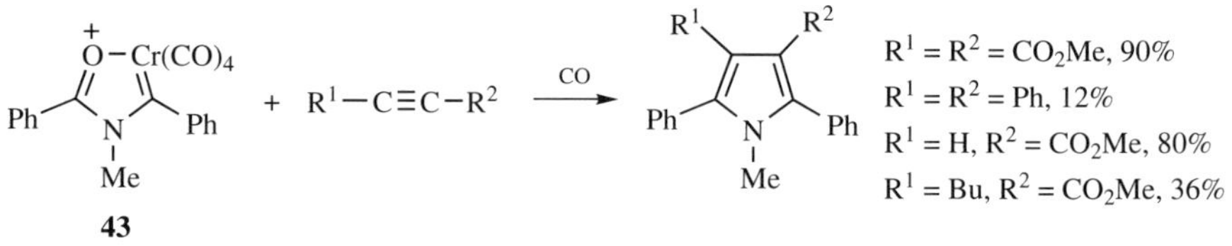

$R^1 = R^2 = CO_2Me$, 90%
$R^1 = R^2 = Ph$, 12%
$R^1 = H, R^2 = CO_2Me$, 80%
$R^1 = Bu, R^2 = CO_2Me$, 36%

Figure 4.9

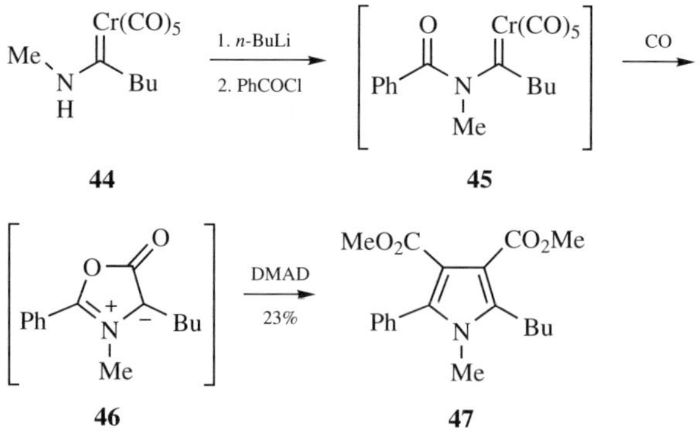

Figure 4.10

4.2.2. Structure and Spectral Properties

4.2.2.1. Structure

Most of the research done on munchnones since the review by Gingrich and Baum[1] has involved synthetic applications. Only one additional X-ray crystal structure has been described in the interim. Toupet and co-workers[29] reported the crystal structure of **48**, concluding that the "mesoionic character is unambiguously established" and that the structure is intermediate between resonance forms **48a** and **48b** (Fig. 4.11). Some bond lengths are shown, and they correspond to those found earlier by Boyd and co-workers[30] for **49**. The dihedral angles in **48** between the munchnone ring and the phenyl ring and the *p*-methoxyphenyl ring are 22.5° and 29.8°, respectively. In **49**, this angle of twist is 44°.

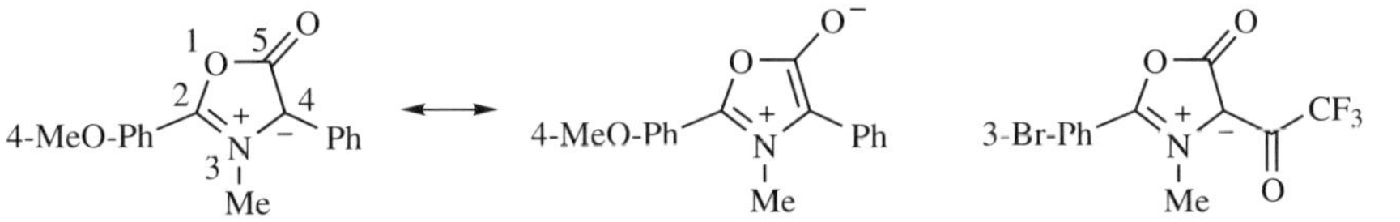

O1-C5 1.454 Å	O1-C5 1.428 Å
O1-C2 1.337 Å	O1-C2 1.299 Å
C5-O1 1.238 Å	C5-O1 1.235 Å
N3-C2 1.332 Å	N3-C2 1.310 Å
N3-C4 1.411 Å	N3-C4 1.376 Å
C4-C5 1.375 Å	C4-C5 1.374 Å

Figure 4.11

4.2.2.2. Spectral Properties

In continuation of earlier work,[31,32] Petride and Raileanu[33] synthesized a series of oxazolo[3,2-*a*]pyridinium-2-olates **50** and recorded their spectral properties (Fig. 4.12). Although these C-3 unsubstituted munchnones were unstable, some spectral data could be obtained.

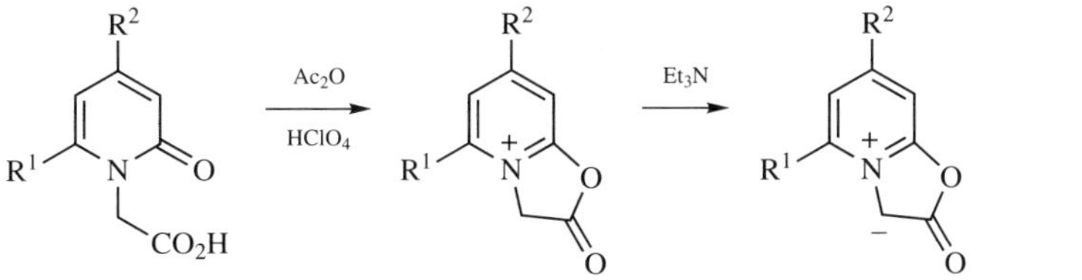

50a R^1 = H, R^2 = Me
50b R^1 = Me, R^2 = H
50c R^1 = Me, R^2 = OAc

Figure 4.12

The proton NMR spectra of **50b** and **50c** revealed the C-3 proton at 6.05 and 5.95 ppm, respectively. These shifts led the authors to conclude that these munchnones are "not aromatic" and that resonance structure **50** is the major contributor to the hybrid. Compound **50a** was too labile to be studied by NMR. The UV and IR spectra of **50a–c** were also recorded, although with difficulty, since decomposition was occurring. The authors believe that the λ_{max} (CH_2Cl_2) at 360 and 353 nm for **50a** and **50b**, respectively, are $\pi \rightarrow \pi^*$ transitions. The IR spectra of **50a–c** all show carbonyl bands at 1700–1708 cm^{-1} and 1726–1747 cm^{-1}, in addition to the acetoxy absorption at 1770 cm^{-1} for **50c**.

Petride and Raileanu[34] also prepared a series of acylated bicyclic munchnones **51** and examined the spectral properties of these stable compounds, which were synthesized by in situ acylation of **50** (Fig. 4.13). Two examples are illustrated of the several compounds prepared in this study. Simple analogs of type **51a** have a coplanar acyl group, whereas those derivatives with a substituent (usually methyl) ortho to the acyl group (**51b**) cause the latter to be twisted out of planarity. The acyl groups in **51a** have more single bond character (1601–1631 cm^{-1}) than the acyl groups in **51b** (1633–1664 cm^{-1}), reflecting greater delocalization of negative charge into the acyl group in **51a** than in **51b**.

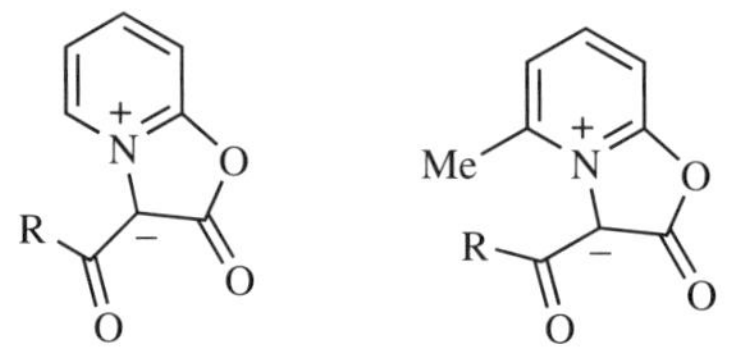

51a (coplanar)   **51b** (nonplanar)

Figure 4.13

4.2.3. Reactions

4.2.3.1. Electrophilic Reactions

As presented in some detail by Gingrich and Baum[1] (see their Table 4.31 and Scheme 4.51) azlactones react with α,β-unsaturated imines to give α-pyridones. Sandhu and co-workers[35–37] continued the studies in this area, and they now view this reaction as involving initial electrophilic attack on the azlactone (munchnone) followed by cyclization to an α-pyridone, rather than prior ring-opening to the corresponding ketene tautomer, as was originally proposed (Table 4.1). With azlactone **52** and *N*-aryl cinnamaldehyde anils **53**, in the presence of acetic anhydride, the 4-substituted azlactones **54** are isolated (Fig. 4.15).[36] Dalla Croce

TABLE 4.1. REACTION OF AZLACTONES WITH α,β-UNSATURATED IMINES

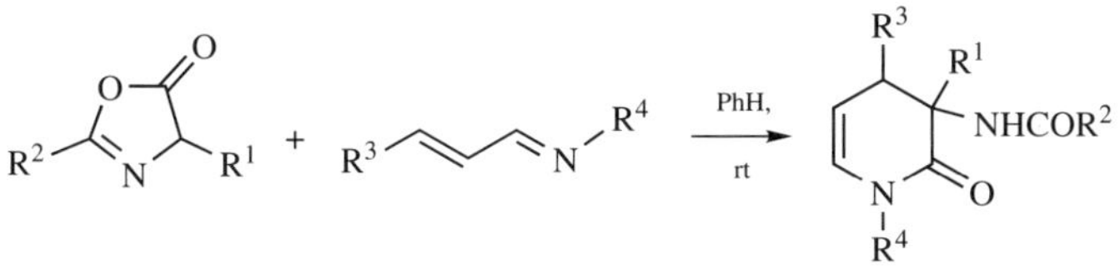

Figure 4.14

R^1	R^2	R^3	R^4	Yield (%)	Reference
Me	Ph	2-furyl	Ph	55	35
Me	Ph	2-furyl	4-Me-Ph	72	35
Me	Ph	2-furyl	4-MeO-Ph	72	35
Me	Ph	2-furyl	4-Br-Ph	62	35
Me	Ph	2-furyl	4-EtO-Ph	63	35
Ph	Ph	Ph	Me	60	36
Ph	Ph	Ph	cyclohexyl	60	36
Ph	Ph	Me	cyclohexyl	78	37
Me	Ph	Me	cyclohexyl	88	37
Me	4-NO_2-Ph	Me	cyclohexyl	80	37
Me	4-Cl-Ph	Me	cyclohexyl	85	37
Me	Bn	Me	cyclohexyl	78	37
Me	4-MeO-Ph	Me	cyclohexyl	90	37
Me	tBu	Me	cyclohexyl	70	37
Ph	Ph	Me	i-Pr	65	37
Me	Ph	Me	i-Pr	72	37

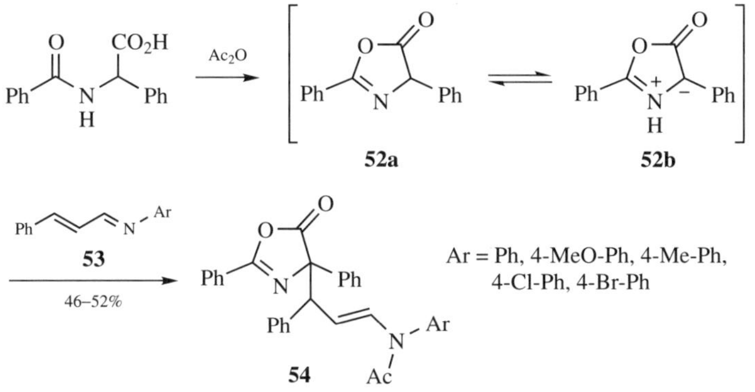

Figure 4.15

and co-workers[38] also found that N-tosyl imine **55** reacts with azlactone **52** to form substituted azlactone **56** as a mixture of diastereomers (Fig. 4.16). At higher temperatures pyrroles and pyridine products are formed (vide infra). Raileanu and Petride[39] studied the site of protonation of various acylated munchnones with

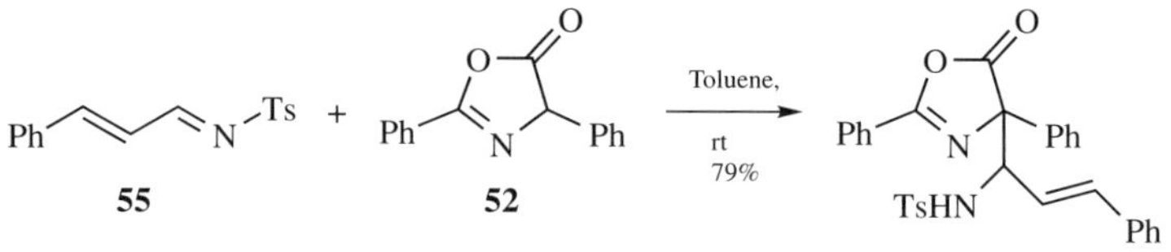

484   Mesoionic Oxazoles

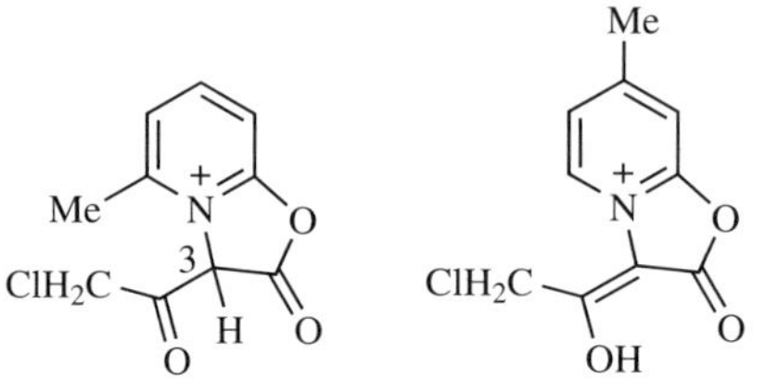

Figure 4.16

perchloric and trifluoroacetic acid (TFA). Protonation of nonplanar acylated munchnones with perchloric acid affords C-3 protonated products **57**, whereas under similar conditions planar acylated munchnones give rise to acyl group oxygen protonation leading to **58** (Fig. 4.17). Using trifluoroacetic acid, these authors were unable to confirm the earlier results of Boyd and Wright[40] that these munchnones undergo C-3 protonation. Thus planar acylated munchnones are not protonated at all by TFA, and the self-acylated munchnone **59** is protonated on the pyridone oxygen to give **60**.

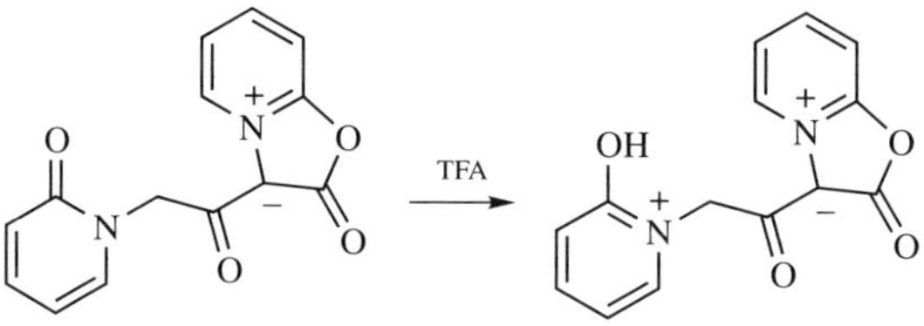

Figure 4.17

Although the acylation of munchnones was the first reaction of these mesoionic compounds to be discovered (and during their synthesis),[41] Kawase and co-workers[42–49] have greatly exploited the synthetic potential of this chemistry. These chemists have found that N-acylated prolines **61** are transformed into 5-trifluoromethyloxazoles **62** with trifluoroacetic acid anhydride (TFAA) (Fig. 4.18).[42,43] Acid hydrolysis furnishes the alcohols **63** in good overall yield. The corresponding N-CHO, N-Ac, N-i-Bu prolines failed to react in this fashion.

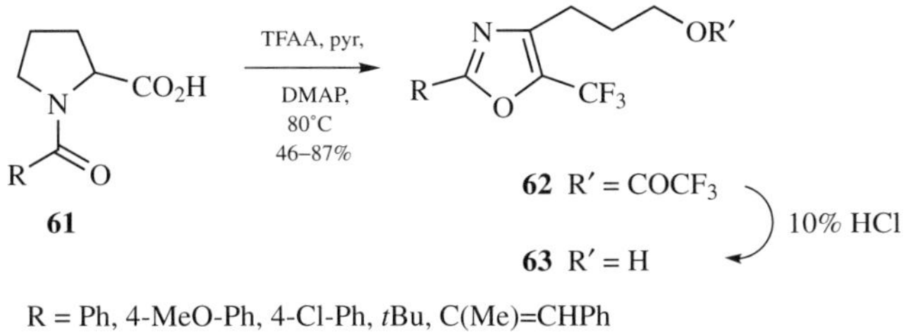

1,3-Oxazolium 5-Oxides (Munchnones)

Figure 4.18

R = Ph, 4-MeO-Ph, 4-Cl-Ph, *t*Bu, C(Me)=CHPh

A suggested mechanism, which is related to the Dakin-West reaction,[50] is shown in Scheme 4.5. Evidence in support of this mechanism is seen with the isolation of **64** and **65** with bulky R groups, the latter structure of which was supported by X-ray crystallography (Fig. 4.19).

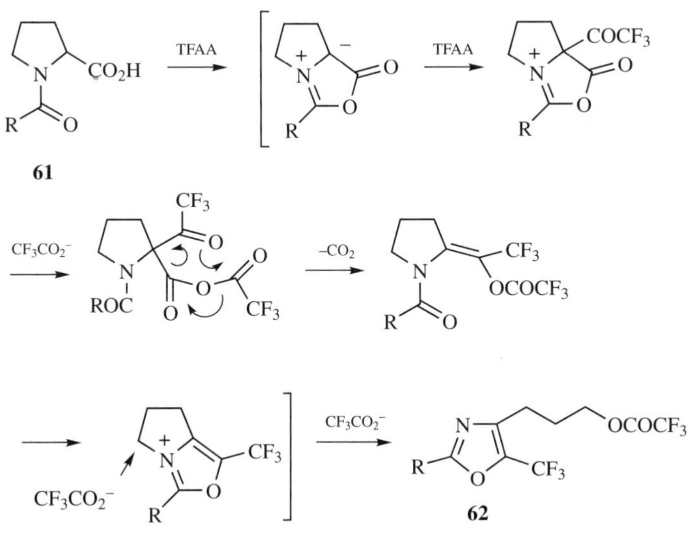

Scheme 4.5

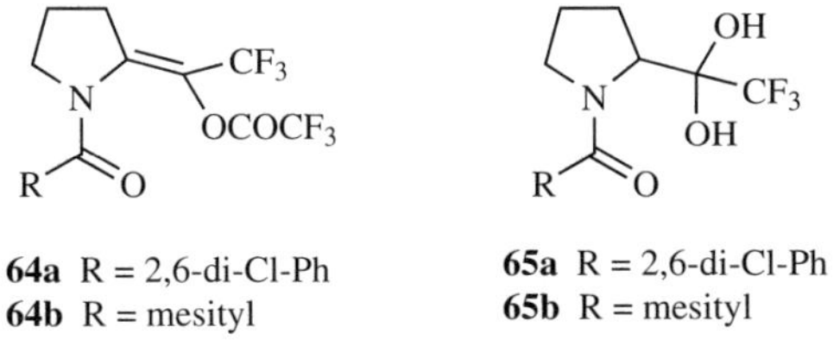

64a R = 2,6-di-Cl-Ph
64b R = mesityl

65a R = 2,6-di-Cl-Ph
65b R = mesityl

Figure 4.19

TABLE 4.2. REACTIONS OF MUNCHNONES WITH PERFLUOROALKYLCARBOXYLIC ACID ANHYDRIDES[a]

Figure 4.20

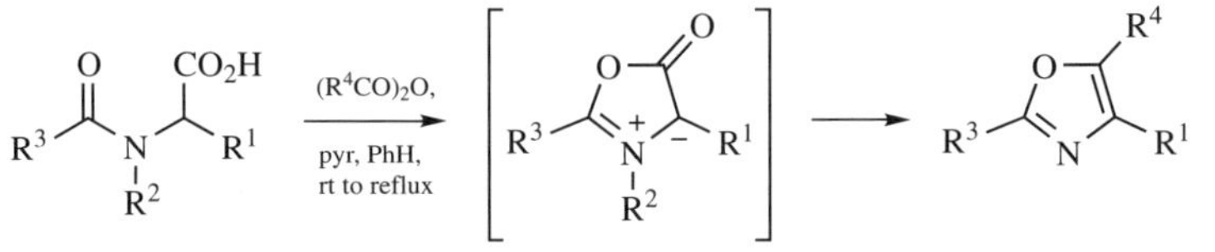

R^1	R^2	R^3	R^4	Yield (%)
Bn	Me	Ph	CF_3	13
Bn	Et	Ph	CF_3	16
Bn	Bn	Ph	CF_3	88
Bn	Bn	tBu	CF_3	83
Bn	Bn	PhCH=CMe	CF_3	61
Ph	Bn	Ph	CF_3	92
Ph	Bn	2-thienyl	CF_3	93
Me	Bn	Ph	CF_3	46
sec-Bu	Bn	Ph	CF_3	51
Ph	Bn	Ph	C_2F_5	92
Ph	Bn	Ph	n-C_3F_7	98

[a] Data from Ref. 44.

Kawase[43,44] has now extended this to other munchnones (Table 4.2), making this a general preparation of 5-perfluoroalkyloxazoles (but mainly 5-trifluoromethyloxazoles). In all cases, the munchnones were generated in situ from the corresponding N-acylamino acids.

At higher temperatures in the absence of base, however, N-acylprolines **61** undergo conversion to acyloin amides **66** (Fig. 4.21).[45] Other N-acylamino acids behave similarly, for example, N-benzoyl-N-methylphenylalanine **67** affords acyloin **68** and only a small amount of oxazole **70**, in addition to some benzoate **69**. Kawase[45] proposed that the initially formed munchnone undergoes transformation to an oxazolium species, which, depending on the site of attack by trifluoroacetate, is converted either to an oxazole, an acyloin, or a benzoate (Scheme 4.6; see Scheme 4.5).

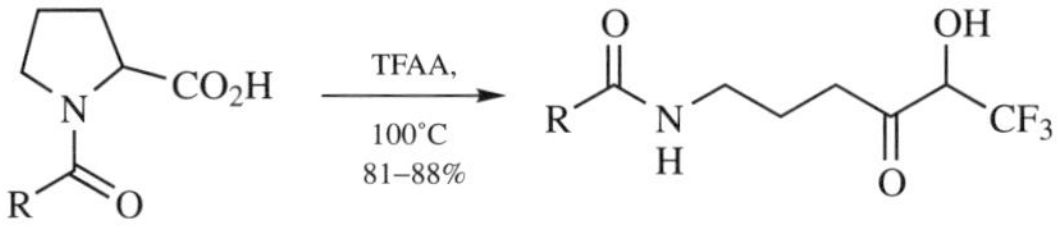

R = Ph, 4-Me-Ph, 4-Cl-Ph

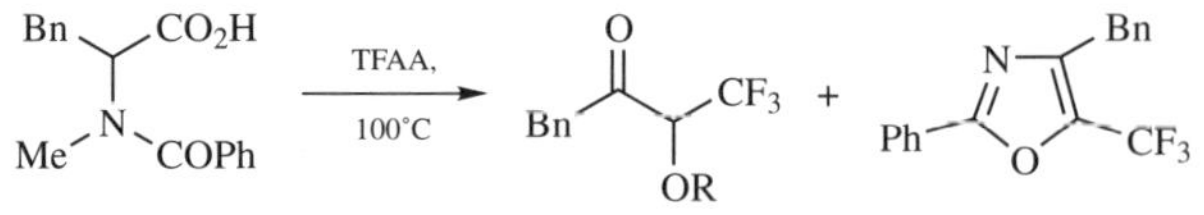

68 R = H, 54%
69 R = COPh, 11%
70 5%

Figure 4.21

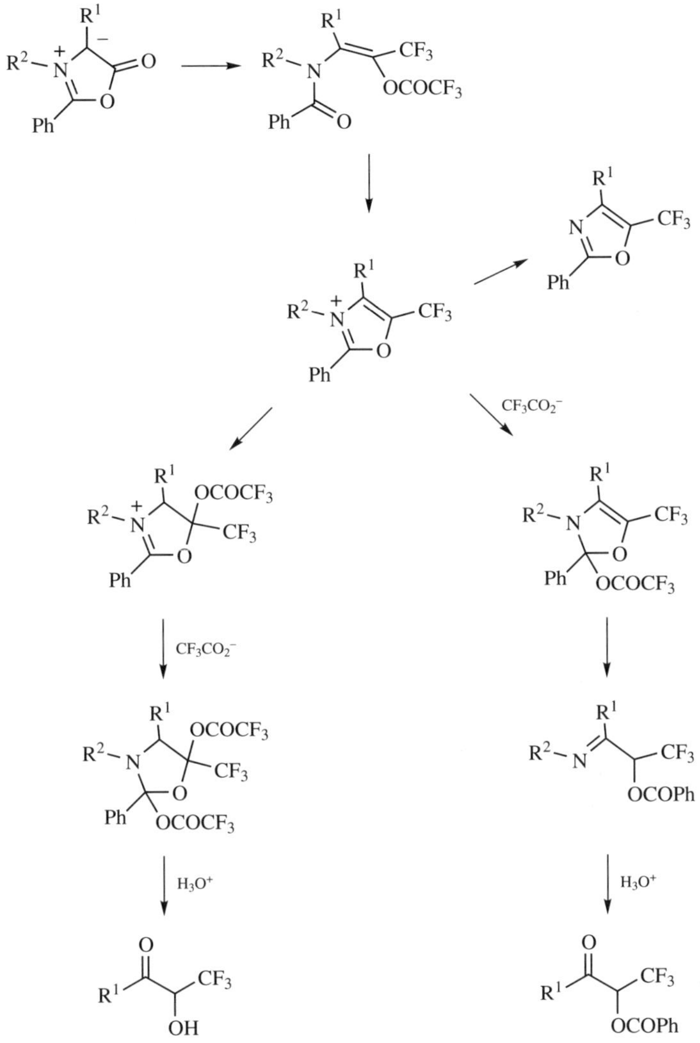

1,3-Oxazolium 5-Oxides (Munchnones)

Scheme 4.6

Furthermore, Kawase[46] found that *N*-acetylamino acids **71** react with TFAA in the presence of pyridine to form 1,4-oxazin-3-ones **75**, as summarized below (Fig. 4.22). The only unprecedented step in this proposed mechanism is the final transformation of **74** to **75**.

In yet another variation of TFAA reactions of in situ–generated munchnones, Kawase[47] described a novel synthesis of 2-trifluoromethyltetrahydro-3-benzazepinones **77** from tetrahydroisoquinoline-1-carboxylic acids **76** (Table 4.3). This novel conversion is thought to involve an oxazolium intermediate, which is formed from the corresponding munchnone (see Scheme 4.5). Addition of TFA followed by a

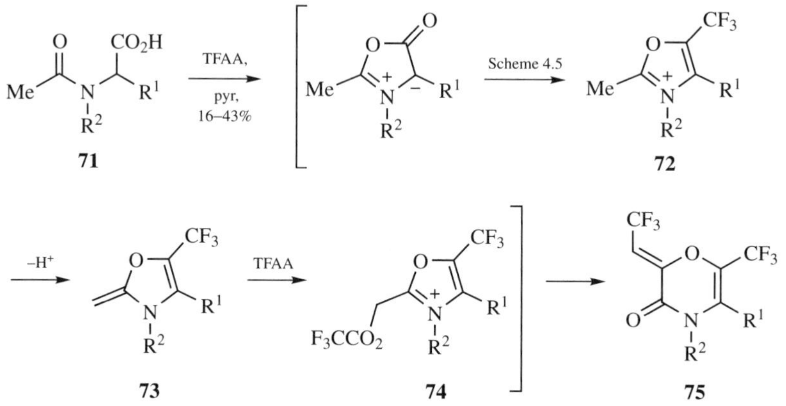

Figure 4.22

R¹ = Bn, Ph, *i*-Pr, Me, *i*-Bu; R² = Bn, Me

ring-opening, ring-closing sequence affords the final product **77** (Fig. 4.24). In support of this mechanism is the isolation of some of the Dakin-West product **78**, and **79** when R = Me (28%). Moreover, when R = OEt only the corresponding Dakin-West product **78** is isolated (61%).

Kawase and co-workers[48] found *N*-alkoxycarbonylprolines **80** undergo a novel conversion to 3-trifluoroacetyl dihydropyrroles **81** (Table 4.4). This reaction is believed to involve formation of the *N*-alkoxycarbonyl dihydropyrroles, which are subsequently acylated (supported by control experiments). However, formation of the dihydropyrrole ring remains curious but presumably an alkoxymunchnone is

TABLE 4.3. REACTIONS OF TETRAHYDROISOQUINOLINE-1-CARBOXYLIC ACIDS WITH TRIFLUOROACETIC ANHYDRIDE

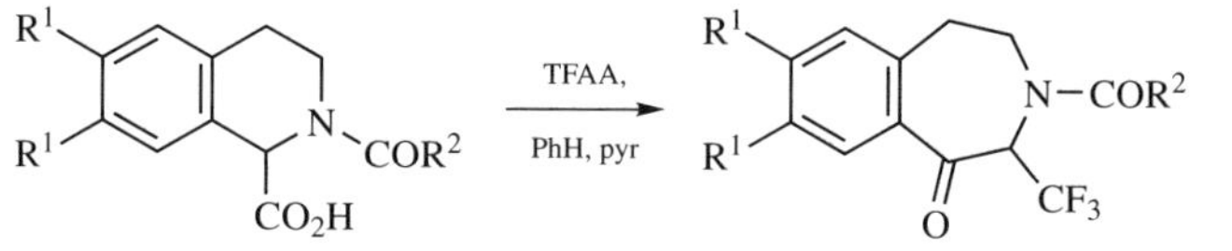

Figure 4.23

R¹	R²	Yield **77** (%)
H	*t*Bu	98
OMe	*t*Bu	92
H	Ph	61
H	4-Cl-Ph	58
H	4-MeO-Ph	52

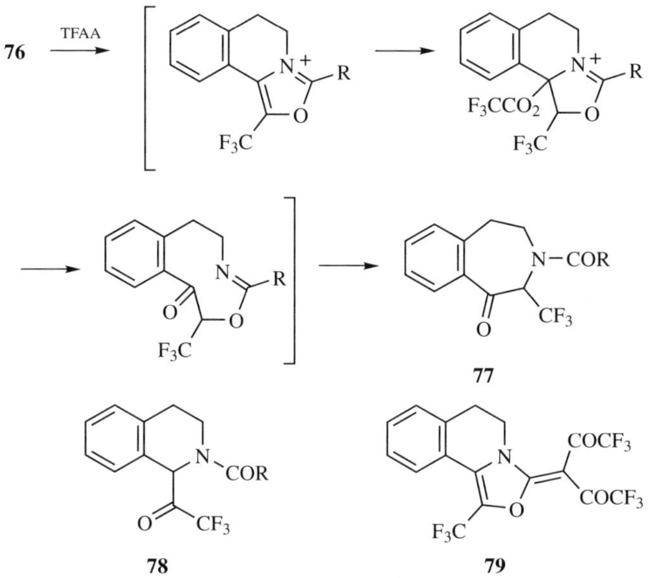

1,3-Oxazolium 5-Oxides (Munchnones) 489

Figure 4.24

TABLE 4.4. REACTIONS OF *N*-ALKOXYCARBONYLPRO-LINES WITH TRIFLUOROACETIC ANHYDRIDE

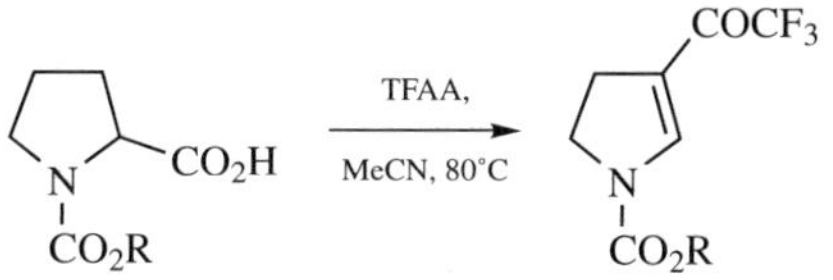

80 → (TFAA, MeCN, 80°C) → **81**   Figure 4.25

R	Solvent	Yield **81** (%)
Me	MeCN	66
Et	MeCN	61
Bu	DMF	68
Ph	MeCN	32
Fluoren-9-yl	MeCN	34
Allyl	MeCN	30[a]
Bn	MeCN	0[b]

[a] *N*-allylacetamide was isolated in 54% yield.
[b] The only products isolated were *N*-benzylacetamide (94%) and proline (45%).

490   Mesoionic Oxazoles

produced at the outset. Indoline **82** undergoes a similar transformation to indole **83**, but trifluoromethyl-acetylation does not occur (Fig. 4.26).

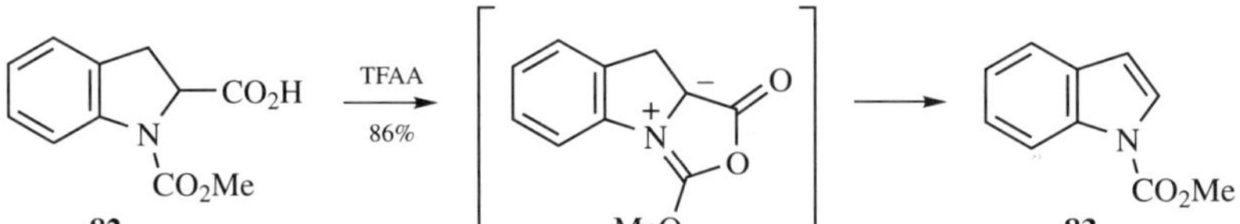

Figure 4.26

This group[49] also employed the Dakin-West reaction to prepare a series of amino alcohols (not shown) by reduction of the trifluoromethyl ketones **85**, resulting from treatment of *N*-alkoxycarbonylamino acids **84** with TFAA (Fig. 4.27).

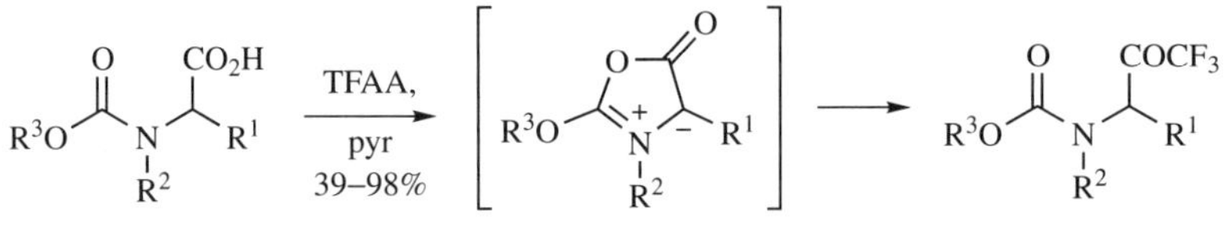

R^1 = H, Me, Bn, Ph, *i*-Bu; R^2 = Bn, Me, 3,4-di-MeO-Bn; R^3 = Me, Et, Bn

Figure 4.27

4.2.3.2. Nucleophilic Additions

The hydrolysis of munchnones is well known. Petride[33,51–53] studied the kinetics and mechanism of the reaction of water and other nucleophiles with bicyclic munchnones. For example, **86** reacts with water to give the expected product **87** (Fig. 4.28). Ethanol affords ester **88**, and **90** is formed when aniline is allowed to

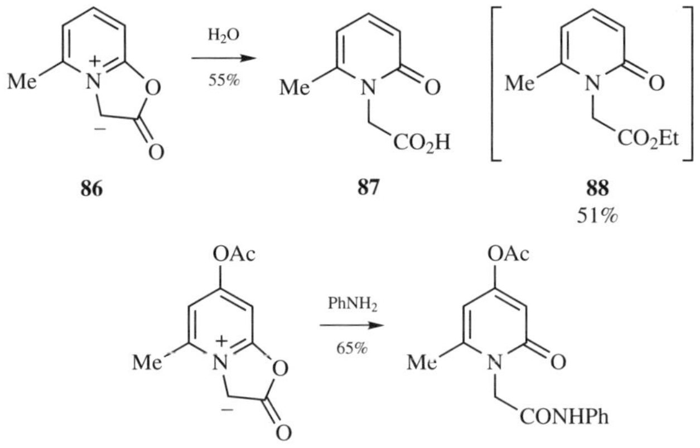

Figure 4.28

react with munchnone **89**. The nonplanar bicyclic munchnones (e.g., **51b**) are much more reactive with nucleophiles than are the coplanar analogs (e.g., **51a**).[51–53] Slightly different mechanisms are proposed to account for this difference.[53]

Kawase and co-workers[54] employed ^{18}O-labeled munchnones **91** to support the earlier notion[55] that hydrolysis occurs via attack at C-5 and not C-2. Moreover, $H_2^{18}O$ is not incorporated into the amide carbonyl of **92** (Fig. 4.29). Molecular orbital calculations (PM 3) are consistent with attack at C-5 for the hard nucleophile water, but attack is predicted to occur at C-2 for softer nucleophiles such as amines. Indeed, as discussed below, this is the case. Thus whereas there is a higher positive charge on C-5 than on C-2, the LUMO coefficient is larger on C-2 than on C-5.

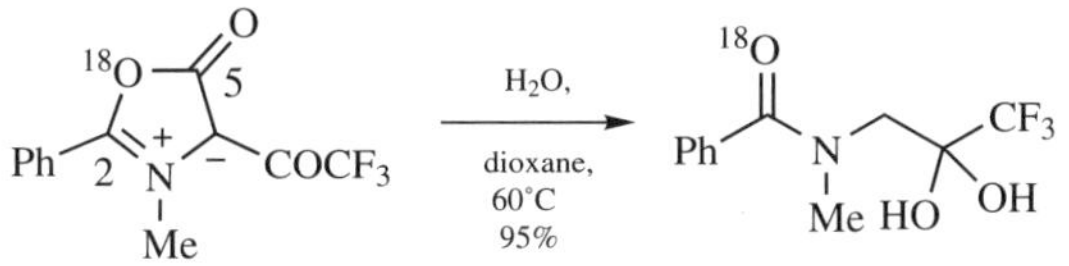

Figure 4.29

Kawase and co-workers[56,57] found, as predicted, that amidines react with munchnones to give imidazoles **93** and dihydroimidazole alcohols **94** (Table 4.5). A proposed mechanism is shown in Scheme 4.7.

The release of ammonia in the proposed mechanism for the formation of **93** suggested to Kawase and co-workers[58] that ammonia might serve as a nucleophile

TABLE 4.5. NUCLEOPHILIC REACTIONS OF MUNCHNONES WITH AMIDINES[a]

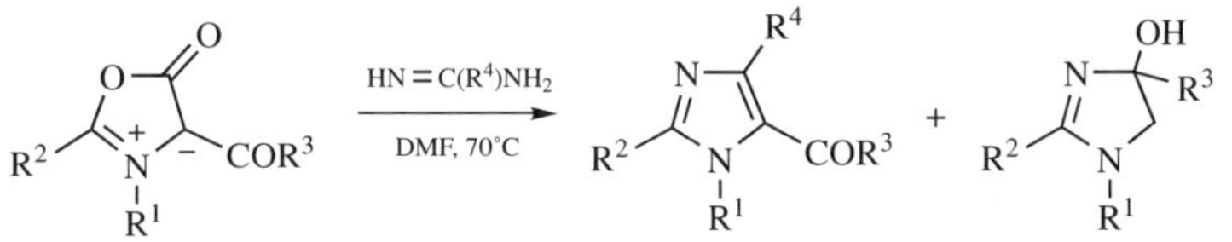

R^1	R^2	R^3	R^4	Yield **93** (%)	Yield **94** (%)
Me	Ph	CF_3	H	63	19
Ph	Ph	CF_3	H	54	—
Ph	Me	CF_3	H	55	14
Me	tBu	CF_3	H	49	15
Me	Ph	CF_3	Me	46	16
Me	Ph	CF_3	Ph	55	25
Me	Ph	C_2F_5	H	60[a]	14
Me	Ph	$n-C_3F_7$	H	56[b]	20

[a] Data from refs. 56 and 57.
[b] Yield refers to products using in situ-generated munchnones.

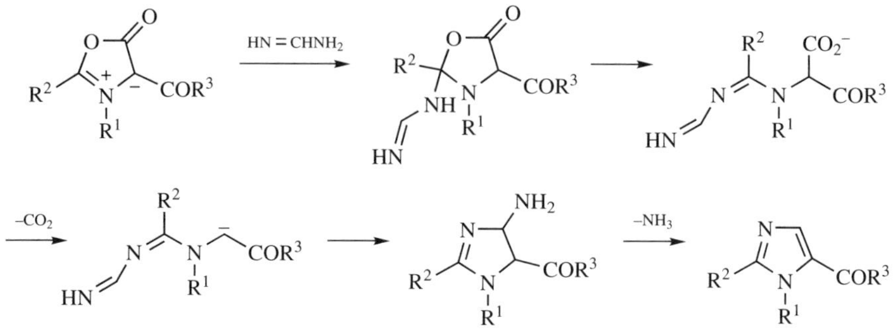

Scheme 4.7

with munchnones. Indeed, ammonium acetate reacts smoothly with munchnones to form **94** and imidazoles **95** upon dehydration (Table 4.6). A mechanism similar to that give in Scheme 4.7 can be envisaged. Thus imine **96** can cyclize to the product **94** (Fig. 4.32).

Kawase and co-workers[59] demonstrated the remarkable ability of phenylhydrazine to react with 4-trifluoroacetylmunchnones at three different positions, depending solely on solvent. Thus, in DMF, C-2 attack is preferred, leading to 1,2,4-triazines **97** (Fig. 4.33). But, in benzene, C-5 attack occurs affording pyrazoles

TABLE 4.6. NUCLEOPHILIC REACTIONS OF IN SITU–GENERATED MUNCHNONES WITH AMMONIA[a]

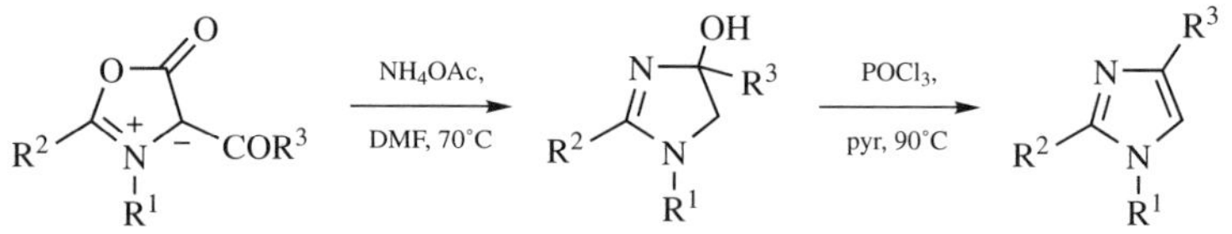

R^1	R^2	R^3	Yield **94** (%)	Yield **95** (%)
Me	Ph	CF_3	94	88
Ph	Ph	CF_3	91	93
Ph	Me	CF_3	91[b]	96
Me	tBu	CF_3	78[b]	88
Me	Bn	CF_3	98	91
Me	Ph	C_2F_5	93	99
Me	Ph	n-C_3H_7	87	93

[a] Munchnones were generated from the N-acylamino acids with the appropriate perfluoro carboxylic acid anhydride.
[b] Yield refers to product obtained using isolated munchnone.

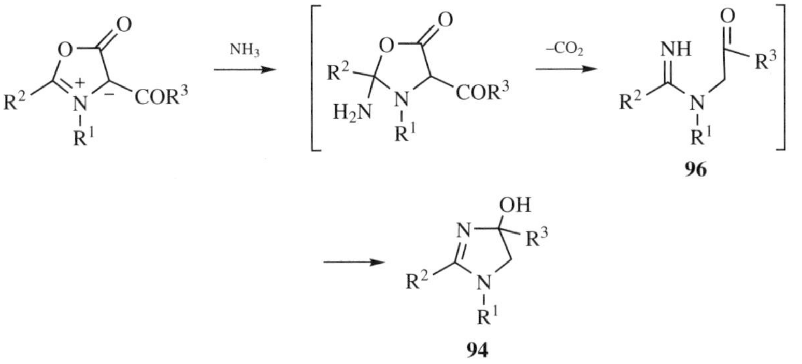

1,3-Oxazolium 5-Oxides (Munchnones) 493

Figure 4.32

98. Finally, in 1,2-dichloroethane, attack on the acyl group leads to pyrazolones **99**. Pyrazoles **98** may derive from ring-opened ketene, but detailed mechanisms for the formation of these compounds were not proposed.

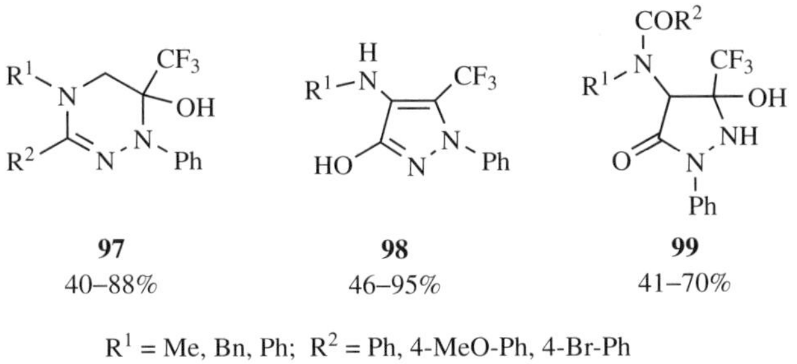

97
40–88%

98
46–95%

99
41–70%

R^1 = Me, Bn, Ph; R^2 = Ph, 4-MeO-Ph, 4-Br-Ph

Figure 4.33

Kawase and co-workers[60] also examined aminomalonate as a nucleophile with munchnones. In this reaction, initial attack at C-5 is followed by enolate cyclization onto the trifluoroacetyl group, leading to pyrrolidin-2-ones **100** (Fig. 4.34). The less enolic ethyl glycinate and N-methylglycinate failed to react.

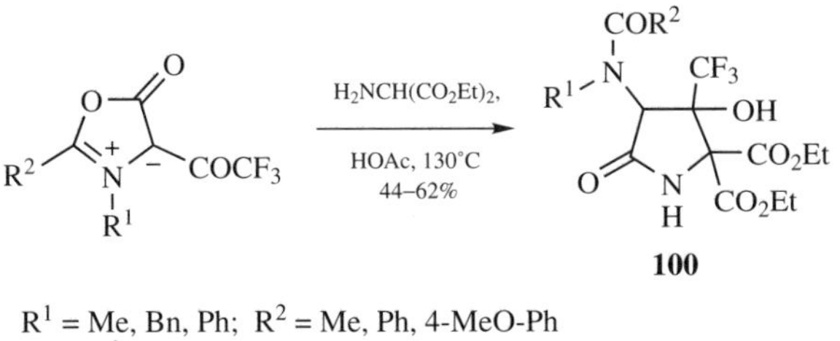

$H_2NCH(CO_2Et)_2$, HOAc, 130°C, 44–62%

100

R^1 = Me, Bn, Ph; R^2 = Me, Ph, 4-MeO-Ph

Figure 4.34

Sain and Sandhu[61] found that azlactones react with dienamine **101** to give cyclohexenone **102** in good yield (Fig. 4.35). Because these reactions occur at or below room temperature, the authors favor a pathway involving C-5 attack on a munchnone rather than thermal equilibration to the ketene tautomer and subsequent

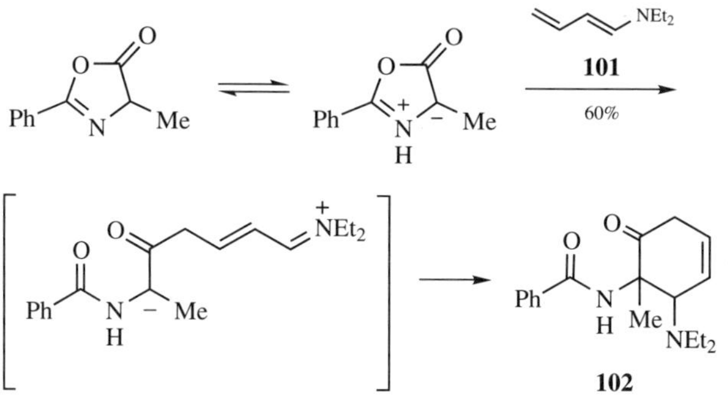
NEt2
101
60%
O
Ph
N
Me
H
NEt2
102

Figure 4.35

reaction with the dienamine. Märkl and Dormeister[62] observed that tris(trimethylsilyl)phosphine **104** reacts with munchnones **42** and **103** to give 1,3-azaphospholes **106** after hydrolysis of the intermediates **105** (Fig. 4.36).

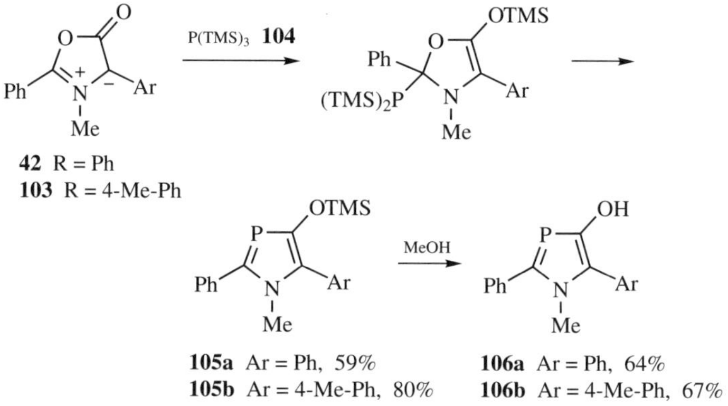
P(TMS)3 104
OTMS
(TMS)2P
Ph
Ar
Me
42 R = Ph
103 R = 4-Me-Ph
MeOH
OH
105a Ar = Ph, 59%
105b Ar = 4-Me-Ph, 80%
106a Ar = Ph, 64%
106b Ar = 4-Me-Ph, 67%

Figure 4.36

4.2.3.3. Cycloaddition Reactions

As Gingrich and Baum[1] said in their review, "the most important reactions (of munchnones) from a synthetic point of view are 1,3-dipolar cycloaddition reactions." If anything, as will be seen, this is even more true today. Although the factors governing the regioselectivity of unsymmetrical munchnone cycloadditions are not completely understood, and clearly involve several competing effects, the synthetic utility of this chemistry is enormous and indisputable.

4.2.3.3.1. Acetylenic Dipolarophiles

The archetypal acetylenic dipolarophile is DMAD, and this remarkably cooperative compound continues to find employment in the munchnone industry. Anderson and co-workers[63,64] used the munchnone generation DMAD trapping protocol to synthesize various pyrrolizines as potential antileukemic agents, such as **107** and the pyrrolo[1,2-c]thiazoles **108** (Fig. 4.37).

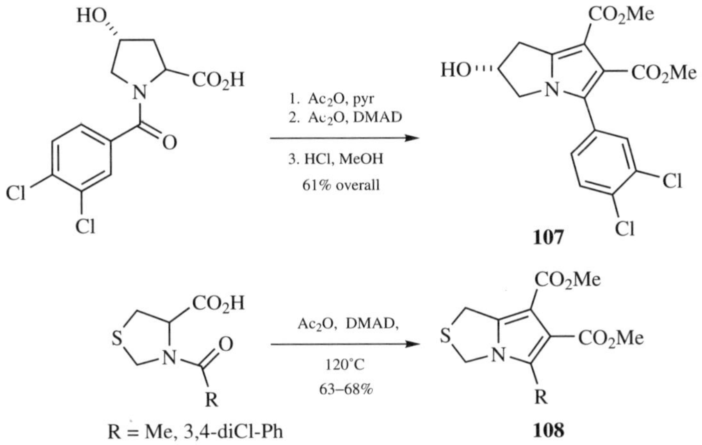

Figure 4.37

Padwa and co-workers[65] used pyrrolothiazole **108** (R = Me)[66] to prepare sulfone **109** (Fig. 4.38). However, attempts to extrude sulfur dioxide from **109** to form the novel dipole **110** were unsuccessful. Györgydeák and colleagues[67] converted a series of 2-substituted 3-acyl-1,3-thiazolidine-4-carboxylic acids **111** into the corresponding pyrrolo[1,2-c]thiazoles **113** via munchnones **112**, which were trapped with DMAD (Fig. 4.39).

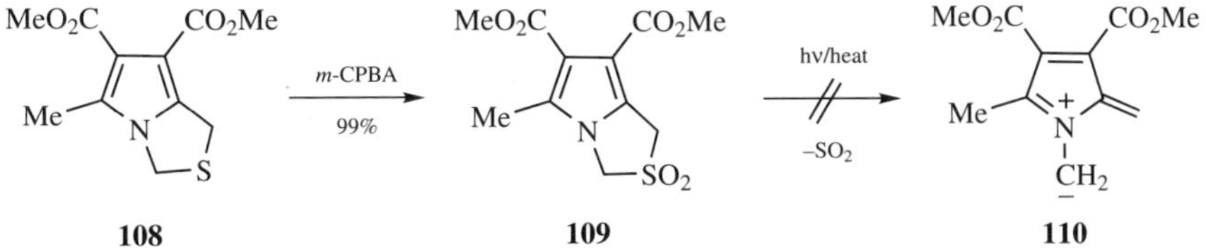

Figure 4.38

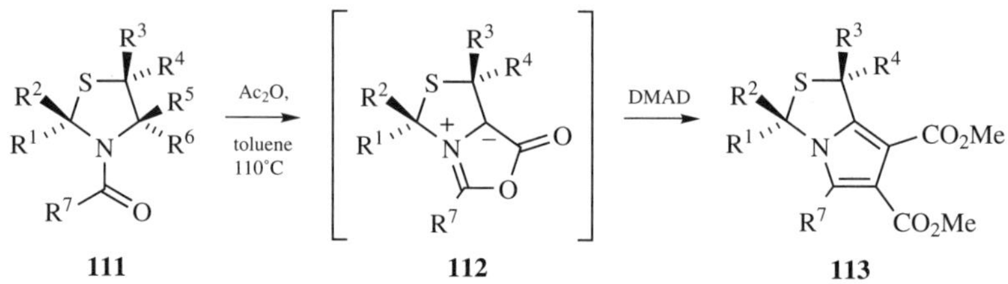

496   Mesoionic Oxazoles

Figure 4.39

R^1, R^2 = H, Ph, Ar, sugar; R^3, R^4 = H, Me; R^7 = Me, Et, tBu

Pinho e Melo and co-workers[68] heated a diastereomeric mixture of (2R,4R)- and (2S,4R)-2-phenylthiazolidine-2-carboxylic acids **114** with acetic anhydride in the presence of DMAD to afford pyrrolothiazole **115** in good yield and 99% ee (Fig. 4.40). Robba and co-workers[69] synthesized the 2-azapyrrolo[1,2-a]indole ring system via a munchnone cycloaddition strategy. Trapping the munchnone derived from proline derivative **116** gave pyrrole **117** in 75% yield (Fig. 4.41). Further elaboration yielded the desired **118** and subsequent target compounds.

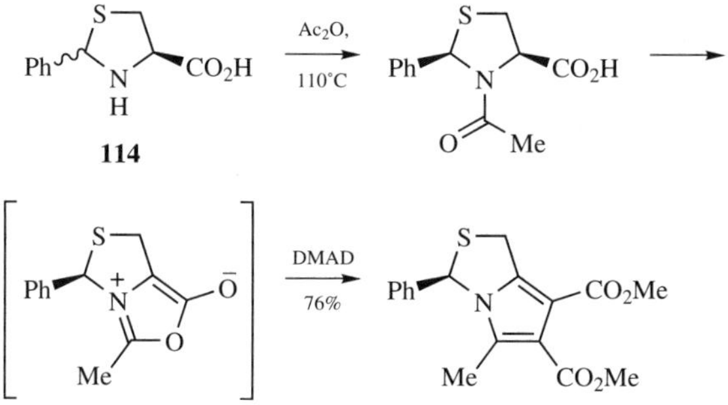

Figure 4.40

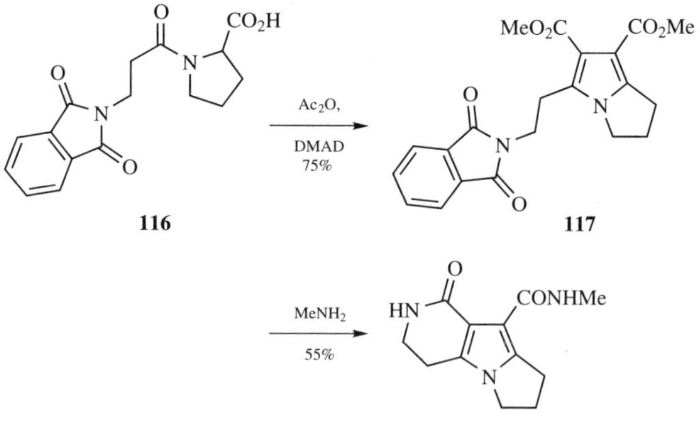

Figure 4.41

Vasella and co-workers[70] employed munchnone chemistry in the synthesis of several pyrrolopyridines and imidazopyridines as novel inhibitors of β-D-glucosidases. Thus treatment of lactam glycine **119** with acetic anhydride in the presence

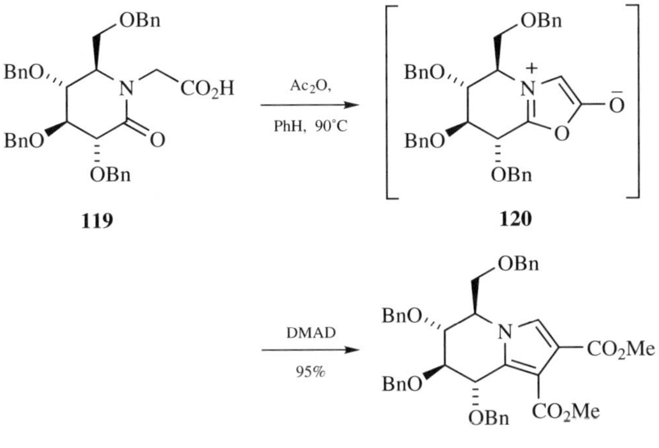

119 **120** **121**

Figure 4.42

of DMAD affords pyrrole **121** via munchnone **120** in 95% yield (Fig. 4.42). These authors also generated munchnone **120** in the presence of methyl propiolate using EDC (Fig. 4.43).[70] This reaction gives pyrroles **122** and **123** in the ratio of 4:3 (77% yield). The imidazopyridine **125** was synthesized by treating amino acid **124** with

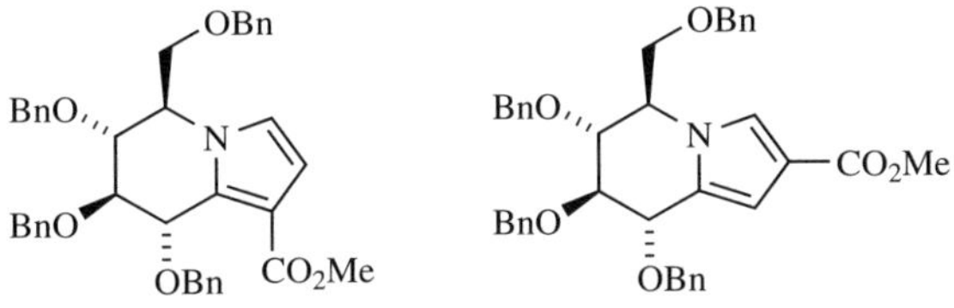

122 **123**

Figure 4.43

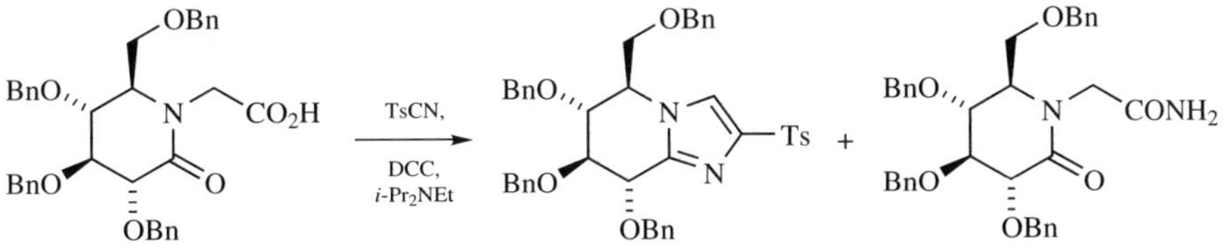

124 **125** 53% **126** 23%

Figure 4.44

p-toluenesulfonyl cyanide and DCC (Fig. 4.44). The yield of **125** was lower (38%) when the munchnone was generated with mesyl chloride.[70] The minor amide product **126** is proposed to arise from the diastereoisomeric munchnone adduct **127**, which undergoes fragmentation rather than decarboxylation (Fig. 4.45).[70]

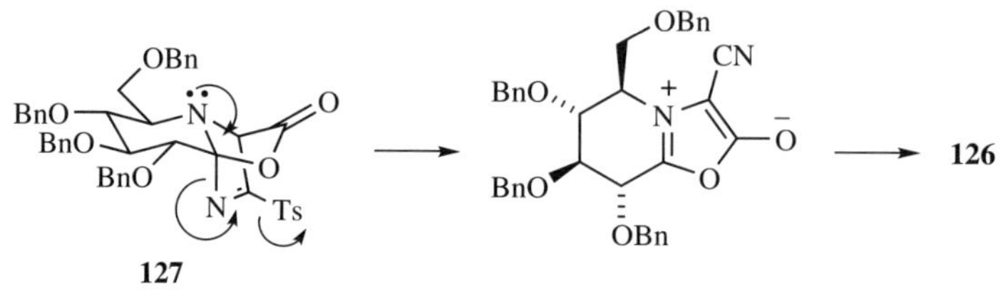

Figure 4.45

Pinho e Melo and co-workers[68] explored similar reactions of the **114**-derived munchnone with methyl propiolate and methyl vinyl ketone to afford pyrrolothiazoles **128** (55%) and **129a** (25%), respectively (Fig. 4.46). The latter reaction also gave spiro lactone **130a** in 9% yield; its structure was established by X-ray crystallography.[71] The regiochemistry in these reactions is consistent with that described earlier. Similar products **129b** (10%) and **130b** (8%) are produced with ethyl vinyl ketone. Again, the structure of the spirolactone **130b** was clinched by X-ray crystallography.

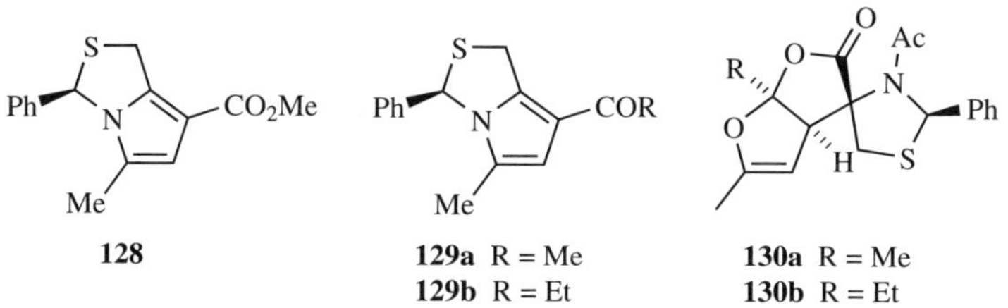

Figure 4.46

The authors propose a pathway to these novel spirolactones as involving munchnone ring opening to ketene **131** followed by condensation with acetic anhydride and alkyl vinyl ketone in some fashion to give intermediate **132** (Fig. 4.47). Loss of acetic acid leads to spirolactone **130**. Pinho e Melo and colleagues[68] also found that acrylonitrile reacts with this munchnone to afford a 1:1 mixture of epimers **133** in very low yield (3%) (Fig. 4.48). Oxidation with DDQ gave the aromatic pyrrole **134**.

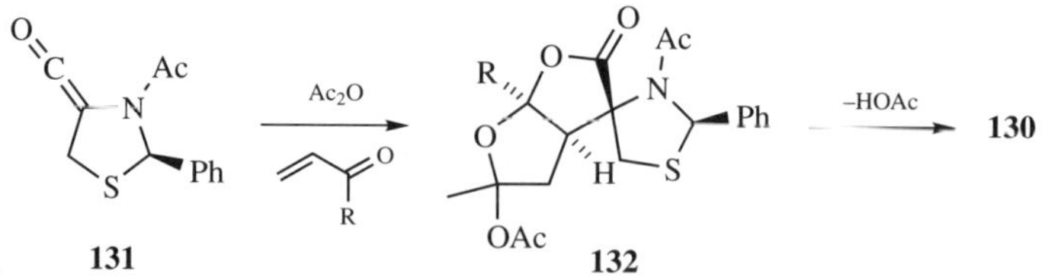

Figure 4.47

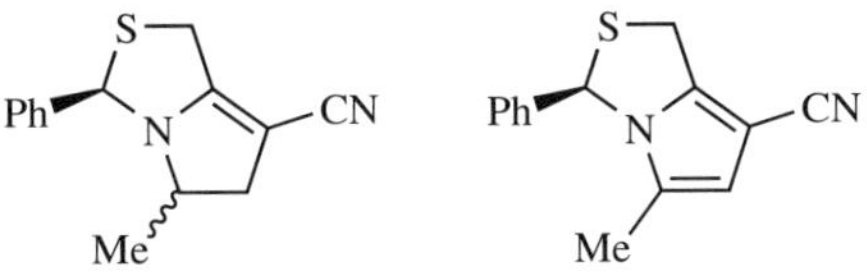

1,3-Oxazolium 5-Oxides (Munchnones)

Figure 4.48

Chan and co-workers[72] generated the novel munchnone–sydnone hybrids **137–138** by cyclodehydration of the sydnone glycine **135** and alanine **136**, respectively (Fig. 4.49). Trapping with DMAD gave **139** and **140**. Maleic anhydride and dimethyl maleate failed to ambush these munchnone–sydnone hybrids. Exposure of **135** to trifluoroacetic anhydride gave acylated munchnone **141** (by NMR). Workup affords the hydrolysis product **142**.

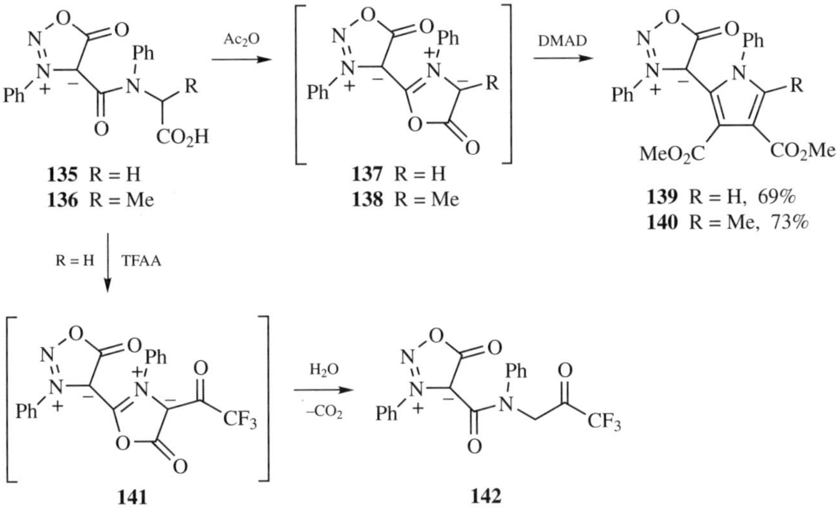

135 R = H
136 R = Me

137 R = H
138 R = Me

139 R = H, 69%
140 R = Me, 73%

141

142

Figure 4.49

Yamanaka and co-workers[73] effected the 1,3-dipolar cycloaddition between munchnones and polyfluoro-2-alkynoic acid esters to afford the corresponding 4-(polyfluoroalkyl)pyrrole-3-carboxylates **144** (Fig. 4.50). The reaction proceeds

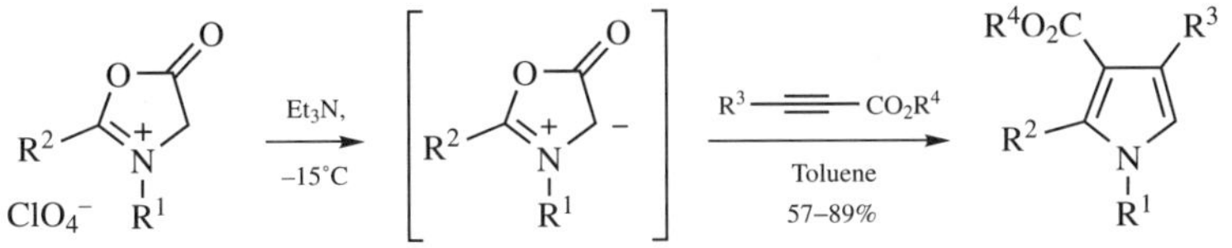

143

144

R^1 = Ph, Me (one case); R^2 = Ph, 4-NO_2-Ph, 4-Cl-Ph, 4-Me-Ph, 4-MeO-Ph, Me; R^3 = CHF_2, CF_3, $H(CF_2)_3$, CO_2Me; R^4 = Me, n-C_6H_{11} (one case)

Figure 4.50

rapidly at low temperature, and the various munchnones were generated from 1,3-oxazolium perchlorates **143** using the method of Boyd and Wright.[13,14]

Under these reaction conditions methyl 2-butynoate failed to react with the munchnone from **143**, and DMAD gave the corresponding pyrrole in 75% yield (Fig. 4.51). Interestingly, changing the solvent from toluene to acetonitrile promoted Michael addition products (e.g., **145**). In this case, only a 24% yield of pyrrole **144** was realized. The rapid reaction of these polyfluorinated esters in this 1,3-dipolar cycloaddition reaction is attributed to a lowering of LUMO energy as a result of the powerful electron-withdrawing ability of the CHF_2 group.

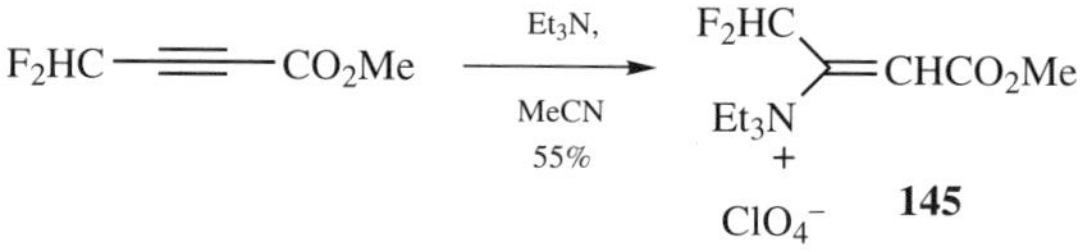

Figure 4.51

Padwa and co-workers[74] found that unsymmetrical munchnone **146**, which was generated from N-acetyl-N-benzylglycine and refluxing acetic anhydride, reacts with methyl propiolate to give an 8:1 mixture of pyrroles **147** and **148** (Fig. 4.52). The same product ratio is obtained from the reaction of methyl propiolate and the azomethine ylide derived from N-benzyl-N-(α-cyanoethyl)-N-[(trimethylsilyl)-methyl]amine.

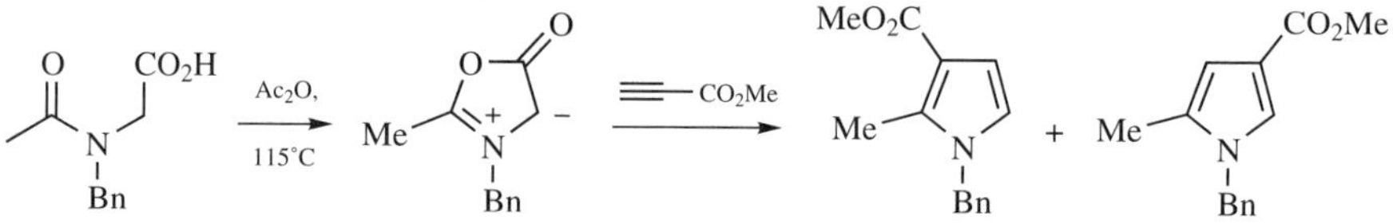

Figure 4.52

Similarly, Toupet and co-workers[75] determined the structures of the pyrrole adducts **150** from the cycloaddition of munchnones **149** with methyl phenylpropionate by X-ray crystallography (Fig. 4.53). These highly regioselective cycloadditions are in accord with FMO predictions. A detailed study by this same team[76] involved cycloaddition reactions of munchnones with methyl propiolate and methyl

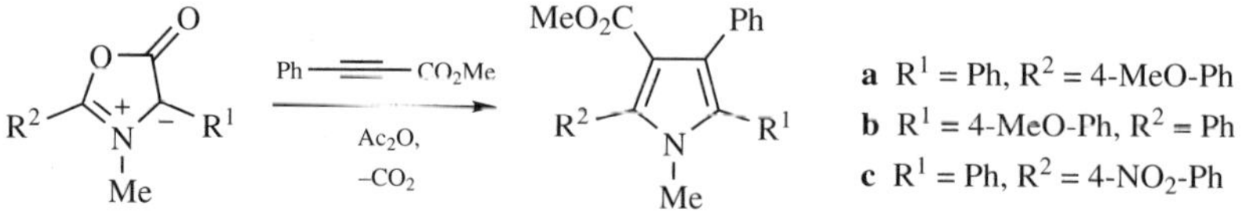

a R^1 = Ph, R^2 = 4-MeO-Ph
b R^1 = 4-MeO-Ph, R^2 = Ph
c R^1 = Ph, R^2 = 4-NO_2-Ph

Figure 4.53

TABLE 4.7. 1,3-DIPOLAR CYCLOADDITION REACTIONS OF IN SITU–GENERATED MUNCHNONES AND METHYL PROPIOLATE AND METHYL 3-PHENYLPROPIOLATE[a]

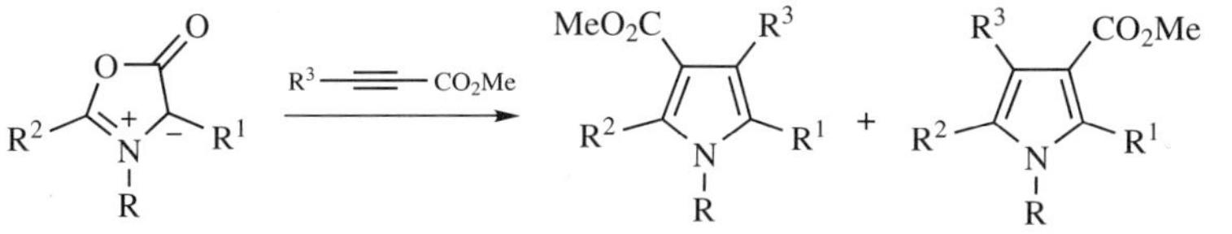

R	R^1	R^2	R^3	Yield (%)	151 (%)	152 (%)
Me	Ph	Me	H	100	48	52
Me	Me	Ph	H	100	65	35
Ph	Ph	Me	H	100	35	65
Ph	Me	Ph	H	100	65	35
Me	Ph	4-NO$_2$-Ph	H	100	50	50
Me	Ph	Me	Ph	65	95	5
Me	Me	Ph	Ph	55	95	5
Ph	Ph	Me	Ph	40	90	10
Ph	Me	Ph	Ph	47	90	10
Me	Ph	4-MeO-Ph	Ph	40	100	0
Me	4-Me-Ph	Ph	Ph	60	100	0
Me	Ph	4-NO$_2$-Ph	Ph	70	100	0
Me	3-NO$_2$-Ph	Ph	Ph	82	100	0

[a] Data from ref. 76.

Figure 4.54

3-phenylpropiolate to give mixtures of the regioisomeric pyrroles **151** and **152** (Table 4.7).

Yebdri and Texier[77] studied cycloaddition reactions of the well-known munchnone **153** (derived from proline and acetic anhydride) with methyl propiolate and methyl 3-phenylpropiolate to give mixtures of pyrrolizines **154** and **155**, respectively (Fig. 4.55).

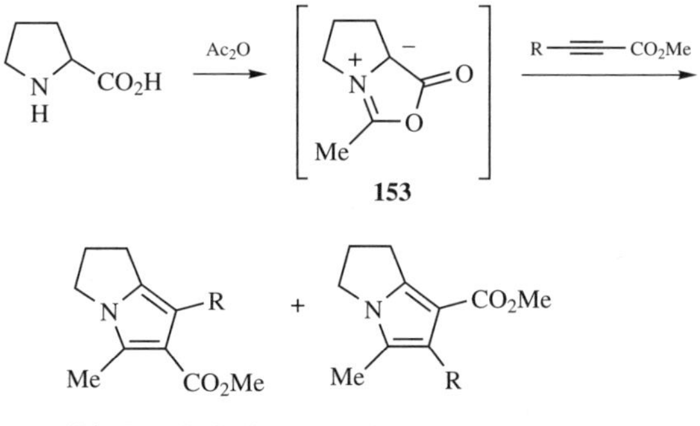

154a R = H, 21%
155a R = Ph, 40%

154b R = H, 42%
155b R = Ph, 21%

Figure 4.55

TABLE 4.8. 1,3-DIPOLAR CYCLOADDITION REACTIONS OF IN SITU–GENERATED MUNCHNONES AND TERMINAL ALKYNES[a,b]

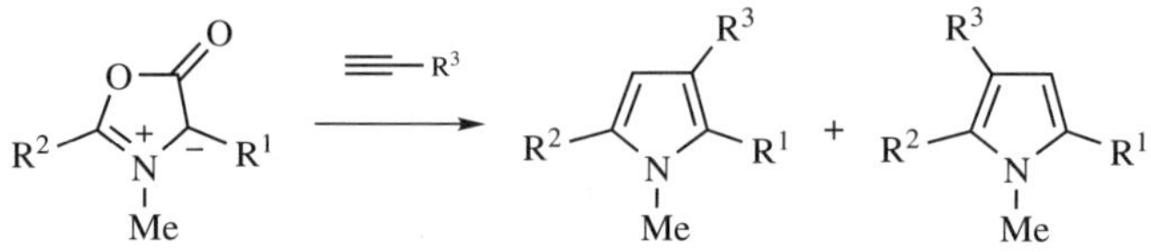

Figure 4.56

R^1	R^2	R^3	Yield (%)	156 (%)	157 (%)
Ph	H	CO_2Et	77	75	25
H	Ph	CO_2Et	30	14	86
Ph	H	COPh	80	75	25
H	Ph	COPh	58	0	100
Ph	H	Ph	34	100	0
Me	H	CO_2Et	51	84	16
H	Me	CO_2Et	24	25	75
Me	H	COPh	85	80	20
H	Me	COPh	11	0	100
Me	H	Ph	4	100	0
Ph	Me	CO_2Et	87	38	62
Me	Ph	CO_2Et	99	43	57
Ph	Me	COPh	29	2	98
Me	Ph	COPh	83	18	82
Ph	Me	Ph	50	98	2
Me	Ph	Ph	44	99	1

[a] Data from ref. 78.
[b] Munchnones were generated from the corresponding N-acylamino acids using acetic anhydride in toluene at 80°C.

Dalla Croce and La Rosa[78] examined the 1,3-dipolar cycloaddition reactions of unsymmetrical munchnones with terminal alkynes to give pyrroles 156 and 157 (Table 4.8). The reaction is generally regioselective; the major pyrrole isomer from the monosubstituted munchnones have adjacent hydrogens, irrespective of the munchnone substituent. With the disubstituted munchnones, the major regioisomer is derived from attachment of C-4 of the munchnone and the β-carbon of the alkyne, except for phenylacetylene, which shows the opposite regiochemistry. The authors interpreted this behavior as a consequence of the electron-rich nature of phenylacetylene as a dipolarophile with a larger LUMO coefficient on the α carbon.

Gribble and co-workers[79] observed the same behavior with N-benzylmunchnones 158 and 161 to give pyrroles 159 and 160, respectively, in a highly regioselective fashion (Fig. 4.57).

Coppola and co-workers[80] studied extensively the dipolar cycloaddition of methyl propiolate with unsymmetrical munchnones. In addition to their own results,

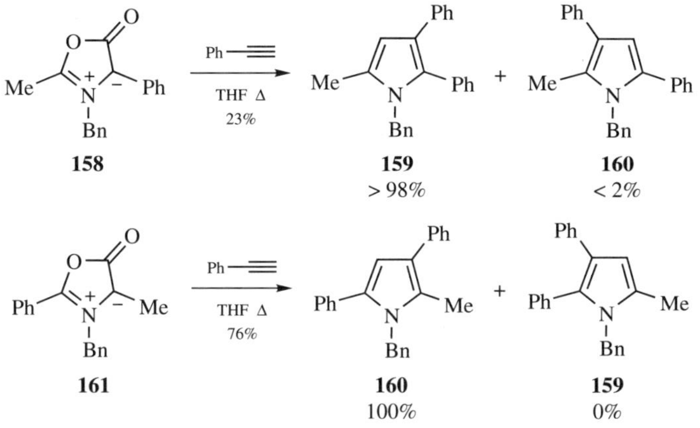

Figure 4.57

these investigators summarized much previous data on these cycloadditions: "No single criterion can successfully be used to correlate the experimental observations regarding the regioselectivity in munchnone cycloaddition reactions." Steric and electronic effects must be considered in analyzing these cycloadditions. One revealing result from this study is the use of C-13 labeled munchnones **162** and **163**, which exhibit essentially no regiochemical bias in reacting with methyl propiolate (Fig. 4.58; Table 4.9).

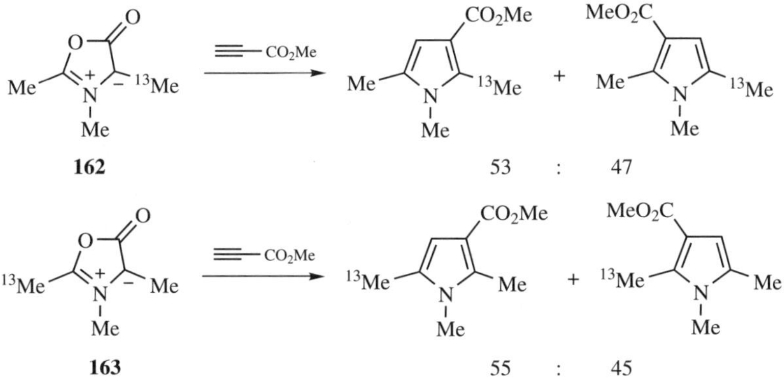

Figure 4.58

These workers also examined the influence of thio-substitution on the cycloaddition of munchnones with methyl propiolate (Table 4.10). The results are consistent with an unsymmetrical transition state in which bond formation between the two-least encumbered positions occurs first. This leads to pyrrole products, e.g., **166** (R = H), with adjacent hydrogens.

TABLE 4.9. 1,3-DIPOLAR CYCLOADDITION REACTIONS OF IN SITU–GENERATED MUNCHNONES AND METHYL PROPIOLATE[a]

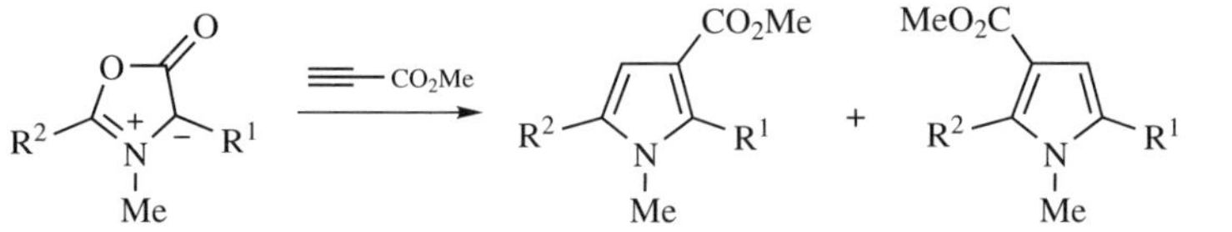

Figure 4.59

R^1	R^2	Yield (%)	164 (%)	165 (%)
i-Pr	H	56	57	43
H	i-Pr	92	25	75
Me	i-Pr	88	67	33
i-Pr	Me	84	23	77
Me	Et	64	55	45
tBu	H	53	67	33
Bn	H	95	83	17
H	Bn	91	16	84

[a] Data from ref. 80.

Coppola and co-workers[81] reexamined the reaction of munchnone **168**, derived from N-formylproline, with methyl propiolate (Fig. 4.61). In contrast to an earlier study that reported complete regioselectivity,[82] they found both regioisomers **169** and **170** in this reaction.

Despite the uncertainties regarding regiochemistry, the reaction of propiolates with munchnones has found use in synthesis. Kane and co-workers[83] synthesized

TABLE 4.10. 1,3-DIPOLAR CYCLOADDITION REACTIONS OF IN SITU–GENERATED MUNCHNONES AND METHYL PROPIOLATE[a]

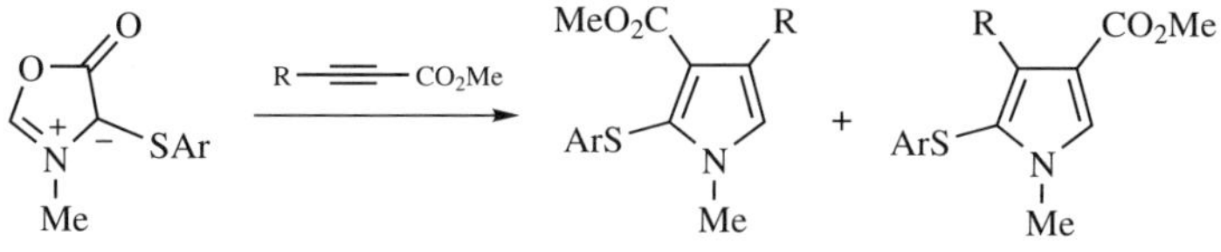

Figure 4.60

Ar	R	Yield (%)	166 (%)	167 (%)
4-MeO-Ph	H	91	80	20
Ph	H	93	82	18
3-CF$_3$-Ph	H	94	84	16
4-NO$_2$-Ph	H	93	84	16
Ph	Me	90	25	75
Ph	Ph	94	0	100

[a] Data from ref. 80.

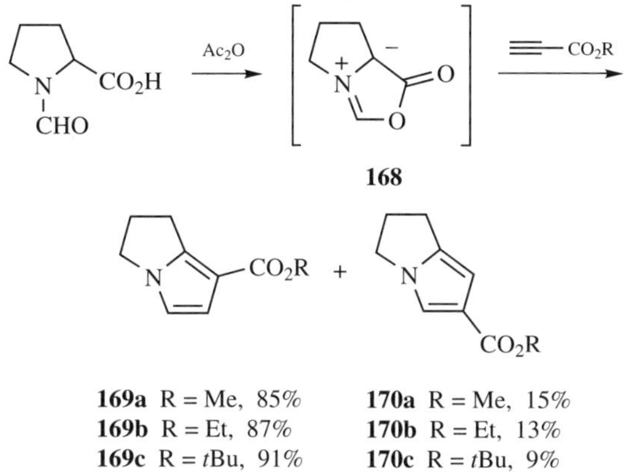

169a R = Me, 85%   **170a** R = Me, 15%
169b R = Et, 87%   **170b** R = Et, 13%
169c R = tBu, 91% **170c** R = tBu, 9%

Figure 4.61

the calcium channel activator FPL 64176 **174** using a munchnone cycloaddition protocol (Fig. 4.62). Thus reaction of amino acid **171** with acetic anhydride in the presence of acetylenic dipolarophile **172** gave pyrrole **173** in 49% yield. Base-induced elimination of the 4-nitrophenethyl protecting group afforded FPL 64176 **174** in 85% yield. The other *N*-protecting groups benzyl and 4-methoxybenzyl could not be removed from the pyrrole products.

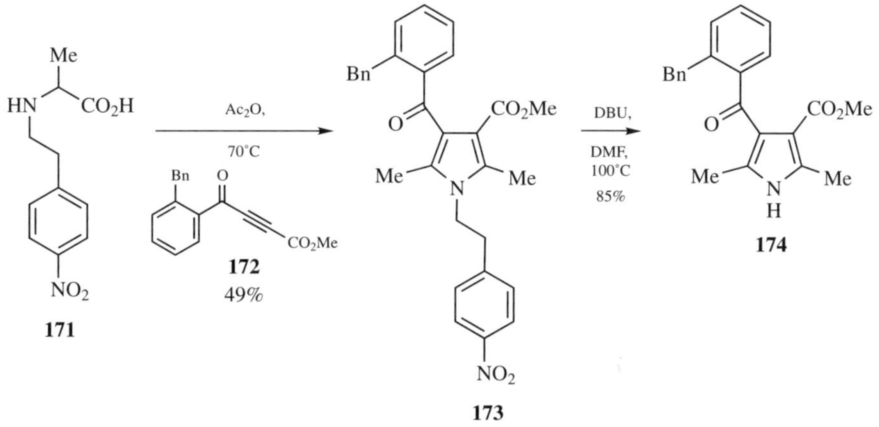

Figure 4.62

Similar to the work of Anderson[63,64] and in continuation of their proline-derived munchnone generation and DMAD trapping[69] (**116 → 117**), Ladurée and co-workers[84] also used the reaction between the munchnones derived from cyclic *N*-acylamino acids **175** and **177** and DMAD to prepare a series of potential new antileukemic agents derived from cycloadducts **176** and **178** (Fig. 4.63).

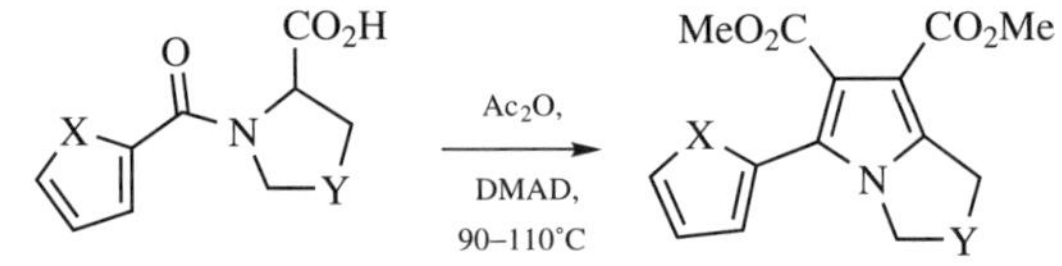

506   Mesoionic Oxazoles

[Figure 4.63 showing reaction schemes]

175a X = O, Y = CH
175b X = S, Y = CH
175c X = Y = S

176a, 45%
176b, 80%
176c, 82%

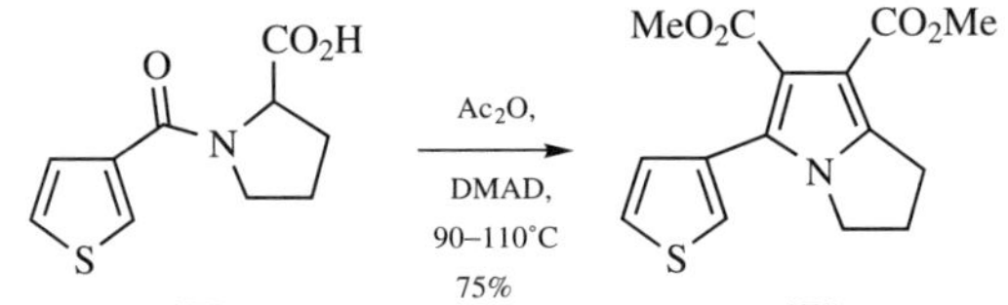

177 → **178** (75%)

Figure 4.63

La Rosa and co-workers[85] found that munchnones **179** react with arylsulfonyl alkynes **180** to furnish 3-arylsulfonyl pyrroles in good yields (Table 4.11). In the case of unsymmetrical munchnones, e.g., **179** (R^1 = Ph, R^3 = Me), only one regioisomer **181** is observed. Pyrroles were identified using two-dimensional NMR techniques. The regioselectivity is explained in terms of a preferred HOMO–

TABLE 4.11. 1,3-DIPOLAR CYCLOADDITION REACTIONS OF IN SITU–GENERATED MUNCHNONES AND ARYLSULFONYL ALKYNES[a]

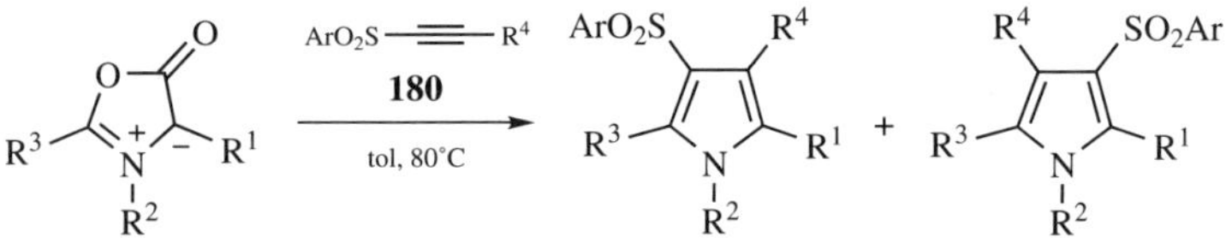

Figure 4.64   **179**   **181**   **182**

R^1	R^2	R^3	R^4	Ar	Yield (%)	**181** (%)	**182** (%)
Ph	Me	Ph	H	Ph	72	—	—
Ph	Me	Ph	Me	Ph	63	—	—
Ph	Me	Ph	Ph	4-Me-Ph	55	—	—
H	Ph	H	H	Ph	61	—	—
H	Ph	H	Me	Ph	62	—	—
H	Ph	H	Ph	4-Me-Ph	68	—	—
Ph	Me	Me	H	Ph	62	100	0
Ph	Me	Me	Me	Ph	67	100	0
Ph	Me	Me	Ph	4-Me-Ph	72	100	0
Me	Me	Ph	H	Ph	86	75	25
Me	Me	Ph	Me	Ph	92	90	10
Me	Me	Ph	Ph	4-Me-Ph	83	88	12

[a] Data from ref. 85.

munchnone/LUMO-alkyne interaction, one of the few cases in which FMO theory satisfactorily explains the regiochemistry of munchnone 1,3-dipolar cycloadditions.

Intramolecular 1,3-dipolar cycloadditions represent a powerful synthetic tool. Kato and co-workers[86] were apparently the first to report an intramolecular munchnone–alkyne cycloaddition. Thus munchnones **184**, as generated from N-acylamino acids **183**, yield the corresponding benzopyrano[4,3-b]pyrroles **186** after extrusion of carbon dioxide from adduct **185** (Fig. 4.65). The yields shown are for high-dilution reaction conditions. Under normal conditions of concentration, the yields are still about 60%. Interestingly, attempts to divert the intramolecular cycloaddition by the addition of N-phenylmaleimide had no effect on the reaction pathway.

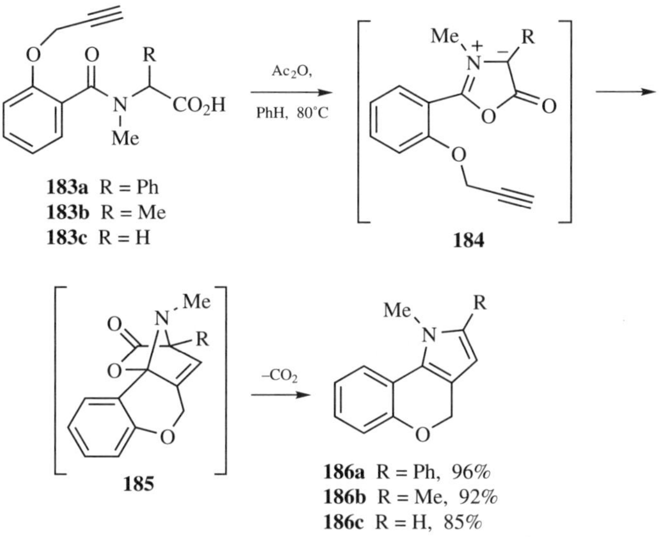

183a R = Ph
183b R = Me
183c R = H

186a R = Ph, 96%
186b R = Me, 92%
186c R = H, 85%

Figure 4.65

Sainsbury and co-workers[87] used the intramolecular 1,3-dipolar cycloaddition of tetrahydroquinolines **187** to construct 11-acetoxycarbonyl-1,2,3,4,5,10-hexahydroindeno[2,3-a]isoquinoline **188** (Fig. 4.66). The presence of an ester functionality on the alkyne is essential for the cycloaddition to occur. With one less methylene group in the tether, the yield of the corresponding pentaleno[2,3-a]isoquinoline is 37%.

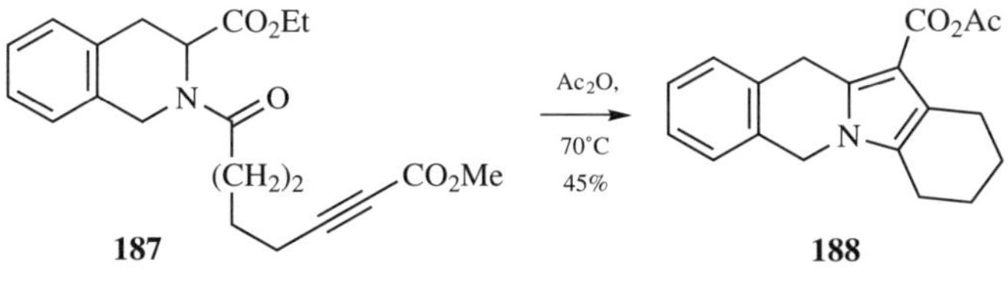

Figure 4.66

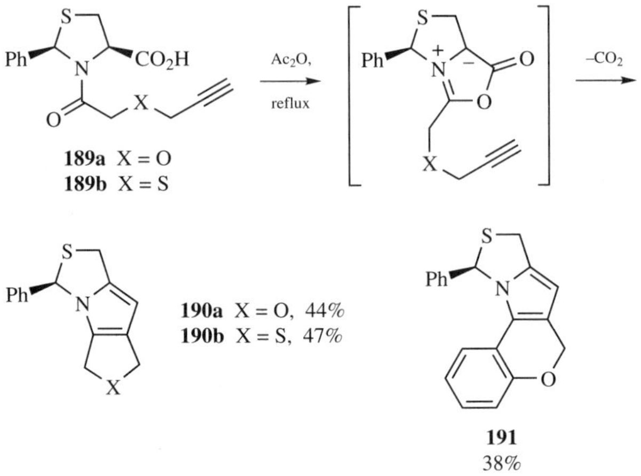

Figure 4.67

Pinho e Melo and co-workers[88] employed an intramolecular munchnone cycloaddition to construct several 1H-pyrrolo[1,2-c]thiazole derivatives. Thus the N-acylthiazolidines **189** on heating in acetic anhydride yielded **190** (Fig. 4.67). The structure of **190a** was confirmed by X-ray crystallography. Analog **191** was also synthesized in this study.

Martinelli and co-workers[89,90] employed an intramolecular munchnone cycloaddition to craft a series of 4-keto-4,5,6,7-tetrahydroindoles **192–195** in two steps (Fig. 4.68). The requisite acetylenic precursors were prepared from glutaric

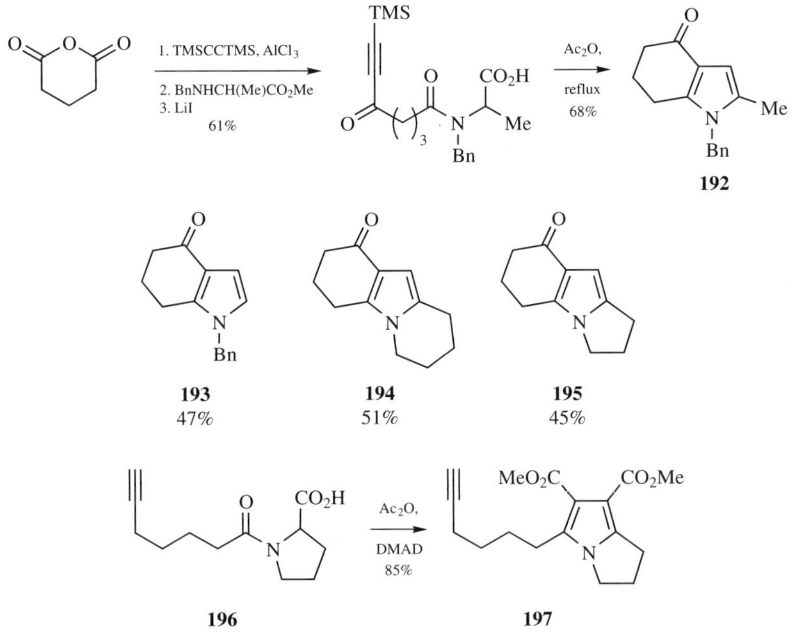

Figure 4.68

anhydride (or 3-methylglutaric anhydride). The overall sequence is illustrated for the synthesis of **192**. An electrophilic acetylenic unit appears to be necessary for successful intramolecular 1,3-dipolar cycloaddition. For example, the munchnone from **196** failed to yield any tetrahydroindole but could be trapped with DMAD to give **197** (85% yield). Trapping with ethyl propiolate was also achieved to afford the corresponding pyrroles as a 2:1 regioisomeric mixture (not shown).

Jursic[91] studied the cycloaddition reaction of a munchnone with acetylene from several theoretical standpoints using density functional theory on AM1 geometries. The predicted activation energy for the 1,3-dipolar cycloaddition is 11.49 kcal/mol and the elimination of carbon dioxide from the cycloadduct to give a pyrrole is 5.82 kcal/mol. Both reactions are extremely exothermic as observed experimentally.

4.2.3.3.2. Aryne Dipolarophiles

Despite its obvious attractiveness as a potential route to isoindoles and benz-fused isoindoles, no new examples of the dipolar cycloaddition of munchnones and arynes have been reported since the first report by Kato and co-workers.[92] This area is ripe for further exploration, particularly because many new mild methods for generating arynes and hetarynes have been developed since 1976.

4.2.3.3.3. Olefinic Dipolarophiles

The 1,3-dipolar cycloaddition reactions of munchnones with olefinic dipolarophiles continues to be of enormous interest in regard to both mechanisms and synthetic applications. Unlike the comparable cycloadditions with acetylenic dipolarophiles that yield only pyrroles, reactions of munchnones with olefinic dipolarophiles can lead to a variety of interesting products.

Nan'ya and co-workers[93,94] reexamined the double cycloaddition of munchnone **42** with two equivalents of several maleimides **198** to give cycloadducts **199** and **200** (Fig. 4.69). Whereas *N*-phenylmaleimide **198a** gives only the exo, endo adduct

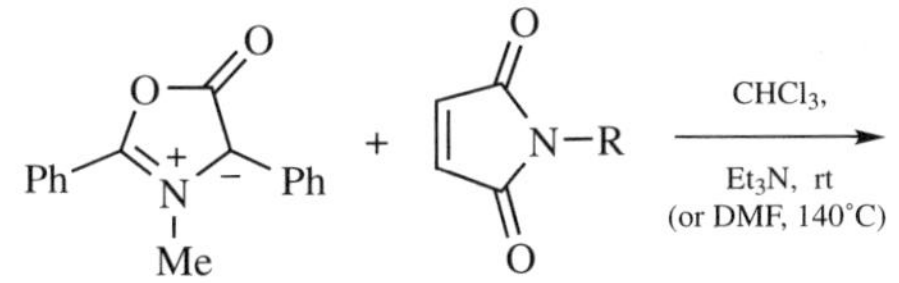

42

198a R = Ph
198b R = Me
198c R = H

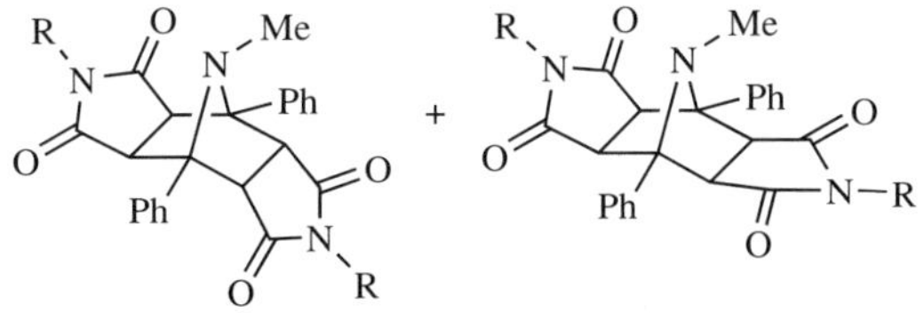

199a R = Ph, 75%
199b R = Me, 50%
199c R = H, 44%

200b R = Me, 22%
200c R = H, 22%

Figure 4.69

TABLE 4.12. 1,3-DIPOLAR CYCLOADDITION REACTIONS OF IN SITU–GENERATED MUNCHNONES AND 1,4-QUINONES[a]

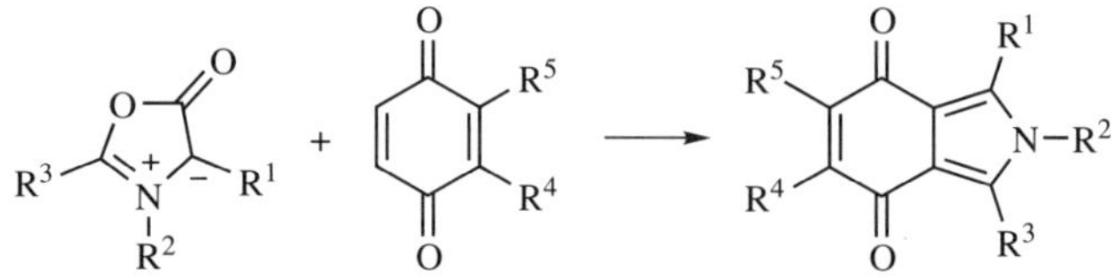

Figure 4.70   201

R^1	R^2	R^3	R^4	R^5	Yield (%)
Ph	Me	Ph	Me	H	35
Ph	Me	Ph	Ph	H	35
-(CH$_2$)$_3$-	Ph	Me	H	}	30
Ph	-(CH$_2$)$_3$-	Me	H		
-(CH$_2$)$_3$-	Ph	Ph	H	}	32
Ph	-(CH$_2$)$_3$	Ph	H		
Ph	Me	Ph	H	Br	43
Ph	-(CH$_2$)$_3$-	H	Br	}	27
-(CH$_2$)$_3$-	Ph	H	Br		

[a] Data from ref. 97.

199a (75%), *N*-methylmaleimide **198b** yields a mixture of exo, endo adduct **199b** (50%) and exo, exo adduct **200b** (22%). Similarly, maleimide **198c** itself affords exo, endo adduct **199c** (44%) and exo, exo adduct **200c** (22%). Huisgen and co-workers[95,96] earlier reported that munchnone **42** and *N*-phenylmaleimide **198a** gave the exo, exo adduct **200** (R = Ph).

In continuation of earlier work,[1] Nan'ya and co-workers[97] reported the synthesis of isoindolediones **201** by the reaction of munchnones with 1,4-benzoquinones (Table 4.12). Reactions with the unsymmetrical munchnone were not regioselective.

Fabre and co-workers[98] found that the proline derivatives **202** react with acetic anhydride and 2-chloroacrylonitrile to form the unexpected bridged adducts **204** (Fig. 4.71). In addition to the minor formation (17%) of the regioisomer corresponding to **204a**, only 5% of the expected munchnone-derived product **205** is obtained (from **202a**). Apparently, the pyridine ring is preventing munchnone formation and instead leads to mesoionic species **203**, which reacts with 2-chloroacrylonitrile to give the major products.

The proline–derived munchnone **153** reacts with electron-deficient alkenes to afford pyrrolizines **207** (Fig. 4.72).[77] The intermediate adducts **206** can be isolated.

Reaction of munchnone **153** with fumaronitrile followed by workup with methanol affords the novel products **211** and **212**,[77] the latter of which was

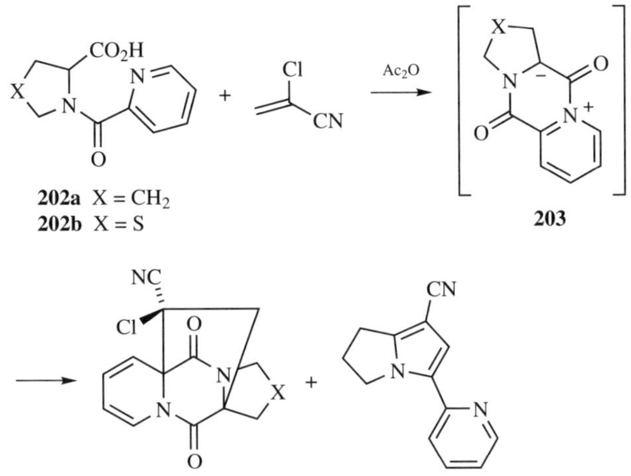

1,3-Oxazolium 5-Oxides (Munchnones)   511

202a X = CH$_2$
202b X = S

203

204a X = CH$_2$, 52%
204b X = S, 44%

205, 5%

Figure 4.71

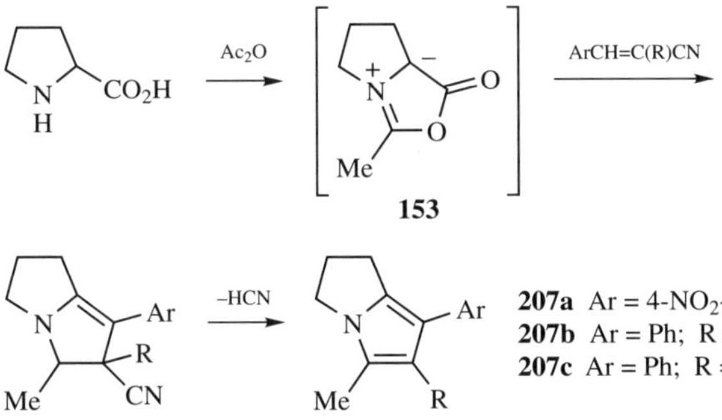

153

206

207a Ar = 4-NO$_2$-Ph; R = CO$_2$Me, 65%
207b Ar = Ph; R = CO$_2$Me, 40%
207c Ar = Ph; R = CN, 42%

Figure 4.72

confirmed by X-ray crystallography (Fig. 4.73). Presumably, the initial cycloadduct **208** undergoes rearrangement to **209** followed by product formation. Texier and coworkers[99] reported the cycloaddition reactions of several munchnones with methyl α-cyanocinnamate and α-cyanocinnamonitrile to give mixtures of 2-pyrrolines **213** and **214** (Table 4.13). Further heating with acetic anhydride gives the expected pyrroles. In some cases, the 2-pyrrolines are not isolated. All of the munchnones were generated by the cyclodehydration of the appropriate N-acylamino acids and acetic anhydride. The regioselectivity is very high in most cases and is rationalized by a combination of electronic and steric effects.

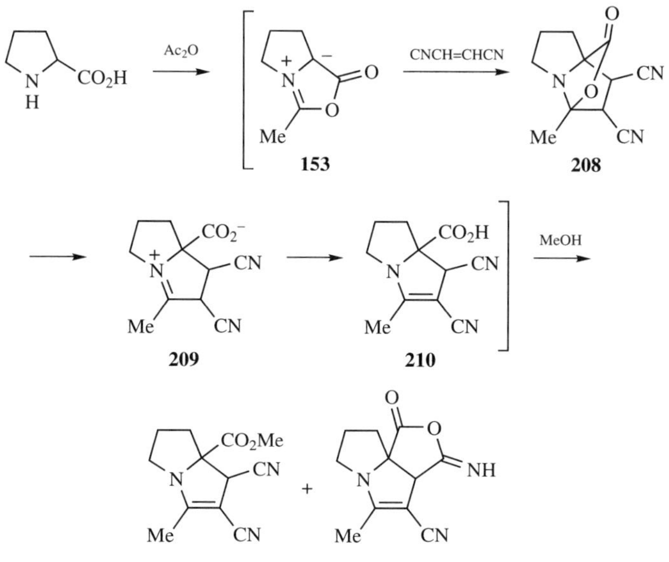

512 Mesoionic Oxazoles

Figure 4.73

TABLE 4.13. 1,3-DIPOLAR CYCLOADDITION REACTIONS OF IN SITU–GENERATED MUNCHNONES AND METHYL α-CYANOCINNAMATE AND α-CYANOCINNAMONITRILE[a]

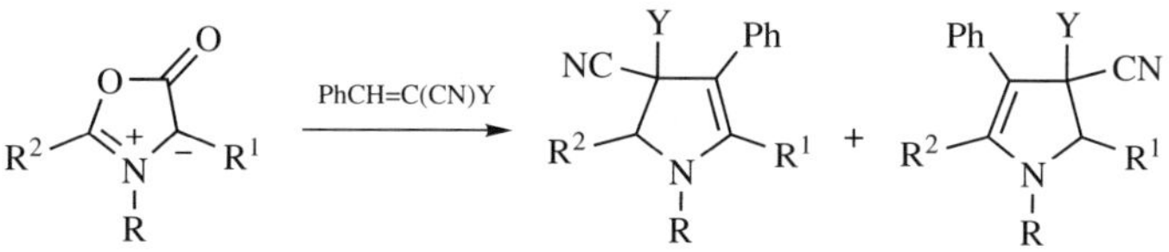

Figure 4.74

R	R^1	R^2	Y	Yield (%)	213 (%)	214 (%)
Me	Ph	Me	CO_2Me	95	0	100
Me	Me	Ph	CO_2Me	50	100	0
Ph	Ph	Me	CO_2Me	92	0	100
Ph	Me	Ph	CO_2Me	95	100	0
Me	Ph	4-MeO-Ph	CO_2Me	86	35	65
Me	4-MeO-Ph	Ph	CO_2Me	90	0	100
Me	Ph	4-NO_2Ph	CO_2Me	85	0	100
Me	Ph	Me	CN	92	0	100
Me	Me	Ph	CN	86[b]	60	40
Ph	Ph	Me	CN	85	0	100
Ph	Me	Ph	CN	84	65	35
Me	Ph	4-MeO-Ph	CN	95[b]	50	50
Me	4-MeO-Ph	Ph	CN	82[b]	0	100

[a] Data from ref. 99.
[b] The yield of the corresponding pyrroles resulting from loss of HCN.

Eguchi and co-workers[100] studied the cycloaddition of several munchnones with electron-deficient trifluoromethylated olefins. Thus munchnones **216** react with alkenes **217** to give the corresponding 4-(trifluoromethyl)pyrroles **218** (Fig. 4.75). Yields of pyrroles from alkene **217a** are higher (56–89%) than those from alkene **217b** (9–33%). The regiochemistry in **218** was easily established by the spin–spin coupling ($^4J_{HF} = 1.2$ Hz) between the CF_3 group and the C-5H in the ^{1}H-NMR spectrum. This cycloaddition sequence was also applied to the synthesis of bicyclic pyrroles **219** and **220**, although the regiochemistry is less selective than that observed earlier (Fig. 4.76).

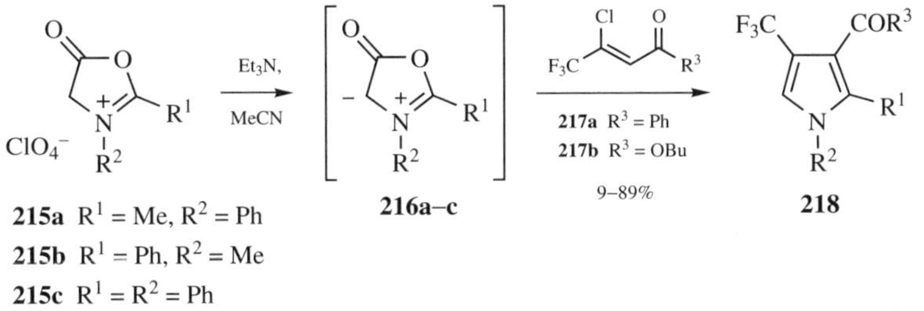

Figure 4.75

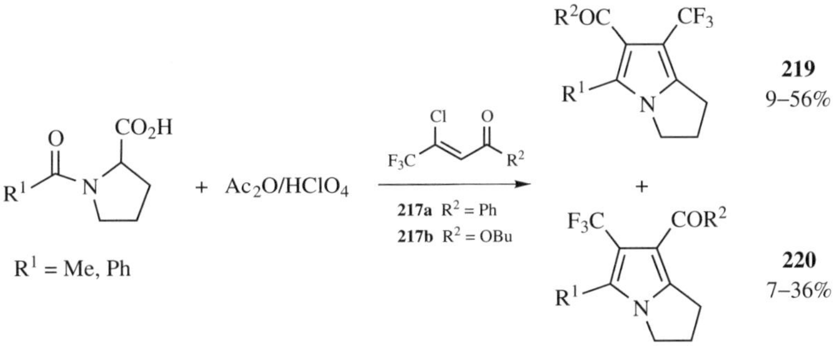

Figure 4.76

Similarly, N-acylisoquinolines also furnish the corresponding fused pyrroles **221** and **222** (Fig. 4.77).[100] The regiochemistry was established by virtue of the five-bond H-F coupling in the major isomer.

Gelmi and co-workers[101] found that vinyl phosphonium salts serve as alkyne synthetic equivalents in reacting with munchnones to give pyrroles **223** (Table 4.14). The munchnones were generated in situ by cyclodehydration of the corresponding N-acylamino acids with acetic anhydride. In the case of 1-propenyltriphenylphosphonium bromide a single regioisomer was obtained. The regiochemistry apparently obtains from a strong interaction between the phosphonium group and the carbonyl group that overwhelms the simple polarization of the vinyl group.

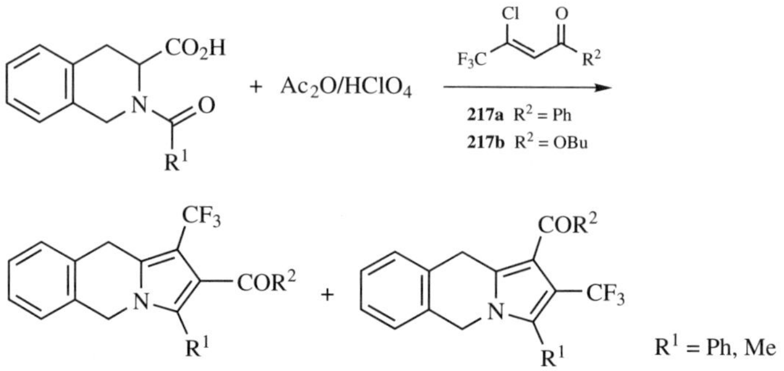

514 Mesoionic Oxazoles

Figure 4.77

221 (50–71%) and 222 (0–24%), R¹ = Ph, Me
217a R² = Ph
217b R² = OBu

This group[101] also found that azlactones react with vinyl phosphonium salts to give N-unsubstituted pyrroles **224** (Table 4.15). Further support for the regiochemistry comes from the reaction of 2-carboxyvinyltriphenylphosphonium bromide **225** with munchnones **149** to give the pyrroles shown (Fig. 4.80).[101] With azlactone **226**, the dihydropyrrole **227** is isolated.

The N-tosyl imine **55**, which reacts with azlactone **52** as an electrophile (presented earlier), also reacts with **52** in a 1,3-dipolar cycloaddition reaction to give pyrroles **228** and **229** (Fig. 4.81).[38] The pyridine **230** and pyridone **231** apparently arise by addition of the azlactone to the imine and subsequent cyclization. The product distribution depends highly on experimental conditions, and α,β-unsaturated N-alkyl- and N-arylimines undergo nucleophilic attack only by the azlactone to give pyridines. Erba and co-workers[102] observed a novel formation of

TABLE 4.14. 1,3-DIPOLAR CYCLOADDITION REACTIONS OF IN SITU–GENERATED MUNCHNONES AND VINYL PHOSPHONIUM SALTS[a]

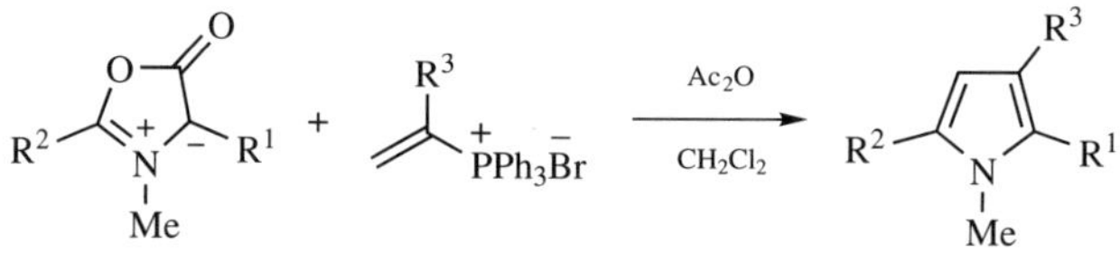

Figure 4.78   223

R¹	R²	R³	Yield (%)
Ph	Ph	H	53
4-Cl-Ph	Ph	H	49
Ph	4-Cl-Ph	H	51
4-MeO-Ph	Ph	H	46
Ph	4-Cl-Ph	Me	34
4-MeO-Ph	Ph	Me	35

[a] Data from ref. 101.

TABLE 4.15. 1,3-DIPOLAR CYCLOADDITION REACTIONS OF MUNCHNONES (AZLACTONES) AND VINYL PHOSPHONIUM SALTS[a]

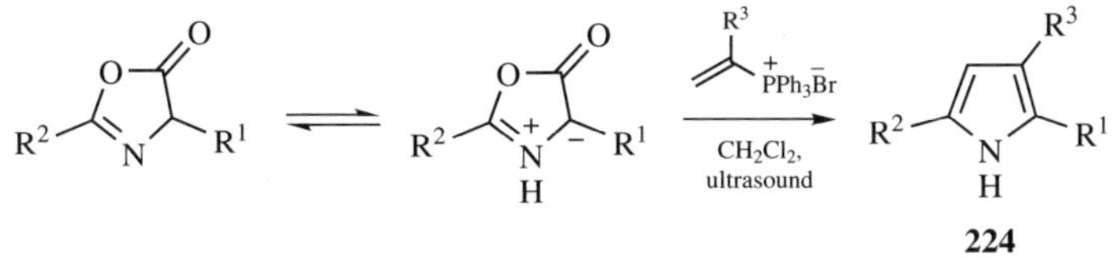

Figure 4.79

R^1	R^2	R^3	Yield (%)
Ph	Ph	H	41
4-Cl-Ph	Ph	H	35
Ph	4-Cl-Ph	H	32
4-Me-Ph	Ph	H	38
Ph	4-Me-Ph	Me	30[b]
Ph	i-Pr	H	48
Ph	i-Pr	Me	41

[a] Data from ref. 101.
[b] Reaction run in refluxing THF:DMF.

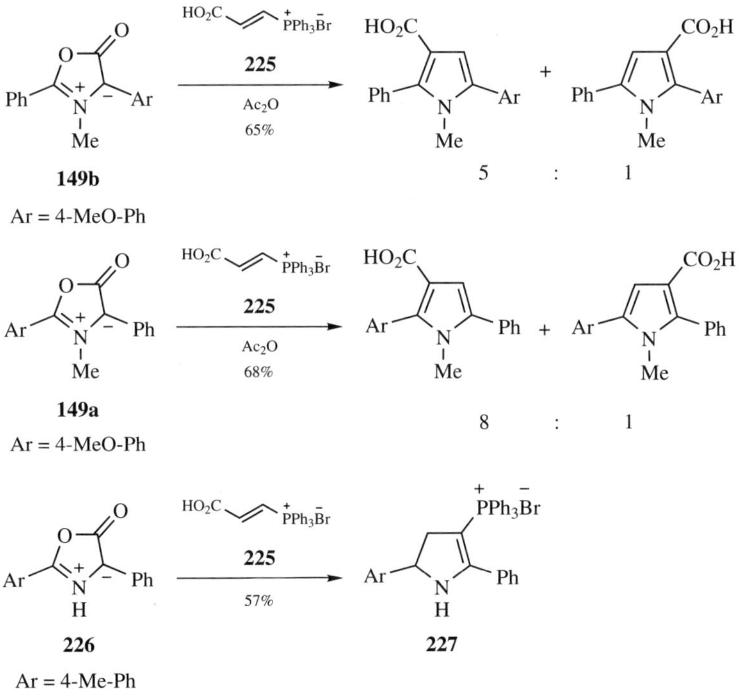

Figure 4.80

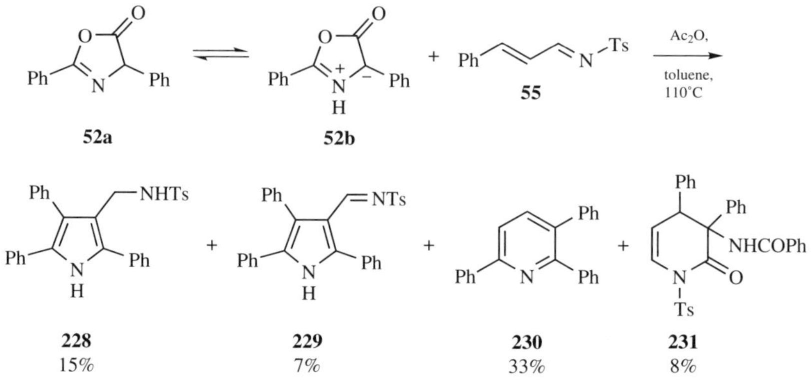

Figure 4.81

pyrrole imines **234** from the 1,3-dipolar cycloaddition of munchnones **232** and 5-amino-1-aryl-4,5-dihydro-4-methylene-1,2,3-triazoles **233** (Table 4.16). Treatment with benzaldehyde yielded 3-formylpyrroles **235**. The reaction presumably involves loss of carbon dioxide, nitrogen, and morpholine from the initial cycloadduct. Unsymmetrical munchnones behave regioselectively and furnish products derived from bonding between C-2 of the munchnone and the methylene terminus of **233**. The pyrrole isomers were identified by NMR spectroscopy. This study featured the synthesis of several new munchnones.

TABLE 4.16. 1,3-DIPOLAR CYCLOADDITION REACTIONS OF MUNCHNONES AND METHYLENE TRIAZOLES[a]

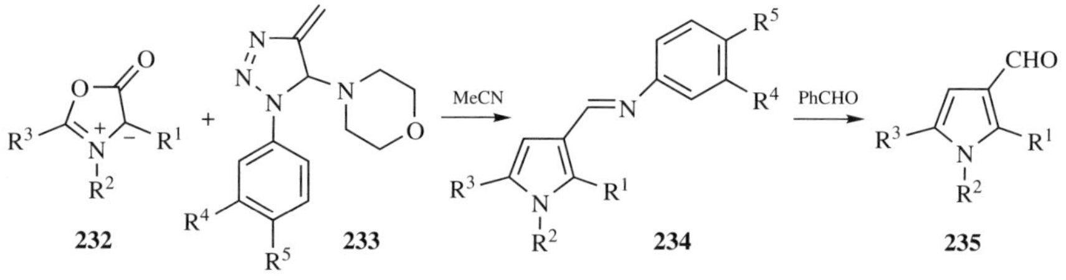

Figure 4.82

R^1	R^2	R^3	R^4	R^5	Yield **234**[b] (%)	Yield **235** (%)
Ph	Me	Ph	CF_3	H	10	65
Ph	Me	4-MeO-Ph	CF_3	H	—	68
Ph	Me	2-Cl-Ph	CF_3	H	—	81
4-MeO-Ph	Me	4-MeO-Ph	CF_3	H	—	69
Ph	Et	Ph	CF_3	H	—	68
Ph	Ph	Ph	H	NO_2	—	89
4-MeO-Ph	Me	Ph	CF_3	H	—	52
Ph	Me	Me	H	NO_2	—	78
Me	Me	Ph	H	NO_2	23	45

[a] Data from ref. 102.

[b] In most cases, the imine **234** was not isolated but directly converted to **235**.

This same group[103,104] described the reactions of munchnones with 4-methylene-4,5-dihydroisoxazoles **236** to give cycloadducts **237** and with benzylideneisoxazolones **238** to give cycloadducts **239** as a mixture of isomers (Fig. 4.83). The latter undergo facile loss of benzonitrile to afford pyrroles **240**. This transformation, which also results in the interconversion of the isomeric **239**, involves ring opening to a zwitterionic species. The spiro compounds are isomeric at the spiro center and exist as a mixture of regioisomers.

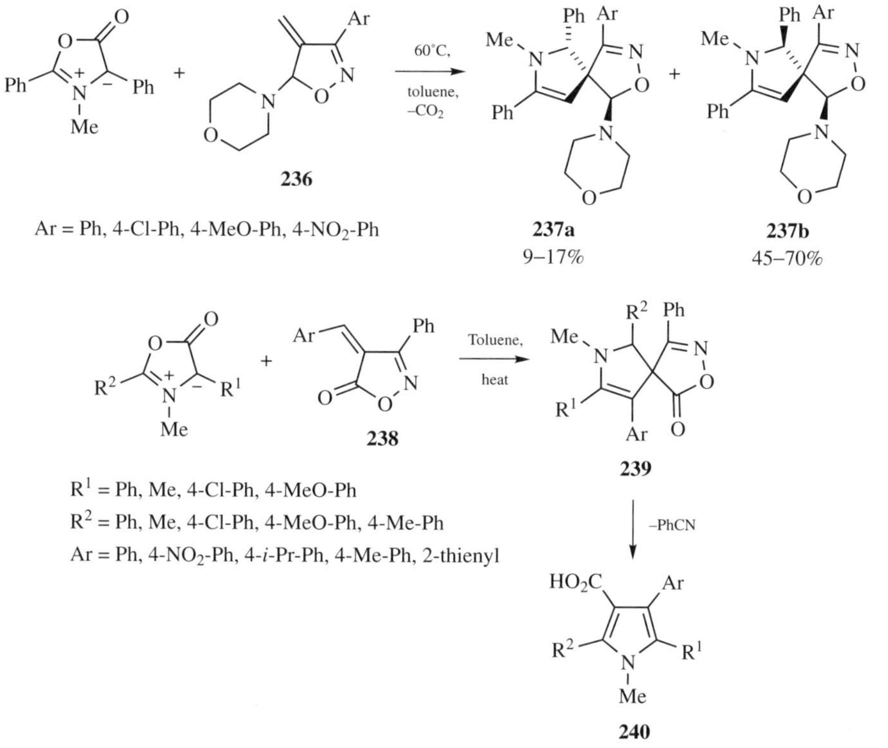

Figure 4.83

Clerici and co-workers[105] investigated the behavior of munchnones with isothiazole 1,1-dioxide **241**. As shown in Table 4.17, the products **242** are formed in good yields, along with minor amounts of the tautomeric imine (not shown). The azalactones–munchnones were generated from the corresponding *N*-acylamino acids and acetic anhydride.

This cycloaddition works equally well with *N*-methylmunchnones (Table 4.18). Once again, the regioselectivity favors products **243**, resulting from an apparent strong repulsion between the carbonyl group of the munchnone and the sulfone group, leading to the observed (exclusive) products. Adducts **243** are converted to the corresponding pyrroles **244** on heating to their melting points (180°–220°C), e.g., **245** → **246** (Fig. 4.86).

TABLE 4.17. 1,3-DIPOLAR CYCLOADDITION REACTIONS OF IN SITU–GENERATED MUNCHNONES AND ISOTHIAZOLE 1,1-DIOXIDE 241[a]

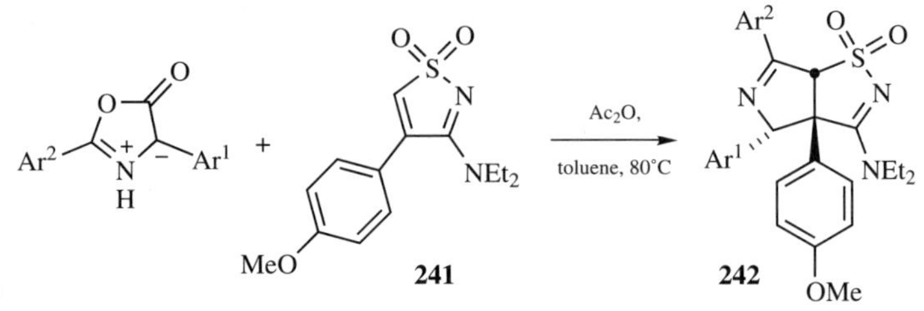

Figure 4.84

Ar[1]	Ar[2]	Yield 242 (%)	Yield[b] (%)
Ph	Ph	60	(10)
4-Cl-Ph	Ph	65	(0)
Ph	4-Cl-Ph	74	(13)
4-Me-Ph	Ph	70	(5)
Ph	4-Me-Ph	70	(0)

[a] Data from ref. 105.
[b] Refers to the imine tautomer.

TABLE 4.18. 1,3-DIPOLAR CYCLOADDITION REACTIONS OF IN SITU–GENERATED MUNCHNONES AND ISOTHIAZOLE 1,1-DIOXIDE 241[a]

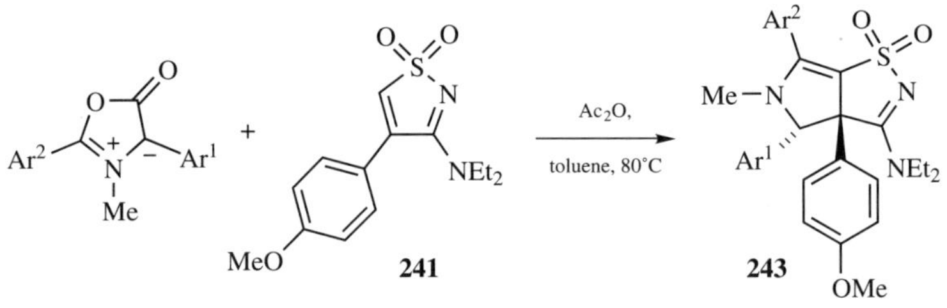

Figure 4.85

Ar[1]	Ar[2]	Yield 243 (%)	Yield 244[b] (%)
Ph	Ph	90	45
Ph	4-Me-Ph	91	68
4-Me-Ph	Ph	90	84
Ph	4-MeO-Ph	75	70
4-MeO-Ph	Ph	90	64
Ph	4-F-Ph	90	25
Ph	4-Br-Ph	63	73
Ph	4-NO$_2$-Ph	70	75

[a] Data from ref. 105.
[b] Refers to the corresponding pyrroles 244 formed on heating 243 at 180°–220°C.

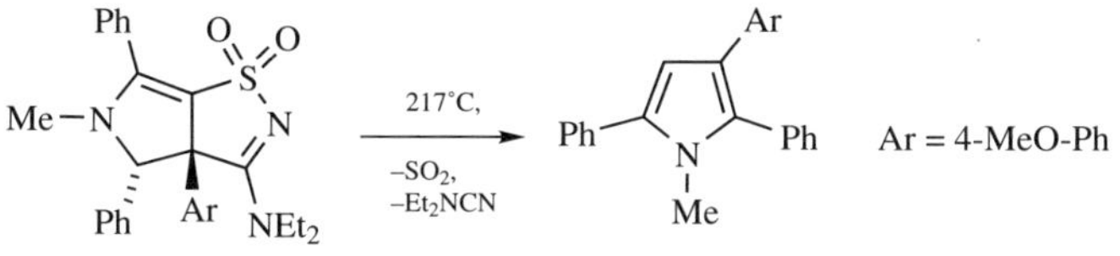

1,3-Oxazolium 5-Oxides (Munchnones) 519

Figure 4.86

This same group[106] described the interesting reaction between munchnones and vinylisothiazole 1,1-dioxide **247** (Fig. 4.87). The initial cycloadducts **248**, which form at room temperature, are converted at higher temperature into nitriles **250** via pyrrole tautomers **249**, which could also be isolated.

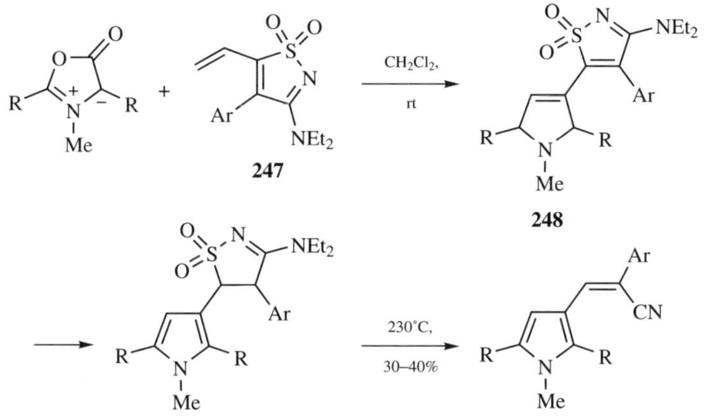

R = Ph, 4-Me-Ph, 4-MeO-Ph; Ar = 4-MeO-Ph

Figure 4.87

The ability of the nitro group to serve both as a powerful electron-withdrawing group and a leaving group in the form of nitrous acid has led to its employment in munchnone cycloaddition chemistry. Thus Jiménez and co-workers[107,108] studied the reaction of munchnone **149c** with sugar derivatives **251** and **252** (Fig. 4.88). These nitroalkenes react in a highly regioselective fashion to give pyrroles **253** and **254** in 69 and 65% yield, respectively. Deacylation and periodate cleavage afford pyrrole aldehyde **255**. A series of NOE experiments established the structure of **255** and the regioselectivity. No other regioisomers were found. This cycloaddition, like many involving munchnones, is opposite of that predicted by FMO theory. However, the authors carried out an ab initio molecular orbital (MP2/6-31B) calculation on a simpler system **256** + **257** → **258**, which does agree with experimental observation (Fig. 4.89). Furthermore, from these calculations the authors conclude that this 1,3-dipolar cycloaddition is concerted but slightly asynchronous.

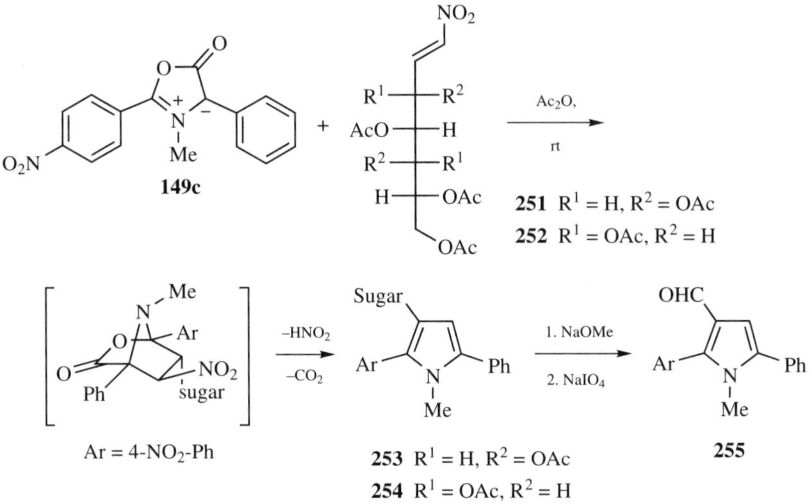

Figure 4.88

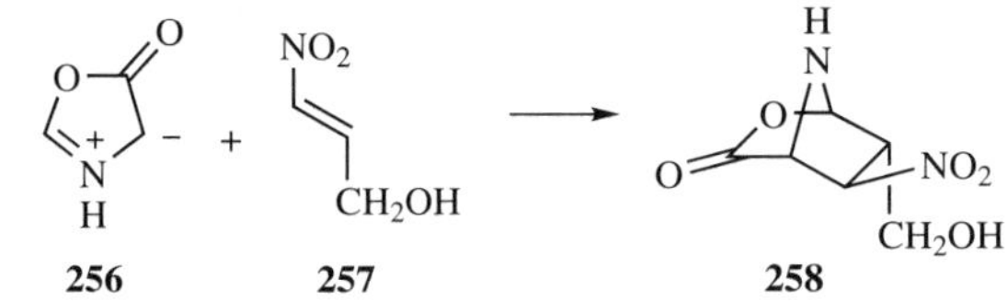

Figure 4.89

Gribble and co-workers[79] also observed apparent anti-FMO regiochemistry in the dipolar cycloadditions of munchnones **158** and **161** with β-nitrostyrene to give pyrroles **159** and **160** (Fig. 4.90). The isomers were distinguished by NMR and NOE experiments.

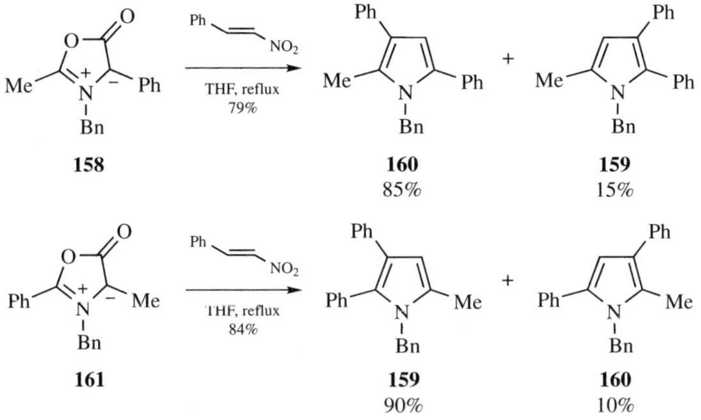

Figure 4.90

The idea of employing the reaction of a nitroarene or nitroheterocycle with a munchnone to synthesize a fused pyrrole ring system was developed by two groups. Nesi and co-workers[109] found that munchnone **42** reacts with 3-methyl-4-nitro-isoxazole **259** and 4-nitro-3-phenylisoxazole **260** to give the corresponding 5*H*-pyrrolo[3,4-*d*]isoxazoles **261** and **262**, respectively, in good yield (Fig. 4.91). Presumably loss of carbon dioxide follows that of nitrous acid.

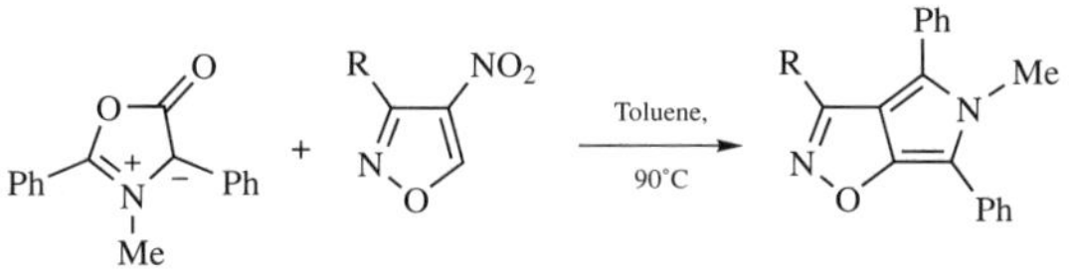

42 **259** R = Me **260** R = Ph **261** R = Me, 52% **262** R = Ph, 69%

Figure 4.91

Gribble and co-workers[110,111] found that munchnones react smoothly with 2- and 3-nitroindoles to afford pyrrolo[3,4-*b*]indoles **263** and **264** (Table 4.19). In the

TABLE 4.19. 1,3-DIPOLAR CYCLOADDITION REACTIONS OF IN SITU–GENERATED MUNCHNONES AND 2- AND 3-NITROINDOLES[a]

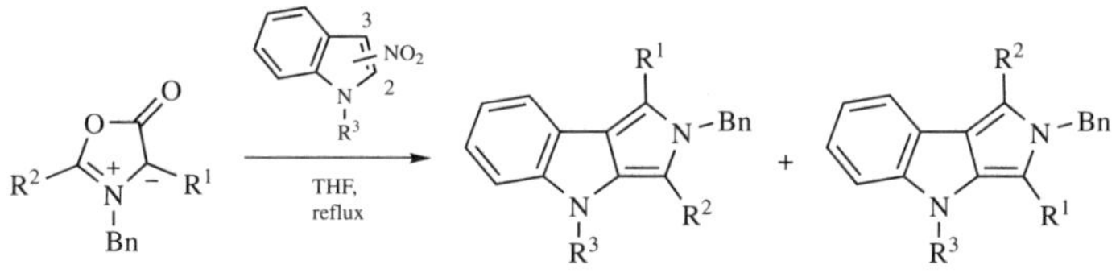

263 **264**

Figure 4.92

R^1	R^2	R^3	NO_2	Yield (%)	**263** (%)	**264** (%)	Reference
Ph	Ph	CO_2Et	2	94	—	—	110
Ph	Ph	CO_2Et	3	60	—	—	110
Me	Me	CO_2Et	2	53	—	—	110
Me	Me	CO_2Et	3	39	—	—	110
Ph	Ph	SO_2Ph	2	76	—	—	110
Ph	Ph	SO_2Ph	3	65	—	—	110
Me	Me	SO_2Ph	2	17	—	—	110
Me	Me	SO_2Ph	3	67	—	—	110
Ph	Me	CO_2Et	2	85	10	90	111
Me	Ph	CO_2Et	2	88	10	90	111
Ph	Me	CO_2Et	3	65	95–100	0–5	111
Me	Ph	CO_2Et	3	89	40–50	50–60	111
Ph	Me	SO_2Ph	3	74	94–100	0–6	111
Me	Ph	SO_2Ph	3	76	10–30	70–90	111
Me	Ph	Me	3	<1%[b]	—	—	111

[a] Munchnones were generated from the corresponding *N*-acylamino acids using DIPC in refluxing THF.
[b] Reflux in diglyme for 24 h.

522 Mesoionic Oxazoles

case of unsymmetrical munchnones, the regiochemistry does not follow simple FMO theory. The structures of regioisomers were established using a combination of NOE techniques, independent synthesis, and X-ray crystallography. This novel pyrrole annulation reaction was also employed to synthesize benzo[b]furo[2,3-c]pyrroles **265** and benzo[b]thieno[2,3-c]pyrroles **266** (Fig. 4.93).[110]

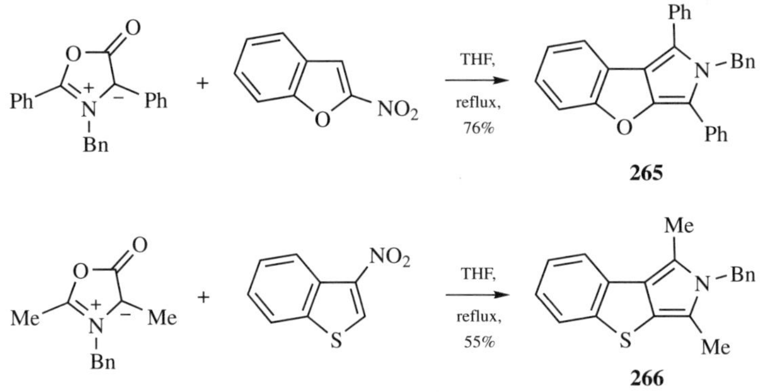

Figure 4.93

Unsymmetrical munchnones **158** and **161** react regioselectively with 2-nitrobenzofuran to give predominantly benzo[b]furo[2,3-c]pyrroles **267** and **268**, respectively, resulting from bond formation between C-2 of the munchnone and C-3 of the nitrobenzofuran, analogous to pyrroloindole **264** (Fig. 4.94; Table 4.19).[79]

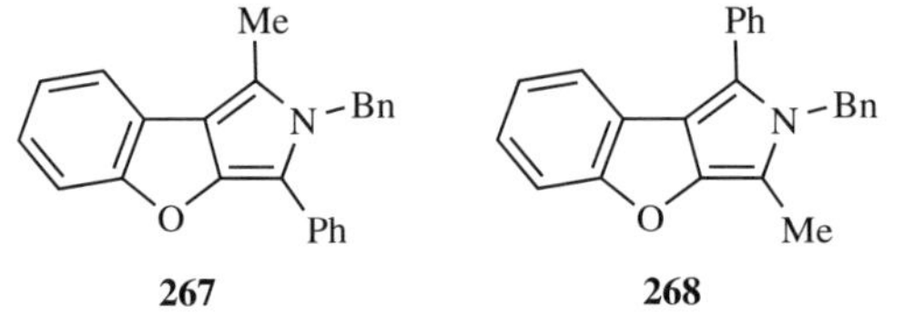

Figure 4.94

The higher reactivity of ring-strained olefins has been exploited by several workers in 1,3-dipolar cycloaddition reactions of munchnones. Thus Kato and co-workers[112] reported that munchnone **42** reacts with 1,2,3-triphenyl-1H-phosphirene **269** in boiling acetonitrile to give 1-methyl-2,3,4,5-tetraphenylpyrrole **270** (45%

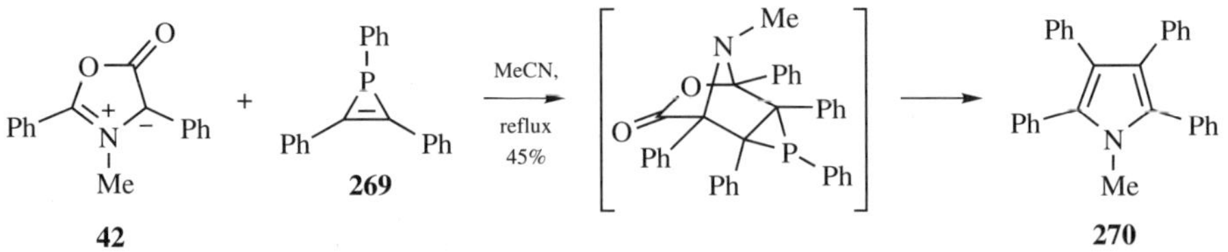

Figure 4.95

yield) (Fig. 4.95). The yield of **270** is only 15% when **42** is generated in situ. Control experiments demonstrated that phosphirene **269** does not decompose to diphenylacetylene appreciably under the reaction conditions. Moreover, the reaction of diphenylacetylene and munchnone **42** afforded only a 21% yield of pyrrole **270** (acetonitrile, reflux, 10 h).

Unfortunately, this same group[113] found that munchnone **42** gave complex product mixtures only on reaction with benzocyclopropene **271**, in an unsuccessful

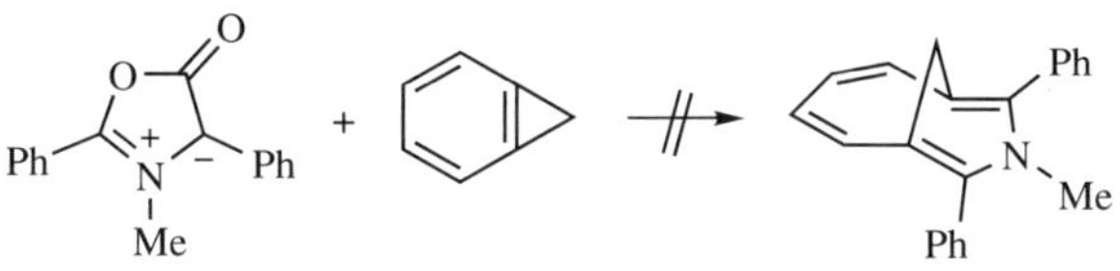

Figure 4.96

attempt to synthesize methanoazonine **272** (Fig. 4.96). As discussed later, a similar reaction was successful with an isomunchnone. Likewise, Kato and co-workers[114] were unable to induce munchnones **42** or **149c** to react cleanly with benzocyclobutadiene **273**, although several other mesoionic heterocycles do afford well-

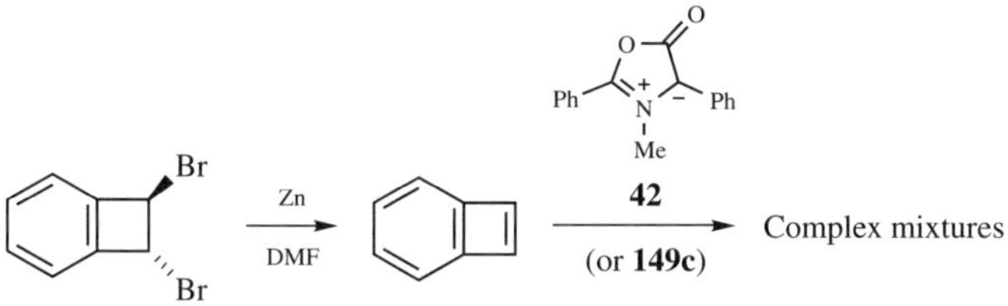

Figure 4.97

defined cycloadducts in high yield (Fig. 4.97). Martin and co-workers[115] found that munchnone **42** reacts with isopropylidenecyclobutenone **274** to form dihydroazepine **275** (Fig. 4.98). At room temperature, the two bis-adducts **276** and **277** were isolated, although the regiochemistry of the cycloaddition has not been established.

Maryanoff and co-workers[116] continued their studies of the reaction of 1,2-dicyanocyclobutene **279** with munchnones and found that 3-(4-chlorophenyl)alanine **278** reacts with **279** in the presence of acetic anhydride to give imino acid **281** (Fig. 4.99). Esterification of **281** afforded ester **284**, which structure was confirmed by X-ray crystallography. Further heating of **281** yielded the originally expected dihydroazepine **283**. The isolation of acid **281** suggests that decarboxylation of the munchnone primary adducts need not be concerted.

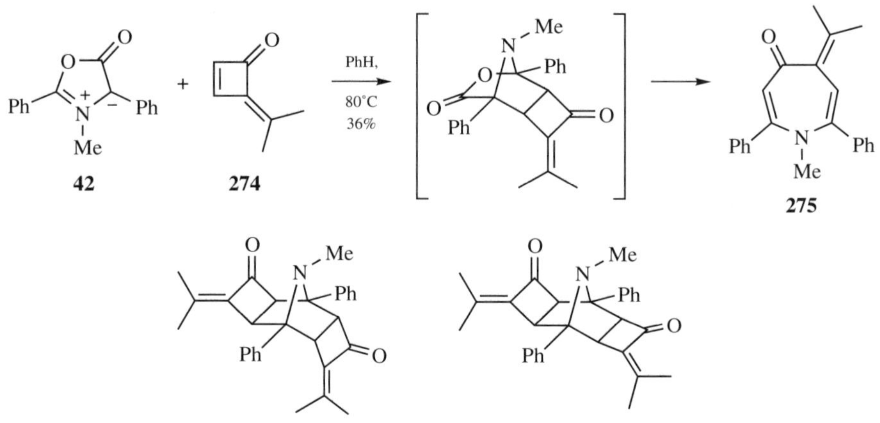

42
274
275
276
20% (exo, endo)
277
4% (exo, exo)

Figure 4.98

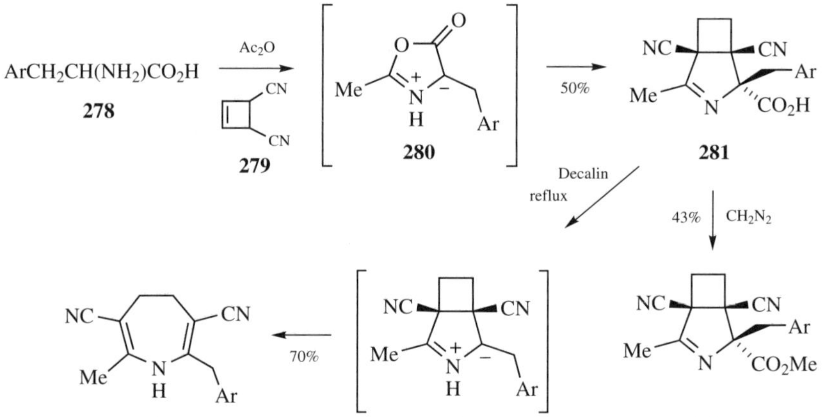

278
279
280
281
283
282
284

Figure 4.99

Maryanoff and Turchi[117] pursued a detailed theoretical study of the reaction between 1,2-dicyanocyclobutene **279** and munchnone **280**, prepared by cyclodehydration of **278** and acetic anhydride (Fig. 4.100). The results from these AM1 molecular orbital calculations led to the conclusions that the transition state leading to the exo cycloadduct **281** is favored electrostatically and that azomethine ylide **282** is a discrete intermediate in the formation of dihydroazepine **283**. More recently, Turchi[118] reported cycloaddition reactions between munchnone **285** and **279** to afford dihydroazepine **286** in high yield. Further cyclization of **286** gave tricycle **287**. Likewise, diester **288** reacts with munchnone **42** to give dihydroazepine **289**.

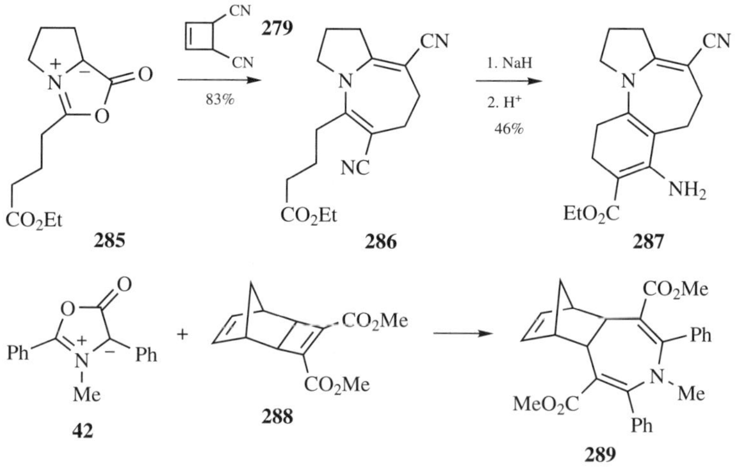

Figure 4.100

Kato and co-workers[119] explored reactions of fulvenes with a variety of mesoionic heterocycles. Unfortunately, reactions of munchnone **42** with fulvenes **290–292** afforded complex mixtures in each case, and no identifiable products were reported (Fig. 4.101). However, Friedrichsen and co-workers[120] reported the reaction between **42** and 6,6-dimethylfulvene **292** in the presence of manganese dioxide to give **293** in low yield.

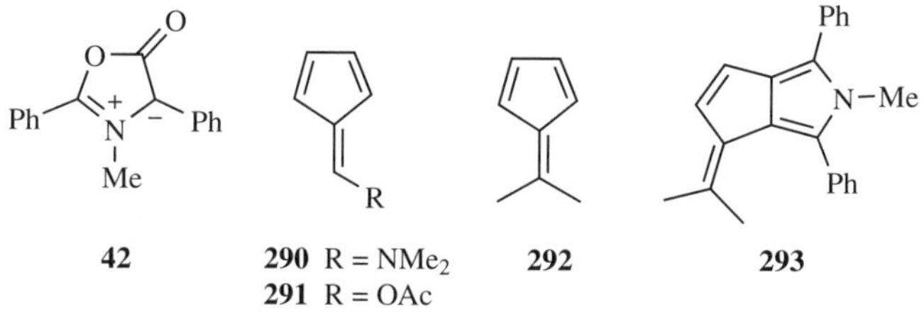

Figure 4.101

Kato and co-workers[121] also studied the cycloaddition reactions of tropone **294** with several mesoionic heterocycles. Unfortunately, despite heroic efforts, the reaction of **294** with munchnone **42** was complex and could not be unraveled (Fig. 4.102). However, as described later, the reaction of tropone with isomunchnones was successful.

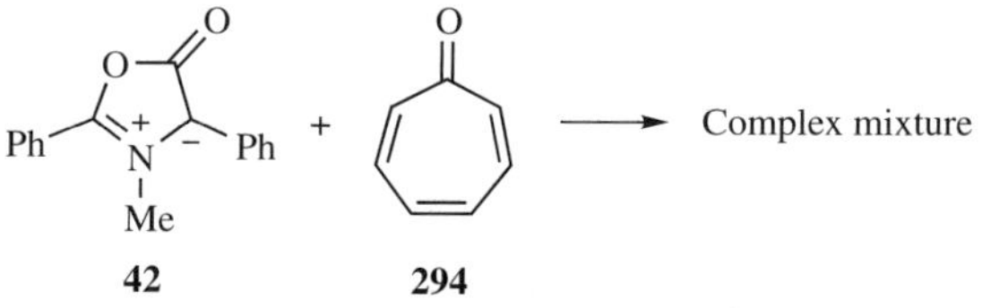

Figure 4.102

Wu and co-workers[122] effected the cycloaddition between azlactone **295**, presumably via munchnone **296**, and fullerene-60 (C_{60}) to give dihydropyrrole **297** in excellent yield (Fig. 4.103).

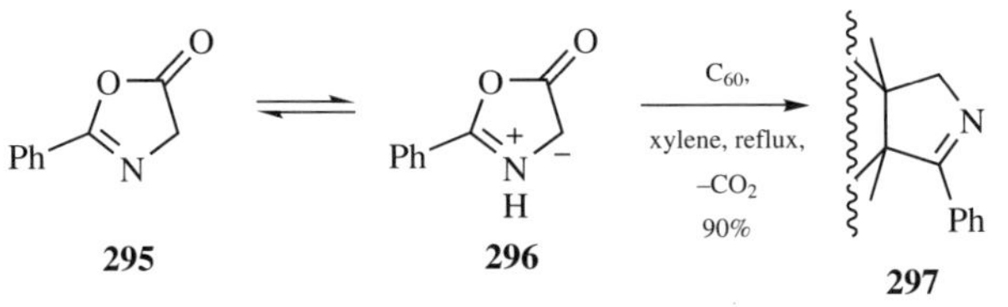

Figure 4.103

Gribble and co-workers[123] found that munchnones **42** and **299** react in a novel tandem fashion with 1,5-cyclooctadiene **298** to give the caged compounds 10-methyl- **300a** and 10-benzyl-9,10-diphenyl-10-azatetracyclo[6.3.0.0.$^{4,11}0^{5,9}$]undecane **300b** (Fig. 4.104). The latter compound was characterized by X-ray crystallography. A similar reaction of munchnone **299** with 1,3,5,7-cyclooctatetraene afforded cycloadduct **301** in low yield. Photolysis of **301** gave azahomopentaprismane **302** in good yield. Gribble and co-workers[124] reported that **300a** and **300b** not only exhibit the expected restricted nitrogen inversion (ring strain) ($\Delta G^{\neq} = 12.2$ and 10.6 kcal/mol, respectively) but also display remarkably slow bridge-head phenyl rotation ($\Delta G^{\neq} = 9.8$ kcal/mol).

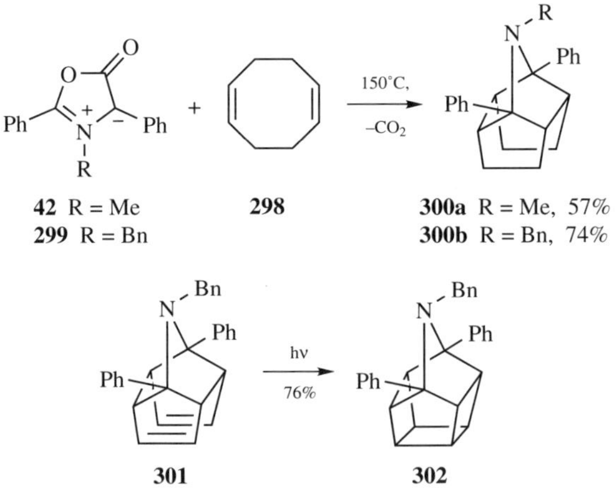

42 R = Me
299 R = Bn

300a R = Me, 57%
300b R = Bn, 74%

Figure 4.104

Padwa and co-workers[125] applied the intramolecular munchnone–olefin cycloaddition reaction to craft a series of novel caged compounds **304, 305, 307** (Fig. 4.105). Whereas the alanine-derived munchnone from **303a** affords only **304a**, the 2-phenylglycine-derived munchnone from **303b** gives **305** as the major regioisomer. Cyclization of the munchnones derived from the glycine derivatives

306 is also highly regioselective to afford **307**. Several *o*-vinylaryl systems were also examined in this study.[125] For example, treatment of **308** with acetic anhydride led to a mixture of regioisomers **309** and **310** (Fig. 4.106).

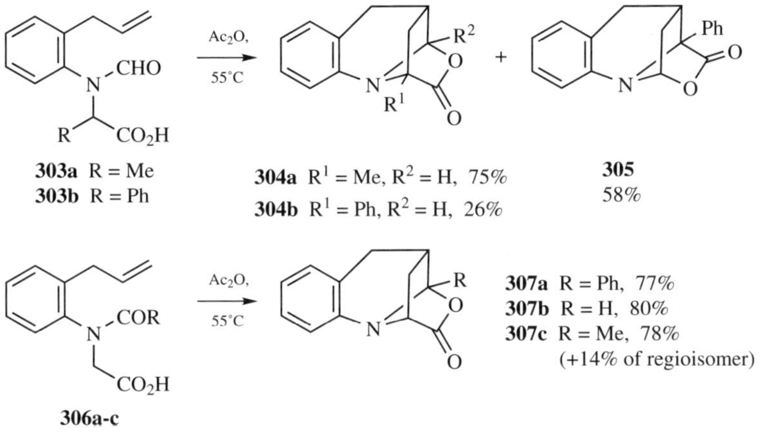

303a R = Me
303b R = Ph

304a R^1 = Me, R^2 = H, 75%
304b R^1 = Ph, R^2 = H, 26%

305 58%

306a-c

307a R = Ph, 77%
307b R = H, 80%
307c R = Me, 78%
(+14% of regioisomer)

Figure 4.105

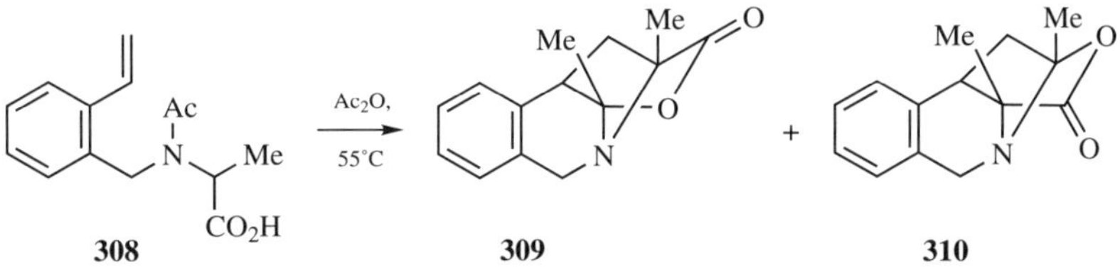

308

309 34%

310 51%

Figure 4.106

4.2.3.3.4. Carbonyl Compounds

No new examples of 1,3-dipolar cycloaddition reactions between munchnones and carbonyl compounds have been reported in the interim since the review by Gingrich and Baum.[1] The original work by Huisgen and co-workers[126,127] remains the only study of this interesting reaction.

4.2.3.3.5. Thiocarbonyl Compounds

No additional examples of the 1,3-dipolar cycloaddition reactions between munchnones and thiocarbonyl compounds have been reported since the review by Gingrich and Baum.[1]

4.2.3.3.6. Oxygen

Kawase[128,129] reexamined the reaction of munchnones with oxygen. This is a powerful method for the synthesis of imides (e.g., **311** → **312**), and, based on

^{18}O labeling experiments, the mechanism of this autoxidation is different from that originally proposed by Huisgen and co-workers.[19] This reaction is particularly useful for the preparation of tetrahydroisoquinolones **314** and tetrahydrocarbolones **316** (Fig. 4.107). Based on experiments with $^{18}O_2$, the mechanism shown in Scheme 4.8 appears to operate in the autoxidation of munchnones, as proposed by Kawase.[128,129]

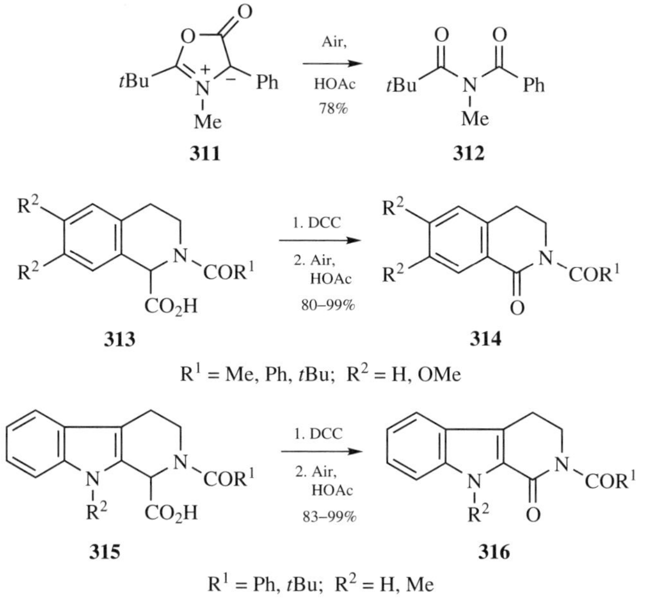

Figure 4.107

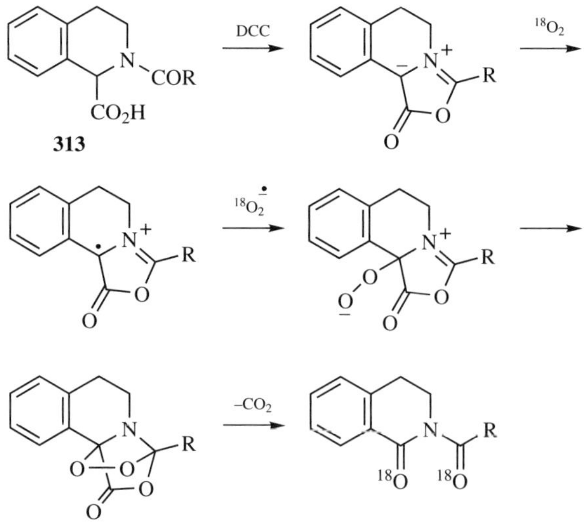

Scheme 4.8

4.2.3.3.7. Nitrogen-Containing Dipolarophiles

The father of munchnone chemistry, Rolf Huisgen,[130] reported the 1,3-dipolar cycloaddition of munchnone **42** with 4-nitrophenyldiazonium salt **317** to give the triazolium salt **318** (Fig. 4.108). This appears to be an important new route to these heterocycles.

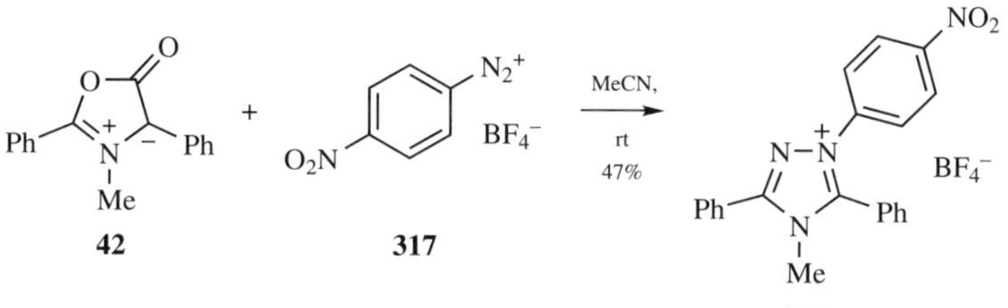

Figure 4.108

Ferraccioli and co-workers[131] employed the reaction of munchnones with N-(phenylsulfonyl)imines as a general synthesis of imidazoles **319** (Table 4.20). Regioselectivity is very high and was determined using NOE measurements. The munchnones were prepared in situ using DCC in toluene (25°–60°C).

A subsequent study by this group[132] uncovered a small amount of the regioisomer **323** in the reaction of munchnone **320** with imine **321** (Fig. 4.110). The

TABLE 4.20. 1,3-DIPOLAR CYCLOADDITION REACTIONS OF IN SITU–GENERATED MUNCHNONES AND N-(PHENYLSULFONYL)IMINES[a,b]

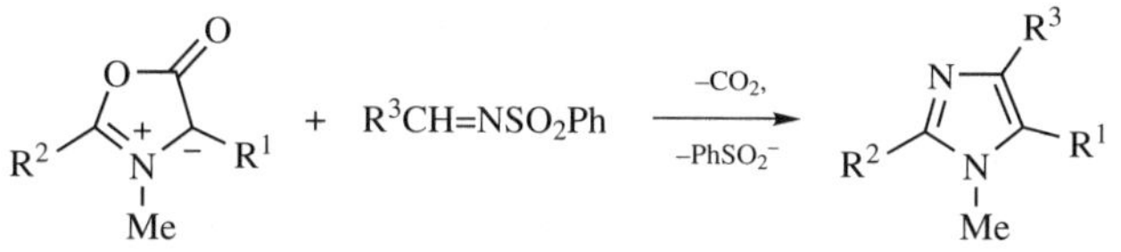

Figure 4.109

R^1	R^2	R^3	Yield **319** (%)
Ph	Ph	Ph	64
Me	Ph	Ph	29
Ph	Me	Ph	40
Ph	Ph	H	20
Ph	Ph	4-NO$_2$Ph	65
Me	Ph	4-NO$_2$-Ph	50
Ph	Me	4-NO$_2$-Ph	30
Ph	Ph	4-MeO-Ph	45
Me	Ph	4-MeO-Ph	55
Ph	Me	4-MeO-Ph	42

[a] Data from ref. 131.
[b] The munchnones were generated in situ by the cyclodehydration of the N-acylamino acid precursors with DCC in toluene (25°–60°C).

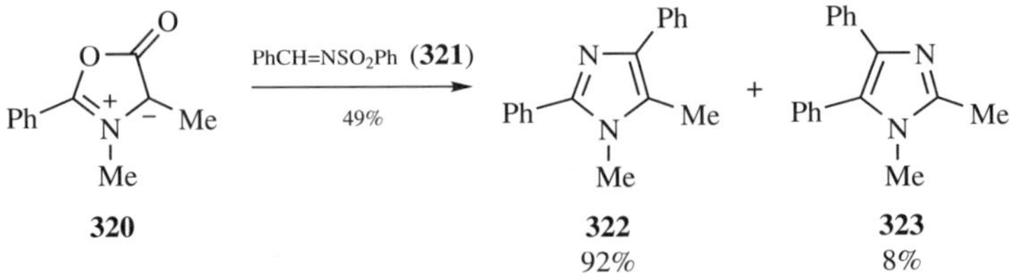

Figure 4.110

authors use the perturbation MO treatment, which is less approximate than the FMO method, to predict correctly the observed regiochemistry. Moreover, they conclude that, based on Mulliken total atomic charges, the resonance form shown for munchnones in this chapter (e.g., **320**) is the best representation. Interestingly, the reaction between imine **321** and munchnone **324** also produces a small amount (6%) of amidine **325**, presumably formed by carbon dioxide loss from the initial cycloadduct (Fig. 4.111).[132]

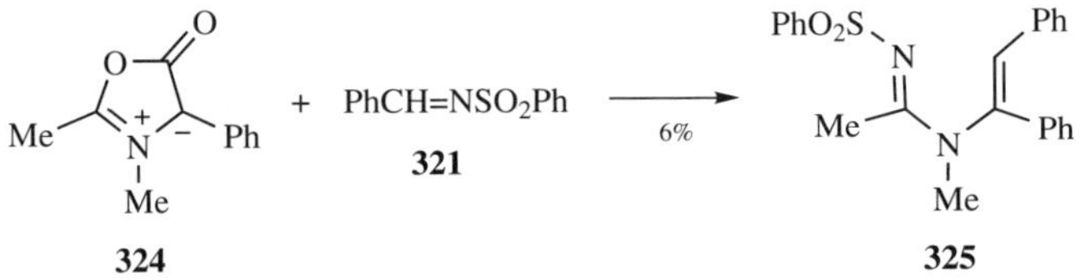

Figure 4.111

Bilodeau and Cunningham[133] effected the solid-phase synthesis of a series of imidazoles **329** via munchnone generation and trapping regioselectively with *N*-tosylimines (Scheme 4.9). The resin bound imidazoles **328** were first treated

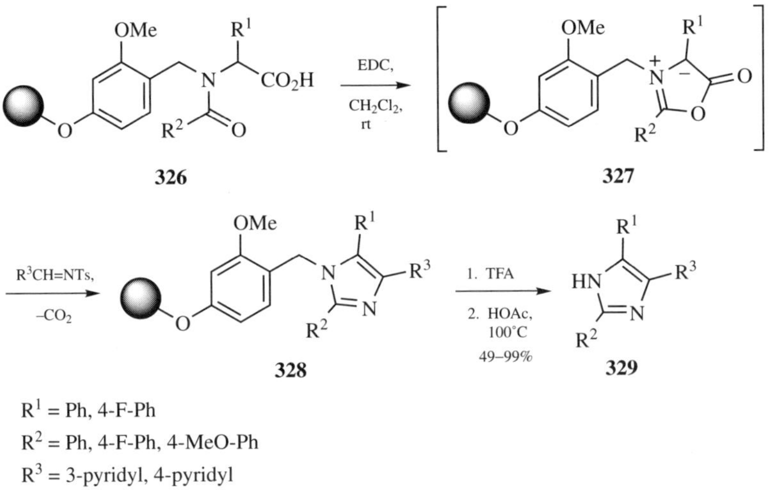

R^1 = Ph, 4-F-Ph
R^2 = Ph, 4-F-Ph, 4-MeO-Ph
R^3 = 3-pyridyl, 4-pyridyl

Scheme 4.9

1,3-Oxazolium 5-Oxides (Munchnones)   531

with aqueous trifluoroacetic acid (TFA) to remove impurities on the resin and then with hot acetic acid to actually cleave the resin.

Dalla Croce and co-workers[134] pursued the chemistry of unsaturated *N*-(phenylsulfonyl)imines **330** and munchnone **42** (Fig. 4.112). While the major products are the imidazoles **331**, pyrroles **332** and amides **333** also are formed in this reaction. The structure of amide **333a** was confirmed by X-ray crystallography.

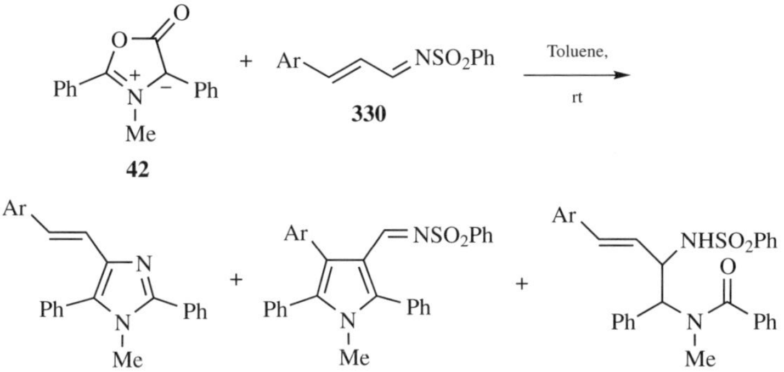

331a, 38%   **332a**, 11%   **333a**, 7%
331b, 30%   **332b**, 8%   **333b**, 6%
331c, 35%   **332c**, 13%   **333c**, 9%

a: Ar = Ph; **b**: Ar = 4-NO$_2$-Ph; **c**: Ar = 4-MeO-Ph

Figure 4.112

This same group[135] generated bicyclic munchnones (e.g., **334**) and trapped them with imines to afford either imidazoles **335** or spirocyclic β-lactams **336**, depending on conditions, although mixtures are produced and yields of the β-lactams are invariably low (Fig. 4.113). The latter can arise by nucleophilic attack on the munchnone itself or the ring-opened ketene tautomer (vide infra).

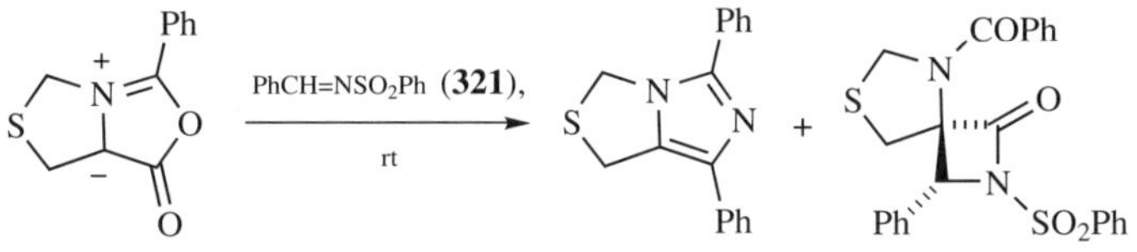

335 52%   **336** 26%

Figure 4.113

Hamper and co-workers[136] employed a solid-phase protocol for the generation of a large library of 5-(trifluoroacetyl)imidazoles (200 compounds) using the reaction between munchnones and benzamidines. Storr and co-workers[137] employed the cycloaddition of munchnones **338** and **339** with 2-phenylbenzazete **337** to craft 3*H*-1,3-benzodiazepines **340** and **341** (Scheme 4.10). Upon heating and

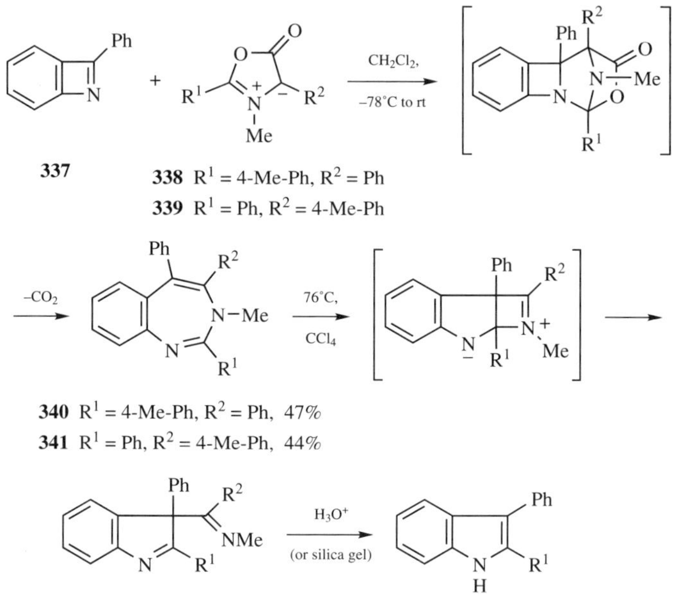

532 Mesoionic Oxazoles

Scheme 4.10

337

338 R^1 = 4-Me-Ph, R^2 = Ph
339 R^1 = Ph, R^2 = 4-Me-Ph

340 R^1 = 4-Me-Ph, R^2 = Ph, 47%
341 R^1 = Ph, R^2 = 4-Me-Ph, 44%

342 R^1 = 4-Me-Ph, R^2 = Ph, 92%
343 R^1 = Ph, R^2 = 4-Me-Ph, 88%

344 R^1 = 4-Me-Ph
345 R^1 = Ph, 83%

subsequent hydrolysis, these benzodiazepines undergo conversion to 2,3-diarylindoles **344** and **345**. The observed regiochemistry is consistent with that seen with other electron-deficient dipolarophiles.

Rodríguez and co-workers[138] found that munchnone **52** reacts with nitrosobenzene in hot xylene to give N-benzoyl-N'-phenylbenzamidine **346** (Fig. 4.114). At room temperature, the intermediate oxadiazoline **347** can be isolated.

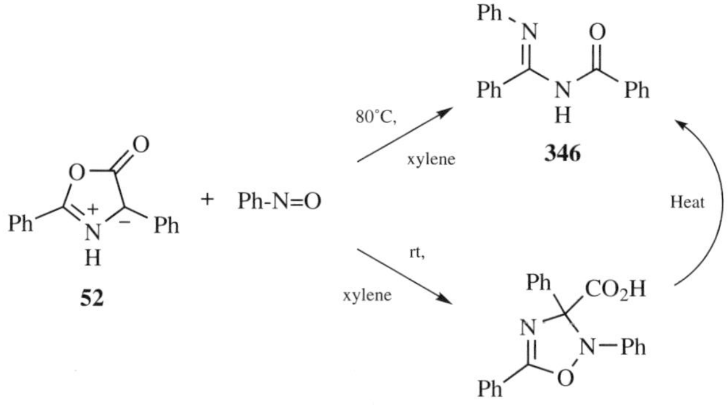

Figure 4.114

The regiochemistry of the initial cycloadduct was established from the reaction between munchnone **348** and nitrosobenzene (Fig. 4.115). The structure of **350** was confirmed by basic hydrolysis to *p*-toluic acid and *N*-phenylbenzamidine.

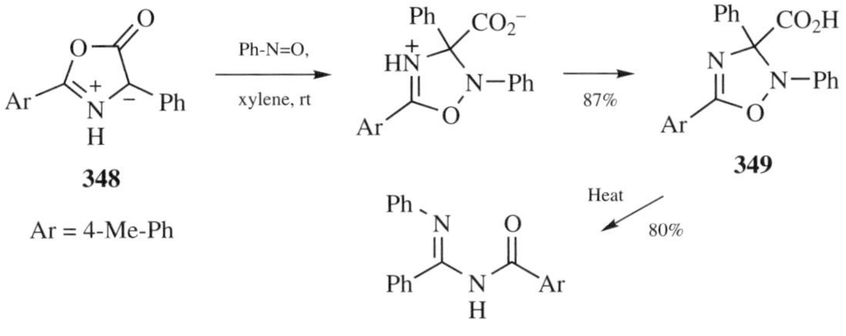

Figure 4.115

4.2.3.3.8. Heterocumulenes

No new examples of this 1,3-dipolar cycloaddition have been reported since the review by Gingrich and Baum.[1]

4.2.3.3.9. Phosphorus-Containing Dipolarophiles

Apparently independently, Märkl[139] and Regitz[140–142] discovered that 1,3-dipolar cycloaddition reactions of munchnones and phosphaalkenes or phosphaalkynes provide a direct synthesis of 1,3-azaphospholes **353** (Table 4.21). The intermediate cycloadducts cannot be isolated. The various phosphaalkynes were generated from phosphaalkenes or, in the case of methylidynephosphane **352**

TABLE 4.21. 1,3-DIPOLAR CYCLOADDITION REACTIONS OF MUNCHNONES AND PHOSPHAALKYNES

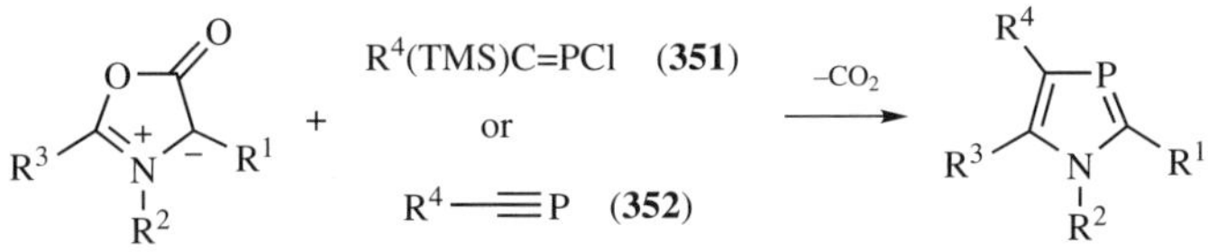

Figure 4.116

Dipolarophile	R^1	R^2	R^3	R^4	Yield **353** (%)	Reference
351	Ph	Ph	Ph	Ph	50	139
351	4-Cl-Ph	Ph	Ph	Ph	40	139
351	4-MeO-Ph	Ph	Ph	Ph	43	139
351	Ph	Me	Ph	Ph	33	139
351	4-MeO-Ph	Me	Ph	Ph	55	139
352	Ph	Me	Ph	*t*Bu	63	140
352	Ph	Me	Ph	H	23	141
352	Ph	Me	Ph	Mes	46	142

534   Mesoionic Oxazoles

($R^4 = H$), by flash vacuum pyrolysis (FVP) of either **352** ($R^4 = t$Bu) or dichloromethylphosphine.

4.2.3.4. Ring-Chain Valence Tautomerism

As alluded to in earlier sections, munchnones can undergo ring opening to ketene tautomers, which can then engage in chemical processes. A few clear examples of this pathway have been described since the review by Gingrich and Baum.[1]

The reaction of in situ generated munchnone **42** with unsaturated imine **354** gives β-lactam **355**, in contrast to the results described earlier for other imines (e.g., **42** + **330** → **331**, **332**, **333**) (Fig. 4.117).[134] Presumably, this reaction involves ring-opened ketene **42a**.

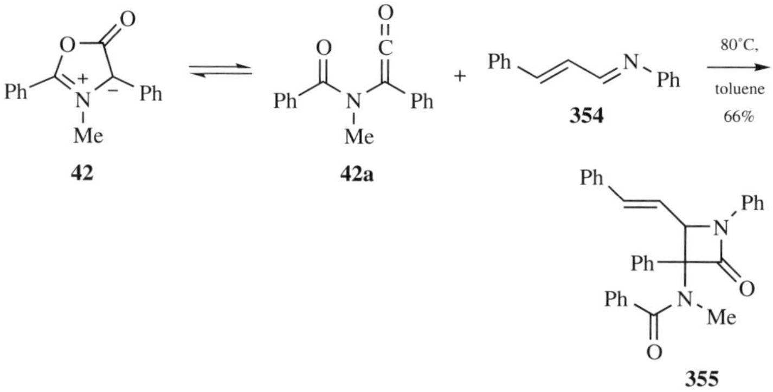

Figure 4.117

Regitz and co-workers[143] found that the stable 2,3,4-tri-*tert*-butylazete **356** reacts with munchnone **42** to afford a mixture of unseparable *E/Z* isomers of oxaazabicyclo[2.2.0]hexene **357** (Fig. 4.118). Presumably, munchnone **42** is reacting through the ring-opened tautomer **42a**. The structure of *Z*-**357** was established by X-ray crystallography.

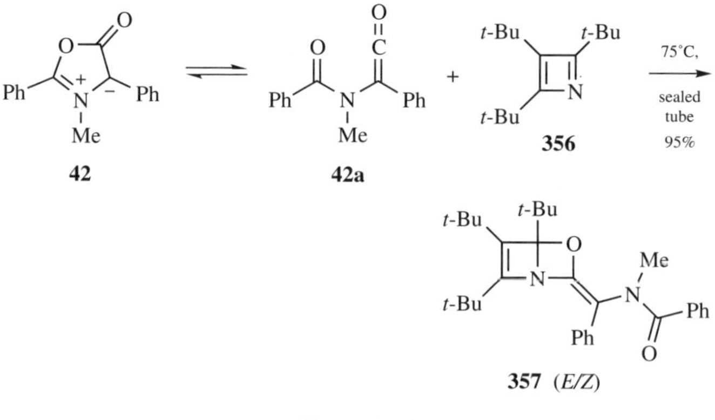

Figure 4.118

4.3. 1,3-OXAZOLIUM 5-IMINES (MUNCHNONE IMINES)

Because all of the work with munchnone imines that has been reported since the review by Gingrich and Baum[1] was described by one research group and involves only cycloaddition chemistry, it is collected in this section. In a series of papers, Laude and co-workers[144–149] examined 1,3-dipolar cycloaddition reactions of munchnone imines derived from Reissert compounds. For example, treatment of the readily assembled Reissert compounds **358** and **360** with HBF_4 forms the munchnone imines **359** and **361**, respectively (Fig. 4.119).[144] Both **359** and **361** undergo smooth intramolecular 1,3-dipolar cycloaddition with the tethered alkyne unit to afford pyrroles **362** and **363**, respectively, after extrusion of HNCO (Fig. 4.120).[144]

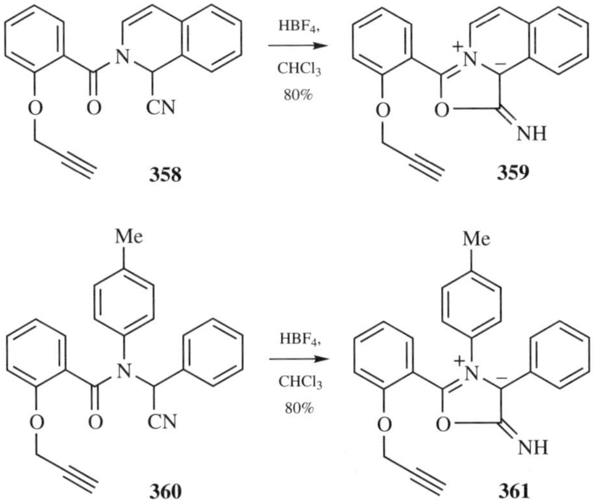

Figure 4.119

This group[145] also diverted the usual Diels-Alder cycloaddition pathway of Reissert salts with olefinic esters to a 1,3-dipolar cycloaddition pathway by the addition of triethylamine. Thus treatment of munchnone imine **364** with ethyl acrylate and triethylamine affords the 1,3-dipolar cycloaddition product **366** (30%) as the major product, formed by fragmentation of cycloadduct **365** (Fig. 4.121). The Diels-Alder product (not shown) is formed in 15% yield. Similar products to **366** are formed with dimethyl and diethyl maleate and fumarate. Laude and co-workers[146] also were able to trap munchnone imine **367** with dipolarophiles to furnish **368** (Fig. 4.122). No Diels-Alder cycloadducts derived from the oxazolium salt were detected. In contrast, fumarate and acrylate esters give only Diels-Alder cycloadducts from the tautomeric oxazolium salt (not shown). However, benzoquinones and 1,4-naphthoquinone react in a 1,3-dipolar fashion with munchnone imine **372** derived from Reissert compound **369** to give **373** (Scheme 4.11).[147] Diels-Alder cycloadducts derived from oxazolium salt **371** were not observed. In a

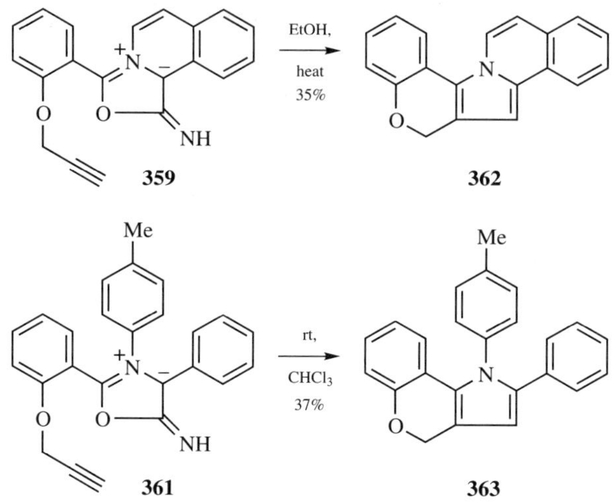

Figure 4.120

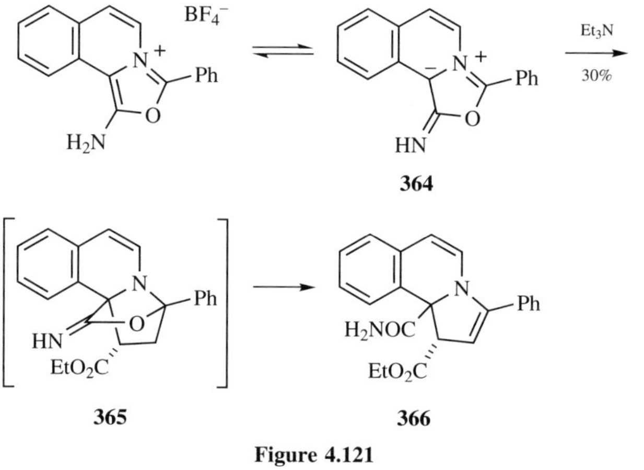

Figure 4.121

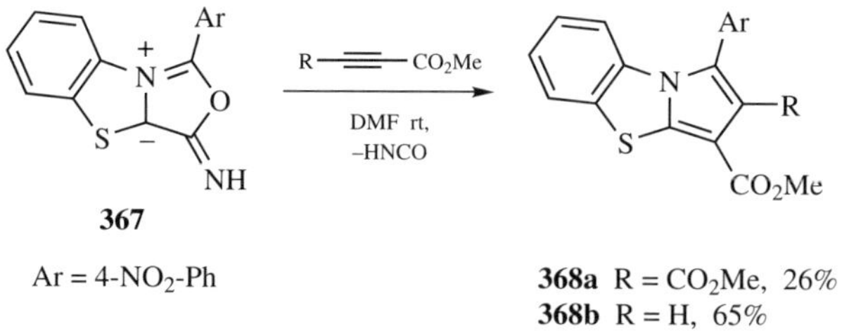

Figure 4.122

368a R = CO$_2$Me, 26%
368b R = H, 65%

Ar = 4-NO$_2$-Ph

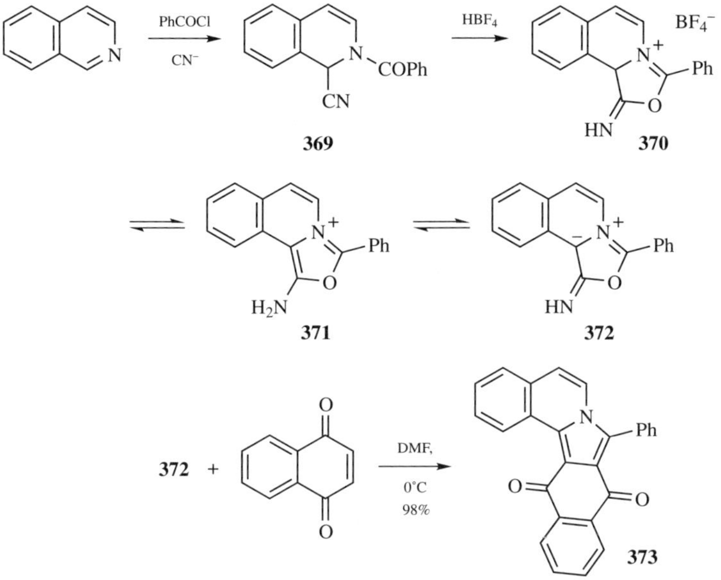

Scheme 4.11

reinvestigation of earlier work by McEwen and co-workers,[150] Laude and co-workers[148,149] showed that the product from the reaction between munchnone imine **374** and several acrylates affords 2-pyridones (e.g., **377**) and not a 3-carboethoxypyrrole as originally claimed. Thus, once again, a 1,3-dipolar cycloaddition reaction wins out over a Diels-Alder cycloaddition (Scheme 4.12).[148,149] In some cases, cycloadduct **375** and dihydropyridone **376** could be isolated.

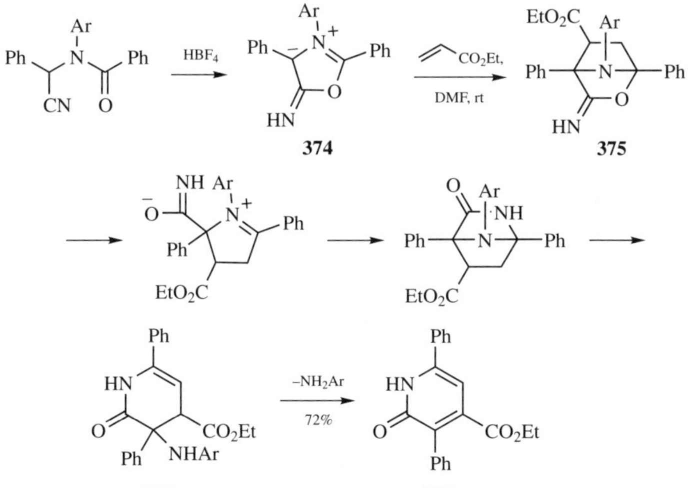

Scheme 4.12

4.4. 1,3-OXAZOLIUM 4-OXIDES (ISOMUNCHNONES)

Although mesoionic 1,3-oxazolium 4-oxides, "isomunchnones," occupied only a few pages in Gingrich and Baum,[1] in the intervening years this ring system has exploded in popularity, largely due to the efforts of Padwa and co-workers. Padwa summarized his isomunchnone work in several reviews.[151–156] Although isomunchnones are rarely isolable, these carbonyl ylide 1,3-dipoles undergo 1,3-dipolar cycloaddition reactions with alacrity.

4.4.1. Synthesis

Only a few methods are available for the synthesis of isomunchnones, and the original methods are summarized by Gingrich and Baum.[1] Padwa and co-workers[154] provided a more recent historical account of the synthesis of isomunchnones. As will be seen, the major new development is the generation of isomunchnones using the rhodium(II)-catalyzed decomposition of α-diazo carbonyl compounds. In continuation of earlier work,[157] Haddadin and Tannus[158] extended their new synthesis of isomunchnones **383** from *N*-benzoylphenylglyoxanilides **382** (Scheme 4.13) to a series of new compounds.

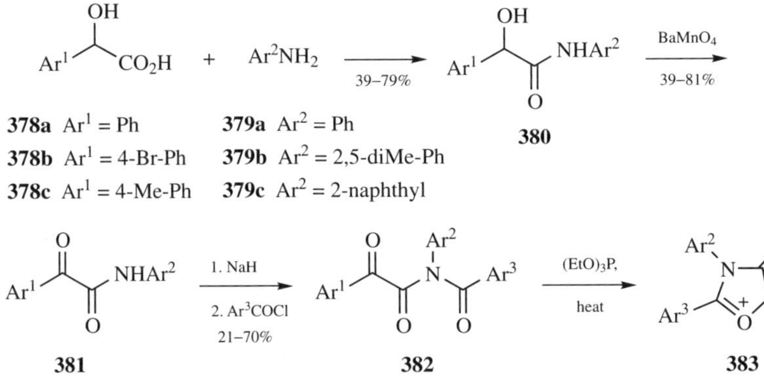

Scheme 4.13

In some cases these highly colored isomunchnones were quite stable for weeks (**383a–c**), moderately stable for 2–10 min (**383d–n**), or unstable within seconds (**383o**). Indeed, isomunchnones **383a–c** crystallized out of the reaction mixture (Fig. 4.123).

Only a few of these isomunchnones **383** reacted with *N*-phenylmaleimide (NPM) to give stable adducts. Thus, although **383a** gave a mixture of exo and endo adducts as reported earlier,[157] **383b** and **383c** gave mainly endo adducts **384** and **385**,

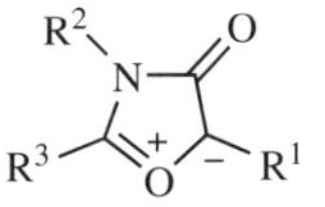

1,3-Oxazolium 4-Oxides (Isomunchnones)

383a $R^1 = R^2 = R^3 = Ph$
383b $R^1 = R^2 = Ph, R^3 = 4\text{-}NO_2\text{-}Ph$
383c $R^1 = 4\text{-}Br\text{-}Ph, R^2 = Ph, R^3 = 4\text{-}NO_2\text{-}Ph$
383d $R^1 = R^2 = Ph, R^3 = 4\text{-}MeO\text{-}Ph$
383e $R^1 = R^2 = Ph, R^3 = t\text{-}Bu$
383f $R^1 = R^2 = Ph, R^3 = CH=CHPh$
383g $R^1 = R^3 = Ph, R^2 = 2,5\text{-}diMe\text{-}Ph$
383h $R^1 = Ph, R^2 = 2,5\text{-}diMe\text{-}Ph, R^3 = 4\text{-}NO_2\text{-}Ph$
383i $R^1 = R^3 = Ph, R^2 = 2\text{-naphthyl}$
383j $R^1 = Ph, R^2 = 2\text{-naphthyl}, R^3 = 4\text{-}NO_2\text{-}Ph$
383k $R^1 = 4\text{-}Br\text{-}Ph, R^2 = Ph, R^3 = 4\text{-}Me\text{-}Ph$
383l $R^1 = 4\text{-}Br\text{-}Ph, R^2 = Ph, R^3 = CH=CHPh$
383m $R^1 = 4\text{-}Me\text{-}Ph, R^2 = R^3 = Ph$
383n $R^1 = R^2 = Ph, R^3 = 1\text{-naphthyl}$
383o $R^1 = R^2 = Ph, R^3 = Me$

Figure 4.123

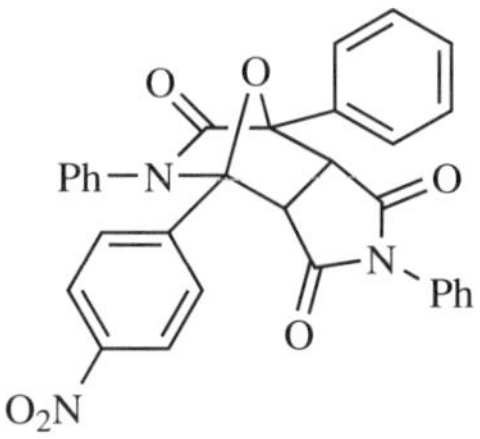

384 36%

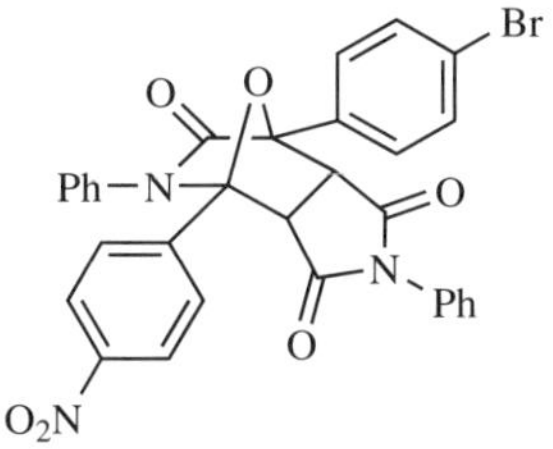

385 46%

Figure 4.124

respectively (Fig. 4.124). The authors were unable to isolate cycloadducts of the other isomunchnones with NPM, the problem apparently being that the latter chemical reacts with the reagent (and solvent) triethyl phosphite.

In a series of papers, Mathias and Moore[159–162] described a new synthesis of isomunchnones **387** via the thermal cyclization of *N*-(chloroacetyl)lactams **386** (Fig. 4.125). Compound **386d**, which would afford a 5-5 fused ring system, is stable up to 150°C. These isomunchnones can be captured by NPM to give fused 2-pyridones in moderate yields. The reaction with DMAD affords **389** in much lower yields ($\leq 17\%$), and other olefinic dipolarophiles (fumarate, maleate, acrylate, and dicyanocyclobutene) are unreactive. Reaction of *N*-(chloroacetyl)benzamide in the presence of NPM gave **391** in low yield.

The major thrust of this work is the synthesis of poly(oxyvinylene)lactams **395** by heating **386b** in the absence of a trapping agent (Scheme 4.14). This appears to be the first report of the self-addition of a mesoionic compound. These low molecular weight polymers form at ambient temperature, and a proposed pathway is illustrated. The oxazolone salt **392** is attacked by isomunchnone **387b** to give dimer **393**. This is followed by nucleophilic ring opening by another isomunchnone to afford **394** and hence polymer **395**.

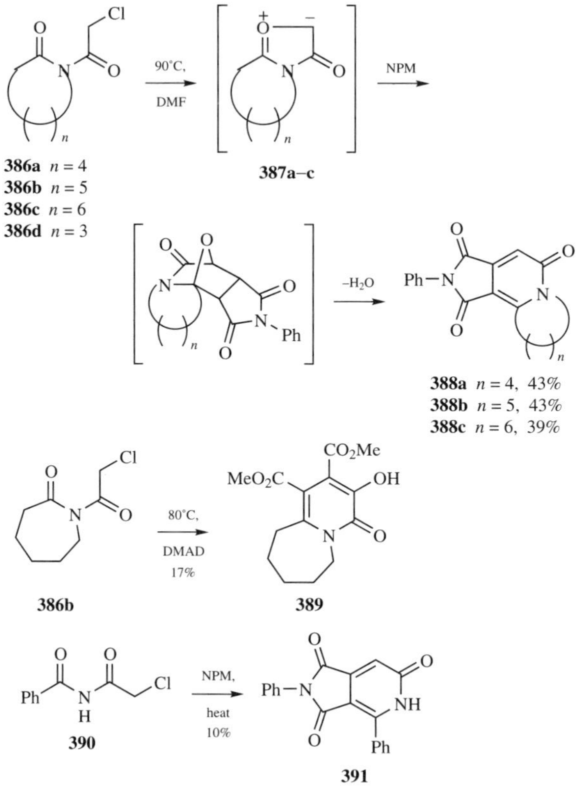

Figure 4.125

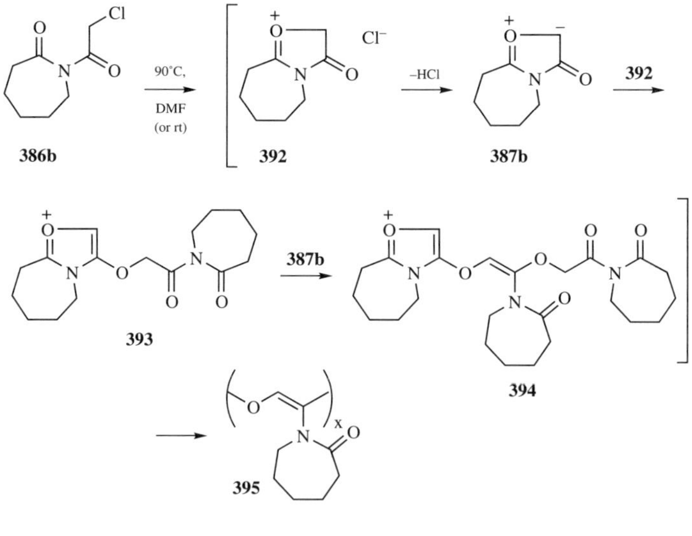

Scheme 4.14

Doyle and co-workers[163] were the first to generate isomunchnones from diazo imides using rhodium(II) catalysis. For example, isomunchnone **397** was produced from diazo imide **396** (Fig. 4.126), but attempts to trap this species with ethyl acrylate were unsuccessful. The only material identified was the isomunchnone hydrolysis product.

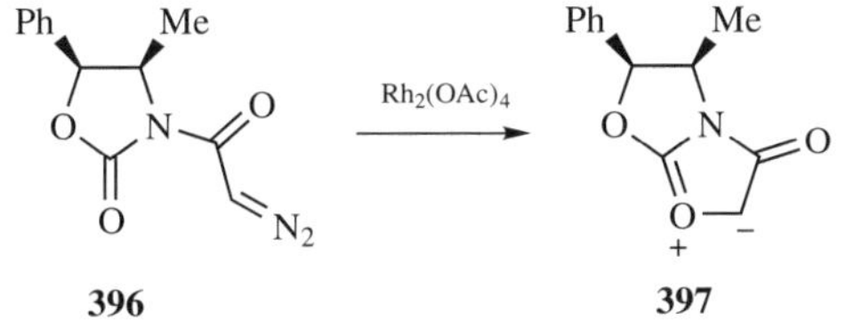

Figure 4.126

Doyle's use of rhodium(II) to generate a rhodium-carbenoid species from an α-diazo carbonyl compound is reminiscent of the first successful synthesis of isomunchnones by Hamaguchi and Ibata,[164] which involved copper(II) to generate a copper–carbenoid species from an α-diazo imide that cyclized to an isomunchnone.

The first successful generation and trapping of isomunchnones using this strategy was described independently by Maier[165,166] and Padwa.[167,168] Maier and Evertz[165] were the first workers to report the intramolecular dipolar cycloaddition of isomunchnones to alkenes, the reaction that Padwa would later exploit so spectacularly. Thus diazo imide **399** was readily prepared from amide **398** by acylation and diazo transfer (Fig. 4.127). Reaction of **399** with rhodium acetate generates isomunchnone **400** which smoothly cyclizes to afford tricycle **401**. Several other examples are described in this Chapter. Reductive ring opening of **401** led to 2-piperidone **402**.

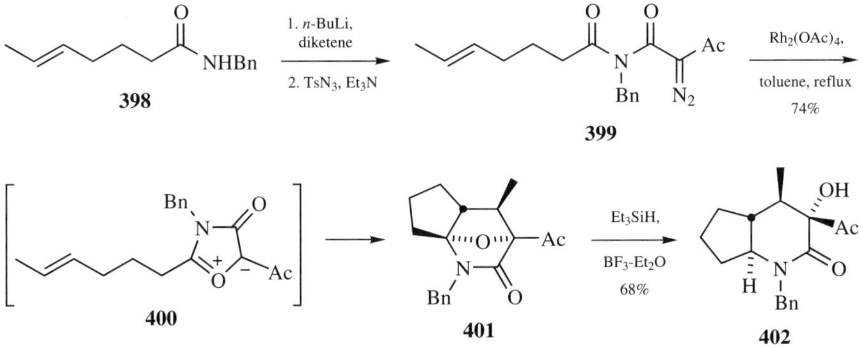

Figure 4.127

Maier and Schöffling[166] have extended this intramolecular isomunchnone cycloaddition to a synthesis of fused furans by employing an alkyne dipolarophile. Thus the diazo acetylenes **403** are smoothly converted to furans **404** with catalytic rhodium acetate (Fig. 4.128).

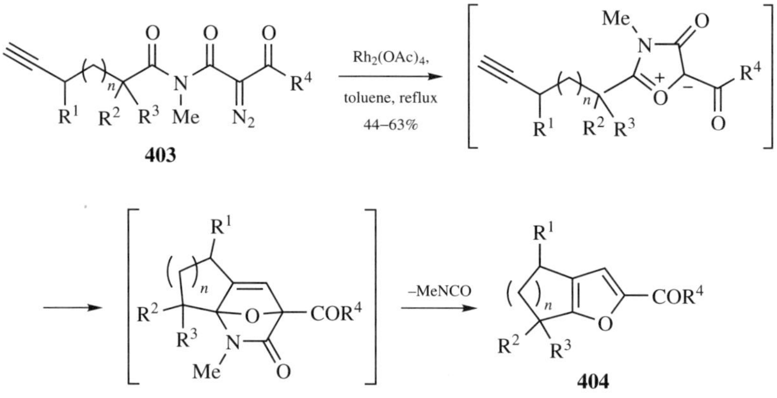

R^1 = H, OCOPh; R^2 = H, Me; R^3 = H, Me; R^4 = Me, OMe; n = 1, 2

Figure 4.128

Padwa and co-workers[167] also explored the rhodium-catalyzed reaction of diazo imides to form isomunchnones. Subsequent in situ trapping affords aza-substituted polycyclic compounds. Thus **405** smoothly forms isomunchnones **406**, which can be intercepted in high yield with DMAD to give furans **407**, following loss of methyl isocyanate from the cycloadducts (Fig. 4.129). Likewise, the bicyclic

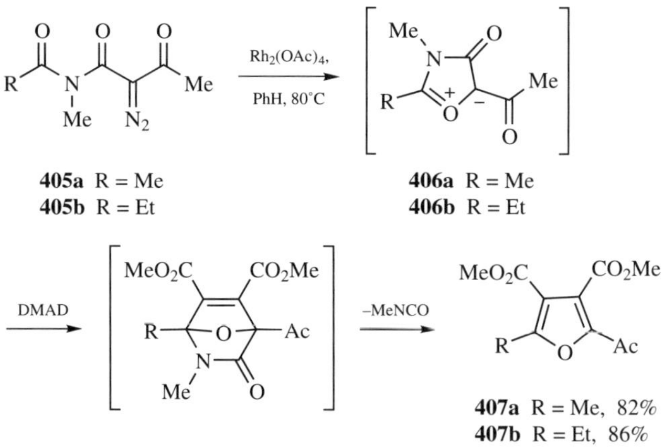

405a R = Me
405b R = Et

406a R = Me
406b R = Et

407a R = Me, 82%
407b R = Et, 86%

Figure 4.129

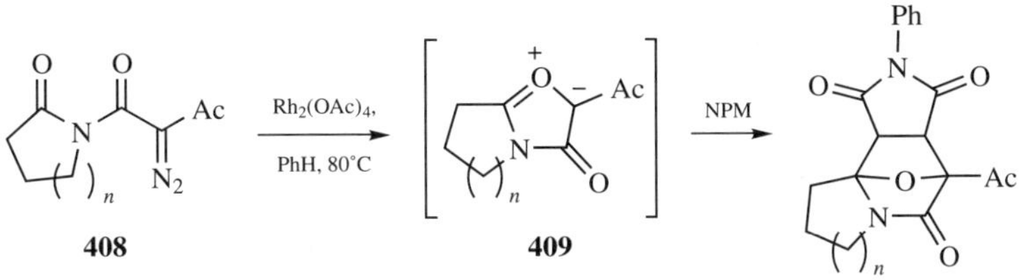

408 **409**

410a $n = 0$, 61%
410b $n = 1$, 78%
410c $n = 2$, 85%
410d $n = 3$, 75%

Figure 4.130

isomunchnones **409** can be trapped with NPM to afford adducts **410** (Fig. 4.130).[167] Trapping isomunchnone **409** ($n = 2$) with DMAD led to furan isocyanate **411**, after a retro-Diels-Alder reaction (Fig. 4.131).

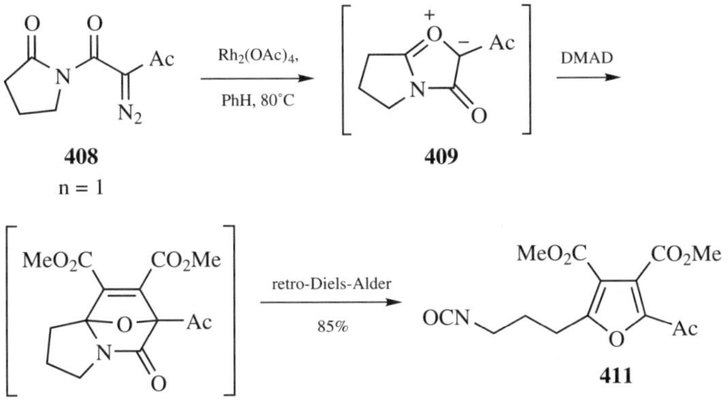

408 n = 1 **409** **411**

Figure 4.131

In a collaborative effort, Doyle and co-workers[168] expanded the rhodium-catalyzed generation of isomunchnones from diazoacetoacetamides and subsequent trapping with dipolarophiles.[167] In the case of diazoacetoacetyl urea **412**, the derived isomunchnone **413** reacts with methyl propiolate to give a 2:1 mixture of cycloadducts **414** (Fig. 4.132). The resulting regiochemistry is successfully rationalized using FMO theory as being isomunchnone–HOMO controlled. This result represents one of the few reactions in which the cycloadducts are stable. The full synthetic power of the rhodium(II)-catalyzed generation of isomunchnones from α-diazo carbonyl compounds is elaborated on in Section 4.4.3.2.

The newest method for generating isomunchnones was reported by Padwa and co-workers.[169,170] Thus Kuethe and Padwa developed an exciting new application of the venerable Pummerer reaction of imidosulfoxides to generate and trap isomunchnones with alkenes. For example, the readily prepared imidosulfoxide

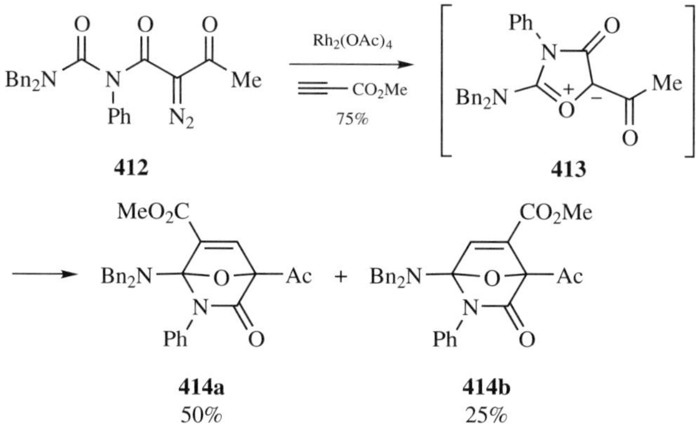

Figure 4.132

415 on exposure to acetic anhydride and a trace of *p*-toluenesulfonic acid affords isomunchnone **416** (Fig. 4.133). Trapping with NPM or maleic anhydride yields cycloadducts **417**.

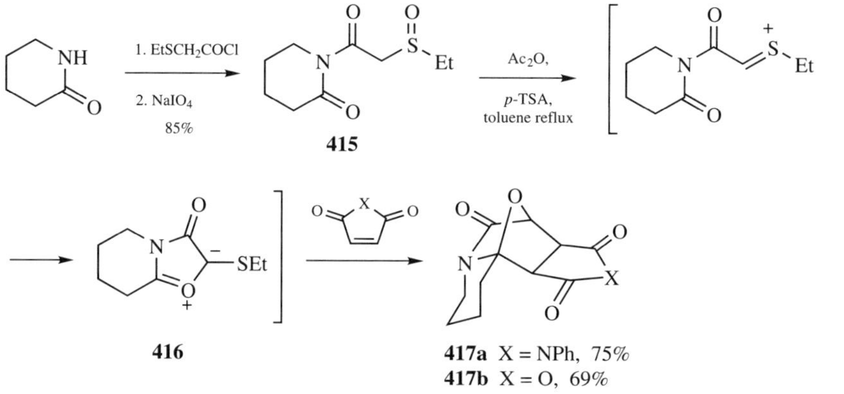

Figure 4.133

This strategy is a powerful route to bicyclic pyridones and their transformation products. Thus these workers[169,170] applied the methodology to formal syntheses of the lupinine alkaloids (±)-lupinine and (±)-anagyrine **422** (Scheme 4.15). Thus imidosulfoxide **415** is converted to the corresponding isomunchnone, which is trapped with methyl acrylate to give **418**. Oxidation, ring opening, and triflate formation affords **419**. A Stille cross-coupling installs the pyridine unit and further manipulation leads to **421**, which has previously been converted to (±)-anagyrine **422**.

In the full account of this work, Padwa and co-workers[170] demonstrated that the 1,3-dipolar cycloaddition is an endo cycloaddition and the regiochemistry is

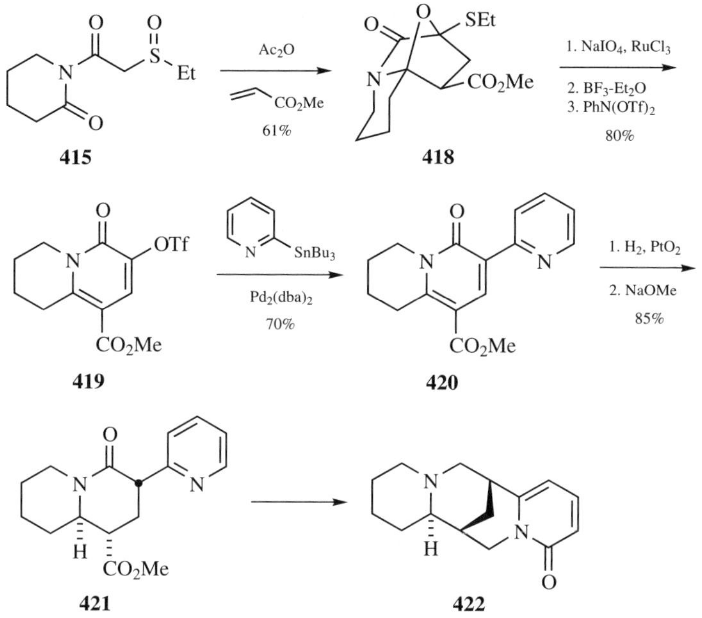

1,3-Oxazolium 4-Oxides (Isomunchnones) 545

Scheme 4.15

consistent with that of a HOMO-dipole controlled process as judged from the products **424** and **425**, which arise from the reaction between isomunchnone **423** and methyl propiolate and phenyl vinyl sulfone, respectively (Fig. 4.134). Isomunchnone **423** is trapped with DMAD to give the expected furan in 41% yield.

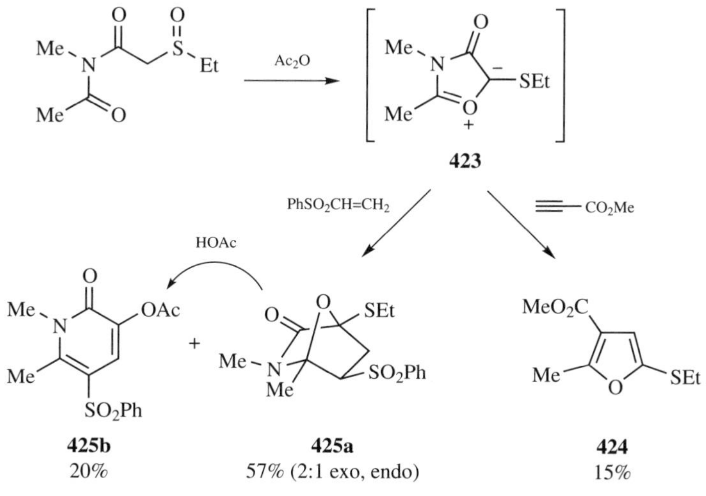

Figure 4.134

4.4.2. Structure and Spectra Properties

Only one new example of the isolation and spectral data of isomunchnones has been reported since the cases presented by Gingrich and Baum.[1] Regitz and co-workers[171] prepared a series of remarkably stable isomunchnones **426** (Fig. 4.135). These yellow solids can often be obtained in analytical purity and have melting points in the range of 153°–207°C. Their synthesis and 1,3-dipolar cycloaddition reactions are described in Section 4.4.3.2.

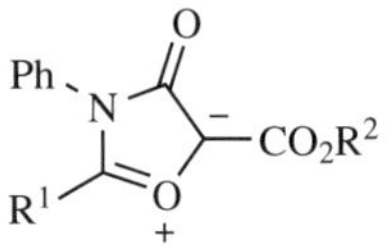

R^1 = 1- and 2-naphthyl, mesityl, 4-MeO-Ph
R^2 = Me, Et

426

Figure 4.135

4.4.2.1. Infrared and Mass Spectra

Isomunchnones **426** exhibit a single infrared carbonyl absorption at v 1686–1690 cm^{-1}. The mass spectra reveal a significant parent ion in each case, but the base peak in the mass spectrum of most isomunchnones **426** corresponds to the nitrilium ion R^1-C≡N$^+$-Ph resulting from ring cleavage.

4.4.2.2. NMR Spectra

The proton NMR spectra of isomunchnones are unexceptional, but the carbon spectra indicate the mesoionic ring system. Thus C-2 and C-4 appear at 150–160 ppm and C-5 resonates at 116–120 ppm.

4.4.2.3. Molecular Orbital Calculations

The limited number of molecular orbital calculations, which deal with the regioselectivities of isomunchnone 1,3-dipolar cycloaddition reactions, are covered in Section 4.4.3.2.

4.4.3. Reactions

4.4.3.1. Nucleophiles

The only significant reaction of isomunchnones with nucleophiles is hydrolysis. However, because this is normally observed only as a side reaction during 1,3-dipolar cycloaddition reactions, the few examples of isomunchnone hydrolysis are covered in the next section.

4.4.3.2. Cycloaddition Reactions

As presented by Gingrich and Baum,[1] the isomunchnone ring system—a masked carbonyl dipole—is exceptionally reactive as a 1,3-dipole in 1,3-dipolar cycloaddition reactions. In the intervening years, the major research efforts in isomunchnone chemistry have entailed synthetic applications to specific targets such as alkaloids.

Kato and co-workers[113] had much better success in performing unusual 1,3-dipolar cycloadditions with isomunchnones than with munchnones (vide supra). Thus the room temperature union of isomunchnone **383a** with benzocyclopropene **271** leads to a syn cycloadduct (Fig. 4.136). The latter is remarkably stable, is recovered unchanged on heating to 300°C, and is impervious to the action of tributylphosphine, in an abortive attempt to excise the bridging oxygen, which would have led to a methanooxonine.

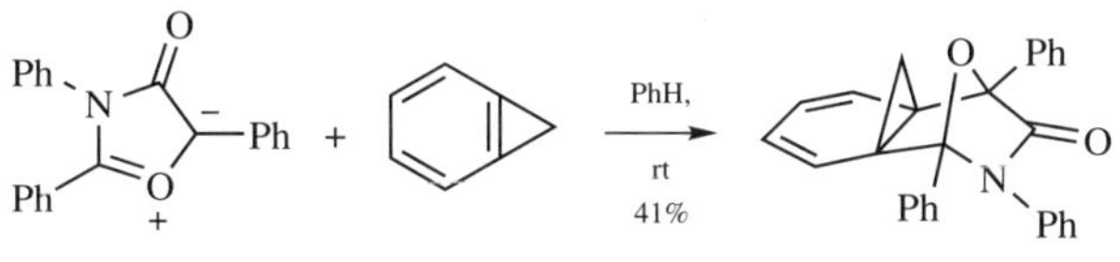

Figure 4.136

Kato and co-workers[121] found that isomunchnone **383a** reacts with tropone **294** to afford **427**, which is apparently the first example of a [4π + 6π] cycloadduct both involving a mesoionic heterocycle and a carbonyl ylide. We saw earlier that tropone did not react cleanly with a munchnone. The one-pot reaction of **428** gave **427** in somewhat higher yield. Whereas heating the latter in bromobenzene affords *o*-benzoylmandelanilide (69%), heating **427** in refluxing toluene in the presence of DMAD leads to furan **429** (Fig. 4.137).

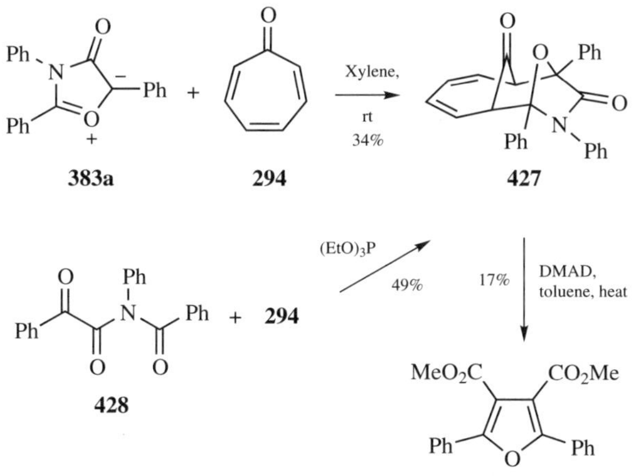

Figure 4.137

548   Mesoionic Oxazoles

Kato and co-workers[172,173] explored the chemistry of 2-*tert*-butylfulvenes with isomunchnones, as well as with several other mesoionic compounds, in a novel approach to pseudo-hetero-azulenes. Thus the isomunchnone **383a**, generated as before in situ from *N*-benzoylphenylglyoxyanilide (**428**) with triethylphosphite, reacted with 2-*tert*-butyl-6-(dimethylamino)fulvene (**430**) to give the [4π + 6π] adduct diphenylcyclopenta[*c*]pyran **432** in low yield (Fig. 4.138). Likewise, reaction of **383a** with dimethylfulvene **433** gave a mixture of two adducts, **434** and **435**, the latter arising from a [4π + 2π] cycloaddition. Moreover, this reaction, run in the presence of air, afforded a small amount of tricyclic oxygenated dimer **436**, established by X-ray crystallography. The regiochemistry leading to **434** and **435** is unknown and is shown arbitrarily.

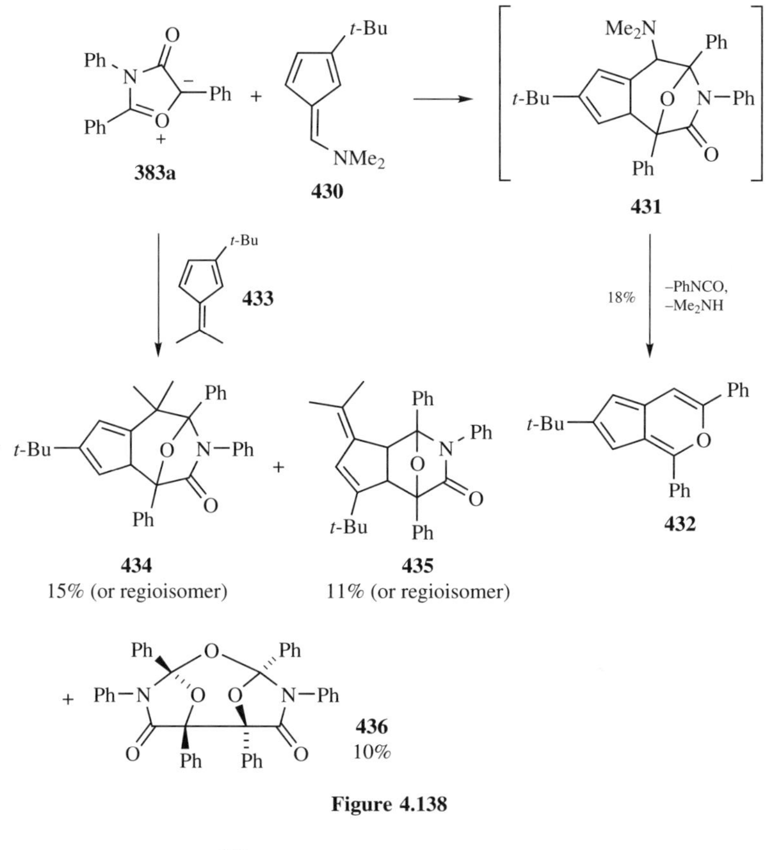

Figure 4.138

Regitz and co-workers[143] also found that azete **356** reacts with isomunchnones. In contrast to the chemistry observed with munchnone **42** (vide supra), isomunchnones **437** react with **356** to afford regioselectively the primary cycloadducts **438** (Fig. 4.139). These compounds undergo facile hydrolysis to **439**, and an X-ray crystal structure of **439b** confirmed its identity. Thermolysis of **439** in toluene yields pyrrole **440**.

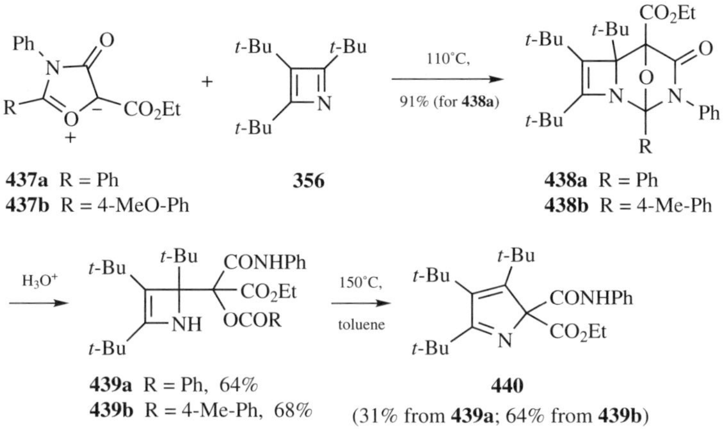

437a R = Ph
437b R = 4-MeO-Ph

356

438a R = Ph
438b R = 4-Me-Ph

439a R = Ph, 64%
439b R = 4-Me-Ph, 68%

440 (31% from **439a**; 64% from **439b**)

Figure 4.139

Details regarding the preparation of isomunchnones **437** were not disclosed in this paper but presumably involved rhodium(II)-catalyzed decomposition of α-diazo imides as described subsequently.[171] Thus Regitz and co-workers[171] employed the cycloaddition of isomunchnones **426** with phosphaalkynes **442** to prepare 1,3-oxaphospholes **443** (Fig. 4.140). The isomunchnones can either be generated and trapped in situ or isolated. Both procedures are equally successful. This sequence is clearly the method of choice for the synthesis of the relatively little investigated 1,3-oxaphospholes. The requisite diazo carbonyl compounds **441** are readily prepared from 2-diazomalonic ester chlorides and N-phenylcarboxamides (27–36% yield). The presumed bicyclic intermediates could not be detected by NMR. The authors note that the high regioselectivity, which was established by X-ray crystallography, is not "under charge control" in view of the polarization of the C≡P triple bond.

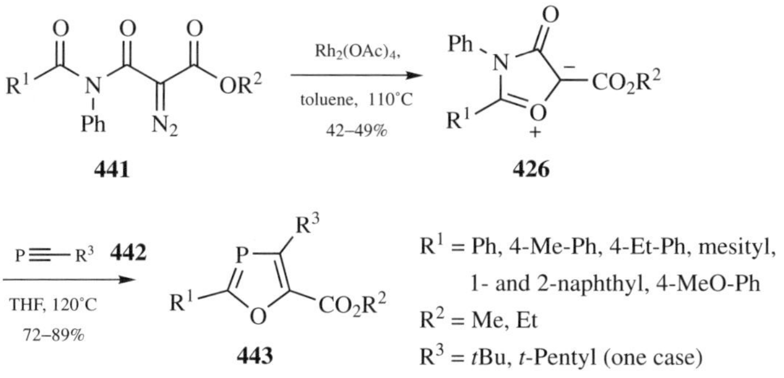

441 → **426**

442 → **443**

R^1 = Ph, 4-Me-Ph, 4-Et-Ph, mesityl, 1- and 2-naphthyl, 4-MeO-Ph
R^2 = Me, Et
R^3 = tBu, t-Pentyl (one case)

Figure 4.140

Padwa and co-workers are the consummate practitioners of the rhodium(II)-catalyzed decomposition of α-diazo carbonyl compounds leading to isomunchnones

550 Mesoionic Oxazoles

and their subsequent 1,3-dipolar cycloaddition reactions. These workers have made elegant use of both intermolecular and intramolecular versions of this chemistry.

Padwa and Hertzog[174] described intermolecular cycloaddition reactions of isomunchnones with both electron-rich and electron-deficient dipolarophiles. The resulting regiochemistry is in accord with FMO theory. Diazo imide **444** is readily prepared from 2-pyrrolidinone, and under the usual conditions it reacts via **445** with N-phenylmaleimide to give the expected cycloadducts (86% yield; endo:exo, 2.4:1), and also with DMAD to give a furan after ring opening (Fig. 4.141). Interestingly, isomunchnone **445** could not be generated from chloroacetyl lactam **386d**, as described earlier.

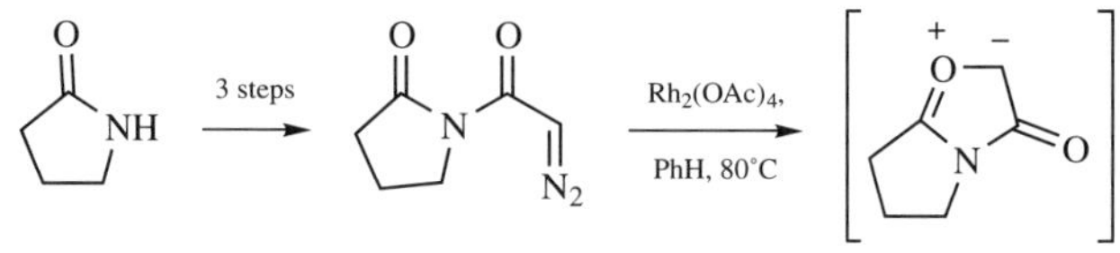

Figure 4.141

Whereas **445** does not react with electron-rich dipolarophiles, the more delocalized isomunchnone **446** does react with both electron-rich and electron-deficient dipolarophiles.[174] A detailed FMO analysis is consistent with these observations and with the regiochemistry exhibited by diethyl ketene acetal and methyl vinyl ketone (Scheme 4.16). The reaction of **446** with the ketene acetal to give **447** is LUMO-dipole HOMO-dipolarophile controlled (so-called Type III process). In

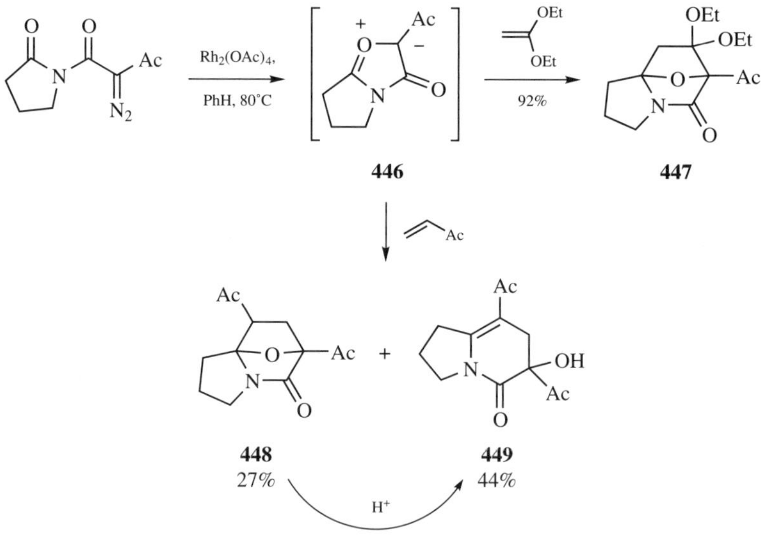

Scheme 4.16

contrast, the reaction of **446** with methyl vinyl ketone to give **448** is HOMO-dipole LUMO-dipolarophile controlled (so-called Type I process). In competition, experiments using a mixture of N-phenylmaleimide and ketene acetal only a cycloadduct from the former was isolated. This result is consistent with a smaller energy gap for this Type I process than for the Type III reaction ($\Delta E = 7.05$ eV vs. 8.69 eV). The difference in reactivity between isomunchnones **445** and **446** is also manifest in their behavior with methyl propiolate (not shown).

In a careful study of rhodium catalysts for the decomposition of α-diazo imide **450**, Padwa and co-workers[175,176] found that perfluorinated ligands greatly favor isomunchnone formation, whereas acetate leads to the generation of a six-membered carbonyl ylide. Thus **450** is converted to isomunchnone **451** with either rhodium perfluorobutyroamidate ($Rh_2(pfm)_4$), rhodium perfluorobutyrate ($Rh_2(pfb)_4$), or rhodium trifluoroacetate ($Rh_2(tfa)_4$) but is converted to **454** with $Rh_2(OAc)_4$ (Scheme 4.17). Neither 1,3-dipole can be isolated, but isomunchnone

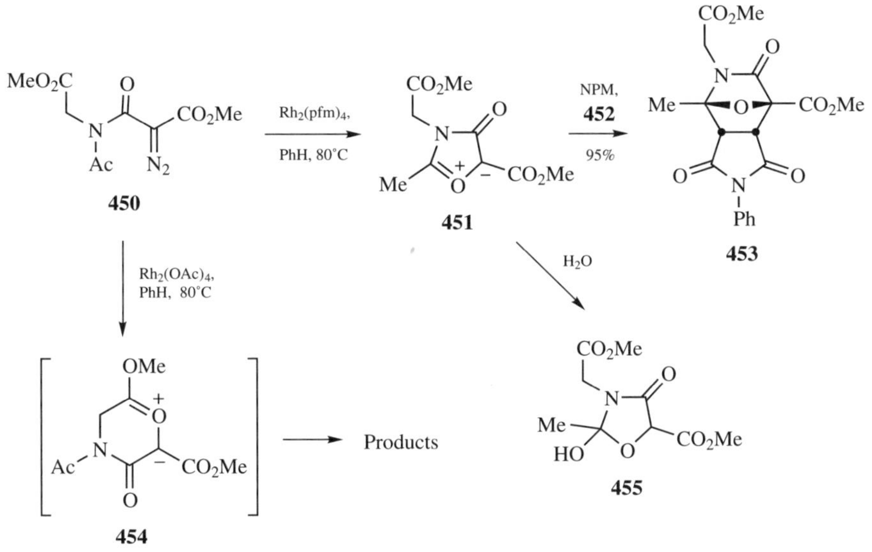

Scheme 4.17

451 can be hydrolyzed to hemiketal **455** in 90% yield and trapped with N-phenylmaleimide (**452**) to give endo cycloadduct **453**. The mechanistic basis for this dramatic ligand effect is not understood. Interestingly, the addition of the strong Lewis acid $Sc(OTf)_3$ to the $Rh_2(OAc)_4$ reaction mixture diverts the pathway to an isomunchnone mode.

Padwa and Prein[177] presented an extensive experimental and theoretical study of the 1,3-dipolar cycloaddition reactions of isomunchnones with olefinic dipolarophiles. The α-diazo carbonyl isomunchnone precursors were synthesized in the usual fashion from amides and diazoethylmalonyl chloride. For example, isomunchnone **457** was readily generated from **456** using rhodium catalysis to form the corresponding carbenoid (Fig. 4.142). Trapping with *N*-phenylmaleimide affords exo-anti-adduct **458** in 90% yield.

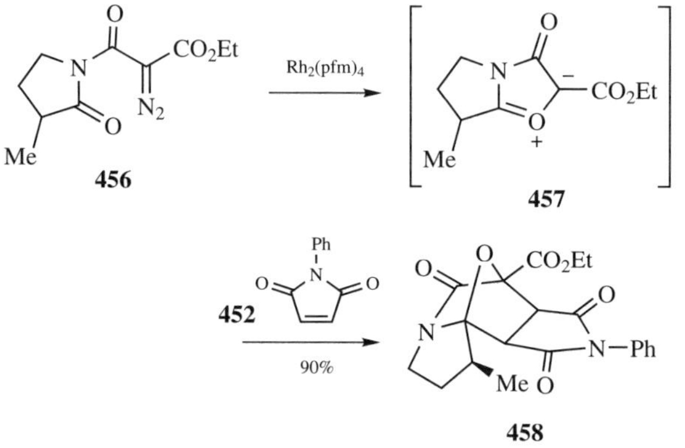

Figure 4.142

Wudl and co-workers[178] found that several isomunchnones **460** react with C_{60} reversibly to form cycloadducts **461** (Fig. 4.143). These workers suggested that such cycloadducts could function as isomunchnone repositories. For example, heating **461** (R^1 = Ph, R^2 = Me, R^3 = Me) in the presence of *N*-phenylmaleimide (NPM) afforded C_{60} and the known isomunchnone-NPM cycloadduct in 94% yield.

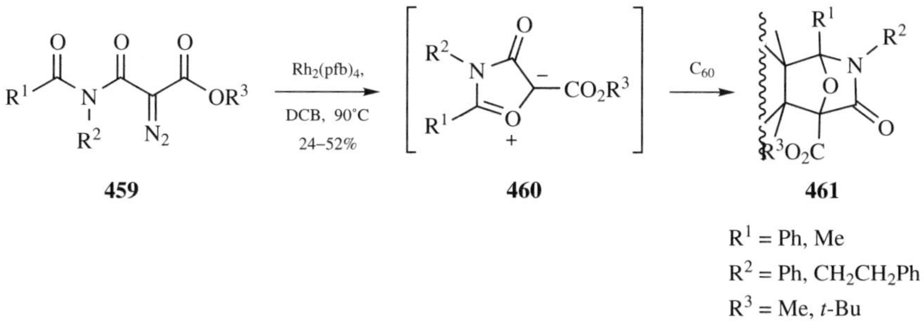

R^1 = Ph, Me
R^2 = Ph, CH_2CH_2Ph
R^3 = Me, *t*-Bu

Figure 4.143

The dipolar cycloaddition chemistry of isomunchnones is a powerful and concise route to polycyclic azaheterocycles, and Padwa has been the pioneer in this effort. Sheehan and Padwa[179] employed the rhodium-catalyzed isomunchnone generation and subsequent trapping to a synthesis of 2-pyridones and the alkaloid (±)-ipalbidine (**465**) (Fig. 4.144). Thus α-diazo imide **462** was readily constructed from 2-pyrrolidinone and allowed to react with rhodium acetate in the presence of *cis*-1-(phenylsulfonyl)-1-propene to afford 2-pyridone **464** after loss of phenylsulfinic acid. Further manipulation, featuring a Stille coupling, gave (±)-ipalbidine **465**.

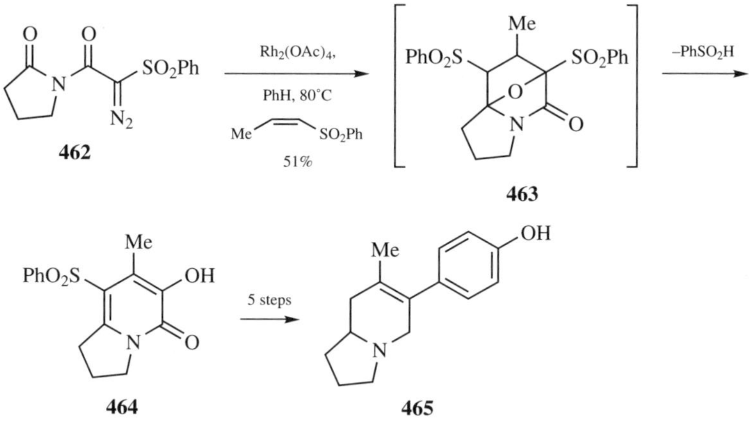

Figure 4.144

Padwa and co-workers[180,181] expanded on their isomunchnone cycloaddition technology to encompass the synthesis of several other highly substituted 2-pyridones, such as the angiotensin-converting enzyme inhibitor (−)-A58365A **469**, which was synthesized using this protocol (Scheme 4.18). Pyroglutamic acid **466** was readily converted to isomunchnone precursor **467** in 49% overall yield. Treatment with rhodium acetate in the presence of methyl vinyl ketone under the usual conditions afforded 2-pyridone **468** in high yield. Further manipulation yielded (−)-A58365A **469**. Similar selection of starting materials allowed for the synthesis of the indolizidine alkaloids δ-coniceine and a formal synthesis of (±)-septicine.

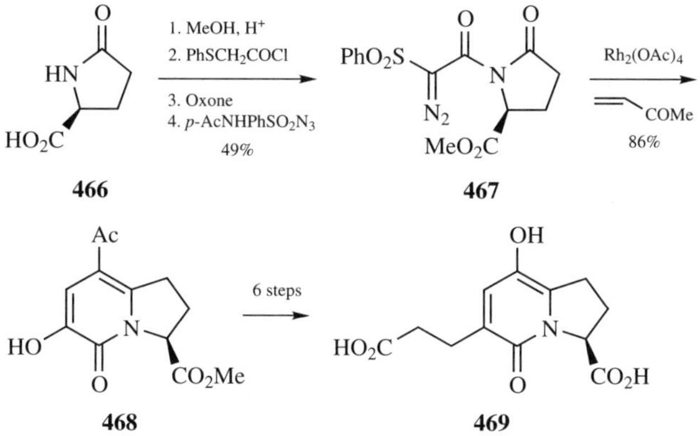

Scheme 4.18

Padwa and Prein[182] generated several chiral isomunchnones, using the rhodium-catalyzed deamination of chiral α-diazo imides and trapped them with various dipolarophiles. The best results were obtained with isomunchnones derived

554 Mesoionic Oxazoles

from phenylalanine methyl ester, apparently due to π-stacking and shielding of one face. Two of these reactions are illustrated (Fig. 4.145). This work and that of Harwood[183–185] (vide infra) represent the first examples of acyclic stereocontrol in the 1,3-dipolar cycloaddition reactions of mesoionic heterocycles. The isomunchnone derived from diazo imide **470** also reacts smoothly with NPM and 1,4-naphthoquinone. In all cases the syn:anti ratio is ≥95:5, and only exo adducts were observed. Isomunchnones derived from alanine and leucine were less stereocontrolling. The results from this study reinforce the notion that the actual 1,3-dipole is an isomunchnone-rhodium carbenoid species and not the free isomunchnone.

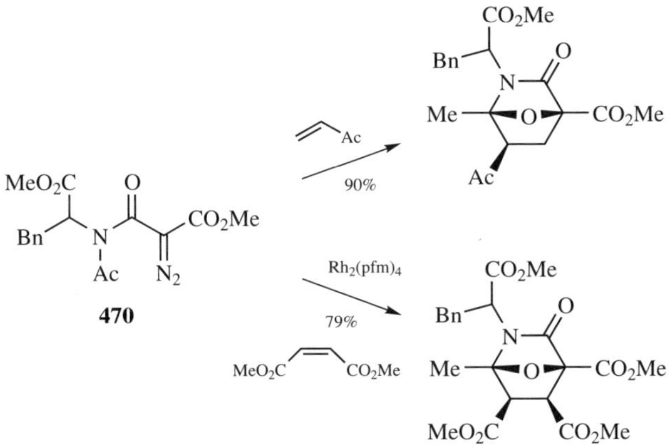

Figure 4.145

Independently, Harwood[183–185] also demonstrated the role of chiral-templated isomunchnones in 1,3-dipolar cycloaddition reactions. Thus using the rhodium(II)-catalyzed decomposition of diazo carbonyl compounds, Harwood and co-workers[183] explored cycloadditions of isomunchnone derivatives of (5R)- and (5S)-phenyloxazin-2,3-dione. Along with the work of Padwa (vide supra), these reactions appear to represent the first examples of chirally templated isomunchnone 1,3-dipolar cycloadditions. For example, reaction of **471** under standard rhodium acetate conditions in the presence of NPM affords a mixture of *endo*-**472** and *exo*-**473** adducts (Fig. 4.146). *N*-Methylmaleimide and DMAD react with **471** similarly.

Additional work by the Harwood team[184] with diazo imides **474**, and related ones, revealed that the resulting isomunchnones react with methyl propiolate to form adducts **475** (Fig. 4.147). Diazo imides **474** (R = CO$_2$Et, Ac, H) also react with maleimides and DMAD to afford adducts with high endo, exo selectivities and moderate diastereofacial selectivities.

This group[185] also found that the isomunchnones derived from these diazo imides react with aromatic aldehydes with excellent diastereofacial- and exo-selectivity. Thus 4-nitrobenzaldehyde reacts with the isomunchnone derived from **476** to give adduct **477** (Fig. 4.148). Hydrolysis leads to α,β-hydroxy acid **478** in optically pure form.

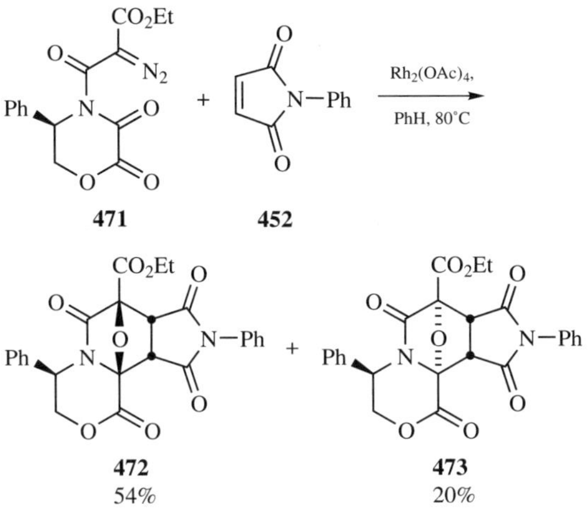

1,3-Oxazolium 4-Oxides (Isomunchnones)

Figure 4.146

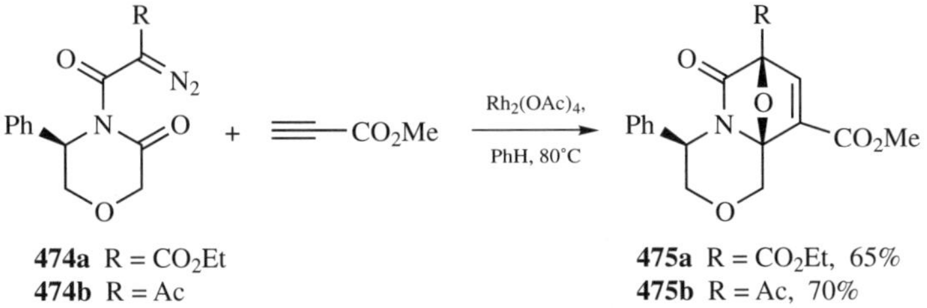

Figure 4.147

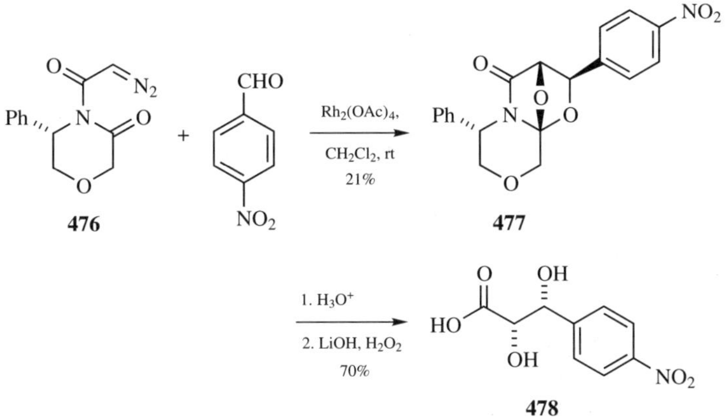

Figure 4.148

556 Mesoionic Oxazoles

Kappe and co-workers[186] employed an isomunchnone generation-trapping sequence to access conformationally restricted dihydropyrimidine derivatives as novel calcium channel modulators. The stable isomunchnone **481** was prepared from dihydropyrimidone **479** by the standard N-malonylacylation and azide transfer to give **480**, and then treatment of the latter with rhodium acetate (Scheme 4.19). Isomunchnone **481**, which is stable in the open air for months, reacts with N-methylmaleimide and methyl vinyl ketone to give adducts **482** and **483**, respectively, the latter arising by rearrangement of the primary adduct.

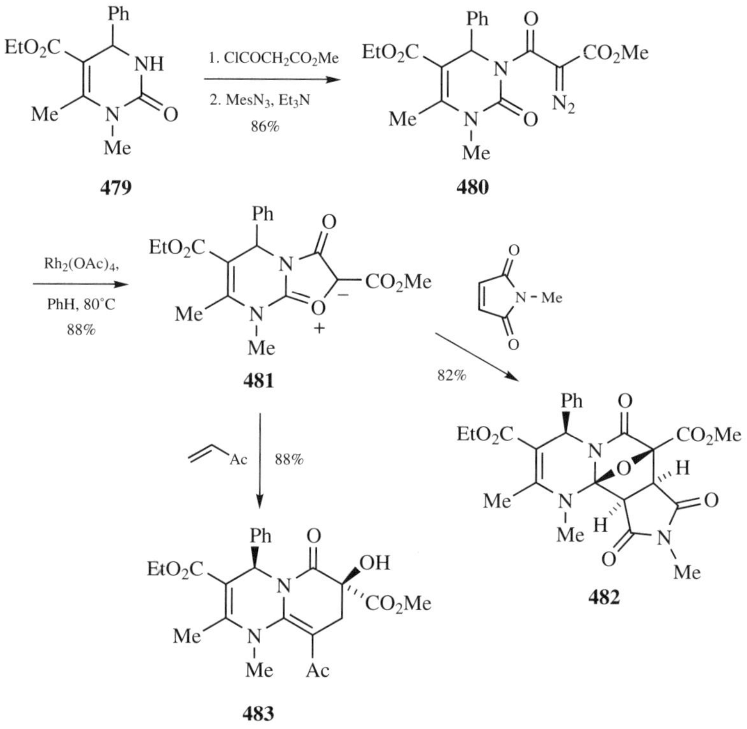

Scheme 4.19

In similar fashion,[186] the conformationally restricted analogs **485** were prepared via intramolecular cycloadditions from the isomunchnones generated from α-diazo imides **484** (Fig. 4.149). The structures of these cycloadducts were established by X-ray crystallography.

Kappe[187] investigated further the generation and trapping of aminoisomunchnones. Interestingly, when diazoacetylurea **486** is treated with rhodium acetate, the ammonium ylide **488** is obtained in good yield (Scheme 4.20). The structure of this unanticipated product was established by X-ray crystallography. However, when this reaction is performed in the presence of DMAD, then 2-dimethylaminofuran **491** is isolated, presumably indicating the generation of isomunchnone **490** via the

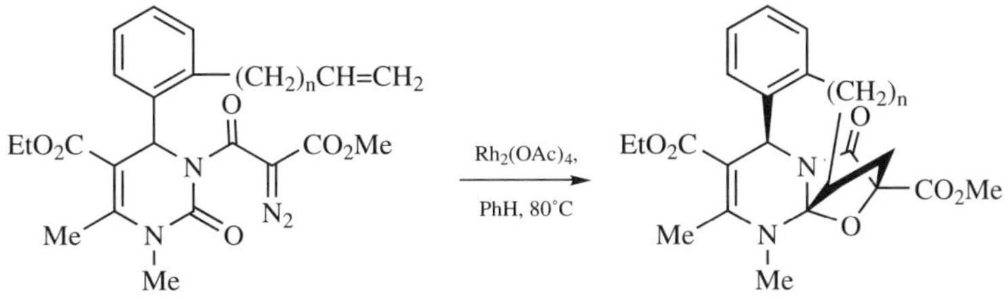

1,3-Oxazolium 4-Oxides (Isomunchnones) 557

Figure 4.149

484a $n = 0$
484b $n = 1$

485a $n = 0$, 92%
485b $n = 1$, 88%

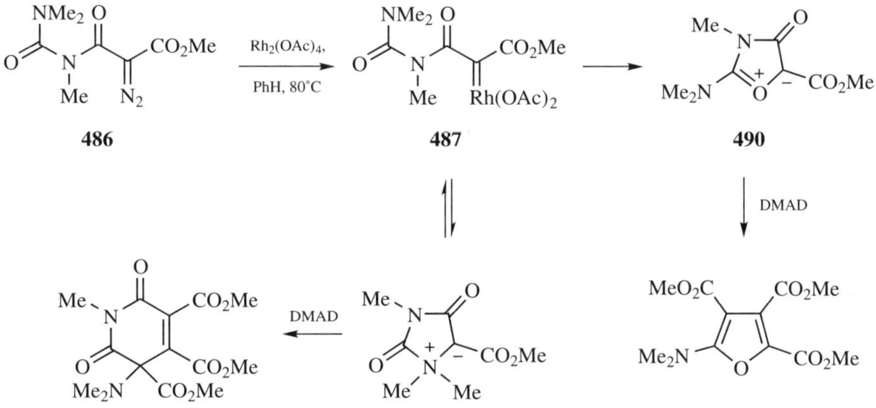

Scheme 4.20

rhodium–carbenoid intermediate **487**. The reversibility of the sequence was shown by the fact that ylide **488** reacts with DMAD in the presence of rhodium acetate to give a small amount (10–15%) of 2-aminofuran **491**, presumably via an equilibrium concentration of isomunchnone **487**.

In a continuation of these observations, Padwa and co-workers[188] examined in depth the rhodium-catalyzed chemistry of α-diazo esters leading either to ammonium or carbonyl ylides, depending on the reaction conditions. However, the major thrust of their paper involves nonmesoionic carbonyl ylides and ammonium ylides.

Gowravaram and Gallop[189] adapted the rhodium-catalyzed generation of isomunchnones from diazo imides to the solid-phase synthesis of furans, following a 1,3-dipolar cycloaddition reaction with alkynes. A variety of furans **492** were prepared in this fashion (Fig. 4.150). With unsymmetrical electron-deficient alkynes (e.g., methyl propiolate), the anticipated regiochemistry is observed, e.g., HOMO-dipole LUMO-dipolarophile, as seen previously.

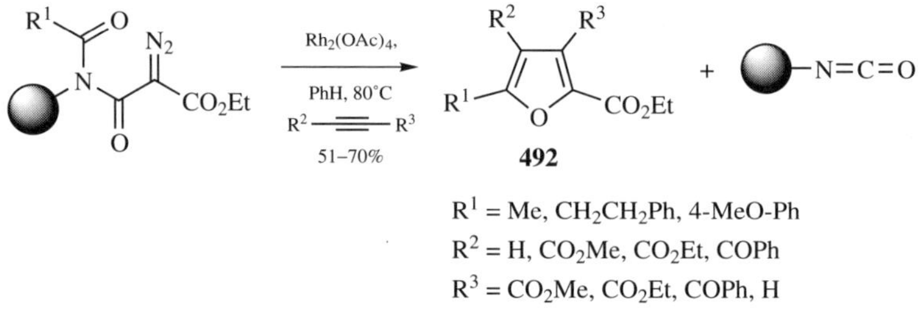

558   Mesoionic Oxazoles

Figure 4.150

R^1 = Me, CH_2CH_2Ph, 4-MeO-Ph
R^2 = H, CO_2Me, CO_2Et, COPh
R^3 = CO_2Me, CO_2Et, COPh, H

Independently, Austin and co-workers[190,191] also adopted the isomunchnone generation and trapping protocol to the solid-phase synthesis of furans. Model studies revealed that rhodium perfluorobutyroamidate ($Rh_2(pfm)_4$) afforded none of the by-product **495**, which forms via a tandem cyclopropanation-Cope rearrangement. Moreover, this catalyst is more soluble in organic solvents than is rhodium trifluoroacetate (Scheme 4.21). The cycloadducts **494** and **497** readily fragment to the corresponding furans (e.g., **498**) on heating in benzene, which was found to be superior as a solvent to methanol or chloroform.

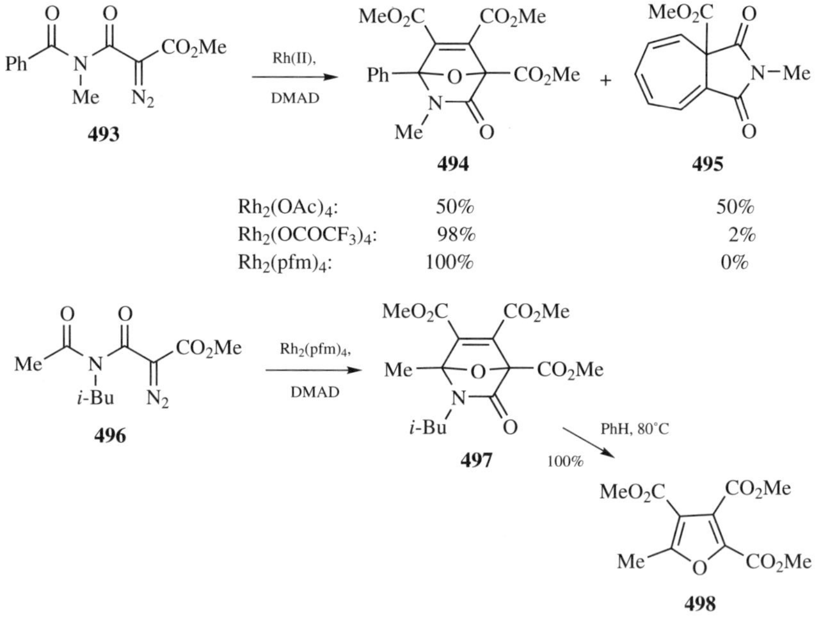

	494	495
$Rh_2(OAc)_4$:	50%	50%
$Rh_2(OCOCF_3)_4$:	98%	2%
$Rh_2(pfm)_4$:	100%	0%

Scheme 4.21

Subsequent application of this methodology to the solid-phase synthesis of furans was straightforward, as shown for **499→500** using a Wang resin (Fig. 4.151).[190,191] The resin-bound cycloadducts could be isolated.

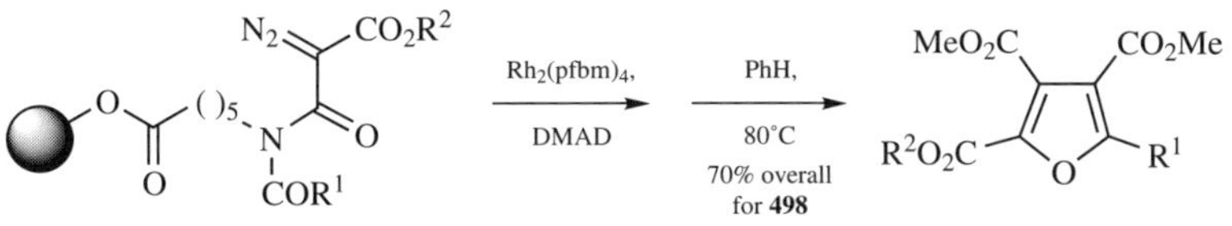

1,3-Oxazolium 4-Oxides (Isomunchnones) 559

Figure 4.151

In related work, this same group[199] found that isomunchnones **502** and **503** also react with electron-rich enol ethers to give the corresponding cycloadducts **504** in high yield (Fig. 4.152). In each case, a single diastereomer is isolated and the order of dipolarophilicity is $CH_2=CHOR > (E)\text{-}RCH=CHOR > (Z)\text{-}RCH=CHOR > CR_2=CHOR \gg CH_2=CR(OR)$.

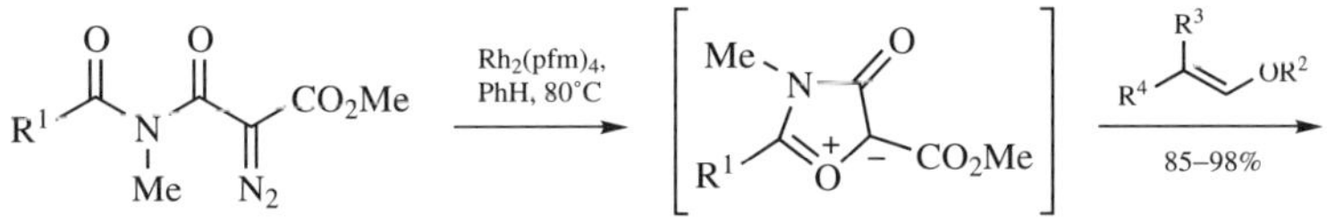

501 R^1 = Me
493 R^1 = Ph

502 R^1 = Me
503 R^1 = Ph

504
R^1 = Me, Ph
R^2 = Et, Bn, Ph, c-Hex, TMS, Me, t-Bu, i-Pen
R^3 = H, Me, n-Hex
R^4 = H, Me, n-Hex

Figure 4.152

A competition experiment involving isomunchnone **502** and ethyl acrylate and ethyl vinyl ether afforded both cycloadducts, **505** and **506**, although the former predominated (Fig. 4.153). The authors concluded that similar FMO energetics are operating for both electron-rich and electron-deficient dipolarophiles in their 1,3-dipolar cycloaddition reactions with isomunchnones.

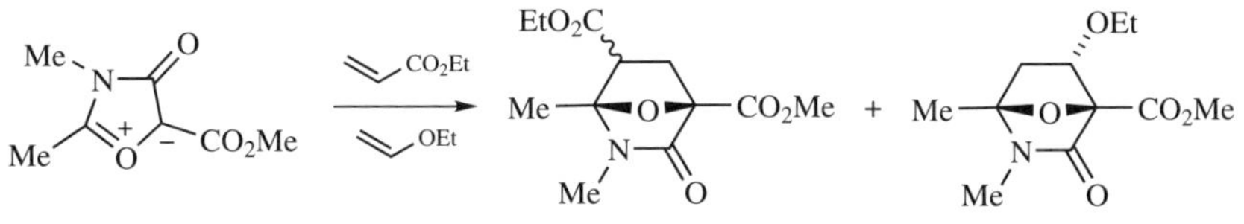

502

505 76%

506 19%

Figure 4.153

560   Mesoionic Oxazoles

As Moody and co-workers[193] discovered, isomunchnones can occasionally form even when they are not the desired product! Thus these workers inadvertently obtained oxazolidinedione **509** via isomunchnone **508** rather than the desired oxoindoline when diazo compound **507** was treated with rhodium(II) perfluorobutyramide, leading ultimately to the synthesis of the marine alkaloid convolutamydine C (Fig. 4.154). This unexpected transformation was readily circumvented by replacing the N-BOC protecting group with N-benzyl or N-p-methoxybenzyl.

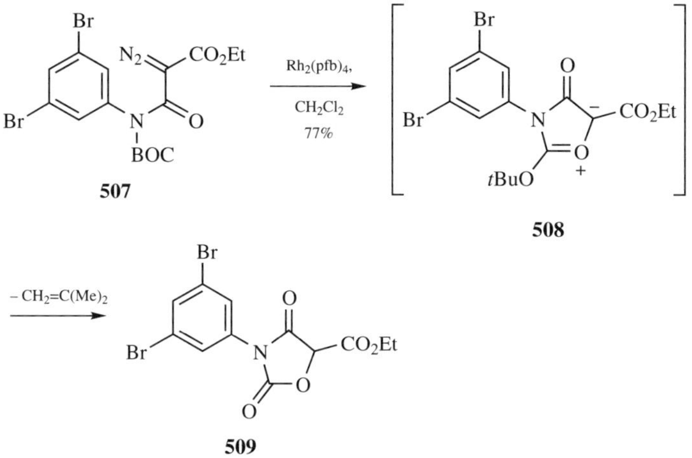

Figure 4.154

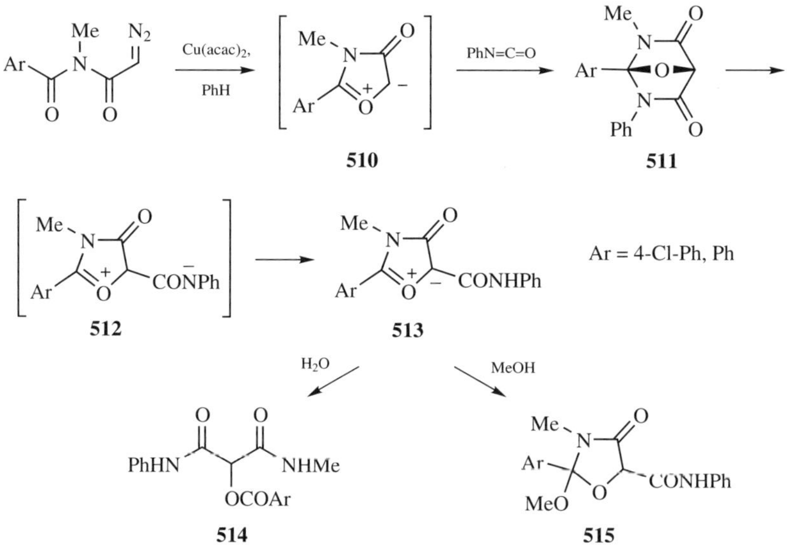

Scheme 4.22

Hamaguchi and Nagai[194] revised the structures for the 1:1 adduct of various mesoionic heterocycles and isocyanates, which were originally proposed by Potts and co-workers.[195–197] Thus the acylated isomunchnone structure **513** is now proposed for the reaction product of isomunchnone **510** and aryl isocyanates and aryl thioisocyanates (Scheme 4.22).

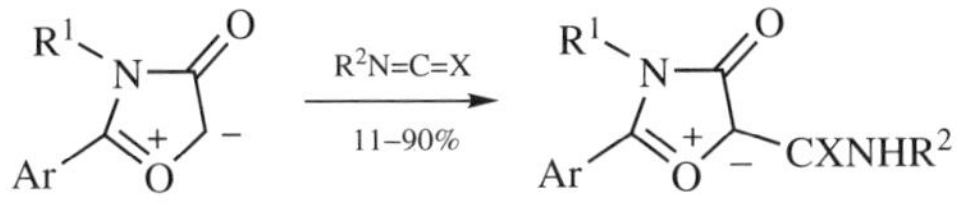

Figure 4.155

Similar acylated isomunchnones were obtained with benzoyl isocyanate, phenyl isothiocyanate, and benzoyl isothiocyanate, as summarized for **516 → 517** (Fig. 4.155).[194] The authors[194] suggested that the NMR and IR spectra are more consistent with acylated isomunchnones than with bicyclic adducts **511**. Thus the presence of an exchangeable proton at 9–11 ppm in the ^{1}H-NMR spectrum and the amide carbonyl absorption in the IR spectrum at 1670 cm^{-1} are not consistent with bicyclic lactam **511** but rather with isomunchnone amide **513**. Likewise, the NMe absorption at 3.72 ppm in the ^{1}H-NMR spectra of these adducts is at too low field to be expected for **511**. When the authors follow the reaction in an NMR tube, they observe peaks for the 1:1 adduct **511** after 10 s, at 5.02 ppm (bridgehead methine) and 2.97 ppm (*N*-Me), which disappear after 30 min to give the NMR spectrum of **513**.

Although Maier achieved the first intramolecular 1,3-dipolar cycloaddition reaction of an isomunchnone, it was Padwa who unleashed the synthetic utility of this reaction. Thus Padwa and co-workers[198,199] also found that isolated π-bonds can successfully and efficiently capture the in situ–generated isomunchnones, as shown by the examples **518 → 519** (Fig. 4.156). The alkene can also be tethered adjacent to the nitrogen atom (not shown). The indole double bond in **520** intercepts an isomunchnone 1,3-dipole to give the single diastereomer **521**, the structure of which is supported by X-ray crystallography.

The alkyne-tethered diazo imide **522** leads to furan **524**, resulting from a retro-Diels-Alder loss of benzyl isocyanate from adduct **523** (Fig. 4.157).[199]

The initial cycloadducts, such as **519**, can be ring opened with acid to *N*-acyliminium ions, which can then either deprotonate to an enamide or be reduced to a hydroxy enamide (Fig. 4.158). An example of the latter pathway is the reductive cleavage of **525** to bicyclic piperidone **526**.[199]

Padwa and co-workers[200,201] effected intramolecular 1,3-dipolar cycloaddition reactions of isomunchnones tethered with other examples of π-systems. Several substrates with tethers of varying lengths were examined in this study. For example, reaction of diazo imide **527**, readily assembled from the appropriate *o*-alkenyl

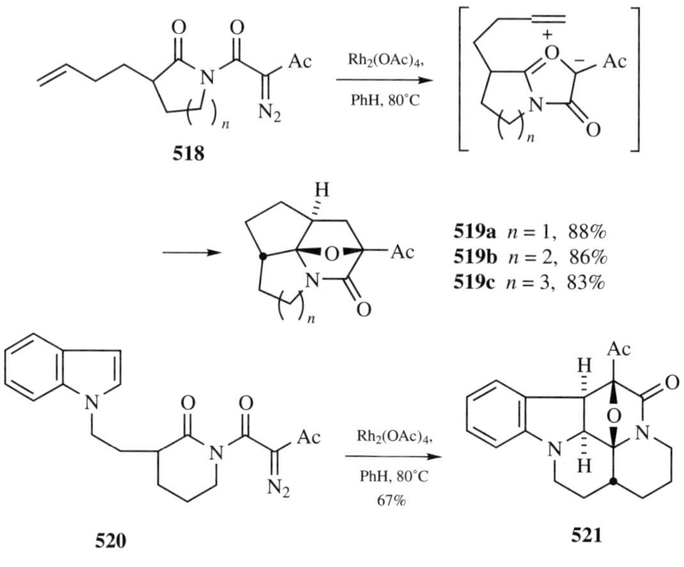

Mesoionic Oxazoles

519a $n = 1$, 88%
519b $n = 2$, 86%
519c $n = 3$, 83%

Figure 4.156

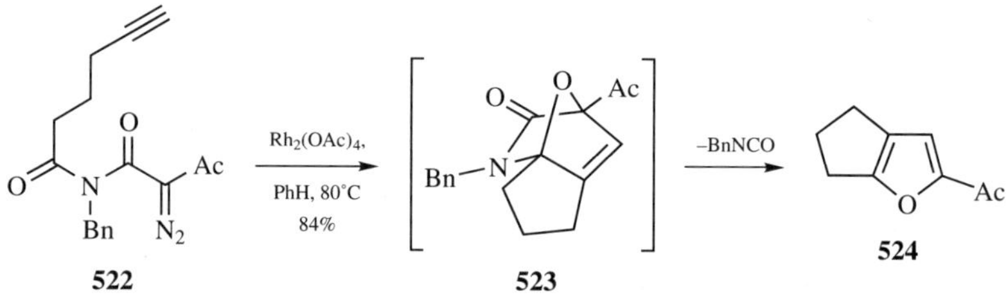

Figure 4.157

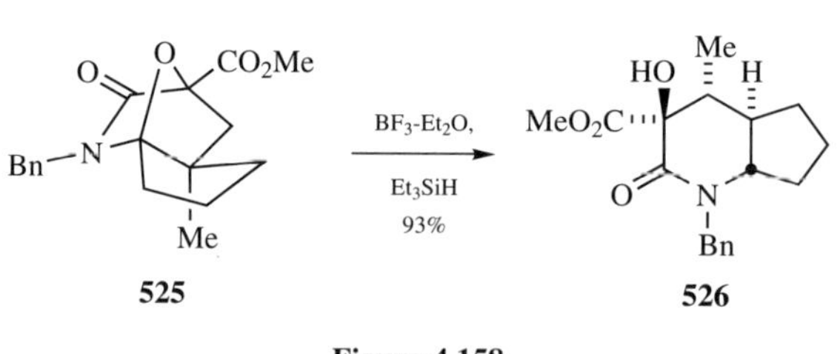

Figure 4.158

aniline, with catalytic rhodium perfluorobutyrate affords cycloadduct **529** in high yield via the intermediate isomunchnone **528** (Fig. 4.159).

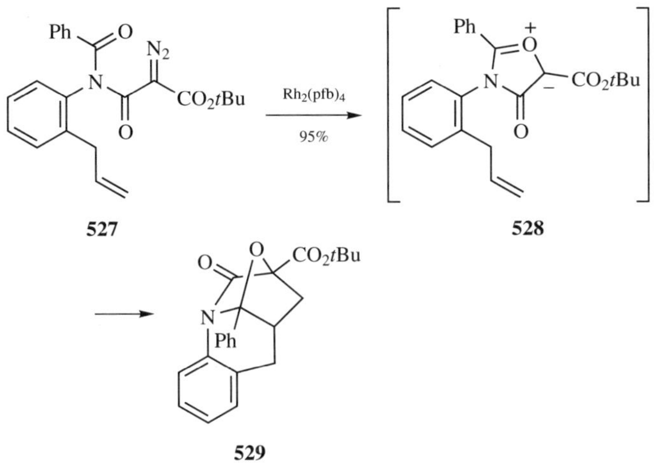

Figure 4.159

Further studies by this group[201] revealed that the preferred exo cycloaddition mode that is observed experimentally is supported by molecular mechanics calculations. In cases in which ring strain prevents an intramolecular cycloaddition, the isomunchnone can nevertheless be intercepted by *N*-phenylmaleimide in an intermolecular reaction.

Padwa and co-workers[202] extended this methodology to intramolecular 1,3-dipolar cycloaddition reactions of isomunchnones to tethered heterocycles. Thus the furanyl diazo imide **530** undergoes the usual conversion to isomunchnone **531**, which cyclizes to **532** in good yield (Fig. 4.160). The analogous compound with one less methylene group in the tether fails to yield a cycloadduct with the furan ring, although the isomunchnone can be trapped with DMAD (not shown). The thienyl diazo imide corresponding to **530** did not undergo a similar cycloaddition.

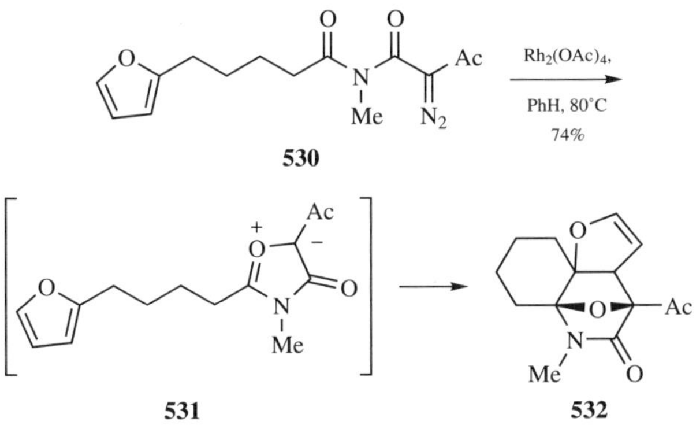

Figure 4.160

As indicated previously, in similar fashion, indole analog **520** undergoes a smooth cycloaddition, via an isomunchnone intermediate, to adduct **521** (Fig. 4.156). Unfortunately, attempts to apply this chemistry to a synthesis of the alkaloid vallesamidine **534** were not successful (Fig. 4.161). Thus diazo imides **533** failed to cyclize onto the indole double bond via the corresponding isomunchnone.

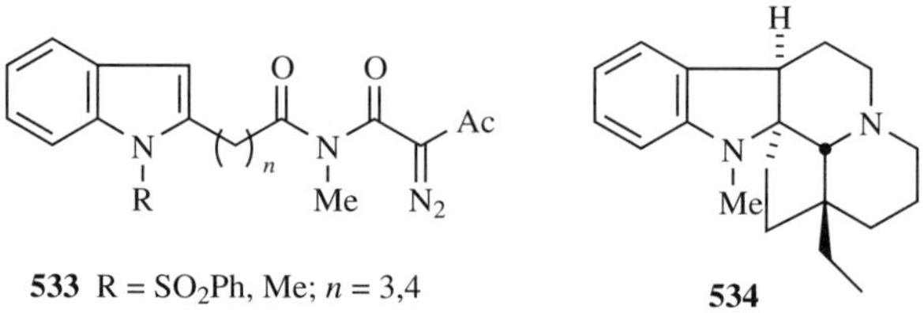

533 R = SO$_2$Ph, Me; n = 3,4

534

Figure 4.161

However, subsequent work by Padwa and co-workers[203] led to a successful formal synthesis of vallesamidine. Thus, following the results from a closely related model study, these workers subjected diazo imide **535** to the standard rhodium-catalyzed carbenoid generation and cyclization to give **537** via isomunchnone **536** (Fig. 4.162). Conversion of **537** to a previously synthesized precursor to vallesamidine **538** was uneventful.

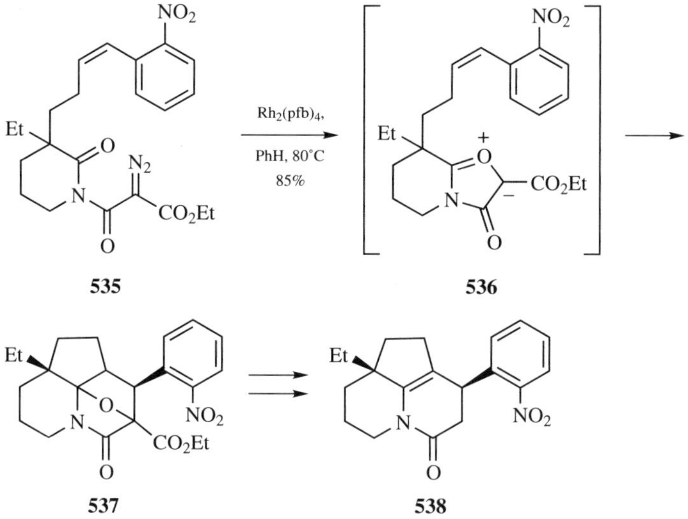

535

Rh$_2$(pfb)$_4$, PhH, 80°C
85%

536

537

538

Figure 4.162

Padwa and co-workers[204] adapted their tandem cyclization-cycloaddition methodology to culminate in a π-cyclization reaction enroute to B-ring homologs of erythrinane alkaloids. Thus, for example, treatment of diazo imide **539** with Rh$_2$(pfb)$_4$ affords the expected cycloadduct **540** in nearly quantitative yield

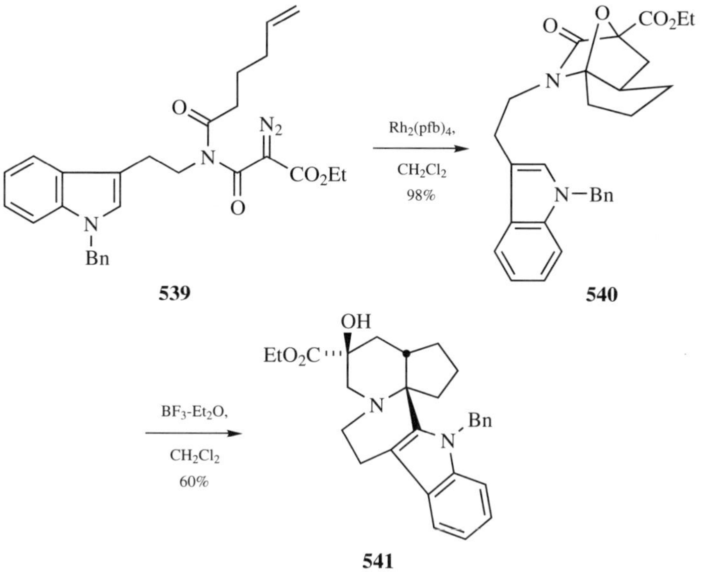

1,3-Oxazolium 4-Oxides (Isomunchnones) 565

Figure 4.163

(Fig. 4.163). Exposure of the latter to boron trifluoride etherate induces formation of the N-acyliminium ion, which is presumed to be in equilibrium with the corresponding enamide. Finally, the nucleophilic indole ring captures the N-acyliminium ion to give **541** in good yield. The π-nucleophile in this sequence can also be alkenes and electron-rich benzene rings.

This group[205] also engineered a clever approach to lysergic acid using an intramolecular isomunchnone cycloaddition strategy. Although ultimately unsuccessful, this study provided much interesting chemistry, including several examples of intramolecular isomunchnone 1,3-dipolar cycloadditions. After the completion of model studies, these workers adopted the sequence of reactions illustrated (Scheme 4.23). The desired cycloaddition reaction **543** → **544** proceeded in very high yield. However, the double bond in **545** could not be isomerized as required for a synthesis of lysergic acid.

A successful synthesis of the *Lycopodium* alkaloid (±)-lycopodine (**552**) was achieved by Padwa and colleagues.[206] Following the guidance provided by several model studies, these workers adopted the final route shown in Scheme 4.24. The isomunchnone precursor **547** was assembled from **546**, which was crafted from 5-methylcyclo-hex-2-en-1-one. Reaction of diazo imide **547** under the usual conditions afforded a 3:2 mixture of cycloadducts **548** and **549**. Acid treatment afforded the tetracyclic lactam **550**. Further manipulation gave **551**, which had been previously converted to (±)-lycopodine (**552**).

In the full account of this rhodium-catalyzed isomunchnone generation and intramolecular 1,3-dipolar cycloaddition reaction, followed by a terminal Mannich

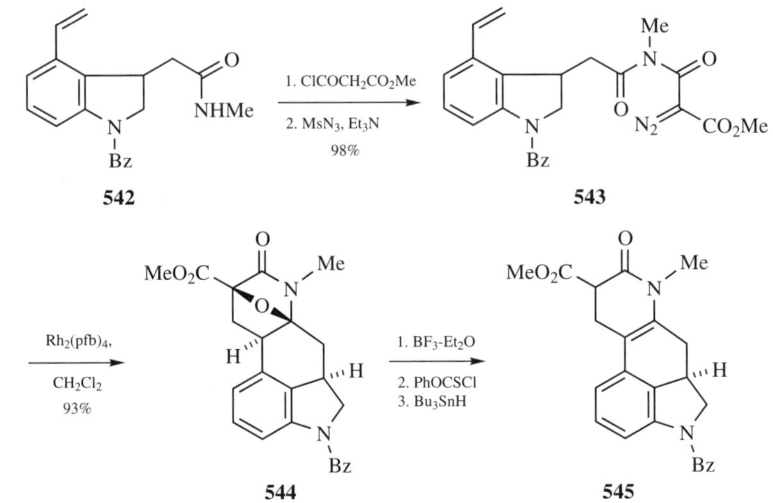

Scheme 4.23

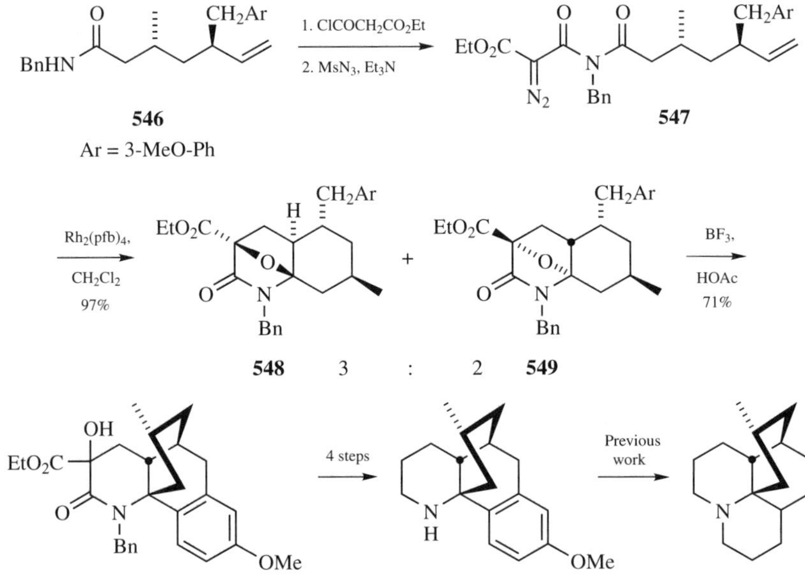

Scheme 4.24

cyclization, Padwa and co-workers[207] revealed the full synthetic power of this methodology. For example, exposure of α-diazo imide **553** to rhodium perfluorobutyrate (Rh$_2$(pfb)$_4$) affords cycloadduct **554** in nearly quantitative yield, via the corresponding isomunchnone (Fig. 4.164). Acid-induced ring opening of the ether

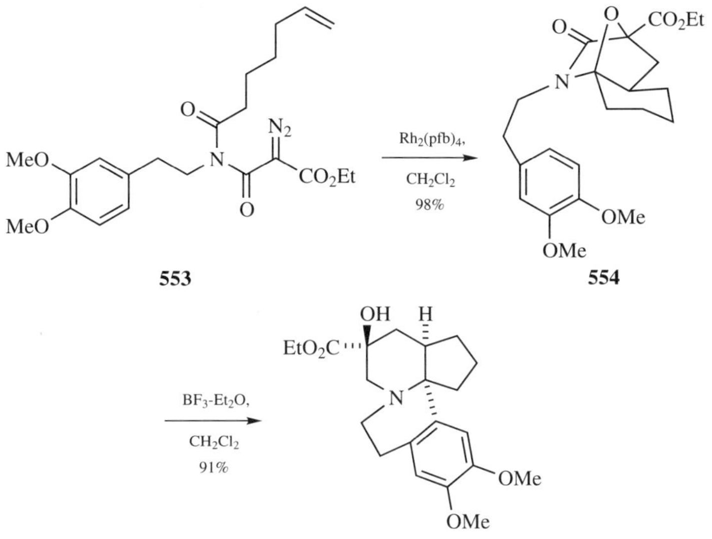

1,3-Oxazolium 4-Oxides (Isomunchnones) 567

Figure 4.164

bridge provides an *N*-acyliminium ion, which is captured by the pendant electron-rich benzene ring to give **555**. Both a tethered indole ring and olefinic units similarly serve as Mannich reaction nucleophiles.

Padwa[208] has concisely summarized his "domino cycloaddition/*N*-acyliminium ion cyclization cascade" process—a tactic akin to a two-move chess combination—involving the generation of an isomunchnone 1,3-dipole, intramolecular 1,3-dipolar cycloaddition reaction, *N*-acyliminium ion formation, and Mannich cyclization.

Kappe and co-workers[209] used Padwa's cyclization-cycloaddition cascade methodology to construct several rigid compounds (e.g., **557**) that mimic the putative receptor-bound conformation of dihydropyridine type calcium channel modulators (Fig. 4.165).

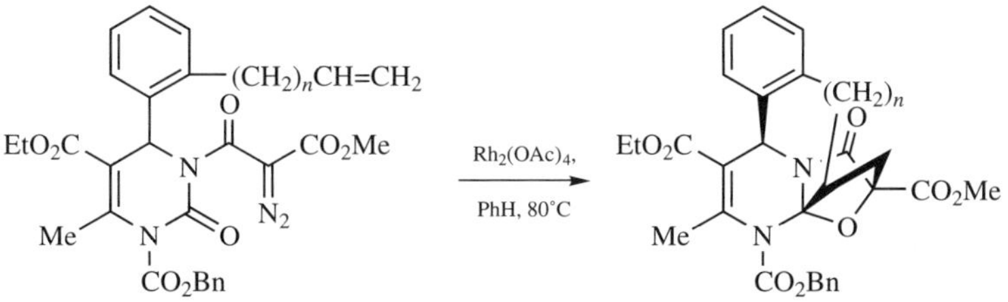

557a $n = 0$, 71%
557b $n = 1$, 56%

Figure 4.165

568   Mesoionic Oxazoles

As mentioned earlier, Padwa and co-workers employed the Pummerer reaction to generate and trap isomunchnones. This group[210,211] also adapted the intramolecular version of this tactic to the synthesis of several alkaloids of the pyridine, quinolizidine, and clavine classes. In each case, a 2-pyridone serves as the keystone intermediate. For example, Kuethe and Padwa[210] employed the Pummerer reaction of imidosulfoxides that contain tethered π-bonds in a formal synthesis of the frog alkaloid (±)-pumiliotoxin C (Scheme 4.25). Treatment of the easily assembled imidosulfoxide **558** with acetic anhydride and a trace of *p*-toluenesulfonic acid afforded a mixture of pyridones **561** and **562** via isomunchnone **559** and cycloadduct **560**. Both compounds were separately converted to **563**, which was transformed into a known precursor to pumiliotoxin C **564**.

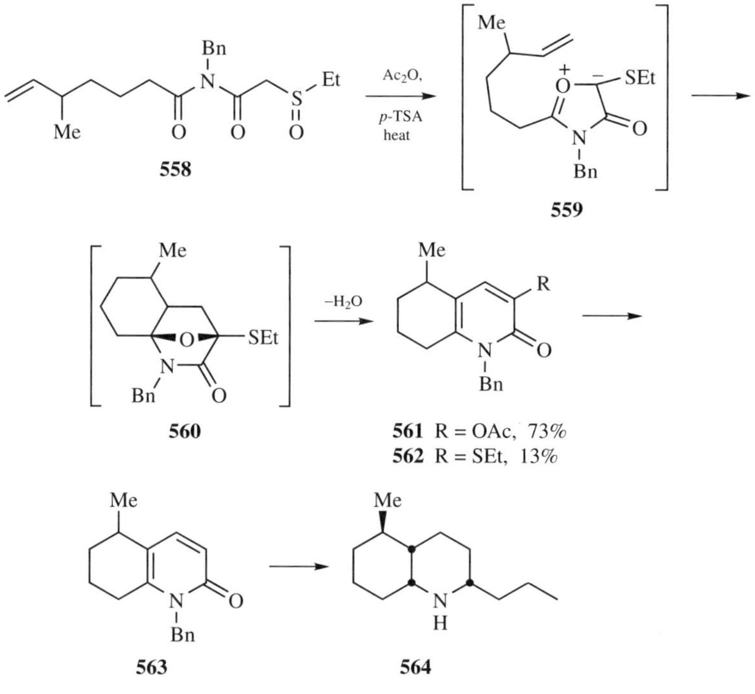

Scheme 4.25

Padwa and co-workers[211] also used this methodology to synthesize the azafluorenone alkaloid onychine **570** (Scheme 4.26). The sulfoxide **565** was prepared from 2-(2-butenyl)benzoic acid in four steps. Generation of the thionium ion **566** under standard Pummerer reaction conditions was followed by cyclization to isomunchnone **567** and then to cycloadduct **568**, which loses water to form α-pyridone **569**. Subsequent manipulation involving deoxygenation and debenzylation completed the synthesis.

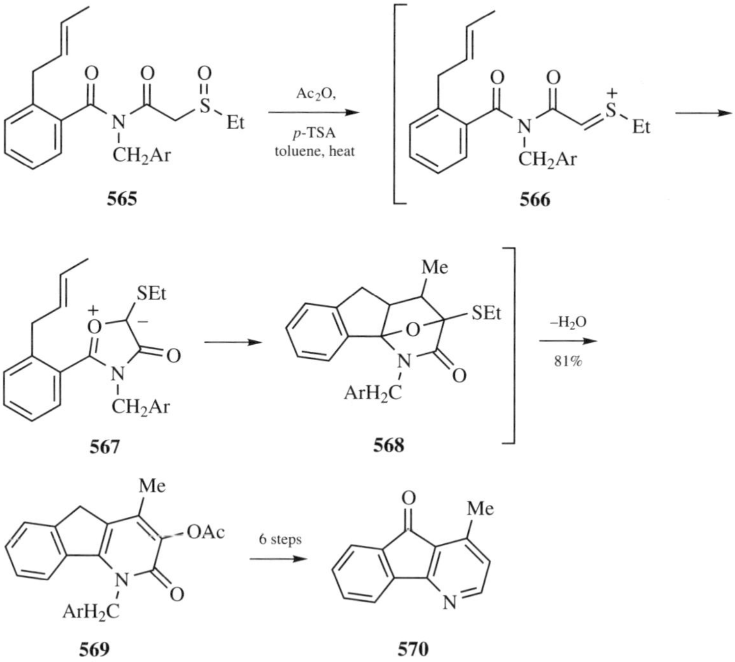

Scheme 4.26

In similar fashion, the azaanthraquinone alkaloid dielsiquinone **571** was synthesized for the first time. Also, the quinolizidine alkaloids (±)-lupinine **572** and (±)-anagyrine **573**, and the ergot alkaloid (±)-costaclavine **574** were synthesized using this Pummerer cyclization-cycloaddition cascade of imidosulfoxides and isomunchnones (Fig. 4.166).

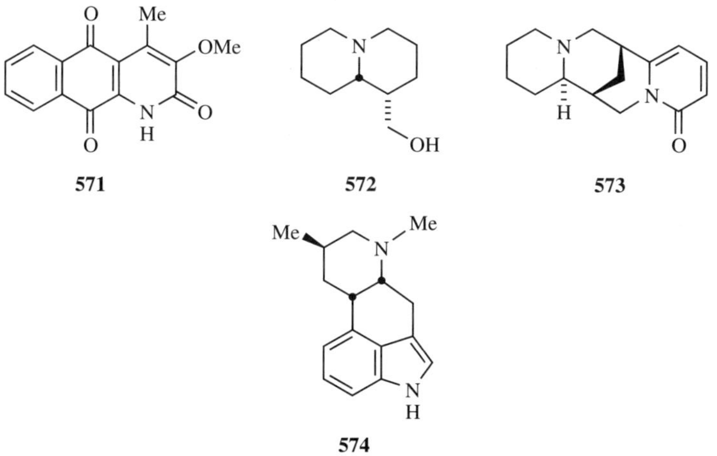

Figure 4.166

4.5. 1,3-OXAZOLIUM 4-IMINES (ISOMUNCHNONE IMINES)

As noted also by Gingrich and Baum,[1] "no example of the mesoionic ring system **575** has been reported" (Fig. 4.167).

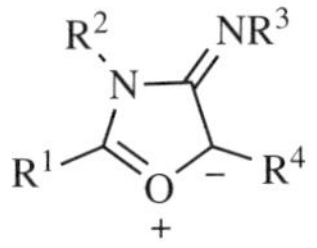

575

Figure 4.167

4.6. SUMMARY

The fascinating mesoionic 1,3-oxazolium oxides—munchnones and isomunchnones—have emerged from the relative obscurity of theoretical fascination and simple chemical transformations to become powerful and versatile synthetic tools for the organic chemist.

Munchnones undergo a range of electrophilic, nucleophilic, and 1,3-dipolar cycloaddition reactions. In particular, the latter reactions, following loss of carbon dioxide from the initial cycloadduct, lead to a diverse array of pyrroles, fused pyrroles, and derived ring systems. Sophisticated molecular orbital calculations are beginning to unravel the interesting regiochemistry that is observed in 1,3-dipolar cycloaddition reactions involving unsymmetrical munchnones.

Until recently, isomunchnones languished in popularity relative to that enjoyed by munchnones. Fortunately, this situation has been rectified by the discovery that diazo imides undergo an efficient cyclization reaction with copper(II) or, notably, rhodium(II) to generate isomunchnones. The resulting isomunchnones can be trapped intermolecularly in a 1,3-dipolar cycloaddition reaction with alkynes to afford furans, after loss of an isocyanate from the initial cycloadduct. Furthermore, as demonstrated elegantly through numerous examples by Padwa, these mesoionic heterocycles can be generated and captured intramolecularly by pendant dipolarophiles to yield complex ring systems and derived natural products.

The imagination and ingenuity of the organic chemist ensure a promising future in synthesis for these mesoionic oxazolium oxides.

Acknowledgments

The author wishes to thank Dartmouth College for a Senior Faculty Fellowship to initiate work on this manuscript during a sabbatical leave at the University of California at Santa Cruz, 1999–2000. The author also thanks Dr. Phil Crews and his students and colleagues for their hospitality during this sabbatical.

REFERENCES

1. Gingrich, H. L.; Baum, J. S. In Turchi, I. J., ed. *Oxazoles, The Chemistry of Heterocyclic Compounds*, Vol. 45, Wiley: New York, 1986, pp. 731–961.
2. Huisgen, R. In Padwa, A., ed. *1,3-Dipolar Cycloaddition Chemistry*, Vol. 1, Wiley-Interscience: New York, 1984, pp. 1–176.
3. Potts, K. T. In Padwa, A., Ed. *1,3-Dipolar Cycloaddition Chemistry*, Vol. 2, Wiley-Interscience: New York, 1984, pp. 1–82.
4. Ollis, W. D.; Stanforth, S. P.; Ramsden, C. A. *Tetrahedron* **1985**, *41*, 2239–2329.
5. Huisgen, R. In Curran, D. P., ed. *Advances in Cycloaddition*, Vol. 1 JAI Press: Greenwich, CT, 1988, pp. 1–32.
6. Kato, H.; Kobayashi, T. *J. Synth. Org. Chem. Jpn.* **1990**, *48*, 672–680 [*Chem. Abstr.* **1990**, *113*, 171912e].
7. Gupta, R. R.; Kumar, M.; Gupta, V. *Heterocyclic Chemistry, II*, Springer: Berlin, 1999, pp. 584–597.
8. Fortt, S. M. In Sainsbury, M., ed. *Rodd's Chemistry of Carbon Compounds*, Vol. IV C/D, Suppl. 2 to 2nd ed., Elsevier: Amsterdam, 1998, pp. 49–57.
9. Hartner, F. W. Jr. In Katritzky, A. R.; Rees, C. W.; Scriven, E. F. V., eds. *Comprehensive Heterocyclic Chemistry II*, Vol. 3, Elsevier: Oxford, UK 1996, pp. 261–318.
10. Potts, K. T.; Yao, S. *J. Org. Chem.* **1979**, *44*, 977–979.
11. Anderson, W. K.; Heider, A. R. *Synth. Commun.* **1986**, *16*, 357–364.
12. Slebioda, M. *Pol. J. Chem.* **1997**, *71*, 1045–1048.
13. Boyd, G. V.; Wright, P. H. *J. Chem. Soc. Perkin Trans. 1* **1972**, 909–913.
14. Boyd, G. V.; Wright, P. H. *J. Chem. Soc. Perkin Trans. 1* **1972**, 914–918.
15. Hershenson, F. M.; Pavia, M. R. *Synthesis* **1988**, 999–1001.
16. Wilde, R. G. *Tetrahedron Lett.* **1988**, *29*, 2027–2030.
17. Takagaki, H.; Yasuda, N.; Asaoka, M.; Takei, H. *Chem. Lett.* **1979**, 183–186.
18. Huisgen, R.; Funke, E.; Schaefer, F. C.; Knorr, R. *Angew. Chem. Int. Ed. Engl.* **1967**, *6*, 367–368.
19. Bayer, H. O.; Huisgen, R.; Knorr, R.; Schaefer, F. C. *Chem. Ber.* **1970**, *103*, 2581–2597.
20. Funke, E.; Huisgen, R. *Chem. Ber.* **1971**, *104*, 3222–3228.
21. Keating, T. A.; Armstrong, R. W. *J. Am. Chem. Soc.* **1995**, *117*, 7842–7843.
22. Keating, T. A.; Armstrong, R. W. *J. Am. Chem. Soc.* **1996**, *118*, 2574–2583.
23. Keating, T. A.; Armstrong, R. W. *J. Org. Chem.* **1996**, *61*, 8935–8939.
24. Strocker, A. M.; Keating, T. A.; Tempest, P. A.; Armstrong, R. W. *Tetrahedron Lett.* **1996**, *37*, 1149–1152.
25. Dömling, A.; Ugi, I. *Angew. Chem. Int. Ed. Engl.* **2000**, *39*, 3168–3210.
26. Ugi, I.; Lohberger, S.; Karl, R. In Heathcock, C. H., ed. *Comprehensive Organic Synthesis*, Vol. 2, Pergamon: Oxford, UK, 1991, pp. 1083–1109.
27. Mjalli, A. M. M.; Sarshar, S.; Baiga, T. J. *Tetrahedron Lett.* **1996**, *37*, 2943–2946.
28. Merlic, C. A.; Baur, A.; Aldrich, C. C. *J. Am. Chem. Soc.* **2000**, *122*, 7398–7399.
29. Toupet, L.; Texier, F.; Carrié, R. *Acta Crystallogr. Sect. C Cryst. Struct. Commun.* **1991**, *C47*, 328–330.
30. Boyd, G. V.; Davies, C. G.; Donaldson, J. D.; Silver, J.; Wright, P. H. *J. Chem. Soc. Perkin Trans. 2* **1975**, 1280–1282.
31. Tighineanu, E.; Chiraleu, F.; Raileanu, D. *Tetrahedron Lett.* **1978**, 1887–1890.
32. Tighineanu, E.; Chiraleu, F.; Raileanu, D. *Tetrahedron* **1980**, *36*, 1385–1397.
33. Petride, H.; Raileanu, D. *Rev. Roum. Chim.* **1988**, *33*, 729–739.

34. Petride, H.; Raileanu, D. *Rev. Roum. Chim.* **1989**, *34*, 1251–1261.
35. Prajapati, D.; Sandhu, J. S.; Baruah, J. N.; Kametani, T.; Nagase, H.; Kawai, K.; Honda, T. *Heterocycles* **1984**, *22*, 287–293.
36. Sain, B.; Baruah, J. N.; Sandhu, J. S. *J. Chem. Soc. Perkin Trans. 1* **1985**, 773–777.
37. Sain, B.; Sandhu, J. S. *J. Heterocycl. Chem.* **1986**, *23*, 1007–1010.
38. Dalla Croce, P.; Ferraccioli, R.; La Rosa, C. *J. Chem. Soc. Perkin Trans. 1* **1994**, 2499–2502.
39. Raileanu, D.; Petride, H. *Rev. Roum. Chim.* **1990**, *35*, 459–466.
40. Boyd, G. V.; Wright, P. H. *J. Chem. Soc. C* **1970**, 1485–1490.
41. Lawson, A.; Miles, D. H. *Chem. Ind.* [London] **1958**, 461–462.
42. Kawase, M.; Miyamae, H.; Narita, M.; Kurihara, T. *Tetrahedron Lett.* **1993**, *34*, 859–862.
43. Kawase, M.; Miyamae, H.; Kurihara, T. *Chem. Pharm. Bull.* **1998**, *46*, 749–756.
44. Kawase, M. *Heterocycles* **1993**, *36*, 2441–2444.
45. Kawase, M. *Tetrahedron Lett.* **1994**, *35*, 149–152.
46. Kawase, M.; Saito, S.; Kikuchi, H.; Miyamae, H. *Heterocycles* **1997**, *45*, 2185–2195.
47. Kawase, M. *J. Chem. Soc. Chem. Commun.* **1992**, 1076–1077.
48. Kawase, M.; Hirabayashi, M.; Koiwai, H.; Yamamoto, K.; Miyamae, H. *Chem. Commun.* **1998**, 641–642.
49. Kawase, M.; Hirabayashi, M.; Kumakura, H.; Saito, S.; Yamamoto, K. *Chem. Pharm. Bull.* **2000**, *48*, 114–119.
50. Buchanan, G. L. *Chem. Soc. Rev.* **1988**, *17*, 91–109.
51. Petride, H. *Rev. Roum. Chim.* **1990**, *35*, 747–755.
52. Petride, H. *Rev. Roum. Chim.* **1991**, *36*, 1113–1122.
53. Petride, H. *Rev. Roum. Chim.* **1991**, *36*, 1299–1306.
54. Kawase, M.; Koiwai, H.; Saito, S.; Kurihara, T. *Tetrahedron Lett.* **1998**, *39*, 6189–6190.
55. Singh, G.; Singh, S. *Tetrahedron Lett.* **1964**, 3789–3793.
56. Kawase, M. *J. Chem. Soc. Chem. Commun.* **1994**, 2101–2102.
57. Kawase, M.; Saito, S. *Chem. Pharm. Bull.* **2000**, *48*, 410–414.
58. Kawase, M.; Saito, S.; Kurihara, T. *Heterocycles* **1995**, *41*, 1617–1620.
59. Kawase, M.; Koiwai, H.; Yamano, A.; Miyamae, H. *Tetrahedron Lett.* **1998**, *39*, 663–666.
60. Kawase, M.; Miyamae, H.; Saito, S. *Heterocycles* **1999**, *50*, 71–74.
61. Sain, B.; Sandhu, J. S. *Indian J. Chem. Sect B* **1992**, *31B*, 153–155.
62. Märkl, G.; Dorfmeister, G. *Tetrahedron Lett.* **1987**, *28*, 1089–1092.
63. Anderson, W. K.; Chang, C.-P.; McPherson, H. L. Jr. *J. Med. Chem.* **1983**, *26*, 1333–1338.
64. Anderson, W. K.; Mach, R. H. *J. Med. Chem.* **1987**, *30*, 2109–2115.
65. Padwa, A.; Fryxell, G. E.; Gasdaska, J. R.; Venkatramanan, M. K.; Wong, G. S. K. *J. Org. Chem.* **1989**, *54*, 644–653.
66. Kane, J. M. *J. Org. Chem.* **1980**, *45*, 5396–5397.
67. Györgydeák, Z.; Szilágyi, L.; Kajtár, J.; Argay, G.; Kálmán, A. *Monatsh. Chem.* **1994**, *125*, 189–208.
68. Pinho e Melo, T. M. V. D.; Soares, M. I. L.; Barbosa, D. M.; Gonsalves, A. M. D. R.; Beja, A. M.; Paixão, J. A.; Silva, M. R.; Da Veiga, L. A. *Tetrahedron* **2000**, *56*, 3419–3424.
69. Ladurée, D.; Lancelot, J.-C.; Robba, M. *Tetrahedron Lett.* **1985**, *26*, 1295–1296.
70. Granier, T.; Gaiser, F.; Hintermann, L.; Vasella, A. *Helv. Chim. Acta* **1997**, *80*, 1443–1456.
71. Silva, M. R.; Beja, A. M.; Paixão, J. A.; Alte Da Veiga, L.; Barbosa, D. M.; Soares, M. I. L.; Pinho e Melo, T. M. V. D.; Gonsolves, A. M. d. R. *Acta Crystallogr. Sect. C Cryst. Struct. Commun.* **1999**, C55, 1094–1096.

72. Lo, C. W.; Chan, W. L.; Szeto, Y. S.; Yip, C. W. *Chem. Lett.* **1999**, 513–514.
73. Funabiki, K.; Ishihara, T.; Yamanaka, H. *J. Fluorine Chem.* **1995**, *71*, 5–7.
74. Padwa, A.; Chen, Y.-Y.; Dent, W.; Nimmesgern, H. *J. Org. Chem.* **1985**, *50*, 4006–4014.
75. Toupet, L.; Mazari, M.; Texier, F.; Carrié, R. *Acta Crystallogr. Sect. C Cryst. Struct. Commun.* **1991**, *C47*, 1528–1531.
76. Texier, F.; Mazari, M.; Yebdri, O.; Tonnard, F.; Carrié, R. *Bull. Soc. Chim. Fr.* **1991**, 962–967.
77. Yebdri, O.; Texier, F. *J. Heterocycl. Chem.* **1986**, *23*, 809–812.
78. Dalla Croce, P.; La Rosa, C. *Heterocycles* **1988**, *27*, 2825–2832.
79. Gribble, G. W.; Simon, W. M.; Pelkey, E. T.; Trujillo, H. A. Personal communication, 2000.
80. Coppola, G. P.; Noe, M. C.; Schwartz, D. J.; Abdon, R. L. II; Trost, B. M. *Tetrahedron* **1994**, *50*, 93–116.
81. Coppola, B. P.; Noe, M. C.; Hong, S. S.-K. *Tetrahedron Lett.* **1997**, *38*, 7159–7162.
82. Pizzorno, M. T.; Albonico, S. M. *J. Org. Chem.* **1974**, *39*, 731.
83. Santiago, B.; Dalton, C. R.; Huber, E. W.; Kane, J. M. *J. Org. Chem.* **1995**, *60*, 4947–4950.
84. Ladurée, D.; Lancelot, J.-C.; Robba, M.; Chenu, E.; Mathé, G. *J. Med. Chem.* **1989**, *32*, 456–461.
85. Dalla Croce, P.; Gariboldi, P.; La Rosa, C. *J. Heterocycl. Chem.* **1987**, *24*, 1793–1797.
86. Kato, H.; Wang, S.-Z.; Nakano, H. *J. Chem. Soc. Perkin Trans. 1* **1989**, 361–363.
87. Sainsbury, M.; Strange, R. H.; Woodward, P. R.; Barsanti, P. A. *Tetrahedron* **1993**, *49*, 2065–2076.
88. Pinho e Melo, R. M. V. D.; Barbosa, D. M.; Ramos, P. J. R. S.; Gonsalves, A. M. D. R.; Gilchrist, T. L.; Beja, A. M.; Paixão, J. A.; Silva, M. R.; Alte Da Veiga, L. *J. Chem. Soc. Perkin Trans. 1* **1999**, 1219–1223.
89. Hutchison, D. R.; Nayyar, N. K.; Martinelli, M. J. *Tetrahedron Lett.* **1996**, *37*, 2887–2890.
90. Nayyar, N. K.; Hutchison, D. R.; Martinelli, M. J. *J. Org. Chem.* **1997**, *62*, 982–991.
91. Jursic, B. S. *Theochem* **1996**, *365*, 55–61.
92. Kato, H.; Nakazawa, S.; Kiyosawa, T.; Hirakawa, K. *J. Chem. Soc. Perkin Trans. 1* **1976**, 672–675.
93. Nan'ya, S.; Goto, S.; Butsugan, Y. *J. Heterocycl. Chem.* **1990**, *27*, 1519–1520.
94. Nan'ya, S.; Kurachi, Y.; Butsugan, Y. *J. Heterocycl. Chem.* **1991**, *28*, 1853–1855.
95. Gotthardt, H.; Huisgen, R.; Schaefer, F. C. *Tetrahedron Lett.* **1964**, 487–491.
96. Gotthardt H.; Huisgen, R. *Chem. Ber.* **1970**, *103*, 2625–2638.
97. Nan'ya, S.; Tange, T.; Maekawa, E.; Ueno, Y. *J. Heterocycl. Chem.* **1986**, *23*, 1267–1271.
98. Fabre, J. L.; Farge, D.; James, C.; Lavé, D. *Tetrahedron Lett.* **1985**, *26*, 5447–5450.
99. Texier, F.; Mazari, M.; Yebdri, O.; Tonnard, F.; Carrié, R. *Tetrahedron* **1990**, *46*, 3515–3526.
100. Okano, T.; Uekawa, T.; Morishima, N.; Eguchi, S. *J. Org. Chem.* **1991**, *56*, 5259–5262.
101. Clerici, F.; Gelmi, M. L.; Trimarco, P. *Tetrahedron* **1998**, *54*, 5763–5774.
102. Erba, E.; Gelmi, M. L.; Pocar, D.; Trimarco, P. *Chem. Ber.* **1986**, *119*, 1083–1089.
103. Dalla Croce, P.; La Rosa, C.; Gelmi, M. L.; Ballabio, M. *J. Chem. Soc. Perkin Trans. 2* **1988**, 423–425.
104. Clerici, F.; Erba, E.; Mornatti, P.; Trimarco, P. *Chem. Ber.* **1989**, *122*, 295–300.
105. Baggi, P.; Clerici, F.; Gelmi, M. L.; Mottadelli, S. *Tetrahedron* **1995**, *51*, 2455–2466.
106. Clerici, F.; Gelmi, M. L.; Soave, R.; Valle, M. *Tetrahedron* **1998**, *54*, 11285–11296.
107. Avalos, M.; Babiano, R.; Bautista, I.; Fernandez, J. I.; Jimenez, J. L.; Palacios, J. C.; Plumet, J.; Rebolledo, F. *Carbohydr. Res.* **1989**, *186*, C7–C8.
108. Avalos, M.; Babiano, R.; Cabanillas, A.; Cintas, P.; Jiménez, J. L.; Palacios, J. C.; Aguilar, M. A.; Corchado, J. C.; Espinosa-García, J. *J. Org. Chem.* **1996**, *61*, 7291–7297.
109. Nesi, R.; Giomi, D.; Turchi, S.; Tedeschi, P.; Ponticelli, F. *Gazz. Chim. Ital.* **1993**, *123*, 633–635.
110. Gribble, G. W.; Pelkey, E. T.; Switzer, F. L. *Synlett* **1998**, 1061–1062.

111. Gribble, G. W.; Pelkey, E. T.; Simon, W. M.; Trujillo, H. A. *Tetrahedron* **2000**, *56*, 10133–10140.
112. Kobayashi, T.; Minemura, H.; Kato, H. *Heterocycles* **1995**, *40*, 311–317.
113. Kato, H.; Toda, S.; Arikawa, Y.; Masuzawa, M.; Hashimoto, M.; Ikoma, K.; Wang, S. Z.; Miyasaka, A. *J. Chem. Soc. Perkin Trans. 1* **1990**, 2035–2040.
114. Kato, H.; Kobayashi, T.; Horie, K.; Oguri, K.; Moriwaki, M. *J. Chem. Soc. Perkin Trans. 1* **1993**, 1055–1059.
115. Martin, H.-D.; Mais, F.-J.; Mayer, B.; Hecht, H.-J.; Hekman, M.; Steigl, A. *Monatsh. Chem.* **1983**, *114*, 1145–1147.
116. Maryanoff, C. A.; Karash, C. B.; Turchi, I. J.; Corey, E. R.; Maryanoff, B. E. *J. Org. Chem.* **1989**, *54*, 3790–3792.
117. Maryanoff, C. A.; Turchi, I. J. *Heterocycles* **1993**, *35*, 649–657.
118. Turchi, I. J. personal communication, 2000.
119. Kato, H.; Aoki, N.; Kawamura, Y.; Yoshino, K. *J. Chem. Soc. Perkin Trans. 1* **1985**, 1245–1247.
120. Friedrichsen, W.; Schröer, W.-D.; Debaerdemaeker, T. *Liebigs Ann. Chem.* **1981**, 491–501.
121. Kato, H.; Kobayashi, T.; Tokue, K.; Shirasawa, S. *J. Chem. Soc. Perkin Trans. 1* **1993**, 1617–1620.
122. Wu, S.-H.; Sun, W.-Q.; Zhang, D.-W.; Shu, L.-H.; Wu, H.-M.; Xu, J.-F.; Lao, X.-F. *J. Chem. Soc. Perkin Trans. 1* **1998**, 1733–1738.
123. Gribble, G. W.; Sponholtz, W. R. III; Switzer, F. L.; D'Amato, F. J.; Byrn, M. P. *Chem. Commun.* **1997**, 993–994.
124. Gribble, G. W.; Switzer, F. L.; Bushweller, J. H.; Jewett, J. G.; Brown, J. H.; Dion, J. L.; Bushweller, C. H.; Byrn, M. P.; Strouse, C. E. *J. Org. Chem.* **1996**, *61*, 4319–4327.
125. Padwa, A.; Lim, R.; MacDonald, J. G.; Gingrich, H. L.; Kellar, S. M. *J. Org. Chem.* **1985**, *50*, 3816–3823.
126. Huisgen, R.; Funke, E. *Angew. Chem., Int. Ed. Engl.* **1967**, *6*, 365–366.
127. Huisgen, R.; Funke, E.; Gotthardt, H.; Panke, H.-L. *Chem. Ber.* **1971**, *104*, 1532–1549.
128. Kawase, M. *J. Chem. Soc. Chem. Commun.* **1990**, 1328–1329.
129. Kawase, M. *Chem. Pharm. Bull.* **1997**, *45*, 1248–1253.
130. Bronberger, F.; Huisgen, R. *Tetrahedron Lett.* **1984**, *25*, 65–68.
131. Consonni, R.; Dalla Croce, P.; Ferraccioli, R.; La Rosa, C. *J. Chem. Res. Synop.* **1991**, 188–189.
132. Bonati, L.; Ferraccioli, R.; Moro, G. *J. Phys. Org. Chem.* **1995**, *8*, 452–462.
133. Bilodeau, M. T.; Cunningham, A. M. *J. Org. Chem.* **1998**, *63*, 2800–2801.
134. Dalla Croce, P.; Ferraccioli, R.; La Rosa, C.; Pilati, T. *J. Chem. Soc. Perkin Trans. 2* **1993**, 1511–1515.
135. Dalla Croce, P.; Ferraccioli, R.; La Rosa, C. *Tetrahedron* **1995**, *51*, 9385–9392.
136. Hamper, B. C.; Jerome, K. D.; Yalamanchili, G.; Walker, D. M.; Chott, R. C.; Mischke, D. A. *Biotechnol. Bioeng.* **2000**, *71*, 28–37.
137. Manley, P. W.; Rees, C. W.; Storr, R. C. *J. Chem. Soc. Chem. Commun.* **1983**, 1007–1008.
138. Rodríguez, H.; Pavez, H.; Márquez, A.; Navarrete, P. *Tetrahedron* **1983**, *39*, 23–27.
139. Märkl, G.; Dorfmeister, D. *Tetrahedron Lett.* **1986**, *27*, 4419–4422.
140. Rösch, W.; Richter, H.; Regitz, M. *Chem. Ber.* **1987**, *120*, 1809–1813.
141. Fuchs, E. P. O.; Hermesdorf, M.; Schnurr, W.; Rösch, W.; Heydt, H.; Regitz, M.; Binger, P. *J. Organomet. Chem.* **1988**, *338*, 329–340.
142. Mack, A.; Pierron, E.; Allspach, T.; Bergsträsser, U.; Regitz, M. *Synthesis* **1998**, 1305–1313.
143. Bach, P.; Bergsträsser, U.; Leininger, S.; Regitz, M. *Bull. Soc. Chim. Fr.* **1997**, *134*, 927–936.
144. Schmitt, G.; An, N. D.; Vebrel, J.; Laude, B. *Bull. Soc. Chim. Belg.* **1986**, *95*, 215–216.
145. Schmitt, G.; Laude, B.; Vebrel, J.; Rodier, N.; Theobald, F. *Bull. Soc. Chim. Belg.* **1989**, *98*, 113–123.

146. Berrabah, M.; Schmitt, G.; An, N. D.; Laude, B. *Bull. Soc. Chim. Belg.* **1991**, *100*, 613–616.
147. Monnier, K.; Schmitt, G.; Laude, B.; Theobald, F.; Rodier, N. *Tetrahedron Lett.* **1992**, *33*, 1609–1610.
148. Perrin, S.; Monnier, K.; Laude, B.; Kubicki, M. M.; Blacque, O. *Tetrahedron Lett.* **1998**, *39*, 1753–1754.
149. Perrin, S.; Monnier, K.; Laude, B.; Kubicki, M.; Blacque, O. *Eur. J. Org. Chem.* **1999**, 297–303.
150. McEwen, W. E.; Grossi, A. V.; MacDonald, R. J.; Stamegna, A. P. *J. Org. Chem.* **1980**, *45*, 1301–1308.
151. Padwa, A. *Acc. Chem. Res.* **1991**, *24*, 22–28.
152. Padwa, A.; Hornbuckle, S. F. *Chem. Rev.* **1991**, *91*, 263–309.
153. Padwa, A.; Krumpe, K. E. *Tetrahedron* **1992**, *48*, 5385–5453.
154. Osterhout, M. H.; Nadler, W. R.; Padwa, A. *Synthesis* **1994**, 123–141.
155. Padwa, A.; Weingarten, M. D. *Chem. Rev.* **1996**, *96*, 223–269.
156. Padwa, A. *Top. Curr. Chem.* **1997**, *189*, 121–158.
157. Haddadin, M. J.; Kattan, A. M.; Freeman, J. P. *J. Org. Chem.* **1982**, *47*, 723–725.
158. Haddadin, M. J.; Tannus, H. T. *Heterocycles* **1984**, *22*, 773–778.
159. Mathias, L. J.; Moore, D. R. *J. Am. Chem. Soc.* **1985**, *107*, 5817–5818.
160. Mathias, L. J.; Moore, D. R. *Polym. Prep.* **1986**, *27*, 118–119.
161. Moore, D. R.; Mathias, L. J. *Macromolecules* **1986**, *19*, 1530–1536.
162. Moore, D. R.; Mathias, L. J. *J. Org. Chem.* **1987**, *52*, 1599–1601.
163. Doyle, M. P.; Dorow, R. L.; Terpstra, J. W.; Rodenhouse, R. A. *J. Org. Chem.* **1985**, *50*, 1663–1666.
164. Hamaguchi, M.; Ibata, T. *Tetrahedron Lett.* **1974**, 4475–4476.
165. Maier, M. E.; Evertz, K. *Tetrahedron Lett.* **1988**, *29*, 1677–1680.
166. Maier, M. E.; Schöffling, B. *Chem. Ber.* **1989**, *122*, 1081–1087.
167. Padwa, A.; Hertzog, D. L.; Chinn, R. L. *Tetrahedron Lett.* **1989**, *30*, 4077–4080.
168. Doyle, M. P.; Pieters, R. J.; Taunton, J.; Pho, H. Q.; Padwa, A.; Hertzog, D. L.; Precedo, L. *J. Org. Chem.* **1991**, *56*, 820–829.
169. Kuethe, J. T.; Padwa, A. *J. Org. Chem.* **1997**, *62*, 774–775.
170. Padwa, A.; Heidelbaugh, T. M.; Kuethe, J. T. *J. Org. Chem.* **1999**, *64*, 2038–2049.
171. Ruf, S. G.; Bergsträsser, U.; Regitz, M. *Tetrahedron* **2000**, *56*, 63–70.
172. Kato, H.; Kobayashi, T.; Ciobanu, M.; Iga, H.; Akutsu, A.; Kakehi, A. *Chem. Commun.* **1996**, 1011–1012.
173. Kato, H.; Kobayashi, T.; Ciobanu, M.; Kakehi, A. *Tetrahedron* **1997**, *53*, 9921–9934.
174. Padwa, A.; Hertzog, D. L. *Tetrahedron* **1993**, *49*, 2589–2600.
175. Prein, M.; Padwa, A. *Tetrahedron Lett.* **1996**, *37*, 6981–6984.
176. Prein, M.; Manley, P. J.; Padwa, A. *Tetrahedron* **1997**, *53*, 7777–7794.
177. Padwa, A.; Prein, M. *J. Org. Chem.* **1997**, *62*, 6842–6854.
178. González, R.; Knight, B. W.; Wudl, F.; Semones, M. A.; Padwa, A. *J. Org. Chem.* **1994**, *59*, 7949–7951.
179. Sheehan, S. M.; Padwa, A. *J. Org. Chem.* **1997**, *62*, 438–439.
180. Straub, C. S.; Padwa, A. *Org. Lett.* **1999**, *1*, 83–85.
181. Padwa, A.; Sheehan, S. M.; Straub, C. S. *J. Org. Chem.* **1999**, *64*, 8648–8659.
182. Padwa, A.; Prein, M. *Tetrahedron* **1998**, *54*, 6957–6976.
183. Angell, R.; Drew, M. G. B.; Fengler-Veith, M.; Finch, H.; Harwood, L. M.; Jahans, A. W.; Tucker, T. T. *Tetrahedron Lett.* **1997**, *38*, 3107–3110.

184. Angell, R.; Fengler-Veith, M.; Finch, H.; Harwood, L. M.; Tucker, T. T. *Tetrahedron Lett.* **1997**, *38*, 4517–4520.
185. Drew, M. G. B.; Fengler-Veith, M.; Harwood, L. M.; Jahans, A. W. *Tetrahedron Lett.* **1997**, *38*, 4521–4524.
186. Kappe, C. O.; Peters, K.; Peters, E.-M. *J. Org. Chem.* **1997**, *62*, 3109–3118.
187. Kappe, C. O. *Tetrahedron Lett.* **1997**, *38*, 3323–3326.
188. Padwa, A.; Snyder, J. P.; Curtis, E. A.; Sheehan, S. M.; Worsencroft, K. J.; Kappe, C. O. *J. Am. Chem. Soc.* **2000**, *122*, 8155–8167.
189. Gowravaram, M. R.; Gallop, M. A. *Tetrahedron Lett.* **1997**, *38*, 6973–6976.
190. Whitehouse, D. L.; Nelson, K. H. Jr.; Savinov, S. N.; Austin, D. J. *Tetrahedron Lett.* **1997**, *38*, 7139–7142.
191. Whitehouse, D. L.; Nelson, K. H. Jr.; Savinov, S. N.; Löwe, R. S.; Austin, D. J. *Bioorg. Med. Chem.* **1998**, *6*, 1273–1282.
192. Savinov, S. N.; Austin, D. J. *Chem. Commun.* **1999**, 1813–1814.
193. Miah, S.; Moody, C. J.; Richards, I. C.; Slawin, A. M. Z. *J. Chem. Soc. Perkin Trans. 1* **1997**, 2405–2412.
194. Hamaguchi, M.; Nagai, T. *J. Chem. Soc. Chem. Commun.* **1985**, 726–728.
195. Potts, K. T.; Husain, S. *J. Org. Chem.* **1972**, *37*, 2049–2050.
196. Potts, K. T.; Baum, J.; Houghton, E.; Roy, D. N.; Singh, U. P. *J. Org. Chem.* **1974**, *39*, 3619–3627.
197. Potts, K. T.; Baum, J.; Datta, S. K.; Houghton, E. *J. Org. Chem.* **1976**, *41*, 813–818.
198. Hertzog, D. L.; Austin, D. J.; Nadler, W. R.; Padwa, A. *Tetrahedron Lett.* **1992**, *33*, 4731–4734.
199. Padwa, A.; Hertzog, D. L.; Nadler, W. R.; Osterhout, M. H.; Price, A. T. *J. Org. Chem.* **1994**, *59*, 1418–1427.
200. Padwa, A.; Austin, D. J.; Price, A. T. *Tetrahedron Lett.* **1994**, *35*, 7159–7162.
201. Padwa, A.; Austin, D. J.; Price, A. T.; Weingarten, M. D. *Tetrahedron* **1996**, *52*, 3247–3260.
202. Padwa, A.; Hertzog, D. L.; Nadler, W. R. *J. Org. Chem.* **1994**, *59*, 7072–7084.
203. Padwa, A.; Harring, S. R.; Semones, M. A. *J. Org. Chem.* **1998**, *63*, 44–54.
204. Marino, J. P. Jr.; Osterhout, M. H.; Price, A. T.; Semones, M. A.; Padwa, A. *J. Org. Chem.* **1994**, *59*, 5518–5520.
205. Marino, J. P. Jr.; Osterhout, M. H.; Padwa, A. *J. Org. Chem.* **1995**, *60*, 2704–2713.
206. Padwa, A.; Brodney, M. A.; Marino, J. P. Jr.; Sheehan, S. M. *J. Org. Chem.* **1997**, *62*, 78–87.
207. Padwa, A.; Brodney, M. A.; Marino, J. P. Jr.; Osterhout, M. H.; Price, A. T. *J. Org. Chem.* **1997**, *62*, 67–77.
208. Padwa, A. *Chem. Commun.* **1998**, 1417–1424.
209. Jauk, B.; Belaj, F.; Kappe, C. O. *J. Chem. Soc. Perkin Trans. 1* **1999**, 307–314.
210. Kuethe, J. T.; Padwa, A. *Tetrahedron Lett.* **1997**, *38*, 1505–1508.
211. Padwa, A.; Heidelbaugh, T. M.; Kuethe, J. T. *J. Org. Chem.* **2000**, *65*, 2368–2378.

Author Index

The entries in this index refer to the chapter number (boldface) and the reference number.

Aakermark, B. **1** 81
Abbotto, A. **2** 14
Abdon, R. L. II **4** 80
Abraham, W. **1** 227
Abramo, K. H. **2** 153
Abronin, I. A. **2** 2
Ackermann, E. **1** 133
Adachi, K. **1** 260
Adachi, M. **1** 117
Adamczeski, M. **1** 350, 351 **2** 72
Adams, G. W. **2** 107
Adams, J. K. **1** 119
Addie, M. S. **1** 171
Addo, M. F. **1** 213
Agal'tsov, A. M. **2** 136
Aguilar, E. **1** 55
Aguilar, M. A. **4** 108
Aitzhanova, M. A. **1** 578 **2** 2 3 21
Akaji, K. **1** 482, 490, 491, 492
Åkermark, B. **1** 365, 466, 540
Akullian, V. **1** 225
Akutsu, A. **4** 172
Alberola, A. **1** 382
Albinsson, B. **2** 151
Albonico, S. M. **4** 82
Aldrich, C. C. **4** 28
Allspach, T. **4** 142
Almendros, P. **1** 132
Alte Da Veiga, L. **4** 71, 88
Alvarez, M. M. **1** 315
Alvarez-Ibarra, C. **2** 66, 86
Amat-Guerri, F. **2** 121, 143, 149
Amato, J. S. **1** 486
Amiel, P. **1** 552
Amos, R. A. **1** 475
An, N. D. **4** 144, 146
Ana, G. **1** 64
Anderson, B. A. **1** 169, 364, 374

Anderson, O. P. **1** 501
Anderson, W. K. **4** 11, 63, 64
Andrade-Gorden, P. **1** 213
Ang, K. H. **1** 72
Angell, R. **4** 183, 184
Ansari, A. J. **1** 145
Ansell, M. F. **3** 82
Anz, S. J. **1** 315
Aoki, N. **4** 119
Aoki, T. **1** 577
Aoyama, T. **2** 53
Arai, N. **1** 428
Araki, E. **3** 36
Arao, H. **1** 12, 65, 294, 295, 296, 297, 298 **2** 31
Araynos, A. **1** 534
Arbain, D. **1** 160 **3** 69
Arbogast, J. W. **1** 315
Arcadi, A. **1** 562
Argay, G. **4** 67
Arikawa, Y. **4** 113
Armstrong, R. W. **1** 411, 493, 494, 508
    4 21, 22, 23, 24
Artyukhov, A. N. **2** 125
Asano, T. **1** 518, 519
Asaoka, M. **1** 15 **4** 17
Aso, Y. **1** 101
Austin, D. J. **4** 190, 191, 192, 198, 200, 201
Avalos, M. **4** 107, 108
Awad, R. W. **1** 138, 140, 141

Babaev, E. V. **1** 392, 394, 395, 396, 397, 398, 399, 400, 401, 402
Babiano, R. **4** 107, 108
Babii, S. B. **1** 572
Baccar, B. **1** 152
Bacescu, M. **2** 123
Bach, A. **2** 163
Bach, P. **4** 143

Baggi, P. **4** 105
Bagley, M. **1** 84, 85, 86, 89, 90, 91, 93, 94
Baiga, T. J. **4** 27
Bailey, T. R. **1** 225
Baizman, E. **1** 210
Baker, G. A. S. **1** 34
Baker, R. **1** 355, 356
Baker, W. R. **1** 368
Balaban, A. T. **1** 64 **2** 71, 123
Balaban, M-C. **1** 64
Bald, E. **1** 470
Ballabio, M. **4** 103
Ballantyne, L. **1** 426
Bally, I. **2** 123
Balšánek, V. **1** 581
Barbe, J. **1** 552
Barbieri, G. **1** 248
Barbosa, D. M. **4** 68, 71, 88
Barbosa, J. **1** 504, 505
Barda, D. A. **1** 525
Barni, E. **1** 193
Barreau, M. **1** 170
Barrett, A. G. M. **1** 151, 501, 517, 564
Barrish, J. C. **1** 28, 46, 47
Barsanti, P. A. **4** 87
Barsukov, I. N. **2** 97
Barton, D. H. R. **1** 353
Baruah, J. N. **4** 35, 36
Baughn, C. **1** 387
Baum, J. **4** 196, 197
Baum, J. S. **4** 1
Baumer, V. N. **1** 246 **2** 112, 141
Baumgaertel, H. **2** 145
Baur, A. **4** 28
Bautista, I. **4** 107
Bayer, A. **1** 59
Bayer, H. O. **4** 19
Beau, J.-M. **1** 508
Becke, L. M. **1** 169
Behnke, J. **2** 73
Beidrzycka, Z. **2** 89
Beisswenger, T. **2** 50
Beja, A. M. **4** 68, 71, 88
Belaj, F. **4** 209
Belen'kii, L. I. **1** 10, 241, 242 **2** 119, 120, 144
Belikov, A. B. **3** 43
Belshaw, P. J. **1** 549 **2** 103
Belyashova, A. I. **2** 110
Benassi, R. **1** 248
Benayahu, Y. **1** 443
Bender, S. **2** 162, 163
Bengston, G. **3** 84
Beppu, K. **1** 196

Beresis, R. T. **1** 108, 109
Bergdahl, M. **1** 366, 466
Berger, R. **1** 554
Bergsträsser, U. **4** 142, 143, 171
Berliner, M. A. **1** 515
Berrabah, M. **4** 146
Bertounesque, E. **1** 505
Bertram, A. **1** 432
Beuhler, A. J. **2** 146
Bhat, S. V. **3** 59
Bhatt, M. V. **3** 104
Biard, J. F. **1** 444
Biedrzycka, Z. **2** 88
Bienayme, H. **1** 582, 584
Bilodeau, M. T. **4** 133
Binger, P. **4** 141
Birladeanu, L. **2** 123
Blacque, O. **4** 148, 149
Blanchard, S. G. **2** 61
Blum, C. A. **1** 184, 185 **3** 115, 116
Blume, E. **2** 6
Boche, G. **1** 348, 349 **2** 87
Bödeker J. **2** 17
Boell, W. **3** 49
Bogdanov, V. S. **1** 578 **2** 2 **3** 21, 27, 54, 55
Boger, D. L. **3** 1, 2, 3
Bonati, L. **4** 132
Booher, R. N. **1** 169, 373
Borda, A. **1** 193
Borowski, E **1** 434
Bosold, F. **1** 348, 349 **2** 87
Bossio, R. **1** 156, 157, 164, 251 **2** 39
Boswell, G. E. **2** 61
Boto, A. **1** 204
Bouchoux, G. **2** 101
Boulos, J. **2** 47
Bowie, J. H. **2** 3, 107
Boyce, R. J. **1** 30, 31
Boyd, G. V. **1** 3, 4, 13, 18 **3** 9, 10 **4** 13, 14, 30, 40
Boykin, D. W. **2** 93
Bozhenko, S. V. **1** 397
Bradamante, S. **2** 14
Bradsher, C. K. **1** 393
Brain, C. T. **1** 218
Braña, M. F. **2** 49
Branch, S. K. **1** 137 **2** 98
Brazhnikova, M. G. **1** 434
Bredereck, H. **2** 118
Brennan, C. J. **1** 467
Breusova, E. G. **2** 133
Bringmann, G. **3** 46
Britton, T. C. **1** 511
Brocklehurst, B. **2** 128

Brodney, M. A. **3** 70 **4** 206, 207
Bronberger, F. **4** 130
Brookes, L. M. **3** 122
Brooks, D. A. **1** 57, 360, 515
Brooks, M. **1** 59
Broom, N. J. P. **1** 216, 217
Brovarets, V. S. **1** 573, 575, 576
Brown D. J. **2** 5
Brown, B. R. **1** 45, 46, 47
Brown, J. H. **4** 124
Brown, R. G. **2** 122
Brunner, H. **1** 192
Buchanan, G. L. **1** 180 **4** 50
Buchta, V. **1** 581
Buchwald, S. L. **1** 534
Buck, K. **1** 227
Buck, R. T. **1** 90, 91
Buhro, W. E. **1** 532
Bull J. R. **2** 7
Burger, K. **1** 116, 262, 263, 264, 265, 266, 267, 268, 269, 270 **2** 77, 78, 81, 82, 83, 85, 96
Burgess, E. M. **1** 49
Burgett, A. W. G. **1** 531
Burmester, K. **2** 17
Burri, K. **3** 51
Buscemi, S. **1** 76
Bushweller, C. H. **4** 124
Bushweller, J. H. **4** 124
Butsugan, Y. **4** 93, 94
Buttke, K. **2** 145
Butula, I. **3** 50
Bychkov, N. N. **2** 161
Bykanova, N. V. **3** 100
Byrn, M. P. **4** 123, 124

Cabana, E. Q. **1** 351
Cabanillas, A. **4** 108
Cacchi, S. **1** 562
Caesar, J. C. **3** 99
Cahill, P. A. **2** 146
Camoutsis, C. C. **1** 122
Campagne, J-M. **1** 467
Campos, E. **1** 114
Carabateas, P. M. **1** 225
Cardwell, K. S. **1** 153, 569
Carlsen, Per. H. J. **1** 506
Carlson, R. P. **1** 105
Carlsson, C. **2** 151, 152
Carmeli, S. **1** 439
Carpignano, R. **1** 193
Carreira, E. M. **1** 468
Carrié, R. **2** 52 **4** 29, 75, 76, 99
Cascia, L. **1** 562

Casimir, J. **1** 420
Casiraghi, G. **1** 334
Cassels, R. **1** 217
Cassidy, M. **1** 355
Castellano, J. M. **2** 49
Castellucci, E. **2** 108
Castro, B. **1** 473
Catalan, J. **2** 126, 127, 149
Caton, M. P. L. **3** 82
Cee, V. J. **1** 363, 430
Celatka, C. A. **1** 106, 107, 108
Cermola, F. **2** 56 **3** 142
Cermola, F. M. **3** 85
Cervantes, H. **1** 114
Ceulemans, E. **1** 563 **2** 40
Cha, J. K. **3** 147, 148, 149
Chamberlin, J. W. **1** 435
Chan, W. L. **4** 72
Chang, C.-P. **4** 63
Chang, S. **1** 477
Chang, Y. C. **1** 304 **3** 138
Chao, I. **1** 311
Chapman, T. **1** 569
Charifson, P. S. **2** 61
Chattopadhyay, S. K. **1** 56, 541, 542, 544, 545
Chauhan, P. M. S. **1** 309
Chen, B. **1** 177 **2** 21 **3** 97
Chen, B. C. **1** 24 **2** 91
Chen, H-C. **1** 310
Chen, J. **2** 130
Chen, Q. **3** 97
Chen, S. **1** 435
Chen, T. **1** 207 **2** 55
Chen, T-F. **1** 580 **3** 151
Chen, X. **1** 371, 530, 531, 535
Chen, Y.-Y. **4** 74
Chen, Y.-X. **2** 74
Cheng, H-Y. **1** 217
Cheng, Z. **1** 516
Chenu, E. **4** 84
Cheon, S. H. **1** 508
Chepeleva, L. V. **2** 112
Chernega, A. N. **1** 576
Cheskis, M. A. **1** 241, 242
Chiacchio, U. **1** 404
Chien, J. Y. **1** 119
Chinn, R. L. **4** 167
Chiraleu, F. **4** 31, 32
Chittari, P. **1** 352
Chizhikova, Z. A. **2** 144
Chizhov, O. S. **3** 21
Chong, L. S. **1** 493
Chott, R. C. **4** 136

Chou, T-S. **1** 310, 311, 312
Christ, W. J. **1** 508
Christmann, O. **1** 244
Chuang, J. M. **1** 276
Chuche, J. **1** 135, 136
Chung, Y. M **1** 561
Chuvylkin, N. D. **1** 10
Cicchi, S. **1** 547
Cimminiello, G. **3** 139, 142
Cintas, P. **4** 108
Ciobanu, M. **4** 172, 173
Cioranescu, E. **2** 123
Clapham, B. **1** 560, 579
Clardy, J. **1** 448
Clark, A. D. **1** 74
Clark, D. A. **1** 201 **2** 59
Clauss, K-U. **1** 227
Clerici, A. **2** 100
Clerici, F. **4** 101, 104, 105, 106
Cobb, J. E. **2** 61
Coburn, S. P. **3** 47
Cohen, L. A. **1** 271
Colens, A. **1** 487
Colletti, S. L. **1** 554
Collins, J. L. **2** 61
Colombo, L. **1** 334
Connell, R. **1** 540
Connell, R. D. **1** 81
Connolly, C. J. **2** 11
Consonni, R. **4** 131
Cooke, J. W. B. **1** 569
Coppola, G. P. **4** 80, 81
Corbett, T. H. **1** 439
Corchado, J. C. **4** 108
Cordero, F. M. **1** 547
Corey, E. R. **4** 116
Cornforth, J. W. **1** 149, 465
Cornforth, R. H. **1** 465
Corrao, S. **1** 345
Cottard, M. **1** 466
Covel, J. A. **1** 430
Craig, T. **3** 119
Cramp, S. M. **1** 564
Crank, G. **1** 252, 253, 254, 255, 256, 257 **3** 56
Crast, L. B. **1** 276
Cremer, D. **2** 157
Crew, A. P. A. **1** 309
Crews, P. **1** 350, 351, 442 **2** 72
Crimmin, M. J. **1** 206
Crow, F. W. **2** 90
Crowe, E. **1** 347 **2** 10
Crozet, M. P. **1** 170
Cuadrado, P. **1** 382

Cullen, T. G. **1** 336
Cundy, D. J. **1** 229
Cunico, R. F. **1** 221, 222, 223 **2** 46, 63
Cunningham, A. **2** 60
Cunningham, A. M. **4** 133
Curran, D. P. **4** 5
Curran, T. T. **2** 99
Curtin, D. Y. **1** 303
Curtis, E. A. **4** 188
Curtis, M. D. **2** 147, 148
Cutcliffe, D. **1** 225

D'Auria, M. **1** 1
D'Amato, F. J. **4** 123
Da Veiga, L. A. **4** 68
Dahn, H. **2** 95
Dai, L-X. **1** 167
Dale, J. A. **1** 533
Dalko, M. **1** 571
Dall'Occo, T. **1** 376, 377
Dalla Croce, P. **4** 38, 78, 85, 103, 131, 134, 135
Dalton, C. R. **4** 83
Dambek, P. J. **2** 59
Dandarova, M. **3** 83
Dansereau, S. M. K. **3** 67
Dardenne, G. **1** 420
Darmanyan, A. P. **1** 315
Das, J. **1** 175
Datta, A. **1** 219
Datta, S. K. **4** 197
Davidson, D. **1** 166
Davidson, J. G. **1** 532
Davies, C. G. **4** 30
Davis, M. **1** 529
Dax, S. L. **1** 408
de Díaz, R. G. **2** 35
de la Fuente, G. **1** 421
De Meijere, A. **3** 84
de Miguel, P. **2** 49
De Proft, F. **1** 313
Deady, L. W. **1** 25
Debaerdemaeker, T. **4** 120
Defterdarović, N. **1** 245 **2** 62
Dehaen, W. **1** 563 **2** 40, 75
del Valle, J. C. **2** 126, 127
Delpierre, G. R. **1** 436
DeMattei, J. A. **1** 493
Demillequand, M. **1** 466
deMontigny, P. **1** 554
Denisko, O.V. **1** 232
Dent, W. **4** 74
DeSimone, R. W. **1** 184, 185, 288, 289 **3** 115, 116
Dess, D. B. **1** 211

Dettori, G. **1** 214
Devos, A. **1** 487
Dewar, M. J. S. **3** 7
Dey, J. K. **2** 150
Dhar, T. G. M. **1** 567
Dherbomez, M. **1** 444
Diana, G. D. **1** 225
DiDonato, G. C. **1** 47
Diederich, F. **1** 315
Dion, J. L. **4** 124
Diwu, Z. **2** 134
Doerr, J. **3** 86
Dogan, I. **2** 159
Dogra, S. K. **2** 150
Dolby, L. J. **1** 423
Doleschall, G. **1** 75 **2** 51
Dominguez, X. A. **1** 421
Dömling, A. **1** 583 **4** 25
Donaghue, P. F. **2** 3
Donaldson, J. D. **4** 30
Dondoni, A. **1** 5, 7, 372, 372a, 376, 377, 378, 379 **2** 33 **3** 60
Dorfmeister, D. **4** 139
Dorfmeister, G. **4** 62
Dormoy, J. R. **1** 473
Doroshenko, A. O. **2** 112, 125, 137, 138, 139, 140, 141, 142
Dorow, R. L. **4** 163
Douglas, A. W. **1** 378
Douhal, A. **2** 121, 143
Downing, S. V. **1** 55
Doyle, K. J. **1** 11, 80, 87, 88, 92 **2** 54
Doyle, M. P. **1** 532 **4** 163, 168
Drach, B. S. **1** 572, 573, 575, 576
Dreizler, H. **2** 157
Drew, M. G. B. **4** 183, 185
Drobinskaya, N. A. **3** 32
Drushlyak, T. G. **1** 246
Duan, J. J.-W. **1** 502, 504, 505
Duarte, M. P. **1** 568
Dull, D. L. **1** 533
Dumats, J. **1** 571
Dumic, M. **3** 50
Dunlop, R. B. **1** 512, 513
Dvorak, C. A. **1** 475
Dyall, L. K. **1** 563 **2** 40

Eastwood, F. W. **1** 66, 436
Edmunds, J. J. **1** 151, 501
Effenberger, F. **1** 259 **2** 50
Efimov, A. V. **1** 392, 395, 400
Egdell, R. G. **2** 160
Egi, R. **1** 343

Eguchi, S. **1** 99, 130, 314 **4** 100
Ehrler, J. **1** 478
Eissenstat, M. A. **1** 224, 225
El'yanov, B.S. **3** 44
El-Azhary, A. A. **2** 109
Elder, J. S. **1** 216, 217
Elliott, M. C. **1** 88, 92
Elliott, R. C. **1** 532
Ellis, K. O. **1** 182
Ellman, J. A. **1** 511
Entwhistle, D. A. **1** 463, 464
Erba, E. **4** 102 104
Ermolenko, I. G. **1** 247
Espinosa-García, J **4** 108
Esser, L. **1** 530, 535
Evans, D. A. **1** 363, 497, 498, 499, 511
Evans, D. L. **1** 27
Evertz, K. **4** 165
Evin, G. **1** 473

Fabero, F. **2** 149
Fabre, J. L. **4** 98
Fabrizi, G. **1** 562
Falick, A. M. **1** 200
Falorni, M. **1** 214
Fan, W-Q **1** 231
Fantin, G. **1** 5, 372, 372a, 376, 379 **2** 33
Farge, D. **4** 98
Farina, C. **3** 52
Farooq, S. **1** 478
Fate, G. D. **2** 90
Faulkner, D. J. **1** 452
Fedyunyaeva, I. A. **1** 247 **2** 113, 125
Fengler-Veith, M. **4** 183, 184, 185
Fenical, W. **1** 448
Ferkous, F. **3** 95
Fernandes, P. S. **1** 121
Fernandez, J. I. **4** 107
Fernández, R. **1** 444
Ferraccioli, R. **4** 38, 131, 132, 134, 135
Fiege, M. **1** 415
Fields, S. **1** 327 **3** 125
Fieser, L. F. **1** 367
Fieser, M. **1** 367
Filippova, T. M. **1** 434
Filla, S. A. **1** 500
Finch, H. **4** 183, 184
Findlay, R. H. **2** 160
Firestone, R. A. **3** 31, 37
Firl, J. **2** 77
Fischer, B. **1** 2, 240, 320, 328, 389, 390 **3** 11, 124, 143
Fischer, J. **1** 333

Fischer, M. **1** 133
Fisher, M. H. **1** 554
Flammang, R. **1** 68 **2** 8, 101
Flaugh, M. E. **1** 169, 373
Fleury, J-P. **2** 52
Fliege, E. R. L. **2** 156
Florent'ev, V. L. **3** 5, 32, 126
Floyd, D. **1** 28, 47
Floyd, M. B. **3** 136
Fogagnolo, M. **1** 5, 372, 372a, 376, 379 **2** 33 **3** 60
Fominykh, N. G. **2** 162
Foote, C. S. **1** 315
Fortt, S. M. **4** 8
Frangopol, P. T. **2** 123
Franke, R. **2** 162, 163
Freedman, J. **1** 125
Freedman, S. B. **1** 355
Freeman, F. **1** 207, 208 **2** 55
Freeman, J. P. **4** 157
Frenna, V. **1** 76
Fresneda, P. M. **1** 132
Friebe, T. **1** 466
Friebe, T. L. **1** 366
Friedrichsen, W. **1** 187 **4** 120
Friestad, G. K. **1** 504, 505
Frisque-Hesbain, A.-M. **1** 487
Fryxell, G. E. **4** 65
Fuchs, E. P. O. **4** 141
Fujieda, H. **1** 318, 323, 324 **3** 132, 134
Fujii, T. **1** 159, 160 **3** 69
Fujioka, H. **1** 508
Fujita, A. **1** 196
Fujita, E. **1** 460
Fujita, S. **1** 440
Fujita, T. **1** 37, 181 **2** 41
Fukumoto, T. **1** 101
Fukushima, K. **1** 83 **2** 30, 37
Fukuyama, T. **1** 483
Funabashi, M. **1** 150
Funabiki, K. **4** 73
Funahashi, T. **2** 23
Funke, E. **4** 18, 20, 126, 127
Furihata, K. **1** 450
Furukawa, J. **1** 429
Furuya, T. **1** 440
Fuse, H. **1** 536
Fusetani, N. **1** 440, 441, 453, 454, 455, 509

Gaa, K **1** 266, 267, 268, 269 **2** 96
Gabriel, S. **1** 179
Gage, J. R. **1** 497, 498, 499
Gaiser, F. **4** 70
Gala, D. **2** 9
Galanin, M. D. **2** 144
Galliani, G. **1** 377
Gallop, M. A. **4** 189
Gambale, R. J. **1** 286
Ganem, B. **1** 461
Ganesan, A. **1** 173, 174
Gangloff, A. R. **1** 81, 365, 366, 466
Gardelli, C. **1** 51
Garey, D. **1** 61
Gariboldi, P. **4** 85
Garner, P. **1** 546
Garst, M. E. **3** 96
Gasdaska, J. R. **4** 65
Gebhard, R. **1** 226
Geerlings, P. **1** 313
Gehring, A. M. **1** 549
Geith, K. **1** 262, 263, 264, 265, 267, 268, 269
   2 78, 81, 82, 83, 85, 96
Gelmi, M. L. **4** 101 102 103 105 106
Genereux, P. E. **1** 201
George, M. V. **1** 280
Gerhart, F. **1** 154
Gerothanassis, I. P. **2** 94
Gerritz, S. W. **1** 500
Gertitschke, P. **1** 116
Gheorghiu, M. D. **2** 71
Ghosez, L. **1** 487
Ghosh, A. K. **1** 472
Ghosh, P. B. **2** 5
Ghosh, S. **2** 20
Giacomelli, G. **1** 214
Giardina, G. A. M. **3** 52
Gibbs, E. M. **1** 201
Gilbertson, T. J. **2** 90
Gilchrist, T. L. **1** 548 **4** 88
Ginanneschi, M. **2** 108
Gingrich, H. L. **4** 1, 125
Giomi, D. **1** 547 **2** 42, 84 **3** 61, 62 **4** 109
Giori, P. **1** 228
Giraud, L. **1** 170
Girault P. **2** 29
Gisselbrecht, J. P. **3** 86
Glaser, K. **1** 105
Glunčić, B. **1** 258 **3** 58
Goddard, C. J. **1** 142
Gogonas, E. P. **1** 95
Gol'dfarb, Ya. L. **1** 242
Goldenberg, H. J. **1** 45, 46
Goldstein, S. W. **2** 59
Golik, J. **1** 434
Gollnick, K. **1** 290, 291 **2** 19 **3** 140, 141
Gololobov, Y. G. **1** 131
Golovinsky, E. V. **1** 555

Gompper, R. **1** 244, 259 **2** 118
Gonsalves, A. M. D. R. **4** 68, 71, 88
Gontar, A. F. **2** 97
Gonzales, S. **1** 61, 426
Gonzalez, A. G. **1** 421
Gonzalez, A. M. **1** 382
Gonzalez, J. **3** 28
Gonzalez, R. **4** 178
Good, R. H. **2** 16
Gopalan, R. **1** 385
Gordon, T. D. **1** 209, 210
Gorelik, V. S. **2** 136
Goth H. **2** 77
Goto, S. **4** 93
Gotoh, H. **2** 38
Gotthardt, H. **4** 95, 96, 127
Gowravaram, M. R. **4** 189
Grabowski, Z. **1** 26 **2** 89
Graf, F. **3** 76
Gramer, C. J. **1** 364
Granier, T. **4** 70
Grasselli, J. G. **1** 23
Gravalos, D. G. **1** 446
Gray-Nunez, Y. **2** 61
Graziano, L. **3** 85
Graziano, M. L. **2** 56 **3** 139, 142
Gream, G. E. **1** 436
Gribble, G. W. **4** 79, 110, 111, 123, 124
Griesbeck, A. G. **1** 415
Griffiths, D. V. **3** 99
Griffiths, P. A. **3** 99
Grigg R. **2** 13 **3** 71, 72
Grimes, D. **1** 105
Grissom, J. W. **1** 405, 406, 407
Gromova, G. P. **1** 242
Gronsdemange-Pale, C. **1** 135, 136
Gross, M. **3** 86
Grossi, A. V. **4** 150
Grubbs, R. H. **1** 477
Gruen, D. M. **1** 316
Grueva, E. S. **1** 555
Grugni, M. **3** 52
Gryczynski, I. **2** 135
Gu, H. H. **1** 567
Gu, Y-G. **1** 57
Guallar, V. **2** 121
Guarneri, M. **1** 228
Guesten, H. **2** 106
Gunar, V. I. **3** 43, 44
Gunawan, I. **1** 105
Gunzner, J. L. **1** 527
Gupta, R. R. **4** 7
Gupta, V. **4** 7

Györgydeák, Z. **4** 67

Hachiya, S-I. **1** 501
Hacksell, U. **1** 123 **2** 64
Haddadin, M. J. **4** 157, 158
Hadjiarapoglou, L. P. **1** 95
Hagen, T. J. **1** 559
Hagihara, N. **1** 98
Hagiwara, S. **1** 130
Hajjem, B. **1** 152
Hall, J. H. **1** 119
Hall, S. E. **1** 28, 46, 47
Hallinan, E. A. **1** 559
Hallmeier, K. H. **2** 163
Ham, W.-H. **1** 508
Hamada, Y. **1** 32, 33, 163, 277, 352, 495, 496, 507, 510, 516
Hamaguchi, M. **4** 164, 194
Hamana, H. **1** 96
Hamano, K. **1** 424
Hammond, P. R. **2** 124
Hamper, B. C. **4** 136
Han, W-C. **1** 46, 47
Hannan, P. C. T. **1** 216, 217
Hansen, P. E. **1** 209, 210
Hara, O. **1** 163
Harada, K. **1** 292, 293
Harley, E. A. **1** 355
Harms, K. **1** 348, 349 **2** 87
Harn, N. K. **1** 169, 364, 374
Harran, P.G. **1** 371, 530, 531, 535
Harreus, A. **1** 250
Harring, S. R. **4** 203
Harris, D. N. **1** 45, 46, 47
Harris, E. **3** 37
Harris, E. E. **3** 31
Hartman, D. A. **1** 105
Hartner, F. W. **1** 17
Hartner, F. W Jr. **4** 9
Härtner, M. W. **1** 527
Haruta, J. **1** 570
Harwood, L. M. **4** 183, 184, 185
Hasan, N. M. **3** 41
Hasegawa, H. **1** 383, 384
Hashimoto, H. **1** 570
Hashimoto, K. **1** 440, 441, 453, 454, 455
Hashimoto, M. **4** 113
Hashimoto, N. **2** 24
Hashimoto, R. **1** 424
Hassner, A. **1** 2, 240, 320, 328, 389, 390 **3** 11, 124, 143
Haugland, R. P. **2** 134
Hawkins, L. D. **1** 508

Hay, D. W. P. **3** 52
Hayashi, Y. **1** 492
Hayes R. **3** 71
Hayes, R. N. **2**, 107
He, X. **2** 104
Heathcock, C. H. **1** 195, 479, 485 **4** 26
Hecht, H.-J. **4** 115
Hecht, S. M. **1** 27
Heckman, M. **4** 115
Hedberg, A. **1** 44
Heidelbaugh, T. M. **4** 170, 211
Heider, A. R. **4** 11
Heimgartner, H. **2** 57
Hell, S. W. **2** 135
Helmreich, B. **1** 270
Helquist, P. **1** 81, 82, 365, 366, 466, 540
Hénaff, N. **1** 565, 566
Hendrix, J. A. **1** 151, 501
Henke, B. R. **2** 61
Henkel, T. **1** 427
Hennessy, A. J. **1** 564
Hennig, C. **2** 163
Henry, R. A. **1** 119
Herdeis, C. **1** 226
Hermann, H. **1** 349
Hermesdorf, M. **4** 141
Hermitage, S. A. **1** 153, 569
Herrmann, A. **3** 86
Hershenson, F. M. **4** 15
Hertzog, D. L. **4** 167, 168, 174, 198, 199, 202
Hett, R. **1** 366, 466
Hetzheim, A. **1** 403
Heydt, H. **4** 141
Hibi, S. **1** 292, 293, 294, 298
Hidaka, T. **1** 450
Hiemstra, H. **2** 70
Higa, T. **1** 446, 447
Higuchi, N. **1** 380, 381
Hilf, C. **1** 348 **2** 87
Hill, K. W. **3** 87
Hind, S. L. **1** 90, 91, 93
Hinkle, B. **1** 426
Hintermann, L. **4** 70
Hirabayashi, M. **1** 550 **4** 48, 49
Hirakawa, K. **4** 92
Hirata, K. **1** 474
Hirotsu, Y. **1** 323 **3** 134
Hixson, S. S. **1** 388
Ho, M. S. **3** 89
Ho, S. M. S. **3** 90
Ho, W-B. **1** 288
Hodges, J. C. **2** 11

Hoekstra, J. W. **1** 532
Hoekstra, W. J. **1** 213
Hoess, E. **1** 264, 265, 266
Hoferichter, R. **3** 63
Hoffmann, H. M. R. **1** 52
Höfle, G. **1** 425, 438
Hojo, M. **1** 147 **2** 22
Hollins, R. A. **1** 119
Holmes, J. L. **2** 102
Honda, T. **4** 35
Hong, S. S.-K. **4** 81
Hong, T. **1** 111
Honma, T. **2** 76
Hoogenboom, B. E. **1** 168
Hop, C. E. C. A. **2** 102
Hoppe, I. **2** 58
Hoppilliard, Y. **2** 101
Horak, V. **1** 34
Horie, K. **4** 114
Horita, K. **1** 501
Hormes, J. **2** 162, 163
Hornbuckle, S. F. **4** 152
Höß, E. **2** 82, 83
Hossner, F. **1** 347 **2** 10
Hottenroth, J. M. **1** 521
Houghton, E. **4** 196, 197
Houk, K. N. **3** 28
Houwing, H. A. **2** 70
Howe, R. K. **1** 77
Howell, R. **1** 105
Huang, C.-H. **3** 15, 34, 35
Huang, W. **1** 177
Huang, W-S. **1** 155
Huber, E. W. **1** 125
Huber, E. W. **4** 83
Hübl, D. **1** 262, 263 **2** 78, 81, 85
Huebl, D. **1** 116
Huff, B. E. **1** 200
Hughes, M. J. **1** 347 **2** 10
Hughlett, R. K. **1** 283
Huisgen, R. **4** 2, 5, 18, 19, 20, 95, 96, 126, 127, 130
Hulin, B. **1** 201 **2** 59
Hull, K. G. **1** 502, 504, 505
Hull-Ryde, E. A. **2** 61
Hulshizer, B. L. **1** 213
Humbert, S. **2** 101
Hungate, R. W. **1** 197, 198, 199
Hunsmann, G. **1** 425, 438
Husain, S. **4** 195
Hutchison, D. R. **4** 89, 90
Huth, A. **1** 194
Hutton, J. **3** 77

Author Index

Ibata, T. **1** 83, 100, 317, 318, 319, 321, 322, 323, 324, 325, 326, 329, 330, 331, 332, 551 **2** 30, 36, 37 **3** 22, 23, 24, 128, 129, 130, 131, 132, 133, 134, 135 **4** 164
Ichiba, T. **1** 446
Ichikawa, Y. **3** 88
Ichimura, K. **1** 299, 300, 301, 302
İçli, S. **2** 159
Iddon, B. **1** 14
Iesce, M. R. **2** 56 **3** 85, 139, 142
Iga, H. **4** 172
Ihlenfeldt, H. G. **1** 59
Ikai, K. **1** 325, 326 **3** 135
Ikeda, K. **1** 260
Ikeda, T. **1** 299, 300
Ikoma, K. **4** 113
Ilisiu, A.-M. **2** 75
Inanaga, J. **1** 474
Ionova, L. V. **3** 32
Iqbal, J. **1** 366, 466
Irie, M. **1** 150, 462
Ishibashi, H. **1** 159, 160 **3** 69, 144
Ishibashi, M. **1** 437 **2** 34
Ishibashi, Y. **1**, 41
Ishihara, M. **3** 107
Ishihara, T. **4** 73
Ishikawa, F. **3** 16, 17, 18, 25
Iso, Y. **1** 96
Isobe, M. **3** 88
Isoda, S. **3** 16
Isogami, Y. **1** 100, 317, 318, 319, 331, 332 **3** 22, 23, 24, 128
Itagaki, F. **1** 433
Ito, A. **1** 39, 40 **2** 44
Ito, H. **1** 299, 300, 301, 302
Ito, S. **2** 23
Ito, Y. **1** 380, 381
Itoh, I. **3** 25
Itoh, M. **1** 99
Itoh, T. **1** 383, 384
Ivanova, Z. N. **3** 39, 73, 74, 75
Iversen, L. L. **1** 355
Iwabuchi, J. **1** 428
Iwanami, M. **3** 33
Iwanowicz, E. J. **1** 567
Iwasaki, S. **1** 429
Iwasaki, T. **1** 158 **2** 25
Iwase, K. **1** 577
Iwashima, M. **1** 504, 505
Iyer, R. P. **1** 385, 386
Izuta, R. **1** 161 **3** 145, 146

Jackson, J. L. **2** 13 **3** 71, 72
Jackson-Mülly, M. **1** 126
Jacobi, P. A. **1** 184, 185, 189, 190 **3** 108, 110, 111, 112, 113, 114, 115, 116, 118, 119, 120, 121
Jacobsen, N. W. **1** 144
Jadrijevic-Mladar Takac, M. **3** 50
Jaetsch, T. **1** 488
Jahans, A. W. **4** 183, 185
James, C. **4** 98
Janda, K. D. **1** 560
Janowski, W. K. **1** 74
Jansen, R. **1** 425, 438
Januszewaski, H. **1** 26
Janvier, P. **1** 584
Jaret, R. S. **2** 9
Jauk, B. **4** 209
Jebaratnam, D. J. **1** 48
Jelling, M. **1** 166
Jenkins, G. **1** 309
Jenny, C. **2** 57
Jeong, S. **1** 371
Jerome, K. D. **4** 136
Jesdapaulpaan, S. **1** 256
Jewett, J. G. **4** 124
Jiménez, J. L. **4** 107, 108
Jiménez, R. **1** 114
Jin, H. **1** 508
Jin, S. **3** 148
Jinguji, M. **3** 107
Jones, B. G. **1** 137 **2** 98
Jones, G. **2** 16
Jones, M. F. **1** 484
Jones, P. G. **1** 418
Jonsson M. **2** 151, 152
Jordan, S. I. **1** 463, 464
Jordis, U. **1** 27
Jug, K. **1** 392
Jung, G. **1** 58, 59
Jung, M. E. **3** 67
Jurasek, A. **3** 81
Jurkiewicz, E. **1** 425, 438
Jursic, B. S. **3** 29, 30, 127 **4** 91
Jüttner, F. **1** 431

Kabotianskaya, E. B. **3** 54, 55
Kaczmarek, C. S. R. **1** 189 **3** 112, 113
Kada, R. **3** 83
Kagemoto, T. **1** 150
Kahn, H. R. **1** 252, 255, 256
Kaiser, D. **1** 58, 59
Kajitani, H. **1** 445
Kajtár, J. **4** 67
Kakegawa, H. **1** 128
Kakehi, A. **2** 23 **4** 172, 173

Kálmán, A. **4** 67
Kaluzny, B. D. **2** 90
Kamat, P. V. **1** 280
Kametani, T. **4** 35
Kamitori, Y. **1** 147 **2** 22
Kampf, J. W. **2** 148
Kane, A. A. **2** 110
Kane, J. M. **4** 66, 83
Kang, S. H. **1** 508
Kann, N. **1** 466
Kanzelberger, M. **1** 105
Kao, C. **2** 130
Kappe, C. O. **4** 186, 187, 188, 209
Kappeler, W. H. **2** 59
Karady, S. **1** 486
Karash, C. B. **4** 116
Karl, R. **4** 26
Karolak-Wojciechowska, J. **1** 552
Karpeiskii, M. Ya. **3** 5, 32
Kasai, P. H. **2** 158
Kascheres, A. **2** 48
Kasha, M. **2** 126, 127
Kashima, C. **1** 12, 65, 292, 293, 294, 295, 296, 297, 298 **2** 31
Kashiuchi, M. **2** 36
Kashman, Y. **1** 443
Kasukhin, L. F. **1** 131
Kataoka, M. **3** 42
Kato, H. **4** 6, 86, 92, 112, 113, 114, 119, 121, 172, 173
Kato, Y. **1** 440, 441
Katritzky, A. R. **1** 188, 229, 230, 231, 232, 233, 234, 235 **2** 74 **4** 9
Katsuki, T. **1** 307, 474, 506
Kattan, A. M. **4** 157
Katunuma, N. **1** 128
Kauffman, J. A. **1** 213
Kauffman, J. M. **1** 119, 186
Kaun, C. P. **1** 221, 222, 223
Kawada, K. **3** 98
Kawai, A. **1** 163
Kawai, K. **4** 35
Kawai, N. **1** 419
Kawamatsu, Y. **1** 37 **2** 41
Kawamura, Y. **2** 32 **4** 119
Kawanishi, K. **1** 343, 344 **2** 69
Kawase, M. **1** 202, 203, 550 **2** 79, 80 **4** 42, 43, 44, 45, 46, 47, 48, 49, 54, 56, 57, 58, 59, 60, 128, 129
Kazmierski, W. M. **2** 61
Keating, T. A. **4** 21 22 23 24
Keck, G. E. **1** 430, 457
Keefe, K. **1** 61

Keehn, P. M. **2** 67
Kellar, S. M. **4** 125
Kelleher, N. L. **2** 103
Kelly, T. R. **1** 104, 371a
Kempson, J. **1** 543, 544
Kempter, C. **1** 58
Kernan, M. R. **1** 452
Keyaniyan, S. **3** 84
Khalafy, J. **1** 556, 557
Khan, H. R. **3** 56
Kharlanov, V. A. **2** 122
Khavtasi, N. S. **1** 281
Khuhara, M. **1** 342
Kidwai, M. **1** 120
Kiguchi, T. **1** 414
Kikuchi, H. **4** 46
Kilpert, C. **1** 333
Kim, D. S. H. L. **1** 208
Kim, H. S. **1** 561
Kim, J. N. **1** 561
Kim, K. **1** 128, 249
Kim, T. Y. **1** 561
Kimel, W. **3** 20
King, M. D. **1** 183
Kingston, D. G. I. **1** 436
Kirichenko, A. V. **2** 137
Kirstgen, R. **1** 250
Kishi, Y. **1** 508
Kiso, Y. **1** 482, 490, 491, 492
Kissick, T. P. **1** 43
Kitagawa, O. **3** 98
Kitagawa, T. **1** 341, 342, 343, 344 **2** 68, 69
Kitamura, A. **1** 447
Kiyomori, A. **1** 534
Kiyosawa, T. **4** 92
Klasinc, L. **2** 106
Klaubert, D. H. **2** 134
Klein, R. F. X. **1** 34
Klein, U. **1** 333
Kneen, C. **1** 355
Knight D. W. **1** 537
Knight, B. W. **4** 178
Knorr, R. **4** 18 19
Kobayashi, H. **1** 429
Kobayashi, J. **1** 433, 437, 536
Kobayashi, S. **1** 117
Kobayashi, T. **4** 6, 112, 114, 121, 172, 173
Kobayashi, Y. **3** 98
Kobrakov, K. I. **1** 9
Köckritz, A. **1** 272, 273, 274, 275
Kocy, O. **1** 46
Koegler, S. **1** 290, 291 **2** 19 **3** 140, 141

Koehler, U. **1** 250
Koenig, H. **3** 49, 76
Kohrt, J. T. **1** 517
Koide, N. **1** 314
Koiwai, H. **4** 48, 54, 59
Kojanni, E. **1** 141
Kolodina, N. S. **2** 110
Kolter, R. **1** 456
Kondo, K. **1** 437
Kondo, Y. **1** 370
Kondrat'eva, G. Ya. **1** 578 **3** 13, 14, 15, 21, 27, 34, 35, 39, 40, 55, 73, 74, 75, 79, 100
Kong, Y. C. **1** 249
Konopelski, J. P. **1** 521
Koren, R. **3** 83
Koreshkov, Yu. D. **3** 54, 55
Korzhenevskaya, N. G. **1** 282
Kosc, T. Z. **2** 146
Koseki, K. **1** 441, 453, 454
Koshelev, V. M. **2** 97
Koskinen, A. M. P. **1** 44, 60
Kotani, E. **1** 117
Koumbis, A. E. **1** 526
Kovac, J. **3** 81
Kowalczyk, P. J. **1** 225
Kowalczyk, T. C. **2** 146
Koyama, T. **1** 447
Koyama, Y. **1** 423
Kozikowski, A. P. **3** 41
Krafka, E. **2** 157
Krasovitskii, B. M. **1** 247 **2** 129
Kraus, K. G. **1** 338
Kress, T. J. **1** 169
Krishnamachari, S. L. N. G. **1** 239
Krishnamurthy, A. **1** 176
Krishnamurthy, D. **1** 430
Kronenthal, D. R. **1** 43, 175
Kruger, J. **1** 468
Krumpe, K. E. **4** 153
Kuan, C. P. **2** 46, 63
Kubicki, M. M. **4** 148, 149
Kubo, H. **1** 159, 160 **3** 69 144
Kubota, H. **1** 335
Kudinova, M. K. **1** 434
Kuethe, J. T. **4** 169, 170, 210, 211
Kujundžić, N. **1** 258 **3** 58
Kulkarni, B. A. **1** 173, 174
Kulkarni, S. P. **1** 385, 386
Kumakura, H. **4** 49
Kumar, A. **2** 154, 155
Kumar, D. **1** 113
Kumar, M. **4** 7
Kumar, R. **1** 120

Kuneš, J. **1** 581
Kuno, A. **1** 103
Kunze, B. **1** 425, 438
Kuo, Y.-C **2** 53
Kurachi, Y **4** 94
Kurihara, T. **1** 203 **2**, 79 **4** 42, 43, 54, 58
Kuriyama, N. **1** 490, 492
Kuriyama, S. **1** 284
Kurizono K. **3** 36
Kuznetsova, R. T. **2** 133
Kyrychenko, A. V. **2** 138, 140, 141

L'abbé, G. **2** 75
L'vova, S. D. **3** 43, 44
La Rosa, C. **4** 38, 78, 85, 103, 131, 134, 135
Ladurée, D. **4** 69, 84
Lafontaine, J. A. **1** 51
Laguna, M. A. **1** 382
Lahmani, F. **2** 143
Lake, D. H. **2** 61
Lakhan, R. **1** 220, 529 **3** 6
Lakowicz, J. R. **2** 135
LaMattina, J. L. **1** 337
Lamphere, C. H. **2** 59
Lan, X. **1** 230, 231, 232
Lancelot, J.-C. **4** 69, 84
Landi, J. J. **1** 417
Lang, F. **1** 104, 371a
Langridge, D. C. **1** 388
Lantos, I. **1** 306
Lao, X.-F. **4** 122
Larina, L. I. **1** 237, 238
Larsson, A. **2** 151, 152
Laude, B. **1** 391 **4** 144, 145, 146, 147, 148, 149
Lautens, M. **1** 134
Lavé, D. **4** 98
LaVerne, J. A. **2** 128
Lawesson, S.-O. **1** 346
Lawhorn, D. E. **1** 373
Lawson, A. **4** 41
Lawson, J. P. **1** 243, 459a, 471, 475
Lazo, J. S. **2** 60
Leahy, J. W. **1** 51
Lee, G. C. M. **3** 96
Lee, J. C. **1** 102, 111 **3** 147, 148, 149
Lee, K. **1** 190 **3** 120, 121
Lee, V. Y. **1** 375
Lee, W. S. **1** 128
Leesnitzer, L. M. **2** 61
LeGourrierec, D. **2** 122
Lehmann, J. **2** 61
Leighton, J. L. **1** 497, 498, 499
Leimgruber, W. **3** 20

Leininger, S. **4** 143
Lenhard, J. M. **2** 61
Lesko, J. **3** 83
Letourneux, Y. **1** 444
Leung, Y. **3** 93
Levin, J. I. **1** 29 **3** 64, 65, 66
Levy, D. E. **1** 488
Lewis, D. M. **1** 201 **2** 59
Lewis, J. R. **1** 19, 20, 21, 22
Lex, J. **1** 415 **3** 86
Li, G. **1** 48
Li, J. **1** 531
Li, S. **2** 21
Li, W. S. **3** 117
Li, W-S. **1** 215
Li, Y. **2** 21
Li, Y-J. **1** 469
Li, Y-M. **1** 456
Li, S. **3** 150
Liebscher, J. **1** 416, 418, 528
Lim, R. **4** 125
Lim, S. **1** 514
Lin, C. C. **3** 47
Lin, X. **2** 104
Lin, Y-R. **1** 167
Linden, A. **1** 431
Linderman, R. J. **1** 459a
Lindquist, N. **1** 448
Ling, M. **1** 204
Liotta, D. **3** 78
Lipshutz, B. H. **1** 197, 198, 199, 200 **3** 4
Litak, P. T. **1** 119, 186
Liu, B. **3** 70, 123
Liu, K. C. **1** 77
Liu, L. **1** 476
Liu, P. **1** 106, 107, 110, 362
Liu, W. **1** 472
Lluch, J. M. **2** 121
Lo, C. W. **4** 72
Lobo, A. M. **1** 568
Lohberger, S. **4** 26
Lohrenz, J. C. W. **1** 349
Lokshin, A. I. **1** 246 **2** 114, 131
López-Ranz, M. M. **2** 86
López-Sánchez, M. I. **2** 86
Lopyrev, V. A. **1** 237, 238
Loughlin, W. A. **1** 183
Loupy, A. **1** 278
Lowder, P. D. **1** 57
Löwe, R. S. **4** 191
Lu, H-F. **1** 311
Lu, T-J. **1** 285, 308
Lu, Y.-Q **2** 37

Luchetta, L. M. **1** 358 **2** 12
Lue, P. **2** 74
Lunn, G. **1** 279
Luttmann, M. **3** 52
Lykke, K. R. **1** 316

MacDonald, J. G. **4** 125
Mach, R. H. **4** 64
Macielag, M. J. **1** 345
Mack, A. **4** 142
MacLeod, A. M. **1** 355
Macomber, D. **1** 188
Mader, M. M. **1** 466
Madison, L. L. **1** 456
Maeda, S. **1** 158
Maekawa, E. **4** 97
Magnus. P. **1** 524
Mahieu, C. **1** 278
Mahuren, J. D. **3** 47
Maiboroda, D. A. **1** 392, 398
Maier, G. V. **2** 133
Maier, M. E. **4** 165, 166
Mais, F.-J. **4** 115
Majo, V. J. **1** 115
Mak, T. C. **3** 91
Makarov, M. G. **3** 79
Malak, H. **2** 135
Malamas, M. S. **1** 105
Malashkhiya, M. V. **1** 281, 282
Malecha, J. W. **1** 151, 501
Mallamo, J. P. **1** 225
Mancuso, A. J. **1** 367a
Manley, P. J. **4** 176
Manley, P. W. **4** 137
Manoharan, M. **1** 313
Maquestiau, A. **1** 68 **2** 8
Marcaccini, S. **1** 156, 157, 164, 251 **2** 39
Marić, D. **2** 106
Marinelli, F. **1** 562
Marinić, Ž. **1** 245 **2** 62
Marino, J. P. Jr. **4** 204, 205, 206, 207
Märkl, G. **4** 62, 139
Marlier, M. **1** 420
Márquez, A. **4** 138
Marr, J. G. **2** 90
Marsch, M. **1** 348, 349 **2** 87
Marsden, S. P. **1** 97
Marshakova, S. A. **2**, 161
Martin, D. C. **2**, 148
Martin, G. E. **2**, 90
Martin, H.-D. **4** 115
Martin, J. C. **1** 211
Martin, V. S. **1** 506

Martinelli, M. J. **1** 508 **4** 89, 90
Maruyama, T. **1** 12, 292 **3** 36
Maryanoff, B. E. **1** 213 **4** 116
Maryanoff, C. A. **2** 1 **4** 116, 117
Masamune, S. **1** 488, 500
Mashraqui, S. H. **2** 67
Masson, N. **1** 217
Mastellari, A. **1** 5, 377, 379 **3** 60
Masuda, R. **1** 147 **2** 22
Masuzawa, M. **4** 113
Mathé, G. **4** 84
Mathias, L. J. **4** 159, 160, 161, 162
Matsui, T. **1** 163
Matsumoto, K. **1** 158, 335 **2** 25, 43 **3** 23
Matsumura, K. **2** 24
Matsunaga, S. **1** 440, 441, 453, 454, 455, 509
Matsuo T. **3** 26, 38
Matsushita, M. **1** 570
Matsushita, T. **1** 78, 79
Matsuzaki, K. **2** 38
Matthews, T. R. **1** 351
Mayer, B. **4** 115
Mazari, M. **4** 75 76 99
Mazzu, A. Jr. **1** 27
McCarthy, K. E. **1** 6, 197, 198, 199, 200, 288 **3** 137
McCleary, M. A. **1** 183
McClymont, E. L. **1** 361
McCombie, S. W. **1** 353
McComsey, D. F. **1** 213
McCracken, J. **3** 87
McDermott, R. E. **2** 59
McDonald, R. J. **4** 150
McEwen, W. E. **1** 388 **4** 150
McGarvey, G. J. **1** 63
McGown, L. B. **2** 153
McIver, E. G. **1** 524
Mclaws, M. D. **1** 430
McLeod, Jr., D. **2** 158
McNeil, A. **1** 544
McPhail, A. T. **1** 580 **3** 151
McPherson, H. L. Jr. **4** 63
McWhorter, W. J., Jr. **1** 508
Medhurst, A. D. **3** 52
Medici, A. **1** 5, 372, 372a, 376, 377, 379 **2** 33
Medvedskaya, L. B. **2**, 2 **3** 39, 40, 73, 74, 75, 79, 100
Meek, G. **1** 204
Meguro, K. **1** 37, 181 **2** 41
Mehicic, M. **1** 23
Meinhardt, M. B. **2** 146
Meinke, P. T. **1** 554
Mekonnen, B. **1** 253, 254

Mena, E. **2** 149
Mendoza, M. **2** 66
Menjo, Y. **1** 462 **2** 26
Merchant, K. J. **1** 355
Merke, I. **2** 157
Merlic, C. A. **4** 28
Merslavic, M. **1** 146
Methot, J.-L. **1** 523
Meyer, K. G. **1** 360
Meyers, A. I. **1** 27, 35, 36, 55, 112, 459a, 471, 475
Miah, S. **4** 193
Michel, I. M. **1** 45, 46
Migallón, A. S. **2** 49
Mikami, Y. **1** 536
Miki, T. **3** 26, 38
Miles, D. H. **4** 41
Miller, C. P. **1** 50, 62, 489
Miller, L. **1** 303
Miller, M. M.**1** 476
Miller, R. D. **1** 375
Miller, T. A. **1** 200
Milne, J. C. **1** 456, 549
Milstein, D. **1** 520
Minemura, H. **4** 112
Minster, D. K. **1** 27
Mircea, D. **1** 64
Mischke, D. A. **4** 136
Misra, R. N. **1** 28, 45, 46, 47
Mitina, V. G. **2** 112, 137, 142
Mitsunobu, O. **1** 42
Miura, M. **2** 32
Miyamae, H. **1** 203 **2** 7 **4** 42, 43, 46, 48, 59, 60
Miyasaka, A. **4** 113
Miyashita, O. **2** 24
Miyatake, R. **3** 107
Miyaura, N. **1** 96a
Miyazaki, H. **2** 34
Miyoshi, M. **2** 25
Mizugaki, M. **2** 34
Mizuno, M. **1** 508
Mizuno, Y. **1** 260
Mjalli, A. M. **4** 27
Mohapatra, D. K. **1** 219
Mohr, R. **3** 63
Molina, P. **1** 132
Molinski, T. F. **1** 172, 354, 359, 449, 452 **2** 18
Monahan, S. D. **1** 357
Mondal, S. **1** 422
Monnier, K. **1** 391
Monnier, K. **4** 147, 148, 149
Monshizadegam, H. **1** 45
Montgomery, J. **1** 463, 464

Moody, C. J. **1** 11, 80, 85, 86, 87, 88, 89, 90, 91, 92, 93, 94 **2** 54 **4** 193
Moore, D. R. **4** 159, 160, 161, 162
Moore, R. E. **1** 439
Moreno, M. **2** 121
Morgan, B. A. **1** 209, 210
Mori, T. **2** 43
Morisawa, Y. **3** 42
Morishima, N. **4** 100
Morison, R. A. **1** 47
Morita, T. **1** 158
Moriwaki, M. **4** 114
Moriya, T. **1** 335 **2** 43
Mornatti, P. **4** 104
Moro, G. **4** 132
Mosher, H. S. **1** 533
Motoyoshiya, J. **2** 38
Mottadelli, S. **4** 105
Mowlem, T. J. **1** 88, 92
Moylan, C. R. **1** 375
Mrozek, A. **1** 552
Muderawan, I. W. **1** 183
Muellar, R. H. **1** 43, 175
Muir, J. C. **1** 458
Mukaiyama, T. **1**, 71, 470
Mukkarram, S. M. J. **1** 200
Mularski, C. J. **1** 337
Muller-Bottischer, H. **1** 417
Mulqueen, G. C. **1** 30, 31
Muneer, M. **1** 280
Muniz-Miranda, M. **2** 115
Murakami, M. **1** 380, 381 **3** 33
Murata, I. **3** 107

Nabil, M. **1** 444
Nadin, A. **1** 527
Nadler, W. R. **4** 154, 198, 199, 202
Nagai, S. **2** 38
Nagai, T. **4** 194
Naganawa, A. **3** 88
Nagao, Y. **1** 128, 460
Nagarajan, K. **1** 24 **2** 91
Nagase, H. **4** 35
Nagata, H. **1** 370
Nagata, K. **1** 383, 384
Nagatsu, A. **1** 445
Nagayoshi, K. **1** 118
Nahm, S. **1** 191
Naito, T. **1** 414 **3** 16, 17, 18, 19, 25
Nakagawa, A. **1** 424
Nakagawa, H. **1** 342
Nakagawa, Y. **2** 34
Nakahara, Y. **1** 196

Nakamura, K. **1** 117
Nakamura, T. **1** 437
Nakamura, Y. **1** 38, 39, 40 **2** 44
Nakano, H. **4** 86
Nakano, S. **1** 317, 319 **2** 36 **3** 22
Nakata, M. **1** 508
Nakawa, H. **1** 317, 318, 319 **2** 36 **3** 22, 23
Nakazawa, S. **4** 92
Nakazawa, T. **3** 107
Namikoshi, M. **1** 429
Nan'ya, S. **4** 93, 94, 97
Narita, M. **4** 42
Narsaiah, B. **1** 148
Natalie, K. J. Jr. **1** 215
Navarrete, P. **4** 138
Nayyar, N. K. **4** 89, 90
Negrini, E. **1** 5
Nelson, J. A. **1** 105
Nelson, K. H. Jr. **4** 190, 191
Nesi, R. **2** 42, 84 **3** 61, 62 **4** 109
Neville, M. **1** 257
Newton, L. S. **1** 201
Newton, R. **1** 569
Niclas, H. J. **2** 145
Nicolaides, D. N. **1** 138, 139, 140, 141
Nicolaou, K.C. **1** 526, 527 **3** 117
Nie, X. **2** 21
Nieto, R. M. **1** 442
Nikitina, A. N. **2** 144
Nilsson, B. M. **1** 123 **2** 64
Nimmesgern, H. **4** 74
Ninomiya, I. **1** 414
Nishimoto, K. **1** 78, 79
Nishiyama, S. **1**, 41
Nissinen, M. J. **1** 60
Noda, T. **1** 500
Noe, M. C. **4** 80, 81
Noguchi, H. **1** 453
Noguchi, M. **3** 68
Noma, M. **1** 453, 454
Nomura, M. **2** 32
North, P. C. **3** 82
Nuber, B. **1** 192
Nyce, P. **2** 9

O'Hanlon, P. J. **1** 206, 216, 217
O'Neil, M. P. **1** 316
Oberdorf, K. **1** 250
Obukhov, A. E. **2** 116, 117, 119, 120
Odawara, A. **1** 158 **2** 43
Odeh, I. M. A. **3** 110
Odom, J. D. **1** 512, 513
Ogawa, A. K. **1** 493, 494

Ogawa, T. **1** 196
Ogletree, M. L. **1** 46, 47
Ogura, F. **1** 101
Oguri, K. **4** 114
Oguri, T. **1** 419
Ohashi, T. **1** 99
Ohba, M **1** 159, 160, 161 **3** 69, 144, 145, 146
Ohba, S. **1** 41, 261
Ohkubo, M. **1** 103
Ohlsen, S. R. **3** 80
Ohno, M. **1** 99, 314
Ohsawa, A. **1** 383, 384
Okada, M. **1** 383
Okada, R. **1** 297
Okano, T. **4** 100
Oksenberg, D. **1** 213
Okuda, K. **2** 38
Okuda, S. **1** 429
Okumura, K. **1** 38, 39, 40 **2** 25, 44
Old, D. W. **1** 534
Ollis, W. D. **4** 4
Olschewski, G. **1** 192
Oltra, H. M. **1** 521
Oluwadiya, J. O. **1** 124
Omote, Y. **1** 292, 293, 294
Omura, S. **1** 428
Omura, Y. **3** 16, 17, 25
Oppenhuizen, M. **1** 532
Orband-Miller, L. A. **2** 61
Orellana, G. **2** 66, 86
Orlova, I. A. **1** 399, 401
Ortiz, P. **2** 86
Osamura, Y. **1** 78
Osterhout, M. H. **4** 154, 199, 204, 205, 207
Ostrovskii, V. A. **2** 132
Ostval'd, G. V. **3** 27, 55
Ostvald, G. V. **3** 54
Ota, Y. **2** 23
Otsubo, T. **1** 101
Ott, W. **3** 78
Ottlinger, R. **2** 77
Owens, K. A. **1** 554
Ozaki, T. **1** 130
Ozaki, Y. **1** 158

Padwa, A. **1** 271, 404 **3** 70, 123 **4** 2, 3, 65, 74, 125, 151, 152, 153, 154, 155, 156, 157, 167, 168, 169, 170, 174, 175, 176, 177, 178, 179, 180, 181, 182, 188, 198, 199, 200, 201, 202, 203, 204, 205, 206, 207, 208, 210, 211
Padyukova, N. Sh. **3** 126
Pagani, G. A. **2** 14
Paget, C. J., Jr. **1** 373

Paixão, J. A. **4** 68 71 88
Pajpanova, T. I. **1** 555
Pal, S. K. **1** 340
Palacios, J. C. **4** 107, 108
Palmer, M. H. **2** 160
Pan, J. **2** 130
Panek, J. S. **1** 106, 107, 108, 109, 110, 362, 370a
Panke, H.-L. **4** 127
Panomarev, O. A. **2** 113
Pansegrau, P. D. **1** 175
Papageorgiou, G. K. **1** 139, 141
Papaleo, S. **3** 61
Park, J. M. **1** 546
Parkinson, C. J. **1** 151, 501
Parks, D. J. **2** 61
Parrilli, M. **3** 139, 142
Parsons, R. L. **1** 195
Parsons, R. L., Jr. **1** 479
Partyka, R. A. **1** 276
Pascual, A. **1** 70 **2** 45
Pasichnichenko, K. Yu. **1** 398
Passerini, M. **1** 165
Patel, H. V. **1** 121
Patel, M. M. **1** 46, 47
Patsenker, L. D. **1** 246, 247 **2** 112, 114, 131
Patt, W. C. **2** 11
Pattenden, G. **1** 30, 31, 56, 204, 432, 458, 463, 464, 484, 537, 538, 541, 542, 543, 544, 545
Patterson, G. M. L. **1** 439
Paul, J. M. **1** 218
Pavez, H. **4** 138
Pavia, M. R. **4** 15
Pavlopoulos, T. G. **2** 124
Pavlychev, A. A. **2** 162
Pederson, B. S. **1** 346
Pedrini, P. **1** 5, 372, 372a, 376, 379 **2** 33 **3** 60
Pei, J. **1** 177 **2** 21 **3** 150
Pei, W. **1** 177 **2** 21 **3** 150
Peisach, J. **3** 87
Pelkey, E. T. **4** 79, 110, 111
Pellin, M. J. **1** 316
Peña, M. R. **1** 61, 426
Peng, J. **1** 513
Penton, H. R., Jr. **1** 49
Pepino, R. **1** 156, 157, 164, 251 **2** 39
Pepper, A. C. **1** 94
Pérez, J. D. **1** 69 **2** 35
Perlmutter, P. **1** 66
Perrin, S. **1** 391 **4** 148, 149
Perrocheau, J. **2** 52
Perronnet, J. **2** 29
Perumal, P. T. **1** 115
Pesa, F. A. **1** 23

Peters, E.-M. **4** 186
Peters, K. **4** 186
Petit, A. **1** 278
Petride, H. **4** 33, 34, 39, 51, 52, 53
Pevear, D. C. **1** 225
Pfister, K. III **3** 37
Philippides, A. **1** 144
Philipsborn, W.V. **1** 24
Phillips, A. J. **1** 553
Pho, H. Q. **4** 168
Pickett, J. A. **1** 305
Pierron, E. **4** 142
Pieters, R. J. **4** 168
Pihko, P. M. **1** 44, 60
Pil'o, S. G. **1** 573, 575, 576
Pilati, T. **4** 134
Pinho e Melo, T. M. V. D. **4** 68, 71, 88
Piotrowski, D. W. **1** 409, 410
Pitner, J. B. **2** 153
Pittalis, A. **1** 334
Pitts, W. J. **1** 567
Pivnenko, N. S. **2** 113
Pivnichny, J. V. **1** 554
Pivsa-Art, S. **2** 32
Pizzorno, M. T. **4** 82
Plisnier, M. **2** 101
Plumet, J. **4** 107
Plunkett, K. D. **2** 61
Plüss, T **1** 431
Pocar, D. **4** 102
Pohl, M. **1** 418
Poli, T. **1** 228
Politis J. K. **2** 147, 148
Polo, C. **1** 251
Ponomarev, O. A. **2** 112, 137, 142
Pons, J. E. **1** 216
Ponticelli, F. **4** 109
Poon, D. J. **1** 475
Pope, A. **1** 217
Popova, N. A. **1** 247 **2** 113
Popovich, T. P. **1** 575
Posada, P. **2** 49
Posokhov, E. A. **2** 142
Possel, O. **2** 70
Potapova, N. P. **1** 434
Potts B. **3** 77
Potts, K. T. **4** 3, 10, 195, 196, 197
Pouchert, C. J. **2** 73
Pour, M. **1** 581
Prabhakar, S. **1** 568
Prabhu, A. V. **1** 385
Prager, R. H. **1** 72, 73, 74, 556, 557 **2** 15
Prajapati, D. **4** 35

Precedo, L. **4** 168
Prein, M. **4** 175, 176, 177, 182
Price, A. T. **4** 199, 200, 201, 204, 207
Pridgen, L. N. **1** 306, 369
Pridzun, L. **1** 425
Procopiou, A. **1** 564
Provencal, D. P. **1** 51
Prowse, K. S. **1** 6, 143, 287, 288 **3** 137
Pryde, D. C. **1** 475
Ptyagina, L. M. **2** 141
Puar, M. S. **2** 9
Puk, E. **1** 68 **2** 8
Pulido, F. J. **1** 382

Qi, M. **1** 233
Qui, Y. **1** 504, 505
Quiñoà, E. **1** 350 **2** 72
Quiroga, M. L. **2** 66, 86

Rachwal, B. **1** 188
Rachwal, S. **1** 188
Racoveanu-Schiketanz, A. **2** 71
Radspieler, A. **1** 528
Rahm, A. **3** 95
Rahman, L. T. **1** 127
Raileanu, D. **4** 31, 32, 33, 34, 39
Rajaniakova, O. **3** 81
Rakhmatullaev, I. A. **2** 136
Ram, B. **1** 176
Rama Rao, A. V. **3** 102, 103
Ramachandran, K. **1** 580 **3** 151
Ramirez, M. **1** 61
Ramos, P. J. R. S. **4** 88
Ramsden, C. A. **4** 4
Rao, A. V. R. **3** 101
Ratnam, K. R. **1** 385, 386
Raveglia, L. F. **3** 52
Rawlinson, D. J. **1** 53, 54
Ray, S. **1** 340 **2** 20
Reader, M. **1** 544
Rebolledo, F. **4** 107
Reck, S. **1** 187
Rector, S. R. **1** 127
Redchenko, V. V. **2** 161
Reddy G. S. **3** 104
Reddy, D. R. **3** 101
Reddy, P. V. **3** 59
Rees, C. W. **4** 137
Regitz, M. **4** 140, 141, 142, 143, 171
Reich, F. **2** 118
Reichenbach, H. **1** 425, 438
Reid, J. A. **1** 175
Rein, T. **1** 466

Reina, M. **1** 421
Reisch, J. **1** 124
Remion, J. **1** 487
Reno, M. J. **1** 553
Rettig, W. **2** 122
Reuter, W. **3** 31
Rheingold, A. L. **3** 87
Rice, R. L. **2** 60
Richards, I. C. **4** 193
Richter, H. **4** 140
Rickborn, B. **2** 4, 65 **3** 105, 106
Ridgewell, R. E. **1** 47
Rigolio, R. **3** 52
Rippon, D. E. **1** 537, 544
Rissanen, K. **1** 60
Rizzi, J. P. **2** 59
Robba, M. **4** 69 84
Roberts, J. D. **2** 92
Roberts, R. S. **1** 564
Robinson, R. **1** 178
Rodda, H. J. **2** 3
Rodenhouse, R. A. **4** 163
Rodier, N. **4** 145, 147
Rodin, O. G. **2** 161
Rodríguez, H. **4** 138
Rodriguez, J. **1** 442
Roesener, J. A. **1** 451
Rogers, J. W. **1** 283
Rogers, N. H. **1** 206
Röhr, G. **1** 272, 273
Ronda, L. G. **2** 148
Rösch, W. **4** 140, 141
Rosenberg, D. **1** 194
Roth, G. P. **1** 417
Roy, A. **1** 134
Roy, D. N. **4** 196
Roy, R. S. **1** 549
Rtishchev, N. I. **2** 132
Rubin, Y. **1** 315
Rudi, A.; **1** 443
Ruf, S. G. **4** 171
Rybakov, V. B. **1** 395, 400, 401
Ryden, R. **1** 257

Sá, M. C. M. **2** 48
Saalfrank, R. W. **1** 133
Sabuco, J. F. **1** 170
Sachs, G. **3** 96
Sadighi, J. P. **1** 534
Saeki, H. **1** 474
Sagi, M. **2** 34
Saha, C. K. **1** 340
Saigo, K. **1** 470

Sain, B. **4** 36, 37, 61
Saindane, M. **3** 78
Sainsbury, M. **4** 8, 87
Saito, H. **1** 40 **2** 44
Saito, S. **1** 550 **4** 46, 49, 54, 57, 58, 60
Sakai, H. **1** 103
Sakakibara, J. **1** 445
Sakamoto, T. **1** 261, 370 **2** 34
Salgado-Zamora, H. **1** 114
Salvagno, A. M. **1** 417
Salvatore, B. A. **1** 502, 503, 504, 505
Sánchez-Pavon, E. **1** 114
Sandhu, J. S. **4** 35, 36, 37, 61
Sankawa, U. **1** 453
Sano, M. **3** 25
Sano, S. **1** 128
Santiago, B. **4** 83
Santiago, K. J. **1** 363
Sanz, C. R. **2** 49
Sarau, H. M. **3** 52
Sargent, M. V. **1** 160 **3** 69
Sarin, P. S. **1** 436
Sarma, C. R. **1** 176
Sarshar, S. **4** 27
Sasaki, H. **1** 341, 342, 343, 344 **2** 28, 68, 69
Sasaki, Y. **1** 158
Saski, T. **1** 536
Satake, K. **3** 70
Sato, H. **1** 314
Sato, Y. **1** 118
Satoh, T. **2** 32
Saunders, G. A. **1** 356
Saunders, J. **1** 355
Savage, M. A. **1** 559
Savarino, P. **1** 193
Savin, K. A. **1** 430, 457
Savinov, S. N. **4** 190, 191, 192
Sawabe, A. **1** 488, 500
Sbrana, G. **2** 108
Scarborough, R. M. **1** 213
Scarlato, G. R. **1** 411, 493
Scarpati, R. **2** 56 **3** 85, 139, 142
Scavo, F. **1** 540
Schaefer, F. C. **4** 18, 19, 95
Schaltenbrand, W. E. **3** 47
Schaub, B. **1** 413
Schaus, J. M. **1** 373
Schaus, J. V. **1** 370a
Scheibye, S. **1** 346
Scheidecker, S. **1** 274
Scherrer, V. **1** 126
Scheuer, P. J. **1** 446, 451
Schierlinger, C. **1** 266 **2** 96

Schiketanz, I. **1** 64 **2** 71
Schlessinger, R. H. **1** 469
Schleyer, M. **1** 443
Schlosser, M. **1** 412, 413
Schmatz, D. M. **1** 554
Schmickler, H. **3** 86
Schmid, H. **1** 126
Schmidt, D. B. **3** 52
Schmitt, G. **4** 144, 145, 146, 147
Schmitz, W. D. **1** 475
Schneider, M. **2** 145
Schneider, S. **3** 46
Schnell, M. **1** 272, 273, 274, 275
Schnurr, W. **4** 141
Schöffling, B. **4** 166
Schöllkopf, U. **1** 154 **2** 6, 58
Schröer, W.-D. **4** 120
Schulman, J. **2** 47
Schulte-Frohlinde, D. **2** 159
Schumacher, W. A. **1** 46, 47
Schumann, I. **1** 194
Schwan, T. J. **1** 182
Schwartz, D. J. **4** 80
Seager, C. H. **2** 146
Searle, P. A. **1** 449
Sedishev, I. P. **1** 578
Seitz, G. **3** 63
Seki, M. **2** 43
Selnick, H. G. **3** 114, 118, 122
Selve, C. **1** 473
Semeria, D. **1** 278
Semones, M. A. **4** 178, 203, 204
Sen, A. K. **1** 339
Sen, P. K. **1** 212
Sengupta, D. K. **1** 339
Sengupta, S. **1** 422
Seres, P. **1** 75 **2** 51
Seto, H. **1** 450
Seto, S. **1** 159
Sewald, N. **1** 266
Shafer, C. M. **1** 172, 354, 359 **2** 18
Shafiullah, A. J. A. **1** 145
Shah, U. **1** 105
Shanholtz, C. E. **1** 63
Shapiro, R. **1** 236
Sharpless, K. B. **1** 307, 506
Sheehan, S. M. **4** 179, 181, 188, 206
Shemelev, V. V. **2** 162
Shephard, R. G. **1** 387
Sher, P. M. **1** 28, 46, 47
Sheridan, J. **2** 154, 155
Shershukov, V. M. **2** 112, 125, 129, 138, 142
Shi, G. **1** 84 **2** 27 **3** 48

Shi, X. **1** 318, 321, 322, 324, 330, 332 **2**, 111 **3** 128, 130, 131, 132, 133
Shibanuma, T. **1** 71
Shibata, M. **1** 507
Shigemori, H. **1** 433
Shilcrat, S. C. **1** 306
Shilo, O. P. **2** 112
Shimada, S. **3** 45, 57
Shimazu, H. **2** 24
Shimizu, E. **1** 161 **3** 145
Shimizu, H. **1** 128
Shimonishi, Y. **1** 433
Shin, C. G. **1** 38, 39, 40
Shin, C.-G **2** 44
Shindo, H. **1** 96
Shingu, T. **1** 343, 344 **2** 69
Shinkai, I. **1** 378
Shinose, M. **1** 428
Shioiri, T. **1** 32, 33, 163, 277, 352, 419, 495, 496, 507, 510, 516, 518, 519 **2** 53
Shiomi, K. **1** 428
Shiono, M. **1** 71
Shiraiwa, M. **1** 370
Shirasawa, S. **4** 121, 122
Shiro, M. **1** 128
Shkumat, A. P. **2** 114
Shmelev, L. V. **3** 73, 75
Shoop, W. L. **1** 554
Short, K. M. **1** 129
Showell, G. A. **1** 356
Shridhar, D. R. **1** 176
Shu, L.-H. **4** 122
Shul'ts, E. E. **3** 53
Shvaika, O. P. **1** 281, 282
Shvekhgeimer, G. A. **1** 9
Siahann, T. J. **1** 200
Sicinska, W. **2** 89
Siderius, H. **1** 168
Silks, L. A. III **1** 512, 513
Silva, M. R. **4** 68, 71, 88
Silver, J. **4** 30
Silverberg, L. J. **1** 215
Sime, F. M. **1** 206
Simon, W. M. **4** 79, 111
Simonsen, K. B. **1** 526
Simpson, J. H. **1** 215
Šindler-Kulyk, M. **1** 245 **2** 62
Singer, K. D. **2** 146
Singh, A. K. **1** 215
Singh, G. **4** 55
Singh, J. **1** 43, 175, 209
Singh, R. L. **1** 220
Singh, S. **4** 55

Singh, U. P. **4** 196
Sircar, J. C. **1** 112
Sivaprasad, A. **1** 148
Sjolin, A. **1** 153
Slawin, A. M. Z. **1** 91 **4** 193
Slebioda, M. **4** 12
Smith, A. B. III **1** 502, 503, 504, 505
Smith, J. A. **1** 72, 73 **2** 15
Smith, T. E. **1** 363
Smith, T. P. **1** 188
Snagoshchenko, L. P. **1** 282
Snow, R. J. **1** 355, 356
Snyder, J. P. **4** 188
Snyder, S. A. **1** 526
Soares, M. I. L. **4** 68, 71
Soave, R. **4** 106
Somoza, F. B. **2** 148
Sonaseth, M. S. **1** 385, 386
Song, I-G. **1** 102
Song, Z. Z. **3** 90, 91, 92
Sonogashira, K. **1** 98
Sosnovsky, G. **1** 53, 54
Southern, P. F. **3** 77
Spada, A. P. **1** 288
Spanka, C. **1** 560
Spergel, S. H. **1** 43
Spoerel, U. **2** 157
Sponholtz III, W. R. **4** 123
Spoors, P. **1** 504, 505
Sprague, P. W. **1** 47
Srzić, D. **1** 245 **2** 62, 106
Stahl, W. **2**, 157
Stamegna, A. P. **4** 150
Stanchev, M. S. **1** 555
Stanforth, S. P. **4** 4
Stanovnik, B. **1** 146
Stapelfeld, A. **1** 559
Stashina, G. A. **1** 578
Steenken, S. **2** 159
Stefaniak, L. **1** 26 **2** 92
Steglich, W. **1** 333
Steigl, A. **4** 115
Stein, P. D. **1** 28
Stepanova, S. V. **3** 43, 44
Stephanidou-Stephanatou, J. **1** 139
Stetinova, J. **3** 81, 83
Stiefvater, O. L. **2** 154, 155
Stille, J. K. **1** 520
Storr, C. **1** 309
Storr, R. C. **4** 137
Strange, R. H. **4** 87
Straub, C. S. **4** 180, 181
Strocker, A. M. **4** 24

Strouse, C. E. **4** 124
Stutz, A. E. **1** 508
Subramanyam, C. **3** 68
Sudakova, G. N. **2** 110
Suga, H. **1** 318, 321, 322, 323, 324, 325, 326, 329, 330, 332, 551 **3** 128, 129, 130, 131, 132, 133, 134, 135
Sugihara, Y. **3** 107
Sugiyama, Y. **1** 37 **2** 41
Sujino, K. **1** 150
Sun, J. **1** 167
Sun, W.-Q. **4** 122
Sun, X. **1** 582, 584
Sund, E. H. **1** 283
Surov, Yu. N. **2** 113, 114
Sutherland, A. J. **1** 579
Suzuki, A. **1** 96a
Suzuki, M. **2** 25
Suzuki, T. **1** 424
Svensson, C. E. **1** 556
Swaminathan, S. **1** 215
Swern, D. **1** 367a
Switzer, F. L. **4** 110, 123, 124
Szargan, R. **2** 163
Szeto, Y. S. **4** 72
Szilágyi, L. **4** 67

Tada, Y. **2** 76
Taddei, F. **1** 248
Takagaki, H. **4** 17
Takahashi, M. **1** 447
Takahashi, S. **1** 424
Takahashi, T. **1** 147 **2** 22
Takahashi, Y. **1** 428
Takamura, I. **3** 16
Takao, T. **1** 433
Takashima, K. **2** 43
Takasugi, H. **1** 103
Takayama, H. **2** 34
Takei, H. **1** 15 **4** 17
Takeuchi, H. **1** 130
Takeyama, S. **2** 43
Talamas, F. X. **1** 508
Tamai, Y. **1** 491
Tamburlin, I. **1** 170
Tamura, H. **1** 317, 318, 331, 332 **3** 24, 128
Tanada, Y. **1** 510
Tanaka, C. **1** 284
Tanaka, H. **1** 78, 79
Tanaka, J. **1** 447
Tanaka, Y. **1** 428
Tang, J. **1** 162
Tange, T. **4** 97

Taniguchi, M. **1** 508
Tanimoto, N. **1** 500
Tanke, R. S. **1** 476
Tannus, H. T. **4** 158
Taunton, J. **4** 168
Tavares, F. X. **1** 35, 36, 471
Tawada, H. **1** 37 **2** 41
Taylor, E. A. **1** 49
Taylor, E. C. **3** 28
Taylor, R. J. K. **1** 171
Tebbe, M. **1** 81
Tebby, J. C. **3** 99
Tedeschi, P. **4** 109
Tellew, J. E. **1** 493
Tempest, P. A. **4** 24
Terlouw, J. K. **2** 102
Ternai, B. **1** 529 **3** 6
Terpstra, J. W. **4** 163
Terui, Y. **2** 76
Texier, F. **4** 29 75 76 77 99
Thapar, G. S. **1** 176
Theilig, G. **1** 205
Theobald, F. **4** 145, 147
Thielert, K. **1** 194
Thieme, M. **1** 416, 418
Thom, S. M. **1** 484
Thomas, J. P. **1** 387
Thomas, R. M. **1** 458
Threadgill, M. D. **1** 37 **2** 98
Tian, Z. Q. **3** 87
Tighineanu, E. **4** 31 32
Timón, I. **1** 421
Tino, J. A. **1** 508
Tišler, M. **1** 146
Tith, S. **1** 61
Tobinaga, S. **1** 117
Toda, S. **4** 113
Todd, Lord, A. R. **1** 436
Todorova, A. K. **1** 431
Togo, H. **1** 150, 462 **2** 26
Tohda, Y. **1** 98
Tojo, T. **3** 45
Tokue, K. **4** 121
Tolstikov, G. A. **3** 53
Tong, W.-Q **2** 61
Tonnard, F. **4** 76, 99
Toppet, S. **2** 75
Tori, K. **2** 76
Torroba, T. **1** 251
Touchette, K. M. **3** 118
Toupet, L. **4** 29, 75
Toyoda, J. **1** 319 **3** 22
Traldi, P. **2** 100

Traylor, T. G. **3** 87
Trifonov, R. E. **2** 132
Trimarco, P. **4** 101, 102, 104
Trost, B. M. **1** 585 **4** 80
Trujillo, H. A. **4** 79, 111
Tsai, C. Y. **1** 310, 312
Tsisevich, A. A. **1** 394, 396, 402
Tsoleridis, C. A. **1** 141
Tsuda, M. **1** 536
Tsujioka, T. **1** 117
Tsveniashvili, V. Sh. **1** 281, 282
Tsymbalov, S. **1** 559
Tucci, F. C. **1** 410
Tucker, T. T. **4** 183, 184
Tuinman, A. **2** 7
Tullis, J. S. **1** 82
Tupitsyn, I. F. **2** 110
Turchi, I. J. **1** 8, 336, 345 **2** 1 **3** 7, 8, 12 **4** 1, 116, 117, 118
Turchi, S. **2**, 42, 84 **3** 61, 62 **4** 109
Turner, S. **3** 80

Ubukata, M. **1** 462
Uchida, T. **1** 447
Ududong, U. E. S. **1** 184, 185, 189 **3** 112, 113, 115, 116
Ueda, K. **1** 508
Ueda, Y. **1** 276
Uekawa, T. **4** 100
Uenishi, J. **1** 508
Ueno, K. **3** 25
Ueno, Y. **4** 97
Ugi, I. **1** 583 **4** 25, 26
Ugozzoli, F. **3** 60
Ung-Truong, M-N. **2** 95
Usifoh, C. O. **1** 124
Uto, Y. **1** 553

Vaccaro, H. A. **1** 488
Vaccaro, W. D. **1** 200
Vafina, G. F. **3** 53
Valasinas, L. **1** 580 **3** 151
Valeriote, F. A. **1** 439
Valle, M. **4** 106
Valluri, M. **3** 102, 103
Van der Linden, J. B. **1** 207 **2** 55
Van Duyne, G. D. **1** 448
van Leusen, A. M **1** 168 **2** 70
Vanelle, P. **1** 170
Vankevich, A. V. **2** 112
VanSant, K. A. **1** 243
VanSickle, A. **1** 501
Varella, E. A. **1** 138, 140

Varie, D. L. **1** 169
Varma, R. S. **1** 113
Vasella, A. **4** 70
Vasil'ev N. V. **2** 97
Vasil'eva, I. A. **2** 144
Veal, C. J. **1** 212
Vebrel, J. **4** 144, 145
Vecchietti, V. **3** 52
Vedejs, E. **1** 327, 357, 358, 405, 406, 407, 408, 409, 410, 525 **2** 12 **3** 125
Veliz, E. A. **1** 521
Venit, J. J. **1** 215
Venkataratnam, R. V. **1** 148
Venkatasubramanian, R. **1** 239
Venkatraman, S. **1** 459, 480, 481
Venkatramanan, M. K. **1** 404 **4** 65
Verbist, J. F. **1** 444
Verezubova, A. A. **2** 141
Verkade, J. G. **1** 162
Veronese, A. C. **1** 228
Vettori U. **2** 100
Vicentini, C. B. **1** 228
Videnov, G. **1** 58, 59
Vieira, E. **1** 416
Vinick, F. J. **1** 304, 305 **3** 138
Vinkovic, M. **3** 50
Vinogradov, N. N. **2** 161
Vinogradova, T. K. **1** 576
Viscardi, G. **1** 193
Vivona, N. **1** 76
Vogel, E. **3** 86
Vojnović, D. **1** 245 **2** 62
Volkov, V. L. **2** 129
Volter, K. E. **1** 183
Von Philipsborn, W. **1** 431 **2** 91
Voronkov, M. G. **1** 237, 238
Vydzhak, R. N. **1** 573, 575

Wada, M. **1** 147 **2** 22
Wager, C. A. **1** 430
Wager, T. T. **1** 430
Waite, D. **1** 544
Wälchli M. R. **1** 437
Walgenbach, P. **3** 86
Walker, D. G. **1** 459a **3** 110, 111
Walker, D. M. **4** 136
Walker, F. J. **1** 338
Walker, G. **1** 206, 216
Walker, M. A. **1** 485
Walker, S. M. **1** 309
Walsh, C. T. **1** 456, 549 **2** 103
Waluk, J. **2** 140
Wang, S. Z. **4** 113

Wang, S.-Z. **4** 86
Wang, Y. **1** 558
Ward, S. **1** 210
Wargowski, D. A. **2** 146
Warmke, L. M. **1** 554
Warner, P. M. **1** 48
Wasielewsky, M. R. **1** 316
Wasserman, H. H. **1** 6, 143, 285, 286, 287, 288, 289, 304, 305, 308 **3** 136, 137, 138
Watanabe, M. **1** 462 **2** 26
Watanabe, T. **3** 42
Watterson, S. H. **1** 567
Weaver, J. D., III. **1** 224
Weaver, R. E. **1** 215
Webb, G. A. **2** 89, 92
Webb, H. **1** 200
Webb, M. L. **1** 45, 46, 47
Weber, B. **1** 72, 73 **2** 15
Weberndoerfer, V. **3** 76
Weglarz, M. A. **1** 457
Weingarten, M. D. **4** 155, 201
Weinreb, S. M. **1** 29, 191 **3** 3, 64, 65, 68
Weinstock, L. M. **1** 486
Weiss, M. **1** 166
Weng, L. **3** 151
Weng, L-I. **1** 580
Wenkert, D. **1** 580 **3** 151
Wentrup, C. **2** 101
Wepsiec, J. P. **1** 169
Wertsching, A. **1** 61
Wessels, F. L. **1** 182
Whetten, R. J. **1** 315
White, J. B. **1** 508
White, R. E. **1** 47
White, R. L. Jr. **1** 182
Whitehouse, D. L. **4** 190, 191
Whiting, A. **1** 565, 566
Whitney, S. E. **2** 4, 65 **3** 105, 106
Wilde, R. G. **4** 16
Wilkinson, R. C. **1** 387
Williams, C. M. **1** 72, 73, 556, 557 **2** 15
Williams, D. H. **1** 436
Williams, D. R. **1** 57, 360, 361, 515, 553
Williams, E. L. **1** 67
Wilman, D. E. V. **1** 137 **2** 98
Wilson, J. **1** 216
Wilson, J. M. **1** 217
Wilson, K. J. **1** 63
Wingert, H. **1** 250
Winters, M. **2** 4 **3** 106
Wipf, P. **1** 50, 62, 127, 459, 480, 481, 489, 514, 522, 523, 553 **2** 60
Wirth, U. **1** 133

Witanowski, M. **1** 26 **2** 88, 89, 92
Wolbers, P. **1** 52
Wolfe, J. P. **1** 534
Wong, G. S. K. **4** 65
Wong, H. N. C. **3** 89, 90, 91, 92, 93, 94, 109
Wong, M. K. **3** 93
Wood, R. D. **1** 461
Woodall, P. **1** 216
Woodward, P. R. **4** 87
Worsencroft, K. J. **4** 188
Wright, P. H. **4** 13, 14, 30, 40
Wu, H. **1** 235
Wu, H.-M. **4** 122
Wu, J. **2** 21, 104
Wu, S.-H. **4** 122
Wu, T. **3** 70
Wu, Y **1** 366, 466
Wudl, F. **4** 178
Wunderlin, D. A. 1 69
Wyvratt, M. J. **1** 554

Xie, L. **1** 235, 483
Xu, J.-F. **4** 122
Xu, M. **1** 84 **2** 27 **3** 48
Xu, Y. **1** 84 **2** 27 **3** 48

Yadav, J. S. **3** 102, 103
Yalamanchili, G. **4** 136
Yamada, S. I. **1** 419, 460
Yamaguchi, M. **1** 474
Yamakawa, M. **2** 76
Yamamoto, I. **2** 38
Yamamoto, K. **4** 48, 49
Yamamura, S. **1** 41
Yamanaka, H. **1** 261, 370 **2** 34 **4** 73
Yamane, T. **1** 574
Yamano, A. **4** 59
Yamashita, T. **2** 36
Yanagida, S. **1** 130
Yang Y. **3** 94, 109
Yang, J. Z. **1** 232
Yang, Q. **1** 66
Yang, Z. **1** 229
Yang, Z. C. **1** 521
Yannakopoulou, K. **2** 74
Yao, S. **4** 10
Yarmolenko, S. N. **2** 112
Yasuda, N. **4** 17
Yasumatu, M. **3** 36
Yasumuro, K. **1** 455
Ye, X. **1** 177 **2** 21 **3** 97, 150

Yebdri, O. **4** 76, 77, 99
Yelland, M. **1** 309
Yip, C. W. **4** 72
Yokokawa, F. **1** 32, 33, 495, 496, 510, 518, 519, 522
Yokose, K. **1** 423
Yokoyama, M. **1** 150, 462 **2** 26
Yonaga, M. **1** 508
Yoo, J. U. **1** 289
Yoo, S.-K. **1** 539
Yoshida, H. **1** 428
Yoshida, W. Y. **1** 446
Yoshikawa, T. **3** 16, 17, 18, 19
Yoshino, K. **4** 119
You, D. **2** 105
Young, D. W. **1** 212
Yranzo, G. I. **2** 35
Yu, Z. **2** 104
Yuan, C-Y. **1** 155
Yun, B.-S. **1** 450
Yuumoto, Y. **1** 414

Zakson-Aiken, M. **1** 554
Zaparaucha, A. 1 278
Zatsepina, N. N. **2** 110
Zdravkovski, Z. **3** 30, 127
Zeeck, A. **1** 427
Zehnacker-Rentien, A. **2** 143
Zelinski, Y. **1** 434
Zhang, C. **2** 134
Zhang, D.-W. **4** 122
Zhang, W. **2** 105
Zhang, X. **1** 426 **2** 105
Zhang, Y-X. **1** 155
Zhao, G. **1** 582, 584
Zhao, Z. **1** 411
Zhmurova, I. N. **1** 131
Zhou, X-T. **1** 167
Zhou, Z. Y. **3** 91
Zhu, H. **2** 105
Zhu, J. **1** 582, 584
Zhu, S. **1** 558
Zhukov, S. G. **1** 395, 400, 401
Zhulin, V. M. **3** 27, 54, 55
Ziegler, C. B. Jr. **1** 129
Zimmermann, H. **1** 133
Zinn, M. F. **1** 393
Zsindely, J. **1** 126
Zvak, V. **3** 81
Zvolinskii, V. I. **1** 9
Zyabrev, V. S. **1** 572
Zyuz', K. V. **1** 573

Subject Index

General oxazole substitution patterns are described in the text although entries in an associated table may be more specific in accordance to the examples from the original literature. For instance, a table of 4,5-disubstituted-2-methyloxazoles can be referred to in the text as 2,4,5-trisubstituted oxazoles. Tables of oxazoles that incorporate multiple substitution patterns or functional groups are titled with a general designation as substituted, disubstituted, or trisubstituted. The individual entries in a table are not included in the index. *Italicized* page numbers refer to tables.

Alphabetized lists of the general classes of oxazoles and exact names of oxazoles follow the subject index. The list of general classes of oxazoles includes all classes of oxazoles described in the text or tables as either starting materials or products. The entries in these lists are sorted and arranged in accordance with the sort order rules defined in Microsoft Word. The subject index is cross-referenced wherever possible to facilitate locating a general substitution pattern, a synthetic method or a reaction.

For example, if one is interested in using an isocyanide as a starting material, the general classes of oxazoles that can be prepared from an isocyanide, i.e. **2-(aminomethyl)-4-benzyl-5-morpholinooxazoles, 2-acyl-5-alkoxyoxazoles, 2-acyl-5-ethoxyoxazoles, 2,4-diaryl-5-oxazolecarboxamides, 2,4,5-trisubstituted oxazoles, 5-(aminomethyl)oxazoles, chiral *N*-Boc-, 5-(aminomethyl)oxazoles, chiral *N*-protected-, 5-(dialkylamino)-2,4-disubstituted oxazoles, 5-(*N*-alkyl-*N*-phenylamino)-2-(arylthio)oxazoles, 5-(*N*-alkyl-*N*-phenylamino)-2,4-(diarylthio)oxazoles, 5-(substituted 4-nitrophenyl)oxazoles, 5-alkoxy-2-(arylthio)oxazoles, 5-aryl(heteroaryl)-4-oxazolecarboxylic acid methyl esters, 5-aryl(heteroaryl)oxazoles, 5-aryloxazoles, 5-substituted-4-tosyloxazoles, 5-substituted-4-oxazolecarboxylic acid esters, 5-substituted oxazoles, and *N*-cyclohexyl-2,4-diaryl-5-oxazolecarboxamides**, are identified as bolded entries.

Additionally, all of these bolded entries are included alphabetically in the list of general classes of oxazoles that follows the subject index. Here, for example, the entry for 5-aryloxazoles lists all pages that discuss 5-aryloxazoles and will direct one not only to the pages that use isocyanides as a starting material but to other synthetic methods or reactions as well.

A table index follows the subject index.

A factor, 442
Acetamide-BF$_3$·OEt$_2$, 92–93, *93*
Acetic anhydride, 481, 483, 496, 497, 498, 499, 500, 501, 505–508, 510, 511, 513, 517, 523, 524, 527
Acetylation, oxazole reactions, 131, 137
Acrylic acid, 424
Acrylonitrile, 212, 367, 421, 428
1,3-Acyl transfer, 462

N-Acylisoxazol-5-ones (2-acylisoxazol-5-ones), oxazole synthesis, 23–25, 359–360
2-(1-aminoalkyl)-4-oxazolecarboxylic acid esters, 359
2-(1-aminoalkyl)-5-oxazolecarboxylic acid esters, 359
2,4-disubstituted oxazoles, 23, *23*, 24
2,4,5-trisubstituted oxazoles, *23*
2,5-disubstituted oxazoles, 23, *23*, 24

N-Acylisoxazol-5-ones (2-acylisoxazol-5-ones),
    oxazole synthesis   (Continued)
  bis-oxazoles, 24
  tris-oxazoles, 24
  almazole A and B, 359–360
  bis-oxazoles, 24
  FVP, 23
  photolysis, 23, 359, 360
  tris-oxazoles, 24
  N-acyl-1H-azirine, 25
N-Acylamino ketones (2-acylamino ketones),
    94–118
  cyclodehydration of, 94–97, 99–100, 106
  Dakin-West reaction, 94
  N-phenacylamino ketones, 106
N-Acylaziridines, oxazole synthesis, 16–18, *18*
  2-phenyl-4-substituted oxazoles, *18*
  2-styryl-4-substituted oxazoles, *18*
  2,4-disubstituted oxazoles, 16, 17
  unsymmetrical aziridines, regiochemistry, 16
  halichondramide, 16
  β-iodoazides, 17
N-Acyl-1,2,3-triazoles, oxazole synthesis,
    17–20, *18*
  2-alkyl(aryl)oxazoles, *18*
  2-substituted oxazoles, 17
  2-trimethylsilyl-1,2,3-triazole, 17
  bis-oxazoles, 20
N-Acyl-1,2,4-triazoles, oxazole synthesis, 17,
    20, *20*
  5-alkyl(aryl)oxazoles, *20*
  5-aryloxazoles, 20
  5-substituted oxazoles, 20
  thermal rearrangement, 20
  flash vacuum pyrolysis, 20
  [1,5]-acyl shift, 20
Acylation of, 19, 53, 78, 80–82, 108–109,
    120–121, 126, 128, 137, 138, 195, 201,
    204, 205, *208*, 222, 224, 231, 247,
    482–484
  2-(trimethylsilyl)-1,2,4-triazole, 19
  2-(trimethylsilyl)oxazoles, *222*, 247
  5-(1,3-dithian-2-yl)oxazole, 201, *204*
  5-(acylamino)-4-methyloxazoles, 138
  aminomalononitrile tosylate, 108–109
  bimetallic oxazoles, 205, *208*
  bis-oxazolyl methanol, 195
  dimethyl [(aminophenylthio)methyl]malonate,
    126
  enolates of O-trimethylsilyl acyltrimethylsilane
    cyanohydrins, 120–121
  ethyl isocyanoacetate, 78
  Friedel-Crafts, 128, 137

  isocyanides, 78, 80–82
  isocyanoacetic acid ester, 82
  lithiomethyl isocyanide, 81, 82
  metalated isocyanides, 80
  methyl isocyanoacetate, 80
  munchnones, 482–484
  nitrogen heterocycles to Reissert salts, 224
  oxazole, 224, 231
  propargyl amines, 53
Acyloins, 486
α-Acyloxyketones, 92–94, *93*, 366
  2,4-diaryloxazoles, 92, *93*
  2,4,5-trisubstituted oxazoles, 366
  **N-cyclohexyl-2,4-diaryl-
    5-oxazolecarboxamides**, 94
  cyclization of with
    ammonium acetate, 92, 366
    ammonium formate, 94
    thiourea, 366
    urea, 92
    acetamide-$BF_3 \cdot OEt_2$, 92–93, *93*
Agrochemicals, 29
Aldehydes, 68, 83–84, *85*, 86, 87, 88, 89, 90, 110,
    116, 117, 119, *119*, 122, 124, 133, 134,
    135, 136, *137*, 150–151, 178, *179*,
    179–180, 181, 195–196, 200, *201*, *203*,
    207–208, *209*, 210, 222, *222*, 224–225,
    227–228, 248–250, 251, 252, *254*, 268,
    302, 361–362, 364, 458–459
  **2-(hydroxymethyl)-4-oxazolecarboxylic acid
    ethyl esters**, 207–208, *209*
  2-alkenyloxazoles, trans, 249–250, 251
  2-alkyl-4-amino-5-aryloxazoles, 118–119,
    119
  **2-amino-5-(hydroxymethyl)-
    4-methyloxazoles**, 135–136, *137*
  4-alkenyloxazoles, trans, 248–249
  4-amino-2,5-diaryloxazoles, 118–119, *119*
  4-methyl-2-oxazolemethanols, 222, *222*
  **4-(methoxymethyl)-2′-methyl-5-substituted-
    2,4′-bis-oxazoles**, *203*
  5-aryl(heteroaryl)oxazoles, 90
  5-aryloxazoles, 87, 88, 90
  5-heteroaryl-sustituted oxazoles, 116
  5-substituted oxazoles, 86, 200, *201*, 361–362,
    364
  bis-oxazoles, 250, 251
  oxazole N-oxides, 68, 69
  **N-cyclohexyl-2,4-diaryl-5-
    oxazolecarboxamides**, 83–84, *85*
  tris-oxazoles, 250, 251
  5-oxazolecarboxaldehyde, 195–196
  2-acylamino, 117

Subject Index

[3+2] cycloaddition reaction with
  5-alkoxyoxazoles, 178–179, *179*, 180, 181, 458–459
  aldol reactions, 227–228, 275, 322
  arylglyoxals, 83–84, *85*
  benzaldeydes and α-hydroxyimino aromatic ketones, 68–69
  Bucherer-Bergs reaction, 122, 124
  heterodienophiles, 178–179, *179*, 180–181, 458–459
  hydroxymethylation, 135–136, *137*
  cis- and *trans*-2-oxazolines, 178–179, *179*, 180, 181, 227–228, 458–459
  oxetanes, 253, 254
  oxidation of (chloromethyl)oxazoles, 133, 134
  Paternò-Büchi photocycloaddition, 252, 253, *254*
  reaction with
    (trimethylsilyl)oxazoles, 222, *222*, 227
    (trimethylsilyl)thiazole, 224
    2-amino-4-methyloxazole, 135–136, *137*
    aroyl cyanides, 119, *119*
    lithiated oxazoles, 196, 200, *201*, 202, *203*, 207
    oxazole Reformatzsky type reagent, 207, *209*
    oxazole Wittig and Horner-Emmons reagents, 248–250, 251, 302
    TosMIC, 86, 88–89, 90, 361–362, 362
  serine-derived, 110
Aldol reactions, 227–228, 275, 322
Alkenyl-sulfonyl fluorides, 425
Alkylation, oxazole reactions, 138, 139
  2-alkyl-5-(diacylamino)-4-methyloxazoles, 138
  5-(acylamino)-2-alkyl-4-methyloxazoles, 138
  5-(diacylamino)-4-methyloxazoles, 138
  via acylation of oxazoles, 138
2-Alkynoic acids, polyfluoro, 499
Alzheimer disease, 197
Amidines, 491, *491*, 530
Amino acids, 141, 143–145, *145*, 178, 183–184, 227, 486, *486*, 487, 497, 505, 507, 511, 513, 517
  N-acyl, 486, *486*, 487, 497, 505, 507, 511, 513, 517
  C-glycosyl, 183–184
  β-hydroxy, 227
  β-hydroxy, *erythro*, 178
  α-substituted α-(trifluoromethyl), 141, 143–145, *145*
Amino polyols, *erythro*, chiral, 179–181
1-Aminobenzotriazole, 443
Aminomalonate, 493

Aminomalononitrile tosylate (AMNT), 108, *110*
  acylation of with carboxylic acids, 108, *110*
  diazonamide, 214, 341
Ammonium ylides, 557
Amphimedine, 434
Anagyrine, 544, 569
Andrographis paniculata, 446
14,15-Anhydropristinamycin II$_B$, 268–271
Anils, 482
Annuloline, 60
Antialgicidals, 257
Antibacterial agents, 50, 116, 147, 247, 358–359
Antibiotics, 10, 100, 209, 256, 257, 262, 263, 264, 265, 273, 438
  A23187 (calcimycin), 100
  peptide, 10
  streptogramin A, 257
  thiopeptide, 262
Anticyanobacterial agents, 257
Antidiabetic agents, 8, 42
Antiemetic agents, 216
Antiflea/antitick agents, 358
Antifungal activity/agents, 256, 345, 371
Antihelminthic agents, 195, 260
Antiinflammatory agents, 440
Antileukemic activity/agents, 255, 345, 495, 505
Antimycin A3, 156
(+)-Antimycin A3, 157, 158
Antineoplastic alkaloids, 250
Antipicornaviral agents, 121
Antiprotozoal agents, 45
Antitubercular agent, 231
Antitumor agents, 125, 255, 442, 447
Arbuzov reaction, 248, 249
Aromatic acyl(aroyl) cyanides, 118–119, *119*, 340, 341
Aroylation, oxazole reactions, 134, *134*, 137, *139*
  2-(acylamino)-4-methyloxazoles, 137
  4-acyloxazoles, 134
  4-aroyl-2-(arylthio)-5-ethoxyoxazoles, 134, *134*
  5-acyl-2-amino-4-methyloxazoles, 137
  5-acyl-4-alkyl-2-aminooxazoles, *139*
  2-(arylthio)-5-ethoxyoxazoles, 134
  2-aminooxazoles, 137, *139*
Arylazooxazoles, 136–137, *138*
Ascidian, 433
Asymmetric α-alkylation of 2-aminomethyloxazoles, 253, 254, 255, 256
  natural products, 254, 255
Autooxidation, 528
Aza-Claisen rearrangement, 163, 181–182
Aza-Cope rearrangement, 182

Azaanthraquinone alkaloids, 568
Azafluorenone alkaloids, 568
Azahomopentaprismane, 526
Azaphenanthrene alkaloids, 431
1,3-Azaphospholes, 494, 533
2-Azapyrrolo[1,2-*a*]indoles, 496
Azepines, 523, 524
 dihydro, 523, 524
Azetes, 534
Azides, 59–62, *61*, *63*, 122, 124–125
 2-alkyl(aryl)amino-4-cyano-5-methoxyoxazoles, 61
 2,4,5-trisubstituted oxazoles, 59, *61*
 2,5-disubstituted oxazoles, 59, *61*
 ***N*-(2,4-dialkyloxazol-5-yl)oxazolidinones**, 62
 (Z)-β-(acyloxy)vinyl azides, 59, *61*
 acyl azides, 122, 124–125
 Aza-Wittig reaction, intramolecular, 59, *61*
 iminophosporanes, 60
 methyl 3,3-diazidocyanoacrylate, 61, *63*
 oxazole alkaloids, 60
Azide transfer, 556
Azidothiazoles, 361, 363
 4-cyanooxazoles, 361
 thermolysis, 361
Aziridinomitosine, 244–247
Azirines, 21, 27, 62
 C-aroylamino, 27
 N-acyl, 25
Azlactones, 476, 482, *483*, 483, 494, 514, 526
Azo dyes, 136, 138
Azomethine ylides, 229, 235, 240, 241, 242–246, 500
Azulenes, 548

Barton-McCombie deoxygenation, 195
Baeyer-Villiger reaction (rearrangement), 152–153, 462
Beckmann rearrangement, base catalyzed, 25–26
Bengazoles, 195–197, 258, 261, 264
Benzamidines, 531
3-Benzazepinones, 487
Benzazetes, 531
Benzo[*b*]furo[2,3-*c*]pyrroles, 522
Benzo[*b*]thieno[2,3-*c*]pyrroles, 522
Benzo[*h*]1,6-naphthyridines, 432
Benzocyclobutadiene, 523
Benzocyclopropene, 523, 547
1,3-Benzodiazepines, 531, 532
Benzoxazoles, 429
4(5*H*)-Benzoxazolones, 36
Benzopyrano[4,3-*b*]pyridines, 432
Benzopyrano[4,3-*b*]pyrroles, 507

N-Benzoylphenylglyoxanilides, 538
Berninamycin/berninamycin A, 8
Bis-heteroannulation, 444, 449, 462
1,4-Bis-(imidazol-5-yl)benzenes, 189–190
Bis-oxazoles, 20, 24, 31, 33, 110, *112*, 189, 195, 196, 198, 200, *203*, 211, 212, 250, 258, 259, 260, 261, 262, 263, 316, 317, 318, 322, 324, 326, 329, 330, 331, 334, 340, 347, 348, 349, 350, 354, 405, 408
 2,2'-, 198
 2,4'-, *112*, 200, *203*, 212, 260, 316, 317, 318, 327, 347
Bistratamide D, 13–14, 264
Blood platelet aggregation inhibitors, 80
BMS 89,180, 291
Boroxines, 439
2-Bromoleptoclinidinone, 433
Bromoacetonitrile, 359, 361
Burcherer-Bergs reaction, 122
Burgess reagent, 10, 111, 114, 116, *117*, 278, 279, 281, 282, 285, 287, 288, 292, 293, 310–311, 312, 313, 319, 354, 356, 358, 368, 361
Butenolides, 440, 441, 451
 α-face reduction of, 453
 hydroxy, 440, 441
 hydrolysis to, 452
 keto, 446
 methyl, 451

^{13}C NMR, 3, *4*, 152, 154–155, 193, 195, 227, 232, 392, *396, 397, 398, 399*, 399–400
 2-lithiooxazole and isocyanovinyllithium alkoxide equilibrium, 193–194
 2-(trimethylsilyl)oxazole and formaldehyde, 227
 2-zincated benzoxazole, 195
 2,5-oxazole endoperoxide, 152, 154–155
 azafulvene, 232
Calcimycin (antibiotic A23187), 100
Calcium channel activators/modulators, 505, 556, 567
Calyculins, 14, 74, 75, 127, 248, 258, 260, 263, 289, 297, 295–316
 ent-calyculin A, 302
 (−)-calyculin A, 304–308
 (+)-calyculin A, 310, 312–313
 calyculin A, 74, 75, 295, 296
 (−)-calyculin B, 312
 calyculin C, 263, 289, 295, 296, 298, 299, 301–302, 310–312
 C(34)-*epi*-calyculin C, 299
 2,4-disubstituted oxazoles, 299–300

2-substituted-4-oxazolecarboxylic acid methyl ester, 309
4-(chloromethyl)-2-substituted oxazole, 298
absolute stereochemistry revised, 305
base and olefination selectivity, 303, 305, 315
  n-BuLi, 315
  KHMDS, 303, 305
  LiHMDS, 315
Burgess reagent, 310–311, 312–313
C(26)-C(32) oxazole fragment, 299–300, 302–303, 310–312
  Cu(II)OAc/*tert*-butylperbenzoate, 312
  CuBr$_2$/HMTA/DBU, 312
  diphenylsulfoxonium triflate, 300
  KHMDS/I$_2$, 311–312
  MnO$_2$, 312
  NiO$_2$, 300, 302–303, 312
C(26)-C(37) oxazole fragment, 296–299, 301–302, 304–308, 309–310, 312–313
  ethylbromopyruvate, 305, 307
  NiO$_2$, 302–303, 312–313
  TFAA, 307
  Wittig and Horner-Emmons reagents, 297, 298, 299, 302–303, 303–305, 309, 314, 315, 316
  Cornforth methodology, 309
  D-galactose, 307
  gulonolactone, 305, 307
  oxazolidine-2-seleonone, 312–313
  reductive elimination, 308
  Silks-Odom protocol, 312–313
  spiroketal β-elimination, 303, 305
Camptothecin, 442
Carbonic anhydrase II, 453
Carbon monoxide, 480
Cerebral protective agents, 42
Ciglitazone, 104
Chan-type rearrangement of imides, 330, 333
  diazonamide fragment, 330, 333
  polyoxazoles, 330, 333
Chloroacetonitrile, 39
γ-Chlorobutyryl chloride, 450
Chlorodecarboxylation, 331, 334–335, 336
Chromium complexes, 480
Chromium ketenes, 479, 480
Cinchomeronic anhydrides, 418
Clavans, 247
Clavine alkaloids, 569
Claisen rearrangement, 139, 141, 182, 183
(−)-Colchicine, 465
Coniceine, 553
Convolutamydine C, 560
Cope rearrangement, 558

Copper-carbenoids, 541
Copper salts, 6, 9–11, 91, 112, 281–282, 312, 319, 321
  2-alkyl(aryl)-4-oxazolecarboxylic acid esters, 9
  2-alkyl(aryl)-4-oxazolecarboxylic acid amides, 9
  2,5-diaryl-4-oxazoleacetic acids, 10
  2,4,5-trisubstituted oxazoles, 11
  calyculins, 312
  CuBr$_2$/DBU/HMTA, 9–11, 91, 112, 281–282, 312, 319, 321
  CuBr$_2$/LiBr/CaCO$_3$, 10–11
  hennoxazoles, 319, 321
  (−)-madumycin II, 281–282
  phorboxazole fragment, 10
  rhizoxin fragment, 10
Cornforth reaction, 73–78, 266, 267, 269, 309
Cornforth rearrangement, 185, 191, 192, *193*, 201, 361
Costaclavine, 569
Coumermycin analogues, 147
Crenatine, 104–105
Cross-Coupling reactions, transition-metal-catalyzed, 211–221, *212, 216, 218, 220*, 369
  2-(heteroaryl)oxazoles, 216
  2-alkenyl(aryl)-5-substituted oxazoles, *220*
  2-alkenyl(aryl)oxazoles, *220*
  2-alkyl(aryl)-4,5-diphenyloxazoles, 211, *212*
  2-alkyl(aryl)oxazoles, 211
  2-aryl-5-phenyloxazoles, *220*
  2-aryl(heteroaryl)-4-methyloxazoles, 216, *218*
  2-aryloxazoles, 216, *220*
  2-substituted-5-(3-methyl-5-isoxazolyl)oxazoles, 216–217
  2-substituted oxazoles, 217, 218
  2,5-diphenyl-4-substitued oxazoles, 369
  2,5-disubstituted oxazoles, 218
  4-(*E*)-alkenyl-2-phenyloxazoles, 214–215, *216*
  4-substituted oxazoles, 214
  vinyloxazoles, 214
  Ni(dppe)Cl$_2$-catalyzed, 211–212, *212*
    Grignard reagents, 211–212
  Palladium-catalyzed, 212–221, 369, 370
    2,4′-bis-oxazoles, 211, 212, 213
    2-(stannyl)oxazoles, 215, 216, 217, *218*
    bromooxazoles, 212, 213, 215, 217, 221, 369
    diazonamide, 215, 216, 217
    dimethylsulfomycinamate, 213–214
    organostannanes, 214, 370
    oxazole triflates, 213, 214, 215, *216*
    oxazolezinc reagents, 218–219, *220*, 221
    phorboxazole, 215, 216

Cumulenes, 55
Cyanomethylimidates, 75–76, 77
Cyclizations, oxazole synthesis, 51–62, 54, 58, 59, 61, 63, 125, 126, 126, 361
   2-alkyl(aryl)amino-4-cyano-5-methoxyoxazoles, 61
   2,4,5-trialkyloxazoles, 54
   2,4,5-trisubstituted oxazoles, 52, 59, 61
   2,5-dialkyloxazoles, 54
   2,5-diaryloxazoles, 361
   2,5-disubstituted-4-oxazolecarboxylic acid esters, 56–57
   2,5-disubstituted-4-oxazolecarboxylic acid ethyl esters, 58
   2,5-disubstituted oxazoles, 52, 59, 61
   4-aryl-2,5-dialkyloxazoles, 54
   5-[(phenylsulfonyl)methyl]-2-substituted oxazoles, 57, 59
   5-methyleneoxazoles, 53
   ***N*-(2,4-dialkyloxazol-5-yl)oxazolidinones**, 62
   1,2-bis(1H-benzotriazol-1-yl)-1,2-ethanediol, 125–126, 126
   internal diazonium salts, 125
   propargylic amides, 51–59
   azides, 59–62
Cycloaddition reactions, 163–183, 171, 172, 173, 174, 177, 179, 180, 252–254, 254, 348, 357, 418–436, 422–430, 436–455, 457, 459, 461, 463, 466
   [2 + 2], 152, 154, 241, 247, 461
   [2 + 3], 241, 243
   [3 + 2], 34, 128, 163, 168–174, 171, 172, 173, 174, 177, 178, 179, 180, 348, 357, 455–460, 461
   [4 + 2], 152, 165, 234, 443
   [4 + 3]-oxyallyl, 466
   [4 + 4], 165
   4-phenyl-1,2,4-triazoline-3,5-dione (PTAD), 169–171, 171, 457, 461
   activation energy, 420
      oxazole-olefin, 418
      oxazole-alkyne, 436
   5-alkoxyoxazole-acetylene, 441
   amino acids, 178, 183–184
      C-glycosyl, 183–184
      β-hydroxy, 178
   amino polyols, *erythro*, chiral, 179–181
   concerted process, 443
   D-AB1, 179, 180
   1-deoxynorjirimycin, 179, 181
   diethylazodicarboxylate (DEAD), 170–172, 172, 456–457, 461
   dioxetane intermediate, 461

1,3-dipolar, 28, 229, 369
1,5-dipolar, 191
electron transfer mechanism, 453
   electrochemical studies, 453
   electron acceptors, 453
   radical cations, 453
   single electron reduction, 453
exo-diene, 163
fullerene, 166, 169
FVP, 164
heterodienophiles, 168–181, 455–463, 461
intramolecular, 174–176, 430–436, 444–455, 462–463
intermolecular, 422, 418–430, 437–444, 455–462
inverse electron demand, 429, 453
mechanisms,
   dipolar, 455, 457
   diradical, 455
   nitrile ylide with PTAD, 457
   dipolar mechanism, 457
nitroso compounds, 171–172, 173, 457–458, 461
1,2,4-oxadiazoles, 171–172, 173, 457–458, 461
oxazole-alkyne, 436–455
   steric affects, 437
oxazole-isothiazolone, 425
oxazole-olefin, 418–436
   4-methyloxazole, 418
   regiochemistry, 424
oxazole as a dienophile, 422, 429
oxazolo-*o*-quinodimethane, 164–166
2-oxazolines, 175–180, 181, 177, 179, 457–459, 460, 461
oxazolones, 181–184
photocycloaddition, 252–254, 254
phytosphingosine, 179, 181
pyridines, 367, 369
pyrroles, 166–169
sealed tube, 429, 436, 438, 439, 440, 459
sphingosine, 179, 181
tetracyanoethylene (TCNE), 166–169
3-thiazolines, 171, 173, 174, 459–460, 461
thioaldehydes, 168, 171, 173, 174, 459–460, 461
1,2,4-triazolines, 169–171, 171, 172, 456–457, 461
Cyclodehydrating agents, 94–118, 185, 288, 290, 319, 332, 335, 336, 338, 350, 361, 362
   (Boc)$_2$O, 117–118
   (C$_6$H$_5$)$_3$P, Cl$_3$CCCl$_3$, (C$_2$H$_5$)$_3$N, 184, 332, 335, 336, 338
   (C$_6$H$_5$)$_3$P/Br$_2$, 109

Subject Index 605

(C₆H₅)₃P/BrCCl₂CCl₂Br, 319, 323, 350
(C₆H₅)₃P/CCl₄/DBU, 109, 116
(C₆H₅)₃P/I₂/(C₂H₅)₃N, 109–111, *112*, 114, 327, 329, 357
(C₆H₅)₃PCl₂, 360, 362
Acetyl chloride, 106
Cl₃CCOCl, 106
H₂SO₄, 94
P₂O₅, 94, 98–99, 109
PCl₅, 109
POCl₃, 94–97, 109
PPA, 94, 98–100, 109
PPE, 185
p-toluenesulfonic acid, 94, 100, 288, 290
SOCl₂, 94, 96–98, 109
TFAA and TFAA/TFA (cat), 94, 101–105, *107*, 108
Cyclooxygenase 2 inhibitor, 365
Cyclohexenones, 494
Cyclopentenone, 428
Cyclopropanation, 558
Cysteine protease inhibitors, 56
Cytotoxic activity/agents, 256, 258, 261, 283, 292, 324

D-AB1, 179, 180
Dakin-West reaction, 94, 106, 357, 485, 488
DAST, 274–275, 324, 357
DBN, 432, 434, 435
DBU, 436, 439
DDQ, 453
DEAD, 170–172, *172*, 456, 457
Decomposition, 421, 422
    Diels-Alder adducts, acid catalyzed, 422
    oxazoles, oxidative, 421
4,5-Dehydrotropone, 444
Deoxo-fluor, 322, 357
1-Deoxynorjirimycin, 179, 181
Dendroamide A, 257, 258
Dess-Martin periodinane, 110, 111, *112*, 113, 114, 207, 271, 276, 277, 288, 290, 293, 319, 339, 340, 346, 350, 351, 357
Deuteration, oxazole reactions, 128
    2-deuterooxazole, 128
Deuterioformylation, 161
Deutero-L-alanine, 423
α-Diazo compounds, 27–39, *30, 31, 34, 37, 39*, 245, 264, 273, 324, 327, 328, 348, 359, 360, 538, 541, 543, 547–567
    acetates, 36
    acetoacetamides, 543
    acetoacetates, 33, 359
    acetoacetyl urea, 543, 556

acetophenones, 28–29, 38, 264
acetylindole, 32, 326
carbonyl compounds, 27, 29, 324, 327, 538, 541, 543, 549, 551, 554
ethyl diazobenzoylacetate, 28
ethyl formyldiazoacetate, 273
ethyl phenylsulfonyldiazoacetate, 30
ethyl (trimethylsilyl)diazoacetate, 36–37
ethylmalonylchloride, 551
esters, 36, 38, *39*, 557
formylacetates, 273
imides, 541, 542, 547–567
ketones, 34, 37–39, *37, 39*, 245
    α-(trimethylsilyl), 34–36, *37*
    polymer bound, 34–36
    solid-phase synthesis, 34–36
ketoesters, 38, 328, 360
    polymer bound, 360
malonates, 27, 31, 33, *34*, 348, 549, 551
methyl cyanodiazoacetate, 31
nitriles, 31
phosphonates, 32
propanoates, 359, 361
propionates, 28
sulfones, 30–31, *30, 31*
Diazonamide(s), 30, 32, 106, 215, 216, 217, 260, 261, 263, 323–344, 345, 346
    Diazonamides A and B, 323–344
    2-chloroindole, 328, 330
    2-methyl-5-oxazolecarbonylchloride, 329, 331
    2-vinyloxazole, 336, 338
    3-(oxazol-5-yl)indoles, 324
    3,4-bridged indole, 326
    4-chloro-2,5-disubstituted oxazoles, 340, 341
    4-chlorooxazole, 335, 337
    4-cyanooxazole, 341, 343
    5-bromo-2-phenyloxazole, 328, 331
    (C₆H₅)₃P, Cl₃CCCl₃, (C₂H₅)₃N, 332, 335, 336, 338
    (C₆H₅)₃P, I₂, (C₂H₅)₃N, 327, 329
    N-acylamino ketone, 329, 335, 344
    aromatic acyl cyanides, 340, 341
    atropisomers, 335, 340
        NMR, 335
    benzofuranyloxazole, 328
    Chan-type rearrangement of imides, 330, 333
    chlorodecarboxylation, 331, 334–335, 336
    DDQ, 327, 331, 333, 334, 335, 336, 338, 344
    Doyle's protocol, 328, 331
    fragments,
        A-C-D-E-F-G-H-I, 344, 346
        A-F-G-I, 340, 342, 343, 344

Diazonamide(s) (*Continued*)
  B-C-D-E-G-H-I, 335, 329, 331
  B-C-D-F-G, 327–328
  B-C-D-F-G-H, 327, 329, 331
  C-D-E, 324, 326, 328, 329, 330, 331, 332, 333, 334, 335, 337, 338, 340, 341
  C-D-E-F, 326, 329, 330, 331, 333, 334–335, 336
  C-D-E-G, 335, 337
  C-D-E-G-H, 326, 329, 330, 332
  C-D-E-G-H-I, 336, 338
  C-D-E-G-I, 334, 335, 337
  C-D-F-G, 327
  C-D-G, 327, 328, 329
  F-G-H, 328
  Hantzch cyclization, 328, 331
  Heck coupling, 328, 329, 341, 343
  Horner-Wadsworth-Emmons reaction, 336, 338, 339, 340
  IBX, 335, 338
  indole bis-oxazole, 326–327, 329, 330, 331, 333
    dichloro, 330, 332, 334, 335, 336
  iodoindolyloxazole, 329, 332
  lithiomethylisocyanide, 334, 337
  macrocyclization, 339
  macrolactamization, 327
  Mosher amides, 330
  rhodium carbene methodology, 324, 326–329, 330
  2-chloroindole, 328, 330
  α-diazo-β-ketoester, 328, 329, 330
  α-diazoacetylindole, 326
  α-diazocarbonyl compounds, 324, 327
  Schöllkopf methodology, 334, 337
  Stille coupling, 328, 329, 335, 341, 343
  Suzuki coupling, 328, 335, 337
  triaryl acetaldehyde, 343, 344, 345
  Ullman coupling, 329
Diazonium salts, oxazole reactions, 125, 136–138, *138*
  4-alkyl(aryl)-2-amino(acetamido)-5-(arylazo)oxazoles, *138*
  pyrazolo[3,4-*d*]oxazoles, 125
  azo dyes, 136–138, *138*
  aryl diazonium tetrafluoroborates, 136, 137, 138, *138*
  internal, 125
Dichloroacetonitrile, 76–77, 364–365
Diels-Alder, 28, 121, 127, 163, 165, 166, 167, 174, 175, 182, 231, 232, 234, 235, 252, 367, 370, 371, 372, 417–472, *431*, *445*, *452*, *461*

adducts (cycloadducts), 174, 175, 252, 418, 419, 420, 422, 424, 425, 426, 428, 429, 435, 437, 442, 443, 456, 459, 460, 462
  bis(benzyne), 443
  dehydroformylation, 428
  endo, 419, 425, 426, 429
  2,5-endoperoxide, 252, 460
  exo, 419, 425, 426, 429
  isolable, 442
azabutadiene-olefin, 420
  forming bond lengths, 420
  transition state charge transfer, 420
Lewis acid catalysis, 428, 432, 433, 451, 458, 463
mechanism, 420–421, 453, 455, 457, 459, 460
oxazole-alkyne, 436–455, *445*, 465–467
  intermolecular, 437–444
  intramolecular, 444–455
oxazole-heterodieneophile, 455–463, *461*
  chiral catalysts, 459
  computational study, 437
  intermolecular, 455–462
  intramolecular, 462–463
oxazole-olefin, 418–436, *431*, 463–465
  intermolecular, 418–430
  intramolecular, 430–436
oxazolium salts, 231–233, 234, 235–236, 369, 370, 467
reactions
  intermolecular, 437–444, 418–430, 455–462
  intramolecular, 418–430, 444–455, 462–463
  rate, 422, 426, 437, 464, 465
regiochemistry, 420–421, 424, 425, 427–428, 432, 433, 434–435, 436–437, 438, 448, 451, 458, 459, 463, 464–465
  Brønstead acids, 438
  electronic or substituent affects, 420–421, 424, 425, 427–428, 434–435, 436–437, 448, 459, 464–465
  Lewis acids, 428, 432, 433, 451, 458, 459, 463, 464
  pressure affects, 426–427, 458
  solvent affects, 432, 433
  steric affects, 421
Diels-Alder retro-Diels-Alder sequence, 436, 437, 438, 443–444, 444–455
  furans from oxazoles and alkynes, 436, 437
  furans from furans and alkynes, 438
  intermolecular, 443–445
  intramolecular
    bis-heteroannulation, 444
    oxazole-alkyne, 444–455
Dielsiquinone, 569

Subject Index

Dienamines, 494
Dienes, 418, 419, 420, 422, 424, 425, 426, 427, 428, 429, 430, 435, 436, 437, 438, 439, 440, 441, 442, 443, 444, 456, 457, 458, 459
  1,1-dimethoxy-tetrachlorocyclopentadiene, 429
  2,3-dimethyl-1,3-butadiene, 429
  2-aminooxazoles, 427, 429, 442
  2-aryl-substituted 4-phenyloxazole, 440
  3,6-bis(trifluoromethyl)-1,2,4,5-tetrazine, 430
  4-(4-methoxyphenyl)oxazole, 443
  4-(4-nitrophenyl)oxazoles, 426, 443
  4-methyloxazole, 418, 424, 440
  4-phenyloxazole, 425, 437, 438, 439, 440, 443, 444
  5-alkoxyoxazoles, 426, 427, 436, 441, 457, 458, 459
  5-ethoxy-4-methyloxazole, 419, 425, 428, 441
  alkoxyoxazoles, 426, 427, 436, 441, 457, 458, 459
  all-carbon, 422, 429, 437
  azadienes, 420, 429, 435
  o-chloranil, 429
  cyclohexadiene, 429
  ethoxyoxazoles, 419, 441, 456, 458
  hexachlorocyclopentadiene, 429
  N-methylmaleimide, 443
  quinones, 426
Dienophiles, 418, 419, 420, 421, 423, 424, 425, 426, 427, 428, 429, 432, 434, 436, 437, 438, 439, 440, 441, 442, 443, 447, 448, 449, 451, 455–463, 466, 535
  acetylenic, 447
    2-butyn-1,4-diol diacetate, 440
    3,3,3-trifluoropropyne, 440
    acetylenic diimidazole, 437
    activated alkynes, 437
    bis(tributylstannyl)acetylene, 440
    bis(trimethylsilyl)acetylene, 438
    diethylacetylene, 437
    dimethylacetylene dicarboxylate, 442, 466
    ethyl 2-butynoate, 438
    ethyl phenyl-prop-1-ynoate, 440, 441
    trimethylsilyl groups, 448
    trimethylsilylacetylene, 439
    trimethylsilylpropyne, 439
    unactivated alkynes, 451
  acrylonitrile, 421, 428
  aldehydes, 458, 459
    α-benzyloxy aldehydes, 459
    2-(*S*)-(benzyloxy)propanal, 459
    cinnamaldehyde, 459
    propanal, 459
  alkoxy dihydrofuran, 423
  allyl *t*-butyl ether, 421
  benzyne, 443
  cyclohexenones, 428
  cyclopentenone, 428
  diethyl fumarate, 419, 535
  diethyl maleate, 427, 535
  cis-2,5-dimethoxy-2,5-dihydrofurans, 419, 423
  dimethyl fumarate, 426, 427, 535
  dimethyl maleate, 426, 535
  ethyl acrylate, 535
  heterodienophiles, 455–463
  isothiazolone, 425
  maleic acid, 425
  maleic anhydride, 418, 419, 427, 449
  mono-activated olefins, 432
  nitro olefins, 424
  4-nitrooxazoles, 429
  oxazoles, 429
  N-phenylmaleimide, 424, 426
  singlet oxygen, 460
  sulfonyl alkenes, 424, 425
Diethoxyphosphonoglycine amides, 146, *147*
Diethylazodicarboxylate (DEAD), 170–172, *172*, 456, 457
Diethyloxomalonate (diethyl ketomalonate), 175–176, *177*, 457, 458
Diimidazole copper ligands, 437
Diimidazolo-[3^2]metacyclophanes, 190–191
Dimethylsulfomycinate, 42, 213–214
1,4,2-Dioxazole, 363, 365
1,3-Dipolar cycloaddition reactions, 28, 229, 234, 494, 499, 500, *501*, 502, 502, *504*, *506*, 507, *510*, *512*, 514, *514*, *515*, *516*, 516, *518*, *521*, 522, *529*, 533, *533*, 535, 537, 538, 544, 547, 554, 559
  Intramolecular, 507, 508, 509, 526, 541, 561, 563–565
Dipolarophiles,
  1,3,5,7-cyclooctatetraene, 526
  1,4-naphthoquinone, 535, 554
  1,5-cyclooctadiene, 526
  2-chloroacrylonitrile, 510
  acetylenic, 502, 505, 506, *506*, 507, 508, 509
  acrylonitrile, 498
  alkynes, 557
  arynes, 509
  azetes, 534, 548
  benzazetes, 531
  benzocyclobutadiene, 523
  benzoquinone, 535
  cyclobutenes, 523, 524, 539
  α-cyanocinnamonitrile, 511, *512*

Dipolarophiles (Continued)
diazonium salts, 529
diethyl ketene acetal, 550
dimethyl maleate, 499, 539
dimethyl acetylenedicarboxylate, 241, 475, 476, 478–480, 495, 496, 497, 499, 500, 505, 509, 539, 542, 543, 545, 547, 550, 554, 556, 557, 563
diphenylacetylene, 523
enol ethers, 559
ethyl acrylate, 241, 535, 539, 559
ethyl propiolate, 241, 475, 509
ethyl vinyl ketone, 498
fulvenes, 525
fumaronitrile, 510
isocyanates, 561
maleic anhydride, 499, 544
maleimide, 367, 369
methyl acrylate, 544
methyl propiolate, 241, 475, 497, 498, 500, *501*, 501, 502, 503, *504*, 504, 543, 545, 551, 554, 557
methyl vinyl ketone, 498, 550, 551, 553, 556
methyl α-cyanocinnamate, 511, *512*
nitroalkenes, 519, 520
nitroarenes, 521
nitrobenzofuran, 522
nitroheterocycles, 521, 522
nitroindoles, 521, *521*
nitrosobenzene, 532, 533
N-methylmaleimide, 510, 556
N-phenylmaleimide, 241, 476, 507, 509, 510, 538, 539, 543, 544, 550, 551, 552, 554, 563
olefinic, 509, 513, 539, 551
oxygen, 527
phenyl vinyl sulfone, 241, 545
phosphaalkenes, 533
phosphaalkynes, 533, *533*, 549
thiocarbonyls, 476
thioisocyanates, 561
tropone, 525
vinyl phosphonium salts, 513, 514, *514*, *515*
Disubstituted oxazoles, 393–395, 397
^{1}H NMR, *393–395*
^{13}C NMR, *397*
DMAD, 28, 74, 232, 234, 235, 241, 242, *243*
DNA gyrase inhibitor, 262
Doyle's protocol, 328, 331

Electrophilic reactions, oxazole reactions, 128–138, *134*, *136*, *137*, *138*, *139*
 2-(acylamino)-4-methyloxazoles, 137

2-alkyl-5-(diacylamino)-4-methyloxazoles, 138
2-amino-5-(hydroxymethyl)-4-methyloxazoles, *137*
4-alkyl(aryl)-2-amino(acetamido)-5-(arylazo)oxazoles, *138*
4-alkyl(aryl)-2-amino-5-oxazolethiocarboxamides, 135, *136*
4-acyloxazoles, 134
4-aroyl-2-(arylthio)-5-ethoxyoxazoles, 134, *134*
5-(acylamino)-2-alkyl-4-methyloxazoles, 138
5-(diacylamino)-4-methyloxazoles, 138
5-acyl-2-amino-4-methyloxazoles, 137
5-acyl-4-alkyl-2-aminooxazoles, *139*
pyrazolo[3,4-*d*]oxazoles, 125
acetylation, 131, 137, 138
acylation, 131–132
alkylation, 138, 139
aroylation, 134, *134*, 137
bromination, 129–130, *131*
deuteration, 128
diazonium salts, 136–137, *138*
diazotization, 136–137
formylation, 130–133
Friedel-Crafts reaction, 137–138, *139*
hydroxymethylation, 135–136, *137*
isothiocyanates, 135, *136*
nitration, 130
Ellipticine, 421
Enamino esters, 62–63, *64*
acyclic, 63
cyclic, 62–63, *64*
Ergolines, 87–88
Ergot alkaloids, 569
Erythrinane alkaloids, 564
ESR spectroscopy, 409
argon matrices, 409
dihydrooxazole radicals, 409
Ethambutol, 231
Ethyl acrylate, 212, 241
Ethyl isocyanoacetate, 78–79, 88, 134
synthesis of oxazole, 88
5-alkoxy-2-(arylthio)oxazoles, 78
Eupolauramine, 5, 431, 432
Evodone, 444, 446

Flash-lamp pumped laser dyes, 49, 95
Flash vacuum pyrolysis (FVP), *20*, 20–21, 23–24, 120, 121, 164–165, 182
Fluorescence microscopy stains, 96
^{19}F NMR, 401, *401*, *402*, *403*
Formylation, 130–133, 159–162, *161*

N-formylformamide, 159–162, *161*
  of oxazoles, 130–133
    regiochemistry, 130, 132
    MO calculations, 132
    ring vs. substituent, 131–132
    Vilsmeier-Haack, 132, 133
  via ozonolysis of oxazole, 159–162
Frangulanine, 100
Fredericamycin, 441
Friedel-Crafts reaction, 137–138, *139*, 438
  2-(acylamino)-4-methyloxazoles, 137
  2-alkyl-5-(diacylamino)-4-methyloxazoles, 138
  5-(acylamino)-2-alkyl-4-methyloxazoles, 138
  5-(diacylamino)-4-methyloxazoles, 138
  5-acyl-2-amino-4-methyloxazoles, 137
  5-acyl-4-alkyl-2-aminooxazoles, *139*
  acylation, 438
  low temperature acetylation, aroylation, 137
  Lewis acid, 137
Fullerene(s), 166, 169, 526, 552
Fulvenes, 525, 548
Fungicides, 170
Furanosesquiterpenes, 447
Furanoterpenes, 444
Furanothiopyran, 453
Furans, 418, 422, 436, 437, 438, 439, 440, 441, 443, 446, 449, 451, 452, 453, 463, 467, 542, 543, 547, 550, 558
  2,4-disubstituted, 439
  2-ethoxy, 436, 441
  3,4-bis(trimethylsilyl), 438
  3,4-diethyl, 437
  3,4-disubstituted, 437, 439
  ethoxy, 449
  fused to 5- and 7-membered rings, 453
  highly substituted, 418, 441, 463
  isobenzo, 443
  metalation of, 439
  methoxy, 446
  mono-silylated, 439
  rosefuran, 439
  stannyl, 440
  synthesis, 542, 543, 547, 550, 558
  thio, 453
  tetrahydro, 422
  tricyclic, 451, 452, 453
  trifluoromethyl, 440
  trisubstituted, 440, 442, 467
Furo[3,4-*d*]tropones, 444

G. aspera Harv., 449
Garner's acid, 354, 355

Geigerin, 449, 450
Ghosez's reagent, 289
β-Glucosidase inhibitors, 497
Glycinates, 493
  ethyl, 493
  N-methyl, 493
Gnidia latifolia, 447
Gnididione, 447
Gnididione ketal, 447
Grigg, 436, 455
(−)-Griseoviridin, 282–283, 284, 285
Griseoviridin, 257, 259
Group A streptogramin antibiotics, oxazole natural products, 257, 259, 264–283, 284, 285
  (−)-griseoviridin, 282–283, 284, 285
  (−)-madumycin II, 278–282
  (−)-virginiamycin M$_2$, 275–277
  14,15-anhydropristinamycin II$_B$, 268–271
  griseoviridin, 257, 259
  madumycin, 207
  madumycin I, 272–274
  madumycin II, 257, 259
  virginiamycin M$_2$, 257, 259, 268, 270
  2-methyloxazoles, regioselective functionalization, 265–269
    2-(phenylsulfonylmethyl)-4-oxazolecarboxylic acid *t*-butyl ester, 265–266, 268, 270
    2-methyl-4-oxazolecarboxylic acid ester, 265, 267–268
    2-methyl-5-(trimethylsilyl)-4-oxazolecarboxylic acid, 265–266
    4-(hydroxymethyl)-2-methyloxazole, 269, 271
    Cornforth reaction, 266, 267–268, 269
    lateral lithiation, 267–268, 269, 271
  Burgess reagent, 278–279, 281–282
  CuBr/Cu(OAc)$_2$/*t*-butylperbenzoate, 278–279
  CuBr$_2$/DBU/HMTA, 281–282
  DAST, 274–275
  lactoxime *O*-vinylethers, 268, 270
  macrocyclization, 271
  macrolactamization, 272, 273, 277, 279
  macrolactonization, 279, 282
  oxazole Reformatsky-type reagents, 272–273, 277
  N-acylamino acid ester, 274–275
  NiO$_2$ oxidation, 274–275
  ring-closing olefin metathesis, 282, 285
  Stille coupling, 268
Grubbs's catalyst, 283, 285
GW475151, 76, 365

^{1}H NMR, 3, *4*, 53, 130, 152, 154–155, 193, 221, 227, 231, 234, 335, 392, *393, 394, 395, 396*, 461, 482, 506, 513, 546
  2-(trimethylsilyl)oxazole and formaldehyde, 227
  2-lithiooxazole and isocyanovinyllithium alkoxide equilibrium, 193
  2,5-oxazole endoperoxide, 152, 154–155, 461
  allene intermediates, 53
  atropisomers of diazonamide, 335
  bromination of oxazoles, 130
  oxazole, 3, 392
  oxazolium salts, 231, 234
    bis-Reissert salt, tautomer, 231
    trimethylsilyl enol ether of isocyanovinylalkoxide, 221
Halichondramide(s), 16, 248, 262, 345, 348
Halishigamides, 345
Halogenation of oxazoles, 129–130, *131*, 199, *199*, 331, 332, 334–335, 336, 367
  bromination, 129–130, *131*
    aromatic substitution vs. addition, 129
    ring opened products, 129–130, *131*
    4,5- vs. 2,5- addition products, 129–130
    2-oxazolines, 129–131, *131*
    3-oxazolines, 129–131, *131*
  chlorination, 331, 332, 334–335, 336, 337, 367
    chlorodecarboxylation, 331, 334–335, 336
    NCS, 332, 334–335, 336, 337
    side chain, 367
  iodination, 199, *199*
    2-lithiooxazoles, 199
Hantzch reacton, 106, 348–351
Harmonic oscillator model of aromaticity (HOMA), 357
Hassner, 457, 462
Hennoxazole A, 260
Hennoxazoles, oxazole natural products, 260, 261, 316–323, 324, 325
  $(C_6H_5)_3P$, $BrCl_2CCCl_2Br$, 319, 323
  2,4'-bis-oxazoles, 316, 317, 318, 322
  2-(5-bromo-3-penten-1-yl)-4-oxazolecarboxylic acid methyl ester, 318–319
  2-(stannyl)oxazole, 318
  2-lithiooxazole, 318
  2-phenyl-4-(trimethylstannyl)oxazole, 317–318
  2-styryl-4-oxazolecarboxylic acid methyl ester, 320–321
  4-acetyl-2-styryloxazole, 321–322
  Burgess reagent, 319
  $CBrCl_3$/DBU, 322, 324
  $CuBr_2$/DBU/HMTA, 319

DAST, 324
Deoxo-fluor, 322
Dess-Martin periodinane, 319
  dithiane substituted oxazole, 320–321
  Robinson-Gabriel methodology, 317–318
  Stille coupling, 317, 318
Herbicides, 27
Herpes simplex virus (HSV-1) inhibitor, 260, 316
Heterocycles from oxazole rearrangements, oxazoles reactions, 183–191, 367–368, 371
  1,3,4-oxadiazoles, 367–368
  1,4-bis-(imidazol-5-yl)benzenes, 189–190
  Cornforth rearrangement, 191, 192
  diimidazolo-[3^2]metacyclophanes, 190–191
  α-diketo amidine, 188
  hydantoins, 191, 192
  imidazoles, 187–189, *189*
  molecular orbital calculations, 191
  pyrazoles, 185, 187
  pyrazolones, 185
  pyrimidines, 187
  pyrimidinols, 187–189, *189*
  pyrroles, 185–186, *186*
    β-amino, 185
  tetrahydrofuran-2-ones, 371
  thiazoles, 191–192, *193*
  Vilsmeier-Haack reaction, 185
Hetero-Cope rearrangement, 141
Heterodienophiles, 168–181, *171, 172, 174, 179*, 455–463, *461*
  C=N, 174–175, 176, 457, 462–463
    hydantoin, 457
    imine, 174–175, 176, 462–463
  C=O, 168, 175–176, *177*, 178–179, *179*, 180–181, 457–459, 460, *461*
    aldehydes, 178–179, 180–181, 458, 459
    α-benzyloxy aldehydes, 459
    2-(S)-(benzyloxy)propanal, 459
    diethyloxomalonate (diethyl ketomalonate), 175–176, *177*, 457, 458
  C=S, 168, 171, 173, *174*, 459–460, *461*
    intramolecular reactions, 174–175, 176, 462–463
  N=N, 168–172, 171, 456, 457, *461*
    DEAD, 170–172, 456, 457
    PTAD, 169–171, 457
  N=O, 168, 171–172, 457
Hippadine, 453
HIV-1 inhibitor, 258, 283
Holographic media, 152
5-HT$_{1A}$-Serotonin agonists, 87
Human neutrophil elastase inhibitor, 76–77

Hydantoins, 191, 192, 457
Hydrolysis, oxazole reactions, 146–149, *147*
  alcoholysis, 148–149, 150
  coumermycin analogues, 147
  diethoxyphosphonoglycine amides, 146, *147*
  masked dipeptides, 148–149
  masked α-amino ketone, 147
  mugineic acid, 147
  oxazolocoumarins 146, 147
β-Hydroxyamides, 8, *112*, 121, 123, 357
  cyclodehydration with
    DAST, 357
    Deoxo-fluor, 357
    diphenylsulfoxonium triflate, 8
    thermal, 121, 123
  precursors to *N*-acylamino ketones, 112
Hydroxymethylation, oxazole reactions, 135–136, *137*
  2-amino-5-(hydroxymethyl)-4-methyloxazoles, *137*
  aliphatic aldehydes, 136
  aromatic aldehydes, 135–136
  Schiff base, 135
  x-ray crystal structure, 135
Hydroxypyridines, 420, 421, 422, 423–427, 428, 432, 433
Hypoglycemic agent, 94, 104
Hypolipemic agents, 94
Hypolipidemic agents, 92, 185

Ibata, 457, 458
Icthyotoxic activity/agents, 255, 345
Ifetroban sodium, 112, 114–115
Imerubrine, 466
Imidates/Thioimidates-Cornforth reaction, 73–78, *75*, *77*, 363, 365
  2-alkyl-4-oxazolecarboxylic acid methyl esters, 74
  2-substituted-4,5-oxazoledicarboxylic acid dimethyl esters, 73–75, *75*
  2,5-disubstituted oxazoles, 75, 76, *77*
  4-oxazolecarboxylic acid esters, 73
  calyculin A intermediate, 74
  cyanomethylimidates, 75–76, *77*
  dichloroacetonitrile, 76–77
  C-formylated imidates, 73
  β-ketoimidates, 75–76
  lactoxime *O*-vinyl ethers, thermal cyclization, 73–74, *75*, 363, 365
  organozinc reagents, 75–76, *77*
Imidazo[2,1-*b*][1,2,4]triazines, 238
Imidazo[1,2-*a*]imidazoles, 238
Imidazo[1,2-*a*]pyridines, 238, 240

Imidazo[5,1-*b*]oxazoles, 45–47, *47*
Imidazoles, 187–189, *189*, 238, 491, *491*, 492, *492*, 529, *529*, 530, 531
Imidazolines, 463
Imidazolinones, 238
Imidazopyridines, 497
Imides, synthesis, 527
Imidosulfoxides, 543, 569
Imines, 482, 483, *483*, 514, *516*, 529, *529*, 530, 531, 534
  N-alkyl, 514
  N-aryl, 514, *516*, 534
  N-tosyl, 483, 514, 529, *529*, 530, 531
  α,β-unsaturated, 482, *483*
Iminium ions, 565, 567
  N-acyl, 565, 567
Iminium salts, 467
Iminophosphoranes, 56, 59, 60
Immunomodulating agents, 183
IMPDH antagonist, 363
Indeno[2,3-*a*]isoquinolines, 507
Indoles, 490, 508, 532
Indolizidine alkaloids, 553
Integressine, 104–105
Iodobenzene diacetate, 44–45, *46*, 462
Iodonium salts, 34–35, *36*
  benzoxazolones, *36*
Ipalbidine, 552
Ishikawa, 418
Isoclavans, 247
Isocyanides, 78–89, *79*, *83*, *85*, *87*, 334, 337, 361, 362, 363–364, 370–372
  2-(aminomethyl)-4-benzyl-5-morpholinooxazoles, 371–372
  2-acyl-5-alkoxyoxazoles, 78
  2-acyl-5-ethoxyoxazoles, *79*
  2,4-diaryl-5-oxazolecarboxamides, *83*
  2,4,5-trisubstituted oxazoles, 371
  5-(aminomethyl)oxazoles, *N*-Boc, chiral, 80
  5-(aminomethyl)oxazoles, *N*-protected, chiral, 80
  5-(dialkylamino)-2,4-disubstituted oxazoles, 84, *87*
  5-(*N*-alkyl-*N*-phenylamino)-2-(arylthio)oxazoles, *79*
  5-(*N*-alkyl-*N*-phenylamino)-2,4-(diarylthio)oxazoles, *79*
  5-(substituted 4-nitrophenyl)oxazoles, 363
  5-alkoxy-2-(arylthio)oxazoles, 78
  5-aryl(heteroaryl)-4-oxazolecarboxylic acid methyl esters, 80
  5-aryl(heteroaryl)oxazoles, 89
  5-aryloxazoles, *87*, 88

Isocyanides (*Continued*)
  5-substituted-4-tosyloxazoles, 87
  5-substituted 4-oxazolecarboxylic acid esters, 82, *83*
  5-substituted oxazoles, 87, 361–362, 364
  N-cyclohexyl-2,4-diaryl-5-oxazolecarboxamides, 84, *85*
  arylsulfenyl chlorides, 78–79
  combinatorial libraries, 371
  desulfonylation of tosyloxazole, 87–88, 90
  dipeptide, 372
  ethyl isocyanoacetate, 78, 88
  β-hydroxy-α-amino acid synthons, 82
  isocyanoacetamides, 371–373
  isocyanoacetic acid esters and α-alkoxyacids, 82–83, *83*
  isothiocarbamoyl chlorides, 78–79
  α-ketoimidoyl chlorides, 78, *79*
  lithiated isocyanides, 80–81
  lithiomethylisocyanide, 81, 82, 334, 337, 370
  macrocyclic depsipeptides, 371–372
  methyl isocyanoacetate, 80, 82, 84
  N,N-dialkylisocyanoacetamides, 84, *87*
  N-alkylisocyanoacetanilides, 79
  N-tosylimines, 85, *87*
  Passerini reaction, 83–84, *85*
  phthaloxazolin, 362, 364
  TosMIC, 78, 84, 86–90, 361, 362, 363–364
    polystyrene, 88, 90
    ROMPgel, 361, 364
    tentagel, 88
  Ugi reaction, 371–373
α-Isocyanoesters, 227
Isognididone, 447
Isoindoles, 428
Isoquinolines, 426, 441, 443, 467, 487, *488,* 507, 513
Isoquinolinequinones, 426
Isoquinolones, 528
  tetrahydro, 528
Isothiazole, 1,1-dioxide, 517, *518,* 519
Isothiazolones, 425
Isothiocyanates, 135, *136*
  4-alkyl(aryl)-2-amino-5-oxazolethiocarboxamides, 135, *136*
  alkyl- and arylthioureas, 135
Isoxazoles, 20–22, *22,* 25–27, 62, 517, 521
  2-(alkylamino)-4-cyanooxazoles, 21, *22*
  2-(aroylamino)-5-methyloxazoles, 26
  2,5-diaryloxazoles, 27
  5-alkyl-2-(alkylamino)-4-cyanooxazoles, 21, *22*

1-azirine, 21
Beckman rearrangement, base catalyzed, 25–26
C-aroylamino azirine, 27
FVP, 20
nitrile ylide, 21
photolysis, 27
thioureas, 21–22, *22*
Isoxazolones, benzylidene, 517

Jacobi, 444, 446, 448, 449, 450
JandaJel™, 360, 362
Jones oxidation, 295

Kabiramide C, 201, 206, 348, 350
Kabiramides, 248, 262–263, 345
Keramamides, 257, 258
Ketenes, 224, 478, 479, 482, 493, 494, 531, 534
  tautomers, 482, 494, 531, 534
Ketones, 34, 37–39, *39,* 41–51, *42, 45, 46, 47, 50,* 65–68, 68, 92–94, *93,* 174, 207, *209,* 237, 245, 249, 361, 363, 366, 437, 446, 447, 448, 449
  2-alkyl(aryl)-5-fluoro-4-(trifluoromethyl)oxazoles, 47
  2-alkyl-5-aryloxazoles, 43, 44, *45, 46*
  2-(alkylthio)steroidal oxazoles, 51
  2-amino-4-substituted oxazoles, 41, 50
  2-aryl-5-(4-hydroxyphenyl)oxazoles, 36, *37*
  2-aryl-5-methyl-4-oxazoleacetic acid methyl esters, 42
  2-phenyl-4-substituted oxazoles, *42*
  2,4-diaryloxazoles, 92, *93*
  2,4-disubstituted oxazoles, 41
  2,4,5-trisubstituted oxazoles, 49–50, *50*
  4,5-disubstituted 2-phenyloxazoles, *42*
  5-alkyl-2,4-diaryloxazoles, 39
  5-aryl-2-(chloromethyl)oxazoles, 38, *39*
  5-aryl-4-oxazolecarboxaldehydes, 45
  imidazo[5,1-*b*]oxazoles, 45–46, *47*
  oxazolo[5,4-*c*]pyrazoles, 50
  oxazolo[5,4-*c*]pyridazines, 50
  steroidal oxazoles, 69
  tris-oxazoles, 43
  acetylenic, 437, 446, 448, 449
  aromatic, 39–41, *42,* 43–45, *45, 46,* 68–69
  azidoacetophenones, 45, 48
  desoxybenzoins, 361, 363
  dimethylsulfomycinate, 42
  furano, 448, 449
  α-hydroximino, 68–69
  iodobenzene diacetate/trifluoromethanesulfonic acid, 44–45, *46*

Subject Index 613

kinetic deprotonation, 447
α-(methoxyimino), 65–68, *68*
α-methyl, 43–45, *45*, *46*
oxidation of with
 iron solvates of nitriles, 49–50, *50*
 copper(II) triflate, 49–50
phenacyl halides, 45, 50, 92
3,3,3-trifluoroalanine, 47–48
α-acyloxyketones, 92
microwave accelerated reaction with urea, 50
reaction with
 oxazole Reformatsky reagent, 207, *209*
 Vilsmeier reagent, 45, 48
Ritter-like reaction, 361, 363
steroidal oxazoles from reaction with KOCN, 50–51, 52
α-substituted, 34, *34*, 39, 41–51, 92–94, 174, 237, 245, 366
 acetoxy, 49–51, 366
 acyloxy, 92–94, *93*, 366
 azido, 45, 48, 60
 diazo, 34, *34*, 37–38, *39*, 245
 halo, 42–44, 45, 50–51, *47*, 249
 phenacyl-2-pyridone, 237
 sulfides, 174
 triflates, 43–45, *45*, *46*
thallium triflate, 43–44, *45*, 361
tris-oxazoles, 43
Kharash-Sosnovsky reaction, 12–13
2-cyclohexyl-4-oxazolecarboxylic acid esters, 12
2-isopropyl-4-oxazolecarboxylic acid esters, 12
2-substituted-4-oxazolecarboxylic acid esters, 13
Cu(I)/Cu(II)-*tert* butylperbenzoate, 12–13, 114, 278–279, 285, 287
ulapualide A intermediate, 13
Kondrat'eva, 367, 418, 427, 431, 436
Kondrat'eva pyridine synthesis
 intramolecular, 431, 434, 435
 intermolecular, 418, 427
Kozikowski, 421

Lactams, 497, 531, 534, 539
β-, synthesis, 531, 534
glycine, 497
N-(chloroacetyl), 539
poly(oxyvinylene), 539
Lawesson's reagent, 191
Lactoxime *O*-vinyl ethers, 73–74, *75*, 268, 270, 363, 365
Levin, 430

Lewis acids, 38, 115, 175, *177*, 178–181, *179*, 227–228, 428, 432, 433, 451, 458, 459, 463
 aldol reactions, 227–228
 amide acetal cyclization, 115
 catalyst for α-diazoacetophenones with nitriles, 38
 Diels-Alder catalysts, 428, 432, 433, 451, 458, 463
 oxazoles and heterodienophiles, 175, *177*, 178–181, *179*, 458, 459, 463
Leukotriene A-E fragments, 158–159, 160
Lignans, 440, 466
Ligularone, 446
Lipoxygenase inhibitors, 43
Lithiation, 87, 90, 191–203, *198*, *199*, *201*, *203*, 215, 221, 265, 266–267, 269, 271
2-iodo-5-substituted oxazoles, 199, *199*
2-substituted oxazoles, 197–198, *198*
2,2′-bis-oxazoles, 198
2,4-diiodo-5-substituted oxazoles, 199
2,5-disubstituted oxazoles, *198*
4-(methoxymethyl)-2′-methyl-5-substiuted-2,4′-bis-oxazoles, 200, *203*
4-iodo-5-substituted oxazoles, 199, *199*
4-iodooxazoles, 199
5-(1,3-dithian-2-yl)oxazoles, 201, 203, *204*, 205
5-substituted oxazoles, 200, *201*
bis-oxazolyl methanols, 195
2-(lithiomethyl)oxazole, 202, 206–207
2-lithiooxazole, 193–199, *199*, 201
5-lithiooxazole, 200, 201
2-lithiooxazole acyclic isocyanovinyllithium alkoxide equilibrium, 193–195
 ^{1}H NMR and ^{13}C NMR, 193–195
 molecular orbital calculations, 194–195
 x-ray crystal structures, 194–195
2-lithiooxazole-borane complex, 197–198, *198*
2-(methylthio)oxazole, 200–201, *201*
2-methyl-4-oxazolecarboxylic acid, 265–267
2-methyl-4-substituted oxazoles, regioselective, 202
2-methyl-5-(trimethylsilyl)-4-oxazolecarboxylic acid, 265
4-(hydroxymethyl)-2-methyloxazole, 269, 271
4-(methoxymethyl)-2′-methyl-2,4′-bis-oxazole, regioselective, 200, *203*
4-methyloxazole, 197, 215, 221
4-oxazolecarboxylic acid, 266–267
5-(3-methyl-5-isoxazolyl)oxazole, 193–194
5-methyl-4-tosyloxazole, 87, 90
5-phenyloxazole, 205, 207–208, *208*, 221

Lithiation (*Continued*)
  dithiane containing oxazoles, 201, 203–205, *204*, 206
  homoenolate equivalent, 210
  lithium diethylamide, 202
  natural products, 195–197, 201, 202–203, 265–271
    bengazoles, 195–195
    kabiramide C, 201
    phorboxazoles, 202–203, 207
    group A streptogramin antibiotics, 265–271
  oxazoles, 192–207
  silyl protected 4-oxazolemethanol, 195
  transmetalation, 205, 210–211
Lipoxygenase inhibitors, 43
Lupinine, 544
Lupinine alkaloids, 544, 569
Lycopodine, 565
Lysergic acid, 565

Macrocyclic depsipeptides, 371–373
Macrocyclization, 271, 352–353, 354, 372
Macrolactamization, 272, 273, 277, 279, 356
Macrolactonization, 279, 282, 352, 372
(−)-Madumycin II, 278–282
Madumycin, 207
Madumycin I, 272–274
Madumycin II, 257, 259
Manganese dioxide, 8–9, 11, 196, 269, 312, 442
  2-substituted-4-oxazolecarboxylic acid methyl ester, 9
  2,4-disubstituted oxazoles, 8
  bengazoles, 196
  berninamycin A fragment, 8
  microcin B17 fragments, 10, 14
Mannich reactions, 565, 567
Marine alkaloids, 434
Masamune's protocol, 289
Mass spectrometry, 403–405
  electrospray MS/MS, 405
  fragmentation patterns, 403–404
    benzylic cleavage, 403
    carbon monoxide, loss of, 403
    HCN, loss of, 403
    methyl group cleavage, 403
    nitriles, loss of, 403
    phenyl migration, 405
  McLafferty rearrangement, 403–404
  Microcin B17, 405
  negative ions, 405
  oxazole anions, 405
  oxirene radical cations, 403

  radical cations, 403
  ring cleavage, 403, 404, 405
  substituent affects, 403–404
Melanin biosynthesis inhibitors, 256
Melanoxadin, 256, 257
Melanoxazal, 256, 257
Mesoionic 1,3-oxazolium-5-olate, 105
Methanoazonine, 523
Methanooxonine, 547
Methyl acrylate, 165, 231
Methyl triflate, 229, *243*, 370, 478
O-Methylhalfordinol, 60
Methylidynephosphanes, 533
Michael addition reactions, 500
Microcin B17, 10, 14, 262, 264, 357, 405
Microwave assisted (accelerated) reactions, 39, 40, 41, *42*, 50, 51, 116, *117*
Microwave spectroscopy, 409
  dipole moment, 409
  Van der Waals complex, 409
    centrifugal distortion constants, 409
    nitrogen quadrupole coupling constants, 409
    rotational constants, 409
Mirbazoles, 287
Miscellaneous oxazoline oxidations, 13–16, 322, 324, 357, 358, 359
  2-alkyl-4-oxazolecarboxylic acid esters, 16
  2-substituted-4-oxazolecarboxylic acid esters, 14
  2,5-disubstituted oxazoles, 15
  5-bromooxazoles, 15
  calyculin fragment, 14
  DBU/BrCCl$_3$, 13–14, 322, 324, 357, 358
  DBU/CCl$_4$, 14, 359
  halogenation/dehydrohalogenation, 13
  iodination, 14–15
  NBS/AIBN, 15–16
  nodulosporic acid, 358
  nonproteinogenic oxazole amino acid esters, 358–359
  side-chain bromination, 16
  silyl amides, 14
Miscellaneous oxazole reactions, 250–255
  2-(α-aminoalkyl)oxazoles, optically active, 253–255, *255*
  asymmetric α-alkylation, 253, 254, 255, 256
  2-(aminomethyl)oxazoles, 254, *255*, 256
  natural products, 254, 256
  oxazoles as masked carboxylic acids, 254
  Paternò-Büchi photocycloaddition, 252–253, *254*
  α-amino-β-hydroxy methyl ketones, 253, *254*

latent α-aminoketone or α-aminoaldehyde, 253
oxetanes, 253
pseudodistomin A and B, 250–251, 252
reductive cyclization, 251–252
Miscellaneous oxazole syntheses, 118–127, *119, 122, 123, 126, 127*
   2-alkyl-4-amino-5-aryloxazoles, *119*
   2-alkyl(aryl)oxazoles, 121–122, 123, *123*
   2-substituted-4-oxazolecarboxylic acid esters, 126
   2-substituted-4-oxazolecarboxylic acid methyl esters, *127*
   2-substituted oxazoles, 121
   2,4,5-trisubstituted oxazoles, 120
   4-amino-2,5-diaryloxazoles, *119*
   4-amino-2,5-disubstituted oxazoles, 118
   4,5-dialkyl-2-methyloxazoles, 121
   4,5-disubstituted-2-methyloxazoles, *122*
   5-(aroylamino)-2-aryloxazoles, 125, *126*
   pyrazolo[3,4-*d*]oxazoles, 125
   acylation of enolates of *O*-trimethylsilyl acyltrimethylsilane cyanohydrins, 120–121, *122*
   aroyl cyanides, aldehydes, and ammonium acetate, 118–119, *119*
   Bucherer-Bergs reaction, 122
   cyclization of internal diazonium salts, 125
   cyclization of 1,2-bis(1*H*-benzotriazol-1-yl)-1,2-ethanediol, 125–126, *126*
   from acyl azides, 122, 124–125
   from dimethyl amino[(phenylthio)methyl]malonate, 126, *127*
   retro Diels-Alder reaction, 121–123, *123*
   α-oxoamidines, 119
Mitsunobu dehydration, 8
Molecular orbital calculations (theory), 128, 132, 166, 191, 194, 195, 200, 235, 420, 436, 437, 455, 457, 491, 500, 502, 506, 507, 509, 519, 520, 522, 524, 530, 543, 545, 546, 550, 551, 557, 559, 563
Monic acid, 106
Monooximes, 65–69
   10-(methoxyimino)phenanthrene-9-one, 65–66, 68
   1-phenyl-1,2-propanedione-2-oxime, 68–69
   butanedione, 68–69
   orthoquinone, 66–67
Monosubstituted oxazoles,
   ^{1}H NMR, *393*
   ^{13}C NMR, *396*
Monoterpene alkaloids, 463

Mosher amides or esters, 309, 330
Mugineic acid, 147
Mukaiyama
   aldol condensation, 275, 322
   amide coupling, 277
   reagent, carbodiimides, 22
Munchnones, *491, 492, 501, 502, 504, 506, 510, 512, 514, 516, 518, 521, 529, 533*
   1,3-dipolar cycloaddition reaction with
     1,4-quinones, *510*
     2-nitroindoles, *521*
     3-nitroindoles, *521*
     arylsulfonyl alkynes, *506*
     isothiazole 1,1-dioxide, *518*
     methyl 3-phenylpropiolate, *501*
     methyl propiolate, *501, 504*
     methyl α-cyanocinnamate, *512*
     methylene triazoles, *516*
     N-(phenylsulfonyl)imines, *529*
     phosphaalkynes, *533*
     terminal alkynes, *502*
     vinyl phosphonium salts, *514, 515*
     α-cyanocinnamonitrile, *512*
   reaction with
     amidines, *491*
     ammonia, *492*
     perfluoroalkylcarboxylic acid anhydrides, *486*
Muscarinic cholinergic receptor ligands, 197
Muscoride A, 33, 111, 260, 264
Mycalolides, 262, 263, 345

Naphthyridines, 432, 436
Neurokinin-3 receptor antagonists, 425
Ni(dppe)Cl$_2$, 211–212, *212*
Nickel peroxide, 5–8, 9, 16–17, 114, 274, 275, 285, 288, 300, 302, 303, 312, 313, 347, 354, 356
   2-alkyl(aryl)-4-oxazolecarboxylic acid esters, 8
   2-aryl-4-oxazolecarboxylic acid esters, 8
   2-aryloxazoles, 5
   2,4-disubstituted oxazoles, *6–7*
   natural products, 274, 275, 285, 288, 300, 302, 303, 312, 313, 347, 354, 356
Nitration of oxazoles, oxazole reactions, 130, 132
   ring substitution vs. side chain, 130, 132
   non-protonating reaction conditions, 130
   N-nitropicolinium tetrafluoroborate, 130
Nitrile ylides, 21, 62, 84, 86, 182, 191, 457, 459
   acyl, intermediates in rhodium carbene reactions, 28
   alkoxyoxazoles to 1-azirines, 182

Nitrile ylides (*Continued*)
azides to oxazoles, 62
cycloaddition with PTAD, 457
isoxazoles to oxazoles, 21
N,N-dialkylisocyanoacetamides and *N*-tosylbenzaldimine, 84, 86
oxazoles to thiazoles, 191
Nitriles, 27–32, *30, 31,* 34, 35, *36, 37,* 38, *38, 39,* 39–41, *40, 42,* 43–45, *45, 46,* 49–50, *50,* 245, 264, 273, 324, 326, 348, 359, 360, 361, 363, 462, 519
2-alkyl-5-aryloxazoles, 43–44, *45, 46*
2-alkyl(alkenyl)-5-methoxy-4-oxazolecarboxylic acid methyl esters, 27
2-alkyl(aryl)-5-ethoxy-4-(phenylsulfonyl)oxazoles, 30, *30*
2-alkyl(aryl)-5-ethoxyoxazoles, 36
2-alkyloxazoles, fused (benzoxazolones), 34, *36*
2-aralkyloxazoles, fused (benzoxazolones), 34, *36*
2-aryl-4-substituted oxazoles, 360
2-aryl-5-methoxy-4-oxazolecarboxylic acid methyl esters, 27
2-aryl-5-(4-hydroxyphenyl)oxazoles, 36, *37*
2-aryloxazoles, fused (benzoxazolones), 34, *36*
2-phenyl-4-substituted oxazoles, *42*
2,4-diaryloxazoles, 39
2,4-disubstituted oxazoles, 359
2,4,5-trisubstituted oxazoles, 38, 49, *50*, 361, 363
2,5-disubstituted-4-oxazolecarboxylic acid methyl ester, 33, *34*
3-(oxazol-5-yl)indoles, 32, 324
4-cyanooxazoles, 31
4,5-disubstituted-2-methyloxazoles, 38–39, *40*
4,5-disubstituted-2-phenyloxazoles, *42*
5-(perfluoroalkyl)-2-substituted 4-oxazolecarboxylic acid esters, 359
5-alkyl-2,4-diaryloxazoles, 39
5-aryl-2-(chloromethyl)oxazoles, 38, *39*
5-aryl-2-phenyloxazoles, 28
5-ethoxy-2-substituted 4-(triethylsilyl)oxazoles, 36–37
5-methoxy-2-substituted 4-oxazolecarboxylic acid methyl esters, 348
benzoxazolones, *36*
bis-oxazoles, 31
tris-oxazoles, 348
Lewis acid catalyzed reactions, 38, *39*, 245, 264

reaction with
acetylenes, 39–40, *40*
α-diazo compounds, 27–32, *30, 31, 37, 38, 39,* 245, 264, 273, 324, 326, 348, 359–361
iodonium salts, 34, 35, *36*
ketones, 39–41, *42,* 43–45, *44, 45, 46,* 49–50, *50,* 361, 363
^{14}N and ^{15}N NMR, 400
Nitroso compounds, 171–172, *173*, 457, 532
NMR spectroscopy, 392–403
^{13}C 392, 399–400, *396, 397, 398, 399*
2-metalated oxazoles, 399
charge distribution mapping, 400
coupling constants, 400
deshielding affects, 392
long-range coupling, 400
shielding affects, 392
solvent affects, 399, 400
substituent affects, 392, 399, 400
^{19}F, 401, *401, 402, 403*
coupling constants, *401, 402, 403*
fluorotrichloromethane (standard), 401
^{1}H, 392, *393, 394, 395, 396*
coupling constants, 392
cyclophanes, 392
deshielding affects, 392
order of, 392
protonated oxazoles, 392
solvent affects, 392
variable-temperature, 392
^{14}N and ^{15}N, 400
charge distribution mapping, 400
coupling constants, 400
linewidth, 400
solvent affects, 400
substituent affects, 400
^{17}O, 401
furan, 401
sydnones, 401
Nodulisporic acid, 358
Nonlinear optical chromophore, 219
Non-steroidal anti-inflammatory agents, 425
(−)-Normalindine, 81, 435
Normalindine, 463
(+)-Norsecurinine, 95, 449
(−)-Norsecurinine, 95, 449
Norsecurinine, 448, 449
antipodes, 448
(+)-Nostocyclamide, 30, 34, 111, 257, 258, 264
Novobiocin, 147
Nucleophilic substitution, oxazole reactions, 138–146, *142, 145*
2-alkoxy-4,5-diphenyloxazoles, 145

2-aryl-5-substituted 4-
    (trifluoromethyl)oxazoles, *142*
4-diethylphosphono-5-substituted oxazoles,
    146
  [3,3]-sigmatropic rearrangements, 145
  2-halooxazoles, 140–141, 145
  4-halooxazoles, 140–141
  5-halooxazoles, 140–146
  bis-nucleophiles, 140, 142
  carbanions of active methylene compounds,
    140, 141
  Claisen rearrangement, 141
  hetero-Cope rearrangement, 141
  α-substituted α-(trifluoromethyl) amino acids,
    141, 143–145, *145*
  ultrasonic mediated cyclization, 145

Ohba, 463
Onychine, 568
Optical spectroscopy, 406–409
  Infrared and Raman, 406–407
    ring stretch, 406
    POPOP, 406
    Surface enhanced Raman (SERS), 407
  Ultraviolet and Visible, 407–408
    benzoxazole photophysics, 408
    bis-oxazoles, 408
    excited states, 407
    ground states, 407
    luminescence 407, 408
    oxazole yellow fluorescent dyes, 408
    pH affects, 408
    POPOP, 408
    polymeric oxazoles, 408
    solvent affects, 408
    UV-amplified spontaneous emission (ASE)
      laser spike spectroscopy, 407–408
Organomercury reagents, oxazole synthesis,
    39–41, *42*
  2-phenyl-4-substituted oxazoles, *42*
  2,4-diaryloxazoles, 39
  4,5-disubstituted-2-phenyloxazoles, *42*
  5-alkyl-2,4-diaryloxazoles, 39
  microwave assisted reaction, 39
  mercury tosylate, 39–40
Organometallic reactions, 27–41, 75–76, *77*,
    191–211, *198*, *199*, *201*, *203*, *208*, *209*,
    220, 324, 348, 359
  oxazole reactions, 191–211
  **2-(2-hydroxyethyl)-4-oxazolecarboxylic acid
    esters**, 207
  **2-(2-hydroxyethyl)-4-oxazolecarboxylic acid
    ethyl esters**, *209*

2-acyl-5-phenyloxazoles, 205, 208, *208*
2-alkenyl(aryl)-5-substituted oxazoles, *220*
2-alkenyl(aryl)oxazoles, *220*
2-aryl-5-phenyloxazoles, *220*
2-aryloxazoles, *220*
2-iodo-5-substituted oxazoles, 199, *199*
2-substituted oxazoles, 197–198, *198*,
    218–219
2,2'-bis-oxazoles, 198
2,4-diiodo-5-substituted oxazoles, 199
2,5-disubstituted oxazoles, 75–76, *77*, *198*,
    218–219
4-(methoxymethyl)-2'-methyl-5-substiuted-
    2,4'-bis-oxazoles, 200, *203*
4-iodo-5-substituted oxazoles, 199, *199*
4-iodooxazoles, 199
5-(1,3-dithian-2-yl)oxazoles, 201, 203, *204*,
    205
5-substituted oxazoles, 200, *201*
bis-oxazolyl methanols, 195
  lithiation of oxazoles, 191–203
  oxazole-2-ylzinc reagents, 194, 195, 205,
    208, *208*
  oxazole Reformatsky-type reagent, 205, 207,
    208, 209, *209*, 210
  (trimethylstannyl)ethyloxazole, 210–211
oxazole synthesis, 27–41, *30*, *31*, *34*, *36*, *37*, *38*,
    39, *40*, *42*, 324, 348, 359
**2-alkyl(alkenyl)-5-methoxy-4-
    oxazolecarboxylic acid methyl esters**, 27
**2-alkyl(aryl)-5-ethoxy-4-
    (phenylsulfonyl)oxazoles**, 30, *30*
2-alkyl(aryl)-5-ethoxyoxazoles, 36
2-alkyloxazoles, fused (benzoxazolones), 34,
    36
2-aralkyloxazoles, fused (benzoxazolones),
    34, *36*
2-aryl-4-substituted oxazoles, 360
**2-aryl-5-methoxy-4-oxazolecarboxylic acid
    methyl esters**, 27
2-aryl-5-(4-hydroxyphenyl)oxazoles, 36, *37*
2-aryloxazoles, fused (benzoxazolones), 34,
    36
2-phenyl-4-substituted oxazoles, *42*
2,4-diaryloxazoles, 39
2,4-disubstituted oxazoles, 359
**2,5-disubstituted-4-oxazolecarboxylic acid
    methyl esters**, 33, *34*
3-(oxazol-5-yl)indoles, 32, 324
4-cyanooxazoles, 31
4,5-disubstituted-2-methyloxazoles, 38–39,
    40
4,5-disubstituted-2-phenyloxazoles, *42*

618   Subject Index

Organometallic reactions (*Continued*)
  5-(perfluoroalkyl)-2-substituted 4-oxazolecarboxylic acid esters, 359
  5-alkyl-2,4-diaryloxazoles, 39
  5-aryl-2-phenyloxazoles, 28
  5-ethoxy-2-substituted 4-(triethylsilyl)oxazoles, 36
  benzoxazolones, *36*
  bis-oxazoles, 31
  tris-oxazoles, 348
    organomercury reagents, 39–41
    organotellurium reagents, 38–39
    rhodium carbene additions, 27–38
Organotellurium reagents, oxazole synthesis, 38–39, *40*
  4,5-disubstituted-2-methyloxazoles, 38–39, *40*
  amidotellurinylation, 38–39
  internal acetylene, 38–39
  benzenetellurinyl triflate, 39, *40*
Organozinc and Reformatsky-type reagents, 75–76, *77*, 87, 194, 195, 205, 207–209, *208, 209*, 211, 217, 218–219, *220*, 221, 224, 272–273, 277, 319
  2-(2-hydroxyethyl)-4-oxazolecarboxylic acid esters, 207
  2-(2-hydroxyethyl)-4-oxazolecarboxylic acid ethyl esters, *209*
  2-acyl-5-phenyloxazoles, 205, 208, *208*
  2-alkenyl(aryl)-5-substituted oxazoles, *220*
  2-alkenyl(aryl)oxazoles, *220*
  2-aryl-5-phenyloxazoles, *220*
  2-aryloxazoles, *220*
  2-substituted oxazoles, 218–219
  2,5-disubstituted oxazoles, 75–76, *77*, 218–219
  alkenylzinc bromide, 75–76
  cross-coupling, 211, 217, 218–220, *220*, 221
  natural products, 207, 209, 272–273, 277, 319
  oxazole Reformatsky reagent, 205, 207, 208, *209*, 272–273, 277
  reaction with cyanomethylimidates, 76, *77*
  transmetalation, 205
Oxaazabicyclo[2.2.0]hexene, 534
1,2,4-Oxadiazoles, 171–172, *173*, 457, *461*
1,3,4-Oxadiazoles, 367, 368
Oxadiazolines, 532
1,3-Oxaphospholes, 549
1,4-Oxazin-3-ones, 487
Oxazoles
  Diels-Alder reactions, 367, 369, 370, 417–472
    heterodienophiles, 455–463
    oxazole-alkyne, 436–455, 465–467
    oxazole-olefin, 418–436, 463–465

  oxazolium salts, 367, 369, 370, 467
  ESR spectroscopy, 409
  from
    α-acyloxyketones, 92–94
    α-substituted ketones, 41–51
    azides, 59–62
    aziridines, 16–17
    enamino esters, 62–65
    hydrazones, 65–73
    imidates, 73–78
    isocyanides, 78–89
    isoxazoles, 20–22, 25–27
    isoxazolones, 23–25
    ketones, 33–34, *34*, 39, 41–51
    lactoxime *O*-vinyl ethers, 73–74, 268, 270, 363, 365
    N-acylamino acids/esters, 94–118
    N-acylamino ketones (2-acylamino ketones), 94–118
    N-acylamino nitriles, 94–118
    N-acylpeptides, 94–118
    nitriles, 27–32, *30, 31*, 34, 35, *36, 37*, 38, *38, 39*, 39–41, *40, 42*, 43–45, *44, 45, 46*, 49–50, *50*, 245, 264, 273, 324, 326, 348, 359–361, 363
    oxazolines, 4–16
    oximes, 65–73
    propargylic amides, 51–59, *54, 59*, 361, 363
    thioimidates, 73–78
    triazoles, 17–20
    vinyl bromides, 89–92, 114, 366, 367, 368
    vinylogous amides, 62–65
  functional group equivalents, 101–102, 147, 148, 157, 159, 161, 162, 210, 221, 222, 253, 254, 265, 359, 371
    activating group, 162, 371
    α-amino aldehydes, 253
    α-amino ketones 147, 253
    carbanions, 221, 222, 265
    carbonyl 1,1 dipole, 156–157
    carboxylic acids, 162, 254
    diacylhydrazide, 359
    dipeptides, 101–102, 148, 254
    ω-esters, 159
    formic anhydride, 161
    homoenolates, 210
    protected carboxylic acids, 157
  mass spectrometry, 403–405
  microwave spectroscopy, 409
  natural products, 5, 8–9, 10, 12, 13–14, 15, 16, 30, 32, 33, 34, 60, 74, 75, 95–96, 109, 111, 127, 195–197, 202, 203, 207, 214, 215, 248, 255–357, 405

mono-oxazoles
  annuloline, 60
  berninamycin/berninamycin A, 8
  bistratamide D, 13–14, 264
  calyculins, 14, 74, 75, 127, 248, 258, 260, 263, 289, 297, 295–316
  dendroamide A, 257, 258
  group A streptogramin antibiotics, 264–283, 284, 285
    (−)-griseoviridin, 282–283, 284, 285
    (−)-madumycin II, 278–282
    (−)-virginiamycin M_2, 275–277
    14,15-anhydropristinamycin II_B, 268–271
    griseoviridin, 257, 259
    madumycin, 207
    madumycin I, 272–274
    madumycin II, 257, 259
    virginiamycin, 207
    virginiamycin M_2, 257, 259, 268, 270
  keramides, 257, 258
  melanoxadin, 256, 257
  melanoxazal, 256, 257
  O-methylhalfordinol, 60
  (+)-nostocyclamide, 30, 34, 111, 257, 258, 264
  phenoxan, 8–9, 15, 33, 256, 257, 263
  phorbazoles, 95–96, 264
  phthaloxazolins, 256, 257, 362, 364
  pimprinethine, 32, 256, 257, 263
  pimprinine, 32, 256, 257, 263
  rhizoxin, 10, 256, 257, 264
  rhizoxin D, 264
  tantazoles, 258, 260, 283–295, 296
  texaline, 256, 257
  texamine, 256, 257, 264
  theonezolide A, 257, 259
  thinagazole(s), 5, 8, 12, 109, 258, 260, 283–295, 296
bis-oxazoles
  bengazoles, 195–197, 258, 261, 264
    A, 195–197
    C, 196
    D, 196
  diazonamides, 260, 261, 263, 324–344, 345, 346
    A, 261
  hennoxazoles, 260, 261, 316–323, 324, 325
  muscoride A, 33, 111, 260, 264
  phorboxazoles, 10, 202, 203, 207, 214, 215, 261, 262, 264
  promothiocin A, 30, 34, 262, 264

tris-oxazoles and tetra-oxazoles
  microcin B17, 10, 14, 262, 264, 357, 405
  ulapualides, 13, 262, 263, 344, 345, 347, 348, 350, 353, 344–357
  halichondramides, 16, 248, 262, 345
  halishigamides, 345
  kabiramides, 248, 262–263, 345
  mycalolides, 262, 263, 345
NMR spectroscopy, 3–4, 392–403
optical spectroscopy, 406–409
photoelectron spectra, 409
proton acidities, 3
reactions, 127–255
  conversion to other heterocycles, 183–191,
  cycloaddition reactions, 163–183
  electrophilic reactions, 128–138
  miscellaneous reactions, 250–255
  nucleophilic reactions and hydrolysis, 138–149
  organometallic reactions, 191–211
  oxazolium salts, 229–248
  oxidation, 152–163
  reduction, 149–152
  sigmatropic rearrangements, 163–183
  transition metal-catalyzed cross-coupling reactions, 211–221
  trimethylsilyloxazoles, 221–228
  Wittig reactions, 248–250
synthesis of, 4–127
  from α-acyloxyketones, 92–94
  from α-substituted ketones, 41–51
  from imidates and thioimidates-Cornforth reaction, 73–78
  from isocyanides, 78–89
  from oxidation of oxazolines, 4–16
  from vinyl bromides, 89–92
  from vinylogous amides and enamino esters, 62–65
  via cyclizations, 51–62
  via miscellaneous reactions, 118–127
  via organometallic reactions, 27–41
  via rearrangements, 16–27
  via Robinson-Gabriel and related reactions, 94–118
triflates, 213, 214, 215, *216*
x-ray absorption near-edge structure technique, 409–410
Oxazolidinediones, 560
Oxazoline oxidation, oxazole synthesis, 4–16, 6–7, 322, 324, 357, 358–359
2-alkyl(aryl)-4-oxazolecarboxylic acid esters, 8, 9

Oxazoline oxidation, oxazole synthesis
  (*Continued*)
  2-alkyl(aryl)-4-oxazolecarboxylic acid amides, 9
  2-alkyl-4-oxazolecarboxylic acid esters, 16
  2-aryl-4-oxazolecaboxylic acid esters, 8
  2-aryloxazoles, 5
  2-cyclohexyl-4-oxazolecarboxylic acid esters, 12
  2-isopropyl-4-oxazolecarboxylic acid esters, 12
  2-substituted-4-oxazolecarboxylic acid esters, 13, 14
  2-substituted-4-oxazolecarboxylic acid methyl esters, 9
  2,4-disubstituted oxazoles, *6–7*, 8
  2,5-diaryl-4-oxazoleacetic acids, 10
  2,5-disubstituted oxazoles, 15
  2,4,5-trisubstituted oxazoles, 11
  5-bromooxazoles, 15
  copper salts, 9–11
  Kharasch-Sosnovsky reaction, 12–13
  manganese dioxide, 8–9
  miscellaneous oxidations, 13–16, 358–359
  nickel peroxide, 5–8
2- and 3-Oxazolines, 8, 121, 123, 175–180, *177, 179*, 181, 227–228, 357, 429, 457–459, 460, *461*
  from
  2-aminooxazoles, 429
  2-(trimethylsilyl)oxazoles and aldehydes, 227–228
  5-alkoxy- and 5-alkyloxazoles, 175–180, *177, 179*, 181, 458–459, 460, *461*
  β-hydroxyamides, 8, 121, 123, 357
  DAST, 357
  Deoxo-fluor, 357
  diphenylsulfoxonium triflate, 8
  thermal, 121, 123
  α-isocyanoacetate esters and aldehydes, 227
1,3-Oxazolium 4-oxides:
  with acetylenic dipolarophiles, 539, 540, 542–545, 550, 555–559, 562
  with carbonyl dipolarophiles, 554, 555
  1,3-dipolar cycloadditions:
    with azetes, 548, 549
    with fullerenes, 552
    with fulvenes, 548
    with furan, 563
    intramolecular dipolarophiles, 541, 542, 556, 557, 561–569
    isolable primary cycloadducts, 546, 548, 549, 556, 561
    with olefinic dipolarophiles, 538–540, 543–545, 547, 550–569

  with phosphorus dipolarophiles, 549
  regiochemistry, 543–545, 557–559
  theoretical studies, 546, 550, 551, 563
  with tropone, 547
  hydrolysis of, 546, 551, 560
  infrared spectra, 546, 561
  mass spectra, 546
  NMR spectra, 546, 561
  synthesis:
    from *N*-acyl-*N*-arylglyoxanilides, 538
    from diazoimides, 541–544, 550–567
    from *N*-haloacetyl lactams, 539, 540
    from imidosulfoxides, 543–545, 568, 569
    isolable derivatives of, 538
    of alkaloids, 552, 553, 562, 564–569
    of chiral examples, 553, 554
    on solid supports, 558
  X-ray crystallography, 548, 549, 556, 561
1,3-Oxazolium 5-imines:
  Diels-Alder reactions, 535–537
  1,3-dipolar cycloadditions, 535–537
  synthesis from Reissert compounds, 535
1,3-Oxazolium 5-oxides:
  autoxidation, 527, 528
  cycloadditions *via* the presumed ketene valence tautomer:
    with azetes, 534
    with enones, 498
    with imines, 531, 534
  1,3-dipolar cycloadditions:
    with acetylenic dipolarophiles, 475–480, 495–509
    with arynes, 509
    with azetes, 531, 532, 534
    with carbonyl dipolarophiles, 527
    with 1,5-cyclooctadiene, 526
    with 1,3,5,7-cyclooctatetraene, 526
    with fullerenes, 526
    with fulvenes, 525
    with heterocycles, 521–523
    intramolecular acetylenic dipolarophiles, 507–509
    with nitrogen dipolarophiles, 529–533
    with olefinic dipolarophiles, 509–527
    with oxygen, 527, 528
    with phosphorus dipolarophiles, 533
    regiochemistry of, 500–507, 511–522
    theoretical studies, 509, 519, 524, 530
    with tropone, 525
  electrophilic reactions:
    with acids, 483, 484
    with imines, 482, 483
  infrared spectra, 476, 482

Subject Index   621

NMR spectra, 482
nucleophilic reactions:
 amidine additions, 491
 amine additions, 490–493
 dienamine additions, 494
 hydrazine additions, 492
 phosphine additions, 494
 water and alcohol additions, 490, 491
protonation, 483, 484
reaction with acylating agents, 484–490
synthesis:
 from N-acylamino acids, 105, 108, 474, 476
 of N-acyl analogues, 476
 from chromium carbene complexes, 479, 480
 in situ generation, 475, 481, 482
 isolable derivatives, 476, 481, 482
 from oxazolonium perchlorates, 476, 499, 500
 on solid supports, 479, 530, 531
 via Ugi condensation, 479
ultraviolet spectra, 482
X-ray crystallography, 481, 485, 498, 500, 508, 511, 522, 523, 531
N-Oxazolium methylides, non stabilized, 240, 241
Oxazolium salts, 105, 108, 229–248, 370, 467
 [2 + 3] cycloadditions, 241, 243
 2-amino-3-phenacyloxazolium salts, 238
 addition vs ring opening, 229–230
 Diels-Alder reactions, 231–233, 234, 235–236
 from enol trifluoroacetates, 105, 108
 from oxazoles and
  1-chloroethyl chloroformate, 231
  alkylating agents, 229–230, 239–243, *243*, 245–247
  ketenes, 247–248
 intermediates for aziridinomitosine, 244–247
 intramolecular, 234–235, 245
 intramolecular Diels-Alder reaction, 370
 pyrroles, 231, 231–233, 240–241, *243*
 reaction with
  amines, 237–240
  cuprates, 229
  cyanide, 234–235,
  Grignard reagents, 229
  hydride, 230
  organolithium reagents, 229
  phenylsilane/cesium fluoride, 241, 242, 244, 245
  sodium borohydride, 246–247
  tributyltin hydride, 242
  trimethylsilylcyanide, 242–243, 245
 bis-Reissert salts, 231–233
  Diels-Alder reactions, 231–233

Ethambutol, 231
 pyrroles, 231
 bis-pyrroles, 231–233
fused-Reissert salts, 235–240
 Diels-Alder reactions, 235–236
 pyrroles, 236–237
 imidazo[1,2-*a*]pyridines, 238, 240
 pyrido[2,1-*c*][1,2,4]triazines, 238, 240
 trimethylsilyl, 240–241
 trimethylsilylmethyl, 239, 241
  non stabilized N-oxazolium methylides, 239, 241
 pyrroles, 240–241
Oxazolo[3,2-*a*]pyridinium-2-olates, 481, 482, 506, 513, 546
Oxazolo[4,5-*b*]pyridines, 100, 429
Oxazolo[5,4-*b*]pyridines, 98
Oxazolo[5,4-*c*]pyrazoles, 50
Oxazolo[5,4-*c*]pyridazines, 50, 69, 70
 IR spectral data, 482, 546
 NMR spectral data, 482, 506, 513, 546
 UV spectral data, 482
2(3H)-Oxazolones, 121, 181, 182
2(5H)-Oxazolones, 181
4(5H)-Oxazolones, 36, 214, 318
5(2H)-Oxazolones, 182, 184
5(4H)-Oxazolones, 53, 117, 122, 141, 145, 182, 183, 184, 476
Oxidation, oxazole reactions, 133–134, 152–163, 460–462
 ozonolysis, 159–162
  activating group, 162
  deuterioformylation, 161
  formylation, 162, *162*
  N-formylformamide, 161
  oxazole 4,5-epoxide, 162–163
  peptide coupling, 162
  positive-working photoresists, 162, 164
  side chain oxidation, 133–134
  singlet oxygen, 152–159, 162, 164, 166, 168, 254, 256, 460–462
   [2 + 2] cycloaddition, 152, 461
   [4 + 2] cycloaddition, 152–154, 460
   ^{13}C NMR, 152
   ^{1}H NMR, 152, 154–155, 461
   activating group, 371
   carbonyl 1,1 dipole, 156
   natural products, 156–157, 158, 159, 254, 256
   oxazole 2,5-endoperoxides, 152–156, 460, 462
   oxygen-18 labeling studies, 152–153, 460–461

622 Subject Index

Oxidation, oxazole reactions (*Continued*)
  photooxidation, 152, 162, 164
  triamides, 152–155, 462
Oximes and hydrazones, oxazole synthesis, 65–73, *68, 72*
  2-arylphenanthro[9,10-*d*]oxazoles, 65–67, *68*
  2-heteroarylphenanthro[9,10-*d*]oxazoles, 65–67, *68*
  2,4,5-trisubstituted oxazoles, 68
  4-alkyl(alkenyl)-5-(trifluoromethyl)oxazoles, 72
  4-aryl-5-(trifluoromethyl)oxazoles, 71–72, *72*
  oxazolo[5,4-*c*]pyridazines, 69
  steroidal oxazoles, 69
  aldehyde *tert*-butyl(methyl)hydrazones, 71–72, *72*
  aromatic ketones, α-hydroximino, 68–69
  aromatic methylated hydrocarbons, 65–68
  formamidoxime, *N*-heteroaryl, 69–70
  heterocyclic methylated hydrocarbons, 65–68
  monooximes, 65–69
    10-(methoxyimino)phenanthrene-9-one, 65–66, *68*
    1-phenyl-1,2-propanedione-2-oxime, 68–69
    butanedione, 68–69
    orthoquinone, 66–67
  radical reaction, 65
  silica gel in toluene, 71
  steroidal, 69–70
(+)-Oxerine, 82
Oxerine, 463
Oxicam, 425
^{17}O NMR, 401
Oxy-Cope rearrangement, 447, 449
Oxygen-18 labeling, 152–153, 460–461, 491, 528
Ozonolysis, 159–162
  activating group, 162
  deuterioformylation, 161
  formylation, 162, *162*
  N-formylformamide, 161
  mixed anhydride, 160
  oxazole 4,5-epoxide, 162–163
  peptide coupling, 162

Padwa, 453
Pandamine, 100
(±)-Paniculide A, 98
Paniculide A 446, 447
Passerini reaction, 83–84, *85*
Paternò-Büchi photocycloaddition, 252–253, *254*
  α-amino-β-hydroxy methyl ketones, *erythro*, 253, *254*

latent α-aminoketone or α-aminoaldehyde, 253
oxetanes, 253
Pentaleno[2,3-*a*]isoquinoline, 507
Peptide antibiotic, 10
Peptidomimetics, 111
Perchloric acid, 476, 478
Peripheral analgesic agent, 260, 316
Petasalbine, 446
PGE_2 antagonists, 359
Phenoxan, 8–9, 15, 33, 256, 257, 263
  NADH ubiquinone oxidoreductase inhibitor, 256
  oxazole-γ-pyrone, 356
4-Phenyl-1,2,4-triazoline-3,5-dione (PTAD), 169–171, *171*, 457
Phenylhydrazines, 492
Phenyloxazin-2,3-diones, 554
Phenylsilane/cesium fluoride, 241, 242, 244, 245
Phorbazoles, 95–96, 264
Phorboxazole(s), 10, 202, 203, 207, 214, 215, 261, 262, 264
  A, 261, 262
  B, 261, 262
Phosphaalkenes, 533
Phosphaalkynes, 533, *533*, 549
Phosphazene superbases, 82
Phosphines, tris(trimethylsilyl), 494
Phosphirenes, 522
Photoelectron spectra, 409
  ionization potentials, 409
Photooxidation, 152, 162, 164
Photoresists, 152, 162–163, 164
Phthaloxazolin, 256, 257, 362, 364
  oxazole trienes, 256
Phytosphingosine, 179, 180
Pimprinethine, 32, 256, 257, 263
Pimprinine, 32, 256, 257, 263
Plant growth regulators, 27
(−)-Plectrodorine, 82, 463, 464
Plectrodorine, 463
Polymer (resin) bound, 35, 88–89, 90, 116, *117*, 360, 361, 364
  α-diazoacetophenone, 35
  α-diazo-β-ketoesters, 360
  Burgess reagent, 116, *117*
  quaternary ammonium hydroxide, 89
  TosMIC, 88–89, 90, 361, 364
Polythiazolines, 287–288, 289
POPOP, 406, 408
Proline, 448, 449, 484, 486, 488, *489*, 496, 504, 510
  D-, 448, 449
  L-, 448, 449
  N-acyl, 484, 486, 488, *489*, 496, 504, 510

Promothiocin A, 30, 34, 262, 264
Propargylic amides, oxazole synthesis, 51–59, *54, 58, 59*, 361
  2,5-dialkyloxazoles, *54*
  2,5-diaryloxazoles, 361
  2,5-disubstituted-4-oxazolecarboxylic acid esters, 56–57
  2,5-disubstituted-4-oxazolecarboxylic acid ethyl esters, *58*
  2,5-disubstituted oxazoles, 52
  2,4,5-trialkyloxazoles, *54*
  2,4,5-trisubstituted oxazoles, 52
  4-aryl-2,5-dialkyloxazoles, *54*
  5-[(phenylsulfonyl)methyl]-2-substituted oxazoles, 57, *59*
  5-methyleneoxazoles, 53
  3-methyl-1,4-benzoxazepin-5(4*H*)-one, 53
  α-acetamido-α-alkynylmalonates, 56–57
  allenes, 53, 55–56
  allenyl esters, 56
  cumulene intermediates, 55
  diethyl α-alkynylmalonates, *58*
  intramolecular 5-endo cyclization, 57
  palladium catalyzed cyclization, 361, 363
  porcine liver esterase, 56–58
  precursors to (*E*)-β-iodo(vinyl)sulfones, 52, 57
  iminophosporanes, 56, 59
Protein phosphatase inhibitors, 295
(±)-Pyrenolide C, 68, 156–157, 159
Pseudodistomins, 250
Pseudomonic acid, 116
Pteridines, 428
Pumiliotoxin C, 568
Pummerer reaction, 543, 568, 569
Pyrans, 548
Pyrazoles, 185, 187, 492, 493
Pyrazolo[3,4-*d*]oxazoles, 125
Pyrazolones, 185, 493
Pyridazines, 430
Pyridine alkaloids, 568
Pyridines, 367, 369, 418, 420, 421, 422, 423–427, 428, 429, 430, 431, 432, 433, 434, 435, 441, 463, 464, 483, 514, 567
  3-alkoxy, 418
  3-hydroxy, 420, 422, 424, 427, 428, 432, 433
  3-isopropyl, 421
  3-substituted, 421
  4-carboxylic acid, 425
  4-cyano, 421
  4-isopropyl, 421
  4-substituted, 421
  amino, 428
  anilino, 428
  benzopyrano[4,3-*b*], 432
  carbamoyl, 428
  cyclopenta[*c*], 464
  dihydro, 567
  highly substituted, 418, 430, 463
  hydroxy, 418, 419–421, 423–427
  synthesis, 483, 514
  tricyclic, 432–433
  trifluoromethyl, 424
Pyrido[2,1-*c*][1,2,4]triazines, 238, 240
α-Pyridones, synthesis, 482, 514, 537, 544, 552, 553, 568
Pyridoxal, [CD$_3$], 423
Pyridoxine antagonist, 188, 190
Pyridoxol, 422, 423–424
Pyrimidines, 187, 428, 556
  dihydro, 556
Pyrimidinols, 187–188, *189*, 190
L-Pyroglutamic acid, 453, 553
Pyrolysis, 62–63, *64*
Pyrroles, 28, 166–169, 185–186, *186*, 231, 231–233, 240–241, 242–243, *243*, 422, 428, 476, 479, 480, 483, 488, *489*, 496, 497, 499, 500, *501, 502*, 502, *504*, 505, 506, *506*, 509, 511, *512*, 513, 514, *514, 515*, 516, 517, 519, 520, 522, 526, 531, 535, 548
  1-methyl-2,3,4,5-tetraphenyl, 522
  3-formyl, 516
  4-(polyfluoroalkyl)-3-carboxylates, 499
  4-(trifluoromethyl), 513
  β-amino, 185–186, *186*
  bis-, 231–233
  dihydro, 166–169, 488, *489, 512*, 514, 526
  from nonstabilized *N*-oxazolium methylides and dipolarophiles, 239–241
  from azomethine ylides, 242–243, *243*
  N-acetyl, 476
  synthesis, 28, 166–169, 185–186, *186*, 231, 231–233, 240–241, 242–243, *243*, 479, 480, 483, 496, 497, 499, 500, *501, 502*, 502, *504*, 505, 506, *506*, 509, 511, 513, 514, *514, 515*, 517, 519, 520, 531, 535, 548
Pyrrolidin-2-ones, 493
Pyrrolidines, 424, 449, 456, 465, 467
2-Pyrrolines, 511
Pyrrolizidines, 495
Pyrrolizines, 501, 510
Pyrrolo[1,2-*c*]thiazoles, 495, 496, 498, 508
Pyrrolo[3,4-*b*]indoles, 521, 522
Pyrrolo[3,4-*b*]pyridines, 371–372
Pyrrolo[3,4-*d*]isoxazoles, 521

Pyrrolophenanthridones, 453
Pyrrolopyridines, 497

Quinolines, 507
  tetrahydro, 507
Quinolizidine alkaloids, 569
Quinone-oxazole bioreductive alkylating agent, 87, 89
1,4-Quinones, *510*, 535
Quinones, 426

Radical reactions, 5, 9, 12–13, 16, 65
Radiosensitizer, 45
Rearrangements, oxazole reactions, 183–191
  imidazoles, 187–191, *189*
  pyrazoles, 185, 187
  pyrimidines, 187
  pyrimidinols, 187–188, *189*, 190
  pyrroles, 185–186, *186*
Rearrangement reactions, oxazole synthesis, 16–27, *18, 20, 22*, 62, 359–360
  2-(1-aminoalkyl)-4-oxazolecarboxylic acid esters, 359
  2-(1-aminoalkyl)-5-oxazolecarboxylic acid esters, 359
  2-(alkylamino)-4-cyanooxazoles, 21, *22*
  2-(aroylamino)-5-methyloxazoles, 26
  2-alkyl(aryl)oxazoles, *18*
  2-phenyl-4-substituted oxazoles, *18*
  2-styryl-4-substituted oxazoles, *18*
  2-substituted oxazoles, 17
  2,4-disubstituted oxazoles, 16, 17, 23, *23*, 24
  2,4,5-trisubstituted oxazoles, *23*
  2,5-diaryloxazoles, 27
  2,5-disubstituted oxazoles, 23, *23*, 24
  5-alkyl-2-(alkylamino)-4-cyanooxazoles, 21, *22*
  5-alkyl(aryl)oxazoles, *20*
  5-aryloxazoles, 20
  5-substituted oxazoles, 20
  bis-oxazoles, 24
  tris-oxazoles, 24
  N-acylaziridines, 16–17
  N-acylisoxazolones, 23–25
  N-acyltriazoles, 17–20
  isoxazoles, 20–21, 21–22
Reduction, oxazole reactions, 149–152, 348
  anion-pi radicals, 152
  methoxy group cleavage of 5-methoxyoxazoles, 348
    LiAlH$_4$, 348
    Li(C$_2$H$_5$)$_3$BH, 348
  polarographic data, 152

cis-oxazolines, catalytic hydrogenation, 152
  ring cleavage, 149–152
    catalytic hydrogenation, high catalyst loads, 152
    dissolving metal, 151–152
    NaBH$_4$, 150–151
    nickel-aluminum alloy, 149–150
    Red-Al$^®$, 150
Reformatsky-type reagents, 205, 207, 208, 209, *209*, 210
Reissert compounds and salts, 224–225, *226*, 231–233, 235–240, 535–537
Resins, 479, 558
  Rink, 479
  Wang, 479, 558
Retro Diels-Alder reactions, 121, *123*
Rhizoxin, 10, 256, 257, 264
Rhizoxin D, 264
Rhodium carbene reactions, 27–38, *30, 31, 34, 36, 37, 38, 39*, 111, 273, 324, 326, 327, 348, 359–361, 362, 538, 541–543, 549–560, 563–567
  2-alkyl(alkenyl)-5-methoxy-4-oxazolecarboxylic acid methyl esters, 27
  2-alkyl(aryl)-5-ethoxy-4-(phenylsulfonyl)oxazoles, 30, *30*
  2-alkyl(aryl)-5-ethoxyoxazoles, 36
  2-alkyloxazoles, fused (benzoxazolones), 34, *36*
  2-aralkyloxazoles, fused (benzoxazolones), 34, *36*
  2-aryl-4-substituted oxazoles, 360
  2-aryl-5-methoxy-4-oxazolecarboxylic acid methyl esters, 27
  2-aryl-5-(4-hydroxyphenyl)oxazoles, 36, *37*
  2-aryloxazoles, fused (benzoxazolones), 34, *36*
  2,4-disubstituted oxazoles, 359
  2,5-disubstituted-4-oxazolecarboxylic acid methyl esters, 33, *34*
  3-(oxazol-5-yl)indoles, 32, 324
  4-cyanooxazoles, 31
  5-(perfluoroalkyl)-2-substituted 4-oxazolecarboxylic acid esters, 359
  5-aryl-2-phenyloxazoles, 28
  5-ethoxy-2-substituted 4-(triethylsilyl)oxazoles, 36
  benzoxazolones, *36*
  bis-oxazoles, 31
  tris-oxazoles, 348
  2-(bromomethyl)-4-oxazolecarboxylic acid ethyl ester, 273, 359, 361

Subject Index 625

α-diazo-β-ketoesters, 38, 328, 360, 362
  polymer bound, 360, 362
α-diazoacetates, 273
  formyl, 273
α-diazoacetoacetamides, 543
α-diazoacetoacetates, 33, 359
  methyl, 33
  perfluoroalkyl, 359
α-diazoacetoacetyl urea, 543, 556
α-diazoacetophenones, 28–29
α-diazoacetylindole, 32, 326
α-diazocarbonyl compounds, 27, 29, 324, 327, 538, 541, 543, 549, 551, 554
α-diazoesters, 28, 36, 38, *39*, 359, 361, 557
α-diazoimides, 538, 541–543, 549–560, 563–567
α-diazoketones, 33–34, *34*, 37–38, *39*
α-diazomalonates, 27, 31, 33, *34*, 348, 549, 551
α-diazonitriles, 31
α-diazophosphonates, 32
α-diazosulfones, 30–31, *30, 31*
α-(trimethylsilyl)diazoketones, 34–36, *37*
  polymer bound, 34–36
  solid-phase synthesis, 34–36
α-(triethylsilyl)diazoesters, 36–37
acyl nitrile ylides, 28
bromoacetonitrile, 273, 359, 361
ethyl diazobenzoylacetate, 28
ethyl α-formyldiazoacetate, 273
ethyl phenylsulfonyldiazoacetate, 30
ethyl (trimethylsilyl)diazoacetate, 36–37
ethylmalonylchloride, 551
JandaJel™, 360, 362
ligand effects, *31*
iodonium salts, 34–36, *36*
methyl cyanodiazoacetate, 31
natural products, 32, 33, 34, 111, 273, 324, 326, 327, 328, 329, 348
N-H insertion reactions, 33–34, *34*, 111, 326–327, 328, 360, 362
  diazonamide, 324, 326–327, 328, 329
  (+)nostocyclamide, 34, 111
  promothiocin A, 34
  muscoride A, 33,
  phenoxan, 33
  solid-phase synthesis, 360, 362
nitriles, 27–32, *30, 31,* 34, *36, 37, 38, 39,* 273, 324, 326, 348, 359–361
Rickborn, 443
Ring-closing olefin metathesis, 282, 285
Robinson-Gabriel and related reactions, 94–118, *107, 110, 112, 117,* 185, 318, 361, 362
  2-alkyl(aryl)-4-oxazoleacetic acids, 94

2-aryl-5-(dialkylamino)-4-oxazoleacetic acid ethyl esters, 185
2-substituted oxazoles, *117*
2,4′-bis-oxazoles, *112*
2,4-disubstituted-5-(t-butoxycarbonyl)oxazoles, 117–118
2,4-disubstituted-5-(trifluoromethyl)oxazoles, 104, *107*
2,4,5-trisubstituted oxazoles, 94, 95, 96, *112, 117*
2,5-dialkyl(aryl)-4-oxazoleacetic acids, 94
2,5-disubstituted 4-oxazolecarboxylic acid esters, *112*
2,5-disubstituted oxazoles, 94, 95, *117*
2,5-disubstituted-4-[2-(4-nitrophenoxy)ethyl]oxazoles, 94
4-(3-hydroxypropyl)-2-substituted 5-(trifluoromethyl)oxazoles, 105
5-(acylamino)-2,4-disubstituted oxazoles, 101, 103
5-(trifluoroacetamido)oxazoles, 104
5-amino-4-cyano-2-substituted oxazoles, 108, *110*
5-aryl-bis-oxazoles, 110
5-heteroaryl-substituted oxazoles, 116
5-methoxy-4-methyl-2-substituted oxazoles, 95
benzofuranyloxazole, 106
naphtho[2,3-*d*]oxazoles, 98
oxazolo[4,5-*b*]pyridines, 100
oxazolo[5,4-*b*]pyridines, 98
phorbazoles, 95–96
N-acylamino acids/esters, 94, 97–99, *107,* 109, 117, 274–275, 360, 362
N-acylamino β-ketoesters, 111
N-acylamino ketones (2-acylamino ketones), 94–97, 99–100, 106, 110–111, *112, 117*
N-acylamino nitriles, 94, 101–104
N-acylpeptides, 94, 101–102, 104
N-phenacylamino ketone, 106
aminomalononitrile tosylate (AMNT), 108, *110*
azole peptide mimetic, 109
benzofuranyloxazole, 106, 108
calcimycin (A23187), 100–101
ciglitazone, 104
cyclodehydrating agents, 94–118
Dakin-West reaction, 94, 106, 357
flash-lamp pumped laser dyes, 95
fluorescence microscopy stains, 96
Hantzch reaction, 106
Ifetroban sodium, 112, 114–115
Lewis acid, amide acetal cyclization, 115
mesoionic 1,3-oxazolium-5-olate, 105

Robinson-Gabriel and related reactions
(*Continued*)
   microwave accelerated reactions, 116, *117*
   natural products, 95, 98, 100, 104–105, 106, 111, 116, 274–275, 317–318
      (−)-norsecurine, 95
      (−)-tantazole, 100
      (+)-norsecurinine, 95
      (+)-nostocyclamide, 111
      (±)-paniculide A, 98
      (±)-stemonamide, 98
      A23187 (calcimycin), 100
      crenatine, 104–105
      diazonamide A, 106, 108
      frangulanine, 100
      integressine, 104–105
      monic acid, 106
      muscoride, 111
      pandamine, 100
      phorbazoles, 95
      pseudomonic aicd, 116
   oxazolium salts, 105, 108
   oxazolophane, 102, 104
   peptidomimetic, 109, 111
Ritter-like reaction, 361, 363
ROMPgel TosMIC reagent, 361, 364
Rosefuran, 439

Schlosser-Wittig reaction, 249, 350–351
Scintillating polymers, 369
Securinega alkaloids, 448
^{77}Se NMR, 312
Selnick, 453
Senile dementia, 197
Septicine, 553
Serine-threonine phosphatase inhibitors, 258
Sesquiterpenes, 446, 449
Shimada, 430
Sigmatropic reactions, rearrangements, and shifts, 20, 143, 145, 180, 181, 182, 237, 371
   [1,3]-, 143
   [1,5]-, 20, 237
   [2,3]-, 371
   [3,3]-, 145, 180, 181, 182
Silicon tetrafluoride, 475
Silks-Odom protocol, 312–313
Singlet oxygen, 152–159, *161*, 162, 164, 166, 168, 254, 256, 438, 440, 460–462
   [2+2] cycloaddition, 152, 461
   [4+2] cycloaddition, 152–154, 460
   ^{13}C NMR, 152
   ^{1}H NMR, 152, 154–155, 461
   activating group, 371

Bayer-Villiger rearrangement, 152–153, 462
   carbonyl 1,1 dipole, 156–157
   dioxetanes, 152–153, 461–462
   formylation, 161–162, *161*
   imino anhydrides, 152–153, 462
   natural products, 156–157, 158, 159
   oxazole 2,5-endoperoxides, 152–156, 460
   oxygen-18 labeling studies, 152–153, 460–461
   photooxidation, 152, 162, 164
   triamides, 152–155, 460
   natural products, 156–157, 158, 159, 254, 256
      4,5-dihydroxypipecolinic acid, 254, 256
      antimycin A$_3$, 156, 158
      (±)-pyrenolide C, 156–157, 159
Skeletal muscle relaxants, 94
Solid phase synthesis, 34, 78, 88, 360, 371, 479, 530, 531, 558
Sonogoshira coupling, 38
Sphingosine, 179, 181
Spirolactones, 498
Stannyloxazoles, 193, 210–211, 215, 216, *218*, 221, 222, 224, 318, 362, 370
Staudinger reaction, 56, 59
Stemona alkaloids, 450
   (−)-Stemoamide, 450, 453
   (±)-Stemoamide, 98, 450–453
Stille coupling, 213, 215, 268, 317, 318, 328, 329, 330, 335, 341, 343, 369, 370, 544, 552
Streptogramin A antibiotics, 263–283
Strychnos alkaloids, 435
Substance P antagonist, 109, 428
Sulfones, 495
Suzuki coupling, 36, 328, 335, 337, 439
Swern oxidation, 209, 268, 269, 273, 274, 282, 284, 307, 348, 364, 446
Sydnones, 499

(−)-Tantazole, 100, 109
(−)-Tantazole B, 294, 295, 296
Tantazole(s), 258, 260
Tautomerism, 476, 478
Tautomycin, 438
Tetrabutylammonium fluoride (TBAF), 228, 464
Tetracyanoethylene (TCNE), 166–169
Tetraoxaporphyrin, 437
Texaline, 256, 257
Texamine, 256, 257, 264
Thallium triflate, 43–44, *45*
Theonezolide A, 257, 259
(−)-Thiangazole, 12
Thiangazole(s) and tantazole, oxazole natural products, 5, 8, 12, 109, 258, 260, 283–295, 296

(−)-tantazole, 100, 109
(−)-tantazole B, 294, 295, 296
(−)-thiangazole, 12
thiangazole(s), 5, 8, 109, 258, 260, 283–295, 296
  2-chloro-1,3-dimethylimidazolium hexafluorophosphate, 293
  2-methylcysteine esters, 284, 285, 287, 290, 293
  2-methylserines, 290
  bis-thiazoline nitrile, 284, 286
  Burgess reagent, 285, 287, 288
  Cu(I)Br/t-butylperbenzoate, 285, 287, 288
  Dess-Martin periodinane, 288
  Ghosez's reagent, 289
  β-lactones, 294–295
  Masamune's protocol, 289
  NBS/benzoyl peroxide, 288
  NiO_2 oxidation, 285, 288
  oxazole tripeptide, 291
  thiazolines, 284, 287–288, 289, 292, 294
  threonine amides/esters, 285, 287, 290, 293
  $TiCl_4$, 287–288, 289, 292
  tris-oxazolines, 289, 292
  TsOH cyclodehydration of β-ketoamides, 288, 290
Thiazoles, 191–192, *193*
1,3-Thiazolidines, 495, 496, 508
  N-acyl, 508
Thiazolines, 171, 173, *174*, 284–286, 287–288, 289, 292, 294, 295, 459, *461*
  3-thiazolines, 171, 173, *174*, 459, *461*
  natural products,
    from 2-methylcysteine esters, 284
    from ammonium thiol esters, 294
    from oxazolines, 289, 292
    from $TiCl_4$ mediated cyclizations, 287–288, 289, 292
  bis-, 295
  tris-, 288, 289, 295
  poly-, 287
Thioaldehydes, 171, 173, *174*, 459, *461*
Thiobenzophenone, 478
Thioureas, 135
Thrombin receptor (PAR-1) antagonists, 111
Thromboxane A_2 receptor antagonist, 10, 89
Tosylmethylisocyanide (TosMIC), 78, 84, 86–90, 169, 361, 362, 363, 364
  solid-phase synthesis, 88–89, 90
  partial ergoline 5-HT_{1A} serotonin agonists, 87–88
  5-substituted-4-tosyloxazoles, 87, 90
  ROMPgel supported, 361, 364
Triamides, oxazole oxidations, 152–163, 462

1,2,4-Triazines, 492
1,2,3-Triazoles, 516, *516*
1,2,4-Triazolines, 169–171, *171, 172*, 456–457, *461*
Triazolium salts, synthesis, 529
Trifluoroacetic acid, 128, 438, 439, 443, 479, 484, 487, 531
Trifluoroacetic anhydride, 65, 94, 101–105, *107*, 108, 156, 289, 291, 305, 307, 349, 484, 485, *486*, 486, 487, *488*, 488, 489, *489*, 490, 499
Trimethylsilyloxazoles, oxazole reactions, 221–228, *222, 223, 226*
  2-(heteroaryl)oxazoles, 224, *226*
  2-acyl-4-methyloxazoles, *223*, 224
  2-acyl-5-aryloxazoles, *223*
  2-acyl-5-phenyloxazoles, 224
  2-acyloxazoles, 222
  2-substituted oxazoles, 222
  4-methyl-2-oxazolemethanol derivatives, 222, *222*
  5-aryl-2-(trimethylsilyl)oxazoles, 221–222
  mechanism for reaction with aldehydes, 224–225, 227, 227–228
  oxazolines, 227–228
  reaction with
    aldehydes, *222*, 227–228
    acylating agents, 222–223, *223*, 224
    ketenes, 224
    Reissert salts, 224–225, *226*
    soft carbanion equivalents, 222
  TBAF, 228
  thiazolium ylides, 224, 225, 227
Trisubstituted oxazoles, *395–396, 398–399*
  ^{1}H NMR, 395–396
  ^{13}C NMR, *398–399*
Tris-oxazoles, 24, 43, 127, 211, 248, 249, 250, 262–263, 264, 344–357, 345, 347, 348, 349, 350, 353, 354
Tropone, 525, 547
Turchi, 430

Ugi reaction, 371–373, 479
Ulapualide A, 13, 262–263, 344, 345, 347, 348, 350, 353
Ulapualides, 13, 127, 248, 262–263, 344–357
  Ulapualide A, 13, 262–263, 344, 345, 347, 348, 350, 353
  2-(acetoxymethyl)-4-oxazolecarbonyl chloride, 354
  2-methyl-4-oxazolecarboxylic acid ethyl ester, 355
  bis-oxazoles, 347, 348, 349, 350, 354

Ulapualides (Continued)
  biogenetic synthesis, 345, 347
  2,4′-bis-oxazoles, 347
  2-methyl-4-oxazolecarboxylic acid ethyl ester, 345
  NiO_2 oxidation, 347
  Wittig reagents, 347
  Burgess reagent, 356
  Garner's acid, 354, 355
  Hantzsch condensation, 348–351
  $(C_6H_5)_3P/BrCCl_2CCl_2Br$, 350
  2-styryl-4-oxazolecarboxylic acid ethyl ester, 349
  Dess-Martin periodinane, 350
  Schlosser-Wittig olefination, 350–351
  Horner-Wadsworth-Emmons reaction, 353, 354, 356
  macrocyclization, 352–353, 354
  macrolactamization, 356
  macrolactonization, 352
  NiO_2 oxidation, 356
  phosphonium salts, 347, 350–351, 352–353, 355
  rhodium mediated [3 + 2] cycloaddition, 348
    4-(hydroxymethyl)oxazoles, 348
    4-cyanooxazoles, 348
    5-methoxy-2-substituted-4-oxazolecarboxylic acid methyl ester, 348
    diazomalonates, 348
  tris-oxazoles, 345, 347, 348, 349, 350, 351, 353, 354, 356

Vallesamidine, 564
Vilsmeier-Haack formylation, 132, 133
Vilsmeier reagent, 45, 48, 185–186, *186*
Vinyl bromides/chlorides, oxazole synthesis, 89–92, 114, 366, 367, 368
  2-alkyl(aryl)-5-hydrazino-4-phosphorylated-oxazoles, 367
  2-aryl-4-cyano-5-hydrazinooxazoles, 367, 368
  4,5-(diarylthio)-2-phenyloxazoles, 366
  2-(acylamino)-3,3-dichloroacetonitrile, 367, 368
  BMS 180291, 89, 91
  β,β-dichloro-α-(toluenesulfonyl)benzamide, 366, 367, 368
Vinylogous amides/enamino esters, oxazole synthesis, 62–65, *64*

2-alkyloxazoles, 62, *64*
2-alkyl-5-aryloxazoles, 62, *64*
2,4-dialkyloxazoles, 62, *64*
5-(trifluoromethyl)oxazole, 63
  2-(trifluoromethyl)histidine, 65
  acyclic enamino esters, 63
  cyclic enamino esters, 62, *64*
  pyrolysis, 62–63, *64*
  retro ene reaction, 63
  trifluoroacetic anhydride, 65
  vinylogous β-ketoamide, 62
(−)-Virginiamycin M_2, 275–277
Virginiamycin, 207
Virginiamycin M_2, 257, 259, 268, 270
Vitamin B_6, 130, 423

Wasserman, 460
Weinreb, 430
Weinreb amides, 98, 278
Wipf's protocols, 15, 111, 114, 322
Wittig and Horner-Emmons reagents, 248–250, 296, 297, 298, 299, 302–303, 303–305, 309, 314, 315, 316, 347, 350–351
  2-alkenyloxazoles, *trans*, 248, 249, 250
  4-alkenyloxazoles, *trans*, 249–250, 251
  bis-oxazoles, 250, 251
  tris-oxazoles, 250, 251
  Arbuzov reaction, 248, 249
  natural product syntheses, 248, 249–250, 251, 296–299, 302–303, 303–305, 309, 314, 315, 316, 347, 350–351, 352, 353, 355
  phosphonium salts, 249–250, 251, 350–351, 352, 353, 355
  reactivity, 249, 250
  Schlosser modification, 249, 350–351
  selectivity, 249–250, 251, 303, 305, 315

X-ray absorption near-edge structure technique, 409
  electron density, 410
X-ray crystal structures, 135, 170, 194, 195, 238, 312, 367, 457, 481, 485, 498, 500, 508, 511, 522, 523, 526, 531, 534, 548, 549, 556, 561

Yamaguchi conditions, 282

Oxazoles – General Classes

2-(1-aminoalkyl)-4-oxazolecarboxylic acid esters, 359–360
2-(1-aminoalkyl)-5-oxazolecarboxylic acid esters, 359–360
2-(2-hydroxyethyl)-4-oxazolecarboxylic acid esters, 207
2-(2-hydroxyethyl)-4-oxazolecarboxylic acid ethyl esters, 207–209, *209*
2-(acylamino)-4-methyloxazoles, 137
2-(alkylamino)-4-cyanooxazoles, 21–22, *22*
2-(alkylthio)steroidal oxazoles, 51
2-(α-aminoalkyl)oxazoles, optically active, 253–255, *255*
2-(aminomethyl)-4-benzyl-5-morpholinooxazoles, 371–372
2-(aroylamino)-5-methyloxazoles, 26
2-(arylthio)-5-ethoxyoxazoles, 134, *134*
2-(dichloroacetyl)oxazoles, 247–248
2-(heteroaryl)oxazoles, 216, 224–225, *226*
2-(methylthio)oxazoles, 200, *201*
2-(nitrophenyl)oxazolo[4,5-*b*]pyridines, 98
2-(nitrophenyl)oxazolo[5,4-*b*]pyridines, 98
2-(stannyl)oxazoles, 318
2-(trifluoromethyl)oxazoles, *401*
2-(trimethylsilyl)oxazoles, 221–228
2-acetamidooxazoles, 453
2-acyl-4-methyloxazoles, *223*
2-acyl-5-alkoxyoxazoles, 78
2-acyl-5-aryloxazoles, *223*
2-acyl-5-ethoxyoxazoles, *79*
2-acyl-5-phenyloxazoles, 205, 208, *208*, 224
2-acyloxazoles, 222–223
2-alkenyl(aryl)-5-substituted oxazoles, *220*
2-alkenyl(aryl)oxazoles, *220*
2-alkenyloxy-4,5-diphenyloxazoles, 182, 183
2-alkenyloxazoles, *trans*, 249, 250, 251
2-alkoxy-4,5-diphenyloxazoles, 145, 180
2-alkoxyoxazoles, 461
2-alkyl(alkenyl)-5-methoxy-4-oxazolecarboxylic acid methyl esters, 27
2-alkyl(aryl)-4,5-diphenyloxazoles, 211–212, *212*
2-alkyl(aryl)-4-oxazoleacetic acids, 94
2-alkyl(aryl)-4-oxazolecarboxylic acid amides, 9
2-alkyl(aryl)-4-oxazolecarboxylic acid esters, 8, 9
2-alkyl(aryl)-5-aryloxazoles, 169
2-alkyl(aryl)-5-ethoxy-4-(phenylsulfonyl)oxazoles, 30, *30*
2-alkyl(aryl)-5-ethoxy-4-oxazolecarboxamides, 191, 192, *193*
2-alkyl(aryl)-5-ethoxy-4-oxazolethiocarboxamides, 192, *193*
2-alkyl(aryl)-5-ethoxyoxazoles, 36–37
2-alkyl(aryl)-5-fluoro-4-(trifluoromethyl)oxazoles, 47–48, 140
2-alkyl(aryl)-5-hydrazino-4-phosphorylatedoxazoles, 367
2-alkyl(aryl)amino-4-cyano-5-methoxyoxazoles, 61, *63*
2-alkyl(aryl)oxazoles, *18,* 121, 122, 123, *123*, 211–212, *212*
2-alkyl-4-amino-5-aryloxazoles, *119*
2-alkyl-4-oxazolecarboxylic acid esters, 16
2-alkyl-4-oxazolecarboxylic acid methyl esters, 74
2-alkyl-5-(diacylamino)-4-methyloxazoles, 138
2-alkyl-5-aryloxazoles, 43, 44, *45, 46*, 62–64, *64*
2-alkyl-5-methoxy-4-(4-nitrophenyl)oxazoles, 167, 168
2-alkyloxazoles, fused (benzoxazolones), 34, *36*
2-alkyloxazoles, 62–64, *64*, 458, 463
2-amino-3-phenacyloxazolium salts, 238
2-amino-4,5-disubstituted oxazoles, 152
2-amino-4-alkyloxazoles, 443
2-amino-4-methyloxazoles, 136, *137*
2-amino-4-substituted oxazoles, 41, 50
2-amino-5-(hydroxymethyl)-4-methyloxazoles, *137*
2-amino-5-aroyloxazoles, 188, *189*
2-aminooxazoles, 367, 369, 427, 429, 442
2-aralkyloxazoles, fused (benzoxazolones), 34, *36*
2-aroyl-5-aryloxazoles, 132–133
2-aryl(heteroaryl)-4-methyloxazoles, 216–217, *218*
2-aryl-4-phenyloxazoles, 466
2-aryl-4-cyano-5-hydrazinooxazoles, 367, 368
2-aryl-4-oxazolecarboxylic acid esters, 8
2-aryl-4-substituted oxazoles, 360, 362
2-aryl-5-(4-hydroxyphenyl)oxazoles, 36, *37*
2-aryl-5-(aryloxadiazolyl)phenyloxazoles, 408
2-aryl-5-(dialkylamino)-4-oxazoleacetic acid ethyl esters, 185, *186*
2-aryl-5-fluoro-4-(trifluoromethyl)oxazoles, *142*
2-aryl-5-methoxy-4-oxazolecarboxylic acid methyl esters, 27
2-aryl-5-methyl-4-oxazoleacetic acid methyl esters, 42

2-aryl-5-methyl-7-nitroimidazo[5,1-*b*]oxazoles,
   45–46, *47*
2-aryl-5-phenyloxazoles, *220*, 405, 406
2-aryl-5-substituted-4-(trifluoromethyl)oxazoles,
   142
2-aryl-substituted-4-phenyloxazoles, 440
2-aryloxazoles, fused (benzoxazolones), 34, *36*
2-aryloxazoles, 5, 129, 130, 216, *220*, 407, 458
2-arylphenanthro[9,10-*d*]oxazoles, 65–67, *68*
2-carboethoxyoxazoles, 419
2-cyclohexyl-4-oxazolecarboxylic acid esters, 12
2-ethoxyoxazoles, 456
2-halooxazoles, 212
2-heteroarylphenanthro[9,10-*d*]oxazoles, 65–67,
   68
2-iodo-5-substituted oxazoles, 199, *199*
2-isopropyl-4-oxazolecarboxylic acid esters, 12
2-lithiooxazole-borane complexes, 197–198, *198*
2-lithiooxazoles, 193–195, 199, *199*, 318
2-metalated oxazoles, 399
2-methyl-4-oxazolecarboxylic acid esters, 265
2-methyl-4-substituted oxazoles, 202, 203, 206–
   207
2-phenyl-4-substituted oxazoles, *18, 42*
2-phenyloxazoles, 154–155, 431, 432
2-styryl-4-substituted oxazoles, *18*
2-substituted 4,5-oxazoledicarboxylic acid
   dimethyl esters, 73–75, *75*
2-substituted 4-oxazolecarboxylic acid esters,
   13–14, 126, 309, 357
2-substituted 4-oxazolecarboxylic acid methyl
   esters, 9, *127*
2-substituted 5-(3-methyl-5-isoxazolyl)oxazoles,
   216, 219
2-substituted oxazoles, 17–20, 117, *117*, 121, 196,
   197–198, *198*, 217, 218, 222, 465
2-unsubstituted oxazoles, 154
2-vinyloxazoles, 336, 338
2,2′-bis-oxazoles, 198
2,4′-bis-oxazoles, *112*, 200, 202, 260, 316–320,
   322, 323, 324, 347–350
2,4-dialkyloxazoles, 62–64, *64*
2,4-diaryl-5-oxazolecarboxamides, 83
2,4-diaryloxazoles, 39, 92, *93*
2,4-diiodo-5-substituted oxazoles, 199
2,4-dimethyl substituted oxazoles, 130
2,4-disubstituted-5-(*tert*-butoxycarbonyloxy)oxa-
   zoles, 117–118
2,4-disubstituted-5-(trifluoromethyl)oxazoles,
   104–106, *107*, 108
2,4-disubstituted oxazoles, *6–7*, 8, 16, 17, 23–24,
   23, 41, 152, 261–262, 264, 296, 298,
   299–300, 345, 359, 361

2,4,5-trialkyloxazoles, *54*
2,4,5-triaryloxazoles, 408
2,4,5-trisubstituted oxazoles, *6–7*, 11, 23, *23*, 38,
   49, *50*, 52–53, *54*, 59, *61*, 68, 94, 95, *112*,
   117, 120, 152, 153, 240, 361, 363, 366,
   371, 457
2,5-bis(biphenylyl)oxazoles, 408
2,5-dialkyl(aryl)-4-oxazoleacetic acids, 94
2,5-dialkyloxazoles, *54*, 62
2,5-diaryl-4-(trimethylsilyl)oxazoles, 34
2,5-diaryl-4-oxazoleacetic acids, 10
2,5-diaryloxazoles, 27, 361, 363, 399, 407
2,5-diphenyl-4-(trialkylstannyl)oxazoles, 370
2,5-diphenyl-4-substituted oxazoles, 369
2,5-diphenyloxazoles, 408
2,5-disubstituted-4-[2-(4-nitrophenoxy)ethyl]oxa-
   zoles, 94
2,5-disubstituted-4-nitrooxazoles, 21
2,5-disubstituted-4-oxazolecarboxylic acid esters,
   56–57, *112*
2,5-disubstituted-4-oxazolecarboxylic acid ethyl
   esters, *58*
2,5-disubstituted-4-oxazolecarboxylic acid
   methyl esters, 33, *34*
2,5-disubstituted oxazoles, 15, 23–24, *23*, 52–3,
   59, *61*, 75, 76, *77*, 94, 95, *117*, 152, *198*,
   218, 219, 240, 242, *243*
3-(2-alkyloxazol-5-yl)indoles, 256
3-(oxazol-5-yl)indoles, 32, 324
4-(3-hydroxypropyl)-2-substituted 5-(trifluoromethyl)oxazoles, 105–106
4-(4-nitrophenyl)oxazoles, 426, 443
4-(chloromethyl)-2-substituted oxazole, 298
4-(*E*)-alkenyl-2-phenyloxazoles, 214–215, *216*
4-(hydroxymethyl)oxazoles, 348
4-(iodomethyl)-2-substituted oxazoles, 248, 249
4-(methoxymethyl)-2′-methyl-5-substituted-2,4′-
   bis-oxazoles, 200, *203*
4-(trifluoromethyl)oxazoles, *402*
4-acyloxazoles, 134, 187
4-alkenyloxazoles, *trans*, 249–250, 251
4-alkyl(alkenyl)-5-(trifluoromethyl)oxazoles, 72
4-alkyl(aryl)-2-amino(acetamido)-5-(arylazo)oxa-
   zoles, *138*
4-alkyl(aryl)-2-amino(acetamido)oxazoles, *138*
4-alkyl(aryl)-2-amino-5-oxazolethiocarboxa-
   mides, 135, *136*
4-alkyl(aryl)-2-aminooxazoles, 135, *136*
4-alkyl-2-aminooxazoles, 136, *139*
4-alkyloxazoles, 437
4-amino-2,5-diaryloxazoles, *119*
4-amino-2,5-disubstituted oxazoles, 118
4-aminooxazoles, 427

Subject Index

4-aroyl-2-(arylthio)-5-ethoxyoxazoles, 134, *134*
4-aryl-2,5-dialkyloxazoles, *54*
4-aryl-5-(trifluoromethyl)oxazoles, 71–72, *72*
4-bromo-2,5-diaryloxazoles, 36
4-chloro-2,5-disubstituted oxazoles, 340, 341
4-chlorooxazoles, 335, 337
4-cyanooxazoles, 22, 31, 341, 343, 348, 361, 363
4-diethylphosphono-5-substituted oxazoles, 146
4-formyloxazoles, 187
4-iodo-5-substituted oxazoles, 199, *199*
4-iodooxazoles, 199
4-methyl-2-oxazolemethanol derivatives, 222, *222*
4-oxazoleacetic acid esters, 185
4-oxazolecarboxylic acid esters, 73
4-oxazolecarboxylic acids, 273
4-phenyl-5-(propargyloxy)-2-substituted oxazoles, 182, 184
4-phenyloxazoles, 130, 444
4-substituted oxazoles, 214, 318
4,5-(diarylthio)-2-phenyloxazoles, 366
4,5-bis(chloromethyl)-2-substituted oxazoles, 165, 166, 167
4,5-dialkyl-2-methyloxazoles, 121
4,5-diaryl-2-substituted oxazoles, 406
4,5-disubstituted 2-methyloxazoles, 38–39, *40*, *122*
4,5-disubstituted 2-phenyloxazoles, *42*
4,5-disubstituted oxazoles, 201, 203, *204*, 205
5-[(phenylsulfonyl)methyl]-2-substituted oxazoles, 57–58, *59*
5-(1,3-dithian-2-yl)oxazoles, 201, 203, *204*, 205
5-(acylamino)-2,4-disubstituted oxazoles, 101, 103
5-(acylamino)-2-alkyl-4-methyloxazoles, 138
5-(acylamino)-4-methyloxazoles, 138
5-(aminomethyl)oxazoles, *N*-Boc, chiral, 80
5-(aminomethyl)oxazoles, *N*-protected, chiral, 80
5-(aroylamino)-2-aryloxazoles, 125–126, *126*
5-(diacylamino)-4-methyloxazoles, 138
5-(dialkylamino)-2,4-disubstituted oxazoles, 84, *87*
5-(dialkylamino)-2-substituted oxazoles, 84
5-(*N*-alkyl-*N*-phenylamino)-2-(arylthio)oxazoles, 79
5-(*N*-alkyl-*N*-phenylamino)-2,4-(diarylthio)oxazoles, 79
5-(perfluoroalkyl)-2-substituted 4-oxazolecarboxylic acid esters, 359–360
5-(perfluoroalkyl)oxazoles, *403*, 486, *486*
5-(substituted 4-nitrophenyl)oxazoles, 363
5-(trifluoroacetamido)oxazoles, 104
5-(trifluoromethyl)oxazoles, 63, 357, 484, 486

5-acyl-2-amino-4-methyloxazoles, 137
5-acyl-4-alkyl-2-aminooxazoles, *139*
5-acyl-4-methyloxazoles, 188, 190
5-acyloxazoles, 187
5-alkoxy-2-(arylthio)oxazoles, 78
5-alkoxy-2-aryloxazoles, 175
5-alkoxy-2-methyloxazoles, 175
5-alkoxyoxazoles, 170, 171, *177*, 178, 357, 422, 426, 427, 436, 437, 441, 457, 458, 459
5-alkyl-2-(alkylamino)-4-cyanooxazoles, 21–22, *22*
5-alkyl-2,4-diaryloxazoles, 39
5-alkyl(aryl)oxazoles, *20*
5-alkyloxazoles, 403
5-amino-4-cyano-2-substituted oxazoles, 108–109, *110*
5-amino-4-diethylphosphonooxazoles, 146, *147*
5-aminooxazoles, 427
5-aryl-2-(4-pyridyl)oxazoles, 408
5-aryl-2-(chloromethyl)oxazoles, 38, *39*
5-aryl-2-(trimethylsilyl)oxazoles, 221–222, *223*
5-aryl-2-methyloxazoles, 406
5-aryl-2-phenyloxazoles, 28
5-aryl-4-oxazolecarboxaldehydes, 45, 48
5-aryl-bis-oxazoles, 110
5-aryl(heteroaryl)-4-oxazolecarboxylic acid methyl esters, 80
5-aryl(heteroaryl)oxazoles, 89
5-aryloxazoles, 20, 87, 88, 169, 170
5-bromo-4-cyano-2-substituted oxazoles, 215, 217
5-bromooxazoles, 15
5-ethoxy-2-substituted-4-(trimethylsilyl)oxazoles, 36–37
5-ethoxy-4-substituted oxazoles, 436
5-ethoxyoxazoles, 456
5-fluoro-2-heteroaryl-4-(trifluoromethyl)oxazoles, 140–141
5-heteroaryl-substituted oxazoles, 116
5-methoxy-2-(*t*-butyldimethylsilyl)oxazoles, 227
5-methoxy-2-substituted 4-oxazolecarboxylic acid methyl ester, 348
5-methoxy-4-methyl-2-substituted oxazoles, 95
5-methoxyoxazoles, 348
5-methyleneoxazoles, 53
5-perfluoroalkyloxazoles, 486
5-phenyloxazoles, 162
5-siloxyoxazoles, 476, 478
5-substituted-4-oxazolecarboxylic acid esters, 78, 82–83, *83*
5-substituted-4-tosyloxazoles, 87
5-substituted oxazoles, 20, 86, 87, 200–201, *201*, 357, 361–362, 364, 465, 466, 467

5-unsubstituted oxazoles, 357
alkoxyoxazoles, 426, 427, 436, 441, 457, 458, 459
alkyloxazoles, 418
alkyl substituted oxazoles, 418
allyloxyoxazoles, 183, 184
amidooxazoles, 435
aminooxazoles, 427, 442
aryl-2,5'-bis-oxazoles, 408
aryloxazoles, 407, 408
benzoxazolones, 34, *36*
benzofuranyloxazole, 106
bis-oxazoles, 20, 24, 31, 110, *112*, 198, 200, *203*, 211, 212, 213, 250, 251, 316, 317, 318, 322, 347, 348, 349, 350, 354, 405, 408
bis-oxazolyl ketones, 196
bis-oxazolyl methanols, 195
bromooxazoles, 212–213
diaryloxazoles 399, 401, 405, 406, 407–408
disubstituted oxazoles, 392, *393–395, 397*
ethoxyoxazoles, 419, 436, 441, 456, 458
fluoroalkyl-substituted oxazoles, 401
fluoroinated oxazoles, 401
imidazo[5,1-*b*]oxazoles, 45–46, *47*
methyloxazoles, 403, 406, 409
monoaryloxazoles, 407
monosubstituted oxazoles, 392, *393, 396*
N-(2,4-dialkyloxazol-5-yl)oxazolidinones, 62

N-cyclohexyl-2,4-diaryl-5-oxazolecarboxamides, 84, *85*, 94
naphtho[2,3-*d*]oxazoles, 98
ortho-phenolic-2,5-diaryloxazoles, 408
oxazole-2-ylzinc reagents, 218–220, *220*
oxazole 2,5-endoperoxides, 152–154
oxazole 4,5-epoxides, 162, 163
oxazole-acetylenes, 465
oxazole-ureas, 428
oxazole-ynones, 448
oxazolo[3,2-*a*]pyridinium perchlorates, 237–240
oxazolo[4,5-*b*]pyridines, 100
oxazolo[5,4-*b*]pyridines, 98
oxazolo[5,4-*c*]pyrazoles, 50
oxazolo[5,4-*c*]pyridazines, 50, 69
phenolic 2-aryloxazoles, 407
pyrazolo[3,4-*d*]oxazoles, 125
steroıdal oxazoles, 50–51, 52, 69–70
substituted oxazoles, *131*, 162, *171, 172, 173, 174, 177*
trialkyloxazoles, 457
triaryloxazoles, 405, 408
trimethylsiloxyoxazoles, 476
tris-oxazoles, 24, 43, 250, 251, 348
trisubstituted oxazoles, 392, *395–396, 398–399*, 443, 456
vinyloxazoles, 214

Oxazoles

(1*E*, 3*E*)-1-dialkylamino-4-[5-(4-nitrophenyl)oxazol-2-yl]-1,3-butadiene, 237
(*Z*)-4-[1,3-bis(benzamide)prop-1-enyl]-2-phenyl-5-(trifluoromethyl)oxazole, 65
1-[(1*E*, 3*E*)-4-(5-phenyloxazol-2-yl)]-1,3-butadienylpiperidine, 237
1-amino-4-(5-phenyloxazol-2-yl)-1,3-butadiene, 237
1,3-bis-(4-methyloxazol-5-yl)benzene, 189, 190
1,3-bis-[2-phenyl-4-(trifluoromethyl)-5-oxazolyloxy]benzene, 140
1,4-bis-(4-methyloxazol-5-yl)benzene, 189, 190
1,4-bis-(5-phenyloxazol-2-yl)benzene, 406
1,4-bis-[2-phenyl-4-(trifluoromethyl)-5-oxazolyloxy]benzene, 140
2-(2-benzenesulfonic acid)-5-(dimethylaminophenyl)oxazole, sodium salt, 408
2-(2-furanyl)-5-phenyloxazole, 131, 406, 408
2-(2-hydroxyphenyl)-5-methyloxazole, 53
2-(2-propenyl)-4-(2-pyridyl)oxazole, 42
2-(2-pyridyl)naphtho[2,3-*d*]oxazole, 98

2-(2-thienyl)-5-phenyloxazole, 406
2-(3-chloropropyl)-5-methoxy-4-methyloxazole, 98
2-(4-bromobutyl)-5-ethoxyoxazole, 234
2-(4-fluorophenyl)-4-oxazolecarboxaldehyde, 133
2-(4-methoxyphenyl)-4-(2-pyridyl)oxazole, 42
2-(4-methoxyphenyl)-4-oxazole triflate, 213, 214
2-(4-methoxyphenyl)-5-(4-pyridyl)oxazole, 96
2-(4-methoxyphenyl)oxazole, 219, 221
2-(5-bromo-3-penten-1-yl)-4-oxazolecarboxylic acid methyl ester, 318–319
2-(acetoxymethyl)-4-oxazolecarbonyl chloride, 354
2-(aminomethyl)-4,5-diphenyloxazole, 253, 254, *255*
2-(bromomethyl)-4-oxazolecarboxylic acid ethyl ester, 207, *209*, 359, 361
2-(chloromethyl)-4-oxazolecarboxylic acid methyl ester, 76, 364–365
2-(dimethylamino)-4-phenyloxazole, 130

Subject Index

2-(dimethylamino)-5-nitro-4-(4-nitrophenyl)oxazole, 130
2-(iodomethyl)-4-oxazolecarboxylic acid ethyl ester, 250, 251
2-(lithiomethyl)oxazole, 202
2-(methylamino)-4-(trifluoromethyl)oxazole, 129
2-(methylthio)oxazole, 200–201, *201*
2-(*N*-benzyl-*N*-methylamino)-4,5-dimethyloxazole, 248
2-(*N*-benzyl-*N*-methylamino)-4-methyloxazole, 248
2-(*N*-benzyl-*N*-methylamino)oxazole, 248
2-(phenylsulfonylmethyl)-4-oxazolecarboxylic acid *t*-butyl ester, 265–266, 268, 270
2-(trimethylsilyl)oxazole, 127, 221
2-[(2-butenyl)oxy]-4,5-diphenyloxazole, 181
2-acetamido-4-methyloxazole, 136
2-allyloxazole, 231
2-allyloxy-4,5-diphenyloxazole, 181
2-amino-4-(5-benzyloxy-2-pyridyl)-5-hydroxyoxazole, 122
2-amino-4-(trifluoromethyl)oxazole, 129
2-amino-4-methyloxazole, 135, 136, 427
2-amino-4-phenyloxazole, 136
2-amino-4,5-dimethyloxazole, 137
2-amino-5-bromo-4-(trifluoromethyl)oxazole, 129
2-aminooxazole, 136, 442
2-benzoyl-4-methyloxazole, 222, 223
2-benzyloxazole, 437
2-benzyloxy-4,5-diphenyloxazole, 182
2-bromooxazole, 221
2-chloro-4,5-diphenyloxazole, 140, 145
2-cyclohexyl-4-oxazoleacetic acid, 42
2-deuterooxazole, 128
2-ethoxy-5-phenyloxazole, 175
2-ethyl-4-oxazolecarboxylic acid, 267
2-lithiooxazole, 194–195
2-mercapto-4-methyloxazole, 453
2-methyl-4-oxazolecarboxylic acid, 267
2-methyl-4-oxazolecarboxylic acid ethyl ester, 10, 265, 345, 355
2-methyl-4-oxazolecarboxylic acid methyl ester, 14, 266
2-methyl-4-phenyl-5-(phenylethynyl)oxazole, 212, 213
2-methyl-4-phenyloxazole, 156
2-methyl-5-(trimethylsilyl)-4-oxazolecarboxylic acid, 265–266
2-methyl-5-(trimethylsilyl)-4-oxazolecarboxylic acid *t*-butyl ester, 265
2-methyl-5-oxazolecarbonylchloride, 329, 331
2-methyl-5-palmityl-4-oxazolecarboxylic acid dimethylamide, 149
2-methyl-5-palmityl-4-oxazolecarboxylic acid ethyl ester, 148
2-methyl-5-palmityl-4-oxazolecarboxylic acid methyl ester, 149
2-methyl-5-phenyl-4-(phenylethynyl)oxazole, 212, 213
2-methyl-5-phenyloxazole, 63, 160, 409
2-methyloxazole, 205, 403, 409
2-oxazolecarboxaldehyde, 132
2-phenyl-4-(trimethylstannyl)oxazole, 317–318
2-phenyl-4-oxazole triflate, 214, 215, *216*
2-phenyl-4-oxazolecarbonyl chloride, 250, 252
2-phenyl-4-oxazolecarboxylic acid methyl ester, 73, 363
2-phenyl-5-oxazole triflate, 214
2-phenyl-5-oxazolecarboxaldehyde, 131
2-phenyl-5-vinyloxazole, 55
2-phenylbenzoxazole, 152
2-phenyloxazole, 63, 129, 130, 154, 431, 432
2-phenyloxazolo[3,2-*a*]pyridinium perchlorate, 237–240
2-styryl-4-oxazolecarboxylic acid ethyl ester, 349
2-styryl-4-oxazolecarboxylic acid methyl ester, 8, 320
2,4-dimethyl-5-ethoxyoxazole, 98
2,4-diphenyl-5-methyloxazole, 40, 41, 151
2,4-diphenyloxazole, 24, 466
2,4,5-trimethyloxazole, 129, 150, 253, *254*, 406
2,4,5-triphenyloxazole *N*-oxide, 162
2,4,5-triphenyloxazole, 50, 122, 151, 162, 366, 405, 466
2,5-bis-[2-phenyl-4-(trifluoromethyl)-5-oxazolylthio]-1,3,4-thiadiazole, 140
2,5-dimethyl-4-oxazolecarboxylic acid, 267
2,5-dimethyl-4-oxazolecarboxylic acid ethyl ester, 24
2,5-dimethyl-4-phenyloxazole, 50, 52
2,5-diphenyl-4-methyloxazole *N*-oxide, 68
2,5-diphenyl-4-oxazolecarboxylic acid ethyl ester, 28
2,5-diphenyloxazole, 24, 44, 63, 68, 122, 129, 157, 160, 401, 408
4-(1-methylethyl)-2-phenyloxazole, 12
4-(4-methoxyphenyl)oxazole, 443
4-(4-nitrophenyl)oxazole, 443
4-(chloromethyl)-2-(4-fluorophenyl)oxazole, 133
4-(chloromethyl)-2,5-diphenyloxazole, 68
4-(chloromethyl)-5-methyl-2-phenyloxazole, 367
4-(hydroxymethyl)-2-methyloxazole, 269
4-(methoxycarbonyl)-2-methyl-5-oxazoleacetic acid methyl ester, 96

4-(methoxymethyl)-2'-methyl-2,4'-bis-oxazole, 200, 202, *203*
4-(nitrophenyl)-2-phenyl-5-oxazolecarboxylic acid ethyl ester, 42
4-acetyl-2-styryloxazole, 321–322
4-benzoyl-5-ethoxy-2-phenyloxazole, 28
4-bromo-2,5-diphenyloxazole, 129, 140, 369
4-bromo-5-methyl-2-phenyloxazole, 25
4-*t*-butyloxazole, 194
4-chloro-2,5-diphenyloxazole, 140
4-cyano-2,5-dimethyloxazole, 25
4-cyano-5-methoxy-2-phenyloxazole, 31
4-cyclohexyl-5-(4-methanesulfonylphenyl)-2-methyloxazole, 365, 366
4-diethylphosphono-5-ethylthiooxazole, 146
4-ethyl-5-methoxy-2-methyloxazole, 426
4-iodo-5-(2-phenylethyl)oxazole, 211, 213
4-methyl-2-(trimethylsilyl)oxazole, 221, 222, *222*, *223*, *226*
4-methyl-2-(trimethylstannyl)oxazole, 215, 216, 217, *218*, 222
4-methyl-2-oxazolecarboxaldehyde, 130
4-methyl-2-phenyloxazole, 63, 155
4-methyl-5-[(phenylthio)methyl]oxazole, 164
4-methyl-5-oxazolecarboxaldehyde, 130
4-methyloxazole, 11, 130, 197, 215, 221, 403, 409, 418, 421, 424, 440
4-nitro-2-phenyloxazole, 429
4-oxazolecarboxylic acid, 266–267
4-phenyloxazole, 194, 239, 241, 425, 437, 438, 439, 440, 443, 444
4,5-dibromo-2-phenyloxazole, 129
4,5-dimethyl-2-oxazolecarboxaldehyde, 130–131
4,5-dimethyl-2-phenyloxazole, 122, 367
4,5-dimethyloxazole, 130
4,5-diphenyl-2-[(3-methyl-2-butenyl)oxy]oxazole, 181
4,5-diphenyl-2-(3-methacryloyloxypropyl)oxazole, 162
4,5-diphenyl-2-(methylthio)oxazole, 211, *212*
4,5-diphenyl-2-(propynyloxy)oxazole, 181
4,5-diphenyl-2-[(2-butenyl)oxy]oxazole, 181
4,5-diphenyl-2-[(2-trimethylstannyl)ethyl]oxazole, 210
4,5-diphenyl-2-methanesulfonyloxazole, 140
4,5-diphenyl-2-methyloxazole, 50, 130, 151, 155, 156, 157, 158
4,5-diphenyloxazole, 130, 466
4,6-dihydrothieno[3,4-*d*]oxazole 5,5-dioxide, 164
5-(1,3-dithian-2-yl)oxazole, 201, 203, *204*, 205
5-(2-methoxy-4-nitrophenyl)oxazole, 363, 364

5-(2,5-dimethoxy-3,4,6-trimethylphenyl)oxazole, 87, 89
5-(3-methyl-5-isoxazolyl)oxazole, 193
5-(3-tri-*n*-butylstannylallyl)oxazole, 362, 364
5-(3,4-dimethoxyphenyl)-2-isopropyloxazole, 100
5-(3,4-dimethoxyphenyl)-2-methyloxazole, 100
5-(3,4,5-trimethoxyphenyl)-4-oxazolecarboxylic acid methyl ester, 82
5-(4-dimethylaminophenyl)-2-(4-nitrophenyl)oxazole, 132
5-(4-hydroxyphenyl)-4-(4-methylphenyl)-2-phenyloxazole, 36
5-(4-methylphenyl)-5'-(2-phenylethyl)-2,4'-bis-oxazole, 212, 213
5-(4-methylphenyl)oxazol-2-ylzinc chloride, 211, 213
5-(4-pentenyl)oxazole, 370
5-(4-methoxyphenyl)-2-(4-pyridyl)oxazole, 95
5-(5-ethoxy-4-methyloxazol-2-yl)pentanal, 174
5-(5-hexenyl)oxazole, 370
5-[(5-hexenyl)oxy]-2-phenyloxazole, 182
5-[[(p-chlorobenzoyl)oxy]methyl]-4-methyloxazole, 164
5-acetyl-2-aminooxazole, 187, 188, *189*
5-bromo-2-(methylamino)-4-(trifluoromethyl)oxazole, 129
5-bromo-2-phenyloxazole, 129, 328, 331
5-bromo-4-methyl-2-phenyloxazole, 25
5-chloro-2,4-diphenyloxazole, 140
5-chloro-4-diethylphosphonooxazole, 146
5-deutero-2-methyl-4-oxazolecarboxylic acid-D, 266–267
5-deutero-4-oxazolecarboxylic acid-D, 266–267
5-ethoxy-2-ethyl-4-(phenylsulfonyl)oxazole, 31–32, *31*
5-ethoxy-2-phenyl-4-(trifluoromethyl)oxazole, 47
5-ethoxy-2-phenyloxazole, 175, 457
5-ethoxy-3-ethyl-2-phenyloxazolium tetrafluoroborate, 235
5-ethoxy-4-(trifluoromethyl)-2-oxazolecarboxylic acid, 424
5-ethoxy-4-(trifluoromethyl)-2-oxazolecarboxylic acid ethyl ester, 28
5-ethoxy-4-iodo-2-phenyloxazole, 38
5-ethoxy-4-methyloxazole, 419, 425, 428, 441
5-ethoxy-4-oxazoleacetic acid ethyl ester, 419
5-ethoxy-β-oxo-2-phenyl-4-oxazolepropanoic acid ethyl ester, 185, 187
5-ethyl-2-methyloxazole, 21
5-ethyl-2-phenyl-4-(2-propynyl)oxazole, 104

5-fluoro-2-phenyl-4-(trifluoromethyl)oxazole, 47, 140, *145*
5-isopropoxy-4-methyloxazole, 422
5-methoxy-2-(4-methoxyphenyl)oxazole, 178, 458
5-methoxy-2-methyl-4-cyanooxazole, 400
5-methoxy-2-phenyl-4-oxazolecarboxylic acid methyl ester, 27
5-methoxy-2-phenyloxazole, 459
5-methyl-2-(4-nitrophenyl)oxazolo[3,2-*a*]pyridinium perchlorate, 237
5-methyl-2-phenyloxazole, 129, 160, 459
5-methyl-4-[(phenylthio)methyl]oxazole, 164
5-methyl-4-tosyloxazole, 87
5-methyloxazole, 403, 409
5-nitro-2-(3-nitrophenyl)oxazole, 130
5-nitro-2-(4-nitrophenyl)oxazole, 130
5-nitro-2-phenyloxazole, 130
5-oxazolecarboxaldehyde, 195
5-phenyl-2-(2-thienyl)oxazole, 406
5-phenyl-4-oxazolecarboxylic acid methyl ester, 82
5-phenyloxazol-2-yllithium, 205
5-phenyloxazol-2-ylzinc chloride, 205, *208*
5-phenyloxazole, 162, 221, 239, 241, 422
5,7-di-*t*-butyl-2-phenylbenzoxazole, 66–67
ethyl 4-cyano-5-methyloxazol-2-yl acetate, 25
N-cyclohexyl-2-(2'-chlorophenyl)-4-phenyloxazolecarboxamide, 85
N-ethyl-4-carboethoxy-2,5-dimethyloxazolium tetrafluoroborate, 230
N-ethyl-4-cyano-2,5-dimethyloxazolium tetrafluoroborate, 229
N,N-diethyl-2-phenylnaphth[1,2-*d*]-5-oxazolesulfonamide, 98
oxazole *o*-quinodimethane, 164–166
α,2,4-triphenyl-5-oxazoleacetonitrile, 140
α,2,5-triphenyl-4-oxazoleacetonitrile, 140
α,4,5-triphenyl-2-oxazoleacetonitrile, 140
α,4,5-triphenyl-2-oxazolepropanol, 210

Tables

TABLE 1.1: 2,4-DISUBSTITUTED- AND 2,4,5-TRISUBSTITUTED OXAZOLES VIA OXIDATION OF OXAZOLINES, *6*

TABLE 1.2: 2-PHENYL- AND 2-STYRYL-4-SUBSTITUTED OXAZOLES FROM REARRANGEMENT OF *N*-ACYLAZIRIDINES, *18*

TABLE 1.3: 2-ALKYL(ARYL)OXAZOLES BY REARRANGEMENT OF *N*(1)-ACYL-1,2,3-TRIAZOLES, *18*

TABLE 1.4: 5-ALKYL(ARYL)OXAZOLES BY FVP OF *N*(1)-ACYL-1,2,4-TRIAZOLES, *20*

TABLE 1.5: 2-(ALKYLAMINO)- AND 5-ALKYL-2-(ALKYLAMINO)-4-CYANOOXAZOLES FROM REARRANGEMENT OF ISOXAZOLE THIOUREAS, *22*

TABLE 1.6: 2,4-DISUBSTITUTED-; 2,5-DISUBSTITUTED-; AND 2,4,5-TRISUBSTITUTED OXAZOLES FROM REARRANGEMENT OF 2-ACYLISOXAZOL-5-ONES, *23*

TABLE 1.7: 2-ALKYL(ARYL)-5-ETHOXY-4-(PHENYLSULFONYL)OXAZOLES FROM α-DIAZOSULFONES, *30*

TABLE 1.8: EFFECT OF RHODIUM(II) CATALYSTS ON THE SYNTHESIS OF 5-ETHOXY-2-ETHYL-4-(PHENYLSULFONYL)OXAZOLE, *31*

TABLE 1.9: 2,5-DISUBSTITUTED-4-OXAZOLECARBOXYLIC ACID METHYL ESTERS FROM RHODIUM-CARBENOID NH INSERTION REACTIONS, *34*

TABLE 1.10: 2-ALKYL(ARYL)-6,7-DIHYDRO-6,6-DIMETHYL-4(5*H*)-BENZOXAZOLONES FROM IODONIUM SALTS AND NITRILES, *36*

TABLE 1.11: 2-ARYL-5-(4-HYDROXYPHENYL)OXAZOLES FROM RESIN-BOUND α-(TRIMETHYLSILYL)DIAZOKETONES AND NITRILES, *37*

TABLE 1.12: 5-ARYL-2-(CHLOROMETHYL)OXAZOLES FROM $BF_3 \cdot OEt_2$ CATALYZED REACTION OF DIAZOKETONES AND DIAZOESTERS WITH CHLOROACETONITRILE, 39

TABLE 1.13: 4,5-DISUBSTITUTED-2-METHYLOXAZOLES FROM INTERNAL ACETYLENES, ACETONITRILE, AND BENZENETELLURINYL TRIFLUOROMETHANESULFONATE, 40

TABLE 1.14: MICROWAVE ASSISTED SYNTHESIS OF 2-PHENYL-4-SUBSTITUTED OXAZOLES AND 4,5-DISUBSTITUTED 2-PHENYLOXAZOLES FROM ARYL KETONES, BENZONITRILE, AND MERCURY(II)TOSYLATE, 42

TABLE 1.15: 2-ALKYL-5-ARYLOXAZOLES FROM AROMATIC α-METHYL KETONES, NITRILES, AND THALLIUM TRIFLATE, 45

TABLE 1.16: 2-ALKYL-5-ARYLOXAZOLES FROM AROMATIC α-METHYL KETONES, NITRILES, AND IODOBENZENE DIACETATE, 46

TABLE 1.17: 2-ARYL-5-METHYL-7-NITROIMIDAZO[5,1-b]OXAZOLES, 47

TABLE 1.18: 2,4,5-TRISUBSTITUTED OXAZOLES FROM OXIDATION OF KETONES WITH IRON(III) SOLVATES OF NITRILES, 50

TABLE 1.19: 2,5-DIALKYL, 2,4,5-TRIALKYL, AND 4-ARYL-2,5-DIALKYLOXAZOLES FROM N-PROPARGYLAMIDES, 54

TABLE 1.20: 2,5-DISUBSTITUTED-4-OXAZOLECARBOXYLIC ACID ETHYL ESTERS FROM DIETHYL α-ALKYNYLMALONATES (DAM), 58

TABLE 1.21: 5-[(PHENYLSULFONYL)METHYL]-2-SUBSTITUTED OXAZOLES FROM PROPARGYL AMIDES VIA (E)-β-IODO(VINYL)SULFONES, 59

TABLE 1.22: 2,5-DISUBSTITUTED AND 2,4,5-TRISUBSTITUTED OXAZOLES FROM INTRAMOLECULAR AZA-WITTIG REACTION OF (Z)-β-(ACYLOXY)VINYL AZIDES, 61

TABLE 1.23: 2-ALKYL(ARYL)AMINO-4-CYANO-5-METHOXYOXAZOLES FROM METHYL-3,3-DIAZIDOCYANOACRYLATE, 63

TABLE 1.24: 2-ALKYL-, 2,4-DIALKYL-, AND 2-ALKYL-5-ARYLOXAZOLES FROM PYROLYSIS OF CYCLIC ENAMINOESTERS, 64

TABLE 1.25: 2-ARYL- AND 2-HETEROARYLPHENANTHRO[9,10-d]OXAZOLES FROM 10-(METHOXYIMINO)PHENANTHRENE-9-ONE, 68

TABLE 1.26: 4-ARYL-5-(TRIFLUOROMETHYL)OXAZOLES FROM ALDEHYDE N-METHYL-N-tert-BUTYLHYDRAZONES, 72

TABLE 1.27: 2-SUBSTITUTED-4,5-OXAZOLEDICARBOXYLIC ACID DIMETHYL ESTERS FROM CYCLIZATION OF LACTOXIME O-VINYL ETHERS, 75

TABLE 1.28: 2,5-DISUBSTITUTED OXAZOLES FROM CYANOMETHYLIMIDATES AND ORGANOZINC REAGENTS, 77

TABLE 1.29: 2-ACYL-5-ETHOXYOXAZOLES FROM CYCLIZATION OF α-KETOIMIDOYL CHLORIDES, 79

TABLE 1.30: 5-SUBSTITUTED 4-OXAZOLECARBOXYLIC ACID ESTERS FROM α-ALKOXYACIDS AND ISOCYANOACETIC ACID ESTERS, 85

TABLE 1.31: N-CYCLOHEXYL-2,4-DIARYL-5-OXAZOLECARBOXAMIDES VIA THE PASSERINI REACTION, 83

TABLE 1.32: 5-(DIALKYLAMINO)-2,4-DISUBSTITUTED OXAZOLES FROM N-TOSYLIMINES AND N, N-DIALKYLISOCYANOACETAMIDES, 87

Subject Index 637

TABLE 1.33: 2,4-DIARYLOXAZOLES FROM CYCLIZATION OF α-ACYLOXYKETONES WITH ACETAMIDE-$BF_3 \cdot OEt_2$, *93*

TABLE 1.34: 2,4-DISUBSTITUTED-5-(TRIFLUOROMETHYL)OXAZOLES FROM *N*-ACYLATED AMINO ACIDS AND TFAA, *107*

TABLE 1.35: 5-AMINO-4-CYANO-2-SUBSTITUTED OXAZOLES FROM AMINOMALONONI-TRILE TOSYLATE AND CARBOXYLIC ACIDS, *110*

TABLE 1.36: 2,5-DISUBSTITUTED 4-OXAZOLECARBOXYLIC ACID ESTERS, 2,4,5-TRISUBSTITUTED OXAZOLES, AND 2,4′-BIS-OXAZOLES FROM DESS-MARTIN PERIODINANE OXIDATION AND $(C_6H_5)_3P/I_2/(C_2H_5)_3N$ CYCLIZATION, *112*

TABLE 1.37: 2-SUBSTITUTED-; 2,5-DISUBSTITUTED-; AND 2,4,5-TRISUBSTITUTED OXAZOLES VIA MICROWAVE-ACCELERATED CYCLODEHYDRATION OF 2-ACYLAMINO KETONES USING BURGESS REAGENT, *117*

TABLE 1.38: 4-AMINO-2,5-DIARYLOXAZOLES AND 2-ALKYL-4-AMINO-5-ARYLOXAZOLES FROM CONDENSATION OF AROYL CYANIDES, ALDEHYDES, AND AMMONIUM ACETATE, *119*

TABLE 1.39: 4,5-DISUBSTITUTED-2-METHYLOXAZOLES FROM O-TRIMETHYLSILYLFORMYLTRIMETHYLSILANE CYANOHYDRIN, *122*

TABLE 1.40: 2-ALKYL(ARYL)OXAZOLES VIA RETRO-DIELS-ALDER REACTION, *123*

TABLE 1.41: 5-(AROYLAMINO)-2-ARYLOXAZOLES FROM 1,2-BIS(1*H*-BENZOTRIAZOL-1-YL)-1,2-ETHANEDIOL, *126*

TABLE 1.42: 2-SUBSTITUTED 4-OXAZOLECARBOXYLIC ACID METHYL ESTERS FROM DIMETHYL AMINO[PHENYLTHIO)METHYL]MALONATE, *127*

TABLE 1.43: 2-OXAZOLINES AND 3-OXAZOLINES FROM SUBSTITUTED OXAZOLES AND BR_2/CH_3OH, *131*

TABLE 1.44: 4-AROYL-2-(ARYLTHIO)-5-ETHOXYOXAZOLES FROM AROYLATION OF 2-(ARYLTHIO)-5-ETHOXYOXAZOLES, *134*

TABLE 1.45: 4-ALKYL(ARYL)-2-AMINO-5-OXAZOLETHIOCARBOXAMIDES FROM 4-ALKYL(ARYL)-2-AMINOOXAZOLES AND ISOTHIOCYANATES, *136*

TABLE 1.46: 2-AMINO-5-(HYDROXYMETHYL)-4-METHYLOXAZOLES FROM 2-AMINO-4-METHYLOXAZOLES AND ALDEHYDES, *137*

TABLE 1.47: 4-ALKYL(ARYL)-2-AMINO(ACETAMIDO)-5-(ARYLAZO)OXAZOLES FROM 4-ALKYL(ARYL)-2-AMINO(ACETAMIDO)OXAZOLES AND ARYL DIAZONIUM TETRAFLUOROBORATE SALTS, *138*

TABLE 1.48: 5-ACYL-4-ALKYL-2-AMINOOXAZOLES FROM FRIEDEL-CRAFTS REACTIONS OF 4-ALKYL-2-AMINOOXAZOLES, *139*

TABLE 1.49: 2-ARYL-5-SUBSTITUTED 4-(TRIFLUOROMETHYL)OXAZOLES FROM NUCLEOPHILIC SUBSTITUTION OF 2-ARYL-5-FLUORO-4-(TRIFLUOROMETHYL) OXAZOLES, *142*

TABLE 1.50: *N*-BENZOYL-α-(TRIFLUOROMETHYL)AROMATIC α-AMINO ACIDS FROM 5-FLUORO-2-PHENYL-4-(TRIFLUOROMETHYL)OXAZOLE, *145*

TABLE 1.51: DIETHYLPHOSPHONOGLYCINE AMIDES FROM HYDROLYSIS OF 5-AMINO-4-DIETHYLPHOSPHONOOXAZOLES, *147*

TABLE 1.52: FORMYLATION OF AMINES AND ALCOHOLS VIA *N*-FORMYLFORMAMIDE, *161*

TABLE 1.53: *N*-PHENYL-5-ACYL-Δ³-1,2,4-TRIAZOLINE-1,2-DICARBOXIMIDES FROM SUBSTITUTED OXAZOLES AND PTAD, *171*

TABLE 1.54: 1,2-DICARBETHOXY-3,5-DISUBSTITUTED- AND 3,5,5-TRISUBSTITUTED Δ³-1,2,4-TRIAZOLINES FROM SUBSTITUTED OXAZOLES AND DEAD, *172*

TABLE 1.55: 2,5-DIHYDRO-1,2,4-OXADIAZOLES FROM OXAZOLES AND ARYLNITROSO COMPOUNDS, *173*

TABLE 1.56: 3-THIAZOLINES FROM OXAZOLES AND THIOALDEHYDES, *174*

TABLE 1.57: 2-OXAZOLINE-4,5,5-TRICARBOXYLATES AND 3-OXAZOLINE-2,5,5-TRICARBOXYLATES FROM [3 + 2] CYCLOADDITION OF OXAZOLES AND DIETHYLOXOMALONATE, *177*

TABLE 1.58: *CIS*-2-OXAZOLINE-4-CARBOXYLATES AND *TRANS*-2-OXAZOLINE-4-CARBOXYLATES FROM LEWIS ACID-CATALYZED [3 + 2] CYCLOADDITION OF OXAZOLES AND ALDEHYDES, *179*

TABLE 1.59: 5-ARYL-2-(DIALKYLAMINOCARBONYL)-4-(DIMETHYLAMINO)-3-PYRROLECARBOXYLIC ACID ETHYL ESTERS FROM 2-ARYL-5-(DIALKYLAMINO)-4-OXAZOLEACETIC ACID ETHYL ESTERS UNDER VILSMEIER-HAACK CONDITIONS, *186*

TABLE 1.60: 1(*H*)-5-ACETYL-2-AMINOIMIDAZOLES AND 2-AMINO-4-METHYL-5-PYRIMIDINOLS FROM 5-ACETYL-2-AMINOOXAZOLE AND AMINES, *189*

TABLE 1.61: 1(*H*)-5-AROYL-2-(DIMETHYLAMINO)IMIDAZOLES AND 4-ARYL-2-(DIMETHYLAMINO)-5-PYRIMIDINOLS FROM 2-AMINO-5-AROYLOXAZOLES AND DIMETHYLAMINE, *189*

TABLE 1.62: 2-ALKYL(ARYL)-5-AMINO-4-CARBOETHOXYTHIAZOLES VIA CORNFORTH REARRANGEMENT OF 2-ALKYL(ARYL)-5-ETHOXY-4-OXAZOLETHIOCARBOXAMIDES, *193*

TABLE 1.63: 2-SUBSTITUTED AND 2,5-DISUBSTITUTED OXAZOLES FROM 2-LITHIOOXAZOLE-BORANE COMPLEXES AND ELECTROPHILES, *198*

TABLE 1.64: 4-IODO-5-SUBSTITUTED AND 2-IODO-5-SUBSTITUTED OXAZOLES FROM 2-LITHIOOXAZOLES AND IODINE, *199*

TABLE 1.65: 5-SUBSTITUTED OXAZOLES VIA LITHIATION OF 2-(METHYLTHIO)OXAZOLE AND REDUCTIVE DESULFURIZATION, *201*

TABLE 1.66: 4-(METHOXYMETHYL)-2'-METHYL-5-SUBSTITUTED-2,4'-BIS-OXAZOLES VIA REGIOSELECTIVE LITHIATION OF 4-(METHOXYMETHYL)-2'-METHYL-2,4'-BIS-OXAZOLE, *203*

TABLE 1.67: ALKYLATION AND ACYLATION OF 5-(1,3-DITHIAN-2-YL)OXAZOLE, *204*

TABLE 1.68: 2-ACYL-5-PHENYLOXAZOLES FROM ACYLATION OF 5-PHENYLOXAZOL-2-YLZINC CHLORIDE, *208*

TABLE 1.69: 2-(2-HYDROXYETHYL)-4-OXAZOLECARBOXYLIC ACID ETHYL ESTERS FROM 2-(BROMOMETHYL)-4-OXAZOLECARBOXYLIC ACID ETHYL ESTER REFORMATSKY REAGENT AND ALDEHYDES OR KETONES, *209*

TABLE 1.70: 2-ALKYL(ARYL)-4,5-DIPHENYLOXAZOLES FROM Ni(dppe)Cl$_2$-CATALYZED CROSS-COUPLING OF 4,5-DIPHENYL-2-(METHYLTHIO)OXAZOLE AND GRIGNARD REAGENTS, *212*

TABLE 1.71: 4-(*E*)-ALKENYL-2-PHENYLOXAZOLES FROM PALLADIUM-CATALYZED CROSS-COUPLING REACTIONS OF 2-PHENYL-4-OXAZOLE TRIFLATE, *216*

Subject Index 639

TABLE 1.72: 2-ARYL(HETEROARYL)-4-METHYLOXAZOLES FROM PALLADIUM-CATALYZED CROSS-COUPLING OF 4-METHYL-2-(TRIMETHYLSTANNYL)OXAZOLE, *218*

TABLE 1.73: 2-ALKENYL(ARYL)OXAZOLES OR 2-ALKENYL(ARYL)-5-SUBSTITUTED OXAZOLES FROM PALLADIUM-CATALYZED CROSS-COUPLING OF OXAZOL-2YLZINC CHLORIDES, *220*

TABLE 1.74: 2-ARYLOXAZOLES AND 2-ARYL-5-PHENYLOXAZOLES FROM PALLADIUM-CATALYZED CROSS-COUPLING OF OXAZOL-2YLZINC CHLORIDES, *220*

TABLE 1.75: 4-METHYL-2-OXAZOLEMETHANOL DERIVATIVES FROM 4-METHYL-2-(TRIMETHYLSILYL)OXAZOLE AND ALDEHYDES, *222*

TABLE 1.76: 2-ACYL-4-METHYLOXAZOLES AND 2-ACYL-5-ARYLOXAZOLES FROM 4-METHYL-2-(TRIMETHYLSILYL)OXAZOLE OR 5-ARYL-2-(TRIMETHYLSILYL) OXAZOLES AND ELECTROPHILES, *223*

TABLE 1.77: 2-(HETEROARYL)OXAZOLES FROM 4-METHYL-2-(TRIMETHYLSILYL) OXAZOLE AND REISSERT SALTS, *226*

TABLE 1.78: 2-ACYL-1-METHYL-5-SUBSTITUTED PYRROLE-3,4-DICARBOXYLIC ACID DIMETHYL ESTERS FROM 2,5-DISUBSTITUTED OXAZOLES, METHYLTRIFLATE, AND DMAD, *243*

TABLE 1.79: *ERYTHRO* α-AMINO-β-HYDROXY METHYL KETONES FROM PATERNÒ-BÜCHI PHOTOCYCLOADDITION OF 2,4,5-TRIMETHYLOXAZOLE WITH ALDEHYDES OR α-KETOESTERS, *254*

TABLE 1.80: OPTICALLY ACTIVE 2-(α-AMINOMETHYL)OXAZOLES FROM ASYMMETRIC α-ALKYLATION OF 2-(AMINOMETHYL)-4,5-DIPHENYLOXAZOLE, *255*

TABLE 2.1: ^{1}H NMR OF MONOSUBSTITUTED OXAZOLES, *393*

TABLE 2.2: ^{1}H NMR OF DISUBSTITUTED OXAZOLES, *393–395*

TABLE 2.3: ^{1}H NMR OF TRISUBSTITUTED OXAZOLES, *395–396*

TABLE 2.4: ^{13}C NMR OF MONOSUBSTITUTED OXAZOLES, *396*

TABLE 2.5: ^{13}C NMR OF DISUBSTITUTED OXAZOLES, *397*

TABLE 2.6: ^{13}C NMR OF TRISUBSTITUTED OXAZOLES, *398–399*

TABLE 2.7: ^{19}F NMR OF 2-(TRIFLUOROMETHYL)OXAZOLES, *401*

TABLE 2.8: ^{19}F NMR OF 4-(TRIFLUOROMETHYL)OXAZOLES, *402*

TABLE 2.9: ^{19}F NMR OF 5-(PERFLUOROALKYL)OXAZOLES, *403*

TABLE 3.1: INTERMOLECULAR OXAZOLE-OLEFIN DIELS-ALDER REACTIONS, *431*

TABLE 3.2: INTERMOLECULAR OXAZOLE-ALKYNE DIELS-ALDER REACTIONS, *445*

TABLE 3.3: INTRAMOLECULAR OXAZOLE DIELS-ALDER REACTIONS, *452*

TABLE 3.4: INTERMOLECULAR OXAZOLE-HETERODIENOPHILE DIELS-ALDER REACTIONS, *461*

TABLE 4.1: REACTION OF AZLACTONES WITH α,β-UNSATURATED IMINES, *483*

TABLE 4.2: REACTIONS OF MUNCHNONES WITH PERFLUOROALKYLCARBOXYLIC ACID ANHYDRIDES, *486*

TABLE 4.3: REACTIONS OF TETRAHYDROISOQUINOLINE-1-CARBOXYLIC ACIDS WITH TRIFLUOROACETIC ANHYDRIDE, *488*

TABLE 4.4: REACTIONS OF *N*-ALKOXYCARBONYLPROLINES WITH TRIFLUOROACETIC ANHYDRIDE, *489*

TABLE 4.5: NUCLEOPHILIC REACTIONS OF MUNCHNONES WITH AMIDINES, *491*

TABLE 4.6: NUCLEOPHILIC REACTIONS OF IN-SITU-GENERATED MUNCHNONES WITH AMMONIA, *492*

TABLE 4.7: 1,3-DIPOLAR CYCLOADDITION REACTIONS OF IN-SITU-GENERATED MUNCHNONES AND METHYL PROPIOLATE AND METHYL 3-PHENYLPROPIOLATE, *501*

TABLE 4.8: 1,3-DIPOLAR CYCLOADDITION REACTIONS OF IN-SITU-GENERATED MUNCHNONES AND TERMINAL ALKYNES, *502*

TABLE 4.9: 1,3-DIPOLAR CYCLOADDITION REACTIONS OF IN-SITU-GENERATED MUNCHNONES AND METHYL PROPIOLATE, *504*

TABLE 4.10: 1,3-DIPOLAR CYCLOADDITION REACTIONS OF IN-SITU-GENERATED MUNCHNONES AND METHYL PROPIOLATE, *504*

TABLE 4.11: 1,3-DIPOLAR CYCLOADDITION REACTIONS OF IN-SITU-GENERATED MUNCHNONES AND ARYLSULFONYL ALKYNES, *506*

TABLE 4.12: 1,3-DIPOLAR CYCLOADDITION REACTIONS OF IN-SITU-GENERATED MUNCHNONES AND 1,4-QUINONES, *510*

TABLE 4.13: 1,3-DIPOLAR CYCLOADDITION REACTIONS OF IN-SITU-GENERATED MUNCHNONES AND METHYL α-CYANOCINNAMATE and α-CYANOCINNAMONITRILE, *512*

TABLE 4.14: 1,3-DIPOLAR CYCLOADDITION REACTIONS OF IN-SITU-GENERATED MUNCHNONES AND VINYL PHOSPHONIUM SALTS, *514*

TABLE 4.15: 1,3-DIPOLAR CYCLOADDITION REACTIONS OF MUNCHNONES (AZLACTONES) AND VINYL PHOSPHONIUM SALTS, *515*

TABLE 4.16: 1,3-DIPOLAR CYCLOADDITION REACTIONS OF MUNCHNONES AND METHYLENE TRIAZOLES, *516*

TABLE 4.17: 1,3-DIPOLAR CYCLOADDITION REACTIONS OF IN-SITU-GENERATED MUNCHNONES AND ISOTHIAZOLE 1,1-DIOXIDE, *518*

TABLE 4.18: 1,3-DIPOLAR CYCLOADDITION REACTIONS OF IN-SITU-GENERATED MUNCHNONES AND ISOTHIAZOLE 1,1-DIOXIDE, *518*

TABLE 4.19: 1,3-DIPOLAR CYCLOADDITION REACTIONS OF IN-SITU-GENERATED MUNCHNONES AND 2-AND 3-NITROINDOLES, *521*

TABLE 4.20: 1,3-DIPOLAR CYCLOADDITION REACTIONS OF IN-SITU-GENERATED MUNCHNONES[a] AND *N*-(PHENYLSULFONYL)IMINES, *529*

TABLE 4.21: 1,3-DIPOLAR CYCLOADDITION REACTIONS OF MUNCHNONES AND PHOSPHAALKYNES, *533*